Ready for student success?

Set your goals high with *Geometry* from Holt, Rinehart and Winston.

READY

Holt mathematics goes the extra distance to help you and your students achieve more. *Geometry* provides traditional math instruction with frequent practice while including options for students to communicate and explore content in ways that illuminate the transitions between concrete and abstract thinking.

With traditional learning in focus, and flexible support on demand, an important range of educational demands can be mastered with confidence:

SET

* Reaching all learners

* Teaching for understanding

* Integrating with technology

GO FARTHER...

Need more than just an assortment of teaching tools?

You'll find clear and accessible goals for support in *Geometry*.

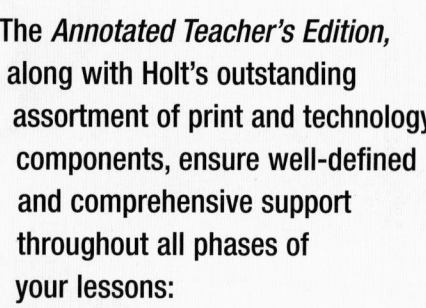

PREPARE

TEACH

ASSESS

Even when teaching with traditional tools, it's important to find clarification and support for alternative resources. Whether you use activity notes, teaching suggestions, content connections, or technology tips for your lessons, Holt's *Geometry* carefully incorporates your options so that you can find them and use them most efficiently when and if you need them.

The *Annotated Teacher's Edition,* along with Holt's outstanding assortment of print and technology components, ensure well-defined and comprehensive support throughout all phases of your lessons:

✳ Preparing with ease

✳ Teaching with options

✳ Assessing with confidence

Geometry

TABLE OF CONTENTS

Table of Contents

READY...

Each chapter in *Geometry* includes important features that get students ready to learn. Familiar contexts and clear entry points help students to focus and gain motivation as they proceed from basic skills towards understanding of new concepts.

Features that get your students ready to learn:

Quick Warm-Up helps students review previously taught and prerequisite skills.

NAME _____ CLASS _____ DATE _____

Quick Warm-Up: Assessing Prior Knowledge
4.5 *Proving Quadrilateral Properties*

How are the quadrilaterals in each pair alike? How are they different?

1. a parallelogram and a square _____

2. a rhombus and a square _____

Each manageable **Objective** helps students focus by breaking down the skills and concepts to be learned in each lesson.

Objective

- Prove quadrilateral conjectures by using triangle congruence postulates and theorems.

Proving Quadrilateral Properties

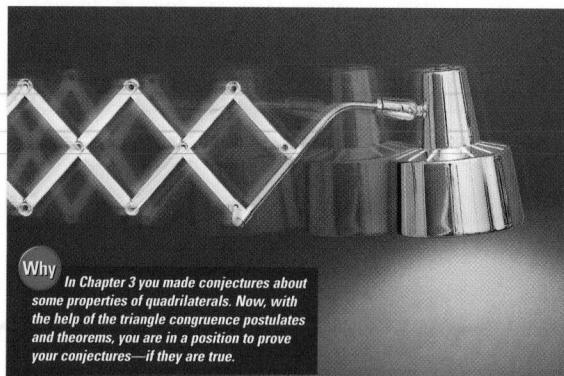

Objective

- Prove quadrilateral conjectures by using triangle congruence postulates and theorems.

Why *In Chapter 3 you made conjectures about some properties of quadrilaterals. Now, with the help of the triangle congruence postulates and theorems, you are in a position to prove your conjectures—if they are true.*

APPLICATION
ENGINEERING

A property of rhombuses explains why this lamp assembly stays perpendicular to the wall as it moves. You learned this property as a conjecture in Lesson 3.2. Do you know which property it is?

An Important Conjecture and Proof

In the Activity below, you will make a conjecture about an important property of parallelograms. The proof of the conjecture is given on the following page.

Activity
Rotational Symmetry in Parallelograms

1. Trace the parallelogram at right onto two sheets of tracing paper.

2. Place one figure over the other so that they match. Place the point of your pencil at point *C* and hold it firmly in place. Rotate the top piece of paper 180°. Describe the result.

3. Does △*PLM* seem be congruent to △*GML*? Fill in the blank to complete the conjecture below.

Conjecture:
A diagonal of a parallelogram divides the parallelogram into ____?____.

CHECKPOINT ✔

Why answers your students' question, *"When am I ever going to use this?"*

WHY?

Activity features get students practicing, thinking, and making connections.

Every **Application** motivates and helps your students connect with math by demonstrating how concepts and principles are used in the real world.

CRITICAL THINKING In the diagram below, the diagonal $\overleftrightarrow{ML}$ is a transversal to two different pairs of parallel lines. Name those parallel lines. List the alternate interior angles for each pair of lines.

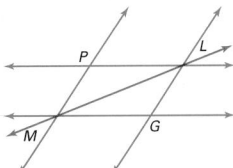

EXAMPLE

PROOFS

Given: parallelogram *PLGM* with diagonal $\overline{LM}$

Prove: $\triangle LGM \cong \triangle MPL$

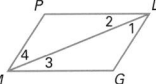

SOLUTION

Proof A (paragraph format):

$\overline{PL}$ and $\overline{GM}$ are parallel, according to the definition of a parallelogram. Therefore, $\angle 3$ and $\angle 2$ are congruent alternate interior angles. Similarly, $\angle 1$ and $\angle 4$ are congruent because $\overline{PM}$ and $\overline{GL}$ are also parallel. Finally, diagonal $\overline{LM}$ is congruent to itself. Thus, two angles and the included side are congruent in $\triangle LGM$ and $\triangle MPL$. Therefore, the triangles are congruent by the ASA Congruence Postulate.

Proof B (two-column format):

Statements	Reasons
1. Parallelogram *PLGM* has diagonal $\overline{LM}$.	Given
2. $\overline{PL} \parallel \overline{GM}$	Def. of a parallelogram
3. $\angle 3 \cong \angle 2$	‖ lines ⇒ alt. int. angles ≅
4. $\overline{PM} \parallel \overline{GL}$	Def. of a parallelogram
5. $\angle 1 \cong \angle 4$	‖ lines ⇒ alt. int. angles ≅
6. $\overline{LM} \cong \overline{LM}$	Reflexive Prop. of Congruence
7. $\triangle LGM \cong \triangle MPL$	ASA Congruence Postulate

This result is stated as a theorem on page 247. In Exercises 24-30, you will be asked to complete a flowchart proof of this theorem.

CRITICAL THINKING How can you use the result from the Example to prove that opposite angles of a parallelogram are congruent? You will be asked to write this proof in the exercise set.

🦅 *Look Back*

75. Suppose that the angles in one triangle are congruent to the angles in another triangle. Are the triangles necessarily congruent? Why or why not? *(LESSON 4.3)*

Write a congruence statement for each pair of triangles below, and name the postulate or theorem to justify it. *(LESSONS 4.2 AND 4.3)*

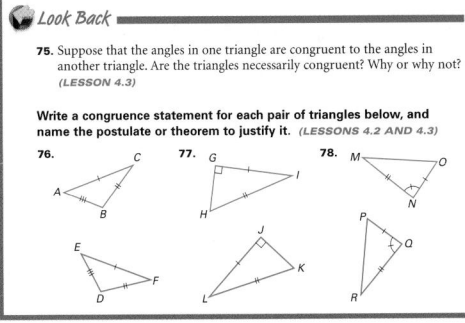

Each chapter in *Geometry* supports both skills-based and concept-building instruction— targeting traditional lesson requirements and including options for changing demands in your classroom:

Example-Solution-Try This format helps to segment lessons into manageable bites.

Skills practice abounds, beginning with **Communicate** and **Guided Skills Practice** in the chapter and extending to **Extra Practice** in the **Info Bank**.

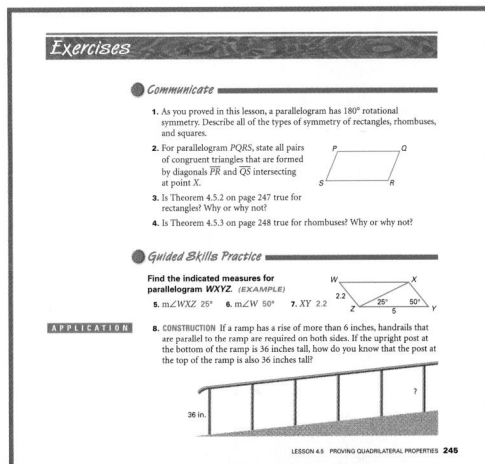

Extra Practice

Critical Thinking questions develop problem-solving and higher-order thinking skills.

Look Back and **Look Beyond** exercises offer practice, reinforce learning, and foreshadow upcoming topics or challenge students to stretch their knowledge.

For Students

HOLT, RINEHART AND WINSTON (A5)

GO FARTHER!

Geometry offers students ample opportunities to review, extend, and apply what they've learned. Holt's outstanding integration fosters lasting comprehension—with quick access to an assortment of unique and optional tools.

Engaging and practical options for ensuring comprehension:

Real-world integration includes the compelling stories and actual news events of **Eyewitness Math.**

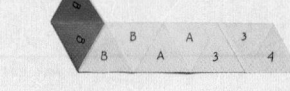

REAL-WORLD

CHAPTER PROJECT FOUR

FLEXAGONS

In 1939, a university mathematics student named Arthur H. Stone was playing with a strip of paper trimmed from a notebook. He discovered something interesting—flexagons.

Flexagons are polygons made from folded paper that show different faces when "flexed." The instructions below will give you a chance to play with two of these unique figures yourself.

For your first flexagons, it may be helpful to use paper with one color on the front and another on the back, as shown. Later, you can get creative by drawing designs on the flexagons, which have a kaleidoscopic effect when flexed.

Activity 1

FLEXAGON A hexaflexagon has three faces, but only two are visible [at a] time. Be sure to make your cuts and folds as precisely as possible [to ensure] that your flexagon will flex smoothly.

[First] cut a strip of paper divided into 10 [equi]lateral triangles. Label the front and [bac]k of the strip as shown.

Front

Back

2. Fold the strip so that the triangles labeled 1 face each other.

3. Fold the strip so that the triangles labeled 2 face each other.

4. Fold the strip so that the triangles marked 3 face each other. Carefully glue together the triangles labeled 4, and let the glue dry.

5. Now you are ready to flex your hexaflexagon. Pinch together two triangles and push in the opposite side so that the flexagon looks like a **Y** shape when viewed from above. Open the flexagon from the center. Repeat. How does the arrangement of the faces change as you flex the flexagon?

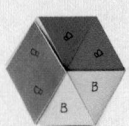

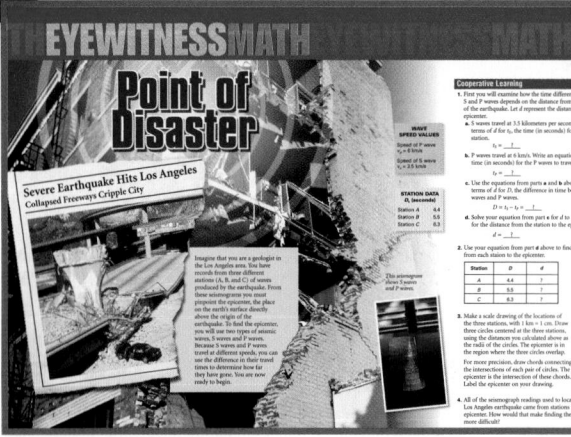

For Students

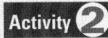

HEXAHEXAFLEXAGON

A hexahexaflexagon has twice as many faces as the hexaflexagon. It is also more complicated to construct, so follow the directions below carefully.

1. First cut a strip of paper divided into 19 equilateral triangles. Label the front and back of the strip as shown.

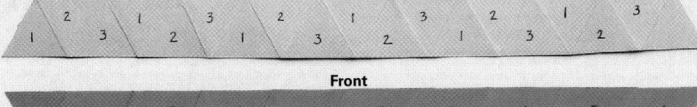

Front

Back

2. Fold the strip so that each pair of adjacent 4s, 5s, and 6s face each other. The strip should coil around itself and look like the one shown below.

3. Fold the strip so that each pair of adjacent 3s face each other.

4. Tuck one end of the strip under the other so that the remaining pair of 3s face each other. Fold down the flap, and carefully glue the unlabeled triangles together.

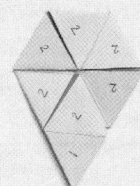 

5. Flex your hexahexaflexagon in the same way as you did the first flexagon. Can you get all six faces of this flexagon to show?

Extension

1. What are the front-back face combinations?

2. Are any combinations of faces not possible?

3. Is there a pattern to the order in which the faces are revealed?

Integration of **Multiple Representations** includes verbal, symbolic, and graphical representations.

Performance assessment in **Geometry** encourages your students' independence and confidence in math and helps to increase the probability that they will remember what they learn.

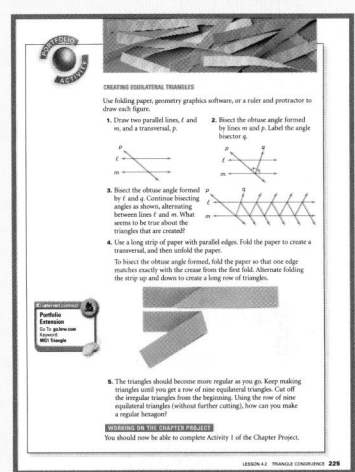

Assessment options include:

❊ **Portfolio Activity**

❊ **Chapter Project**

❊ **College Entrance Exam Practice**

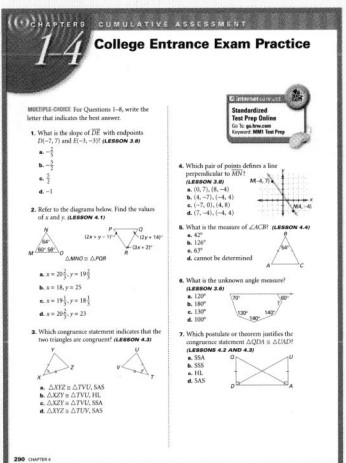

For Students

HOLT, RINEHART AND WINSTON (A7)

PREPARE WITH EASE

No matter what your teaching preference may be, the *Annotated Teacher's Edition* for **Geometry** saves you time and offers you superior guidance in lesson preparation.

Chapter and lesson support includes a variety of well-organized traditional tools and strategies, including:

* **Background Information**
* **Chapter Resources**
* **Chapter Objectives**

About Chapter 4

Background Information
This chapter introduces students to the triangle congruence postulates. First, students develop intuitive notions of triangle congruence through exploration. Later in the chapter, students apply the congruence postulates in formal proofs.

CHAPTER RESOURCES

* Block-Scheduling Handbook
* Writing Activities for Your Portfolio
* Tech Prep Masters
* Long-Term Project
* Assessment Resources: Mid-Chapter Assessment Chapter Assessments Alternative Assessments
* Test and Practice Generator
* Tech___y Handbook

Triangle Congruence

ALL AROUND YOU—IN NATURE, ART, AND HUMAN technology—you find things that are the same shape and size. Such things are said to be congruent. In the photos on these pages, notice that there are many congruent triangles.

Triangles have the property of being rigid, which makes them useful in building bridges and other structures. Also, since any polygon can be divided into a number of triangles, the properties of triangles can be used to study polygons in general.

*Buckminster Fuller
(1895–1983)*

Lessons

4.1 ● Congruent Polygons

4.2 ● Triangle Congruence

4.3 ● Analyzing Triangle Congruence

4.4 ● Using Triangle Congruence

4.5 ● Proving Quadrilateral Properties

4.6 ● Conditions for Special Quadrilaterals

4.7 ● Compass and Straightedge Constructions

4.8 ● Constructing Transformations

Chapter Project Flexagons

About the Photos

Many functional structures were created by using triangles in their design. Domes constructed with congruent triangles are called *geodesic domes* and are used as solar greenhouses, pet and hay shelters, pool covers, and astronomical observatories. Some of the strongest bridges in the world, such as the beam bridge shown above, use the geometric principle of triangle rigidity to maximize the strength of their structures.

Proving Quadrilateral Properties

Objective
* Prove quadrilateral conjectures by using triangle congruence postulates and theorems.

APPLICATION
ENGINEERING

Why In Chapter 3 you made conjectures about some properties of quadrilaterals. Now, with the help of the triangle congruence postulates and theorems, you are in a position to prove your conjectures—if they are true.

A property of rhombuses explains why this lamp assembly stays perpendicular to the wall as it moves. You learned this property as a conjecture in Lesson 3.2. Do you know which property it is?

An Important Conjecture and Proof

In the Activity below, you will make a conjecture about an important property of parallelograms. The proof of the conjecture is given on the following page.

Activity
Rotational Symmetry in Parallelograms

1. Trace the parallelogram at right onto two sheets of tracing paper.
2. Place one figure over the other so that they match. Place the point of your pencil at the point C and hold it firmly in place. Rotate the top piece of paper 180°. Describe the result.
3. Does $\triangle PLM$ seem to be congruent to $\triangle GML$? Fill in the blank to complete the conjecture below.

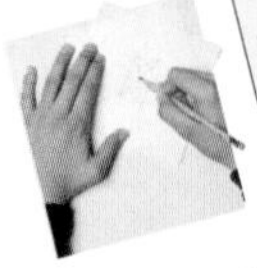

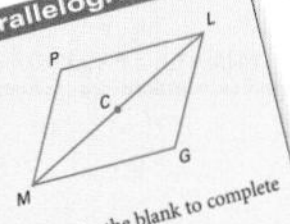

CHECKPOINT ✓

Conjecture:
A diagonal of a parallelogram divides the parallelogram into _____ ?

Alternative Teaching Strategy
TECHNOLOGY Have students use geometry graphics software to make the parallelograms for this lesson. Students should construct figures that retain their properties when vertices or segments are dragged.

Inclusion Strategies
COGNITIVE STRATEGIES Some students may be more comfortable writing two-column proofs with words, phrases, and sentences rather than with symbols and abbreviations. Others may prefer writing proofs as flowcharts. Let students experiment with different methods.

LESSON 4.5 **243**

Prepare

jectives
polygons.

using [4.1]

ty. [4.2]

ence

—SSS,

SAS,
stulates

.3]

ve

to

QUICK WARM-UP

How are the quadrilaterals in each pair alike? How are they different?

1. a parallelogram and a square Alike: opposite sides parallel and congruent Different: square has four right angles and four congruent sides

2. a rhombus and a square Alike: four equal sides, opposite angles equal, diagonals perpendicular Different: square has four right angles

Also on Quiz Transparency 4.5

Teach

Why A diagonal in any quadrilateral creates two triangles. Because of this, students' experiences with triangle congruency will help them prove conjectures about properties of quadrilaterals.

☞ For Activity Notes and the answer to the Checkpoint, see page 244.

---

* **Lesson Resources**
* **Lesson Objectives**
* **Quick Warm-Up**

For Teachers

Manhattan Bridge,
New York

- Use congruence of triangles to conclude congruence of corresponding parts. [4.4]
- Develop and use the Isosceles Triangle Theorem. [4.4]
- Prove quadrilateral conjectures by using triangle congruence postulates and theorems. [4.5]
- Develop conjectures about special quadrilaterals—parallelograms, rectangles, and rhombuses. [4.6]
- Construct congruent copies of segments, angles, and triangles. [4.7]
- Construct an angle bisector. [4.7]
- Translate, rotate, and reflect figures by using a compass and straightedge. [4.8]
- Prove that translations, rotations, and reflections preserve congruence and other properties. [4.8]
- Use the Betweenness Postulate to establish the Triangle Inequality Theorem. [4.8]

About the Chapter Project

In 1939 an American graduate student at Princeton made an interesting discovery. He folded strips of paper to form an object called a flexagon.

Flexagons have an interesting mathematical property. The hexaflexagon, for example, has three faces, but only two are visible at any given time. All flexagons have at least one hidden face. To find the hidden face, the flexagon must be folded, or "flexed," a certain way.

After completing the Chapter Project, you will be able to do the following:

- Create a hexaflexagon and a hexahexaflexagon.
- Describe the patterns in the order of faces of a flexagon.

About the Portfolio Activities

Throughout the chapter, you will be given opportunities to complete Portfolio Activities that are designed to support your work on the Chapter Project.

The theme of each Portfolio Activity and of the Chapter Project is congruent polygons.

- In the Portfolio Activity on page 225, you will use a strip of paper to fold a series of triangles. The more triangles you fold, the closer they become to a set of congruent equilateral triangles.
- Tessellations involve covering a surface with congruent shapes that fit together without gaps or overlapping. In the Portfolio Activity on page 252, you will explore tessellations with congruent, nonregular quadrilaterals.
- In the Portfolio Activity on page 281, you will explore tessellations with congruent hexagons that are not regular but that have one pair of parallel and congruent opposite sides.

Portfolio Activities appear at the end of Lessons 4.2, 4.5, and 4.8. Each serves as preparation for the Chapter Project. The Portfolio Activities, as well as the Chapter Project Activities, are appropriate for inclusion in the student's portfolio. Students should be encouraged to include in their portfolios any other work in which they feel a sense of pride or a sense of accomplishment.

Alternatives for teaching preparation are conveniently located when you need them, including:

- ❋ **About the Chapter Project**
- ❋ **About the Portfolio Activities**
- ❋ **Internet Connect** (links to go.hrw.com)
- ❋ **In-text references to Lesson Keywords**
- ❋ **Online Technology Updates**
- ❋ *One-Stop Planner® CD-ROM*

ONE-STOP PLANNER: THE ULTIMATE PLANNING TOOL

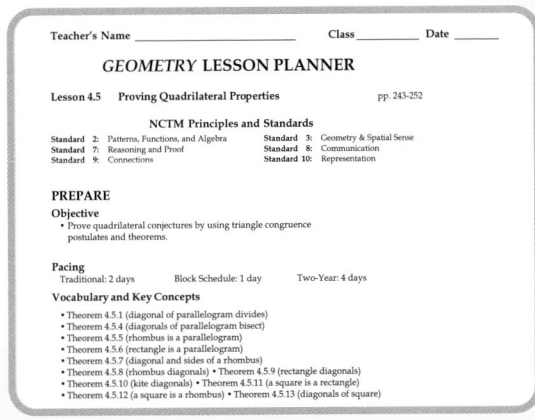

Teacher's Name _____   Class _____ Date _____

GEOMETRY LESSON PLANNER

Lesson 4.5   Proving Quadrilateral Properties   pp. 243–252

NCTM Principles and Standards

Standard 2: Patterns, Functions, and Algebra   Standard 3: Geometry & Spatial Sense
Standard 7: Reasoning and Proof   Standard 8: Communication
Standard 9: Connections   Standard 10: Representation

PREPARE

Objective
- Prove quadrilateral conjectures by using triangle congruence postulates and theorems.

Pacing
Traditional: 2 days   Block Schedule: 1 day   Two-Year: 4 days

Vocabulary and Key Concepts
- Theorem 4.5.1 (diagonal of parallelogram divides)
- Theorem 4.5.4 (diagonals of parallelogram bisect)
- Theorem 4.5.5 (rhombus is a parallelogram)
- Theorem 4.5.6 (rectangle is a parallelogram)
- Theorem 4.5.7 (diagonal and sides of a rhombus)
- Theorem 4.5.8 (rhombus diagonals) • Theorem 4.5.9 (rectangle diagonals)
- Theorem 4.5.10 (kite diagonals) • Theorem 4.5.11 (a square is a rectangle)
- Theorem 4.5.12 (a square is a rhombus) • Theorem 4.5.13 (diagonals of square)

IT'S EASY

For Teachers

HOLT, RINEHART AND WINSTON A9

TEACH WITH OPTIONS

With *Geometry* you can expect full teaching support organized for maximum convenience throughout the wrap. Separation between traditional and alternative options makes for easy selection.

Options include:

❋ **Additional Examples**

❋ **Teaching Tips**

❋ **Critical Thinking references**

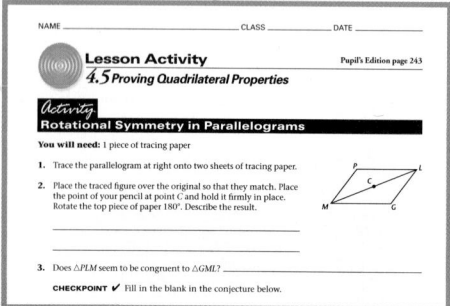

NAME _____ CLASS _____ DATE _____

Lesson Activity   Pupil's Edition page 243

4.5 *Proving Quadrilateral Properties*

Activity
Rotational Symmetry in Parallelograms

You will need: 1 piece of tracing paper

1. Trace the parallelogram at right onto two sheets of tracing paper.
2. Place the traced figure over the original so that they match. Place the point of your pencil at point C and hold it firmly in place. Rotate the top piece of paper 180°. Describe the result.

_____

3. Does △PLM seem to be congruent to △GML? _____

CHECKPOINT ✔ Fill in the blank in the conjecture below.

Additional teaching support includes:

❋ **Applications**

❋ **Math Connections**

❋ **Alternative Teaching Strategies**

❋ **Inclusion Strategies**

❋ **Enrichment**

❋ **Reteaching the Lesson**

❋ **Reduced ancillaries at point-of-use**

Activity **Notes**

In this Activity, students discover and make a conjecture about the rotational symmetry of parallelograms by rotating and tracing a figure.

For a student worksheet of this Activity and detailed Teacher Notes, see page 60 in the Lesson Activities booklet.

CHECKPOINT ✔

3. A diagonal of a parallelogram divides the parallelogram into two congruent triangles.

CRITICAL THINKING

Lines $\overrightarrow{PL}$ and $\overrightarrow{MG}$: ∠PLM and ∠LMG

Lines $\overrightarrow{MP}$ and $\overrightarrow{GL}$: ∠PML and ∠MLG

ADDITIONAL EXAMPLE

Given: parallelogram *PLGM*
Prove: ∠POL ≅ ∠GOM

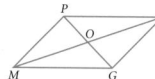

Statements
1. $\overline{PL} \cong \overline{MG}$
2. $\overline{PO} \cong \overline{GO}, \overline{MO} \cong \overline{LO}$
3. △POL ≅ △GOM
4. ∠POL ≅ ∠GOM

Reasons
1. Prop. of parallelogram
2. Def. of bisector
3. SSS
4. CPCTC

CRITICAL THINKING

You can prove that ∠P is congruent to ∠G by CPCTC. To prove that ∠PLG ≅ ∠GMP, use substitution to show that m∠1 + m∠2 = m∠3 + m∠4.

244 LESSON 4.5

CRITICAL THINKING   In the diagram below, the diagonal $\overleftrightarrow{ML}$ is a transversal to two different pairs of parallel lines. Name those parallel lines. List the alternate interior angles for each pair of lines.

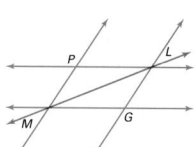

EXAMPLE   **Given:** parallelogram *PLGM* with diagonal $\overline{LM}$.

PROOFS   **Prove:** △LGM ≅ △MPL

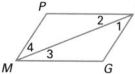

SOLUTION

Proof A (paragraph format):

$\overline{PL}$ and $\overline{GM}$ are parallel, according to the definition of a parallelogram. Therefore, ∠3 and ∠2 are congruent alternate interior angles. Similarly, ∠1 and ∠4 are congruent because $\overline{PM}$ and $\overline{GL}$ are also parallel. Finally, diagonal $\overline{LM}$ is congruent to itself. Thus, two angles and the included side are congruent in △LGM and △MPL. Therefore, the triangles are congruent by the ASA Congruence Postulate.

Proof B (two-column format):

Statements	Reasons
1. Parallelogram *PLGM* has diagonal $\overline{LM}$.	Given
2. $\overline{PL} \parallel \overline{GM}$	Def. of a parallelogram
3. ∠3 ≅ ∠2	∥ lines ⇒ alt. int. angles ≅
4. $\overline{PM} \parallel \overline{GL}$	Def. of a parallelogram
5. ∠1 ≅ ∠4	∥ lines ⇒ alt. int. angles ≅
6. $\overline{LM} \cong \overline{LM}$	Reflexive Prop. of Congruence
7. △LGM ≅ △MPL	ASA Congruence Postulate

This result is stated as a theorem on page 247. In Exercises 24-30, you will be asked to complete a flowchart proof of this theorem.

CRITICAL THINKING   How can you use the result from the Example to prove that opposite angles of a parallelogram are congruent? You will be asked to write this proof in the exercise set.

Interdisciplinary Connection

PHYSICS   Parallelograms are used in physics to show vector addition. For example, the figure at right shows the movement of an object that is being pulled in the direction of $\overrightarrow{AB}$ with a force of 50 pounds and in the direction of $\overrightarrow{AD}$ with a force of 100 pounds. Have students look through their physical science textbooks for examples of these parallelogram diagrams.

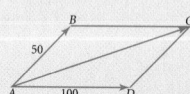

For Teachers

A System of Geometry Knowledge

With the theorems, postulates, and definitions you now know, you can prove all of the conjectures you have made about the properties of parallelograms and the other special quadrilaterals. It is best to start with the simplest ones first. As you progress, you will find that it is often possible to use a previously proven result as part of a proof of a more complicated theorem. In this way, you are building a system of knowledge.

The exercises in this lesson and the next will guide you through a series of proofs of the conjectures about quadrilateral properties that you made in the Activities in Lessons 3.3 and 3.4. The earlier exercises will give you the most guidance, but in the later exercises you will be on your own.

Exercises

● Communicate

1. As you proved in this lesson, a parallelogram has 180° rotational symmetry. Describe all of the types of symmetry of rectangles, rhombuses, and squares.

2. For parallelogram *PQRS*, state all pairs of congruent triangles that are formed by diagonals $\overline{PR}$ and $\overline{QS}$ intersecting at point *X*.

3. Is Theorem 4.5.2 on page 247 true for rectangles? Why or why not?

4. Is Theorem 4.5.3 on page 248 true for rhombuses? Why or why not?

● Guided Skills Practice

Find the indicated measures for parallelogram WXYZ. *(EXAMPLE)*

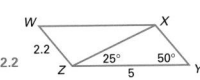

5. m∠*WXZ* 25°   6. m∠*W* 50°   7. *XY* 2.2

APPLICATION

8. **CONSTRUCTION** If a ramp has a rise of more than 6 inches, handrails that are parallel to the ramp are required on both sides. If the upright post at the bottom of the ramp is 36 inches tall, how do you know that the post at the top of the ramp is also 36 inches tall?

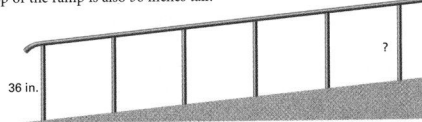

Reteaching the Lesson

COOPERATIVE LEARNING Have students draw the diagonals of an isosceles trapezoid and look for pairs of triangles that appear congruent. Have them work in small groups to write proofs of their conjectures.

LESSON 4.5 **245**

Assess

Selected Answers
Exercises 5–8, 9–71 odd

ASSIGNMENT GUIDE

In Class	1–8
Core	9–23 odd, 24–41
Core Plus	42–66, 67–73 odd
Review	75–79
Preview	80–81

✐ Extra Practice can be found beginning on page 818.

Error Analysis

When students write paragraph proofs such as those in Exercises 61–63, they should first write a conditional such as "If a figure is a square, then it is a rectangle." Point out that the "if" part of the conditional is the given, and the "then" part is what they are to prove.

8. The ramp, handrail, and the two upright posts form a parallelogram. Opposite sides of a parallelogram are congruent.

Interactive teaching options feature:

❋ *Mathepedia*®—an interactive mathematics encyclopedia on CD-ROM!

❋ **go.hrw.com**—our award-winning Web sit that helps you "click" with technology

❋ *Geometry Investigations* CD-ROM—a dynamic addition to the study of geometry!

MATHEPEDIA

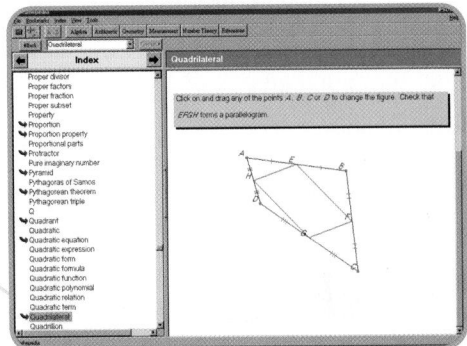

GEOMETRY INVESTIGATIONS

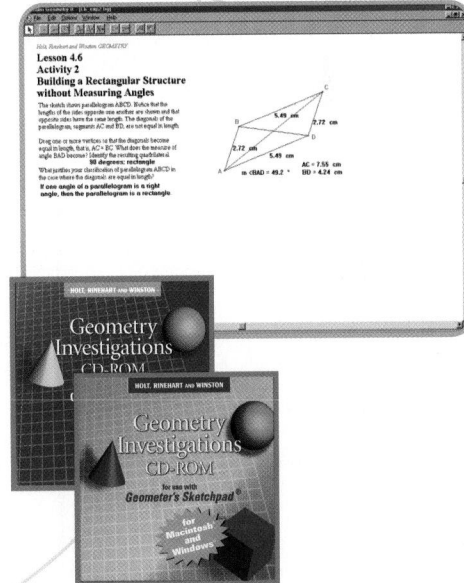

ASSESS
WITH CONFIDENCE

Geometry covers a wealth of potential assessment goals for your classroom. Tools address skill-oriented assessment goals and additional needs that challenge students to solve problems, reason critically, understand concepts, and better communicate mathematically.

Skill-oriented assessment includes:

✳ **Practice and Apply** with attention given to algebra skills

✳ **Leveled practice** with Practice Masters, Levels A, B, C

✳ **Chapter Test**

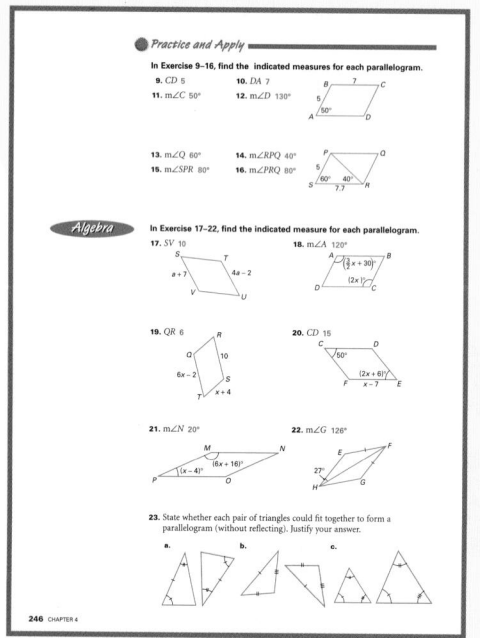

 **Chapter Test**

In the diagram below, *ABCD ≅ EFGH*. Complete the following statements about congruence.

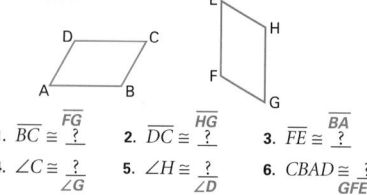

1. $\overline{BC} \cong \underset{FG}{\underline{\ ?\ }}$
2. $\overline{DC} \cong \underset{HG}{\underline{\ ?\ }}$
3. $\overline{FE} \cong \underset{BA}{\underline{\ ?\ }}$
4. $\angle C \cong \underset{\angle G}{\underline{\ ?\ }}$
5. $\angle H \cong \underset{\angle D}{\underline{\ ?\ }}$
6. $CBAD \cong \underset{GFEH}{\underline{\ ?\ }}$

For Exercises 7–12, are the triangles in each pair congruent? State the postulate or theorem that supports your answer.

7. yes; SSS

8. yes; ASA

9. yes; SAS

10. yes; HL

Find each indicated measure.

11. $m\angle Y$ 50°

12. $m\angle Q$ 60°

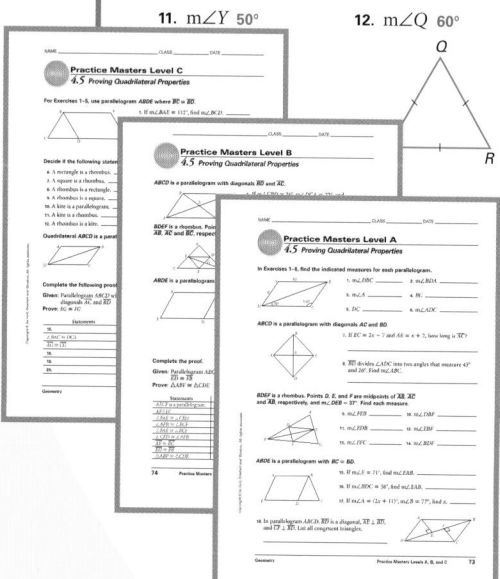

15. SURVEYING A surveyor needs to measure the distance across a field from point *J* to point *K*. What is the distance? Justify your answer.

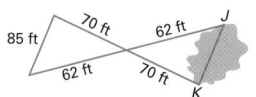
85 ft, 70 ft, 62 ft, 62 ft, 70 ft

Find the indicated measure for each parallelogram.

16. *RU* 44

17. $m\angle XYZ$ 110°

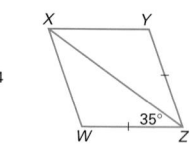
$2x + 10$, $4x - 24$, 15

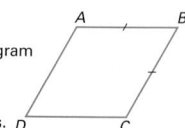
35°

Given: *ABCD* is a parallelogram and $\overline{AB} \cong \overline{BC}$

Prove: *ABCD* is a rhombus.

Statements	Reasons
ABCD is a parallelogram.	Given
$\overline{AB} \cong \overline{CD}$, $\overline{AD} \cong \overline{BC}$	18. __?__
$\overline{AB} \cong \overline{BC}$	19. __?__
$\overline{AB} \cong \overline{BC} \cong \overline{CD} \cong \overline{AD}$	20. __?__
ABCD is a rhombus.	21. __?__

Construct each of the following with a compass and straightedge.

22. a trapezoid

23. a rhombus

Is each of the following triangles possible? If not, why not?

24. $AB = 17, BC = 20, AC = 22$ yes

25. $DE = 25, EF = 10, DF = 40$ no; $DE + EF < DF$

26. $GH = 9, HI = 14, GI = 9$ yes

27. $XY = 13, YZ = 11, XZ = 24$ no; $XY + YZ = XZ$

College Entrance Exam Practice

MULTIPLE-CHOICE For Questions 1–8, write the letter that indicates the best answer.

1. What is the slope of $\overline{DE}$ with endpoints $D(-7, 7)$ and $E(-3, -3)$? **(LESSON 3.8)**

a. $-\dfrac{2}{5}$

b. $-\dfrac{5}{2}$

c. $\dfrac{5}{2}$

d. -1

2. Refer to the diagrams below. Find the values of x and y. **(LESSON 4.1)**

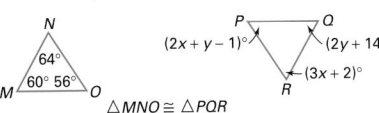

$\triangle MNO \cong \triangle PQR$

a. $x = 20\dfrac{2}{3}, y = 19\dfrac{2}{3}$

b. $x = 18, y = 25$

c. $x = 19\dfrac{1}{3}, y = 18\dfrac{1}{3}$

d. $x = 20\dfrac{2}{3}, y = 23$

3. Which congruence statement indicates that the two triangles are congruent? **(LESSON 4.3)**

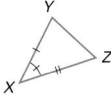

a. $\triangle XYZ \cong \triangle TVU$, SAS

b. $\triangle XZY \cong \triangle TVU$, HL

c. $\triangle XZY \cong \triangle TVU$, SSA

d. $\triangle XYZ \cong \triangle TUV$, SAS

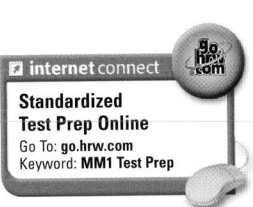

4. Which pair of points defines a line perpendicular to $\overline{MN}$? **(LESSON 3.8)**

a. $(0, 7), (8, -4)$

b. $(4, -7), (-4, 4)$

c. $(-7, 0), (4, 8)$

d. $(7, -4), (-4, 4)$

5. What is the measure of $\angle ACB$? **(LESSON 4.4)**

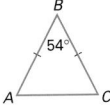

a. $42°$

b. $126°$

c. $63°$

d. cannot be determined

6. What is the unknown angle measure? **(LESSON 3.6)**

a. $120°$

b. $180°$

c. $130°$

d. $100°$

7. Which postulate or theorem justifies the congruence statement $\triangle QDA \cong \triangle UAD$? **(LESSONS 4.2 AND 4.3)**

a. SSA

b. SSS

c. HL

d. SAS

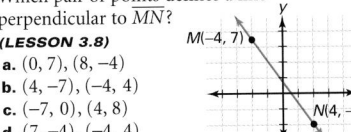

internet connect

Standardized Test Prep Online

Go To: **go.hrw.com**
Keyword: **MM1 Test Prep**

Alternative Assessment

The following suggest alternative assessments for students who may benefit from a different type of assessment than the regular chapter quizzes and the mid-chapter/end-of-chapter test. Visit the HRW web site to get additional Alternative Assessment material.

internet connect

Alternative Assessment
Go To: **go.hrw.com**
Keyword: MG1 Alt Assess

Performance Assessment

1. Given: $\triangle ACX$ and $\triangle BDX$;
$\overline{AC} \cong \overline{DB}$;
$\overline{AC}$ is parallel to $\overline{DB}$.

Prove: $\triangle ACX \cong \triangle BDX$

2. Given: Isosceles trapezoid $ABCD$

Prove: $\triangle AXD \cong \triangle BXC$

3. Prove the following conjecture: The diagonals of a rectangle are congruent.

4. Prove the following conjecture: The base angles of an isosceles triangle are congruent.

Portfolio Project

Suggest that students choose one of the following projects for inclusion in their portfolios.

1. Using geometry software, draw an irregular quadrilateral. Show that joining the midpoints of the sides always forms a parallelogram in the interior of the quadrilateral. Prove that this is true using the triangle midsegment theorem.

2. Create three irregular figures that tessellate. Use a different method for forming each one: rotations, translations, and/or reflections.

3. Draw a parallelogram. Using a compass and protractor to perform a construction, translate the parallelogram 2 inches to the left on your paper. Draw your translation vector on your paper.

internet connect

The table below identifies the pages in this chapter that contain internet and technology information.

Content Links	
Activities Online	pages 213, 257, 264
Portfolio Extensions	pages 225, 252
Homework Help Online	pages 213, 222, 232, 240, 248, 258, 269, 276

Resource Links	
Parents can go online and find concepts that students are learning–lesson by lesson–and questions that pertain to each lesson, which facilitate parent-student discussion.	

Go To: **go.hrw.com**
Keyword: MG1 Parent Guide

Technical Support

The following may be used to obtain technical support for any HRW software product.

Online Help: **www.hrwtechsupport.com**
e-mail: **tschrw@hbtechsupport.com**

HRW Technical Support Center: (800)323-9239
7 AM to 10 PM Monday through Friday CST

Visit the HRW math web site at: **www.hrw.com/math**

208C CHAPTER 4 INTERLEAF

Alternative assessment includes:

* **College Entrance Exam Practice**

* **Performance Assessment**

* **Portfolio Project**

* **Internet Connect with Portfolio Links**

* *One-Stop Planner CD-ROM with Dynamic Test Generator*

TEST GENERATOR

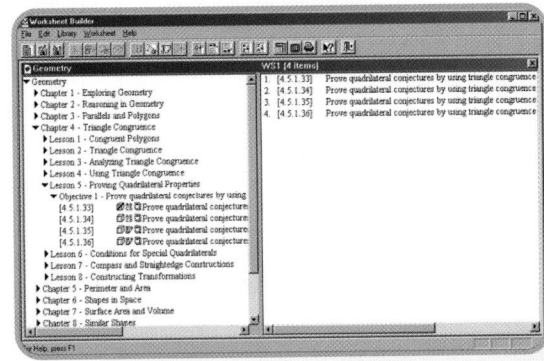

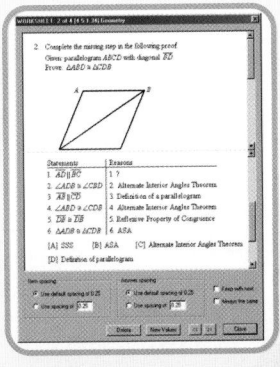

For Teachers

HOLT, RINEHART AND WINSTON A13

TEACHING RESOURCES

OVERHEAD TRANSPARENCIES

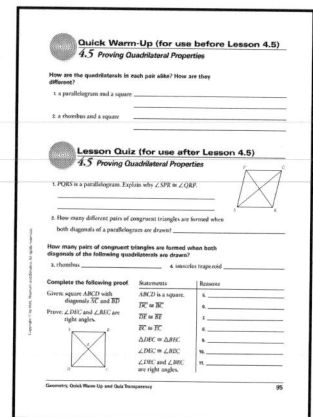

Quiz Transparencies

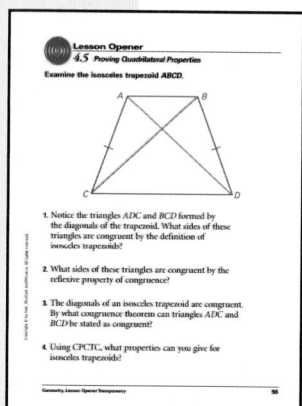

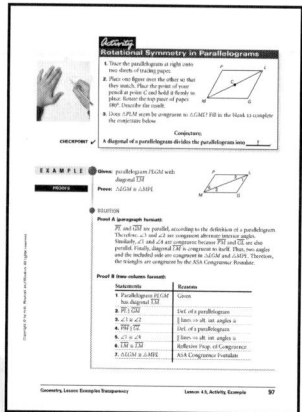

Lesson Presentation Transparencies

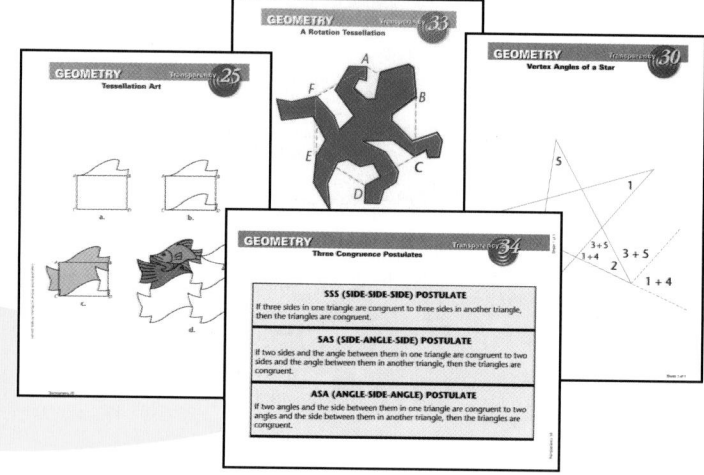

Teaching Transparencies

PRACTICE & ASSESSMENT

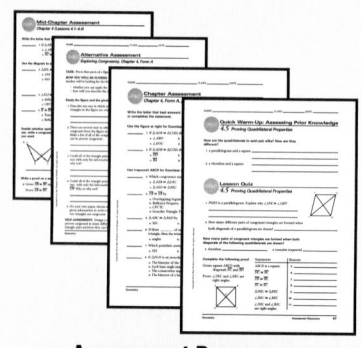

Assessment Resources
Quick Warm-Up, Lesson Quiz, Mid-Chapter Assessment,
Chapter Assessments A & B, Alternative Assessments A & B

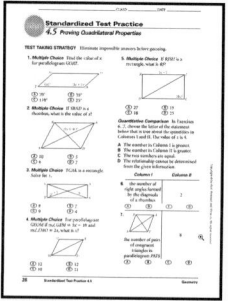

Standardized Test Practice

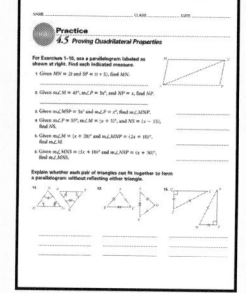

Test and Practice Generator Item Listing

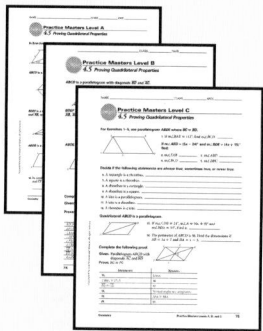

Practice Masters Levels A, B, C

Teaching Resources

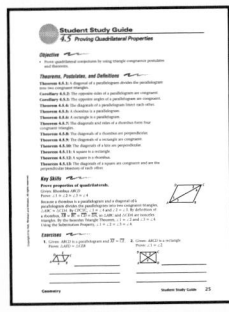

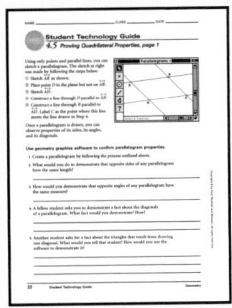

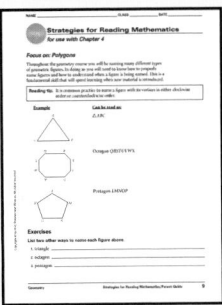

- **Homework Help Online**
- **Standardized Test Prep Online**

Student Study Guide

Student Technology Guide

Spanish Resources

Building Success in Mathematics

ACTIVITIES & EXTENSIONS

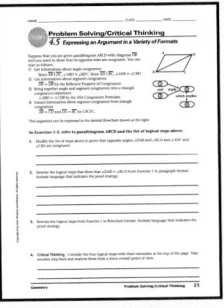

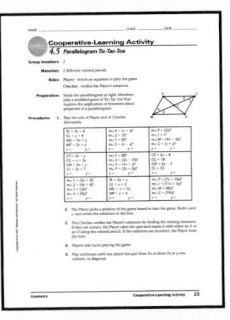

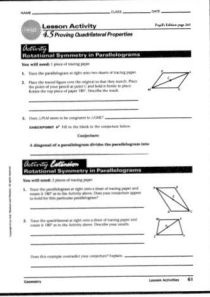

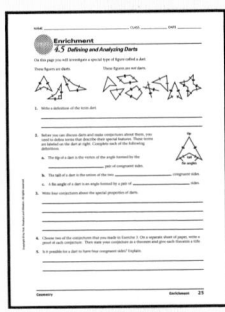

Problem Solving/ Critical Thinking Masters

Cooperative-Learning Activities

Lesson Activities

Enrichment Masters

Long-Term Projects

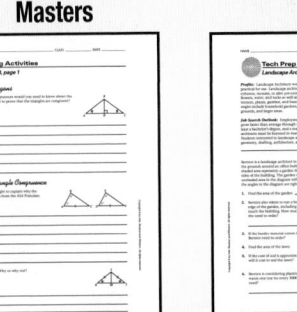

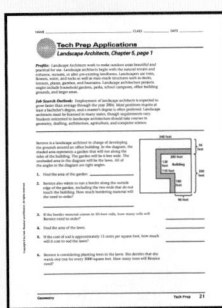

Writing Activities for Your Portfolio

Tech Prep Masters

TECHNOLOGY

Geometry One-Stop Planner CD-ROM

Mathepedia CD-ROM

Dynamic Test Generator

Lesson Presentations on CD-ROM

Geometry Investigations CD-ROM

Activities Online

TEACHER'S TOOLS

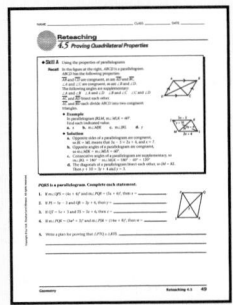

HRW Technology Handbook

Block-Scheduling Handbook

Reteaching Masters

Teaching Resources

HOLT, RINEHART AND WINSTON A15

NOW YOU'RE READY

Integrate technology with **Internet Connect**— a unique resource that helps you to enrich and expand lessons and strengthen students' computer skills. In-text references lead you and your students to **go.hrw.com** where resources help you review, extend, apply, and assess learning.

NOW YOU'RE SET

Teachers, students, and parents can benefit from these exceptional resources— all available online:

* Portfolio Extension
* Alternative Assessment
* Homework Help Online
* Activities Online
* Standardized Test Prep Online

go.hrw.com also features:

* Calculator Keystroke Guides
* Parent Guides
* State-specific Resources

GO FARTHER WITH

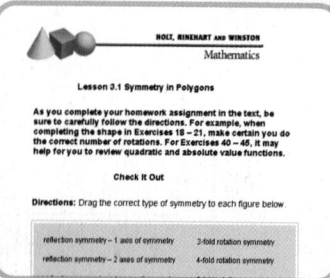

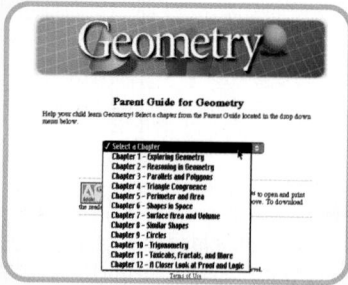

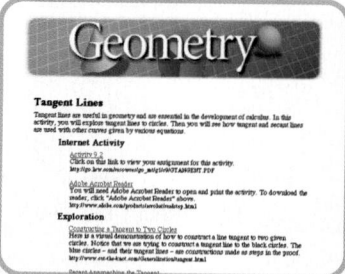

Internet

ANNOTATED TEACHER'S EDITION

Geometry

HOLT, RINEHART AND WINSTON

A Harcourt Education Company

Austin • Orlando • Chicago • New York • Toronto • London • San Diego

For permission to reprint copyrighted material, grateful acknowledgment is made to the following sources:

Associated Press: From "Astronomers Dispute NASA Gauge of Universe's Age," June 2, 1999. Copyright © 1999 by Associated Press.

Discover Magazine: From "Egg Over Alberta" by Paul Hoffman, photos by Annette Del Zoppo, from Discover, May 1988. Text copyright © 1988 by Paul Hoffman. Photos copyright © 1988 by Annette Del Zoppo.

Dover Publications, Inc.: "6 Accomplishments" and "31 Digits Are Symbols" from *My Best Puzzles in Logic and Reasoning* by Hubert Phillips. Copyright © 1961 by Dover Publications, Inc.

The Griffith Institute, Ashmolean Museum, Oxford: Adapted excerpt from Howard Carter's Notebook.

National Council of Teachers of Mathematics: Adapted from "Activities: Spatial Visualization" by Glenda Lappan, Elizabeth A. Phillips, and Mary Jean Winter from *Mathematics Teacher*, vol. 77, no. 8, November 1984. Copyright © 1984 by the National Council of Teachers of Mathematics.

The New York Times Company: From "Math Problem, Long Baffling, Slowly Yields" by Gina Kolata and map, "The Efficient Traveling Salesman," from *The New York Times*, March 12, 1991. Copyright © 1991 by The New York Times Company.

Oxford University Press, London: Figures 2.5 and 3.7 from *Chinese Mathematics: A Concise History* by Lǐ Yǎn and Dù Shirán, translated by John N. Crossley and Anthony W.-C. Lun. Translation copyright © 1987 by John N. Crossley and Anthony W.-C. Lun.

Greg Stec: From "Message of the Maya in Modern Translation" by Greg Stec from *The Christian Science Monitor*, June 22, 1989. Copyright © 1989 by Greg Stec.

Printed in the United States of America

ISBN 0-03-070056-6

1 2 3 4 5 6 7 048 07 06 05 04 03

James E. Schultz, *Senior Series Author*

Dr. Schultz has over 30 years of experience teaching at the high school and college levels and is the Robert L. Morton Professor of Mathematics Education at Ohio University. He helped to establish standards for mathematics instruction as a co-author of the *NCTM Curriculum and Evaluation Standards for School Mathematics* and A Core Curriculum: *Making Mathematics Count for Everyone.*

Kathleen A. Hollowell, *Senior Author*

Dr. Hollowell is an experienced high school mathematics and computer science teacher who currently serves as Director of the Mathematics & Science Education Resource Center, University of Delaware. Dr. Hollowell is particularly well versed in the special challenge of motivating students and making the classroom a more dynamic place to learn.

Wade Ellis, Jr.

Professor Ellis has co-authored numerous books and articles on how to integrate technology realistically and meaningfully into the mathematics curriculum. He was a key contributor to the landmark study *Everybody Counts: A Report to the Nation on the Future of Mathematics Education.*

Paul A. Kennedy

A professor in the Department of Mathematics at Southwest Texas State University, Dr. Kennedy is a leader in mathematics education reform. His research focuses on developing algebraic thinking by using multiple representations and technology. He has been the author of numerous publications and he is often invited to speak and conduct workshops on the teaching of secondary mathematics.

CONTRIBUTING AUTHORS

Martin Engelbrecht

A mathematics teacher at Culver Academies, Culver, Indiana, Mr. Engelbrecht also teaches statistics at Purdue University, North Central. An innovative teacher and writer, he integrates applied mathematics with technology to make mathematics accessible to all students.

Kenneth Rutkowski

A mathematics teacher at St. Stephen's Episcopal School, Austin, Texas, Mr. Rutkowski is an innovative geometry teacher. He sponsors his school's mathematics honor society, serves on various professional committees, and conducts creative teacher workshops.

Table of Contents

1 EXPLORING GEOMETRY 2

MATH CONNECTIONS

Algebra 15, 16, 20, 21, 22, 23, 32, 48, 65, 66, 75

APPLICATIONS

Science
Archaeology 44
Computer Programming 67
Engineering 74
Geology 24

Language Arts
Communicate 13, 21, 29, 39, 45, 54, 63

Business and Economics
Construction 48

Life Skills
Navigation 20, 21, 33, 60
Travel 74

Sports and Leisure
Hobbies 14
Scuba Diving 33

REASONING IN GEOMETRY

MATH CONNECTIONS

Algebra 81, 83, 84, 85, 87, 107, 108, 110, 112, 113, 122, 124

Patterns in Data 98

APPLICATIONS

Science
Biology 90, 103, 105, 130
Botany 98
Chemistry 103
Computer Programming 130
Geology 103
Weather 94

Language Arts
Communicate 82, 94, 102, 111, 120

Life Skills
Navigation 116

Sports and Leisure
Entertainment 120
Fine Arts 97
Games 86
Hobbies 111
Humor 96
Music 97
Photography 115

PARALLELS AND POLYGONS

MATH CONNECTIONS

Algebra 145, 152, 160, 174, 181, 187, 191, 192, 195

Technology 174

APPLICATIONS

Science
Agriculture 142, 143
Archaeology 185, 186
Architecture 159
Biology 180, 189
Botany 141
Cartography 176
Civil Engineering 166
Engineering 188, 204
Gemology 182

Language Arts
Communicate 142, 151, 159, 164, 173, 180, 186, 193

Social Studies
Geography 146, 172

Life Skills
Carpentry 161, 167
Construction 195, 196
Drafting 166
Furniture 145
Navigation 161, 167, 173, 175
Traffic Safety 146
Transportation 182

Sports and Leisure
Art 204
Crafts 204
Gardening 153
Music 189
Optical Illusions 151
Painting 188
Quilting 147
Recreation 145
Sports 176, 184, 186

MATH CONNECTIONS

Algebra 215, 234, 238, 240, 246, 278   **Coordinate Geometry** 216, 242

APPLICATIONS

Science
Archaeology 233
Astronomy 229, 230
Engineering 220, 233, 243
Forestry 226, 230

Language Arts
Communicate 213, 220, 230, 238, 245, 257, 264, 275

Life Skills
Carpentry 215, 222, 255
Construction 223, 238, 239, 245, 260
Design 279, 288
Fashion 216
Navigation 233, 269
Road Signs 242
Surveying 242, 288

Sports and Leisure
Art 251, 260
Marching Band 279
Quilting 215, 223
Recreation 223
Sports 234, 251

MATH CONNECTIONS

Algebra 297, 299, 309, 317, 319, 326, 327, 329, 331, 332, 333, 343, 344
Coordinate Geometry 309
Maximum/Minimum 310, 352
Technology 311, 359
Number Theory 327
Probability 354, 357

APPLICATIONS

Science
Agriculture 297, 301, 325
Astronomy 319
Automobile Engineering 319
Aviation 298
Civil Engineering 337, 345, 352
Environmental Protection 345
Farming 311
Irrigation 318
Meteorology 347, 358
Solar Energy 300

Language Arts
Communicate 298, 307, 317, 326, 335, 342, 349, 356

Life Skills
Construction 300
Drafting 337
Home Improvement 302
Landscaping 301, 306
Meal Planning 318
Public Safety 329
Safety 323, 366
Travel 329
Transportation 358

Sports and Leisure
Gardening 297
Recreation 298, 359
Skydiving 358
Sports 329, 366
Theater Arts 304

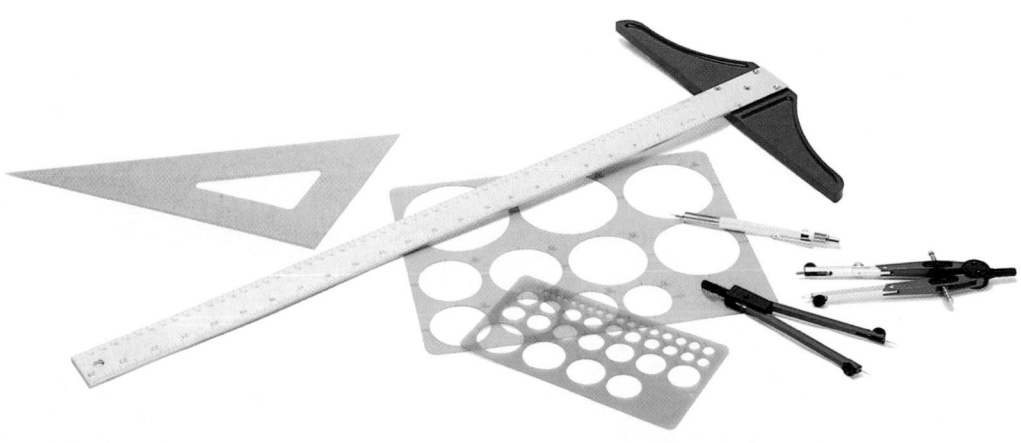

SHAPES IN SPACE

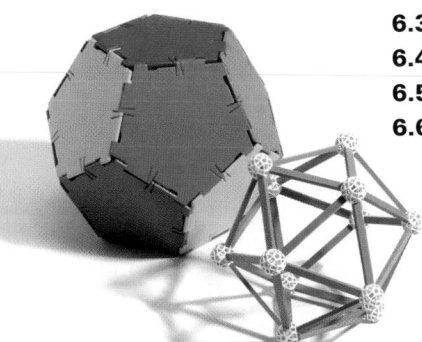

MATH CONNECTIONS

Algebra 377, 393, 400, 401, 403, 405, 406, 408

Coordinate Geometry 394, 407
Technology 425

APPLICATIONS

Science
Archaeology 378
Architecture 377
Astronomy 401
Chemistry 383, 390, 391, 394
Geology 424
Optics 394
Physical Science 401

Language Arts
Communicate 375, 383, 391, 399, 405, 413

Business and Economics
Packaging 393
Fund Raising 424

Life Skills
Design 424
Graphics 400
Hobbies 407
House Painting 406
Mechanical Drawing 372
Navigation 385, 407

Sports and Leisure
Recreation 377

MATH CONNECTIONS

Algebra 431, 434, 442, 444, 450, 456, 457, 459, 475

Coordinate Geometry 444

Maximum/Minimum 432, 434, 457, 466, 468, 483

Technology 466

APPLICATIONS

Science
Archaeology 448
Architecture 444
Biology 430, 433, 435
Botany 435
Cartography 471
Chemistry 435
Engineering 453, 455
Geology 460, 463
Marine Biology 458
Optics 484
Physics 492
Physiology 435

Language Arts
Communicate 433, 441, 448, 456, 464, 473, 482

Social Studies
Geography 475

Business and Economics
Coins 454, 456
Manufacturing 433, 434, 443, 458, 467
Printing 492
Product Packaging 458
Small Business 467

Life Skills
Carpentry 459
Construction 449, 451
Cooking 435
Food 475
Metalwork 475

Pottery 483
Public Health 458
Transportation 484

Sports and Leisure
Aquariums 439
Hobbies 434, 484
Hot-Air Ballooning 469, 473
Recreation 443, 444, 467
Scale Models 468
Sports 474, 475, 492

MATH CONNECTIONS

Algebra 499, 504, 513, 516, 524, 529, 551   **Coordinate Geometry** 506, 530, 532

APPLICATIONS

Science
Architecture 510, 511
Astronomy 501, 549
Biology 531, 546
Earth Science 505
Engineering 505, 532, 533
Optics 505, 541
Paleontology 540
Wildlife Management 514

Language Arts
Communicate 502, 511, 520,
   528, 536, 547

**Business and
Economics**
Business 550
Packaging 550
Storage 550

Life Skills
Food 550
Graphic Design 523
Interior Decorating 515
Landscaping 515
Photography 498

Sports and Leisure
Fine Art 515
Music 531
Quilting 504
Sports 550

Other
Indirect Measurement 516

CIRCLES

MATH CONNECTIONS

Algebra 570, 575, 578, 579, 594, 595, 597, 601, 602, 604, 606, 607, 609, 610, 611, 614, 615, 617

Coordinate Geometry 612
Technology 611, 613
Trigonometry 572

APPLICATIONS

Science
Agriculture 624
Archaeology 597
Astronomy 616
Communications 578, 597, 624
Cartography 587
Civil Engineering 571
Engineering 579, 608
Geology 600, 604
Lunar Exploration 608
Optics 583
Space Flight 578
Wildlife Management 603

Social Studies
Demographics 571
Language Arts
Communicate 569, 576, 585, 592, 605, 613

Business and Economics
Package Design 608

Life Skills
Carpentry 580, 584, 586
Computer Graphics 616
Design 578, 597

Landscaping 571
Navigation 588, 591, 593, 603, 624
Photography 585
Stained Glass 586
Structural Design 617
Surveying 624

Sports and Leisure
Games 567
Sports 571

TRIGONOMETRY

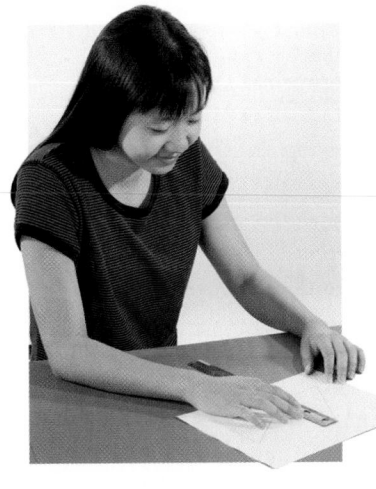

MATH CONNECTIONS

Algebra 635, 645, 667, 679, 684　　　　**Polar Coordinates** 662
Coordinate Geometry 669, 679, 685　　**Trigonometry** 669, 685
Technology 644

APPLICATIONS

Science
Agriculture 673
Architecture 661, 684
Astronomy 641, 652, 653, 692
Aviation 678
Engineering 636, 692
Forestry 645, 661
Indirect Measurement 636, 668, 692

Physics 673
Rotary Motion 649
Wildlife Management 660

Language Arts
Communicate 634, 643, 651, 657, 666, 676, 683

Life Skills
Construction 645
Landscaping 668

Navigation 657, 668, 679, 692
Public Safety 685
Surveying 636, 661

Sports and Leisure
Recreation 640, 645, 673
Scuba Diving 678
Sports 668, 675, 676

TAXICABS, FRACTALS, AND MORE 696

MATH CONNECTIONS

Algebra 700, 704, 710, 727   **Trigonometry** 719
Coordinate Geometry 705   **Knot Theory** 728

APPLICATIONS

Science
Architecture 762
Genetics 721
Telecommunications 741
Wildlife Management 719

Language Arts
Communicate 702, 709, 716,
  724, 733, 742, 752

Social Studies
Geography 739, 762

**Business and
Economics**
Fund Raising 719
Manufacturing 727
Marketing 705

Life Skills
Commuting 710
Food 728
Law Enforcement 718
Photography 705
Public Safety 710

Sports and Leisure
Fine Arts 705
Hobbies 744

MATH CONNECTIONS

Algebra 796, 804   Number Theory 795

APPLICATIONS

Science
Computer Architecture 798
Computer Databases 781
Electricity 803
Environmental Science 804
Physics 788
Telecommunications 796

Language Arts
Communicate 772, 779, 787, 794, 801

Business and Economics
Advertising 780
Criminal Law 784
Law 795

Life Skills
Academics 790
Landscaping 774

Sports and Leisure
Sports 774

INFO BANK TABLE OF CONTENTS 817

Bringing Math *to Life*

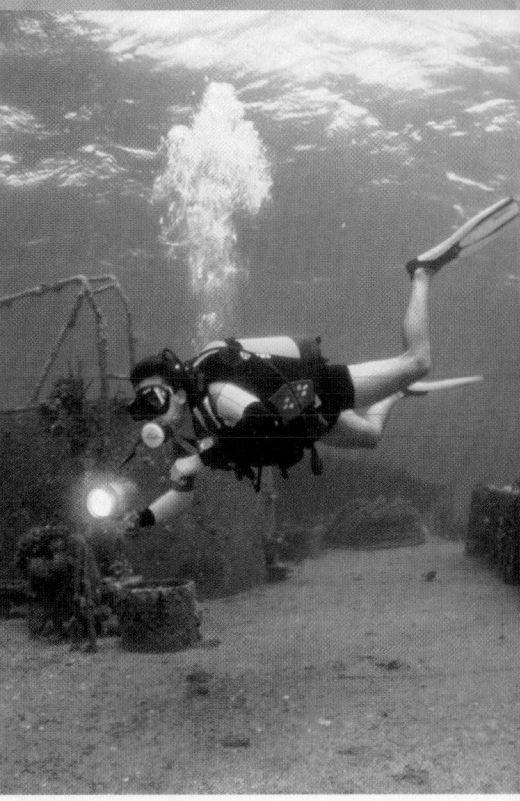

Our lives are touched daily by mathematics. Computers, with circuitry and software that applies mathematical logic, control many of the dimensions of the infrastructure around us. Traffic lights, power grids, and automated teller machines are among the many things that depend on computer-based mathematical logic structures to work properly.

The role of drill and practice in mathematics has long been a debated topic. Evidence of the drill-and-practice approach to teaching mathematics goes back to papyri from ancient Egypt and even further to clay tablets from ancient Babylon. The drill-and-practice approach may be valid at certain points in the instructional sequence, but when used as the primary or sole means of instruction, it decontextualizes the mathematics and invites the impression that mathematics is an abstraction that has no practical value. Most teachers agree that a combination of drill and practice and an instructional context consisting of applications, connections, and problem solving is the best way to teach mathematics.

APPLICATIONS

Applications provide evidence that mathematics is important in our world and that the geometry learned in the classroom today does have practical uses. Throughout Holt, Rinehart and Winston's *Geometry*, students are introduced to concepts through real-world applications. These applications provide connections to domains of human activity such as business, consumer economics, science, life skills, and leisure activities.

Some applications are very close to students' daily lives. For example, in Lesson 5.3, students are asked to compare the value of two different-sized pizzas with different prices.

Other applications suggest exciting possibilities for leisure activity or adventure, such as airplane piloting or scuba diving. In Lesson 10.6, students

determine the course of a student pilot soloing from Austin to Llano, Texas.

Students may also recognize the power of mathematics

to increase their number of options in career choices. For example, in Lesson 3.6, students compute the angles used in cutting diamonds.

Mathematics is the main tool of science, providing tools to quantify, verify, and describe results and theories In Lesson 8.6, students investigate how a change in linear dimensions can affect the volume of an object, and how this explains the limitations in size of various animals .

CONNECTIONS

Holt, Rinehart and Winston's *Geometry* emphasizes the connections of geometry with other mathematical

disciplines, in particular with algebra. Students are given many opportunities to hone their algebra skills throughout *Geometry*, by solving linear, quadratic, radical, and systems of equations, proportions, and by studying the properties of equality. Places where algebra skills are used in the course of geometry instructions are clearly marked by the following icon, which you will see throughout this book:

Students use algebra to understand geometric concepts, as in coordinate proofs and derivations of formulas. Students also use geometry to understand algebraic concepts, such as the sum of a geometric series.

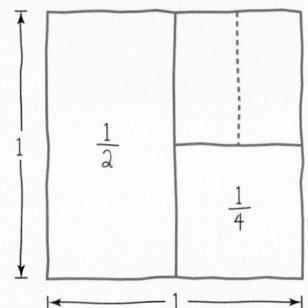

In Chapter 11, students are introduced to other mathematical disciplines that are connected with geometry, such as topology, graph theory, and fractals.

Activities

These words of the philosopher and mathematician Bertrand Russell apply remarkably well to the pedagogy of Holt, Rinehart and Winston's *Geometry*. After a very few preliminaries, students begin to explore geometric figures by using paper folding to construct segment and angle bisectors. When applied to triangles, these paper folding techniques lead to the observation, spoken of by Bertrand Russell above, of "three or more lines that meet in a point"—a surprising and attractive result.

By working cooperatively and observing figures created by their classmates, students realize that a given result may be true, not just for an isolated case, but for any figure they might create. If geometry graphics software is available, students can make the same observation dynamically by dragging a figure through an infinite number of shapes—always with the same result. Holt, Rinehart and Winston's

Geometry supports the use of geometry graphic software by suggesting optional technology approaches to Activities, when appropriate.

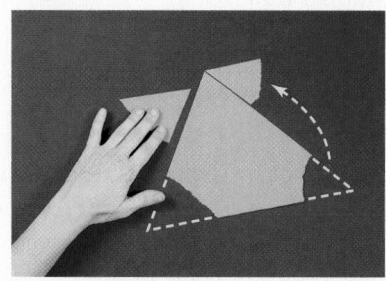

Understanding and retention of theorems and formulas are motivated by actual student experiences in the Activities and, in many cases, by student conjectures arising out of them. In Lesson 3.5, for example, the Triangle Sum Theorem is first demonstrated by having students tear off corners of a triangle and reassemble them to form a straight line.

This *physical* demonstration of the Triangle Sum Theorem is immediately related to a line drawing, and students fill in a table by using postulates and theorems they have previously discovered about parallel lines and transversals.

Finally, the procedure is given its most *abstract* presentation as a proof, still closely tied to the students' experience of tearing the corners off of a triangle.

Proving the Triangle Sum Theorem

To prove the theorem, begin by drawing line ℓ through a vertex of the triangle so that it is parallel to the side opposite the vertex.

TWO-COLUMN PROOF

Given: $\triangle ABC$

Prove: $m\angle 1 + m\angle 2 + m\angle 3 = 180°$

Plan: Study the illustration, which is related to the Activity in which you tore off the corners of a triangle. You can use what you discovered in the Activity to write a two-column proof of the Triangle Sum Theorem.

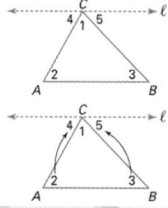

Proof:

Statements	Reasons
1. $\ell \parallel \overline{AB}$	As drawn (justification: the Parallel Postulate)
2. $m\angle 1 + m\angle 4 + m\angle 5 = 180°$	The angles fit together to form a straight line.
3. $\angle 2 \cong \angle 4$ ($m\angle 2 = m\angle 4$) $\angle 3 \cong \angle 5$ ($m\angle 3 = m\angle 5$)	Alternate Interior Angles Theorem
4. $m\angle 1 + m\angle 2 + m\angle 3 = 180°$	Substitution in Step 2

An increasing number of educators are realizing that more learning takes place when students construct knowledge for themselves. Instead of presenting students with rules, theorems, principles, and worked-out examples, this approach calls for students to be given questions to investigate, problems to explore, and conjectures to verify or disprove. A teacher is no longer the source of all information. Instead, he or she acts as a facilitator, presenting questions to explore and pointing to areas that need further discussion and clarification.

STUDENT OWNERSHIP

It may be hard, especially today, to justify the teaching of some of the traditional theorems of geometry, such as theorems about segments and angles related to circles. (Indeed, some teachers may choose to de-emphasize them in favor of theorems more obviously applicable to the real world.) But when students construct their own knowledge of these traditional theorems, the result is personal ownership—and consequent *pride* of ownership. In fact, a number of teachers have reported that the material on circles in Chapter 9 provided some of the most exciting and valuable experiences of their geometry courses. The thrill of genuine intellectual discovery, and the realization, perhaps for the first time, of personal creative potential within oneself are benefits that will extend far beyond a student's classroom encounter with geometry.

CHOOSING THE DIRECT APPROACH

In many classroom situations, a key teacher decision will be when to use a discovery approach and when to use direct instruction. With Holt, Rinehart and Winston's *Geometry*, both approaches are available to the teacher. Through the use of examples and lists of theorems and postulates provided at the end of each chapter and in the back matter of this book, a teacher can choose to present materials in a more traditional way.

VAN HIELE'S THEORIES OF THE LEVELS OF MATHEMATICAL LEARNING

If one goal of mathematics teaching is to lead students to ever higher levels of abstraction in their thinking, as was suggested by Van Hiele in his groundbreaking work *Structure and Insight*, it is certainly true that students do not start with abstract thinking. Rarely will a student be instinctively drawn to abstract results and demonstrations without prior grounding in concrete experiences. As Van Hiele's research showed, there must be a progression from one level of thinking to another. Holt, Rinehart and Winston's *Geometry* embodies this progression.

Holt, Rinehart and Winston's *Geometry* begins with purely visual and descriptive materials (Van Hiele Levels 1 and 2) in the introductory lessons, often calling upon students' earlier life experiences of the subject matter. When proofs are introduced (Van Hiele Level 3), they remain connected as closely as possible to student experiences.

Logic is touched upon briefly in Chapter 2, as a prelude to the idea of logical demonstration. The presentation of logic is rooted in concrete realities with which students can be comfortable. In Chapter 12, after having gained considerable experience with geometry proofs, students are taken further into the field of formal logic (Van Hiele Level 4), where they learn of formal arguments, truth tables, smart machines, and computer logic gates. The highest level of abstract thinking— the contemplation of the nature of logical laws (Van Hiele Level 5)—is achieved by relatively few people and is beyond the scope of this course. But the material presented here will provide students food for thought for the rest of their lives.

Proof in Holt, Rinehart and Winston's *Geometry*

In simplest language, a proof is a convincing argument that uses logical reasoning to prove that something is true. To the professional mathematician, proof may be the very "heart and soul" of mathematics. But why should the average person be concerned with proof? In fact, a mathematical proof is more than a guarantee that a certain result is true. Much more importantly, *a proof is a way of understanding why something is true.* Proofs give students an understanding of the theorems and formulas they learn to use.

Reasoning and proofs are themes that are infused throughout Holt, Rinehart and Winston's *Geometry*, with the concept of a proof being developed gradually. In the first three chapters, students begin by exploring figures, creating definitions, and looking for geometric relationships. As students formulate different conjectures, a genuine need for proof arises.

FORMAL PROOFS—AND MORE

In geometry, the expression "formal proof" usually refers to a proof that is presented in the traditional two-column or statement-reason form. *Every* statement in the argument of a formal proof is given a separate justification in terms of a system of axioms, rules, definitions, given information, and previously established results.

In the past, formal two-column proofs were used almost exclusively in the geometry classroom. Today, however, there is a growing realization among educators that exclusive use of formal proofs can be tedious and frustrating to the student, as well as limit his or her creativity. For this reason, Holt, Rinehart and Winston's *Geometry* offers students opportunities to experience several styles of proof.

INFORMAL PROOFS

Proof in Holt, Rinehart and Winston's *Geometry* is presented gradually, beginning in Chapter 2 with informal proofs. To whet the student's appetite for the kind of understanding that is made possible by proofs, the first examples of proofs the students see are amusing, elegant—and entirely informal. For example, a proof is given of the fact that a chessboard can not be completely covered by 31 dominoes, two squares at a time, if two opposite corner squares are removed. In the proof, there is no systematic listing of statements and no appeal to an axiomatic system.

In some cases, a proof may even be a "proof without words," as in the classical example of the proof of the algebraic formula for the sum of the first *n* odd counting numbers.

PROBLEM SOLVING

Part II

How could you find the sum of the first *n* odd counting numbers without actually adding them? **Make a table** like the one below, and see if you can discover the answer.

n	First n odd numbers	Sum of the first n odd numbers
1	1	1
2	1, 3	4
3	1, 3, 5	9
4	1, 3, 5, 7	16
5	?	?
6	?	?
n	?	?

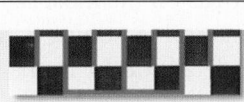

The pathway through the maze suggests a proof.

CHECKPOINT ✔ The diagram to the right of the table is a "proof without words" of the algebraic result that you may have discovered. Explain in your own words how the diagram proves the result.

Part III

You are given the figure at right, which is built entirely of squares. The area of square *C* is 64, and the area of square *D* is 81.

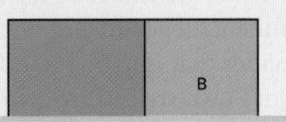

ALGEBRAIC AND TWO-COLUMN PROOFS

Formal proofs have their place within axiomatic mathematical systems, such as Euclid's system of axioms, definitions, and common notions. In Holt, Rinehart and Winston's *Geometry*, formal proofs are introduced in terms of the assumptions and rules of algebra, with which the students are already familiar.

It is an easy step from such algebraic reasoning to simple two-column proofs like the one at right. Two-column proofs establish the mental discipline of ensuring that all bases are covered by an argument so that there will be no logical loopholes that would render the argument invalid.

When students are first exposed to proofs in Holt, Rinehart and Winston's *Geometry*, they are not required to complete whole proofs on their own. Instead, they are asked to fill in the blanks of proofs that are partially completed for them. This gives beginning students help in getting started. Another benefit of this approach is that students are led to focus on the most significant mathematical reasoning within the proofs—not the details of the form.

PARAGRAPH PROOFS

Paragraph proofs are introduced at the same time as two-column proofs. Paragraph proofs are conversational. The proof is written out in much the same way that a person would conduct a verbal argument, sometimes omitting reasons for steps that can safely be assumed to be obvious. Many students and teachers find paragraph proofs more "friendly" and easier to follow than two-column proofs, though perhaps more difficult to create. The philosophy of Holt, Rinehart and Winston's *Geometry* program is not to favor any one form of proof over another; in many cases, the choice of form is left up to the student, depending upon the nature of the matter that is to be proved and the student's own stylisic preferences.

Algebraic Properties of Equality

The importance of Euclid's work lies not so much in what he discovered as in *the way he organized the existing knowledge of geometry* of his time. Starting from simple beginnings, he built up a large system of geometry knowledge.

Euclid began *The Elements* with five basic postulates, or statements that are accepted as true without proof. In addition to the postulates, Euclid included twenty-three definitions and five statements he called "common notions."

If equals are added to equals, then the wholes are equal.

Algebra

You may recognize this from your study of algebra as the Addition Property of Equality. This property is used to solve equations, as in the example below.

$$x - 3 = 5$$
$$x - 3 + 3 = 5 + 3 \quad \text{Addition Property of Equality}$$
$$x = 8 \quad \text{Simplify.}$$

Notice that *equals* (the 3s) are added to *equals* (the sides of the equation) to give two *wholes* (the sides of the new equation), which are themselves equal.

Two-Column Proofs

The proof in Example 1 might be written out as shown below. A **two-column** format has been used. This format is especially convenient for many of the proofs you will do in your study of geometry.

As you may notice, the steps in the final form of a proof may not be in the

TWO-COLUMN PROOF

You can use boxes and colors to make the proof easier to understand.

	Statements	Reasons
1.	$AB = CD$	Given
2.	$AB + BC = BC + CD$	Addition Property of Equality
3.	$AB + BC = AC$	Segment Addition Postulate
4.	$BC + CD = BD$	Substitution Property of Equality
5.	$AC = BD$	Substitution Property of Equality

TWO-COLUMN PROOF Complete the proof below of Theorem 4.5.3.

Theorem

Opposite angles of a parallelogram are congruent.   4.5.3

Given: parallelogram $ABCD$ with diagonals $\overline{BD}$ and $\overline{AC}$

Prove: $\angle BAD \cong \angle DCB$ and $\angle ABC \cong \angle CDA$

Proof:

Statements	Reasons
$ABCD$ is a parallelogram.	Given
$\triangle ABD \cong$ **35.** ___?___	**36.** ___?___
$\triangle BAC \cong$ **37.** ___?___	**38.** ___?___
39. ___?___ and **40.** ___?___	CPCTC

Paragraph Proofs

An alternative to a two-column proof is a **paragraph proof**. An advantage of a paragraph proof is that you have a chance to explain your reasoning in your own words. A paragraph proof of the Overlapping Segments Theorem might

PROOF

You are given $AB = CD$. Add BC to both sides of the equation, resulting in $AB + BC = BC + CD$. In the figure, $AB + BC = AC$ and $BC + CD = BD$ by the Segment Addition Postulate. The expressions on the left of these equations match the expressions in the previous equations, so you can substitute the equivalent expressions, AC and BD. The result is $AC = BD$.

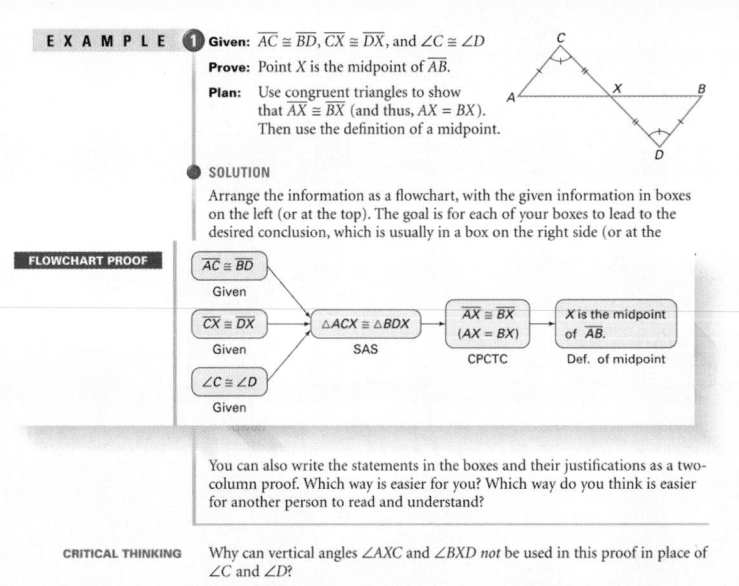

EXAMPLE **1** **Given:** $\overline{AC} \cong \overline{BD}$, $\overline{CX} \cong \overline{DX}$, and $\angle C \cong \angle D$

Prove: Point X is the midpoint of $\overline{AB}$.

Plan: Use congruent triangles to show that $\overline{AX} \cong \overline{BX}$ (and thus, $AX = BX$). Then use the definition of a midpoint.

● **SOLUTION**

Arrange the information as a flowchart, with the given information in boxes on the left (or at the top). The goal is for each of your boxes to lead to the desired conclusion, which is usually in a box on the right side (or at the

FLOWCHART PROOF

$\overline{AC} \cong \overline{BD}$
Given

$\overline{CX} \cong \overline{DX}$
Given

$\angle C \cong \angle D$
Given

$\triangle ACX \cong \triangle BDX$
SAS

$\overline{AX} \cong \overline{BX}$
$(AX = BX)$
CPCTC

X is the midpoint of $\overline{AB}$.
Def. of midpoint

You can also write the statements in the boxes and their justifications as a two-column proof. Which way is easier for you? Which way do you think is easier for another person to read and understand?

CRITICAL THINKING Why can vertical angles $\angle AXC$ and $\angle BXD$ *not* be used in this proof in place of $\angle C$ and $\angle D$?

FLOWCHART PROOFS

Another form of proof, and one that is rapidly gaining in popularity among teachers and students alike, is the flowchart form. In a flowchart proof, individual statements and their supporting reasons appear in or near "boxes" that are connected by pathways. The pathways through the flowchart reveal the structure of the reasoning.

The flowchart form is particularly helpful as an aid to understanding the logical relations among the steps of a proof. Flowchart proofs may be more labor-intensive and require more time, but students generally will find them to be intellectually easier, not harder, to produce.

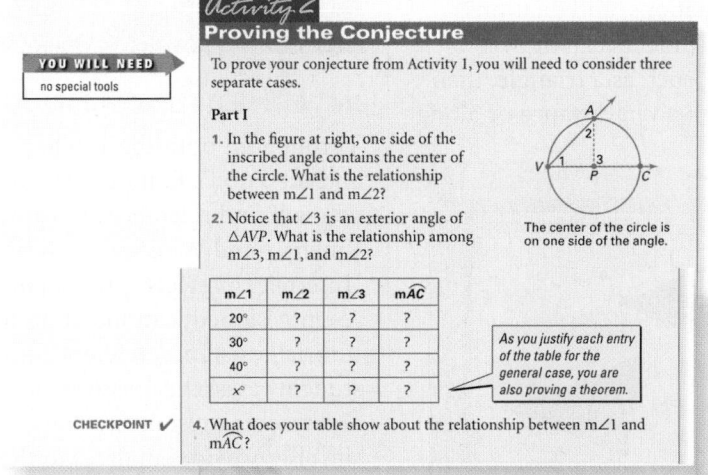

Activity 2
The Diagonals of a Parallelogram

YOU WILL NEED
graph paper

1. Use the coordinates of the parallelograms given in the table below to test the following theorem: The diagonals of a parallelogram bisect each other. (Theorem 4.5.4) Draw the first figure in the coordinate plane. Three vertices of a parallelogram are given. Find the fourth vertex and fill in the blanks of the table.

Three vertices of a parallelogram	Fourth vertex	Midpoint of $\overline{BD}$	Midpoint of $\overline{AC}$
$A(0, 0)$, $B(2, 6)$, $D(10, 0)$	$C(?, ?)$	$(?, ?)$	$(?, ?)$

CHECKPOINT ✔

PROOF

2. The figure at right represents the general case of the theorem. Use the coordinates of the figure to fill in the blanks of the table. Based on this information, what have you proven about the diagonals of a parallelogram? Explain your answer.

CRITICAL THINKING How can you prove this theorem without using coordinate geometry? Which proof seems easier to you? Explain why.

COORDINATE PROOFS

A coordinate proof uses algebra to prove theorems about figures that are represented in a coordinate plane or space. In additon to being, in some cases, a more simple and straightforward method for proving a given geometry theorem, coordinate proofs give students a chance to experience the interconnected nature of mathematical knowledge—plus occasions and motivation for practicing their algebra skills.

Coordinate geometry appears throughout Holt, Rinehart and Winston's *Geometry*. Actual proofs using coordinate geometry first occur in Chapter 5, after students have been introduced to the distance and midpoint formulas.

Activity 2
Proving the Conjecture

YOU WILL NEED
no special tools

To prove your conjecture from Activity 1, you will need to consider three separate cases.

Part I

1. In the figure at right, one side of the inscribed angle contains the center of the circle. What is the relationship between m$\angle$1 and m$\angle$2?

2. Notice that $\angle$3 is an exterior angle of $\triangle AVP$. What is the relationship among m$\angle$3, m$\angle$1, and m$\angle$2?

The center of the circle is on one side of the angle.

m$\angle$1	m$\angle$2	m$\angle$3	m$\overarc{AC}$
20°	?	?	?
30°	?	?	?
40°	?	?	?
$x°$	?	?	?

As you justify each entry of the table for the general case, you are also proving a theorem.

CHECKPOINT ✔

4. What does your table show about the relationship between m$\angle$1 and m$\overarc{AC}$?

TABLE PROOFS

Table proofs are a unique feature of Holt, Rinehart and Winston's *Geometry* program. In a table proof, students contemplate a figure and fill in blanks in a table, first for specific cases of the figure, and then for the general case (represented by one or more variables). Then they are asked to justify their entries for the general case by displaying their reasoning, typically algebraically. The completed result is a proof of a geometrical proposition.

Technology

TECHNOLOGY GOALS

To meet changing curriculum requirements, geometry teachers are continually searching for the best way to integrate technology into their classroom. They know that technology allows their students to become active participants in the learning process.

Through the use of technology, today's mathematics classrooms are being transformed into laboratories where students explore and experiment with mathematical concepts rather than just memorize isolated facts. Students make generalizations and reach conclusions about mathematical concepts and relationships and apply them to real-world situations. Holt, Rinehart and Winston's *Geometry* supports instruction that utilizes technology with activities and examples that encourage students to make and test conjectures and to confirm mathematical ideas for themselves.

Teachers also know that technology enhances cooperative learning. Cooperative-learning groups allow students to compare results, brainstorm, and reach conclusions based on group results. As in real life, where scientists and financial analysts often consult with each other, students in the technology-oriented classroom learn to communicate and consult with each other. Students learn that such consultations are not "cheating," but are rather a method of sharing information that will be used to solve a problem.

Many lessons in Holt, Rinehart and Winston's *Geometry* utilize the power of technology as an exploration tool. For example, in Lesson 1.5, students investigate special points in triangles. Students can use geometry graphics software to draw a triangle and construct its circumcenter and incenter, then investigate different cases by "dragging" the vertices of the triangle to different locations.

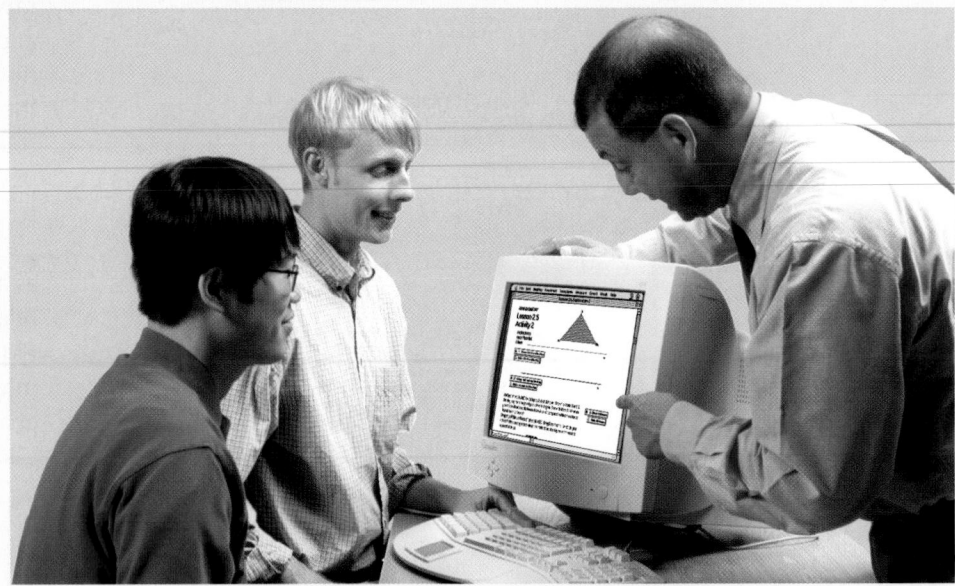

GEOMETRY GRAPHICS SOFTWARE

The most prominent technology in Holt, Rinehart and Winston's *Geometry* is geometry graphics software. This software is commonly used in geometry courses because it is a powerful tool for visualizing geometric concepts. Holt, Rinehart and Winston's *Geometry* carefully incorporates geometry graphics software in its instruction so that its use is appropriate and promotes mathematical reasoning.

Geometry graphics software performs many geometric constructions, as well as transformations such as translations, reflections, rotations, and dilations. For example, students can draw a general figure, such as a triangle, then translate it by a given distance or along

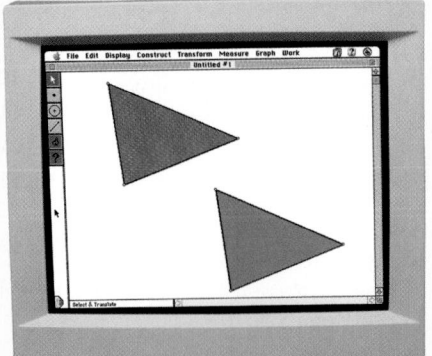

a given vector. One advantage to using software is that students can then explore different cases by dragging the vertices of the original triangle, and the vertices of the translated triangle will move accordingly. This type of exploration allows students to observe many different cases without having to redraw figures.

INTERNET TEACHING AIDS

Throughout both the Teacher's Edition and the Pupil's Edition of *Geometry*, you will find references to Holt, Rinehart and Winston's web site. These "Internet Connects" provide math resources of educational links for tutorial assistance, references for student research, classroom activities, and teaching resources. The HRW web site also provides updates to the technology used in the text.

Assessment

ASSESSMENT GOALS

An essential aspect of any learning environment, such as a geometry classroom, is the process of assessing or evaluating what students have learned. Informally, this has been done with paper-and-pencil tests given by the teacher on a regular basis to measure students' knowledge of the material. Formal evaluations with standardized tests are generally conducted over a period of years to establish performance records for both individuals and groups of students within a school or school district. Both types of tests are very good at measuring the ability of a student to perform a certain compuation or to recall a specific fact. They fall short, however, in evaluating other key goals of learning mathematics, such as being able to solve problems, to reason critically, to understand concepts, and to communicate mathematically, both verbally and in writing. Other techniques, usually referred to as alternative assessment, are needed to evaluate students' performance on these goals of instruction.

The goals of an alternative-assessment program are to provide a means of evaluating students' progress in non-computational areas of mathematical learning. Thus, the design and structure of alternative assessment techniques must be quite different from those of the computation-oriented, paper-and-pencil tests of the past.

TYPES OF ALTERNATIVE ASSESSMENT

In the world outside school, a person's work is evaluated by what that person can do (that is, by the results the person achieves) and not by taking a test. For example, a musician may demonstrate skill by making music, a pilot by flying an airplane, a writer by writing a book, and a surgeon by performing an operation. Students learn to think

mathematically and to solve problems on a continuous basis over a long period of time as they study mathematics at many grade levels. Students, too, can demonstrate what they have learned by collecting a representative sample of their best work in a portfolio. A portfolio should illustrate achievements in problem solving, critical thinking, writing about mathematics, mathematical applications or connections, and any other activity that demonstrates an understanding of both concepts and skills.

Specific examples of the kinds of work that students can include in their portfolio are solutions to nonroutine problems, graphs, tables or charts, computer printouts, group reports or reports of individual research, simulations, and artwork or models. Each entry should be dated and should be chosen to show the student's growth in mathematical competence and maturity.

A portfolio is just one way for students to demonstrate their performance on a mathematical task. Performance assessment can also be achieved in other ways, such as by asking students questions and evaluating their answers, by observing their work in cooperative-learning

groups, by having students give verbal presentations, by working on extended projects and investigations, and by keeping journals.

Peer assessment and self-evaluation are also valuable methods of assessing students' performance. Students should be able to critique their classmates' work and their own work against standards set by the teacher. In order to evaluate their work, students need to know the teacher's goals of instruction and the criteria (scoring rubrics) that have been established for evaluating performance against the goals. Students can help to design their own self-assessment forms that they then fill out on a regular basis and give to the teacher. They can also help to construct test items that are incorporated into tests given to their classmates. This work is ideally done in small groups of four students. The teacher can then choose items from each group to construct the test for the entire class. Another alternative testing technique is to have students work on take-home tests that pose more open-ended and nonroutine questions and problems. Students can devote more time to such tests and, in doing so, demonstrate their understanding of concepts and skills and their ability to do mathematics independently.

SCORING

The use of alternative assessment techniques implies the need to have a set of standards against which students' work is judged. Numerical grades are difficult to assign because growth in understanding and problem solving cannot be measured by a single number or letter grade. Instead, scoring rubrics or criteria can allow the teacher more flexibility to recognize and comment on all aspects of a student's work, pointing out both strengths and weaknesses that need to be corrected.

A scoring rubric can be created for each major instructional goal, such as being able to solve problems or communicate mathematically. A rubric generally consists of four or five short descriptive paragraphs that can be used to evaluate a piece of work. For example, if a five-point paragraph scale is used, a rating of 5 may denote that the student has completed all aspects of the assignment and has a comprehensive understanding of problems. The content of the fifth paragraph specifies the details of what constitutes the rating of highly satisfactory. On the other hand, a rating of 1 designates an essentially unsatisfactory performance, and the first paragraph would specify the details of what constitutes an unsatisfactory rating. The other three paragraphs provide an opportunity for the teacher to recognize significant accomplishments by the student as well as aspects of the work that need improvement. Thus, scoring rubrics are a far more realistic and educationally substantive way to evaluate a student's performance than a single grade, which is usually determined by an answer being either right or wrong.

The Holistic Scoring Rubric illustrated on the next page is an excellent and effective guide for use by math teachers who are practicing performance assessment in their

classrooms. The scorer gathers evidence about a student's mathematical ability. The rubric is then used to assign a single performance rating based on an overall view of the full contents of the student's portfolio. The Holistic Scoring Rubric lists the types of work and tools for entries that are appropriate for a student to place in his or her portfolio.

Holt, Rinehart and Winston's *Geometry* provides numerous opportunities in each chapter of the Pupil's Edition for portfolio assessment. Investigation and discovery are encouraged in all of the activities in the student lessons of the program. Interdisciplinary topics from astronomy to zoology and applications from arts to transportation are designated throughout the lessons. Projects are found at the end of every chapter, and Portfolio Activities are found within the chapter. Communicate exercises offer excellent writing opportunities. Tools such as geometry graphics software and algebra tiles can be used to support the activities that students include in their portfolios.

ASSESSMENT AND *GEOMETRY*

Throughout the textbook, students are asked to explain their work; describe what they are doing; compare and contrast different approaches; analyze a problem; make sketches, graphs, tables, and other models; hypothesize, conjecture, and look for counterexamples; and make and prove generalizations.

All of these activities, including the more traditional responses to routine problems, provide the teacher with a wealth of assessment opportunities to see how well students are progressing in their understanding and knowledge of geometry. The assessment task can be aligned with the major process goals of instruction, with scoring rubrics established for each goal. For example, a teacher may decide to organize his or her assessment tasks based on the following general skills of mathematics: problem solving, reasoning, communicating, and connecting.

Within each of these areas, specific goals can be written and shared with students. In this way, the assessment process becomes an integral part not only of evaluating students' progress, but also of the instructional process itself. The results of assessment can be used to modify the instructional approach in order to enhance learning for all students.

In addition to the many opportunities for performance assessment found in the lessons and chapter tests of *Geometry*, a variety of assessment types are integrated into the chapter-end material. The Chapter Review and Assessment and the Cumulative Assessment include both traditional and alternative assessment. The Cumulative Assessments are formatted in the style of college entrance exams. In addition to multiple-choice and free-response items, each Cumulative Assessment contains student-produced response questions with solutions recorded in answer grids.

Holistic Scoring Rubric

PERFORMANCE GOALS AND CRITERIA				
Mathematical Reasoning Selecting and using appropriate types of reasoning and methods of proof through inductive and deductive reasoning	**Problem Solving** Solving of problems through the use of exploration, appropriate strategies, and a systematic approach	**Communication** Communicating ideas, thoughts, and approaches through the use of everyday language, mathematical language and symbols, graphs, tables, charts, and diagrams	**Mathematical Connections** Recognizing connections between different mathematical ideas or between mathematics and other disciplines	**Use of Tools** Using technology (calculators, computers, etc.) and/or manipulatives

	Mathematical Reasoning	Problem Solving	Communication	Mathematical Connections	Use of Tools
LEVEL FOUR: SUPERIOR	• Uses sophisticated mathematical reasoning • Provides strong supporting arguments • Includes examples and counterexamples	• Shows thorough understanding of the problems' mathematical ideas and processes • Uses and synthesizes multiple strategies that lead to correct solutions	• Contains a complete response with clear, precise, and appropriate language • Uses effective diagrams such as graphs, tables, or charts	• Demonstrates a comprehensive knowledge of connections to other mathematical topics or other disciplines	• Makes appropriate use of technology and manipulatives to demonstrate mathematical concepts
LEVEL THREE: COMPETENT	• Uses sound mathematical reasoning • Includes some supporting arguments	• Shows basic understanding of the problems' mathematical ideas and processes • Uses appropriate strategies that lead to correct solutions	• Contains a solid response but is expressed less elegantly and less completely • Uses accurate diagrams	• Demonstrates some knowledge of connections to other mathematical topics or other disciplines	• Uses some technology and manipulatives to demonstrate solutions to problems
LEVEL TWO: MARGINAL	• Uses somewhat appropriate mathematical reasoning • Includes incomplete or faulty arguments	• Indicates a partial understanding of the problem • Selects some appropriate strategies that lead to partially correct solutions	• Contains a fairly complete response but uses unclear language • Uses inappropriate and/or unclear diagrams	• Demonstrates few connections to other mathematical topics or other disciplines	• Occasionally uses technology and manipulatives appropriately
LEVEL ONE: LIMITED	• Uses limited mathematical reasoning • Includes no arguments	• Shows little understanding of the problems • Uses poor or inappropriate strategies that lead to incorrect solutions	• Uses some appropriate mathematical language • Uses few, if any, diagrams	• Does not demonstrate or demonstrates inappropriate connections to other mathematical topics or other disciplines	• Rarely uses technology and manipulatives appropriately

NCTM Standards for Geometry

NUMBERS AND OPERATIONS

- **Understand numbers, ways of representing numbers, relationships among numbers, and number systems**

  Lessons 2.2, 2.4, 8.2, 8.4, 8.5, 9.1, 10.1, 10.2, 10.3, 10.4, 10.6, 11.1, 11.6; Chapters 10, 12 Eyewitness Math; Chapters 10, 12 Chapter Project

- **Understand meanings of operations and how they relate to one another**

  Lessons 2.5, 10.7, 11.1, 12.1

- **Compute fluently and make reasonable estimates**

  Lessons 2.1, 3.7, 5.1, 5.3, 5.4, 5.6, 7.4, 10.1, 10.2, 10.4, 10.5, 10.6, 11.1; Chapter 10 Eyewitness Math

ALGEBRA

- **Understand patterns, relations, and functions**

  Lessons 2.1, 2.2, 2.5, 4.4, 4.5, 5.4, 6.5, 7.1, 8.1, 11.1

- **Represent and analyze mathematical situations and structures using algebraic symbols**

  Lessons 1.2, 1.3, 1.7, 3.6, 5.2, 5.3, 5.4, 6.4, 6.5, 8.1, 10.2, 10.4, 10.6, 11.1, 11.4, 11.5, 12.4; Chapter 10 Eyewitness Math

- **Use mathematical models to represent and understand quantitative relationships**

  Lessons 1.2, 1.5, 2.2, 2.5, 3.5, 3.6, 3.7, 3.8, 4.1, 4.2, 4.3, 5.1, 5.4, 5.5, 6.2, 6.3, 7.1, 7.2, 7.3, 7.4, 7.5, 7.6, 8.2, 8.3, 9.1, 9.2, 9.3, 9.4, 10.3, 10.4, 10.5, 10.7, 11.1, 11.2, 11.3, 11.4, 11.7, 12.4; Chapters 9, 10 Eyewitness Math; Chapters 6, 10, 11, 12 Chapter Project

- **Analyze change in various contexts**

  Lessons 1.6, 1.7, 3.5, 3.8, 5.6, 5.7, 8.1, 8.2, 9.6

GEOMETRY

- **Analyze characteristics and properties of two- and three-dimensional geometric shapes and develop mathematical arguments about geometric relationships**

  Lessons 1.1, 1.3, 1.4, 1.5, 2.1, 2.3, 3.1, 3.2, 3.3, 3.4, 3.5, 3.6, 3.7, 3.8, 4.1, 4.2, 4.3, 4.4, 4.5, 4.6, 5.1, 5.2, 5.4, 5.5, 5.6, 6.2, 6.3, 7.1, 7.2, 7.3, 8.1, 8.2, 8.3, 8.4, 8.5, 8.6, 10.1, 10.2, 10.3, 11.1, 11.2, 11.3, 11.4, 11.5, 11.6, 12.4; Chapter 5 Eyewitness Math; Chapters 3, 7, 8 Chapter Project

- **Specify locations and describe spatial relationships using coordinate geometry and other representational systems**

  Lessons 1.2, 1.6, 3.8, 5.6, 6.4, 6.5, 9.6, 10.4, 10.5, 10.6, 10.7, 11.1; Chapter 2 Eyewitness Math; Chapters 5, 9 Chapter Project

- **Apply transformations and use symmetry to analyze mathematical situations**

  Lessons 1.6, 1.7, 2.2, 2.4, 2.5, 3.1, 3.2, 4.8, 5.7, 7.7, 8.1, 10.7, 11.7

- **Use visualization, spatial reasoning, and geometric modeling to solve problems**

  Lessons 1.1, 1.3, 1.4, 1.5, 4.1, 4.2, 4.3, 4.7, 5.3, 5.8, 6.1, 6.6, 7.1, 7.4, 7.5, 7.6, 8.1, 9.1, 9.2, 9.3, 9.4, 9.5, 12.1, 12.2, 12.3, 12.5; Chapters 3, 5, 7 Eyewitness Math; Chapters 1, 4, 6 Chapter Project

MEASUREMENT

- **Understand measurable attributes of objects and the units, systems, and processes of measurement**

  Lessons 1.2, 1.3, 1.5, 1.6, 1.7, 3.2, 3.5, 3.6, 5.1, 5.3, 5.6, 7.1, 11.2; Chapter 7 Eyewitness Math; Chapter 8 Chapter Project

- **Apply appropriate techniques, tools, and formulas to determine measurements**

  Lessons 1.3, 1.4, 1.6, 2.1, 2.5, 3.2, 3.3, 3.6, 3.7, 3.8, 4.2, 4.3, 4.4, 4.5, 4.7, 4.8, 5.1, 5.2, 5.3, 5.4, 5.5, 5.6, 5.7, 5.8, 6.3, 7.1, 7.2, 7.4, 7.5, 7.6, 7.7, 8.1, 8.2, 8.3, 8.4, 8.5, 8.6, 9.1, 9.2, 9.5, 10.1, 10.2, 10.4, 11.1, 11.5, 11.6; Chapters 7, 10 Eyewitness Math; Chapters 5, 7, 8 Chapter Project

DATA ANALYSIS AND PROBABILITY

- **Formulate questions that can be addressed with data and collect, organize, and display relevant data to answer them**

  Lessons 5.7, 5.8; Chapter 2 Eyewitness Math; Chapter 5 Chapter Project

- **Select and use appropriate statistical methods to analyze data**

  Lessons 5.1, 5.8; Chapter 5 Chapter Project

- **Develop and evaluate inferences and predictions that are based on data**

  Lessons 2.1, 2.5, 3.5, 5.7, 5.8; Chapter 10 Chapter Project

- **Understand and apply basic concepts of probability**

  Lesson 5.8

PROBLEM SOLVING

- **Build new mathematical knowledge through problem solving**

  Lessons 1.1, 1.3, 1.4, 1.5, 1.6, 1.7, 2.1, 2.4, 4.8, 5.1, 5.2, 6.1, 6.2, 6.3, 6.6, 9.1, 10.2, 10.3, 11.3; Chapters 2, 3 Eyewitness Math; Chapters 1, 4, 5 Chapter Project

- **Solve problems that arise in mathematics and in other contexts**

  Lessons 1.2, 2.2, 2.5, 3.1, 3.2, 5.4, 5.5, 5.7, 5.8, 7.7, 9.2, 9.3, 11.3; Chapters 10, 12 Eyewitness Math; Chapters 2, 7, 9, 10, 11, 12 Chapter Project

- **Apply and adapt a variety of appropriate strategies to solve problems**

  Lessons 1.7, 2.1, 3.3, 3.6, 3.7, 3.8, 4.1, 4.3, 4.6, 5.1, 5.2, 5.3, 5.6, 6.3, 7.1, 7.3, 7.4, 8.1, 8.5, 8.6, 9.4, 9.5, 10.1, 10.4, 10.5, 10.6, 11.1, 11.2, 11.3, 11.4, 11.5, 11.6, 11.7; Chapters 5, 7, 9 Eyewitness Math; Chapters 3, 6, 8, 10, 11, 12 Chapter Project

- **Monitor and reflect on the process of mathematical problem solving**

  Lessons 3.5, 3.7, 3.8, 4.2, 4.7, 8.3, 11.3, 11.5, 12.1, 12.4; Chapters 10, 12 Eyewitness Math; Chapters 1, 10, 11 Chapter Project

REASONING AND PROOF

● **Recognize reasoning and proof as fundamental aspects of mathematics**

Lessons 1.1, 1.4, 2.1, 2.2, 2.3, 2.4, 2.5, 3.4, 5.7, 11.5, 12.1, 12.2, 12.4; Chapter 2 Eyewitness Math; Chapter 12 Chapter Project

● **Make and investigate mathematical conjectures**

Lessons 1.1, 1.2, 1.4, 1.5, 1.6, 1.7, 2.1, 2.2, 2.4, 2.5, 3.2, 3.3, 3.5, 3.7, 4.1, 4.3, 4.5, 4.6, 5.5, 5.6, 6.2, 7.2, 8.2, 8.3, 8.4, 8.6, 9.1, 9.2, 9.3, 9.4, 10.1, 10.2, 10.4, 10.7, 11.2, 11.3, 11.6, 11.7, 12.1, 12.3, 12.4; Chapters 10, 12 Eyewitness Math; Chapters 10, 11, 12 Chapter Project

● **Develop and evaluate mathematical arguments and proofs**

Lessons 2.1, 3.5, 4.1, 4.2, 4.3, 4.6, 8.3, 9.5, 9.6, 10.2, 10.3, 10.4, 10.5, 10.7, 11.1, 11.2, 11.3, 11.4, 11.5, 11.6, 11.7, 12.1, 12.2, 12.3, 12.5; Chapters 11, 12 Chapter Project

● **Select and use various types of reasoning and methods of proof**

Lessons 2.1, 2.2, 2.4, 2.5, 3.3, 3.4, 3.5, 3.6, 3.7, 3.8, 4.1, 4.2, 4.3, 4.4, 4.5, 4.6, 4.7, 4.8, 5.1, 5.2, 5.4, 5.6, 5.7, 7.3, 7.4, 7.5, 7.7, 8.3, 8.4, 8.5, 9.1, 9.2, 9.3, 9.4, 9.5, 10.4, 10.5, 12.1, 12.2, 12.3, 12.4; Chapter 7 Eyewitness Math; Chapters 2, 12 Chapter Project

COMMUNICATION

● **Organize and consolidate their mathematical thinking through communication**

Lessons 1.0, 1.1, 1.4, 1.6, 2.1, 2.4, 3.1, 3.2, 3.5, 3.8, 4.1, 4.2, 4.3, 4.6, 5.7, 6.1, 8.3, 9.5, 10.1, 10.2, 10.3, 10.4, 10.5, 10.6, 10.7, 11.1, 11.2, 11.3, 11.4, 11.5, 11.6, 11.7, 12.1, 12.2, 12.3, 12.4, 12.5; Chapters 10, 11, 12 Chapter Project

● **Communicate their mathematical thinking coherently and clearly to peers, teachers, and others**

Lessons 1.0, 1.5, 2.3, 2.5, 4.1, 4.8, 8.1, 8.3, 8.4, 8.5, 8.6, 9.1, 9.6, 10.1, 10.2, 10.3, 10.4, 10.5, 10.6, 10.7, 11.1, 11.2, 11.3, 11.4, 11.5, 11.6, 11.7, 12.1, 12.2, 12.3, 12.4, 12.5; Chapters 8, 10, 11, 12 Chapter Project

● **Analyze and evaluate the mathematical thinking and strategies of others**

Lessons 1.0, 1.2, 1.4, 1.7, 2.2, 2.4, 3.5, 3.7, 3.8, 4.2, 4.4, 4.5, 4.7, 5.8, 8.3, 9.1, 9.2, 9.3, 9.4, 11.3, 12.1, 12.4; Chapters 10, 12 Eyewitness Math; Chapters 9, 10, 11 Chapter Project

● **Use the language of mathematics to express mathematical ideas precisely**

Lessons 1.0, 1.1, 1.3, 3.1, 3.3, 3.4, 3.6, 3.8, 4.1, 5.1, 5.2, 5.3, 5.4, 5.5, 5.6, 6.2, 6.3, 6.4, 6.5, 6.6, 7.1, 7.2, 7.3, 7.4, 7.5, 7.6, 7.7, 8.2, 8.3, 9.5, 10.1, 10.2, 10.3, 10.4, 10.5, 10.6, 10.7, 11.1, 11.2, 11.3, 11.4, 11.5, 11.6, 11.7, 12.1, 12.2, 12.3, 12.4, 12.5; Chapters 10, 11, 12 Chapter Project

CONNECTIONS

● **Recognize and use connections among mathematical ideas**

Lessons 1.0, 2.2, 2.3, 3.7, 3.8, 4.1, 4.4, 4.5, 4.7, 5.8, 6.3, 6.5, 9.1, 9.2, 9.3, 10.1, 10.2, 10.3, 11.1, 11.3, 11.4, 11.6; Chapters 5, 12 Eyewitness Math; Chapter 10 Chapter Project

● **Understand how mathematical ideas interconnect and build on one another to produce a coherent whole**

Lessons 1.0, 1.2, 1.5, 1.7, 2.1, 2.4, 2.5, 3.5, 3.6, 4.8, 5.2, 5.8, 6.1, 6.2, 6.5, 7.7, 8.2, 8.3, 8.4, 8.5, 9.6, 10.1, 10.2, 10.5, 10.6, 10.7, 11.1, 11.2, 11.4, 11.6, 11.7, 12.4, 12.5

- **Recognize and apply mathematics in contexts outside of mathematics**

  Lessons 1.0, 1.1, 1.3, 1.6, 3.1, 3.2, 3.3, 3.4, 4.2, 4.3, 4.6, 5.1, 5.3, 5.4, 5.5, 5.6, 5.7, 6.4, 6.6, 7.1, 7.2, 7.3, 7.4, 7.5, 7.6, 8.1, 8.5, 8.6, 9.4, 9.5, 9.6, 10.4, 11.1, 11.6, 12.4, 12.5; Chapters 2, 3, 9, 10, 12 Eyewitness Math; Chapter 1 Chapter Project

REPRESENTATION

- **Create and use representations to organize, record, and communicate mathematical ideas**

  Lessons 1.1, 1.2, 1.5, 2.1, 2.2, 2.3, 3.1, 3.2, 3.3, 3.6, 3.7, 4.1, 4.5, 5.1, 6.1, 6.6, 9.2, 9.5, 10.1, 10.2, 10.3, 10.4, 10.5, 10.6, 10.7, 11.1, 11.2, 11.3, 11.4, 11.5, 11.6, 11.7, 12.1, 12.2, 12.3, 12.4, 12.5; Chapters 8, 9 Chapter Project

- **Select, apply, and translate among mathematical representations to solve problems**

  Lessons 3.8, 4.1, 4.3, 5.2, 6.3, 7.1, 7.7, 9.4, 10.1, 10.6, 10.7, 11.3; Chapter 9 Eyewitness Math; Chapters 6, 8, 10, 11, 12 Chapter Project

- **Use representations to model and interpret physical, social, and mathematical phenomena**

  Lessons 3.8, 4.8, 6.2, 6.5, 7.1, 7.3, 7.4, 7.7, 9.1, 10.7, 11.3; Chapter 9 Eyewitness Math; Chapters 6, 9 Chapter Project

Exploring Geometry

Lesson Presentation CD-ROM
Power Point® presentations for each lesson 1.1-1.7

CHAPTER PLANNING GUIDE

Lesson	1.1	1.2	1.3	1.4	1.5	1.6	1.7	Project and Review
Pupil's Edition Pages	9–16	17–24	25–34	35–42	43–49	50–58	59–67	68–77
Practice and Assessment								
Extra Practice (Pupil's Edition)	818	818	819	819	820	820	821	
Practice Workbook	1	2	3	4	5	6	7	
Practice Masters Levels A, B, and C	1–3	4–6	7–9	10–12	13–15	16–18	19–21	
Standardized Test Practice Masters	1	2	3	4	5	6	7	8
Assessment Resources	1	2	3	4	6	7	8	5, 9–14
Visual Resources								
Lesson Presentation Transparencies Vol. 1	1–4	5–8	9–12	13–16	17–20	21–24	25–28	
Teaching Transparencies	1–3	4	5–8		9–10	11	12	
Answer Key Transparencies	1–3	4–6	7–9	10–13	14–18	19–22	23–32	33–39
Quiz Transparencies	1.1	1.2	1.3	1.4	1.5	1.6	1.7	
Teacher's Tools								
Reteaching Masters	1–2	3–4	5–6	7–8	9–10	11–12	13–14	
Make-Up Lesson Planner for Absent Students	1	2	3	4	5	6	7	
Student Study Guide	1	2	3	4	5	6	7	
Spanish Resources	1	2	3	4	5	6	7	
Block Scheduling Handbook								2–3
Activities and Extensions								
Lesson Activities	1–3	4–5		6–10	11–12	13–16	17–21	
Enrichment Masters	1	2	3	4	5	6	7	
Cooperative-Learning Activities	1	2	3	4	5	6	7	
Problem-Solving/ Critical Thinking	1	2	3	4	5	6	7	
Student Technology Guide	1–2	3	4	5	6	7–8	9	
Long Term Projects								1–4
Writing Activities for Your Portfolio								1–3
Tech Prep Masters								1–4
Building Success in Mathematics								1–3

LESSON PACING GUIDE

Lesson	1.1	1.2	1.3	1.4	1.5	1.6	1.7	Project and Review
Traditional	1 day	1 day	1 day	1 day	1 day	2 days	2 days	2 days
Block	$\frac{1}{2}$ day	$\frac{1}{2}$ day	$\frac{1}{2}$ day	$\frac{1}{2}$ day	$\frac{1}{2}$ day	1 day	1 day	1 day
Two-Year	2 days	2 days	2 days	2 days	2 days	4 days	4 days	4 days

CONNECTIONS AND APPLICATIONS

Lesson	1.1	1.2	1.3	1.4	1.5	1.6	1.7	Review
Algebra	15, 16	20–23	32		48		65, 66	75
Geometry	9–16	17–24	25–34	35–42	43–49	50–58	59–67	70–77
Business and Economics					48			
Life Skills		20, 21	33				60	74
Science and Technology		24			44		67	74
Sports and Leisure	14		33					
Cultural Connection: Africa		23						
Cultural Connection: Asia			34					

BLOCK-SCHEDULING GUIDE

Day	Lesson	Teacher Directed: Lesson Examples, Teaching Transparencies	Student Guided: Activity, Try This	Cooperative-Learning Activity, Lesson Activity, Student Technology Guide	Practice: Practice & Apply, Extra Practice, Practice Workbook	Assessment: Quiz, Mid-Chapter Assessment	Problem Solving, Reteaching
1	1.0	5 min					
	1.1	5 min	15 min	20 min	45 min	15 min	15 min
2	1.2	8 min	8 min	10 min	25 min	8 min	8 min
	1.3	7 min	7 min	10 min	20 min	7 min	7 min
3	1.4	8 min	8 min	10 min	25 min	8 min	8 min
	1.5	7 min	7 min	10 min	20 min	7 min	7 min
4	1.6	15 min	15 min	20 min	55 min	15 min	15 min
5	1.7	15 min	15 min	20 min	55 min	15 min	15 min
6	Assess.	50 min	90 min	90 min	65 min	30 min	
		PE: Chapter Review	PE: Chapter Project, Writing Activities	Tech Prep Masters	PE: Chapter Assessment, Test Generator	Chap. Assess. (A or B), Alt. Assess. (A or B), Test Generator	

Alternative Assessment

The following suggest alternative assessments for students who may benefit from a different type of assessment than the regular chapter quizzes and the mid-chapter/end-of-chapter test. Visit the HRW web site to get additional Alternative Assessment material.

internet connect

Alternative Assessment
Go To: **go.hrw.com**
Keyword: **MG1 Alt Assess**

Performance Assessment

1. Build a three dimensional figure consisting of at least one plane, two segments or lines, four angles, and two rays. On a sheet of paper give the dimensions of the figure you build, including the lengths of segments and the measures of angles.

2. Write and graph the equations of two parallel lines and two perpendicular lines. Give the rules for slope that make lines parallel and perpendicular.

3. Plot the points (0, 6), (0, 1), and (5, 0) on a coordinate grid and connect the points. Rotate the figure 180 degrees. Then reflect the figure across the *x*-axis. Then translate the figure left 5 units and up 3 units. Show the result of each transformation on the coordinate grid.

4. List words that, when reflected spell the same word. Then list words that, when reflected, spell a different word. For each draw the line of reflection.

Portfolio Project

Suggest that students choose one of the following projects for inclusion in their portfolios.

1. Using geometry software, demonstrate the Angle Addition Postulate and the Segment Addition Postulate. As the measurements appear on the screen, point out how they change (or stay the same) as you drag a figure to change its size. After the demonstration print the screen and give a hard copy to your teacher.

2. Create an art work using translations, reflections, and rotations. Work can be done on posterboard or a computer. Also include a paragraph explaining how transformations were used. If the art work is done on a computer, print a hard copy to give to your teacher.

internet connect

The table below identifies the pages in this chapter that contain internet and technology information.

Content Links

Activities Online	pages 16, 34, 57
Portfolio Extensions	pages 42, 58
Homework Help Online	pages 14, 22, 31, 40, 46, 54, 64

Resource Links

Parents can go online and find concepts that students are learning–lesson by lesson–and questions that pertain to each lesson, which facilitate parent-student discussion.

Go To: **go.hrw.com**
Keyword: **MG1 Parent Guide**

Technical Support

The following may be used to obtain technical support for any HRW software product.

Online Help: **www.hrwtechsupport.com**

e-mail: **tschrw@hbtechsupport.com**

HRW Technical Support Center: **(800)323-9239**

7 AM to 10 PM Monday through Friday CST

Visit the HRW math web site at: **www.hrw.com/math**

Technology

Technology Objectives and Suggestions

Lesson 1.1 The Building Blocks of Geometry

This lesson focuses on understanding and identifying the undefined terms point, line, and plane. This lesson is a good place to introduce students to geometry graphics software such as Geometer's Sketchpad or Cabri Geometry. Students can use programs such as these to draw points, lines, and angles as well as other geometric figures. As they drag a point, line, or segment, they can see how these figures change.

Lesson 1.2 Measuring Length

In this lesson students are asked to create their own geometry ruler to study measuring length. Students can use geometry graphics software to construct rulers of different sizes, constructing one ruler of a certain size using geometry software, and shrinking or enlarging it using the appropriate tool in the program.

This lesson also focuses on using the Segment Addition Postulate to solve problems. Students will use this postulate and algebra to solve equations. Students can check their work using graphics calculators or a calculator or software that solves equations, such as the TI-92.

Lesson 1.3 Measuring Angles

In this lesson, students can use geometry graphics software to explore the angle addition postulate. Using the software students should construct an angle with a ray in the center, forming two adjacent angles.

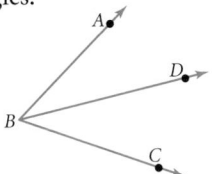

Using the measuring tool, students should find the measure of ∠ABC, ∠ABD, and ∠DBC. These measures should be displayed on the screen. Have students drag point D to change the measures of the three angles. Students can then see that the sum of m∠ABD and m∠DBC is equal to m∠ABC.

Lesson 1.4 Exploring Geometry by Using Paper Folding

The focus of this lesson is to use paper folding to construct segments and lines such as perpendicular lines, parallel lines, and segment and angle bisectors. All of the activities and exercises in this lesson that call for the use of paper folding can be done using geometry software. The activities and Exercises 8–20 on page 40 are very good places to use such technology. You may want to teach a lesson by first having students use paper folding and conclude by having students use geometry software.

Lesson 1.5 Special Points in Triangles

In this lesson students discover points of concurrency in triangles and draw inscribed and circumscribed circles of triangles. One way for students to learn the different points of concurrency is to draw all of them on a single triangle using geometry software. Angle bisectors, perpendicular bisectors, medians, and altitudes can all be different colors. Students can drag the triangle, changing its shape, to see if a special point will ever lie outside the triangle. If students wish to see only one point of concurrency, all other special segments can be "hidden" using the appropriate tool in the software. Students can draw circles using each special point as its center to try to determine which point is the incenter and which point is the circumcenter.

Lesson 1.6 Motion in Geometry

Students are to identify the three basic rigid transformations in this lesson. Teachers can use a computer overhead projector display to demonstrate each of these transformations. Geometry software includes tools that rotate objects a given angle about a point, and reflect objects over a given line. To translate objects, many programs allow you to highlight the object and drag it to another location. (Usually if the object is not highlighted, dragging on a point or a line changes the shape of the object.) Have students try transformations on their own computers once the class demonstration is complete.

Lesson 1.7 Motion in the Coordinate Plane

Some geometry programs, such as Geometer's Sketchpad, include a coordinate grid. Students can plot points on the grid by using the mouse to point and click, or by entering coordinates and allowing the program to plot them. Try using the coordinate grid both ways.

Background Information

This chapter introduces students to basic notions of geometry and geometric objects. The chapter provides immediate experience with geometry through paper folding, exploratory lessons, and options for the use of geometry graphics software.

Chapter Objectives

- Begin to construct a geometry portfolio that will help you to organize your work throughout this course. [**1.0**]

- Understand and identify the undefined terms *point*, *line*, and *plane*. [**1.1**]

- Define *segment*, *ray*, *angle*, *collinear*, *intersect*, *intersection*, and *coplanar*. [**1.1**]

- Investigate postulates about points, lines, and planes. [**1.1**]

- Construct a geometry ruler. [**1.2**]

- Define *length* and *congruent*. [**1.2**]

- Identify and use the Segment Addition Postulate. [**1.2**]

Exploring Geometry

GEOMETRY IS BOTH ANCIENT AND MODERN. From its traditional beginnings as a systematic study in the works of Euclid, through its develpoment in the works of the French philosopher and mathematician Rene Descartes, to its present-day study using sophisticated computers and calculators, geometry has an unbroken tradition in the West that spans well over two thousand years.

This first chapter is for you to get acquainted with geometry. You will be introduced to a number of tools for exploring geometry, such as paper folding, compass and straightedge, and geometry drawing software. In this way, you will discover what geometry is by experiencing it.

Lessons

About the Photos

Modern technological advances, such as the computer and graphics calculator, can help us re-create and explore the mathematical ideas developed by important mathematicians in both ancient and modern times. In his treatise entitled *The Elements*, Euclid (ca. 325–265 B.C.) established the properties of triangles, parallelograms, and circles. In more recent times, French philosopher, mathematician, and scientist René Descartes (1596–1650) founded the analytic geometry, which he based on the coordinate system that now bears his name—the "Cartesian" coordinates. He also made important contributions to algebra, including the conventions of exponential notation.

- Measure angles with a protractor. [**1.3**]
- Identify and use the Angle Addition Postulate. [**1.3**]
- Use paper folding to construct perpendicular lines, parallel lines, segment bisectors, and angle bisectors. [**1.4**]
- Define and make geometry conjectures. [**1.4**]
- Discover points of concurrency in triangles. [**1.5**]
- Draw the inscribed and circumscribed circles of triangles. [**1.5**]
- Identify and draw the three basic rigid transformations: translation, rotation, and reflection. [**1.6**]
- Review the algebraic concepts of *coordinate plane*, *origin*, *x*- and *y-coordinates*, and *ordered pair*. [**1.7**]
- Construct translations, reflections across axes, and rotations about the origin on a coordinate plane. [**1.7**]

PORTFOLIO ACTIVITIES PROJECT

As you begin your study of geometry, set up your own portfolio. The first lesson of this book, Lesson 1.0, tells you how to do this. Throughout this book you will find many suggestions of things to include in your portfolio.

About the Chapter Project

Origami is an ancient art form that originated in Japan. Using a few basic folds, masters of the art have created elegant and intriguing forms that have become the classics of the tradition.

One of the most popular and enduring of the origami classics is the paper crane, which you will study in the Chapter Project. As you fashion the bird, notice how paper folding is used to create the symmetry of the finished form.

After completing the Chapter Project, you will be able to do the following:

- Fold the classic origami crane.
- Analyze the patterns formed on the paper by the folds of the crane.

About the Porfolio Activities

Throughout the chapter, you will be given hands-on projects to do that will enhance and extend your understanding of the material in the lessons.

- In the Portfolio Activity on page 42, you will learn how to create a regular hexagon by folding paper.
- In the Portfolio Activity on page 49, the center of mass of a triangle is found by using the midpoints of the sides of the triangle. The triangle can be balanced on this point.
- Two of the basic rigid transformations are used to construct "snowflake" patterns in the Portfolio Activity on page 58.

Portfolio Activities appear at the end of Lessons 1.4, 1.5, and 1.6. Each serves as preparation for the Chapter Project. The Portfolio Activities, as well as the Chapter Project Activities, are appropriate for inclusion in the student's portfolio. Students should be encouraged to include in their portfolios any other work in which they feel a sense of pride or a sense of accomplishment.

internet connect

Chapter Internet Features and Online Activities

LESSON	KEYWORD	PAGE	LESSON	KEYWORD	PAGE
1.1	MG1 Homework Help	14	1.5	MG1 Homework Help	46
	MG1 Flatland	16	1.6	MG1 Homework Help	54
1.2	MG1 Homework Help	22		MG1 Rigid	57
1.3	MG1 Homework Help	31		MG1 Snow	58
	MG1 Latitude	34	1.7	MG1 Homework Help	64
1.4	MG1 Homework Help	40			
	MG1 Origami	42			

Lesson 1.0 is a unique lesson in that it does not follow the format of the other lessons in the text. This lesson does not contain instructional examples or exercises. The main objective of the lesson is to introduce teachers and students to the nature and possible contents of a student's portfolio. The portfolio is used for assessment and as a means for students to learn about the geometry of the real world.

Students using this textbook are not required to create portfolios. However, there are several reasons why a teacher may wish to use portfolios for assessment. The main reason is to document learning over a period of time. Other advantages include better communication between student and teacher, positive reinforcement for students, and an increase in student awareness of evaluation standards.

Building Your Geometry Portfolio

Objective

● Begin to construct a geometry portfolio that will help you to organize your work throughout this course.

Why *Artists and other professionals often keep portfolios of their work. Although you are probably not yet a professional in any area, the work you do in school may help you decide on your future work and career.*

Building a portfolio will help you organize and display your work. Design it to show your work in a way that reflects your interests and your strengths. You should concentrate on the things your enjoy; these will probably be the things you do best. You may want to create geometric constructions on a computer, study the geometry of beehives and spider webs, or explore the geometry found in works of art.

Geometry in Nature

People have long been attracted to geometric figures in nature, such as the spiral shell of the chambered nautilus. The larger the shell grows, the more closely its proportions approach the value of the golden ratio, which is a very important number in mathematics. The underlying geometric principles of natural objects often seem to be the reason for their visual appeal. As you look around yourself, you will find many examples of geometric beauty in nature.

Geometry in Art

Many artists create works of art by using pure geometric forms. The work of art at right, by a famous artist of the Bauhaus School in Germany, uses rectangular solids and flat surfaces. As the work illustrates, pure geometric forms have their own beauty. Make your own collection of works of art that use pure geometric forms.

Herbert Bayer, *Structure with Three Squares*. 1967. Painted aluminum. 62.9 cm h × 62.9 cm w × 5.7 cm d. Gift of Jan van der Marck in loving memory of his wife Ingeborg. Photograph © 1999 The Detroit Institute of Arts.

Geometry in Architecture

The dimensions of the Parthenon reveal the ancient Greeks' fascination with geometry. The ratio of the height of the original structure to its width is very close to the golden ratio. Geometry is still an important element in architecture today. By applying principles from geometry and physics, architects design structures that are both strong and beautiful.

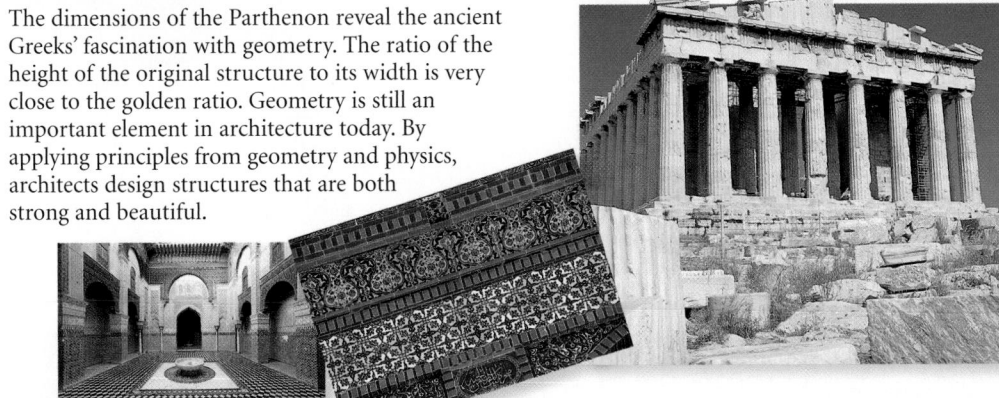

Your Notebook and Journal

Mathematicians keep records of their theories and discoveries. You, too, should keep a notebook of your work, including tests, homework activities, and special activities such as research projects. Your teacher might also want you to keep a journal.

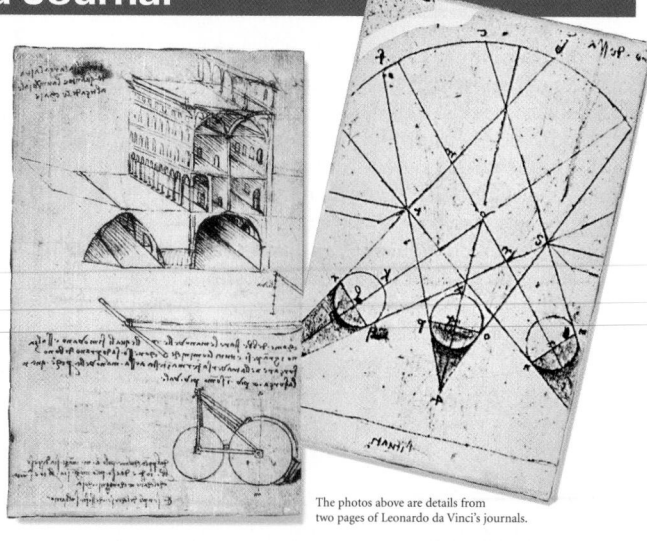

The photos above are details from two pages of Leonardo da Vinci's journals.

Putting Your Portfolio Together

Various containers can be used for your portfolio. File folders, accordion files, and even cereal boxes will work very well. You may have a number of string designs or physical models that will not fit into your portfolio container. These should still be considered a part of your portfolio.

On the following pages are five different things to include in your portfolio. You should begin right away and continue to add to your portfolio throughout this course—don't wait until the last minute.

You Can Begin Now

1. You can create a star with almost any number of points by using the following method: Start by drawing a circle. Draw the desired number of points spaced evenly around the circle. Connect the points, skipping the same number of points each time. Experiment by skipping different numbers of points. You may wish to color your stars or draw two or more stars in the same circle.

   Try to determine when the star can be drawn without picking up your pencil.

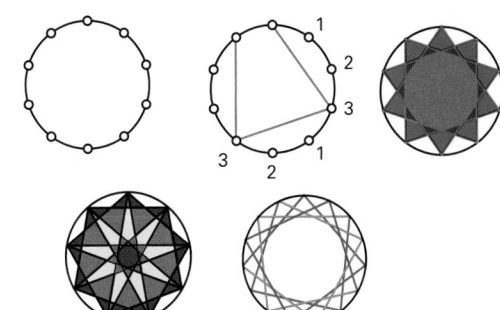

2. Circular designs known as *mandalas* (from the Sanskrit word for "circle" or "center") appear in the Hindu and Buddhist traditions as symbols of the wholeness of the universe. The Aztec calendar, a huge carved stone from ancient Mexico, bears striking resemblances to the mandalas of the East. Write a report on mandalas and their history, and then try creating mandalas of your own.

Aztec calendar (with color added)

3. Interesting designs can be created by using only straight lines. One type of line design is made from string and is known as "string art." Make your own design, using either string or pencil, paper, and a straightedge.

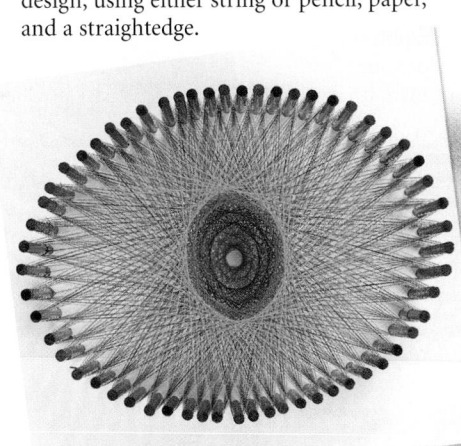

Student project

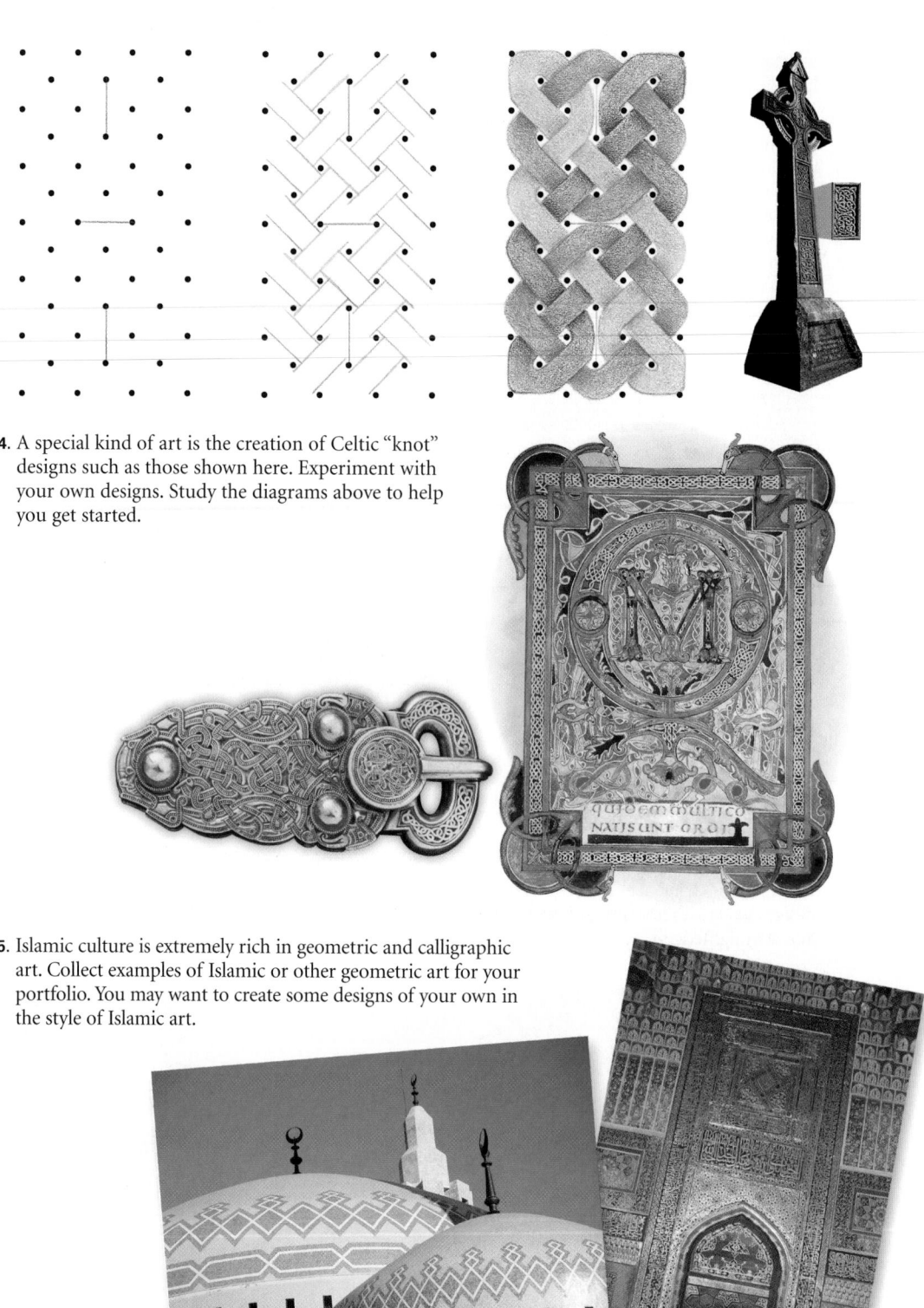

4. A special kind of art is the creation of Celtic "knot" designs such as those shown here. Experiment with your own designs. Study the diagrams above to help you get started.

5. Islamic culture is extremely rich in geometric and calligraphic art. Collect examples of Islamic or other geometric art for your portfolio. You may want to create some designs of your own in the style of Islamic art.

The Building Blocks of Geometry

Why *Geometric figures such as points, lines, and planes can be used to make mathematical models of physical objects. The models can be used to solve real-world problems.*

The spiral galaxy M31 in the constellation Andromeda is the companion galaxy of our galaxy, the Milky Way. Just as a galaxy is made up of stars, a geometric figure is composed of points.

Objectives

- Understand and identify the undefined terms *point, line,* and *plane.*

- Define *segment, ray, angle, collinear, intersect, intersection,* and *coplanar.*

- Investigate postulates about points, lines, and planes.

Prepare

QUICK WARM-UP

Find three examples of each of these in the classroom.

1. point
2. line segment
3. angle
4. line
5. plane

Answers will vary.

Also on Quiz Transparency 1.1

Basic Geometric Figures—Undefined Terms

The most basic figures of geometry are *points, lines,* and *planes.* They can be thought of as building blocks for other geometric figures. Because they are so basic, they are not defined in terms of other figures. In fact, they are often referred to as *undefined terms.* But even though points, lines, and planes are undefined, they can be explained.

The first thing to realize about geometric figures is that they are not real-world objects. Lines and planes have no thickness, and points have no size at all. This book contains many illustrations, but the illustrations are not the same as the geometric figures they represent. Theoretically, geometric figures exist "only in the mind."

Points When you look at the night sky and see the stars, the tiny dots of light seem like points. Points are often shown as dots, but unlike physical dots, geometric points have no size. Points are named by capital letters such as A or X.

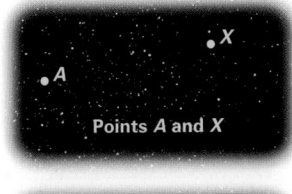

Points A and X

Lines A geometric line has no thickness, is perfectly straight, and extends forever. A line can be named by two points on the line, with a double-headed arrow ($\leftrightarrow$) over the two letters, or by a single lowercase letter.

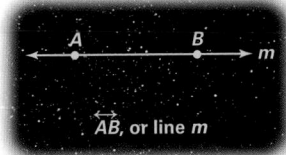

$\overleftrightarrow{AB}$, or line m

Teach

Why For some students, the distinction between a real-world object and a geometric figure may be unclear. Point out that sometimes it is impossible to find concrete objects to illustrate certain geometric definitions. For instance, explain that in geometry, a line has infinite length and no width. Compare this with objects such as a fishing line or cotton thread whose lengths are not infinite and whose widths, although minimal, do exist.

Alternative Teaching Strategy

TECHNOLOGY Have students explore basic geometric figures with geometry graphics software. They can draw a variety of lines, line segments, rays, angles, and planes. Each figure should be labeled with appropriate letters. Have students draw designs or pictures with these basic geometric figures. They can change the shapes of the figures by dragging points or lines across the screen.

Planes A geometric plane extends infinitely in all directions along a flat surface. You can think of any flat surface, such as the top of your desk or the front of this book, as representing a portion of a plane.

In the figure at right, the flat surface represents a portion of a plane. A plane can be named by three points that lie in the plane, such as *M*, *N*, and *O*, and that are not on the same line. A plane can also be renamed by a script capital letter, such as *R*.

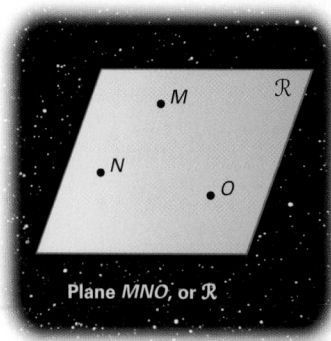

Plane *MNO*, or *R*

Points are said to be **collinear** if a single line can contain them all. (Any *two* points are collinear.) In the figure at right, *A*, *B*, and *C*, but not *D*, are collinear. Points are said to be **coplanar** if a single plane can contain them all. (Any *three* points are coplanar.)

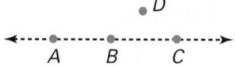

Points *A*, *B*, and *C* are collinear.

Defining Figures in Terms of the Basics

It is now possible to define three important geometric figures. Notice how each of the following definitions builds on the undefined terms *point*, *line*, and *plane*.

Definition: Segment

A **segment** is a part of a line that begins at one point and ends at another. The points are called the **endpoints** of the segment.

1.1.1

The light-gathering capacity of the large mirror of the Mount Palomar telescope makes it possible to see distant stars as points of light.

A segment is named by its endpoints. A bar (‾) is drawn over the two letters representing the endpoints.

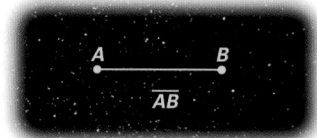

Definition: Ray

A **ray** is a part of a line that starts at a point and extends infinitely in one direction. The point is called the **endpoint** of the ray.

1.1.2

A ray is named by its endpoint and one other point that lies on the ray. The endpoint is named first. An arrow (→) is drawn over the two letters representing the points.

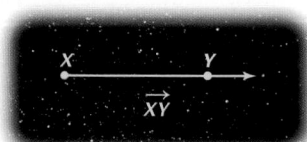

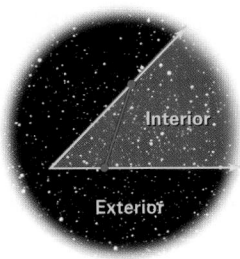

Definition: Angle

An **angle** is a figure formed by two rays with a common endpoint. The common endpoint is called the **vertex of the angle**, and the rays are the **sides of the angle.**

An angle divides a plane into two regions: the **interior** and the **exterior** of the angle. If two points, one from each side of an angle, are connected by a segment, the segment passes through the interior of the angle.

1.1.3

An angle can be named with the angle symbol ($\angle$) and three letters: one point from each side of the angle plus the vertex, with the letter for the vertex in the middle. If there is only one angle with a given vertex, the angle can be named with the angle symbol and the single letter that represents the vertex. Angles can also be named with a number shown in the angle's interior.

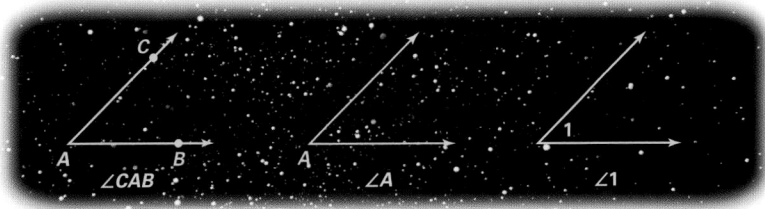

EXAMPLE Name each figure.

a.    b.    c.    d. • Y

e. 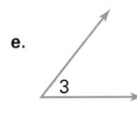   f. 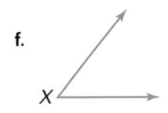   g. 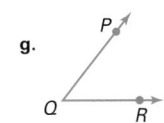   h.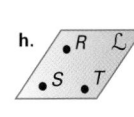

SOLUTION

a. $\overleftrightarrow{XY}$, $\overleftrightarrow{YX}$, or line m b. $\overline{PQ}$ or $\overline{QP}$ c. $\overrightarrow{MN}$ d. point Y

e. $\angle 3$ f. $\angle X$ g. $\angle PQR$, $\angle RQP$, or $\angle Q$

h. plane RST, plane RTS, plane SRT, plane STR, plane TRS, plane TSR, or plane $\mathcal{L}$

Intersections of Geometric Figures

When geometric figures have one or more points in common, they are said to **intersect**. The set of points that they have in common is called their **intersection**. In the Activity that follows, you will discover some fundamental geometry ideas, or **postulates**, involving intersections of geometric figures. Postulates are statements that are accepted as true without proof. Postulates, like undefined terms, are building blocks of geometry. (Postulates are also known as *axioms*.)

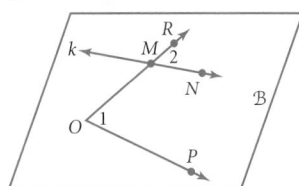

Activity Notes

In this Activity, students use
the given illustration to
complete five postulates.
For Postulates 1.1.4–1.1.8,
encourage students to use a
model of the illustration,
such as a box or a wooden
block, so that they can see
how points, lines, and
planes intersect in three-
dimensional space.

For a student worksheet of this
Activity and detailed Teacher
Notes, see page 1 in the Lesson
Activities booklet.

Cooperative Learning

In pairs, have students build a
model of the figure on page 12 by
using index cards and tape. In
Steps 2–5, students must be
allowed to manipulate the model
before they formulate their
answers.

Checkpoint questions provide an
opportunity for ongoing assessment. The answers are provided
in the Teacher's Edition side copy.

CHECKPOINT ✔

1. point

2. line

3. line

4. plane

5. is in the plane

Activity
Discovering Geometry Ideas in a Model

YOU WILL NEED

no special tools

The illustration at left may be thought of as a model of a real-world object
such as a box or a classroom. Complete each postulate below.

1. Examine the illustration. Identify the places where lines intersect each
   other. What kind of geometric figure is the intersection of two lines?

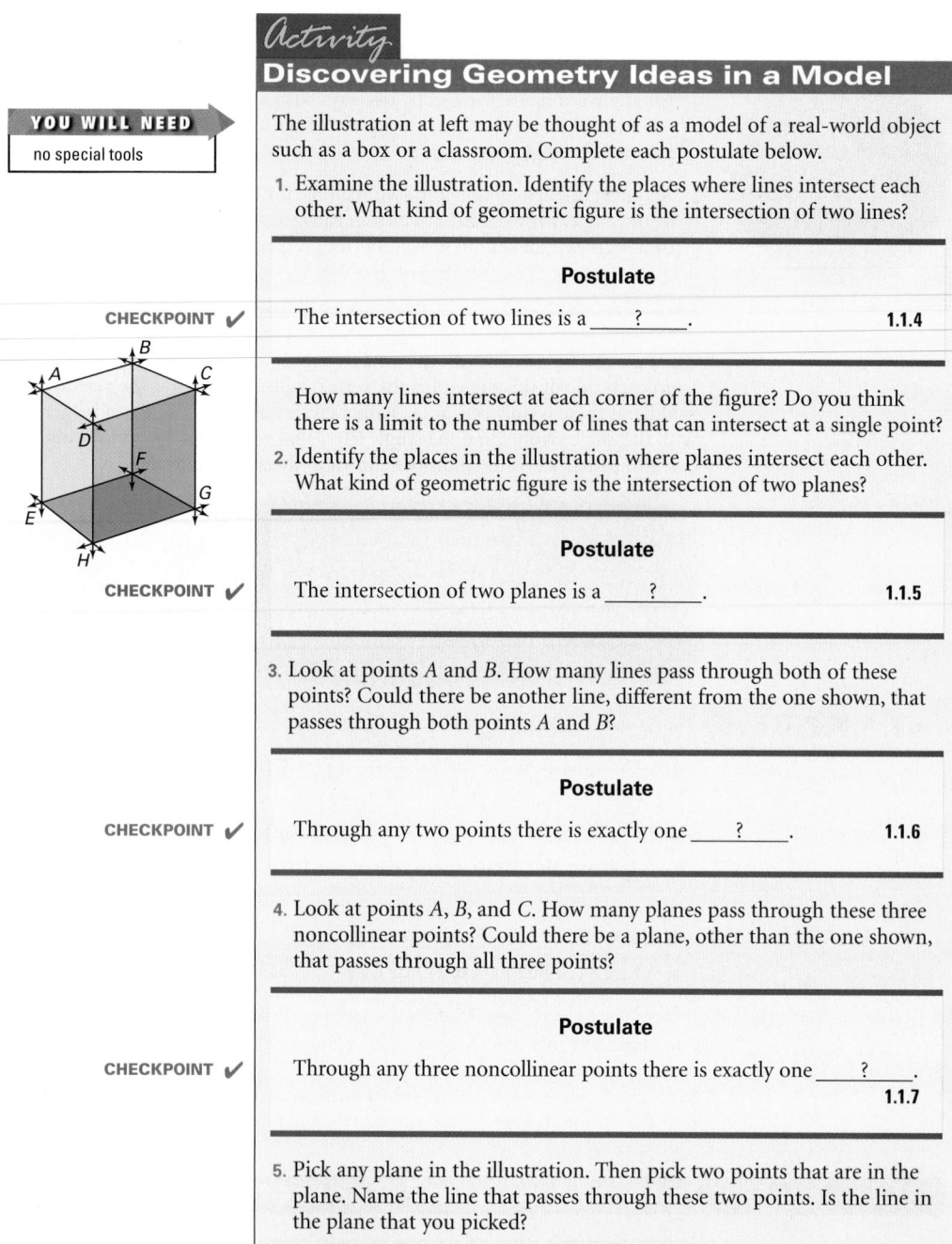

Postulate

CHECKPOINT ✔   The intersection of two lines is a ____?____.   **1.1.4**

How many lines intersect at each corner of the figure? Do you think
there is a limit to the number of lines that can intersect at a single point?

2. Identify the places in the illustration where planes intersect each other.
   What kind of geometric figure is the intersection of two planes?

Postulate

CHECKPOINT ✔   The intersection of two planes is a ____?____.   **1.1.5**

3. Look at points *A* and *B*. How many lines pass through both of these
   points? Could there be another line, different from the one shown, that
   passes through both points *A* and *B*?

Postulate

CHECKPOINT ✔   Through any two points there is exactly one ____?____.   **1.1.6**

4. Look at points *A*, *B*, and *C*. How many planes pass through these three
   noncollinear points? Could there be a plane, other than the one shown,
   that passes through all three points?

Postulate

CHECKPOINT ✔   Through any three noncollinear points there is exactly one ____?____.   **1.1.7**

5. Pick any plane in the illustration. Then pick two points that are in the
   plane. Name the line that passes through these two points. Is the line in
   the plane that you picked?

Postulate

CHECKPOINT ✔   If two points are in a plane, then the line containing them ____?____.   **1.1.8**

Reteaching the Lesson

HANDS-ON STRATEGIES   Have students build a
three-dimensional model that contains at least
one plane. The plane must intersect some segments or lines, thus creating angles, points, and
rays. Students can bring their completed models
to class, or they can complete them in class. One
example is a kite, which is a plane with wooden
supports as intersecting segments. The kite's tail
and string represent rays. After students have
completed their models, use them to review
Postulates 1.1.4–1.1.8 in the Activity.

Exercises

Communicate

1. Explain how geometric figures are different from real-world objects.

2. Examine the room you are in. Name some objects that could be represented by points, lines, and planes.

3. Explain why one point is not enough to name a line.

4. Explain why two points are not enough to name a plane.

5. Why is the order of the letters important in the name of a ray? You may wish to illustrate your reasoning with a diagram.

Automotive designers often create mathematical models on computers before building actual physical models.

Guided Skills Practice

6. Refer to the figure at right. Name a point, a line, a segment, and a ray in the figure. *(EXAMPLE)*
Sample answer: point A, $\overleftrightarrow{AB}$, $\overline{AB}$, $\overrightarrow{AB}$

7. Give four names for the angle in the figure at right. *(EXAMPLE)*
$\angle 1$, $\angle Q$, $\angle PQR$, or $\angle RQP$

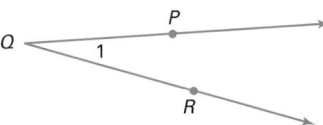

8. Give three names for the plane in the figure at right. *(EXAMPLE)*
Sample answer: plane ONM, plane NOM, or plane $\mathcal{R}$

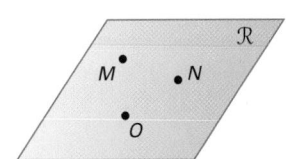

9. $\overline{AB}$, $\overline{BC}$, $\overline{AC}$

10. $\angle A$, $\angle 1$, $\angle BAC$; $\angle B$, $\angle 2$, $\angle ABC$; $\angle C$, $\angle 3$, $\angle ACB$

11. $\angle A$: $\overrightarrow{AB}$, $\overrightarrow{AC}$; $\angle B$: $\overrightarrow{BA}$, $\overrightarrow{BC}$; $\angle C$: $\overrightarrow{CA}$, $\overrightarrow{CB}$

12. plane ABC

18. False. Lines are infinite, never ending in both directions.

19. False. Planes are infinite, extending without bound and having no edges.

20. False. Two intersecting lines are contained in exactly one plane. Three intersecting lines may be contained in two planes.

21. True. For example, two opposite sides of a box intersect the bottom plane of a box, but they don't intersect each other.

22. True. For example, the side, front, and bottom of a box intersect at a point, the corner of the box.

23. False. There are an infinite number of planes through any two points.

● *Practice and Apply*

In Exercises 9–12, refer to the triangle below.

9. Name all of the segments in the triangle.

10. Name each of the angles in the triangle in three different ways.

11. Name the rays that form each side of the angles in the triangle.

12. Name the plane that contains the triangle.

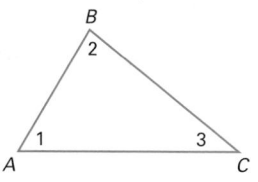

APPLICATION

HOBBIES In Exercises 13–17, refer to the aquarium shown below. State whether each object in the aquarium is best modeled by a point, a line, or a plane.

13. an edge of the aquarium line

14. a grain of sand point

15. a side of the aquarium plane

16. the surface of the water plane

17. a corner of the aquarium point

✓ **internet** connect

Homework Help Online
Go To: **go.hrw.com**
Keyword:
MG1 Homework Help
for Exercises 18–25

In Exercises 18–25, classify each statement as true or false, and explain your reasoning.

18. Lines have endpoints.

19. Planes have edges.

20. Three lines that intersect at the same point must all be in the same plane.

21. Two planes may intersect a third plane without intersecting each other.

22. Three planes may all intersect each other at exactly one point.

23. Any two points are contained in exactly one plane.

24. Any three points are contained in exactly one plane.

25. Any four points are contained in exactly one plane.

24. False. Through any three *noncollinear* points there is exactly one plane. If the points are collinear, the line that contains them is contained in an infinite number of planes.

25. False. Three noncollinear points are contained in exactly one plane, but the fourth point might be in a different plane.

Refer to the figure below for Exercises 26–30.

26. Name a line in the figure. Give three other names for the line.

27. Name a point on line *m*.

28. Name the intersection of lines *m* and *n*.

29. Name an angle in the figure. Name the vertex of this angle and the two rays that form the sides of the angle.

30. Can an angle in the figure be named ∠*A*? Why or why not?

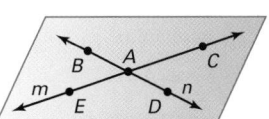

In Exercises 31–40, refer to the figures below.

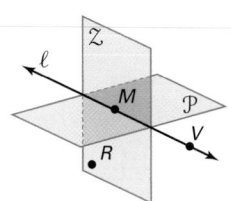

31. Name the intersection of planes 𝒫 and 𝒵.

32. Name a line in plane 𝒫.

33. Name a point in plane 𝒫.

31. line ℓ

32. line ℓ

33. point *M*

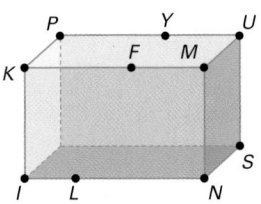

34. Name a point on $\overline{KM}$.

35. Name the intersection of $\overline{MU}$ and $\overline{MN}$.

36. Name three collinear points in the figure.

37. Name two coplanar segments in the figure.

34. point *K, M,* or *F*

35. point *M*

36. Sample answer: points *I, L,* and *N*

37. Sample answer: $\overline{KP}$ and $\overline{MU}$

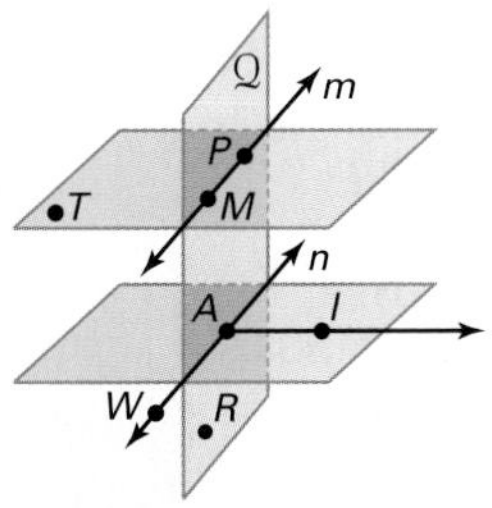

38. Name the intersection of line *n* and $\overline{AI}$.

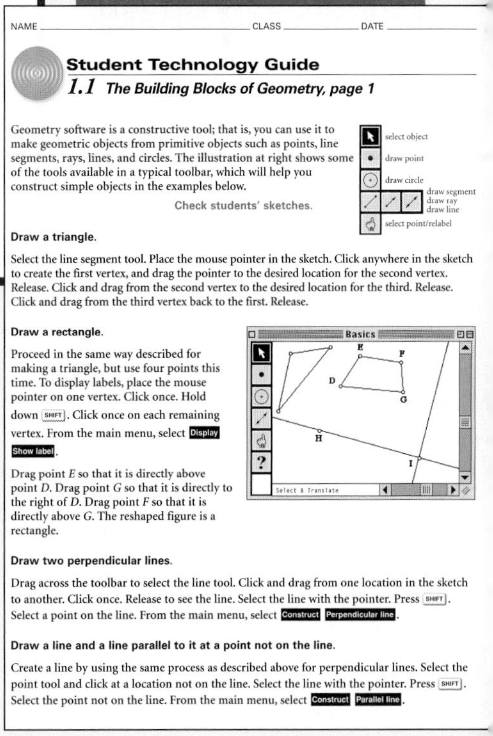

39. Name the intersection of planes 𝒬 and *MPT*.

40. Name three coplanar points in the figure.

38. point *A*

39. line *m*

40. Sample answer: points *M, A,* and *R*

How many different segments can be named in each figure below? Name each segment.

41.

42.
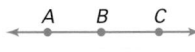

43.
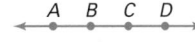

Algebra

44. Write a general rule or formula for finding the number of segments that can be named by a given number of points on a line. Can you explain why the rule works?

44. If there are *n* points, there are $\frac{n(n-1)}{2}$ segments. From each of the *n* points, *n* – 1 segments may be drawn, giving *n*(*n* – 1) segments. But each segment is counted twice, so divide by 2 to get the formula.

26. Sample answer: line *m*, $\overleftrightarrow{AE}$, $\overleftrightarrow{EA}$, or $\overleftrightarrow{CE}$

27. Sample answer: point *E*

28. point *A*

29. Sample answer: ∠*DAC*; *A*; $\overrightarrow{AD}, \overrightarrow{AC}$

30. No. There are 4 angles with a vertex at *A*.

41. 1; $\overline{AB}$

42. 3; $\overline{AB}, \overline{AC}, \overline{BC}$

43. 6; $\overline{AB}, \overline{AC}, \overline{AD}, \overline{BC}, \overline{BD}, \overline{CD}$

Look Beyond

Exercises 58 and 59 are nonroutine problems. Encourage students to draw a diagram, as part of the goal is to find a visual representation of the problems. These exercises introduce students to proofs and logic, which will reappear in later chapters.

45. 1; ∠AVB

46. 3; ∠AVC, ∠AVB, and ∠BVC

47. 6; ∠AVD, ∠AVC, ∠BVD, ∠AVB, ∠BVC, and ∠CVD

48. Each of the n points is contained in $n − 1$ angles. By reasoning as in Ex. 44, there are $\frac{n(n-1)}{2}$ distinct angles.

58. 6 exchanges

In the figure, the points represent the 4 people. Each exchange between two people can be represented by a segment connecting the points that represent those two people. There are 6 segments required to connect all the points in the figure.

How many different angles can be named in each figure below? Name each angle.

45.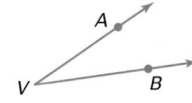

46.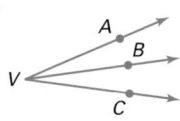

47.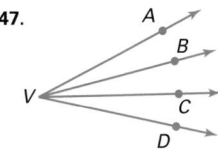

 Algebra

48. Write a general rule or formula for finding the number of angles that can be named by a given number of rays with the same endpoint. Can you explain why the rule works? (Assume that all rays lie on one side of a straight line, as shown.)

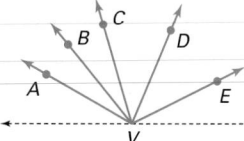

 **Look Back**

Simplify each expression. You may wish to draw a number line.

49. $22 + (−6)$ **16**

50. $7 + 15$ **22**

51. $11 − (−4)$ **15**

52. $−81 − (−30)$ **−51**

53. $|−14 + (−35)|$ **49**

54. $|13 − 10|$ **3**

55. $−123 − 41$ **−164**

56. $|21 + (−35)|$ **14**

57. $|−54 + (−20)|$ **74**

 **Look Beyond**

58. You can **make a diagram** to find the answer to the following problem. Suppose that 4 people are exchanging cards. Each person exchanges one card with each of the other 3 people. How many exchanges are made? (Hint: Draw 4 noncollinear points. Then draw line segments between the points. How many segments did you draw? How does this relate to the exchanges of cards?)

59. If 5 people exchanged cards as described above, how many exchanges would there be? Explain how to determine the number of exchanges when n people exchange cards.

internet connect
Activities Online
Go To: go.hrw.com
Keyword: **MG1 Flatland**

CHALLENGE

 *Algebra*

59. 10 exchanges. For n people, imagine a diagram with n points. If each point is connected to the other $n − 1$ points, this would give $n(n − 1)$ segments. However, this method counts each segment twice, so the number of segments needed is half that amount. Thus, the number of exchanges is $\frac{n(n-1)}{2}$.

Measuring Length

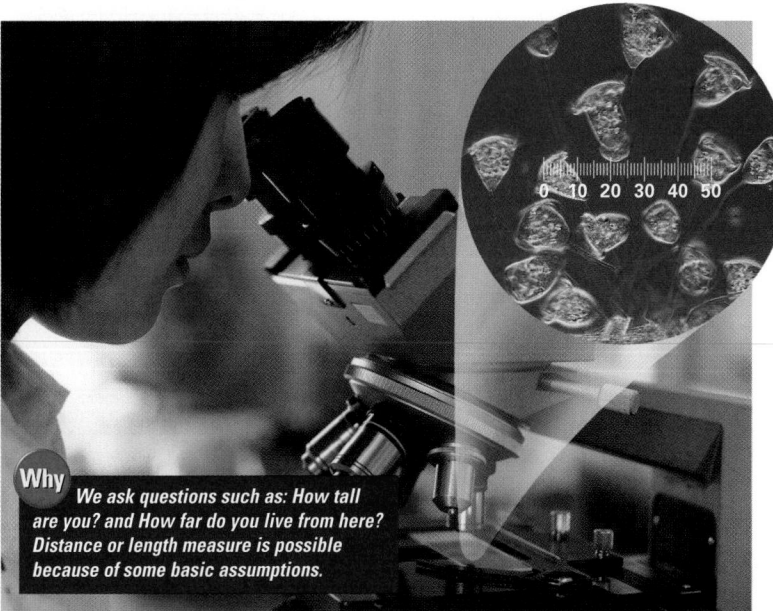

Objectives

● Construct a geometry ruler.

● Define *length* and *congruent*.

● Identify and use the Segment Addition Postulate.

Why *We ask questions such as: How tall are you? and How far do you live from here? Distance or length measure is possible because of some basic assumptions.*

A ruler in the eyepiece of the microscope makes it possible to measure in units of one-millionth of a meter; these units are known as microns.

Prepare

QUICK WARM-UP

Identify an angle, segment, ray, line, and point in the figure below.

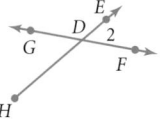

Answers will vary. Sample answer: $\angle EDF, \overline{EH}, \overrightarrow{HE}, \overleftrightarrow{GF}$, point *E*.

Also on Quiz Transparency 1.2

The Length of a Segment

In defining the length of a segment, we will use a *number line*, which is like a ruler. A **number line** is a line that has been set up to correspond with the real numbers.

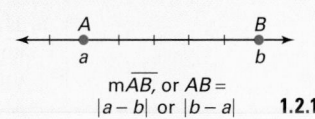

A geometry "ruler"

The **coordinate of a point** on the number line is a real number. In the illustration, −3 is the coordinate of point *A* and 4 is the coordinate of point *B*.

How would you find the distance between points *A* and *B*? Try $|-3 - 4|$. Also try $|4 - (-3)|$. What do you notice about the absolute values? This leads to a definition of the distance between two points.

Definition: Length of $\overline{AB}$

Let *A* and *B* be points on a number line, with coordinates *a* and *b*. Then the measure of $\overline{AB}$, which is called its **length**, is $|a - b|$ or $|b - a|$.

A — a ⋯ B — b

$m\overline{AB}$, or $AB =$ $|a - b|$ or $|b - a|$ **1.2.1**

The measure, or length, of $\overline{AB}$ is written as $m\overline{AB}$ or, more commonly, as just *AB*.

Teach

Why Although many aspects of geometry can be studied without using either length or angle measures, most real-world applications involve measurable quantities. Concepts of measurement are introduced early in the book so that students will be able to apply these concepts to real-world problems.

Alternative Teaching Strategy

TECHNOLOGY Geometry graphics software can be used to create fair rulers. Students can complete the Activity in this lesson by using geometry graphics software instead of a compass and straightedge. Have students create one ruler that can then be made larger or smaller by using the appropriate feature of the software.

ADDITIONAL
EXAMPLE ①

Find RT. $RT = 4$

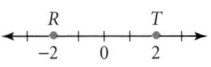

Try This questions provide an opportunity for ongoing assessment. The answers are provided in the Teacher's Edition side copy as well as in the Pupil's Edition Selected Answers.

TRY THIS

$ST = 7$

Activity Notes

In this Activity, students use a compass and straightedge to construct a fair ruler. The concept presented here is that a fair ruler is a line with equal divisions. Explain to students that inches and centimeters are not the only fair ruler units and that other units of measure are accepted as standard so that people can communicate about length.

For a student worksheet of this Activity and detailed Teacher Notes, see page 4 in the Lesson Activities booklet.

Cooperative Learning

Different groups will create different unit divisions on their rulers. Discuss which rulers would be the most appropriate to measure objects of different sizes and how the size of the unit length affects precision of measurement.

EXAMPLE ① Find the measures (lengths) of $\overline{AB}$, $\overline{AX}$, and $\overline{XB}$ on the number line below.

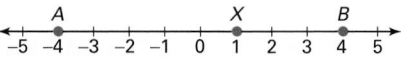

● **SOLUTION**

AB (or m$\overline{AB}$) $= |-4 - 4| = |-8| = 8$ or $AB = |4 - (-4)| = |8| = 8$

AX (or m$\overline{AX}$) $= |-4 - 1| = |-5| = 5$ or $AX = |1 - (-4)| = |5| = 5$

XB (or m$\overline{XB}$) $= |1 - 4| = |-3| = 3$ or $XB = |4 - 1| = |3| = 3$

TRY THIS   Find ST.

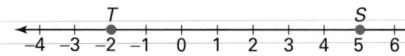

Fair Rulers

The rulers you will use in this book—with a few interesting exceptions (see Exercises 47–49 and the Portfolio Project on page 75)—are "fair" rulers. That is, they have equal intervals from one integer to the next. In the Activity below, you will use a compass and straightedge to construct a simple fair ruler.

Activity
Constructing and Using a Fair Ruler

YOU WILL NEED

compass and straightedge

1. Use your straightedge to draw a line. Choose any point on the line and label it 0. Adjust your compass to an appropriate spacing, and set the point of the compass on the point labeled.

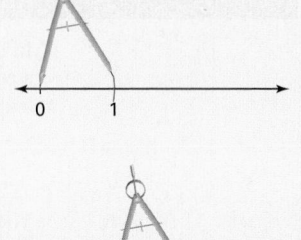

Use the pencil part of the compass to draw a short mark that crosses the number line to the right of 0. Label the point of intersection 1.

2. Set the point of the compass on the new point, and draw another mark to its right. Label the new intersection 2.

3. Repeat the previous step as many times as desired, adding 1 to the label each time.

4. Construct the negative numbers to the left of 0.

5. The distance from 0 to 1 on a ruler is known as the **unit length**. Two common unit lengths are *inches* and *centimeters*. Make up a name for your unit length.

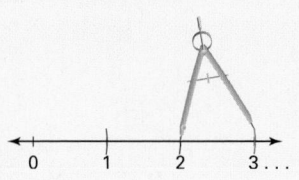

CHECKPOINT ✔

6. Use your ruler to measure an object in your classroom. Estimate the fractional part of the measurement. Compare your measurement with those of your classmates. What do you observe?

Interdisciplinary Connection

SOCIAL STUDIES   Different countries have different standards of measurement. For instance, while European countries use the metric system, the U.S. Standard system of measurement (also known as the Imperial system) is favored in the United States. Ancient cultures had their own systems of measurement (see the Cultural Connection on page 23). Have students research different systems of measurement and present their findings to the class.

Congruent Segments

Congruent figures are figures that are the same size and shape. If you move one of them onto the other, they will match exactly, like the segments in the figure at right. The segments on the ruler that you constructed in the Activity were all congruent because the same compass setting was used for each one.

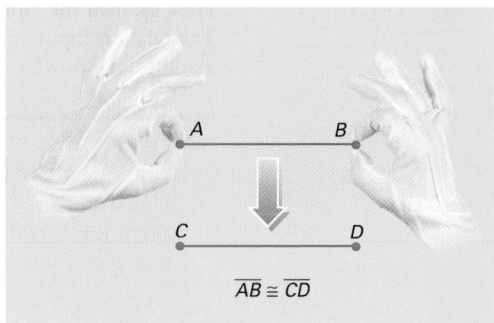

$\overline{AB} \cong \overline{CD}$

The symbol for congruence is $\cong$. $\overline{XY} \cong \overline{YZ}$ is read as "Segment $\overline{XY}$ is congruent to segment $\overline{YZ}$."

In geometry, tick marks are used to show which segments are known to be congruent. Within a given illustration, segments that have a single tick mark are congruent. Similarly, segments that have two tick marks are congruent, and so on.

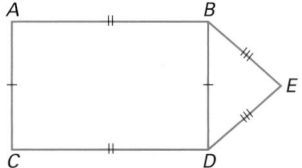

Which segments are congruent?

From your exploration of a fair ruler, the following important principle should be obvious:

Segment Congruence Postulate

If two segments have the same length as measured by a fair ruler, then the segments are congruent. Also, if two segments are congruent, then they have the same length as measured by a fair ruler.　　**1.2.2**

EXAMPLE ② Use the Segment Congruence Postulate to complete the following:

If $XY = YZ$, then ___?___.　　　　　　If $\overline{XY} \cong \overline{YZ}$, then ___?___.

● **SOLUTION**

If $XY = YZ$, then $\overline{XY} \cong \overline{YZ}$.　　　　If $\overline{XY} \cong \overline{YZ}$, then $XY = YZ$.

CRITICAL THINKING　　What could happen if the intervals on a ruler were not evenly spaced? Could you be sure that segments with the same measure were congruent or that congruent segments had equal measures? Explain your answer.

Inclusion Strategies

ENGLISH LANGUAGE DEVELOPMENT　Students who are acquiring English as a second language may have difficulty with terms such as *congruence* and *unit length*. These students can benefit from developing a vocabulary list as part of their portfolio. Students can illustrate each definition to help them learn the term.

ADDITIONAL
E X A M P L E ②

Use the Segment Congruence Postulate to complete the following:

If $AB = CD$, then _____.
$\overline{AB} \cong \overline{CD}$
If $\overline{AB} \cong \overline{CD}$, then _____.
$AB = CD$

Critical Thinking questions provide students an opportunity to analyze, assimilate, and expand their understanding of lesson concepts. The answers are provided in the Teacher's Edition side copy.

CRITICAL THINKING
No; the rulers are not fair.

Teaching Tip

Explain to students that the congruence sign and the equal sign are different. Point out that the equal sign compares two quantities, such as the lengths of two segments. The congruence sign, on the other hand, compares two objects, such as two segments with the same length or two polygons with the same size and shape.

Segment Addition

Look again at the number line in Example 1. Notice that X is between A and B. The relationships among the lengths AX, XB, and AB depend on an important assumption known as the Segment Addition Postulate.

Segment Addition Postulate

If point R is between points P and Q on a line, then
$PR + RQ = PQ$.

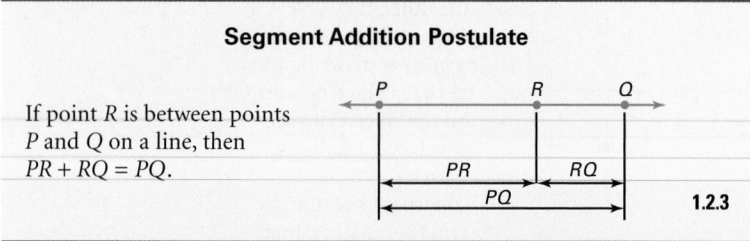

1.2.3

CRITICAL THINKING What is being added in this postulate? In geometry, addition and other arithmetic operations are defined for numbers, not for geometric figures.

Which of these statements make sense?

$AB + CD = 5$
$\overline{AB} + \overline{CD} = 5$
$m\overline{AB} + m\overline{CD} = 5$

E X A M P L E ③

APPLICATION
NAVIGATION

Algebra

The towns of Dyersberg, Newton, and Saint Thomas are located along a straight portion of Ventura Highway. Newton is between Saint Thomas and Dyersberg. The distance from Dyersberg to Saint Thomas is 25 miles. The distance from Dyersberg to Newton is 1 mile more than 3 times the distance from Newton to Saint Thomas. Find the distance from Dyersberg to Newton and from Newton to Saint Thomas.

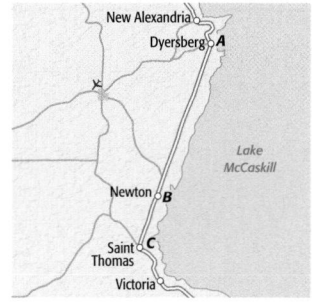

● **SOLUTION**

PROBLEM SOLVING **Write an equation.** First represent each town as a point on a line segment. Let Dyersberg be A, let Saint Thomas be B, and let Newton be C. Let x be the distance in miles from C to B, or CB. Then the distance from A to C, or AC, is $3x + 1$.

Since C is between A and B, $AC + CB = AB$.

$$AC + CB = AB$$
$$(3x + 1) + x = 25$$
$$4x + 1 = 25$$
$$4x = 24$$
$$x = 6 \text{ miles, the distance from Newton to Saint Thomas}$$

The distance from Dyersberg to Newton is found as follows:

$$AC = 3x + 1 = 3(6) + 1 = 19 \text{ miles}$$

As a check, note that the total distance is $6 + 19 = 25$.

Exercises

Communicate

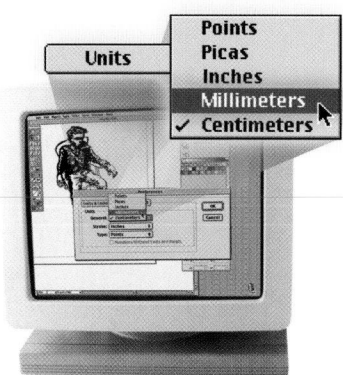

Computer software often allows you to select the unit of measure.

1. The unit length on a ruler can be any desired size. Give some commonly used unit lengths.

2. Explain why it is important for a ruler to have equal intervals.

3. Suppose that the centimeter were the only unit of measure for length. What problems would this create? Discuss why it is useful to have different units for measuring length.

4. Once you have constructed a ruler, why might you want to divide the unit length into smaller intervals?

5. Explain why each of the following statements does or does not make sense:

   a. $\overline{MN} + OP = 30$ **b.** $MN + OP = 30$ **c.** $m\overline{MN} + m\overline{OP} = 30$

Guided Skills Practice

In Exercises 6–8, find the lengths of the segments. *(EXAMPLE 1)*

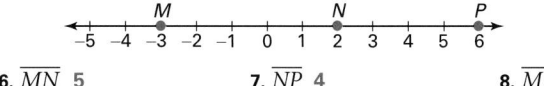

6. $\overline{MN}$ 5   **7.** $\overline{NP}$ 4   **8.** $\overline{MP}$ 9

9. Complete the statements below. *(EXAMPLE 2)*

     a. If $\overline{AB} \cong \overline{CD}$, then ___?___.   **b.** If $AB = CD$, then ___?___.

APPLICATION

10. NAVIGATION The cities of Bloomington, Forsyth, and Decatur are located on a straight road from Bloomington to Decatur, with Forsyth between them. The distance from Bloomington to Decatur is 40 miles. The distance from Bloomington to Forsyth is 7 times the distance from Forsyth to Decatur. Find the distance from Bloomington to Forsyth and from Forsyth to Decatur. *(EXAMPLE 3)*

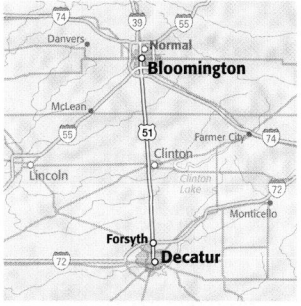

Selected Answers

Exercises 6–10, 11–45 odd

ASSIGNMENT GUIDE

In Class	1–10
Core	11–17 odd, 18–25, 27–37 odd
Core Plus	23–38
Review	39–46
Preview	47–49

✎ Extra Practice can be found beginning on page 818.

Answers to odd-numbered Extra Practice exercises can be found immediately after Selected Answers in the Pupil's Edition.

Error Analysis

Some students may need help with the algebra required in Exercises 23–25 and 26–27. Suggest that they make a three-column table. In the first column, they should write the three segment names, $\overline{PR}$, $\overline{PQ}$, and $\overline{QR}$. In the second column they should write the lengths of segments in terms of x. In the third column they should enter the numerical value for each length.

9. a. $AB = CD$
    b. $\overline{AB} \cong \overline{CD}$

10. $BF = 35$ miles and $FD = 5$ miles

17. $AB = 2$, $BC = 4$, $AC = 6$. The order of the coordinates does not matter in subtraction, because of the absolute value signs. For example, $|-3 - (-1)| = |-1 - (-3)| = 2$.

18. $\overline{AC} \cong \overline{CE} \cong \overline{BD} \cong \overline{DF}$; $\overline{AB} \cong \overline{CD} \cong \overline{EF}$

19. $\overline{AF} \cong \overline{ED}$; $\overline{FG} \cong \overline{EG}$; $\overline{BG} \cong \overline{CG}$; $\overline{FC} \cong \overline{BE}$

In Exercises 11–16, find the length of *AB*.

11. 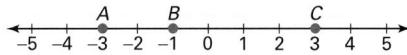 4

12. 2

13. 3

14. 5

15. 5

16. 2

17. Find the length of each segment determined by points *A*, *B*, and *C* on the number line below. Show that the order of the coordinates does not matter.

In Exercises 18 and 19, name all congruent segments.

18.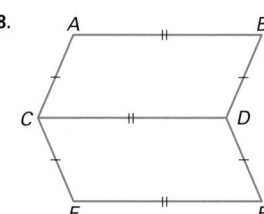

19.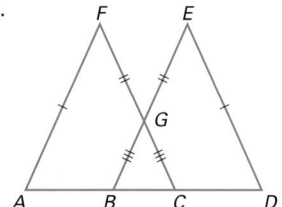

20.

21.

22.

In Exercises 20–22, point *A* is between points *M* and *B* on $\overline{MB}$. Sketch each figure and find the missing lengths.

20. $MA = 30$ $AB = 15$ $MB = $ _?_ **45**

21. $MA = 15$ $AB = $ _?_ **85** $MB = 100$

22. $MA = $ _?_ **16.3** $AB = 13.3$ $MB = 29.6$

Algebra

Find the indicated value in Exercises 23–25.

23. $PR = 25$ $x = $ _?_ **5**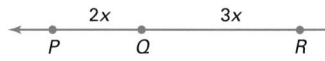

24. $PQ = 25$ $PR = $ _?_ **40**

25. $PQ = 25$ $PR = $ _?_ **37**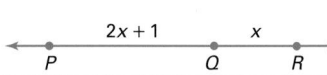

Towns *A*, *B*, *C*, and *X* are located along a straight highway. Town *B* is between *A* and *C*, and the distance from *A* to *C* is 41 miles. *BC* is 2 miles more than twice *AB*.

26. Write an equation and solve it to find *AB* and *BC*.

27. Town *X* is between *A* and *B*, 6 miles from *A*. Find *XC*.

In Exercises 28–33, explain why each statement does or does not make sense.

28. $XY = 5000$ yd

29. $\overline{PQ} = 32$ in.

30. $m\overline{ST} = 6$ cm

31. $XY + XZ = 32$ cm

32. $m\overline{PR} = 46$ cm

33. $\overline{XY} - \overline{XZ} = 12$ cm

CULTURAL CONNECTION: AFRICA The Egyptian *royal cubit* (1550 B.C.E.) was subdivided into 28 units known as *digits* or *fingers*. From this basic unit, a number of others were created:

4 digits = 1 palm

5 digits = 1 handsbreadth

6 digits = 1 fist

8 digits = 1 double palm

12 digits = 1 small span

14 digits = 1 great span

16 digits = 1 foot (t'eser)

24 digits = 1 short cubit

34. A digit is approximately 1.9 centimeters. Draw a line and use a centimeter ruler to mark a unit length of 1 digit. Then use a compass to construct a digit ruler. **Check students' rulers.**

35. A cubit is the distance from the elbow to the tip of the middle finger. Measure your own cubit with your digit ruler. How does it compare with the Egyptian royal cubit?
Sample answer: The Egyptian royal cubit is longer.

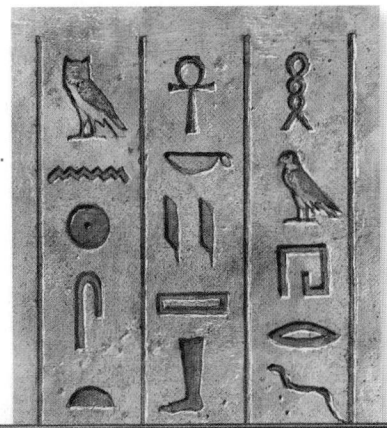

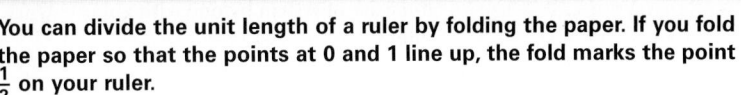

You can divide the unit length of a ruler by folding the paper. If you fold the paper so that the points at 0 and 1 line up, the fold marks the point $\frac{1}{2}$ on your ruler.

36. What point would be marked if you folded the paper again so that the points at 0 and $\frac{1}{2}$ line up? $\frac{1}{4}$

37. Construct a ruler with a fairly large unit length. Divide the interval from 0 to 1 to find the points for $\frac{1}{2}$, $\frac{1}{4}$, $\frac{3}{4}$, and $\frac{1}{8}$. **Check students' rulers.**

38. What are some points on your ruler that you *cannot* find by using this method? **Sample answers:** $\frac{1}{3}, \frac{5}{6}, \frac{3}{5}, \frac{4}{7}$

26. $x + (2 + 2x) = 41$
$AB = 13$ miles;
$BC = 28$ miles

27. $XC = 35$ miles

28. makes sense; both XY and 5000 are numbers

29. does not make sense; $\overline{PQ}$ is not a number

30. does not make sense; numbers do not have measures

31. makes sense; XY, XZ, and 32 are all numbers

32. makes sense; $m\overline{PR}$ and 46 are numbers

33. does not make sense; $\overline{XY}$ and $\overline{XZ}$ are not numbers

Exercises 47–49 expand on the concept of a ruler by presenting the Richter scale. Point out that this scale is not used to measure distance. Instead, the Richter scale is used to convert the amount of ground motion during an earthquake to a number that describes the intensity of the earthquake.

39. positive

40. positive; the result is a positive number

41. negative; the result is a positive number

42. The absolute value is the same, regardless of the order in which the numbers are subtracted.

43. negative

44. positive; the result is a positive number

45. negative; the result is a positive number

46. The absolute values are the same, regardless of the order in which the numbers are subtracted.

Look Back

Choose any two different positive numbers.

39. Add the two numbers together. Is the sum positive or negative?

40. Subtract the smaller number from the larger number. What kind of number do you get? Now take the absolute value of this number.

41. Subtract the larger number from the smaller number. What kind of number do you get? Now take the absolute value of this number.

42. Compare your answers from Exercises 40 and 41. Explain why they are the same or why they are different.

Choose any two different negative numbers.

43. Add the two numbers together. Is the sum positive or negative?

44. Subtract the smaller number from the larger number. What kind of number do you get? Now take the absolute value of this number.

45. Subtract the larger number from the smaller number. What kind of number do you get? Now take the absolute value of this number.

46. Compare your answers from Exercises 44 and 45. Explain why they are the same or why they are different.

Look Beyond

APPLICATION

GEOLOGY Here is one example of a ruler that does not have evenly spaced divisions. The *Richter scale* is used to measure the intensity of earthquakes. An increase of 1 unit on the Richter scale indicates an increase in ground motion by a factor of 10. You can represent this with an "unfair" ruler.

47. Check students' rulers. 1000 mm, 10,000 mm

47. Draw part of a ruler to represent values on the Richter scale. Start with the distance from 0 to 1 as 1 mm. The distance from 1 to 2 will be 10 mm, the distance from 2 to 3 will be 100 mm, and so on. (You will quickly run out of room on your paper!) How far would the distance be from 3 to 4? from 4 to 5?

48. How much more ground motion is caused by an earthquake with an intensity of 8 than by one with an intensity of 6? **100 times**

49. Earthquakes can also be measured by the amount of energy released. An increase of 1 unit on the Richter scale indicates an increase by a factor of 32 in the amount of energy released. How much more energy is released by an earthquake with an intensity of 8 than by one with an intensity of 6? **1024 times**

Two of Earth's continental plates meet at the San Andreas fault. When the plates slip, the resulting shock waves may be felt all over the world.

Measuring Angles

Air speed indicator
Altitude indicator
Altimeter
Automatic direction finder
Vertical speed indicator
Directional gyro

Objectives

● Measure angles with a protractor.

● Identify and use the Angle Addition Postulate.

Why *Angle measure is used in many professions. For example, pilots use an angle measure known as the "heading" of an airplane to navigate safely through the skies.*

The heading of this airplane, as indicated by the directional gyro in the photo above, is 166, a direction that is a little east of due south. (The readings on the dial must be multiplied by 10.)

Defining Angle Measure

A protractor is used to measure angles. As on a ruler, the intervals on a protractor must be equal. Then you can be sure that if two angles have the same measure, they are congruent, and vice versa.

To understand how a protractor is used, study the following example:

EXAMPLE ① Use a protractor to find the measure of $\angle CAB$.

● **SOLUTION**

1. Put the center of the protractor at the vertex.

2. Align the protractor so that $\overrightarrow{AB}$ passes through 0 on the protractor.

3. Read the measure of $\angle CAB$ (in degrees) at the point where $\overrightarrow{AC}$ intersects the scale on the protractor.

The measure of $\angle CAB$ is 121°, or $m\angle CAB = 121°$.

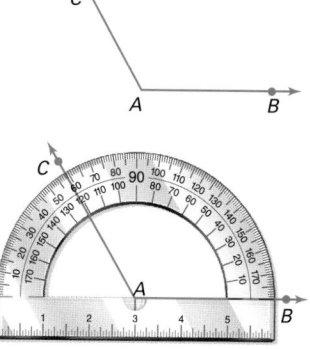

CRITICAL THINKING   Why isn't the measure of $\angle CAB$ 59° instead of 121°?

Alternative Teaching Strategy

TECHNOLOGY Geometry graphics software can be used to explore the Angle Addition Postulate. Have students draw any three rays, all with the same vertex. They should use the "measure" command to find the measure of the three angles. The angle measures should be recorded in a table. Then they can drag on the rays to change the angle measures. They will see that the Angle Addition Postulate holds true regardless of the angle measures.

CRITICAL THINKING
The starting ray $\overrightarrow{AB}$ points to the right where the bottom scale is zero, so the bottom scale must be used.

Draw four angles with obviously different measures on the board or overhead.

1. Order the angles from largest to smallest.
Answers will vary.

2. Which angles are larger than a right angle? Which are smaller?
Answers will vary.

3. Does changing the length of the rays that form an angle change the measure of the angle? no

Also on Quiz Transparency 1.3

Teach

Why Angle measures have many applications. Students can find examples of angle measures in architecture, mechanics, kinesiology, and many other fields.

ADDITIONAL
EXAMPLE ①

Have students draw angle *ABC* with a measure of 67° and angle *XYZ* with a measure of 108°.

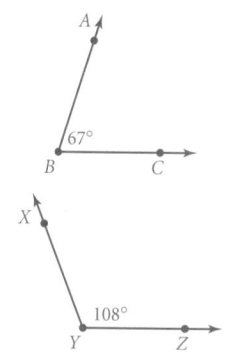

Angles, like segments, are measured in standard units. The most common unit of angle measure is the **degree**. This is the unit of measure that results when a half-circle is divided into 180 equal parts.

A protractor may be thought of as another type of geometry ruler. This "ruler" is a half-circle with coordinates from 0° to 180°. You can use a protractor to define the measure of an angle such as $\angle AVB$.

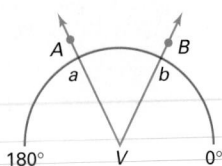

Definition: Measure of an Angle

Suppose that the vertex, V, of $\angle AVB$ is placed on the center point of a half-circle with coordinates from 0° to 180° so that $\vec{VA}$ and $\vec{VB}$ intersect the half-circle. Let a and b be the coordinates of the intersections.

Then the **measure of the angle**, written as $m\angle AVB$, is $|a - b|$ or $|b - a|$.

1.3.1

CRITICAL THINKING

No; the measure of an angle is based on a fraction of a circle; any size circle would give the same circle fraction.

CRITICAL THINKING

In creating a protractor, does the size of the half-circle make a difference? Why or why not?

EXAMPLE ② Use a protractor to find the measures of $\angle BAC$, $\angle CAD$, and $\angle BAD$. What is the value of $m\angle BAC + m\angle CAD$?

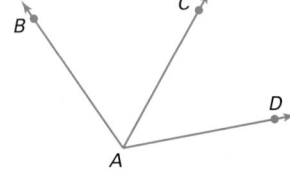

SOLUTION

1. To measure $\angle BAC$, find the points where $\vec{AB}$ and $\vec{AC}$ pass through the scale of the protractor (55° and 120°). Using these coordinates, $m\angle BAC = |55 - 120| = |-65| = 65°$.

2. The coordinate of $\vec{AD}$ is 170°, so $m\angle CAD = |120 - 170| = |-50| = 50°$.

3. Similarly, $m\angle BAD = |55 - 170| = |-115| = 115°$.

4. Using the answers for Steps 1 and 2, $m\angle BAC + m\angle CAD = 65 + 50 = 115°$. Notice that this agrees with the answer for Step 3.

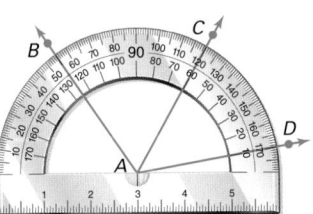

Interdisciplinary Connection

SPORTS Have students research the responsibilities of a navigator in a boat race. What types of tools does the navigator use? What factors affect the accuracy of his or her predictions?

Congruent Angles

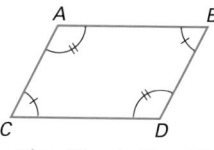

$\angle A \cong \angle D$ and $\angle B \cong \angle C$

Angles, like segments, are congruent if one can be moved onto the other so that they match exactly. Tick marks are used to show which angles are known to be congruent.

Note: Angles are said to match if their sides match. The length of the sides, which are rays that go on forever, does not matter.

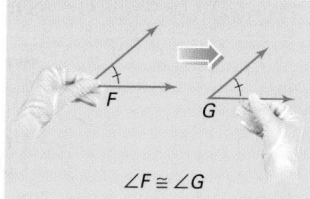

$\angle F \cong \angle G$

From the discussion of a protractor, the following important geometry principle should be obvious:

Angle Congruence Postulate

If two angles have the same measure, then they are congruent. If two angles are congruent, then they have the same measure.

1.3.2

EXAMPLE 3 Complete the following statements by using the Angle Congruence Postulate:

If $m\angle ABC = m\angle DEF$, then ___?___.

If $\angle ABC \cong \angle DEF$, then ___?___.

SOLUTION

If $m\angle ABC = m\angle DEF$, then $\angle ABC \cong \angle DEF$.

If $\angle ABC \cong \angle DEF$, then $m\angle ABC = m\angle DEF$.

Angle Addition

Look again at Example 2. Step 4 shows that $m\angle BAC + m\angle CAD = m\angle BAD$. This suggests the following postulate:

Angle Addition Postulate

If point S is in the interior of $\angle PQR$, then $m\angle PQS + m\angle SQR = m\angle PQR$.

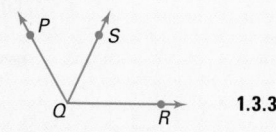

1.3.3

CRITICAL THINKING

As with the Segment Addition Postulate, you should ask yourself, What is really being added—angles or measures of angles?

Which statement makes sense?

$\angle A + \angle B = 180°$

$m\angle A + m\angle B = 180°$

Enrichment

Real numbers and measures of segments and angles satisfy the Trichotomy Property. For every x and y, one and only one of the following statements can be true: $x < y$, $x > y$, or $x = y$. Have students list some real-world relationships or situations that satisfy this property.

Inclusion Strategies

HANDS-ON STRATEGIES Learning to measure angles with a protractor is difficult for many students. Show them how to extend the rays by using index cards or stick-on notes. Have students try protractors of various sizes until they find one that is comfortable and easy to use.

X A M P L E **4**

Give the measure of a complementary and of a supplementary angle to an angle that is 47°.

complementary: 43°
supplementary: 133°

CRITICAL THINKING

Two intersecting lines form four linear pairs: angles 1 and 2, angles 2 and 4, angles 4 and 3, and angles 3 and 1. Each pair satisfies the definition of a linear pair because each line has a ray that intersects it at one point.

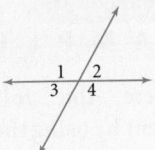

Pairs of Angles

Sometimes it is important to consider pairs of angles.

Special Angle Pairs

Complementary angles are two angles whose measures have a sum of 90°. Each angle is called the **complement** of the other.

Supplementary angles are two angles whose measures have a sum of 180°. Each angle is called the **supplement** of the other.

1.3.4

E X A M P L E **4** Name all complementary and supplementary angle pairs below.

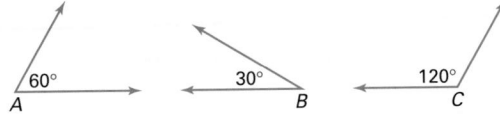

● **SOLUTION**

∠A and ∠B are complementary angles.

∠A is the complement of ∠B. ∠B is the complement of ∠A.

∠A and ∠C are supplementary angles.

∠A is the supplement of ∠C. ∠C is the supplement of ∠A.

If the endpoint of a ray falls on a line so that two angles are formed, then the angles are known as a **linear pair**.

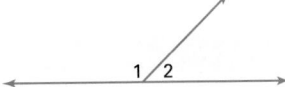

CRITICAL THINKING If two lines intersect at a point, how many linear pairs are formed? Explain how this situation fits the definition of a linear pair.

The definitions above lead to the following property:

Linear Pair Property

If two angles form a linear pair, then they are supplementary.

1.3.5

You will use the Linear Pair Property to prove an important conjecture in Lesson 2.5.

Reteaching the Lesson

COOPERATIVE LEARNING Each student should create a simple dissection puzzle by dividing a square into six regions by using straight line segments. Students should measure all of the angles formed and write several hints for the puzzle solver. For example, a hint might be that one corner of the square is made up of a 15° angle and a 75° angle. Students then should exchange puzzles with a partner and try to reassemble the original square. If necessary, more hints can be provided.

Classification of Angles

Angles can be classified according to their measure.

Definitions: Three Types of Angles

A **right angle** is an angle whose measure is 90°.
An **acute angle** is an angle whose measure is less than 90°.
An **obtuse angle** is an angle whose measure is greater than 90° and less than 180°.

1.3.6

The symbol for a right angle is a small square placed at the vertex of the angle.

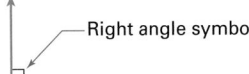
Right angle symbol

Extending Angle Measure

You may have noticed that so far the angle measures studied have been greater than 0° and less than 180°. One reason for this is that the definition of the interior and exterior of an angle doesn't apply outside this range. (Try it and see!)

However, it is sometimes necessary to talk about angles of 0° and less or of 180° and greater. Compass headings, for example, range from 0° to 360°. You can think of them as being based on a 360°, or circular, protractor. Angles with a measure of 180° are called *straight angles*, and angles with a measure greater than 180° are called *reflex angles*.

In Chapter 10, angles are studied in terms of *rotation*, in which case their measures can be greater than 360° and can even be negative.

Exercises

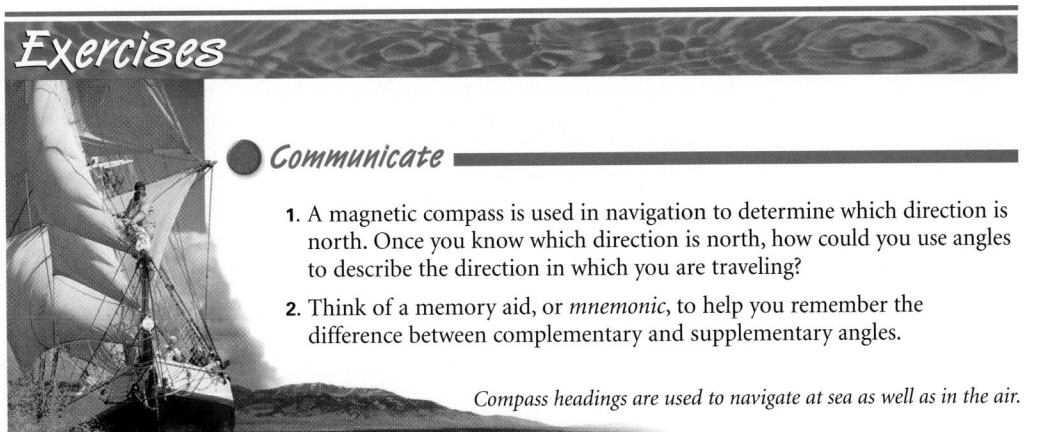

● Communicate

1. A magnetic compass is used in navigation to determine which direction is north. Once you know which direction is north, how could you use angles to describe the direction in which you are traveling?

2. Think of a memory aid, or *mnemonic*, to help you remember the difference between complementary and supplementary angles.

Compass headings are used to navigate at sea as well as in the air.

ASSIGNMENT GUIDE

In Class	1–13
Core	15–47 odd
Core Plus	27–57 odd, 58
Review	59–65
Preview	66–67

✎ Extra Practice can be found beginning on page 818.

Answers to odd-numbered Extra Practice exercises can be found immediately after Selected Answers in the Pupil's Edition.

---

3. Discuss the similarities and differences between segment length and angle measure.

4. Use an example to explain why the Angle Addition Postulate fails if point *S* is not in the interior of ∠*PQR*.

5. Classify each of the following statements as true or false, and explain your reasoning:
   a. All right angles are congruent.
   b. All acute angles are congruent.
   c. All obtuse angles are congruent.

6. Explain why one of these statements makes sense and the other does not:
   a. m∠*A* + m∠2 = 190°
   b. ∠*X* + ∠*Y* = 150°

7. Explain why the definitions of interior and exterior of an angle do not apply to straight angles and reflex angles.

● *Guided Skills Practice*

8. Use the protractor in the figure at right to find the measure of ∠*WVX*.
(EXAMPLE 1) 60°

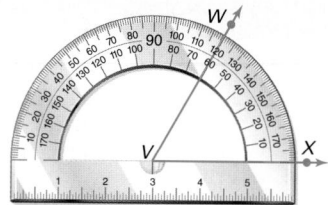

In Exercises 9–11, use the protractor in the figure at right to find the measures below. (EXAMPLE 2)

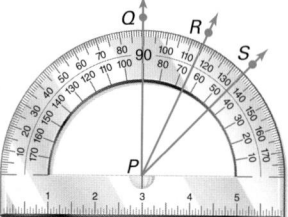

9. m∠*QPR* 25°

10. m∠*RPS* 20°

11. m∠*QPR* + m∠*RPS* 45°

12. Complete each statement by applying the Angle Congruence Postulate. **(EXAMPLE 3)**
   a. If m∠*UVW* = m∠*XYZ*, then _____. ∠*UVW* ≅ ∠*XYZ*
   b. If ∠*UVW* ≅ ∠*XYZ*, then _____. m∠*UVW* = m∠*XYZ*

13. Name a complementary angle pair and a supplementary angle pair in the figure below. **(EXAMPLE 4)**

Complementary angle pair:
∠*CAD* and ∠*DAE*

Supplementary angle pair:
∠*BAD* and ∠*DAE* or
∠*BAC* and ∠*CAE*

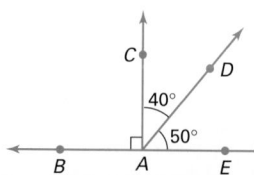

Practice and Apply

Refer to the figure below. Find the measure of each angle.

14. m∠AVB   45°

15. m∠AVC   85°

16. m∠AVD   120°

17. m∠BVC   40°

18. m∠BVD   75°

19. m∠CVD   35°

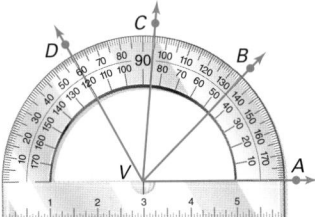

Use a protractor to find the measures of the indicated angles. You may trace the figures or use a piece of paper to extend the rays if necessary.

20. m∠X   35°

21. m∠Y   120°

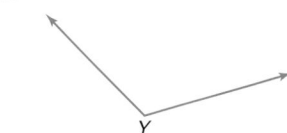

22. m∠ABC   90°

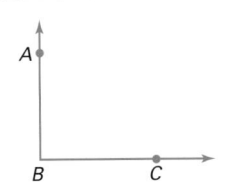

23. m∠2   30°

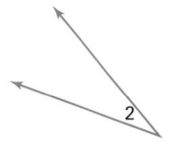

24. m∠RST   93°

25. m∠SRT   32°

26. m∠STR   55°

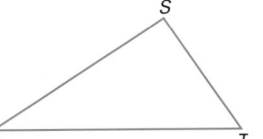

In the figure below, m∠CED = 25° and ∠AEB and ∠BED form a linear pair. Find the following:

27. m∠BEC   25°

28. m∠AEB   130°

29. m∠AEC   155°

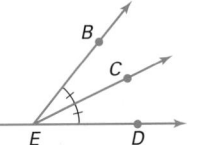

In the figure, m∠EDG = 70° and m∠FDH = 60°. Find the following:

30. m∠2   30°

31. m∠3   30°

32. m∠1   40°

33. m∠4   40°

34. m∠EDJ   140°

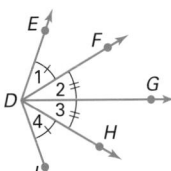

Error Analysis

In Exercises 20–26, remind st[...]
dents to check the degree mark [...]
both rays of each angle. Unles[...]
one ray of the angle is aligned
with the 0 degree mark, students
will need to subtract to find the
measure of the angle.

ECTION

u- use algebra and the
t ddition Postulate to solve
s in Exercises 39–44. Some
ents may need to review
thods for solving multistep
quations.

47. A gradian is smaller than 1 degree since one gradian is $\frac{9}{10}$ of one degree.

48. m∠*TVM* and m∠*OVB* = 20°; m∠*BVM*, m∠*RVT*, and m∠*TVO* = 45°; m∠*BVT* and m∠*MVR* = 65°

Practice

NAME ______ CLASS ______ DATE ______

Practice 1.3 Measuring Angles

Find the measure of each angle in the diagram at right.

1. m∠*SVR* 30°
2. m∠*SVQ* 60°
3. m∠*SVP* 120°
4. m∠*RVQ* 30°
5. m∠*PVR* 90°

6. Name all sets of congruent angles in the diagram below.

∠*AVS* and ∠*BVP*;
∠*AVB*, ∠*WVS*, and ∠*SVP*;
∠*AVW* and ∠*BVS*

Find the missing angle measures.

7. m∠*BTE* = 40°, m∠*ETM* = 60°, m∠*BTM* = 100°
8. m∠*BTE* = 112°, m∠*ETM* = 56°, m∠*BTM* = 168°
9. m∠*BTE* = 45°, m∠*ETM* = 47°, m∠*BTM* = 92°

In the figure at right, m∠*CED* = 39°, m∠*CEL* = (3x − 6)°, and m∠*LED* = (x + 25)°.

10. What is the value of *x*? 5
11. What is m∠*CEL*? 9

In the diagram at right, m∠*DSF* = (45 + x)°. Find the value of x, and then give each indicated angle measure.

12. m∠*DSF* 50° 13. m∠*DSE* 29° 14. m∠*ESF* 21°

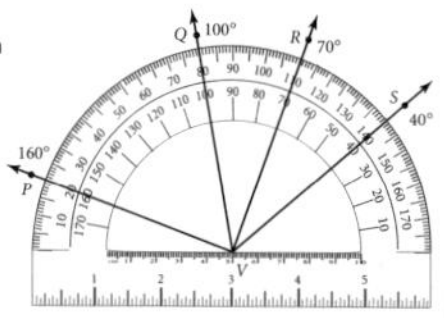

35. In the figure at right, m∠*ALE* = 31° and m∠*SLE* = 59°. What is the measure of ∠*SLA*? What is the relationship between ∠*SLE* and ∠*ALE*? **m∠*SLA* = 90°; ∠*SLE* and ∠*ALE* are complementary angles.**

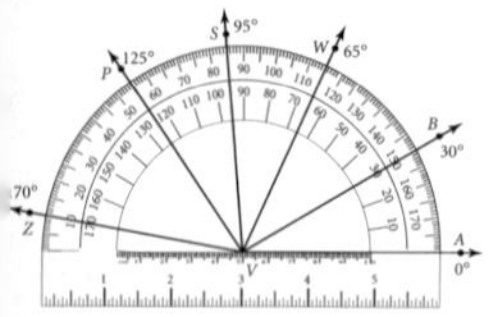

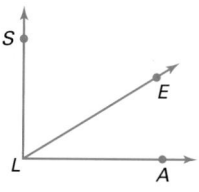

Find the missing angle measures in Exercises 36–38. Refer to the figure below.

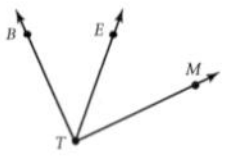

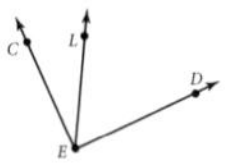

36. m∠*WZX* = 80° m∠*XZY* = 34° m∠*WZY* = ? **114°**

37. m∠*WZX* = 21° m∠*XZY* = ? **22°** m∠*WZY* = 43°

38. m∠*WZX* = ? **18°** m∠*XZY* = 34° m∠*WZY* = 52°

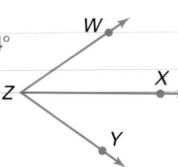

Algebra

In the diagram below, m∠*KNM* = 87°, m∠*LNM* = (2*x* − 8)°, and m∠*KNL* = (*x* + 50)°.

39. What is the value of *x*? **15**

40. What is m∠*KNL*? **65°**

41. What is m∠*LNM*? **22°**

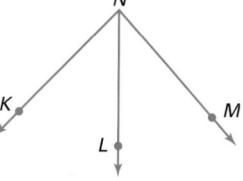

Algebra

In the diagram below, m∠*ADC* = (43 + *x*)°. Find the value of *x*, and then find each indicated angle measure.

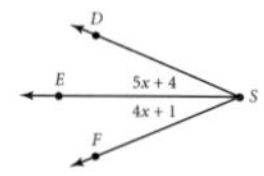

42. m∠*ADC* **49°**

43. m∠*ADB* **30°**

44. m∠*BDC* **19°**

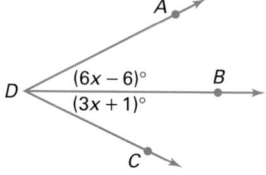

90° **45.** What is the angle between the minute and hour hands on a clock at 3:00?

150° **46.** What is the angle between the minute and hour hands on a clock at 5:00?

47. Another unit of angle measure, used primarily in engineering, is called a *gradian.* There are 100 gradians in a right angle. Is 1 gradian smaller or larger than 1 degree? Why?

CHALLENGE

48. Name all sets of congruent angles in the figure below.

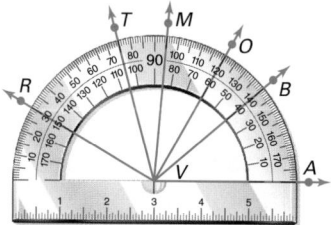

NAVIGATION Exercises 49–56 involve compass headings.
Headings are numbered in the clockwise direction from 000 to 360, starting at north. The headings from 000 to 180 are equal to the measure of the angle formed by the compass needle and a ray that points in the direction of travel. The headings from 180 to 360 are found by subtracting the measure of this angle from 360.

Find the heading for each of the following compass directions:

49. N 000

50. E 090

51. S 180

52. W 270

53. NE 045

54. SW 225

55. NNE 022.5

56. SSW 202.5

57. NAVIGATION A pilot is flying to Chicago on a heading of 355. Refer to the diagram at left. Would the pilot's heading be greater or less than 355 if the plane was traveling to Chicago from point *X*?

58. SCUBA DIVING Scuba divers often navigate in a square pattern at a constant depth. Examine the figure at right. If the diver starts out at a heading of 315, what compass headings are needed to navigate around the rest of the square? Other than using the headings, what else must a diver do in order to navigate the square?

Underwater compasses, like this programmable computer, are used by divers to navigate.

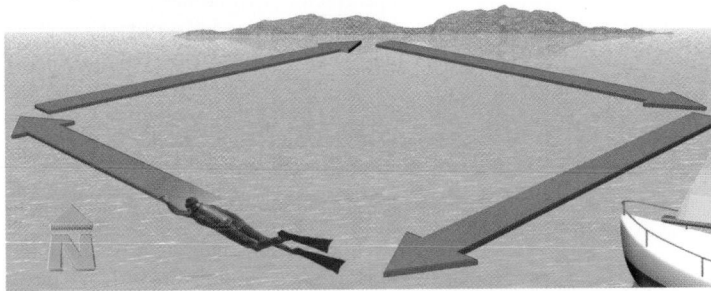

Technology
Students can use geometry graphics software in Exercises 49–56 to construct the appropriate diagram and find the angle measures.

57. Less than 355. The pilot would have to fly more to the west.

58. 45, 135, 225; The diver must also move the same distance in each direction in order to move in a square.

Student Technology Guide

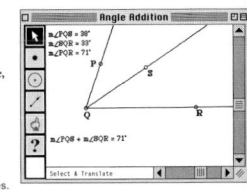

Look Back

Refer to the figure below for Exercises 59–62. *(LESSON 1.1)*

59. Give a different name for ∠1.

60. Identify a line that is coplanar with ∠1.

61. Name all lines that are formed by the intersection of two planes.

62. Identify three collinear points.

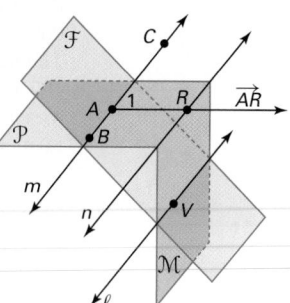

For each figure below, find the length of the indicated segment.
(LESSON 1.2)

63.

64.

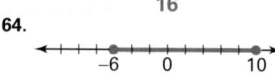

65. If $AB = 27$, find AC and BC. *(LESSON 1.2)* $AC = 21$, $BC = 6$

Look Beyond

66. CULTURAL CONNECTION: ASIA Our unit measure of degrees comes from the ancient Babylonians. The Babylonians based the measurement of an angle on a circle divided into 360 equal parts.

What are some advantages to using this unit of measurement? (Hint: List all of the factors of 360.)

67. Just as the units on a ruler are subdivided, so are the units on a protractor. Astronomers, architects, and surveyors often need angle measurements that are more precise than degrees. A degree is often divided into 60 units called *minutes*, and a minute is divided further into 60 units called *seconds*.

The ancient Babylonians were sophisticated mathematicians.

  a. How many seconds are in 1 degree? 3600
  b. How many seconds are in 1.5 minutes? 90
  c. How many minutes are in 1.75 degrees? 105

Geometry Using Paper Folding

Objectives

- Use paper folding to construct perpendicular lines, parallel lines, segment bisectors, and angle bisectors.

- Define and make geometry conjectures.

Why *Paper folding can be used to create precise geometric figures without the use of special drawing instruments.*

Origami, the ancient Japanese art of paper folding, relies on properties of geometry to produce fascinating and often beautiful shapes.

QUICK WARM-UP

Identify the following in the figure below: Answers will vary. Sample answers are given in Exercises 1–4.

1. perpendicular line segments $\overline{AC}$ and $\overline{BF}$

2. a right angle $\angle ACE$

3. a 45° angle $\angle ACG$

4. a line segment that divides a right angle in half $\overline{CG}$

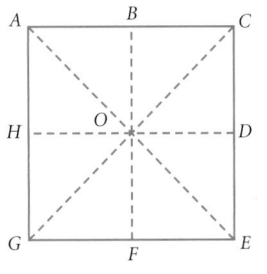

Also on Quiz Transparency 1.4

Paper Folding: The Basics

When created according to certain rules, a diagram is considered to be mathematically precise. These types of diagrams are called **constructions**. In this book, you will learn to construct geometric figures with paper folding, a compass and straightedge, and geometry software.

The term *folding paper* refers to any paper that is suitable to use for folding geometric figures. You should use paper you can see through, such as wax paper, so that you can match figures precisely when you fold. "Patty paper" is ideal because it makes white creases when folded and is easy to write on.

In the following activities, work with a partner. Be sure both you and your partner understand *why* the constructions work. Learning why things work the way they do is one of the most important objectives of this course.

The terms defined below will be used in the Activities in this lesson.

Definitions: Perpendicular and Parallel Lines

Perpendicular lines are two lines that intersect to form a right angle.
Parallel lines are two coplanar lines that do not intersect. **1.4.1**

Teach

Why For many students, paper folding activities are more instructive than paper-and-pencil drawings because the folding actions involve students' kinesthetic and tactile senses as well as their visual abilities.

Alternative Teaching Strategy

TECHNOLOGY Students can explore the Activities in this lesson with geometry graphics software. Many programs allow students to draw parallel and perpendicular lines and to bisect angles. They can also measure angles and segments with the same programs.

Activity 1 **Notes**

Students investigate a special case of the following theorem: If two lines are cut by a transversal in such a way that corresponding angles are congruent, then the two lines are parallel.

For a student worksheet of this Activity and detailed Teacher Notes, see page 6 in the Lesson Activities booklet.

Cooperative Learning

Students working in small groups can generate examples for each Activity. Each group should make several different examples of each figure to provide a basis for conjectures and conclusions.

CHECKPOINT ✔

3. 90°; perpendicular lines
6. perpendicular; parallel; the lines are parallel

Teaching Tip

Students can use the corner of an index card as a right angle "tester."

A **conjecture** is a statement that you think is true. It is an "educated guess" based on observations. Mathematical discoveries often start out as conjectures. In the Activity that follows, you will make conjectures about perpendicular and parallel lines.

Activity 1
Perpendicular and Parallel Lines

YOU WILL NEED

folding paper, and a marker or pencil that will write on folding paper

1. Fold the paper once to make a line. Label the line ℓ.

2. Draw a point on line ℓ and label it A. Fold the paper through A so that line ℓ matches up with itself. Label the new line m.

 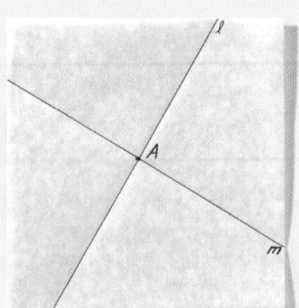

CHECKPOINT ✔

3. Measure the angles formed by lines ℓ and m. What kind of lines are formed?

4. Mark a new point on line ℓ, and label it B.

5. Fold the paper through B so that line ℓ matches up with itself again. Label the new line n.

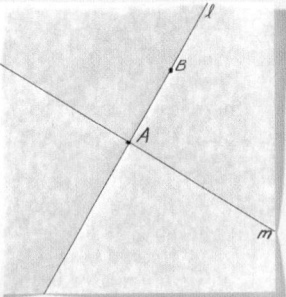 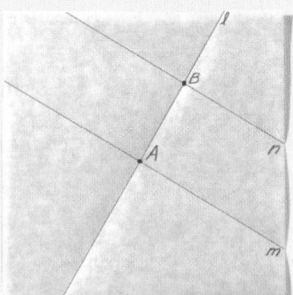

6. What kind of lines are ℓ and n? What kind of lines are m and n? Complete the following conjecture about parallel and perpendicular lines:

CHECKPOINT ✔

If two coplanar lines are perpendicular to the same line, then ___?___.

Interdisciplinary Connection

PHYSICS Airplanes made by paper folding are sometimes entered in contests. Have students use library resources to investigate what types of rules the contests follow and what factors affect the speeds and distances flown.

Measuring the Distance From a Point to a Line

How would you measure the distance from the tree to the fence in the photo at right? Would you use the segment $\overline{XA}$, $\overline{XB}$, or $\overline{XC}$?

In geometry, the distance from a point to a line is the length of the perpendicular segment from the point to the line. Thus, the distance from the tree (point X) to the fence ($\overleftrightarrow{AC}$) is the length of $\overline{XB}$.

CRITICAL THINKING   Why do you think the distance from a point to a line is defined along a perpendicular segment?

You can use paper folding to help you measure the distance from a point to a line.

Activity 2
Finding the Distance From a Point to a Line

1. Fold a line and label it ℓ. Choose a point not on the line and label it P.

2. Fold the paper over so that ℓ lines up with itself, but do not crease the paper yet.

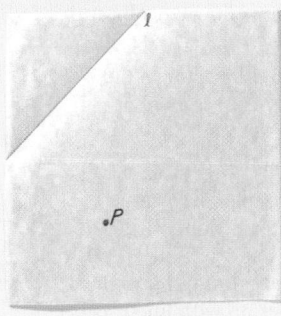

3. Slide the top edge of the paper, keeping ℓ lined up on itself, until P is on the fold. Carefully crease the paper, making sure that P is on the crease.

CHECKPOINT ✔   4. Use a ruler to measure the distance from P to ℓ along the creased fold.

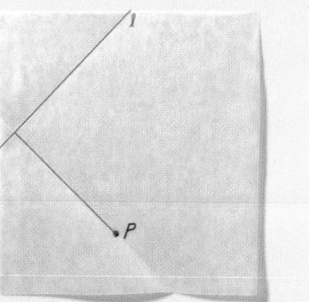

Enrichment

Constructing the perpendicular bisector of a segment automatically gives the *midpoint*. Have students use paper folding to justify this statement: If point M is the midpoint of $\overline{AB}$, then $AM = \frac{1}{2}AB$ and $MB = \frac{1}{2}AB$.

Inclusion Strategies

ENGLISH LANGUAGE DEVELOPMENT   Although the paper-folding activities in this lesson are simple, students who have difficulty reading may need to work with a partner or a group as they follow the steps.

Activity 2 Notes

In this Activity, students find the shortest distance from a point to a line by folding a perpendicular segment. Students should fold their papers to create a perfect match between line 1 and itself. This produces a fold line that is perpendicular to line 1.

For a student worksheet of this Activity and detailed Teacher Notes, see page 6 in the Lesson Activities booklet.

CHECKPOINT ✔
4. Answers will vary.

For a student worksheet of this Activity and detailed Teacher Notes, see page 6 in the Lesson Activities booklet.

CHECKPOINT ✔
2. equal
4. equal

Teaching Tip

In Step 2, have students connect the points on line *m* to points *A* and *B* in order to create a set of "nested" isosceles triangles. This may help students make an appropriate conjecture.

Before students try to make a conjecture in Step 4, remind them that *distance* refers to the distance along a perpendicular segment. Thus, the segments that are measured should be perpendicular to the sides of the angle.

Segment and Angle Bisectors

In Activity 3 below, you will use the following definitions:

Definitions

A **segment bisector** is a line that divides a segment into two congruent parts. The point where a bisector intersects a segment is the **midpoint** of the segment. A bisector that is perpendicular to a segment is called a **perpendicular bisector**. An **angle bisector** is a line or ray that divides an angle into two congruent angles.　　**1.4.2**

Activity 3
Exploring Segment and Angle Bisectors

YOU WILL NEED
folding paper, a marker, and a ruler

Part I　Segment Bisectors

1. Fold line ℓ. Choose points *A* and *B* on ℓ. Fold the paper so that *A* matches up with *B*. Label the resulting line *m*. What is the relationship between *m* and $\overline{AB}$?

2. Choose a point on *m* and label it *C*. Measure $\overline{AC}$ and $\overline{BC}$. Repeat, choosing several different locations for *C*. What do you notice? Complete the following conjecture:

CHECKPOINT ✔

The distances from a point on the perpendicular bisector to the endpoints of the segment are ____?____.

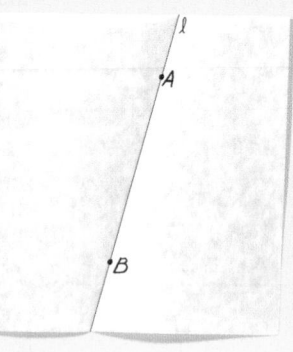

Part II　Angle Bisectors

3. Fold intersecting lines *j* and *k*. Label the intersection *P*. Label a point *Q* on *k* and a point *R* on *j*. Fold the paper through *P* so that line *j* matches up with line *k*. Label the new line *n*. What is the relationship between *n* and ∠*QPR*?

4. Choose a point on *n* and label it *S*. Measure the distance from *S* to *j* and the distance from *S* to *k*. Repeat, choosing several different locations for *S*. What do you notice? Complete the following conjecture:

CHECKPOINT ✔

The distances from a point on the angle bisector to the sides of the angle are ____?____.

Reteaching the Lesson

HANDS-ON STRATEGIES Have each student create a 4-inch square design for a quilt on a piece of paper. Each square must include one or more examples of perpendicular lines, parallel lines, segment bisectors, and angle bisectors. Display the quilt designs on a bulletin board for discussion and comparison.

Exercises

Communicate

1. When you folded line ℓ onto itself in Step 2 of Activity 1, which pairs of angles matched up? Use this to state an alternative definition of perpendicular lines.

2. When you constructed parallel lines m and n in Step 5 of Activity 1, how many right angles were formed? Make a conjecture about how you can determine whether two lines are parallel.

3. When you folded A onto B in Step 1 of Activity 3, how could you tell (without measuring) that the new line divided $\overline{AB}$ into two congruent segments?

4. Explain how you measured the distance from S to j and S to k in Step 4 of Activity 3.

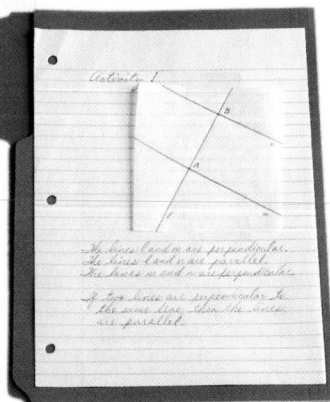

Tape, staple, or glue your folded papers to your work. Label the papers in case they come off.

ASSIGNMENT GUIDE

In Class	1–8
Core	9–21, 23–25 odd
Core Plus	22–32
Review	33–42
Preview	43–44

✐ Extra Practice can be found beginning on page 818.

Answers to odd-numbered Extra Practice exercises can be found immediately after Selected Answers in the Pupil's Edition.

Guided Skills Practice

For Exercises 5–8, use the figures below and your conjectures from the Activities to complete the following statements:

5. Lines ℓ and m are ___?___.
   (ACTIVITY 1) parallel

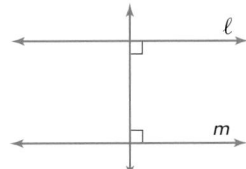

6. The distance from P to ℓ is ___?___. *(ACTIVITY 2)* PB

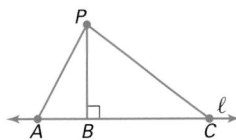

7. AC and BC are ___?___.
   (ACTIVITY 3) equal

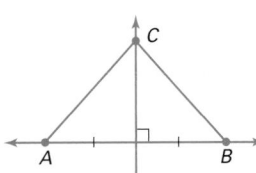

8. BX and CX are ___?___.
   (ACTIVITY 3) equal

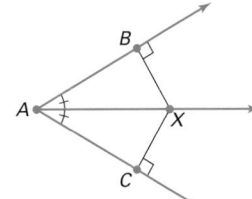

Technology

Exercises 9–21 can be done with geometry graphics software. You may wish to have students do the exercises twice, first by paper folding and then with geometry graphics software.

Error Analysis

For Exercises 22–24, make sure students do not create three lines that intersect at a single point. Explain that the three lines should form a triangle.

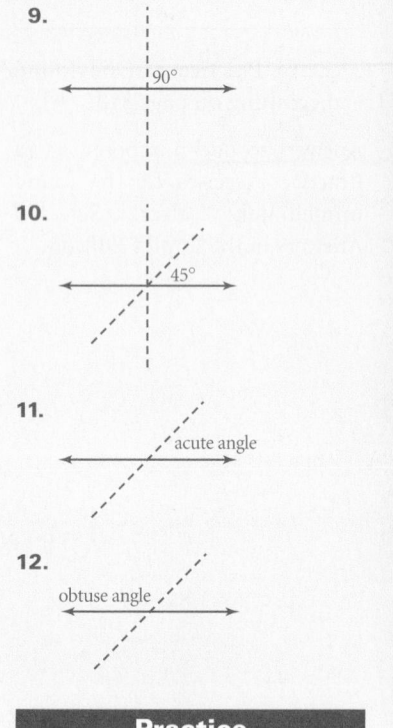

9.

10.

11.

12.

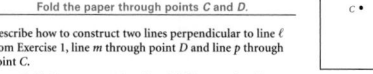
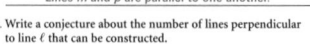

Use a separate piece of folding paper for each exercise. Attach the folded papers to your homework, and be sure to label each piece.

Fold each figure below. Do not use a ruler or protractor. Trace over each figure and label all relevant parts.

9. a 90° angle
10. a 45° angle
11. an acute angle
12. an obtuse angle
13. two complementary angles
14. two supplementary angles
15. two parallel lines
16. a rectangle
17. a square
18. a triangle
19. a right triangle (a triangle with one angle of 90°)
20. a triangle with two equal sides
21. a right triangle with two equal sides

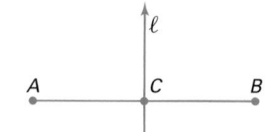
✔ internet connect
Homework Help Online
Go To: go.hrw.com
Keyword:
MG1 Homework Help
for Exercises 22–24

Fold a line and label two points *A* and *B* on the line. Construct the perpendicular bisector of $\overline{AB}$ and label it ℓ. Label a point *C* on ℓ, and fold or draw $\overline{AC}$ and $\overline{BC}$.

22. Using the first conjecture you made in Activity 3, what can you conclude about AC and BC?

23. Write a conjecture about triangles in which the vertex of one angle is on the perpendicular bisector of the side opposite that angle.

24. How could you use the conjecture from Exercise 23 to construct a triangle with three congruent sides? (Hint: Start by tracing $\overline{AB}$ onto another piece of folding paper.)

25. Suppose that ℓ is the perpendicular bisector of $\overline{AB}$ and that $AB = 10$.
    Find AC and BC.
    AC = BC = 5

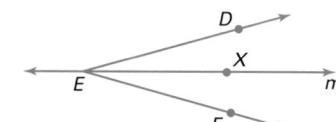

26. Suppose that *m* is the angle bisector of ∠DEF and that m∠DEX = 15°.
    Find m∠FEX and m∠DEF.
    m∠FEX = 15°, m∠DEF = 30°

---

13. ∠1 and ∠2 are complementary

14. ∠1 and ∠2 are supplementary

15. Lines ℓ and *m* are parallel.

16.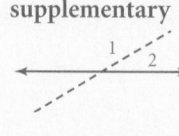

17.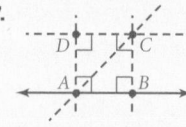

18.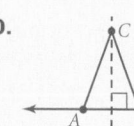

19.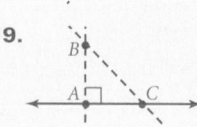

20.

21.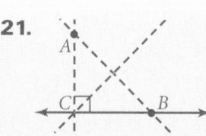

22. The lengths are equal.

Fold two intersecting lines, and label them as shown in the figure.

27. Fold the bisector of ∠*XVY* and label it ℓ. What is the relationship between ℓ and ∠*WVZ*?

28. Fold the bisector of ∠*WVX* and label it *m*. What is the relationship between ℓ and *m*?

29. Repeat Exercises 27 and 28 with several different pairs of intersecting lines. Make a conjecture about the bisectors of angles formed by intersecting lines.

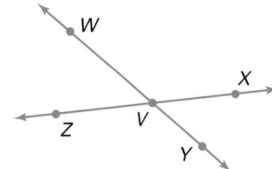

For Exercises 30–32, fold or draw two perpendicular lines on a piece of folding paper and label them ℓ and *m*. Label the intersection of ℓ and *m* point *A*.

30. Fold the angle bisector of one of the angles formed by ℓ and *m*. Choose a point on the angle bisector and label it *B*. Fold two lines through *B*, one perpendicular to ℓ and one perpendicular to *m*.

31. Make a conjecture about the shape you formed in Exercise 30. Measure any sides or angles necessary to confirm your conjecture.

32. $\overline{AB}$ is called a *diagonal* of this shape. Fold the other diagonal of the shape. Make a conjecture about the diagonals of the shape you constructed.

 **Look Back**

Name the geometric figure that each item suggests. *(LESSON 1.1)*

33. the edge of a table

34. the wall of a classroom

35. the place where two walls meet

36. the place where two walls meet the ceiling

Name each figure, using more than one name when possible. *(LESSON 1.1)*

37. 38. 39.

 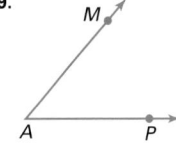

40. 41. 42.

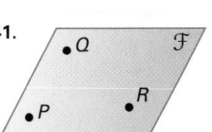

---

27. bisector

28. Lines ℓ and *m* are perpendicular.

29. The bisectors of the angles formed by intersecting lines are perpendicular.

30.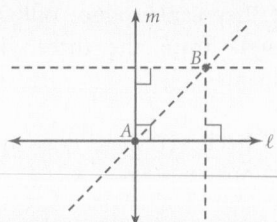

31. The shape appears to be a square.

32. Sample answer: The diagonals of a square are of equal length, and are perpendicular bisectors of each other.

33. line   34. plane

35. line   36. point

37. $\overleftrightarrow{XY}$, $\overleftrightarrow{YX}$, or line ℓ

38. $\overline{MN}$ or $\overline{NM}$

39. ∠*MAP*, ∠*PAM*, or ∠*A*

40. $\overrightarrow{PS}$

41. plane ℱ, plane *PQR*, plane *QRP*, or plane *RQP*

42. point *A*

In Exercises 43 and 44, students use line segments drawn on tracing paper to explore the Side-Side-Side Postulate for triangle congruence. Students should note that they can produce only one triangle from the three given segments.

43. Check student constructions.

44. No; any triangle created with the segments will have the same size and shape.

Assessment

Portfolio Activity

The Portfolio Activity can be used as preparation for the Chapter Project or as a separate activity. The Portfolio Activity on this page shows how to make a hexagon by connecting a series of six 60° angles.

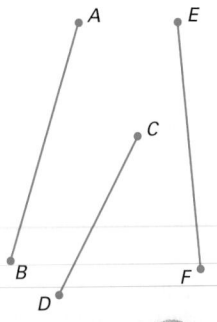

Look Beyond

In Chapter 4, you will examine properties of triangles. In Lesson 4.7, you will use a compass and straightedge to construct a triangle with given side lengths. You can also use tracing paper to construct such a triangle, as directed below.

43. Trace each segment at left onto a separate piece of tracing paper.
Place the paper with $\overline{AB}$ on top of the paper with $\overline{CD}$, and line up A and C. You may want to poke a pin through A and C to keep them aligned.

Place the paper with $\overline{EF}$ on top of the other two, line up E and B, and then line up F with D.

Trace the complete triangle on the top piece of paper.

44. Can you create a different triangle from these segments? Why or why not?

internet connect

Portfolio Extension
Go To: **go.hrw.com**
Keyword:
MG1 Origami

A *regular hexagon* is a six-sided figure with equal sides and equal angles. You can create one by following the steps below.

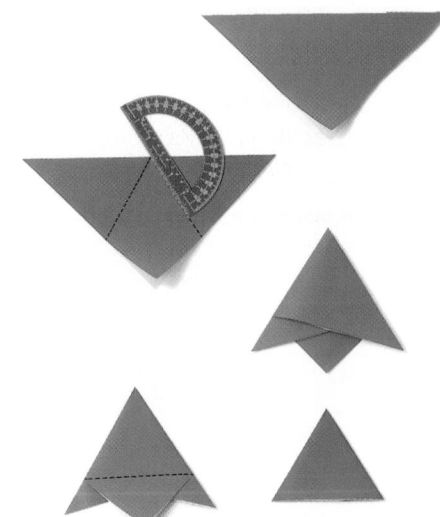

1. Fold a square piece of paper along one diagonal.

2. Place your protractor on the center point of the fold and mark off dashed lines at 60° angles.

3. Fold the right and left portions along the dashed lines.

4. Turn the figure over and mark a new dashed line. Cut along this line.

5. Open the figure. The finished product is a hexagon.

You can also create a snowflake pattern from these folds. Do not open your figure in Step 5. Instead, fold the figure vertically through the center another time. Cut notches into each side of the folded figure. When you unfold the figure, you will have a snowflake.

Special Points in Triangles

Objectives

- Discover points of concurrency in triangles.
- Draw the inscribed and circumscribed circles of triangles.

Why *Special points in triangles have many applications, from art to telecommunications.*

Archaeologists at this site in Belize have uncovered part of a circular structure. In this lesson, you will learn a method for finding the center of a circle when only a part is given.

Triangles and Circles

In this lesson, you will discover some interesting facts about triangles and circles. When you study proofs, you will understand why they are true.

Activity 1

Some Special Points in Triangles

YOU WILL NEED

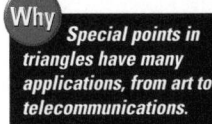

folding paper
OR
geometry software

In this Activity, you will discover some special points in triangles. Save your triangles for the next Activity. If you are using geometry software, you can "drag" the vertices of the triangles to explore different triangles.

1. Draw or fold a triangle. Then construct the perpendicular bisector of each side of the triangle. What do you notice about the perpendicular bisectors?

2. Draw or fold another triangle. Then construct the angle bisector of each angle of the triangle. What do you notice about the angle bisectors?

3. Share your results with your class. Complete the conjectures below.

CHECKPOINT ✔

The perpendicular bisectors of a triangle ___?___ at a single ___?___ .

The angle bisectors of a triangle ___?___ at a single ___?___ .

Alternative Teaching Strategy

TECHNOLOGY AND HANDS-ON STRATEGIES In this lesson, teachers have the choice of using paper folding, a compass, or geometry graphics software for completion of the Activities and exercises. Students should be allowed to choose a method once the teacher has demonstrated all methods. Many students may not have access to geometry graphics software at home and will need to use paper folding or a compass to complete the exercises.

Enrichment

A quadrilateral is cyclic if its four vertices lie on a circle. Challenge students to discover a property of the perpendicular bisectors of the four sides of a cyclic quadrilateral. **The bisectors are concurrent.**

Prepare

QUICK WARM-UP

Use paper and pencil to draw an example of each kind of triangle—right, equilateral, isosceles, acute, obtuse, and scalene. **Answers will vary.**

Also on Quiz Transparency 1.5

Teach

Why Using the incenter and circumcenter of triangles to inscribe and circumscribe circles has many important applications in art and industry.

Activity 1 Notes

In Activity 1, make sure students do not use equilateral or isosceles triangles as their only cases. Emphasize that their conjectures should apply to triangles of any shape.

For a student worksheet of this Activity and detailed Teacher Notes, see page 11 in the Lesson Activities booklet.

CHECKPOINT ✔
3. intersect, point intersect, point

incenter—the intersection point of the angle bisectors of a triangle

circumcenter—the intersection point of the perpendicular bisectors of a triangle

---

Special Circles Related to Triangles

For any triangle, you can draw an *inscribed circle* and a *circumscribed circle*. An **inscribed circle**, as the name suggests, is *inside* the triangle and just touches its three sides. A **circumscribed circle** is outside the triangle (*circum-* means "around") and contains all three vertices.

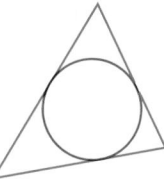
Inscribed circle

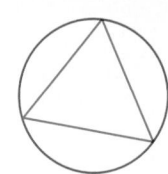
Circumscribed circle

Activity 2 Notes

In the previous Activity, students found that all four types of lines are concurrent. In this Activity, they test whether each intersection point is the center of an inscribed or circumscribed circle.

For a student worksheet of this Activity and detailed Teacher Notes, see page 11 in the Lesson Activities booklet.

CHECKPOINT ✔
2. inscribed; circumscribed

YOU WILL NEED

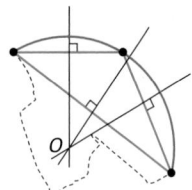

folding paper and compass
OR
geometry software

CHECKPOINT ✔

Activity 2
Constructing the Special Circles

1. Use the triangles you created in Activity 1. Draw circles with centers at the special points you discovered in Activity 1. Can you draw the special circles defined above?

2. Complete the conjectures below.

Inscribed and Circumscribed Circles

The intersection point of the angle bisectors of the angles of a triangle is the center of the ____?____ circle of the triangle.

The intersection point of the perpendicular bisectors of the sides of a triangle is the center of the ____?____ circle of the triangle.

EXAMPLE

APPLICATION
ARCHAEOLOGY

An archaeologist wants to find the original diameter of a broken plate. How can she do this by applying one of the conjectures from Activity 2?

SOLUTION

The archaeologist draws an outline of the broken plate. Then she selects three points on the circumference of the plate and connects them to form a triangle. The intersection of the perpendicular bisectors of the sides of the triangle is the center of the plate. (Only two bisectors are needed, however, the third acts as a check.)

The radius of the circle is the distance from the circle's center, *O*, to the edge of the plate. By measuring this distance and doubling it, the archaeologist can find the diameter of the plate.

The Metropolitan Museum of Art, Rogers Fund, 1.

Inclusion Strategies

USING VISUAL MODELS Students may gain more benefit from using geometry graphics software by making the drawings larger or bolder than usual.

Reteaching the Lesson

COOPERATIVE LEARNING Have students work in pairs. Each student should create one drawing of a triangle showing the altitudes, medians, angle bisectors, and perpendicular bisectors. Segments should be labeled with letters. Students should then exchange drawings and identify the four different types of segments.

Exercises

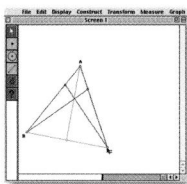

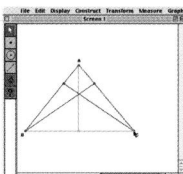

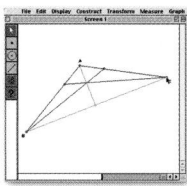

Using geometry graphics software, you can "drag" the points in a figure to examine different cases.

Communicate

1. The lines you constructed in Activity 1 are said to be **concurrent**. The word comes from the Latin words *con* ("together") and *currere* ("to run"). Why is this word appropriate for these lines?

2. The center of an inscribed circle is called the **incenter** of a triangle. Can the incenter be outside the triangle? Explain why or why not.

3. The center of a circumscribed circle is called the **circumcenter** of a triangle. Can the circumcenter be outside the triangle? Explain why or why not.

Geometry graphics software has many useful features for creating constructions.

Guided Skills Practice

4. What is the result when you fold or draw the perpendicular bisectors of the three sides of a triangle? *(ACTIVITY 1)*

5. What is the result when you fold or draw the angle bisectors of the three angles of a triangle? *(ACTIVITY 1)*

For Exercises 6 and 7, use your conjectures about inscribed and circumscribed circles to complete the table. *(ACTIVITY 2)*

	Intersecting lines	Type of circle formed	Name of center
6.	perpendicular bisectors	?	?
7.	angle bisectors	?	?

8. Trace the portion of a circle shown at right. Choose three points on the circle and draw a triangle to connect them. Then construct the circumscribed circle around the triangle to complete the figure. *(EXAMPLE)*

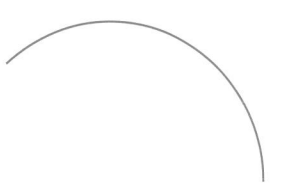

	Intersecting lines	Type of circle formed	Name of center
6.	perpendicular bisectors	circumscribed	circumcenter
7.	angle bisectors	inscribed	incenter

8.

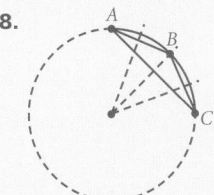

Cooperative Learning

Students can use geometry graphics software to explore the relationship between the center of a circle and the triangle inscribed in the circle.

ADDITIONAL EXAMPLE ①

Draw a curve that is smaller than a semicircle. How can you find the rest of the circle by connecting three points on the curve as a triangle? Find the intersection of the perpendicular bisectors of the triangle (circumcenter). Use the circumcenter and a point on the curve as a radius of the original circle. Draw the circle.

Assess

Selected Answers
Exercises 4–8, 9–37 odd

ASSIGNMENT GUIDE

In Class	1–8
Core	9–23
Core Plus	9–28
Review	29–37
Preview	38–40

✎ Extra Practice can be found beginning on page 818.

Answers to odd-numbered Extra Practice exercises can be found immediately after Selected Answers in the Pupil's Edition.

4. The perpendicular bisectors meet in a single point.

5. The angle bisectors meet in a single point.

9.
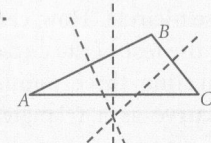

10.
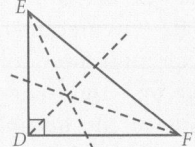

11.
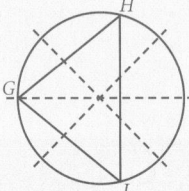

12.
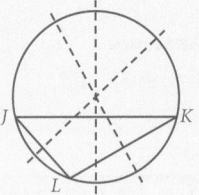

13.
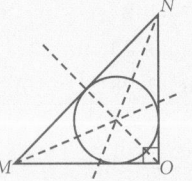

14.
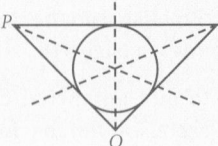

15.
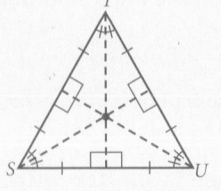

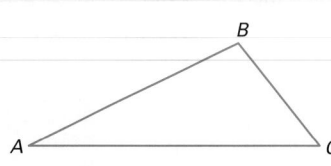

Practice and Apply

internet connect

Homework Help Online

Go To: **go.hrw.com**
Keyword:
MG1 Homework Help
for Exercises 9–23

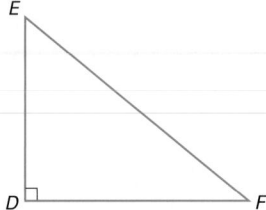

Draw each of the triangles below onto a separate piece of folding paper, or draw them by using geometry software. Triangles are named by the symbol △ and the names of the vertices, such as △*ABC*.

Find the following:

9. the perpendicular bisectors of each side of △*ABC*

10. the angle bisectors of each angle of △*DEF*

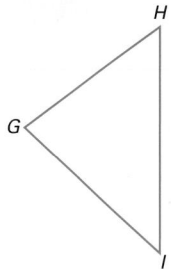

11. the circumscribed circle of △*GHI*

12. the circumscribed circle of △*JKL*

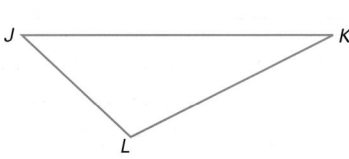

13. the inscribed circle of △*MNO*

14. the inscribed circle of △*PQR*

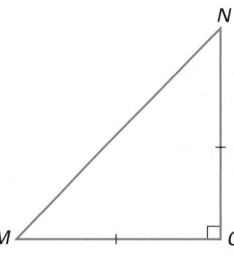

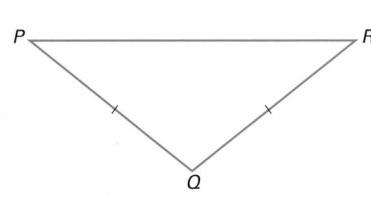

15. the perpendicular bisectors of △*STU*

16. the angle bisectors of △*STU*

17. the circumscribed circle of △*STU*

18. the inscribed circle of △*STU*

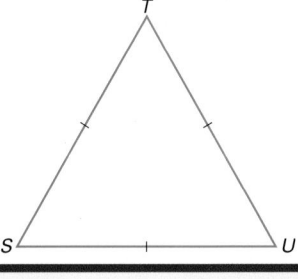

16.
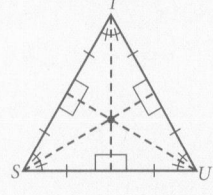

17.
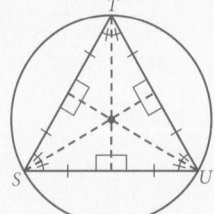

18.
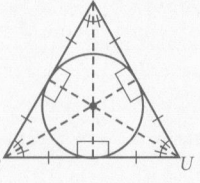

In Exercises 19–23, draw or fold three triangles:
- **an acute triangle—a triangle with all angles less than 90°**
- **an obtuse triangle—a triangle with one angle greater than 90°**
- **a right triangle—a triangle with one angle equal to 90°**

19. Construct the circumcenter of the acute triangle. Is it inside or outside the triangle?

20. Construct the circumcenter of the obtuse triangle. Is it inside or outside the triangle?

21. Where do you think the circumcenter of a right triangle should be? Construct the circumcenter of the right triangle to test your conjecture.

22. Use the circumcenter of the right triangle to draw the circumscribed circle. How does the longest side of the triangle divide the circle?

23. Based on your answer to Exercise 22, make a conjecture about the circumscribed circle of a right triangle. Draw several different right triangles and test your conjecture.

Draw or fold two acute triangles and construct the midpoint of each side. Use one in Exercise 24 and the other in Exercise 25.

24. Connect the midpoints to form another triangle. Your drawing should now contain four small triangles inside the original one. Cut out the four small triangles and compare them. What do you observe?

25. Draw a segment from each midpoint to the opposite vertex. These segments, called the *medians*, are shown in the figure at right. Label point *C* where the medians intersect. Measure the following segments in your triangle and complete the table:

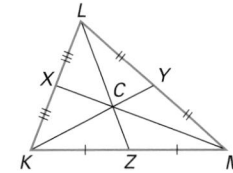

CK = ?	CL = ?	CM = ?
CY = ?	CZ = ?	CX = ?

What is the relationship between the lengths in the first row of the table and the lengths in the second row? Make a conjecture about point *C*, which is known as the *centroid*.

CHALLENGE

26. In the triangle at right, ℓ is the perpendicular bisector of $\overline{AB}$. Suppose that *X* is a point on ℓ. What can you say about *AX* and *BX*?

What can you say about any point on the perpendicular bisector of $\overline{AC}$? of $\overline{BC}$? Explain why the intersection of the perpendicular bisectors is the center of the circumscribed circle.

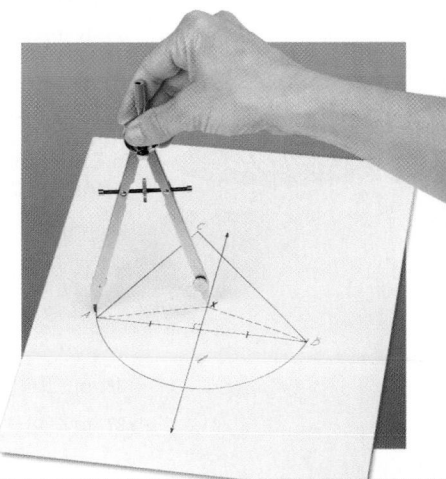

19. inside the triangle

20. outside the triangle

21. on the triangle

22. The longest side of the triangle divides the circle into two equal parts.

23. The center of the circumscribed circle of a right triangle is the midpoint of the longest side of the triangle.

24. The four triangles are exactly the same.

25. Sample answer: The distance from a vertex to the centroid is twice the distance from the centroid to the opposite side.

26. $AX = BX$. For any point, *Y*, on the perpendicular bisector of $\overline{AC}$, $AY = CY$. For any point, *Z*, on the perpendicular bisector of $\overline{BC}$, $BZ = CZ$. The intersection of the perpendicular bisectors is the same distance from each vertex of the triangle.

27. Y is the same distance from $\overline{DE}$ as it is from $\overline{EF}$. Any point on the angle bisector of $\angle D$ is the same distance from $\overline{DF}$ as it is from $\overline{DE}$. Any point on the perpendicular bisector of $\angle F$ is the same distance from $\overline{EF}$ as it is from $\overline{DF}$. The intersection of angle bisectors is the same distance from all 3 sides of the triangle.

28.

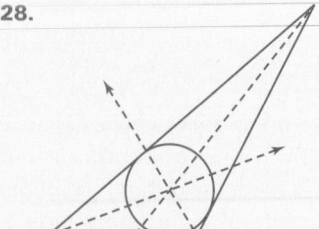

31. No; through any two points there is exactly one line.

32. No; through any three non-collinear points there is exactly one plane. If the points are collinear, they lie in an infinite number of planes.

33. Yes; if two non-parallel lines do not intersect, they are not in the same plane.

---

27. In the triangle at right, m is the angle bisector of $\angle E$. Suppose that Y is a point on m. What can you say about the distance from Y to sides $\overline{DE}$ and $\overline{EF}$?

What can you say about any point on the angle bisector of $\angle D$? of $\angle F$? Explain why the intersection of the angle bisectors is the center of the inscribed circle.

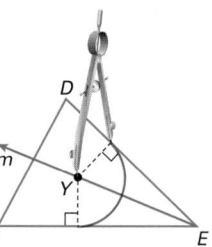

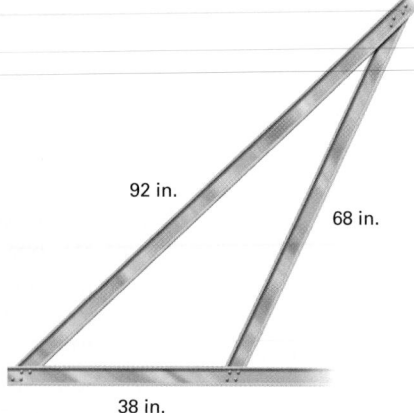

28. CONSTRUCTION
A contractor is installing a round air-conditioning duct. The duct must pass through a triangular opening, as shown in the figure at right. Trace the figure onto folding paper and construct the largest circle that will fit inside the triangle.

92 in.

68 in.

38 in.

Look Back

Complete the statements below. *(LESSON 1.1)*

29. Points that lie on the same line are said to be ____?____ . collinear

30. Points or lines that lie in the same plane are said to be ____?____ .coplanar

In Exercises 31–33, you may wish to include a diagram to illustrate your answer. *(LESSON 1.1)*

31. Is it possible for two points to be noncollinear? Why or why not?

32. Is it possible for three points to be noncoplanar? Why or why not?

33. Is it possible for two lines to be noncoplanar? Why or why not?

Algebra

In Exercises 34–37, refer to the figures below, in which $WY = 48$ and $\angle BAC \cong \angle CAD$. Find the measures listed below. *(LESSONS 1.4 AND 1.5)*

34. XY 23
35. WX 25

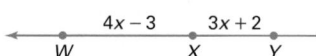

36. $m\angle BAC$ 13°
37. $m\angle CAD$ 13°

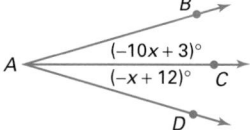

Another special point in a triangle is formed by the altitudes of the sides. An *altitude* is a perpendicular line segment from a vertex of a triangle to the line containing the opposite side.

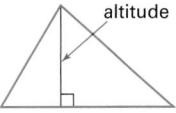

altitude

38. Draw or fold an acute triangle and label the vertices *A*, *B*, and *C*. How would you find the distance from *A* to $\overline{BC}$?

39. Construct the altitude for each vertex of △*ABC*. What do you notice?

40. Draw or fold an obtuse triangle. Construct the altitudes of this triangle. (Hint: Extend the sides of the triangle.) What do you notice? How is this different from the construction you did in Exercise 39?

A *median* of a triangle is a line segment from a vertex to the midpoint of the opposite side (see Exercise 25). In addition to the property of medians that you already examined, there is another interesting feature of medians.

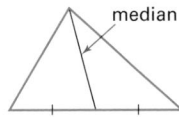

median

1. Cut out a triangle from stiff cardboard. Draw one median of the triangle. Try to balance the triangle on the line you drew. Make a conjecture about the part of the triangle on each side of the median.

2. Draw the other two medians of your triangle. Try to balance the triangle on the point where the medians intersect, known as the *centroid*. Explain in your own words why the centroid is called the *center of mass*.

3. According to a law of physics, a free-falling body should rotate around its center of mass. Test this theory by tossing your triangle like a Frisbee. What do you observe?

You will confirm your conjecture from Step 1 when you examine areas of triangles in Chapter 5.

In Exercises 38–40, students are asked to find the orthocenter of a triangle by using paper folding. Students will find that the orthocenter of an obtuse triangle lies outside the triangle.

38. Fold a line through *A* and perpendicular to $\overline{BC}$. Measure the length of the perpendicular.

39. All three altitudes meet in a single point inside the triangle.

40. If extended, all three altitudes meet in a single point. The point is outside the triangle.

ALTERNATIVE
Assessment

Portfolio Activity

The Portfolio Activity can be used as preparation for the Chapter Project or as a separate activity. The Portfolio Activity on this page has students find the centroid of a triangle, or the center of mass. Students are asked to balance a triangle by centering it on the centroid.

Student Technology Guide

In Lesson 1.5 of the textbook, you explored the behavior of altitudes, medians, perpendicular bisectors, and angle bisectors in triangles. Did you ever wonder how these special lines and segments behave in other figures—or whether they even exist in other figures? The exercises below explore these questions.

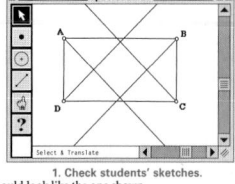

1. **a.** Draw four segments joined at their endpoints. Adjust their positions to form a rectangle like the one shown at right. Label the vertices *A*, *B*, *C*, and *D*, as shown.
1. Check students' sketches.
2–6. Answers may vary. Sample responses are given.
   b. Bisect each angle. The resulting figure should look like the one shown.
2. Unlike the angle bisectors of a triangle, these angle bisectors do *not* intersect at one point. What is the relationship among the angle bisectors in this figure?
   The four angle bisectors form a square; bisectors of consecutive angles are perpendicular; bisectors of opposite angles are parallel.
3. Drag $\overline{CD}$ closer to $\overline{AB}$, then farther away. Be careful that *ABCD* remains a rectangle. Do the angle bisectors ever intersect at one point? Use your observations to write a conjecture. They intersect at one point when *ABCD* is a square. When the length of *ABCD* is twice its width, two vertices of the square formed by the angle bisectors are at midpoints of the longer sides of *ABCD*.
4. Drag $\overline{CD}$ so that *ABCD* is no longer a rectangle. Do the angle bisectors ever intersect at one point? Use your observations to write a conjecture.
   No; they form a rectangle that is not a square.
5. Position $\overline{CD}$ so that *ABCD* is a rectangle again. Drag point *B* to different positions. Do the angle bisectors ever intersect at one point? Use your observations to write a conjecture. The angle bisectors form a non-rectangular quadrilateral; they intersect at one point for several positions of point *B* on the same line.
6. Make a new sketch of rectangle *ABCD*. Use the midpoint and perpendicular line features to construct the four perpendicular bisectors. What can you say about them?
   When *ABCD* is a rectangle, the perpendicular bisectors intersect at one point. If you drag $\overline{CD}$, the perpendicular bisectors always form a parallelogram. If you begin again with *ABCD* as a rectangle and drag point *B*, the perpendicular bisectors intersect at one point whenever $\overline{AB}$ and $\overline{BC}$ are perpendicular.

Motion in Geometry

> Every motion can be modeled in terms of three mathematical "motions," or transformations.

Objective

- Identify and draw the three basic rigid transformations: translation, rotation, and reflection.

This photo illustrates the basic mathematical "motion" or transformation known as translation. *The skier is moving in a straight line, without twisting or turning.*

Teach

Why A key topic in any geometry course is the study of congruence. If two figures can be made to coincide through a rigid motion, then the figures are congruent. Thus, rigid transformations provide a way of studying congruent figures.

Rigid Motion

Compare the two "snapshots" above of the same skier in motion. The picture on the left has been *translated* in the direction of the arrow. Notice that the two pictures are the same size and shape. If you cut one of them out, it would fit onto the other exactly.

As with segments and angles, planar figures that match exactly are said to be congruent. Transformations that do not change the size or shape of a figure are known as **rigid transformations**.

In geometry, the terms **preimage** and **image** refer to a shape that undergoes a motion or transformation. Points on the image of an object are usually named by adding a prime symbol (′) to the original name of the point.

You can easily draw a rigid transformation by tracing the same object in two different positions on a piece of paper. Label the first tracing as the preimage and the second one as the image. The only real trick in drawing the basic transformations is to locate the image properly. You will learn how to do this in the Activities that follow.

Alternative Teaching Strategy

TECHNOLOGY Geometry graphics software can be used to explore rigid motions. Have students use the reflection and rotation features of the program to create transformations of various figures. Students' results can be printed and used for discussion during the lesson.

Translations

In a **translation**, every point of a figure moves in a straight line, and all points move the same distance and in the same direction. The paths of the points are parallel.

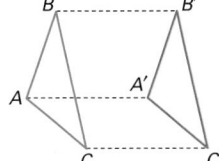

Activity 1
Drawing a Translation

1. Trace both the object and the line at right onto a piece of tracing paper. Do not move the paper yet.

2. Slide your paper so that your drawing of the line stays on the line in the diagram. Trace the object a second time.

CHECKPOINT ✔ 3. Label your first and second drawings of the object as *preimage* and *image*, respectively.

Rotations

In a **rotation**, every point of a figure moves around a given point known as the **center of rotation**. All points move the same angle measure.

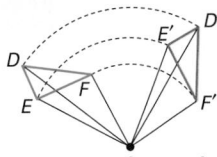

Center of rotation

Activity 2
Drawing a Rotation

1. Trace both the object and the point at right onto a piece of tracing paper. Do not move the paper yet.

2. Place the point of your pencil on the point, and press down to hold the tracing paper in place at that point. Turn the paper around the point as far as you wish. Trace the object a second time.

CHECKPOINT ✔ 3. Label your first and second drawings of the curved object as *preimage* and *image*, respectively.

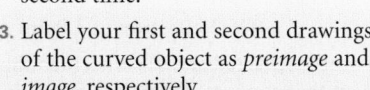

Interdisciplinary Connection

PHYSICS The study of objects reflected by mirrors or other surfaces can provide interesting experiments for students. Have them investigate what happens to an object when it is reflected in a convex or concave surface.

Activity 1 Notes

Students construct a translation and differentiate between an image and a preimage. Students will need tracing paper for this Activity.

For a student worksheet of this Activity and detailed Teacher Notes, see page 13 in the Lesson Activities booklet.

CHECKPOINT ✔
3. The preimage is the original object and the image is the object after it has been translated. Check students' work.

Activity 2 Notes

Students use tracing paper to draw the rotation of an object about a point or center of rotation.

For a student worksheet of this Activity and detailed Teacher Notes, see page 13 in the Lesson Activities booklet.

CHECKPOINT ✔
3. The preimage is the original object and the image is the object after it has been rotated. Check students' work.

Part I

Students reflect a single point across a line and complete the definition of a reflection of a point.

CHECKPOINT ✔

4. They are perpendicular and ℓ bisects $\overline{PP'}$; perpendicular bisector.

Teaching Tip

A Mira reflecting device is a good tool for performing reflections because students can see through the reflecting device to draw figures behind it. Another way to perform reflections is to place a piece of carbon paper inside the folded paper in Part I, Step 2 of Activity 3.

Reflections

Hold a pencil in front of a mirror and focus on the pencil point, P. The reflection of that point, called the reflection image of the point, P', will appear to be on the opposite side side of the mirror at the same distance from the mirror as P.

In a mathematical **reflection**, a line plays the role of the mirror, and every point in a geometric figure is "flipped" across the line. In Part I of Activity 3, paper folding is used to produce the appropriate motion of a reflected point.

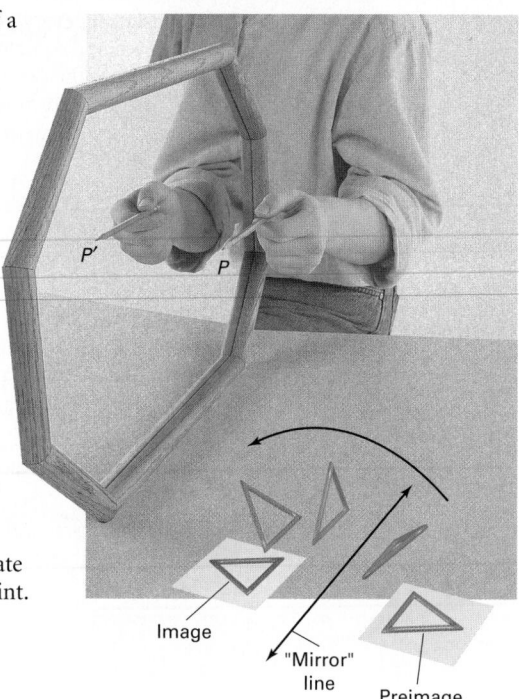

Image

"Mirror" line

Preimage

Activity 3
Drawing a Reflection

> **YOU WILL NEED**
>
> a pointed object, a ruler, a protractor, and scissors

Part I   The Reflection of a Point

1. Draw line ℓ and a point, P, not on line ℓ.

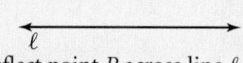

2. Reflect point P across line ℓ as follows: Fold your paper along line ℓ. Use a pointed object such as a sharp pencil or a compass point to punch a hole through the paper at point P. Unfold the paper and label the new point P'.

3. Draw $\overline{PP'}$ and label the intersection of $\overline{PP'}$ and line ℓ as point X.
   a. Measure the length of $\overline{XP}$ and $\overline{XP'}$.
   b. Measure the angles formed by $\overline{PP'}$ and line ℓ.

CHECKPOINT ✔ 4. What do you notice about the relationship between $\overline{PP'}$ and line ℓ? Complete the statement below, and then use your results from the Activities in Lesson 1.5 to explain why it is true.

The Reflection of a Point Across a Line

If a point is reflected across a line, then the line is the ____?____ of the segment that connects the point with its image.

Enrichment

A single-plane mirror can be used to demonstrate a reflection. Interesting effects are produced by using two mirrors hinged together with tape. By varying the angle between the mirrors, different numbers of reflected images are produced. Challenge students to explore the relationship between the hinge angle and the number of images.

Inclusion Strategies

ENGLISH LANGUAGE DEVELOPMENT   The terms in this lesson may be difficult for some students. Illustrate the ideas of *preimage*, *image*, and *reflection* by using a mirror. *Translation* and *rotation* can each be demonstrated by having students physically perform the movements.

Part II   The Reflection of a Triangle

1. Draw line ℓ with △ABC on one side.

2. Fold your paper along line ℓ and punch holes through points A, B, and C to obtain image points A′, B′, and C′. Connect the image points to form △A′B′C′, the reflection of △ABC.

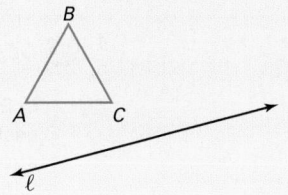

3. Cut out △ABC. Place it face down on △A′B′C′. Do the two triangles match exactly?

4. Form a conjecture about a triangle and its reflected image. Complete the statement below.

The Reflection of a Triangle

CHECKPOINT ✔   If a triangle is reflected across a line, then the reflected image of the triangle is _____?_____ to the original triangle.

Part II

In this Activity, students reflect a triangle across a reflection line. They complete a conjecture about the congruence of triangles after a reflection.

For a student worksheet of this Activity and detailed Teacher Notes, see page 13 in the Lesson Activities booklet.

CHECKPOINT ✔
4. congruent

TRY THIS   Reflect each figure below across line ℓ. Do your conjectures about the reflection of triangles and segments seem to be true when the figures touch the reflection line?

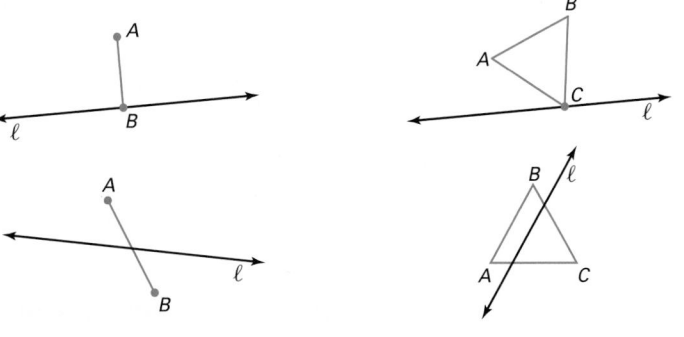

Teaching Tip

Point out that Part II of Activity 3 covers only the case in which the figure being reflected is completely on one side of the reflection line. The Try This problems involve two other cases: the preimage has one point on the reflection line and the preimage crosses the reflection line. Ask students to predict whether their conjecture from Step 4 will be true for the other cases.

TRY THIS
Yes; check students' reflections.

Summary: Three Rigid Transformations

The facts that you discovered by doing hands-on translations, rotations, and reflections in the Activities can be developed into the formal mathematical definitions given below.

A **translation** is a transformation in which every point of the preimage is moved the same distance in the same direction.   **1.6.1**

A **rotation** is a transformation in which every point of the preimage is moved by the same angle through a circle centered at a given fixed point known as the *center of rotation*.   **1.6.2**

A **reflection** is a transformation in which every point of the preimage is moved across a line known as the *mirror line* so that the mirror is the perpendicular bisector of the segment connecting the point and its image.   **1.6.3**

Reteaching the Lesson

COOPERATIVE LEARNING   Some students may find the steps involved in rotations difficult to follow. Provide students with a cardboard square, equilateral triangle, and regular hexagon. Have them create designs by rotating the figures about center points and vertex points. Students can work in groups and choose the number of degrees used in the various rotations.

Selected Answers

Exercises 8–10, 11–37 odd

ASSIGNMENT GUIDE

In Class	1–10
Core	11–22, 23–33 odd
Core Plus	23–33
Review	34–38
Preview	39–42

✎ Extra Practice can be found beginning on page 818.

Answers to odd-numbered Extra Practice exercises can be found immediately after Selected Answers in the Pupil's Edition.

Error Analysis

It may be helpful for students to perform transformations by using carbon paper to transfer the images to their notebook paper.

8.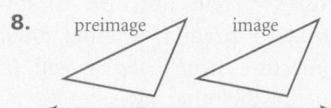

9.

10.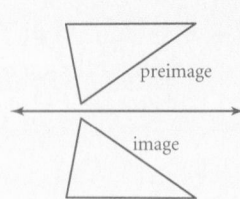

Exercises

● Communicate

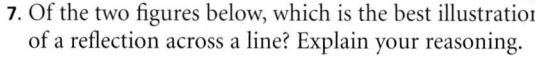

In Exercises 1–6, determine whether each description represents a translation, a rotation, a reflection, or none of these. Explain why. What is the direction of motion, center of rotation, or line of reflection in each case?

1. a canoe drifting straight ahead

2. a ball rolling down a hill

3. the image of a building in a lake

4. hands moving on a clock

5. a pair of scissors opening and closing

6. a slide projected onto a screen

7. Of the two figures below, which is the best illustration of a reflection across a line? Explain your reasoning.

● Guided Skills Practice

internet connect

Homework Help Online
Go To: go.hrw.com
Keyword:
MG1 Homework Help
for Exercises 8–23

In Exercises 8–10, trace the figures onto folding paper.

8. Translate the figure along the given line. *(ACTIVITY 1)*

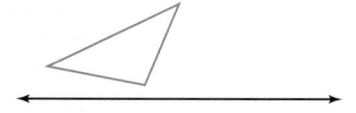

9. Rotate the figure about the given point. *(ACTIVITY 2)*

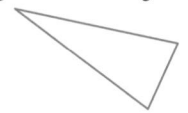

10. Reflect the figure across the given line. *(ACTIVITY 3)*

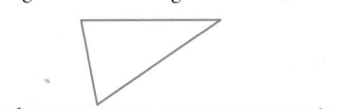

● *Practice and Apply* ━━━━━━━━━━━━

In Exercises 11–19, trace the figures onto folding paper.

Translate each figure along the given line.

11. 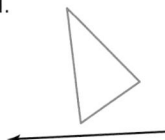   12. 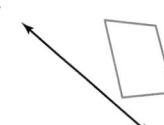   13.

Rotate each figure about the given point.

14. 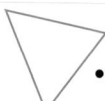   15. 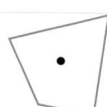   16.

Reflect each figure across the given line.

17.    18. 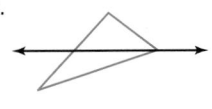   19.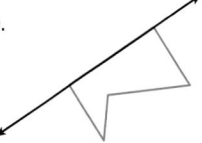

Copy the figures below onto tracing paper. Fold the paper along the line and trace the reflection of the figure on the back of the paper.

20.    21.

22. The word MOM has some unique characteristics when reflected across different lines. Reflect the word across each line.

a.    b.    c.

23. Write your name in capital letters on a piece of folding paper. Reflect your name across a vertical line. Do any of the letters stay the same? Reflect your name across a horizontal line. Do any of the letters stay the same?

24. In the diagram at right, △ABC has been translated to △A'B'C'. Copy the figures and draw $\overline{AA'}$, $\overline{BB'}$, and $\overline{CC'}$. What two things can you say about the relationship among these segments?

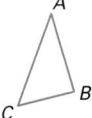 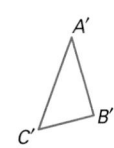

20.

21. 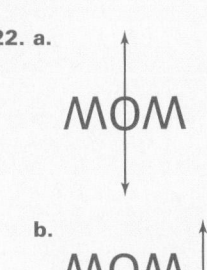 22. a.

b.

c.
MOM
WOW

23. The letters that stay the same in a reflection across a vertical line are A, H, I, M, O, T, U, V, W, X, and Y. The letters that stay the same in a reflection across a horizontal line are B, C, D, E, H, I, K, O, and X.

24. Each image point is the same distance from its preimage point. The segments are parallel and the same length.

11.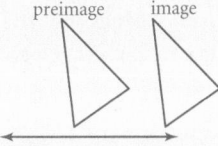
preimage   image

12.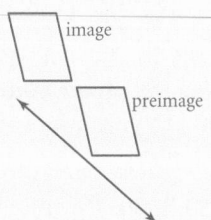
image   preimage

13.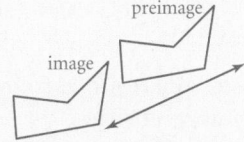
preimage   image

14.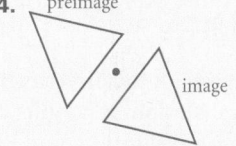
preimage   image

15.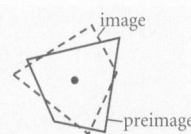
image   preimage

16.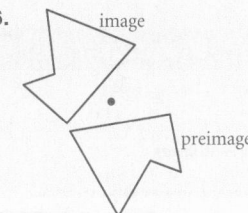
image   preimage

17.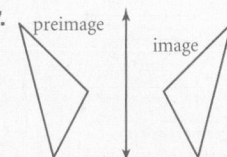
preimage   image

18.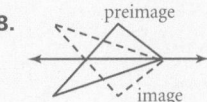
preimage   image

19.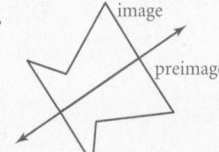
image   preimage

27. a. footprints made by the right and left foot, the pair on the left

b. footprints made by the same foot, the pair on the right

28. Reflection; the preimage is the wood block on the left and the image is the print on the right.

29. Translate square *ABCD* from its preimage position one unit up, one unit down, and one unit to the left. Then translate the square twice to the right by one unit.

30. Rotate *ABCD* around each of its vertices by 90° counter-clockwise. Then rotate *ABEF* 180° clockwise around the point *E*.

31. Reflect square *ABCD* from its preimage position through $\overline{CB}$, $\overline{CD}$, $\overline{DA}$, and $\overline{BA}$. Then reflect the square at *BEFA* through $\overline{EF}$.

32. The arrow in the direction of the translation is parallel to the line of reflection.

---

25. In the figure at right, △*DEF* has been rotated to △*D′E′F′*. Copy the figure and draw ∠*DCD′* and ∠*ECE′*. Measure these angles. What do you think is the measure of ∠*FCF′*?
The angles all measure 70°.

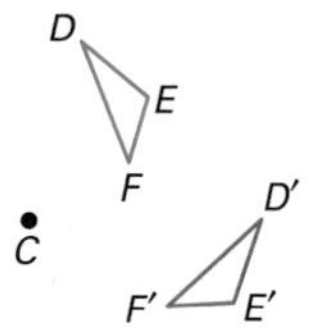

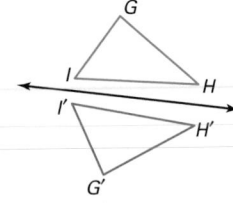

26. In the figure at right, △*GHI* has been reflected to △*G′H′I′*. Copy the figure and draw $\overline{GG'}$, $\overline{HH'}$, and $\overline{II'}$. What is the relationship between these segments and the line of reflection?
The line of reflection is the perpendicular bisector of $\overline{GG'}$, $\overline{HH'}$, and $\overline{II'}$.

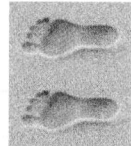

27. While visiting the beach, Teresa saw the two pairs of footprints shown at right.
a. Which pair represents a reflection?
b. Which pair represents a translation?

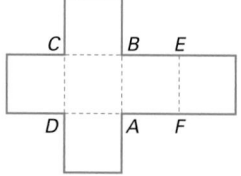

28. What transformation describes the relationship between the picture on the wood block and the one on the print? Which is the preimage and which is the image?

Wood block   *Print*

The shape below, called a *net*, can be cut out and folded to create a cube. Refer to the figure for Exercises 29–31.

29. Explain how to draw the net by starting with one square and applying translations.

30. Explain how to draw the net by starting with one square and applying rotations.

31. Explain how to draw the net by starting with one square and applying reflections.

```
        C   B  E
        D   A  F
```

A fourth type of transformation is a *glide reflection*. A glide reflection is a combination of a translation and a reflection.

32. The footprints shown below represent a glide reflection. Copy the figure, draw an arrow in the direction of the translation, and draw the line of reflection. What do you notice?

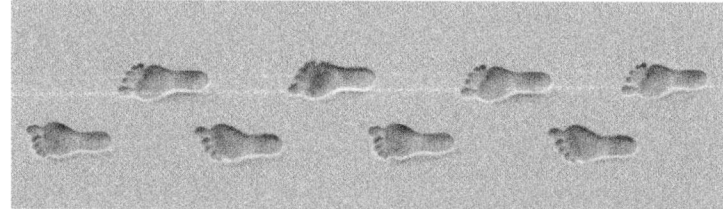

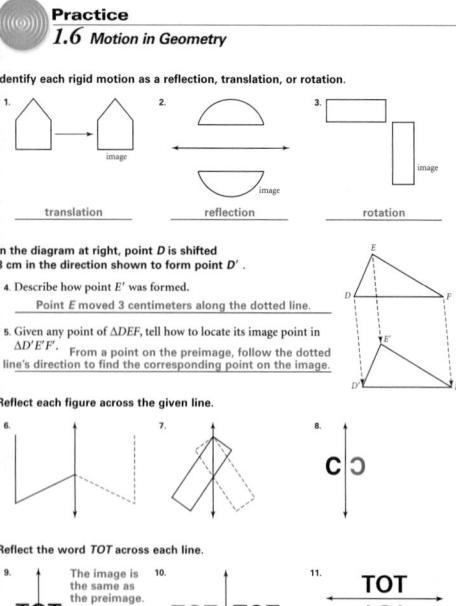

33. The figure below shows a shape that has been transformed by two glide reflections. How could you continue applying glide reflections to form a pattern? Copy the figure and draw two more images of the shape.

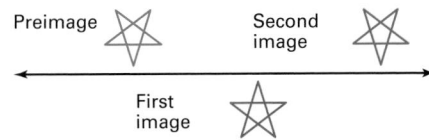

Preimage   Second image

First image

 *Look Back*

For each triangle in Exercises 34–36:
    a. Name all sides and angles of the triangle. *(LESSON 1.1)*
    b. Measure the lengths of the sides in centimeters. *(LESSON 1.2)*
    c. Measure the angles of the triangle. *(LESSON 1.3)*

34.   **35.**   **36.**

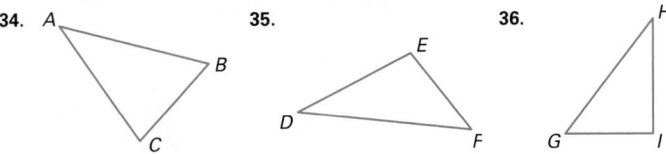

37. In the triangle at right, two angle bisectors have been constructed. Trace the figure and draw the third angle bisector. *(LESSON 1.5)*

38. In the triangle at right, two perpendicular bisectors have been constructed. Trace the figure and draw the third perpendicular bisector. *(LESSON 1.5)*

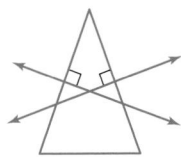

 *Look Beyond*

internet connect

Activities Online

Go To: go.hrw.com
Keyword:
MG1 Rigid

Sometimes you may want to translate a figure a given distance or rotate a figure by a given degree measure. This can be done with tracing paper, a ruler, and a protractor.

39. Trace the figure and the line below onto a piece of tracing paper. Then translate the figure so that point *T* on your paper matches up with point *S* in the diagram. What distance has the figure been translated?

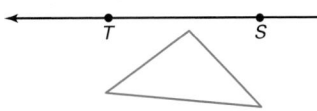

40. Use the method from Exercise 39 and a ruler to translate the figure at right 5 centimeters along the given line.

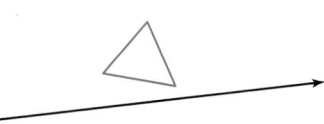

37.

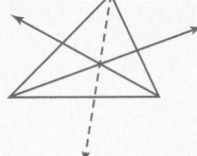

38.

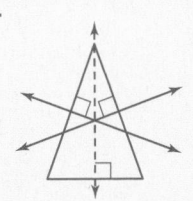

39. *TS* = 2.6 cm;

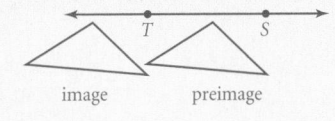

image   preimage

40.

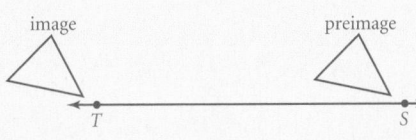

image   preimage

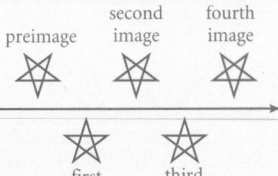

41. m∠WVX = 55°

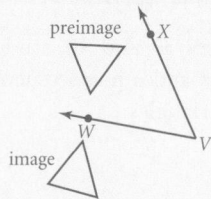

preimage ·X

W

image V

42.

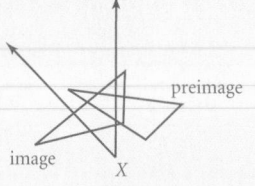

preimage

image
X

41. Trace the figure and the angle at right onto a piece of tracing paper. Then rotate the figure so that $\overrightarrow{VX}$ on your paper matches up with $\overrightarrow{VW}$ in the diagram. By what angle has the figure been rotated?

42. Use the method from Exercise 41 and a protractor to rotate the triangle at right counterclockwise by 45° about point X.

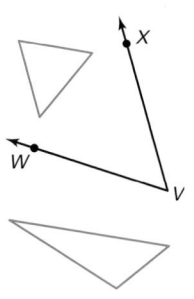

📶 internet connect 🔵 Go.hrw.com

Portfolio Extension

Go To: **go.hrw.com**
Keyword:
MG1 Snow

PORTFOLIO ACTIVITY

Snowflakes You may wish to use geometry software to complete this activity.

A ——————— B

1. Draw $\overline{AB}$.

2. Draw several more segments attached to $\overline{AB}$, all on the same side of the segment, as shown.

3. Reflect the segments across $\overline{AB}$.

4. Rotate the entire figure 60° about A. Then rotate the image 60° about A. Repeat until you have 6 images, including the preimage.

5. Create other snowflakes with this method. If you are using geometry graphics software, try dragging the points of the snowflake to see how your design changes.

Motion in the Coordinate Plane

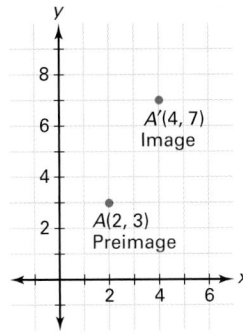

The downtown area of Provo, Utah, is laid out as a coordinate grid. Each block is one hundred units on a side.

Objectives

● Review the algebraic concepts of *coordinate plane, origin, x-* and *y-coordinates,* and *ordered pair.*

● Construct translations, reflections across axes, and rotations about the origin on a coordinate plane.

Why *A coordinate plane enables you to describe geometry ideas by using algebra. For example, you can use algebra to find the transformation of a figure without using special construction tools.*

Map © by RMC, R.L. 99-S-151, www.randmcnally.com

Teach

Why In this lesson, transformations are treated as functions, that is, as rules that are applied to a set of points on a plane to produce a unique new point for every point in the original set.

Operations on the Coordinates of a Point

By applying algebraic operations to the coordinates of a point, you can relocate it on the coordinate plane. For example, if you add 2 to the *x*-coordinate and 4 to the *y*-coordinate of the point $A(2, 3)$, the result is a new point, $A'(4, 7)$.

Preimage $A(2, 3)$ → **Transformation** → Image $A'(4, 7)$

This operation, which is known as a transformation, can be expressed as a rule by using transformation notation.

$$T(x, y) = (x + 2, y + 4)$$

The letter T stands for the transformation, but any other letter could be used. As you may notice, the transformation in the example above is a *function.*

Alternative Teaching Strategy

TECHNOLOGY The Activities and exercises in this lesson can be done with a computer program or a graphics calculator. Set the range for *x* and *y* at −10 to 10.

Activity 1 Notes

In this Activity, students discover that changing the coordinates of a set of points results in horizontal and vertical translations.

For a student worksheet of this Activity and detailed Teacher Notes, see page 17 in the Lesson Activities booklet.

Cooperative Learning

Have students create cases that cover all quadrants in the coordinate plane.

CHECKPOINT ✔

5. To translate a figure a given number of units horizontally, add the same number, h, to the x-coordinate of each point in the figure. To translate a figure a given number of units vertically, add the same number, v, to the y-coordinate of each point in the figure.
Horizontal: $H(x, y) = (x + h, y)$
Vertical: $V(x, y) = (x, y + v)$

Teaching Tip

An alternative way of explaining how to translate a point in the coordinate plane is given below.

horizontally: If point (a, b) is moved 1 unit to the right, the coordinates become $(a + 1, b)$.

vertically: If point (a, b) is moved 1 unit up, the coordinates become $(a, b + 1)$.

In the Activities that follow, you will discover rules for the three basic rigid transformations of geometry in a coordinate plane.

Activity 1
Translations

YOU WILL NEED
graph paper, a ruler, and a protractor

1. Copy $\triangle ABC$ onto graph paper.

2. Pick a number between -10 and $+10$ (except 0). Then choose either x or y.

3. Depending on your choice, add the number you picked to either the x- or the y-coordinate of points A, B, and C. The other coordinate remains unchanged in each point.

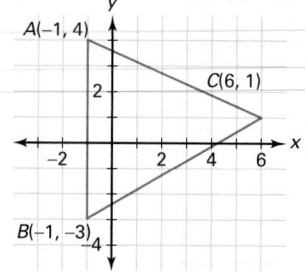

Plot each of your new points, and then connect them to form a new triangle. Does your new triangle seem to be congruent to the original one? Measure the sides and the angles of each triangle.

PROBLEM SOLVING

4. **Look for a pattern.** The original figure has been translated to a new position. In what direction has it moved and by how much? Draw an arrow to show the translation. Compare your results with those of your classmates.

CHECKPOINT ✔

5. Explain how to translate a figure horizontally or vertically in the coordinate plane. What would you do to the coordinates of each point in the figure? Express your method as a set of rules for moving points in the original figure h units horizontally or v units vertically.

Horizontal and Vertical Coordinate Translations

Horizontal translation of h units: $H(x, y) = (\underline{?}, \underline{?})$
Vertical translation of v units: $V(x, y) = (\underline{?}, \underline{?})$ **1.7.1**

APPLICATION
NAVIGATION

Refer to the map at right. Suppose that you are driving from point A at the intersection of 500 W and 200 S to point B at the intersection of 300 E and 400 N. The rule for the horizontal movement is $H(x, y) = (x + 800, y)$. The rule for the vertical movement is $V(x, y) = (x, y + 600)$. These rules can be combined to show both the horizontal and vertical movements: $T(x, y) = (x + 800, y + 600)$.

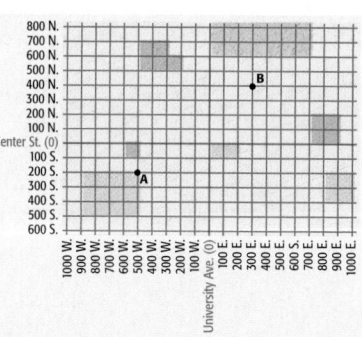

Interdisciplinary Connection

ART Rigid transformations are used frequently in art and design, particularly for wallpaper, tile, and fabric patterns. Have students collect examples and critique some of the patterns. When is repetition boring? When is variety confusing? Encourage students to use the terminology of transformations in their written work.

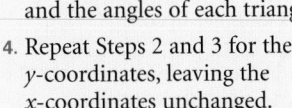

Activity 2
Reflections Across the *x*- or *y*-axis

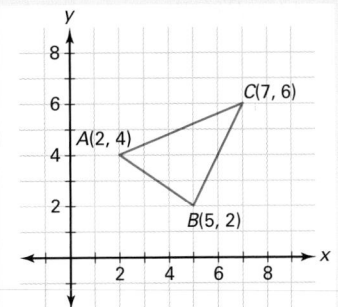

YOU WILL NEED

graph paper, a ruler, and a protractor

1. Copy △*ABC* onto graph paper.

2. Multiply the *x*-coordinate of each vertex by −1. Leave the *y*-coordinate unchanged.

3. Plot each of the new points, and connect them to form the new triangle. Does the new triangle seem to be congruent to the original one? Measure the sides and the angles of each triangle.

4. Repeat Steps 2 and 3 for the *y*-coordinates, leaving the *x*-coordinates unchanged.

5. The original triangle is reflected in each case. What are the lines of reflection, or the "mirrors" of the reflections?

CHECKPOINT ✔ 6. Explain how to reflect a figure across the *x*- or *y*-axis. What would you do to each point in the figure? Express your method as a set of rules for reflecting points in the original figure across the *x*- or *y*-axis.

Reflection Across the *x*- or *y*-axis

Reflection across the *x*-axis: $M(x, y) = (\underline{?}, \underline{?})$
Reflection across the *y*-axis: $N(x, y) = (\underline{?}, \underline{?})$

1.7.2

TRY THIS Experiment with figures in other positions, such as those below. Do your rules seem to work for all positions?

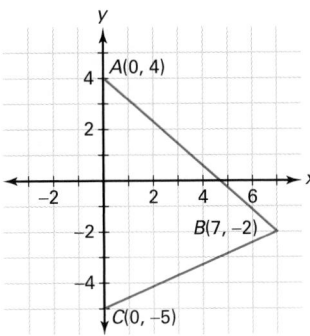 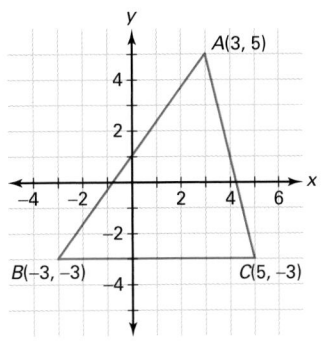

CRITICAL THINKING How would you write a rule for translating a point and then reflecting it across an axis?

Enrichment

If a line can be drawn through a geometric figure so that the two halves are reflections of each other, then the line of reflection is also a line of symmetry. Challenge students to find figures that have more than one line of symmetry. Examples include squares, equilateral triangles, and all the regular polygons.

Inclusion Strategies

USING COGNITIVE STRATEGIES Associating ordered pairs of the form (*x*, *y*) with points in a coordinate plane is difficult for some students. For example, a common mistake is to graph (0, 3) instead of (3, 0). Provide extra practice in plotting points, and ask students to double-check all plotted points.

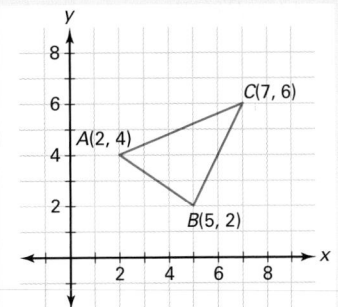

Activity 2 Notes

Have students do this Activity with a triangle drawn in each of the four quadrants. If students are working in groups of four, each student can do one of the quadrants.

For a student worksheet of this Activity and detailed Teacher Notes, see page 17 in the Lesson Activities booklet.

CHECKPOINT ✔

6. To reflect a figure across the *x*-axis, multiply the *y*-coordinate of each point in the figure by −1. To reflect a figure across the *y*-axis, multiply the *x*-coordinate of each point in the figure by −1.
   Reflection across the *x*-axis: $M(x, y) = (x, -y)$
   Reflection across the *y*-axis: $N(x, y) = (-x, y)$

TRY THIS

Yes; the students' methods should work for figures positioned anywhere in the coordinate plane.

CRITICAL THINKING

Combine the rules for translations and reflections. For instance, if the point (*a*, *b*) were translated 1 unit to the right and then reflected across the *x*-axis, the image would be (*a* + 1, −*b*).

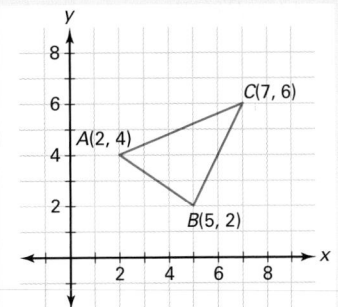

Teaching Tip

Students will find that geometry graphics software is useful for exploring a wide variety of cases before making conjectures.

Activity 3 Notes

In this Activity, students discover a way to manipulate the numbers in ordered pairs in order to rotate a figure 180° about the origin. This is equivalent to reflecting the figure twice, once across each axis.

For a student worksheet of this Activity and detailed Teacher Notes, see page 17 in the Lesson Activities booklet.

CHECKPOINT ✔

5. If the preimage is (x, y), then the image after a 180° rotation about the origin is $(-x, -y)$.

CRITICAL THINKING

Yes, a rotation of 180° about the origin can be thought of as a reflection across the x-axis followed by a reflection across the y-axis or vice-versa. Not all rotations can be thought of in this way. However, rotations that are multiples of 90°, such as 360°, 180°, and 270° can be thought of as a series of reflections across the x- and y-axes.

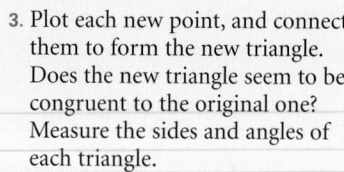

YOU WILL NEED

graph paper, a ruler, and a protractor

Activity 3
180° Rotations About the Origin

1. Copy $\triangle ABC$ onto graph paper.

2. Multiply the x- and y-coordinates of each vertex by −1.

3. Plot each new point, and connect them to form the new triangle. Does the new triangle seem to be congruent to the original one? Measure the sides and angles of each triangle.

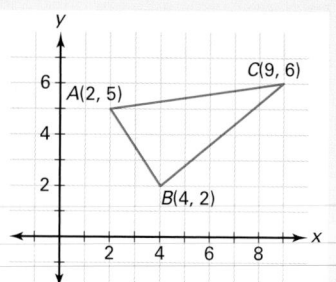

4. The original figure has been rotated. Where is the center of the rotation? By how many degrees has the figure been rotated? Draw a curved arrow to show the rotation.

CHECKPOINT ✔

5. Explain how to rotate a figure 180° about the origin. Express your method as a rule for rotating points in the original figure about the origin.

180° Rotation About the Origin

$$R(x, y) = (\underline{\ ?\ }, \underline{\ ?\ })$$ **1.7.3**

TRY THIS Use your rule from Activity 3 to rotate the triangle with vertices at $(1, 5)$, $(-2, 4)$, and $(-3, -2)$ by 180° about the origin.

CRITICAL THINKING You can think of the figure at right as being formed by two reflections of the upper left panel. First the original panel is reflected across the y-axis. Then the resulting figure is reflected across the x-axis.

Explain why a 180° rotation can be thought of as a combination of these two reflections. Do you think this idea could be extended to describe any rotation as a combination of two reflections?

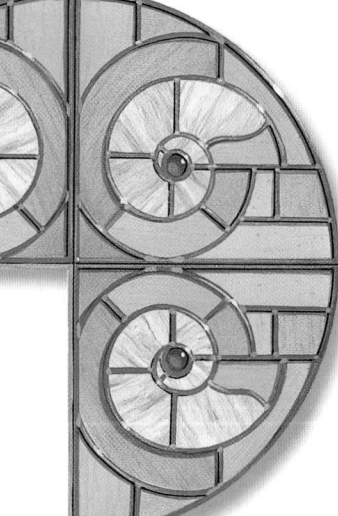

Reteaching the Lesson

COOPERATIVE LEARNING Students can play a spin-off of the game Battleship. A student plots several points and then transforms them all in the same way (for instance, 4 units to the left). The student then gives his or her partner a coordinate plane with the transformed points plotted on it. The partner gets 10 guesses to find information about the original, preimage points in order to figure out how the points have been transformed. The player who finds the correct transformation in the fewest number of guesses wins.

Exercises

Communicate

1. Explain how you would find the *x*- and *y*-coordinates of points *A* and *B* on the coordinate plane at right.

2. Plot the points (1, 5) and (5, 1). Do you think that the order of the coordinates matters? Why do you think the coordinates of a point are called *ordered pairs*?

3. What are some advantages to using the coordinate plane for transformations?

4. What happens when you reflect a figure across the *x*-axis and then reflect the image across the *y*-axis? Explain the result in terms of a single transformation in a coordinate plane.

Guided Skills Practice

5. Plot the points (2, 4), (−1, 4), and (2, 1), and connect them to form a triangle. What rule would you use to translate the triangle 5 units to the right? Write your rule in the form $H(x, y) = (\underline{?}, \underline{?})$. Draw the translated figure and label its coordinates. **(ACTIVITY 1)**

6. Plot the points (1, 3), (−2, 5), and (0, 0), and connect them to form a triangle. What rule would you use to reflect the figure across the *y*-axis? Write your rule in the form $N(x, y) = (\underline{?}, \underline{?})$. Draw the reflected figure and label its coordinates. **(ACTIVITY 2)**

7. Plot the points (2, 2), (4, 4), and (2, 4), and connect them to form a triangle. What rule would you use to rotate the figure 180° about the origin? Write your rule in the form $R(x, y) = (\underline{?}, \underline{?})$. Draw the rotated figure and label the coordinates. **(ACTIVITY 3)**

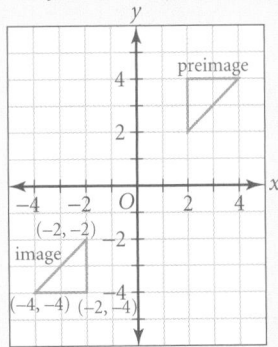

Rainbow over Provo, Utah.

Map © by RMC. R.L. 99-S-151. www.randmcnally.com

Error Analysis

For exercises involving reflections, make sure that students do not become confused. When a point is reflected across the *x*-axis, the change is in the *y*-coordinate: (a, b) becomes $(a, -b)$. The reverse applies when a point is reflected across the *y*-axis. The change is in the *x*-coordinate: (a, b) becomes $(-a, b)$.

5. $H(x, y) = (x + 5, y)$;

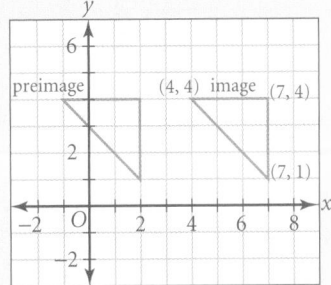

6. $N(x, y) = (-x, y)$;

7. $R(x, y) = (-x, -y)$;

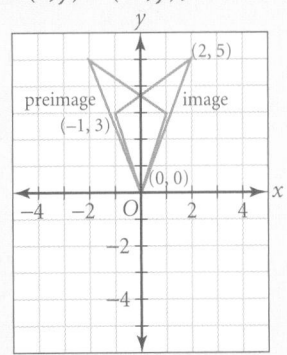

Technology

Exercises 8–17 can be done with geometry graphics software on an *xy*-coordinate plane or with a graphics calculator.

8. horizontal translation;

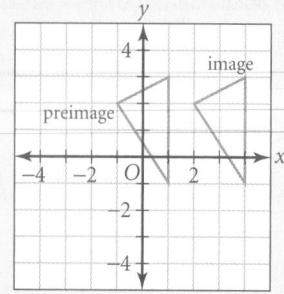

9. horizontal translation;

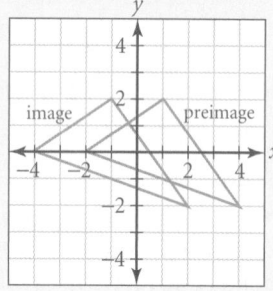

10. vertical translation;

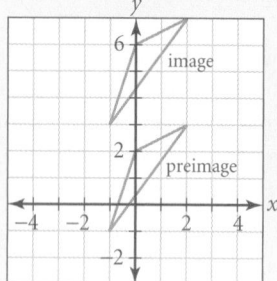

11. vertical translation;

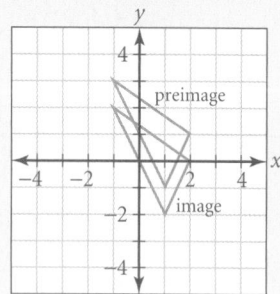

12. reflection across the *x*-axis;

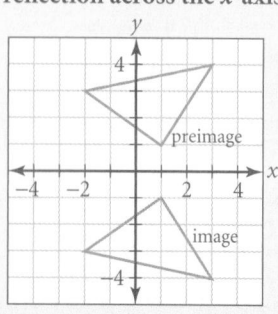

● *Practice and Apply*

In Exercises 8–17, copy each figure onto graph paper, and then use the given rule to transform the figure. Identify the type of transformation given by each rule.

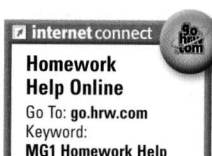

internet connect

Homework Help Online

Go To: **go.hrw.com**
Keyword:
MG1 Homework Help
for Exercises 8-30

8. $H(x, y) = (x + 3, y)$

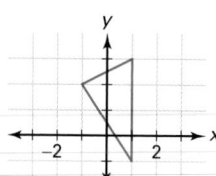

9. $H(x, y) = (x - 2, y)$

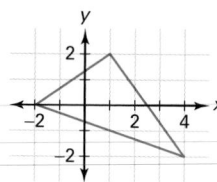

10. $V(x, y) = (x, y + 4)$

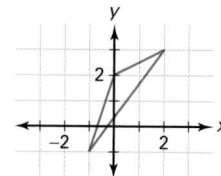

11. $V(x, y) = (x, y - 1)$

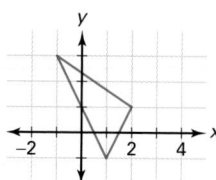

12. $M(x, y) = (x, -y)$

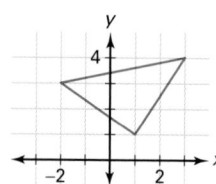

13. $M(x, y) = (x, -y)$

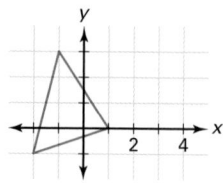

14. $N(x, y) = (-x, y)$

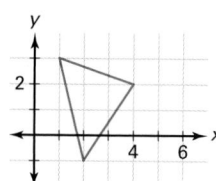

15. $N(x, y) = (-x, y)$

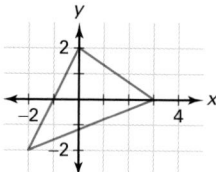

16. $R(x, y) = (-x, -y)$

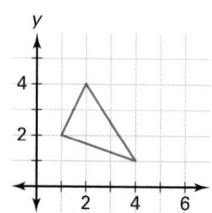

17. $R(x, y) = (-x, -y)$

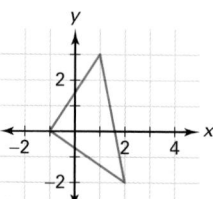

13. reflection across the *x*-axis;

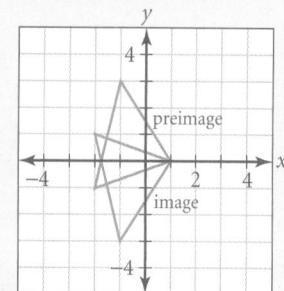

14. reflection across the *y*-axis;

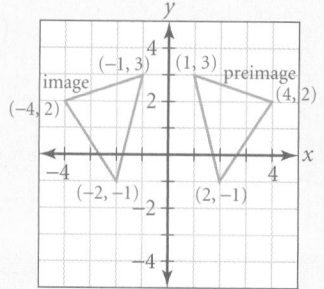

64 LESSON 1.7

In Exercises 18–22, use graph paper to draw the transformation(s) of the figure as indicated. Write the rule(s) you used below each transformed figure.

18. Reflect the figure across the *x*-axis.

19. Reflect the figure across the *y*-axis.

20. Rotate the figure 180° about the origin.

21. Reflect the figure across the *x*-axis, and then reflect the image across the *y*-axis. How does this figure relate to the transformed figure from Exercise 20?

22. Reflect the figure across the *y*-axis, and then reflect the image across the *x*-axis. How does this figure relate to the transformed figures from Exercises 20 and 21?

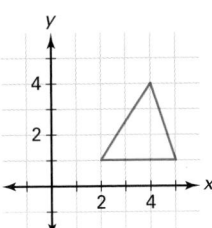

In Exercises 23–30, describe the result of applying each rule to a figure in a coordinate plane.

23. $F(x, y) = (x + 7, y)$

24. $T(x, y) = (-x, y)$

25. $A(x, y) = (x - 6, y + 7)$

26. $P(x, y) = (x, y - 4)$

27. $C(x, y) = (x, y + 7)$

28. $W(x, y) = (-x, -y)$

29. $Z(x, y) = (x - 7, y)$

30. $K(x, y) = (x, y + 2)$

31. Copy the figure below onto graph paper. Apply Rule 1 to the figure, then Rule 2 to its image, and so on. Draw each image that results.

Rule 1: $A(x, y) = (x, -y)$

Rule 2: $B(x, y) = (-x, y)$

Rule 3: $C(x, y) = (x, -y)$

Rule 4: $D(x, y) = (-x, y)$

What is the final result of applying Rules 1–4?

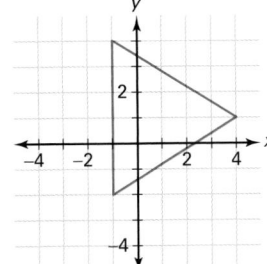

Algebra

In Exercises 32–35, you will use the idea of the *slope* of a line. Recall from algebra that the slope of a line is the change in the *y*-coordinates divided by the change in the *x*-coordinates for any two points on the line. Graph the triangle with vertices *K*(0, 0), *L*(1, 2), and *M*(0, 3).

32. Apply the transformation $T(x, y) = (x + 4, y + 3)$ to the triangle. Label the new vertices *K′*, *L′*, and *M′*.

33. Draw a line through *K* and *K′*. What is its slope? Draw a line through *L* and *L′*. What is its slope? Draw a line through *M* and *M′*. What is its slope?

34. How do the slopes of the lines relate to the transformation rule $T(x, y)$?

35. Based on your results from Exercises 32–34, make a conjecture about the slope of the line through *K* and *K′* when the transformation $T(x, y) = (x + h, y + k)$ is applied to point *K*.

Math
CONNECTION

ALGEBRA In Exercises 32–35, students use the concept of slope to analyze a transformed image. In Exercises 36–49, students use *x*- and *y*-coordinates to transform images.

18.

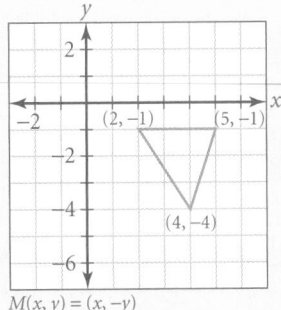

$M(x, y) = (x, -y)$

19.

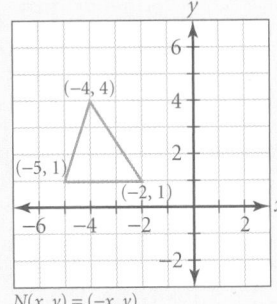

$N(x, y) = (-x, y)$

20.

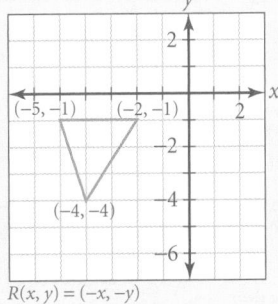

$R(x, y) = (-x, -y)$

21. The result is a 180° rotation about the origin as in Ex. 20.

22. The result is a 180° rotation about the origin as in Ex. 20 and 21.

23. translation, 7 units to the right

24. reflection across the *y*-axis

25. translation, 6 units to the left and 7 units up

26. translation, 4 units down

27. translation, 7 units up

28. rotation of 180° about the origin

29. translation, 7 units to the left

30. translation, 2 units up

31.

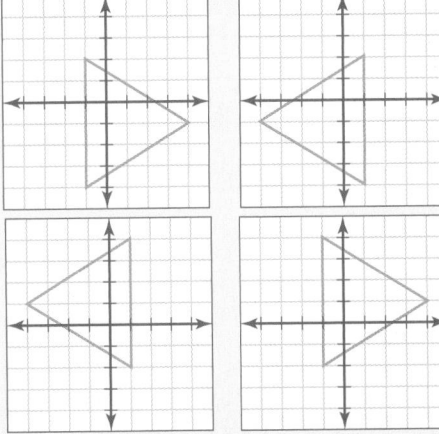

32.

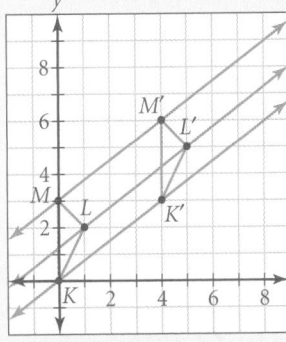

33. $\frac{3}{4}, \frac{3}{4}, \frac{3}{4}$

34. The slope is the ratio of the amount of the vertical translation over the amount of the horizontal translation.

36.

x	y
0	0
2	2
−1	−1

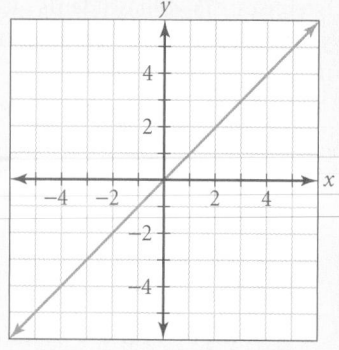

37. 45°

38.–39.

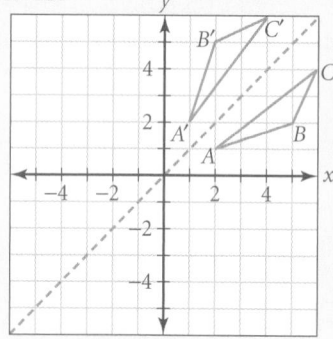

40. The two figures are reflections of each other across the line $y = x$.

41. $R(x, y) = (y, x)$

Practice

AME _____ CLASS _____ DATE _____

Practice

1.7 Motion in the Coordinate Plane

Use the given rule to translate each triangle on the grid provided.

1. $H(x, y) = (x, y + 3)$ 2. $H(x, y) = (x, -y)$

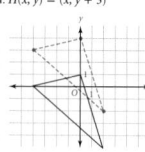

3. $H(x, y) = (x - 4, y)$ 4. $H(x, y) = (x + 2, y + 1)$

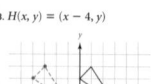

Describe the result of applying each rule below to a figure in a coordinate plane.

5. $G(x, y) = (x + 6, y)$ 6. $F(x, y) = (x, y - 1)$
   translation right 6 units   translation down 1 unit

7. $P(x, y) = (x, -y)$ 8. $H(x, y) = (-x, y)$
   reflection across the x-axis   reflection across the y-axis

9. $T(x, y) = (x - 4, y - 5)$ 10. $R(x, y) = (x + 2, y + 2)$
   translation left 4 units and down 5 units   translation right 2 units and up 2 units

11. $M(x, y) = (x + 5, y + 2)$ 12. $N(x, y) = (-x, -y)$
   translation right 5 units and up 2 units   rotation through 180°

Algebra

In Exercises 36–43, you will discover another rule for transforming an object.

36. On graph paper, graph the line $y = x$. Complete the table of values at right to get started.

x	y
0	?
2	?
−1	?

37. What angle does the line form with the x-axis?

38. Plot each of the following points on the graph, and connect them to form a triangle:

    $A(2, 1)$ $B(5, 2)$ $C(6, 4)$

39. Reverse the x- and y-coordinates of points A, B, and C to create new points A', B', and C'. Plot these points on the graph with A, B, and C, and connect them to form a new triangle.

40. What is the relationship between the figures you drew?

41. Write a rule in the form $R(x, y) = (\underline{?} , \underline{?})$ for the transformation. (Hint: The rule should use only the letters x and y, with no signs or numbers.)

42. Draw segments connecting A to A', B to B', and C to C'. What is the relationship between these segments and the line $y = x$?

43. Choose a point (x, y), where $x < y$. Is this point above or below the line $y = x$? Apply the transformation rule you wrote in Exercise 41 to this point. Is the image of the point above or below the line $y = x$? Repeat this exercise for points where $x > y$ and $x = y$.

CHALLENGES

In Exercises 44–47, you will explore glide reflections in a coordinate plane.

44. What two transformations would you apply to $\triangle ABC$ to get $\triangle DEF$? Express these transformations with a single rule of the form $T(x) = (\underline{?} , \underline{?})$.

45. Apply the transformation rule you found in Exercise 44 to $\triangle DEF$. What is the result? Apply the transformation to the resulting image three or four more times. Describe what happens.

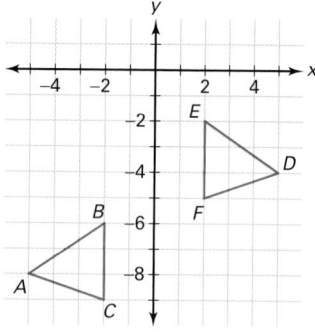

46. Write a general rule for a glide reflection with the y-axis as the line of reflection and with a vertical motion of v.

47. Write a general rule for a glide reflection with the x-axis as the line of reflection and a horizontal motion of h.

---

42. The line $y = x$ is the perpendicular bisector of each segment.

43. above the line; below the line
below the line; above the line
on the line; on the line

44. Reflect $\triangle ABC$ across the y-axis and then translate the image 4 units up. $T(x, y) = (-x, y + 4)$

45. It flips back and forth across the y-axis and moves up to form a pattern.

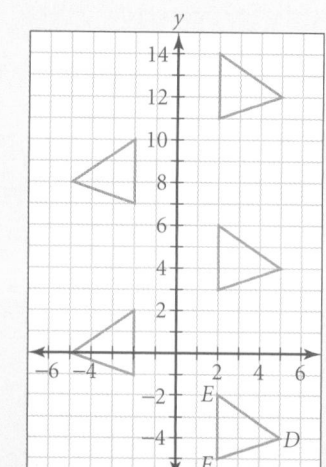

46. $G(x, y) = (-x, y + v)$

47. $G(x, y) = (x + h, -y)$

COMPUTER PROGRAMING The PostScript computer language is designed to communicate between computers and printers. A page in the PostScript language is laid out as a coordinate plane with the origin at the lower left-hand corner of the page. The following PostScript code draws a segment from a point located at (144, 72) to (144, 432).

```
newpath
    144 72 moveto
    144 432 lineto
stroke
showpage
```

48. Each unit in the PostScript coordinate system is $\frac{1}{72}$ in. long. What are the coordinates, in inches, of the endpoints of the segment drawn by the code above? Sketch the segment on a graph with axes labeled in inches.

49. Rewrite the code above to draw a segment between points whose coordinates, in inches, are (1, 1) and (3, 4.5).

 Look Back

Give the compass heading for each direction below. *(LESSON 1.3)*

50. W 270

51. NW 315

52. SW 225

53. SSW 202.5

Tell how you would find each point of a triangle. *(LESSON 1.5)*

54. circumcenter

55. incenter

Describe or draw an example of each type of translation below. If you make a drawing, label the preimage and image. *(LESSON 1.6)*

56. translation

57. rotation

58. reflection

59. glide reflection

 **Look Beyond**

60. Plot the points (1, 2), (4, 2), and (1, 8) in a coordinate plane, and connect them to form a triangle. Apply the transformation $Q(x, y) = (-y, x)$ to the triangle. What type of transformation is this? Try applying the same transformation to several different triangles to confirm your answer.

61. Based on your results from Exercise 60, write a transformation, $R(x, y)$, for the rotation of a figure 90° about the origin in the clockwise direction.

57. In a rotation, a figure turns around a given point called the center of rotation.

58. In a reflection, a figure is flipped over a line.

59. In a glide transformation, a figure is reflected across a line, while being translated in a direction parallel to that line.

60. A 90° counterclockwise rotation about the origin;

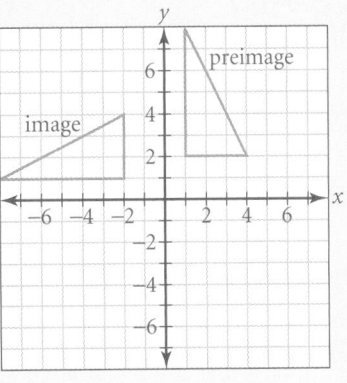

61. $Q(x, y) = (y, -x)$

In Exercises 60 and 61, students apply an unfamiliar transformation rule to the points in a triangle and discover that the transformation is a rotation.

48. (2, 1) and (2, 6)

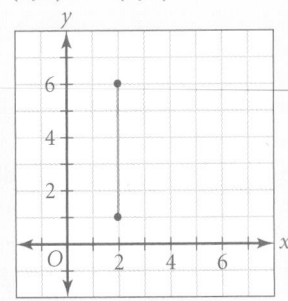

49. newpath
    72 72 moveto
    216 324 lineto
stroke
showpage

54. Find the point where the perpendicular bisectors of the sides meet.

55. Find the point where the angle bisectors meet.

56. In a translation, all the points of a figure move a given distance in a given direction.

Student Technology Guide

NAME _____ CLASS _____ DATE _____

Student Technology Guide
1.7 Motion in the Coordinate Plane

The rule $T(x, y) = (x + h, y + k)$, where h and k are fixed real numbers, gives a translation of $P(x, y)$ to $P'(x + h, y + k)$. If a geometric figure, such as a line, is defined by an equation, then you can use the translation rules to write and graph an equation for the geometric object after the translation is applied. Using a graphics calculator, you can explore lines and translations of them.

Example: The equation $y = 2x$ describes a line. Write an equation for the new line after you apply the rule $T(x, y) = (x + 3, y + (-2))$. Then graph both equations. How are they related?
• original equation: $y = 2x$
  new equation: $y = 2(x + 3) + (-2)$
• Press [Y=]. For Y1 and Y2, press 2 [X,T,θ,n] [ENTER]
  2 [( ] [X,T,θ,n] [+] 3 [ )] [+] [(-)] 2 [ENTER].
• Press [WINDOW]. Enter the ranges $-9 \le x \le 9$ and $-6 \le y \le 6$. Press [GRAPH].

The graphs of the equations are parallel lines.

Apply the translation $T(x, y) = (x + 3, y + (-2))$ to each equation. Write the new equation. Graph the original equation along with its translation on the same calculator display.   1–3. See Answer Key.

1. $y = 2.5x$ 2. $y = -2x$ 3. $y = 4x$
   $y = 2.5x + 5.5$ $y = -2x - 8$ $y = 4x + 10$

4. When you apply a translation to a given line, do a line and its image ever meet? Explain your response.

   The new line and the original line are parallel. If both h and k are nonzero, the lines will never meet.

The rule $T(x, y) = (-x, -y)$ describes a transformation of $P(x, y)$ to $Q(-x, -y)$.

In Exercises 5–8, apply the rule above to the given equation. Graph the original equation along with its image on the same calculator display.

5. $y = 2.5x$ 6. $y = -2x$ 7. $y = -x$ 8. $y = 4x$

9. Suppose that a line is described by $y = mx$. What can you say about the line that results when the rule $T(x, y) = (-x, -y)$ is applied? Justify your response.

   The resulting line coincides with the original line because mx translated to $m(-x)$ gives $-[m(-x)]$, or mx.

Focus

Students will follow the steps given to create an origami crane. The crane is made up of several geometric shapes and provides a physical example for some of the concepts discussed in Chapter 1.

Motivate

Origami is an ancient Japanese art in which figures are made from folded paper. The paper folding must be precise in order to create a visually pleasing figure. Books about origami are available at city and school libraries. You may wish to have some of these books available for students when they do the project.

1.–9. Students will need to follow the instructions for making the crane.

CHAPTER ONE

PROJECT

Origami
Paper Folding

Origami, the ancient Japanese art of paper folding, produces intriguing figures from simple paper folds. In this project, you will use paper folding to create a paper crane. First follow Steps 1–6 to make the crane base. Then continue with the final folds to finish your crane.

Use a 6- to 8-inch square of paper.

1. Fold the paper in half as shown.

2. Fold point *A* forward and down to point *B*. Fold point *C* backward and down to point *B*.

3. Open the paper at point *B* and lay it flat.

4. Crease the paper by folding points *D* and *B* inward to point *F* on the center line.

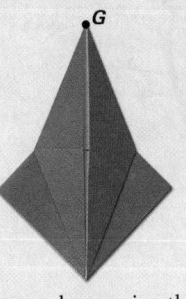
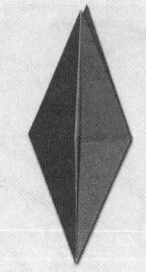

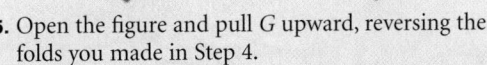

5. Open the figure and pull *G* upward, reversing the folds you made in Step 4.

6. Turn the paper over and repeat Steps 4 and 5.

You have now completed the crane base. Continue with the steps below to complete the crane.

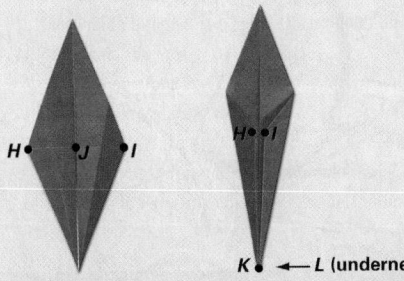

7. From the crane base, fold points *H* and *I* upward into point *J* on the center line. Turn the paper over and repeat.

8. Pull points *K* and *L* outward and reverse the folds. These will become the tail and neck of the crane.

9. Fold the neck to form the head, and bend the wings down as shown. You can blow into the base of the figure to inflate its body.

Your crane is finished.

10. Unfolding a paper crane will reveal a pattern of creases. Trace over and label the geometric figures formed by the creases. Are any of the figures congruent with one another?

11. The crane base can be used to create many origami figures. Can you figure out how to fold a frog or a fish?

12. Find an origami book at your library and expand your collection of origami figures.

Cooperative Learning

Origami instructions are sometimes difficult to follow. Students working in groups will be able to consult with each other about the steps. Provide each group with an origami book from the library and encourage them to try other shapes. Each group should create a variety of origami figures. Students can add these figures to their portfolio.

Discuss

Have students unfold some of their figures and study the lines created by the creases. Discuss the various geometric shapes that appear. Introduce the concept of congruence by pointing out shapes that appear to be congruent. Ask students if there is something about the way they folded the paper that would ensure that the shapes are congruent.

10.
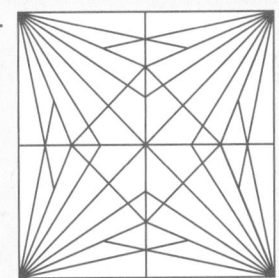

11. Answers may vary. Students will use trial and error to make their own origami structure.

Chapter Review

Chapter Test, Form A

NAME ______________ CLASS ________ DATE ________

Chapter Assessment
Chapter 1, Form A, page 1

Write the letter that best answers the question or completes the statement.

d 1. What plane contains points M, N, and O?
a. plane B
b. plane X
c. plane MN
d. plane MNO

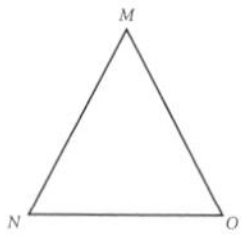

a 2. How many rays in the figure have A as an endpoint?
a. 5 b. 10
c. 3 d. 15

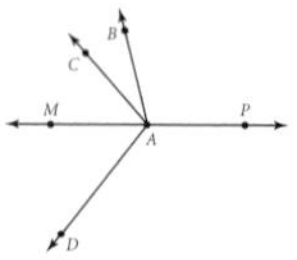

c 3. Every point on the ________ of a segment is equidistant from the endpoints of the segment.
a. angle bisector b. median
c. perpendicular bisector d. plane

a 4. Which segments are congruent?

a. $\overline{DE}$ and $\overline{FG}$ b. $\overline{DE}$ and $\overline{EF}$
c. $\overline{EF}$ and $\overline{FG}$ d. $\overline{DG}$ and $\overline{EG}$

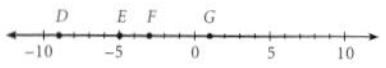

c 5. H is between points Q and R. $QH = 23$ and $HR = 12$. What is the length of QR?
a. 12 b. 11 c. 35 d. 48

b 6. At how many points do the perpendicular bisectors of the sides of a triangle intersect?
a. 0 b. 1 c. 2 d. 3

NAME ______________ CLASS ________ DATE ________

Chapter Assessment
Chapter 1, Form A, page 2

Use the figure to answer Exercises 7–9.

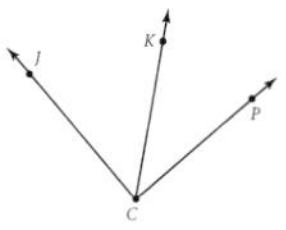

b 7. If $m\angle JCK = 58°$ and $m\angle KCP = 35°$, what is $m\angle JCP$?
a. 180° b. 93°
c. 90° d. 83°

a 8. If $m\angle JCK = 92°$ and $m\angle JCP = 116°$, what is $m\angle KCP$?
a. 24° b. 208° c. 90° d. 92°

d 9. If $m\angle KCP = 35°$ and $m\angle JCP = 145°$, what is $m\angle JCK$?
a. 35° b. 45° c. 190° d. 110°

b 10. What kind of transformation changes $\overline{ED}$ to $\overline{E'D'}$?
a. glide
b. translation
c. rotation
d. reflection

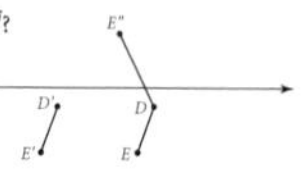

d 11. What kind of transformation is shown in the figure at right?
a. slide
b. translation
c. rotation
d. reflection

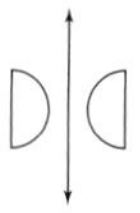

a 12. What are the coordinates of $\triangle XYZ$ at right?
a. $(1, -1), (2, -3), (3, -1)$
b. $(-1, 1), (-3, 2), (-1, 3)$
c. $(1, 1), (2, 2), (3, 3)$
d. $(-1, 1), (-2, -2), (-3, -3)$

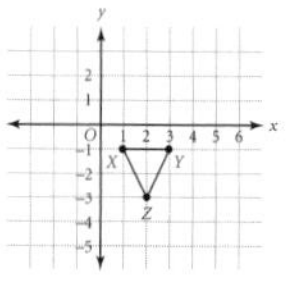

d 13. What would be the coordinates of $\triangle X'Y'Z'$, the image of $\triangle XYZ$ reflected across the y-axis?
a. $(1, 1), (2, 3), (3, 1)$
b. $(1, -1), (2, 1), (3, -1)$
c. $(1, -1), (-1, -1), (0, -3)$
d. $(-1, -1), (-3, -1), (-2, -3)$

1 Chapter Review and Assessment

VOCABULARY

- acute angle 29
- angle 11
- angle bisector 38
- center of rotation 51
- circumcenter 45
- circumscribed circle 44
- collinear 10
- complement 28
- complementary angles 28
- concurrent 45
- congruent 19
- conjecture 36
- construction 35
- coordinate of a point 17
- coplanar 10
- degree 26
- endpoint 10
- exterior 11
- image 50
- incenter 45
- inscribed circle 44
- interior 11
- intersect 11
- intersection 11
- length 17
- line 9
- linear pair 28
- measure of an angle 26
- midpoint 38
- number line 17
- obtuse angle 29
- parallel lines 35
- perpendicular bisector 38
- perpendicular lines 35
- plane 10
- point 9
- postulate 11
- preimage 50
- ray 10
- reflection 52
- right angle 29
- rigid transformation 50
- rotation 51
- segment 10
- segment bisector 38
- sides of an angle 11
- supplement 28
- supplementary angles 28
- translation 51
- unit length 18
- vertex of an angle 11

POSTULATES

Lesson	Section	Postulate or Theorem
1.1	1.1.4 Postulate	The intersection of two lines is a point.
	1.1.5 Postulate	The intersection of two planes is a line.
	1.1.6 Postulate	Through any two points there is exactly one and only one line.
	1.1.7 Postulate	Through any three noncollinear points there is exactly one plane.
	1.1.8 Postulate	If two points are in a plane, then the line containing them is in the plane.
1.2	1.2.2 Segment Congruence Postulate	If two segments have the same length as measured by a given fair ruler, then the segments are congruent. Also, if two segments are congruent, then they have the same length as measured by a given ruler.
	1.2.3 Segment Addition Postulate	If point R is between points P and Q on a line, then $PR + RQ = PQ$.
1.3	1.3.2 Angle Addition Postulate	If point S is in the interior of $\angle PQR$, then $m\angle PQS + m\angle SQR = m\angle PQR$.
	1.3.3 Angle Congruence Postulate	If two angles have the measure, then they are congruent. If two angles are congruent, then they have the same measure.
	1.3.5 Linear Pair Property	If two angles form a linear pair, then they are supplementary.

Key Skills & Exercises

LESSON 1.1

Key Skills

Identify and name geometric figures.

In the figure below, A, B, and C are points, $\overleftrightarrow{AC}$ is a line, $\overline{AB}$ and $\overline{AC}$ are segments, $\overrightarrow{AB}$, $\overrightarrow{AC}$, and $\overrightarrow{CA}$ are rays, and $\angle BAC$ is an angle.

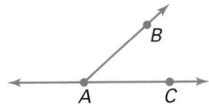

Exercises

Refer to the figure below.

1. Name all segments in the figure.
2. Name all angles in the figure.

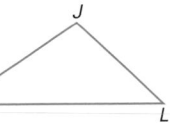

Refer to the figure below.

3. Name all lines in the figure.
4. Name all rays in the figure.

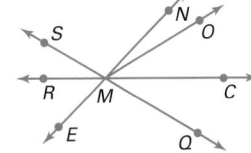

LESSON 1.2

Key Skills

Determine the length of a given segment.

In the figure below, find the lengths of $\overline{CD}$, $\overline{DE}$, and $\overline{EF}$.

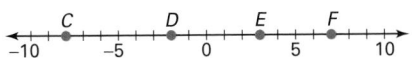

$CD = |-8 - (-2)| = 6$
$DE = |-2 - 3| = 5$
$EF = |3 - 7| = 4$

Determine whether segments are congruent.

In the figure above, are $\overline{CD}$, $\overline{DE}$, and $\overline{EF}$ congruent segments?

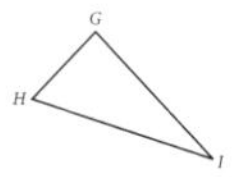

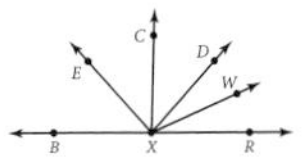

None of the segments are congruent.

Add the lengths of segments.

In the figure below, $AB = 7$ and $BX = 15$. Find AX.

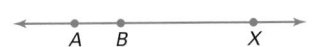

$AX = AB + BX = 7 + 15 = 22$

Exercises

Refer to the figure below.

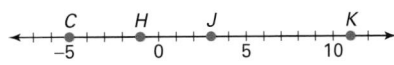

5. Find the length of every segment in the figure.
6. Name all congruent segments in the figure.

In Exercises 7 and 8, point A is between points R and P on a line. Sketch a figure for each exercise and find the missing measure.

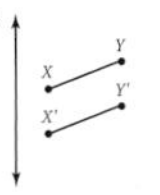

7. $RA = 25, AP = 13, RP = \underline{?}$
8. $RA = \underline{?}, AP = 7, RP = 13$

---

1. $\overline{KJ}, \overline{JL}, \overline{LK}$
2. $\angle J, \angle K, \angle L$
3. $\overleftrightarrow{SQ}, \overleftrightarrow{RC}, \overleftrightarrow{EN}$
4. $\overrightarrow{MR}, \overrightarrow{MC}, \overrightarrow{MQ}, \overrightarrow{MO}, \overrightarrow{MN}, \overrightarrow{CR}, \overrightarrow{MS}, \overrightarrow{ME}, \overrightarrow{QS}, \overrightarrow{EN}, \overrightarrow{RC}, \overrightarrow{SQ}, \overrightarrow{NE}$
5. $CH = 4; CJ = 8; CK = 16; HJ = 4; HK = 12; JK = 8$
6. $\overline{CJ} \cong \overline{JK}; \overline{CH} \cong \overline{HJ}$

7.

$RP = 38$

8.

$RA = 6$

Chapter Test, Form B

NAME ______ CLASS ______ DATE ______

Chapter Assessment
Chapter 1, Form B, page 1

1. Name the segments in the triangle at right. $\overline{GH}, \overline{HI}, \overline{IG}$
2. Name the angles in the triangle at right. $\angle G, \angle H, \angle I$
3. Name the plane that contains the points G, H, and I. plane GHI

4. Name the rays in the figure which have X as an endpoint.

$\overrightarrow{XB}, \overrightarrow{XE}, \overrightarrow{XC}, \overrightarrow{XD}, \overrightarrow{XW}, \overrightarrow{XR}$

5. Describe the relationship between lines ℓ and m.

They are perpendicular

S is between points B and D. Sketch each figure in the space provided and find the missing measure.

6. $BS = 49$, $SD = 5$, $BD = \underline{54}$
7. $BS = \underline{16.1}$, $SD = 13.9$, $BD = 30$
8. $BS = 115$, $SD = \underline{65}$, $BD = 180$

9. For which of the above exercises, if any, are $\overline{BS}$ and $\overline{SD}$ congruent? none
10. Explain the difference between the inscribed circle and circumscribed circle of a triangle.
An inscribed circle touches each side of the triangle and is located inside the triangle. A circumscribed circle passes through each vertex and is located outside the circle.

NAME ______ CLASS ______ DATE ______

Chapter Assessment
Chapter 1, Form B, page 2

In Exercises 11–13, find the missing measures.

11. $m\angle FGL = 17°$, $m\angle LGM = 32°$, $m\angle FGM = \underline{49°}$
12. $m\angle FGL = 48°$, $m\angle LGM = \underline{48°}$, $m\angle FGM = 96°$
13. $m\angle FGL = \underline{30°}$, $m\angle LGM = 65°$, $m\angle FGM = 95°$
14. For which of the above exercises, if any, are $\angle FGL$ and $\angle LGM$ congruent? Exercise 12

Identify each transformation as a translation, rotation, or reflection.

15.

translation

16.

reflection

17.

rotation

18. Using the rule $T(x, y) = (x + 3, y + 3)$, transform $\triangle ABC$. Draw $\triangle A'B'C'$ on the grid provided, and give the coordinates of its vertices. (3, 3), (5, 6), (7, 3)
19. What type of transformation is shown in Exercise 18? translation
20. Write the rule to transform the new triangle back to $\triangle ABC$'s original coordinates. $R(x, y) = (x - 3, y - 3)$
21. Use the rule you wrote in Exercise 20 to transform $\triangle ABC$. Give the coordinates of the new vertices. (−3, −3), (−1, 0), (1, −3)

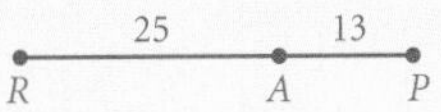

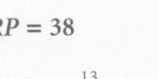

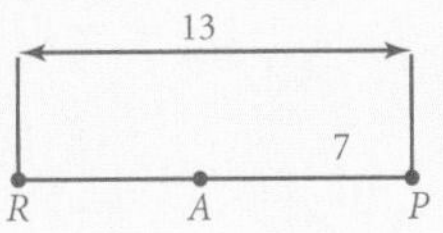

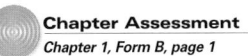

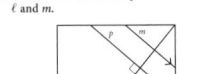

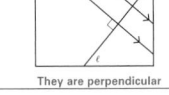

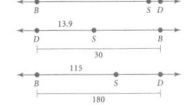

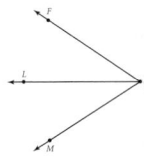

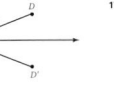

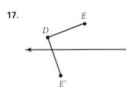

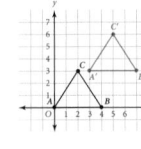

9. $m\angle PVT = 125°$;
$m\angle QVT = 95°$;
$m\angle RVT = 65°$;
$m\angle SVT = 30°$;
$m\angle PVS = 95°$;
$m\angle QVS = 65°$;
$m\angle RVS = 35°$;
$m\angle PVR = 60°$;
$m\angle QVR = 30°$;
$m\angle PVQ = 30°$

10. $\angle QVT \cong \angle PVS$;
$\angle SVT \cong \angle PVQ \cong \angle QVR$;
$\angle QVS \cong \angle RVT$

11. $m\angle LKN = 38°$

12. $m\angle LKM = 15°$

13.
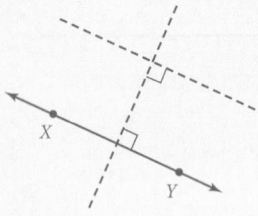

14.
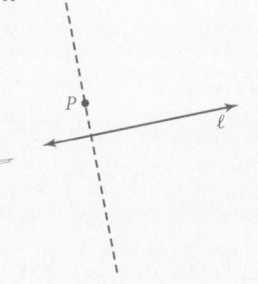

15.
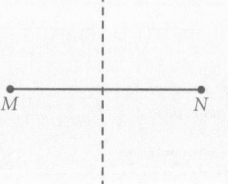

16.
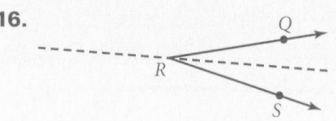

Key Skills

Determine the measure of a given angle.

For the figure below, find the measure of $\angle AVB$ and $\angle BVC$.

$m\angle AVB = |40° - 20°| = 20°$
$m\angle BVC = |60° - 40°| = 20°$

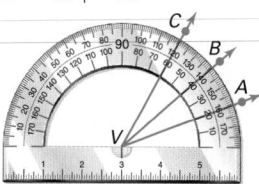

Add measures of angles.

For the figure above, find $m\angle AVC$.

$m\angle AVC = m\angle AVB + m\angle BVC = 20° + 20° = 40°$.

Determine whether angles are congruent.

In the figure above, are $\angle AVB$ and $\angle BVC$ congruent angles?

Yes; $\angle AVB \cong \angle BVC$ because they have the same measure.

Exercises

Refer to the figure below.

9. Find the measure of every angle in the figure.

10. Name all congruent angles in the figure.

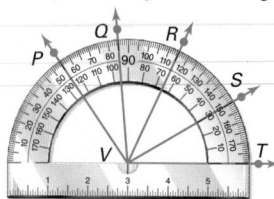

In Exercises 11 and 12, find the missing measures.

11. $m\angle LKM = 12°$
$m\angle MKN = 26°$
$m\angle LKN = \underline{?}$

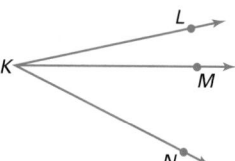

12. $m\angle LKM = \underline{?}$
$m\angle MKN = 75°$
$m\angle LKN = 90°$

Key Skills

Use paper folding to construct geometric figures.

For the figure below, construct the perpendicular bisector of $\overline{AB}$, the angle bisector of $\angle ABC$, and a line parallel to $\overleftrightarrow{AB}$ that passes through C.

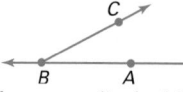

To construct the perpendicular bisector of $\overline{AB}$, fold the paper so that A matches up with B.

To construct the angle bisector of $\angle ABC$, fold the paper so that $\overrightarrow{BA}$ matches up with $\overrightarrow{BC}$.

To construct a line parallel to $\overleftrightarrow{AB}$ through C, fold the perpendicular bisector of $\overline{AB}$ onto itself so that C is on the fold.

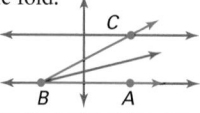

Exercises

Trace each figure onto folding paper and construct the given figure.

13. a line parallel to $\overleftrightarrow{XY}$

14. a line through P perpendicular to ℓ

15. the perpendicular bisector of $\overline{MN}$

16. the angle bisector of $\angle QRS$

LESSON 1.5
Key Skills

Construct the circumscribed circle of a triangle.

The center of the circumscribed circle of a triangle is the intersection of the perpendicular bisectors of the triangle.

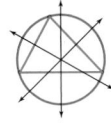

Construct the inscribed circle of a triangle.

The center of the inscribed circle of a triangle is the intersection of the angle bisectors of the triangle.

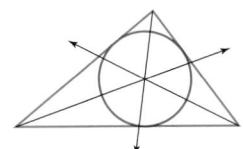

Exercises

Use geometry software or folding paper and a compass to construct the following:

17. the angle bisectors of an acute triangle

18. the perpendicular bisectors of an obtuse triangle

19. a right triangle circumscribed by a circle

20. a triangle with two congruent sides and its inscribed circle

LESSON 1.6
Key Skills

Identify translations, rotations, and reflections.

  *Translation*

  *Rotation*

  *Reflection*

Translate a figure along a line.

Rotate a figure about a point.

Reflect a figure across a line.

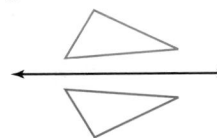

Exercises

21. Identify the following as being best represented by a translation, rotation, or reflection.
    a. geometry ʎɹʇǝɯoǝƃ
    b. a child on a slide
    c. a Ferris wheel

22. Translate the figure along the line.

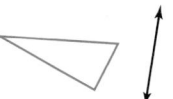

23. Rotate the figure about the point.

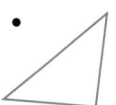

24. Reflect the figure across the line.

17. Sample answer:

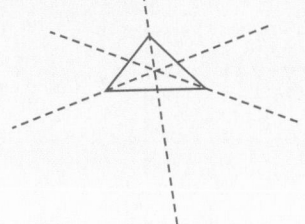

18. Sample answer:

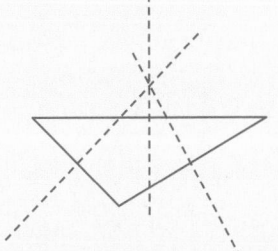

19.

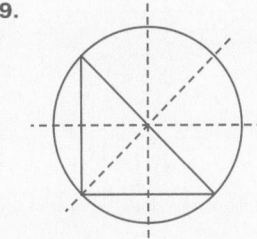

20.

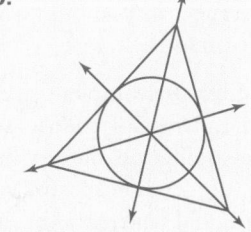

21. a. reflection
b. translation
c. rotation

22.

23.

24.

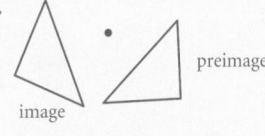

25.

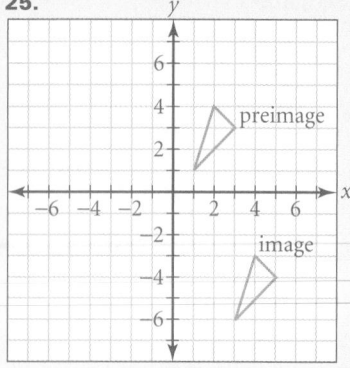

26. $T(x, y) = T(x - 3, y)$

27.

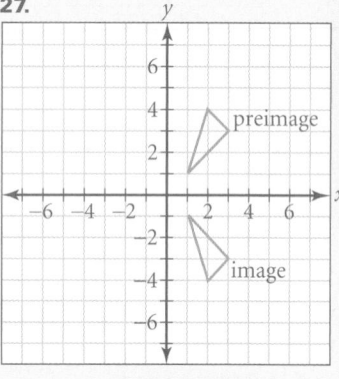

28. $R(x, y) = (-x, -y)$

29. It is 49 km from Smithville to LaGrange and 71 km from Bastrop to LaGrange.

30. To draw the complete circle, the student can choose any three points on the existing part of the circle, draw a triangle, and construct the circumscribed circle.

Key Skills

Use the coordinate plane to transform geometric figures.

Plot the points $(2, 3)$, $(-1, 2)$, and $(0, 0)$, and connect them to form a triangle.

Give the rules for translating the figure 5 units down and for reflecting the figure across the y-axis, and draw the transformed figures.

To translate 5 units down, use the rule $V(x, y) = (x, y - 5)$.

To reflect across the the y-axis, use the rule $N(x, y) = (-x, y)$.

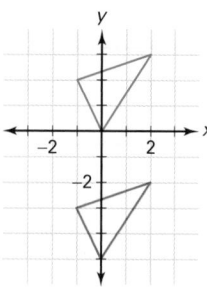

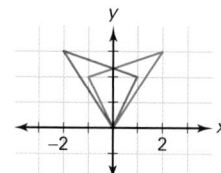

Exercises

Plot the points **(1, 1), (3, 3),** and **(2, 4),** and connect them to form a triangle.

25. Transform the triangle by using the rule $T(x, y) = (x + 2, y - 7)$.

26. What rule would you use to translate the triangle 3 units to the left?

27. Transform the triangle by using the rule $S(x, y) = (x, -y)$.

28. What rule would you use to rotate the triangle $180°$ about the origin?

Applications

29. TRAVEL Smithville is between Bastrop and LaGrange on a straight highway. The distance from Bastrop to Smithville is 22 kilometers. The distance from Bastrop to LaGrange is 5 kilometers more than 3 times the distance from Bastrop to Smithville. Write an equation for the distances, and then solve it to find the distance from Smithville to LaGrange and from Bastrop to LaGrange.

30. ENGINEERING Jenny has a piece of a broken gear from an antique clock. She needs to find the original size of the gear in order to get a replacement part. Trace the part of the gear shown at right, and construct the complete circle.

Chapter Test

Refer to the figure below.

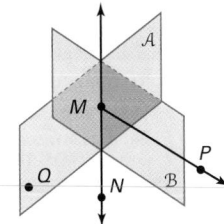

$\overleftrightarrow{MN}$ **1.** Name the intersection of planes A and B.

2. Name three coplanar points in the figure.
M, N, and P or **M, N, and Q**

3. **NAVIGATION** The towns of Limon, Rocky Ford, and Timpas are located along a straight road. Rocky Ford is between Limon and Timpas. The distance from Limon to Rocky Ford is 13 miles more than 5 times the distance from Rocky Ford to Timpas. The distance from Limon to Timpas is 103 miles. Find the distance from Limon to Rocky Ford and from Rocky Ford to Timpas. **88 mi; 15 mi**

Point C is between points B and F on $\overline{BF}$. Sketch each figure and find the missing measure.

4. $BC = 42$ $CF = \underline{?}$ $BF = 60$ **18**

5. $BC = \underline{?}$ $CF = 23$ $BF = 51$ **28**

In the figure below, m∠RPS = 32° and ∠QPS and ∠SPT form a linear pair. Find the measure of each angle.

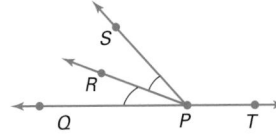

6. m∠QPR **32°**   **7.** m∠SPT **116°**   **8.** m∠RPT **148°**

Use a separate piece of paper to fold each figure below. Do not use a ruler or protractor. Trace over each figure and label all relevant parts.

9. two parallel lines

10. a right triangle

11. Use folding paper to construct the angle bisectors of a right triangle.

Refer to the figure below for Exercises 12–13. Suppose that a is the bisector of ∠XWZ and that m∠YWZ = 18° Find the measure of each angle.

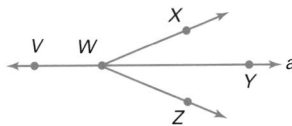

12. m∠XWY **18°**   **13.** m∠XWZ **36°**

14. **ART** A potter wants to reconstruct a broken ceramic base. He needs to find the original size of the base. Trace the part of the base shown below and construct the complete circle.

Trace each figure below onto folding paper.

15. Translate the figure along the line.

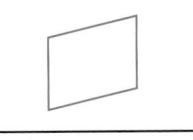

16. Rotate the figure about the point.

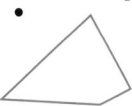

17. Reflect the figure across the line.

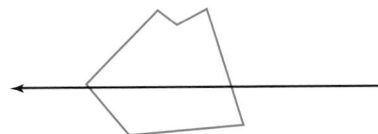

Plot the points (5, 1), (7, –1), and (3, –3), and connect them to form a triangle.

18. Transform the triangle by using the rule $T(x, y) = (x - 2, y + 3)$.

19. What rule would you use to translate the triangle 4 units to the right? $S(x, y) = (x + 4, y)$

9.

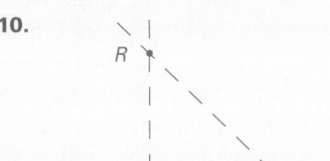

10.

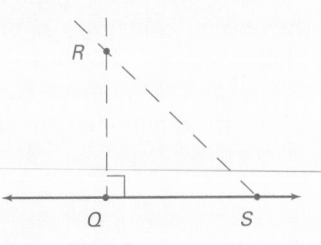

11.

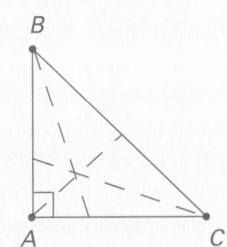

14.

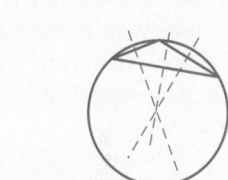

15.   preimage   image

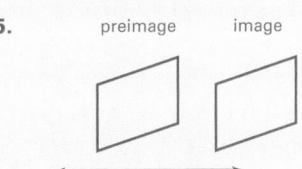

16.   image

    preimage

17.   preimage

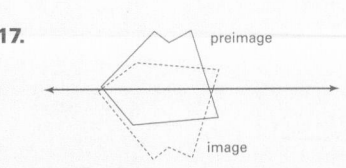
    image

18.

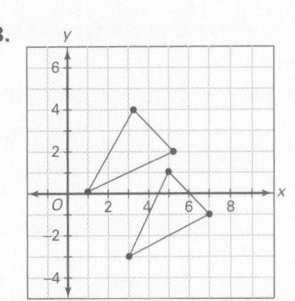

CHAPTER 1 CUMULATIVE ASSESSMENT

College Entrance Exam Practice

MULTIPLE-CHOICE For Questions 1–3, write the letter that indicates the best answer.

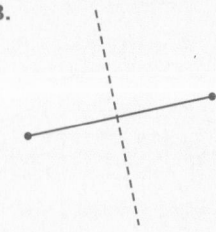

1. Find *NB* on the line below. ***(LESSON 1.2)***

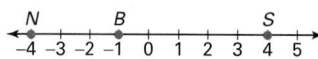

a. 5
b. 3
c. 4
d. 2

2. Which expression states that $\overline{EF}$ is congruent to $\overline{HG}$? ***(LESSON 1.2)***
a. $EF \cong HG$
b. $\overline{EF} = \overline{HG}$
c. $\overline{EF} \cong \overline{HG}$
d. $EF = HG$

3. Refer to the figure below. If $\overline{WZ}$ bisects $\angle XWY$, which of the following statements is true? ***(LESSON 1.3)***

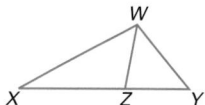

a. $m\angle XWZ > m\angle YWZ$
b. $m\angle XWZ < m\angle YWZ$
c. $m\angle XWZ = m\angle YWZ$
d. $m\angle XWZ \neq m\angle YWZ$

internet connect
Standardized Test Prep Online
Go To: **go.hrw.com**
Keyword: **MM1 Test Prep**

Name each figure. ***(LESSON 1.1)***

4.

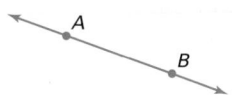

5.

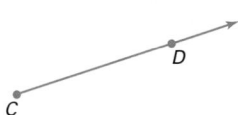

6.

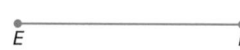

7.

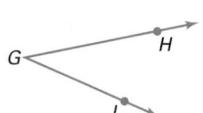

For Items 8–11, trace each diagram onto folding paper and construct the given figure. ***(LESSONS 1.4 AND 1.5)***

8. the perpendicular bisector

College Entrance Exam Practice

Multiple-Choice Samples
The first half of the Cumulative Assessment contains a multiple-choice section. This part of the Cumulative Assessment consists of items commonly found on standardized tests.

Free-Response Grid Samples
The second half of the Cumulative Assessment consists of a free-response section. This part requires student-produced response items like those commonly found on college entrance exams. These questions require the use of machine-scored answer grids. You may wish to have students practice answering these items in preparation for standardized tests.

1. b
2. c
3. c
4. $\overleftrightarrow{AB}$ or $\overleftrightarrow{BA}$
5. $\overrightarrow{CD}$
6. $\overline{EF}$ or $\overline{FE}$
7. $\angle HGI$, $\angle IGH$, or $\angle G$
8.

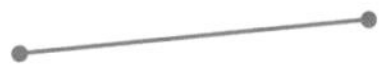

9. the angle bisector

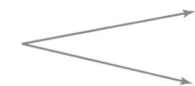

10. the circumcenter
(Draw the circumscribed circle.)

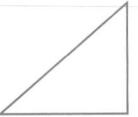

11. the incenter
(Draw the inscribed circle.)

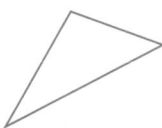

For Items 12–16, identify each transformation as a translation, rotation, or reflection.
(LESSONS 1.6 AND 1.7)

12.

13.

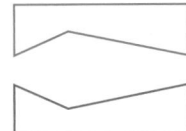

14.

15. $S(x, y) = (x - 1, y)$

16. $T(x, y) = (x, -y)$

For Items 17 and 18, refer to the figure below. *(LESSON 1.2)*

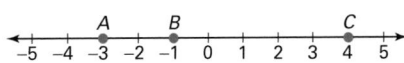

17. $AB =$ ___?___

18. $AC =$ ___?___

19. Refer to the figure below. If $m\angle AVD = 85°$, what is $m\angle AVC$? *(LESSON 1.3)*

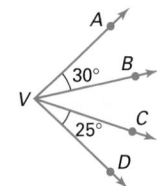

20. Refer to the figure below. What is $m\angle MON$? *(LESSON 1.5)*

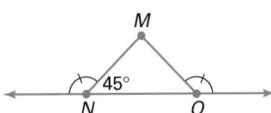

9.

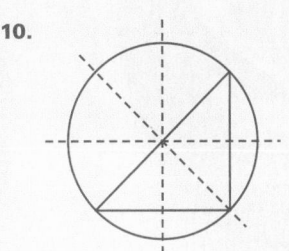

10.

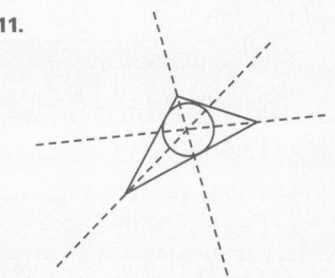

11.

12. rotation

13. reflection

14. translation

15. translation

16. reflection

17. $AB = 2$

18. $AC = 7$

19. $m\angle AVC = 60°$

20. $m\angle MON = 45°$

Reasoning in Geometry

Lesson Presentation CD-ROM
Power Point® presentations for each lesson 2.1-2.5

CHAPTER PLANNING GUIDE

Lesson	2.1	2.2	2.3	2.4	2.5	Project and Review
Pupil's Edition Pages	80–87	90–98	99–106	107–116	117–125	88–89, 126–135
Practice and Assessment						
Extra Practice (Pupil's Edition)	821	822	822	823	823	
Practice Workbook	8	9	10	11	12	
Practice Masters Levels A, B, and C	22–24	25–27	28–30	31–33	34–36	
Standardized Test Practice Masters	9	10	11	12	13	14
Assessment Resources	15	16	17	19	20	18, 21–26
Visual Resources						
Lesson Presentation Transparencies Vol. 1	29–32	33–36	37–40	41–44	45–48	
Teaching Transparencies	13–15		16–17	18		
Answer Key Transparencies	40–47	48–55	56–61	62–65	66–73	74–79
Quiz Transparencies	2.1	2.2	2.3	2.4	2.5	
Teacher's Tools						
Reteaching Masters	15–16	17–18	19–20	21–22	23–24	
Make-Up Lesson Planner for Absent Students	8	9	10	11	12	
Student Study Guide	8	9	10	11	12	
Spanish Resources	8	9	10	11	12	
Block Scheduling Handbook						4–5
Activities and Extensions						
Lesson Activities	22–25		26–28		29–31	
Enrichment Masters	8	9	10	11	12	
Cooperative-Learning Activities	8	9	10	11	12	
Problem-Solving/ Critical Thinking	8	9	10	11	12	
Student Technology Guide	10		11	12–13		
Long Term Projects						5–8
Writing Activities for Your Portfolio						4–6
Tech Prep Masters						5–8
Building Success in Mathematics						4–6

LESSON PACING GUIDE

Lesson	2.1	2.2	2.3	2.4	2.5	Project and Review
Traditional	2 days	2 days	2 days	2 days	2 days	2 days
Block	1 day	1 day	1 day	1 day	1 day	1 day
Two-Year	4 days	4 days	4 days	4 days	4 days	4 days

CONNECTIONS AND APPLICATIONS

Lesson	2.1	2.2	2.3	2.4	2.5	Review
Algebra	81, 83, 84, 85, 87			107, 108, 110, 112, 113	122, 124	
Geometry	80–87	90–98	99–106	107–116	117–125	128–136
Patterns in Data		98				
Life Skills				116		
Science and Technology		90, 94, 98	103, 105			130
Sports and Leisure	86	96, 97		111, 115	120	
Cultural Connection: Asia	87					

BLOCK-SCHEDULING GUIDE

Day	Lesson	Teacher Directed: Lesson Examples, Teaching Transparencies	Student Guided: Activity, Try This	Cooperative-Learning Activity, Lesson Activity, Student Technology Guide	Practice: Practice & Apply, Extra Practice, Practice Workbook	Assessment: Quiz, Mid-Chapter Assessment	Problem Solving, Reteaching
1	2.1	15 min	15 min	20 min	55 min	15 min	15 min
2	2.2	15 min	15 min	20 min	55 min	15 min	15 min
3	2.3	15 min	15 min	20 min	55 min	15 min	15 min
4	2.4	15 min	15 min	20 min	55 min	15 min	15 min
5	2.5	15 min	15 min	20 min	55 min	15 min	15 min
6	Assess.	50 min	90 min	90 min	65 min	30 min	
		PE: Chapter Review	**PE:** Chapter Project, Writing Activities	Tech Prep Masters	**PE:** Chapter Assessment Test Generator	Chap. Assess. (A or B), Alt. Assess. (A or B), Test Generator	

PE: Pupil's Edition

Alternative Assessment

The following suggest alternative assessments for students who may benefit from a different type of assessment than the regular chapter quizzes and the mid-chapter/end-of-chapter test. Visit the HRW web site to get additional Alternative Assessment material.

internet connect

Alternative Assessment
Go To: **go.hrw.com**
Keyword: **MG1 Alt Assess**

Performance Assessment

1. Write three different definitions that could fit some or all of the objects shown below. In each case, state which objects fit the definition, and which do not.

2. Using geometry graphics software, create an electronic page showing a proof of any of the conjectures in this chapter. For instance a proof the Vertical Angle Theorem can be accompanied by a diagram showing vertical angles which the viewer can drag to change the measures displayed on the screen.

Portfolio Project

Suggest that students choose one of the following projects for inclusion in their portfolios.

1. Make up or describe an existing game that uses logic to win. Examples are Mastermind, Clue, or Nim (featured in Chapter 2). Describe how the player must use inductive or deductive reasoning, and what strategies are necessary to win the game.

2. Use the Vertical Angle Theorem and the Overlapping Angles theorem to prove the following statement:
Given: $\angle RLS$ and $\angle MLN$ are
    vertical angles
    $m\angle MLP = m\angle QLN$
Prove: $m\angle QLP = m\angle RLS$

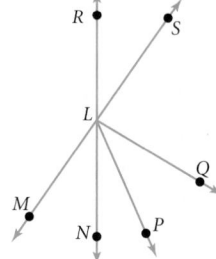

internet connect

The table below identifies the pages in this chapter that contain internet and technology information.

Content Links	
Activities Online	pages 82, 94, 120
Portfolio Extensions	pages 87, 116
Homework Help Online	pages 84, 94, 103, 112, 122

Resource Links

Parents can go online and find concepts that students are learning–lesson by lesson–and questions that pertain to each lesson, which facilitate parent-student discussion.

Go To: **go.hrw.com**
Keyword: **MG1 Parent Guide**

Technical Support

The following may be used to obtain technical support for any HRW software product.

Online Help: **www.hrwtechsupport.com**
e-mail: **tschrw@hbtechsupport.com**

HRW Technical Support Center: **(800)323-9239**
7 AM to 10 PM Monday through Friday CST

Visit the HRW math web site at: **www.hrw.com/math**

Technology

Technology Objectives and Suggestions

Lesson 2.1 An Introduction to Proofs

This lesson introduces proofs and logic by providing three non-routine problems in the activity. In Part 2 of the activity students are asked to continue a sequence. Students can use a computer spreadsheet to complete this part of the activity. In Exercise 24 on page 84, students can use a spreadsheet to complete this exercise as well.

Lesson 2.2 An Introduction to Logic

In this lesson students use conditionals and their converses in logical arguments, and create logical chains from conditionals. Geometry graphics software can be used to test and create conditional statements about geometric figures, such as triangles.

Give students a conditional such as "If the triangle is obtuse, then the medians intersect at one point inside the triangle." Students can use the software to draw an obtuse triangle with the three medians. Students can draw the figure to see that the medians always intersect inside the triangle, no matter what its shape is.

Do the same activity with altitudes by giving students this conditional: "If the triangle is obtuse, then the altitudes intersect at one point inside the triangle." By manipulating the figure students will see that this conditional is false because they can find at least one counterexample.

Lesson 2.3 Definitions

In this lesson, students study definitions of objects and use principles of logic to create definitions of objects. In groups of two or three have students write a definition of an imaginary object that can be drawn using geometry graphics software. An example is given: a *house-pentagon* is a five-sided figure in which the bottom is made up of three sides of a rectangle and the top is made up of two sides of an isosceles triangle.

Students then trade definitions, and the next group tries to draw the figure based on the definition. Students will find that the more complex the definition is, the harder the figure will be to draw. Students should come up with something like the figure shown below.

Lesson 2.4 Building a System of Geometry Knowledge

This lesson looks at Algebraic and Equivalence Properties of Equality. In the lesson and in the exercises students look at the Overlapping Segments Theorem and the Overlapping Angles Theorem. Students can use geometry software to test these conjectures before they prove them.

Have students draw the figure on page 107 and use the measuring tool to measure the distance between AB, BC, CB, AC, and BD. These measurements can remain on the screen. As students drag a point, they can watch as the measurements change. They should be able to see that if AB and CD have equal measures, then AC will equal BD.

Continue this with the Overlapping Angles Theorem. By drawing the figure on page 110, measuring all the angles, and dragging a point on any of the rays students can watch the measures of angles change.

Lesson 2.5 Conjectures That Lead to Theorems

In this lesson students develop theorems from conjectures. Students can use geometry graphics software to test the conjectures presented in the activities.

In Activity 1, have students draw the figure at the top of page 116 and use the measuring tool to measure angles 1 through 4. As they drag any of the lines, they will be able to see that the measures of vertical angles are equal. They can also add the measures of the adjacent angles to see that the sum of the measures of adjacent angles is 180 degrees.

In Activity 2, students can perform the reflection across two parallel lines using geometry software to discover the result of such a transformation. Have students measure the distance between points on the transformed image and the pre-image and come up with a proof for the theorem appearing on page 117.

This chapter introduces students to the ideas and concepts that provide a basis for understanding the deductive nature of geometry.

CHAPTER RESOURCES

- Block-Scheduling Handbook
- Writing Activities for Your Portfolio
- Tech Prep Masters
- Long-Term Project
- Assessment Resources:
  Mid-Chapter Assessment
  Chapter Assessments
  Alternative Assessments
- Test and Practice Generator
- Technology Handbook

Chapter Objectives

- Investigate some interesting proofs of mathematical claims. [**2.1**]
- Understand the meaning of the term *proof*. [**2.1**]
- Define *conditionals* and model them with Venn diagrams. [**2.2**]
- Use conditionals in logical arguments. [**2.2**]
- Form the converses of conditionals. [**2.2**]
- Create logical chains from conditionals. [**2.2**]

Reasoning in Geometry

REGARDLESS OF WHO YOU ARE OR WHAT YOU DO, there are times when you must reason clearly. Whether you are scaling a cliff, or playing a game of nim with a friend, or—like Sherlock Holmes—trying to solve a mystery, the basic principles underlying your reasoning processes are similar.

Geometry presents a unique opportunity for studying the processes of reasoning. Since ancient times, geometry has been for many people the foremost example of a fully reasoned system of knowledge.

Lessons

About the Photos

The photos above appear to have nothing in common, except for one thing: to do what the people in these pictures are doing, you must use your ability to reason. When Alice talks to the cat (from Lewis Carroll's *Alice in Wonderland*), when a judge deliberates before making a final decision, when investigator Sherlock Holmes looks for evidence and proofs, when a rock climber decides on his or her next move, or when you play nim or chess with a partner, the use of logical reasoning maximizes the chance of getting a successful result or outcome.

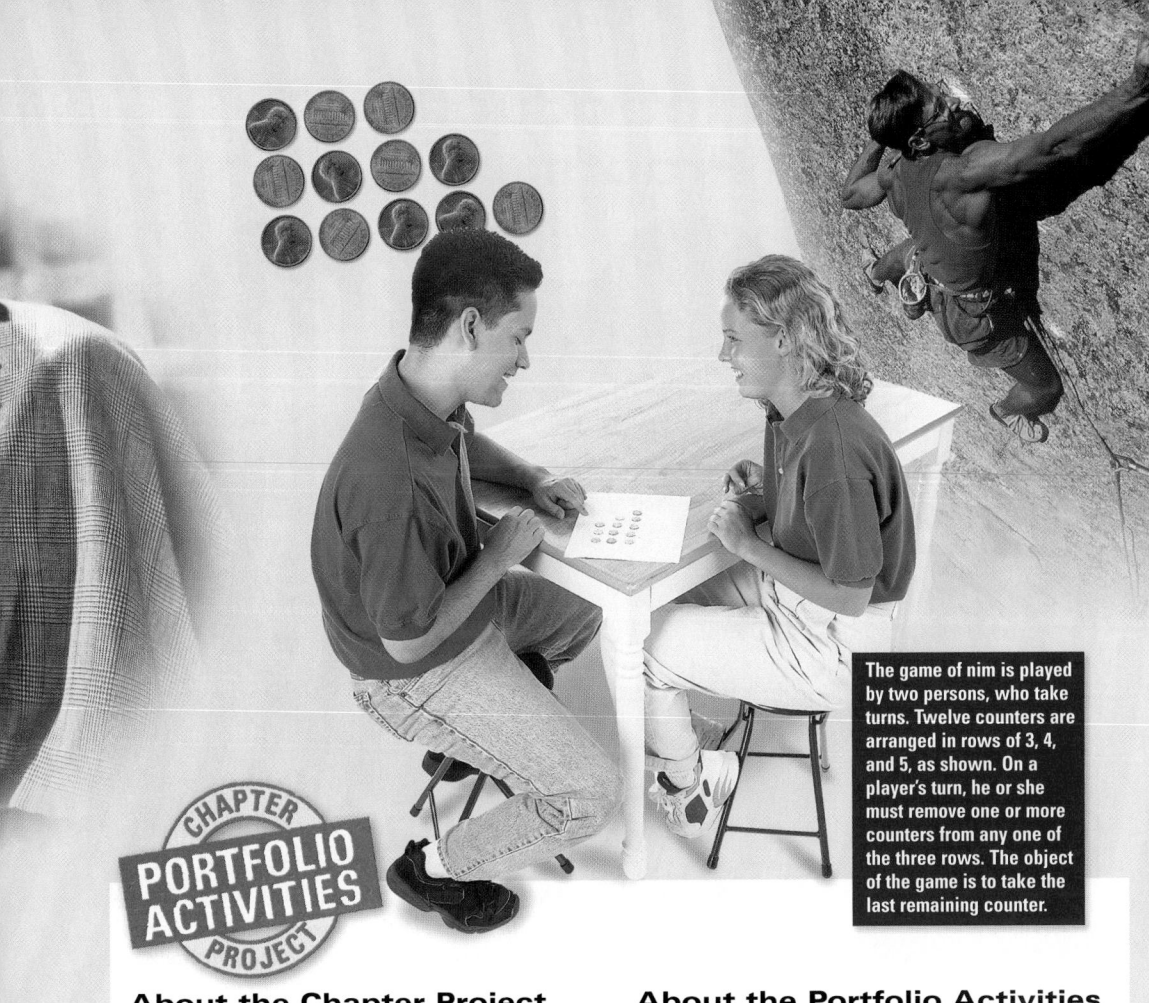

The bulleted objectives (top right):

- Use Euler diagrams to study definitions of objects. [2.3]
- Use principles of logic to create definitions of objects. [2.3]
- Identify and use the Algebraic Properties of Equality. [2.4]
- Identify and use the Equivalence Properties of Equality and of Congruence. [2.4]
- Link the steps of a proof by using properties and postulates. [2.4]
- Develop theorems from conjectures. [2.5]
- Write two-column and paragraph proofs. [2.5]

The game of nim is played by two persons, who take turns. Twelve counters are arranged in rows of 3, 4, and 5, as shown. On a player's turn, he or she must remove one or more counters from any one of the three rows. The object of the game is to take the last remaining counter.

Portfolio Activities appear at the end of Lessons 2.1 and 2.4. Each serves as preparation for the Chapter Project. The Portfolio Activities, as well as the Chapter Project Activities, are appropriate for inclusion in the student's portfolio. Students should be encouraged to include in their portfolios any other work in which they feel a sense of pride or a sense of accomplishment.

About the Chapter Project

In this chapter, you will investigate the principles of logic and apply them to proofs of geometry conjectures. But logic is useful for more than mathematical proofs. As you will see, it is also useful in everyday life and in recreational games and puzzles. In the Chapter Project, *Logic Puzzles and Games*, you will try your hand at the mathematical game known as "sprouts" and two different logic puzzles.

After completing the Chapter Project, you will be able to do the following:

- Analyze certain games and devise strategies for play.
- Solve logic puzzles by using reasoning.
- Solve a type of logic puzzle by setting up a table that displays information in an orderly, structured way.

About the Portfolio Activities

Throughout this chapter, you will be given opportunities to complete Portfolio Activities that are designed to support your work on the Chapter Project.

The rules of the game of nim are given above. The game can be analyzed to develop a strategy for winning. After you have played the game a few times, you will be ready to begin your analysis.

- One of the basic winning patterns for the game of nim is given in the Portfolio Activity on page 87. A second version of the game is also introduced.
- Three different patterns that lead to the winning pattern of the first Portfolio Activity are given in the Portfolio Activity on page 114.

internet connect

Chapter Internet Features and Online Activities

LESSON	KEYWORD	PAGE	LESSON	KEYWORD	PAGE
2.1	MG1 Alice	82	2.3	MG1 Homework Help	103
	MG1 Homework Help	84	2.4	MG1 Homework Help	112
	MG1 Nim1	87		MG1 Nim2	116
2.2	MG1 Logic	94	2.5	MG1 Proof	120
	MG1 Homework Help	94		MG1 Homework Help	122

1. The length of a square is 5 cm. What is the measure of its width? 5 cm

2. What is the sum of the first 5 counting numbers? 15

3. Solve the equation $2x + 1 = 5$. $x = 2$

Also on Quiz Transparency 2.1

Teach

Why When answering a question requires further explanation, the explanation given may be a proof. Ask students to suggest an example of an answer that they were asked to prove. Then ask how they proved it. Encourage students to discuss whether the argument used was logical or not.

An Introduction to Proofs

2.1

Objectives

- Investigate some interesting proofs of mathematical claims.

- Understand the meaning of the term *proof*.

Why *Have you ever needed to prove something you said? If you used a logical argument, you were probably able to make your case. Logical arguments that make a definite point are known as proofs.*

A chessboard has 64 squares, so it can be completely covered by 32 appropriately sized dominoes. This setup is the basis of two famous mathematical questions.

Proving Your Point

PROOF

Suppose that two squares are cut from opposite corners of a chessboard. Can the remaining squares be completely covered by 31 dominoes? If your answer is yes, can you offer a method for showing that it is possible? If your answer is no, can you explain why it cannot be done? In either case, you would be giving a *proof* of your answer.

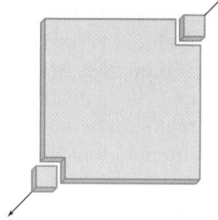

Here is a proof that the altered chessboard cannot be covered by 31 dominoes:

Because of the pattern of a chessboard, each domino must cover one dark square and one light square. Thus, any arrangement of dominoes must cover the same number of dark squares as light squares.

Notice that the squares which were cut off are the *same color*, leaving more squares of one color than the other. Therefore, it is not possible to cover the altered board with the 31 dominoes.

Alternative Teaching Strategy

USING VISUAL MODELS Students can explore the use of logical arguments in advertisements for products. Bring some examples from newspapers and magazines to class for discussion, or suggest that students find their own examples of logical arguments in product advertisements. Ask students whether each advertisement proves its case.

Activity
Three Challenges

CHECKPOINT ✔

Algebra

PROBLEM SOLVING

Part I

Suppose that you change the chessboard problem so that you cut off one dark square and one light square anywhere on the board.

Use the diagram at right and your own explanation to prove that the altered board can be covered by 31 dominoes.

The pathway through the maze suggests a proof.

Part II

How could you find the sum of the first *n* odd counting numbers without actually adding them? **Make a table** like the one below, and see if you can discover the answer.

n	First *n* odd numbers	Sum of the first *n* odd numbers
1	1	1
2	1, 3	4
3	1, 3, 5	9
4	1, 3, 5, 7	16
5	?	?
6	?	?
n	?	?

CHECKPOINT ✔

The diagram to the right of the table is a "proof without words" of the algebraic result that you may have discovered. Explain in your own words how the diagram proves the result.

Part III

You are given the figure at right, which is built entirely of squares. The area of square *C* is 64, and the area of square *D* is 81.

CHECKPOINT ✔

Use the above information to determine whether the overall figure is a square. (Recall that all four sides of a square must be the same length.)

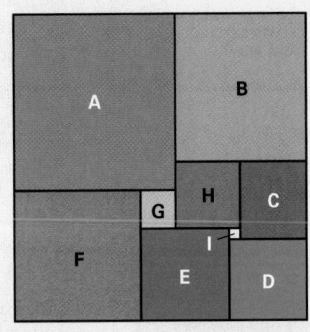

Interdisciplinary Connection

LANGUAGE ARTS Authors of detective stories, such as Arthur Conan Doyle, use logical arguments to solve mysteries. Ask students to read Conan Doyle's tales about Sherlock Holmes and find logical arguments that are used to solve mysteries.

Enrichment

Ask students to list the first 10 even integers. Then ask them to prove informally that if *n* is any positive integer, then 2*n* is always an even integer.

Activity **Notes**

In this Activity, students make conjectures based on specific experiment results and then use their conjectures to prove a generalization.

For a student worksheet of this Activity and detailed Teacher Notes, see page 22 in the Lesson Activities booklet.

Teaching Tip

Bring a chessboard and a set of dominoes to class so that students can experiment with the challenge in Part I. After students have found an answer to Part II, suggest that they extend the values of *n* beyond 6 to check their answers.

CHECKPOINT ✔

Part I

By laying dominoes end-to-end along the maze pattern, all 62 remaining squares can be covered because every light square will be paired with a dark square no matter what squares are removed. There is no problem turning corners because the dominoes can lie vertically or horizontally.

Part II

The diagram shows that the sum of the first *n* odd counting numbers is n^2 for values of *n* from 1 to 8; this would hold true for any value of *n*. Each band of dots represents an odd counting number in the sequence. When *n* = 8, the sum of the odd numbers in each band is 64.

Part III

Length of $A + B = 18 + 15 = 33$
Length of $A + F = 18 + 14 = 32$
The sides are not the same length, so the figure is not a square.

ALGEBRA In Part II of the Activity on page 81, students can find the sum of the first *n* odd counting numbers by observing the pattern in the chart. They can then use the diagram to check their answers.

Cooperative Learning

Students can work together in groups of three or four to explore the three challenge activities. They can test their ideas together and arrive at the best answer. When each group has agreed on the answer for each challenge, the solutions can be discussed by the entire class.

Assess

Selected Answers

Exercises 6–8, 9–45 odd

ASSIGNMENT GUIDE	
In Class	1–8
Core	9–32
Core Plus	16–39
Review	40–45
Preview	46–49

✐ Extra Practice can be found beginning on page 818.

What Is a Proof?

A **proof** is a convincing argument that something is true. But before you allow yourself to be convinced by a supposed proof, you should make sure that it is sound.

In mathematics, a proof starts with things that are agreed on (called *postulates* or *axioms*). Then logic is used to reach a conclusion.

There are many different styles of proofs in mathematics. Some proofs follow a prescribed form and are called *formal proofs*. For example, the calculations below are a formal proof that, for the given equation, $x = 4$.

$5x + 4 = 24$	*Given*
$5x = 20$	*Subtraction Property of Equality*
$x = 4$	*Division Property of Equality*

As you can see, a definite form is followed in this proof. Each statement on the left is given a *justification* in the column on the right. But this is not always the case. The proofs you did in the Activity did not follow any particular form, but they are just as mathematically sound (if they are correct!) as formal proofs.

In addition to the free-form proofs used in this lesson, you will learn the following styles for proofs:

- two-column proofs, Lesson 2.4
- paragraph proofs, Lesson 2.4
- flowchart proofs, Lesson 4.4
- coordinate proofs, Lesson 5.7
- table proofs, Lesson 9.3

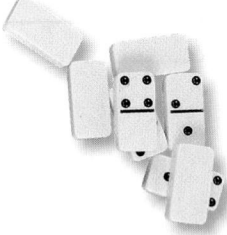

Exercises

Communicate

☑ **internet** connect
Activities Online
Go To: **go.hrw.com**
Keyword:
MG1 Alice

1. In your own words, describe what a proof is.

2. If the indicated squares were removed from the chessboard at right, explain why you could not cover the remaining squares with 31 dominoes.

3. In Part II of the Activity, a geometric solution is given for an algebraic problem. What do you think might be a advantage for a geometric solution over an algebraic one?

4. Consider the following argument for proving that the figure in Part III of the Activity is a square: I measured the sides and they were the same, so it is a square. Is this a proof? Why or why not?

5. Explain how you proved whether the overall figure in Part III of the Activity was a square. Could anyone find a flaw in your argument?

Inclusion Strategies

ENGLISH LANGUAGE DEVELOPMENT Students may need to be reminded that words can take on different meanings in different contexts. This is particularly important when students are learning new mathematical terms, such as *argument,* which is commonly used in everyday life. Discuss the everyday meaning of this word along with its mathematical meaning.

Reteaching the Lesson

USING PATTERNS Have students continue the following sequence for five more numbers: 1, 3, 5, 7, 9, . . . Ask them to describe the rule they devised to continue the sequence. Then have the students develop an informal proof for the following statement: If *n* is any positive integer, then $2n + 1$ is an odd integer.

Guided Skills Practice

Algebra

6. Two squares of each color need to be removed.

PROOF

7. 100; 10,000

8. $B = 36$, $C = 16$, $D = 16$, $E = 36$, $F = 16$, $G = 4$, $H = 4$; overall area = 196

6. Suppose that four squares were removed from a chessboard. What would need to be true about the colors of these squares in order for the remaining squares to be covered with 30 dominoes? *(ACTIVITY)*

7. According to your answer in Part II of the Activity, what is the sum of the first 10 odd numbers? of the first 100 odd numbers? *(ACTIVITY)*

8. The diagram at right is built entirely of squares. The overall figure is also a square. If the area of square A is 64 and the area of square I is 4, what are the areas of the other squares? What is the area of the overall square? *(ACTIVITY)*

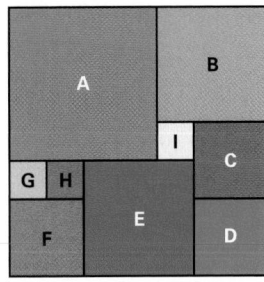

Practice and Apply

For Exercises 9–11, consider a variation of the chessboard problem (pages 80–81) in which the chessboard is covered with square tiles, each the size of 4 squares on the board. The board can then be completely covered with 16 tiles. You may wish to draw diagrams on graph paper.

9. If one square were removed from each corner of the board, could the altered board be covered completely with 15 tiles? Why or why not?

10. If four squares were removed from the top row of the board, could the altered board be covered completely with 15 tiles? Why or why not?

PROOFS

11. State a rule for removing four squares so that the altered board can be covered by 15 tiles. How would you prove your rule?

Algebra

12. Copy the diagram at right and label each area. Use the diagram and your own explanation to prove the following:

$$(x + a)(x + b) = x^2 + ax + bx + ab$$

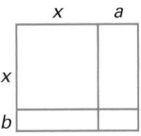

The following procedure is a shortcut for squaring a number that ends in 5: Multiply the number formed by the first digits (all digits except the final 5) by the next consecutive number. Then put 25 at the end of the product. For example, to square 35, multiply 3×4 and then put 25 at the end of the product, 12, to get 1225.

13. Use the shortcut to square 25, 75, and 105. Does it work? **625, 5625, 11,025; yes**

14. Explain how the diagram at right illustrates how the shortcut works for squaring 35.

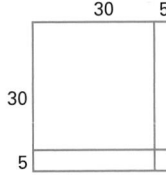

PROOF

15. Use the diagram and your own explanation to prove that this shortcut works for squaring any number ending in 5.

Error Analysis

For Exercise 16, students might make simple arithmetic errors when finding the cubes of successive integers. Using a calculator to find the cubes can eliminate such errors.

Math
CONNECTION

ALGEBRA In Exercise 7, students use a formula from Part II of the Activity to find the sum of the first 10 and of the first 100 odd numbers. In Exercise 12, students must use the FOIL method to multiply binomials. Exercises 16–26 involve infinite sequences.

Technology

Calculators will be helpful to students in Exercises 16–26 and 33–36.

9. No

10. No

11. Sample answer: Remove squares in groups the same size and shape as the tile.

12. Answer should include that the area of the diagram can either be found by multiplying the length and the width or by adding the areas of the four regions.

14. Sample answer: The shortcut says that $35^2 = 3 \cdot 4 \cdot 100 + 25 = 1225$. The diagram can be rearranged as a rectangle which is 30 by 40 and a square which is 5 by 5.

15. Sample answer: The shortcut says that $(10n + 5)^2 = n(n+1) \cdot 100 + 25$. The diagram can be rearranged as a rectangle which is $10n$ by $10(n+1)$ and a square which is 5 by 5.

16. Continuing to fill in the table until one billion is reached is too time-consuming.

18. Column B: $2^3, 5^3, 8^3, \ldots$; Column C: $3^3, 6^3, 9^3, \ldots$

19. Every entry in column C is the cube of a multiple of 3. No, columns A and B do not contain any cubes of multiples of 3.

20. The entry for 999^3 occurs in column C, because 999 is a multiple of 3.

21. Because 999^3 will occur in column C, the cube of the number immediately following 999 must occur in column A. Therefore 1000^3 must occur in column A.

23. The numbers in the sequence are getting smaller. The numerator of the fraction remains 1, while the numbers in the denominators are getting larger, meaning that 1 is being divided into more and more parts. The value of the sequence fraction is therefore decreasing.

24. $\frac{1}{2}, \frac{1}{4}, \frac{1}{8}, \frac{1}{16}, \frac{1}{32}; \frac{31}{32}$, or 0.96875

$\frac{1}{2}, \frac{1}{4}, \frac{1}{8}, \frac{1}{16}, \frac{1}{32}, \frac{1}{64}; \frac{63}{64}$, or 0.984375

$\frac{1}{2}, \frac{1}{4}, \frac{1}{8}, \frac{1}{16}, \frac{1}{32}, \frac{1}{64}, \frac{1}{128}; \frac{127}{128}$, or 0.9921875

$\frac{1}{2}, \frac{1}{4}, \frac{1}{8}, \frac{1}{16}, \frac{1}{32}, \frac{1}{64}, \frac{1}{128}, \frac{1}{256}; \frac{255}{256}$, or 0.99609375

25. The sums seem to be approaching 1. If the sum were continued infinitely, the sum would be 1.

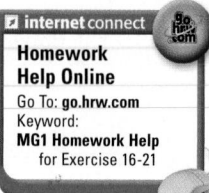
internet connect

Homework Help Online

Go To: **go.hrw.com**
Keyword:
MG1 Homework Help
for Exercise 16-21

Use the table below to answer Exercises 16–21. The numbers in the table are perfect cubes: $1^3 = 1$, $2^3 = 8$, $3^3 = 27$, $4^3 = 64$, . . .

A	B	C
1	8	27
64	125	216
?	?	?
?	?	?

16. Suppose that you wanted to know which column contains the number 1,000,000,000 (one billion). One method is to fill in the table until you reach 1,000,000,000. What is one disadvantage of this method?

17. Another method is to analyze the pattern of the numbers in the table. One billion (1,000,000,000) is the cube of what number? (Hint: $10^3 = 1000$, $100^3 = \underline{\quad ? \quad}, \ldots$) **1000**

18. Look at the columns in the table. Notice that column A contains the numbers $1^3, 4^3, 7^3, \ldots$ What numbers occur in column B? in column C?

19. What is true of every number in column C? Is this true of any number in column A or B?

20. In what column does the number 999^3 occur?

21. Prove that 1,000,000,000 occurs in column A.

Exercises 22–28 involve partial sums of the infinite sequence $\frac{1}{2}, \frac{1}{4}, \frac{1}{8}, \frac{1}{16}, \ldots$, in which the denominator of each term is doubled to obtain the next term.

22. What are the next four terms of the sequence? $\frac{1}{32}, \frac{1}{64}, \frac{1}{128}, \frac{1}{256}$

23. Are the terms getting larger or smaller? Explain your answer.

24. Complete the table below, in which terms of the sequence are added together.

Number of terms	Terms	Sum of terms
1	$\frac{1}{2}$	$\frac{1}{2}$, or 0.5
2	$\frac{1}{2}, \frac{1}{4}$	$\frac{3}{4}$, or 0.75
3	$\frac{1}{2}, \frac{1}{4}, \frac{1}{8}$	$\frac{7}{8}$, or 0.875
4	$\frac{1}{2}, \frac{1}{4}, \frac{1}{8}, \frac{1}{16}$	$\frac{15}{16}$, or 0.9375
5	?	?
6	?	?
7	?	?
8	?	?

25. What number do the sums in the table seem to be approaching?

26. Because the sequence has infinitely many terms, you cannot add them all with your calculator. However, it is still possible to find the sum of the infinite sequence. Start by drawing a square with an area of 1 square unit. Label the lengths of the sides of the square.

27. Divide your square into pieces with areas equal to the terms of the sequence. Use the diagram at right as a guide.

28. Explain how the diagram indicates the sum of the infinite sequence. What is the sum of the sequence?

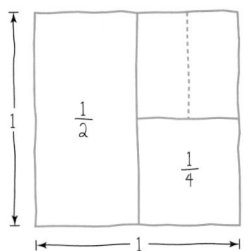

26–27.

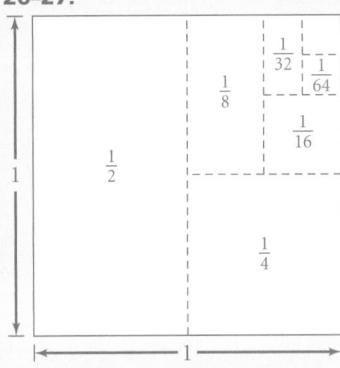

Suppose that you know the square of a positive integer n and you want to find the square of $n + 1$. In Exercises 29–32, you will discover a method.

Algebra

29. Draw a square array of dots with 5 rows of 5 dots, as shown at right. Explain how the array represents the square of 5, or 5^2.

30. Increase the side lengths of your square to represent the square of the number $(5 + 1)$, or 6^2. Use your diagram to show that $6^2 = 5^2 + 5 + 6$.

PROOF

31. Use your diagram to prove that the square of the number $n + 1$ is found by adding n and $n + 1$ to the square of n.

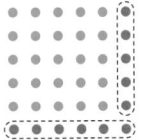

32. Use this method to find the square of 21, given that the square of 20 is 400. **441**

Algebra

Exercises 33–37 refer to the following conjecture:

$1 + 2 + 3 + 4 + \cdots + n = \dfrac{n(n + 1)}{2}$, where n is any positive integer

33. Verify that the conjecture is true for values of n from 1 to 5.

34. The triangle at right represents the sum $1 + 2 + 3 + 4 + 5 + 6$. Count the dots to find the sum. **21**

35. Find the sum of the integers 1 through 6 by using the formula in the conjecture. Does the result agree with your answer to Exercise 34? **21; yes**

36. A second triangle is added to the original figure, as shown at right. Find the number of dots in the resulting rectangle. **42**

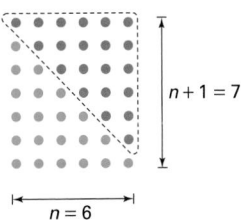
$n + 1 = 7$
$n = 6$

PROOF

37. How does the number of dots in one of the triangles relate to the number of dots in the rectangle? Express this relationship as a formula in terms of n and $n + 1$. Explain how this proves the conjecture.

28. The area of the square is 1 unit^2. If this pattern of dividing the square is infinitely continued, the sum of the areas of all the rectangles is 1, which is the sum of the given sequence.

29. The number of dots is found by multiplying the number of rows (5) by the number of columns (5), so the number of dots represents the number 5^2.

30. The number of the number of dots added equals $5 + 6$, so the total number of dots is $5^2 + 5 + 6$.

31. A square array of dots with n rows of n dots represents the number n^2. When the square is increased to $n + 1$ rows of $n + 1$ dots $n + n + 1$ dots will have been added to the number of dots in the original square.

The diagram shows n^2 dots (as in Exercise 29) with the new rows divided so that one contains n dots and the other contains $n + 1$ dots. The total number of dots is: $(n + 1)^2 = n^2 + n + (n + 1)$.

33. For $n = 1$: $1 = 1$ and $\dfrac{1(1 + 1)}{2} = \dfrac{1(2)}{2} = 1$.

For $n = 2$: $1 + 2 = 3$ and $\dfrac{2(2 + 1)}{2} = \dfrac{2(3)}{2} = 3$.

For $n = 3$: $1 + 2 + 3 = 6$ and $\dfrac{3(3 + 1)}{2} = \dfrac{3(4)}{2} = 6$.

For $n = 4$: $1 + 2 + 3 + 4 = 10$ and $\dfrac{4(4 + 1)}{2} = \dfrac{4(5)}{2} = 10$.

For $n = 5$:
$1 + 2 + 3 + 4 + 5 = 15$ and $\dfrac{5(5 + 1)}{2} = \dfrac{5(6)}{2} = 15$.

37. The number of dots in the triangle is one-half the number of dots in the rectangle. The diagram suggests that the sum of the integers from 1 to n can be related to the pattern of the dots in the triangle. This triangle encloses half the number of dots in the rectangle with dimensions n and $n + 1$, therefore it contains $\dfrac{n(n + 1)}{2}$ dots.

38. The conjecture gives the formula for the sum of the first n integers: $\frac{n(n+1)}{2}$. Notice that if n is the row number, we get the following:

For $n = 1$, $\frac{1(1+1)}{2} = 1$

For $n = 2$, $\frac{2(2+1)}{2} = 3$

For $n = 3$, $\frac{3(3+1)}{2} = 6$

For $n = 4$, $\frac{4(4+1)}{2} = 10$

These are the last numbers in each of the first four rows. Using this pattern, on the 45th row, the last number will be $\frac{45(45+1)}{2} = 1035$. Since there will be 45 numbers on the 45th row, the row begins with 991 and ends with 1035, so 1000 must occur on the 45th row.

CHALLENGE

PROOF

APPLICATION

The murder was committed in the study with the candlestick by Colonel Mustard.

A disjunctive syllogism can often be used when at least one of two things must be true. If one of the two can be shown to be false, then the other must be true.

38. The positive integers are written in a triangular array, as shown at right.

Use the conjecture given for Exercises 33–37 to prove that the number 1000 occurs in the 45th row.

```
        1
      2   3
    4   5   6
  7   8   9   10
11 ...
```

39. GAMES In a popular mystery game, players try to solve a murder. The "murder" is represented by three cards: the murderer, the place, and the weapon. The cards are placed in an envelope and kept concealed until the end of the game.

To solve the murder, you must use logic. Suppose you have determined that the crime was committed in the study with the candlestick by either Colonel Mustard or Professor Plum. Then you learn that one of the other players is holding the Professor Plum card, which means that the murderer cannot be Professor Plum. You should now be able to solve the murder. What is your full conclusion?

Your conclusion about the identity of the murderer involves a special kind of argument known as a *disjunctive syllogism*. Explain in your own words when you can use this kind of an argument.

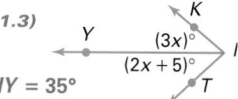

Look Back

40.–41.

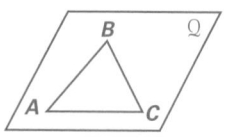

a triangle

40. Draw a plane and label it Q. Draw three noncollinear points in plane Q, and label them A, B, and C. *(LESSON 1.1)*

41. In the figure you drew for Exercise 40, draw $\overline{AB}$, $\overline{BC}$, and $\overline{AC}$. What shape is formed? *(LESSON 1.1)*

Use the diagram below for Exercises 42 and 43. *(LESSON 1.2)*

42. If $LN = 18$, find the value of x. **3**

43. If $LM = 7$, find LN. **23**

In the diagram at right, m∠KIT = 80°. *(LESSON 1.3)*

44. Find the value of x. **15**

45. Find m∠KIY and m∠TIY. m∠KIY = 45°, m∠TIY = 35°

 *Algebra*

The triangular array of numbers shown below is commonly known as *Pascal's triangle,* named after Blaise Pascal (1623–1662). Pascal's triangle has applications in geometry (see Lessons 11.1 and 11.7), algebra, and probability.

46. How would you find the entries for the next row in the triangle?

47. Write out the first eight rows of Pascal's triangle.

48. For each of the first five rows, add the entries in the row. What is the pattern in the sums?

49. What other interesting patterns can you find in Pascal's triangle?

```
            1
          1   1
        1   2   1
      1   3   3   1
    1   4   6   4   1
  1   5  10  10   5   1
```

CULTURAL CONNECTION: ASIA
The illustration at right is the earliest known version of "Pascal's triangle." It is from a Chinese book printed around 1303 C.E.

Portfolio Extension
Go To: **go.hrw.com**
Keyword:
MG1 Nim1

 *PORTFOLIO ACTIVITY*

In the game of nim (see page 79), the object is to take the last remaining counter. In a second version of the game, the object is to force your opponent to take the last counter.

- Play a few games of both versions of nim with a friend. How is your strategy for the second version different from your strategy for the original version of nim?

PROOF

- If you are able to leave two rows with two counters each, you can always win. Show that this is true for both versions of the game.

46. The first and last entry in each row are 1. All other entries are found by adding the numbers in the row above the entry, to the left and right.

47.
```
                1
              1   1
            1   2   1
          1   3   3   1
        1   4   6   4   1
      1   5  10  10   5   1
    1   6  15  20  15   6   1
  1   7  21  35  35  21   7   1
```

48. 1st row: 1
2nd row: $1 + 1 = 2$
3rd row: $1 + 2 + 1 = 4$
4th row: $1 + 3 + 3 + 1 = 8$
5th row: $1 + 4 + 6 + 4 + 1 = 16$
The pattern of the sums appears to be given by powers of 2:
$2^0 = 1, 2^1 = 2, 2^2 = 4, 2^3 = 8, 2^4 = 16.$

49. Answers may vary. Students should check carefully to make sure that their pattern applies to all rows, columns, or diagonals.

In Exercises 46–49, students look at patterns in Pascal's triangle. They will study this further in Lessons 11.6.

ALTERNATIVE
Assessment

Portfolio Activity

The Portfolio Activity can be used as preparation for the Chapter Project or as a separate activity. The Portfolio Activity on this page shows a variation of the game of nim, which was introduced on page 79.

Student Technology Guide

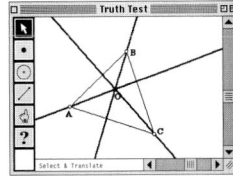

NAME _____ CLASS _____ DATE _____

Student Technology Guide
2.1 An Introduction to Proofs

One day, Jamie made the statement that if you draw a triangle with lines through each vertex perpendicular to the side not containing that vertex, then the lines intersect inside the triangle. Michael decided to make this dynamic sketch to test the truth of the statement.

Is Jamie's claim true when other triangles are examined?

Use geometry graphics software in each of the following exercises:

1. **a.** Using the software, draw a triangle and show the labels for its vertices.   Check students' sketches.
   b. Select a vertex and the side that does not contain it. Construct a perpendicular line. Display it as a heavy line by selecting the pointer, clicking once on the line, and selecting Display Line Weight Thick .
   c. Repeat part **b** for the other two vertices and sides. Label the point O where the heavy lines intersect.

2. Is point O inside the triangle you sketched? Drag a vertex to change the triangle's shape. Continue to drag vertices to view many different triangles and perpendicular lines. Does point O remain inside the triangle? What can you conclude?
   No. The perpendicular bisectors can meet in the interior or the exterior. They can even meet at a point on the triangle itself.

In Exercises 3–5, use geometry software to test the truth of each statement.

3. Four points A, B, C, and D lie along the same horizontal line. Point B is to the right of A, C is to the right of B, and D is to the right of C. Then $AB + BC + CD = AD$.   The statement is always true by the Segment Addition Postulate.

4. Four points A, B, C, and D, lie in the same plane and $AB = BC = CD = DA$. The figure determined must be a square.   The statement is not always true. The sides can have the same length, but the angles are not necessarily right angles.

5. Four points A, B, C, and D, lie in the same plane. These points determine a four-sided figure.   The statement is false. If the four points lie along the same line, for example, they do not determine a polygon at all.

6. If you believe that the statement in Exercise 3 is true, try to write a proof. If you believe that it is false, justify your position.
   $AB + BD = AD$, but $BD = BC + CD$. Therefore, by substitution,
   $AB + (BC + CD) = AB + BC + CD = AD$.

THE EYEWITNESS MATH

TOO TOUGH FOR COMPUTERS

Focus

Students explore how the number of possible routes in a tour grows factorially with the number of cities.

Motivate

After students read the article, ask "Why would a solution to a traveling salesman problem be useful in manufacturing stereo equipment and other devices using electronic circuits?" (A circuit board travels from point to point under a laser drill just as a salesman travels from town to town.)

Math Problem, Long Baffling, Slowly Yields

by Gina Kolata, *New York Times*

A century-old math problem of notorious difficulty has started to crumble. Even though an exact solution still defies mathematicians, researchers can now obtain answers that are good enough for most practical applications.

The traveling salesman problem asks for the shortest tour around a group of cities. It sounds simple–just try a few tours out and see which one is shortest. But it turns out to be impossible to try all possible tours around even a small number of cities.

Companies typically struggle with traveling salesmen problems involving tours of tens of thousands or even hundreds of thousands of points.

For example, such problems arise in the fabrication of circuit boards, where lasers must drill tens to hundreds of thousands of holes in a board. What happens is that the boards move and the laser stays still as it drills the holes. Deciding what order to drill these holes is a traveling salesman problem.

Very large integrated circuits can involve more than a million laser-drilled holes, leading to a traveling salesman problem of more than a million "cities."

In the late 1970s, investigators were elated to solve 50-city problems, using clever methods that allow them to forgo enumerating every possible route to

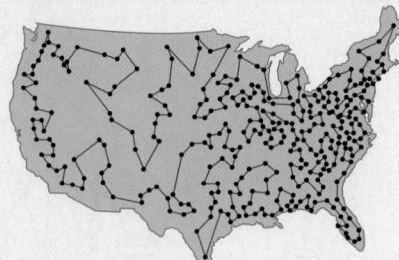

As of 1968, the computer solution for the shortest route connecting 532 cities with AT&T central offices looked like this, It was the biggest such problem calculated up to that time. Now, a route for 2,392 destinations has been computed, and mathematicians are working on a 3,038-city problem. (Source: New York University and Institute for Systems Analysis)

find the best one. By 1980, they got so good that they could solve a 318-city problem, an impressive feat but not good enough for many purposes.

Dr. David Johnson and Dr. Jon Bentley of AT&T Bell Laboratories are recognized by computer scientists as the world champions in solving problems involving about 100,000 cities. By running a fast computer for two days, they can get an answer that is guaranteed to be either the best possible tour or less than 1 percent longer than the best one.

In most practical situations, an approximate solution is good enough, Dr. Johnson said. By just getting to within about 2 percent of the perfect solution of a problem involving drilling holes in a circuit board, the time to drill the holes can usually be cut in half, he said.

The researchers break large problems into many smaller ones that can be attacked one by one, and give these fragments to fast computers that can give exact answers.

For example, Dr. Bentley said, "If I ask you to solve a traveling salesman problem for 1,000 cities in the U.S., you would do it as a local problem. You might go from New York toward Trenton and then move to Philadelphia," he explained. Then the researchers would repeat this process from other hubs, like Chicago, and combine the results. "We end up calculating only a few dozen instances per point," Dr. Bentley said. "if you have a million cities, you might do only 30 million calculations."

Cooperative Learning

The most obvious way to solve a "traveling-salesman" problem is to try all possible routes and determine which is the shortest. How many routes would you need to consider? One way to find the number of possible routes is to multiply the number of choices for each city along the route. Suppose that a route contains four cities: A, B, C, and D. One of the cities will be the starting and ending point. From that city, there are 3 possible choices for the first city to visit, then from that city there are 2 remaining possible choices for the second city to visit, then from that point there is 1 remaining possible choice for the third city to visit. Thus, the number of possible routes is

One way to visualize the number of routes is by using a tree diagram like the one below.

1	×	3	×	2	×	1	=	6
starting city (given)		choices to visit first		choices to visit second		choices to visit third		number of possible routes

In order to reduce the number of computations needed to find the shortest route, notice that a route traveled in reverse order has the same length.

ABCDA has the same length as *ADCBA*.
ABDCA has the same length as *ACDBA*.
ACBDA has the same length as *ADBCA*.

Thus, it is necessary to compute the lengths of only three routes.

1. Copy and complete the table below.

Number of cities	Multiply	Factorial form	Evaluate	Divide by 2
4	$3 \times 2 \times 1$	3!	6	3
5	?	?	?	?
6	?	?	?	?
7	?	?	?	?
n	?	?	?	?

Based on the table above, complete the following formula:
  The number of routes, N, that must be calculated to determine the shortest route for visiting n cities is $N = $ ___?___.

2. Use the formula above to determine the following: How many routes must be calculated to determine the shortest route through 10 cities? through 15 cities? through 20 cities? through 50 cities?

3. Suppose that a computer could calculate the length of 1 billion routes per second. How long would it take to find the shortest route through 20 cities? through 21 cities? through 50 cities? Give your answers in seconds, days, and years.

4. Compare your answers to Exercise 3 with the estimated age of the universe, about 15 billion years. Do you think the traveling-salesman problem is really "too tough for computers"? Why or why not?

An Introduction to Logic

Prepare

QUICK WARM-UP

For Exercises 1–3, classify each statement as true or false.

1. A statement is a sentence that is either true or false.   **true**

2. If Tom lives in Boston, then he lives in Massachusetts.   **true**

3. If Tom lives in Massachusetts, then he lives in Boston.   **false**

4. What is an Euler diagram? a diagram that is used to show logical relationships

Also on Quiz Transparency 2.2

Teach

Why The foundation of logical reasoning is some form of class organization. To explain this, use examples of classes of things that include other classes. From scientific discoveries to personal decisions, the application of logical reasoning facilitates understanding.

Objectives

● Define *conditionals* and model them with Euler diagrams.

● Use conditionals in logical arguments.

● Form the converses of conditionals.

● Create logical chains from conditionals.

APPLICATION
BIOLOGY

Why *A proof of any form requires logical reasoning. Logical reasoning ensures that the conclusions you reach are true—if the rest of the statements in the argument are true.*

Organisms can be classified according to their structure. For example, all of the flowers shown above belong to the orchid family. Class organization, which also extends to manufactured things, is the basis of logical reasoning.

Drawing Conclusions From Conditionals

The force of logic comes from the way information is structured. For example, all Corvettes are Chevrolets, a fact which can be represented by an Euler (pronounced "oiler") diagram like the one at right. Note: Euler diagrams are often called Venn diagrams.

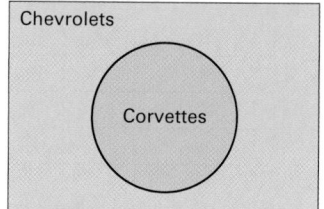

From the Euler diagram, it is easy to see that the following statement is true:

If a car is a Corvette, then it is a Chevrolet.

"If-then" statements like this one are called **conditionals**. In logical notation, conditionals are written as follows:

If *p* then *q*

or

$p \Rightarrow q$ (Read as "*p* implies *q*.")

In a conditional, the part following the word *if* is the **hypothesis**. The part following the word *then* is the **conclusion**.

If <u>a car is a Corvette</u>, then <u>it is a Chevrolet</u>.
     Hypothesis   *Conclusion*

Alternative Teaching Strategy

COOPERATIVE LEARNING Working in small groups, have students write one statement per group that requires logical reasoning to prove. Groups can then exchange statements and try to prove the new statement they received. The statements and proofs can then be shared with the class. Mathematical and nonmathematical statements are acceptable.

Now consider the following statement:

Susan's car is a Corvette.

By placing Susan's car into the Euler diagram, you can see that it is a Chevrolet. (Anything inside the circle is also inside the rectangle.)

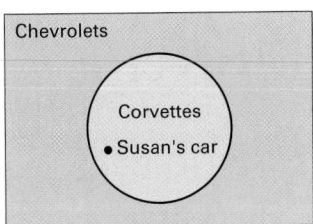

The complete process of drawing the conclusion that Susan's car is a Chevrolet can be written as a *logical argument*. This particular argument, which is known as a *syllogism* (see Lesson 12.1), has three parts.

1. If a car is a Corvette, then it is a Chevrolet.
2. Susan's car is a Corvette.
3. Therefore, Susan's car is a Chevrolet.

The process of drawing *logically certain* conclusions by using an argument is known as **deductive reasoning,** or **deduction**.

E X A M P L E **1** Recall the following definitions from your earlier studies:

An *equilateral triangle* is a triangle with three congruent sides.
An *isosceles triangle* is a triangle with at least two congruent sides.

PROBLEM SOLVING

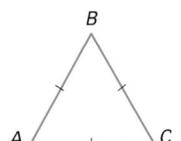

a. Draw an Euler diagram that conveys the following information:

If a triangle is equilateral, then the triangle is isosceles.

Triangle *ABC* is equilateral.

b. What conclusion can you draw about triangle *ABC*?

● **SOLUTION**

a.

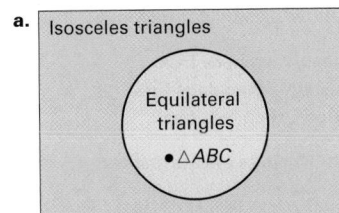

b. Triangle *ABC* is isosceles.

Interdisciplinary Connection

HISTORY Throughout history, people have been challenged to prove various hypotheses. Students can consider the following questions: What events proved that Earth was round, not flat? How can you prove who won the World Series in 1945? Is it possible to prove the exact date that Europeans arrived on the shores of North America?

ADDITIONAL
E X A M P L E ②

Write a conditional with the hypothesis "the polygon has four congruent sides" and the conclusion "the polygon is a square." Write the converse. Is the conditional true? Is the converse true?

Conditional: If the polygon has four congruent sides, then the polygon is a square. False.

Converse: If the polygon is a square, then the polygon has four congruent sides. True.

TRY THIS

Conditional: If an animal is a snake, then the animal is a reptile. True.

Converse: If the animal is a reptile, then the animal is a snake. False.

ADDITIONAL
E X A M P L E ③

Given:

1. If there is a parade, then fireworks will go off.

2. If it is July 4, then flags are flying.

3. If flags are flying, then there is a parade.

Prove: If it is July 4, then fireworks will go off.

If it is July 4, then flags are flying. If flags are flying, then there is a parade. If there is a parade, then fireworks will go off. Therefore, if it is July 4, then fireworks will go off.

Reversing Conditionals

When you interchange the hypothesis and the conclusion of a conditional, the new conditional is called the **converse** of the original conditional.

Conditional: If a car is a Corvette, then it is a Chevrolet.
Converse: If a car is a Chevrolet, then it is a Corvette.

The original conditional is true. But what about its converse? If there is an example of a Chevrolet that is not a Corvette—and there certainly is—then the converse is false. An example which proves that a statement is false is called a **counterexample**.

E X A M P L E ② Write a conditional with the hypothesis "a triangle is equilateral" and the conclusion "the triangle is isosceles." Then write the converse of your conditional. Is the conditional true? Is the converse true?

● **SOLUTION**

Conditional: If a triangle is equilateral, then it is isosceles.
Converse: If a triangle is isosceles, then it is equilateral.

The original conditional is true, according to the definition of an equilateral triangle. (If a triangle has three congruent sides, then it has at least two congruent sides.)

The converse is false, however, as the counterexample at right shows. (△*MNO* is isosceles but not equilateral.)

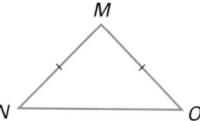

TRY THIS Write a conditional with the hypothesis "an animal is a snake" and the conclusion "the animal is a reptile." Then write the converse of your conditional. Is the conditional true? Is the converse true?

Logical Chains

Conditionals can be linked together. The result is a logical chain. In the next example, three different conditionals are linked together to form a **logical chain**. (It does not matter whether the conditionals are actually true.)

E X A M P L E ③ Consider the following silly conditionals:

If cats freak, then mice frisk.
If sirens shriek, then dogs howl.
If dogs howl, then cats freak.

PROOF Prove that the following conditional follows logically from the three given conditionals:

If sirens shriek, then mice frisk.

Enrichment

Some students may wish to consider the symbolic form of the logical reasoning presented in this lesson. Essentially, the reasoning is as follows: If a statement of the form If p, then q is assumed to be true and if p is known to be true, then q must be true. Symbolically, this can be written in two ways:

$$p \Rightarrow q \qquad \text{If } p \text{ then } q$$
$$\underline{p} \qquad\qquad \underline{p}$$
$$\therefore q \qquad\qquad \text{Therefore, } q$$

Inclusion Strategies

USING COGNITIVE STRATEGIES Students may wish to explore further the following statement from page 92: "It does not matter whether the conditionals are actually true." By analyzing the logical arguments given in this lesson, students should be able to discern that it is the form or structure of the statements that is the key in applying logic to arrive at correct conclusions.

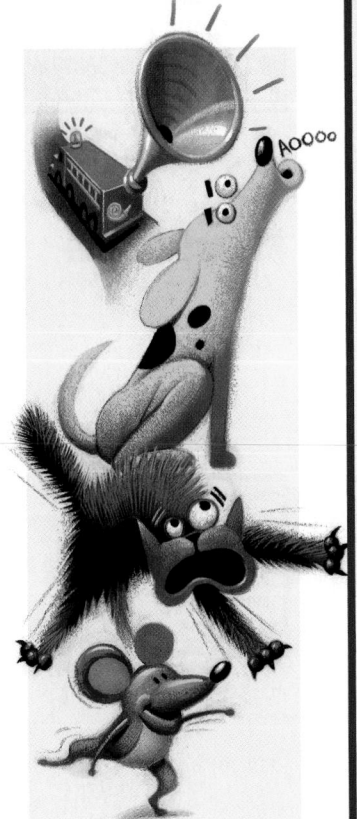

● **SOLUTION**

Identify the hypothesis of the conditional you are trying to prove:

> **If sirens shriek,** . . .

Look for a conditional that begins with "if sirens shriek."

> **If sirens shriek,** **then dogs howl.**

Look for a conditional that begins with "if dogs howl."

> **If dogs howl,** **then cats freak.**

Look for a conditional that begins with "if cats freak out."

> **If cats freak,** **then mice frisk.**

There is a zigzag pattern in the steps of the logical chain.

> **If sirens shriek,** **then dogs howl.**
>
> **If dogs howl,** **then cats freak.**
>
> **If cats freak,** **then mice frisk.**

Finally, by linking the first hypothesis and the last conclusion, you can conclude:

> **If sirens shriek,** **then mice frisk.**

TRY THIS   Working individually or in groups, create logical chains of your own.

CRITICAL THINKING   Notice that Example 3 does not prove that mice frisk. What is proven instead? What is lacking in the argument that is necessary to prove that mice actually frisk?

The proof of the conditional in Example 3 relied on the following property:

If-Then Transitive Property

Given: You can conclude:

If *A* then *B*, and   If *A* then *C*.

if *B* then *C*.   **2.2.1**

Notice that the same property is used repeatedly in long chains of conditionals.

Reteaching the Lesson

USING COGNITIVE STRATEGIES   The following statements can be used to help students think about drawing conclusions from conditionals.

If John wishes to travel outside the United States, he needs to apply for and be granted a passport.

John is now in France.

What conclusions can be drawn from these two statements?

You can conclude that John has applied for, and has been granted, a passport.

TRY THIS

A sample logical chain is given:

1. When it's Tonya's night to cook, she always makes hamburgers.

2. When Tonya makes hamburgers, she burns them.

3. If the hamburgers are burned, we order pizza.

Therefore, when it's Tonya's night to cook, we order pizza.

CRITICAL THINKING

What is actually proven is that "mice are frisking *if* sirens are shrieking." What is missing is a statement that says sirens are in fact shrieking.

Selected Answers

Exercises 6–8, 9–45 odd

ASSIGNMENT GUIDE

In Class	1–8
Core	9–16, 17–33 odd
Core Plus	29–35 odd, 36–38
Review	39–46
Preview	47–53

✐ Extra Practice can be found beginning on page 818.

Error Analysis

A common mistake is to assume that if a conditional is true, then its converse is true also. Also, it is possible for a conditional to be false and its converse to be true.

6. If a worker is a United States Postal worker, then the person is a federal employee. John is a United States Postal worker. Therefore, John is a federal employee.

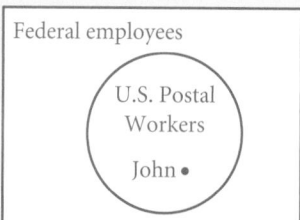

Federal employees

U.S. Postal Workers

John •

Exercises

● Communicate

APPLICATION

1. WEATHER Look at the satellite photo at right, which shows a cloud cover over Amsterdam, the capital of the Netherlands. Can you conclude that it was raining in Amsterdam at the time the satellite photo was taken?

✦ internet connect

Activities Online
Go To: **go.hrw.com**
Keyword:
MG1 Logic

2. Draw an Euler diagram to illustrate the conditional "If it rains, then it is cloudy."

3. What is the converse of the conditional given in Exercise 2? Use your Euler diagram to illustrate whether the converse is true.

4. Explain how to write the converse of a given conditional.

5. Explain how to disprove a given conditional.

Clouds over the Netherlands, looking south. The North Sea is on the right.

● Guided Skills Practice

6. Consider the following statements:

> All United States Postal workers are federal employees.
> John is a United States Postal worker.

Write the first statement as a conditional and use deduction to show that John is a federal employee. Draw an Euler diagram to illustrate the logic of your deduction. *(EXAMPLE 1)*

✦ internet connect

Homework Help Online
Go To: **go.hrw.com**
Keyword:
MG1 Homework Help
for Exercises 7, 17–20

7. Write a conditional with the hypothesis "two lines are parallel" and the conclusion "the two lines do not intersect." Then write the converse of your conditional. Is the conditional true? Is the converse true? *(EXAMPLE 2)*

8. Consider the three conditionals below. *(EXAMPLE 3)*

> If a number is divisible by 2, then the number is even.
> If a number is even, then the last digit is 0, 2, 4, 6, or 8.
> If a number is divisible by 4, then the number is divisible by 2.

Prove that the following conditional follows logically from the three conditionals above:

> If a number is divisible by 4, then the last digit is 0, 2, 4, 6, or 8.

---

7. Conditional: If two lines are parallel, then the two lines do not intersect.
Converse: If two lines do not intersect, then the two lines are parallel.
The conditional is true. The converse is not true.

8. Form a logical chain:
If a number is divisible by 4, then the number is divisible by 2.
If a number is divisible by 2, then the number is even.
If a number is even, then the last digit is 0, 2, 4, 6, or 8.
Therefore, if a number is divisible by 4, then the last digit is 0, 2, 4, 6, or 8, by the If-Then Transitive Property.

● *Practice and Apply*

For Exercises 9–12, refer to the following statement:

All people who live in Ohio live in the United States.

9. Rewrite the statement as a conditional.

10. Identify the hypothesis and the conclusion of the conditional.

11. Draw an Euler diagram that illustrates the conditional.

12. Write the converse of the conditional.

For Exercises 13–16, use the Euler diagram to write a conditional.

13.
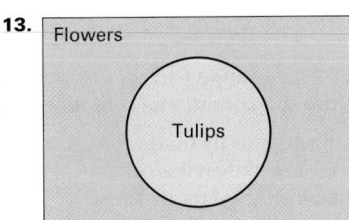

14.
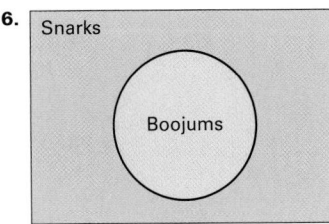

Illustration from The Hunting of the Snark *by Lewis Carroll. A boojum seems to have been a particularly dangerous kind of snark. (See Exercise 16.)*

15.
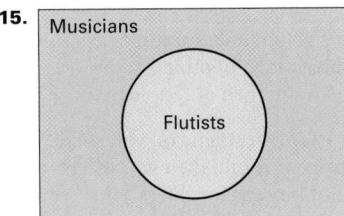

16.

For Exercises 17–20, identify the hypothesis and conclusion of each conditional. Write the converse of each conditional. If the converse is false, give a counterexample to show that it is false.

17. If it is snowing in Chicago, then it is snowing in Illinois.

18. If two angles are complementary, then the sum of their measures is 90°.

19. If the measure of each angle in a triangle is less than 90°, then the triangle is acute.

20. If a figure is rotated, then its size and shape stay the same.

For Exercises 21–23, refer to the diagram below, and write a conditional with the given hypothesis and conclusion.

21. Hypothesis: $\angle AXB$ and $\angle BXD$ form a linear pair.
Conclusion: $\angle AXB$ and $\angle BXD$ are supplementary.

22. Hypothesis: $\angle AXB$ and $\angle BXD$ are supplementary.
Conclusion: $m\angle AXB + m\angle BXD = 180°$

23. Hypothesis: $m\angle BXC + m\angle CXD = 90°$
Conclusion: $m\angle AXB = 90°$

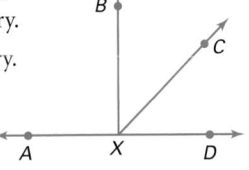

19. Hypothesis: The measure of each angle in a triangle is less than 90°.
Conclusion: The triangle is acute.
Converse: If a triangle is acute, then the measure of each angle is less than 90°.
The converse is true.

20. Hypothesis: A figure is rotated.
Conclusion: Its size and shape stay the same.
Converse: If a figure's size and shape stay the same, then the figure is rotated.

The converse is false. Sample counter-example: A figure which is translated has the same size and shape, but it is not rotated.

21. If $\angle AXB$ and $\angle BXD$ form a linear pair, then $\angle AXB$ and $\angle BXD$ are supplementary.

22. If $\angle AXB$ and $\angle BXD$ are supplementary, then $m\angle AXB + m\angle BXD = 180°$.

23. If $m\angle BXC + m\angle CXD = 90°$ then $m\angle AXB = 90°$.

Technology

It may be helpful for students to use calculators in Exercises 47–50.

9. If a person lives in Ohio, then the person lives in the United States.

10. Hypothesis: a person lives in Ohio
Conclusion: the person lives in the United States

11.

| People who live in the USA |
| People who live in Ohio |

12. Converse: If a person lives in the United States, then the person lives in Ohio.

13. If a plant is a tulip, then it is a flower.

14. If two angles form linear pairs, then the angles are supplementary angles.

15. If a person is a flutist, then the person is a musician.

16. If a thing is a boojum, then it is a snark.

17. Hypothesis: It is snowing in Chicago.
Conclusion: It is snowing in Illinois.
Converse: If it is snowing in Illinois, then it is snowing in Chicago.
The converse is false.
Sample counterexample: It is snowing in Springfield, Illinois, but not in Chicago, Illinois.

18. Hypothesis: Two angles are complementary.
Conclusion: The sum of the angle measures is 90°.
Converse: If the sum of the measures of two angles is 90°, then the angles are complementary.
The converse is true.

24. Conclusion: "Mikey" is a rodent.

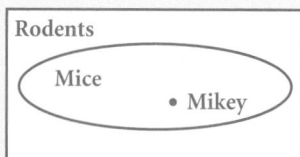

25. Conclusion: Socrates is a mortal.

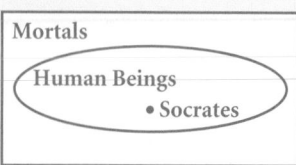

26. Conclusion: Jennifer will get a sunburn.

27. Conclusion: Ingrid lives in Scandinavia.

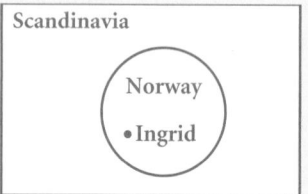

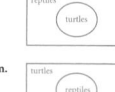
In Exercises 24–29, use the given statements to draw a conclusion. Then draw an Euler diagram to illustrate your conclusion.

24. If an animal is a mouse, then the animal is a rodent. Mikey is a mouse.

25. If someone is human, then he or she is mortal. Socrates is human.

26. If Jennifer goes to the beach, then she will get a sunburn. Jennifer is going to the beach tomorrow.

27. If someone lives in Norway, then he or she lives in Scandinavia. Ingrid lives in Norway.

28. If a figure is a square, then the figure is a rectangle. Figure *ABCD* is a square.

29. If two points are in plane $\mathcal{P}$, then the line containing them is in plane $\mathcal{P}$. Points S and T are in plane $\mathcal{P}$.

In Exercises 30–33, arrange each set of statements to form a logical chain. Then write the conditional that follows from the logical chain.

30. If it is cold, then birds fly south.
If the days are short, then it is cold.
If it is winter, then the days are short.

31. If the police catch Tim speeding, then Tim gets a ticket.
If Tim drives a car, then Tim drives too fast.
If Tim drives too fast, then the police catch Tim speeding.

32. If quompies plaun, then romples gleer.
If ruskers bleer, then homblers frain.
If homblers frain, then quompies plaun.

33. If you go to a movie, then you will spend all of your money.
If you clean your room, then you will go to a movie.
If you cannot buy gas for the car, then you will be stranded.
If you spend all of your money, then you cannot buy gas for the car.

APPLICATION

34. **HUMOR** Can logic be used to prove the impossible? Consider the following logical chain:

> The independent farmer is disappearing.
> That man is an independent farmer.
> Therefore, that man is disappearing.

Write the above argument using a conditional statement. How would you criticize the argument?

Is this man disappearing? (See Exercise 34.)

28. Conclusion: Figure *ABCD* is a rectangle.

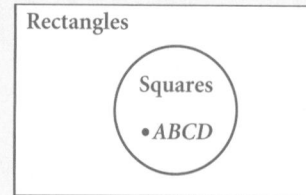

29. Conclusion: The line containing points S and T is in the plane.

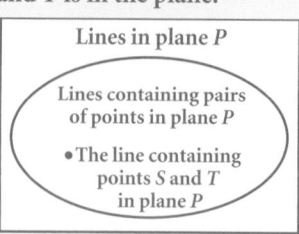

35. The following is an old saying dating back to at least the fifteenth century:

> For want of a nail, the shoe was lost.
> For want of a shoe, the horse was lost.
> For want of a horse, the rider was lost.
> For want of a rider, the battle was lost.
> For want of the battle, the war was lost.

Rewrite the saying as a logical chain of conditionals. Then write the conclusion that follows from the logical chain.

FINE ARTS People have different opinions about what constitutes a work of art. Consider the objects below. Use one of the conditional statements below or one of your own to present a logical proof that each work is or is not a work of art.

If an object displays form, beauty, and unusual perception on the part of its creator, then the object is a work of art.

If an object displays creativity on the part of the person who made it, then the object is a work of art.

36.

Two Open Modular Cubes/Half Off,
Sol Lewitt, 1972

37.

Fond Jaune,
Alexander Calder, 1972

38. MUSIC Write a conditional with the given hypothesis and conclusion.

Hypothesis: A person performs classical music.

Conclusion: The person dislikes jazz.

Write the converse of the conditional.

Yo-Yo Ma is a famous classical cellist who also loves jazz. Is this a counterexample to the original conditional or to the converse? Is either the conditional or the converse true?

35. If a nail is lost, then the shoe is lost.
If a shoe is lost, then the horse is lost.
If a horse is lost, then the rider is lost.
If a rider is lost, then the battle is lost.
If a battle is lost, then the war is lost.
Conclusion: If a nail is lost, then the war is lost.

36.–37.
Answers will vary. Using the first conditional to lay the foundations, student arguments should include statements that support the claim that the object displays form, beauty, and unusual perception on the part of its creator. Using the second conditional to lay the foundations, student arguments should include statements that support the claim that the object displays creativity on the part of its creator. Check student arguments.

38. Conditional: If a person performs classical music, then the person dislikes jazz.
Converse: If a person dislikes jazz, then the person performs classical music.
The counterexample disproves the original conditional. Neither the conditional nor the converse is true.

In Exercises 47–53, students explore inductive reasoning and the Fibonacci sequence. They are asked to recognize and continue the pattern in the sequence.

39. a plane

40. Fold the paper in half lengthwise, unfold it, and then fold it in half widthwise. The lengthwise crease is the perpendicular bisector of the widthwise crease, and vice versa.

41. Fold over one side of the angle until it meets the edge of the opposite side of the angle. The line formed is the bisector of the angle.

42. circumcenter, or center of the circumscribed circle of the triangle

43. incenter, or center of the inscribed circle of the triangle

44. obtuse; acute; right

45. reflection, rotation, and translation

46.

YAM | MAY

52. Every third number in the sequence is even. Pattern: odd, odd, even, odd, odd, even, …
This pattern occurs because the sum of an even and an odd is odd. To get an even number, two odds (or two evens) must be added.

53. Every fourth number in the sequence is divisible by 3. Every fifth number is divisible by 5.

39. A floor is best modeled by what geometric figure? *(LESSON 1.1)*

40. With folding paper, construct a segment and its perpendicular bisector. *(LESSON 1.4)*

41. With folding paper, construct an angle and its bisector. *(LESSON 1.4)*

Complete the statements below. *(LESSON 1.5)*

42. The perpendicular bisectors of a triangle meet at the ___?___.

43. The angle bisectors of a triangle meet at the ___?___.

44. The center of the circumscribed circle of a triangle is outside the triangle if the triangle is ___?___, is inside the triangle if the triangle is ___?___, and is on the triangle if the triangle is ___?___.

45. Name the three basic types of rigid transformations. *(LESSON 1.6)*

46. Reflect the word **YAM** across a vertical line. *(LESSON 1.6)*

 *Look Beyond*

CONNECTION

PATTERNS IN DATA In this lesson, you learned about deductive reasoning. Another kind of reasoning is called inductive reasoning. Inductive reasoning is based on the recognition of patterns.

For Exercises 47–50, use inductive reasoning to find the next number in each sequence.

47. 5, 8, 11, 14, _?_ 17

48. 20, 27, 36, 47, 60, _?_ 75

49. 2, 6, 18, 54, _?_ 162

50. 3, 1, $\frac{1}{3}$, $\frac{1}{9}$, _?_ $\frac{1}{27}$

51. The beginning of the Fibonacci sequence given below. The numbers of this sequence are known as Fibonacci numbers. What are the next five Fibonacci numbers?
1, 1, 2, 3, 5, 8, . . . **13, 21, 34, 55, 89**

52. Which terms in the Fibonacci sequence are even? Describe the pattern of even and odd numbers in the sequence, and explain why this pattern occurs.

53. Which terms in the Fibonacci sequence are divisible by 3? by 5? Guess a pattern for the multiples of 3 and of 5, and check your answer by examining the sequence.

APPLICATION

BOTANY *In artichokes and other plants, the number of spirals in each direction are often Fibonacci numbers. This artichoke has 5 clockwise spirals and 8 counterclockwise spirals.*

Definitions

Objectives

- Use Euler diagrams to study definitions of objects.
- Use principles of logic to create definitions of objects.

Why *Have you ever disagreed with someone, only to find out that you and the other person had different definitions of the same terms? In mathematics it is especially important to know the definitions of the terms you are studying.*

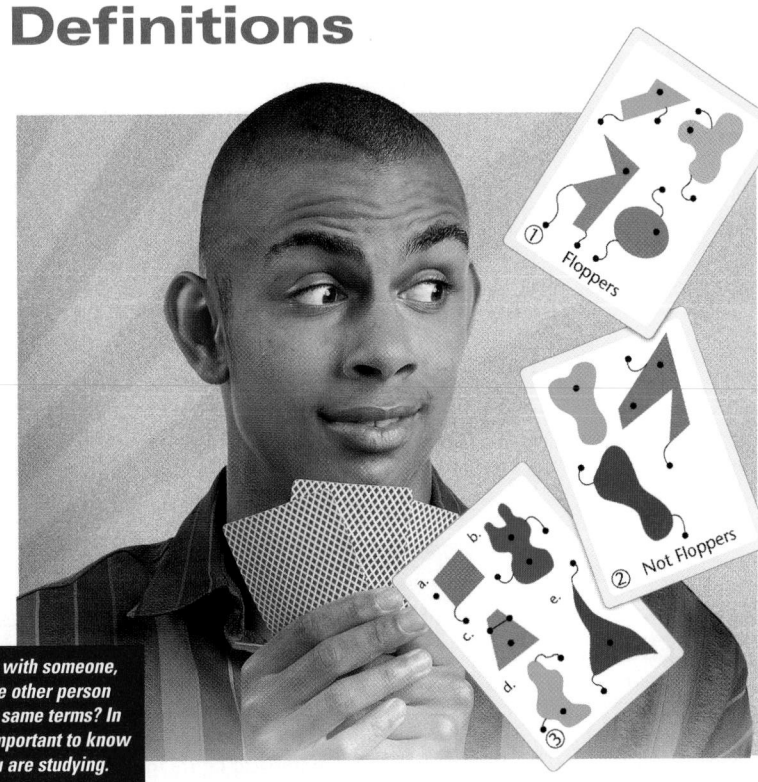

Prepare

QUICK WARM-UP

1. Write the definition of a triangle as a conditional.
   If a figure is a triangle, then it has three sides.

2. What is the converse of "If A, then B"?
   If *B*, then *A*.

3. If a conditional is a true statement, will its converse be true?
   Not necessarily.

Also on Quiz Transparency 2.3

Definitions and Euler Diagrams

Which of the figures in card 3 above are floppers? Simply by observing the differences between the figures in the card 1 and card 2, it is possible to write a definition of a flopper.

A flopper is a figure with one "eye" and two "tails."

Using this definition, you can see that figures **d** and **e** are floppers.

Definitions have a special property when they are written as conditional statements. For example:

If a figure is a flopper, then it has one eye and two tails.

You can also write the **converse** of the conditional by interchanging the hypothesis and conclusion:

If a figure has one eye and two tails, then it is a flopper.

Notice that both the original conditional and its converse are true. *This special property is true for all definitions.* The two true conditionals can be combined into a compact form by joining the hypothesis and the conclusion with the phrase "if and only if," which is represented by $p \Leftrightarrow q$.

$$p \text{ if and only if } q \qquad or \qquad p \Leftrightarrow q$$

The resulting "if-and-only-if" statement is known as a **biconditional.**

Teach

Why Disagreements can arise when different meanings are assigned to the same term. Discuss with students why it is important in mathematics to know the definitions of the terms used.

Alternative Teaching Strategy

USING DISCUSSION Use an object that is available to all students in class, such as a pencil, and ask students to define the object. As students suggest various characteristics of the object, write them on the board. Have students agree on the least number of characteristics that can be used in a definition of the object.

This Activity will help students understand that the essential nature of a geometric figure must be incorporated into its definition. They also learn that the name given to an object can be arbitrarily chosen. However, once an object has been named and defined, the definition must be able to distinguish that object from all others unambiguously.

For a student worksheet of this Activity and detailed Teacher Notes, see page 26 in the Lesson Activities booklet.

Cooperative Learning

Students can do Activity 1 in groups of 2 or 3. Each student should think about at least one characteristic for the definition of the object depicted. Then have groups compare and evaluate their definitions in a class discussion.

Teaching Tip

Students need to understand that *p if and only if q* is a compact form of writing two conditionals: *If p then q* and *if q then p*. An abbreviated form of this is *p iff q*. Students can also read the symbolic expression $p \Leftrightarrow q$ as *p implies q* and *q implies p*.

By combining the conditional and its converse, you create the following definition of a flopper, expressed in logical terms:

A figure is a flopper if and only if it has one eye and two tails.

Euler diagrams can be used to represent the two parts of the definition:

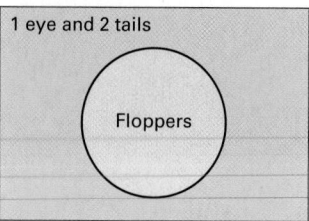 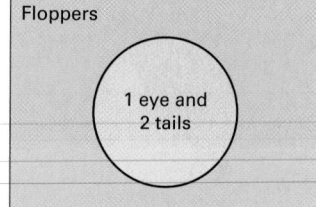

CRITICAL THINKING

Can you create a diagram to represent the fact that both the original statement and its converse are true? How would the parts of the diagram be related?

Activity 1
Capturing the "Essence" of a Thing

YOU WILL NEED

no special tools

1. Look at the figure at right. Suppose that it is a geometrical object you want to study. Make up your own name for the object. Then answer the question, What must be true of a geometrical figure in order for it to be a __(your name for the object)__?

2. According to your concept of a __(your name for the object)__, which of the objects below would you consider to be one? (There are no set rules for this. The conditions are up to you.)

CHECKPOINT ✔

3. Write your own definition of a __(your name for the object)__. Base your definition on your answer to Step 1 or any other conditions you place on the object. Test your definition of the object to be sure that it is actually true.

Interdisciplinary Connection

SOCIAL STUDIES Discuss the need for members of Congress to use precise language when writing laws. What could happen if one person's interpretation of a particular law is different from another person's interpretation? You can also mention that in our system of government, such disputes are usually settled in the judicial system. Many legal disputes are followed closely in the media. Ask students to find newspaper or magazine articles describing legal disputes that occurred over differing interpretations of a law.

Enrichment

Have students design and make games in which imaginary and geometric figures must be identified or drawn according to definitions. Students can add to the games throughout the year and play them for lesson reinforcement and review.

CHECKPOINT ✔

3. Answers will vary. A sample answer is given: A house-pentagon is a five-sided polygon with a base made up of three sides of a rectangle and a roof made up of two sides of an isosceles triangle.

TRY THIS Create your own object, and then give it a name and a definition. Your object does not have to be geometrical or even mathematical, but it should be something you can draw. Test your definition to be sure it is valid. Share your definition with others.

Activity 2
Adjacent Angles

YOU WILL NEED

no special tools

An important concept in geometry is that of *adjacent angles*. By examining the figures in the boxes below, you should be able to form an idea about what adjacent angles are—and also what they are not. This information will enable you to write your own definition of adjacent angles.

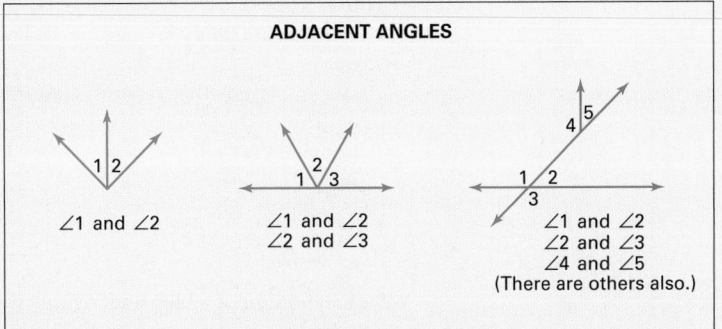

ADJACENT ANGLES

∠1 and ∠2

∠1 and ∠2
∠2 and ∠3

∠1 and ∠2
∠2 and ∠3
∠4 and ∠5
(There are others also.)

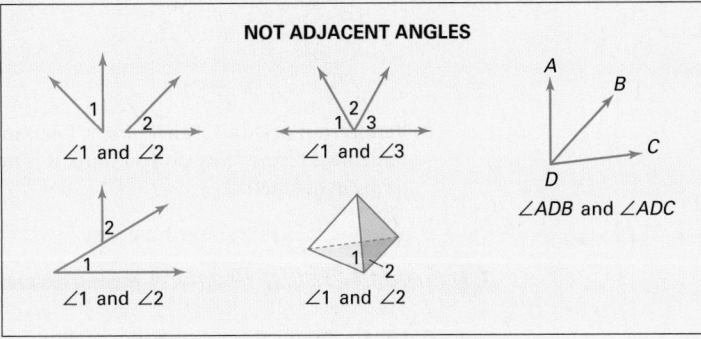

NOT ADJACENT ANGLES

∠1 and ∠2

∠1 and ∠3

∠ADB and ∠ADC

∠1 and ∠2

∠1 and ∠2

1. List the adjacent angles in the figure at right.

2. What do adjacent angles have in common? Can they overlap?

CHECKPOINT ✔
3. Angles that do not overlap have *no interior points in common*. Use this fact and your answers from Steps 1 and 2 to write a definition of adjacent angles.

Definition: Adjacent Angles

Adjacent angles are angles in a plane that have their __?__ and one __?__ in common but __?__ .

Activity 2 **Notes**

In this Activity, students write a definition of adjacent angles by listing the characteristics that they have in common and by comparing them with angles that are not adjacent.

For a student worksheet of this Activity and detailed Teacher Notes, see page 26 in the Lesson Activities booklet.

CHECKPOINT ✔

3. Vertices; side; no interior points in common

Cooperative Learning

After completing Activities 1 and 2, put students into groups and give them sets of attribute blocks. Tell them to divide all the blocks into two categories and write a definition for each category. Then have them group the blocks into three categories with three definitions, and so on. Ask students if it is easier to write definitions when there are fewer categories and have them explain their reasoning.

Inclusion Strategies

USING COGNITIVE STRATEGIES Students who are having difficulty with the activities in this lesson can enhance their understanding of definitions by referring to Chapter 1 and making a list of the definitions of geometric terms introduced in that chapter.

Reteaching the Lesson

USING PATTERNS The idea of a transformation can be used to reteach the lesson. For example, ask students how they would define a *rigid motion*. Have them write a definition of a rigid motion as a conditional statement, as the converse of the statement, and in "if and only if" form. Students can also find definitions in other subject areas, such as science and language arts, and write them as conditionals.

ASSIGNMENT GUIDE

In Class	1–7
Core	9–29 odd
Core Plus	18–29
Review	30–38
Preview	39–42

✐ Extra Practice can be found beginning on page 818.

Error Analysis

Some students may have difficulty writing the sentences in Exercises 8–16 as conditional statements. Ask students to write several examples on the board, so they can refer to them as they do the exercises.

6. Sample answer: A glosh is a figure which has 6 sides, 2 of which are parallel, and the remaining 4 sides are equal in length. Figures a, c, d, and f are gloshes.

7. ∠1 and ∠3; ∠1 and ∠2; ∠2 and ∠4; ∠3 and ∠4

Exercises

● Communicate

1. Explain how a definition is different from a conditional statement.

2. Choose a definition from a dictionary and write it as a biconditional. Is it a valid definition?

3. Explain why the following statement is not a definition:

   A tree is a plant with leaves.

4. The following are blops: The following are not blops:

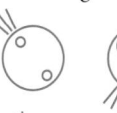  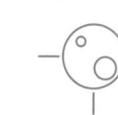 

Write a definition of a blop and identify which of the following are blops:

  a.   **b.**   **c.**   **d.**

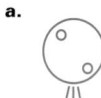 

5. Recall from Lesson 1.1 that *points*, *lines*, and *planes* are referred to as undefined terms. Why do you think it is necessary to have undefined terms in geometry?

● Guided Skills Practice

6. Use the figure at right to write a definition of a glosh. Write your definition as precisely as possible.

   Examine the shapes below. According to your definition, which (if any) of the figures are gloshes? *(ACTIVITY 1)*

   a.   **b.**   **c.**   **d.**   **e.**   **f.**

   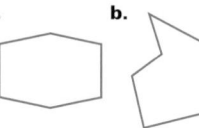   

7. List all pairs of adjacent angles in the figure at right. *(ACTIVITY 2)*

   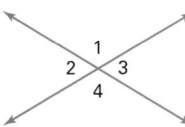

Practice and Apply

In Exercises 8–16, use the following steps to determine whether the given sentence is a definition.
- **a.** Write the sentence as a conditional statement.
- **b.** Write the converse of the conditional.
- **c.** Write a biconditional statement.
- **d.** Decide whether the sentence is a definition, and explain your reasoning.

8. A teenager is a person who is 13 years old or older.

9. A teenager is a person from 13 to 19 years old.

10. Zero is the integer between −1 and 1.

11. An even number is divisible by 2.

12. An angle is formed by two rays.

13. A right angle has a measure of 90°.

14. GEOLOGY Granite is a very hard, crystalline rock.

15. CHEMISTRY Hydrogen is the lightest of all known substances.

16. BIOLOGY An otter is a small furry mammal with webbed feet that are used for swimming.

17. Name all pairs of adjacent angles in the figure at right.
∠WVX and ∠XVY;
∠XVY and ∠YVZ;
∠WVY and ∠YVZ;
∠WVX and ∠XVZ

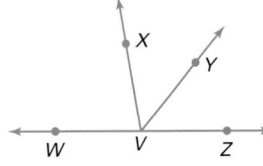

internet connect

Homework Help Online
Go To: go.hrw.com
Keyword:
MG1 Homework Help
for Exercises 8–16

APPLICATIONS

There are 13 species of otters in the world, inhabiting 5 continents. (See Exercise 16.)

For Exercises 18–22, explain why the indicated angles are *not* adjacent.

18.
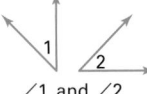
∠1 and ∠2

19.
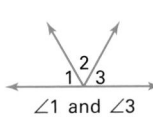
∠1 and ∠3

20.
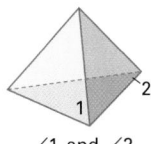
∠1 and ∠2

21.
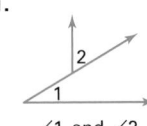
∠1 and ∠2

22.
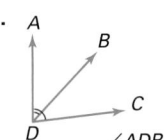
∠ADB and ∠ADC

23. The following are flishes:
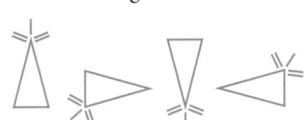

The following are not flishes:

Which of the following are flishes? **shapes b and d**

a.

b.

c.

d.

11. a. Conditional: If a number is even, then it is divisible by 2.
- **b.** Converse: If a number is divisible by 2, then it is even.
- **c.** Biconditional: A number is even if and only if it is divisible by 2.
- **d.** The statement is a definition because the conditional and the converse are both true.

12. a. Conditional: If something is an angle, then it is formed by two rays.
- **b.** Converse: If something is formed by two rays, then it is an angle.
- **c.** Biconditional: Something is an angle if and only if it is formed by two rays.
- **d.** The statement is not a definition because the converse is not true. Two rays form an angle only if they have a common endpoint.

8. a. Conditional: If a person is a teenager, then the person is 13 years old or older.
- **b.** Converse: If a person is 13 years old or older, then the person is a teenager.
- **c.** Biconditional: A person is a teenager if and only if the person is 13 years old or older.
- **d.** The statement is not a definition because the converse is not true. A person who is 20 years old or older is 13 years old or older but is not a teenager.

9. a. Conditional: If a person is a teenager, then the person is from 13 to 19 years old.
- **b.** Converse: If a person is from 13 to 19 years old, then the person is a teenager.
- **c.** Biconditional: A person is a teenager if and only if the person is from 13 to 19 years old.
- **d.** The statement is a definition because the conditional and the converse are both true.

10. a. Conditional: If an integer is zero, then it is between −1 and 1.
- **b.** Converse: If an integer is between −1 and 1, then it is zero.
- **c.** Biconditional: An integer is zero if and only if it is between −1 and 1.
- **d.** The statement is a definition because the conditional and the converse are both true.

27. Sample answer: A polygon is a regular polygon if and only if all of its sides and angles are congruent.

24. The following are zobbles:

The following are not zobbles:

Which of the following are zobbles? **shape a**

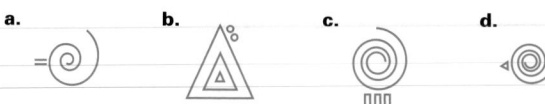

a. b. c. d.

25. The following are parallelograms:

The following are not parallelograms:

Which of the following are parallelograms? **shapes b and c**

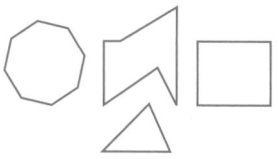 

a. b. c. d.

26. The following are polygons:

The following are not polygons:

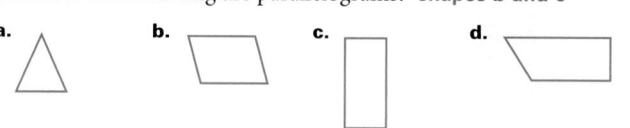

Write a definition of a polygon.

Sample answer: A figure is a polygon if and only if it is a closed plane figure consisting of line segments that intersect exactly two other line segments, one at each endpoint.

27. The following are regular polygons:

The following are not regular polygons:

Write a definition of a regular polygon.

28. Examine the figures below. Choose two or more figures with some features in common. Then choose a name and create a definition for the figures you chose. Tell which of the other figures below are examples of the object you defined.

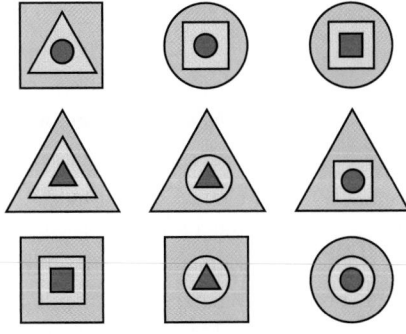

APPLICATION

Ciconiiformes:
   Great Blue Heron;
   Roseate Spoonbill

Cuculiformes:
   Greater Roadrunner;
   Yellow-Billed Cuckoo

Apodiformes:
   White-Throated Swift;
   Ruby-Throated Hummingbird

29. BIOLOGY Scientists who study birds are known as ornithologists. They classify birds into 29 different orders. Each order has a number of defining characteristics. Read the descriptions of the three orders below, and classify each of the six birds shown according to their order.

White-Throated Swift

Greater Roadrunner

Ruby-Throated Hummingbird

Great Blue Heron

Roseate Spoonbill

Yellow-Billed Cuckoo

Ciconiiformes: long-legged, long-necked birds that wade in shallow water or, in some cases, feed on open ground

Cuculiformes: small-to medium-sized, slender, usually long-tailed birds with zygodactyl feet (*Zygodactyl* means that the toes are arranged in pairs, with two in front and two in back.)

Apodiformes: small or very small birds with tiny feet, extremely short humeri, and long bones in the outer portion of the wing (*Humeri* are the bones of the upper arm part of the wing extending from the "shoulder" to the "elbow.")

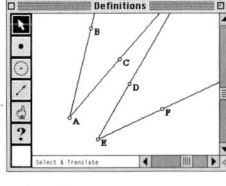

Student Technology Guide

Exercises 39–42 build on what students know about reflections from Chapter 1. Students are asked to reflect a figure across more than one reflection line. Geometry graphics software can be used to perform these exercises.

31.–32.

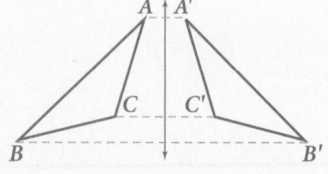

35.–36.

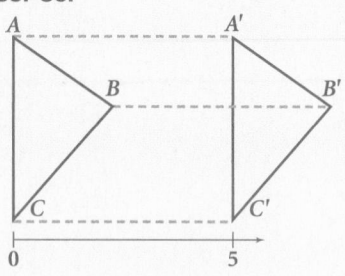

30. Point *B* is the midpoint of $\overline{AC}$. Find *x* and *AB*. *(LESSON 1.2)*

$x = -3$; *AB* = 7

Refer to the figure below for Exercises 31–34. *(LESSON 1.6)*

31. Reflect the triangle across the given line.

32. Draw $\overline{AA'}$, $\overline{BB'}$, and $\overline{CC'}$.

33. They are parallel.

33. What is the relationship among the three segments you drew in Exercise 32?

34. The line of reflection is the perpendicular bisector of the line segments.

34. What is the relationship between the line of reflection and each of the segments you drew in Exercise 32?

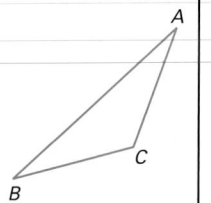

Refer to the figure below for Exercises 35–38. *(LESSON 1.6)*

35. Translate the triangle 5 units along the given line.

36. Draw $\overline{AA'}$, $\overline{BB'}$, and $\overline{CC'}$.

37. They are parallel and the same length.

37. What is the relationship among the three segments you drew in Exercise 36?

38. They are parallel.

38. What is the relationship between the given line and each of the segments you drew in Exercise 36?

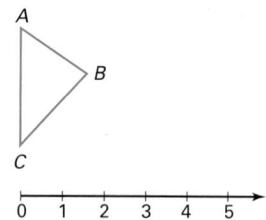

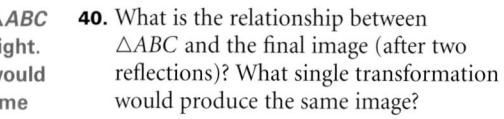

39. Copy the figure at right onto folding paper. Reflect △*ABC* across line ℓ, and then reflect the image across line *m*.

40. The image is △ABC moved to the right. A translation would produce the same image as the two reflections.

40. What is the relationship between △*ABC* and the final image (after two reflections)? What single transformation would produce the same image?

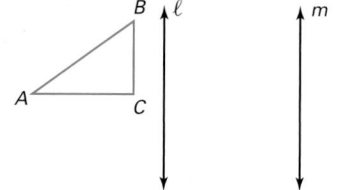

41. Copy the figure at right onto folding paper. Reflect △*DEF* across line *n*, and then reflect the image over line *p*.

42. The image is △DEF rotated. A rotation of 180° around the intersection of lines *n* and *p* produces the same image.

42. What is the relationship between △*DEF* and the final image (after two reflections)? What single transformation would produce the same image?

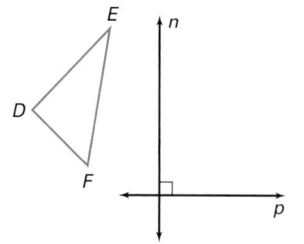

39.

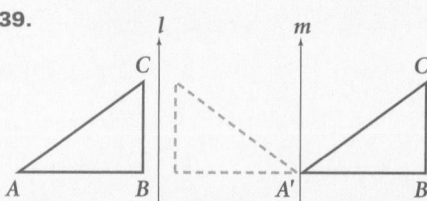

41.

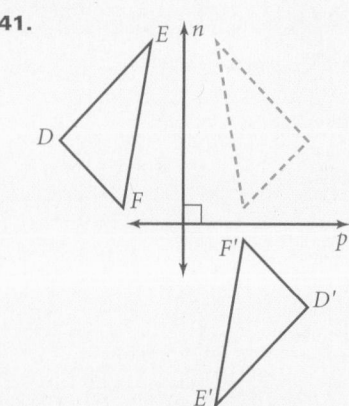

Building a System of Geometry Knowledge

Objectives

- Identify and use the Algebraic Properties of Equality.

- Identify and use the Equivalence Properties of Equality and of Congruence.

- Link the steps of a proof by using properties and postulates.

Why In algebra, you used the Properties of Equality to link the steps of a proof. Each link is a justification which guarantees that each statement you write in a proof is true—assuming that the statements you start with are true.

The first geometry "textbook," The Elements, was written by Euclid, a Greek mathematician who lived in Alexandria, Egypt, around 300 B.C.E.

QUICK **WARM-UP**

1. How many endpoints does a line have? none

2. Solve the equation $x - 6 = 14.$ $x = 20$

3. Does the symbol $\overline{AB}$ refer to the segment itself or its measure? the segment itself.

4. What is another way to write $m\overline{AB} = 5$ units? $AB = 5$

Also on Quiz Transparency 2.4

Algebraic Properties of Equality

The importance of Euclid's work lies not so much in what he discovered as in *the way he organized the existing knowledge of geometry* of his time. Starting from simple beginnings, he built up a large system of geometry knowledge.

Euclid began *The Elements* with five basic postulates, or statements that are accepted as true without proof. In addition to the postulates, Euclid included twenty-three definitions and five statements he called "common notions." The second common notion reads as follows:

If equals are added to equals, then the wholes are equal.

You may recognize this from your study of algebra as the Addition Property of Equality. This property is used to solve equations, as in the example below.

$$x - 3 = 5$$
$$x - 3 + 3 = 5 + 3 \qquad \textit{Addition Property of Equality}$$
$$x = 8 \qquad \textit{Simplify.}$$

Notice that *equals* (the 3s) are added to *equals* (the sides of the equation) to give two *wholes* (the sides of the new equation), which are themselves equal.

Teach

Why The first geometric ideas were based on what people observed to be true. By 300 B.C.E., a large body of geometric knowledge existed, but apparently no one had yet sorted out the basic postulates to prove theorems. In this lesson, students will see that a proof is a logical argument that links an initial true statement with other true statements in order to prove a new fact.

Alternative Teaching Strategy

USING MODELS To introduce proofs, give students several steps of a two-column proof that are in the wrong order. Students must use logical reasoning to reorder the steps correctly. Next give students copies of a proof with some steps deleted. Have students write in the missing steps. After students have completed two-column proofs successfully on their own, have them rewrite some of the proofs as paragraph proofs. Then have them rewrite some paragraph proofs as two-column proofs.

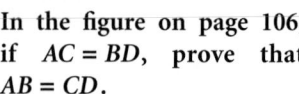

ADDITIONAL EXAMPLE 1

In the figure on page 106, if $AC = BD$, prove that $AB = CD$.

$AC = BD$	Given
$AC - BC = BD - BC$	Subtraction Property
$AC - BC = AB$	Segment Addition
$BD - BC = CD$	Segment Addition
$AB = CD$	Substitution

Teaching Tip

This lesson introduces students to formal geometric proofs. Some students may benefit from using a standard two-column format.

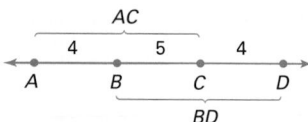

Algebraic Properties of Equality

Let a, b, and c be real numbers or expressions representing real numbers.

Addition Property	If $a = b$, then $a + c = b + c$.	2.4.1
Subtraction Property	If $a = b$, then $a - c = b - c$.	2.4.2
Multiplication Property	If $a = b$, then $ac = bc$.	2.4.3
Division Property	If $a = b$ and $c \neq 0$, then $\frac{a}{c} = \frac{b}{c}$.	2.4.4
Substitution Property	If $a = b$, you may replace a with b in any true equation containing a and the resulting equation will still be true.	2.4.5

Algebra

In the figure, the lengths of $\overline{AB}$ and $\overline{CD}$ are equal. What can you conclude about the lengths of the two overlapping segments, $\overline{AC}$ and $\overline{BD}$?

It is easy to see the following:

$AC = 4 + 5 = 9$
$BD = 5 + 4 = 9$
Thus, $AC = BD$.

CRITICAL THINKING How does the conclusion illustrate the Addition Property of Equality? In terms of Euclid's second common notion, what *equals* are added to what other *equals*?

Linking Steps to Prove a Theorem

The following example illustrates how each step in a proof can be linked to the information that is given. The proof is a continuation of the example about overlapping segments. The result is called a *theorem*.

EXAMPLE 1 In the figure, $AB = CD$. Prove that $AC = BD$.

SOLUTION

PROOF

You are interested in the lengths AC and BD. Because B is between A and C and C is between B and D, the Segment Addition Postulate can be used to write an equation for each length.

$$AB + BC = AC \qquad BC + CD = BD$$

You are given $AB = CD$. Notice what happens when you add BC to both sides of the equation.

> The two sides of the new equation express the lengths **AC** and **BD**.

$AB = CD$	Given
$\underbrace{AB + BC}_{AC} = \underbrace{BC + CD}_{BD}$	Addition Property of Equality
	Segment Addition Postulate

Now you can substitute AC and BD into the equation to get the conclusion you want: $AC = BD$.

Interdisciplinary Connection

TRADE OCCUPATIONS People working in trade occupations, such as carpenters, plumbers, and electricians, frequently use the fact "if equals are added to equals, then the wholes are equal". Students can probably suggest examples of situations when they also used this common notion, for example, dividing candy among siblings, etc.

The result that was proven in Example 1, along with its converse, can be stated as a theorem. You will be asked to prove the converse in Exercise 21. In mathematics, a **theorem** is a statement that has been proved deductively. In your future work in geometry, you can use a theorem to justify a statement without writing out the whole proof.

Overlapping Segments Theorem

Given a segment with points A, B, C, and D arranged as shown, the following statements are true:

1. If $AB = CD$, then $AC = BD$.
2. If $AC = BD$, then $AB = CD$.

2.4.6

Two-Column Proofs

The proof in Example 1 might be written out as shown below. A **two-column** format has been used. This format is especially convenient for many of the proofs you will do in your study of geometry.

As you may notice, the steps in the final form of a proof may not be in the same order as the steps you followed to discover the ideas.

TWO-COLUMN PROOF

> You can use boxes and colors to make the proof easier to understand.

Statements	Reasons
1. $AB = CD$	Given
2. $AB + BC = BC + CD$	Addition Property of Equality
3. $AB + BC = AC$	Segment Addition Postulate
4. $BC + CD = BD$	Segment Addition Postulate
5. $AC = BD$	Substitution Property of Equality

Paragraph Proofs

An alternative to a two-column proof is a **paragraph proof**. An advantage of a paragraph proof is that you have a chance to explain your reasoning in your own words. A paragraph proof of the Overlapping Segments Theorem might read as follows:

PROOF

You are given $AB = CD$. Add BC to both sides of the equation, resulting in $AB + BC = BC + CD$. In the figure, $AB + BC = AC$ and $BC + CD = BD$ by the Segment Addition Postulate. The expressions on the left of these equations match the expressions in the previous equations, so you can substitute the equivalent expressions, AC and BD. The result is $AC = BD$.

Inclusion Strategies

ENGLISH LANGUAGE DEVELOPMENT Some students may have difficulty reading and understanding mathematics. You can help these students by suggesting that they read the material in this lesson prior to discussing it in class. They can then write down their questions and note aspects of the lesson that are not clear. If they still have questions after the lesson is discussed in class, you can meet individually with them.

Teaching Tip

Additional Example 1 on page 108 is a proof of Part II of the Overlapping Segment Theorem on page 109.

Initially, students may show a strong preference for two-column proofs over paragraph proofs. Two-column proofs allow students to see the steps of the proof more easily. Have students try writing both types of proofs.

Cooperative Learning

In groups of three or four, give students a statement to prove in two-column form. One student in the group should write the first step of the proof and pass it to the second student. The second student should write the second step of the proof, and so on. The groups can discuss the completed proof and then compare their results with the answer given by you.

Math
CONNECTION

ALGEBRA The Equivalence Properties of Equality on page 110 and the Algebraic Properties of Equality on page 108, connect geometric figures to the real numbers. Thus, all properties of real numbers are now assumed to be true and thus can be used to prove geometric statements.

CRITICAL THINKING

The Transitive Property of Congruence guarantees that the segments are congruent. $\overline{AB}$ is congruent to the segment connecting the points of the compass, and this is congruent to $\overline{BC}$. Thus, $\overline{AB} \cong \overline{BC}$.

The Equivalence Properties of Equality

In addition to the Algebraic Properties of Equality, there are three important properties known as the *Equivalence Properties of Equality*. They are so obvious that you probably don't think of them when you use them. In geometry, however, they are often used to justify steps in a proof.

Equivalence Properties of Equality

Reflexive Property	For any real number a, $a = a$.	2.4.7
Symmetric Property	For all real numbers a and b, if $a = b$, then $b = a$.	2.4.8
Transitive Property	For all real numbers a, b, and c, if $a = b$ and $b = c$, then $a = c$.	2.4.9

There are also equivalence properties for relations other than equality, as you will learn in the next section.

The Equivalence Properties of Congruence

Congruence, like the relation of equality, satisfies the equivalence properties. The congruent shapes below illustrate this fact. Any relation that satisfies these three equivalence properties is called an **equivalence relation**.

Equivalence Properties of Congruence

Reflexive Property
figure $A \cong$ figure A
2.4.10

Symmetric Property
If figure $A \cong$ figure B,
then figure $B \cong$ figure A.
2.4.11

Transitive Property
If figure $A \cong$ figure B
and figure $B \cong$ figure C,
then figure $A \cong$ figure C.
2.4.12

CRITICAL THINKING

In Lesson 1.4 you used a compass to draw congruent segments on a line. What property of congruence guarantees that the segments are actually congruent? (Hint: Imagine a segment that connects the points of the compass.)

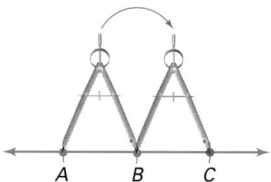

Enrichment

Students can explore the following phrases to determine whether they represent equivalence relations:

1. is greater than or equal to

2. is less than

3. is approximately equal to

4. is about the same size

EXAMPLE ② 2

The first stamp pictured at right measures 3 centimeters by 6 centimeters. The second stamp is congruent to the first. The third stamp is congruent to the second.

APPLICATION
HOBBIES

From the given information, what can you conclude about the first and the third stamps? State your conclusions by using geometry terms. What property discussed in this lesson justifies your conclusion?

Using a template or ruler, draw three triangles with the same dimensions. If the second triangle has exactly the same dimensions as the first and the third triangle has exactly the same dimensions as the second, what can you conclude about the first and third triangles? **The first triangle is congruent to the third triangle by the Transitive Property of Congruence.**

● **SOLUTION**

The first stamp is congruent to the third stamp (and so the measurements of the stamps are equal). The justification for this conclusion is the Transitive Property of Congruence.

Russian stamps commemorating the Russian American Company

TRY THIS   Create examples of your own that illustrate the Equivalence Properties of Congruence. Share your answers with your classmates.

TRY THIS
Sample answer: Line segments $\overline{AB}$, $\overline{BC}$, and $\overline{CD}$ such that $AB = 10$, $\overline{AB} \cong \overline{BC}$, and $\overline{BC} \cong \overline{CD}$. $\overline{AB} \cong \overline{AB}$ by the Reflexive Property. Since $\overline{AB} \cong \overline{BC}$, $\overline{BC} \cong \overline{AB}$ by the Symmetric Property. Since $\overline{AB} \cong \overline{BC}$ and $\overline{BC} \cong \overline{CD}$, then $\overline{AB} \cong \overline{CD}$ by the Transitive Property.

Exercises

● *Communicate*

1. Why is it necessary to use postulates in geometry?

2. Explain the difference between a postulate and a theorem.

3. Rephrase Euclid's second common notion in your own words. Give an example that uses this notion.

4. Euclid's first common notion reads as follows:

Things which are equal to the same thing are also equal to one another.

Rephrase this statement in your own words. How is Euclid's first common notion related to the Equivalence Properties of Equality?

● *Guided Skills Practice*

For Exercises 5 and 6, refer to the diagram below. *(EXAMPLE 1)*

W   X   Y   Z

5. $\overline{WX} \cong \overline{YZ}$
$WX = 8$
$WY = 19$
Find XZ. **19**

6. $\overline{WY} \cong \overline{XZ}$
$WX = 30$
$WY = 75$
Find YZ. **30**

Reteaching the Lesson

COOPERATIVE LEARNING   Arrange the class into small groups. Have each group prepare to teach part of the lesson to the rest of the class. When the groups are finished, have each group quiz their classmates over part of the lesson they taught.

Selected Answers

Exercises 5–8, 9–59 odd

ASSIGNMENT GUIDE

In Class	1–8
Core	9, 11, 13–20, 21–29 odd
Core Plus	21–29 odd, 30–46, 47–53 odd
Review	54–60
Preview	61–64

✐ Extra Practice can be found beginning on page 818.

Error Analysis

For exercises in which students are asked to write their own proofs, recommend that they model their proofs by following the examples shown on page 109.

9. Subtraction

10. Addition; Division

11. Transitive

12. Symmetric; Transitive

13. m∠MLN + m∠NLP

14. m∠NLP + m∠PLQ

15. Angle Addition Postulate

For Exercises 7 and 8, refer to the diagram below, in which △CDE ≅ △C′D′E′ and △C′D′E′ ≅ △C″D″E″. *(EXAMPLE 2)*

7. What conclusion can you draw about △CDE and △C″D″E″?

8. What property justifies your conclusion?

7. △CDE ≅ △C″D″E″

8. Transitive Property of Congruence

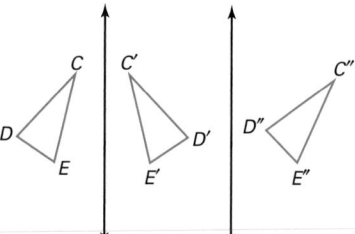

Practice and Apply

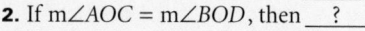

Algebra

PROOFS

In Exercises 9–12, identify the Properties of Equality that justify the indicated steps.

9.
$$x + 6 = 14 \quad \textit{Given}$$
$$x + 6 - 6 = 14 - 6 \quad \underline{?}\textit{ Property}$$
$$x = 8$$

10.
$$2x - 3 = 17 \quad \textit{Given}$$
$$2x - 3 + 3 = 17 + 3 \quad \underline{?}\textit{ Property}$$
$$2x = 20$$
$$2x \div 2 = 20 \div 2 \quad \underline{?}\textit{ Property}$$
$$x = 10$$

11.
$$a + b = c + d \quad \textit{Given}$$
$$c + d = e + f \quad \textit{Given}$$
$$a + b = e + f \quad \underline{?}\textit{ Property}$$

12.
$$AB + CD = XY \quad \textit{Given}$$
$$CD + DE = XY \quad \textit{Given}$$
$$XY = CD + DE \quad \underline{?}\textit{ Property}$$
$$AB + CD = CD + DE \quad \underline{?}\textit{ Property}$$

PROOF

In Exercises 13–19, you will complete and prove the Overlapping Angles Theorem.

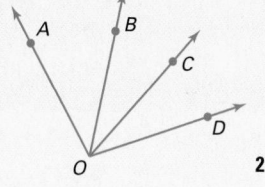

Overlapping Angles Theorem

Given ∠AOD with points B and C in its interior as shown, the following statements are true:

1. If m∠AOB = m∠COD, then __?__.
2. If m∠AOC = m∠BOD, then __?__.

2.4.13

☐ internet connect

Homework Help Online

Go To: **go.hrw.com**
Keyword:
MG1 Homework Help
 for Exercises 13–20
 and 30–46

Refer to the figure at right to answer Exercises 13–19.

13. ___?___ + ___?___ = m∠MLP

14. ___?___ + ___?___ = m∠NLQ

15. What postulate justifies your answers to Exercises 13 and 14?

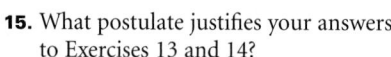

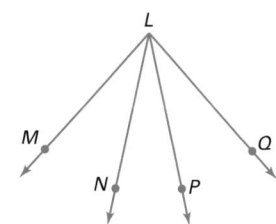

Part 1: Let m∠MLN = m∠PLQ.

16. Use the Addition Property of Equality to add m∠NLP to both sides of the equation.

17. Use the Substitution Property of Equality, the Angle Addition Postulate, and your conclusions to Exercises 13, 14, and 16 to write the conclusion for the first part of the Overlapping Angles Theorem.

Part 2: Let m∠MLP = m∠NLQ.

18. Use the Subtraction Property of Equality to subtract m∠NLP from both sides of the equation.

19. Use the Substitution Property of Equality, the Angle Addition Postulate, and your conclusions to Exercises 13, 14, and 18 to write the conclusion for the second part of the Overlapping Angles Theorem.

20. Prove the second part of the Overlapping Segments Theorem (see page 109), which is the converse of the first part. Draw and label your own diagram. (Hint: Use the Subtraction Property of Equality.)

Algebra

Refer to the diagram at right, in which WX = YZ. Use the Overlapping Segments Theorem to complete the following:

21. $WX = a$, $WY = 15$, $XZ = \underline{\ ?\ }$ 15

22. $WY = 9$, $XY = 2a + 1$, $YZ = a^2$,

$a = \underline{\ ?\ }$, $WZ = \underline{\ ?\ }$
　　　2　　　　　13

Algebra

Refer to the diagram at right, in which m∠EDF = m∠HDG. Use the Overlapping Angles Theorem to complete the following:

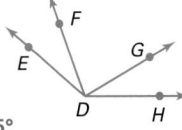

23. m∠EDF = 25°, m∠EDG = 85°, m∠HDF = $\underline{\ ?\ }$ 85°

24. m∠EDG = $(9x - 8)°$, m∠FDG = $(6x + 8)°$, m∠GDH = $(2x - 4)°$,

$x = \underline{\ ?\ }$, m∠HDF = $\underline{\ ?\ }$, m∠HDG = $\underline{\ ?\ }$
　　12　　　　　　100°　　　　　　20°

Refer to the three triangles below for Exercises 25–28.

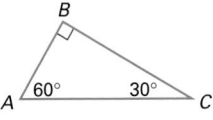 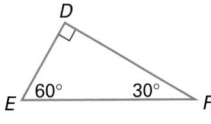 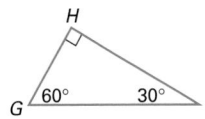

25. Complete the conclusion below.

$AB = DE$

$DE = GH$

$AB = ?$

26. Complete the conclusion below.

m∠ABC = m∠EDF

m∠EDF = m∠GHI

m∠ABC = ?

27. Which property justifies your answer to Exercise 25?

28. Which property justifies your answer to Exercise 26?

Math
C O N N E C T I O N

ALGEBRA In Exercises 9–12, students use the Algebraic Properties of Equality and the Equivalence Properties of Equality to prove solutions to equations. In Exercises 21–24, students use the same properties to prove the Overlapping Segments Theorem and the Overlapping Angles Theorem.

16. m∠MLN + m∠NLP = m∠PLQ + m∠NLP

17. m∠MLP = m∠NLQ

18. m∠MLP − m∠NLP = m∠NLQ − m∠NLP

19. m∠MLN + m∠PLQ

20. Answers will vary depending on letters chosen.

A　B　　C　D

Statements	Reasons
$AC = BD$	Given
$AC = AB + BC$	Segment Addition Postulate
$BD = CD + BC$	Segment Addition Postulate
$AB + BC$ $= CD + BC$	Substitution Property
$AB = CD$	Subtraction Property

25. GH

26. m∠GHI

27. Transitive Property

28. Transitive Property

34. m∠1 + m∠4 + m∠5 = 180°

35. Substitution Property

36. m∠3

37. m∠1 + m∠2 + m∠3 = 180°

39. 90°

(See Exercise 29.)

PROOFS

CHALLENGE

29. JoAnn wears the same size hat as April. April wears the same size hat as Lara. Will Lara's hat fit JoAnn? What property of congruence justifies your answer? **Yes; Transitive Property**

Given: m∠BAC + m∠ACB = 90°
m∠DCE + m∠DEC = 90°
m∠ACB = m∠DCE

Prove: m∠BAC = m∠DEC

Proof:

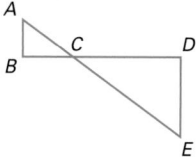

Statements	Reasons
m∠BAC + m∠ACB = 90°	Given
m∠DCE + m∠DEC = 90°	**30.** __?__ Given
m∠BAC + m∠ACB = m∠DCE + m∠DEC	**31.** __?__ Transitive Prop.
m∠ACB = m∠DCE	Given
m∠BAC + m∠ACB = m∠ACB + m∠DEC	**32.** __?__ Substitution Prop.
33. __?__ m∠BAC = m∠DEC	Subtraction Property

Given: m∠1 + m∠4 + m∠5 = 180°
m∠4 = m∠2
m∠5 = m∠3

Prove: m∠1 + m∠2 + m∠3 = 180°

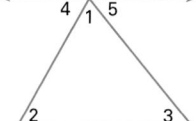

Proof: Since m∠4 = m∠2, you can replace m∠4 with m∠2 in the equation **34.** ____?____ by **35.** ____?____. Similarly, you can replace m∠5 with **36.** ____?____, giving **37.** ____?____.

Given: m∠CBD = m∠CDB
m∠ABD = 90°
m∠EDB = 90°

Prove: m∠ABC = m∠EDC

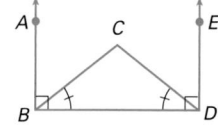

Proof:

Statements	Reasons
m∠ABD = 90°	Given
m∠ABC + m∠CBD = m∠ABD	**38.** __?__ Angle Addition Postulate
39. m∠ABC + m∠CBD = __?__ 90°	Transitive (or Substitution) Property
m∠EDB = 90°	**40.** __?__ Given
41. __?__ m∠CDE + m∠CDB = m∠EDB	Angle Addition Postulate
m∠CDE + m∠CDB = 90°	**42.** __?__ Trans. (or Subs.) Prop.
m∠ABC + m∠CBD = m∠CDE + m∠CDB	**43.** __?__ Trans. (or Subs.) Prop.
m∠CBD = m∠CDB	**44.** __?__ Given
m∠ABC + m∠CBD = m∠CDE + m∠CBD	**45.** __?__ Substitution Property
46. __?__ m∠ABC = m∠CDE	Subtraction Property

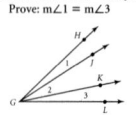
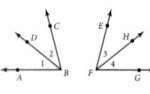

Examine the relationships below. Determine which of the following properties each relationship satisfies: the Reflexive Property, the Symmetric Property, and the Transitive Property. Then state whether the relationship is an equivalence relation. If a property is not satisfied by a given relationship, give a counterexample.

47. $a < b$, where a and b are real numbers.

48. $a \leq b$, where a and b are real numbers.

49. a is divisible by b, where a and b are real numbers.

50. Figure A is a reflection of figure B.

51. Figure A is a rotation of figure B.

52. A is a sister of B, where A and B are people.

53. A has the same last name as B, where A and B are people.

 *Look Back*

Recall the five postulates about points, lines, and planes that you discovered on page 12 and refer to the figure below to answer Exercises 54–59. *(LESSON 1.1)*

54. Name three points that determine plane $\mathcal{R}$.

55. Which postulate justifies your answer to Exercise 54?

56. Name two points that determine line ℓ.

57. Which postulate justifies your answer to Exercise 56?

58. Name the intersection of plane $\mathcal{R}$ and plane $\mathcal{M}$.

59. Which postulate justifies your answer to Exercise 58?

60. PHOTOGRAPHY A *tripod* is a three-legged stand used by photographers to hold their cameras steady. Why do you think tripods provide more stability than four-legged stands? Which of the five postulates justifies your answer? *(LESSON 1.1)*

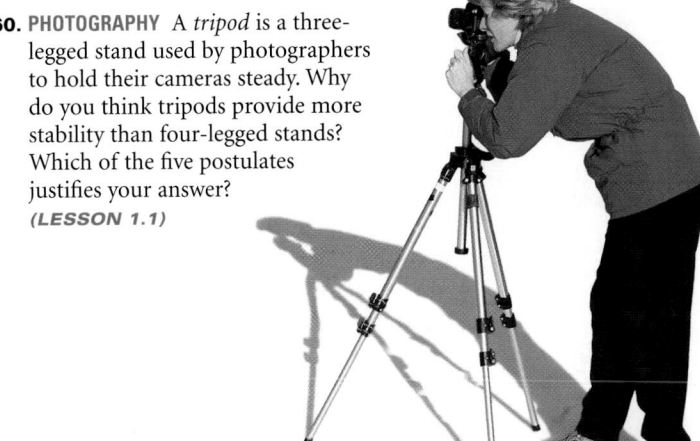

47. Transitive; the relation is not an equivalence relation. Counterexample to the Reflexive Property: $5 \not< 5$ Counterexample to the Symmetric Property: $2 < 3$ but $3 \not< 2$.

48. Reflexive and Transitive; the relation is not an equivalence relation. Counterexample to the Symmetric Property: $1 \leq 2$ but $2 \not\leq 1$.

49. Reflexive and Transitive; the relation is not an equivalence relation. Counterexample to the Symmetric Property: 10 is divisible by 5, but 5 is not divisible by 10.

50. Symmetric; the relation is not an equivalence relation. Counterexample to the Reflexive Property: letter **b** is not the reflection of itself (it is the reflection of letter **d**). Counterexample to the Transitive Property: letter **b** is the reflection of letter **d**, and letter **d** is the reflection of letter **b**, but letter **b** is not the reflection of letter **b**.

51. Reflexive, Symmetric, and Transitive; it is an equivalence relation. (Note: Any figure is a rotation of itself, by an amount of 360°.)

52. The relation is not an equivalence relation. Counterexample to the Reflexive Property: Cassie is not a sister of herself. Counterexample to the Symmetric Property: Cassie is a sister of John, but John is not a sister of Cassie. Counterexample to the Transitive Property: Cassie is a sister of Diana, and Diana is a sister of Cassie, but Cassie is not a sister of herself.

53. Reflexive, Symmetric, and Transitive; it is an equivalence relation.

54. Sample answer: S, E, I

55. Postulate 1.1.7 Through any three non-collinear points there is exactly one plane.

56. Sample answer: F and Z

57. Postulate 1.1.6 Through any two points there is exactly one line.

58. line l

59. Postulate 1.1.5 The intersection of two planes is a line.

60. Sample answer: A tripod has three "points" touching the ground. These "points" touch the ground, which is a type of plane. Since only 3 points are necessary to determine a plane, the 3 rest evenly on the ground. A four-legged stand may not rest evenly because the fourth "point" may not lie on the same plane as the others. This answer is justified by Postulate 1.1.7.

Exercises 60–63 examine non-Euclidean geometry. Students are asked to look at a globe and determine whether postulates that are true in Euclidean geometry are also true in spherical geometry.

Portfolio Activity

The Portfolio Activity can be used as preparation for the Chapter Project or as a separate activity. The Portfolio Activity on this page shows strategies for winning the game of Nim, which was first mentioned on page 79.

61. lines of longitude and the equator

62. Sample answer: Lines are defined this way because when any two points on a sphere are connected by the shortest path, which is extended in each direction, a circle is formed which divides the sphere into two halves.

63. Sample answer: On Earth, the north and south poles are two "points" through which many lines of longitude pass.

64. Sample answer: The route would appear to curve on a traditional map: the pilot would fly north from Washington, D.C., then fly south to get to London.

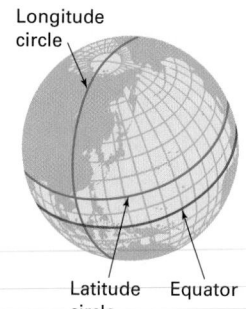

Longitude circle

Latitude circle   Equator

Because the system of logic we use in geometry is based on the postulates of Euclid, it is commonly known as *Euclidean geometry*. It is possible to have geometries in which one or more of Euclid's postulates do not hold. Geometry on a sphere, such as Earth's surface, is known as *spherical geometry*. In spherical geometry, a *line* is defined as a circle that divides the sphere into two equal halves.

61. Using this definition of a line, which of the following are lines?
- lines of latitude
- lines of longitude
- the equator

62. Explain why lines are defined this way in spherical geometry. (Hint: Pick two points on a globe and stretch a string between them to find the shortest distance from one point to the other.)

63. Explain why the following postulate is not true in spherical geometry:

Through any two points, one and only one line passes.

APPLICATION

64. NAVIGATION Pilots navigating long distances often travel along the lines of spherical geometry, called *great circles*. Using a globe and string, determine the shortest route for a pilot traveling from Washington, D. C., to London, England. What do you notice?

🖥 **internet** connect

Portfolio Extension
Go To: **go.hrw.com**
Keyword:
MG1 Nim2

These are winning combinations.

In a game of nim, if you leave two rows of two counters each, then you can always win, in either version of the game. (See pages 79 and 87.)

The following strategies allow you to achieve this winning combination—or an even quicker win, depending on your opponent's moves.

- On your turn, leave two rows with the same number of counters in each row.

or

- On your turn, leave three rows with one, two, and three counters, in any order.

Explain how these combinations can be converted to eventual wins, no matter what your opponent does, in either version of the game.

In either version of nim, the person who moves first can always win. Try to discover the first move that will allow you to always win. Explain the strategy behind this move.

Conjectures That Lead to Theorems

Objectives

- Develop theorems from conjectures.
- Write two-column and paragraph proofs.

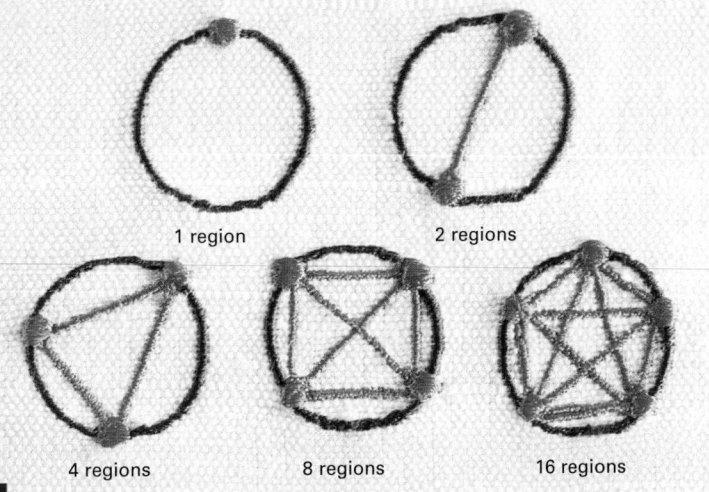

1 region   2 regions

4 regions   8 regions   16 regions

Do you see a pattern in the number of regions in the circles? Do you think the pattern will continue?

Why *When you make predictions based on patterns, you are using inductive reasoning.*

A Need for Proof

From the pattern of circles above, you might make this conjecture:

True or false?

The number of regions doubles each time a point is added.

This conjecture can be tested by drawing pictures. What do you notice about the figure with 6 points?

6 points
? regions

As you can see, the number of regions for 6 points is not 32 but 31. Thus, the conjecture is false. For a conjecture to be considered true by mathematicians it must first be proven deductively, as in Activity 1.

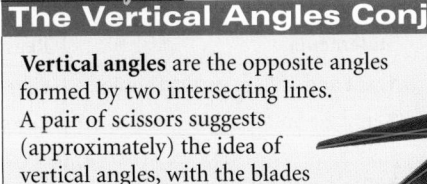
Activity 1
The Vertical Angles Conjecture

YOU WILL NEED

ruler and protractor
OR
geometry software

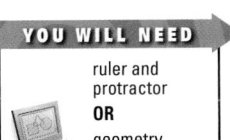

Vertical angles are the opposite angles formed by two intersecting lines. A pair of scissors suggests (approximately) the idea of vertical angles, with the blades making one of the angles and the handles forming the other.

Alternative Teaching Strategy

USING MODELS Have students examine the sums of odd integers, make conjectures about the sums, and prove their conjectures. Sample conjecture: The sum of two odd integers is even. Proof: If n is any integer, then $2n + 1$ is an odd integer; $2n + 1 + 2n + 1 = 4n + 2 = 2(2n + 1)$; the integer $2(2n + 1)$ has a factor of 2 and therefore is even.

Prepare

QUICK WARM-UP

Classify each statement as true or false.

1. A conjecture is a statement that you think is true but that may or may not be true.   true

2. The sum of the measures of the angles that form a linear pair is 90 degrees.   false

3. A reflection of a figure across a line changes the size of the figure but not its shape.   false

Also on Quiz Transparency 2.5

Teach

Why Since inductive reasoning is based on an examination of specific examples, it can never be used to prove a generalization. Methods that use deductive reasoning are needed to prove that a generalization is always true.

Activity 1 Notes

In Activity 1, students measure several pairs of vertical angles, make a conjecture, and develop a proof for their conjecture. The purpose of this Activity is to strengthen students' understanding of the relationship between conjectures and proofs.

For a student worksheet of this Activity and detailed Teacher Notes, see page 29 in the Lesson Activities booklet.

Cooperative Learning

The two Activities in this lesson are appropriate for small-group work. Have students work in groups of two or three, using either geometry graphics software or a ruler and protractor with folding paper. Have students discuss and share their ideas as they complete their work.

CHECKPOINT ✔

7. Subtraction Property of Equality

1. Draw several pairs of intersecting lines. In the figure at right, ∠1 and ∠2 form one pair of vertical angles, and ∠3 and ∠4 form another.

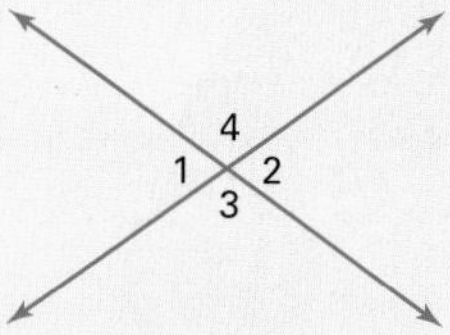

2. Measure each pair of vertical angles. What do you notice?
3. Make a conjecture about vertical angles.
4. What is the relationship between ∠3 and ∠1? between ∠3 and ∠2?

PROOF

5. Complete the following:

$m\angle 1 + m\angle 3 = \underline{\quad ? \quad}$ $\qquad m\angle 2 + m\angle 3 = \underline{\quad ? \quad}$

6. What Property of Equality leads to the following conclusion?

$$m\angle 1 + m\angle 3 = m\angle 2 + m\angle 3$$

CHECKPOINT ✔

7. What Property of Equality leads to the conclusion that $m\angle 1 = m\angle 2$?

Inductive and Deductive Reasoning

Inductive reasoning is the process of forming conjectures that are based on observations. As you have seen, a conjecture can turn out to be false; thus, inductive reasoning is not accepted in mathematical proofs.

Activity 1 began with inductive reasoning and concluded with deductive reasoning. In Steps 1–3, you used inductive reasoning to make a conjecture based on observations. In Steps 4–7, you use deductive reasoning to complete an informal proof. The conjecture you made in Activity 1 is stated and proved formally below.

Vertical Angles Theorem

If two angles form a pair of vertical angles, then they are congruent.

2.5.1

TWO-COLUMN PROOF

Given: ∠1 and ∠2 are vertical angles.
Prove: $\angle 1 \cong \angle 2$

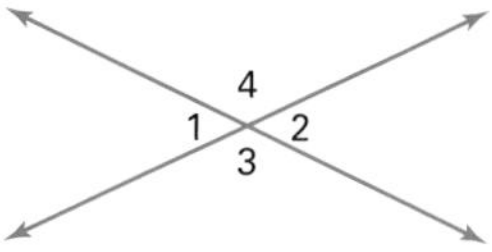

Proof:

Statements	Reasons
1. ∠1 and ∠2 are vertical angles.	Given
2. $m\angle 1 + m\angle 3 = 180°$ $m\angle 2 + m\angle 3 = 180°$	Linear Pair Property
3. $m\angle 1 + m\angle 3 = m\angle 2 + m\angle 3$	Substitution Property of Equality
4. $m\angle 1 = m\angle 2 \quad (\angle 1 \cong \angle 2)$	Subtraction Property of Equality

Interdisciplinary Connection

LAW Lawyers use deductive reasoning when trying to prove that a person is innocent or guilty in a trial. They begin by assembling the known facts that they will use to reason logically to a conclusion.

In Activity 2, you will explore another example of an inductive result that is proven as a theorem.

Activity 2
Reflections Across Parallel Lines

YOU WILL NEED

paper, ruler, and protractor

OR

geometry software

1. Draw two parallel lines, ℓ_1 and ℓ_2, and $\triangle ABC$ as shown in the figure at right.

2. Reflect $\triangle ABC$ across ℓ_1. Label its image $\triangle A'B'C'$.

3. Reflect $\triangle A'B'C'$ across ℓ_2. Label its image $\triangle A''B''C''$.

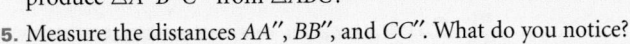

4. Study the relationship between $\triangle ABC$ and $\triangle A''B''C''$. What single transformation do you think would produce $\triangle A''B''C''$ from $\triangle ABC$?

5. Measure the distances AA'', BB'', and CC''. What do you notice?

6. Measure the distance between ℓ_1 and ℓ_2. How does this distance relate to the distances you measured in Step 5?

7. Do all of the points on the triangle seem to have moved in the same direction? Explain your answer.

CHECKPOINT ✔ 8. Complete the following theorem, which is proven below:

Theorem

Reflection across two parallel lines is equivalent to a ___?___ of ___?___ the distance between the lines and in a direction ___?___ to the lines.

2.5.2

CRITICAL THINKING What must you prove about the reflection of a point across two parallel lines in order to show that the theorem you completed above is true?

PROOF The diagram at right suggests a proof of the result in Step 6 of Activity 2.

1. Which of the indicated distances are equal? Why?

2. What is the distance between ℓ_1 and ℓ_2?

3. What does the expression $2D_1 + 2D_2$ represent in the diagram? How does it compare with the distance between ℓ_1 and ℓ_2?

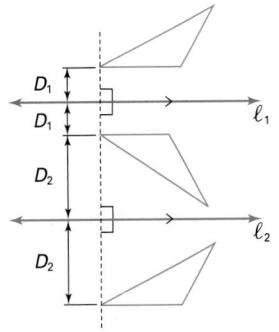

CRITICAL THINKING Do you think the result from Step 6 of Activity 2 is true for any point on the figure? How could you prove that each point moves in the same direction?

Activity 2 Notes

In this Activity, students reflect a triangle across two parallel reflection lines and discover that the result is a translation.

For a student worksheet of this Activity and detailed Teacher Notes, see page 29 in the Lesson Activities booklet.

CHECKPOINT ✔
8. translation; twice; perpendicular

CRITICAL THINKING
Yes. Each point moves in a direction perpendicular to reflection line ℓ_1 (and also ℓ_2) so that the lines of direction are parallel (the directions are the same),

CRITICAL THINKING
Yes. Each point moves in a direction perpendicular to reflection line ℓ_1 (and also ℓ_2), so the lines of direction are parallel (the directions are the same).

Inclusion Strategies

USING DISCUSSION You can enhance students' understanding of mathematical proofs by discussing some examples of everyday uses of logic. Use a chair as an example, and ask students how they know that a chair will not collapse when they sit on it. Inductively, students rely on their previous experience with chairs. Deductively, students can consider the construction and design of the chair.

Enrichment

Some students may wish to write the informal proof that follows Activity 2 as a formal proof with statements and reasons. They can begin by writing down the conjecture and then listing the given information.

ASSIGNMENT GUIDE	
In Class	1–9
Core	11–23 odd, 25–34, 35, 37
Core Plus	25–34, 35, 37 39–43
Review	44–50
Preview	51–55

✐ Extra Practice can be found beginning on page 818.

Error Analysis

Incorrect answers for Exercises 17–20 are probably the result of algebraic errors that occur in the process of solving the equations. Encourage students to review the necessary algebraic concepts.

The Importance of Theorems

Conjectures, like the prediction about the regions of a circle at the beginning of this lesson, may turn out to be false. This is because they are based on a finite number of observations or measurements. You should always ask yourself whether there is a case that might prove your conjecture to be false.

Theorems are different from conjectures. In the Vertical Angles Theorem, the measures of the angles in the diagram do not matter. They could be any size at all, and the proof would still work. This is why the theorem is true for all possible pairs of vertical angles.

Exercises

internet connect

Activities Online

Go To: **go.hrw.com**
Keyword:
MG1 Proof

● **Communicate**

1. Consider the following statement: Every integer is less than 1,000,000,000. Does the statement seem true for integers you use in everyday life? Is it really true?

2. Explain the difference between inductive and deductive reasoning.

3. Why is deductive reasoning the only type of reasoning allowed in proofs?

4. Explain the difference between theorems and conjectures.

5. Describe how you can translate a given figure by using only reflections.

APPLICATION

6. **ENTERTAINMENT** If you have enjoyed all of the films by a certain director that you have seen so far, does this mean that you will enjoy the next film of his that you see? Explain your answer.

Films by one director

Reteaching the Lesson

GUIDED RESEARCH   Students can explore the intersection of two perpendicular lines by measuring all four angles formed at the point of intersection. You can ask them to make a conjecture and prove it. For example: If two lines are perpendicular, then all four angles at the point of intersection are right angles.

7. Use the results of Activity 1 to find the measures of all of the angles in the figure at right. *(ACTIVITY 1)*

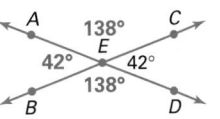

Copy the figure below onto folding paper and use it in Exercises 8 and 9.

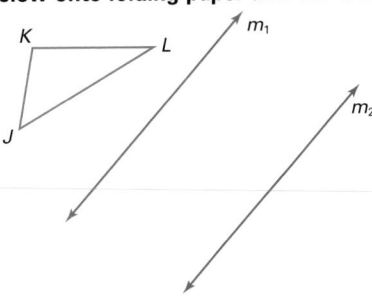

8. Reflect △*JKL* across m_1, and then reflect the image across m_2. *(ACTIVITY 2)*

9. Reflect △*JKL* across m_2, and then reflect the image across m_1. How is the final image different from the final image in Exercise 8? *(ACTIVITY 2)*

Practice and Apply

Refer to the diagram below, which consists of three intersecting lines. For Exercises 10–12, tell which angle is congruent to the given angle.

10. ∠*LEI* ∠*DES*

11. ∠*SEF* ∠*LEP*

12. ∠*PED* ∠*FEI*

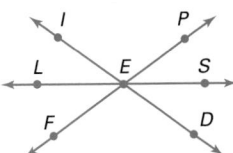

For each pair of intersecting lines, find m∠*ABC*.

13.

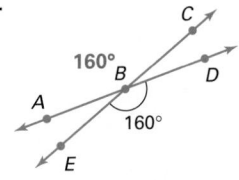

14.

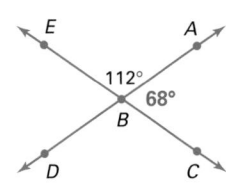

15.

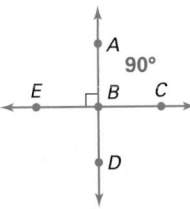

16.

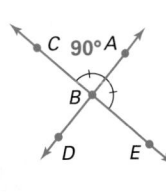

Technology

In Exercises 30–32, students can use geometry graphics software to create the appropriate images.

Math
CONNECTION

ALGEBRA In Exercises 17–20 students use the Vertical Angle Theorem and the definition of a linear pair to set up and solve equations. In Exercises 40–42, students use a pattern of differences to extend a sequence.

8.

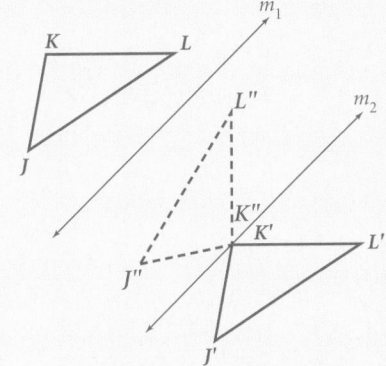

9.

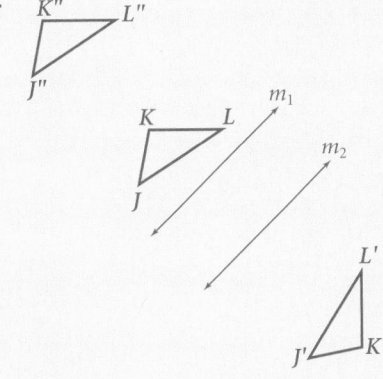

The first set of reflections translates the figure right and downward.

The second set of reflections translates the figure left and upward.

21. inductive reasoning; not a proof; it has not been shown for all cases

22. deductive reasoning; a proof; the hypothesis was given

23. deductive reasoning; a proof; angles in the diagram form a linear pair

24. inductive reasoning; not a proof; it has not been shown for all cases

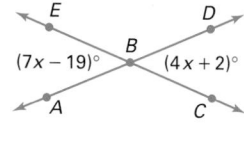
Homework Help Online
Go To: **go.hrw.com**
Keyword:
MG1 Homework Help
for Exercises 17–20

For Exercises 17–20, find the value of x and m∠ABC.

17. $x = 15$, m∠ABC = 60°

18. $x = 7$, m∠ABC = 150°

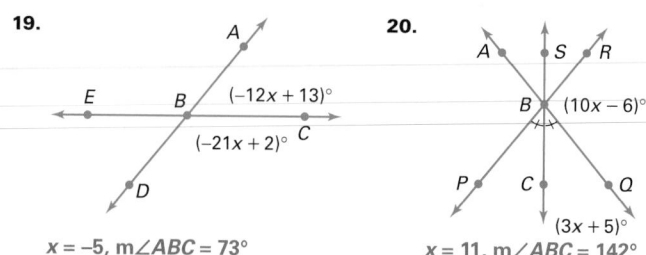

19.

20.

$x = -5$, m∠ABC = 73°

$x = 11$, m∠ABC = 142°

Tell whether each argument is an example of inductive or deductive reasoning. Is the argument a proof? Why or why not?

21. Every time John eats strawberries, he breaks out in hives. Therefore, John is allergic to strawberries.

22. If Erika did not turn in her homework, then she made a bad grade. Erika did not turn in her homework. Therefore, Erika made a bad grade.

23. Angles that form a linear pair are supplementary. Therefore, ∠1 and ∠2 are supplementary.

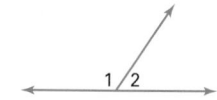

24. The number of black-footed ferrets has increased each year since 1985. Therefore, there will be more black-footed ferrets next year than there are this year.

The black-footed ferret is the rarest mammal in North America. In 1985, fewer than 20 were known to exist.

In Exercise 25–27, complete the two-column proof of the following theorem:

Congruent Supplements Theorem

If two angles are supplements of congruent angles, then the two angles are congruent.

2.5.3

Given: $\angle 1 \cong \angle 3$, $\angle 1$ and $\angle 2$ are supplementary, and $\angle 3$ and $\angle 4$ are supplementary.

Prove: $\angle 2 \cong \angle 4$

Proof:

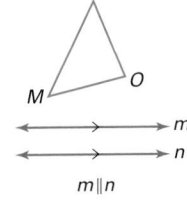

Statements	Reasons
$m\angle 1 + m\angle 2 = 180°$ $m\angle 3 + m\angle 4 = 180°$	Definition of supplementary angles
$m\angle 1 + m\angle 2 = m\angle 3 + m\angle 4$	**25.** ___?___ Transitive (or Substitution) Property
$\angle 1 \cong \angle 3$ $(m\angle 1 = m\angle 3)$	**26.** ___?___ Given (Angle Congruence Postulate)
$m\angle 1 + m\angle 2 = m\angle 1 + m\angle 4$	Substitution Property
$m\angle 2 = m\angle 4$ $(\angle 2 \cong \angle 4)$	**27.** ___?___ Subtraction Property (Angle Congruence Postulate)

28. Write a paragraph proof of the Congruent Supplements Theorem.

In Exercises 29–34, refer to Activity 2 and the proof outlined on page 119. Refer to the figure below. Suppose that $\triangle MNO$ is reflected across line m, and then its image is reflected across line n. What is the distance from $\triangle MNO$ to $\triangle M''N''O''$ if the distance from m to n is

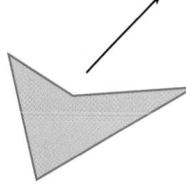

29. 5 cm? **10 cm**

30. 10 cm? **20 cm**

31. x cm? **2x cm**

Refer to the figure below. Suppose that you wish to use two reflections to translate the figure in the direction of the arrow.

32. If you wish to translate the figure 10 cm, how far apart should the parallel lines used in the reflections be?

33. How can you determine the location of the parallel lines?

34. Is there more than one pair of parallel lines that will give a translation of 10 cm? Use diagrams to illustrate your answer.

28. You are given that $\angle 1$ and $\angle 2$ and $\angle 3$ and $\angle 4$ are pairs of supplementary angles. Then $m\angle 1 + m\angle 2 = 180°$ and $m\angle 3 + m\angle 4 = 180°$. So $m\angle 1 + m\angle 2 = m\angle 3 + m\angle 4$ by substitution. Since $\angle 1 \cong \angle 3$ is given, $m\angle 1 = m\angle 3$. Using substitution again, $m\angle 1 + m\angle 2 = m\angle 1 + m\angle 4$. Now, subtracting $m\angle 1$ from both sides, $m\angle 2 = m\angle 4$. Therefore $\angle 2 \cong \angle 4$.

32. The parallel lines should be $\frac{1}{2}(10 \text{ cm}) = 5 \text{ cm}$ apart.

33. The parallel lines should be drawn perpendicular to the direction of the arrow.

34. Yes. Any pair of lines which are 5 centimeters apart and perpendicular to the direction of the arrow will produce the desired translation. Check students' diagrams.

35.

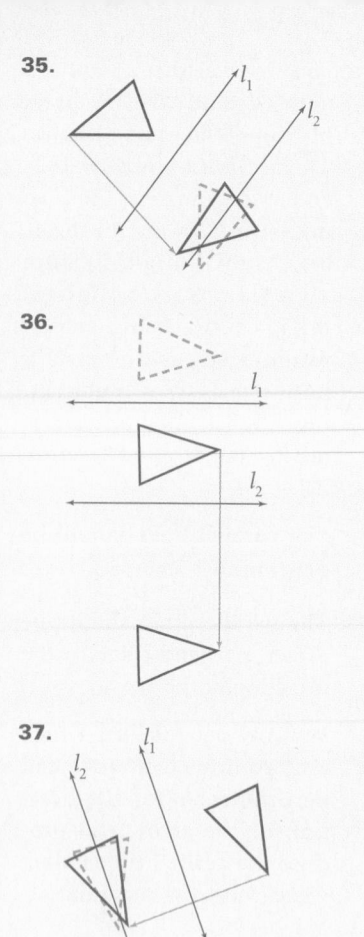

36.

37.

For Exercises 35–38, copy each figure onto folding paper. Reflect each figure across ℓ_1, and then reflect the image across ℓ_2. Draw an arrow that represents the direction and distance that the figure has been translated.

35. 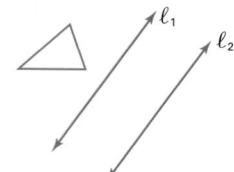   **36.**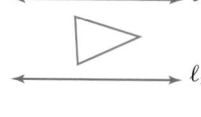

37.    **38.**

CHALLENGE

Exercises 39–43 refer to the pattern in the number of regions in the circles on page 117.

Draw circles with 6 and with 7 points, and connect the points in each circle with segments. To be sure that you have the maximum number of regions, make sure there are no points where three segments intersect inside the circle. (Hint: Do not space the points regularly on the circle.) How many regions are possible for a circle with

You may want to use geometry graphics software for Exercise 40.

39. 6 points? **31**

40. 7 points? **57**

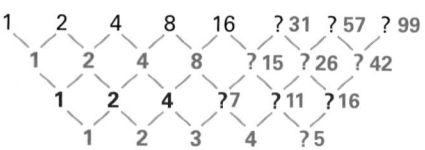

One way to find a pattern in a sequence is by subtracting each term from the next. This gives a new sequence, called the *first differences*. This method can be continued to find second differences, third differences, etc., until the pattern is clear.

Algebra

41. Complete the chart below. (Work from the bottom up to fill in the chart.)

Number of regions	1	2	4	8	16	? 31	? 57	? 99
First differences		1	2	4	8	? 15	? 26	? 42
Second differences			1	2	4	? 7	? 11	? 16
Third differences				1	2	3	4	? 5

42. Using the pattern, what conjecture would you make about the number of regions in a circle with 8 points? with 9 points? with 10 points?

43. If you found that the pattern in the chart above was correct for 100, 1000, or even more points on a circle, would you have proven that the pattern is correct for any number of points? Why or why not?

38.

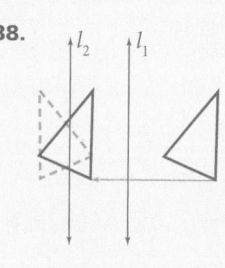

42. 99 regions; 163 regions; 256 regions

43. No; an induction proof cannot be a proof for all cases.

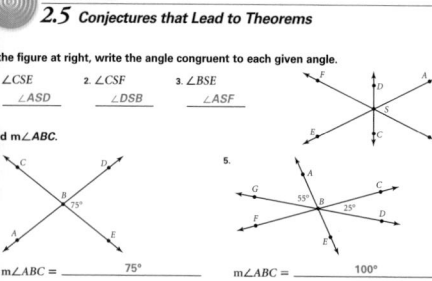

Look Back

Refer to the figure below for Exercises 44–46.
(LESSON 1.1)

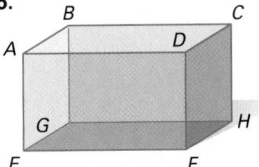

44. Name four points in the figure.

45. Name four segments in the figure.

46. Name four planes in the figure.

Exercises 47 and 48 refer to the statements below. *(LESSON 2.2)*

If I am well rested in the afternoon, then I am in a good mood.
If I sleep until 8:00 A.M., then I am well rested in the afternoon.
If it is Saturday, then I sleep until 8:00 A.M.

47. Arrange the statements to form a logical chain.

48. Write the conditional statement that follows from the argument formed by the logical chain in Exercise 47.

49. Draw an Euler diagram to illustrate the argument below. *(LESSON 2.2)*

Every member of the Culver High School football team goes to Culver High School. Brady is a member of the Culver High School football team. Therefore, Brady goes to Culver High School.

50. Below is a chain of conditionals. Construct an Euler diagram for these statements, and complete the concluding statement. *(LESSON 2.2)*

If you are taking geometry, then you are developing good reasoning skills.

If you are developing good reasoning skills, then you will be able to succeed in many different careers.

If you are able to succeed in many different careers, then you will be able to choose your profession.

Therefore, if you are taking geometry, then ____?____.

(See Exercise 49.)

Look Beyond

51. Copy the figure at right. Reflect △ABC across ℓ_1, and then reflect its image across ℓ_2. Label the final image △A″B″C″.

52. What single transformation would produce the same final image as the two reflections?

53. What seems to be special about the intersection point, D?

54. Use a protractor to measure ∠ADA″, ∠BDB″, and ∠CDC″. What seems to be the relationship between the measures of these angles and the measure of ∠1?

55. Write a conjecture about the reflection of a figure across two intersecting lines. Include your results from Exercise 54 in your conjecture.

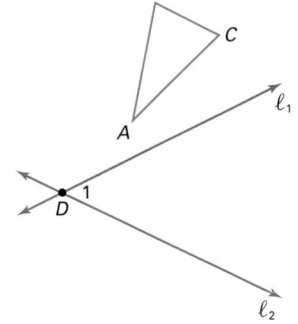

49.

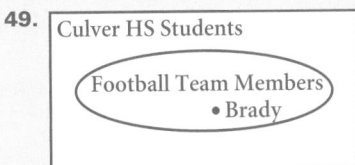

Culver HS Students

Football Team Members
• Brady

50.

People who will be able to choose their profession

People who will be able to succeed in many careers

People who are developing good reasoning skills

People who are taking geometry

you will be able to choose your own profession

Look Beyond

In Exercises 51–55, students build on Activity 2 by reflecting a triangle across two intersecting lines. Then they are asked to make a conjecture.

44. Sample answer: points A, B, C, D

45. Sample answer: $\overline{AB}$, $\overline{BC}$, $\overline{CD}$, $\overline{AD}$

46. Sample answer: planes ABG, DCH, ADE, EGH

47. If it is Saturday, then I sleep until 8:00 A.M.
If I sleep until 8:00 A.M., then I am well rested during the afternoon.
If I am well rested in the afternoon, then I am in a good mood.

48. If it is Saturday, then I am in a good mood.

Student Technology Guide

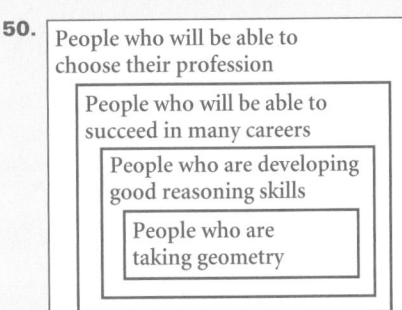

NAME _____ CLASS _____ DATE _____

Student Technology Guide
2.5 *Conjectures That Lead to Theorems, page 1*

Making a conjecture in geometry is similar to taking a sample and drawing a conclusion about a population in statistics. When you look for a conjecture, you select a representative sample from a population and attempt to see if what is true of the sample is true of the entire population.

Consider the population to be the set of all addition problems involving two even numbers. Consider the sample to be the addition problems below. In each of the five addition problems, all of the numbers being added are even.

2 + 2   4 + 12   28 + 36   72 + 100   12,428 + 54,844   1,345,002 + 18,334

Can a conclusion about sums of even numbers be drawn? To find out, begin by finding each of the sums above. Then use the following test: A number *n* is even if $\frac{n}{2}$ is an integer. You can recognize an integer quotient by observing whether its decimal part is 0.

You can explore possible conclusions by using a graphics calculator.

- Using parentheses, enter each sum in the sample and divide by 2. Three of the quotients are shown on the display at right.

Each of the quotients is an integer. You have reason to believe that the sum of two even numbers is another even number.
- Test your conclusion by randomly creating some more addition problems involving only even numbers.
Again, each of the quotients is an integer. Thus, you have more reason to believe that the sum of two even numbers is another even number.

You can approach the same question about even-numbered sums by using a spreadsheet. The formula 2*INT(1000*RAND()) in columns A and B will give a random three-digit even number, and the formula IF(A1+B1)/2=INT((A1+B1)/2),True,False) in column C will test the sum to see if it is even.

Use a spreadsheet to prove your conclusion about each population below.

1. products of even numbers always even; (2k)(2n) = 2(2kn), which is an even number.
2. sums of odd numbers always even; (2k + 1) + (2n + 1) = 2(k + n + 1), which is an even number.
3. squares of even numbers always even because it is the product of two even numbers
4. squares of odd numbers always odd; (2n + 1)² = 4n² + 4n + 1 = 2(2n² + 2n) + 1

Focus

The objective of this project is to have students practice their logical thinking skills.

Motivate

Students have learned a great deal in this chapter about logical thinking and the uses of reasoning processes in geometry. They now know that it is also important to reason clearly and correctly in everyday situations. Logical reasoning skills, however, cannot be developed without studying and practicing them. Logic puzzles of all kinds are an interesting and entertaining way to develop logic skills.

PROJECT TWO
CHAPTER

Logic Puzzles and Games

Activity 1

The game of sprouts was invented by mathematicians John Conway and Michael Paterson. The object is to be the last player to make a legal move.

Rules of the Game

Start by drawing three spots on a piece of paper.
The game is played by two people, who take turns.

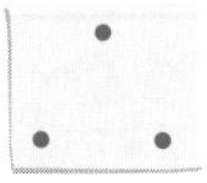

To make a move, draw a curve joining two spots or starting and ending at the same spot, and place a new spot on the curve.
• No curve can cross another curve.
• No spot can have more than three curves coming from it.

Play a few games of sprouts with a friend. Try to figure out a strategy that will enable you to always win if you have the first turn.

You can vary the game by starting with four, five, or any other number of spots.

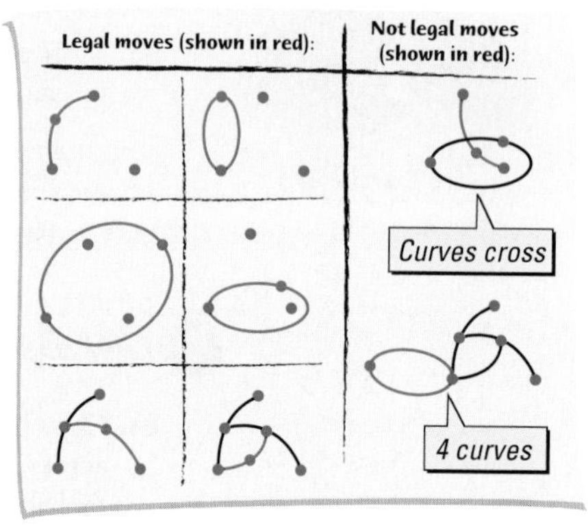

Legal moves (shown in red):	Not legal moves (shown in red):

Curves cross

4 curves

Activity 1

Choose any spot, say *A*. Draw a curve starting at *A*, going around *B*, and back to *A*. This will cause the game to last an odd number of moves, so that player 1 will always win.

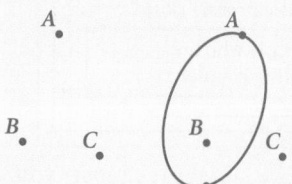

Activity 2

E = 4; O = 2; T = 7; M = 5; V = 3; L = 1; A = 6;
H = 8; I = 0

I L O V E M A T H
0 1 2 3 4 5 6 7 8

I LOVE MATH

Activity 3

Accomplishments: Valerie plays the harp and speaks Italian.

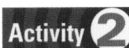

Activity 2

In the following puzzle, each letter represents a different digit. Use the clues given to determine the digits, and then arrange the letters in numerical order to solve the puzzle.

$O + O = E$ HAT $T \times V = OL$

$O \times O = E$ + MAT

            LEVE $M \times M = OM$ $H \times I = I$

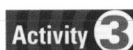

Activity 3

The following puzzle is by the British writer known as "Caliban."

Accomplishments

"My four granddaughters are all accomplished girls." Canon Chasuble was speaking with evident self-satisfaction. "Each of them," he went on, "plays a different musical instrument and each speaks one European language as well as—if not better than—a native."

"What does Mary play?" asked someone.

"The cello."

"Who plays the violin?"

"D'you know," said Chasuble, "I've temporarily forgotten. Anno Domini, alas! But I know it's the girl who speaks French."

The remainder of the facts which I elicited were of a somewhat negative character. I learned that the organist is not Valerie; that the girl who speaks German is not Lorna; and that Mary knows no Italian. Anthea doesn't play the violin; nor is she the girl who speaks Spanish. Valerie knows no French, Lorna doesn't play the harp, and the organist can't speak Italian.

What are Valerie's accomplishments?

	Cello	Violin	Organ	Harp	French	German	Italian	Spanish
Mary								
Valerie								
Lorna								
Anthea								
French								
German								
Italian								
Spanish								

Cooperative Learning

Students can work with a partner to do each logic puzzle. Have them first read each puzzle by themselves before trying to solve it together. In the game in Activity 1, have pairs of students play several rounds of sprouts with 3, 4, and 5 spots. In Activity 2, have students communicate their ideas to their partners as they proceed. In Activity 3, have students work the puzzle alone, consulting the partner only after they have tried by themselves. When helping a partner solve a puzzle, encourage students to explain the logic behind a particular answer, instead of just giving hints.

Discuss

After students in each group complete all three activities, have them compare their answers and the conclusions they arrived at through logical thinking.

Chapter Review and Assessment

VOCABULARY

SUMMARY OF POSTULATES AND THEOREMS

Lesson	Number	Theorem or Postulate
2.2	2.2.1 If-Then Transitive Property	Given: "If A then B, and if B then C." You can conclude: "If A then C."
2.4	2.4.1 Addition Property	If $a = b$, then $a + c = b + c$.
	2.4.2 Subtraction Property	If $a = b$, then $a - c = b - c$.
	2.4.3 Multiplication Property	If $a = b$, then $ac = bc$.
	2.4.4 Division Property	If $a = b$ and $c \neq 0$, then $\frac{a}{c} = \frac{b}{c}$.
	2.4.5 Substitution Property	If $a = b$, you may replace a with b in any true equation containing a and the resulting equation will still be true.
	2.4.6 Overlapping Segments Theorem	Given a segment with points A, B, C, and D (in order) the following statements are true: 1. If $AB = CD$, then $AC = BD$. 2. If $AC = BD$, then $AB = CD$.
	2.4.7 Reflexive Property of Equality	For any real number a, $a = a$.
	2.4.8 Symmetric Property of Equality	For all real numbers a and b, if $a = b$, then $b = a$.
	2.4.9 Transitive Property of Equality	For all real numbers a, b, and c, if $a = b$ and $b = c$, then $a = c$.
	2.4.10 Reflexive Property of Congruence	figure $A \cong$ figure A
	2.4.11 Symmetric Property of Congruence	If figure $A \cong$ figure B, then figure $B \cong$ figure A.
	2.4.12 Transitive Property of Congruence	If figure $A \cong$ figure B and figure $B \cong$ figure C, then figure $A \cong$ figure C.
	2.4.13 Overlapping Angles Theorem	Given $\angle AOD$ with points B and C in its interior, the following statements are true: 1. If m$\angle AOB$ = mCOD, then m$\angle AOC$ = mBOD. 2. If m$\angle AOC$ = mBOD, then m$\angle AOB$ = mCOD.

Chapter Test, Form A

NAME _____ CLASS _____ DATE _____

Chapter Assessment

Chapter 2, Form A, page 1

Write the letter that best answers the question or completes the statement.

b 1. What conjecture can you make about the sequence's relation to zero?

$$\frac{1}{3}, \frac{1}{6}, \frac{1}{12}, \frac{1}{24}, \ldots$$

   a. The denominator is doubled, so the terms get further from zero.
   b. The denominator is increasing, so the terms get closer to zero.
   c. The denominators are multiples of 3, so the terms get closer to zero.
   d. The numerator is always 1, so the terms do not change their distance from zero.

d 2. What is the converse of the statement below?
   If a triangle is equilateral, then it is isosceles.

   a. A triangle is equilateral if and only if it is isosceles.
   b. All equilateral triangles are isosceles.
   c. A triangle is isosceles if and only if it is equilateral.
   d. If a triangle is isosceles, then it is equilateral.

c 3. What conclusion can you draw from the two statements below?
   If an animal is a dog, it has four legs.
   Murphy is a dog.

   a. Murphy is an animal.   b. Murphy is not an animal.
   c. Murphy has four legs.   d. Murphy does not have four legs.

a 4. Choose which of the following is a counterexample for the *converse* of the statement "If an animal is a dog, then it has four legs."

   a. Brandy the cat has four legs.
   b. Spot the dog has four legs.
   c. Murphy the dog is an animal.
   d. Polly the parrot has two legs.

c 5. Which proposed definition is true?

   a. Two lines are parallel if and only if they intersect to form a right angle.
   b. Two lines are perpendicular if and only if they are coplanar and do not intersect.
   c. Two lines are parallel if and only if they are coplanar and do not intersect.
   d. Two lines are parallel if and only if they intersect and do not form right angles.

d 6. Which statement is the biconditional of "All rings are round"?

   a. If it is a ring, then it is round.
   b. If it is a ring, then it is not round.
   c. If it is round, then it is a ring.
   d. It is a ring if and only if it is round.

NAME _____ CLASS _____ DATE _____

Chapter Assessment

Chapter 2, Form A, page 2

c 7. Which angles form an adjacent pair?
   a. $\angle 1$ and $\angle 3$
   b. $\angle 3$ and $\angle 4$
   c. $\angle 2$ and $\angle 3$
   d. $\angle 2$ and $\angle 4$

a 8. What property justifies the conclusion of the statement below?
   If $\overline{BC} \cong \overline{DE}$, then $\overline{DE} \cong \overline{BC}$.

   a. Symmetric Property of Congruence   b. Addition Property of Equality
   c. Reflexive Property of Congruence   d. Transitive Property of Congruence

c 9. What property justifies the conclusion?

   $x + 7 = 21$ ← Given
   $x + 7 - 7 = 21 - 7$
   $x = 14$ ← conclusion

   a. Symmetric Property of Equality   b. Addition Property of Equality
   c. Subtraction Property of Equality   d. Reflexive Property of Equality

b 10. What is the measure of $\angle ABC$?
   a. 105°
   b. 75°
   c. 15°
   d. 90°

d 11. Which of the following is the name given to a statement that has been proven by deductive reasoning?

   a. conjecture
   b. postulate
   c. property
   d. theorem

c 12. What property justifies the conclusion of the statement below?

   m$\angle ABC$ = m$\angle CBA$

   a. Symmetric Property of Equality   b. Addition Property of Equality
   c. Reflexive Property of Equality   d. Transitive Property of Equality

2.5	2.5.1 Vertical Angles Theorem	If two angles form a pair of vertical angles, then they are congruent.
	2.5.2 Theorem	Reflection across two parallel lines is equivalent to a translation of twice the distance between the lines and in a direction perpendicular to the lines.
	2.5.3 Theorem	Reflection across two intersecting lines is equivalent to a rotation about the point of intersection through twice the measure of the angle between the lines.

Key Skills & Exercises

LESSON 2.1
Key Skills

Give a proof of a conjecture.

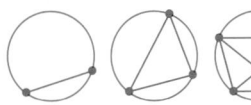

The following table shows the number of points and the corresponding number of segments:

Points	2	3	4	5
Segments	1	3	6	10

Conjecture: For n points on a circle, the number of segments needed to connect all points is $\frac{n(n-1)}{2}$.

Proof: For each point, a segment connects it to every other point, giving $n - 1$ segments. Therefore, the total number of segments should be $n(n-1)$. However, $\overline{AB}$ is same segment as $\overline{BA}$, so each segment is counted twice. Thus, the total must be divided by 2, giving $\frac{n(n-1)}{2}$.

Exercises

The horizontal line intersects the curve at three points and divides it into four pieces.

1. Into how many pieces will the curve be divided by two horizontal lines that each intersect the curve at three points each? by three horizontal lines?

2. Complete the table below, subtracting to find the first differences.

Lines	1	2	3	4
Sections	4	?	?	?
First differences		?	?	?

3. Based on the table, make a conjecture about the pattern in the number of sections.

4. Write a proof of your conjecture.

LESSON 2.2
Key Skills

Draw a conclusion from a conditional.

What conclusion follows from these statements?
  If an animal is a cat, then it has four legs.
  Dinah is a cat.
Conclusion: Dinah has four legs.

State the converse of a conditional.

Converse of the conditional above:
  If an animal has four legs, then it is a cat.

Exercises

Refer to the following statements:
  If a "star" doesn't flicker, then it is a planet.
  The evening star doesn't flicker.

5. Give the conclusion that follows from the statements above.

6. Write the converse of the conditional above. Is the converse true?

---

1. 7; 10

2. 7, 10, 13; 3, 3, 3

3. The curve is divided by n horizontal lines into $1 + 3n$ sections.

4. Proof: The first line divides the figure into 4 pieces, two above and two below. The next line creates regions above, below, and between the lines. There are two pieces above and two below as before, with three pieces between the lines. Each additional line will create a new region between two

lines, with three pieces of the curve in it. Therefore, each additional line creates 3 additional pieces of the curves. Since the curve starts with one section, the number of sections for n lines is $1 + 3n$.

5. The evening star is a planet.

6. Converse: If a "star" is really a planet, then it doesn't flicker. It is a true conditional because a planet does not flicker.

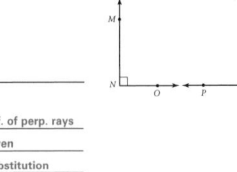
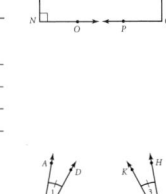

7. *c, b, a*

8. If Darren tells Charles a secret, then Andrew tells Tina the secret.

9. *c* and *d*

10. A torus is a doughnut-shaped object.

11. No; the converse is not true because a figure with four sides is not always a square.

12. Yes; both the conditional and its converse are true.

Arrange statements into a logical chain and draw a conclusion.

   a. If John gets his driver's license, then he will buy a car.
   b. If John buys a car, then he will not have any money.
   c. If John passes his driver's license exam, then he will get his driver's license.

The order of the logical chain is **c, a, b**. The conclusion is "If John passes his driver's license exam, then he will not have any money."

Exercises 7 and 8 refer to the following statements:

   a. If June tells Andrew a secret, then Andrew tells Tina the secret.
   b. If Charles tells June a secret, then June tells Andrew the secret.
   c. If Darren tells Charles a secret, then Charles tells June the secret.

7. Arrange the statements into a logical chain.

8. What conditional statement does the argument formed by this logical chain prove?

LESSON 2.3
Key Skills

Write a definition of an object.

These objects are trapezoids.

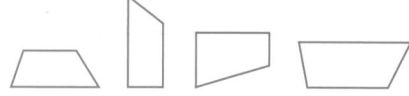

These objects are not trapezoids.

Which of the following objects are trapezoids?

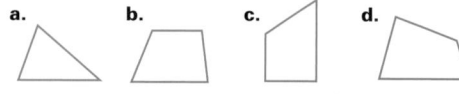

Write a definition of a trapezoid.

Objects **b** and **c** are trapezoids. A trapezoid is a four-sided figure with exactly two parallel sides.

Determine whether a statement is a definition.

Is the following statement a definition?
   A right angle measures 90°.
If a statement is a definition, then the conditional form and its converse are both true.
Conditional: If an angle is a right angle, then it measures 90°.
Converse: If an angle measures 90°, then it is a right angle.
Since both are true, the statement is a definition.

Exercises

These objects are tori (singular, *torus*).

These objects are not tori.

9. Which of the following objects are tori?

   a.   **b.**   **c.**   **d.**

10. Write a definition of *torus*.

Determine whether the following statements are definitions:

11. A square is a figure with four sides.

12. A point that divides a segment into two congruent parts is the midpoint.

LESSON 2.4
Key Skills

Use the Properties of Equality and Congruence to write proofs.

In the figure below, $\angle 1 \cong \angle 3$. Prove that $\angle 2 \cong \angle 4$.

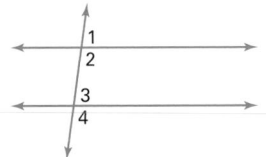

Two-column proof:

Statements	Reasons
$m\angle 1 + m\angle 2 = 180°$	Linear Pair Property
$m\angle 3 + m\angle 4 = 180°$	Linear Pair Property
$m\angle 1 + m\angle 2 = m\angle 3 + m\angle 4$	Substitution Property
$m\angle 1 = m\angle 3$	Given
$m\angle 2 = m\angle 4 \ (\angle 2 \cong \angle 4)$	Subtraction Property

Paragraph proof:

By the Linear Pair Property, $m\angle 1 + m\angle 2 = 180°$ and $m\angle 3 + m\angle 4 = 180°$. By the Substitution Property, $m\angle 1 + m\angle 2 = m\angle 3 + m\angle 4$. You are given $m\angle 1 = m\angle 3$, so by the Subtraction Property, $m\angle 2 = m\angle 4$, or $\angle 2 \cong \angle 4$.

Exercises

In the figure below, $m\angle 1 = 90°$.

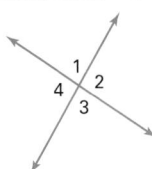

13. What property would you use to prove that $m\angle 2 = 90°$?

14. What property would you use to prove that $m\angle 4 = 90°$?

15. What theorem would you use to prove that $m\angle 3 = 90°$?

16. Use your answers from Exercises 13–15 to write a two-column or paragraph proof of the following conjecture:

> If two lines intersect to form a right angle, then all of the angles formed are right angles.

LESSON 2.5
Key Skills

Use deductive reasoning to prove a conjecture.

Make a conjecture about the angle formed by the angle bisectors of a linear pair, and prove your conjecture.

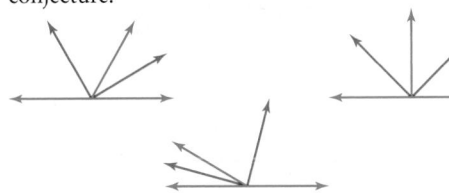

By measuring the angles, you can make the following conjecture: The angle bisectors of a linear pair form a right angle.

Exercises

Three numbers, *a*, *b*, and *c*, are called a *Pythagorean triple* if $a^2 + b^2 = c^2$.

17. Show that the numbers 3, 4, and 5 are a Pythagorean triple and that the numbers 5, 12, and 13 are another Pythagorean triple.

---

16. Two-column proof

Statements	Reasons
$m\angle 1 + m\angle 2 = 180°$	Linear Pair Property
$m\angle 1 = 90°$	Given
$90° + m\angle 2 = 180°$	Substitution Property
$m\angle 2 = 90°$	Subtraction Property
$m\angle 1 + m\angle 4 = 180°$	Linear Pair Property
$90° + m\angle 4 = 180°$	Substitution Property
$m\angle 4 = 90°$	Subtraction Property
$m\angle 1 = m\angle 3$	Vertical Angles Theorem
$m\angle 3 = 90°$	Transitive Property

17. Let $a = 3$, $b = 4$, $c = 5$. Then $a^2 + b^2 = c^2$ gives: $3^2 + 4^2 = 5^2$ which is true so 3, 4, and 5 form a Pythagorean triple.

Let $a = 5$, $b = 12$, $c = 13$. Then $a^2 + b^2 = c^2$ gives: $5^2 + 12^2 = 13^2$ which is true, so 5, 12, and 13 form a Pythagorean triple.

18. Let $a = 3$, $b = 4$, $c = 12$ and $d = 13$. Then $a^2 + b^2 + c^2 = d^2$ gives: $3^2 + 4^2 + 12^2 = 13^2$ which is true, so 3, 4, 12, and 13 form a Pythagorean quadruple.

19. Conjecture: If a, b, and c are a Pythagorean triple such that $a^2 + b^2 = c^2$ and c, d and e are another Pythagorean triple such that $c^2 + d^2 = e^2$, then $a^2 + b^2 + d^2 = e^2$.

20. Paragraph proof: $a^2 + b^2 = c^2$ and $c^2 + d^2 = e^2$ because a, b, c and c, d, e each form Pythagorean triples. Using the Substitution Property, substitute $a^2 + b^2$ for c^2 in the equation $c^2 + d^2 = e^2$: $a^2 + b^2 + d^2 = e^2$.

21. 3^2 4 1 4 = 9 16 = 7. The program would display "NO REAL SOLUTIONS".

22. Yes, it is a mammal because it has hair and produces milk. Since the platypus is a mammal and it lays eggs, it is a monotreme.

Given: the figure below, in which $\vec{OB}$ bisects $\angle AOC$ and $\vec{OD}$ bisects $\angle BOE$

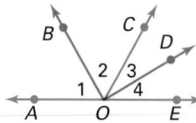

Prove: $m\angle 2 + m\angle 3 = 90°$

Proof:

Statements	Reasons
$m\angle 1 = m\angle 2$	Definition of angle bisector
$m\angle 3 = m\angle 4$	Definition of angle bisector
$m\angle 1 + m\angle 2 + m\angle 3 + m\angle 4 = 180°$	Linear Pair Property
$m\angle 2 + m\angle 2 + m\angle 3 + m\angle 3 = 180°$	Substitution Property
$2(m\angle 2 + m\angle 3) = 180°$	Distributive Property
$m\angle 2 + m\angle 3 = 90°$	Division Property

Four numbers, *a*, *b*, *c*, and *d*, are called a *Pythagorean quadruple* if $a^2 + b^2 + c^2 = d^2$.

18. Show that the numbers 3, 4, 12, and 13 are a Pythagorean quadruple.

19. If a, b, and c are a Pythagorean triple and c, d, and e are another Pythagorean triple, make a conjecture about a, b, d, and e.

20. Use Properties of Equality to prove the conjecture you made in Exercise 19.

Applications

21. **COMPUTER PROGRAMMING** The following is part of a computer program used to evaluate the quadratic formula:

```
INPUT A
INPUT B
INPUT C
IF B² – 4*A*C < 0
THEN
DISP "NO REAL SOLUTIONS"
END
```

Suppose that $A = 1$, $B = 3$, and $C = 4$. Can you determine what the program would do?

22. **BIOLOGY** In biological classification, animals are grouped into categories according to physical characteristics. Consider the following definitions:

Mammals are animals that have hair and produce milk.
Monotremes are mammals that lay eggs.

Suppose that a biologist is examining a platypus and finds that it has hair, produces milk, and lays eggs. Is a platypus a monotreme? Why or why not?

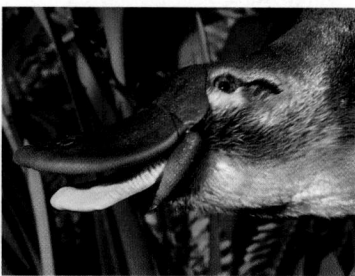

The platypus, about half the size of a house cat, is a mammal that lays eggs. The young are nourished by milk from the mother's mammary glands.

Chapter Test

1. What are the next four terms in the sequence?
$$\frac{1}{1}, \frac{1}{4}, \frac{1}{9}, \frac{1}{16}, \cdots \frac{1}{25}, \frac{1}{36}, \frac{1}{49}, \frac{1}{64}$$

2. What conjecture can you make about the sequence of terms in Exercise 1?

For Exercises 3–4, refer to the following statement:

> If a quadrilateral is a square, then it is a rectangle.

3. Identify the hypothesis and the conclusion of the conditional.

4. Write the converse of the conditional. If the converse is false, give a counterexample to show how it is false.

For Exercises 5–6, refer to the following statements:

> a. If Brett flies his kite, then it is cool.
> b. If it is cool, then it is autumn.
> c. If it is windy, then Brett flies his kite.

5. Arrange the statements to form a logical chain. **c, a, b**

6. What conditional statement results from this logical chain?

For Exercise 7, use the following steps to determine whether the sentence is a definition.

> a. Write the sentence as a conditional statement.
> b. Write the converse of the conditional.
> c. Write a biconditional statement.
> d. Decide whether the sentence is a definition, and explain your reasoning.

7. An obtuse angle has a measure of 100°.

8. Use the figure in which $m\angle ADB = (8x - 31)°$, $m\angle BDC = (3x + 9)°$, and $m\angle ADC = 110°$ to find $m\angle BDC$. **45°**

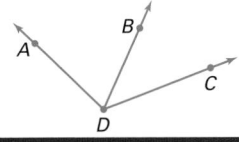

Complete each proof.

Given: $m\angle 1 + m\angle 2 = 90°$
$m\angle 3 + m\angle 4 = 90°$
$m\angle 2 = m\angle 3$

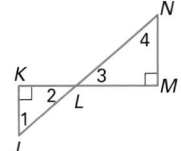

Prove: $m\angle 1 = m\angle 4$

Statements	Reasons	
$m\angle 1 + m\angle 2 = 90°$	9. ?	Given
$m\angle 3 + m\angle 4 = 90°$		
$m\angle 1 + m\angle 2 = m\angle 3 + m\angle 4$	10. ?	Transitive Property
$m\angle 2 = m\angle 3$	11. ?	Given
$m\angle 1 + m\angle 2 = m\angle 2 + m\angle 4$	12. ?	Substitution Property
$m\angle 1 = m\angle 4$	13. ?	Subtraction Property

Given: $\angle 1$ and $\angle 2$ are supplementary
$m\angle 1 = 145°$

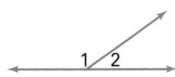

Prove: $m\angle 2 = 35°$

Statements	Reasons	
$\angle 1$ and $\angle 2$ are supplementary	14. ?	Given
$m\angle 1 = 145°$		
$m\angle 1 + m\angle 2 = 180°$	15. ?	Defn. of Supp. Angles
$145° + m\angle 2 = 180°$	16. ?	Substitution Property
$m\angle 2 = 35°$	17. ?	Subtraction Property

For Exercises 18–19, find x and $m\angle ABC$.

18.

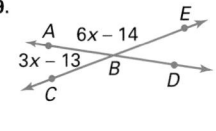

19.

2. The denominators are successive perfect squares.

3. Hypothesis: A quadrilateral is a square.
Conclusion: It is a rectangle.

4. If a quadrilateral is a rectangle, then it is a square. The converse is false. Sample counterexample: A rectangle with length 5 cm and width 4 cm is not a square.

6. If it is windy, then it is autumn.

7. a. If an angle is an obtuse angle, then it has a measure of 100°.

   b. If an angle has a measure of 100°, then it is an obtuse angle.

   c. An angle is an obtuse angle if and only if it has a measure of 100°.

   d. The statement is not a definition because the biconditional is false; not all obtuse angles have a measure of 100°.

18. $x = 14$; $m\angle ABC = 74°$

19. $x = 23$; $m\angle ABC = 56°$

Multiple-Choice Samples

The first half of the Cumulative Assessment contains a multiple-choice section. This part of the Cumulative Assessment consists of items commonly found on standardized tests.

Free-Response Grid Samples

The second half of the Cumulative Assessment consists of a free-response section. This part requires student-produced response items like those commonly found on college entrance exams. These questions require the use of machine-scored answer grids. You may wish to have students practice answering these items in preparation for standardized tests.

internet connect

**Standardized
Test Prep Online**

Go To: **go.hrw.com**
Keyword: **MM1 Test Prep**

MULTIPLE-CHOICE For Questions 1–4, write the letter that indicates the best answer.

1. Refer to the figure below. Which of the following statements is true? *(LESSON 1.3)*

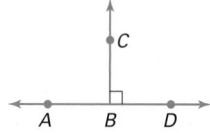

a. $\angle ABC$ is an acute angle.
b. $\angle ABC$ is an obtuse angle.
c. $\angle ABC$ is a right angle.
d. $\angle ABC$ is a complementary angle.

2. What are the first four terms in the sequence for the rule $\frac{n(n+1)}{2}$, where n is a positive integer? *(LESSON 2.1)*

a. 0, 1, 3, 6, ...   **b.** 1.5, 2.5, 3.5, 4.5, ...
c. 1, 2.5, 5, 8.5, ... **d.** 1, 3, 6, 10, ...

3. Which process is used to form conjectures? *(LESSON 2.5)*

a. deductive reasoning
b. inductive reasoning
c. transformation
d. indirect proof

4. Refer to the figure below. Which of the following statements is true? *(LESSON 2.5)*

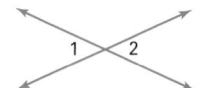

a. $\angle 1$ and $\angle 2$ form a linear pair.
b. $\angle 1$ and $\angle 2$ are parallel.
c. $m\angle 1 = m\angle 2$
d. $m\angle 1 > m\angle 2$

A square is divided into regions by pairs of perpendicular lines as shown. *(LESSON 2.1)*

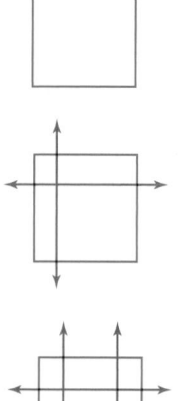

5. How many regions are formed by three pairs of perpendicular lines? by four pairs of perpendicular lines?

6. Write a conjecture about the number of regions formed by n pairs of perpendicular lines.

7. Write a proof of the conjecture you made in Item 6.

---

1. c

2. d

3. b

4. c

5. 16; 25

6. The number of regions formed by n pairs of perpendicular lines is $(n+1)^2$

7. Paragraph proof:
The vertical line of the first pair of perpendicular lines will divide the square into 2 vertical strips and the horizontal line will divide each of these strips in half for a total of $2(2) = 4$ regions. For each additional perpendicular line, the vertical line will add a vertical strip and the horizontal line will divide each vertical strip into one more region. For n perpendicular lines, we have $n + 1$ vertical strips which are divided into $n + 1$ pieces for a total of $(n+1)(n+1)$ or $(n+1)^2$ regions.

A triangle has vertices at (4, 1), (2, 2), and (3, 0). What are the vertices of the image after the following transformations? *(LESSON 1.7)*

8. reflection across the *x*-axis

9. translation 3 units to the left and 1 unit down

10. Write a conditional statement based on the Euler diagram below. *(LESSON 2.2)*

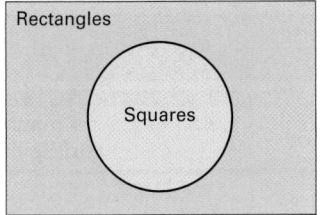

Rectangles

Squares

11. The figures below are splorts.

The figures below are not splorts.

Write a definition of *splort*. *(LESSON 2.3)*

For Items 12–15, refer to the diagram at right. *(LESSON 1.1)*

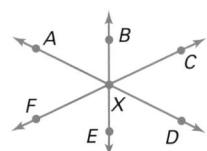

12. Name five points in the figure.

13. Name three lines in the figure.

14. Name four rays in the figure.

15. Name eight angles in the figure.

16. Which of the following is a definition? *(LESSON 2.3)*

 a. The midpoint of a segment is a point that divides the segment into two equal parts.

 b. An acute triangle is a triangle that has an angle of less than 90°.

 c. Two points determine exactly one line.

 d. An equilateral triangle is an isosceles triangle.

FREE-RESPONSE GRID
The following Items may be answered by using a free-response grid such as that commonly used by standardized-test services.

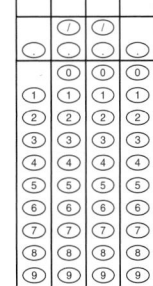

For Items 17 and 18, refer to the figure below. *(LESSONS 1.3 AND 2.5)*

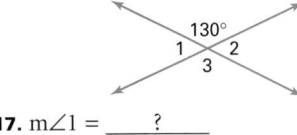

17. m∠1 = _____?_____

18. m∠3 = _____?_____

For Items 19 and 20, refer to the information and the figure below. *(LESSON 1.3)*

m∠QPT = 140°
m∠QPR = (2x − 5)°
m∠RPS = (4x + 10)°
m∠SPT = (3x)°

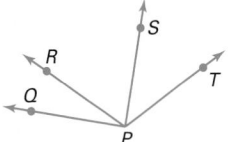

19. x = _____?_____

20. m∠RPS = _____?_____

8. (4, −1), (2, −2), and (3, 0)

9. (1, 0), (−1, 1), and (0, −1)

10. If a figure is a square, then it is a rectangle.

11. Sample answer: A splort is any figure with a 90° rotational symmetry.

12. Sample answer: points *A, B, C, D, E*

13. Sample answer: $\overleftrightarrow{AX}$, $\overleftrightarrow{CF}$, $\overleftrightarrow{BE}$

14. Sample answer: $\overrightarrow{XB}$, $\overrightarrow{XD}$, $\overrightarrow{EB}$, $\overrightarrow{BE}$

15. Sample answer: ∠*AXB*, ∠*CXB*, ∠*DXB*, ∠*FXB*, ∠*AXC*, ∠*CXD*, ∠*DXF*, ∠*FXA*

16. *a*

17. 50°

18. 130°

19. 15

20. 70°

Parallels and Polygons

Lesson Presentation CD-ROM
Power Point® presentations for each lesson 3.1-3.8

CHAPTER PLANNING GUIDE

Lesson	3.1	3.2	3.3	3.4	3.5	3.6	3.7	3.8	Project and Review
Pupil's Edition Pages	138–147	148–154	155–161	162–167	170–176	177–182	183–189	190–197	168–169, 198–207
Practice and Assessment									
Extra Practice (Pupil's Edition)	824	824	825	825	826	826	827	827	
Practice Workbook	13	14	15	16	17	18	19	20	
Practice Masters Levels A, B, and C	37–39	40–42	43–45	46–48	49–51	52–54	55–57	58–60	
Standardized Test Practice Masters	15	16	17	18	19	20	21	22	23
Assessment Resources	27	28	29	30	32	33	34	35	31, 36–41
Visual Resources									
Lesson Presentation Transparencies Vol. 1	49–52	53–56	57–60	61–63	64–67	68–70	71–74	75–78	
Teaching Transparencies	19–22	23–25	26	27		28–30	31	32–33	
Answer Key Transparencies	80–86	87–89	90–95	96–101	102–103	104–106	107–111	112–123	124–127
Quiz Transparencies	3.1	3.2	3.3	3.4	3.5	3.6	3.7	3.8	
Teacher's Tools									
Reteaching Masters	25–26	27–28	29–30	31–32	33–34	35–36	37–38	39–40	
Make-Up Lesson Planner for Absent Students	13	14	15	16	17	18	19	20	
Student Study Guide	13	14	15	16	17	18	19	20	
Spanish Resources	13	14	15	16	17	18	19	20	
Block Scheduling Handbook									6–7
Activities and Extensions									
Lesson Activities	32–34	35–40	41–43		44–45	46–49	50–54		
Enrichment Masters	13	14	15	16	17	18	19	20	
Cooperative-Learning Activities	13	14	15	16	17	18	19	20	
Problem-Solving/ Critical Thinking	13	14	15	16	17	18	19	20	
Student Technology Guide		14–15		16	17	18	19		
Long Term Projects									9–12
Writing Activities for Your Portfolio									7–9
Tech Prep Masters									11–14
Building Success in Mathematics									7–8

LESSON PACING GUIDE

Lesson	3.1	3.2	3.3	3.4	3.5	3.6	3.7	3.8	Project and Review
Traditional	2 days	2 days	2 days	2 days	2 days	2 days	2 days	2 days	2 days
Block	1 day	1 day	1 day	1 day	1 day	1 day	1 day	1 day	1 day
Two-Year	4 days	4 days	4 days	4 days	4 days	4 days	4 days	4 days	4 days

CONNECTIONS AND APPLICATIONS

Lesson	3.1	3.2	3.3	3.4	3.5	3.6	3.7	3.8	Review
Algebra	145	152	160		174	181	187	191, 192, 195	
Geometry	138–147	148–154	155–161	162–167	170–176	177–182	183–189	190–197	200–207
Technology					174				
Life Skills	145, 146		161	166, 167	173, 175	182		195, 196	
Science and Technology	141, 142, 143		159, 161	166	176	180, 182	185, 186, 188, 189		204
Sports and Leisure	145, 147	151, 153			176		184, 186, 188, 189		204
Social Studies	146				172				
Cultural Connection: Africa	145							196	

BLOCK-SCHEDULING GUIDE

Day	Lesson	Teacher Directed: Lesson Examples, Teaching Transparencies	Student Guided: Activity, Try This	Cooperative-Learning Activity, Lesson Activity, Student Technology Guide	Practice: Practice & Apply, Extra Practice, Practice Workbook	Assessment: Quiz, Mid-Chapter Assessment	Problem Solving, Reteaching
1	3.1	15 min	15 min	20 min	55 min	15 min	15 min
2	3.2	15 min	15 min	20 min	55 min	15 min	15 min
3	3.3	15 min	15 min	20 min	55 min	15 min	15 min
4	3.4	15 min	15 min	20 min	55 min	15 min	15 min
5	3.5	15 min	15 min	20 min	55 min	15 min	15 min
6	3.6	15 min	15 min	20 min	55 min	15 min	15 min
7	3.7	15 min	15 min	20 min	55 min	15 min	15 min
8	3.8	15 min	15 min	20 min	55 min	15 min	15 min
9	Assess.	50 min	90 min	90 min	65 min	30 min	
		PE: Chapter Review	**PE:** Chapter Project, Writing Activities	Tech Prep Masters	**PE:** Chapter Assessment, Test Generator	Chap. Assess. (A or B), Alt. Assess. (A or B), Test Generator	

PE: Pupil's Edition

Alternative Assessment

The following suggest alternative assessments for students who may benefit from a different type of assessment than the regular chapter quizzes and the mid-chapter/end-of-chapter test. Visit the HRW web site to get additional Alternative Assessment material.

internet connect

Alternative Assessment
Go To: **go.hrw.com**
Keyword: **MG1 Alt Assess**

Performance Assessment

1. Write the coordinates for the vertices of a pentagon with reflectional symmetry but not rotational symmetry. Plot the figure on grid paper.

2. Write the coordinates of a parallelogram that is not a rectangle or a square. Draw the diagonals. Use the midpoint formula to show that the diagonals of the parallelogram bisect each other.

3. Use the Parallel Postulate and the Triangle Sum Theorem to prove the following:

Given: $\overline{AB}$ is parallel to $\overline{DE}$
Prove: m$\angle ACB$ = m$\angle ECD$

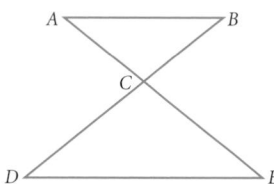

Portfolio Project

Suggest that students choose one of the following projects for inclusion in their portfolios.

1. Explain why an equilateral triangle will tessellate in a plane. How many equilateral triangles can fit around a single point with no overlaps or gaps? Why?

2. Plot a regular hexagon on graph paper. Find the slope of each side of the figure. What pattern do you notice? What other patterns can you find by drawing interior segments in a hexagon?

---

internet connect

The table below identifies the pages in this chapter that contain internet and technology information.

Content Links

Activities Online	pages 143, 151, 173
Portfolio Extensions	pages 147, 196
Homework Help Online	pages 143, 152, 159, 164, 174, 181, 187, 194

Resource Links

Parents can go online and find concepts that students are learning–lesson by lesson–and questions that pertain to each lesson, which facilitate parent-student discussion.

Go To: **go.hrw.com**
Keyword: **MG1 Parent Guide**

Technical Support

The following may be used to obtain technical support for any HRW software product.

Online Help: **www.hrwtechsupport.com**
e-mail: **tschrw@hbtechsupport.com**

HRW Technical Support Center: **(800)323-9239**
7 AM to 10 PM Monday through Friday CST

Visit the HRW math web site at: **www.hrw.com/math**

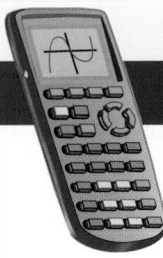

Technology

Technology Objectives and Suggestions

Lesson 3.1 Symmetry in Polygons

Students can explore reflectional and rotational symmetry using geometry software. Have them draw two sets of the polygons described on page 139. In the first set, have them draw the figures with reflectional symmetry but not rotational symmetry. Then have them draw another set with rotational symmetry. Do any of the figures with rotational symmetry have reflectional symmetry?

Lesson 3.2 Properties of Quadrilaterals

In this lesson students identify the properties of quadrilaterals and the relationships among the properties. For instance, students see that the diagonals of a parallelogram bisect each other. All of the activities in this lesson work best with geometry software. Students can draw figures, drag and change them, and see that the properties still hold true.

Lesson 3.3 Parallel Lines and Transversals

In this lesson students look at parallel lines cut by a transversal and the relationships among the special angles formed. Students can explore these relationships by drawing parallel lines using geometry software and measuring the angles formed by a transversal. They will find that changing the position of the transversal does not change the angle relationships, nor does moving the parallel lines closer together and further apart, provided that they are still parallel lines.

Lesson 3.4 Proving That Lines are Parallel

This lesson is an extension of Lesson 3.3 in that students use what they learn about the special angle relationships formed by parallel lines and a transversal to prove that lines are parallel. Have students use geometry software to draw nonparallel lines cut by a transversal.

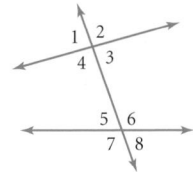

Display the measures of each of the angles formed by the transversal. Have students move the lines until they are sure that the lines are parallel. Ask them how they can know that the two lines are parallel. Their answers should be ones like "Because corresponding angles are congruent," etc.

Lesson 3.5 The Triangle Sum Theorem

In this lesson students identify the Parallel Postulate and the Triangle Sum Theorem. Students can investigate the sum of the measures of the angles of a triangle by using geometry software. Have students draw a triangle, measure the angles, and find the sum of the angle measures. Then have them drag a point on the triangle to change its shape. As the triangle changes, they will be able to see that the sum of the angle measures will always be 180 degrees.

They can extend activity to include the Exterior Angle Theorem. Have them draw an exterior angle on the triangle and measure it. By changing the shape of the triangle, they will be able to see that the sum of the remote interior angles is equal to the measure of the exterior angle.

Lesson 3.6 Angles in Polygons

In this lesson students develop and use formulas for the sums of the measures of interior and exterior angles of a polygon. Geometry software can be used for all activities in this lesson. For Activity 1, draw the figure and have the program calculate the sums of the angles as students drag and change the shape of the polygon.

Lesson 3.7 Midsegments of Triangles and Trapezoids

In this lesson students develop and use formulas from the properties of triangle and trapezoid midsegments. Geometry software can be used for all of the activities. By dragging a base of a trapezoid and making it disappear to a point, use technology to show that the Triangle Midsegment Theorem and the Trapezoid Midsegment Theorem are related.

Lesson 3.8 Analyzing Polygons Using Coordinates

In this lesson students develop and use theorems about equal slopes and the slopes of perpendicular lines. Have students use graphics calculators in this lesson to plot equations of parallel and perpendicular lines to create polygons such as rectangles and rhombuses.

Background Information

The content of this chapter introduces students to the concept of a polygon and explores in detail the properties of quadrilaterals. Parallel lines and transversals are studied, and their properties are then used to develop a number of important facts about the sums of interior and exterior angles of polygons and about midsegments of triangles and trapezoids. The chapter concludes with the use of coordinates to analyze polygons.

CHAPTER RESOURCES

- Block-Scheduling Handbook
- Writing Activities for Your Portfolio
- Tech Prep Masters
- Long-Term Project
- Assessment Resources:
  Mid-Chapter Assessment
  Chapter Assessments
  Alternative Assessments
- Test and Practice Generator
- Technology Handbook

Chapter Objectives

- Define *polygon*. [**3.1**]
- Define and use *reflectional symmetry* and *rotational symmetry*. [**3.1**]
- Define *regular polygon*, *center of a regular polygon*, *central angle of a regular polygon*, and *axis of symmetry*. [**3.1**]
- Define *quadrilateral*, *parallelogram*, *rhombus*, *rectangle*, *square*, and *trapezoid*. [**3.2**]
- Identify the properties of quadrilaterals and the relationships among the properties. [**3.2**]

Parallels and Polygons

RECTANGLES, TRIANGLES, AND HEXAGONS ARE examples of *polygons*. Patterns of polygons are often used for decorative purposes. The tiling pattern at right below is from the Alhambra, a famous Islamic fortress in Spain.

The interplay of parallel beams and polygons in the photograph of the Thorncrown Chapel below suggests the theme of this chapter.

As you will learn, parallel lines and their properties provide a basis for classifying and exploring four-sided polygons known as *quadrilaterals*.

Lessons

3.1 • Symmetry in Polygons

3.2 • Properties of Quadrilaterals

3.3 • Parallel Lines and Transversals

3.4 • Proving That Lines Are Parallel

3.5 • The Triangle Sum Theorem

3.6 • Angles in Polygons

3.7 • Midsegments of Triangles and Trapezoids

3.8 • Analyzing Polygons With Coordinates

Chapter Project String Figures

Thorncrown Chapel, Eureka Springs, Arkansas Designed by E. Fay Jones

About the Photos

Patterns of polygons and parallel lines are frequently found in nature, decorative arts, architecture, and fine arts. The photographs above, from left to right, show the Thorncrown Chapel in Eureka Springs, Arkansas; Moroccan tiles from the Alhambra (an Islamic fortress in Spain); a honeycomb; piano strings; and a detail of Piet Mondrian's *Composition With Red, Yellow and Blue* (1930). In all of these items, sets of polygons and parallel lines provide strength, beauty, and functionality to the objects and structures formed.

Piet Mondrian. *Composition with Red, Blue and Yellow.* 1930. Oil on Canvas. 20" × 20"

About the Chapter Project

In this chapter, you will investigate the properties of polygons, The key that unlocks the study of polygons is found in the properties of parallel lines, which you will explore in Lessons 3.3 and 3.4. In the Chapter Project, *String Figures*, you will learn to make one of the world's most popular string figures, which is known in the United States as "Jacob's ladder." The figure consists of polygons between two (roughly) parallel lines.

After completing the Chapter Project, you will be able to do the following:

- Make popular string figures by following a set of instructions.

- Appreciate the skills and ingenuity of the ancient peoples that created such figures.

About the Portfolio Activities

Throughout the chapter, you will be given opportunities to complete Portfolio Activities that are designed to support your work on the Chapter Project.

The theme of each Portfolio Activity and of the Chapter Project is geometric art.

- In the Portfolio Activity on page 147, you will study artful ways of forming the surface of a quilt with patches. The repetition of the individual shapes often results in intriguing overall designs.

- Tessellation, the art of covering a surface with congruent shapes that fit together without gaps or overlapping, is the subject of the Portfolio Activity on page 154. Translation tessellations are studied in depth.

- Rotation tessellations are the subject of the Portfolio Activity on page 197.

- Define *transversal, alternate interior angles, alternate exterior angles, same-side interior angles,* and *corresponding angles.* [**3.3**]

- Make conjectures and prove theorems by using postulates and properties of parallel lines and transversals. [**3.3**]

- Identify and use the converse of the Corresponding Angles Postulate. [**3.4**]

- Prove that lines are parallel by using theorems and postulates. [**3.4**]

- Identify and use the Parallel Postulate and the Triangle Sum Theorem. [**3.5**]

- Define *interior* and *exterior angles* of a polygon. [**3.6**]

- Develop and use formulas for the sums of the measures of interior and exterior angles of a polygon. [**3.6**]

- Define *midsegment of a triangle* and *midsegment of a trapezoid.* [**3.7**]

- Develop and use formulas based on the properties of triangle and trapezoid midsegments. [**3.7**]

- Develop and use theorems about equal slopes and slopes of perpendicular lines. [**3.8**]

- Solve problems involving perpendicular and parallel lines in the coordinate plane by using appropriate theorems. [**3.8**]

Portfolio Activities appear at the end of Lessons 3.1, 3.2, and 3.8. Each serves as preparation for the Chapter Project. The Portfolio Activities, as well as the Chapter Project Activities, are appropriate for inclusion in the student's portfolio. Students should be encouraged to include in their portfolios any other work in which they feel a sense of pride or a sense of accomplishment.

☑ internet connect

Chapter Internet Features and Online Activities

LESSON	KEYWORD	PAGE	LESSON	KEYWORD	PAGE
3.1	MG1 Reuleaux	143	3.5	MG1 HyperGeo	173
	MG1 Homework Help	143		MG1 Homework Help	174
	MG1 Quilts	147	3.6	MG1 Homework Help	181
3.2	MG1 Bridge	151	3.7	MG1 Homework Help	187
	MG1 Homework Help	152	3.8	MG1 Homework Help	194
3.3	MG1 Homework Help	159		MG1 Escher	196
3.4	MG1 Homework Help	164			

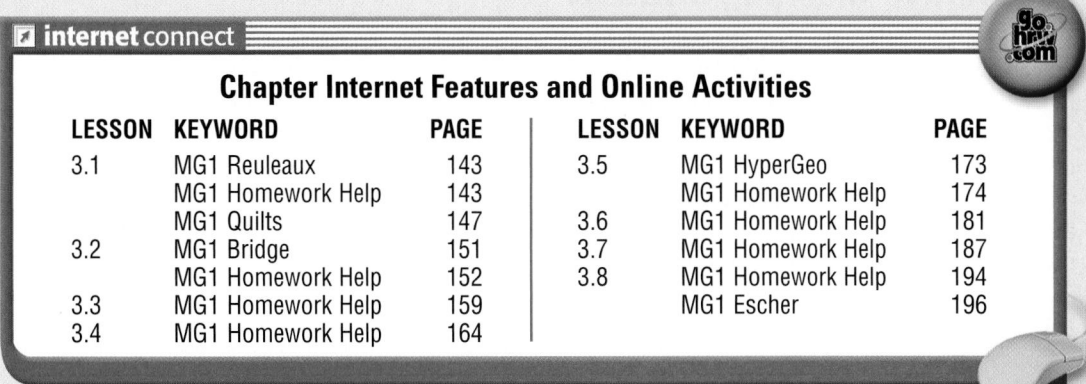

Symmetry in Polygons

QUICK WARM-UP

Refer to the diagram to answer the questions.

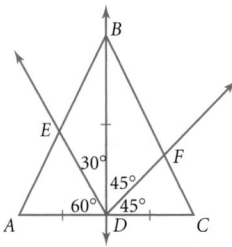

1. Name a ray that bisects $\overline{AC}$. any of these: $\overrightarrow{DE}, \overrightarrow{DB},$ $\overrightarrow{DF}, \overrightarrow{BD}$

2. Name the perpendicular bisector of $\overline{AC}$. $\overrightarrow{BD}$

3. Name the bisector of $\angle CDB$. $\overrightarrow{DF}$

Also on Quiz Transparency 3.1

Objectives

● Define *polygon*.

● Define and use *reflectional symmetry* and *rotational symmetry*.

● Define *regular polygon*, *center of a regular polygon*, *central angle of a regular polygon*, and *axis of symmetry*.

Why Polygons appear all around you in man-made objects. Understanding the mathematical properties of polygons will help you understand how to use them for artistic and practical purposes.

Symmetrical polygons give this Native American blanket design an attractive appearance. They also have interesting mathematical properties.

Defining Polygons

In Lesson 2.3 you learned how to use examples and "nonexamples" to write definitions. Use the following figures to define a *polygon*:

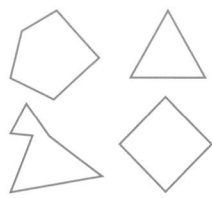 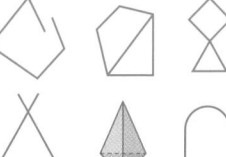

These are polygons.   These are not polygons.

Compare your definition of a polygon with the definition below. Does your definition have all the requirements it needs? Does it have more than it needs?

Polygon

A **polygon** is a plane figure formed from three or more segments such that each segment intersects exactly two other segments, one at each endpoint, and no two segments with a common endpoint are collinear. The segments are called the **sides of the polygon**, and the common endpoints are called the **vertices of the polygon**.   **3.1.1**

Why Students are intuitively familiar with the concept of symmetry from their experiences with symmetry in nature and in real-world objects. Have students suggest examples of objects that have symmetry.

Alternative Teaching Strategy

TECHNOLOGY Students can experiment with reflectional and rotational symmetry by using reflecting devices and their own notebook paper. After they have done this, have them use geometry graphics software to create polygons and regular polygons with the reflection and rotation features of the program. Challenge students to use the software to draw one general polygon and one regular polygon of each type listed on page 139. For each polygon, have students draw any lines of symmetry that exist.

A polygon is named according to the number of its sides. Familiarize yourself with information in the table below.

Polygons Classified by Number of Sides

Triangle	3	Nonagon	9
Quadrilateral	4	Decagon	10
Pentagon	5	11-gon	11
Hexagon	6	Dodecagon	12
Heptagon	7	13-gon	13
Octagon	8	*n*-gon	*n*

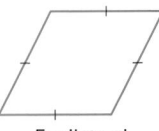
Equiangular

Equilateral

An **equiangular polygon** is one in which all angles are congruent. An **equilateral polygon** is one in which all sides are congruent. A **regular polygon** is one that is both equiangular and equilateral.

The **center of a regular polygon** is the point that is equidistant from all vertices of the polygon. A **central angle** of a regular polygon is an angle whose vertex is the center of the polygon and whose sides pass through two consecutive vertices.

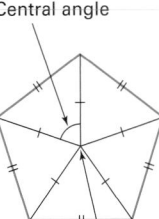
Central angle
Center

Reflectional Symmetry

Lay your pencil on the photograph of the blanket on page 138 so that the part of the photo on one side of your pencil is the mirror image of the part on the other side. The line of your pencil is an *axis of symmetry*.

Is the cat's face symmetrical?

Reflectional Symmetry

A figure has **reflectional symmetry** if and only if its reflected image across a line coincides exactly with the preimage. The line is called an **axis of symmetry**.   3.1.2

Imagine reflecting each of the following figures across the lines shown. As you can see, the reflected image will coincide exactly with the preimage. Therefore, each figure has reflectional symmetry, and each line is an axis of symmetry.

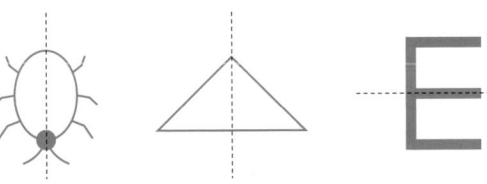

Interdisciplinary Connection

INDUSTRIAL ARTS   Nuts and bolts are used in a great variety of objects, from kitchen tables to the Hubbell Space Telescope. You may wish to discuss why the design of nuts and bolts is frequently based on regular polygons. Discuss the kinds of regular polygons most often used. Ask why the number of sides is usually small.

Teaching Tip

Make sure students are familiar with the terms on page 139 and throughout the lesson. Have students create a vocabulary list for this lesson, including a written definition and, if possible, a picture.

Activity 1 Notes

In this Activity, students determine whether triangles that are classified according to their sides have axes of symmetry. They are expected to complete the triangles symmetry conjecture by analyzing the relationship between each axis of symmetry and the side it intersects.

For a student worksheet of this Activity and detailed Teacher Notes, see page 32 in the Lesson Activities booklet.

Teaching Tip

You may wish to extend the Activity about triangles to polygons with more than three sides. Ask students to make conjectures about the maximum number of lines of symmetry that a given polygon can have. Is it possible for a polygon that has one or more axes of symmetry to have no congruent sides?

CHECKPOINT ✔

4. An axis of symmetry in a triangle is the perpendicular bisector of the side it intersects, and it passes through the vertex of the angle opposite that side of the triangle.

Triangles Classified by Number of Congruent Sides

Three congruent sides:	equilateral
At least two congruent sides:	isosceles
No congruent sides:	scalene

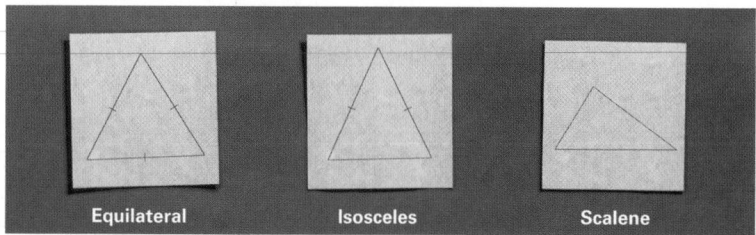

Equilateral   Isosceles   Scalene

Activity 1
Reflectional Symmetry in Triangles

YOU WILL NEED
folding paper, ruler, and protractor

1. Draw an example of each type of triangle shown above, and find any axes of symmetry that exist. How many axes of symmetry does each type of triangle have?

2. Fold each triangle along its axes of symmetry, if it has any. After folding, which angles seem to coincide?

3. Complete a table like the one below.

Type of triangle	Number of axes of symmetry	Number of congruent angles
equilateral	?	?
isosceles	?	?
scalene	?	?

CHECKPOINT ✔

4. Study the relationship between each axis of symmetry and the side it intersects. Use your ruler and protractor to make measurements. Complete the following conjecture:

Triangle Symmetry Conjecture

An axis of symmetry in a triangle is the ____?____ ____?____ of the side it intersects, and it passes through the ____?____ of the angle opposite that side of the triangle.

Enrichment

Have students draw the polygons shown at right on cardboard or tracing paper and cut them out. These cutouts are to be used to make polygons with rotational symmetry. Students should name the polygons thus created.

Rotational Symmetry

Rotational Symmetry

A figure has **rotational symmetry** if and only if it has at least one rotation image, not counting rotation images of 0° or multiples of 360°, that coincides with the original image.

3.1.3

The flower at right has approximate rotational symmetry. Notice that it will coincide with itself 5 times if it is rotated completely about its center of rotation, which is also known as its center of symmetry. Thus, the figure of the flower is said to have *5-fold rotational symmetry*.

Note: *All* geometric figures have 0° and 360° rotational symmetry. A figure that has only 0° and 360° rotational symmetry is said to have only *trivial* rotational symmetry. For such a figure in a plane, *any* point in the plane of the figure is a center of rotation.

Activity 2
Rotational Symmetry in Regular Polygons

1. Draw or trace a regular pentagon such as the one in the illustration. Label the center P and the vertices Q, R, S, T, and U. Draw $\overrightarrow{PQ}$. Then copy the figure onto a sheet of tracing paper.

2. Use a pencil point to anchor the traced figure on top of the original at their centers so that the figures coincide.

3. Rotate the top figure counterclockwise until point Q on the traced figure coincides with point R on the original. Continue to rotate point Q, stopping at each point—S, T, U, and finally Q. How many rotations did you make in all?

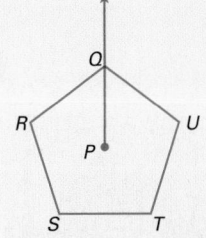

CHECKPOINT ✔

4. Do the central angles of a regular polygon seem to be congruent? (You will prove that they are congruent in Lesson 4.2.)

Assuming that the central angles of a regular polygon are congruent, complete the rule below for finding the measure of a central angle of a regular polygon.

The Central Angle of a Regular Polygon

The measure, θ, of a central angle of a regular polygon with n sides is given by the following formula: $\theta = $ ____?____ .

Inclusion Strategies

HANDS-ON STRATEGIES Students can use cardboard cutouts of polygons to determine which ones have rotational symmetry or reflection symmetry. They can do this by tracing around the polygons and changing the position of the cutout to see what changes of position give a perfect match.

Activity 2 Notes

In this Activity, students use tracing paper to explore rotational symmetry. Students are then asked to write a formula for the measure of a central angle of a regular polygon.

For a student worksheet of this Activity and detailed Teacher Notes, see page 32 in the Lesson Activities booklet.

Teaching Tip

For Activity 2, have a class discussion about how to determine whether a polygon is regular. Students may suggest tracing or using measuring tools such as rulers and protractors. Consider also how to measure the angles in the diagram in Activity 2 by using only a compass.

Cooperative Learning

Students can work in groups of three or four to locate the center points of regular polygons. Students can explore various methods and then decide which ones they think work best. Have groups present their methods to the class. Then discuss how to check the accuracy of the results.

CHECKPOINT ✔

4. The measure, θ, of a central angle of a regular polygon with n sides is given by the following formula: $\theta = \frac{360}{n}$.

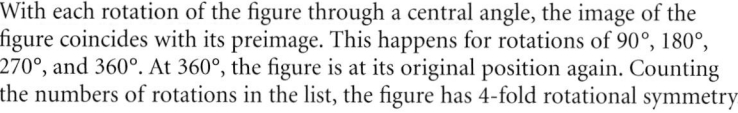

ADDITIONAL EXAMPLE 1

Find the measure of the central angle of each regular polygon below.

a. heptagon ≈ 51.43°

b. decagon   36°

c. 15-gon   24°

Assess

Selected Answers

Exercises 6–8, 9–63 odd

ASSIGNMENT GUIDE

In Class	1–8
Core	9–27 odd, 28–39 41–49 odd
Core Plus	23–32, 33–47 odd 49–58
Review	59–64
Preview	65–68

✎ Extra Practice can be found beginning on page 818.

Error Analysis

In Exercises 18–20, make sure that students preform the number of rotations indicated because some students may stop after the first rotation. In Exercises 39–44, you may need to review how to graph quadratic and absolute-value functions and how to write equations for lines.

EXAMPLE

APPLICATION
AGRICULTURE

A windmill is a device for pumping water by using energy from the wind. If you connected the tips of the blades with segments, what figure would be formed? What is the measure of a central angle of the figure? For what degrees of rotation does the rotational image of the figure coincide with the preimage? What is the rotational symmetry of the figure?

SOLUTION

The windmill has 4 blades, so the segments connecting the tips of the blades will form a regular 4-sided polygon, that is, a square. Use the formula to find the measure of central angle:

$$\theta = 360 \div 4 = 90°$$

With each rotation of the figure through a central angle, the image of the figure coincides with its preimage. This happens for rotations of 90°, 180°, 270°, and 360°. At 360°, the figure is at its original position again. Counting the numbers of rotations in the list, the figure has 4-fold rotational symmetry.

Exercises

Communicate

Taj Mahal, India

1. Describe the position of all the axes of symmetry in the picture at left.

2. Explain how the picture at left illustrates the definition of symmetry. Identify a point and its preimage. What is the relationship between the point and preimage and their axis of symmetry?

3. Why are 0° and 360° rotations not used to define rotational symmetry?

4. What kinds of symmetry does a regular hexagon have? Describe all possible rotations and reflections of a regular hexagon.

5. Identify regions of the Escher woodcut at right that have rotational symmetry. Where are the centers of rotation?

Reteaching the Lesson

USING PATTERNS You can review the main ideas of the lesson by using several cutouts of polygons to illustrate reflectional and rotational symmetry. Which of the figures have reflectional symmetry? Which have rotational symmetry? Discuss which conjectures from Activities 1 and 2 apply to these figures.

6. Draw the axis of symmetry in the triangle at right. Does this triangle confirm your conjecture from Activity 1? Why or why not? *(ACTIVITY 1)*
Yes, the axis of symmetry bisects both the side and the angle through which it passes.

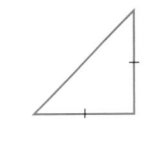

7. Find the measure of a central angle for each regular polygon below. *(ACTIVITY 2)*

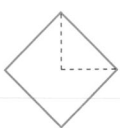   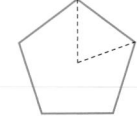   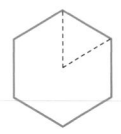

square: 90°   pentagon: 72°   hexagon: 60°

APPLICATION

8. **AGRICULTURE** What is the rotational symmetry of a windmill with 4 blades? with 6 blades? *(EXAMPLE)*
4 blades: 4-fold, 6 blades: 6-fold

● *Practice and Apply*

For Exercises 9–12, copy each figure and draw all of the axes of symmetry.

9.   **10.**   **11.**   **12.**

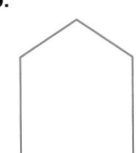   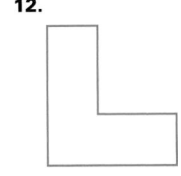

13. How many axes of symmetry does a circle have? Explain your answer.

Each figure below shows part of a shape with reflectional symmetry, with its axis of symetry shown as a black line. Copy and complete each shape.

14.   **15.**   **16.**

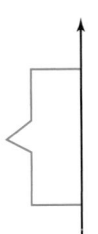      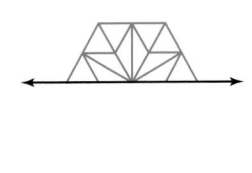

17. Which of the completed shapes from Exercises 14–16 also have rotational symmetry? Exercises 14 and 16

14 .

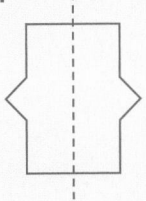

16.

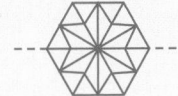

15.

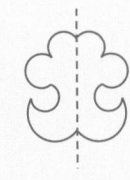

ALGEBRA In Exercises 40–45, students will graph quadratic and absolute-value functions on a graphics calculator or graph paper. They will also write the equation for the axis of symmetry.

6. Yes, the axis of symmetry bisects both the side and the angle through which it passes.

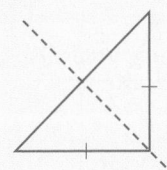

9.

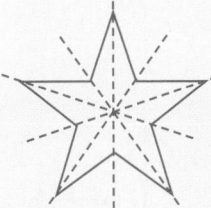

10.

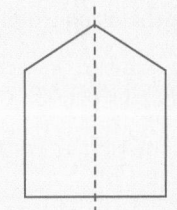

11.

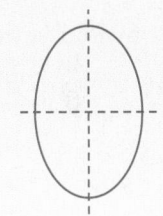

12.

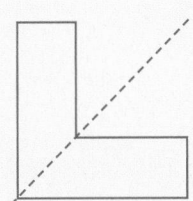

13. A circle has an infinite number of axes of symmetry. Any line through the center of the circle is an axis of symmetry; an infinite number of such lines can be drawn.

18.

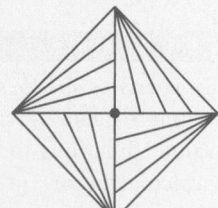

19.

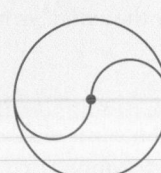

20.

21.

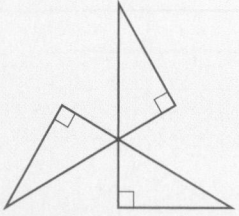

23. Sample answer: an isoceles triangle that is not equilateral

24. Sample answer: a rectangle that is not a square

25. Sample answer: an equilateral triangle

26. Sample answer: a regular pentagon

27. Sample answer: a regular octagon

32. Draw the bisector of the angle.

If a figure has *n*-fold rotational symmetry, then it will coincide with itself after a rotation of $\left(\frac{360}{n}\right)^\circ$. For example, an equilateral triangle has 3-fold symmetry, so it will coincide with itself after a rotation of $\left(\frac{360}{3}\right)^\circ = 120^\circ$.

Each figure below shows part of a shape with the given rotational symmetry. Copy and complete each shape.

18.

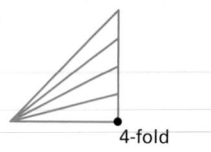

4-fold

19.

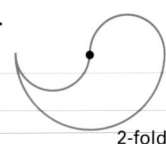

2-fold

20.

6-fold

21.

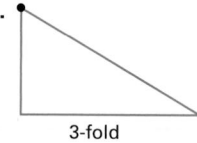

3-fold

22. Which of the completed shapes from Exercises 18–21 also have reflectional symmetry? Exercise 20

23. Draw a figure with exactly 1 axis of symmetry.

24. Draw a figure with exactly 2 axes of symmetry.

25. Draw a figure with exactly 3 axes of symmetry.

26. Draw a figure with 5-fold rotational symmetry.

27. Draw a figure with 8-fold rotational symmetry.

For Exercises 28 and 29, copy the figure at right.

28. The axis of symmetry is the perpendicular bisector of *AB*.

28. Draw an axis of symmetry for $\overline{AB}$ that passes through $\overline{AB}$ at a single point. What is the relationship between $\overline{AB}$ and the axis of symmetry?

29. Reflect $\overline{AB}$ over the axis of symmetry.

What point is the image of *A*? What point is the image of *B*? What segment is the image of $\overline{AB}$? *B* is the image of *A*; *A* is the image of *B*; $\overline{BA}$ is the image of $\overline{AB}$.

For Exercises 30 and 31, copy the figure at right.

30. The axis of symmetry is the angle bisector of *ABC*.

30. Draw an axis of symmetry for $\angle ABC$ that passes through $\angle ABC$ at a single point.

What is the relationship between $\angle ABC$ and the axis of symmetry?

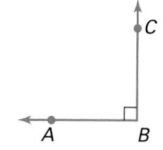

31. Reflect $\angle ABC$ over the axis of symmetry.

What ray is the image of $\overrightarrow{BA}$? $\overrightarrow{BC}$ is the image of $\overrightarrow{BA}$.
What ray is the image of $\overrightarrow{BC}$? $\overrightarrow{BA}$ is the image of $\overrightarrow{BC}$.
What angle is the image of $\angle ABC$? $\angle CBA$ is the image of $\angle ABC$.

32. Describe how to draw an axis of symmetry for any given angle.

Figure *ABCD* below is equilateral. Determine whether each of the following could be the result of a reflection, a rotation, or either:

33. $\overline{DX}$ is the image of $\overline{BX}$. either

34. $\overline{AB}$ is the image of $\overline{AD}$. either

35. $\overline{CB}$ is the image of $\overline{AD}$. rotation

36. $\angle BAX$ is the image of $\angle BCX$. reflection

37. $\angle DCX$ is the image of $\angle BCX$. reflection

38. $\angle DAX$ is the image of $\angle BCX$. rotation

39. $\angle CXD$ is the image of $\angle AXD$. reflection

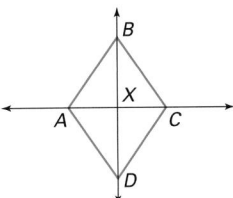

Algebra

Graph each equation below on a graphics calculator or graph paper. Then write an equation for the axis of symmetry of each graph.

40. $y = (x-1)^2 + 3$

41. $y = 2(x-4)^2 + 3$

42. $y = -(x+2)^2 + 3$

43. $y = -2(x+5)^2 + 3$

44. $y = |x| + 3$

45. $y = |x+3|$

46. 0.36°; 0.036°

46. What is the measure of a central angle of a regular chiliagon (1000 sides)? What is the measure of a central angle of a regular myriagon (10,000 sides)?

47. Draw a quadrilateral that is equilateral but not equiangular.

48. Draw a quadrilateral that is equiangular but not equilateral.

49. The extended table is not regular. It is not equilateral but it is equiangular.

49. FURNITURE A table is in the shape of a regular octagon. The table can be extended by putting in a leaf, as shown. Is the extended table regular? equilateral? equiangular?

50. RECREATION The Ferris wheel shown at right has 16 cars. Imagine a polygon formed by connecting the cars. What is the measure of a central angle of this polygon? 22.5°

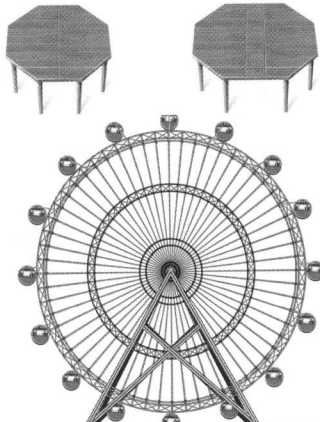

CULTURAL CONNECTION: AFRICA The designs below are taken from Egyptian bowls dating back to 3500 B.C.E. Describe the symmetries in each design.

51. **52.** **53.** **54.**

51. The Egyptian bowl has no reflection symmetry axes. There are nontrivial rotational symmetries for rotations of 72°, 144°, 216°, and 288°.

52. The Egyptian bowl has no reflection symmetry axes. There are nontrivial rotational symmetries for rotations of 120° and 240°.

53. The Egyptian bowl has 2 axes of symmetry—vertically and horizontally through the center. There is a nontrivial rotational symmetry for a rotation of 180°.

54. The Egyptian bowl has 4 axes of symmetry—2 vertically and horizontally through the center—and 2 diagonally through the center of the diamond shaped figures. There are nontrivial rotational symmetries for rotations of 90°, 180°, and 270°.

40.

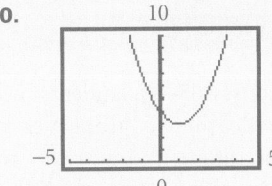

Axis of symmetry: $x = 1$

41.

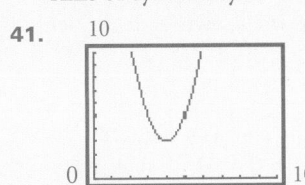

Axis of symmetry: $x = 4$

42.

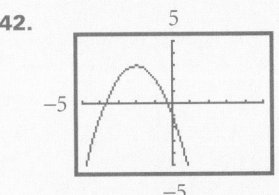

Axis of symmetry: $x = -2$

43.

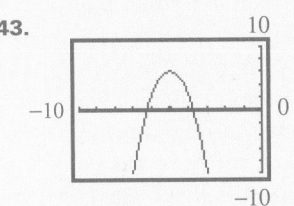

Axis of symmetry: $x = -5$

44.

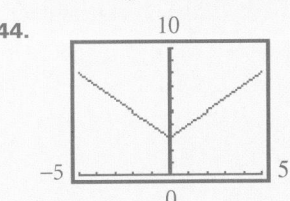

Axis of symmetry: $x = 0$

45.

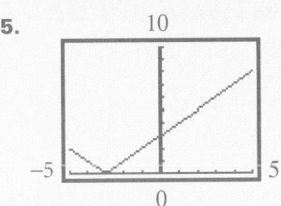

Axis of symmetry: $x = -3$

47. Sample answer:

48. Sample answer:

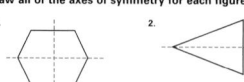
In Exercises 65–68, students look at the geometric properties of flags of different countries. Students can also consider the reflectional and rotational symmetry of their own state or school flag.

55. The sign is a regular octagon with 8 axes of symmetry—4 lines through the center and the midpoint of each side and 4 lines through the center and each vertex. It has nontrivial rotational symmetries for rotations of 45°, 90°, 135°, 180°, 225°, 270°, and 315°.

56. The sign is a regular quadrilateral or square. It has 4 axes of symmetry—2 lines through the center and the midpoint of each side and 2 lines through the center and each vertex. It has nontrivial rotational symmetries for rotations of 90°, 180°, and 270°.

Practice

 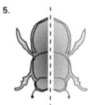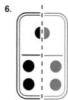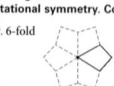 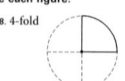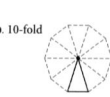
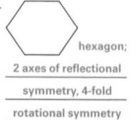
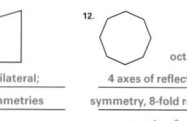

TRAFFIC SAFETY Examine each road sign below. Identify the type of polygon. Is it regular? Describe all of its symmetries (ignore any figures or words on the sign).

55.  56.  57.  58.

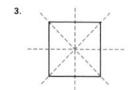 **Look Back**

Refer to the diagram at right for Exercises 59–61.
(**LESSON 1.1**)

59. Name the intersection of and $\overleftrightarrow{AB}$ and $\overleftrightarrow{MN}$. O

60. Sample answer: *BOM* **60.** Name three points that determine plane 𝒯.

61. Name the intersection of planes 𝒮 and 𝒯. $\overleftrightarrow{AB}$

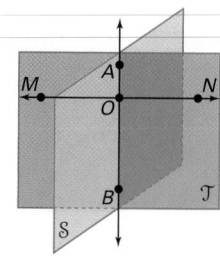

Refer to the diagram at right for Exercises 62–64.
(**LESSON 1.1**)

62. m∠ABD = 80°, m∠ABC = 30°, m∠CBD = ? 50°

63. m∠ABC = 25°, m∠CBD = 60°, m∠ABD = ? 85°

64. m∠ABC = m∠CBD, m∠ABD = 88°, m∠ABC = ? 44°

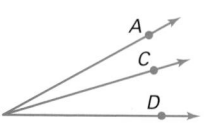

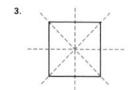 **Look Beyond**

 France  Japan  Syria  United Kingdom  St. Lucia

 Qatar Micronesia  Canada Macedonia  Israel

United States 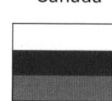 Guyana Russia Ethiopia Iceland

GEOGRAPHY Name all of the countries listed above whose flags have the type of symmetry given below.

65. reflectional, with a horizontal axis of symmetry

66. reflectional, with a vertical axis of symmetry

67. rotational symmetry of 180°

68. no symmetry

---

57. The sign is a regular triangle or equilateral triangle. It has 3 axes of symmetry—3 lines through the center and each vertex. It has nontrivial rotational symmetries for rotations of 120° and 240°.

58. The sign is a equiangular quadrilateral or rectangle. It is not regular. It has 2 axes of symmetry—2 lines through the center and the midpoint of each side. It has a nontrivial rotational symmetry for a rotation of 180°.

65. France, Israel, Japan, Iceland, Macedonia, Micronesia, Qatar, and Guyana

66. Canada, Israel, Japan, Macedonia, Micronesia, Russia, Saint Lucia, Ethiopia, Syria

67. Israel, Japan, Macedonia, Micronesia, United Kingdom

68. United States

Portfolio Extension
Go To: **go.hrw.com**
Keyword:
MG1 Quilts

QUILTING The following designs are called 9-patch quilt blocks because they are based on a grid of 9 congruent squares.

Ohio Star

Shoo Fly

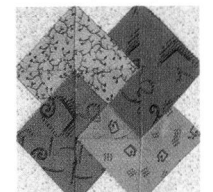

Card Trick

Star Cross

Friendship Star

1. Describe all of the types of symmetry in each quilt block shown above.

2. The quilt design at right consists of Ohio Star blocks. Copy one of the 9-patch blocks shown above or create your own. Design your own quilt by using the block you chose. It may be helpful to use graph paper.

3. Describe the symmetry of your overall quilt design. Did any interesting patterns occur that were not part of the original block?

Quilting is a traditional American social activity.

Properties of Quadrilaterals

QUICK WARM-UP

Refer to the figure to answer the questions below.

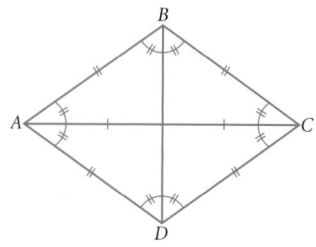

1. Name two angles that are bisected by $\overline{AC}$.
   $\angle BAD$, $\angle BCD$

2. Name a segment that is bisected by $\overline{BD}$. $\overline{AC}$

3. Name four isosceles triangles. triangles ABC, DAB, BCD, and ADC

Also on Quiz Transparency 3.2

Objectives

● Define *quadrilateral, parallelogram, rhombus, rectangle, square,* and *trapezoid.*

● Identify the properties of quadrilaterals and the relationships among the properties.

Why **A baseball diamond is actually a square. The names of three other quadrilaterals also describe the figure.**

The length of each side of a baseball diamond is 90 feet, and the pitcher's mound lies on the diagonal from home plate to second base at a point about 6 feet closer to home plate than to second base.

Teach

Why Point out that properties of quadrilaterals are used widely in determining property lines and in checking measurements for playing fields in sports.

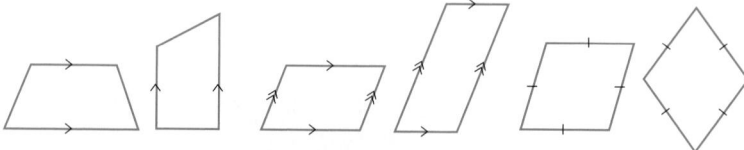

Special Quadrilaterals

Any four-sided polygon is a **quadrilateral**. Quadrilaterals that have certain properties are called *special quadrilaterals*. Study the definitions below. (In the figures, arrowheads are used like tick marks to indicate that two lines, segments, or rays are parallel.)

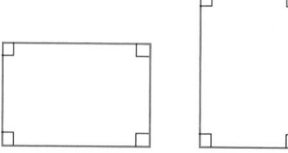

A **trapezoid** is a quadrilateral with one and only one pair of parallel sides.

A **parallelogram** is a quadrilateral with two pairs of parallel sides.

A **rhombus** is a quadrilateral with four congruent sides.

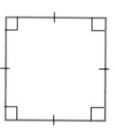

A **rectangle** is a quadrilateral with four right angles.

A **square** is a quadrilateral with four congruent sides and four right angles.

Alternative Teaching Strategy

USING MANIPULATIVES Have students complete the lesson Activities with geoboards. The grid on a geoboard makes it easy to measure segments by counting pegs. Students may also complete the Activities with plastic or cardboard shapes of each type of quadrilateral.

Students can test their conjectures by using a ruler to measure diagonals. Once they are finished with the Activities, have them group the shapes according to the diagram on page 150. Ask students which of the shapes are parallelograms, rhombuses, and rectangles.

Which of the quadrilaterals seem to have rotational symmetry? Which have reflectional symmetry? What can you learn about quadrilaterals from their symmetries? Keep this in mind as you do the Activities that follow.

Activity 1
Parallelograms

1. Draw a parallelogram that is not a rhombus, a rectangle, or a square. Measure the angles and the sides of the figure. Which angles and sides appear to be congruent?

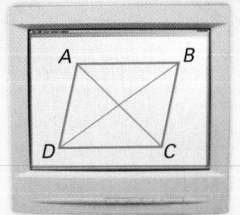

2. Draw diagonals to connect the vertices. Measure the diagonals. Which parts of the figure appear to be congruent? If you are using geometry drawing software, see if the conjecture holds when you vary the shape of your figure by dragging one of the vertices.

3. *Consecutive angles* of a polygon are angles that have a side in common. What do you notice about consecutive angles in a parallelogram?

CHECKPOINT ✔ 4. What conjectures can you make about the sides, angles, or other parts of the figure? Complete the conjectures below.

Conjectures: Properties of Parallelograms

Opposite sides of a parallelogram are _____?_____.

Opposite angles of a parallelogram are _____?_____.

Diagonals of a parallelogram _____?_____.

Consecutive angles of a parallelogram are _____?_____.

Activity 2
Rhombuses

1. Draw a rhombus that is not a square. Draw diagonals to connect the vertices. Make measurements as in Activity 1.

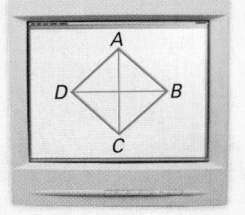

CHECKPOINT ✔ 2. Do the conjectures you made about parallelograms in Activity 1 seem to be true for your rhombus? Discuss why they should or should not be true for rhombuses. Complete the conjecture below.

Conjecture: A Property of Rhombuses

A rhombus is a _____?_____.

3. What new conjecture can you make about rhombuses that is not true for all parallelograms? Fill in the blank in the conjecture below.

Conjecture: A Property of Rhombuses

The diagonals of a rhombus are _____?_____.

Interdisciplinary Connection

INDUSTRIAL DESIGN Parallelograms, unlike triangles, are not rigid figures. Certain relationships between sides and angles of parallelograms evolve from principles that apply to triangles as well. This makes parallelograms suitable for many applications. For example, folding security gates for display windows in stores and doorways often have rhombuses in their design. Invite students to suggest other applications of flexible parallelograms.

Enrichment

Have students write their conjectures for Activities 1–4 as conditional statements. Then have them write the converse of each conjecture. Have students investigate which of the converses seem to be true.

CRITICAL THINKING

Rotational symmetry: parallelogram, rhombus, rectangle, square.
Reflectional symmetry: rhombus, rectangle, square.
By rotating and reflecting quadrilaterals we can determine congruence of angles or sides.

Activity 1 Notes

In this Activity, students measure the sides and angles of a parallelogram and make four conjectures. You may have students use computers or pencil and paper. The purpose of Activities 1–4 is to formulate conjectures about the properties of the parallelograms being studied.

For a student worksheet of this Activity and detailed Teacher Notes, see page 35 in the Lesson Activities booklet.

CHECKPOINT ✔

4. Opposite sides of a parallelogram are congruent.

Opposite angles of a parallelogram are congruent.

Diagonals of a parallelogram bisect each other.

Consecutive angles of a parallelogram are supplementary.

Activity 2 Notes

In this Activity, students measure the sides and angles of a rhombus and make two new conjectures. You may have students use computers or pencil and paper.

For a student worksheet of this Activity and detailed Teacher Notes, see page 35 in the Lesson Activities.

CHECKPOINT ✔

2. A rhombus is a parallelogram.

3. The diagonals of a rhombus are perpendicular.

Activity 3 Notes

In this Activity, students measure the sides and angles of a rectangle and make two new conjectures. You may have students use computers or pencil and paper.

For a student worksheet of this Activity and detailed Teacher Note, see page 35 in the Lesson Activities booklet.

CHECKPOINT ✔

2. A rectangle is a parallelogram.

3. The diagonals of a rectangle are congruent.

Activity 4 Notes

In this Activity, students measure the sides and angles of a square and make two new conjectures. You may have students use computers or pencil and paper.

For a student worksheet of this Activity and detailed Teacher Notes, see page 35 in the Lesson Activities booklet.

CHECKPOINT ✔

2. A square is a parallelogram, a rectangle, and a rhombus.

3. The diagonals of a square bisect each other, are congruent, and are perpendicular.

YOU WILL NEED

ruler and protractor
OR
geometry graphics software

CHECKPOINT ✔

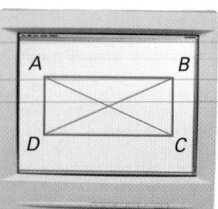

Activity 3
Rectangles

1. Draw a rectangle that is not a square. Draw diagonals to connect the vertices. Make measurements as you did in Activity 1.

2. Do the conjectures you made about parallelograms in Activity 1 seem to be true for rectangles? Discuss why they should or should not be true for rectangles. Fill in the blank in the conjecture below.

Conjecture: Property of Rectangles

A rectangle is a ____?____.

3. What new conjecture can you make about rectangles that is not true for all parallelograms? Fill in the blanks in the conjecture below.

Conjecture: Property of Rectangles

The diagonals of a rectangle ____?____.

Activity 4
Squares

CHECKPOINT ✔

1. Draw a square with diagonals connecting the vertices. Make measurements as in Activity 1.

2. What conjectures that you made in the preceding Activities seem to be true for squares? Discuss why they should or should not be true for squares. Fill in the blank in the conjecture below.

Conjecture: A Property of Squares

A square is a ____?____, a ____?____, and a ____?____.

3. Fill in the blank in the conjecture below.

Conjecture: A Property of Squares

The diagonals of a square ____?____ each other,
are ____?____, and are ____?____.

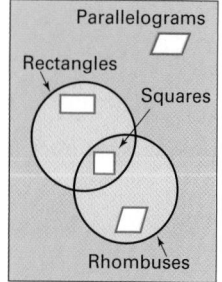

The Inheritance of Properties

Is a square a rectangle? Is a square a rhombus? In the Activities, you may have made some conjectures about these questions. The Euler diagram at left shows relationships among the classes of quadrilaterals. You will be asked to prove these relationships in Chapter 4.

Certain regions of the Euler diagram "inherit" properties from the larger regions in which they are located. Explain what is meant by the *inheritance of properties*. Illustrate your answer with examples from the conjectures you made in the Activities.

Parallelograms
Rectangles
Squares
Rhombuses

Inclusion Strategies

HANDS-ON STRATEGIES Students can use thin strips of tagboard to make quadrilaterals of different sizes. You should precut the tagboard and punch holes in the ends of each strip. Students can put paper clips through the holes at the ends of four strips to create the sides of a quadrilateral. These nonrigid models can aid in the discussion of the Euler diagram on this page.

Reteaching the Lesson

USING VISUAL MODELS Draw congruent acute angles with congruent sides on each of two acetate sheets and use the overhead projector to show how the angles can be positioned to illustrate general parallelograms and rhombuses. Use a pair of right angles to discuss rectangles and squares.

Exercises

Communicate

1. Explain the relationship between a rhombus and a square.

2. Explain the relationship between a rectangle and a parallelogram.

3. **OPTICAL ILLUSIONS** The drawing at right, composed of three rhombuses, forms an optical illusion. What does the drawing appear to be? Can you see the illusion more than one way? Try drawing your own optical illusions with rhombuses, parallelograms, or trapezoids.

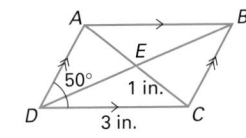

4. An alternative definition of a trapezoid is "a quadrilateral with at least one pair of parallel sides." How would this definition affect the classification of quadrilaterals?

Guided Skills Practice

5. Use your conjectures from Activity 1 and the diagram at right to find the measurements below. **(ACTIVITY 1)**

   a. *AB* **3 in.**   b. m∠*ABC* **50°**

   c. *AE* **1 in.**   d. m∠*BCD* **130°**

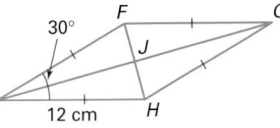

6. Use your conjectures from Activity 2 and the diagram at right to find the measurements below. **(ACTIVITY 2)**

   a. *FG* **12 cm**   b. m∠*FGH* **30°**

   c. m∠*GHI* **150°**   d. m∠*FJG* **90°**

7. Use your conjectures from Activity 3 and the diagram at right, in which *KO* is 2 ft, to find *LN*. **(ACTIVITY 3) 4 ft.**

8. Use your conjectures from Activity 4 and the diagram at right to find m∠*PTQ* and m∠*QTR*. **(ACTIVITY 4)**
   m∠*PTQ* = m∠*QTR* = **90°**

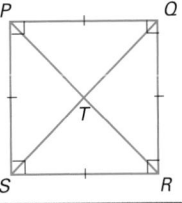

Math
CONNECTION

ALGEBRA For Exercises 33 and 34, students use the properties of parallelograms to set up equations and solve for *x*. Exercise 34 requires some review of solving equations with square roots.

● *Practice and Apply*

For Exercises 9–32, use your conjectures from Activities 1–4 to find the indicated measurements.

In parallelogram *WXYZ*, *WX* = 10, *WZ* = 4, *WY* = 13, and m∠*WZY* = 130°

9. *YZ* 10   **10.** *XY* 4

11. *WV* 6.5   **12.** *VY* 6.5

13. m∠*WXY* 130°   **14.** m∠*XWZ* 50°

15. m∠*XYZ* 50°

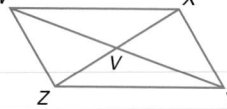

In rhombus *FGHI*, *FG* = 21, *FH* = 15, and m∠*FGH* = 70°.

16. *GH* 21   **17.** *HI* 21

18. *FJ* 7.5   **19.** *JH* 7.5

20. m∠*FIH* 70°   **21.** m∠*GFI* 110°

22. m∠*FJG* 90°   **23.** m∠*HJI* 90°

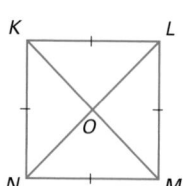

In rectangle *ABCD*, *AB* = 6, *AD* = 8, and *AC* = 10.

24. *CD* 6   **25.** *BC* 8

26. *BD* 10   **27.** *AE* 5

28. *BE* 5

In square *KLMN*, *KL* = 50 and *KM* ≈ 70.7.

29. *LM* 50   **30.** *LN* ≈ 70.7

31. m∠*KOL* 90°   **32.** m∠*LOM* 90°

Algebra

33. *x* = 60,
m∠*P* = m∠*R* = 120°
m∠*Q* = m∠*S* = 60°

33. In parallelogram *PQRS*, m∠*P* = (2*x*)° and m∠*Q* = *x*°. Find *x* and the measure of each angle in *PQRS*.

34. In rectangle *WXYZ*, diagonal *WY* = *x* − 2 and diagonal *XZ* = √*x*. Find *x* and the length of the diagonals of *WXYZ*. *x* = 4, *WY* = *XZ* = 2

☑ internet connect
Homework Help Online
Go To: go.hrw.com
Keyword:
MG1 Homework Help
for Exercises 35–41

Use the definitions of quadrilaterals and your conjectures from Activities 1–4 to decide whether each statement is true or false. If the statement is false, give a counterexample.

35. If a figure is a parallelogram, then it cannot be a rectangle. false; rectangle

36. If a figure is not a parallelogram, then it cannot be a square. true

37. If a figure is a parallelogram, then it cannot be a trapezoid. true

38. If a figure is a trapezoid, then it cannot be a rectangle. true

39. If a figure is a square, then it is a rhombus. true

40. If a figure is a rectangle, then it cannot be a rhombus. false; square

41. If a figure is a rhombus, then it cannot be a rectangle. false; square

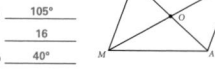
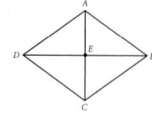
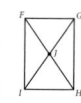

42. In Activity 1, you made the following conjecture: The opposite sides of a parallelogram are congruent. What is the converse of this statement? Do you think that the converse is true?

43. Use your answer from Exercise 42 to explain whether the following statement is true or false: If a figure is a rhombus, then it is a parallelogram.

44. GARDENING Susan is making a shelter for her tomatoes by stretching plastic over a wooden frame. Each wall of the frame is a rectangle, with diagonal braces added for support, as shown. If the brace connecting points A and C has a length of 73 in., how long is the brace connecting points B and D? **73 in.**

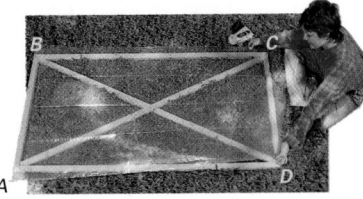

Look Back

Write each statement as a conditional and then write the converse of the conditional. Is the statement a definition? *(LESSON 2.3)*

45. All whales are mammals.

46. All squares are four-sided polygons.

47. All squares are rectangles.

For Exercises 48 and 49, refer to the figure at right. *(LESSON 3.1)*

48. 90°, 180°, 270° **48.** List all nontrivial degrees of the rotational symmetry for the figure.

49. Copy the figure and draw all of its axes of symmetry.

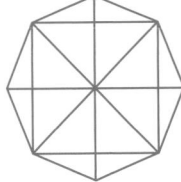

Look Beyond

50. In Activity 2, you made the following conjecture: The diagonals of a rhombus are perpendicular to each other. What is the converse of this statement? Do you think that the converse is true?

51. A figure with two pairs of congruent adjacent sides and opposite sides that are not congruent is called a **kite**. Draw a kite with diagonals connecting the vertices. How does this figure relate to your answer for Exercise 50?

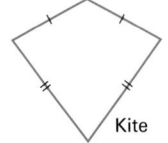
Kite

52. In Activity 3, you made the following conjecture: The diagonals of a rectangle are congruent. What is the converse of this statement? Do you think that the converse is true?

53. A trapezoid in which the nonparallel sides are congruent is called an **isosceles trapezoid**. Draw an isosceles trapezoid and the diagonals connecting the vertices. How does this figure relate to your answer for Exercise 52?

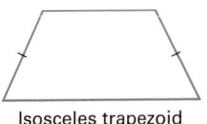
Isosceles trapezoid

---

47. If a figure is a square, then it is a rectangle. Converse: If a figure is a rectangle, then it is a square. This is not a definition because not all rectangles are squares.

49.

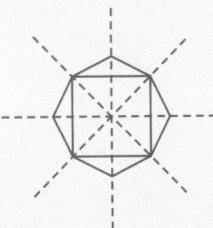

50. If the diagonals of a quadrilateral are perpendicular, then it is a rhombus. The converse is not true.

51.

Any kite has perpendicular diagonals, but is not a rhombus, thus showing the converse in Exercise 50 is not true.

Look Beyond

In Exercises 50–53, students explore the geometric properties of kites and trapezoids to make conjectures. They also establish the converses of two conjectures that they made in the lesson and determine whether they are true.

42. If opposite sides of a quadrilateral are congruent, then it is a parallelogram. The converse is true.

43. If a figure is a rhombus, then its opposite sides are congruent, so it must be a parallelogram. The statement is true.

45. If the creature is a whale, then it is a mammal. Converse: If the creature is a mammal, then it is a whale. This is not a definition because not all mammals are whales.

46. If a figure is a square, then it is a four-sided polygon. Converse: If a figure is a four-sided polygon, then it is a square. This is not a definition because not all four-sided polygons are squares.

Student Technology Guide

Student Technology Guide
3.2 *Properties of Quadrilaterals, page 1*

The sketch at the right shows quadrilateral *ABCD* and the measures of its sides and its angles. Notice that the quadrilateral shown has no particularly noteworthy properties.

When you drag a vertex of quadrilateral *ABCD* to another location to form a different quadrilateral, you *continuously deform* the original figure.

Note: If you need to deform quadrilateral *ABCD* so that two opposite sides, say $\overline{AD}$ and $\overline{BC}$, are parallel, select *B* and $\overline{AD}$. Construct a parallel line through *B*. Then drag *C* so that it is on this line.

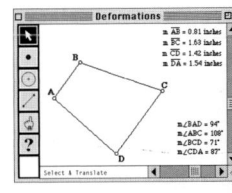

1. Sketch quadrilateral *ABCD*. Find and display the lengths of the sides and the measures of the angles. — Check students' sketches.

Use geometry graphics software to deform quadrilateral *ABCD* so that you know *by definition only* that it illustrates the specified figure. Explain your reasoning. On another paper, illustrate with a sketch.

See note above. Select

2. parallelogram ___ and $\overline{AB}$. Construct a parallel line through *D*. Drag *C* so that it is also on this line. Now opposite sides are parallel, so *ABCD* is a parallelogram by definition.

3. rectangle ___ To change the parallelogram into a rectangle, drag point *B* so that m∠*ABC* = 90°. Maintain all parallels drawn in Exercise 2.

4. rhombus ___ Maintaining all parallels, drag vertices so that $AB = BC = CD = DA$.

5. square ___ Maintaining all parallels, drag vertices so that $AB = BC = CD = DA$ and m∠*ABC* = 90°.

6. Matt wants to deform quadrilateral *ABCD* so that two sides are 1.5 inches long and the other two sides are 1.25 inches long. On another paper, sketch such a quadrilateral. Is the resulting figure unique? — The two illustrations in the answer key meet the given conditions. The quadrilateral is not unique.

Portfolio Activity

The Portfolio Activity can be used as preparation for the Chapter Project or as a separate activity. In the Portfolio Activity on this page, students learn how to draw Escher-type tessellations. Have students complete their tessellation designs on posterboard, and display them in the classroom.

The tessellation pattern by M. C. Escher shown below is an example of a **translation tessellation**. Each of the repeating figures is a translation of other figures in the design. You can make your own translation tessellation by following the steps below. You may have adjust your curves to get a pattern that you like.

Draw your figures on graph paper or tracing paper, or use geometry or tessellation software.

1. Start with a square, rectangle, or other parallelogram. Replace one side of the parallelogram with a curve, as shown.

2. Translate the curve to the opposite side of the parallelogram.

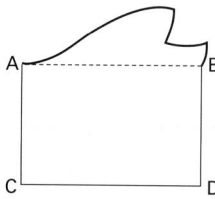

3. Repeat Steps 1 and 2 for the other two sides of your parallelogram.

4. Your figure will now fit together with itself on all sides. You can add details to your figure or divide it into two or more parts, as shown below. Translate the entire figure to create an interlocking design.

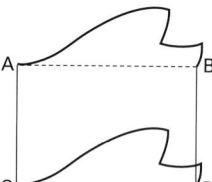

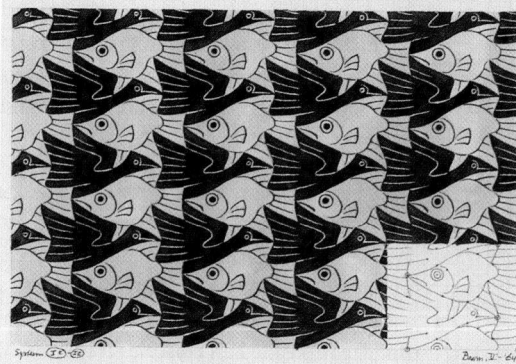

M. C Escher's "Symmetry Drawing E121" ©1999 Cordon Art B. V. -Baarn-Holand. All rights reserved.

Parallel Lines and Transversals

Why For practical as well as artistic reasons, parallel lines are often used in architecture and engineering. The striking appearance of the parallel cables that support the suspended portions of the Golden Gate Bridge contribute to the beauty of the structure.

Objectives

● Define *transversal, alternate interior angles, alternate exterior angles, same-side interior angles,* and *corresponding angles.*

● Make conjectures and prove theorems by using postulates and properties of parallel lines and transversals.

Prepare

QUICK WARM-UP

Refer to the diagram to answer the questions.

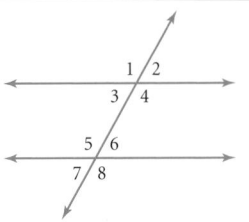

1. Is ∠4 congruent to ∠1? If so, what postulate or theorem explains why?
   Yes; vertical angles are congruent (Vertical Angle Theorem).

2. If ∠5 is congruent to ∠1, what can you conclude about ∠1 and ∠8?
   ∠8 is congruent to ∠1.

3. What is the relationship between ∠3 and ∠4?
   They are a linear pair and are supplementary.

Also on Quiz Transparency 3.3

Teach

Why Relationships between the angles formed when parallel lines are cut by a transversal are used in the construction of roads, railroads and runways to ensure that lines are parallel.

Transversals and Special Angles

When parallel lines are taken by themselves, it is hard to imagine how they can be studied. Look, for example, at the parallel lines at right. What conjectures could you be expected to make about them?

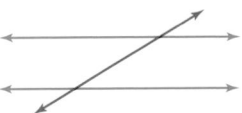

There is not much to study.

Now look at the figure with a third line intersecting the given parallel lines. There are now many discoveries to make. The line that intersects the two parallel lines is known as a *transversal*.

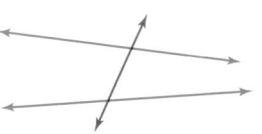

The transversal changes the picture.

Definition: Transversal

A transversal is a line, ray, or segment that intersects two or more coplanar lines, rays, or segments, each at a different point.

3.3.1

Notice that, according to the definition, the lines, rays, or segments that are cut by the transversal do not have to be parallel. This will be important in Lesson 3.4.

Alternative Teaching Strategy

COOPERATIVE LEARNING Have groups of 3 or 4 students draw parallel lines cut by a transversal on posterboard. Have each group investigate the following pairs of angles by measuring them with a protractor: alternate interior, alternate exterior, consecutive interior, and corresponding. For each type of angle pair, have each group develop a conjecture that completes this statement: If two parallel lines are cut by a transversal, then _____. Groups should write their conjectures on their posters.

Activity Notes

In this Activity, students look at pairs of angles formed when two parallel lines are cut by a transversal. After using a protractor or geometry graphics software to measure the angles, students are asked to make four conjectures about four classes of angles.

For a student worksheet of this Activity and detailed Teacher Notes, see page 41 in the Lesson Activities booklet.

Teaching Tip

Ask students to name the type of angles represented by ∠2 and ∠8. Point out that angles such as ∠3 and ∠5 can be called either consecutive interior angles or same-side interior angles.

CHECKPOINT ✔

5. Alternate interior angles are congruent. Alternate exterior angles are congruent. Same-side interior angles are supplementary. Corresponding angles are congruent.

Activity
Special Angle Relationships

YOU WILL NEED

paper, ruler, and protractor

OR

geometry graphics software

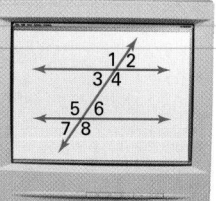

In each step of this Activity, the terms *interior* and *exterior* will be used as shown at right.

Start with two parallel lines. (If you are using lined paper, select two horizontal lines on the paper.) Then draw a third line that intersects both of the lines. Number each of the eight angles that are formed, as shown at left.

Exterior
Interior
Exterior

There are traditional names for certain special pairs of angles in the figure you drew. Measure each of the angles in the special pairs defined below. In Step 5, you will make a conjecture about each pair of angles.

1. Angles 3 and 6 are **alternate interior angles**, as are angles 4 and 5.

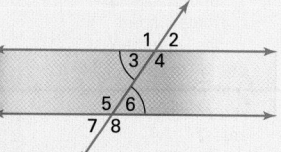

2. Angles 1 and 8 are **alternate exterior angles**. Name another pair of alternate exterior angles.

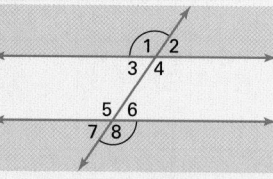

3. Angles 3 and 5 are **same-side interior angles**. Name another pair of same-side interior angles.

4. Angles 1 and 5 are **corresponding angles**. Name three other pairs of corresponding angles.

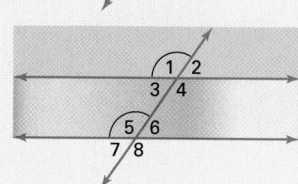

CHECKPOINT ✔

5. Fill in the blanks in the conjectures below using words from this list: congruent, complementary, supplementary.

Conjectures

For two parallel lines cut by a transversal:

Alternate interior angles are ____?____.

Alternate exterior angles are ____?____.

Same-side interior angles are ____?____.

Corresponding angles are ____?____.

Interdisciplinary Connection

PHYSICS Knowledge of the optical properties of different materials and of the angle relationships formed by parallel rays is essential in the design of precision optical instruments such as cameras, binoculars, and telescopes.

EXAMPLE ● Indicate whether the pairs below are alternate interior, alternate exterior, same-side interior, or corresponding angles.

a. ∠1 and ∠8
b. ∠7 and ∠3
c. ∠5 and ∠4
d. ∠3 and ∠5

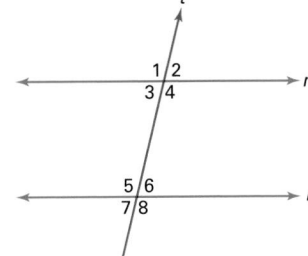

a. alternate exterior angles
b. corresponding angles
c. alternate interior angles
d. same-side interior angles

TRY THIS List three special pairs of angles not mentioned in the Example above.

One Postulate and Three Theorems

In the Activity on the previous page, one of the four conjectures you may have made is that corresponding angles are congruent. Notice what happens if you slide (translate) one of the parallel lines closer to the other. Eventually, the indicated corresponding angles will overlap. Do you think the corresponding angles will match exactly? Does this diagram seem to support your conjecture?

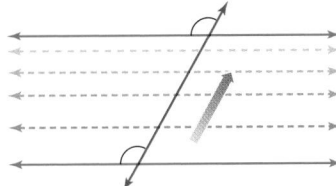

Move the parallel lines together.

The conjecture about corresponding angles will not be proven in this book, but because it seems obvious, it will be given as a postulate.

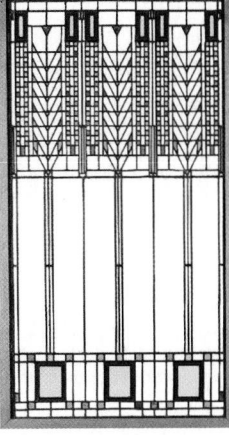

Frank Lloyd Wright, *Tree of Life*, 1904. Leaded glass. 41.5 × 26.75 in. Milwaukee Art Museum, Gift of Frederick Layton Art League in memory of Miss Charlotte Partridge and Miss Miriam Frink. © The Frank Lloyd Wright Foundation, Scottsdale, AZ.

Corresponding Angles Postulate

If two lines cut by a transversal are parallel, then corresponding angles are congruent.   **3.3.2**

Because the corresponding angles conjecture has been given as a postulate, it can be used to prove theorems. In particular, you can use it to prove the other three conjectures that you made in the Activity about parallel lines and transversals on the previous page.

ADDITIONAL
EXAMPLE ❶

Indicate whether the pairs of angles below are alternate interior, alternate exterior, same-side interior, or corresponding.

a. ∠1 and ∠5
corresponding

b. ∠3 and ∠6
alternate interior

c. ∠4 and ∠8
corresponding

d. ∠2 and ∠7
alternate exterior

e. ∠4 and ∠6
same-side interior

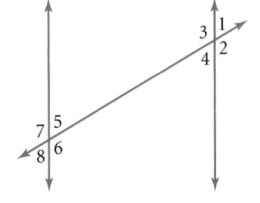

TRY THIS
same-side exterior, vertical angles, linear pair

Teaching Tip

Ask students if there is a pair of corresponding angles, other than ∠1 and ∠3, that they could use to prove that ∠1 is congruent to ∠2.

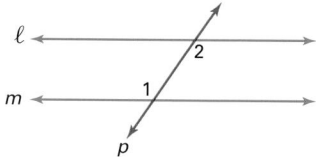

To prove your conjecture about alternate interior angles, begin by drawing a figure in which two parallel lines are cut by a transversal. Show that two alternate interior angles, such as ∠1 and ∠2, are congruent.

> The symbol ∥ means "is parallel to."

Given: ℓ ∥ m
Line p is a transversal.

Prove: ∠1 ≅ ∠2

Plan: Study the figure for ideas, and plan your strategy.

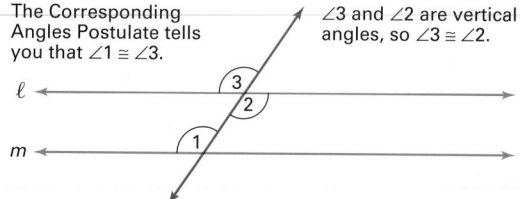

The Corresponding Angles Postulate tells you that ∠1 ≅ ∠3.

∠3 and ∠2 are vertical angles, so ∠3 ≅ ∠2.

From the information in the figure you know the following:

∠1 ≅ ∠3 and ∠3 ≅ ∠2

Thus, ∠1 ≅ ∠2 (the desired result).

What postulate or property allows you draw the final conclusion? (Recall the properties of congruence.)

Now you can write your proof. In proofs of this nature, you may find it convenient to use a two-column format.

> In writing proofs in geometry, you may wish to develop abbreviations for the reasons. For example, an abbreviation for the reason in Step 2 is "∥s ⇒corr. ∠s ≅."

Proof:

Statements	Reasons
1. Line ℓ is parallel to line m. Line p is a transversal.	Given
2. ∠1 ≅ ∠3	If parallel lines are cut by a transversal, then corresponding angles are congruent.
3. ∠3 ≅ ∠2	Vertical angles are congruent.
4. ∠1 ≅ ∠2	Transitive Property of Congruence

The conjecture can now be stated as a theorem.

Alternate Interior Angles Theorem

If two lines cut by a transversal are parallel, then alternate interior angles are congruent.

3.3.3

In the exercise set, you will be asked to prove the two remaining conjectures from the Activity.

Reteaching the Lesson

USING DISCUSSION   Point out that the Corresponding Angles Postulate states that for two parallel lines cut by a transversal, all four pairs of corresponding angles are congruent. Have a class discussion to show that if a single pair of corresponding angles is congruent, then the other three pairs of corresponding angles are also congruent.

Exercises

Communicate

The John Hancock Center, Chicago, Illinois, is 1127 feet tall.

1. ARCHITECTURE Describe all of the transversals that you can find in the photo of the John Hancock building at left.

2. In the diagram at right, what type of angle pair are ∠1 and ∠2? What can you say about them? Name three more angle pairs of this type. What type of angle pair are ∠1 and ∠4? What can you say about them? Name three more angle pairs of this type.

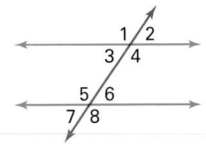

3. In the diagram above, what might you call angles ∠1 and ∠7? What do you think is true of this angle pair? Name another angle pair of this type.

4. For each diagram below, describe all of the transversals or explain why there are no transversals.

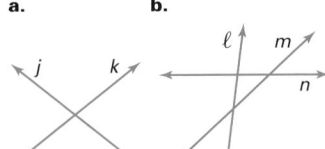

a.   b.   c.   d.

Guided Skills Practice

For Exercises 5–8, refer to the diagram below. List two angle pairs of each indicated type. *(ACTIVITY AND EXAMPLE)*

5. ∠3 and ∠6; ∠4 and ∠5 **5.** alternate interior angles

6. ∠1 and ∠8; ∠2 and ∠7 **6.** alternate exterior angles

7. ∠3 and ∠5; ∠4 and ∠6 **7.** same-side interior angles

9. ∠4, ∠5, ∠8, **8.** corresponding angles

10. ∠3, ∠6, ∠7   **Sample answer:** ∠1 and ∠5; ∠2 and ∠6

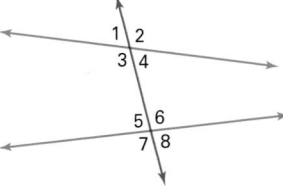

Practice and Apply

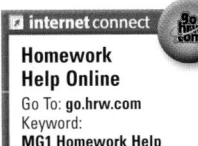
In the figure at right, lines *m* and *n* are parallel.

9. List all angles that are congruent to ∠1.

10. List all angles that are congruent to ∠2.

11. Are there any angles in the figure that are not congruent to ∠1 or to ∠2? Explain.

12. If m∠1 = 130°, find the measure of each angle in the figure.

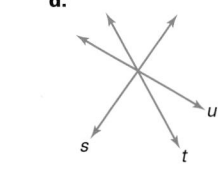

12. m∠1 = m∠4 = m∠5 = m∠8 = 130°;
m∠2 = m∠3 = m∠6 = m∠7 = 50°

Selected Answers
Exercises 5–8, 9–51 odd

ASSIGNMENT GUIDE

In Class	1–8
Core	9–12, 13–33 odd, 34–36
Core Plus	13–35 odd, 37–44
Review	45–51
Preview	52–53

✐ Extra Practice can be found beginning on page 818.

Error Analysis

The transversals in the diagram for Exercises 13–21 may cause some students to make errors in classifying the angle pairs.

11. No; every angle either is vertical to, corresponds to, or is vertical to a corresponding angle of ∠1 or ∠2. Hence every angle is congruent to ∠1 or ∠2.

Technology

Students may use geometry graphics software to complete Exercises 9–27.

13. ∠1 ≅ ∠6 by the Vertical Angles Theorem. ∠1 ≅ ∠9 by the Corresponding Angles Postulate. ∠9 ≅ ∠14 by the Vertical Angle Theorem, so ∠1 ≅ ∠14 by the Transitive Property of Congruence.

14. ∠2 ≅ ∠5 by the Vertical Angles Theorem.

15. ∠3 ≅ ∠4 by the Vertical Angles Theorem. ∠3 ≅ ∠8 by the Corresponding Angles Postulate. ∠8 ≅ ∠11 by the Vertical Angles Theorem, so ∠3 ≅ ∠11 by the Transitive Property of Congruence.

Math
CONNECTION

ALGEBRA For Exercises 28–33, students will use properties of special angles to set up equations and solve for x. They will need to know how to solve multistep equations.

For Exercises 13–17, refer to the diagram below. Lines *p* and *q* are parallel. Name all angles congruent to the given angle, and give the theorems or postulates that justify your answer.

13. ∠1   **14.** ∠2   **15.** ∠3

16. the angle formed by ∠1 and ∠2

17. the angle formed by ∠2 and ∠3

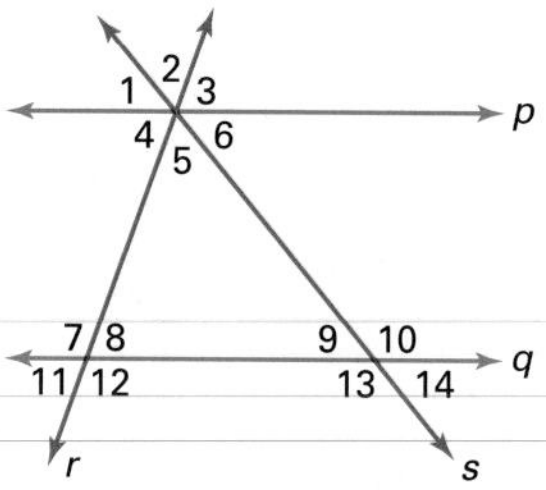

Refer again to the diagram at right. Determine whether the indicated line is a transversal. If so, identify two lines that it intersects.

18. line *p*   **19.** line *q*

20. line *r*   **21.** line *s*

In △*ABC*, $\overline{DE} \parallel \overline{CB}$, and ∠*ADE* ≅ ∠*AED*. Find the indicated angle measures.

22. m∠*ADE* 50°   **23.** m∠*AED* 50°

24. m∠*DEB* 130°   **25.** m∠*BDE* 25°

26. m∠*CDB* 105°   **27.** m∠*ABD* 25°

$\ell_1 \parallel \ell_2$ $\ell \parallel$

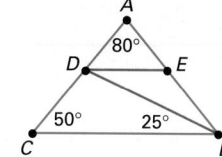

Algebra

In the diagram below, $\ell_1 \parallel \ell_2$, m∠2 = x°, and m∠6 = (3x − 60)°. Find the indicated angle measures.

28. m∠2 30°   **29.** m∠1 150°

30. m∠3 30°   **31.** m∠5 150°

32. m∠7 30°   **33.** m∠8 150°

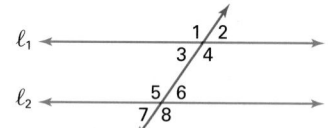

Write a two-column or paragraph proof for each theorem below.

34.
> ### Alternate Exterior Angles Theorem
> If two lines cut by a transversal are parallel, then alternate exterior angles are congruent.   **3.3.4**

35.
> ### Same-Side Interior Angles Theorem
> If two lines cut by a transversal are parallel, then same-side interior angles are supplementary.   **3.3.5**

36. In the diagram at right, lines ℓ and *m* are parallel. Write a two-column proof that the measure of each angle in the figure is 90°.

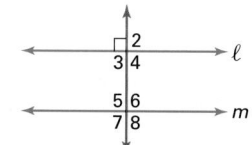

Practice

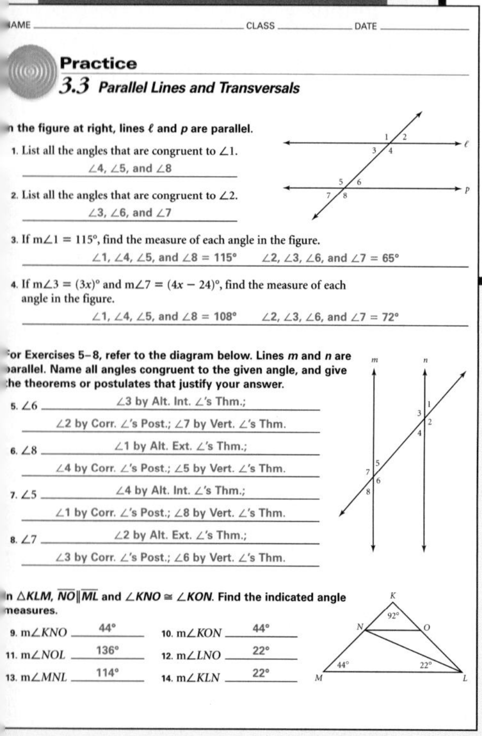

NAME _____ CLASS _____ DATE _____

Practice

3.3 Parallel Lines and Transversals

In the figure at right, lines ℓ and p are parallel.

1. List all the angles that are congruent to ∠1.
   ∠4, ∠5, and ∠8

2. List all the angles that are congruent to ∠2.
   ∠3, ∠6, and ∠7

3. If m∠1 = 115°, find the measure of each angle in the figure.
   ∠1, ∠4, ∠5, and ∠8 = 115°   ∠2, ∠3, ∠6, and ∠7 = 65°

4. If m∠3 = (3x)° and m∠7 = (4x − 24)°, find the measure of each angle in the figure.
   ∠1, ∠4, ∠5, and ∠8 = 108°   ∠2, ∠3, ∠6, and ∠7 = 72°

For Exercises 5–8, refer to the diagram below. Lines m and n are parallel. Name all angles congruent to the given angle, and give the theorems or postulates that justify your answer.

5. ∠6
   ∠3 by Alt. Int. ∠'s Thm.;
   ∠2 by Corr. ∠'s Post.; ∠7 by Vert. ∠'s Thm.

6. ∠8
   ∠1 by Alt. Ext. ∠'s Thm.;
   ∠4 by Corr. ∠'s Post.; ∠5 by Vert. ∠'s Thm.

7. ∠5
   ∠4 by Alt. Int. ∠'s Thm.;
   ∠1 by Corr. ∠'s Post.; ∠8 by Vert. ∠'s Thm.

8. ∠7
   ∠2 by Alt. Ext. ∠'s Thm.;
   ∠3 by Corr. ∠'s Post.; ∠6 by Vert. ∠'s Thm.

In △KLM, $\overline{NO} \parallel \overline{ML}$ and ∠KNO ≅ ∠KON. Find the indicated angle measures.

9. m∠KNO 44°   10. m∠KON 44°

11. m∠NOL 136°   12. m∠LNO 22°

13. m∠MNL 114°   14. m∠KLN 22°

16. By the Vertical Angles Theorem, the angle formed by ∠1 and ∠2 is congruent to the angle formed by ∠5 and ∠6. By the Corresponding Angles Postulate, the angle formed by ∠1 and ∠2 is congruent to ∠7. By the Vertical Angles Theorem, ∠7 ≅ ∠12, so by the Transitive Property of Congruence, ∠12 is congruent to the angle formed by ∠1 and ∠2.

17. By the Vertical Angles Theorem, the angle formed by ∠2 and ∠3 is congruent to the angle formed by ∠4 and ∠5; and by the Corresponding Angles Postulate, is congruent to ∠10. By the Vertical Angles Theorem, ∠10 ≅ ∠13, so by the Transitive Property of Congruence, ∠13 is congruent to the angle formed by ∠2 and ∠3.

18. not a transversal

19. transversal; intersects lines *r* and *s*

20. transversal; intersects lines *p* and *q* and lines *s* and *q*

21. transversal; intersects lines *p* and *q* and lines *r* and *q*

CARPENTRY In the diagram of a partial wall frame, the *ceiling joist*, $\overline{PQ}$, and *soleplate*, $\overline{RS}$, are parallel.

37. How is *corner brace* $\overline{PT}$ related to $\overline{PQ}$ and $\overline{RS}$?

38. How are ∠RTP and ∠QPT related?

39. How is $\overline{PT}$ related to the vertical beams that it crosses?

40. How are ∠1 and ∠2 related?

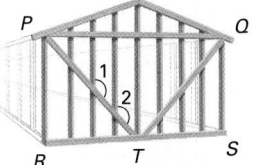

NAVIGATION A *periscope* is an instrument used on submarines to see above the surface of the water. A periscope contains two parallel mirrors that face each other.

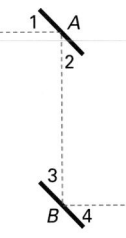

41. Identify a transversal in the diagram at left. What does this line represent?

42. Are ∠1 and ∠4 alternate exterior angles? Why or why not?

43. The angle at which light rays are reflected from a mirror is congruent to the angle they form with the mirror. For example, ∠1 ≅ ∠2. Prove that ∠1 ≅ ∠4.

44. Suppose that m∠1 = 45°. Find the measures of ∠2, ∠3, and ∠4.

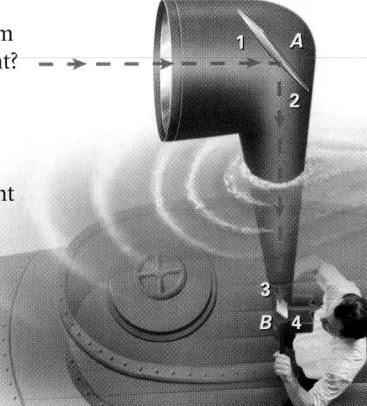

Look Back

45. ray

45. A ____?____ is a part of a line that starts at a point and extends without end in one direction. *(LESSON 1.1)*

46. perpendicular

46. Two lines that intersect to form a right angle are ____?____. *(LESSON 1.4)*

47. parallel

47. Two coplanar lines that do not intersect are ____?____. *(LESSON 1.4)*

48. vertical angles

48. ____?____ are the opposite angles formed by two intersecting lines. *(LESSON 2.5)*

49. polygon

49. A ____?____ is a plane figure formed from three or more segments such that each segment intersects exactly two other segments, one at each endpoint. *(LESSON 3.1)*

50. trapezoid

50. A ____?____ is a quadrilateral with exactly one pair of parallel sides. *(LESSON 3.2)*

51. rhombus

51. A ____?____ is a quadrilateral with four congruent sides. *(LESSON 3.2)*

Look Beyond

In the diagram at right, $\overline{AB} \parallel \overline{CD}$.

52. Write a two-column proof that the sum of the interior angles of trapezoid *ABCD* equals 360°.

53. Write a paragraph proof that the sum of the interior angles of any trapezoid, parallelogram, or rectangle is 360°.

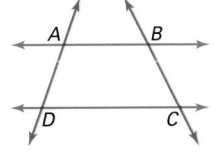

---

1. Write the converse of this conditional: If Rosalia is 18 years old, then she can vote. **If Rosalia can vote, then she is 18 years old.**

2. If two parallel lines are cut by a transversal, how many pairs of corresponding angles are formed? **4 pairs**

3. If two parallel lines are cut by a transversal, then what is the relationship between alternate interior angles? **They are congruent.**

Also on Quiz Transparency 3.4

Teach

Why Architects and contractors make extensive use of parallel lines in their plans and structures and require fail-safe ways of constructing them. Consequently, the geometry of parallel lines and the knowledge of how to prove that lines are parallel is a significant aspect of the work that these people do.

Proving That Lines Are Parallel

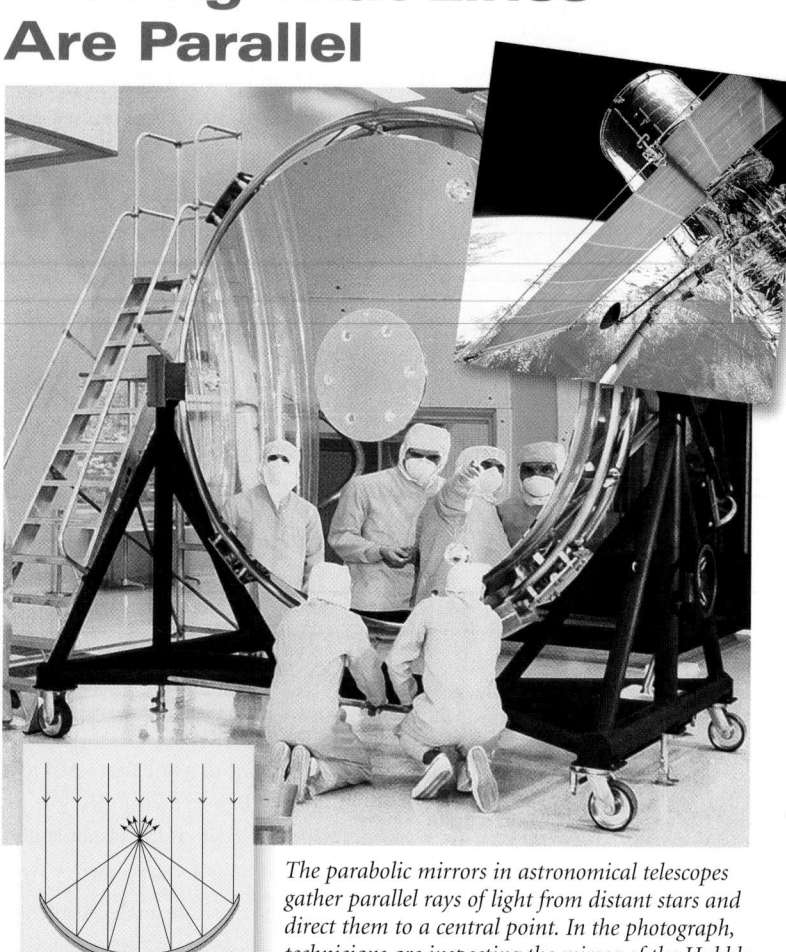

Objectives

● Identify and use the converse of the Corresponding Angles Postulate.

● Prove that lines are parallel by using theorems and postulates.

Why *Suppose that you needed to create a number of parallel lines or that you had to be sure that certain given lines were parallel. The converses of the transversal properties that you learned in Lesson 3.3 will enable you to do these things.*

The parabolic mirrors in astronomical telescopes gather parallel rays of light from distant stars and direct them to a central point. In the photograph, technicians are inspecting the mirror of the Hubble Space Telescope.

The Converses of the Transversal Properties

In Lesson 3.3, you studied a postulate and three theorems about parallel lines and transversals. In each of these, the parallel lines and a transversal were given, and conclusions were drawn about certain special angles. In this lesson, the process is reversed in order to write the converses of the postulate and theorems.

EXAMPLE **1** Write the converse of the Corresponding Angles Postulate.

● **SOLUTION**

Identify the hypothesis and the conclusion of the Corresponding Angles Postulate. Then interchange the hypothesis and the conclusion to form the converse.

Alternative Teaching Strategy

TECHNOLOGY Have students draw parallel lines cut by a transversal on their computer screens. Using the measuring tool of the geometry graphics software, have them measure each of the eight angles formed. Then have them drag the parallel lines, bringing them closer together or farther apart. Have them also change the position of the transversal. Students will see the angle measures change on the screen, but should notice that the angle relationships remain the same (for instance, corresponding angles are still congruent). Have them change the lines so that the angle relationships are not the same as before. They will see that when this happens, the lines are no longer parallel. Finally, ask them to find a case when all of the angles formed by the transversal are congruent. They should discover that this happens when the transversal is perpendicular to the lines.

Original statement: If two lines cut by a transversal are parallel, then corresponding angles are congruent.

Converse: If corresponding angles are congruent, then two lines cut by a transversal are parallel.

Another way of stating the converse is as follows:

Theorem: Converse of the Corresponding Angles Postulate

If two lines are cut by a transversal in such a way that corresponding angles are congruent, then the two lines are parallel.　　　**3.4.1**

Notice that the converse is labeled as a theorem. This is because it can be proved with the theorems and postulates you already know. However, the proof will be given later because it involves a special form of reasoning known as indirect proof, which you have not yet studied (see Lesson 12.4).

Using the Converses

In addition to the Converse of the Corresponding Angles Postulate, the converse of each of the other transversal theorems is also true. You will be asked to prove these converses in Exercises 18–25 by using Theorem 3.4.1.

EXAMPLE **2** Suppose that m∠1 = 64° and m∠2 = 64° in the figure at right. What can you conclude about lines ℓ and m?

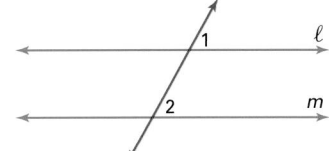

● **SOLUTION**

∠1 and ∠2 are corresponding angles.
The Converse of the Corresponding Angles Postulate states that if corresponding angles are congruent, then the lines are parallel. You can conclude that lines ℓ and m are parallel.

EXAMPLE **3** Given line ℓ and point P not on the line, draw a line through P that is parallel to ℓ.

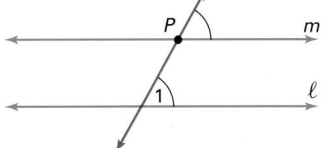

● **SOLUTION**

Draw a line through point P and line ℓ. Label and measure ∠1.

Using a protractor, draw a new line through P such that the new angle corresponds to ∠1 and has the same measure as ∠1. By the Converse of the Corresponding Angles Postulate, the new line, m, is parallel to line ℓ.

Note: You can also do this as a construction with a compass and a straightedge (see Exercises 32 and 33 in Lesson 4.7).

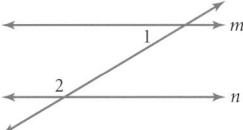

ASSIGNMENT GUIDE

In Class	1–7
Core	9–17 odd, 18–23, 26–37
Core Plus	18–40
Review	41–47
Preview	48–51

✎ Extra Practice can be found beginning on page 818.

Error Analysis

Tell students that citing a theorem or a postulate instead of its converse is, in many instances, an incorrect practice when trying to establish proofs.

5. If alternate interior angles are congruent, then the two lines cut by the transversal are parallel.

6. The indicated angles are corresponding and congruent, so the Converse of the Corresponding Angles Postulate states lines *m* and *n* are parallel.

7.

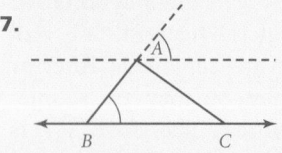

8. Alternate Exterior Angles Theorem

9. Alternate Exterior Angles Theorem

10. Alternate Interior Angles Theorem

11. Alternate Interior Angles Theorem

12. Corresponding Angles Postulate

13. Corresponding Angles Postulate

Exercises

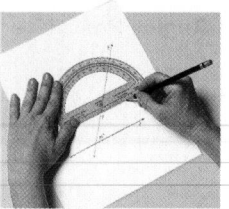

Communicate

For Exercises 1 and 2, state the converse of each theorem.

1. If two lines cut by a transversal are parallel, then corresponding angles are congruent.

2. If two lines cut by a transversal are parallel, then same-side interior angles are supplementary.

3. Explain how a theorem that is used to prove that lines are parallel can be used to develop a method for drawing parallel lines, as in the photo at left.

4. Explain why the lines in the diagram below are not parallel.

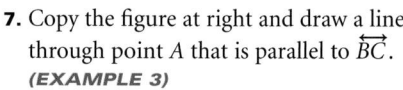

Guided Skills Practice

5. Write the Converse of the Alternate Interior Angles Theorem. *(EXAMPLE 1)*

6. Are lines *m* and *n* in the figure at right parallel? Why or why not? *(EXAMPLE 2)*

7. Copy the figure at right and draw a line through point *A* that is parallel to $\overleftrightarrow{BC}$. *(EXAMPLE 3)*

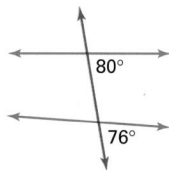

Practice and Apply

For Exercises 8–15, refer to the diagram at left, and fill in the name of the appropriate theorem or postulate.

8. If m∠1 = m∠7, then $\ell_1 \| \ell_2$ by the Converse of the ____?____.

9. If m∠2 = m∠8, then $\ell_1 \| \ell_2$ by the Converse of the ____?____.

10. If m∠4 = m∠6, then $\ell_1 \| \ell_2$ by the Converse of the ____?____.

11. If m∠3 = m∠5, then $\ell_1 \| \ell_2$ by the Converse of the ____?____.

12. If m∠2 = m∠6, then $\ell_1 \| \ell_2$ by the Converse of the ____?____.

13. If m∠3 = m∠7, then $\ell_1 \| \ell_2$ by the Converse of the ____?____.

14. If m∠4 + m∠5 = 180°, then $\ell_1 \| \ell_2$ by the Converse of the ____?____.

15. If m∠3 + m∠6 = 180°, then $\ell_1 \| \ell_2$ by the Converse of the ____?____.

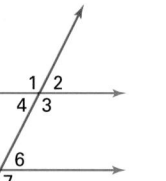

14. Same-Side Interior Angles Theorem

15. Same-Side Interior Angles Theorem

16. Write a two-column proof that $\overline{DF} \parallel \overline{GH}$.

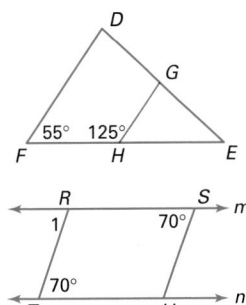

17. Lines m and n in the figure at right are parallel. Write a two-column proof that quadrilateral $RSUT$ is a parallelogram.

Technology

Students may wish to use geometry graphics software to investigate the theorems presented in the exercises.

For Exercises 18–24, complete the two-column proof of the following theorem:

Converse of the Same-Side Interior Angles Theorem

If two lines are cut by a transversal in such a way that same-side interior angles are supplementary, then the two lines are parallel.

3.4.2

Given: $\angle 1$ and $\angle 2$ are supplementary.

Prove: $\ell_1 \parallel \ell_2$

Proof:

Statements	Reasons
$\angle 1$ and $\angle 2$ are supplementary.	**18.** ?
$m\angle 1 + m\angle 2 = 180°$	**19.** ?
$m\angle 1 + m\angle 3 = 180°$	**20.** ?
$m\angle 1 + m\angle 2 = m\angle 1 + m\angle 3$	**21.** ?
$m\angle 2 = m\angle 3 \ (\angle 2 \cong \angle 3)$	**22.** ?
$\ell_1 \parallel \ell_2$	**23.** ?

Write two-column proofs of the indicated theorems.

24. ### Converse of the Alternate Interior Angles Theorem

If two lines are cut by a transversal in such a way that alternate interior angles are congruent, then the two lines are parallel.

3.4.3

25. ### Converse of the Alternate Exterior Angles Theorem

If two lines are cut by a transversal in such a way that alternate exterior angles are congruent, then the two lines are parallel.

3.4.4

16. Given: $m\angle DFH = 55°$; $m\angle GHF = 125°$

$\angle DFH$ and $\angle GHE$ are corresponding angles.

Prove: $\overline{DF} \parallel \overline{GH}$

1. $m\angle DFH = 55°$; $m\angle GHF = 125°$
   $\angle DFH$ and $\angle GHE$ are corresponding angles.
2. $m\angle FHG + m\angle GHE = 180°$
3. $125° + m\angle GHE = 180°$
4. $m\angle GHE = 55°$
5. $\angle DFH \approx \angle GHE$
6. $\overline{DF} \parallel \overline{GH}$

1. Given
2. Linear Pair Property
3. Substitution Property
4. Subtraction Property
5. Angle Congruence Postulate
6. Converse of the Corresponding Angles Postulate.

38. Using the T square as the transversal, each line drawn with the triangle will have congruent corresponding angles to every other line drawn with the triangle. Thus, by the Converse of the Corresponding Angles Postulate, the lines will be parallel.

39. Paint lines that have congruent corresponding angles to the first line. By the Converse of the Corresponding Angles Postulate, each of these lines will be parallel to the first.

For Exercises 26–37, complete the paragraph proofs of the theorems in the boxes below.

Theorem

If two coplanar lines are perpendicular to the same line, then the two lines are parallel to each other. **3.4.5**

PARAGRAPH PROOFS

26. transversal

27. both are right angles

28. line *m*

29. line *p*

30. Converse of Corresponding Angles Postulate

31. ∠3

32. Corresponding Angles Postulate

33. ∠2

34. Corresponding Angles Postulate

35. ∠2

36. Transitive Property of Congruence or Substitution

37. Converse of Corresponding Angles Postulate

APPLICATIONS

Given: $m \perp \ell$ and $p \perp \ell$

Prove: $p \| m$

Proof:

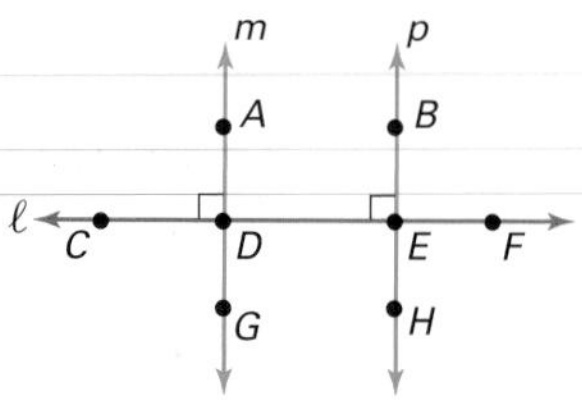

Line ℓ is a **26.** _____?_____ of m and p, by definition. ∠ADC ≅ ∠BED because **27.** _____?_____. Therefore, **28.** _____?_____ is parallel to **29.** _____?_____ by **30.** _____?_____.

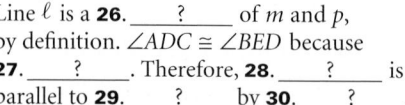

Theorem

If two coplanar lines are parallel to the same line, then the two lines are parallel to each other. **3.4.6**

Given: $\ell \| n$, $m \| n$, and p is a transversal of ℓ, m, and n.

Prove: $\ell \| m$

Proof:

Because $\ell \| n$, ∠1 ≅ **31.** _____?_____ by **32.** _____?_____. Because $m \| n$, then ∠3 ≅ **33.** _____?_____ by **34.** _____?_____. Thus, ∠1 ≅ **35.** _____?_____ by **36.** _____?_____, and $\ell \| m$ by **37.** _____?_____.

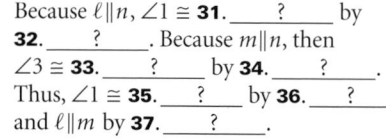

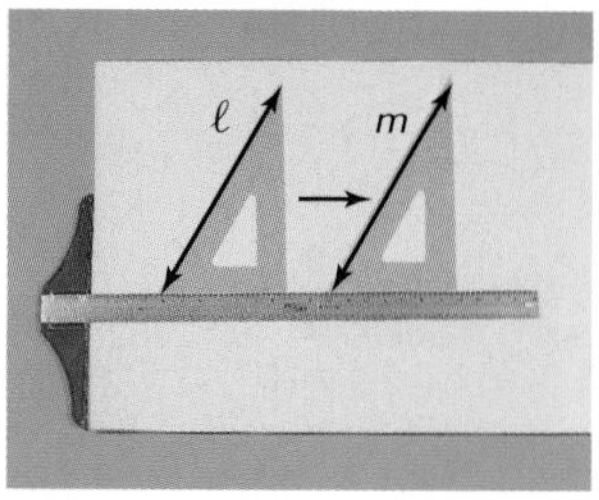

38. DRAFTING A T square and a triangle can be used to draw parallel lines. While holding the T square in place, slide the triangle along the T square as shown. How can you prove that the resulting lines, ℓ and m, are parallel?

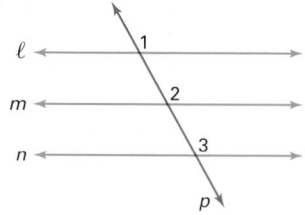

39. CIVIL ENGINEERING Suppose that you are painting lines for angled parking in a parking lot. How could you make sure that the lines are parallel?

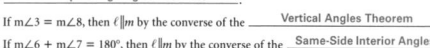
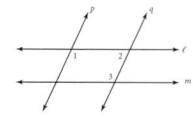

40. CARPENTRY A *plumb bob* is a weight hung at the end of a string, called a *plumb line*. The weight pulls the string straight down so that the plumb line is perfectly vertical. Suppose that the angle formed by the wall and the roof is 120° and the angle formed by the plumb line and the roof is 120°. Explain why this shows that the wall is vertical.

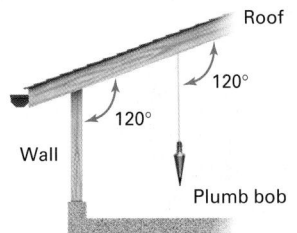

Roof

120°

120°

Wall

Plumb bob

Plumb bob

Look Back

For Exercises 41–45, refer to the following statement: Every rectangle is a parallelogram. *(LESSON 2.2)*

41. Rewrite the statement as a conditional.

42. Identify the hypothesis and conclusion of the statement.

43. Draw an Euler diagram that illustrates the statement.

44. Write the converse of the statement, and construct its Euler diagram.

45. Is the converse true or false? If it is true, write a paragraph proof of it. If it is false, disprove it with a counterexample.

46. NAVIGATION A pilot is flying at a compass heading of 155. What is the heading of a pilot flying in the opposite direction? *(LESSON 1.3)* 335

47. NAVIGATION A pilot is flying at a compass heading of 235. What are the two possible headings of a pilot flying perpendicular to the first pilot? *(LESSON 1.3)* 145 and 325

Look Beyond

Exercises 48 and 49 refer to the photo of railroad tracks shown below.

48. How could you prove that the railroad tracks are parallel?

49. In the photo, the lines appear to meet at the horizon. Why? Some people use the expression "meet at infinity" when referring to parallel lines. What does this mean? Does it make sense?

50. They are parallel because they are opposite edges of a cube.

51. They would intersect at a single point.

50. The drawing at right represents a cube. What is true of the lines that contain the edges shown in red?

51. What would happen if the red edges in the drawing were extended?

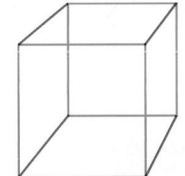

43.

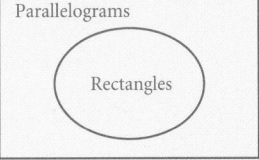

Parallelograms

Rectangles

44. Converse: If a figure is a parallelogram, then it is a rectangle.

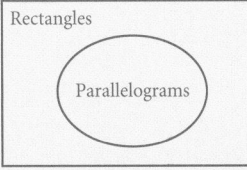

Rectangles

Parallelograms

45. It is false. For example, a rhombus is a parallelogram, but it is not necessarily a rectangle.

48. By showing that both tracks are perpendicular to the railroad ties. This would guarantee that the tracks are parallel.

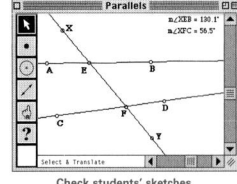

Focus

The design of a 25-foot-long metal egg highlights the difficulty in tessellating curved surfaces and illustrates how mathematical models can be adapted to fit economic and physical constraints.

Motivate

After students read the article, ask questions such as, "What unexpected problems did Resch encounter in designing the egg? How did economic considerations affect Resch's mathematical model for the egg?" As an additional project, have students paint a hollow or boiled egg with a pattern similar to Resch's egg. Have them describe any difficulties they had in copying the pattern onto the egg.

EYEWITNESS MATH

Egg Over Alberta

by Paul Hoffman

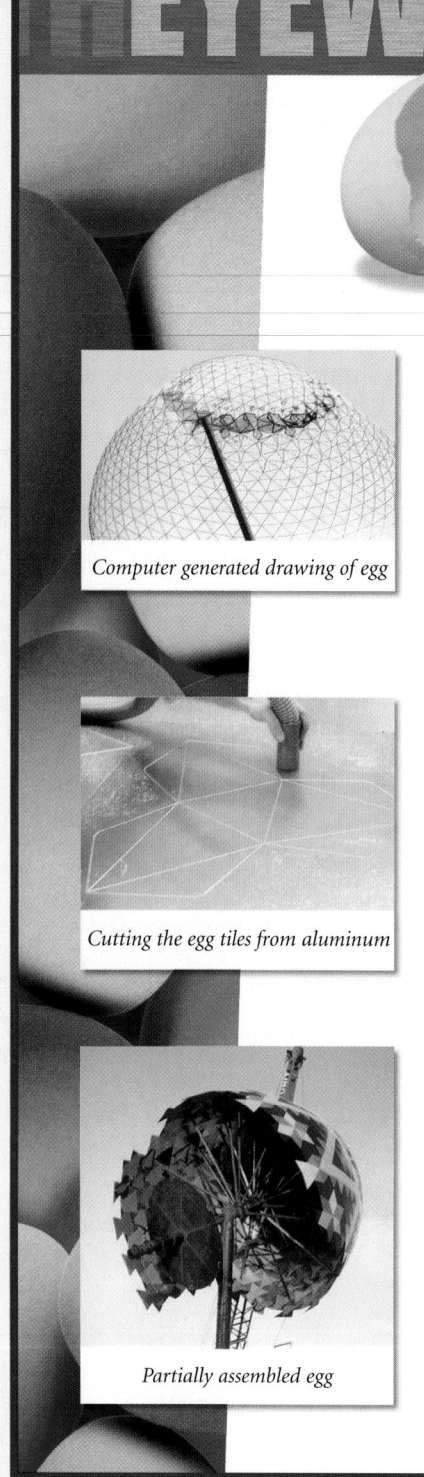

Computer generated drawing of egg

Cutting the egg tiles from aluminum

Partially assembled egg

The town leaders of Vegreville, Alberta, contacted Dale Resch, a computer science professor, for a special project—to build a 31-foot Easter egg.

The problem Resch faced was that no one other than a chicken had ever built a chicken egg. With no formal training in mathematics, Resch relied on his ability to play with geometric abstractions in his mind, then with his hands or a computer, to turn those abstractions into physical reality.

Resch assumed that someone had developed the mathematics of an ideal chicken egg. He soon found, however, that there was no formula for an ideal chicken egg.

After four months of contemplation and simulation, Resch realized that he could tile the egg with 2,208 equilateral triangles and 524 three-pointed stars (equilateral but non-regular hexagons) that varied slightly in width, depending on their position on the egg.

For six weeks, Resch led a team of volunteers in assembling the egg. Residents were afraid it might blow down. Long after the egg was finished, Resch used a computer to analyze the egg's structural integrity and found that it was ten times stronger than it needed to be.

Many Ukrainians still practice the traditional art of painting pysanki, *eggs decorated with colorful geometric patterns.*

Cooperative Learning

Use folding paper and a compass or geometry graphics software for the following explorations.

1. The tiles on the Vegreville egg have two shapes; equilateral triangles, and three-pointed stars, which are equilateral, but not regular, hexagons. The three-pointed stars may be created as follows:

Draw or fold an equilateral triangle. Fold the perpendicular bisectors of the sides to find the center of the triangle. Place the point of your compass at the center of the triangle and draw a circle that completely encloses the triangle. The radius of the circle should be at least twice the distance from the center to a vertex of the triangle.

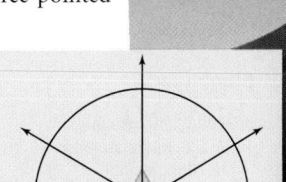

Draw segments from each vertex to the points where the perpendicular bisectors of the sides intersect the circle to form an isosceles triangle on each side of the equilateral triangle. The figure formed by the equilateral triangle plus the three isosceles triangles is a three-pointed star. Explain why the three-pointed star is equilateral.

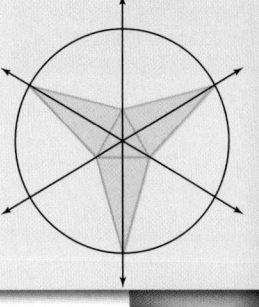

2. What are the axes of symmetry of the three-pointed star that you created in Step 1? What type of rotational symmetry does it have?

3. Draw an equilateral triangle on each side of your three-pointed star. Draw three new three-pointed stars in the gaps between the equilateral triangles. Draw an equilateral triangle on each side of these three-pointed stars, and draw three-pointed stars in the gaps between these triangles. Continue drawing three-pointed stars and equilateral triangles in this way until you understand the pattern of shapes in a plane. How is your pattern like the pattern of tiles on the Vegreville egg? (The pattern of tiles is easiest to see near the top and bottom of the egg.) How is it different?

Cooperative Learning

If students are using a computer to complete the Cooperative Learning activities, have them work in pairs. If they are using folding paper or paper and compass, have them work in groups of no more than four, with students performing the constructions simultaneously.

The activities are sequenced to help students appreciate the difficulty of Resch's task. As each group proceeds through the steps (all students should do the constructions), have them discuss the questions, with one member of the group recording the group's answers.

Discuss

Discuss polygons that can tessellate a plane (such as triangles, rectangles, and regular hexagons) and those that cannot (such as regular pentagons and regular 11-gons). What properties do these polygons have? Do the measures of their angles divide into 360° evenly? Why is this important?

Refer to the following diagram for Exercises 1 and 2.

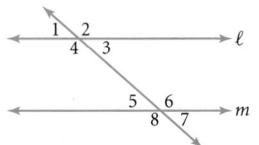

1. If ∠1 and ∠5 have different measures, are lines ℓ and m parallel? Why or why not?

   No; if lines ℓ and m were parallel, then ∠1 ∠5, which are corresponding angles, would have the same measure.

2. If   m∠3 + m∠6 = 180°, what can you conclude about lines ℓ and m?

   They are parallel.

Also on Quiz Transparency 3.5

Teach

Why Euclidean geometry is a useful model of the physical world on a small scale. However, the Euclidean model needs to be modified when considering a geometry that can serve as a model of Earth or the universe.

The Triangle Sum Theorem

Objective

- Identify and use the Parallel Postulate and the Triangle Sum Theorem.

Why Euclidean geometry was founded on five postulates or axioms. One of them, the Parallel Postulate, has been thoroughly investigated by modern mathematicians.

Three different geometries apply to the triangles drawn on the three different surfaces. The differences in the geometries are based on the Parallel Postulate.

The Parallel Postulate

•P

←————————→ ℓ

There is one and only one line through point P that is parallel to line ℓ.

The following postulate is a modern equivalent of Euclid's fifth postulate:

The Parallel Postulate

Given a line and a point not on the line, there is one and only one line that contains the given point and is parallel to the given line.

3.5.1

In Lesson 3.4 you drew a line through a given point and parallel to a given line. You probably never questioned whether such a line actually existed, or whether there could be more than one such line. The *assumption* that there is in fact exactly one such parallel line is known as the Parallel Postulate.

Alternative Teaching Strategy

TECHNOLOGY Students can investigate the sum of the measures of the angles of a triangle by using geometry graphics software. Have students draw a triangle, measure the angles, and find the sum of the angle measures. Then have them drag a point on the triangle to change its shape. As the triangle changes, they will be able to see that the sum of the angle measures will remain 180°. Students can then create a computer model of the drawing at the top of page 172 and use it to study the proof given.

They can extend this activity to include the Exterior Angle Theorem. Have them draw an exterior angle on the triangle and measure it. By changing the shape of the triangle, they will be able to see that the sum of the remote interior angles is equal to the measure of the exterior angle.

The Parallel Postulate is used to prove a theorem about the angles of a triangle. Before you look at the theorem, try the Activity below.

Activity

The Triangle Sum Theorem

1. Cut out a triangle from a piece of paper.

2. Tear two corners off the triangle.

3. Position the two torn-off corners next to the third angle as shown at right.

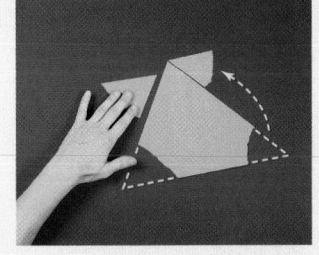

CHECKPOINT ✔ 4. Make a conjecture about the sum of the measures of the angles of a triangle.

5. In the diagram below, line ℓ has been drawn through a vertex of the triangle so that it is parallel to the opposite side. How does the Parallel Postulate relate to this figure?

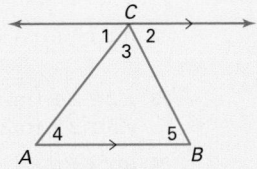

6. Fill in the table below for a figure like the one above. Use geometry theorems, not physical measurements, to find the answers.

m∠1	m∠2	m∠3	m∠4	m∠5	m∠3 + m∠4 + m∠5
40°	30°	?	?	?	?
20°	80°	?	?	?	?
30°	100°	?	?	?	?

7. Does the table support the conjecture you made in Step 4? Explain.

CRITICAL THINKING   Does your work in the Activity prove your conjecture? Why or why not?

The conjecture from the Activity is stated below as a theorem, and it is proven on the following page.

The Triangle Sum Theorem

The sum of the measures of the angles of a triangle is 180°.   **3.5.2**

Interdisciplinary Connection

PHYSICS   Non-Euclidean geometry plays an essential role in relativity theory. For some insights from a recent book for the general public, students may look up information on *geodesics* in *Black Holes and Time Warps: Einstein's Outrageous Legacy*, by Kip S. Thorne (W. W. Norton and Company, 1994).

Enrichment

Students can investigate what happens when they draw a line through each vertex of a triangle parallel to the opposite side. Ask students to formulate conjectures about the sides and angles in the resulting figure.

Proving the Triangle Sum Theorem

To prove the theorem, begin by drawing line ℓ through a vertex of the triangle so that it is parallel to the side opposite the vertex.

TWO-COLUMN PROOF

Given: $\triangle ABC$

Prove: $m\angle 1 + m\angle 2 + m\angle 3 = 180°$

Plan: Study the illustration, which is related to the Activity in which you tore off the corners of a triangle. You can use what you discovered in the Activity to write a two-column proof of the Triangle Sum Theorem.

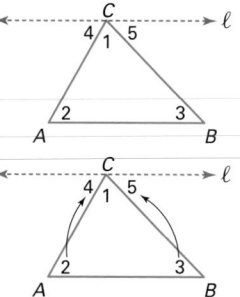

Proof:

Statements	Reasons
1. $\ell \parallel \overline{AB}$	As drawn (justification: the Parallel Postulate)
2. $m\angle 1 + m\angle 4 + m\angle 5 = 180°$	The angles fit together to form a straight line.
3. $\angle 2 \cong \angle 4$ $(m\angle 2 = m\angle 4)$ $\angle 3 \cong \angle 5$ $(m\angle 3 = m\angle 5)$	Alternate Interior Angles Theorem
4. $m\angle 1 + m\angle 2 + m\angle 3 = 180°$	Substitution in Step 2

Another Geometry

It is possible to create geometries in which the Parallel Postulate is not true. On the surface of a sphere, for example, lines are defined differently from the way they are defined on flat surfaces, and there are no parallel lines.

APPLICATION
GEOGRAPHY

On the surface of a sphere, a line is defined as a **great circle**, which is a circle that lies in a plane that passes through the center of the sphere. (A great circle divides a sphere into two equal parts.) The equator is a great circle on the surface of Earth. Lines of longitude, which run north and south, are also great circles. Notice that any two distinct lines (great circles) intersect at two points. Thus, there are no parallel lines on a sphere.

CRITICAL THINKING

Discuss the following statement: On the surface of a sphere, the shortest path between two points is not a straight line. What is the shortest path?

Inclusion Strategies

HANDS-ON STRATEGIES Students may need to actually measure the angles listed in the table on page 171. Using both the geometry theorems and their measurements, students should be convinced that their answers confirm and support their conjectures.

Reteaching the Lesson

USING VISUAL MODELS Redraw the diagram on page 172 that was used in the proof of the Triangle Sum Theorem. This time, however, construct a line through vertex B parallel to $\overline{AC}$. Use this new diagram to prove the theorem. In the construction step, review the Parallel Postulate. Students should see that any vertex can be used to prove the theorem.

Exercises

Assess

Selected Answers
Exercises 5–7, 9–49 odd

Communicate

1. Explain how the torn-triangle Activity is similar to a proof of the Triangle Sum Theorem.

2. Explain why the torn-triangle Activity is not a proof of the Triangle Sum Theorem.

3. What role does the Parallel Postulate play in the proof of the Triangle Sum Theorem?

APPLICATION

4. **NAVIGATION** You can use a globe and a piece of string to approximate the distance of the shortest route between two points on Earth. Estimate the shortest route from New York City to Bangkok, Thailand. What might be some difficulties in traveling along this route?

internet connect

Activities Online
Go To: go.hrw.com
Keyword:
MG1 HyperGeo

ASSIGNMENT GUIDE

In Class	1–7
Core	9–25 odd, 26–40
Core Plus	24–45
Review	46–50
Preview	51–52

✐ Extra Practice can be found beginning on page 818.

Error Analysis

For Exercises 38 and 39, students may need a more detailed description of *exterior angles* and *remote interior angles*.

Guided Skills Practice

For Exercises 5–7, refer to the diagram below. (ACTIVITY)

5. Name two pairs of alternate interior angles in the diagram. ∠1 and ∠4; ∠3 and ∠5

6. What is the sum of m∠1, m∠2, and m∠3? 180°

7. If m∠4 = 65° and m∠5 = 50°, what is m∠2? 65°

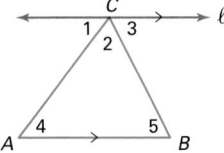

Practice and Apply

For Exercises 8–12, two angle measures of a triangle are given. Find the missing angle measure, or state that the triangle does not exist.

8. m∠1 = 85°, m∠2 = 45°, m∠3 = ___?___ 50°

9. m∠A = 45°, m∠B = ___?___, m∠C = 90° 45°

10. m∠K = ___?___, m∠L = 60°, m∠M = 60° 60°

11. m∠X = 90°, m∠Y = ___?___, m∠Z = 90° No such triangle exists.

12. m∠F = 105°, m∠G = 80°, m∠H = ___?___ No such triangle exists.

For Exercises 13–20, refer to the diagram below, in which $\overline{DF} \parallel \overline{BC}$, $\overline{AB} \parallel \overline{FC}$, m∠ADE = 60°, and m∠ACB = 50° Find the following:

13. m∠B 60°   14. m∠A 70°

15. m∠AED 50°   16. m∠EDB 120°

17. m∠DEC 130°   18. m∠FEC 50°

19. m∠ECF 70°   20. m∠F 60°

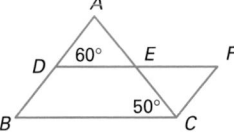

Find the indicated angle measure for each triangle.

21. $m\angle 1$ 30°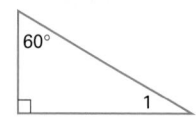

22. $m\angle 2$ 60°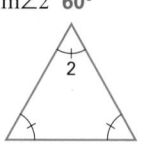

23. $m\angle 3$ 100°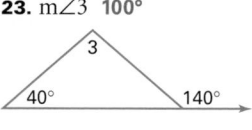

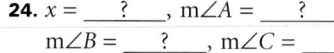

Find x and the measure of each angle of the triangle.

24. $x =$ ___?___ , $m\angle A =$ ___?___ ,   **25.** $x =$ ___?___ , $m\angle D =$ ___?___ ,
      $m\angle B =$ ___?___ , $m\angle C =$ ___?___ $m\angle E =$ ___?___ $m\angle F =$ ___?___

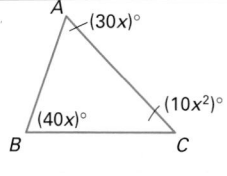

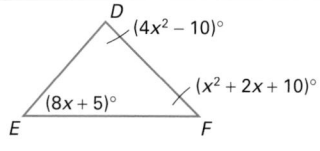

In Exercises 26–44, you will have an opportunity to discover and prove an important geometry theorem related to the figure at right. Begin by copying and completing the following table:

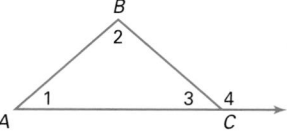

$m\angle 1$	$m\angle 2$	$m\angle 1 + m\angle 2$	$m\angle 3$	$m\angle 4$
30°	70°	**26.** ?	**27.** ?	**28.** ?
30°	80°	**29.** ?	**30.** ?	**31.** ?
40°	80°	**32.** ?	**33.** ?	**34.** ?
40°	90°	**35.** ?	**36.** ?	**37.** ?

26. 100°   **27.** 80°   **28.** 100°

29. 110°   **30.** 70°   **31.** 110°

32. 120°   **33.** 60°   **34.** 120°

35. 130°   **36.** 50°   **37.** 130°

38. An angle such as $\angle 4$ that forms a linear pair with an angle of a polygon is called an *exterior angle* of the polygon. How many exterior angles are possible at each vertex of a given polygon? Are the exterior angles at a given vertex congruent? Explain your reasoning.

39. For an exterior angle of a polygon at a given vertex, the angles of the polygon at the other vertices are called **remote interior angles** of the exterior angle. In the diagram above, $\angle 1$ and $\angle 2$ are remote interior angles of $\angle 4$. Use the table above to complete the following theorem:

Exterior Angle Theorem

The measure of an exterior angle of a triangle is equal to ___?___ .

3.5.3

C O N N E C T I O N

40. TECHNOLOGY Use geometry graphics software to draw the figure shown for Exercises 26–37 so that you can slide point C along the ray, displaying $m\angle 1 + m\angle 2$ and $m\angle 4$. Describe your results.

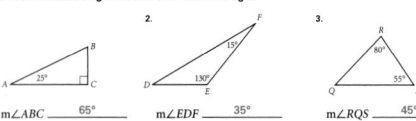
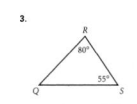
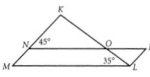

Complete the two-column proof of the Exterior Angle Theorem below.

Given: $\triangle ABC$ with exterior angle $\angle 4$ and remote interior angles $\angle 1$ and $\angle 2$

Prove: $m\angle 4 = m\angle 1 + m\angle 2$

Proof:

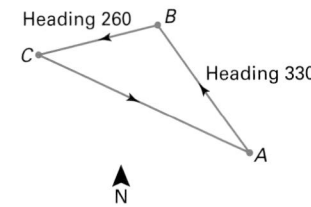

Statements	Reasons
$\triangle ABC$ with exterior angle $\angle 4$	Given
$m\angle 4 + m\angle 3 = 180°$	**41.** ?
$m\angle 1 + m\angle 2 + m\angle 3 = 180°$	**42.** ?
$m\angle 4 + m\angle 3 = m\angle 1 + m\angle 2 + m\angle 3$	**43.** ?
$m\angle 4 = m\angle 1 + m\angle 2$	**44.** ?

41. Linear Pair Property

42. Triangle Sum Theorem

43. Trans or Subs Prop

44. Subtraction Property

45. NAVIGATION Towns A, B, and C form a triangle in which $\angle A \cong \angle C$. A pilot flies from town A to town B at a heading of 330, then to town C at a heading of 260. At what heading should the pilot fly to return to town A? (Hint: Draw vertical (north-south) lines through each of the points. Use same-side interior angles.) **115**

Look Back

46. The set of common points of two figures is their _____?_____. **(LESSON 1.1)**

47. _____?_____ points determine a line. **(LESSON 1.1)**

48. _____?_____ points determine a plane. **(LESSON 1.1)**

49. Adjacent supplementary angles form a _____?_____. **(LESSON 1.3)**

46. intersection

47. 2

48. 3 noncollinear

49. linear pair

In Exercises 51 and 52, students explore the geometry of a Mercator projection. Students see how the great circles on a globe are transformed into straight lines.

APPLICATION

50. SPORTS The diagram at right shows the layout of a baseball diamond. Suppose that the line from home plate to the third baseman forms a 10° angle with the third-base line and that the line from home plate to the shortstop forms a 30° angle with the third-base line.

If the batter hits the ball so that the path of the ball bisects the angle formed by the shortstop, home plate, and the third baseman, what angle does the path of the ball make with the third-base line? *(LESSON 1.3)*
20°

Look Beyond

APPLICATION

CARTOGRAPHY A map that is commonly used in navigation is the Mercator projection, shown below. All straight lines on this map, called *rhumb lines*, have a constant compass heading.

51. yes; the vertical lines and the horizontal line at the equator

52. Answers will vary. It may be helpful to have students trace a rhumb line for a specific direction, such as northwest. They should notice that the path must curve as it approaches the poles.

51. Do any rhumb lines correspond to great circles on a globe? If so, which ones?

52. Suppose that you draw a rhumb line that is not a latitude or longitude line. Trace this line (with your finger) on a globe and describe its path.

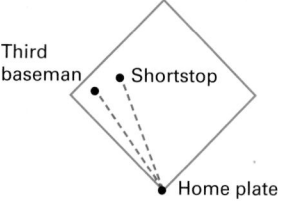

Gerardus Mercator (1512–1594)

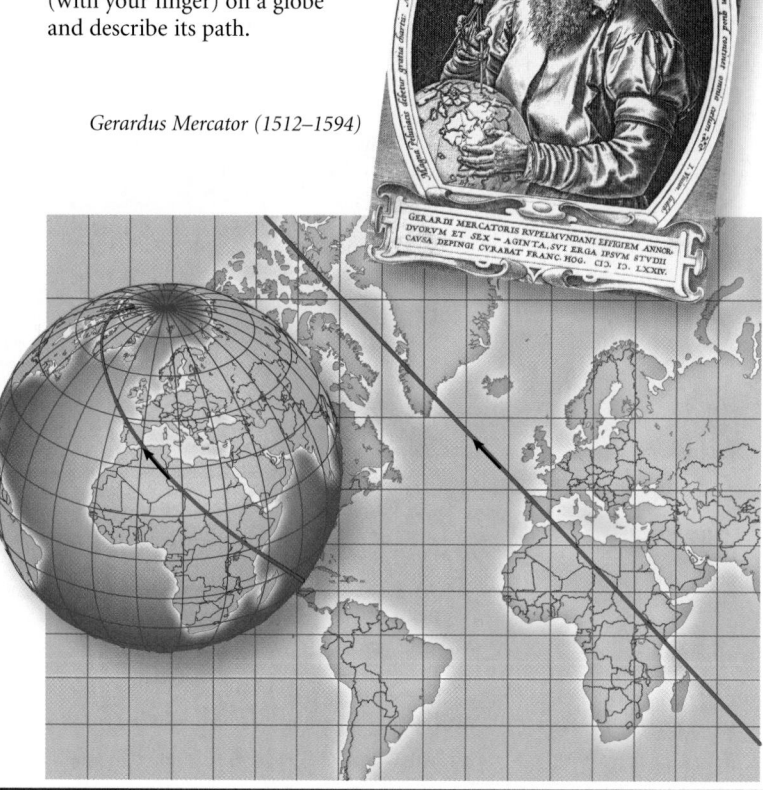

Angles in Polygons

Objective

● Develop and use formulas for the sums of the measures of interior and exterior angles of a polygon.

Why *This human polygonal structure required careful planning and design in order for all of the pieces to fit together properly. How do you think the designers of the figure achieved the final result?*

Prepare

QUICK WARM-UP

1. If ∠*A* in triangle *ABC* has a measure of 70°, what is m∠*B* + m∠*C*?

110°

2. If triangle *XYZ* is a triangle with three congruent angles, what is the measure of ∠*X*?

60°

Also on Quiz Transparency 3.6

Angle Sums in Polygons

A **convex polygon** is one in which no part of a line segment connecting any two points on the polygon is outside the polygon. A **concave polygon** does not have this characteristic. In this book, the word *polygon* will mean a convex polygon unless otherwise stated.

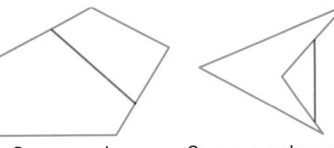

Convex polygon   Concave polygon

Teach

Why Point out to students that when polygons are used to create patterns, it is the angle relationships that are important in determining the kinds of patterns that can be created. Polygonal patterns occur frequently in wallpaper, tiles, linoleum, and many other products for homes and businesses.

Activity 1
Sums of Interior Angles

YOU WILL NEED
calculator (optional)

Pentagon *ABCDE* has been divided into three triangular regions by drawing all possible diagonals from one vertex.

1. Find each of the following:

m∠1 + m∠2 + m∠3 = _____?_____
m∠4 + m∠5 + m∠6 = _____?_____
m∠7 + m∠8 + m∠9 = _____?_____

2. Add the three expressions.

m∠1 + m∠2 + m∠3 + m∠4 + · · · + m∠9 = _____?_____

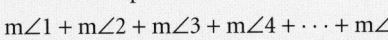

Activity 1 **Notes**

Some students may need help to complete the table in Step 4. You may wish to suggest that students add a column with the heading "Sum of Measures of Angles of Each Triangle" before the last column.

For a student worksheet of this Activity and detailed Teacher Notes, see page 46 in the Lesson Activities booklet.

Alternative Teaching Strategy

TECHNOLOGY Geometry graphics software can be used by students to conduct investigations and make conjectures about the sums of interior and exterior angles of polygons. Students can follow the exercises on pages 177–179 by actually drawing all of the polygons on the computer. Students can find the sum of the interior angle measures and of the exterior angle measures and make conjectures based on their results.

Enrichment

Students can investigate whether the angle-sum formula would work for concave polygons if interior angles were allowed to have measures greater than 180°. Ask what disadvantages such measures might present.

Challenge groups of students to discover a method of finding the sum of the angles of a polygon in terms of the number of sides. Provide hints, if necessary, to stimulate group work. Your hints should lead students through Activity 1. Encourage the groups to write a formula that summarizes their results.

CHECKPOINT ✔

5. The sum of the measures of the interior angles of a polygon with n sides is $180(n-2)$.

CHECKPOINT ✔

$$\frac{180°(n-2)}{n}$$

---

3. Use the diagram and the result from Step 2 to determine the sum of the measures of the interior angles of pentagon *ABCDE* (that is, $m\angle EAB + m\angle B + m\angle BCD + m\angle CDE + m\angle E = \underline{\quad?\quad}$).

4. You can form triangular regions by drawing all possible diagonals from a given vertex of any polygon. Complete the table below. Use sketches to illustrate your answers.

Polygon	Number of sides	Number of triangular regions	Sum of measures of angles
triangle	?	1	180°
quadrilateral	?	?	?
pentagon	?	3	540°
hexagon	?	?	?
n-gon	?	?	?

CHECKPOINT ✔

5. Write a formula for the sum of the interior angles of a polygon in terms of the number of sides, *n*. Complete the formula below.

Sum of the Interior Angles of a Polygon

The sum of the measures of the interior angles of a polygon with *n* sides is ____?____.

3.6.1

Recall that a regular polygon is one in which all the angles are congruent and all the sides are congruent. Equilateral triangles and squares are examples of regular polygons. In an equilateral triangle, each angle has a measure of 60°. In a square, each angle has a measure of 90°.

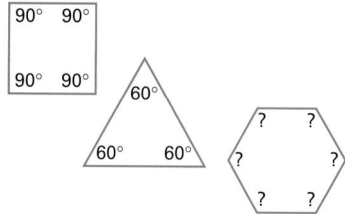

CHECKPOINT ✔ Complete the chart below. Then complete the formula beneath it.

Regular polygon	Number of sides	Sum of measures of interior angles	Measure of one interior angle
triangle	?	180°	?
quadrilateral	?	?	90°
pentagon	?	?	?
hexagon	?	?	?
n-gon	?	?	?

The Measure of an Interior Angle of a Regular Polygon

The measure of an interior angle of a regular polygon with *n* sides is ____?____.

3.6.2

Interdisciplinary Connection

DESIGN Polygons are the basic design element for many patterns in wallpaper, fabrics, tiles, etc. Ask students to collect examples of polygonal patterns from magazines or other publications for inclusion in their portfolio.

Inclusion Strategies

ENGLISH LANGUAGE DEVELOPMENT The geometry vocabulary presented in this lesson may be difficult for some students. Before teaching the lesson, prepare students by discussing the following terms and providing an example of each one: *convex polygon, concave polygon, regular polygon, interior polygon angles, exterior polygon angles, pentagon, hexagon,* and *n-gon*. Students should sketch a picture and write the definition for each term in their portfolio.

Activity 2
Exterior Angle Sums in Polygons

1. Draw a triangle and extend each side in one direction to form an exterior angle at each vertex. Find the sum of the measures of the three exterior angles that you formed. Record your results.

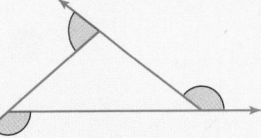

2. Cut out the exterior angles and fit them together. Record your results.

3. Repeat Steps 1 and 2 for a quadrilateral.

4. Repeat Steps 1 and 2 for a pentagon.

5. Make a conjecture about the sum of the measures of the exterior angles of a polygon (one at each vertex). You will prove your conjecture in Steps 6–10.

6. What is the sum of the measures of all interior and exterior angles of the triangle at right?

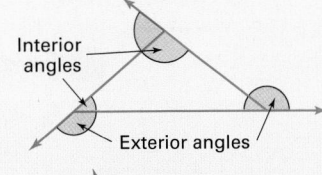

Interior angles

Exterior angles

7. What is the sum of the measures of all interior and exterior angles of the quadrilateral at right?

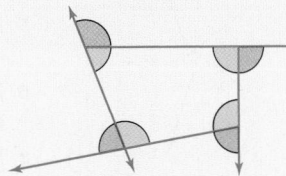

8. Using your results from Steps 6 and 7, write a formula for the sum of the measures of all interior and exterior angles of an n-gon.

9. Complete the table below.

Polygon	Number of sides	Sum of exterior and interior angles	Sum of interior angles	Sum of exterior angles
triangle	3	540°	180°	360°
quadrilateral	?	?	?	?
pentagon	?	?	?	?
hexagon	?	?	?	?
n-gon	?	?	?	?

CHECKPOINT ✔

10. Use the formula from Activity 1 and the formula from Step 8 above to write an expression for the sum of the measures of the exterior angles of an n-gon. Use algebra to simplify the expression. Complete the theorem.

Theorem: Sum of the Exterior Angles of a Polygon

The sum of the measures of the exterior angles of a polygon is ___?___.

3.6.3

Teaching Tip

Exterior angles may not be as obvious as interior angles. To minimize the possibility of adding the incorrect exterior angles, ask students to extend only one ray at each vertex, following the same direction around the polygon. The angles thus formed should be labeled as exterior.

CHECKPOINT ✔

10. The sum of the measures of the exterior angles of a polygon is 360°.

Reteaching the Lesson

USING PATTERNS You can connect an interior point of an n-gon to each vertex to form n triangles. The sum of their angle measures in all of these triangles is $180n$. Subtract 360 for the angles that meet at the interior point. The result is the sum of the interior angle measures.

For the exterior angles, choose an exterior point, P. In one direction around the polygon draw rays with endpoint P parallel to each side. Show how the resulting figure can be used to prove that the exterior angles have a sum of 360°.

ASSIGNMENT GUIDE

In Class	1–7
Core	9–41 odd
Core Plus	15–41 odd, 42, 43
Review	44–47
Preview	48–51

✐ Extra Practice can be found beginning on page 818.

Error Analysis

Students may confuse the theorems for the sum of interior angles and the sum of exterior angles. Have them recall that the sum of exterior angles is always 360°, while the sum of interior angles depends on the number of sides in the polygon.

Practice

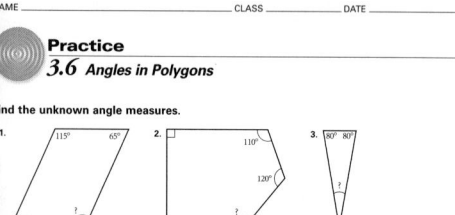

NAME _____ CLASS _____ DATE _____

Practice
3.6 Angles in Polygons

Find the unknown angle measures.

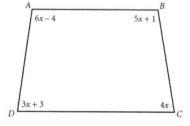

For each polygon, determine the measure of an interior angle and the measure of an exterior angle.

4. a square _interior: 90°; exterior: 90°_
5. a regular nonagon _interior: 140°; exterior: 40°_
6. an equiangular triangle _interior: 60°; exterior: 120°_
7. an equiangular hexagon _interior: 120°; exterior: 60°_

For Exercises 8–11, an interior angle measure of a regular polygon is given. Find the number of sides of the polygon.

8. 120° ___6___
9. 90° ___4___
10. 168° ___30___
11. 144° ___10___

For Exercises 12–15, an exterior angle measure of a regular polygon is given. Find the number of sides of the polygon.

12. 40° ___9___
13. 120° ___3___
14. 18° ___20___
15. 60° ___6___

Find each angle measure of trapezoid *ABCD*.

6. ∠A ___116°___
7. ∠B ___101°___
8. ∠C ___80°___
9. ∠D ___63°___

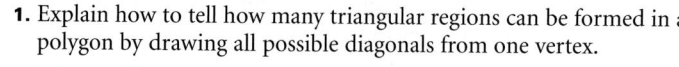
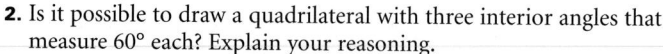
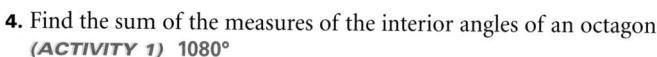
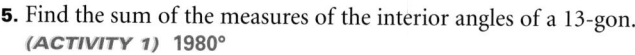

● Communicate

1. Explain how to tell how many triangular regions can be formed in a polygon by drawing all possible diagonals from one vertex.

2. Is it possible to draw a quadrilateral with three interior angles that measure 60° each? Explain your reasoning.

3. The figures at left can be a "proof without words" of the result you may have discovered in Activity 2. A polygon has its sides extended as rays. Imagine that you are looking at the figure from farther and farther away. Explain why the sum of the exterior angles is 360°.

● Guided Skills Practice

4. Find the sum of the measures of the interior angles of an octagon. *(ACTIVITY 1)* **1080°**

5. Find the sum of the measures of the interior angles of a 13-gon. *(ACTIVITY 1)* **1980°**

6. Find the sum of the measures of the exterior angles of a heptagon. *(ACTIVITY 2)* **360°**

7. Find the sum of the measures of the exterior angles of an 11-gon. *(ACTIVITY 2)* **360°**

● Practice and Apply

8. Refer to the figure at right to find the indicated measures.

   a. $x =$ ___?___ **78**

   b. $y =$ ___?___ **102**

   c. $z =$ ___?___ **127**

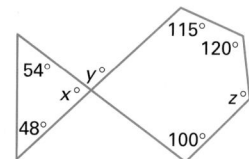

For Exercises 9–14, find the indicated angle measure.

9.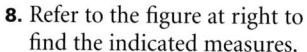

10.

11.

12.

13.

14.

For each polygon, determine the measure of an interior angle and the measure of an exterior angle.

15. a rectangle 90°; 90°

16. an equilateral triangle 60°; 120°

17. a regular dodecagon 150°; 30°

18. an equiangular pentagon 108°; 72°

Algebra

For Exercises 19–21, an interior angle measure of a regular polygon is given. Find the number of sides of the polygon.

19. 135° 8

20. 150° 12

21. 165° 24

For Exercises 22–24, an exterior angle measure of a regular polygon is given. Find the number of sides of the polygon.

22. 60° 6

23. 36° 10

24. 24° 15

For Exercises 25–37, find the indicated angle measure.

25. m∠A 36°

26. m∠B 72°

27. m∠C 108°

28. m∠D 144°

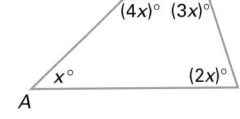

29. m∠E 70°

30. m∠F 110°

31. m∠G 50°

32. m∠H 130°

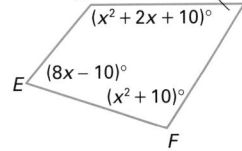

33. m∠I 140°

34. m∠J 121°

35. m∠K 71°

36. m∠L 150°

37. m∠M 58°

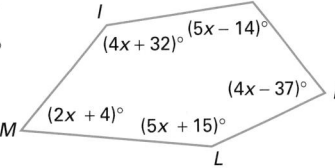

38. What is the maximum possible number of acute angles in a triangle? Can a triangle have no acute angles? Explain your reasoning.

39. What is the maximum possible number of acute angles in a quadrilateral? Can a quadrilateral have no acute angles? Explain your reasoning.

40. What is the maximum possible number of acute angles in a pentagon? Can a pentagon have no acute angles? Explain your reasoning.

41. Find the sum of the measures of the numbered vertex angles of a 5-pointed star polygon. (Hint: First find the measure of the exterior angle indicated by a question mark in each diagram below.) 180°

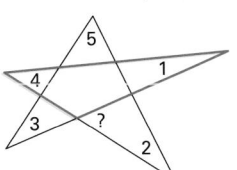 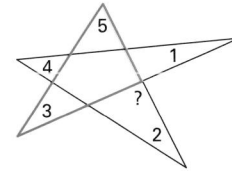 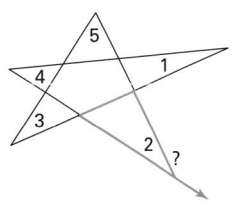

38. 3; Sample answer: an equilateral triangle has three 60° angles. No; if all three angles measure 90° or more, then the sum of the measures of the interior angles would be more than 180°.

39. 3; Sample answer: a quadrilateral with angle measures 60°, 70°, 70°, 160° has 3 acute angles, but if a quadrilateral had 4 acute angles, then the sum of their measures would be less than 360°. Yes; for example, a rectangle has 4 right angles.

40. 3; Sample answer: a pentagon with angle measurements 80°, 80°, 80°, 150°, 150° has 3 acute angles, but if a pentagon had 4 acute angles then the sum of the measures of all 5 angles would be less than 360 + 180 = 540°. Yes; for example, a regular pentagon has 108° interior angles.

Technology

Students may wish to use geometry graphics software to complete Exercises 48–50.

**Math
C O N N E C T I O N**

ALGEBRA In Exercises 25–37, students use the theorems for the sum of exterior angles and the sum of interior angles to set up and solve multistep and quadratic equations.

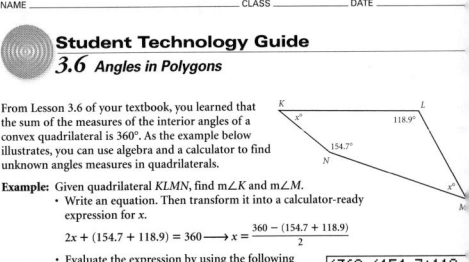

Student Technology Guide

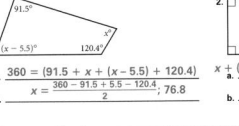

43.

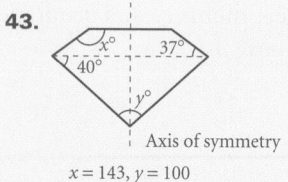

Axis of symmetry

$x = 143, y = 100$

APPLICATIONS

The brilliant cut was developed by the mathematician Marcel Tolkowsky in 1919.

42. GEMOLOGY Precious stones are often cut in a *brilliant cut* to maximize the amount of light reflected by the stone. The angles of the cut depend on the refractive properties of the type of stone. The optimal angles for a diamond are shown in the cross section below. The cut has reflectional symmetry across the axis shown. Find the measures of the indicated angles in the figure. **$x = 145, y = 98$**

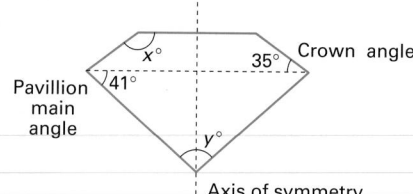

Crown angle

Pavillion main angle

Axis of symmetry

43. GEMOLOGY A brilliant cut topaz should have a pavilion main angle of 40° and a crown angle of 37°. Sketch a cross section of such a gem and find the other angles in the cross section.

Look Back

44. How is the distance from a point to a line determined? *(LESSON 1.4)*

45. List all pairs of supplementary angles in the photo below. What are these types of angles called? *(LESSON 1.3)*

APPLICATION

46. TRANSPORTATION Due to zoning regulations, the measure of an angle at an intersection cannot be less than 75°. If m∠1 = 75°, what is m∠4? *(LESSON 1.3)* **105°**

47. List all pairs of congruent angles in the photo. What are these types of angles called? *(LESSON 2.2)*
∠1 and ∠3, ∠2 and ∠4; vertical angles

Look Beyond

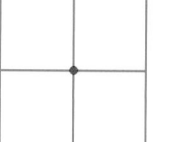

Some regular polygons fit together around a single point with no overlaps or gaps. For example, four squares fit together at a point, as shown at left.

48. What is the measure of each angle at the indicated point? What is the sum of the measures of the angles at this center point? **90°; 360°**

49. For a regular n-gon to form a pattern like the one described above, the measure of its interior angles must be a factor of 360°. Explain why this is true.

50. What other regular n-gons will fit together around a point? How can you be sure that you have found all of the possible n-gons?

APPLICATION

51. BIOLOGY A beehive is constructed from regular hexagons, as shown. What do you think are some advantages of using hexagons?

Midsegments of Triangles and Trapezoids

Why Can you think of a way to estimate the width of the pyramid halfway up the stairs without measuring it at that point? This lesson suggests a way.

The Mayan Temple of the Giant Jaguar in Tikal, Guatemala, has four approximately trapezoidal faces. A stairway rises 100 feet along one side.

Objectives

- Define *midsegment of a triangle* and *midsegment of a trapezoid*.

- Develop and use formulas based on the properties of triangle and trapezoid midsegments.

Midsegments of Triangles

Definition: Midsegment of a Triangle

A midsegment of a triangle is a segment whose endpoints are the midpoints of two sides.   **3.7.1**

Activity 1

Triangle Midsegments: A Conjecture

YOU WILL NEED

ruler and protractor
OR
geometry graphics software

1. Draw △ABC. Find the midpoints, M and N, of sides $\overline{AB}$ and $\overline{AC}$. Then draw $\overline{MN}$, the midsegment.

2. Measure $\overline{MN}$ and $\overline{BC}$. What is the relationship between their lengths?

3. Measure ∠1 and ∠2. Measure ∠3 and ∠4. What do your measurements suggest about $\overline{BC}$ and $\overline{MN}$? What postulate or theorem allows you to draw this conclusion?

CHECKPOINT ✓

4. Complete the conjecture below.

Triangle Midsegment Conjecture

A midsegment of a triangle is _____?_____ to a side of the triangle and has a measure equal to _____?_____ of that side.

Alternative Teaching Strategy

HANDS-ON STRATEGIES You may wish to have students use graph paper for Activities 1 and 2. You may also wish to suggest that they make the bases of the figures horizontal and that they place the vertices at points where horizontal and vertical grid lines intersect.

Prepare

QUICK WARM-UP

1. If M is the midpoint of $\overline{AB}$ and $MA = 7$, then $AB = \underline{\ ?\ }$ and $MB = \underline{\ ?\ }$.
   14; 7

2. What is the average of 15, 27, and 36? 26

3. If the average of x and 20 is 16.25, what is x?
   $x = 12.5$

Also on Quiz Transparency 3.7

Teach

Why Point out that properties of midsegments can be used to solve many practical problems in carpentry, architecture, model building, bridge building, and other construction-related fields. Later in the lesson, you may wish to ask students for examples of such applications from their own experience.

☞ For Activity Notes and the answer to the Checkpoint, see page 184.

In this Activity, students make a conjecture about triangles and the lengths of their midsegments. Students need to be aware that their conjectures in Step 4 cannot be accepted as true statements until they are proved deductively.

For a student worksheet of this Activity and detailed Teacher Notes, see page 50 in the Lesson Activities booklet.

CHECKPOINT ✔

4. A midsegment of a triangle is parallel to a side of the triangle and has a measure equal to one-half of that side.

ADDITIONAL
EXAMPLE ①

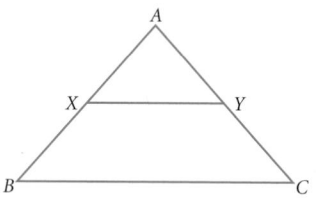

$\overline{XY}$ is the midsegment of triangle ABC. If $BC = 15$, what is XY? **7.5**

Activity 2 Notes

In this Activity, students make a conjecture about the midsegment of a trapezoid. Have students compare this conjecture with the one they made in Activity 1. In what ways are the conjectures similar? In what ways are they different?

For a student worksheet of this Activity and detailed Teacher Notes, see page 50 in the Lesson Activities booklet.

EXAMPLE ①

APPLICATION
SPORTS

Jose is on his school swim team. During the summer, he enjoys training at a small lake near his house. To evaluate his progress, he needs to know the distance across the lake, XY. How can he use the conjecture from Activity 1 to find this distance?

● **SOLUTION**

Jose can select a point, A, from which he can measure segments $\overline{AX}$ and $\overline{AY}$. Then he can find the midpoints of $\overline{AX}$ and $\overline{AY}$ and measure the distance between them. Since this distance is half the length of $\overline{XY}$, Jose can double the distance to find his answer.

Midsegments of Trapezoids

Definition: Midsegment of a Trapezoid

A midsegment of a trapezoid is a segment whose endpoints are the midpoints of the nonparallel sides.

3.7.2

Activity 2
Trapezoid Midsegments: A Conjecture

YOU WILL NEED

ruler and protractor
OR
geometry graphics software

1. Draw trapezoid $ABCD$. Find the midpoints, M and N, of the nonparallel sides. Draw $\overline{MN}$, the midsegment.

2. Measure the length of bases $\overline{AB}$ and $\overline{DC}$ and of midsegment $\overline{MN}$.

3. Find a relationship between MN and the lengths of the bases, AB and DC. (Hint: Find $AB + DC$).

4. Measure $\angle 1$ and $\angle 2$ and then $\angle 4$ and $\angle 5$. What do your measurements suggest about the relationship between $\overline{DC}$ and $\overline{MN}$?

5. Measure $\angle 2$ and $\angle 3$. What do you notice? Also measure $\angle 5$ and $\angle 6$. What do your measurements suggest about the relationship between $\overline{AB}$ and $\overline{MN}$?

6. Complete the conjecture below.

Trapezoid Midsegment Conjecture

CHECKPOINT ✔

A midsegment of a trapezoid is ____?____ to the bases of the trapezoid and has a measure equal to ____?____.

Enrichment

Trapezoidal shapes are prevalent in many ancient pyramids. As a research project, have students find various examples of trapezoids in ancient structures.

Inclusion Strategies

USING DISCUSSION You may wish to remind students about important definitions of polygons. Place a triangle, a trapezoid, and a parallelogram on the board or overhead. Ask students to discuss the differences between the figures and to develop definitions of each polygon for their portfolio. Then introduce the concept of midsegment for the triangle and trapezoid.

EXAMPLE ② The base of the pyramid of the Temple of the Giant Jaguar is a square that is 150 feet on a side. The top is a square that is 40 feet on a side. What is the width of the pyramid at a point midway between the base and the top?

APPLICATION
ARCHAEOLOGY

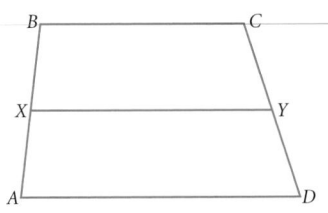
40 feet
95 feet
150 feet

● **SOLUTION**

Use the trapezoid midsegment conjecture:

Length of midsegment $= \frac{1}{2}$(base 1 + base 2)

$\qquad = \frac{1}{2}(40 + 150)$

$\qquad = \frac{1}{2}(190)$

$\qquad = 95$ feet

Making the Connection

YOU WILL NEED

ruler and protractor
OR
geometry graphics software

1. Draw trapezoid *ABCD*, which may be any shape or size, with midsegment $\overline{MN}$. Then fill in a table like the one below by gradually reducing the length of $\overline{AB}$. (Choose your own measurements for *AB* and *DC*.)

DC	AB	MN
6	5	?
6	4	?
6	3	?
6	2	?
6	1	?
6	0.5	?
6	0.1	?

You can use the table feature of a graphics calculator to find the values of MN as AB approaches zero. Enter the function as y = (6 + x) ÷ 2.

2. As *AB* approaches 0, what type of figure does the trapezoid become?

CHECKPOINT ✔ 3. Write a formula for the length of the midsegment of a "trapezoid" with one base length equal to 0. How does your formula relate to the Triangle Midsegment Conjecture?

Reteaching the Lesson

HANDS-ON STRATEGIES You may wish to implement Activity 1 as a paper-folding activity. To compare the length of the triangle midsegment with that of the third side, you can make appropriate marks along the edge of a 3-by-5 card. The conjecture for trapezoid midsegments can be derived from the conjecture for triangles by drawing a diagonal of the trapezoid.

CHECKPOINT ✔

6. A midsegment of a trapezoid is parallel to the bases of the trapezoid and has a measure equal to the average of the base lengths.

ADDITIONAL
EXAMPLE ②

$\overline{XY}$ is the midsegment of trapezoid *ABDC*. If *BC* = 15 and *AD* = 26, what is *XY*?
20.5

Activity 3 **Notes**

In this Activity, students will relate their triangle midsegment conjecture to their trapezoid midsegment conjecture. For the table in Step 1, students should draw each new trapezoid as the length of base *AB* is reduced. As the length of *AB* approaches 0, they will see that the trapezoid begins to resemble a triangle.

For a student worksheet of this Activity and detailed Teacher Notes, see page 51 in the Lesson Activities booklet.

CHECKPOINT ✔

3. $\frac{1}{2}(B_1 + 0) = \frac{1}{2}B_1$; the formula is the same as the one used for a triangle midsegment.

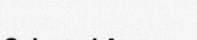
ASSIGNMENT GUIDE	
In Class	1–9
Core	10–20, 21–27 odd
Core Plus	11–15 odd, 16–20, 21–27 odd, 28–32
Review	33–40
Preview	41–45

✐ Extra Practice can be found beginning on page 818.

Error Analysis

If students make mistakes in Exercises 12, 13, and 15, they may be applying the conjecture for triangle midsegment instead of the one for trapezoids. In Exercises 14 and 15, make sure that they write and solve the appropriate equations. In Exercises 16–22, students may need a hint about which lengths to find first.

9. Length of triangle midsegment $= \frac{\text{base}}{2}$; Length of trapezoid midsegment $= \frac{\text{base}_1 + \text{base}_2}{2}$; the formula for the length of the trapezoid midsegment can be used to find the length of a triangle midsegment by letting the short base $= 0$.

Exercises

Communicate

1. In the figure at right, the horizontal segments across the large triangle are all midsegments of smaller triangles. What is the ratio of the base of the red triangle to the lower base of the blue trapezoid? Explain your reasoning.

2. If the length of the lower base of the blue trapezoid in the figure at right is 1, what are the lengths of the midsegments in the figure? Describe the pattern in the sequence of lengths you just found.

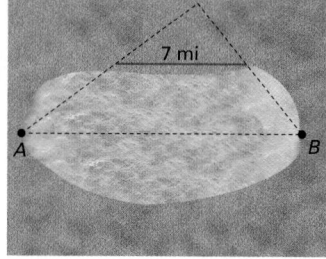

3. In Activity 3, you saw how a triangle can be considered a *limiting case* of a trapezoid. What do you think *limiting case* means?

4. Consider the following method for finding the length of the midsegment of a trapezoid: First subtract the length of the shorter base from the length of the longer base. Then take half the difference, and add it to the length of the shorter base. Does this method work? Why or why not?

Guided Skills Practice

For Exercises 5–9, refer to the conjectures you made in Activities 1–3.

5. What is the length of the midsegment of a triangle with a base of 12? *(ACTIVITY 1)* 6

APPLICATION

6. **SPORTS** A swimmer is practicing in the lake shown at right. The red line is the midsegment of the triangle. What is the distance from point *A* to point *B*? *(EXAMPLE 1)* 14 mi

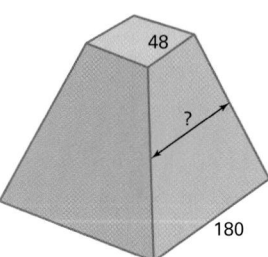

7. What is the length of the midsegment of a trapezoid with bases of 7 and 15? *(ACTIVITY 2)* 11

APPLICATION

8. **ARCHAEOLOGY** The structure shown at right has four congruent sides that are trapezoids with a height of 96 ft and bases of 180 ft and 48 ft. What is the width of the structure at a point midway between the bottom and the top? *(EXAMPLE 2)* 114 ft

9. Write the Triangle Midsegment Conjecture and the Trapezoid Midsegment Conjecture as formulas for the length of a midsegment in terms of the length of the base(s). How are the two formulas related? *(ACTIVITY 3)*

● *Practice and Apply*

Use the conjectures you made in Activities 1–3 to find the indicated measures.

10. *AB* 40

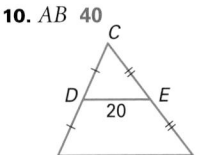

11. *IJ* 25

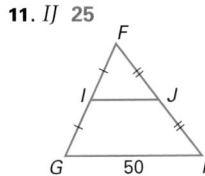

12. *PQ* 45

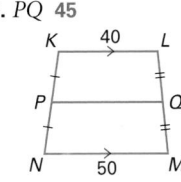

Algebra

13. *RS* 80

14. *DE* 15

15. *FG* 10

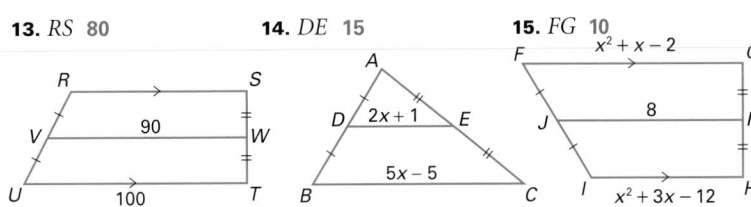

In Exercises 16–20, refer to the diagram of △ABC at right below.

16. *DE* = 10; *FG* = 20; *HI* = 30

17. *DE, FG,* and *HI,* are $\frac{1}{4}$, $\frac{1}{2}$, and $\frac{3}{4}$, respectively, of *BC*.

16. Find *DE, FG,* and *HI* in △*ABC* at right.

17. Describe the relationships among the lengths you found in Exercise 16.

18. The midsegment of a triangle divides two sides of the triangle into two congruent segments. Make a conjecture about the lengths of parallel segments that divide two sides of a triangle into four congruent segments.

19. What do you think is true about parallel segments that divide two sides of a triangle into three congruent segments? into eight congruent segments? into *n* congruent segments? Draw several triangles and test your conjecture.

20. Are the conjectures you made in Exercises 18 and 19 also true for trapezoids? Why or why not?

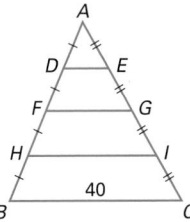

21. Segments $\overline{MN}$ and $\overline{PQ}$ divide two sides of △*JKL* into three congruent segments, as shown. Write a conjecture about $\overline{MN}$ and $\overline{PQ}$. Draw several other triangles and test your conjecture.

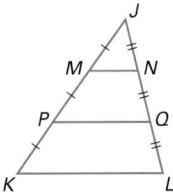

22. In the figure at right, the three red segments are midsegments of the large triangle. Find the length of each midsegment.

Add the side lengths of the outer triangle. Then add the side lengths of the smaller triangle formed by the midsegments. What is the relationship between the two sums?

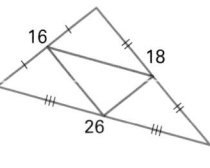

20. No, because both bases of a trapezoid must be taken into account. For example, for *n* = 4, the midsegment divides the trapezoid into two smaller trapezoids, which have midsegments of length $\frac{3b_1 + b_2}{4}$ and $\frac{b_1 + 3b_2}{4}$.

21. By the argument of Exercise 19, each successive segment decreases by $\frac{1}{n} = \frac{1}{3}$ if the base length *KL* is taken as one unit. Therefore $MN = \frac{1}{3}KL$ and $PQ = \frac{2}{3}KL$.

22. Length of midsegments: 8, 9, and 13
perimeter of outer triangle: 60
perimeter of inner traingle: 30
The perimeter of the inner triangle is one-half the perimeter of the outer triangle.

Technology

Exercises 23–26 can be completed with the aid of geometry graphics software.

Math
CONNECTION

ALGEBRA For Exercises 13–15, students use the triangle and trapezoid midsegment conjectures to write and solve multistep and quadratic equations.

18. If parallel segments divide two sides of a triangle into four congruent segments, the length of the shortest segment is $\frac{1}{4}$ of the length of the base, the length of the middle segment is $\frac{2}{4}$ or $\frac{1}{2}$ of the length of the base and the length of longest segment is $\frac{3}{4}$ of the length of the base.

19. If parallel segments divide two sides of a triangle into three congruent segments, the length of the shortest segment is $\frac{1}{3}$ of the length of the base and the length of the other segment is $\frac{2}{3}$ of the length of the base. If parallel segments divide two sides of a triangle into eight congruent segments, the length of the shortest segment is $\frac{1}{8}$ of the length of the base and the length of each of the remaining segments is $\frac{m}{8}$ of the length of the base where *m* = 2, 3, 4, 5, 6, 7. If parallel segments divide two sides of a triangle into n congruent segments, the length of each segment is $\frac{m}{n}$ of the length of the base where *m* is the segment number from 1 to *n* − 1.

23.

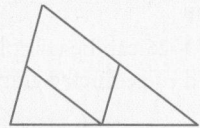

Parallelogram; The two sides of the figure are triangle midsegments making opposite sides parallel, therefore a parallelogram is formed inside the triangle.

24.

Rhombus; Two sides of the figure are triangle midsegments making opposite sides parallel, therefore a parallelogram is formed. Because the triangle is isosceles, the upper adjacent sides are congruent. The lower adjacent sides are midsegments of the congruent triangle sides, and thus half of their length. So the lower sides are congruent to each other and the upper sides. Thus, the figure is a rhombus.

In Exercises 23–26, write informal arguments based on your conjectures about triangle midsegments and quadrilaterals.

23. Draw a scalene triangle. Draw any two midsegments of the triangle. What type of quadrilateral having these midsegments as adjacent sides is formed? Explain why.

24. Draw an isosceles triangle. Draw the midsegments that connect the two equal sides to the third side. What type of quadrilateral having these midsegments as adjacent sides is formed? Explain why.

25. Draw a right triangle that is not isosceles. Draw the two midsegments that connect the legs to the hypotenuse. What type of quadrilateral having these midsegments as adjacent sides is formed? Explain why.

26. Draw an isosceles right triangle. Draw the two misdegments that connect the legs to the hypotenuse. What type of quadrilateral having these midsegments as adjacent sides is formed? Explain why.

27. In Exercises 21–27 of Lesson 2.1, you examined the sum of the infinite sequence $\frac{1}{2} + \frac{1}{4} + \frac{1}{8} + \frac{1}{16} + \cdots$ by using areas of squares and rectangles. Another way of determining this sum involves the Triangle Midsegment Conjecture.

Figure A is a square with a side length of 1, and each horizontal segment is a midsegment of a triangle. Find the length of each midsegment in figure A. Then use figure B to explain why the sum of the infinite sequence is 1.

Figure A	Figure B
1	1

APPLICATIONS

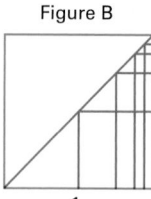

28. ENGINEERING Is $\overline{FC}$ the midsegment of trapezoid $ABDE$ in the figure at left? Explain your answer.   No, the length of the midsegment is 47.5 ft but FC = 45 ft.

PAINTING A painter is using a 20-ft ladder whose base is 5 ft from the wall.

29. Determine the distance from the ladder to the wall at a point halfway up the ladder.   2.5 ft

30. Determine the distance from the ladder to the wall at a point three-quarters of the way up the ladder.   1.25 ft

31. The painter can reach the wall from 2 ft away or less. Estimate the percent of the ladder from which the painter can reach the wall.   40%

Practice

NAME _____ CLASS _____ DATE _____

Practice

3.7 Midsegments of Triangles and Trapezoids

Find the indicated measures.

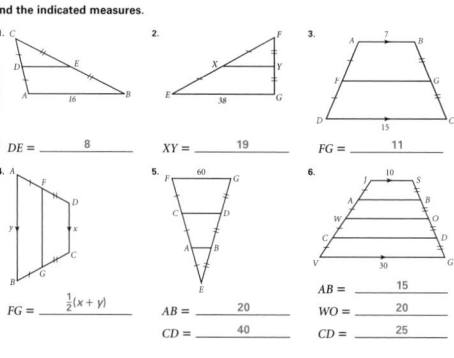

DE = _____ 8	XY = _____ 19	FG = _____ 11

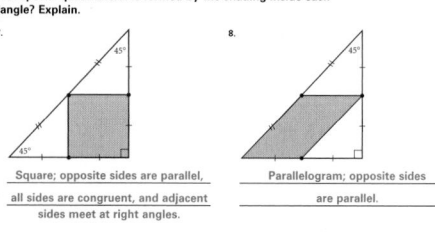

FG = _____ $\frac{1}{2}(x + y)$	AB = _____ 20	AB = _____ 15
	CD = _____ 40	WO = _____ 20
		CD = _____ 25

What special quadrilateral is formed by the shading inside each triangle? Explain.

7.   **8.**

Square; opposite sides are parallel, all sides are congruent, and adjacent sides meet at right angles.

Parallelogram; opposite sides are parallel.

25.

Rectangle; Two sides of the figure are triangle midsegments making opposite sides parallel, therefore a parallelogram is formed. Consecutive interior angles are supplementary, so all angles measure 90°.

32. MUSIC A *hammered dulcimer* is an ancient trapezoidal stringed instrument. The bases of the trapezoid are approximately 17 in. and 38 in. Estimate the length of a string at the center of the dulcimer. **27.5 in.**

 Look Back

APPLICATION

33. BIOLOGY Draw an Euler diagram to illustrate the relationships below. *(LESSON 2.3)*

A squirrel is a rodent.
All rodents are mammals.
All mammals are animals.

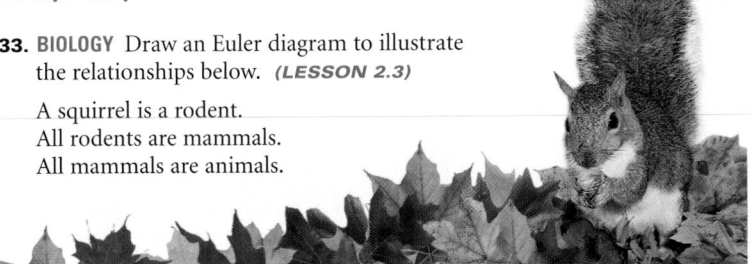

Use the conjectures you made in Lesson 3.2 to classify each statement as true or false. Explain your reasoning. *(LESSON 3.2)*

34. All rhombuses are squares.

35. All rectangles are parallelograms.

36. All squares are rectangles.

37. All parallelograms are rhombuses.

For Exercises 38–40, find the unknown angle measure in each figure. *(LESSON 3.5)*

38.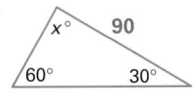

39.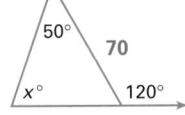

40.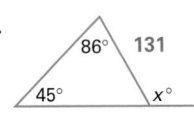

 Look Beyond

41. Draw a triangle and all of its midsegments, as shown at left. Cut out the four triangles that are formed, and compare them with each other. Write a conjecture about them.

42. Test your conjecture from Exercise 41 by using several other triangles. Based on your conjecture, what is the ratio of the area of each of the four triangles to the area of the original triangle? **1:4**

43. The figure at left is formed by drawing the midsegments of the three outer triangles in the figure above. The large outer triangle is equilateral and has a side length of 1. Find and add the side lengths of all of the shaded triangles. $\frac{1}{4}$; **6.75**

44. Is the total area of the shaded triangles in the figure more or less than the area of the outer triangle? **less than**

45. If the midsegments of each shaded triangle were drawn and the resulting center triangle of each was "unshaded," would the shaded areas of the new figure be more or less than the shaded areas above? Would the sum of all the side lengths be more or less? Explain your reasoning.

41. If the midsegments of a triangle are used to divide a triangle into 4 triangles, the four small triangles are congruent.

45. The shaded areas of the new figure would be less than the area of the shaded region in the original figure because triangular areas are being removed from the area. The sum of all the side lengths would be more because triangular borders are being added.

In Exercises 41–45, students explore the fractal called Serpenski's triangle. Students compare the areas of the shaded and unshaded regions.

33.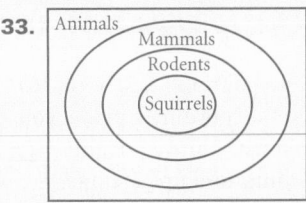

34. False; rhombuses do not have four right angles in general, so they need not be squares.

35. True; all rectangles have two sets of parallel sides, thus all rectangles are parallelograms.

36. True; squares have four right angles and opposite sides congruent, thus all squares are rectangles.

37. False; parallelograms need not have all four sides congruent.

Student Technology Guide

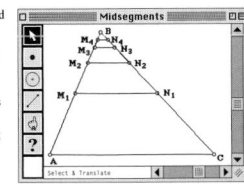

NAME _____ CLASS _____ DATE _____

Student Technology Guide
3.7 Midsegments of Triangles and Trapezoids

When you carry out the procedure described below, you perform an *iteration*.
① Sketch △ABC as shown at right.
② Locate midpoints M_1 and N_1 of sides $\overline{AB}$ and $\overline{BC}$, respectively.
③ Construct segments $\overline{M_1B}$ and $\overline{N_1B}$.
④ Locate midpoints M_2 and N_2 of segments $\overline{M_1B}$ and $\overline{N_1B}$, respectively.
⑤ Continue creating segments and locating midpoints as described in Steps 3 and 4.

Use geometry graphics software in each of the following exercises:

1. Carry out the procedure described above to get a sketch like the one shown. Label the points as shown. — Check students' sketches.

2. Find and display AC, M_1N_1, M_2N_2, M_3N_3, and M_4N_4. — Answers may vary. Each successive length should be one half the length calculated in the previous step of the iteration.

3. Suppose that M_kN_k is the kth segment drawn according to the process described above. Write an equation for M_kN_k in terms of AC. — $M_kN_k = \frac{1}{2^k}AC$

4. Modify △ABC so that $AB = BC = AC$; that is, make △ABC equilateral.
   a. What can you say about △M_kN_kB for each positive integer k? — △M_kN_kB is also equilateral.
   b. How are the lengths of the sides of △M_kN_kB related to those of △$M_{k+1}N_{k+1}B$? — The lengths of the sides of △M_kN_kB are twice as long as those of △$M_{k+1}N_{k+1}B$.

5. Viewing the diagram above as a set of stacked trapezoids, locate and mark the midpoints of $\overline{AM_1}$ and $\overline{CN_1}$. Call the midpoints R_1 and S_1 respectively. Locate and mark the midpoints of $\overline{M_1M_2}$ and $\overline{N_1N_2}$, $\overline{M_2M_3}$ and $\overline{N_2N_3}$, and so on. Call these midpoints R_2 and S_2, R_3 and S_3, and so on, respectively. Find and display R_1S_1, R_2S_2, and so on. — Check students' sketches.

6. Suppose that $\overline{R_kS_k}$ is the kth segment drawn according to the process described in Exercise 5. Write an equation for R_kS_k in terms of AC. — $R_kS_k = \frac{1}{2}(AC + M_kN_k)$; $\frac{1}{2}\left(AC + \frac{1}{2^k}AC\right) = \left(\frac{2^k+1}{2^{(k+1)}}\right)AC$

Prepare

QUICK WARM-UP

1. To go from $(2, 5)$ to $(8, 16)$ in a coordinate plane, you must move right _?_ units and up _?_ units.
6; 11

2. Evaluate $\frac{d-c}{b-a}$ for $a = -3$, $b = 10$, $c = 0$, and $d = 39$.
3

3. If you multiply any non-zero real number by its reciprocal, the product is _?_ . **1**

Also on Quiz Transparency 3.8

Teach

Why The concept of slope has applications in many real-world situations. Examples include finding the steepness of stairs, the pitch of a roof, and the grade of a highway. You may wish to ask students to suggest examples of their own.

Analyzing Polygons With Coordinates

Objectives

● Develop and use theorems about equal slopes and slopes of perpendicular lines.

● Solve problems involving perpendicular and parallel lines in the coordinate plane by using appropriate theorems.

Why *Geometers use a mathematical method to indicate steepness. This method, while not as colorful as the names of roller coasters, is more precise.*

Amusement parks often have descriptive names for their roller coasters. Names such as "Shocker" or "Wild Thing" give riders an idea about the steepness of the falls they will experience.

Slope: A Measure of Steepness

The *slope* of a line or surface tells you how steeply it rises or falls in terms of a ratio. This ratio is found by using a right triangle. Recall that the sides of a right triangle that are adjacent to the right angle are called the legs of the triangle and that the remaining side is the hypotenuse of the triangle.

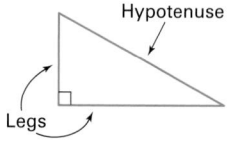

Consider a right triangle in a coordinate plane with horizontal and vertical legs. The slope of the hypotenuse is the ratio of the length of the vertical leg, the *rise*, to length of the horizontal leg, the *run*. If the hypotenuse rises from left to right, the slope is positive; if it falls from left to right, the slope is negative.

EXAMPLE ❶ Find the slope of the segment with endpoints at $(2, 3)$ and $(8, 6)$.

● **SOLUTION**

Draw a right triangle as shown. By counting squares, you can see that the rise is 3 and the run is 6. Thus, the slope is $\frac{3}{6} = \frac{1}{2}$, or 0.5.

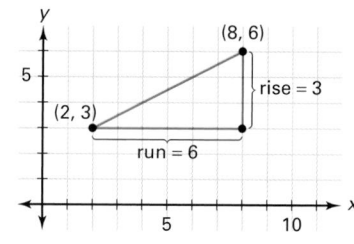

Note: In Example 1, you found that the slope of the segment with the given endpoints is 0.5. To use this method to find the slope of a line, choose any two points on the line and find the slope of the segment with those endpoints.

Alternative Teaching Strategy

HANDS-ON STRATEGIES You can use models of staircases made from wooden or plastic cubes to introduce the concepts of rise and run. Then develop a more formal definition of slope. Place strips of construction paper on two sets of stairs strategically positioned to explore parallelism and perpendicularity.

To find the slope of a line with two given points without drawing a picture, use the definition below.

Definition of Slope

The slope of a nonvertical line that contains the points (x_1, y_1) and (x_2, y_2) is equal to the ratio $\frac{y_2 - y_1}{x_2 - x_1}$.

3.8.1

CRITICAL THINKING In the case of a vertical line or segment, the slope is undefined. Explain why the slope of a vertical line or segment is undefined.

EXAMPLE 2 Find the slope of $\overline{AB}$ with the endpoints $A(5, -3)$ and $B(2, 3)$.

SOLUTION

$$\text{slope} = \frac{y_2 - y_1}{x_2 - x_1} = \frac{3 - (-3)}{2 - 5} = \frac{6}{-3} = -2$$

Parallel and Perpendicular Lines

Algebra

$y = 1.5x + 2$ (0, 2)

-3 3

(0, -3) $y = 1.5x - 3$

Recall from algebra the slope-intercept form of a line:

$$y = mx + b,$$

where m represents the slope and b represents the y-intercept

The two lines in the graph at left have the same slope, 1.5. As you can see, they seem to be parallel. The theorem below, which will not be proved formally in this book, follows from the algebraic concept of slope.

Parallel Lines Theorem

Two nonvertical lines are parallel if and only if they have the same slope. Any two vertical lines are parallel.

3.8.2

The slopes of perpendicular lines also have a special relationship to each other. This relationship is stated in the following theorem, which can be proved using the coordinate proof methods described in Lesson 5.7.

Perpendicular Lines Theorem

Two nonvertical lines are perpendicular if and only if the product of their slopes is -1. Any vertical line is perpendicular to any horizontal line.

3.8.3

Interdisciplinary Connection

INDUSTRIAL ARTS Many construction projects require the implicit or explicit use of ideas related to rise, run, and slope. Students can look for examples in their school buildings and in other familiar buildings. They can find more examples by consulting books on architecture, furniture design, art, etc.

Math CONNECTION

ALGEBRA On this page, students are give equations in slope-intercept form. Remind students that the value of the slope, m, determines whether the lines represented by the equations are parallel or perpendicular.

Have students investigate why two equations of the form $y = mx + b$ that have the same m-values but different b-values cannot have a common solution. They can use the substitution method to see that solving the equation to find a common solution would result in a false algebraic statement such as $0 = -3$.

Teaching Tip

Have students decide which of the two points will be (x_1, y_1) when calculating the slope of a particular line segment. Students will discover that the calculations give the same result for either choice.

The fact that the product of the slopes of perpendicular lines is −1 can be stated another way. In order for the product of two numbers to equal −1, one number must be the negative reciprocal of the other. Thus, if the slope of a line is $\frac{a}{b}$, then the slope of any line perpendicular to that line must be $\frac{-b}{a}$. You can use this relationship to test whether lines are perpendicular without multiplying their slopes.

EXAMPLE ③ Draw quadrilateral *QUAD* with vertices at $Q(1, 4)$, $U(7,8)$, $A(9, 5)$, and $D(3, 1)$. What type of quadrilateral is *QUAD*?

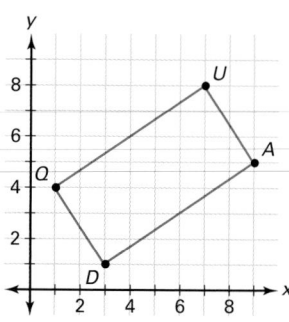

● **SOLUTION**

Based on the figure, it appears that *QUAD* is a parallelogram and perhaps a rectangle. You can test these conjectures by finding the slope of each segment.

slope of $\overline{QU} = \frac{8-4}{7-1} = \frac{4}{6} = \frac{2}{3}$ slope of $\overline{DA} = \frac{5-1}{9-3} = \frac{4}{6} = \frac{2}{3}$

slope of $\overline{QD} = \frac{4-1}{1-3} = \frac{-3}{2}$ slope of $\overline{UA} = \frac{5-8}{9-7} = \frac{-3}{2}$

Because the opposite sides of the quadrilateral have the same slope, the figure is a parallelogram.

Because the slopes of the adjacent sides of the quadrilateral are negative reciprocals, the angles in the figure are right angles. Thus, the figure is a rectangle.

Midsegments in the Coordinate Plane

Coordinates may be used to verify the conjectures about triangle and trapezoid midsegments that you made in the previous lesson.

Recall the following formula from algebra for the midpoint of a segment in a coordinate plane:

Midpoint Formula

The midpoint of a segment with endpoints (x_1, y_1) and (x_2, y_2) has the following coordinates:

$$\left(\frac{x_1 + x_2}{2}, \frac{y_1 + y_2}{2}\right)$$ **3.8.4**

> *Notice that the coordinates of the midpoint are the averages of the coordinates of the endpoints.*

EXAMPLE **4** Draw △ABC with vertices A(2, 6), B(0, 0), and C(4, 0), and use this triangle to test the triangle midsegment conjecture.

● **SOLUTION**

The midpoint of $\overline{AB}$ is $\left(\frac{2+0}{2}, \frac{6+0}{2}\right) = (1, 3)$, and

the midpoint of $\overline{AC}$ is $\left(\frac{2+4}{2}, \frac{6+0}{2}\right) = (3, 3)$.

Thus, the slope of the midsegment is
$\frac{3-3}{3-1} = \frac{0}{2} = 0$.

The slope of $\overline{BC}$ is $\frac{0-0}{4-0} = \frac{0}{4} = 0$, so the
midsegment is parallel to a side of the triangle.

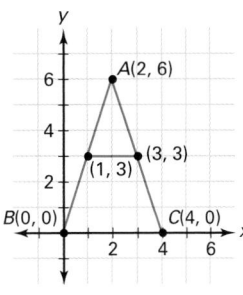

Because $\overline{BC}$ and the midsegment of the triangle are horizontal, you can determine their lengths by counting squares on the grid. The length of $\overline{BC}$ is 4, and the length of the midsegment is 2, so the length of the midsegment is half of the length of $\overline{BC}$. This confirms the Triangle Midsegment Conjecture.

TRY THIS Draw trapezoid DEFG with vertices D(4, 2), E(7, 2), F(9, 0), and G(0, 0), and use this trapezoid to test the Trapezoid Midsegment Conjecture.

CRITICAL THINKING Are the cases above proofs of the Triangle and Trapezoid Midsegment Conjectures? Why or why not?

Exercises

● Communicate

1. Describe lines with the following:
  a. a positive slope　　　　　　**b.** a negative slope
  c. a zero slope　　　　　　　　**d.** an undefined slope

2. Explain the meanings of the terms *rise* and *run*.

3. Suppose that nonvertical lines ℓ_1 and ℓ_2 are perpendicular and that the slope of ℓ_1 is m. What is the slope of ℓ_2? Explain your answer.

4. In the diagram at right, the two lines have slopes of 1 and −1, but they do not appear to be perpendicular. Explain why.

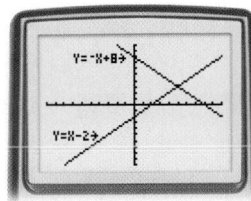

5. In Example 4 and the Try This that follows it, one side of the triangle and trapezoid is on the *x*-axis. What is a possible advantage to placing the figures in this position?

6.

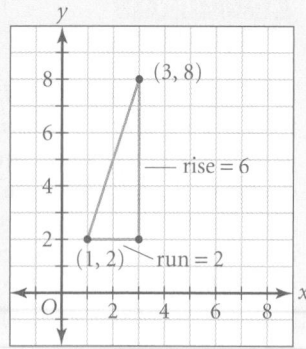

slope = $\frac{6}{2}$ = 3

10.

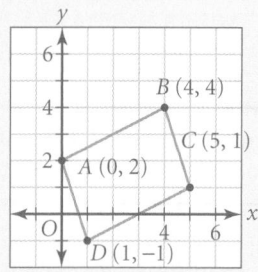

This is a parallelogram because the slopes of opposite sides are equal, so the opposite sides are parallel.

11.

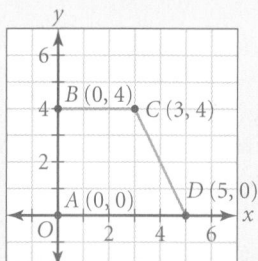

Midsegment from (0, 2) to (4, 2) has length 4. Using the bases, the midpoint length = $\frac{1}{2}$(base 1 + base 2) = $\frac{1}{2}$(3+5). Since the bases and midsegment are all horizontal, they are parallel.

29.

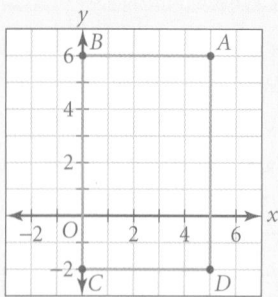

Rectangle; The slopes of $\overline{CD}$ and $\overline{AB}$ are 0 and the slopes of $\overline{DA}$ and $\overline{BC}$ are undefined. Therefore the opposite sides of the polygon are parallel. Vertical lines have undefined slopes and horizontal lines have zero slopes. Thus adja-

Guided Skills Practice

6. A segment has endpoints at (1, 2) and (3, 8). Plot the points and draw a right triangle with this segment as the hypotenuse. Use the right triangle to find the slope of the segment. **(EXAMPLE 1)**

In Exercises 7–9, use the definition of slope to find the slope of the segment with the given endpoints. **(EXAMPLE 2)**

7. (0, 0) and (4, 4) **1**   **8.** (−1, 3) and (4, 5) $\frac{2}{5}$ **9.** (2, 1) and (4, −6) $-\frac{7}{2}$

10. Draw a quadrilateral with vertices at (0, 2), (1, −1), (5, 1), and (4, 4). What type of quadrilateral is this? Explain your answer. **(EXAMPLE 3)**

11. Draw a trapezoid with vertices at (0, 0), (0, 4), (3, 4), and (5, 0), and use this trapezoid to test the Trapezoid Midsegment Conjecture. **(EXAMPLE 4)**

Practice and Apply

In Exercises 12–15, the endpoints of a segment are given. Determine the slope and midpoint of the segment.

12. (0, 0) and (4, 2) $\frac{1}{2}$; (2, 1)   **13.** (−1, 1) and (1, −1) −1; (0, 0)

14. (−3, −1) and (3, 3) $\frac{2}{3}$; (0, 1)   **15.** (−5, 2) and (1, −3) $-\frac{5}{6}$; $\left(-2, -\frac{1}{2}\right)$

In Exercises 16–19, the endpoints of two segments are given. Determine whether the segments are parallel, perpendicular, or neither.

16. (−1, 1) and (2, 3); (2, 2) and (5, 4) **parallel**

17. (−2, 1) and (1, −2); (−1, −1) and (3, 3) **perpendicular**

18. (−2, 2) and (3, 2); (2, −1) and (2, 4) **perpendicular**

19. (−1, 2) and (1, −2); (1, −2) and (2, −1) **neither**

☑ internet connect

Homework Help Online
Go To: go.hrw.com
Keyword:
MG1 Homework Help
for Exercises 20-22

Refer to the diagram at right for Exercises 20–22.

20. Find the slope of $\overline{SA}$. $-\frac{2}{3}$

21. Find the slope of $\overline{TB}$. $\frac{3}{2}$

22. Is ∠1 a right angle? **yes; $\overline{SA}$ is** Explain your answer. **perpendicular to $\overline{TB}$**

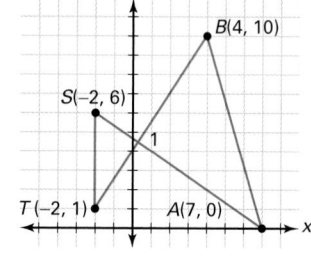

For Exercises 23–28, the vertices of a triangle are given. Use slopes to determine whether each triangle is a right triangle.

23. (−1, 4), (4, 3), (1, 1) **yes**   **24.** (1, 3), (2, 0), (−3, 2) **no**

25. (−2, 3), (3, −1), (−2, −1) **yes**   **26.** (1, 0), (0, 1), (−1, 0) **yes**

27. (1, 2), (3, 3), (4, 0) **no**   **28.** (9, 5), (2, −6), (−1, −1) **yes**

PROOFS

Draw the quadrilateral with the given vertices on a graph. Identify the type of quadrilateral and prove your answer.

29. (0, −2), (5, −2), (5, 6), (0, 6)   **30.** (0, 0), (2, 3), (5, 3), (7, 0)

31. (0, 3), (−1, 1), (2, 0), (3, 2)   **32.** (1, 3), (−2, 1), (−1, −3), (2, −1)

33. (0, 6), (3, 9), (9, 3), (6, 0)   **34.** (1, 0), (5, 1), (−1, 4), (−1, 1)

cent sides are perpendicular and *ABCD* is a rectangle.

30.

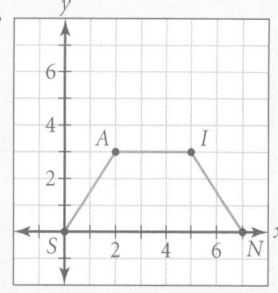

Trapezoid; The slope of $\overline{SA}$ is $\frac{3}{2}$, the slope of

$\overline{IN}$ and $\frac{-3}{2}$, and the slopes of $\overline{AI}$ and $\overline{NS}$ are 0. Therefore $\overline{AI}$ and $\overline{NS}$ are parallel and the other pair is not. Because it has one set of parallel lines, *SAIN* is a trapezoid.

For Exercises 35–37, the vertices of a rhombus are given. Draw the rhombus and use slopes to prove that the diagonals of each rhombus are perpendicular.

35. $(-2, 1), (1, 5), (5, 2), (2, -2)$

36. $(1, 3), (6, 3), (3, -1), (-2, -1)$

Algebra

37. $(0, 0), (0, a), (a, a), (a, 0)$

For Exercises 38–41, the slope and one endpoint of a segment are given. Give the coordinates of the other endpoint. More than one answer is possible. (Hint: Start by drawing a right triangle with the given point as one of its vertices.)

38. slope $= \frac{3}{4}$; $(0, 0)$ **39.** slope $= 2$; $(6, -1)$

40. slope $= -\frac{5}{3}$; $(-1, 4)$ **41.** slope $= -1$; $(2, 5)$

42. The vertices of a triangle are $A(0, 8)$, $B(2, 0)$, and $C(3, 4)$. Find the midpoints of the sides and prove that each midsegment is parallel to a side of the triangle.

43. The vertices of a trapezoid are $K(0, 0)$, $L(0, 7)$, $M(4, 0)$, and $N(4, 9)$. Find the endpoints of the midsegment and prove that it is parallel to the bases.

Algebra

44. Parallelogram $ABCD$ has vertices at $A(-1, x-1)$, $B(x, x+1)$, $C(3, 1)$, and $D(x-2, -1)$. Use the slopes of $\overline{AB}$ and $\overline{CD}$ to find x.   **x = 2**

For Exercises 45–49, draw a quadrilateral that fits the given conditions. Label the vertices, and give the slope of each side. More than one answer may be possible.

45. trapezoid $ABCD$ with vertices $A(3, 5)$ and $B(8, 5)$

46. parallelogram $EFGH$ with vertices $E(3, 2)$ and $F(-1, 5)$

47. rectangle $JKLM$ with vertices $J(1, 6)$ and $K(3, 2)$

48. parallelogram $NPQR$ with vertices $N(-2, 1)$ and $Q(4, -2)$

49. rectangle $STUV$, in which the midpoint of $\overline{ST}$ is $(5, 5)$, the midpoint of $\overline{TU}$ is $(10, 5)$, the midpoint of $\overline{UV}$ is $(7, 1)$, and the midpoint of $\overline{VS}$ is $(2, 1)$

50. CONSTRUCTION According to the Americans With Disabilities Act, a *ramp* is a route with a slope greater than $\frac{1}{20}$. The maximum allowable slope of a ramp is $\frac{1}{12}$, and the maximum rise is 30 in. What are the minimum and maximum runs for a ramp with a rise of 30 in.?   **30 ft, 50 ft**

Math CONNECTION

ALGEBRA In Exercise 37, use the formula for slope to find and compare the slopes of two intersecting lines. In Exercise 44, use the slope formula to solve for x.

35.

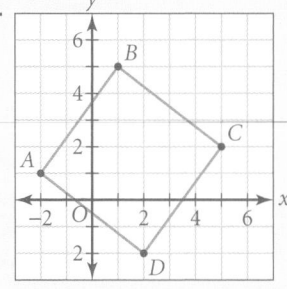

The slope of diagonal $\overline{AC} = \frac{1-2}{-2-5} = \frac{1}{7}$; the slope of diagonal $\overline{BD} = \frac{5-(-2)}{1-2} = -7$.

Since the product of the two slopes is -1, the diagonals are perpendicular.

36.

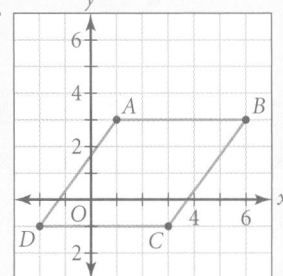

The slope of diagonal $\overline{AC} = \frac{-1-3}{3-1} = \frac{-4}{2} = -2$ and the slope of the diagonal $\overline{BD} = \frac{-1-3}{-2-6} = \frac{-4}{-8} = \frac{1}{2}$.

Since the product of the slopes is -1, the diagonals are perpendicular.

37.

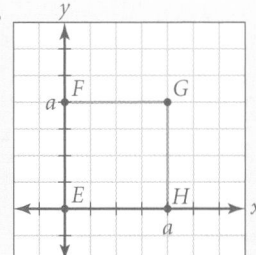

The slope of the diagonal $\overline{EG} = \frac{a-0}{a-0} = \frac{a}{a} = 1$ and the slope of the diagonal $\overline{FH} = \frac{0-a}{a-0} = \frac{-a}{a} = -1$.

Since the product of the slopes is -1, the diagonals are perpendicular.

38. Sample answer: Two possibilities are $(4, 3)$ and $(-4, -3)$.

39. Sample answer: Two possibilities are $(7, 1)$ and $(4, -5)$.

40. Sample answer: Two possibilities are $(-4, 9)$ and $(2, -1)$.

41. Sample answer: Two possibilities are $(3, 4)$ and $(5, 2)$.

42. Midpoints are $E = (1, 4)$, $F = (2.5, 2)$, and $G = (1.5, 6)$. Slope of $\overline{EF} =$ slope of $\overline{AC} = \frac{-4}{3}$, slope of $\overline{EG} =$ slope of $\overline{BC} = 4$, slope of $\overline{FG} =$ slope of $\overline{AB} = -4$. Thus, each midsegment is parallel to a side of the triangle.

43. The endpoints of the midsegment are $(2, 0)$ and $(2, 8)$. The midsegment is vertical, as are the bases, hence it is parallel to them.

In Exercises 58 and 59, students take a retrospective look at the numeration system of cubits, which was used in the mathematics of ancient Egypt.

51. Sample answer: The roof rises $23.0 - 10.5 = 12.5$ ft over a run of $\frac{25}{2} = 12.5$ ft. The slope or pitch is then $\frac{12.5}{12.5} = 1.0$. The house violates the building codes. Adjust the roof so that $\frac{\text{rise}}{12.5} = .7$ rise $= .7(12.5) = 8.75$. Adjust the peak of the roof to (12.5, 19.25).

52. A postulate is accepted without proof. A theorem must be proven using definitions, postulates, or previous theorems.

58.

x	y
0	98
28	95
56	84
84	68
112	41
140	0

Practice

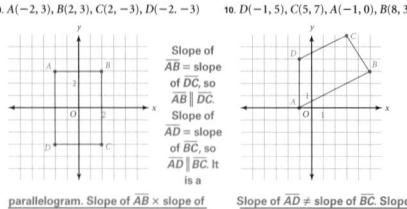

NAME _____ CLASS _____ DATE _____

Practice

3.8 *Analyzing Polygons With Coordinates*

In Exercises 1–4, the endpoints of a segment are given. Determine the slope and midpoint of the segment.

1. $(-1, 1)$ and $(2, 5)$ — slope $= \frac{4}{3}$; midpoint $\left(\frac{1}{2}, 3\right)$

2. $(0, -2)$ and $(3, -2)$ — slope $= 0$; midpoint $\left(\frac{3}{2}, -2\right)$

3. $(4, 3)$ and $(4, -5)$ — slope is undefined. midpoint $(4, -1)$

4. $(-6, 1)$ and $(-3, 0)$ — slope $= -\frac{1}{3}$; midpoint $\left(-\frac{9}{2}, \frac{1}{2}\right)$

In Exercises 5–8, the endpoints of two segments are given. State whether the segments are parallel, perpendicular, or neither.

5. $(2, -4)$ and $(3, 0)$; $(4, -8)$ and $(6, 0)$ — parallel

6. $(-3, 1)$ and $(1, 2)$; $(5, 2)$ and $(4, 6)$ — perpendicular

7. $(7, 2)$ and $(0, 6)$; $(-4, 7)$ and $(3, 5)$ — neither

8. $(-4, 0)$ and $(2, 6)$; $(2, 0)$ and $(-1, -3)$ — parallel

Graph quadrilateral *ABCD* with the given vertices on the grid provided. Justify the type of quadrilateral it is.

9. $A(-2, 3)$, $B(2, 3)$, $C(2, -3)$, $D(-2, -3)$ 10. $D(-1, 5)$, $C(5, 7)$, $A(-1, 0)$, $B(8, 3)$

Slope of $\overline{AB}$ = slope of $\overline{DC}$, so $\overline{AB} \parallel \overline{DC}$. Slope of $\overline{AD}$ = slope of $\overline{BC}$, so $\overline{AD} \parallel \overline{BC}$. It is a parallelogram. Slope of $\overline{AB}$ × slope of $\overline{AD}$ = −1, so $\overline{AB} \perp \overline{AD}$. Slope of $\overline{DC}$ × slope of $\overline{BC}$ = −1, so $\overline{DC} \perp \overline{BC}$. With four right angles, it is a rectangle.

Slope of $\overline{AD}$ ≠ slope of $\overline{BC}$. Slope of $\overline{AB}$ = slope of $\overline{DC}$, so $\overline{AB} \parallel \overline{DC}$. With exactly one pair of parallel sides, it is a trapezoid.

APPLICATION

51. CONSTRUCTION A house is 25 ft wide, and has a peaked roof, as shown at right. City building codes require the pitch (slope) of the roof be at least 0.3 and no greater than 0.7. Use the given coordinates to show that the roof of this house does not meet the building codes. How could the height of the roof be adjusted so that it does meet the codes?

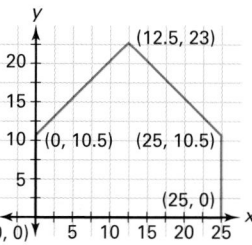

Look Back

52. Explain the difference between a postulate and a theorem. *(LESSON 2.2)*

Refer to the figure at right, in which $\ell_1 \parallel \ell_2$, and find the indicated angle measures. *(LESSON 3.5)*

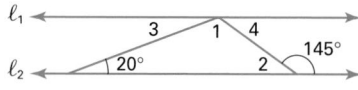

53. $m\angle 3$ 20°   **54.** $m\angle 2$ 35°

55. $m\angle 4$ 35°   **56.** $m\angle 1$ 125°

APPLICATION

57. CONSTRUCTION The diagram at right shows a house whose roof has a pitch of $\frac{1}{3}$. Find the angle that the roof forms with the walls. *(LESSON 3.6)* 108°

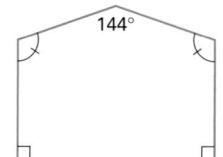

Look Beyond

CULTURAL CONNECTION: AFRICA An ancient Egyptian drawing from 2650 B.C.E. shows a rounded vault. The numbers, which are the marks on the diagram, give the height, y, of the vault at horizontal intervals, x, of 1 cubit.

58. Given that 1 palm is equal to 4 fingers and 1 cubit is equal to 7 palms, copy and complete the table below.

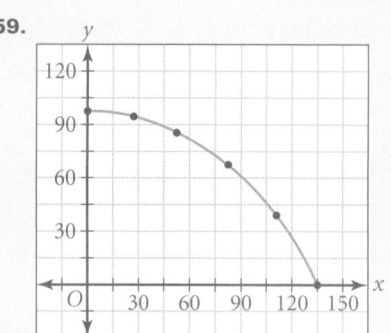

internet connect

Portfolio Extension
Go To: **go.hrw.com**
Keyword:
MG1 Escher

x (in fingers)	Height (from drawing)	y (in fingers)
0 cubit = 0 fingers	3 cubits, 3 palms, and 2 fingers	98 fingers
1 cubit = ?	3 cubits, 2 palms, and 3 fingers	?
2 cubits = ?	3 cubits	?
3 cubits = ?	2 cubits and 3 palms	?
4 cubits = ?	1 cubit, 3 palms, and 1 finger	?
5 cubits = ?	0	?

59. Use the x- and y-coordinates from the table to plot the points on a graph. Then draw a smooth curve through the points.

59.

[Graph with y-axis marked 30, 60, 90, 120 and x-axis marked 30, 60, 90, 120, 150, showing a smooth curve through plotted points.]

In addition to translation tessellations (see page 154), another pattern used by M. C. Escher is known as a **rotation tessellation**. You can make your own rotation tessellation by following the steps below. As in the Portfolio Activity for Lesson 3.2, you may have to make adjustments to your curves in order to get a pattern you like.

Draw your figures on graph paper or tracing paper, or use geometry or tessellation software.

1. Start with a regular hexagon. Replace one side of the hexagon, with a curve, as shown below. Rotate the curve about point B so that point A lies on point C.

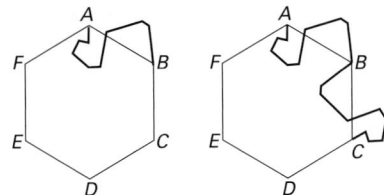

2. Replace side $\overline{CD}$ with a new curve, and rotate it around point D to replace side $\overline{DE}$.

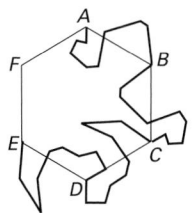

3. Replace side $\overline{EF}$ with a new curve, and rotate it around point F to replace side $\overline{FA}$.

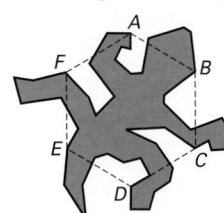

4. Your figure will now fit together with itself on all sides. You can add details to your figure, if desired. Rotate the figure to create an interlocking design.

Portfolio Activity

The Portfolio Activity can be used as preparation for the Chapter Project or as a separate activity. In the Portfolio Activity on this page, students look at another type of tessellation, called a rotation tessellation. Students should follow the steps given to make their own rotation tessellation.

By following the steps, students can produce the elaborate string figure shown.

Motivate

Handmade string figures have been intriguing games for many generations. Many different patterns and combinations are possible. Have students experiment to form their own string figures.

Students will have to experiment, and should expect to make several tries before successfully producing Jacob's ladder.

CHAPTER PROJECT THREE

String Figures

The string activity below and the resulting net have appeared in many parts of the world under different names: Osage diamonds among the Osage Indians of North America, the Calabash net in Africa, and the Quebec bridge in Canada. In the United States, it is commonly known as Jacob's ladder.

1. Start with a piece of string 4 to 5 feet in length. Tie the ends together. Loop the string around your thumbs and little fingers as shown.

2. Use your right index finger to pick up the left palm string from below. In a similar way, use your left index finger to pick up the right palm string.

3. Let your thumbs drop their loop. Turn your hands so that the fingers face out. With your thumbs, reach under all the strings and pull the farthest string back toward you.

4. With your thumbs, go over the near index-finger string, and then reach under and pull back the far index-finger string.

5. Drop the loops from your little fingers. Pass your little fingers over the index-finger string and get the thumb string closest to your little fingers from below.

6. Drop the thumb loops. Pass your thumbs over the index-finger strings, get the near little-finger string from below, and return.

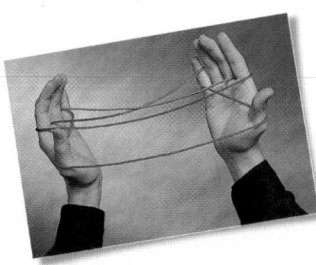
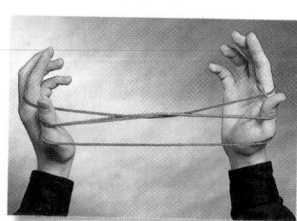
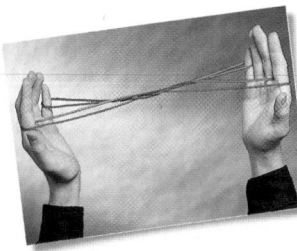

7. Loosen the left index-finger loop with your right hand, and place the loop over your thumb. Do the same with the right index-finger loop.

8. Each thumb now has two loops. Using your right hand, lift the lower loop of the left thumb up and over the thumb. Do the same with the lower loop of the right thumb.

9. Bend your index fingers and insert the tips into the triangles that are near the thumbs.

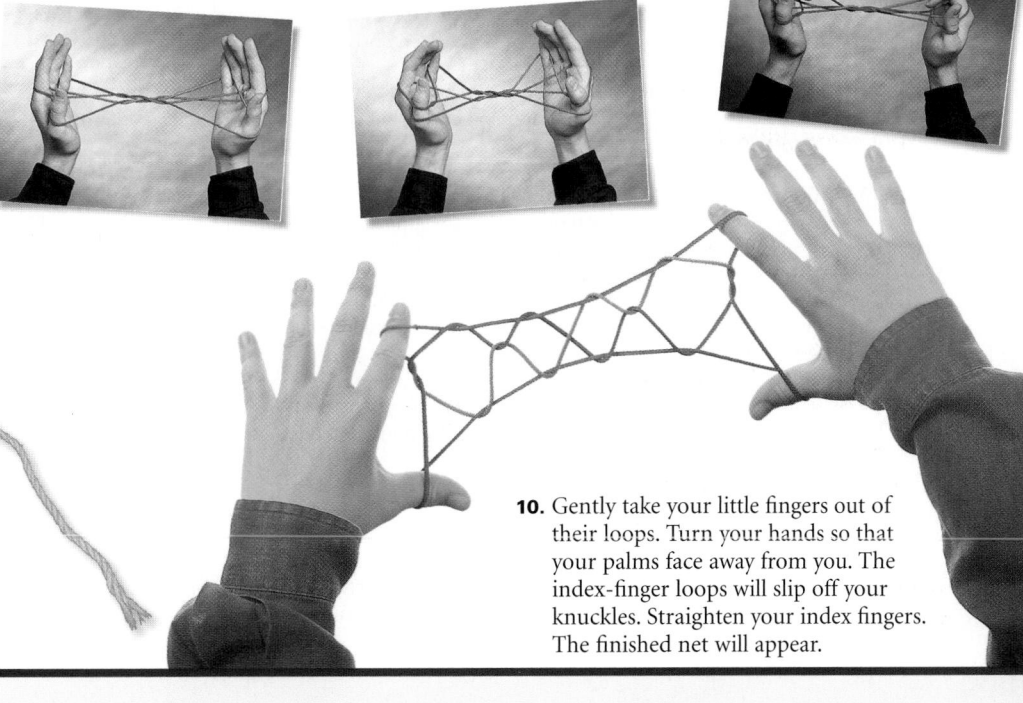

10. Gently take your little fingers out of their loops. Turn your hands so that your palms face away from you. The index-finger loops will slip off your knuckles. Straighten your index fingers. The finished net will appear.

Cooperative Learning

Students can work in small groups to create variations of the string figures. Provide students with books that describe other string figures. Students should sketch the final form of their string figures and identify any polygons that are visible. Students should also identify the lines of symmetry in their figures.

Discuss

Ask students to share their unique string figures with the rest of the class. Ask students why some string figures are very elaborate and yet require relatively few steps to create. Also ask why string figures have intrigued people for so many years.

Chapter Review and Assessment

Chapter Test, Form A

NAME _____ CLASS _____ DATE _____

Chapter Assessment
Chapter 3, Form A, page 1

Write the letter that best answers the question or completes the statement.

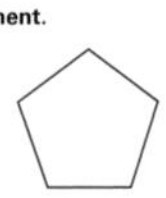

___c___ 1. How many lines of symmetry does the figure at right have?
  a. 0   b. 1   c. 5   d. 10

___c___ 2. What is the angle of rotation for the rotational symmetry of the figure?
  a. 0°   b. 50°   c. 72°   d. 180°

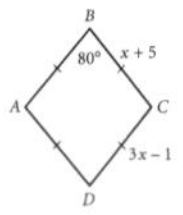

___b___ 3. What type of quadrilateral is $ABCD$?
  a. square   b. rhombus
  c. rectangle   d. trapezoid

___c___ 4. What is the length of side $\overline{AB}$?
  a. 3   b. 6   c. 8   d. 9

___c___ 5. What is the measure of $\angle A$?
  a. 80°   b. 90°   c. 100°   d. 180°

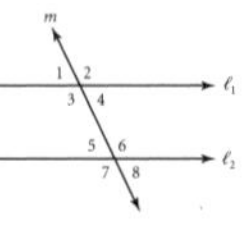

___a___ 6. What type of angles are $\angle 3$ and $\angle 6$?
  a. alternate interior   b. alternate exterior
  c. consecutive interior   d. corresponding

___d___ 7. If $\ell_1 \parallel \ell_2$ and $m\angle 2 = 110°$, then $m\angle 6 =$ _____.
  a. 35°   b. 55°
  c. 70°   d. 110°

___c___ 8. If $\ell_1 \parallel \ell_2$ and $m\angle 5 = 75°$, then $m\angle 3 =$ _____.
  a. 15°   b. 75°   c. 90°   d. 105°

___c___ 9. If $m\angle 5 = 55°$ and $m\angle 4 = 35°$, then ℓ_1 and ℓ_2 _____.
  a. are perpendicular   b. are parallel
  c. intersect at an acute angle   d. intersect at an obtuse angle

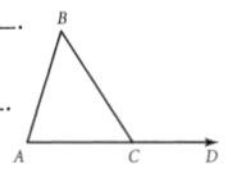

___b___ 10. If $m\angle A = 65°$ and $m\angle BCD = 125°$, then $m\angle B =$ _____.
  a. 55°   b. 60°   c. 65°   d. 185°

___d___ 11. If $m\angle A = 60°$ and $m\angle B = 80°$, then $m\angle BCD =$ _____.
  a. 20°   b. 40°   c. 60°   d. 140°

___a___ 12. If $\overline{AC} \cong \overline{BC}$ and $m\angle BCD = 108°$, then $m\angle A =$ _____.
  a. 54°   b. 72°   c. 36°   d. 90°

NAME _____ CLASS _____ DATE _____

Chapter Assessment
Chapter 3, Form A, page 2

___b___ 13. In $\triangle EFG$, $m\angle E = 85°$ and $m\angle F = 25°$. What is $m\angle G$?
  a. 60°   b. 70°   c. 110°   d. 180°

___d___ 14. What is the sum of the measures of the interior angles of a hexagon?
  a. 180°   b. 360°   c. 540°   d. 720°

___c___ 15. What is the measure of an interior angle of a regular pentagon?
  a. 60°   b. 72°   c. 108°   d. 120°

___c___ 16. If the measure of an exterior angle of a regular polygon is 45°, how many sides does the polygon have?
  a. 5   b. 6   c. 8   d. 10

___b___ 17. If the measure of an interior angle of a regular polygon is 140°, how many sides does the polygon have?
  a. 10   b. 9   c. 8   d. 5

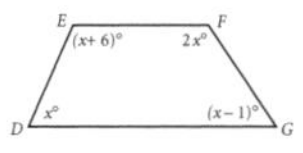

___b___ 18. What is $m\angle G$ in quadrilateral $DEFG$?
  a. 35°   b. 70°
  c. 71°   d. 77°

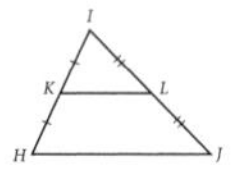

___a___ 19. If $HJ = 26$, then $KL =$ _____.
  a. 13   b. 26
  c. 30   d. 52

___c___ 20. If $KL = 15$, then $HJ =$ _____.
  a. 7.5   b. 15
  c. 30   d. 45

___c___ 21. If $HJ = 3x - 1$ and $KL = x + 1$, then $HJ =$ _____.
  a. 3   b. 4   c. 8   d. 10

$\overline{AB}$ has endpoints $A(-3, -2)$ and $B(-2, 1)$, and $\overline{CD}$ has endpoints $C(2, 1)$ and $D(1, -2)$.

___d___ 22. What is the slope of $\overline{AB}$?
  a. $-\frac{1}{3}$ b. -3 c. $\frac{1}{3}$ d. 3

___a___ 23. What is the relationship between $\overline{AB}$ and $\overline{CD}$?
  a. $\overline{AB} \parallel \overline{CD}$ b. $\overline{AB} \perp \overline{CD}$ c. $AB = CD$ d. $AB = \frac{1}{2}CD$

POSTULATES AND THEOREMS

Lesson	Number	Postulate or Theorem
3.3	3.3.2 Corresponding Angles Postulate	If two lines cut by a transversal are parallel, then corresponding angles are congruent.
	3.3.3 Alternate Interior Angles Theorem	If two lines cut by a transversal are parallel, then alternate interior angles are congruent.
	3.3.4 Alternate Exterior Angles Theorem	If two lines cut by a transversal are parallel, then alternate exterior angles are congruent.
	3.3.5 Same-Side Interior Angles Theorem	If two lines cut by a transversal are parallel, then same-side interior angles are supplementary.
3.4	3.4.1 Theorem: Converse of the Corresponding Angles Postulate	If two lines are cut by a transversal in such a way that corresponding angles are congruent, then the two lines are parallel.
	3.4.2 Converse of the Same-Side Interior Angles Theorem	If two lines are cut by a transversal in such a way that same-side interior angles are supplementary, then the two lines are parallel.
	3.4.3 Converse of the Alternate Interior Angles Theorem	If two lines are cut by a transversal in such a way that alternate interior angles are congruent, then the two lines are parallel.
	3.4.4 Converse of the Alternate Exterior Angles Theorem	If two lines are cut by a transversal in such a way that alternate exterior angles are congruent, then the two lines are parallel.
	3.4.5 Theorem	If two coplanar lines are perpendicular to the same line, then the two lines are parallel.
	3.4.6 Theorem	If two lines are parallel to the same line, then the two lines are parallel.

Lesson	Number	Postulate or Theorem
3.5	3.5.1 The Parallel Postulate	Given a line and a point not on the line, there is one and only one line that contains the given point and is parallel to the given line.
	3.5.2 Triangle Sum Theorem	The sum of the measures of the angles of a triangle is 180°.
	3.5.3 Exterior Angle Theorem	The measure of an exterior angle of a triangle is equal to the sum of the measures of the remote interior angles.
3.6	3.6.1 Sum of the Interior Angles of a Polygon	The sum, s, of the measures of the interior angles of a polygon with n sides is given by $s = (n - 2)180°$.
	3.6.2 The Measure of an Interior Angle of a Regular Polygon	The measure, m, of an interior angle of a regular polygon with n sides is $m = 180° - \frac{360°}{n}$.
	3.6.3 Sum of the Exterior Angles of a Polygon	The sum of the measures of the exterior angles of a polygon is 360°.
3.8	3.8.2 Parallel Lines Theorem	Two nonvertical lines are parallel if and only if they have the same slope.
	3.8.3 Perpendicular Lines Theorem	Two nonvertical lines are perpendicular if and only if the product of their slopes is –1.

Key Skills & Exercises

LESSON 3.1
Key Skills

Identify reflectional symmetry of figures.

Draw all of the axes of symmetry of a regular hexagon.

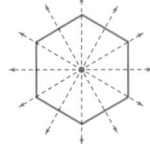

Identify rotational symmetry of figures.

Describe the rotational symmetry of a regular hexagon.

A regular hexagon has 6-fold rotational symmetry. The image will coincide with the original figure after rotations of 60°, 120°, 180°, 240°, 300°, and 360°. After a rotation of 360°, the figure is returned to its original position.

Exercises

Copy each figure below.

1. Draw all axes of symmetry of the figure above.
2. Describe the rotational symmetry of the figure above.

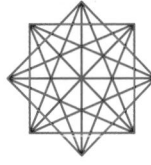

3. Draw all axes of symmetry of the figure above.
4. Describe the rotational symmetry of the figure above.

1.

2. This figure has 5-fold rotational symmetry. The image will coincide with the original figure after rotations of 72°, 144°, 216°, and 288°.

3.
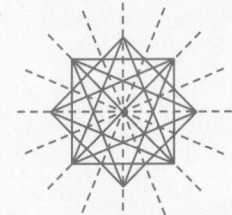

4. This figure has 8-fold rotational symmetry. The image will coincide with the original figure after rotations of 45°, 90°, 135°, 180°, 225°, 270°, and 315°.

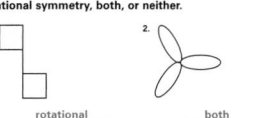
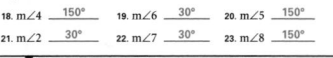
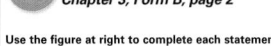
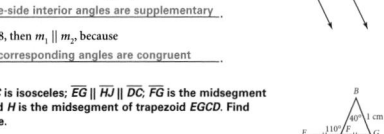
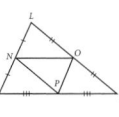
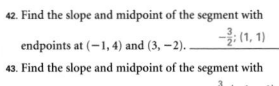

5. 48°

6. 14

7. 90°

8. 13

9. Sample answer: ∠2 and ∠4

10. Sample answer: ∠3 and ∠6

11. ∠3, ∠6, ∠8

Key Skills

Make conjectures about the properties of quadrilaterals.

In a parallelogram, the opposite sides and angles are congruent, and the diagonals bisect each other. In a rhombus, the diagonals are perpendicular to each other. In a rectangle, the diagonals are congruent.

Exercises

ABCD is a parallelogram, *EFGH* is a rhombus, and *JKLM* is a rectangle. Find the indicated measures.

5. m∠*ABC* = 48°   **6.** *AX* = 7
m∠*ADC* = ___?___   *AC* = ___?___

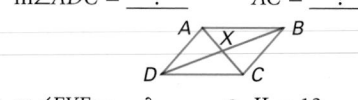

7. m∠*EYF* = ___?___   **8.** *JL* = 13
KM = ___?___

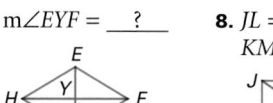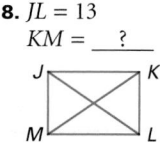

Key Skills

Identify special angle pairs.

In the diagram below, ∠4 and ∠5 are alternate interior angles, ∠2 and ∠7 are alternate exterior angles, ∠4 and ∠6 are same-side interior angles, and ∠2 and ∠6 are corresponding angles.

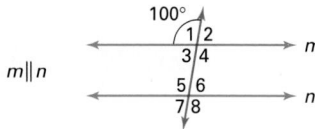

Find angle measures formed by parallel lines and transversals.

For the diagram above,
m∠4 = m∠5 = m∠8 = 100° and
m∠2 = m∠3 = m∠6 = m∠7 = 80°.

Exercises

Refer to the diagram below.

9. Name a pair of corresponding angles.

10. Name a pair of alternate interior angles.

11. Name all angles that are congruent to ∠1.

12. Suppose that m∠1 = 130°. Find the measure of each angle in the figure.

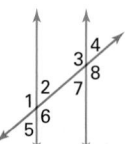

Key Skills

Use the converses of transversal properties to prove that lines are parallel.

Are lines ℓ_1 and ℓ_2 parallel?

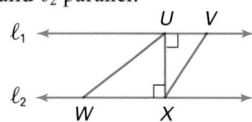

The angles ∠*WXU* and ∠*XUV* are congruent alternate interior angles. By the converse of the Interior Angles Theorem, the lines are parallel.

Exercises

Refer to the diagram below.

13. Is ℓ parallel to *m*? Explain your answer.

14. Is *m* parallel to *n*? Explain your answer.

15. Is *p* parallel to *q*? Explain your answer.

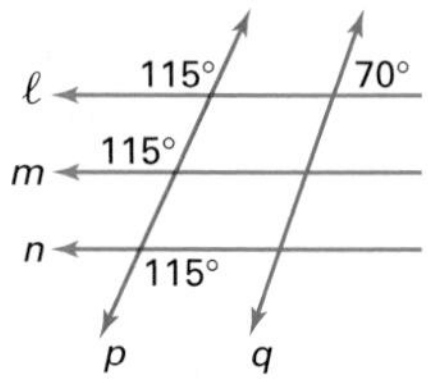

16. Prove that the opposite sides of a rectangle are parallel.

12. m∠2 = m∠5 = m∠4 = m∠7 = 50° and
m∠1 = m∠6 = m∠3 = m∠8 = 130°

13. Yes; the converse of the Corresponding Angles Postulate states they are parallel.

14. Yes; the converse of the Alternate Exterior Angles Theorem states they are parallel.

15. No; because if so, then same-side exterior angles would be supplementary, but 115° + 70° = 185° ≠ 180°.

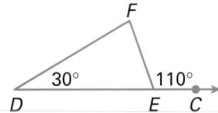
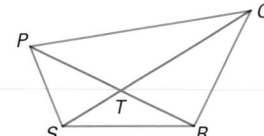
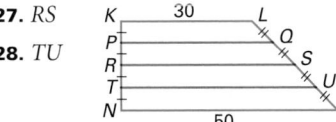

29. Yes

30. b. parallelogram; slopes of opposite sides are equal

31. c. rectangle; slopes of opposite sides are equal and slopes of consecutive sides have product of −1

32. a. trapezoid; one pair of opposite sides has equal slopes

33. The quilt has 2 axes of symmetry—2 lines through the center along the diagonals. It has a nontrivial rotational symmetry for a rotation of 180°.

34. If the frame is a rectangle, then its diagonals will be congruent.

35. 28.6%, 40%

Key Skills

Use slope to determine whether lines and segments are parallel or perpendicular.

A triangle has vertices at $A(0, 0)$, $B(4, 1)$, and $C(3, 5)$. Is $\triangle ABC$ a right triangle?

$\overline{AB}$ has a slope of $\frac{1}{4}$, $\overline{BC}$ has a slope of −4, and $\overline{AC}$ has a slope of $\frac{5}{3}$. Because $\frac{1}{4} \cdot (-4) = -1$, $\overline{AB}$ is perpendicular to $\overline{BC}$. Therefore, $\triangle ABC$ is a right triangle.

Exercises

29. $\triangle FGH$ has vertices at $F(1, 1)$, $G(2, 0)$, and $H(-1, -1)$. Is $\triangle FGH$ a right triangle?

Match each type of special quadrilateral with the correct set of vertices. Explain your reasoning.

    a. trapezoid
    b. parallelogram
    c. rectangle

30. $(0, 2)$, $(2, 3)$, $(5, 2)$, $(3, 1)$

31. $(2, 1)$, $(1, 3)$, $(5, 5)$, $(6, 3)$

32. $(0, 1)$, $(3, 3)$, $(7, 3)$, $(1, -1)$

Applications

33. CRAFTS Ruth is making a quilt with the block design called Bow Ties, shown at right. Describe all lines of symmetry and all rotational symmetries for the block.

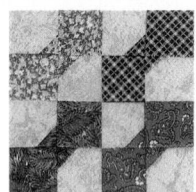

34. ART An artist is making a wooden frame for stretching a canvas. To make sure that the frame is rectangular, he measures the diagonals. If the frame is a rectangle, what will be true of its diagonals?

35. ENGINEERING The *grade* of a road is its slope expressed as a percent. For example, a road that rises 6 ft over a horizontal run of 100 ft has a slope of $\frac{6}{100}$, or a grade of 6%. What is the grade of each section of the road represented by the graph at right?

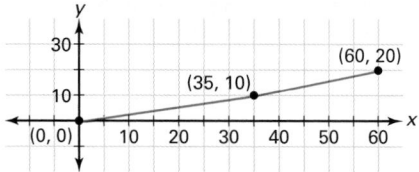

Chapter Test

Copy the figure for Exercises 1 and 2.

1. Draw all axes of symmetry of the figure.

2. Describe the rotational symmetry of the figure.

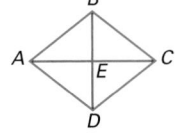

3. **DESIGN** A square table is extended by putting a leaf in the center of the table. Is the extended table regular? equilateral? equiangular? **not regular; not equilateral; equiangular**

For Exercises 4–9, refer to the diagram. In rhombus ABCD, AB = 38, BD = 43, and m∠BCD = 75°. Find the indicated measures.

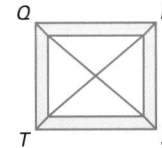

4. *BC* **38**   5. *ED* **21.5**

6. m∠*DEC* **90°**   7. m∠*DAB* **75°**

8. m∠*ADC* **105°**   9. m∠*AEB* **90°**

10. **GRAPHIC ARTS** To set the frame for a rectangular painting, Jason uses diagonal braces as supports. If the brace connecting points *Q* and *S* in the diagram below is 25 inches, how long is the brace connecting *R* and *T*? **25 inches**

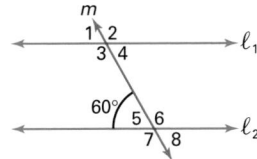

For Exercises 11–13, refer to the diagram below.

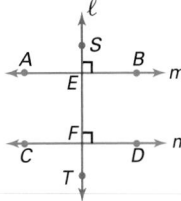

11. Name a pair of alternate exterior angles. **∠1 and ∠8 or ∠2 and ∠7**

12. Name all angles that are congruent to ∠2 **∠3, ∠6, ∠7**

13. Find the measure of each angle in the figure.

Complete the paragraph proof.

Given: *m* ⊥ ℓ and *n* ⊥ ℓ

Prove: *m* ∥ *n*

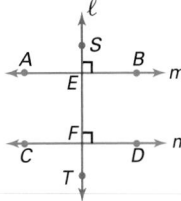

Proof:

Line ℓ is a **14.** __?__ of *m* and *n*, by definition. ∠*SEB* ≅ ∠*EFD* because **15.** __?__ . Therefore, **16.** __?__ is parallel to **17.** __?__ by **18.** __?__ .

For Exercises 19–24, refer to the diagram. QR ∥ PS, PQ ∥ TR, m∠TQR = 70°, and m∠TSU = 50°. Find the indicated measures.

19. m∠*R* **50°**

20. m∠*T* **60°**

21. m∠*SUT* **70°**

22. m∠*QPU* **50°**

23. m∠*PQU* **60°**

24. m∠*PUQ* **70°**

For each polygon, determine the measure of an interior angle and the measure of an exterior angle.

25. a square **90°; 90°**   26. a regular octagon **135°; 45°**

For Exercises 27–29, refer to the diagram of △JKL below. Find the indicated lengths.

27. *OP* **34**

28. *QR* **17**

29. *MN* **51**

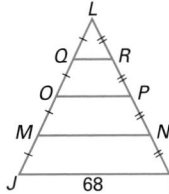

30. Draw a figure with vertices at *A*(−3, 5), *B*(2, 4), *C*(−1, 2). Identify the figure and prove your answer.

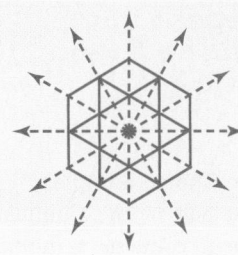

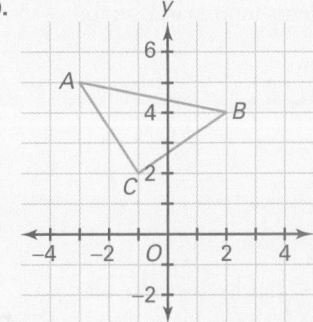

Chapter Test

1.

2. Figure has 6-fold rotational symmetry. The image will coincide with the original figure after rotations of 60°, 120°, 180°, 240°, 300°, and 360°.

13. m∠1 = m∠4 = m∠5 = m∠8 = 60° m∠2 = m∠3 = m∠6 = m∠7 = 120°

14. transversal

15. both are right angles

16. *m*

17. *n*

18. Converse of Corresponding Angles Postulate

30.

Figure ABC is a right triangle. The slope of $\overline{AC}$ is $-\frac{3}{2}$. The slope of $\overline{BC}$ is $\frac{2}{3}$. $\overline{AC}$ is perpendicular to $\overline{BC}$ since the product of their slopes is −1.

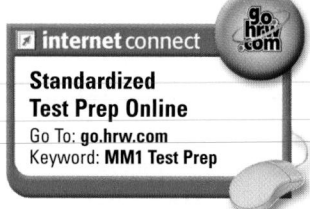

College Entrance Exam Practice

College Entrance Exam Practice

College Entrance Exam Practice

Multiple-Choice Samples

The first half of the Cumulative Assessment contains a multiple-choice section. This part of the Cumulative Assessment consists of items commonly found on standardized tests.

Free-Response Grid Samples

The second half of the Cumulative Assessment consists of a free-response section. This part requires student-produced response items like those commonly found on college entrance exams. These questions require the use of machine-scored answer grids. You may wish to have students practice answering these items in preparation for

1. a

2. d

3. d

4. point; line

5. lies in the same plane

6. perpendicular line segment

7. circumscribed circle

MULTIPLE-CHOICE For Questions 1–3, write the letter that indicates the best answer.

1. Refer to the regular pentagon below. What is m∠1? **(LESSON 3.6)**

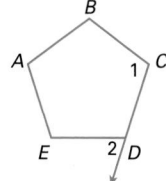

  a. 108°
  b. 120°
  c. 540°
  d. 72°

2. Refer to the figure below. Which of the following statements is true if *l* and *m* are parallel? **(LESSON 3.3)**

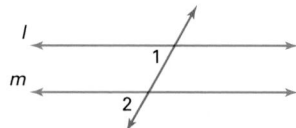

  a. ∠1 and ∠2 are parallel angles.
  b. m∠1 > m∠2
  c. ∠1 and ∠2 are vertical angles.
  d. m∠1 = m∠2

3. Refer to the figure below. Which of the following statements is true? **(LESSON 3.5)**

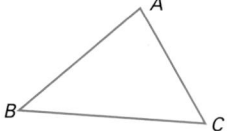

  a. m∠A + m∠B = m∠C
  b. ∠A, ∠B, and ∠C are adjacent angles.
  c. m∠A + m∠B + m∠C = 360°
  d. m∠A + m∠B + m∠C = 180°

Complete the statements in Items 4–7.
(LESSONS 1.1, 1.4, AND 1.5)

4. The intersection of two lines is a ___?___.
   The intersection of two planes is a ___?___.

5. If two points are in a plane, then the line containing them ___?___.

6. The distance from a point to a line is the length of the ___?___ from the point to the line.

7. The intersection of the perpendicular bisectors of a triangle is the center of the ___?___ of the triangle.

For Items 8–11, write the rule for each transformation. *(LESSON 1.7)*

8.

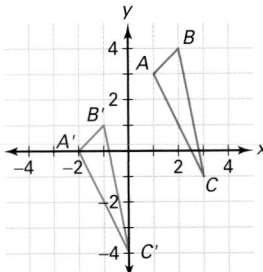

9.

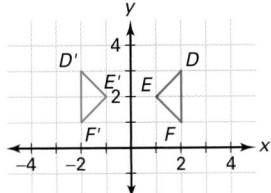

10.

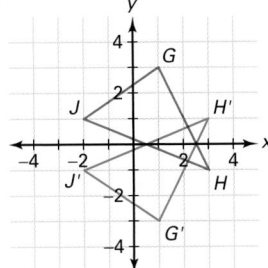

11.

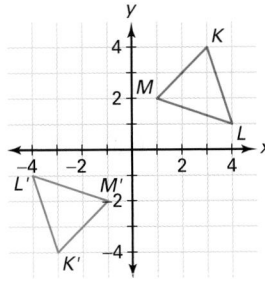

For Items 12–16, refer to the diagram below, in which $\overline{DE} \parallel \overline{FG}$, DE = 3, FG = 8, m∠EGF = 34°, m∠EGD = 30°, and m∠EJH = 42°. *(LESSONS 3.3, 3.5, 3.6, AND 3.7)*

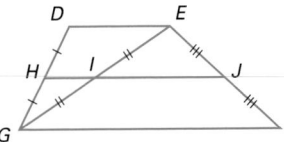

12. Name three angles that are congruent to ∠EGF.

13. What is the ratio of *IJ* to *FG*?

14. *HJ* = _____?_____ , *HI* = _____?_____

15. m∠GFE = _____?_____ , m∠HJF = _____?_____ , m∠GIJ = _____?_____ , m∠GEJ = _____?_____

16. m∠HIE = _____?_____ , m∠GHJ = _____?_____ , m∠GDE = _____?_____ , m∠DHI = _____?_____

FREE-RESPONSE GRID Items 17–20 may be answered by using a free-response grid such as that commonly used by standardized-test services.

17. The measure of an angle is 25°. What is the measure of its complement? *(LESSON 1.3)*

18. The measure of an angle is 75°. What is the measure of its supplement? *(LESSON 1.3)*

19. An angle is its own complement. What is its measure? *(LESSON 1.3)*

20. An angle is its own supplement. What is its measure? *(LESSON 1.3)*

8. $T(x, y) = (x - 3, y - 3)$

9. $T(x, y) = (-x, -y)$

10. $T(x, y) = (x, -y)$

11. $T(x, y) = (-x, -y)$

12. ∠HIG, ∠EIJ, ∠DEG

13. 1:2

14. 5.5; 1.5

15. m∠GFE = 42°, m∠HJF = 138°, m∠GIJ = 146°, m∠GEJ = 104°

16. m∠HIE = 146°, m∠GHJ = 116°, m∠GDE = 116°, m∠DHI = 64°

17. 65°

18. 105°

19. 45°

20. 90°

Triangle Congruence

Lesson Presentation CD-ROM
Power Point® presentations for each lesson 4.1-4.8

CHAPTER PLANNING GUIDE

Lesson	4.1	4.2	4.3	4.4	4.5	4.6	4.7	4.8	Project and Review
Pupil's Edition Pages	210–216	217–225	226–234	235–242	243–252	253–260	261–270	271–281	282–291
Practice and Assessment									
Extra Practice (Pupil's Edition)	828	828	829	829	830	830	831	831	
Practice Workbook	21	22	23	24	25	26	27	28	
Practice Masters Levels A, B, and C	61–63	64–66	67–69	70–72	73–75	76–78	79–81	82–84	
Standardized Test Practice Masters	24	25	26	27	28	29	30	31	32
Assessment Resources	42	43	44	45	47	48	49	50	46, 51–56
Visual Resources									
Lesson Presentation Transparencies Vol. 1	79–82	83–86	87–90	91–94	95–97	98–101	102–105	106–109	
Teaching Transparencies		34	35–36	37–39	40	41–42			
Answer Key Transparencies	128–132	133–136	137–142	143–147	148–155	156–164	165–177	178–184	185–192
Quiz Transparencies	4.1	4.2	4.3	4.4	4.5	4.6	4.7	4.8	
Teacher's Tools									
Reteaching Masters	41–42	43–44	45–46	47–48	49–50	51–52	53–54	55–56	
Make-Up Lesson Planner for Absent Students	21	22	23	24	25	26	27	28	
Student Study Guide	21	22	23	24	25	26	27	28	
Spanish Resources	21	22	23	24	25	26	27	28	
Block Scheduling Handbook									8–9
Activities and Extensions									
Lesson Activities		55–59			60–61	62–65	66–70	71–75	
Enrichment Masters	21	22	23	24	25	26	27	28	
Cooperative-Learning Activities	21	22	23	24	25	26	27	28	
Problem-Solving/ Critical Thinking	21	22	23	24	25	26	27	28	
Student Technology Guide		20–21			22–23	24–25		26–27	
Long Term Projects									13–16
Writing Activities for Your Portfolio									10–12
Tech Prep Masters									15–18
Building Success in Mathematics									9–11

LESSON PACING GUIDE

Lesson	4.1	4.2	4.3	4.4	4.5	4.6	4.7	4.8	Project and Review
Traditional	2 days	2 days	2 days	2 days	2 days	2 days	2 days	2 days	2 days
Block	1 day	1 day	1 day	1 day	1 day	1 day	1 day	1 day	1 day
Two-Year	4 days	4 days	4 days	4 days	4 days	4 days	4 days	4 days	4 days

CONNECTIONS AND APPLICATIONS

Lesson	4.1	4.2	4.3	4.4	4.5	4.6	4.7	4.8	Review
Algebra	215		234	238, 240	246			278	
Geometry	210–216	217–225	226–234	235–242	243–252	253–260	261–270	271–281	284–291
Coordinate Geometry	216			242					
Life Skills	215, 216	222, 223	233	238, 239, 242	245	255, 260	269	279	288
Science and Technology		220	226, 229, 230, 233		243				
Sports and Leisure	215	223	234		251	260		279	
Cultural Connection: Europe							270		

BLOCK-SCHEDULING GUIDE

Day	Lesson	Teacher Directed: Lesson Examples, Teaching Transparencies	Student Guided: Activity, Try This	Cooperative-Learning Activity, Lesson Activity, Student Technology Guide	Practice: Practice & Apply, Extra Practice, Practice Workbook	Assessment: Quiz, Mid-Chapter Assessment	Problem Solving, Reteaching
1	4.1	15 min	15 min	20 min	55 min	15 min	15 min
2	4.2	15 min	15 min	20 min	55 min	15 min	15 min
3	4.3	15 min	15 min	20 min	55 min	15 min	15 min
4	4.4	15 min	15 min	20 min	55 min	15 min	15 min
5	4.5	15 min	15 min	20 min	55 min	15 min	15 min
6	4.6	15 min	15 min	20 min	55 min	15 min	15 min
7	4.7	15 min	15 min	20 min	55 min	15 min	15 min
8	4.8	15 min	15 min	20 min	55 min	15 min	15 min
9	Assess.	50 min	90 min	90 min	65 min	30 min	
		PE: Chapter Review	PE: Chapter Project, Writing Activities	Tech Prep Masters	PE: Chapter Assessment Test Generator	Chap. Assess. (A or B), Alt. Assess. (A or B), Test Generator	

Alternative Assessment

The following suggest alternative assessments for students who may benefit from a different type of assessment than the regular chapter quizzes and the mid-chapter/end-of-chapter test. Visit the HRW web site to get additional Alternative Assessment material.

internet connect

Alternative Assessment
Go To: **go.hrw.com**
Keyword: **MG1 Alt Assess**

Performance Assessment

1. Given: $\triangle ACX$ and $\triangle BDX$;
$\overline{AC} \cong \overline{DB}$;
$\overline{AC}$ is parallel to $\overline{DB}$.

Prove: $\triangle ACX \cong \triangle BDX$

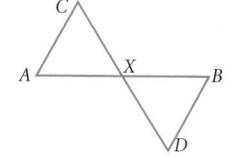

2. Given: Isosceles trapezoid $ABCD$

Prove: $\triangle AXD \cong \triangle BXC$

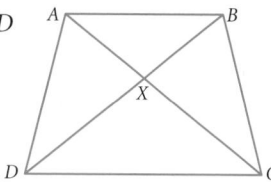

3. Prove the following conjecture: The diagonals of a rectangle are congruent.

4. Prove the following conjecture: The base angles of an isosceles triangle are congruent.

Portfolio Project

Suggest that students choose one of the following projects for inclusion in their portfolios.

1. Using geometry software, draw any irregular quadrilateral. Show that joining the midpoints of the sides always forms a parallelogram in the interior of the quadrilateral. Prove that this is true using the triangle midsegment theorem.

2. Create three irregular figures that tessellate. Use a different method for forming each one: rotations, translations, and/or reflections.

3. Draw a parallelogram. Using a compass and protractor to perform a construction, translate the parallelogram 2 inches to the left on your paper. Draw your translation vector on your paper.

internet connect

The table below identifies the pages in this chapter that contain internet and technology information.

Content Links

Activities Online	pages 213, 257, 264
Portfolio Extensions	pages 225, 252
Homework Help Online	pages 213, 222, 231, 240, 248, 269, 275

Resource Links

Parents can go online and find concepts that students are learning–lesson by lesson–and questions that pertain to each lesson, which facilitate parent-student discussion.

Go To: **go.hrw.com**
Keyword: **MG1 Parent Guide**

Technical Support

The following may be used to obtain technical support for any HRW software product.

Online Help: **www.hrwtechsupport.com**

e-mail: **tschrw@hbtechsupport.com**

HRW Technical Support Center: **(800)323-9239**
7 AM to 10 PM Monday through Friday CST

Visit the HRW math web site at: **www.hrw.com/math**

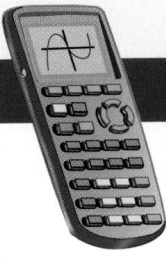

Technology

Technology Objectives and Suggestions

Lesson 4.1 Congruent Polygons

In this lesson students solve problems by using congruent polygons. Students can use geometry software to duplicate the figures in the exercises. Have them use various menu commands to create congruent figures, such as the reflection command and the rotation command.

Lesson 4.2 Triangle Congruence

In this lesson students develop three congruence postulates and explore triangle rigidity. They can construct the triangles in the activities using technology. Students can copy and paste angles and segments, as in Activity 2, and fit the parts together to make a triangle, or the measuring tool may be used and segments can be lengthened or shortened to the correct number of units, as in Activity 1.

Geometry software can also be used to construct the figures in Exercises 14–22. In Exercises 20–22 the instructions may need to be altered a bit to accommodate the use of software instead of a compass and straight edge.

Lesson 4.3 Analyzing Triangle Congruence

In this lesson students use congruence postulates and theorems, such as SSS, SAS, and ASA. They also use counter examples to show that other combinations such as SSA do not work. Have them use geometry software to find a counter-example to the SSA conjecture. To do this, each student should draw a triangle with one 35 degree angle, one 7 centimeter side, and one 4 centimeter side. They can drag the triangle to find that two different triangles, one acute and one obtuse, satisfy the conditions.

Lesson 4.4 Using Triangle Congruence

In this lesson students use congruence of triangles to conclude congruence of corresponding parts (CPCTC). They also develop and use the Isosceles Triangle Theorem. Students can use geometry software to reproduce the figures in the exercise set. Make sure they are constructed in such a way that they do not lose properties, such as parallel lines, when a point or line on the figure is dragged.

Lesson 4.5 Proving Quadrilateral Properties

In this lesson students prove quadrilateral conjectures by using triangle congruence postulates and theorems. Have students use geometry software to make the parallelograms for this lesson. Students should be sure to construct figures that retain their properties, such as parallel segments, when vertices or segments are dragged.

Lesson 4.6 Conditions for Special Quadrilaterals

In this lesson students develop conjectures about special quadrilaterals—parallelograms, rectangles, and rhombuses. Geometry software is a good tool for exploring the conjectures in this lesson. Make sure the figures students construct retain their essential properties when vertices or segments are dragged.

Lesson 4.7 Compass and Straightedge Constructions

If you plan on having students make extensive use of geometry graphics software, you may wish to do this entire lesson with a computer. The steps for most or all of the constructions in this lesson will probably involve drawing an entire circle instead of an arc, but other than that the constructions should use the same steps. However some constructions, such as drawing a perpendicular line or bisecting an angle, may also have their own menu commands.

In addition, have students carry out the construction of a regular polygon such as a hexagon. See the Enrichment section of Lesson 4.7 Teachers' Edition for instructions.

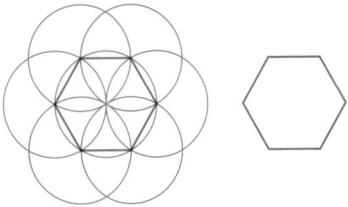

Lesson 4.8 Constructing Transformations

In this lesson students construct and prove that transformations preserve betweeness and congruency. The activities and exercises in this lesson are all appropriate for work with geometry software. Although the software most likely will transform objects with menu commands, have students follow the steps as specified in each activity. For example, in Activity 1, students should construct a translation of segment $\overline{AB}$ by drawing arcs or circles. They should not use a menu command to translate the segment.

Background Information

This chapter introduces students to the triangle congruence postulates. First, students develop intuitive notions of triangle congruence through exploration. Later in the chapter, students apply the congruence postulates in formal proofs.

CHAPTER RESOURCES

- Block-Scheduling Handbook
- Writing Activities for Your Portfolio
- Tech Prep Masters
- Long-Term Project
- Assessment Resources:
  Mid-Chapter Assessment
  Chapter Assessments
  Alternative Assessments
- Test and Practice Generator
- Technology Handbook

Chapter Objectives

- Define *congruent polygons*. [4.1]
- Solve problems by using congruent polygons. [4.1]
- Explore triangle rigidity. [4.2]
- Develop three congruence postulates for triangles—SSS, SAS, and ASA. [4.2]
- Identify and use the SSS, SAS, and ASA Congruence Postulates and the AAS and HL Congruence Theorems. [4.3]
- Use counterexamples to prove that other side and angle combinations cannot be used to prove triangle congruence. [4.3]

Triangle Congruence

ALL AROUND YOU—IN NATURE, ART, AND HUMAN technology—you find things that are the same shape and size. Such things are said to be congruent. In the photos on these pages, notice that there are many congruent triangles.

Triangles have the property of being rigid, which makes them useful in building bridges and other structures. Also, since any polygon can be divided into a number of triangles, the properties of triangles can be used to study polygons in general.

Buckminster Fuller (1895–1983)

Lessons

About the Photos

Many functional structures were created by using triangles in their design. Domes constructed with congruent triangles are called *geodesic domes* and are used as solar greenhouses, pet and hay shelters, pool covers, and astronomical observatories. Some of the strongest bridges in the world, such as the beam bridge shown above, use the geometric principle of triangle rigidity to maximize the strength of their structures.

Manhattan Bridge,
New York

- Use congruence of triangles to conclude congruence of corresponding parts. [4.4]

- Develop and use the Isosceles Triangle Theorem. [4.4]

- Prove quadrilateral conjectures by using triangle congruence postulates and theorems. [4.5]

- Develop conjectures about special quadrilaterals—parallelograms, rectangles, and rhombuses. [4.6]

- Construct congruent copies of segments, angles, and triangles. [4.7]

- Construct an angle bisector. [4.7]

- Translate, rotate, and reflect figures by using a compass and straightedge. [4.8]

- Prove that translations, rotations, and reflections preserve congruence and other properties. [4.8]

- Use the Betweenness Postulate to establish the Triangle Inequality Theorem. [4.8]

About the Chapter Project

In 1939 an American graduate student at Princeton made an interesting discovery. He folded strips of paper to form an object called a flexagon.

Flexagons have an interesting mathematical property. The hexaflexagon, for example, has three faces, but only two are visible at any given time. All flexagons have at least one hidden face. To find the hidden face, the flexagon must be folded, or "flexed," a certain way.

After completing the Chapter Project, you will be able to do the following:

- Create a hexaflexagon and a hexahexaflexagon.

- Describe the patterns in the order of faces of a flexagon.

About the Portfolio Activities

Throughout the chapter, you will be given opportunities to complete Portfolio Activities that are designed to support your work on the Chapter Project.

The theme of each Portfolio Activity and of the Chapter Project is congruent polygons.

- In the Portfolio Activity on page 225, you will use a strip of paper to fold a series of triangles. The more triangles you fold, the closer they become to a set of congruent equilateral triangles.

- Tessellations involve covering a surface with congruent shapes that fit together without gaps or overlapping. In the Portfolio Activity on page 252, you will explore tessellations with congruent, nonregular quadrilaterals.

- In the Portfolio Activity on page 281, you will explore tessellations with congruent hexagons that are not regular but that have one pair of parallel and congruent opposite sides.

Portfolio Activities appear at the end of Lessons 4.2, 4.5, and 4.8. Each serves as preparation for the Chapter Project. The Portfolio Activities, as well as the Chapter Project Activities, are appropriate for inclusion in the student's portfolio. Students should be encouraged to include in their portfolios any other work in which they feel a sense of pride or a sense of accomplishment.

internet connect

Chapter Internet Features and Online Activities

Congruent Polygons

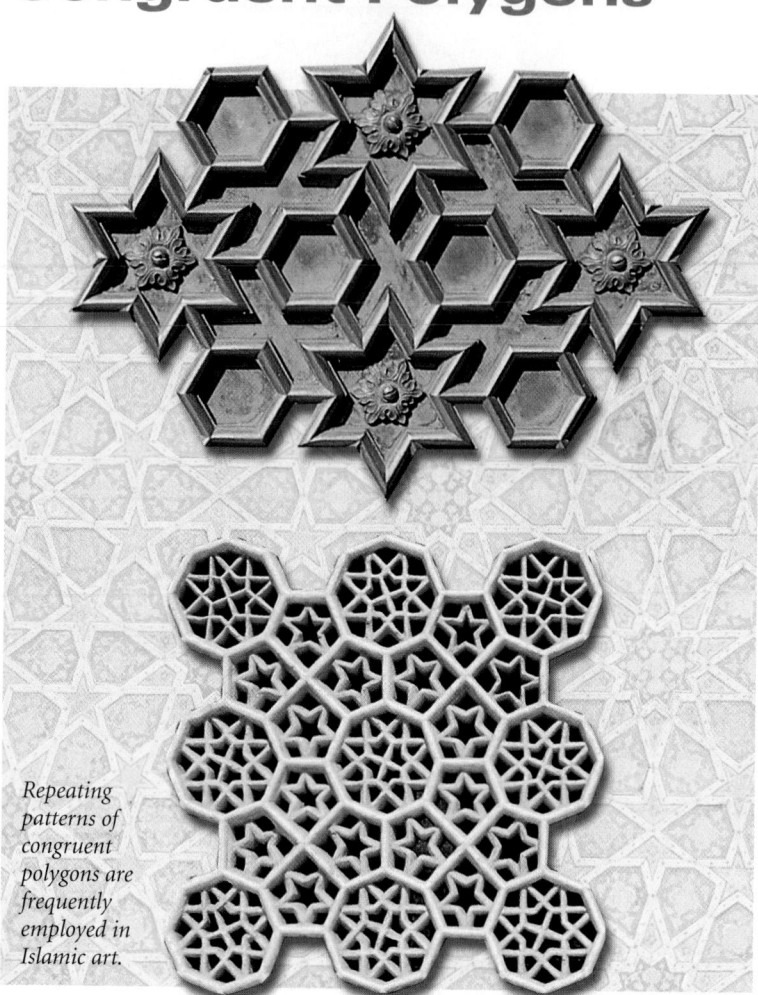

QUICK **WARM-UP**

Display this figure on the overhead or chalkboard.

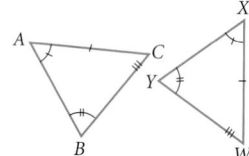

1. Name an angle congruent to ∠Y. **∠B**

2. Name a segment congruent to $\overline{BC}$. **$\overline{YW}$**

3. If ∠A is 50° and ∠Y is 70°, what is the measure of ∠C? (Hint: Use the Triangle Sum Theorem.) **60°**

Also on Quiz Transparency 4.1

Objectives

- Define *congruent polygons.*

- Solve problems by using congruent polygons.

Why *In earlier lessons you learned about congruent segments and angles. In this lesson you will develop a definition of congruent polygons.*

Repeating patterns of congruent polygons are frequently employed in Islamic art.

Teach

Why A definition of congruent polygons will be developed from the ideas that students now have about congruent segments and angles.

Polygon Congruence

Polygons 1 and 2 at right are congruent. If you slide one on top of the other, you will see that they match exactly. Can you think of a way to determine whether two polygons are congruent without actually moving them? What measurements would you need to make?

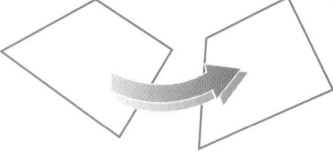

Polygon 1   Polygon 2

If two polygons are congruent, then their respective angles and sides are congruent. The converse is also true: If the respective angles and sides of a polygon match, then the polygons are congruent. These facts will be stated later as a postulate (see page 212), but first you will need to learn some terminology and notation.

Alternative Teaching Strategy

HANDS-ON STRATEGIES Have students use toothpicks, straws, or similar materials to build polygons according to a given set of specifications. For example, build an equilateral triangle with side lengths of 3 inches. When finished, have students compare figures and discuss whether the figures are congruent. Have students label the vertices with letters and write congruence statements for the figures.

Naming Polygons

When naming polygons, the rule is to go around the figure, either clockwise or counterclockwise, and list the vertices in order. It does not matter which vertex you list first.

EXAMPLE **1** **What are all of the possible names for the hexagon at right?**

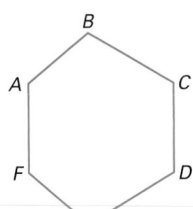

● **SOLUTION**

PROBLEM SOLVING

Make an organized list. You can approach the question systematically, as follows:

Pick a letter from the figure, such as *A*. Then write the letters of each vertex of the figure, going first in one direction and then in the other.

ABCDEF   *AFEDCB*

Then use each of the other letters in the figure as starting points.

BCDEFA   *BAFEDC*
CDEFAB   *CBAFED*
DEFABC   *DCBAFE*
EFABCD   *EDCBAF*
FABCDE   *FEDCBA*

In all, there are 12 possible names for the hexagon.

TRY THIS What are all of the possible names for the pentagon at right?

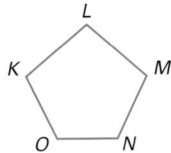

Corresponding Sides and Angles

If two polygons have the same number of sides, it is possible to set up a correspondence between them by pairing their parts. In quadrilaterals *ABCD* and *EFGH*, for example, you can pair angles *A* and *E*, *B* and *F*, *C* and *G*, and *D* and *H*. Notice that you must go in the same order around each of the polygons.

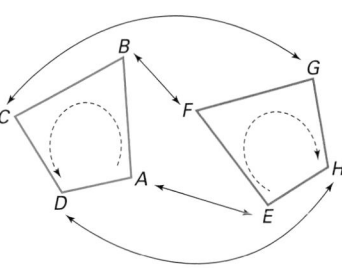

The correspondence of the sides follows from the correspondence of the angles. In this case, side $\overline{AB}$ corresponds to side $\overline{EF}$, and so on.

ADDITIONAL
EXAMPLE **1**

What are all of the possible names for the rectangle below?

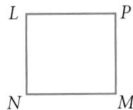

*LPMN, PMNL, MNLP,
NLPM, LNMP, NMPL,
MPLN, PLNM*

TRY THIS
*KLMNO, LMNOK, MNOKL,
NOKLM, OKLMN, ONMLK,
NMLKO, MLKON, LKONM,
KONML*

Teaching Tip

Emphasize that the letters used to name polygons must be listed in either clockwise or counterclockwise order. The letters do not need to be in alphabetical order.

Enrichment

Figures that have the same shape but different sizes are called similar figures. Show students two congruent triangles and two similar triangles. Ask them to list what things must be the same for the triangles to be congruent and what things must be the same for the triangles to be similar. Have them repeat the exercise with other congruent and similar polygons.

Inclusion Strategies

HANDS-ON STRATEGIES Have students create templates to construct different pairs of congruent triangles. The pairs may share various sides, may be rotated or reflected with respect to each other, or may overlap in interesting ways. Each congruency should be described by a statement such as △*ABC* ≅ △*XYZ*.

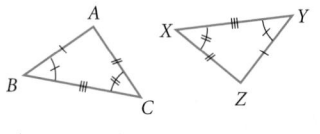
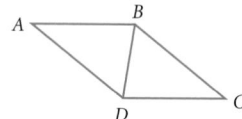
CRITICAL THINKING How many different ways are there of setting up a correspondence between the vertices of the two quadrilaterals on the previous page if they must go in order?

When you write a congruence statement about two polygons, you must write the letters of the vertices in the proper order so that they correspond.

E X A M P L E ② The polygons at right are congruent. Write a congruence statement about them.

● SOLUTION

Write a name for one of the polygons, followed by the congruence symbol. Then imagine moving the other polygon on top of the first one so that they match exactly. Finally, write the name of the second polygon to the right of the congruence symbol, with the corresponding vertices listed in order.

$$ABCD \cong EFGH$$

CRITICAL THINKING There is more than one way to write a congruence statement for polygons $ABCD$ and $EFGH$ above. Complete the congruence statements below, and then write all of the other possibilities as well.

$$BCDA \cong \underline{\quad ? \quad} \qquad\qquad CBAD \cong \underline{\quad ? \quad}$$

You should now be ready to state the Polygon Congruence Postulate.

Polygon Congruence Postulate

Two polygons are congruent if and only if there is a correspondence between their sides and angles such that:

• Each pair of corresponding angles is congruent.
• Each pair of corresponding sides is congruent.

4.4.1

E X A M P L E ③ Prove that $\triangle REX \cong \triangle FEX$.

● SOLUTION

List all of the sides and angles that are given to be congruent.

$\angle R \cong \angle F$	$\overline{RE} \cong \overline{FE}$
$\angle REX \cong \angle FEX$	$\overline{RX} \cong \overline{FX}$
$\angle EXR \cong \angle EXF$	

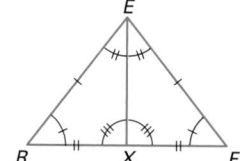

Interdisciplinary Connection

FINE ARTS Many works of art contain repeating patterns that include congruent polygons. Have students bring examples of this kind of artwork to class, and discuss the congruent figures that occur in each work. Then have students use computer graphics, pencil and paper, or some other medium to create artworks that contain congruent polygons.

Six congruences are required for triangles to be congruent—three pairs of angles and three pairs of sides. Thus, one more pair of congruent sides is needed for these triangles.

Notice that $\overline{EX}$ is shared by the two triangles. Use the Reflexive Property of Congruence to justify the statement that $EX \cong EX$. This gives the sixth congruence, so you can conclude that $\triangle REX \cong \triangle FEX$.

ASSIGNMENT GUIDE

In Class	1–9
Core	10–19, 21–33 odd
Core Plus	15–37 odd
Review	39–47
Preview	48–49

✐ Extra Practice can be found beginning on page 818.

Error Analysis

When writing proofs, students may need to review the Properties of Equality and theorems that they have seen in previous lessons.

Exercises

Communicate

1. Sketch a triangle and label the vertices *A*, *B*, and *C*. List all of the possible names for this triangle. Does the order of the vertices matter in naming a triangle? Explain your reasoning.

2. Suppose that quadrilaterals *MNOP* and *QRST* are congruent. List all of the pairs of corresponding sides and angles in the two polygons.

3. When are two segments congruent? Explain.

4. When are two angles congruent? Explain.

5. Explain the difference between the statements $\overline{AB} \cong \overline{CD}$ and $AB = CD$.

6. Does $AB \cong CD$ make sense? Explain.

internet connect

Activities Online
Go To: go.hrw.com
Keyword:
MG1 Pento

Guided Skills Practice

7. Give all possible names for the pentagon at right.
(EXAMPLE 1)

QPTSR	QRSTP
PTSRQ	RSTPQ
TSRQP	STPQR
SRQPT	TPQRS
RQPTS	PQRST

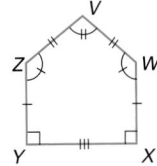

internet connect

Homework Help Online
Go To: go.hrw.com
Keyword:
MG1 Homework Help
for Exercises 9, 30

8. Use notation to write a congruence statement about both pentagons at right.
(EXAMPLE 2)

Sample answer: *PQRST ≅ VZYXW*

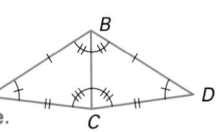

9. Prove that $\triangle ABC \cong \triangle DBC$. *(EXAMPLE 3)*
It is given that $\angle A \cong \angle D$, $\angle ABC \cong \angle DBC$, $\angle ACB \cong \angle DCB$, $\overline{AB} \cong \overline{DB}$, and $\overline{AC} \cong \overline{DC}$. By the Reflexive Property of Congruence, $\overline{BC} \cong \overline{BC}$, so $\triangle ABC \cong \triangle DBC$ by the Polygon Congruence Postulate.

Reteaching the Lesson

USING SYMBOLS Give students statements such as $\triangle ABC \cong \triangle DEF$ and $LMNOP \cong QRSTU$. Students should construct pairs of figures to illustrate the statements.

Technology

Have students create the figures in Exercises 17–25 with geometry graphics software. They must decide whether to use the reflection, rotation, or the copy-and-paste feature to create each figure.

Math
CONNECTION

ALGEBRA In Exercise 29, have students use the Triangle Sum Theorem to set up an equation and then solve for *x*.

14. No; the segments have different lengths.

15. Yes; the segments have the same length.

16. No; the sides have different lengths.

17. △*ABC* ≅ △*DBC*; all corresponding sides and angles are congruent.

18. Yes; all sides have the same length and all angles are right angles.

19. Yes; the measure of the two angles is equal.

---

10. Sample answer:
QUTSR
RSTUQ
UTSRQ

11. ∠QRS, ∠TUQ,
∠RST, ∠UQR,
∠STU

12. ∠QRS ≅ ∠VWX
∠TUQ ≅ ∠YZV
∠RST ≅ ∠WXY
∠UQR ≅ ∠ZVW
∠STU ≅ ∠XYZ

22. a. ∠NMP
b. ∠MNO
c. ∠LQP
d. ∠MPQ

23. a. $\overline{OP}$
b. $\overline{MP}$
c. $\overline{LM}$
d. $\overline{MP}$

Practice and Apply

For Exercises 10 and 11, consider a pentagon, *QRSTU*.

10. Give three other names for pentagon *QRSTU*.

11. Name the interior angles of pentagon *QRSTU*, using three letters for each.

Suppose that *QRSTU* ≅ *VWXYZ*.

12. List all pairs of corresponding angles.

13. Name the segment that is congruent to each segment below.

    a. $\overline{ST}$ $\overline{XY}$ **b.** $\overline{WX}$ $\overline{RS}$ **c.** $\overline{QU}$ $\overline{VZ}$

Determine whether the pairs of figures below are congruent. Explain your reasoning.

14.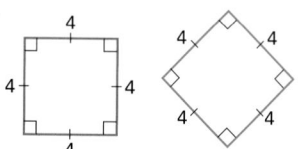

15.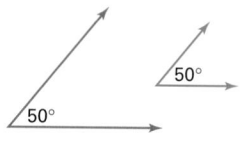

16.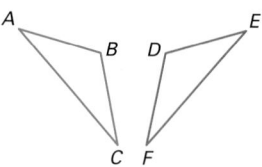

17. △*ABC* and △*DBC*
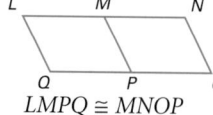

18.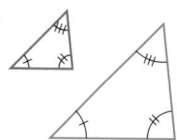

19.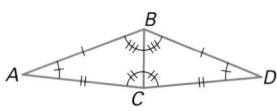

For Exercises 20–25, refer to the figures at right and complete the statements.

20. a. ∠A ≅ __?__ ∠E   **b.** ∠B ≅ __?__ ∠D
    c. ∠C ≅ __?__ ∠F

21. a. $\overline{AB}$ ≅ __?__ $\overline{ED}$ **b.** $\overline{BC}$ ≅ __?__ $\overline{DF}$
    c. $\overline{AC}$ ≅ __?__ $\overline{EF}$

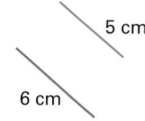

△*ABC* ≅ △*EDF*

22. a. ∠MLQ ≅ __?__   **b.** ∠LMP ≅ __?__
    c. ∠MPO ≅ __?__   **d.** ∠NOP ≅ __?__

23. a. $\overline{PQ}$ ≅ __?__   **b.** $\overline{LQ}$ ≅ __?__
    c. $\overline{MN}$ ≅ __?__   **d.** $\overline{NO}$ ≅ __?__

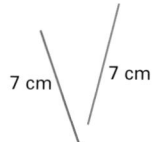

LMPQ ≅ *MNOP*

NAME _____ CLASS _____ DATE _____

Practice
4.1 Congruent Polygons

Determine whether the following pairs of figures are congruent. Explain your reasoning.

1.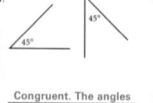
Congruent. The angles have the same measure.

2.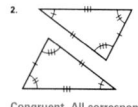
Congruent. All corresponding parts are congruent.

3.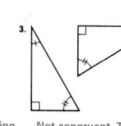
Not congruent. The side lengths could be different.

4.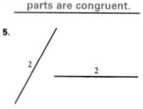
Congruent. All corresponding parts are congruent.

5.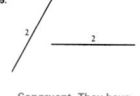
Congruent. They have the same length.

6.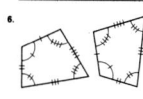
Congruent. All corresponding parts are congruent.

Suppose that hexagon *ABCDEF* ≅ *UVWXYZ*.

7. List all pairs of congruent angles. ____ ∠A ≅ ∠U, ∠B ≅ ∠V, ∠C ≅ ∠W, ∠D ≅ ∠X, ∠E ≅ ∠Y, and ∠F ≅ ∠Z

8. Name the segment that is congruent to each given segment.
a. $\overline{BC}$ ____ $\overline{VW}$ **b.** $\overline{XY}$ ____ $\overline{DE}$ **c.** $\overline{FA}$ ____ $\overline{ZU}$

9. Use the diagram at right to prove that △*ADC* ≅ △*ADB*.
Since ∠C ≅ ∠B, ∠ADC ≅ ∠ADB, and ∠CAD ≅ ∠BAD, all corresponding angles are congruent. Since $\overline{AC}$ ≅ $\overline{AB}$, $\overline{AD}$ ≅ $\overline{AD}$, and $\overline{DC}$ ≅ $\overline{DB}$, all corresponding sides are congruent. Thus, △*ADC* ≅ △*ADB* by the definition of congruent polygons.

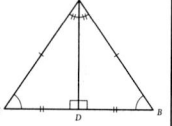

Practice

24. a. $\angle A \cong$ ___?___ $\angle F$ **b.** $\angle E \cong$ ___?___ $\angle J$

c. $\angle C \cong$ ___?___ $\angle H$ **d.** $\angle G \cong$ ___?___ $\angle B$

e. $\angle I \cong$ ___?___ $\angle D$

25. a. $\overline{AB} \cong$ ___?___ $\overline{FG}$ **b.** $\overline{IJ} \cong$ ___?___ $\overline{DE}$

c. $\overline{HI} \cong$ ___?___ $\overline{CD}$ **d.** $\overline{BC} \cong$ ___?___ $\overline{GH}$

e. $\overline{GH} \cong$ ___?___ $\overline{BC}$

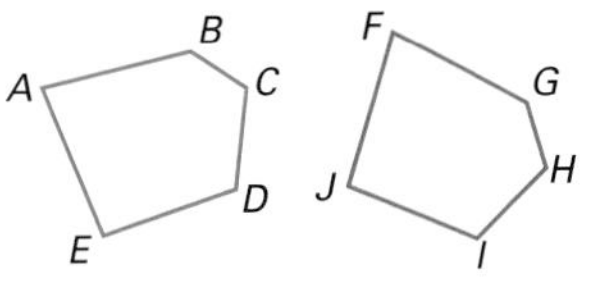

$ABCDE \cong FGHIJ$

Given: $\angle L \cong \angle P$, $\angle M \cong \angle O$, $\angle MQL \cong \angle ORP$, $\overline{LM} \cong \overline{PO}$, $\overline{MQ} \cong \overline{OR}$,
$LR = 5$, $QR = 3$, and $QP = 5$

26. $LQ =$ ___?___ 8

27. $RP =$ ___?___ 8

28. Is $\triangle LMQ \cong \triangle POR$ true? Why or why not?

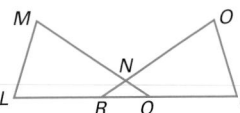

Use the figure of rhombus ABCD at right for Exercises 29 and 30.

Algebra

29. If $m\angle BDA = x°$, what is $m\angle DAB$? What is $m\angle DCB$?

30. Write a two-column proof that $\triangle ABD \cong \triangle CBD$.

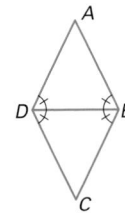

Given: $\triangle XYZ \cong \triangle FGH$ and $\triangle FGH \cong \triangle JKL$

31. What must be true about $\triangle XYZ$ and $\triangle JKL$?

32. Name a property that justifies your conclusion from Exercise 31.

CHALLENGES

33. Given $\triangle ABC \cong \triangle ACB$, what can you say about $\triangle ABC$? Explain your reasoning.

34. Given $\triangle ABC \cong \triangle BCA$, what can you say about $\triangle ABC$? Explain your reasoning.

APPLICATIONS

35. CARPENTRY A shop makes house-shaped mailboxes like the one shown at right. The walls opposite each other are congruent, as are the two slanted portions of the roof.
  a. Name 6 pairs of congruent polygons on this mailbox.
  b. Name 5 pairs of congruent angles on this mailbox.
  c. Name 5 pairs of congruent segments on this mailbox.

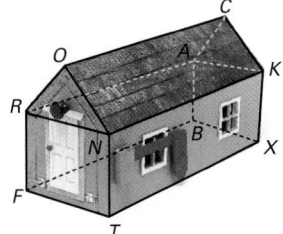

36. QUILTING The quilt shown at right was assembled from 12 identical "Balkan Puzzle" blocks. Make a sketch of one of these blocks. How many different congruent shapes can you find in the quilt? Describe them.

28. The triangles are congruent. It is given that each of the angles are congruent and also two pairs of sides are congruent. From Exercises 26 and 27, the remaining sides $\overline{LQ}$ and $\overline{PR}$ have the property $LQ = PR$; thus, $\overline{LQ} \cong \overline{PR}$.

29. $(180 - 2x)°$
$(180 - 2x)°$

31. $\triangle XYZ \cong \triangle JKL$

32. Transitive Property of Congruence

33. $\triangle ABC$ is isosceles; $\overline{AB}$ and $\overline{AC}$ are corresponding sides, so $\overline{AB} \cong \overline{AC}$.

34. $\triangle ABC$ is equilateral; $\overline{AB}$ and $\overline{BC}$ are corresponding sides, and $\overline{BC}$ and $\overline{CA}$ are corresponding sides, so $\overline{AB} \cong \overline{BC} \cong \overline{CA}$.

35. a. $\triangle RNO \cong \triangle AKC$
$FRNT \cong BAKX$
$FRAB \cong TNKX$
$FBXT \cong RAKN$
$RACO \cong NKCO$
$FRONT \cong BACKX$

  b. Sample answer:
$\angle ORN \cong \angle CAK$
$\angle ONR \cong \angle CKA$
$\angle RON \cong \angle ACK$
$\angle RFT \cong \angle ABX$
$\angle RFB \cong \angle NTX$

  c. Sample answer:
$\overline{FR} \cong \overline{TN}$
$\overline{BA} \cong \overline{XK}$
$\overline{RA} \cong \overline{NK}$
$\overline{FB} \cong \overline{TX}$
$\overline{FT} \cong \overline{BX}$

36.

The quilt contains congruent isosceles right triangles (2 sizes, large and small), parallelograms, rectangles, and squares.

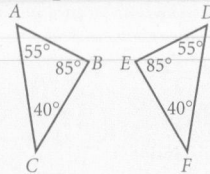
Exercises 48 and 49 require students to transform coordinates according to a given formula. Transformations will be studied in depth in a later chapter.

37. Sample answer:

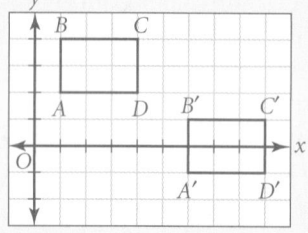

$\triangle ABC \cong \triangle DEF$

38. Yes; Transitive Property of Congruence

44. True; every square is a quadrilateral with four sides of equal length.

45. False; parallelograms have two pairs of parallel sides but trapezoids only have one pair of parallel sides.

46. False; a parallelogram is not required to have sides of equal length.

47. True; every rectangle is a quadrilateral with two pairs of parallel sides.

48.

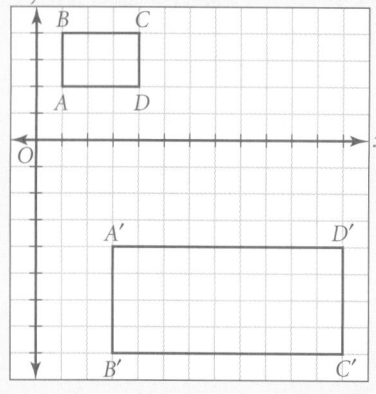

The image is congruent to the preimage because the corresponding sides and angles are congruent.

49.

The image is not congruent to the preimage because the corresponding sides are not congruent.

---

37. FASHION Janet is making a pair of earrings out of polymer clay. She wants the left and right earrings to be congruent. Draw a diagram of both earrings and label the vertices. Write a congruence statement for the two triangles and find the measure of each angle in both earrings.

38. FASHION Janet cuts a triangle out of paper and traces it onto the clay twice to make the two earrings. Will the earrings be congruent? What property of congruence justifies your answer?

**Look Back**

Given: **ABCD** is a parallelogram. Justify each statement below with a theorem, property, or definition. (*LESSONS 3.2, 3.3, AND 3.4*)

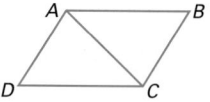

39. $\overline{AB} \parallel \overline{CD}$ **40.** $\overline{AD} \parallel \overline{BC}$ **41.** $\angle BAC \cong \angle DCA$

42. $\angle DAC \cong \angle BCA$ **43.** $\overline{AC} \cong \overline{CA}$

39. definition of parallelogram

40. definition of parallelogram

41. Alternate Interior Angles Theorem

42. Alternate Interior Angles Theorem

43. Reflexive Property of Congruence

Determine whether each of the following statements is true or false. Explain your reasoning. (*LESSON 3.2*)

44. A square is a rhombus.

45. A parallelogram is a trapezoid.

46. A parallelogram is a rhombus.

47. A rectangle is a parallelogram.

**Look Beyond**

COORDINATE GEOMETRY Plot the following points on a graph and connect them to form a rectangle: *A*(1, 2), *B*(1, 4), *C*(4, 4), and *D*(4, 2).

48. Use the rule $F(x, y) = (x + 5, y - 3)$ to transform the figure. Is the image congruent to the preimage? Why or why not?

49. Use the rule $F(x, y) = (3x, -2y)$ to transform the figure. Is the image congruent to the preimage? Why or why not?

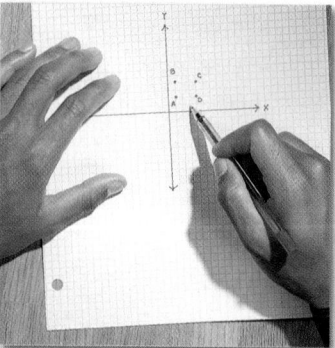

Triangle Congruence

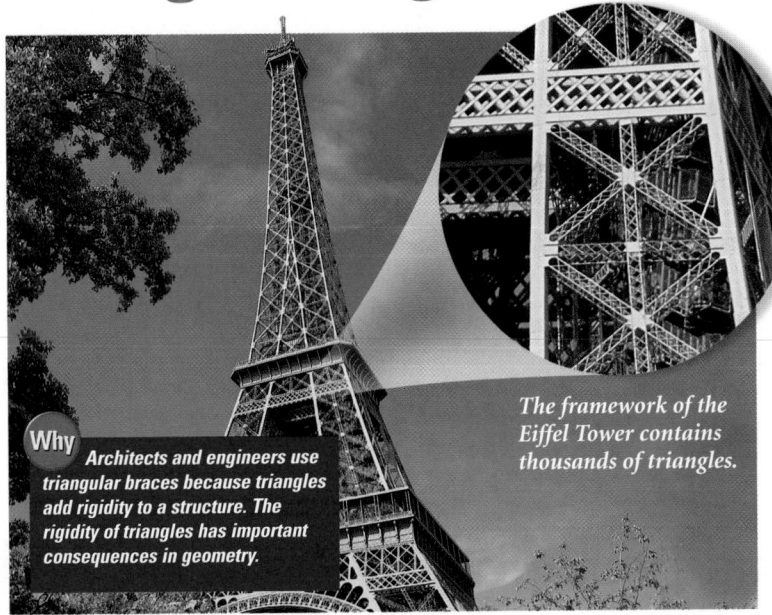

Why Architects and engineers use triangular braces because triangles add rigidity to a structure. The rigidity of triangles has important consequences in geometry.

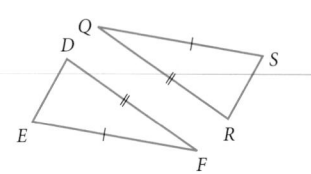

The framework of the Eiffel Tower contains thousands of triangles.

Objectives

- Explore triangle rigidity.

- Develop three congruence postulates for triangles—SSS, SAS, and ASA.

Prepare

QUICK WARM-UP

Triangles *QRS* and *FDE* are congruent. Write all pairs of corresponding parts.

$\overline{DF}$ and $\overline{RQ}$, $\overline{RS}$ and $\overline{DE}$, $\overline{EF}$ and $\overline{SQ}$, $\angle D$ and $\angle R$, $\angle Q$ and $\angle F$, $\angle S$ and $\angle E$

Also on Quiz Transparency 4.2

Triangles in Physical Structures

A physical triangle is rigid. As long as the sides do not change or break loose, the shape of the triangle will not change. The rigidity of triangles contributes a postulate to geometry, which you will investigate in Activity 1 below.

Activity 1
Triangle Rigidity

YOU WILL NEED

drinking straws, string, a ruler, and scissors

1. Construct each triangle described below from drinking straws and string. When you pull the strings tight, your triangles will be rigid.

   $\triangle ABC$:
   $AB = 8$ in., $BC = 9$ in., $CA = 10$ in.

   $\triangle XYZ$:
   $XY = 6$ in., $YZ = 8$ in., $ZX = 10$ in.

2. Compare your triangles with those made by other members of your class. Are your triangles congruent to theirs?

3. Given the lengths of the sides of a triangle, can the triangle have more than one shape? How does your answer to this question relate to the fact that triangles are rigid?

4. Do you need to know the angle measures of a triangle to make a copy of it? Explain.

Teach

Why Triangles are rigid figures and are thus often used as part of structures. Triangular braces can be found in the structures of many bridges, for example. Congruence postulates are shortcuts for showing triangle congruency.

☞ For Activity Notes and the answer to the Checkpoint, see page 218.

Alternative Teaching Strategy

TECHNOLOGY Students can use geometry graphics software to construct the triangles in Activities 1 and 2. Students can copy and paste angles and segments and fit the parts together to make a triangle, or they can alter the lengths of segments by using the measurement feature of the software.

Activity 1 **Notes**

In this Activity, students construct and compare triangles with given side lengths in order to develop an understanding of the SSS Congruence Postulate.

For a student worksheet of this Activity and detailed Teacher Notes, see page 55 in the Lesson Activities booklet.

Teaching Tip

If students are to bring their own materials for constructions, notify them in advance to ensure that they have all the necessary supplies with them at classtime.

CHECKPOINT ✔

5. If the sides of one triangle are congruent to the sides of another triangle, then the two triangles are congruent.

CRITICAL THINKING

No. For example, many different quadrilaterals may be drawn with the same side lengths.

Activity 2 **Notes**

In this Activity, students construct and compare triangles in which different combinations of sides and angles are congruent. Students are expected to develop an understanding of the SAS and ASA Congruence Postulate.

For a student worksheet of this Activity and detailed Teacher Notes, see page 55 in the Lesson Activities booklet.

CHECKPOINT ✔ 5. Your work with the triangle in Steps 1–4 should suggest the important geometry postulate that is partially given below. Complete it by filling in the blanks.

SSS (Side-Side-Side) Postulate

If the ____?____ of one triangle are congruent to the ____?____ of another triangle, then the two triangles are ____?____. **4.2.1**

CRITICAL THINKING Are any polygons other than triangles rigid? If you repeat Activity 1 with quadrilaterals instead of triangles, would your results be similar? If the corresponding sides of two quadrilaterals are congruent, must the quadrilaterals be congruent?

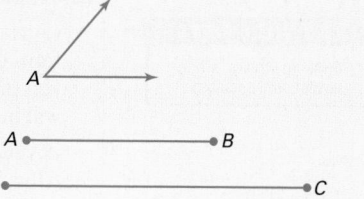

Useful Geometry Tools

If you use the Polygon Congruence Postulate on page 212 to show that two triangles are congruent, you must show that three pairs of sides are congruent and three pairs of angles are congruent. Postulate 4.2.1, which you discovered in Activity 1, provides a shortcut for showing that two triangles are congruent. You need to show only that three pairs of sides are congruent. In Activity 2, you will discover two more shortcuts for proving triangle congruence.

Activity 2
Two More Congruence Postulates

YOU WILL NEED

6 pieces of tracing paper and a straightedge

Part I

1. Trace each figure at right onto a separate piece of tracing paper near the center of the paper.

2. Arrange the pieces of paper on top of each other so that point A of each segment is over the vertex of $\angle A$ and the segments fall along the two rays of the angle.

3. On the top piece of paper, draw a segment connecting points B and C. Trace the rest of $\triangle ABC$.

Interdisciplinary Connection

ARCHITECTURE Have students organize photographs and illustrations of bridges for a bulletin board display. The way triangular elements are used for bracing will be more apparent in bridges than in other structures, and the triangles found in the bridges can be used as examples of congruency.

4. Compare your triangle with those of other members of your class. Are the triangles congruent?

5. In a triangle, an angle formed by two sides of the triangle is called the *included angle* of the two sides. (Informally, it is the angle "between" the two sides.) If two sides and their included angle are fixed in a triangle, is the size and shape of the triangle fixed?

CHECKPOINT ✔

6. Your work with the triangle in Steps 1–5 should suggest the important geometry postulate that is partially given below. Complete the postulate by filling in the blanks.

SAS (Side-Angle-Side) Postulate

If two ____?____ and their ____?____ in one triangle are congruent
to two ____?____ and their ____?____ in another triangle, then the
two triangles are ____?____.   **4.2.2**

Part II

1. Trace each figure at right onto a separate piece of tracing paper near the center of the paper.

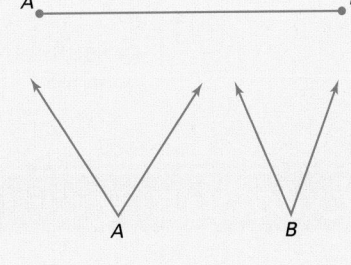

2. Arrange the pieces of tracing paper on top of each other so that the vertices of the angles are over the endpoints of the segments, one side of each angle lies along the segment, and the other sides of the angles are on the same side of the segment.

3. On the top piece of paper, extend the sides of the angles so that they meet to form a triangle. Trace the rest of △*ABC*.

4. Compare your triangle with those of other members of your class. Are the triangles congruent?

5. In a triangle, a side that is part of two angles of the triangle is called the *included side* of the two angles. (Informally, it is the side "between" the two angles.) If two angles and their included side are fixed in a triangle, is the size and shape of the triangle fixed?

CHECKPOINT ✔

6. Your work with the triangle in Steps 1–5 should suggest the important geometry postulate that is partially given below. Complete the postulate by filling in the blanks.

ASA (Angle-Side-Angle) Postulate

If two ____?____ and the ____?____ in one triangle are congruent to
two ____?____ and the ____?____ in another triangle, then the two
triangles are ____?____.   **4.2.3**

Teaching Tip

An alternative to tracing an angle is to make templates from cardboard. Students can trace around a cutout of an angle and then extend the rays of the angle to the lengths specified in the text.

Cooperative Learning

Have students do the Activities in small groups with modeling materials or paper and pencil. Group work promotes student interaction, cooperative learning and efficient task completion.

CHECKPOINT ✔
Part I

6. If two sides and the included angle in one triangle are congruent to two sides and the included angle in another triangle, then the two triangles are congruent.

CHECKPOINT ✔
Part II

6. If two angles and the included side in one triangle are congruent to two angles and the included side in another triangle, then the two triangles are congruent.

Enrichment

Have students check their reasoning about the following questions with geometry graphics software. Imagine triangle *ABC*. Move point *B* toward point *A* along side $\overline{AB}$. Which sides or angles in triangle *ABC* do not change? $\overline{AC}$, ∠*A* Which sides become shorter? $\overline{AB}$, $\overline{BC}$ Which sides become longer? none Which angle decreases in size? ∠*C* Which angle increases in size? ∠*B*

Inclusion Strategies

ENGLISH LANGUAGE DEVELOPMENT Some students may have difficulty understanding the idea of a side that is included between two angles or an angle that is included between two sides. Have students complete statements such as the following: In triangle *ABC*, the angle included between sides *AB* and *BC* is _____.

Assess

Using the New Postulates

The triangle postulates you discovered in the activities of this lesson allow you to save steps in proofs (compare with the proof on pages 212–213). But much more important is the fact that they allow you to determine triangle congruence from limited information.

E X A M P L E   In each pair below, the triangles are congruent. Tell which triangle congruence postulate allows you to conclude that they are congruent, based on the markings in the figures.

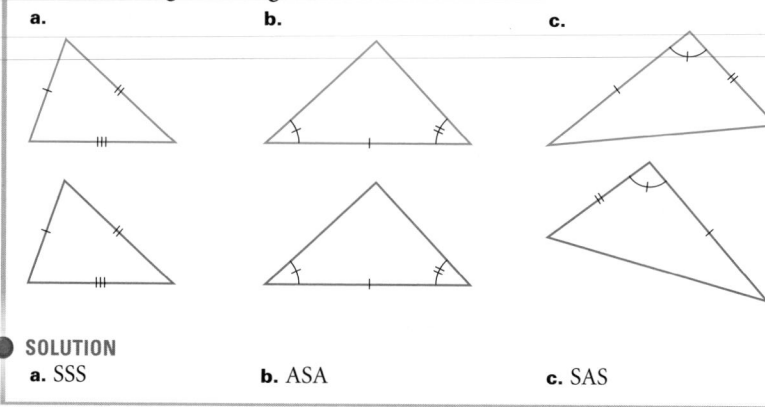

a.   b.   c.

SOLUTION

a. SSS   **b.** ASA   **c.** SAS

Exercises

Communicate

1. What is the advantage of using the SSS, SAS, and ASA Triangle Congruence Postulates instead of the Polygon Congruence Postulate given in Lesson 4.1?

2. Given ∠A at right, describe how you would create △ABC in which AB = 2 inches and AC = 3 inches.

3. When using the ASA Triangle Congruence Postulate, does it matter which two angles are given? Why or why not?

APPLICATION

4. **ENGINEERING** What is meant by triangle rigidity? How does this property of triangles make them especially useful for building structures such as bridges and towers?

Technology

In Exercises 14–19, students can use geometry graphics software to construct the given triangles.

● *Guided Skills Practice* ▬▬▬▬▬▬▬

For each pair below, tell which triangle congruence postulate allows you to conclude that the triangles are congruent. *(EXAMPLE)*

5. SAS
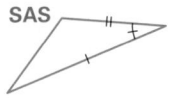

6. SSS
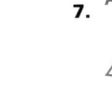

7. ASA
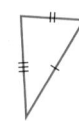

● *Practice and Apply* ▬▬▬▬▬▬▬

Determine whether each pair of triangles can be proven congruent by using the SSS, SAS, or ASA Congruence Postulate. If so, write a congruence statement and identify which postulate is used.

8.
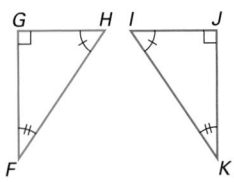
Can't be proven congruent

9.
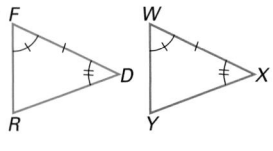
△*FDR* ≅ △*WXY*; ASA

10.
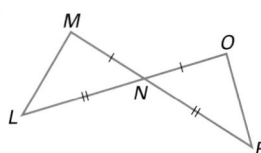
△*LMN* ≅ △*PON*; SAS

11.
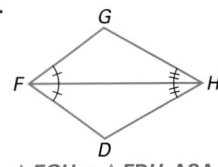
△*FGH* ≅ △*FDH*; ASA

12.
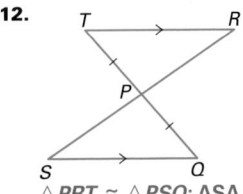
△*PRT* ≅ △*PSQ*; ASA

13.
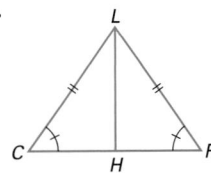
Can't be proven congruent

14. Yes, SSS

15. Yes, SAS

16. No, many triangles can be constructed.

17. Yes, ASA

18. Yes, ASA

19. No, more than one triangle can be constructed.

For Exercises 14–19, some measurements of a triangle are given. Is there exactly one triangle that can be constructed with the given measurements? If so, identify the postulate that applies.

14. △*ABC*: *AB* = 5, *AC* = 7, *BC* = 10

15. △*DEF*: *DE* = 14, *EF* = 12, m∠*E* = 75°

16. △*GHI*: m∠*G* = 60°, m∠*H* = 60°, m∠*I* = 60°

17. △*JKL*: *JK* = 3, m∠*J* = 45°, m∠*K* = 90°

18. △*MNO*: *MN* = 8, m∠*M* = 30°, m∠*N* = 110°

19. △*UVW*: m∠*U* = 40°, *UV* = 10, *VW* = 7

20.

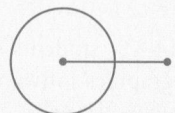

Any segment that connects this endpoint to a point on the circle is congruent to the second segment, because the compass setting used to create the circle was the length of the second segment.

21.

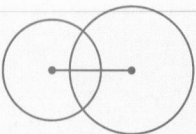

Any segment that connects this endpoint to a point on the circle is congruent to the third segment, because the compass setting used to create the circle was the length of the third segment.

22.

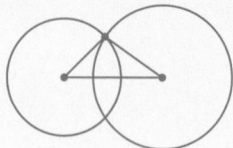

The figure is a triangle with side lengths equal to the measures of the original three line segments.

30. By definition of a regular polygon, all the sides of the polygon are congruent. If a segment is drawn from the center to each vertex, the polygon is divided into triangles. By definition of the center of a regular polygon, the segments from the center to the vertices are all congruent, so the triangles are all congruent by SSS. Thus, the central angles are congruent because CPCTC.

31. Given: Rhombus *ABCD*
Prove: Diagonal $\overline{DB}$ divides rhombus *ABCD* into two congruent triangles.

Statements	Reasons
$\overline{AB} \cong \overline{CD}$	Definition
$\overline{AD} \cong \overline{BC}$	of a rhombus
$\overline{BD} \cong \overline{DB}$	Reflexive Property of Congruence
$\triangle ABD \cong \triangle CDB$	SSS

In Exercises 20–22, you will create a triangle from three given sides.

20. Draw a segment the same length as one of the segments shown at right. Then set your compass to the length of one of the other segments. Place the compass at an endpoint of the first segment you drew, and construct a circle. What is true of any segment that connects this endpoint of the first segment to a point on the circle? Why?

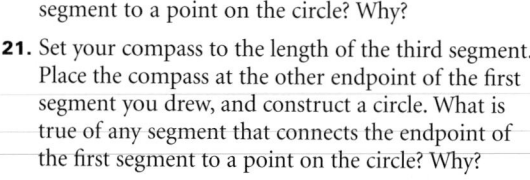

21. Set your compass to the length of the third segment. Place the compass at the other endpoint of the first segment you drew, and construct a circle. What is true of any segment that connects the endpoint of the first segment to a point on the circle? Why?

22. Connect both endpoints of the first segment to one of the points where the circles intersect. What is true of the figure you created? Why?

■ internet connect

Homework Help Online
Go To: go.hrw.com
Keyword:
MG1 Homework Help
for Exercises 23-31

For Exercises 23–29, complete the two-column proof below.

Given: rectangle *ABCD*

Prove: Diagonal $\overline{DB}$ divides rectangle *ABCD* into two congruent triangles.

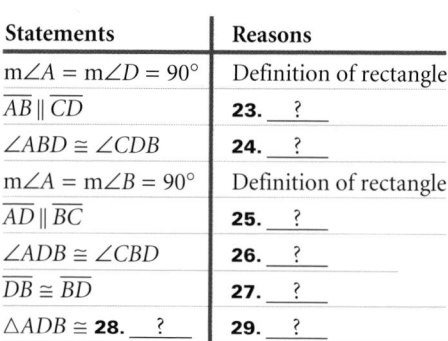

Proof:

Statements	Reasons
$m\angle A = m\angle D = 90°$	Definition of rectangle
$\overline{AB} \parallel \overline{CD}$	**23.** ?
$\angle ABD \cong \angle CDB$	**24.** ?
$m\angle A = m\angle B = 90°$	Definition of rectangle
$\overline{AD} \parallel \overline{BC}$	**25.** ?
$\angle ADB \cong \angle CBD$	**26.** ?
$\overline{DB} \cong \overline{BD}$	**27.** ?
$\triangle ADB \cong$ **28.** ?	**29.** ?

23. Converse of Same-Side Interior Angles Theorem

24. Alternate Interior Angles Theorem

25. Converse of Same-Side Interior Angles Theorem

26. Alternate Interior Angles Theorem

27. Reflexive Property of Congruence

28. $\triangle CBD$

29. ASA

CHALLENGE

APPLICATION

30. Refer to the definition of the center of a regular polygon on page 139. Write a paragraph proof that the central angles of a regular polygon are congruent.

31. Prove that a diagonal of a rhombus divides the rhombus into two congruent triangles.

32. Suppose that two congruent triangles share a common side. Is the figure formed by the two triangles sometimes, always, or never a parallelogram? Explain your reasoning.

33. CARPENTRY A carpenter is building a rectangular bookshelf and finds that it wobbles from side to side. To stabilize the bookshelf, he nails on a board that connects the top left corner to the bottom right corner. Why does this diagonal board stabilize the bookshelf?

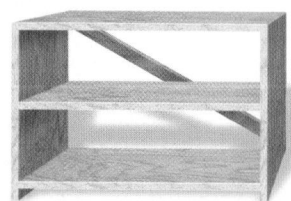

32. Sometimes; two congruent triangles which share a common side will either form a quadrilateral or a triangle. If they form a quadrilateral, they will form a parallelogram or a kite.

33. The diagonal board turns the rectangle into two congruent, rigid triangles.

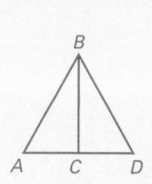

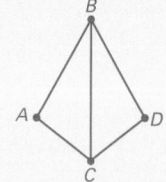

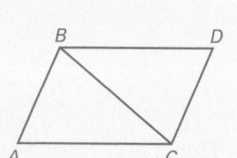

34. CONSTRUCTION In framing a house, certain triangles in the roof structure must be congruent. How can you be sure that the triangles are congruent without measuring any angles? Which triangle congruence postulate would you use?

35. QUILTING A quilter is making a quilt out of regular hexagons that are composed of triangles, as shown at right.

a. Is △FOA ≅ △COD? Why or why not?

b. Is △BOA ≅ △COD? Why or why not?

c. Explain two ways to prove that △FOE ≅ △COB.

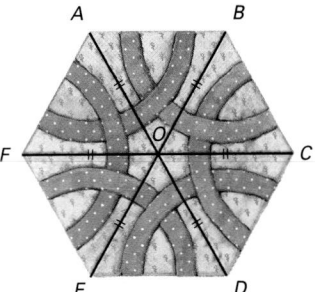

RECREATION In Exercises 36–39, pairs of triangular boat sails are shown. Based on the congruences shown, are the sails in each pair necessarily congruent?

36. Yes, by SAS

37. Yes, by SSS

38. No

39. Yes, by ASA

36.

37.

38.

39.

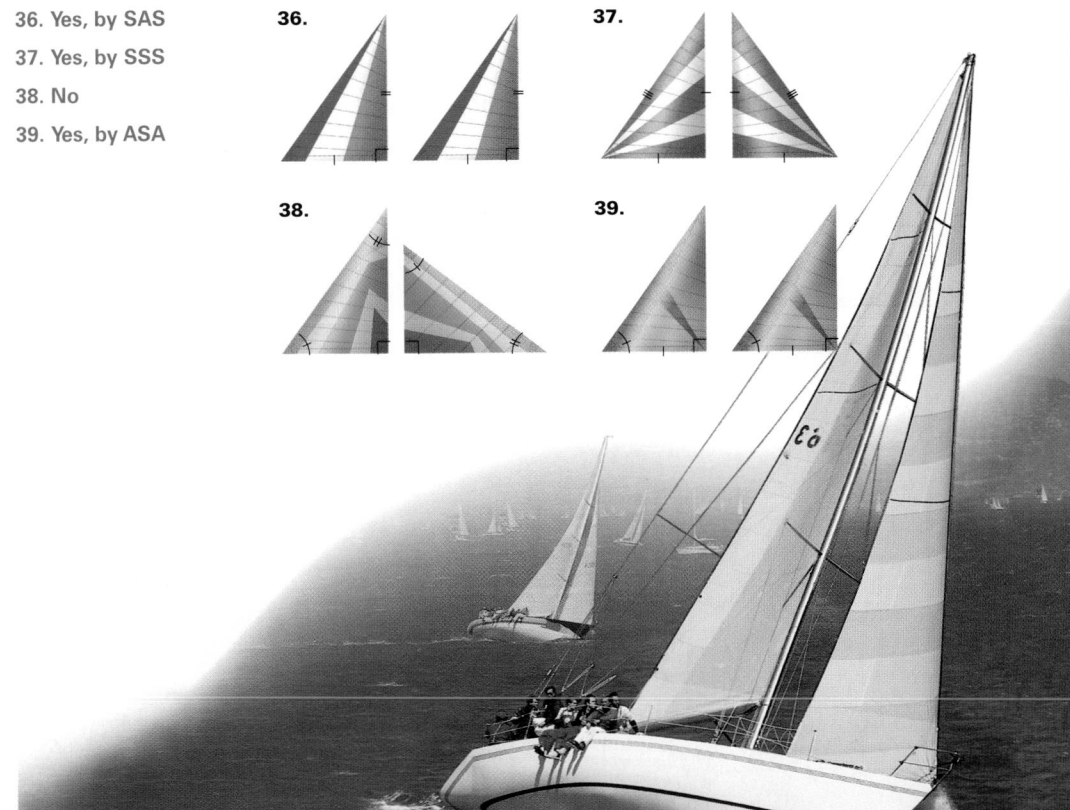

34. Check if the triangles are congruent by measuring the 3 sides of each triangle. By SSS, the triangles are congruent if the lengths of the corresponding sides are the same.

35. a. Yes; because vertical angles are congruent, ∠FOA ≅ ∠COD. It is given that $\overline{FO} \cong \overline{CO}$ and $\overline{OA} \cong \overline{OD}$. By SAS, △FOA ≅ △COD.

b. Yes; because it is a regular hexagon, $\overline{AB} \cong \overline{DC}$. It is given that $\overline{BO} \cong \overline{CO}$ and $\overline{OA} \cong \overline{OD}$. By SSS, △BOA ≅ △COD.

c. △FOE ≅ △COB by either SAS or SSS.

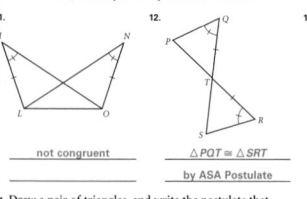
In Exercises 48 and 49, students use geostrips or straws to construct quadrilaterals and decide whether they are rigid. Students also explore special quadrilaterals, which will be studied later in this chapter.

48. No

49. When the congruent segments are opposite each other, it appears that a parallelogram is formed. When the congruent segments are adjacent, it appears that a kite is formed.

 **Look Back**

Complete the two-column proof below. *(LESSON 3.4)*

TWO-COLUMN PROOF

Given: $\ell_1 \parallel \ell_2$ and $\angle 2 \cong \angle 3$

Prove: $\ell_1 \parallel \ell_3$

Proof:

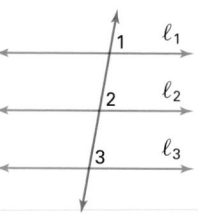

40. Given
41. Given
42. Corresponding Angles Postulate
43. Transitive Property of Congruence
44. Converse of Corresponding Angles Postulate

Statements	Reasons
$\ell_1 \parallel \ell_2$	**40.** ?
$\angle 2 \cong \angle 3$	**41.** ?
$\angle 1 \cong \angle 2$	**42.** ?
$\angle 1 \cong \angle 3$	**43.** ?
$\ell_1 \parallel \ell_3$	**44.** ?

Given: quadrilateral *JKLM* ≅ quadrilateral *WXYZ* *(LESSON 4.1)*

45. Identify an angle that is congruent to $\angle K$. $\angle X$

46. Identify a segment that is congruent to $\overline{KL}$. $\overline{XY}$

47. Is it impossible, possible, or definite that $\angle W \cong \angle L$? **Possible**

 **Look Beyond**

Use straws or geostrips to create four segments—two that are 3 inches long and two that are 5 inches long.

48. Form a quadrilateral from these segments. Is your quadrilateral rigid?

PROBLEM SOLVING

49. Make a model. What type of quadrilateral do you seem to get when the congruent segments are opposite each other? What type of quadrilateral do you seem to get when the congruent segments are adjacent to each other?

CREATING EQUILATERAL TRIANGLES

Use folding paper, geometry graphics software, or a ruler and protractor to draw each figure.

1. Draw two parallel lines, ℓ and m, and a transversal, p.

2. Bisect the obtuse angle formed by lines m and p. Label the angle bisector q.

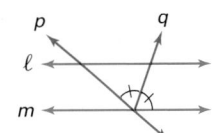

3. Bisect the obtuse angle formed by ℓ and q. Continue bisecting angles as shown, alternating between lines ℓ and m. What seems to be true about the triangles that are created?

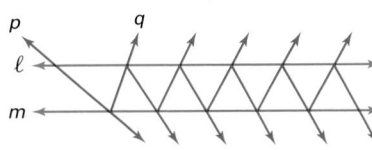

4. Use a long strip of paper with parallel edges. Fold the paper to create a transversal, and then unfold the paper.

To bisect the obtuse angle formed, fold the paper so that one edge matches exactly with the crease from the first fold. Alternate folding the strip up and down to create a long row of triangles.

☑ internet connect

Portfolio Extension
Go To: **go.hrw.com**
Keyword:
MG1 Triangle

5. The triangles should become more regular as you go. Keep making triangles until you get a row of nine equilateral triangles. Cut off the irregular triangles from the beginning. Using the row of nine equilateral triangles (without further cutting), how can you make a regular hexagon?

WORKING ON THE CHAPTER PROJECT

You should now be able to complete Activity 1 of the Chapter Project.

ALTERNATIVE
Assessment

Portfolio Activity

The Portfolio Activity can be used as preparation for the Chapter Project or as a separate activity. In the Portfolio Activity on this page, students make nine equilateral triangles on a strip of paper in preparation for the chapter project, *Flexagons.*

Analyzing Triangle Congruence

Objectives

- Identify and use the SSS, SAS, and ASA Congruence Postulates and the AAS and HL Congruence Theorems.

- Use counterexamples to prove that other side and angle combinations cannot be used to prove triangle congruence.

Why *In the previous lesson, you discovered three different three-part combinations of side and angle measures that determine the shape and size of a triangle. In this lesson, you will learn two more.*

APPLICATION
FORESTRY

In the method of *triangulation*, observers at two different places each take note of their line of sight to an object. The location of the object is the point where their lines of sight cross. By plotting the lines of sight on a map, the location of the object can be found. Why do you think this method is called *triangulation*? Think about this question as you continue your investigation of triangle congruence postulates and theorems.

Observers in towers often work in pairs.

Teach

Why The same information needed to prove that two triangles are congruent can be used to draw or build a triangle. For example, to build a swing set with triangular uprights, you can specify the top angle and the two adjoining lengths (SAS), or you can specify the lengths of the poles and the distance between their bases (SSS). Drawing a triangle by triangulation uses ASA.

Three New Possibilities

You may have realized that the three-part combinations of sides and angles you studied in the previous lesson are not the only combinations that are possible. In fact, there are three others. Which of the following combinations do you think can be used to establish triangle congruence?

Valid or not?

1. **AAA combination**—three angles

2. **AAS combination**—two angles and a side that is not between them

3. **SSA combination**—two sides and an angle that is not between them (that is, an angle opposite one of the two sides)

You can rule out the AAA combination by finding a counterexample, and a simple proof will show that the AAS combination is a valid test for triangle congruence. A counterexample can also be found for the SSA combination.

Alternative Teaching Strategy

TECHNOLOGY Many geometry software packages include activities for the three congruence postulates covered in this lesson. Students can use these computer files to explore the different cases.

E X A M P L E **①** Show that the AAA combination is not a valid test for triangle congruence.

● **SOLUTION**

In order to get a clear idea of what you need to *disprove*, state the combination in the form of a conjecture, as given below.

> Conjecture: If the three angles of one triangle are congruent to the three angles of another triangle, then the triangles are congruent.

Counterexamples to this statement are easy to find. In the triangles at right, there are three pairs of congruent angles, but the two triangles are not congruent. Therefore, the conjecture is *false*.

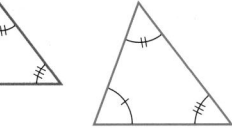

AAA counterexample

E X A M P L E **②** Show that the AAS combination is a valid test for triangle congruence.

● **SOLUTION**

The AAS combination can be converted to the ASA combination.

The triangles at right are an AAS combination. To convert this to an ASA combination, find the measures of the third angle in each triangle. Remember that the sum of the measures of the angles of a triangle is 180°.

> *Notice that the given sides are not between the given angles.*

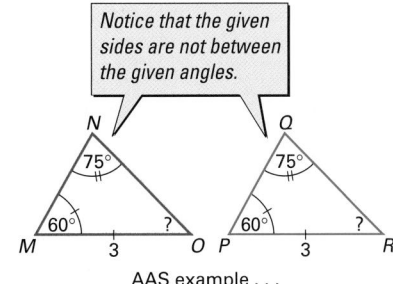

AAS example . . .

$$m\angle O = 180 - (60 + 75) = 45°$$
$$m\angle R = 180 - (60 + 75) = 45°$$

The measure of the third angle is the same in each triangle, so $\angle O \cong \angle R$. The given side, measuring 3 units, is the included side of the 45° and 60° angles in each triangle. Therefore, the two triangles are congruent by the ASA Congruence Postulate.

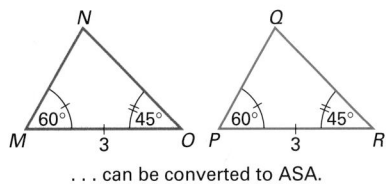

. . . can be converted to ASA.

CRITICAL THINKING

The two triangles at right represent a version of AAS, but the triangles are not congruent. There is an important difference between the two triangles. What is it?

The missing angle measure is 75°.

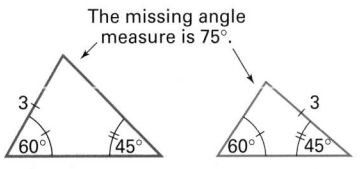

Is this an example of AAS?

Enrichment

It is not necessary to accept all of the conjectures (SAS, SSS, and ASA) as postulates. Challenge students to write the proofs for SSS and ASA, and use them to prove SAS.

Cooperative Learning

Ask students to create a counterexample to the SSA conjecture. To do this, each student should draw a triangle with one 35° angle, one 7-cm side, and one 4-cm side. Students should discover that two different triangles, one acute and one obtuse, satisfy the conditions.

For an AAS combination to be used, *the congruent parts must correspond.* Notice carefully the wording of the following theorem:

AAS (Angle-Angle-Side) Congruence Theorem

If two angles and a nonincluded side of one triangle are congruent to the corresponding angles and nonincluded side of another triangle, then the triangles are congruent.   **4.3.1**

The three combinations you studied in Lesson 4.2 are postulates, but the AAS combination is a theorem. You will be asked to prove this theorem in the exercise set.

TRY THIS Which pairs of triangles below can be proven to be congruent by the AAS Congruence Theorem?

a. 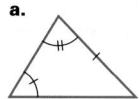   **b.** 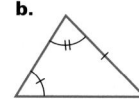   **c.** 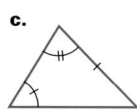   **d.**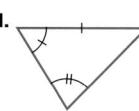

   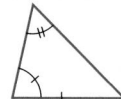      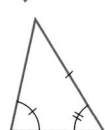

EXAMPLE ③ Show that the SSA combination is not a valid test for triangle congruence.

● **SOLUTION**

It is a little more difficult to find a counterexample for this conjecture, but the figures at right provide one. The sides and angles are congruent, but the triangles are obviously not congruent. Therefore, the conjecture is false.

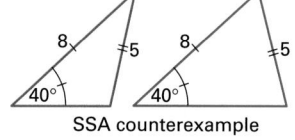

SSA counterexample

When you try to draw a triangle for an SSA combination, the side opposite the given angle can sometimes pivot like a swinging door between two possible positions. This "swinging door" effect show that two triangles are possible for certain SSA information.

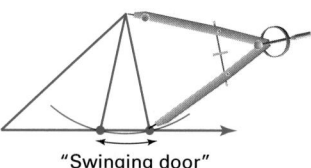

"Swinging door"

The "swinging door" effect, which invalidates certain SSA information as a test for triangle congruence, does not always pose a problem—as you will see in the next section. (See also Exercises 38–39.)

Inclusion Strategies

USING DISCUSSION The concepts introduced in this lesson can be studied by asking the question, "What three pieces of information are necessary and sufficient to enable you to draw or build a unique triangle?" Students will discover that the SSS, ASA, AAS, and SAS combinations work and that the AAA and SSA combinations do not.

A Special Case of SSA

If the given angle in an SSA combination is a right angle, then the "swinging door" side cannot pivot to touch the ray in two different places as in the illustration in Example 3.

Thus, if the given angle is a right angle, SSA can be used to prove congruence. In this case, it is called the Hypotenuse-Leg Congruence Theorem. You will be asked to prove this theorem in the exercise set for Lesson 4.4.

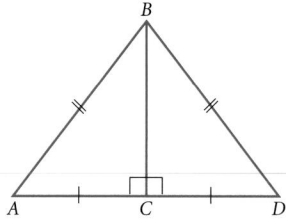

The perpendicular segment touches the line at just one point because it is the shortest segment that can be drawn from the point to the line.

HL (Hypotenuse-Leg) Congruence Theorem

If the hypotenuse and a leg of a right triangle are congruent to the hypotenuse and a leg of another right triangle, then the two triangles are congruent.

4.3.2

EXAMPLE ④ When Venus rises before the Sun it is known as a morning star. When it sets after the Sun, it is known as an evening star.

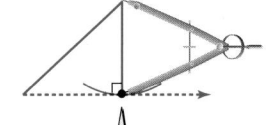
Venus

Venus rising ahead of the Sun as a morning star.

The diagram shows why Venus can be both a morning star and an evening star. Note: The daily rotation of the Earth and the orbits of the planets are both counterclockwise, viewed from above (from the north).

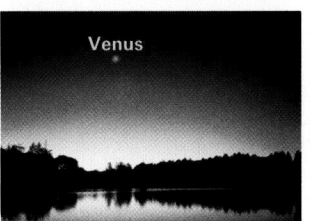

Venus as evening star
Venus as morning star
Sun
Earth

Venus is a morning star when it is to the right of the dashed line in the diagram. It is an evening star when it is on the left.

The orbits of Earth and Venus are nearly circular. The radius of Earth's orbit is about 1.5×10^8 kilometers, while the radius of Venus's orbit is about 1.1×10^8 kilometers. Make a scale drawing of the positions of the Sun, Earth, and Venus when Venus rises as a morning star 30° ahead of the Sun.

● **SOLUTION**

PROBLEM SOLVING

Make a diagram. There are two possible positions for Venus for a given position of Earth. This is an SSA combination, with the Sun, Earth, and Venus at the vertices of a triangle. The side connecting Venus and the Sun is the "swinging door."

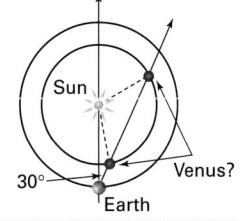

Sun
Venus?
30°
Earth

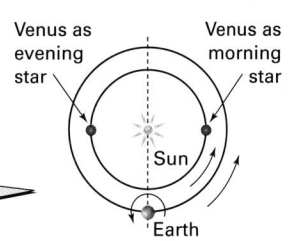
Reteaching the Lesson

USING MODELS To help students remember that SSA does not work, have them construct the figure shown at right. Students start with any isosceles triangle ABC and then they draw segment $\overline{AD}$ so that it meets the line through side $\overline{BC}$. Since $\overline{BA} \cong \overline{CA}$ and side $\overline{AD}$ and $\angle D$ are congruent to themselves, triangles BAD and CAD satisfy SSA. The drawing and the phrase "bad cad" may help students remember that SSA cannot be used to prove triangle congruency.

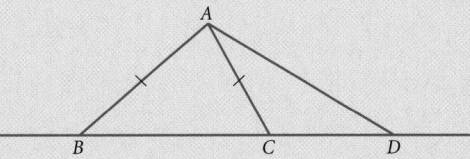

A
B
C
D

Selected Answers

Exercises 6–9, 11–49 odd

ASSIGNMENT GUIDE

In Class	1–9
Core	11–19 odd, 20–24, 25–37 odd
Core Plus	20–24, 25–43 odd
Review	44–50
Preview	51–52

✐Extra Practice can be found beginning on page 818.

Error Analysis

When writing proofs, students may need to review the Triangle Sum Theorem, the properties of parallel lines, and the properties of parallelograms.

Exercises

● Communicate

1. Think of a memory aid, or *mnemonic* device, to help you remember that SSS, SAS, ASA, and AAS are all valid tests for triangle congruence and that AAA and SSA are not valid tests.

For Exercises 2 and 3, is the given information sufficient to determine whether the triangles are congruent? Why or why not?

2. Any two corresponding angles and one corresponding side are congruent.

3. Any two corresponding sides and one corresponding angle are congruent.

4. Explain why any two corresponding sides are sufficient to determine congruence for right triangles.

APPLICATION

5. **FORESTRY** In the method of triangulation, a triangle has one vertex at the object being observed and one at each observer. Which angles and sides are known in this triangle? Which postulate or theorem guarantees that the triangle is "uniquely determined"—that it is the only possible triangle with the given measurements?

● Guided Skills Practice

For Exercises 6–8, is it possible to prove that the triangles are congruent? Explain your reasoning. *(EXAMPLES 1, 2, AND 3)*

6. No, AAA is not a valid test for congruence.

7. Yes, AAS is a valid test for congruence.

8. No, SSA is not a valid test for congruence.

6.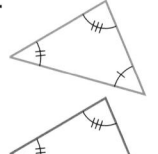

7.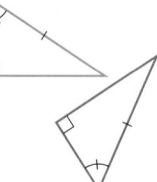

8.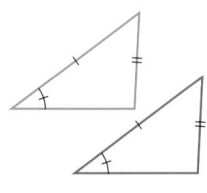

APPLICATION

9. 2×10^8 km and 0.5×10^8 km

9. **ASTRONOMY** Copy the diagram of the positions of the Sun, Earth, and Venus on page 229. Use the diagram to estimate the distance from Earth to Venus for both positions of Venus. *(EXAMPLE 4)*

Practice and Apply

internet connect

Homework Help Online
Go To: **go.hrw.com**
Keyword:
MG1 Homework Help
for Exercises 10-19

In Exercises 10–19, determine whether each pair of triangles can be proven congruent. If so, write a congruence statement and name the postulate or theorem used.

10.

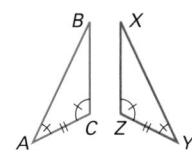

11.

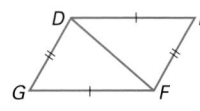

12.

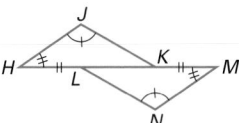

13.

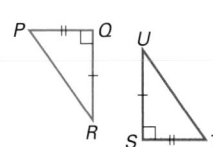

14.

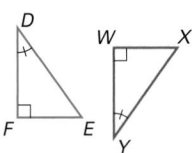

15.

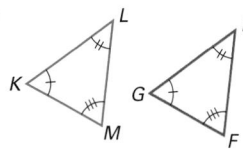

16.

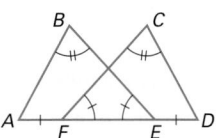

17.

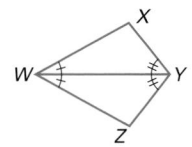

18.

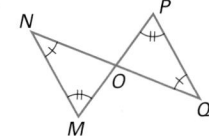

19.

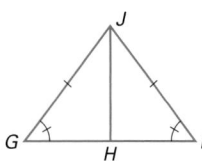

For Exercises 20–24, refer to △*XYZ* below, in which $\overline{XY} \cong \overline{XZ}$. Copy the figure and mark the congruent sides.

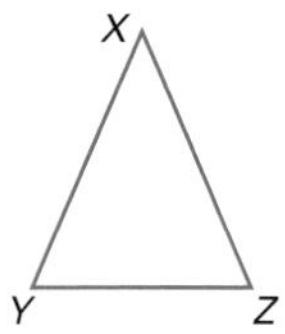

20. Draw a segment perpendicular to $\overline{YZ}$ with one endpoint at *X*. Label the other endpoint *W*.

21. Identify the two right triangles formed by $\overline{WX}$ and name the hypotenuse of each.

22. What can you say about the hypotenuses of the two right triangles? Explain your reasoning.

23. Which property proves that $\overline{WX} \cong \overline{WX}$?

24. What can you say about the two right triangles in the figure? Which theorem or postulate justifies this conclusion?

10. △*ABC* ≅ △*YXZ*; ASA

11. △*DEF* ≅ △*FGD*; SSS

12. △*HJK* ≅ △*MNL*; AAS

13. △*PQR* ≅ △*TSU*; SAS

14. Can't be proven congruent

15. Can't be proven congruent

16. △*ABE* ≅ △*DCF*; AAS

17. △*WXY* ≅ △*WZY*; ASA

18. Can't be proven congruent

19. Can't be proven congruent

Technology

Students can use geometry graphics software to duplicate the segments and angles of the figures in Exercises 10–19. Making these constructions will help reinforce the congruence postulates.

Math CONNECTION

ALGEBRA In Exercise 49, students use the Triangle Sum Theorem to set up an equation and solve for *x*.

20.

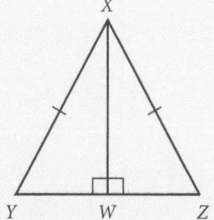

21. △*XWY* has hypotenuse $\overline{XY}$; △*XWZ* has hypotenuse $\overline{XZ}$.

22. $\overline{XY} \cong \overline{XZ}$; Given

23. Reflexive Property of Congruence

24. △*WXY* ≅ △*WXZ* by HL.

25.

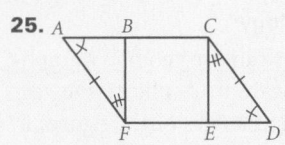

Statements	Reasons
$\angle A \cong \angle D$	Given
$\overline{AF} \cong \overline{DC}$	
$\angle BFA \cong \angle ECD$	
$\triangle AFB \cong \triangle DCE$	ASA

26.

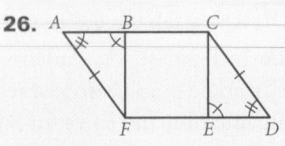

Statements	Reasons
$\angle 1 \cong \angle 4$	Given
$\overline{AF} \cong \overline{DC}$	
$\angle A \cong \angle D$	
$\triangle AFB \cong \triangle DCE$	AAS

35.

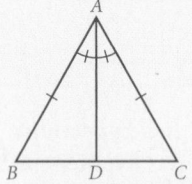

a. $\triangle ABD \cong \triangle ACD$ by SAS.

b. $\angle ABD \cong \angle ADC$ because CPCTC. Because they form a linear pair and are congruent, $\angle ADB$ and $\angle ADC$ are right angles, so $m\angle ADB = m\angle ADC = 90°$.

Practice

For each pair of triangles in Exercises 1–6, is it possible to prove the triangles congruent? If so, write a congruence statement and name the postulate or theorem used.

1.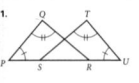

yes; $\triangle PQR \cong \triangle UTS$
by ASA Postulate

2.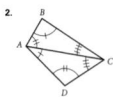

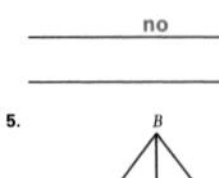

no

3.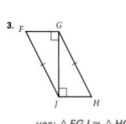

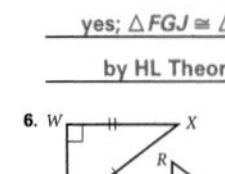

yes; $\triangle FGJ \cong \triangle HGJ$
by HL Theorem

4.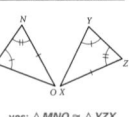

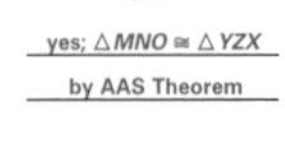

yes; $\triangle MNO \cong \triangle YZX$
by AAS Theorem

5.

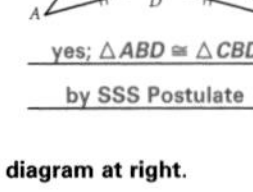

yes; $\triangle ABD \cong \triangle CBD$
by SSS Postulate

6.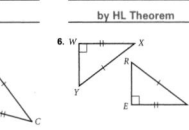

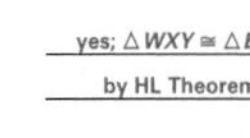

yes; $\triangle WXY \cong \triangle ERG$
by HL Theorem

For Exercises 7–9, refer to the diagram at right.

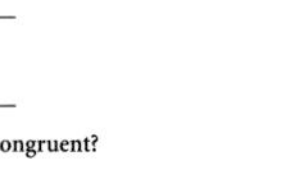

7. $\triangle ABE \cong \triangle CEB$ by _____ AAS

8. $\triangle EDC \cong \triangle CBE$ by _____ ASA

9. $\triangle EDC \cong \triangle AEB$ by _____ Transitive Property of Congruence

10. Of the three triangles described below, which two are congruent?

$\triangle XYZ$: $m\angle X = 40°$, $XY = 9$, and $m\angle Y = 30°$
$\triangle ABC$: $AB = 9$, $m\angle B = 30°$, and $m\angle A = 80°$
$\triangle KLM$: $m\angle L = 30°$, $m\angle M = 110°$, and $KL = 9$

$\triangle XYZ \cong \triangle KLM$

For Exercises 25 and 26, refer to the diagram below.

25. Given: $\angle A \cong \angle D$, $\overline{AF} \cong \overline{DC}$, and $\angle BFA \cong \angle ECD$
Prove: $\triangle AFB \cong \triangle DCE$

26. Given: $\angle 1 \cong \angle 4$, $\overline{AF} \cong \overline{DC}$, and $\angle A \cong \angle D$
Prove: $\triangle AFB \cong \triangle DCE$

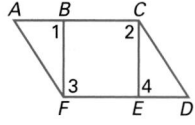

For Exercises 27–29, refer to the diagram below.

27. $\triangle JKL \cong \triangle MKL$ by ___?___ AAS

28. $\triangle MKL \cong \triangle MNL$ by ___?___ SAS

29. $\triangle JKL \cong \triangle MNL$ by ___?___
Transitive Property of Congruence

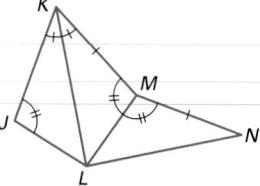

For Exercises 30–34, determine whether the given combination of angles and sides determines a unique triangle. If so, identify the theorem or postulate that supports your answer.

30. $\triangle ABC$: $AB = 6$, $m\angle B = 70°$, and $m\angle A = 40°$ Yes; ASA

31. $\triangle DEF$: $DE = 5$, $EF = 7$, and $m\angle F = 30°$ No

32. $\triangle JKL$: $m\angle J = 50°$, $m\angle K = 75°$, and $m\angle L = 55°$ No

33. $\triangle MNO$: $MN = 8$, $MO = 10$, and $m\angle N = 90°$ Yes; HL

34. $\triangle PQR$: $PQ = 12$, $m\angle P = 45°$, and $m\angle R = 100°$ Yes; AAS

For Exercises 35 and 36, copy equilateral triangle ABC below.

35. Draw the angle bisector of $\angle A$, and label the intersection with $\overline{BC}$ as D.

   a. What can you say about $\triangle ABD$ and $\triangle ACD$? Which postulate or theorem justifies your answer?

   b. What are $m\angle ADB$ and $m\angle ADC$? Explain your reasoning.

CHALLENGE

36. Draw the bisectors of $\angle B$ and $\angle C$, labeling the intersections with the opposite sides as E and F, respectively. Write a paragraph proof that $\overline{AD}$, $\overline{BE}$, and $\overline{CF}$ divide $\triangle ABC$ into six congruent triangles. Hint: Use your results from Exercise 35.

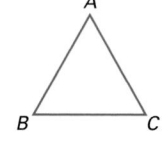

TWO-COLUMN PROOF

37. Write a two-column proof of the AAS Congruence Theorem.

CHALLENGE

In Exercises 38 and 39 you will further explore SSA combinations. If two sides and a nonincluded angle of two triangles are congruent, then the triangles are not necessarily congruent. However, if certain restrictions are placed on the side lengths or angle measure, it is possible to show that the triangles are congruent (see Lesson 10.4).

38. Try to draw $\triangle ABC$ with each set of side and angle measures below. How many different triangles can you draw for each set of measurements?
   a. $AB = 4$, $BC = 1$, and $m\angle A = 75°$
   b. $AB = 3$, $BC = 2$, and $m\angle A = 60°$
   c. $AB = 5$, $BC = 4$, and $m\angle A = 100°$

---

36. **Given:** $\triangle ABC$ is equilateral
   $\overline{AD}$ bisects $\angle A$
   $\overline{BE}$ bisects $\angle B$
   $\overline{CF}$ bisects $\angle C$
Prove: $\triangle AFX \cong \triangle BFX \cong$
   $\triangle BDX \cong \triangle CDX \cong$
   $\triangle CEX \cong \triangle AEX$

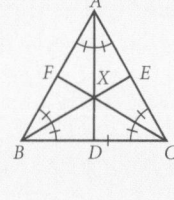

Because $\triangle ABC$ is equilateral, $\angle A \cong \angle B \cong \angle C$. Because $\angle XAE$, $\angle XAF$, $\angle XBF$, $\angle XBD$, $\angle XCD$, and $\angle XCE$ are formed by bisecting the congruent vertices of $\triangle ABC$, they are congruent. From Exercise 35, we know that each angle bisector forms 2 congruent triangles. By CPCTC, we know $\overline{BD} \cong \overline{DC}$, $\overline{CE} \cong \overline{EA}$, and $\overline{AF} \cong \overline{FB}$. Thus, each side of the equilateral triangle has been bisected so $\overline{BD} \cong \overline{DC} \cong \overline{CE} \cong \overline{EA} \cong \overline{AF} \cong \overline{EB}$. From Exercise 35, we also know that the bisector forms two right angles with the side opposite the vertex. Thus, $\angle XEA \cong \angle XEC \cong \angle XDC \cong \angle XDB \cong \angle XFB \cong \angle XFA$. By ASA, $\triangle AFX \cong \triangle BFX \cong \triangle BDX \cong \triangle CDX \cong \triangle CEX \cong \triangle AEX$.

38. a. none   b. none   c. none

39. Try to draw △*ABC* with each set of side and angle measures below. How many different triangles can you draw for each set of measurements?

   a. *AB* = 1, *BC* = 4, and m∠*A* = 75°
   b. *AB* = 2, *BC* = 3, and m∠*A* = 60°
   c. *AB* = 4, *BC* = 5, and m∠*A* = 100°
   d. Complete the following conjecture:

SSA Conjecture

In an SSA combination, if the given side opposite the given angle is ____?____ the other given side, then the triangle is uniquely determined. Under these conditions, SSA can be used to establish congruence.

Test your conjecture by making up some triangles of your own.

APPLICATIONS

40. ARCHAEOLOGY A student is estimating the height of a pyramid. From a certain distance, the angle of elevation of a point on the highest part of the structure is 25°. From a distance of 190 feet closer, the angle of elevation of the point is 30°. Draw a triangle with the point at the top of the structure as one vertex, and the points where the measurements were made as the other vertices. Which postulate or theorem can be used to show that this triangle is uniquely determined?

The Temple of Kukulkan at Chichen Itza. The height of the structure can be estimated by a method that involves creating a triangle as described in Exercise 40. The actual computations involve trigonometry, which you will study in Chapter 10.

41. ARCHAEOLOGY Make a drawing of the triangle for Exercise 40, using the scale 100 feet = 1 centimeter. Use your drawing to estimate the height of the pyramid.

42. ENGINEERING An embankment rises at an angle of 20° from horizontal. On the embankment, two 10-foot vertical poles are each anchored by a guy wire to the ground directly uphill from each pole. If the guy wire of each pole makes an angle of 60° with the embankment, must the wires be the same length? Which postulate or theorem justifies your answer? **Yes, AAS**

43. NAVIGATION Towns *A*, *B*, and *C* are connected by straight roads. For each set of measurements given below, draw a possible map, and determine whether △*ABC* on your map is uniquely determined. If so, state which theorem or postulate supports your answer.

   a. *AB* = 12 miles, *BC* = 7 miles, and *AC* = 8 miles
   b. *AB* = 6 miles, m∠*BAC* = 40°, and m∠*ABC* = 60°
   c. m∠*BAC* = 75°, m∠*ABC* = 50°, and m∠*ACB* = 55°

b.

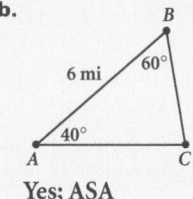

Yes; ASA

c.

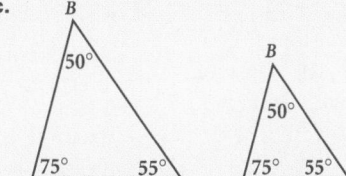

No; AAA is not a valid congruence postulate.

---

39. a. one
   b. one
   c. one
   d. longer than

40.

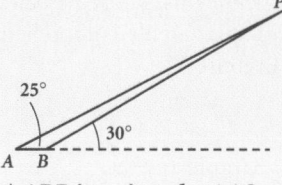

△*ABP* is unique by AAS or ASA.

41.

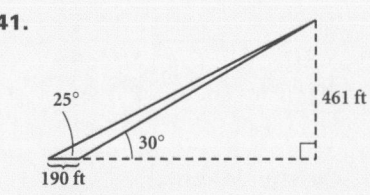

Height ≈ 461 ft

42. Yes; AAS

43. a.

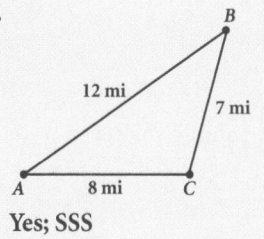

Yes; SSS

Student Technology Guide

Student Technology Guide
4.3 *Analyzing Triangle Congruence, page 1*

Geometry software can be used to help model real-world situations and solve related problems. Consider the following problem:

An embankment makes a 20° angle with the horizontal. This is modeled by ∠*CAB*. Two vertical poles *EF* and *DG*, each 10 feet high, are supported by cables anchored downslope. Each cable makes a 60° angle with the embankment. Are the cables the same length?

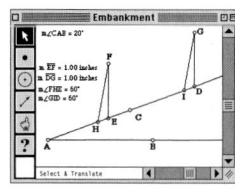

In Exercises 1–3, use geometry graphics software to answer the question above.   Check students' sketches.

1. **a.** Sketch ∠*CAB* such that m∠*CAB* = 20°. Make $\overrightarrow{AB}$ horizontal.
   b. Place two points on $\overrightarrow{AC}$, label them *E* and *D*, and then sketch $\overline{EF}$ and $\overline{DG}$ such that they are vertical and have the same length. (In the sketch above, 1.00 inch represents 10 feet.)
   c. Place two points, *H* and *I*, on $\overrightarrow{AC}$, and then construct $\overline{HF}$ and $\overline{IG}$ such that m∠*FHE* = m∠*GID* = 60°.

2. Does the sketch suggest that *HF* = *IG*? Display *HF* and *IG* to confirm your observation. Then justify your response by using a triangle congruence theorem. *HF* = *IG* = 1.09 inches; $\overline{FE}$ and $\overline{DG}$ are both vertical, so they lie along parallel lines. Thus, ∠*FEH* ≅ ∠*GDI* because they are corresponding angles. △*FEH* ≅ △*GDI* by the AAS Triangle Congruence Theorem. By CPCTC, $\overline{FH}$ ≅ $\overline{GI}$; so *FH* = *GI*.

3. Briefly explain how an accurate computer-aided sketch helps you to be certain about the equality of the cable lengths. In a computer-aided sketch, congruence is much more evident than in a hand-drawn sketch, which may or may not show relative sizes of things.

4. Modify your sketch so that m∠*CAB* = 15°. Keep $\overrightarrow{AB}$ horizontal. Make adjustments so that parts **b** and **c** of Exercise 1 stay the same. Do the support cables still have the same length? What can you conclude about the given information? The adjustment has no effect on the problem. The information about the angle of the embankment can be considered extraneous.

44. Sample answer:

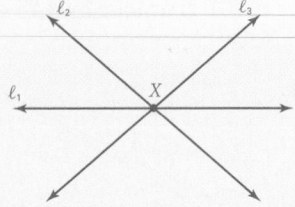

45. Sample answer:

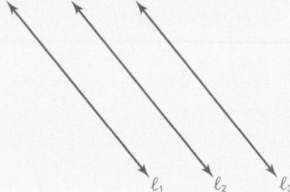

46. Sample answer:

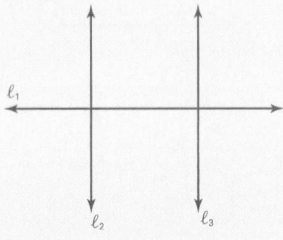

47. $\angle 1 \cong \angle 3 \cong \angle 5 \cong \angle 7$
$\angle 2 \cong \angle 4 \cong \angle 6 \cong \angle 8$

48. Sample answer:
$m\angle 1 + m\angle 4 = 180°$; Same-Side Exterior Angles

51. No; the side lengths would have to be the same.

52. Yes; yes. Each hexagon has three common sides with other hexagons on the soccer ball. Thus, the hexagons must all have the same side length, s. Since each pentagon has five common sides with hexagons on the soccer ball, the pentagons must also have side length s. Thus, all the hexagons are congruent to each other and all the pentagons are congruent to each other.

Look Back

44. Draw three lines that intersect at a point. *(LESSON 1.1)*

45. Draw three lines that do not intersect. *(LESSON 1.1)*

46. Draw two lines that intersect a third line without intersecting each other. *(LESSON 1.1)*

In the diagram at right, $\ell \parallel m$.

47. Identify all congruent angles in the diagram at right. *(LESSON 3.3)*

48. Identify a pair of supplementary angles that are not a linear pair in the diagram at right. What is this type of angle pair called? *(LESSON 3.3)*

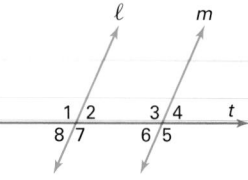

Algebra

49. The angles of a triangle measure $(2x)°$, $(3x)°$, and $(4x)°$. Find each angle measure. *(LESSON 3.5)* 40°, 60°, 80°

50. What can you say about the interior angles of a triangle that has two congruent exterior angles? *(LESSON 3.6)*
Two of the interior angles are congruent.

Look Beyond

51. Consider two regular pentagons. Are they necessarily congruent? What information would you need to have in order to show that two regular polygons are congruent?

APPLICATION

52. SPORTS A soccer ball is composed of regular pentagons and regular hexagons. Are all of the regular pentagons congruent? Are all of the regular hexagons congruent? Explain your reasoning.

Using Triangle Congruence

Objectives

- Use congruence of triangles to conclude congruence of corresponding parts.

- Develop and use the Isosceles Triangle Theorem.

M. C. Escher's "Square Limit" ©1999 Cordon Art B. V. -Baarn-Holand. All rights reserved.

Why *The design elements of this woodcut by M. C. Escher are mathematically related. The size of one part determines the size of another. By using chains of mathematical reasoning, you can deduce things about one part of a figure from information about another part.*

Prepare

QUICK WARM-UP

Given: *ABCD*, shown below, is a rectangle.
Prove: Triangles *ABC* and *CDA* are congruent by ASA.

Since opposite sides of a parallelogram are congruent and parallel, $\overline{AB} \cong \overline{DC}$. $\angle BAC$ is congruent to $\angle DCA$ by the Parallel Postulate. Since m$\angle B = 90°$, $\angle B$ and $\angle D$ are congruent. Therefore, the triangles are congruent by ASA.

Also on Quiz Transparency 4.4

CPCTC in Flowchart Proofs

It follows from the Polygon Congruence Postulate (given in Lesson 4.1) that if two triangles are congruent, then their corresponding parts are congruent. Therefore, if $\triangle ABC \cong \triangle DEF$, you can conclude that $\overline{AB} \cong \overline{DE}$. What other pairs of sides and angles must be congruent?

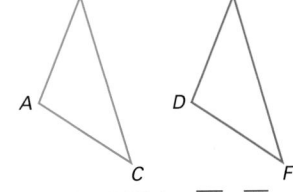

If $\triangle ABC \cong \triangle DEF$, then $\overline{AB} \cong \overline{DE}, \dots$

This idea is often stated in the following form: *Corresponding parts of congruent triangles are congruent*, abbreviated as **CPCTC**. In each proof in this lesson, you will use a triangle congruence postulate or theorem to establish that two triangles are congruent. Then you will use CPCTC.

In the following examples, the **flowchart proof** is introduced. This type of proof will be especially useful for undersanding more complicated proofs that occur later in the book, but it is best to begin with simple examples.

Teach

Why Applications of congruent triangles occur in home building, interior decorating, art, road signs, etc. Knowing how to prove that triangles are congruent enhances students' appreciation of the applicability of congruent triangles in many real-world situations.

Alternative Teaching Strategy

COOPERATIVE LEARNING Give small groups of students a diagram and something to prove about it. Have each group discuss the plan for the proof. Then have each member of the group write at least one step of the proof, according to the group's plan. Have the class compare and evaluate each group's proof.

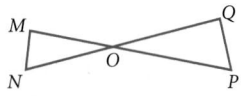

Given: Two segments, $\overline{MP}$ and $\overline{NQ}$, bisect each other at point O.
Prove: $\overline{MN} \cong \overline{PQ}$

Statements
1. $\overline{NO} \cong \overline{QO}$
2. $\overline{OM} \cong \overline{OP}$
3. $\angle NOM \cong \angle QOP$
4. $\triangle MNO \cong \triangle PQO$
5. $\overline{MN} \cong \overline{PQ}$

Reasons
1. Def. of bisector
2. Def. of bisector
3. Vertical Angle Theorem
4. SAS
5. CPCTC

CRITICAL THINKING
If angles C and D were not known to be congruent, the given vertical angles would give SSA information, which does not ensure triangle congruence.

E X A M P L E ❶ **Given:** $\overline{AC} \cong \overline{BD}$, $\overline{CX} \cong \overline{DX}$, and $\angle C \cong \angle D$

Prove: Point X is the midpoint of $\overline{AB}$.

Plan: Use congruent triangles to show that $\overline{AX} \cong \overline{BX}$ (and thus, $AX = BX$). Then use the definition of a midpoint.

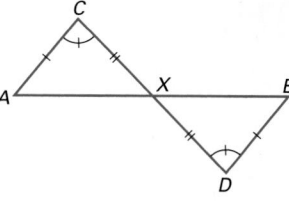

SOLUTION

Arrange the information as a flowchart, with the given information in boxes on the left (or at the top). The goal is for each of your boxes to lead to the desired conclusion, which is usually in a box on the right side (or at the bottom) of the chart. Write the justification for each statement below its box.

FLOWCHART PROOF

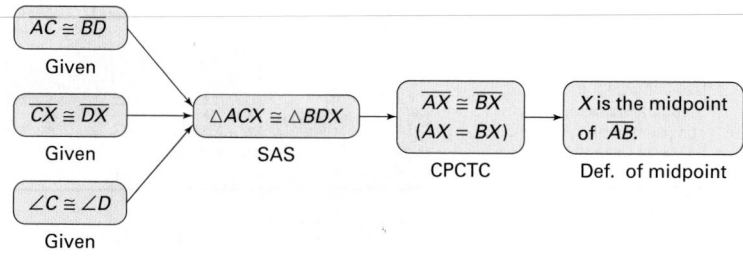

You can also write the statements in the boxes and their justifications as a two-column proof. Which way is easier for you? Which way do you think is easier for another person to read and understand?

CRITICAL THINKING Why can vertical angles $\angle AXC$ and $\angle BXD$ *not* be used in this proof in place of $\angle C$ and $\angle D$?

E X A M P L E ❷ **Given:** $\overline{AB} \cong \overline{CD}$, $\overline{AE} \cong \overline{FD}$, $\angle A \cong \angle D$

Prove: $\overline{EC} \cong \overline{FB}$

Plan: Use the Overlapping Segments Theorem to show that $\overline{AC} \cong \overline{BD}$ and thus, $\triangle ACE \cong \triangle DBF$ by the SAS Congruence Postulate. Then use CPCTC to show that $\overline{EC} \cong \overline{FB}$.

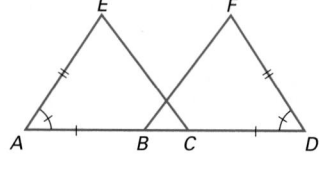

SOLUTION

FLOWCHART PROOF

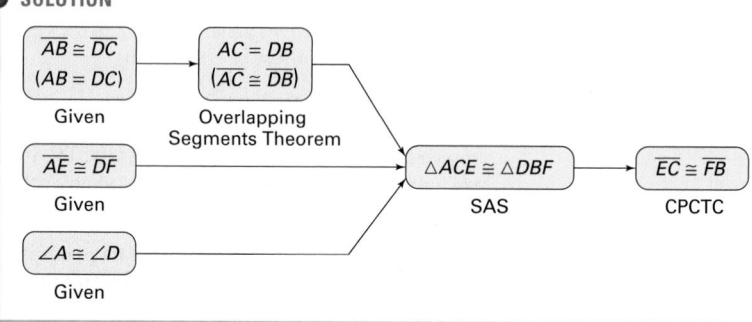

Interdisciplinary Connection

ART Students who are interested in Escher's work may enjoy the following activity: Provide photocopies of some of Escher's drawings and tracing paper. Have students try to find the underlying triangular grid structures used to make the drawings.

The Isosceles Triangle Theorem

An **isosceles triangle** is a triangle with at least two congruent sides. The two congruent sides are known as the **legs** of the triangle, and the remaining side is known as the **base**. The angles whose vertices are the endpoints of the base are known as the **base angles**, and the angle opposite the base is known as the **vertex angle**.

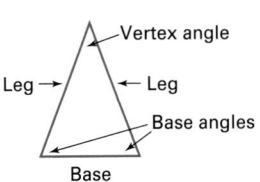

CRITICAL THINKING Is an equilateral triangle an isosceles triangle? Explain.

The following theorems, which you may have already conjectured, are among the great classics of geometry:

Isosceles Triangle Theorem

If two sides of a triangle are congruent, then the angles opposite those sides are congruent.   **4.4.1**

Converse of the Isosceles Triangle Theorem

If two angles of a triangle are congruent, then the sides opposite those angles are congruent.   **4.4.2**

TRY THIS Using the plan provided below, write a flowchart proof of the Isosceles Triangle Theorem.

FLOWCHART PROOF

Given: $\overline{AC} \cong \overline{BC}$

Prove: $\angle A \cong \angle B$

Plan: Draw the bisector of the vertex angle and extend it to the base as shown at right. Show that the two triangles that result are congruent by SAS. Then use CPCTC.

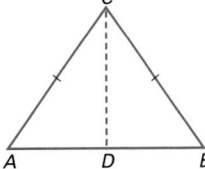

Two Corollaries

A **corollary** of a theorem is an additional theorem that can easily be derived from the original theorem. Once the theorem is known, the corollary should seem obvious. A corollary can be used as a reason in a proof, just like a theorem or a postulate.

There are a number of corollaries to the Isosceles Triangle Theorem that you will be asked to prove. The corollaries on the next page are two of the most important ones. You will be asked to prove them in the exercise set.

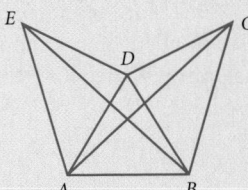
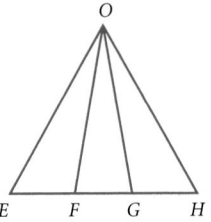
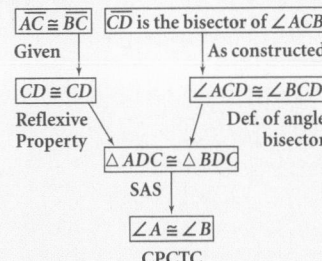

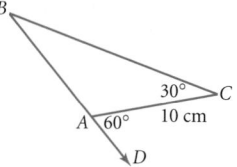

The length of $\overline{AC}$ is 10 cm. What is the length of $\overline{BA}$? Explain your answer.

You know that $\angle B$ is 30° by the Exterior Angle Theorem, so the triangle is isosceles by the converse of the Isosceles Triangle Theorem. Therefore, $BA = 10$ cm.

Math
CONNECTION

ALGEBRA In Example 3, students use the Exterior Angle Theorem to set up a single-step equation and solve for x.

Corollary

The measure of each angle of an equilateral triangle is 60°.

4.4.3

Corollary

The bisector of the vertex angle of an isosceles triangle is the perpendicular bisector of the base.

4.4.4

EXAMPLE ③

APPLICATION
CONSTRUCTION

Algebra

A resort owner plans to install a gondola ride across a small canyon on her property. Study the diagram below. What is the distance across the canyon?

SOLUTION

The 80° angle is an exterior angle.

By the Exterior Angle Theorem:

$$m\angle X + 40° = 80°$$
$$m\angle X = 40°$$

Because two angles of the triangle are congruent, the sides opposite them are congruent, by the Converse of the Isosceles Triangle Theorem. Therefore, the distance across the canyon is 350 feet.

Exercises

Communicate

1. What is CPCTC? How would you use it in a proof?

2. What is a corollary? How is a corollary related to a theorem?

3. If a triangle has three congruent angles, is it necessarily equilateral? Why or why not?

4. Can a right triangle be isosceles? Can a right triangle be equilateral? Explain your reasoning.

Inclusion Strategies

HANDS-ON STRATEGIES Overlapping figures such as those in Example 2 may be difficult for some students to interpret. Have students trace the figures and use two different colored pencils for the segments in each triangle. Students can also redraw the figures as two separate triangles.

Reteaching the Lesson

USING COGNITIVE STRATEGIES In Examples 1 and 2, other results can be proven from the given statements. In Example 1, have students prove that $\angle AXC$ and $\angle BXD$ are congruent angles. In Example 2, have students mark the intersection of $\overline{EC}$ and $\overline{FB}$ as point O. Then have them prove that $\triangle BOC$ is isosceles.

Guided Skills Practice

FLOWCHART PROOFS

Complete the flowchart proofs below. *(EXAMPLES 1 AND 2)*

Given: *ABCDE* is a regular pentagon.

Prove: △*ACD* is isosceles (*AC* = *AD*).

Proof:

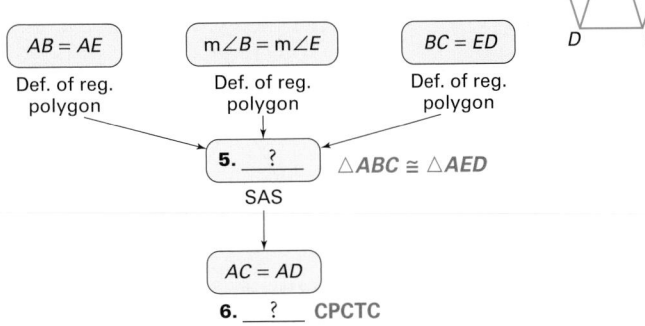

AB = AE	m∠B = m∠E	BC = ED
Def. of reg. polygon	Def. of reg. polygon	Def. of reg. polygon

5. ___?___   △*ABC* ≅ △*AED*

SAS

AC = *AD*

6. ___?___ CPCTC

Given: *QRST* is a rectangle, *TU* = *VS*, and *QV* = *RU*.

Prove: △*QTV* ≅ △*RSU*

Proof:

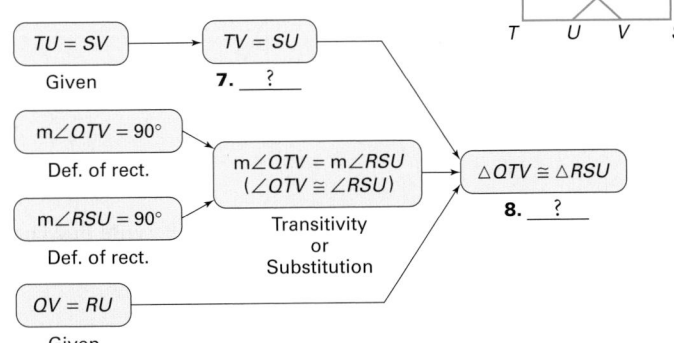

7. Overlapping Segments Theorem

8. HL

TU = SV	→	TV = SU
Given		**7.** ___?___

m∠QTV = 90°
Def. of rect.

m∠RSU = 90°
Def. of rect.

m∠QTV = m∠RSU
(∠QTV ≅ ∠RSU)
Transitivity or Substitution

△QTV ≅ △RSU
8. ___?___

QV = RU
Given

APPLICATION

9. CONSTRUCTION Jerome wants to estimate the distance across a river from point *A* to point *B*. Starting at point *A* and facing directly opposite point *B*, he turns 50° to his right and walks in a straight line to point *C*, where the angle beween the lines of sight to points *A* and *B* is 25°. How does he know that the triangle formed by points *A*, *B*, and *C* is isosceles? If Jerome walked 100 feet from point *A* to point *C*, what is the distance across the river?
(EXAMPLE 3)

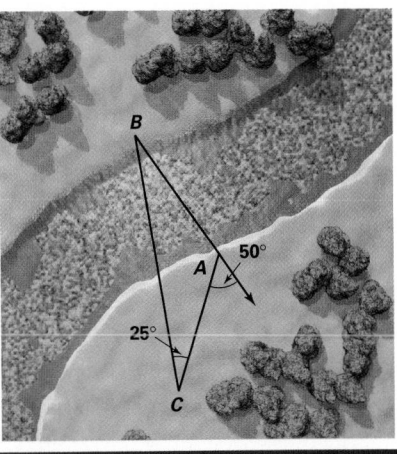

Selected Answers

Exercises 5–9, 11–45 odd

ASSIGNMENT GUIDE

In Class	1–9
Core	11–21 odd, 23–34
Core Plus	20–36
Review	37–43
Preview	44–45

✐ Extra Practice can be found beginning on page 818.

Error Analysis

When writing proofs, students may need to review the properties of parallel lines cut by a transversal, the Exterior Angle Theorem, and the Isosceles Triangle Theorem.

9. ∠*A* is supplementary to an angle measuring 50°, so m∠*A* = 180° − 50° = 130°.
By the Triangle Sum Theorem:
m∠*B* = 180° − m∠*A* − m∠*C*
= 180° − 130° − 25°
= 25°.
By the Converse of the Isosceles Triangle Theorem and the definition of an isosceles triangle, △*ABC* is isosceles; 100 feet.

19.

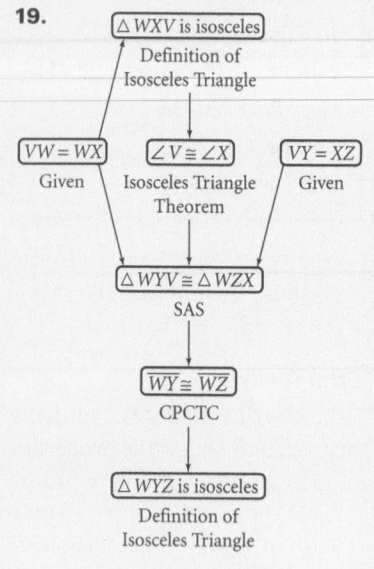

Find each indicated measure.

10. m∠Z **70°**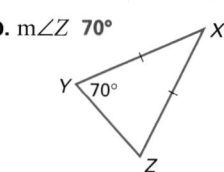

11. KL **23**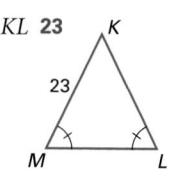

12. QR **7**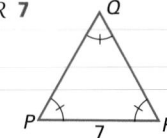

13. m∠F **60°**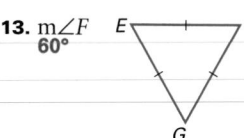

14. m∠ABD **25°**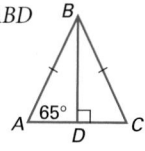

15. GH **24**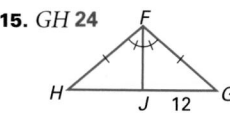

Algebra

16. PR **2**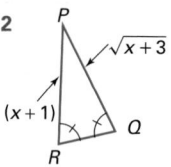

17. m∠L **40°**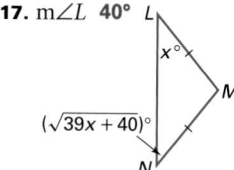

For Exercises 18 and 19, write flowchart proofs.

18. Given: ∠A ≅ ∠D, AB = DE, and AF = DC
Prove: ∠B ≅ ∠E

19. Given: VW = WX and VY = XZ
Prove: △WYZ is isosceles.

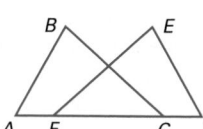

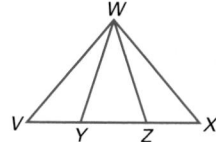

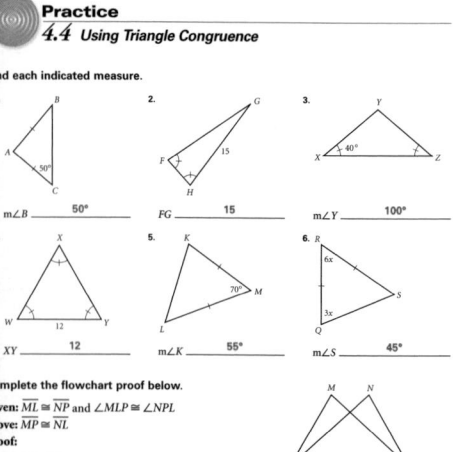
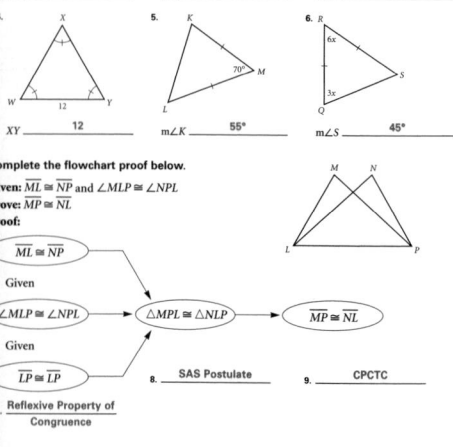
For Exercises 20–22, refer to the diagram below and write a flowchart, paragraph, or two-column proof.

20. Given: $\overline{AB} \parallel \overline{DE}$, and C is the midpoint of $\overline{BE}$.
Prove: $\overline{AB} \cong \overline{DE}$

21. Given: △ABC ≅ △DEC
Prove: $\overline{AB} \parallel \overline{DE}$

22. Given: $\overline{AB} \cong \overline{DE}$ and $\overline{AB} \parallel \overline{DE}$
Prove: C is the midpoint of $\overline{BE}$.

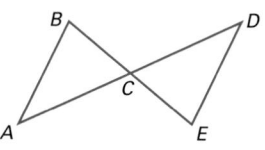

20.

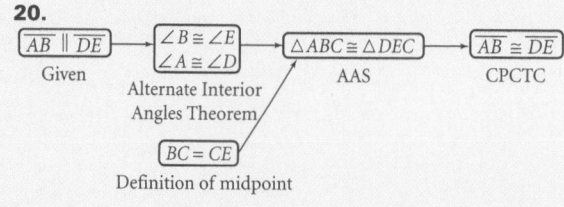

21.

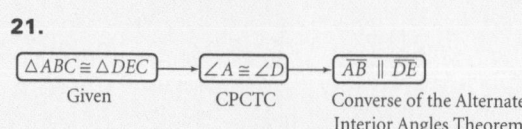

FLOWCHART PROOF

Complete the flowchart proof below of the Converse of the Isosceles Triangle Theorem.

Given: $\angle X \cong \angle Y$ and $\overline{WZ} \perp \overline{XY}$

Prove: $\overline{XZ} \cong \overline{YZ}$

Proof:

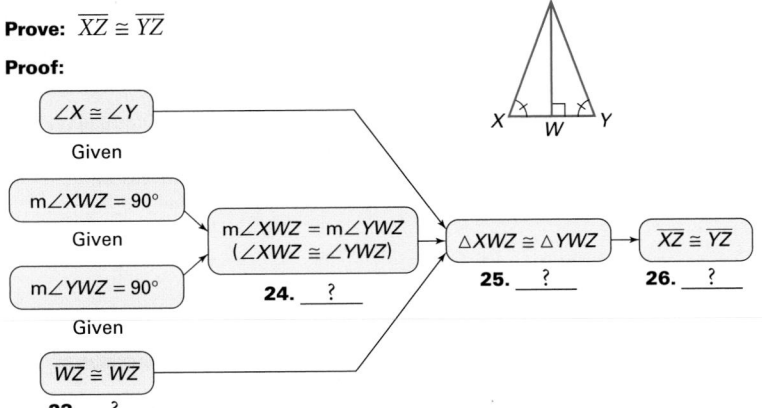

23. Reflexive Property of Congruence

24. Trans. (or Subs.) Property

25. AAS

26. CPCTC

TWO-COLUMN PROOF

Complete the two-column proof below.

Given: $\overline{AB} \cong \overline{BC}$ and $\angle ABX \cong \angle CBX$

Prove: $\overline{AX} \cong \overline{CX}$

Proof:

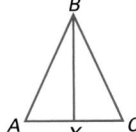

Statements	Reasons
$\overline{AB} \cong \overline{BC}$	Given
$\angle ABX \cong \angle CBX$	Given
$\overline{BX} \cong \overline{BX}$	**27.** ?
$\triangle ABX \cong \triangle CBX$	**28.** ?
$\overline{AX} \cong \overline{CX}$	**29.** ?

27. Reflexive Property of Congruence

28. SAS

29. CPCTC

PROOFS

Based on the two-column proof above, complete the following statement, which you can use in your proof of Corollary 4.4.4 in Exercise 33.

The **30.** _____?_____ of the vertex angle of an isosceles triangle **31.** _____?_____.

32. Write a paragraph proof of Corollary 4.4.3 on page 238.

33. Write a paragraph proof of Corollary 4.4.4 on page 238.

CHALLENGE

34. Use the diagram below to prove that $\overline{AK} \cong \overline{GH}$.

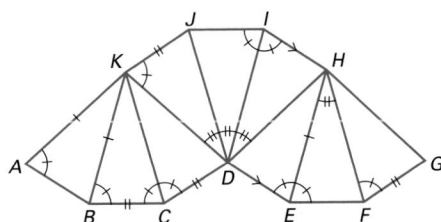

30. bisector

31. bisects the side opposite the angle

32. Let ABC be an equilateral triangle. By the Isosceles Triangle Theorem, $\angle B \cong \angle C$ because $\overline{AC} \cong \overline{AB}$ and $\angle A \cong \angle C$ because $\overline{BC} \cong \overline{BA}$. Hence, $m\angle A = m\angle C = m\angle B$. By the Triangle Sum Theorem, $m\angle A + m\angle B + m\angle C = 180°$. Using substitution, $m\angle A + m\angle A + m\angle A = 180°$. Thus, $m\angle A = 60°$ so $m\angle B = 60°$ and $m\angle C = 60°$.

33. Let ABC be an isosceles triangle with $\angle A \cong \angle C$ and $\overline{BX}$ be the bisector of $\angle B$. By the proof in Exercises 27–29, $\overline{BX}$ bisects $\overline{AC}$ and $\triangle ABX \cong \triangle CBX$. Thus by CPCTC, $\angle AXB \cong \angle CXB$. Since $\angle AXB$ and $\angle CXB$ form a linear pair, $m\angle AXB + m\angle CXB = 180°$. Since $\angle AXB \cong \angle CXB$, $m\angle AXB = m\angle CXB = 90°$. Thus $\overline{BX}$ is the perpendicular bisector of $\overline{AC}$.

34. It is given that $\overline{AK} \cong \overline{BK}$. Since $\angle KBC \cong \angle KCB$, $\triangle KBC$ is isosceles by the Converse of the Isosceles Triangle Theorem, so $\overline{BK} \cong \overline{CK}$. Since $\triangle KBC \cong \triangle KCD$ by SAS, $\overline{CK} \cong \overline{DK}$ because CPCTC. Also, $\triangle KCD \cong \triangle DKJ$ by SAS, so $\overline{DK} \cong \overline{DJ}$. Since $\triangle DKJ$ is isosceles, $\angle DKJ \cong \angle DJK$ by the Isosceles Triangle Theorem, so $\triangle DKJ \cong \triangle DJI$ by AAS, and so $\overline{DJ} \cong \overline{DI}$ because CPCTC. $\triangle DJI \cong \triangle DHI$ by ASA, so $\overline{DH} \cong \overline{DJ}$ because CPCTC. Since $\overline{HI} \| \overline{DE}$, $\angle IHD \cong \angle EDH$ by the Alternate Interior Angles Theorem, so $\triangle DHI \cong \triangle HDE$ by AAS, and so $\overline{DI} \cong \overline{HE}$ because CPCTC. $\angle EHD \cong \angle HDI$ because CPCTC, so $\triangle HDE \cong \triangle HFE$ by ASA, and so $\overline{HD} \cong \overline{HF}$ because CPCTC. Using the Transitive Property of Congruence, $\triangle HFE \cong \triangle DJK$, so $\overline{EF} \cong \overline{JK}$ because CPCTC, and so $\overline{EF} \cong \overline{GF}$ by the Transitive Property of Congruence. Also, $\overline{HE} \cong \overline{HF}$ by the Transitive Property of Congruence, so $\triangle HEF \cong \triangle HFG$ by SAS, and $\overline{HF} \cong \overline{HG}$ because CPCTC. Since it is given that $\overline{AK} \cong \overline{HE}$, $\overline{AK} \cong \overline{HG}$ by the Transitive Property of Congruence.

Look Beyond

In Exercises 44 and 45, students establish another proof of the Isosceles Triangle Theorem and its converse. Have them compare this proof with the one given in this chapter. Ask them which proof seems more complicated and why.

$\triangle ZYX \cong \triangle TBX$

35. The surveyor constructs $\triangle ZYX \cong \triangle TBX$ by SAS. Because CPCTC, $YZ = TB$. Thus, by measuring $\overline{YZ}$, he can determine the distance across the pond.

40. no; no

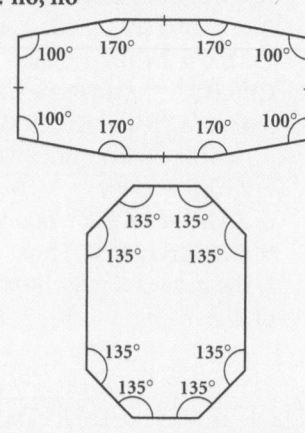

44.

Statements	Reasons
$\overline{AC} \cong \overline{BC}$	Given
$\angle C \cong \angle C$	Reflexive Property of Congruence
$\overline{CB} \cong \overline{CA}$	Given
$\triangle ACB \cong \triangle BCA$	SAS
$\angle A \cong \angle B$	CPCTC

45.

Statements	Reasons
$\angle A \cong \angle B$	Given
$\overline{AB} \cong \overline{BA}$	Reflexive Property of Congruence
$\angle B \cong \angle A$	Given
$\triangle ABC \cong \triangle BAC$	ASA
$\overline{BC} \cong \overline{AC}$	CPCTC
$\triangle ABC$ is isosceles	Definition of isosceles triangle

APPLICATIONS

35. SURVEYING A surveyor needs to measure the distance across a pond from point T to point B. Describe how the measurements shown in the diagram at right enable him to determine the distance.

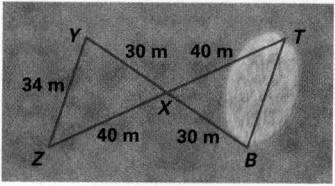

36. ROAD SIGNS A yield sign is an equilateral triangle. What can you say about the angles of the triangle? Explain your reasoning. **By Corollary 4.4.3, all angles are 60°.**

37. interior: 360°
exterior: 360°

38. interior: 720°
exterior: 360°

39. interior: 1800°
exterior: 360°

Look Back

Determine the sum of the interior angles and the sum of the exterior angles for each polygon. *(LESSON 3.6)*

37. square   **38.** hexagon   **39.** dodecagon

APPLICATION

40. ROAD SIGNS A stop sign is a regular (equilateral and equiangular) octagon. Are all equilateral octagons equiangular? Are all equiangular octagons equilateral? Explain your reasoning. *(LESSON 3.1)*

CONNECTION

COORDINATE GEOMETRY Points $A(0, 0)$ and $B(3, 6)$ are endpoints of $\overline{AB}$. Point $C(3, 1)$ is an endpoint of $\overline{CD}$. For each point D given below, determine whether $\overline{AB}$ and $\overline{CD}$ are parallel, perpendicular, or neither. *(LESSON 3.8)*

41. $D(1, 2)$ **perpendicular** **42.** $D(4, 4)$ **neither** **43.** $D(2, -1)$ **parallel**

Look Beyond

PROOFS

Another proof of the Isosceles Triangle Theorem was "discovered" by a computer. This proof had been done by the Greek geometer Pappus of Alexandria (320 C.E.) but had not been known to mathematicians for centuries. It is much simpler than Euclid's proof of the same theorem (which is quite complicated) and even simpler than the one given in this chapter. Discover it for yourself below.

44. Given: $\triangle ABC$ is isosceles, and $\overline{AC} \cong \overline{BC}$.

   Prove: $\angle A \cong \angle B$

   Plan: Show that $\triangle ABC \cong \triangle BAC$.

45. Use the plan given for Exercise 44 to prove the Converse of the Isosceles Triangle Theorem.

Proving Quadrilateral Properties

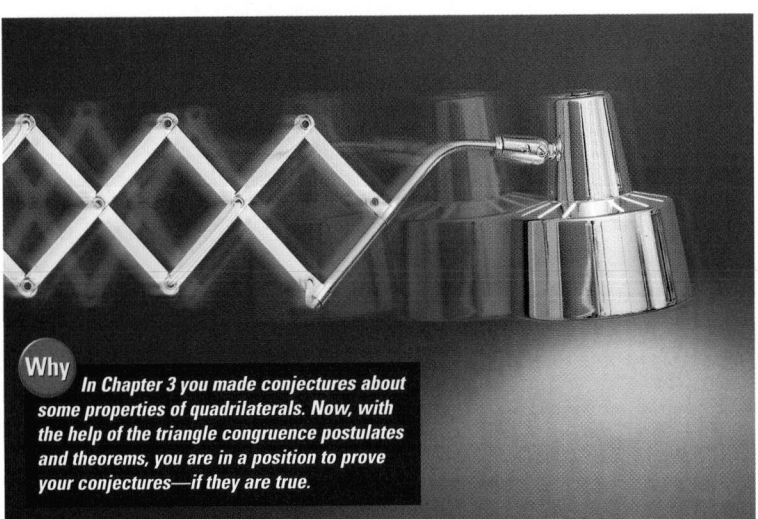

Why *In Chapter 3 you made conjectures about some properties of quadrilaterals. Now, with the help of the triangle congruence postulates and theorems, you are in a position to prove your conjectures—if they are true.*

Objective

● Prove quadrilateral conjectures by using triangle congruence postulates and theorems.

APPLICATION
ENGINEERING

A property of rhombuses explains why this lamp assembly stays perpendicular to the wall as it moves. You learned this property as a conjecture in Lesson 3.2. Do you know which property it is?

An Important Conjecture and Proof

In the Activity below, you will make a conjecture about an important property of parallelograms. The proof of the conjecture is given on the following page.

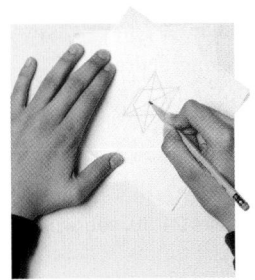

Rotational Symmetry in Parallelograms

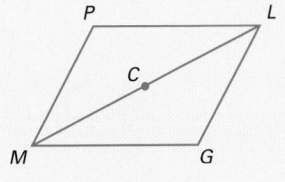

1. Trace the parallelogram at right onto two sheets of tracing paper.

2. Place one figure over the other so that they match. Place the point of your pencil at point *C* and hold it firmly in place. Rotate the top piece of paper 180°. Describe the result.

3. Does △*PLM* seem be congruent to △*GML*? Fill in the blank to complete the conjecture below.

Conjecture:

CHECKPOINT ✔ A diagonal of a parallelogram divides the parallelogram into ____?____.

Prepare

QUICK WARM-UP

How are the quadrilaterals in each pair alike? How are they different?

1. a parallelogram and a square Alike: opposite sides parallel and congruent
   Different: square has four right angles and four congruent sides

2. a rhombus and a square Alike: four equal sides, opposite angles equal, diagonals perpendicular
   Different: square has four right angles

Also on Quiz Transparency 4.5

Teach

Why A diagonal in any quadrilateral creates two triangles. Because of this, students' experiences with triangle congruency will help them prove conjectures about properties of quadrilaterals.

☞ For Activity Notes and the answer to the Checkpoint, see page 244.

Alternative Teaching Strategy

TECHNOLOGY Have students use geometry graphics software to make the parallelograms for this lesson. Students should construct figures that retain their properties when vertices or segments are dragged.

Inclusion Strategies

COGNITIVE STRATEGIES Some students may be more comfortable writing two-column proofs with words, phrases, and sentences rather than with symbols and abbreviations. Others may prefer writing proofs as flowcharts. Let students experiment with different methods.

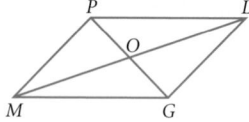
CHECKPOINT ✔
3. A diagonal of a parallelogram divides the parallelogram into two congruent triangles.

CRITICAL THINKING
Lines $\overleftrightarrow{PL}$ and $\overleftrightarrow{MG}$: ∠PLM and ∠LMG

Lines $\overleftrightarrow{MP}$ and $\overleftrightarrow{GL}$: ∠PML and ∠MLG

CRITICAL THINKING
You can prove that ∠P is congruent to ∠G by CPCTC. To prove that ∠PLG ≅ ∠GMP, use substitution to show that m∠1 + m∠2 = m∠3 + m∠4.

CRITICAL THINKING   In the diagram below, the diagonal $\overleftrightarrow{ML}$ is a transversal to two different pairs of parallel lines. Name those parallel lines. List the alternate interior angles for each pair of lines.

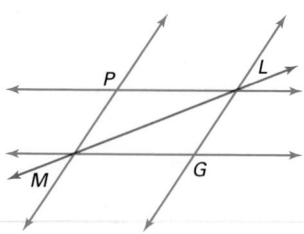

EXAMPLE ●

PROOFS

Given: parallelogram *PLGM* with diagonal $\overline{LM}$

Prove: △LGM ≅ △MPL

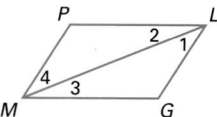

● **SOLUTION**

Proof A (paragraph format):

$\overline{PL}$ and $\overline{GM}$ are parallel, according to the definition of a parallelogram. Therefore, ∠3 and ∠2 are congruent alternate interior angles. Similarly, ∠1 and ∠4 are congruent because $\overline{PM}$ and $\overline{GL}$ are also parallel. Finally, diagonal $\overline{LM}$ is congruent to itself. Thus, two angles and the included side are congruent in △LGM and △MPL. Therefore, the triangles are congruent by the ASA Congruence Postulate.

Proof B (two-column format):

Statements	Reasons
1. Parallelogram *PLGM* has diagonal $\overline{LM}$.	Given
2. $\overline{PL} \parallel \overline{GM}$	Def. of a parallelogram
3. ∠3 ≅ ∠2	∥ lines ⇒ alt. int. angles ≅
4. $\overline{PM} \parallel \overline{GL}$	Def. of a parallelogram
5. ∠1 ≅ ∠4	∥ lines ⇒ alt. int. angles ≅
6. $\overline{LM} \cong \overline{LM}$	Reflexive Prop. of Congruence
7. △LGM ≅ △MPL	ASA Congruence Postulate

This result is stated as a theorem on page 247. In Exercises 24-30, you will be asked to complete a flowchart proof of this theorem.

CRITICAL THINKING   How can you use the result from the Example to prove that opposite angles of a parallelogram are congruent? You will be asked to write this proof in the exercise set.

Interdisciplinary Connection

PHYSICS   Parallelograms are used in physics to show vector addition. For example, the figure at right shows the movement of an object that is being pulled in the direction of $\overrightarrow{AB}$ with a force of 50 pounds and in the direction of $\overrightarrow{AD}$ with a force of 100 pounds. Have students look through their physical science textbooks for examples of these parallelogram diagrams.

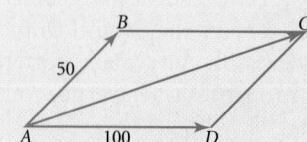

A System of Geometry Knowledge

With the theorems, postulates, and definitions you now know, you can prove all of the conjectures you have made about the properties of parallelograms and the other special quadrilaterals. It is best to start with the simplest ones first. As you progress, you will find that it is often possible to use a previously proven result as part of a proof of a more complicated theorem. In this way, you are building a system of knowledge.

The exercises in this lesson and the next will guide you through a series of proofs of the conjectures about quadrilateral properties that you made in the Activities in Lessons 3.3 and 3.4. The earlier exercises will give you the most guidance, but in the later exercises you will be on your own.

Assess

Selected Answers
Exercises 5–8, 9–71 odd

ASSIGNMENT GUIDE

In Class	1–8
Core	9–23 odd, 24–41
Core Plus	42–66, 67–73 odd
Review	75–79
Preview	80–81

✎ Extra Practice can be found beginning on page 818.

Exercises

Communicate

1. As you proved in this lesson, a parallelogram has 180° rotational symmetry. Describe all of the types of symmetry of rectangles, rhombuses, and squares.

2. For parallelogram *PQRS*, state all pairs of congruent triangles that are formed by diagonals $\overline{PR}$ and $\overline{QS}$ intersecting at point *X*.

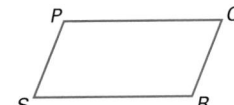

3. Is Theorem 4.5.2 on page 247 true for rectangles? Why or why not?

4. Is Theorem 4.5.3 on page 248 true for rhombuses? Why or why not?

Guided Skills Practice

Find the indicated measures for parallelogram *WXYZ*. *(EXAMPLE)*

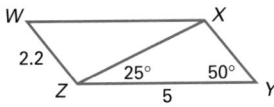

5. m∠*WXZ* **25°**   6. m∠*W* **50°**   7. *XY* **2.2**

APPLICATION

8. **CONSTRUCTION** If a ramp has a rise of more than 6 inches, handrails that are parallel to the ramp are required on both sides. If the upright post at the bottom of the ramp is 36 inches tall, how do you know that the upright post at the top of the ramp is also 36 inches tall?

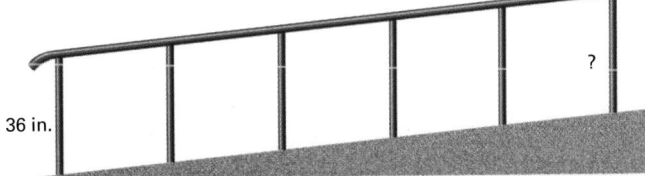

36 in.

Error Analysis

When students write paragraph proofs such as those in Exercises 61–63, they should first write a conditional such as "If a figure is a square, then it is a rectangle." Point out that the "if" part of the conditional is the given, and the "then" part is what they are to prove.

8. The ramp, handrail, and the two upright posts form a parallelogram. Opposite sides of a parallelogram are congruent.

Reteaching the Lesson

COOPERATIVE LEARNING Have students draw the diagonals of an isosceles trapezoid and look for pairs of triangles that appear congruent. Have them work in small groups to write proofs of their conjectures.

Math
CONNECTION

ALGEBRA In Exercises 17–22, students are to use the properties of parallelograms to write and solve multistep equations. In Exercise 72, students write and solve an equation by using the properties of congruent triangles.

Technology

The figures in the exercises can be drawn by using geometry graphics software. Make sure students construct figures that retain their properties when vertices or segments are dragged.

23. a. Yes; the opposite sides are parallel by the Alternate Interior Angles Theorem.
   b. No; the triangles fit together to form a kite.
   c. No; the sides of the triangles will not match up.

● *Practice and Apply*

In Exercise 9–16, find the indicated measures for each parallelogram.

9. *CD* 5   **10.** *DA* 7
11. m∠*C* 50°   **12.** m∠*D* 130°

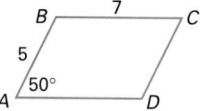

13. m∠*Q* 60°   **14.** m∠*RPQ* 40°
15. m∠*SPR* 80°   **16.** m∠*PRQ* 80°

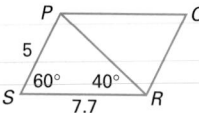

Algebra

In Exercise 17–22, find the indicated measure for each parallelogram.

17. *SV* 10

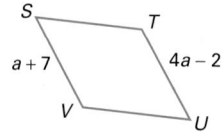

18. m∠*A* 120°

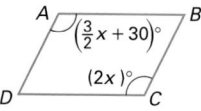

19. *QR* 6

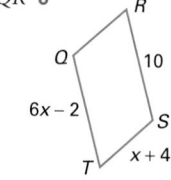

20. *CD* 15

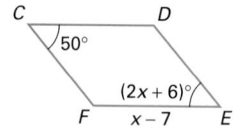

21. m∠*N* 20°

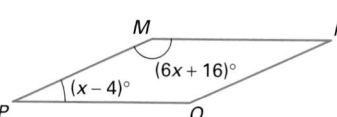

22. m∠*G* 126°

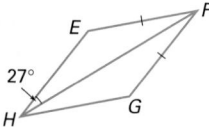

23. State whether each pair of triangles could fit together to form a parallelogram (without reflecting). Justify your answer.

  a.   **b.**   **c.**

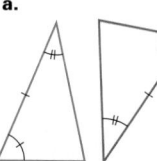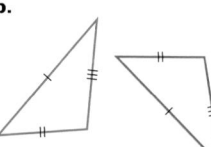

In Exercises 24–71, you will be asked to prove theorems about quadrilaterals.

Fill in the blanks below to complete a flowchart proof of Theorem 4.5.1 (refer back to the Example on page 244).

Theorem

A diagonal of a parallelogram divides the parallelogram into two congruent triangles.
4.5.1

**FLOWCHART PROOF**

Given: parallelogram $ABCD$ with diagonal $\overline{BD}$

Prove: $\triangle ABD \cong \triangle CDB$

Proof:

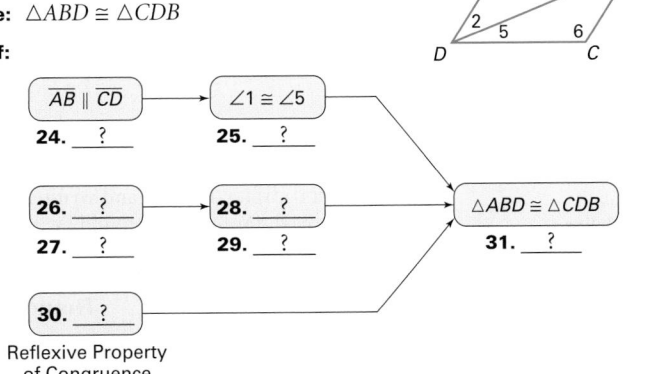

24. definition of parallelogram

25. Alternate Interior Angles Theorem

26. $\overline{AD} \parallel \overline{BC}$

27. definition of parallelogram

28. $\angle 2 \cong \angle 4$

29. Alternate Interior Angles Theorem

30. $\overline{DB} \cong \overline{DB}$

31. ASA

Complete the proof below of Theorem 4.5.2.

Theorem

Opposite sides of a parallelogram are congruent.
4.5.2

TWO-COLUMN PROOF

Given: parallelogram $ABCD$ with diagonal $\overline{BD}$

Prove: $\overline{AB} \cong \overline{CD}$ and $\overline{AD} \cong \overline{CB}$

Proof:

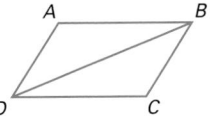

Statements	Reasons
$ABCD$ is a parallelogram.	32. ?
33. ?	A diagonal of a parallelogram divides the parallelogram into two congruent triangles.
$\overline{AB} \cong \overline{CD}$ and $\overline{AD} \cong \overline{CB}$	34. ?

32. Given

33. $\triangle ABD \cong \triangle CDB$

34. CPCTC

41.

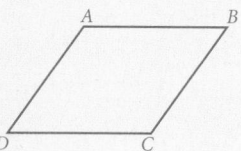

Given: *ABCD* is a parallelogram.

Prove: ∠A and ∠B, ∠B and ∠C, ∠C and ∠D, and ∠D and ∠A are supplementary.

By definition of a parallelogram, $\overline{AB}\|\overline{DC}$ and $\overline{AD}\|\overline{BC}$. By Same-Side Interior Angles Theorem, ∠A and ∠B, ∠B and ∠C, ∠C and ∠D, and ∠D and ∠A are supplementary.

TWO-COLUMN PROOF

35. △*CDB*
36. A diagonal of a parallelogram divides the parallelogram into two congruent triangles.
37. △*DCA*
38. A diagonal of a parallelogram divides the parallelogram into two congruent triangles.
39. ∠*BAD* ≅ ∠*DCB*
40. ∠*ABC* ≅ ∠*CDA*

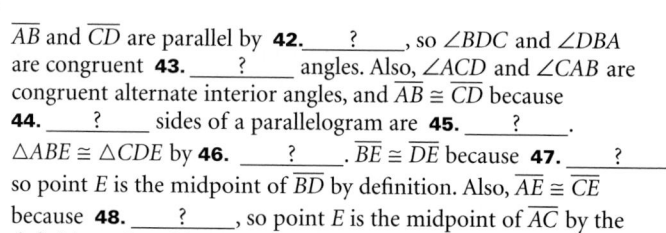

Homework Help Online
Go To: **go.hrw.com**
Keyword:
MG1 Homework Help
for Exercises 41-48, 60-66, 69-71

42. definition of parallelogram
43. alternate interior
44. opposite
45. congruent
46. ASA
47. CPCTC
48. CPCTC

Complete the proof below of Theorem 4.5.3.

Theorem	
Opposite angles of a parallelogram are congruent.	**4.5.3**

Given: parallelogram *ABCD* with diagonals $\overline{BD}$ and $\overline{AC}$

Prove: ∠*BAD* ≅ ∠*DCB* and ∠*ABC* ≅ ∠*CDA*

Proof:

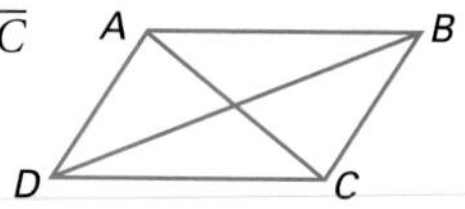

Statements	Reasons
ABCD is a parallelogram.	Given
△*ABD* ≅ **35.** ?	**36.** ?
△*BAC* ≅ **37.** ?	**38.** ?
39. ? and **40.** ?	CPCTC

41. Write a paragraph proof of Theorem 4.5.4. Begin by drawing an appropriate diagram and writing out what is given and what is to be proved. (Your proof should be very short.)

Theorem	
Consecutive angles of a parallelogram are supplementary.	**4.5.4**

Complete the paragraph proof below of Theorem 4.5.5.

Theorem	
The diagonals of a parallelogram bisect each other.	**4.5.5**

Given: parallelogram *ABCD* with diagonals $\overline{AC}$ and $\overline{BD}$ intersecting at point *E*

Prove: Point *E* is the midpoint of $\overline{AC}$ and $\overline{BD}$.

Proof:

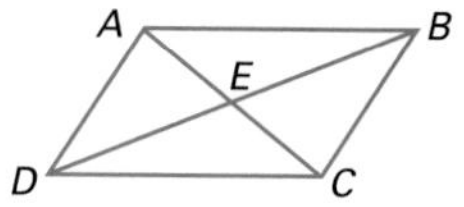

$\overline{AB}$ and $\overline{CD}$ are parallel by **42.** ? , so ∠*BDC* and ∠*DBA* are congruent **43.** ? angles. Also, ∠*ACD* and ∠*CAB* are congruent alternate interior angles, and $\overline{AB} \cong \overline{CD}$ because **44.** ? sides of a parallelogram are **45.** ? . △*ABE* ≅ △*CDE* by **46.** ? . $\overline{BE} \cong \overline{DE}$ because **47.** ? , so point *E* is the midpoint of $\overline{BD}$ by definition. Also, $\overline{AE} \cong \overline{CE}$ because **48.** ? , so point *E* is the midpoint of $\overline{AC}$ by the definition of a midpoint.

TWO-COLUMN PROOFS

In Exercises 49–59, you will complete proofs of the following theorems:

Theorems

A rhombus is a parallelogram. 4.5.6
A rectangle is a parallelogram. 4.5.7

Given: rhombus *EFGH*

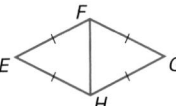

Prove: $\overline{EF} \parallel \overline{GH}$ and $\overline{FG} \parallel \overline{EH}$

Proof:

Statements	Reasons
$\overline{EF} \cong \overline{GH}$	**49.** ?
$\overline{EH} \cong \overline{GF}$	Def. of a rhombus
$\overline{FH} \cong \overline{HF}$	**50.** ?
$\triangle EFH \cong \triangle GHF$	**51.** ?
$\angle EFH \cong \angle GHF$	CPCTC
$\overline{EF} \parallel \overline{GH}$	**52.** ?
$\angle GFH \cong \angle EHF$	**53.** ?
$\overline{FG} \parallel \overline{EH}$	**54.** ?

Given: rectangle *PQRS*

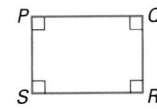

Prove: $\overline{PQ} \parallel \overline{RS}$ and $\overline{PS} \parallel \overline{QR}$

Proof:

Statements	Reasons
$m\angle P = 90°$	Def. of a rectangle
$m\angle S = 90°$	**55.** ?
$m\angle P + m\angle S = 180°$	**56.** ?
$\overline{PQ} \parallel \overline{RS}$	**57.** ?
$m\angle Q = 90°$	**58.** ?
$m\angle P + m\angle Q = 180°$	Add. Prop. of Equality
$\overline{PS} \parallel \overline{QR}$	**59.** ?

49. Definition of a rhombus
50. Reflexive Property of Congruence
51. SSS
52. Converse of the Alternate Interior Angles Theorem
53. CPCTC
54. Converse of the Alternate Interior Angles Theorem
55. Definition of a rectangle
56. Addition Property
57. Converse of the Same-Side Interior Angles Theorem
58. Definition of a rectangle
59. Converse of the Same-Side Interior Angles Theorem

PROOF

Complete the paragraph proof below of Theorem 4.5.8.

Theorem

The diagonals of a rhombus are perpendicular. 4.5.8

---

60. def. of rhombus
61. diagonals of a parallelogram bisect each other
62. Reflexive Property of Congruence
63. SSS
64. CPCTC
65. 180°
66. 90°

Given: rhombus *DEFG* with diagonals $\overline{DF}$ and $\overline{GE}$ intersecting at point *H*

Prove: $\overline{DF}$ and $\overline{GE}$ intersect to form a right angle.

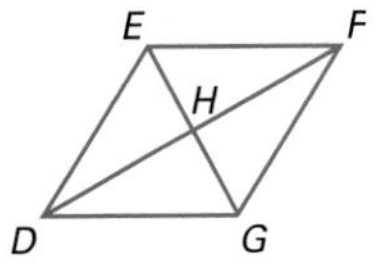

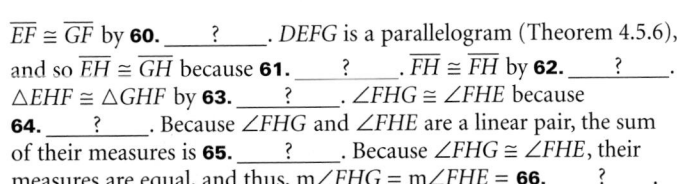

Proof:

$\overline{EF} \cong \overline{GF}$ by **60.** ? . *DEFG* is a parallelogram (Theorem 4.5.6), and so $\overline{EH} \cong \overline{GH}$ because **61.** ? . $\overline{FH} \cong \overline{FH}$ by **62.** ? . $\triangle EHF \cong \triangle GHF$ by **63.** ? . $\angle FHG \cong \angle FHE$ because **64.** ? . Because $\angle FHG$ and $\angle FHE$ are a linear pair, the sum of their measures is **65.** ? . Because $\angle FHG \cong \angle FHE$, their measures are equal, and thus, $m\angle FHG = m\angle FHE =$ **66.** ? .

67. Given: rectangle *RSTU* with diagonals $\overline{RT}$ and $\overline{US}$ intersecting at *V*

Prove: $\overline{RT} \cong \overline{US}$

Statements	Reasons
RSTU is a rectangle.	Given
RSTU is a parallelogram.	Theorem 4.5.7
$ST \cong UR$	Theorem 4.5.2
$TU \cong TU$	Reflexive Property of Congruence
$m\angle RUT = 90°$ $m\angle STU = 90°$	definition of rectangle
$m\angle RUT = m\angle STU$ ($\angle RUT \cong \angle STU$)	Trans. or substitution
$\triangle RTU \cong \triangle SUT$	SAS
$\overline{RT} \cong \overline{SU}$	CPCTC

69. Theorem: A square is a rectangle. A square is a quadrilateral with the property that all of its sides are equal and every angle is a right angle. Since every angle is a right angle , a square is a rectangle by the definition of a rectangle.

67. Write a two-column, flowchart, or paragraph proof of Theorem 4.5.9. (Hint: A rectangle is a parallelogram; therefore, it has the properties of a parallelogram.)

Theorem

The diagonals of a rectangle are congruent.   **4.5.9**

Given: rectangle *RSTU* with diagonals $\overline{RT}$ and $\overline{US}$ intersecting at point *V*

Prove: $\overline{RT} \cong \overline{US}$

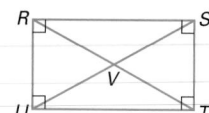

68. Write a two-column, flowchart, or paragraph proof of Theorem 4.5.10. Recall that a kite has two pairs of congruent adjacent sides and that opposite sides are not congruent.

Theorem

The diagonals of a kite are perpendicular.   **4.5.10**

Given: kite *WXYZ* with diagonals $\overline{WY}$ and $\overline{XZ}$ intersecting at point *A*

Prove: $\overline{WY} \perp \overline{XZ}$

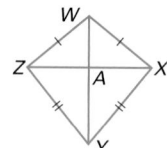

Write a paragraph proof of each theorem below.

Theorem

69. A square is a rectangle.   **4.5.11**

Theorem

70. A square is a rhombus.   **4.5.12**

Theorem

71. The diagonals of a square are congruent and are the perpendicular bisectors of each other.   **4.5.13**

---

70. Theorem: A square is a rhombus. A square is a quadrilateral with the property that all of its sides have the same length. This is the definition of a rhombus.

71. Theorem: The diagonals of a square are congruent and are the perpendicular bisectors of each other.

A square is a rhombus and hence a parallelogram by theorems 4.5.12 and 4.5.6, respectively. Therefore its diagonals are perpendicular bisectors of each other by theorems 4.5.8 and 4.5.5. A square is also a rectangle by theorem 4.5.11 and thus its diagonals are congruent by theorem 4.5.9.

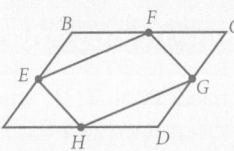

72

CHALLENGE

72. Draw a parallelogram and find the midpoint of each side. Connect the midpoints to form a quadrilateral within the parallelogram. Prove that two pairs of congruent triangles are formed. Use this to prove that the quadrilateral formed by connecting the midpoints is a parallelogram. (Hint: Connect one pair of opposite midpoints in the original figure.)

APPLICATIONS

73. **ART** An artist cuts four congruent right triangles of stained glass. Show how the four triangles can be put together to form each quadrilateral listed below.

    a. a parallelogram that is not a rhombus or a rectangle
    b. a rhombus
    c. a rectangle
    d. a trapezoid
    e. a kite

74. **SPORTS** Because a baseball diamond is a square, it is also a rectangle, a rhombus, and a parallelogram. Which of the following must be true, based on the properties of special quadrilaterals? Explain your reasoning.

    a. The distance from first base to third base equals the distance from second base to home plate.
    b. The center of the pitcher's mound is on the diagonal from first to third base. (Note: The mound is on the diagonal from home plate to second base at a point 3 feet from the midpoint of that diagonal.)
    c. The path from first base to second base is parallel to the path from third base to home plate.

 Look Back

75. Suppose that the angles in one triangle are congruent to the angles in another triangle. Are the triangles necessarily congruent? Why or why not? *(LESSON 4.3)*

Write a congruence statement for each pair of triangles below, and name the postulate or theorem to justify it. *(LESSONS 4.2 AND 4.3)*

76. $\triangle ABC \cong \triangle EDF$, SSS

77. $\triangle GHI \cong \triangle JKL$, HL

78. $\triangle MNO \cong \triangle RQP$, SAS

76.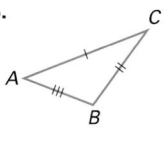

77.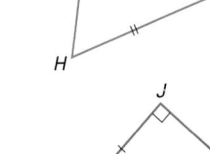

78.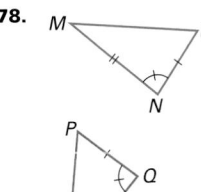

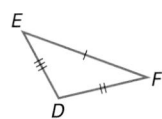

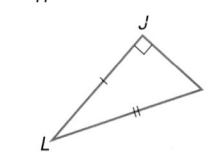

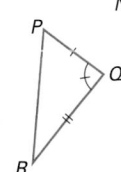

Given: *ABCD* is a parallelogram, and *E*, *F*, *G*, and *H* are the midpoints of sides $\overline{AB}$, $\overline{BC}$, $\overline{CD}$, and $\overline{DA}$.
Prove: *EFGH* is a parallelogram.
Proof: By Theorem 4.5.2, $\overline{AB} \cong \overline{CD}$ and $\overline{BC} \cong \overline{DA}$. By definition of midpoint, $AE = BE = \frac{1}{2}AB$, and $CG = DG = \frac{1}{2}CD$, so $\overline{AE} \cong \overline{CG}$ and $\overline{BE} \cong \overline{DG}$. Also by definition of midpoint, $BF = CF = \frac{1}{2}BC$, and $AH = DH = \frac{1}{2}AD$, so $\overline{BF} \cong \overline{DH}$ and $\overline{CF} \cong \overline{AH}$. By theorem 4.5.3, $\angle A \cong \angle C$ and $\angle B \cong \angle D$, so $\triangle BEF \cong \triangle DGH$ and $\triangle AEH \cong \triangle CGF$ by SAS. $\overline{EF} \cong \overline{GH}$ and $\overline{FG} \cong \overline{HE}$ because CPCTC, so $\triangle EFG \cong \triangle GHE$ by SSS. Because CPCTC, $\angle FEG \cong \angle EGH$, so $\overline{EF} \| \overline{GH}$ by the Converse of the Alternate Interior Angles Theorem. Also because CPCTC, $\angle FGE \cong \angle HEG$, so $\overline{FG} \| \overline{HE}$ by the Converse of

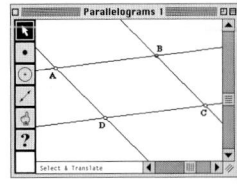

the Alternate Interior Angles Theorem. Thus, EFGH is a parallelogram by the definition of a parallelogram.

73. a.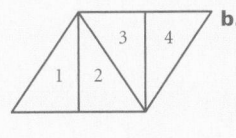

b.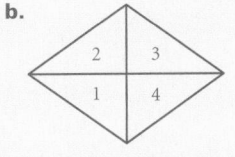

c.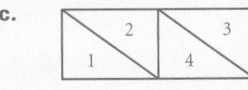

d.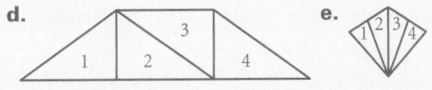

e.

74. a. true, because the diagonals of a rectangle are congruent
    b. false, because the diagonals of a parallelogram bisect each other
    c. true, because opposite sides of a parallelogram are parallel

75. No; AAA does not imply triangles are congruent.

In Exercises 80 and 81, students apply the properties of isosceles trapezoids to write two proofs about angles and diagonals.

Portfolio Activity

The Portfolio Activity can be used as preparation for the Chapter Project or as a separate activity. In the Portfolio Activity on this page, students explore patterns in non-regular polygons. Students create tessellating polygons by rotating quadrilaterals and linking them together.

80.

Statements	Reasons
$\overline{AB} \parallel \overline{CD}$	Definition of a trapezoid
$\overline{AE} \perp \overline{EF}$ $\overline{BF} \perp \overline{EF}$	Given
$\overline{AE} \parallel \overline{BF}$	Converse of Corresponding Angles Postulate
$ABFE$ is a parallelogram.	Definition of parallelogram
$\overline{AE} \cong \overline{BF}$	Theorem 4.5.2
$\overline{AD} \cong \overline{BC}$	Given
$\angle AED \cong \angle BFC$	Right angles are congruent
$\triangle AED \cong \triangle BFC$	HL
$\angle ADE \cong \angle BCE$	CPCTC

81.

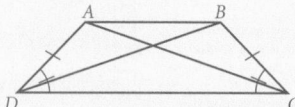

Given: *ABCD* is an isosceles trapezoid.
Prove: $\overline{AC} \cong \overline{BD}$

79. Given $\triangle KLM \cong \triangle RST$, find the values of *x* and *y*. **(LESSONS 3.8 AND 4.4)**

$x = 15$
$y = 20$

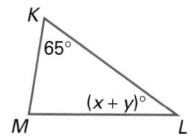 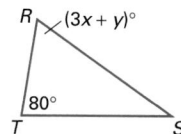

80. Recall that an isosceles trapezoid has congruent legs. In the diagram at right, two segments perpendicular to the bases have been added to form a pair of right triangles.

Use the information given in the diagram to prove that the base angles of an isosceles trapezoid are congruent. (Hint: First prove that *ABFE* is a rectangle and therefore a parallelogram.)

81. Use your result from Exercise 80 to prove that the diagonals of an isosceles trapezoid are congruent.

Portfolio Extension
Go To: go.hrw.com
Keyword:
MG1 Regular

Tessellations With Quadrilaterals

In Chapter 3, you explored several different types of repeating patterns, or tessellations. The only regular polygons that can be used to form tessellations are equilateral triangles, squares, and regular hexagons. (Why?) However, if you use nonregular polygons, there are many more possibilities.

1. Draw a nonregular quadrilateral, such as *ABCD*. Find the midpoint of one of the sides and label it *E*.

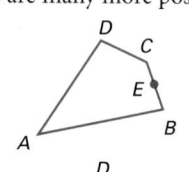

2. Using tracing paper or geometry graphics software, rotate *ABCD* about point *E*.

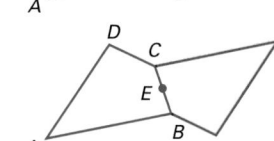

3. Repeat Step 2 for the midpoints of each side of *ABCD*. Then continue the pattern by rotating the images about the midpoints of their sides.

4. Try creating tessellations with different quadrilaterals, including special quadrilaterals and concave quadrilaterals. Do all of the quadrilaterals that you chose work? Can any of them be used to form tessellations in more than one way?

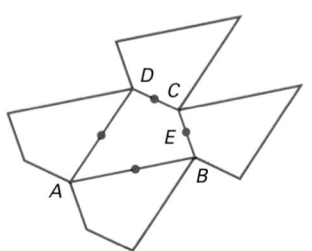

Statements	Reasons
ABCD is an isosceles trapezoid.	Given
$\angle ADC \cong \angle BCD$	Exercise 80
$\overline{AD} \cong \overline{BC}$	Definition of isosceles trapezoid
$\overline{DC} \cong \overline{CD}$	Reflexive Property of Congruence
$\triangle ADC \cong \triangle BCD$	SAS
$\overline{AC} \cong \overline{BD}$	CPCTC

Conditions for Special Quadrilaterals

4.6

Objective

● Develop conjectures about special quadrilaterals—parallelograms, rectangles, and rhombuses.

Why *How can you tell whether a given figure is a parallelogram, a rectangle, a rhombus, or a square? In this lesson, you will discover some theorems that provide answers to this question.*

A perfectly rectangular foundation is required for the proper construction of many modern buildings. Here workers are applying an important geometry principle to ensure that a portion of the foundation is a rectangle.

Visitors to non-western cultures as far away as Mozambique, in southern Africa, have reported observing local workers employing the technique shown above to ensure that a foundation of a house is a rectangle. According to their reports, measurements are made by using long poles (see page 255).

The Conditions That Determine a Figure

If you are given a quadrilateral, how can you tell which of the special quadrilaterals, if any, it is? One way to try is to check the given information against the definitions of the special figures. However, the information you are given may not match the information referred to in the definitions. This does not mean that the figure is not one of the special quadrilaterals.

Alternative Teaching Strategy

TECHNOLOGY Geometry graphics software can be used to explore the conjectures in this lesson. Make sure the figures that students construct retain their essential properties when vertices or segments are dragged.

QUICK WARM-UP

Have students draw quadrilateral *WXYZ* and list all pairs of sides or angles that satisfy each description.

1. opposite sides $\overline{WX}$ and $\overline{ZY}$; $\overline{WZ}$ and $\overline{XY}$

2. opposite angles $\angle X$ and $\angle Z$; $\angle W$ and $\angle Y$

3. adjacent sides $\overline{WX}$ and $\overline{XY}$; $\overline{YZ}$ and $\overline{XY}$; $\overline{YZ}$ and $\overline{ZW}$; $\overline{ZW}$ and $\overline{WX}$

4. consecutive angles $\angle W$ and $\angle X$; $\angle X$ and $\angle Y$; $\angle Y$ and $\angle Z$; $\angle Z$ and $\angle W$

Also on Quiz Transparency 4.6

Teach

Why When building structures that involve quadrilaterals, it is not necessary or convenient to measure four angles and four sides. The theorems covered in this lesson can be applied to simplify projects such as the building foundation shown on this page.

In this Activity, students verify that various properties of parallelograms are true. They can model the parallelograms by using spaghetti or some other material.

For a student worksheet of this Activity and detailed Teacher Notes, see page 62 in the Lesson Activities booklet.

☞ For the answers to the Checkpoints, see Additional Answers, page 893.

Teaching Tip

Have students read all of the conjectures and identify the types of segments or angles specified. They should conclude that the key properties of a quadrilateral involve opposite sides or angles, adjacent sides or angles, and the relationship between the diagonals.

Cooperative Learning

Have students work in small groups to create a chart of the results from Activity 1. For example, the chart may have four columns, with the headings *Property*, *Parallelogram*, *Rectangle*, and *Rhombus*. Under *Property*, students should write items such as "congruent diagonals." Then they should write *always*, *sometimes*, or *never* under the other headings. When all groups are finished, they can compare the chart formats and their content.

YOU WILL NEED

uncooked spaghetti or another material (such as soda straws or toothpicks) to serve as the sides of model quadrilaterals, a ruler, and a protractor

CHECKPOINT ✔

Activity 1
What Does It Take to Make...

In each part of this Activity, decide whether the conjectures listed are true or false. If you believe that a conjecture is false, prove that it is false by giving a counterexample. Make sketches of your counterexamples. If you believe that a conjecture is true, consider how you would prove it.

Part I: What does it take to make a parallelogram?

State whether the following conjectures about parallelograms are true or false:

1. If one pair of opposite sides of a quadrilateral are congruent, then the quadrilateral is a parallelogram.

2. If two pairs of opposite sides of a quadrilateral are congruent, then the quadrilateral is a parallelogram.

3. If one pair of opposite sides of a quadrilateral are parallel and congruent, then the quadrilateral is a parallelogram.

4. If two pairs of sides of a quadrilateral are congruent, then the quadrilateral is a parallelogram.

5. If the diagonals of a quadrilateral bisect each other, then the quadrilateral is a parallelogram.

CHECKPOINT ✔   **Part II: What does it take to make a rectangle?**

State whether the following conjectures about rectangles are true or false:

1. If one angle of a quadrilateral is a right angle, then the quadrilateral is a rectangle.

2. If one angle of a parallelogram is a right angle, then the parallelogram is a rectangle.

3. If the diagonals of a quadrilateral are congruent, then the quadrilateral is a rectangle.

4. If the diagonals of a parallelogram are congruent, then the parallelogram is a rectangle.

5. If the diagonals of a parallelogram are perpendicular, then the parallelogram is a rectangle.

CHECKPOINT ✔   **Part III: What does it take to make a rhombus?**

State whether the following conjectures about rhombuses are true or false:

1. If one pair of adjacent sides of a quadrilateral are congruent, then the quadrilateral is a rhombus.

2. If one pair of adjacent sides of a parallelogram are congruent, then the parallelogram is a rhombus.

3. If the diagonals of a parallelogram are congruent, then the parallelogram is a rhombus.

4. If the diagonals of parallelogram bisect the angles of the parallelogram, then the quadrilateral is a rhombus.

5. If the diagonals of a parallelogram are perpendicular, then the parallelogram is a rhombus.

Interdisciplinary Connection

GEOGRAPHY Parallel dividers are used by navigators when drawing courses on maps. Angle measures are transferred from the compass to make specific course headings. Have students research how parallel dividers are used and give presentations to the class.

The "Housebuilder" Theorem

APPLICATION

CARPENTRY

Carpenters have long used diagonal measurements to determine whether the sides of a house's foundation formed a rectangle. The building method has been used throughout the world, including the African countries Mozambique and Liberia, where people use long poles instead of measuring tapes to compare lengths. In Activity 2 below, you will apply an important theorem, which you may have conjectured in Activity 1.

Activity 2
Building a Rectangular Structure

YOU WILL NEED

uncooked spaghetti or another material (such as soda straws or toothpicks) to serve as the sides of model quadrilaterals

1. Break off two pairs of sides of two different lengths from your modeling material. Let these be the pairs of opposite sides for your model foundation.

2. Model the diagonals of the foundation with two more pieces of your modeling material. Break off the pieces for the diagonals at the correct length.

CHECKPOINT ✔ 3. Compare the lengths of the diagonals. Are they equal? If they are, then your foundation is rectangular, according to one of the conjectures you made in Activity 1. State that conjecture. (This conjecture, when proven, will be called the Housebuilder Theorem.)

4. If the diagonals are not the same length, adjust the sides of your figure until they are. Note: It may be easy to make a reasonably good first guess at a rectangle when you are working with a small model, but it is much more difficult when you are trying to create a large rectangle such as a foundation.

CRITICAL THINKING The head of the lamp in the picture remains vertical as it moves up and down. Which of the conjectures you made in Activity 1 explains why this is true?

Activity 2 Notes

In this Activity, students discover that the diagonals of a rectangle are congruent by forming a rectangular foundation with spaghetti or some other material.

For a student worksheet of this Activity and detailed Teacher Notes, see page 62 in the Lesson Activities booklet.

CHECKPOINT ✔

3. The diagonals are congruent. Conjecture: If the diagonals of a parallelogram are congruent, then the parallelogram is a rectangle.

CRITICAL THINKING

Opposite sides of the lamp assembly are congruent, so the lamp assembly is a parallelogram. Therefore, opposite sides are parallel, which is why the structure can be moved up and down and why the lamp remains parallel to the wall (vertical).

Enrichment

TECHNOLOGY Have students use geometry graphics software to explore these questions about parallelograms, rhombuses, and rectangles.

1. Connect the midpoints of the sides. What figures result?

2. Construct the bisectors of the angles. What figures result?

Inclusion Strategies

USING VISUAL MODELS For Activities 1 and 2, have students place graph paper under their modeling materials. The grid lines will facilitate the construction of parallel and perpendicular lines and congruent segments and angles.

Adding to Your System of Geometry Knowledge

The following theorems are based on your work in Activity 1. You will have the opportunity to prove these theorems in the exercise set.

Theorem

If two pairs of opposite sides of a quadrilateral are congruent, then the quadrilateral is a parallelogram. **4.6.1**

Theorem

If one pair of opposite sides of a quadrilateral are parallel and congruent, then the quadrilateral is a parallelogram. **4.6.2**

Theorem

If the diagonals of a quadrilateral bisect each other, then the quadrilateral is a parallelogram. **4.6.3**

Theorem

If one angle of a parallelogram is a right angle, then the parallelogram is a rectangle. **4.6.4**

The Housebuilder Theorem

If the diagonals of a parallelogram are congruent, then the parallelogram is a rectangle. **4.6.5**

Theorem

If one pair of adjacent sides of a parallelogram are congruent, then the parallelogram is a rhombus. **4.6.6**

Theorem

If the diagonals of a parallelogram bisect the angles of the parallelogram, then the parallelogram is a rhombus. **4.6.7**

Theorem

If the diagonals of a parallelogram are perpendicular, then the parallelogram is a rhombus. **4.6.8**

Reteaching the Lesson

COOPERATIVE LEARNING   Have students work in groups of 2 or 3. Each group should be assigned one class of quadrilateral. Students in each group should write as many true statements as they can about their quadrilateral. Each group should then make a class presentation about the true statements for the class of quadrilateral assigned to them.

Exercises

Communicate

1. The definition of a parallelogram is a quadrilateral with two pairs of parallel sides. Consider the following alternative definition:

   A parallelogram is a quadrilateral whose opposite sides are congruent.

   Does this definition work? How many other definitions for a parallelogram can you think of?

2. Choose a conjecture from Part I of Activity 1 that turned out to be false, and explain why it is false.

3. Choose a conjecture from Part II of Activity 1 that turned out to be false, and explain why it is false.

4. Choose a conjecture from Part III of Activity 1 that turned out to be false, and explain why it is false.

Guided Skills Practice

For Exercises 5–8, determine whether each quadrilateral is necessarily a parallelogram, a rectangle, a rhombus, or none of these. Give all of the names that apply to each quadrilateral. *(ACTIVITY 1)*

5.  none

6. 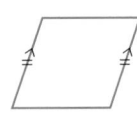 parallelogram

7. 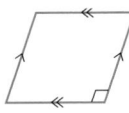 parallelogram, rectangle

8. 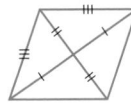 parallelogram, rhombus

9. The figure at right is a parallelogram. What must be true about the figure in order for it to be a rectangle also?
   (ACTIVITY 2)
   The diagonals must be congruent.

Practice and Apply

Exercises 10–13 refer to quadrilateral *ABCD* with diagonals $\overline{AC}$ and $\overline{BD}$ intersecting at point *E*. For each set of conditions given below, determine whether the quadrilateral is necessarily a parallelogram. If so, give the theorem that justifies your answer.

10. $\overline{AB} \cong \overline{CD}$, $\overline{AC} \cong \overline{BD}$ No

11. $\overline{AE} \cong \overline{CE}$, $\overline{BE} \cong \overline{DE}$ Yes, Thm 4.6.3

12. $\overline{AB} \cong \overline{AC}$, $\overline{AB} \parallel \overline{CD}$ No

13. $\overline{AB} \cong \overline{CD}$, $\overline{AB} \parallel \overline{CD}$ Yes, Thm 4.6.2

Assess

Selected Answers

Exercises 5–9, 11–61 odd

ASSIGNMENT GUIDE	
In Class	1–9
Core	11–25 odd, 26–31, 33–43 odd, 45–54
Core Plus	26–31, 33–43 odd, 45–56
Review	57–61
Preview	62–64

✏ Extra Practice can be found beginning on page 818.

Error Analysis

When writing proofs, students may need to review postulates and theorems about triangle congruence, such as the SSS Congruence Postulate.

Technology

Students can use geometry graphics software to construct the figures in the exercises. Make sure that the figures are constructed in such a way that their properties (such as opposite sides parallel) are retained when a vertex or a segment is dragged.

20. The diagonals are perpendicular, so it is a rhombus. ∠KLM is a right angle so KLMN is a rectangle. Therefore KLMN is a square.

21. No, KLMN is a rhombus by Theorem 4.6.6. But suppose ∠KLM > 90° KLMN is not necessarily a square.

22. KLMN is a rhombus by Theorem 4.6.7. KLMN is a rectangle since m∠LKN = 90°. Therefore, KLMN is a square.

23. KLMN is a rhombus by Theorem 4.6.6. KLMN is a rectangle by Theorem 4.6.5. Therefore KLMN is a square.

24. No, KLMN is a rhombus by Theorem 4.6.7. But suppose ∠KLM > 90°. KLMN is not necessarily a square.

---

14. rectangle, Theorem 4.6.5
15. rhombus, Theorem 4.6.6
16. rectangle, Theorem 4.6.4
17. neither
18. rhombus, Theorem 4.6.8
19. neither

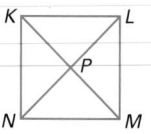

30. Converse of Alternate Interior Angles Theorem

PARAGRAPH PROOF

32. $\overline{FG}$
33. SSS
34. CPCTC
35. same-side interior
36. they are congruent and supplementary
37. opposite angles of a parallelogram are congruent
38. definition of rectangle

PROOFS

Exercises 14–19 refer to a parallelogram *FGHI* with diagonals $\overline{FH}$ and $\overline{GI}$ intersecting at point *J*. For each condition given below, determine whether the parallelogram is a rhombus, a rectangle, or neither. Give the theorem that justifies your answer.

14. $\overline{FH} \cong \overline{GI}$ **15.** $\overline{FG} \cong \overline{GH}$ **16.** m∠FGH = 90°

17. $\overline{FG} \cong \overline{FH}$ **18.** m∠FJG = 90°   **19.** m∠FHG = 90°

For Exercises 20–25, refer to the diagram of parallelogram *KLMN* at left. State whether each set of conditions below is sufficient to prove that *KLMN* is a square. Explain your reasoning.

20. m∠KLM = 90°, m∠KPN = 90°   **21.** $\overline{KL} \cong \overline{LM}, \overline{KL} \cong \overline{MN}$

22. m∠KLN = 45°, m∠LNK = 45°   **23.** $\overline{KN} \cong \overline{MN}, \overline{KM} \cong \overline{LN}$

24. ∠LKM ≅ ∠LMK, ∠LKM ≅ ∠MKN   **25.** △NKL ≅ △KLM

Use the diagram at right to complete the following two-column proof of Theorem 4.6.1:

Given: $\overline{AB} \cong \overline{CD}$ and $\overline{AD} \cong \overline{CB}$

Prove: *ABCD* is a parallelogram.

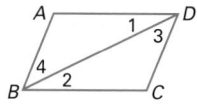

Proof:

Statements	Reasons	
$\overline{AB} \cong \overline{CD}$ and $\overline{AD} \cong \overline{CB}$	**26.** ___?___	Given
$\overline{BD} \cong \overline{BD}$	**27.** ___?___	Reflexive Property
△ABD ≅ △CDB	**28.** ___?___	SSS
∠1 ≅ ∠2 and ∠3 ≅ ∠4	**29.** ___?___	CPCTC
$\overline{AD} \parallel \overline{BC}$ and $\overline{AB} \parallel \overline{CD}$	**30.** ___?___	
ABCD is a parallelogram.	**31.** ___?___	Definition of parallelogram

Use the diagram below to complete the paragraph proof of the Homebuilder Theorem below.

Given: parallelogram *EFGH* with $\overline{EG} \cong \overline{FH}$

Prove: *EFGH* is a rectangle.

Proof:

Opposite sides of a parallelogram are congruent, so $\overline{EH} \cong$ **32.** ___?___ and $\overline{HG} \cong \overline{GH}$ (Reflexive Property). Because $\overline{EG} \cong \overline{FH}$ (given), △EHG ≅ △FGH by **33.** ___?___. ∠EHG ≅ ∠FGH because **34.** ___?___, and ∠EHG and ∠FGH are supplementary because they are **35.** ___?___ angles. ∠EHG and ∠FGH are right angles because **36.** ___?___. Therefore, ∠FEH and ∠EFG are also right angles because **37.** ___?___. Thus, *EFGH* is a rectangle by **38.** ___?___.

Write a two-column, paragraph, or flowchart proof for each of the following theorems:

39. Theorem 4.6.2   **40.** Theorem 4.6.3   **41.** Theorem 4.6.4

42. Theorem 4.6.6   **43.** Theorem 4.6.7   **44.** Theorem 4.6.8

25. KLMN is a rhombus by Theorem 4.6.6, since $\overline{NK} \cong \overline{KL}$ by CPCTC. Also by CPCTC, $\overline{LN} \cong \overline{MK}$, so KLMN is a rectangle by Theorem 4.6.5. Thus, KLMN is a square by definition.

Practice

NAME _____ CLASS _____ DATE _____

Practice

4.6 *Conditions for Special Quadrilaterals*

For Exercises 1–8, refer to quadrilateral *MNOP* with diagonals $\overline{MO}$ and $\overline{NP}$ intersecting at point *Q*. For each set of conditions given, state whether the quadrilateral is a parallelogram. If so, give the theorem that justifies your answer.

1. $\overline{MN} \parallel \overline{PO}$ and $\overline{MN} \cong \overline{PO}$ Yes; if one pair of opposite sides of a quadrilateral are parallel and congruent, then the quadrilateral is a parallelogram.

2. $\overline{PQ} \cong \overline{QN}$ and $\overline{MQ} \cong \overline{QO}$ Yes; if the diag. of a quad. bisect each other, then the quad. is a p-

3. $\overline{MN} \parallel \overline{PO}$ and $\overline{PQ} \cong \overline{ON}$ no

4. $\overline{MN} \parallel \overline{PO}$ and $\overline{NO} \cong \overline{PO}$ no

5. $\overline{MN} \cong \overline{PO}$ and $\overline{MP} \cong \overline{NO}$ Yes; if two pairs of opposite sides of a quadrilateral are congruent, then the quadrilateral is a parallelogram.

6. $\overline{MP} \parallel \overline{NO}$ and $\overline{MP} \cong \overline{PO}$ no

7. $\overline{MP} \parallel \overline{NO}$ and $\overline{NO} \cong \overline{NQ}$ no

8. $\overline{MP} \parallel \overline{NO}$ and $\overline{MP} \cong \overline{NO}$ same theorem as Exercise 1

Exercises 9–16 refer to parallelogram *CLPK* with diagonals $\overline{CP}$ and $\overline{LK}$ intersecting at point *X*. For each condition given below, state whether the parallelogram is a rhombus, a rectangle, or neither. Give the theorem that justifies your answer.

9. m∠K = 90°   rectangle; if one angle of a parallelogram is a right angle, then the parallelogram is a rectangle.

10. $\overline{CL} \cong \overline{LP}$ rhombus; if one pair of adj. sides of a p-gram are cong., then the p-gram is a rhomb

11. m∠KLC = m∠KLP   rhombus; if the diagonals of a parallelogram bisect the angles and m∠PCK = m∠PCL   of the p-gram, then the p-gram is a rhombus.

12. $\overline{KL} \cong \overline{CP}$ rectangle; if the diag. of a parallelogram are congr., then the p-gram is a rectangle

13. $\overline{CL} \cong \overline{KP}$ neither

14. m∠CXL = 90°   rhombus; if the diags. of a parallelogram are perp., then the p-gram is a rhomb

15. $\overline{CK} \cong \overline{CL}$ rhombus; same theorem as Exercise 10

16. $\overline{CK} \cong \overline{LP}$ neither

258 LESSON 4.6

Recall the following conjecture from Lesson 3.7, which you will now prove as a theorem:

The Triangle Midsegment Theorem

A midsegment of a triangle is parallel to a side of the triangle and has a measure equal to half of the measure of that side.   4.6.9

FLOWCHART PROOF

Complete the flowchart proof below.

Given: In $\triangle ABC$, D is the midpoint of $\overline{AC}$, and E is the midpoint of $\overline{BC}$.

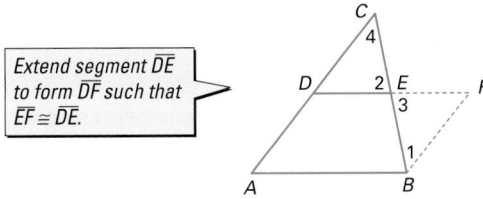

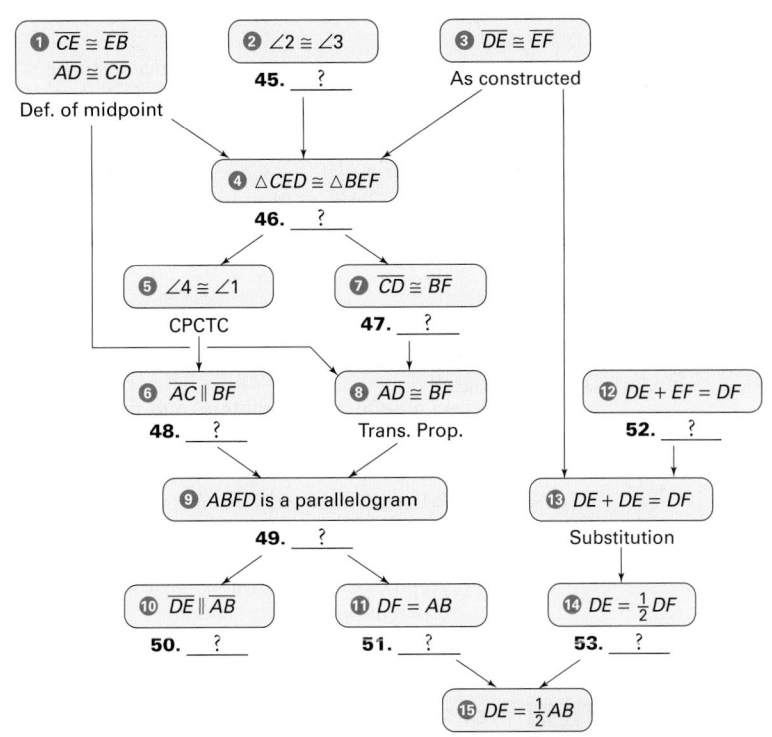

Extend segment $\overline{DE}$ to form $\overline{DF}$ such that $\overline{EF} \cong \overline{DE}$.

Prove: $\overline{DE} \parallel \overline{AB}$ and $DE = \frac{1}{2}AB$

Proof:

45. Vertical Angles Theorem

46. SAS

47. CPCTC

48. Converse of Alternate Interior Angles Theorem

49. Theorem 4.6.2

50. Definition of parallelogram

51. Opposite sides of a parallelogram are congruent.

52. Segment Addition Postulate

53. Division Property

54. Substitution Property

❶ $\overline{CE} \cong \overline{EB}$
   $\overline{AD} \cong \overline{CD}$
   Def. of midpoint

❷ $\angle 2 \cong \angle 3$
   45. ___?___

❸ $\overline{DE} \cong \overline{EF}$
   As constructed

❹ $\triangle CED \cong \triangle BEF$
   46. ___?___

❺ $\angle 4 \cong \angle 1$
   CPCTC

❼ $\overline{CD} \cong \overline{BF}$
   47. ___?___

❻ $\overline{AC} \parallel \overline{BF}$
   48. ___?___

❽ $\overline{AD} \cong \overline{BF}$
   Trans. Prop.

❾ $ABFD$ is a parallelogram
   49. ___?___

⓬ $DE + EF = DF$
   52. ___?___

⓭ $DE + DE = DF$
   Substitution

❿ $\overline{DE} \parallel \overline{AB}$
   50. ___?___

⓫ $DF = AB$
   51. ___?___

⓮ $DE = \frac{1}{2}DF$
   53. ___?___

⓯ $DE = \frac{1}{2}AB$
   54. ___?___

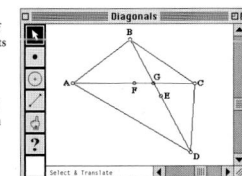

Look Beyond

In Exercises 62–64, students discover an interesting property of quadrilaterals. By connecting the midpoints of a quadrilateral, they will find that the figure formed is always a parallelogram. Have students draw this construction with geometry graphics software. Since the sides of figure *KLMN* become midsegments and a midsegment is parallel to the base of a triangle, this fact can be used to prove that the sides of *KLMN* are parallel.

55. The boards that are 2 feet long must be opposite each other, and the boards that are 3 feet long must be opposite each other. She should also make sure that the diagonals have the same measure.

56. The spacing of the rungs must be the same on both side pieces. In theory, only one corner brace is needed. When one right angle is established, all corresponding angles must also be right angles. By Theorem 4.6.4, each parallelogram must be a rectangle. In practice, more than one brace would be used, because of the flexibility of the wood.

60. Sample answer: A postulate is something that is accepted as true, without proof. A theorem is something you can prove using postulates and definitions. A conjecture is something that you think is true but have not yet proven.

61. Since $\overline{BA} \cong \overline{BC}$, $\triangle ABC$ is isosceles, so $\angle BAC \cong \angle BCA$ by the Isosceles Triangle Theorem. Also, since $\overline{BD} \cong \overline{BE}$, by the Segment Addition Postulate $\overline{AD} \cong \overline{CE}$. By the Reflexive Property of Congruence $\overline{AC} \cong \overline{AC}$, so $\triangle ADC \cong \triangle CEA$ by SAS.

APPLICATIONS

55. ART An artist is making a frame for stretching a canvas. She cuts two boards that are 2 feet long and two boards that are 3 feet long. How can she put the boards together to make sure that the frame is rectangular?

56. CONSTRUCTION In a ladder, all of the rungs are congruent and the two side pieces are congruent. What else is necessary in a ladder in order to ensure that the rungs are parallel?

Corner braces may be used to stabilize the ladder. Explain how corner braces can be placed to ensure that the rungs are perpendicular to the side pieces. How many corner braces are needed? Explain your reasoning.

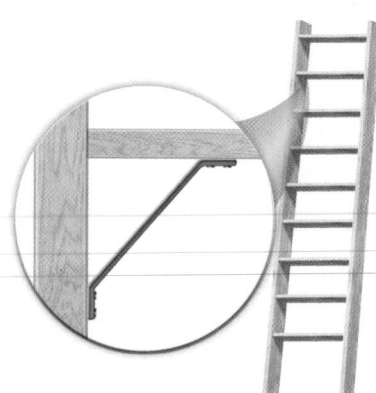

Look Back

For Exercises 57–59, consider a regular polygon with a central angle of 72°. Find each value below. *(LESSONS 3.1 AND 3.6)*

57. the number of sides of the polygon 5

58. the measure of an interior angle 108°

59. the measure of an exterior angle 72°

60. Explain the difference between a postulate, a conjecture, and a theorem. *(LESSONS 1.1, 1.4, AND 2.4)*

PROOF

61. Given: $\overline{BA} \cong \overline{BC}$ and $\overline{BD} \cong \overline{BE}$

Prove: $\triangle ADC \cong \triangle CEA$
(Hint: $\triangle ABC$ is isosceles.)
(LESSON 4.4)

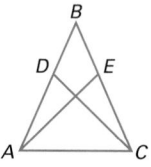

Look Beyond

Draw a general quadrilateral and find the midpoint of each side. Connect the midpoints to form another quadrilateral. Label the figure as shown at right.

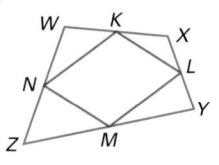

62. What type of quadrilateral appears to be formed by the midpoints? Test your conjecture by drawing several quadrilaterals or, if you are using geometry graphics software, by dragging the vertices to different positions. **parallelogram**

63. Draw diagonal $\overline{WY}$ to form triangles $\triangle WXY$ and $\triangle WZY$. What part of these triangles are segments $\overline{KL}$ and $\overline{MN}$, respectively? **midsegments**

PROOF

64. Use the Triangle Midsegment Theorem to prove that *KLMN* is a parallelogram.

64.

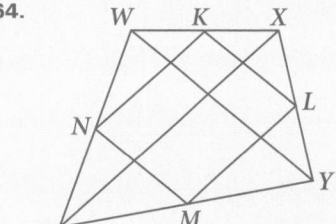

Given: quadrilateral *WXYZ* with *K*, *L*, *M*, and *N* the midpoints of $\overline{WX}$, $\overline{XY}$, $\overline{YZ}$, and $\overline{ZW}$.
Prove: *KLMN* is a parallelogram.
Proof: $\overline{KL}$ is a midsegment of $\triangle WXY$, so by

the Triangle Midsegment Theorem, $\overline{KL} \| \overline{WY}$. $\overline{MN}$ is a midsegment of $\triangle WZY$, so by the Triangle Midsegment Theorem, $\overline{MN} \| \overline{WY}$. Two segments that are parallel to the same segment are parallel to each other, so $\overline{KL} \| \overline{MN}$. Also by the Triangle Midsegment Theorem, $KL = \frac{1}{2}WY$ and $MN = \frac{1}{2}WY$, so by the Transitive Property of Equality $KL = MN$ ($\overline{KL} \cong \overline{MN}$). Since *KLMN* has one pair of opposite sides that are parallel and congruent, *KLMN* is a parallelogram by Theorem 4.6.2.

Compass and Straightedge Constructions

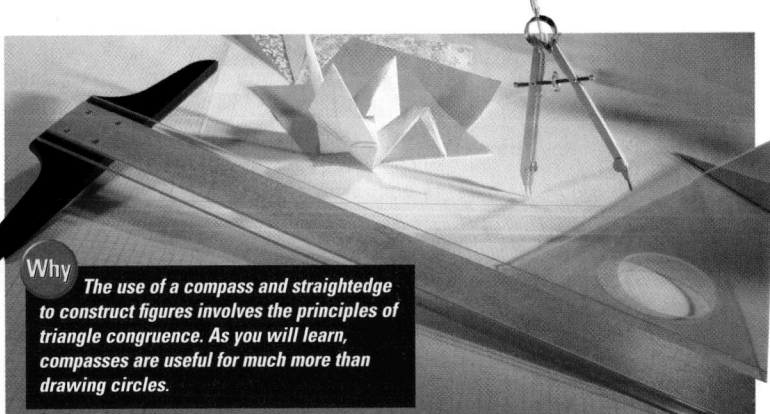

Objectives

- Construct congruent copies of segments, angles, and triangles.
- Construct an angle bisector.

Why *The use of a compass and straightedge to construct figures involves the principles of triangle congruence. As you will learn, compasses are useful for much more than drawing circles.*

A Classical Mathematical Game

For centuries, mathematicians have enjoyed a sort of game. The "game" requires a compass and a straightedge. By using just these tools, you can create precise geometric figures.

Activity 1

CONSTRUCTION
COMPASS and
STRAIGHTEDGE

Note: A straightedge is not a ruler. It has no marks on it and cannot be used to measure distances.

A Segment Congruent to a Given Segment

Given $\overline{AB}$ at right, follow Steps 1–4 to construct $\overline{CD} \cong \overline{AB}$.

$A \bullet \underline{\hspace{2cm}} \bullet B$

1. Using a straightedge, draw line ℓ. $\longleftrightarrow$ ℓ	**2.** Select a point on ℓ and label it C. $\longleftrightarrow$ ℓ $\bullet$ C
3. Set your compass to the distance AB in the given figure.  $A \quad B$	**4.** Place the point of your compass on C and draw an arc that intersects line ℓ. Label the intersection of the arc and the line as point D. 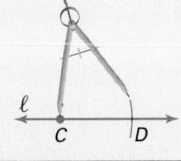 $\ell \longleftrightarrow$ $C \quad D$

Alternative Teaching Strategy

TECHNOLOGY You may wish to teach this lesson with geometry graphics software. The constructions in the Activities and exercises involve similar steps that are accessible from various menu commands.

Activity 1 Notes

Students will use a compass and straightedge to construct a segment that is congruent to a given segment. In this Activity, as well as in Activities 2 and 3, students create precise geometric figures. You may wish to highlight the importance of geometric precision of objects that students like, such as CDs, computer disks, foldable cellphones, etc.

For a student worksheet of this Activity and detailed Teacher Notes, see page 66 in the Lesson Activities booklet.

Teaching Tip

Point out that because every point on a circle is equidistant from the center of the circle, all radii have the same length. All radii in different congruent circles must be congruent to the radii in the original circle. For example, if $\overline{AB}$ and $\overline{CD}$ are radii of two congruent circles, then $\overline{AB}$ and $\overline{CD}$ are congruent.

Activity 2 Notes

Students use the construction from Activity 1 to construct a triangle that is congruent to a given triangle.

For a student worksheet of this Activity and detailed Teacher Notes, see page 66 in the Lesson Activities booklet.

Teaching Tip

Point out that since each side in the constructed triangle is congruent to its corresponding side in the original triangle, SSS ensures congruence. The proof of this is given on page 261.

Three Important Assumptions

Three important assumptions were involved in the construction in Activity 1. Keep these assumptions in mind as you do Activity 2.

1. You can place the point of your compass or the edge of your straightedge precisely on a given point or line (or circle).

2. Given two points, you can set your compass so that the distance between the point and pencil of the compass is equal to the distance between the two given points; you can also set your straightedge precisely on both given points.

3. The distance between the point and pencil of the compass does not change once it has been set—until you reset it. (This is why a compass can be used to draw a circle.)

Copying a Triangle

Activity 2
A Triangle Congruent to a Given Triangle

CONSTRUCTION
COMPASS and STRAIGHTEDGE

Given △ABC at right, follow Steps 1–5 to construct △MNO ≅ △ABC.

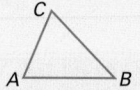

1. Using a straightedge, draw line ℓ. Select a point on line ℓ and label it M. 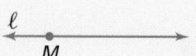	2. Set your compass to the distance AB in the given triangle. Place the the point of your compass on M, and draw an arc that intersects line ℓ. Label the intersection N. 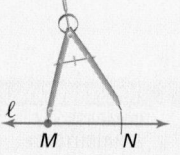
3. Set your compass to the distance AC in the given triangle. Place the point of your compass on M, and draw an arc above $\overline{MN}$. 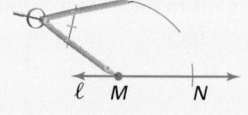	4. Set your compass to the distance BC. Place the point of your compass on N and draw an arc that intersects the first arc you drew. Label the intersection as point O. 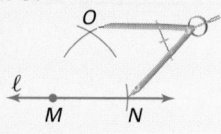

5. Connect points M, N, and O. Result: △MNO ≅ △ABC.

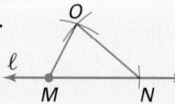

Enrichment

Have students construct a hexagon according to these steps. Begin by drawing a circle. Using any point on the circle as the center, draw another circle with the same radius as the original. Pick one intersection of the two circles, and use this point as the center of another circle with the same radius as the first two. At the point where the third circle intersects the original circle, repeat the procedure by drawing another circle with the same radius. Continue around the original circle until a flower pattern emerges. Connect all intersection points on the original circle. The result is a hexagon.

Justifying a Construction

The steps in the construction of a figure can be justified by combining the assumptions of compass and straightedge constructions with geometry theorems you already know. The proof below justifies the construction in Activity 1.

TWO-COLUMN PROOF

Given: △ABC, with △MNO constructed as in Activity 2.

Prove: △MNO ≅ △ABC

Proof (refer back to Activity 2):

Statements	Reasons
1. $\overline{MN} \cong \overline{AB}$	Compass set at distance AB was used to construct $\overline{MN}$.
2. $\overline{MO} \cong \overline{AC}$	Compass set at distance AC was used to construct $\overline{MO}$.
3. $\overline{NO} \cong \overline{BC}$	Compass set at distance BC was used to construct $\overline{NO}$.
4. △MNO ≅ △ABC	SSS

Bisecting an Angle

Activity 3

Angle Bisector

Given ∠A at right, follow the steps below to construct its bisector.

A

1. Place your compass point at A and draw an arc through the rays of the angle. Label the intersection points as B and C.

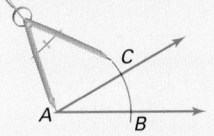

2. Place your compass point first at B and then at C. Using one compass setting, draw arcs that intersect in the interior of ∠A. Label the intersection as F. Draw a ray from A through F. $\overrightarrow{AF}$ is the bisector of ∠BAC.

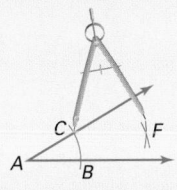

TRY THIS   Write a proof to justify the construction of the angle bisector in Activity 3.

Inclusion Strategies

HANDS-ON STRATEGIES A wide variety of compasses are available, including some without sharp, pointed arms. Others are designed so that the pencil and pointed arm can be retracted for safety. Check a school-supply catalog for alternatives to a traditional compass, particularly for students who lack manual dexterity.

Reteaching the Lesson

USING DISCUSSION Have students work in small groups to complete the following constructions:

1. Copy a segment.
2. Copy a triangle.
3. Copy an angle.
4. Bisect an angle.
5. Construct the perpendicular bisector of a segment.
6. Construct a pair of perpendicular lines.
7. Construct a pair of parallel lines.

Activity 3 Notes

In this Activity, students construct an angle bisector. Before students write the proof in the Try This exercise, have them write a plan for the proof. They should notice that they will be able to justify this construction by using the SSS Congruence Postulate.

For a student worksheet of this Activity and detailed Teacher Notes, see page 66 in the Lesson Activities booklet.

TRY THIS

Given: a diagram of ∠BAC and its bisector, constructed with a compass and straightedge.

Prove: ∠BAF ≅ ∠CAF

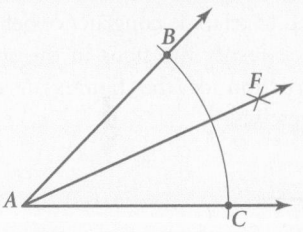

Statements

1. $\overline{AB} \cong \overline{AC}$

2. Draw $\overline{CF}$ and $\overline{BF}$; $\overline{CF} \cong \overline{BF}$.

3. $\overline{AF} \cong \overline{AF}$

4. △BAF ≅ △CAF

5. ∠BAF ≅ ∠CAF

Reasons

1. In the same or ≅ circles, all radii are ≅. (Def. of a circle)

2. In the same or ≅ circles, all radii are ≅. (Def. of a circle)

3. Reflexive Property

4. SSS

5. CPCTC

Selected Answers

Exercises 5–9, 11–55 odd

ASSIGNMENT GUIDE

In Class	1–9
Core	11–37 odd, 38–41
Core Plus	32–46
Review	47–53
Preview	54–55

✐ Extra Practice can be found beginning on page 818.

Error Analysis

Students may need to review postulates and theorems related to parallel lines cut by a transversal and to triangle congruence before they justify the steps in the construction of the figures in the exercises.

Exercises

● Communicate

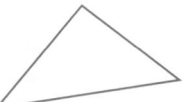

internet connect

Activities Online

Go To: **go.hrw.com**
Keyword:
MG1 American

1. Explain the difference between a straightedge and a ruler. Why is a ruler not used in constructions?

2. In Step 1 of the construction in Activity 3, which of the construction assumptions were used? Explain.

3. In Step 2 of the construction in Activity 3, which of the construction assumptions were used? Explain.

4. Describe how you would do the constructions from the activities with folding paper or on a computer. Name some advantages or disadvantages of one of these methods.

● Guided Skills Practice

Trace the figures below onto paper, and construct a congruent copy of each. *(ACTIVITIES 1 AND 2)*

5.

6.

For Exercises 7–9, refer to the diagram at right. Trace the figure and construct the angle bisector of the indicated angle. *(ACTIVITY 3)*

7. ∠A   **8.** ∠B   **9.** ∠C

● Practice and Apply

Construct a congruent copy of each figure below.

10.

11.

12.

13.

14.

15.

5.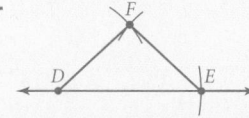

6.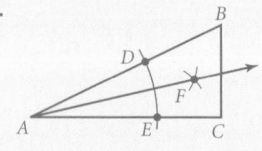

7.

8.

9.

10.

11.

12.

13.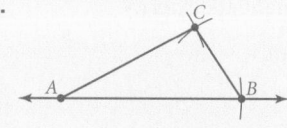

Trace each angle below onto your paper and construct the angle bisector of each.

16.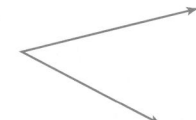

17.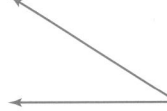

For Exercises 18–21, trace each angle onto paper and follow the steps below to construct a congruent copy.

CONSTRUCTION

You will prove this construction in Exercise 47.

An angle congruent to a given angle

Given ∠R at right, follow these steps to construct ∠B ≅ ∠R.

a. Using a straightedge, draw a ray with endpoint *B*. 	**b.** Place your compass point on *R* and draw an arc. Label the intersection points *Q* and *S*. 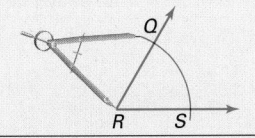
c. Without adjusting your compass, place the compass point at *B*. Draw an arc that crosses the ray, and label the intersection point *C*. 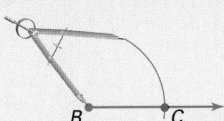	**d.** Set your compass equal to the distance *QS* in ∠R. 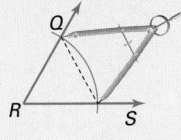
e. Without adjusting your compass, place the compass point at *C*. Draw an arc that crosses the first arc and label the intersection *A*. 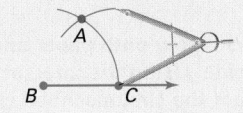	**f.** Draw $\overrightarrow{BA}$ to form ∠B. ∠B is congruent to ∠R. 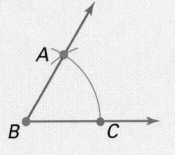

18.

19.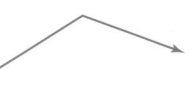

20.

21.

Technology

The constructions in this exercise set may be done with geometry graphics software.

16.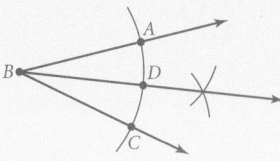

17.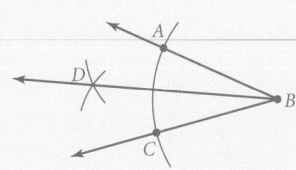

18.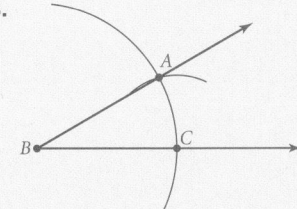

19.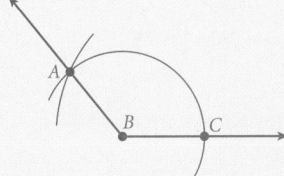

20.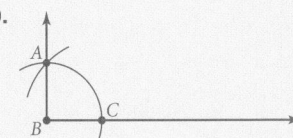

21.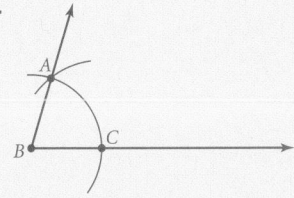

22.

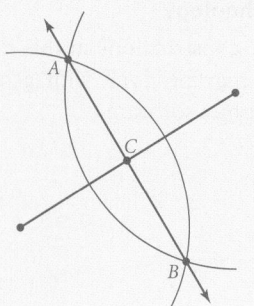

23.

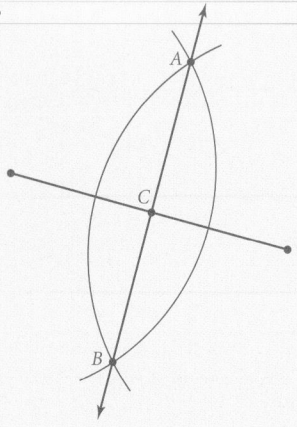

24.

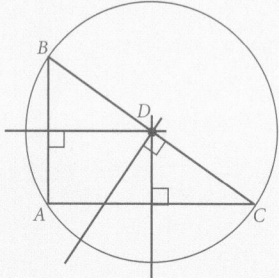

25.

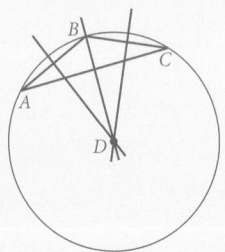

For Exercises 22 and 23, trace each segment onto paper and follow the steps below to construct the perpendicular bisector and midpoint.

You will prove this construction in Exercise 48.

The perpendicular bisector of a given segment and the midpoint of a given segment

Given $\overline{AC}$ at right, follow these steps to construct line the perpendicular bisector of $\overline{AC}$.

a. Set your compass equal to a distance greater than half of AC.

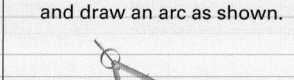

b. Place your compass point on A and draw an arc as shown.

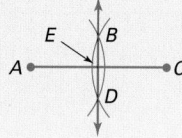

c. Without adjusting your compass setting, place the compass point at C. Draw a new arc as shown and label the intersection points of the two arcs B and D.

d. Use a straightedge to draw $\overleftrightarrow{BD}$. Label the intersection with $\overline{AC}$ as point E. $\overleftrightarrow{BD}$ is the perpendicular bisector of $\overline{AC}$, and E is the midpoint.

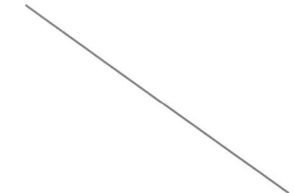

22.

23.

Trace the triangles below onto paper and construct the perpendicular bisector of each side. Using the intersection of the perpendicular bisectors, construct the circumscribed circle of each triangle.

24.

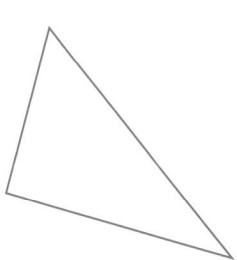

25.

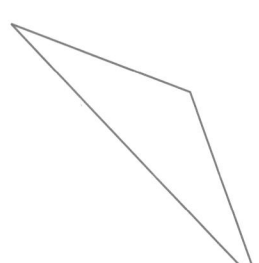

For Exercises 26 and 27, trace each figure onto paper and follow the steps below to construct a line through the given point and perpendicular to each given line.

CONSTRUCTION

You will prove this construction in Exercise 49.

A line through a point perpendicular to a given line

Given point *A* and line *ℓ* at right, follow these steps to construct $\overleftrightarrow{AC} \perp$ line *ℓ*.

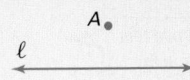

a. Place your compass point on *A* and draw an arc as shown. Label the intersection points *D* and *B*.

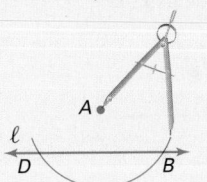

b. Place your compass point on *D* and draw an arc below the line, as shown. The compass does not need to be at the same setting as in the previous step.

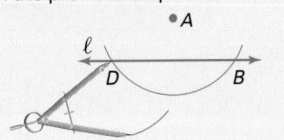

c. Without adjusting your compass setting, place the compass point at *B*. Draw a new arc as shown, and label the intersection point *C*.

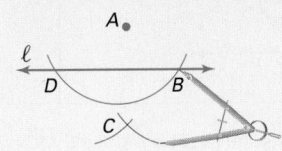

d. Use a straightedge to draw $\overleftrightarrow{AC}$. $\overleftrightarrow{AC}$ is perpendicular to line *ℓ*.

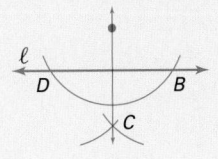

26.

27.

28. A segment from a vertex of a triangle perpendicular to the opposite side is called an *altitude* of the triangle. Trace the triangle at right onto your paper.

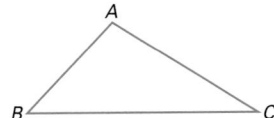

a. Construct the altitude from each vertex.
b. What seems to be true about the three altitudes of a triangle?

29. Draw a large scalene triangle, △*ABC*, on your paper. Follow the steps below to construct the inscribed circle.

a. Construct the angle bisectors to find the incenter, *I*.
b. Construct a line through *I* perpendicular to any side. Label the point of intersection *E*.
c. Draw a circle centered at *I* with radius *IE*. This is the inscribed circle of △*ABC*

26.

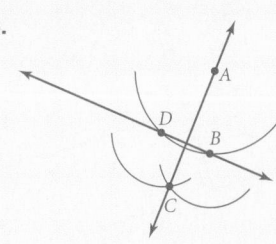

27.

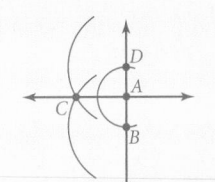

28. a.

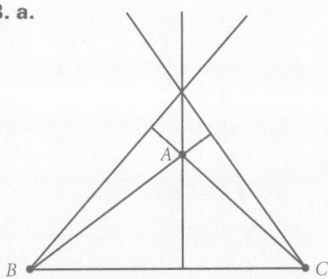

b. The 3 altitudes intersect at one point.

29. a.

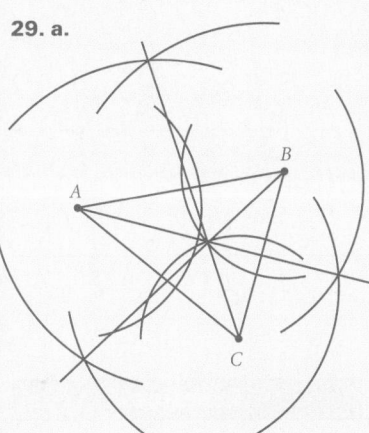

b.

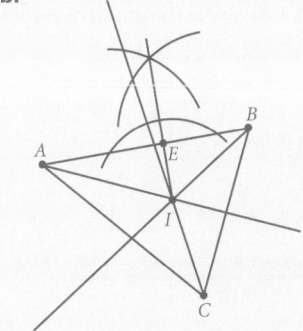

c.

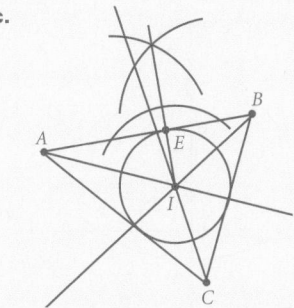

The circle drawn is the inscribed circle of the triangle.

30.

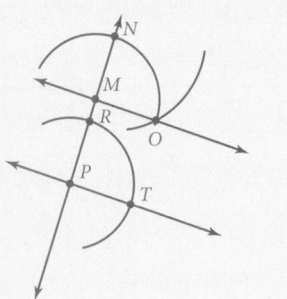

31.

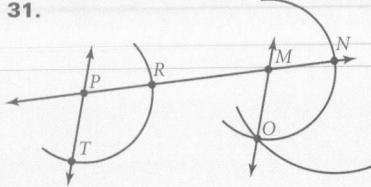

32.

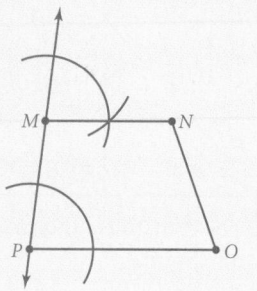

33.

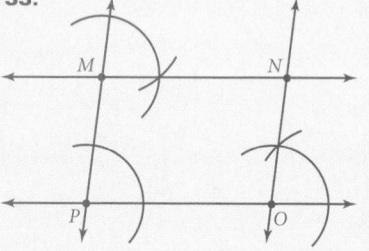

Practice

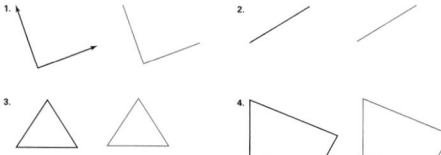
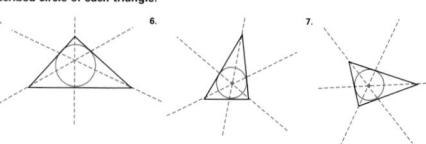
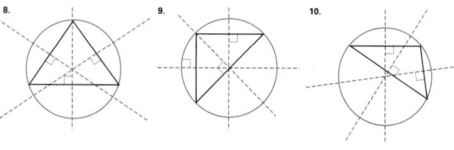
For Exercises 30 and 31, trace each figure onto paper and follow the steps below to construct a line through the given point and parallel to the given line.

CONSTRUCTION

> *You will prove this construction in Exercise 44.*

A line through a point parallel to a given line

Given point *M* and line ℓ at right, follow these steps to construct $\overleftrightarrow{AC} \parallel$ line ℓ.

M•

ℓ

a. Use a straightedge to draw a line through *M* that intersects ℓ. Label the intersection point *P*. 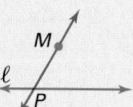	**b.** Place your compass point on *P* and draw an arc as shown. Label the intersection points *R* and *T*. 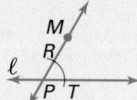
c. Without adjusting your compass setting, place the compass point at *M*. Draw a new arc as shown and label the intersection point *N*. 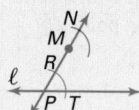	**d.** Set your compass to the distance *RT*. Place your compass point on *N* and draw an arc as shown. Label the point of intersection *O*. 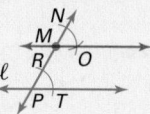

e. Use a straightedge to draw $\overleftrightarrow{MO}$. Line $\overleftrightarrow{MO}$ is parallel to line ℓ.

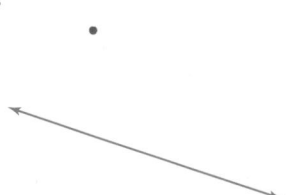

30.   **31.**

Construct each of the following special quadrilaterals with a compass and straightedge:

32. trapezoid   **33.** parallelogram

34. rectangle   **35.** rhombus

36. square   **37.** kite

34.   **35.**

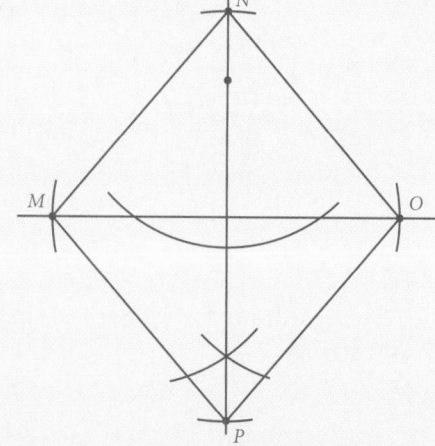

Refer to the construction, shown at right, of an angle congruent to a given angle. Complete the two-column proof that $\angle B \cong \angle R$.

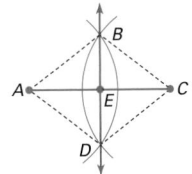

Given: $\angle B$, as constructed

Prove: $\angle B \cong \angle R$

Proof:

Statements	Reasons
$\overline{AB} \cong \overline{QR}$	Same compass setting used
$\overline{BC} \cong \overline{RS}$	**38.** ___?___ Same compass setting used
$\overline{AC} \cong \overline{QS}$	**39.** ___?___ Same compass setting used
$\triangle ABC \cong \triangle QRS$	**40.** ___?___ SSS
$\angle B \cong \angle R$	**41.** ___?___ CPCTC

PROOFS

42. Refer to the construction, shown at right, of the perpendicular bisector of a given segment. Prove that $\overleftrightarrow{BD} \perp \overline{AC}$ and that $\overleftrightarrow{BD}$ bisects $\overline{AC}$.

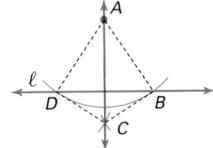

43. Refer to the construction, shown at right, of a line through a point perpendicular to a given line. Prove that $\overleftrightarrow{AC} \perp$ line ℓ.

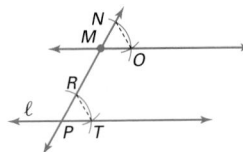

44. Refer to the construction, shown at right, of a line through a given point parallel to a given line. Prove that $\overleftrightarrow{MO} \parallel$ line ℓ.

APPLICATION

45. MAP READING A geologist is at point X near Croton Peak in Big Bend National Park. He wishes to go back to the road by the shortest path. Make your own sketch of the map and use a compass and straightedge to construct his path. (You may assume that the road is perfectly straight.)

Croton Peak

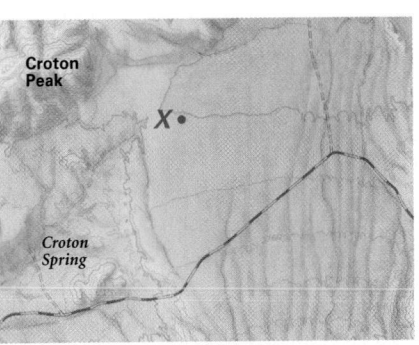
Detail of topographical map

42. The same compass setting was used to create $\overline{AB}$, $\overline{AD}$, $\overline{CB}$, and $\overline{CD}$, so the segments are congruent. Thus, $ABCD$ is a rhombus by definition. By Theorem 4.5.8, the diagonals of a rhombus are perpendicular, so $\overleftrightarrow{BD} \perp \overleftrightarrow{AC}$. By Theorem 4.5.6, $ABCD$ is a parallelogram, so by Theorem 4.5.5, the diagonals of $ABCD$ bisect each other. Thus, $\overleftrightarrow{BD}$ bisects $\overline{AC}$.

43.

Statements	Reasons
$\overline{AB} \cong \overline{AD}$ $\overline{CB} \cong \overline{CD}$	Same compass setting
$ABCD$ is a kite.	Defintion of kite
$\overline{AC} \perp \overline{BD}$	Diagonals of a kite are perpendicular

44.

Statements	Reasons
$\overline{PR} \cong \overline{MN}$ $\overline{PT} \cong \overline{MO}$	Same compass setting
$\overline{RT} \cong \overline{NO}$	Same compass setting
$\triangle PRT \cong \triangle MNO$	SSS
$\angle NMO \cong \angle RPT$	CPCTC
$\overleftrightarrow{MO} \parallel \ell$	Converse of the Corresponding Angles Postulate

45.

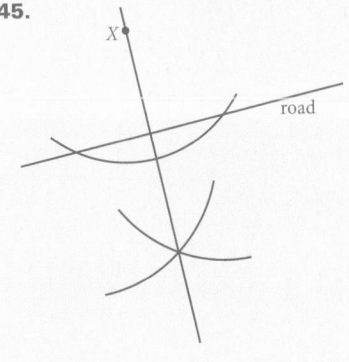

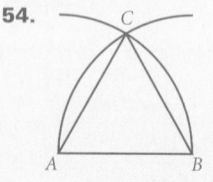
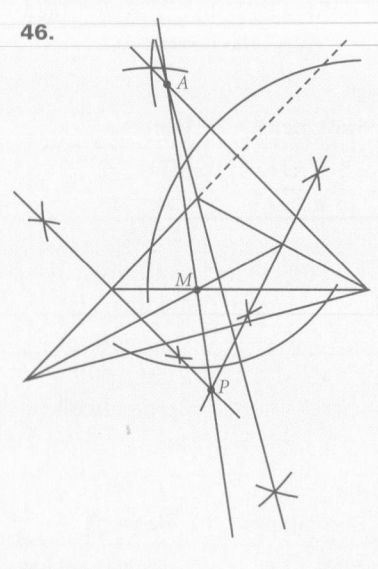
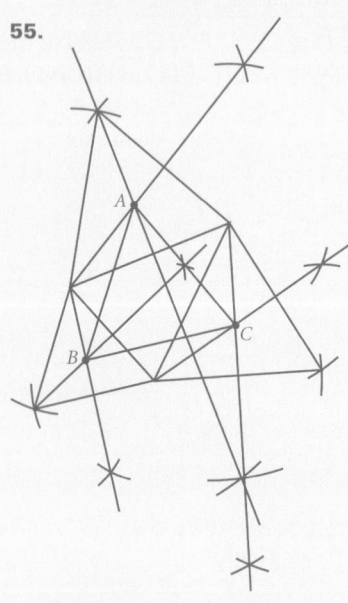
46. Draw a large scalene triangle, and construct at least two of each of the following:

- Perpendicular bisectors: these meet at a point called the *circumcenter*.
- Altitudes: these meet at a point called the *orthocenter*.
- Medians (a segment joining a vertex to the midpoint of the opposite side): these meet at a point called the *centroid*.

If your constructions have been done carefully, these three special points should be collinear. Draw the line through the three points.

Look Back

Identify the type of transformation shown in each figure below.
(LESSON 1.6)

47.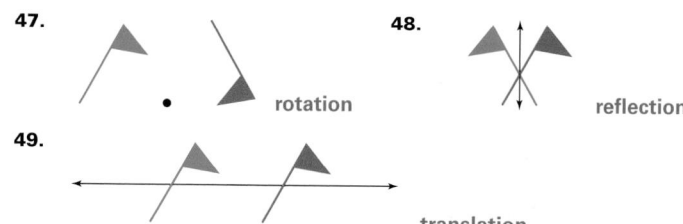

rotation

48.

reflection

49.

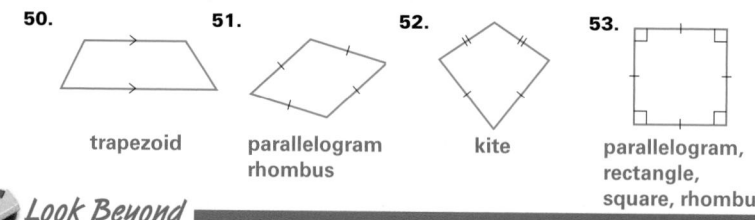

translation

Classify each quadrilateral below as a parallelogram, rectangle, square, rhombus, kite, or trapezoid. List all terms that apply to each figure.
(LESSONS 3.2 AND 3.3)

50. 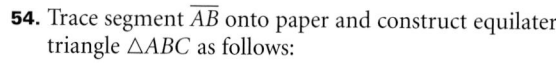 **51.** **52.** **53.**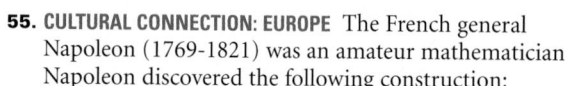

trapezoid   parallelogram, rhombus   kite   parallelogram, rectangle, square, rhombus

Look Beyond

54. Trace segment $\overline{AB}$ onto paper and construct equilateral triangle $\triangle ABC$ as follows:

Place your compass point at A, and adjust the compass so that its pencil point is at B. Draw an arc as shown.

Without adjusting the compass setting, place the compass point at B and draw another arc. Label the intersection point C. Draw segments $\overline{AC}$ and $\overline{BC}$.

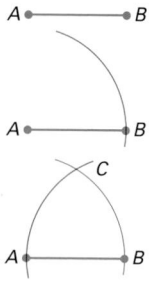

55. CULTURAL CONNECTION: EUROPE The French general Napoleon (1769-1821) was an amateur mathematician. Napoleon discovered the following construction:

Draw any triangle. Construct an equilateral triangle along each side, as shown. Construct the perpendicular bisectors of each equilateral triangle to find the circumcenters, and join the circumcenters to form a triangle. What seems to be true about this triangle?

Napoleon (1769–1821)

Constructing Transformations

Objectives

- Translate, rotate, and reflect figures by using a compass and straightedge.

- Prove that translations, rotations, and reflections preserve congruence and other properties.

- Use the Betweenness Postulate to establish the Triangle Inequality Theorem.

Why All of the transformations you have studied so far have been rigid. That is, the size and shape of the objects that are transformed do not change. You can now prove that these transformations preserve congruence and other properties as well.

Flying in tight formation, these jet planes move as a rigid geometric unit.

Match each transformation with the information needed to perform it.

Transformation
1. rotation   b
2. dilation   d
3. translation   a
4. reflection   c

Needed Information
a. the distance moved, both horizontally and vertically
b. the center and the number of degrees
c. a "mirror" line
d. the center and a scale factor

Also on Quiz Transparency 4.8

Translating Segments and Polygons

Recall from Lesson 1.6 that a translation is a transformation that moves every point of an object the same distance in the same direction.

Activity 1
Translating a Segment

YOU WILL NEED
compass, straightedge, and ruler

Make your own drawing like the one at right. The arrow, known as a translation vector, shows the direction and distance of the translation you are to construct. The distance is the length of the vector.

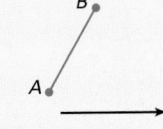

1. Construct a line, ℓ_1, through point A and parallel to the translation vector. Construct another line, ℓ_2, through point B and parallel to the translation vector. Are lines ℓ_1 and ℓ_2 parallel? Explain.

2. Set your compass to the length, x, of the translation vector. On the right side of $\overline{AB}$, construct points A' and B' that are the same distance, x, from points A and B on lines ℓ_1 and ℓ_2, respectively.

CHECKPOINT ✔ 3. Connect points A' and B'. Measure $\overline{AB}$ and $\overline{A'B'}$. Are the two segments congruent?

Teach

Why Most students know intuitively that rigid transformations preserve congruency. Have students discuss the differences between intuitive and scientific knowledge. Intuition, although useful and important, can sometimes lead to incorrect conclusions.

☞ For Activity Notes, see page 272. For the answer to the Checkpoint, see Additional Answers.

Alternative Teaching Strategy

TECHNOLOGY The Activities and exercises in this lesson are appropriate for work with geometry graphics software. Although students can use the software to transform objects with menu commands, have them follow the steps specified in each Activity. For example, in Activity 1, students should construct a translation of segment $\overline{AB}$ by drawing arcs or circles. They should not use a menu command to translate the segment.

The following Example gives a proof that $\overline{AB}$ and $\overline{A'B'}$, as constructed in Activity 1, are congruent:

EXAMPLE

PARAGRAPH PROOF

Given: $\overline{AB}$ and $\overline{A'B'}$, as constructed in Activity 1

Prove: $\overline{AB} \cong \overline{A'B'}$

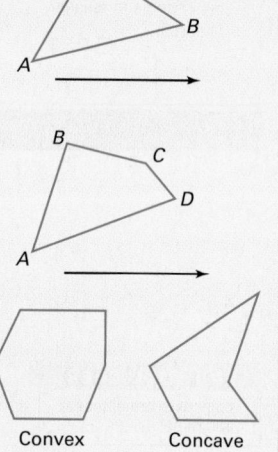

Proof (refer to Activity 1, page 271):

$\overline{AA'} \parallel \overline{BB'}$ because they lie on lines ℓ_1 and ℓ_2, which were constructed to be parallel. (For the method of constructing a line through a given point and parallel to a given line, see page 268.) $\overline{AA'} \cong \overline{BB'}$ because the same compass setting (the distance x) was use to construct them. Quadrilateral $AA'B'B$ is a parallelogram because a pair of opposite sides are congruent and parallel. Therefore, $\overline{AB} \cong \overline{A'B'}$ because opposite sides of a parallelogram are congruent.

Activity 2

Translating Polygons

YOU WILL NEED

no special tools

CHECKPOINT ✓

1. If you translate each of the three sides of $\triangle ABC$ as indicated by the translation vector, will your new figure be congruent to the original one? State the theorem or postulate that justifies your answer.

2. If you translate each of the four sides of quadrilateral $ABCD$ as indicated by the translation vector, will your new figure be congruent to the original one? State the theorem or postulate that justifies your answer. (Hint: Divide the figure into two triangles by drawing a diagonal.)

3. Show how any polygon can be divided into a number of triangles by connecting the vertices with segments. Include both convex and concave polygons in your illustration. What can you conclude about the translation of a polygon?

4. What can you conclude about the translation of an open figure composed of segments, such as the one shown at right? Can you apply your conclusion from Step 3 to this figure?

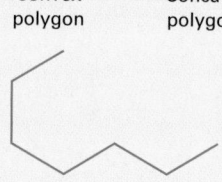

Convex polygon   Concave polygon

CRITICAL THINKING

Do you think your conclusion about translations of segments and polygons can be applied to translations of curves? (Hint: How can you use a number of segments to approximate a curve?)

Preservation of "ABCD"

The rigid transformations, which are also known as *isometries*, preserve *Angles*, *Betweenness*, *Collinearity*, and *Distance* ("ABCD").

CRITICAL THINKING   How do Activities 1 and 2 show that translations preserve angles and distance?

If *A*, *B*, and *X* are collinear points, how can you be sure that their image points, *A′*, *B′*, and *X′*, will be collinear in a translated image of the figure? If *X* is between *A* and *B*, how can you be sure that *X′* will be between *A′* and *B′*?

To answer the above questions, you will need a mathematical way to determine which of three given points is between the other two (if any) from information about the distances between them. The postulate below gives a method. It is the converse of a postulate you have already studied.

Betweenness Postulate
(Converse of the Segment Addition Postulate)

Given three points *P*, *Q*, and *R*, if $PQ + QR = PR$, then *P*, *Q*, and *R* are collinear and *Q* is between *P* and *R*.   **4.8.1**

TRY THIS   You are given the points *R*, *S*, and *T*, where $RS = 9.8$, $TR = 9.6$, and $TS = 19.4$. Assuming that the given distances are exact, are the points collinear? If so, which point is between the other two?

Activity 3
Collinearity and Betweenness

YOU WILL NEED

no special tools

You are given segment *AB* with point *X* between *A* and *B*. Points *A*, *B*, and *X* have been translated as shown at right by the translation vector and connected by segments. Are *A′*, *B′*, and *X′* in the translated image collinear? Does *X′* lie between *A′* and *B′*? How can you prove your answers mathematically? Follow the steps below.

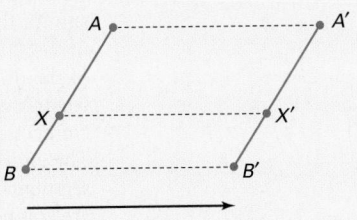

1. What do you know about distances *AB* and *A′B′*? about distances *AX* and *A′X′*? about distances *BX* and *B′X′*? Explain your reasoning.

2. From the Segment Addition Postulate you know that $AX + XB = AB$. What can you conclude about $A′X′ + X′B′$? Explain your reasoning.

CHECKPOINT ✓   3. Is *X′* collinear with *A′* and *B′*? Is *X′* between *A′* and *B′*? Do translations preserve collinearity and betweenness? Explain your reasoning.

Inclusion Strategies

USING COGNITIVE STRATEGIES   To help students remember that the objective of each construction in the Activities is to prove a conjecture, have students complete the following before beginning the proofs:

Conjecture:_____

Given:_____

Prove:_____

Teaching Tip

In Step 3 of Activity 2, remind students of the difference between convex and concave polygons. In a concave polygon, at least one of the interior angles is greater than 180°. A concave polygon "caves" inward.

CRITICAL THINKING

Because a figure is congruent to its image under a translation, the segment lengths and angle measures of the figure are preserved.

TRY THIS

The points are collinear. Point *R* is between *T* and *S*.

Activity 3 Notes

In this Activity, students will discover that translations preserve the collinearity and betweenness of points. Students use the Segment Addition Postulate to show that *X* and *X′* are between their respective points.

For a student worksheet of this Activity and detailed Teacher Notes, see page 71 in the Lesson Activities booklet.

☞ For the answer to the Checkpoint, see Additional Answers.

Cooperative Learning

Have students work in small groups to construct the reflection of a triangle across the line *y* = *x*. After the construction is completed, students should discuss and write a paragraph proof of the construction.

CRITICAL THINKING

$AB + BC$ is apparently larger than AC. If $AB + BC = AC$, then points A, B, and C would be collinear.

The Triangle Inequality Theorem

CRITICAL THINKING   What seems to be true about $AB + BC$ as compared with AC? Suppose that point B is relocated so that $AB + BC = AC$. What happens to $\triangle ABC$?

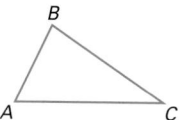

The answers to the questions above suggest the following theorem:

Triangle Inequality Theorem

The sum of the lengths of any two sides of a triangle is greater than the length of the third side.

4.8.2

This theorem can be argued informally by considering the two other possible cases.

Informal Argument

Case 1

Given three segments, if the sum of the lengths of any two of them is **less than** the length of the third segment, then no triangle can be formed by connecting their endpoints. The endpoints of two of the segments cannot be connected.

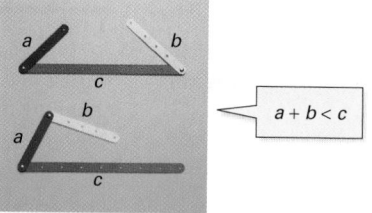

$a + b < c$

The endpoints cannot be connected.

Case 2

Given three segments, if the sum of the lengths of any two of them is **equal to** the length of the third segment, then no triangle can be formed by connecting their endpoints. The endpoints, and hence the segments themselves, are collinear (justification: the Betweenness Postulate).

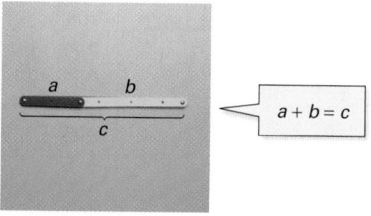

$a + b = c$

The endpoints are collinear.

Thus, if three segments form a triangle, the sum of the lengths of any two segments must be greater than the length of the third. This is the only remaining possibility.

TRY THIS   Which of the following are possible side lengths of a triangle?

a. 14, 8, 25   **b.** 16, 7, 23   **c.** 18, 8, 24

TRY THIS

a. not possible, since
   $14 + 8 < 25$

b. not possible, since
   $16 + 7 = 23$

c. possible, since $18 + 8 > 24$

Reteaching the Lesson

HANDS-ON STRATEGIES   Have students use graph paper and transparencies made from graph paper to translate, rotate, and reflect segments and triangles. They should write two-column proofs to show that congruency is preserved in each case.

Exercises

Communicate

1. How does the Betweenness Postulate justify the translation of the segment in Activity 1?

2. What does "ABCD" refer to? Explain in terms of the translation of a triangle.

3. In the word *isometry*, the root *-metry* means "measure." What do you think the prefix *iso-* means? Check your dictionary. How does this help explain the term *isometry*?

4. In $\triangle ABC$, $AB = 9$ and $AC = 6$. What can you say about BC? Does it have a minimum possible value? Does it have a maximum possible value? Explain your reasoning.

Guided Skills Practice

Trace the figures below. Translate each figure as indicated by the given translation vector. *(ACTIVITIES 1 AND 2 AND EXAMPLE)*

5.

6.

7. Trace the figure at right. Translate the angle as indicated by the given translation vector, and measure both angles. What do you notice? *(ACTIVITY 3)*

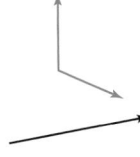

Practice and Apply

7 internet connect

Homework Help Online
Go To: go.hrw.com
Keyword:
MG1 Homework Help
for Exercises 8-19

Trace each figure below and translate it as indicated by the given translation vector.

8.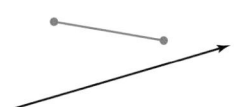

9.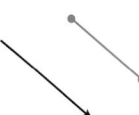

10.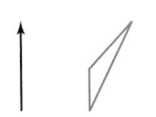

11.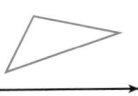

8.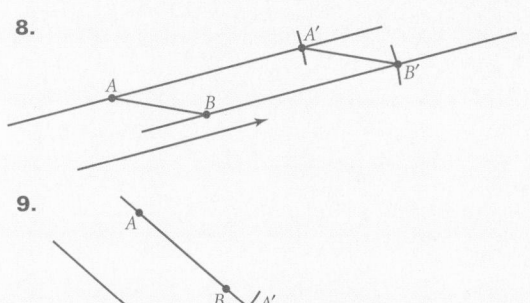

9.

10.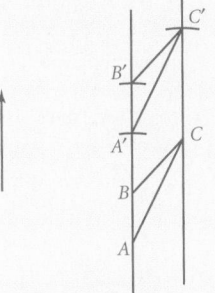

11.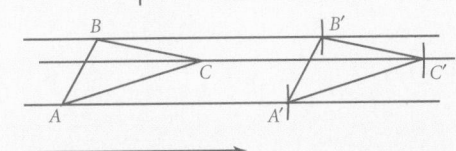

Selected Answers
Exercises 5–7, 9–47 odd

ASSIGNMENT GUIDE	
In Class	1–7
Core	9–25 odd, 26–37
Core Plus	21–25 odd, 26–40
Review	41–48
Preview	49–52

✐ Extra Practice can be found beginning on page 818.

Error Analysis

Students who have difficulty doing the constructions should be given completed figures. Then they should write the proof for each figure.

5.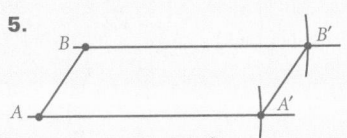

6.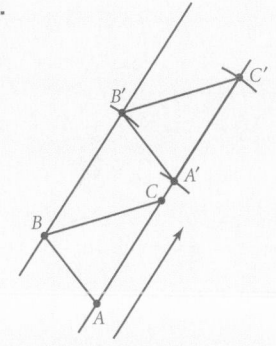

7.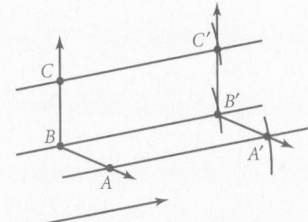

The angles have the same measure.

ALGEBRA In Exercises 24 and 25, students use the Triangle Inequality Theorem to write inequalities about the side lengths of a triangle.

Technology

Most of the exercises in this lesson can be completed by using geometry graphics software. You may wish to explain to students how to adapt the instructions in the exercises for use with your particular geometry graphics software.

12.

13.

14.

15.

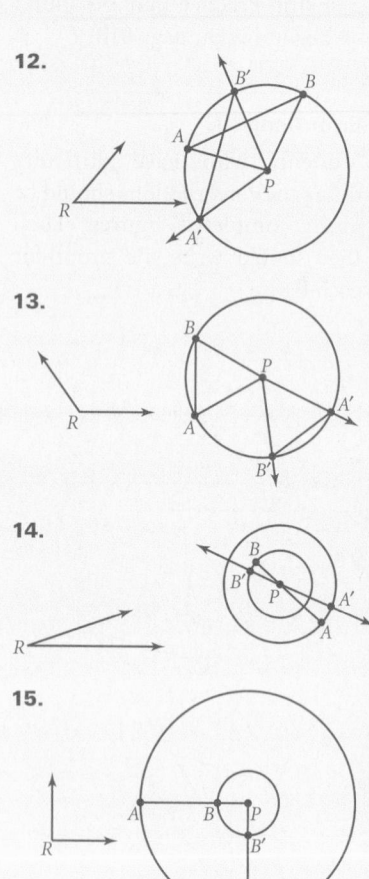

CONSTRUCTION

You will prove this construction in Exercises 26-29.

A rotation about a point by a given angle

Given $\overline{AB}$, point P, and $\angle R$ shown at right, follow the steps below to construct the rotation of $\overline{AB}$ about P by m$\angle R$.

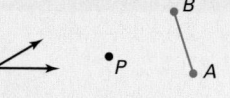

a. Place your compass point on P. Draw an arc through A and an arc through B.

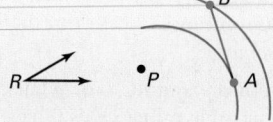

b. Draw $\overline{PA}$. Construct an angle congruent to $\angle R$ with $\overline{PA}$ as one side, as shown. Label the intersection with the arc through A as A'.

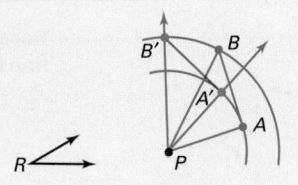

c. Draw $\overline{PB}$. Construct an angle congruent to $\angle R$ with $\overline{PB}$ as one side, as shown. Label the intersection with the arc through B as B'.

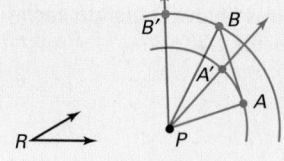

d. Draw $\overline{A'B'}$. $\overline{A'B'}$ is the rotation of $\overline{AB}$ about P by m$\angle R$.

Trace each figure below and follow the steps above to rotate the segment about the given point by the angle below it.

12.

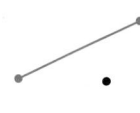

13.

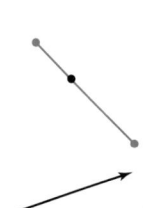

14.

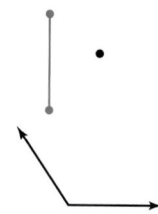

15.

A reflection across a given line

Given $\overline{AB}$ and line ℓ shown at right, follow the steps below to construct the reflection of $\overline{AB}$ across ℓ.

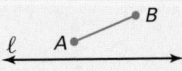

a. Construct a line, m, through A and perpendicular to ℓ, and another line, n, through B and perpendicular to ℓ.

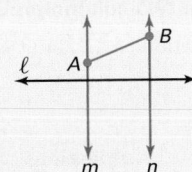

b. Set the compass point on the intersection of ℓ and m and the pencil point at A. Draw an arc that intersects m as shown. Label the intersection point A'.

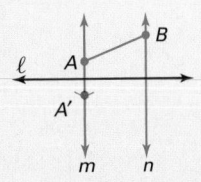

c. Set the compass point on the intersection of ℓ and n and the pencil point at B. Draw an arc that crosses n as shown. Label the intersection point B'.

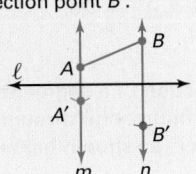

d. Draw $\overline{A'B'}$. $\overline{A'B'}$ is the reflection of $\overline{AB}$ across line ℓ.

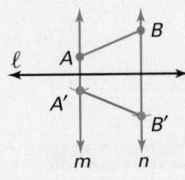

Trace each figure below and follow the steps above to reflect the figure across the given line.

16.

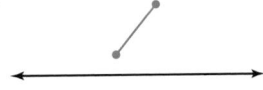

17.

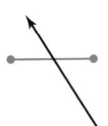

18.

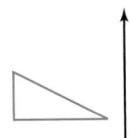

19.

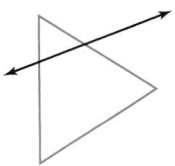

16.

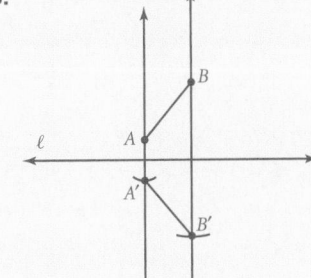

17.

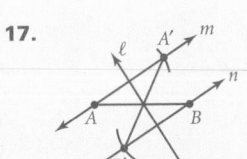

18.

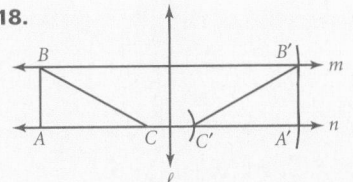

19.

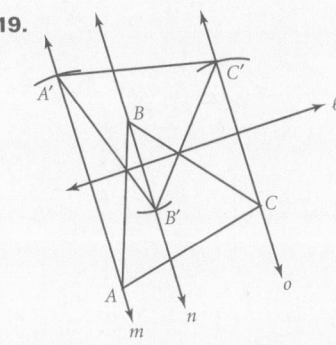

20. $\triangle ABC$ is possible.

21. $\triangle DEF$ is impossible since $DE + EF < DF$.

22. $\triangle GHI$ is impossible since G, H, and I collinear.

23. $\triangle JKL$ is possible.

Which of the following triangles are possible?

20. $AB = 7$, $BC = 10$, $AC = 12$

21. $DE = 6$, $EF = 5$, $DF = 14$

22. $GH = 17$, $HI = 9$, $GI = 8$

23. $JK = 10$, $KL = 10$, $JL = 10$

38. They are the same triangle.

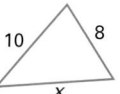

Refer to the diagram below for Exercises 24 and 25.

24. Complete the following inequalities:

   a. $\underline{10\ ?}\ +\ \underline{8\ ?}\ > x$
   b. $x +\ \underline{\ ?\ 8} > 10$
   c. $x +\ \underline{\ ?\ 10} > 8$

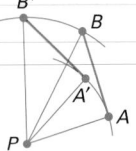

25. Solve the inequalities in Exercise 24 and complete the following statement:
   $\underline{\ ?\ 2} < x < \underline{\ ?\ 18}$

TWO-COLUMN PROOF

Refer to the rotation of a segment about a point by a given angle (page 276). Complete the two-column proof below.

Given: $\overline{A'B'}$ is a rotation of $\overline{AB}$ about point P.

Prove: $\overline{A'B'} \cong \overline{AB}$

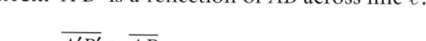

Proof:

26. Same compass setting used

27. Overlapping Angles Theorem

28. SAS

29. CPCTC

Statements	Reasons
$\overline{PA} \cong \overline{PA'}$	Same compass setting used
$\overline{PB} \cong \overline{PB'}$	**26.** ?
$\angle APA' \cong \angle BPB'$	By construction
$\angle APB \cong \angle A'PB'$	**27.** ?
$\triangle APB \cong \triangle A'PB'$	**28.** ?
$\overline{A'B'} \cong \overline{AB}$	**29.** ?

PARAGRAPH PROOF

30. Corresponding Angles Postulate
31. definition of rectangle
32. $A'D$
33. Opposite sides of a rectangle are congruent.
34. Transitive Property of Congruence
35. Subtraction Property
36. SAS
37. CPCTC

Refer to the reflection of a segment across a line (page 277). Two lines are added to the figure: one through *A* and parallel to ℓ and one through *A'* and parallel to ℓ, as shown below. Complete the following proof:

Given: $\overline{A'B'}$ is a reflection of $\overline{AB}$ across line ℓ.

Prove: $\overline{A'B'} \cong \overline{AB}$

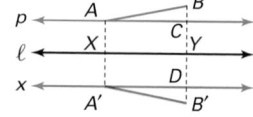

Proof: By construction, $\overline{AA'}$ and $\overline{BB'}$ are perpendicular to ℓ, $\overline{AX} \cong \overline{A'X}$, and $\overline{BY} \cong \overline{B'Y}$ ($BY = B'Y$). $\overline{AA'}$ is perpendicular to p and q, and $\overline{BB'}$ is perpendicular to p and q because **30.** ____?____. Therefore, $ACDA'$ is a rectangle by **31.** ____?____, and $\overline{AC} \cong$ **32.** ____?____ because opposite sides of a rectangle are congruent. By definition, $ACYX$ and $A'DYX$ are rectangles, so $\overline{AX} \cong \overline{CY}$ and $\overline{A'X} \cong \overline{DY}$ because **33.** ____?____. Thus, $\overline{CY} \cong \overline{DY}$ ($CY = DY$) by **34.** ____?____, and $\overline{BC} \cong \overline{B'D}$ ($BC = B'D$) by **35.** ____?____. Then $\triangle ACB \cong \triangle A'DB'$ by **36.** ____?____, and $\overline{A'B'} \cong \overline{AB}$ because **37.** ____?____.

CHALLENGE

38. Use a compass and straightedge or folding paper for the following:
   • Draw a triangle and label the vertices K, L, and M.
   • Rotate the triangle 180° about point K. Label K', L', and M'.
   • Reflect the rotated image, $\triangle K'L'M'$, across $\overline{L'M'}$. Label K'', L'', and M''.
   • Construct the perpendicular bisector of $\overline{L''M''}$, and reflect $\triangle K''L''M''$ across the perpendicular bisector. Label K''', L''', and M'''.
   • Draw a translation vector from M''' to M, and use it to translate $\triangle K'''L'''M'''$. What happens?

39. MARCHING BAND During a marching-band show, a group of band members moves in a triangular formation, as shown at right. Band members *X*, *Y*, and *Z* are the section leaders. What must the section leaders do to ensure that △*XYZ* ≅ △*X'Y'Z'*? What must the other band members in this formation do to ensure that △*XYZ* ≅ △*X'Y'Z'*?

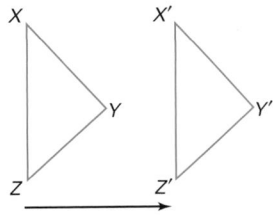

40. DESIGN Trace the design at right and translate it by using $\overline{AB}$ as a translation vector. Translate the image several times, using the same translation vector, in order to create a border pattern. If you were to use this pattern as a border on a page, describe what translation to use to go around a corner.

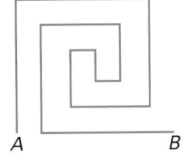

Look Back

Describe all of the types of symmetry in each letter and symbol below.
(LESSON 3.1)

41. A   **42.** N   **43.** ∞   **44.** ⌘

39. Sample answer: they have to walk the same distance in the same direction.

40. Sample answer: rotation of 90° about *B*.

41. reflectional symmetry across a vertical axis

42. rotational symmetry of 180°

43. reflectional symmetry across a vertical and horizontal axis and 180° rotational symmetry

44. reflectional symmetry across a vertical, horizontal, and two diagonal axes and 90°, 180°, and 270° rotational symmetry

Look Beyond

In Exercises 49–52, students construct a geometric figure consisting of equilateral triangles and discover a pattern in the perimeters of the triangles.

45. a, b, and d

46.

Statements	Reasons
$\overline{BX} \perp \overline{AD}$ $\overline{CY} \perp \overline{AD}$ $\overline{AD} \parallel \overline{BC}$	Given
$m\angle BCY = 90°$ $m\angle XBC = 90°$	Same-Side Interior Angles Theorem
$BCYX$ is a rectangle	Defintion of rectangle
$\overline{BX} \cong \overline{CY}$	Opposite sides of a rectangle are congruent.
$\overline{BA} \cong \overline{CD}$	Given
$\angle BXA \cong \angle CYD$	All right angles are congruent
$\triangle BXA \cong \triangle CYD$	HL
$\angle A \cong \angle D$	CPCTC

Practice

45. Which of the following are true for every equilateral triangle? Choose all answers that apply. *(LESSON 4.4)*
  a. A segment joining a vertex to the midpoint of the opposite side divides the triangle into two congruent right triangles.
  b. The angle bisectors are the same as the perpendicular bisectors.
  c. The inscribed circle is the same as the circumscribed circle.
  d. All of the angles measure 60°. **a, b, and d**

PROOFS

Write a two-column, flowchart, or paragraph proof for Exercises 46 and 47. *(LESSONS 4.5 AND 4.6)*

46. Given: $ABCD$ is an isosceles trapezoid; $\overline{BC} \parallel \overline{AD}$; $\overline{AB} \cong \overline{DC}; \overline{BX} \perp \overline{AD}, \overline{CY} \perp \overline{AD}$

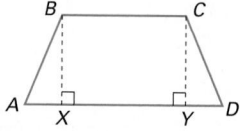

Prove: $\angle A \cong \angle D$

47. Given: $PQRS$ is a parallelogram; $\overline{QS}, \overline{PR},$ and $\overline{XY}$ intersect at Z.

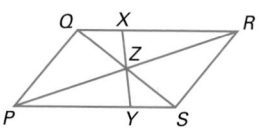

Prove: $\overline{XZ} \cong \overline{YZ}$

APPLICATION

48. NAVIGATION The edges of a parallel ruler, shown at right, always remain parallel as the ruler is opened and closed. Use a geometry theorem to explain why this is true. *(LESSON 4.6)*

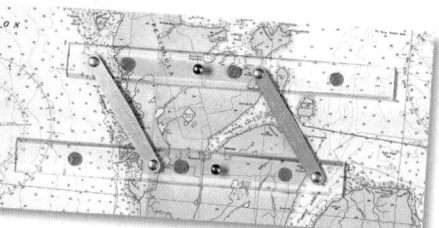

 Look Beyond

For Exercises 49–52, refer to the diagram at right.

49. Draw a large equilateral triangle. Connect the midpoints of the sides to form a second triangle inside the first. Next, draw a third triangle by connecting the midpoints of the sides of the second triangle. Continue this pattern, drawing as many triangles as you can.

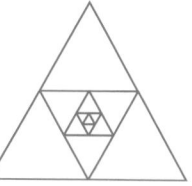

50. Are all of the triangles formed equilateral? Why or why not?

51. Suppose that the length of a side of the original triangle is 1 unit. What is the perimeter of the original triangle? of the second triangle? of the third triangle?

PROBLEM SOLVING

52. Look for a pattern. Add the perimeters of the second, third, fourth, and fifth triangles. How does this compare with the perimeter of the first triangle? What happens as the perimeters of additional triangles are added to this sum?

47.

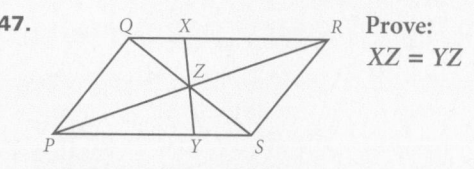

Prove: $XZ = YZ$

Statements	Reasons
$\angle XZQ \cong \angle YZS$	Vertical Angles Theorem
$\angle XQZ \cong \angle YSZ$	Alternate Interior Angles Theorem
$\overline{QZ} \cong \overline{SZ}$	Diagonals of a parallelogram bisect each other.
$\triangle QXZ \cong \triangle SYZ$	ASA
$\overline{XZ} \cong \overline{YZ}$	CPCTC

48. The opposite sides of the quadrilateral formed by the parallel ruler are congruent, so by Theorem 4.6.1 it is a parallelogram. By definition of a parallelogram, the sides are always parallel.

TESSELLATIONS WITH HEXAGONS

Any hexagon with one pair of opposite sides that are parallel and congruent can be used to create a tessellation. Use a compass and straightedge or geometry graphics software to create the following constructions:

1. Construct a hexagon with one pair of opposite sides that are parallel and congruent, and label the vertices *A*, *B*, *C*, *D*, *E*, and *F*, where $\overline{AB}$ and $\overline{ED}$ are the sides that are parallel and congruent.

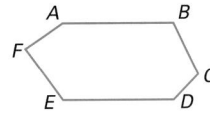

2. Translate the hexagon so that $\overline{A'B'}$ coincides with $\overline{ED}$. (Hint: Draw a translation vector from *A* to *E*.)

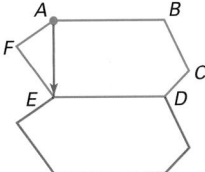

3. Locate the midpoint, *M*, of $\overline{BC}$. Rotate hexagon *ABCDE* 180° about point *M*.

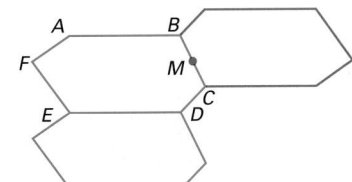

4. By continuing the translations and rotations, you should be able to continue this pattern indefinitely. Try experimenting with different hexagons. If you used geometry graphics software, try dragging the vertices to different locations.

This design was created with concave hexagons.

Student Technology Guide

NAME_____ CLASS_____ DATE_____

Student Technology Guide
4.8 **Constructing Transformations, page 1**

You may be surprised to learn that you can construct a rhombus by using only one segment and reflections across lines. In the sketch at right, you see $\overline{AB}$ and line *k* passing through *B*.

Refer to the sketch at the right to start the construction of a rhombus.

1. a. Make a sketch like the one shown. Label the segment and line as shown.
   b. Select line *k* as the line of reflection and $\overline{AB}$ as the object to be reflected, and then reflect $\overline{AB}$ across line *k*. Label the reflection of point *A* as point *C*. Explain how you know that $\overline{AB} \cong \overline{BC}$.

   _____Reflections preserve congruence._____ Check students' sketches.

   c. After completing part **b**, you will have two adjacent sides of the rhombus left to be constructed by means of transformations. What would you do next to complete the rhombus with one or more reflections? Give your response below and carry out your strategy in your sketch. Draw a line through *A* and *C*. Use this as the line of reflection. Reflect $\overline{AB}$ and $\overline{BC}$ across $\overleftrightarrow{AC}$. The result will be a rhombus because all four sides are congruent.

2. a. The diagram at the right shows square *PQRS* with line *k* passing through points *P* and *R*. Open a new sketch, and sketch $\overline{PQ}$ and line *k* passing through *P*. Place point *X* on *k*.
   b. In order to use line *k* as the line of reflection so that the image of $\overline{PQ}$ will be perpendicular to $\overline{PQ}$, what should m∠*QPX* equal? _____45°_____ Adjust line *k* so that m∠*QPX* equals your response above.
   c. Reflect $\overline{PQ}$ across line *k*. You will have two adjacent sides of the square left to be constructed by means of transformations. How would you complete the square with one or more reflections? Write your strategy below and carry it out in your sketch. Check students' sketches

   Label the image of point *Q* as point *S*. Draw the line through *S* and *Q*.

   _____$\overleftrightarrow{SQ}$ is the next line of reflection._____

   Reflect $\overline{PQ}$ and $\overline{PS}$ across $\overleftrightarrow{SQ}$. Label the image of point *P* as point *R*.

Focus

Flexagons are an interesting mathematical diversion. Each flex of a flexagon reveals different faces. Analysis of the way the faces unfold can be an instructive experiment.

Motivate

Cut off the margins of a few sheets of notebook paper, and use them to construct flexagons. Have students fold two strips, one over the other, in succession, until an accordionlike figure is formed.

Activity 1

1.–5. Check that students follow the instructions carefully. The flexagons will work better if the creases are strong. Students should notice that the flexagon has 3 unique faces which cycle in a repeating pattern. One face has triangles labeled with *B*'s, the next one has triangles labeled with numbers (1, 1, 2, 2, 3, 3) and the last one has triangles labeled with *A*'s.

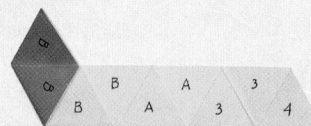

In 1939, a university mathematics student named Arthur H. Stone was playing with a strip of paper trimmed from a notebook. He discovered something interesting—flexagons.

Flexagons are polygons made from folded paper that show different faces when "flexed." The instructions below will give you a chance to play with two of these unique figures yourself.

For your first flexagons, it may be helpful to use paper with one color on the front and another on the back, as shown. Later, you can get creative by drawing designs on the flexagons, which have a kaleidoscopic effect when flexed.

Activity 1

HEXAFLEXAGON A hexaflexagon has three faces, but only two are visible at a time. Be sure to make your cuts and folds as precisely as possible to ensure that your flexagon will flex smoothly.

1. First cut a strip of paper divided into 10 equilateral triangles. Label the front and back of the strip as shown.

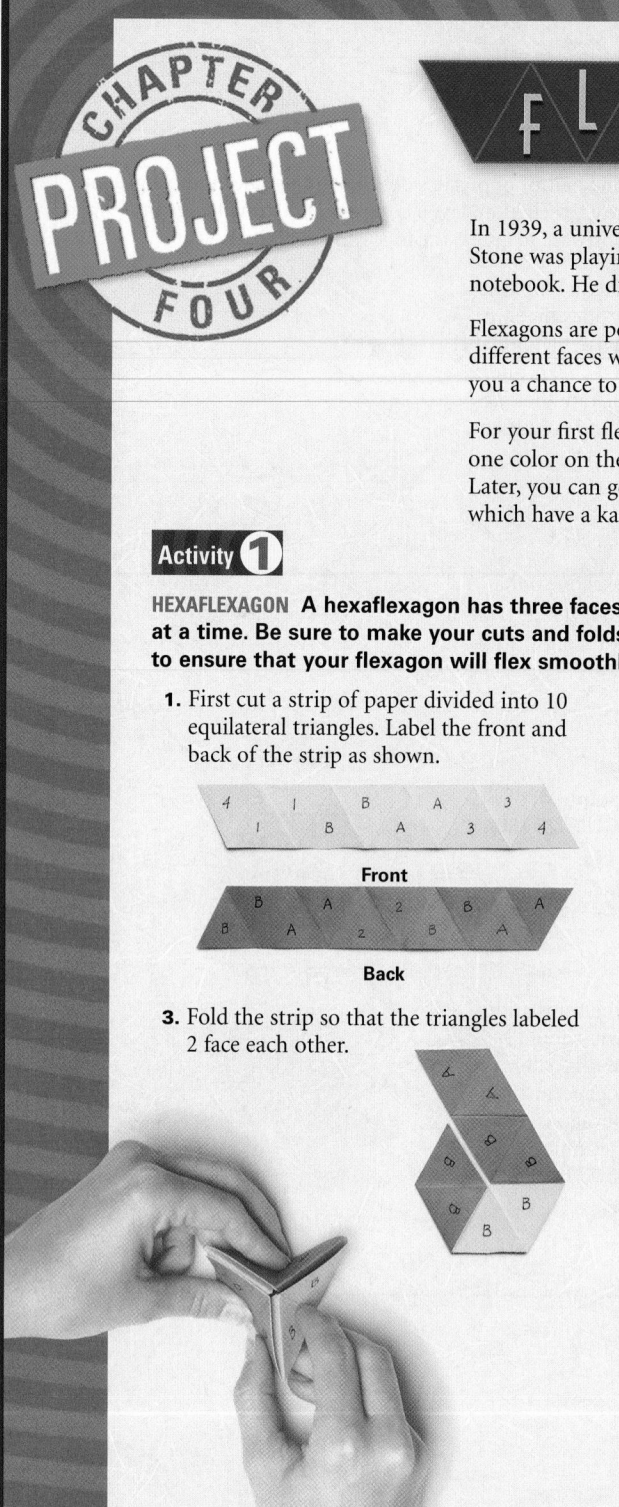

Front

Back

2. Fold the strip so that the triangles labeled 1 face each other.

3. Fold the strip so that the triangles labeled 2 face each other.

4. Fold the strip so that the triangles marked 3 face each other. Carefully glue together the triangles labeled 4, and let the glue dry.

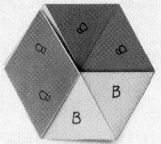

5. Now you are ready to flex your hexaflexagon. Pinch together two triangles and push in the opposite side so that the flexagon looks like a **Y** shape when viewed from above. Open the flexagon from the center. Repeat. How does the arrangement of the faces change as you flex the flexagon?

Activity 2

HEXAHEXAFLEXAGON A hexahexaflexagon has twice as many faces as the hexaflexagon. It is also more complicated to construct, so follow the directions below carefully.

1. First cut a strip of paper divided into 19 equilateral triangles. Label the front and back of the strip as shown.

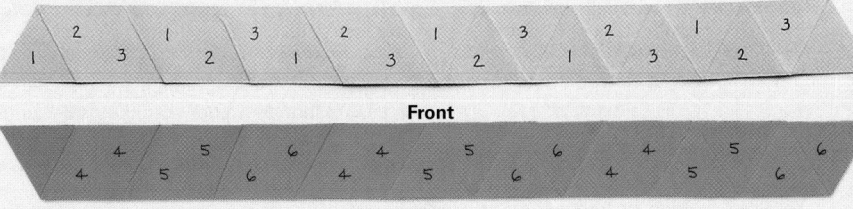

Front

Back

2. Fold the strip so that each pair of adjacent 4s, 5s, and 6s face each other. The strip should coil around itself and look like the one shown below.

3. Fold the strip so that each pair of adjacent 3s face each other.

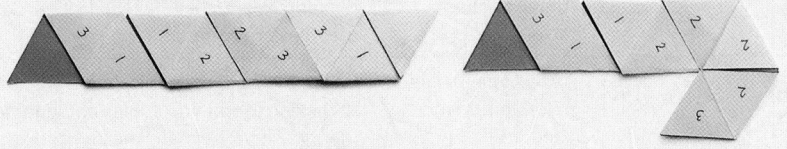

4. Tuck one end of the strip under the other so that the remaining pair of 3s face each other. Fold down the flap, and carefully glue the unlabeled triangles together.

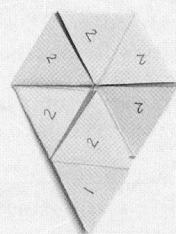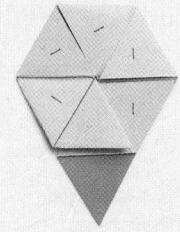

5. Flex your hexahexaflexagon in the same way as you did the first flexagon. Can you get all six faces of this flexagon to show?

Extension

1. What are the front-back face combinations?

2. Are any combinations of faces not possible?

3. Is there a pattern to the order in which the faces are revealed?

Cooperative Learning

Encourage students to do the Activities in groups so that they can help each other interpret the directions. They can also help each other with the constructions. However, each student should make a hexaflexagon and a hexahexaflexagon on his or her own.

Discuss

After all students have finished making their models, ask students to describe how the figures work. For the hexahexaflexagon, if five faces show the number 2, what number is on the other five faces? Which numbers are inside the figure? Talk about why it is important that all of the faces are congruent triangles. (If the faces are not congruent, the model will not flex smoothly.)

Activity 2

1.– 5. Check that students follow the instructions carefully. The hexahexaflexagons will work better if the creases are strong.

Extension

1. Front-back combinations: 2 and 5, 1 and 2, and 5 and 1.

2. Faces labeled 3, 4, and 6 do not appear.

3. Yes, the front cycles through 3 faces and then repeats. With each flex, the face on the front moves to the back. The pattern of the cycle of the front face is face labeled 2, then face labeled 1, and finally face labeled 5.

4 Chapter Review and Assessment

POSTULATES AND THEOREMS

Lesson	Number	Postulate or Theorem
4.1	4.1.1 Polygon Congruence Postulate	Two polygons are congruent if and only if there is a way of setting up a correspondence between their sides and angles, in order, such that (1) all pairs of corresponding angles are congruent, and (2) all pairs of corresponding sides are congruent.
4.2	4.2.1 SSS (Side-Side-Side) Postulate	If the sides of one triangle are congruent to the sides of another triangle, then the two triangles are congruent.
	4.2.2 SAS (Side-Angle-Side) Postulate	If two sides and their included angle in one triangle are congruent to two sides and their included angle in another triangle, then the two triangles are congruent.
	4.2.3 ASA (Angle-Side-Angle) Postulate	If two angles and their included side in one triangle are congruent to two angles and their included side in another triangle, then the two triangles are congruent.
4.3	4.3.1 AAS (Angle-Angle-Side) Congruence Theorem	If two angles and a nonincluded side of one triangle are congruent to the corresponding angles and nonincluded side of another triangle, then the triangles are congruent.
	4.3.2 HL (Hypotenuse-Leg) Congruence Theorem	If the hypotenuse and a leg of a right triangle are congruent to the hypotenuse and corresponding leg of another right triangle, then the two triangles are congruent.
4.4	4.4.1 Isosceles Triangle Theorem	If two sides of a triangle are congruent, then the angles opposite those sides are congruent.
	4.4.2 Converse of the Isosceles Triangle Theorem	If two angles of a triangle are congruent, then the sides opposite those angles are congruent.
	4.4.3 Corollary	The measure of each angle of an equilateral triangle is 60°.
	4.4.4 Corollary	The bisector of the vertex angle of an isosceles triangle is the perpendicular bisector of the base.
4.5	4.5.1 Theorem	A diagonal of a parallelogram divides the parallelogram into two congruent triangles.
	4.5.2 Theorem	The opposite sides of a parallelogram are congruent.
	4.5.3 Theorem	The opposite angles of a parallelogram are congruent.

Chapter Test, Form A

NAME _____ CLASS _____ DATE _____

Chapter Assessment
Chapter 4, Form A, page 1

Write the letter that best answers the question or completes the statement.

Use the figure at right for Exercises 1 and 2.

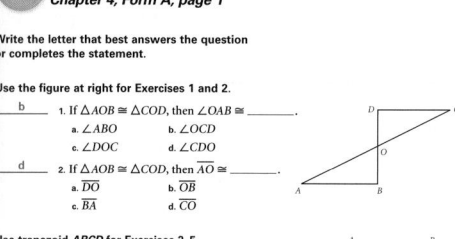

__b__ 1. If △AOB ≅ △COD, then ∠OAB ≅ _____.
  a. ∠ABO   b. ∠OCD
  c. ∠DOC   d. ∠CDO

__d__ 2. If △AOB ≅ △COD, then $\overline{AO}$ ≅ _____.
  a. $\overline{DO}$ b. $\overline{OB}$
  c. $\overline{BA}$ d. $\overline{CO}$

Use trapezoid *ABCD* for Exercises 3–5.

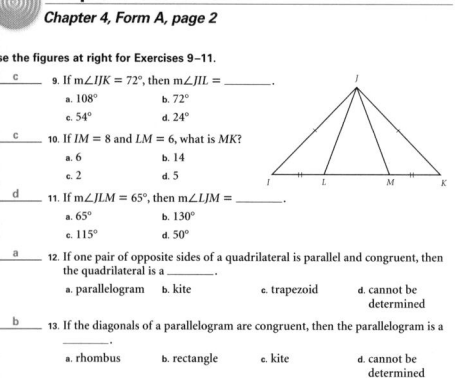

__c__ 3. Which congruence statement is correct?
  a. △AEB ≅ △DEC   b. △AED ≅ △CEB
  c. △AED ≅ △BEC   d. △ABC ≅ △BCD

__b__ 4. $\overline{AB}$ ≅ $\overline{AB}$ by _____.
  a. Overlapping Segments Postulate
  b. Reflexive Property
  c. CPCTC
  d. Isosceles Triangle Theorem

__a__ 5. △ABC ≅ △BAD by _____.
  a. SSS   b. ASA   c. SAS   d. HL

__b__ 6. If three _____ of one triangle are congruent to three _____ of another triangle, then the triangles are congruent.
  a. angles   b. sides   c. angle bisectors   d. altitudes

__d__ 7. Which postulate *cannot* be used to prove that two triangles are congruent?
  a. SSS   b. SAS   c. ASA   d. SSA

__a__ 8. If △FGH is an isosceles triangle, then which of the following is always true?
  a. The bisector of the vertex angle is the perpendicular bisector of the base.
  b. Each base angle measures 50°.
  c. The consecutive angles are congruent.
  d. The bisector of a base angle is the median of the side to which it is drawn.

NAME _____ CLASS _____ DATE _____

Chapter Assessment
Chapter 4, Form A, page 2

Use the figures at right for Exercises 9–11.

__c__ 9. If m∠IJK = 72°, then m∠JIL = _____.
  a. 108°   b. 72°
  c. 54°   d. 24°

__c__ 10. If *IM* = 8 and *LM* = 6, what is *MK*?
  a. 6   b. 14
  c. 2   d. 5

__d__ 11. If m∠JLM = 65°, then m∠LJM = _____.
  a. 65°   b. 130°
  c. 115°   d. 50°

__a__ 12. If one pair of opposite sides of a quadrilateral is parallel and congruent, then the quadrilateral is a _____.
  a. parallelogram   b. kite   c. trapezoid   d. cannot be determined

__b__ 13. If the diagonals of a parallelogram are congruent, then the parallelogram is a _____.
  a. rhombus   b. rectangle   c. kite   d. cannot be determined

__d__ 14. To construct a segment congruent to a given segment, use _____.
  a. the Reflexive Property   b. the definition of congruent segments
  c. CPCTC   d. the Congruent Radii Theorem

__a__ 15. Which segments are congruent?
  a. $\overline{AB}$ and $\overline{CD}$ with A(3, 0), B(8, 0), C(0, −2), and D(0, 3)
  b. $\overline{AB}$ and $\overline{CD}$ with A(3, 0), B(8, 0), C(3, 0), and D(0, 8)
  c. $\overline{AB}$ and $\overline{CD}$ with A(3, 0), B(8, 0), C(3, 0), and D(−8, 0)
  d. $\overline{AB}$ and $\overline{CD}$ with A(3, 0), B(8, 0), C(−3, 0), and D(8, 0)

__a__ 16. What is true about △ABC?
  a. x + 4 > 5   b. x − 5 > 4
  c. 4 + 5 < x   d. 5 − 4 > x

__c__ 17. What is the smallest possible value for *x*?
  a. 1   b. 3
  c. just larger than 1   d. just less than 9

Lesson	Number	Postulate or Theorem
	4.5.4 Theorem	Consecutive angles of a parallelogram are supplementary.
	4.5.5 Theorem	The diagonals of a parallelogram bisect each other.
	4.5.6 Theorem	A rhombus is a parallelogram.
	4.5.7 Theorem	A rectangle is a parallelogram.
	4.5.8 Theorem	The diagonals of a rhombus are perpendicular.
	4.5.9 Theorem	The diagonals of a rectangle are congruent.
	4.5.10 Theorem	The diagonals of a kite are perpendicular.
	4.5.11 Theorem	A square is a rectangle.
	4.5.12 Theorem	A square is a rhombus.
	4.5.13 Theorem	The diagonals of a square are congruent and are the perpendicular bisectors of each other.
4.6	4.6.1 Theorem	If two pairs of opposite sides of a quadrilateral are congruent, then the quadrilateral is a parallelogram.
	4.6.2 Theorem	If one pair of opposite sides of a quadrilateral are parallel and congruent, then the quadrilateral is a parallelogram.
	4.6.3 Theorem	If the diagonals of a quadrilateral bisect each other, then the quadrilateral is a parallelogram.
	4.6.4 Theorem	If one angle of a parallelogram is a right angle, then the parallelogram is a rectangle.
	4.6.5 Housebuilder Theorem	If the diagonals of a parallelogram are congruent, then the parallelogram is a rectangle.
	4.6.6 Theorem	If one pair of adjacent sides of a parallelogram are congruent, then the quadrilateral is a rhombus.
	4.6.7 Theorem	If the diagonals of a parallelogram bisect the angles of the parallelogram, then the parallelogram is a rhombus.
	4.6.8 Theorem	If the diagonals of a parallelogram are perpendicular, then the parallelogram is a rhombus.
	4.6.9 Triangle Midsegment Theorem	A midsegment of a triangle is parallel to a side of the triangle, and its length is equal to half the length of that side.
4.8	4.8.1 Betweenness Postulate	Given three points P, Q, and R, if $PQ + QR = PR$, then Q is between P and R on a line.
	4.8.2 Triangle Inequality Theorem	The sum of the lengths of any two sides of a triangle is greater than the length of the third side.

Chapter Test, Form B

Chapter Assessment
Chapter 4, Form B, page 1

For Exercises 1 and 2, decide whether or not the polygons are congruent. Justify your conclusion.

1.

No; only corresponding angles are congruent.

2.

Yes; when a figure is rotated, it has the same size and shape.

Supply the reasons for each step in the proof below.

Given: $\triangle ABC$ is isosceles and $\overline{BD} \perp \overline{AC}$.
Prove: $\overline{AD} \cong \overline{CD}$

Statements	Reasons
$\triangle ABC$ is isosceles and $\overline{BD} \perp \overline{AC}$.	3. Given
$\overline{AB} \cong \overline{BC}$	4. Definition of isosceles triangle
$\overline{BD} \cong \overline{BD}$	5. Reflexive Property
$\angle ADB$ and $\angle BDC$ are right angles.	6. Definition of perpendicular lines
$\triangle ABD \cong \triangle CBD$	7. Hypotenuse-Leg
$\overline{AD} \cong \overline{CD}$	8. CPCTC

According to the Polygon Congruence Postulate, Pentagon $KLMNO$ is congruent to pentagon $PQRST$ if and only if

9. $\angle K \cong \angle P$, $\angle L \cong \angle Q$, $\angle M \cong \angle R$, $\angle N \cong \angle S$, $\angle O \cong \angle T$ and

10. $\overline{KL} \cong \overline{PQ}$, $\overline{LM} \cong \overline{QR}$, $\overline{MN} \cong \overline{RS}$, $\overline{NO} \cong \overline{ST}$, $\overline{OK} \cong \overline{TP}$

Chapter Assessment
Chapter 4, Form B, page 2

For Exercises 11–13, use the figure at right.

11. How many different triangles are in the figure? ___6___

12. Is $\overline{EI} \cong \overline{HG}$? Why or why not? ___yes; by the Overlapping Segments Theorem___

13. If $\triangle EFG$ is isosceles, why is $\triangle EFH \cong \triangle GFI$? ___SAS___

14. Given: Parallelogram $JKLM$ is a rectangle.

Write a paragraph proof showing that the diagonals are congruent.
Sample proof: By definition of a rectangle, $\overline{KJ} \cong \overline{LM}$ and $\angle KJM$ and $\angle LMJ$ are both right angles. By the Reflexive Property, $\overline{JM} \cong \overline{JM}$. Therefore, $\triangle KJM \cong \triangle LMJ$ by SAS. The diagonals $\overline{KM}$ and $\overline{JL}$ are congruent by CPCTC.

Use a compass and a straightedge for Exercises 15 and 16.

15. Copy $\overline{XY}$.

16. Construct the angle bisector of the given angle.

State whether each set of lengths will form a triangle.

17. $AB = 6$, $BC = 7$, and $CA = 9$ ___yes___

18. $AB = 6$, $BC = 15$, and $CA = 6$ ___no___

19. $AB = \frac{2}{3}$, $BC = \frac{3}{5}$, and $CA = \frac{1}{2}$ ___yes___

1. $\overline{QR}$

2. $\overline{MK}$

3. $\angle K$

4. $\triangle LKM$

5. Yes; SAS

6. Yes; SSS

7. Yes; SSS

8. Yes; ASA

9. Yes; AAS

10. Yes; HL

11. Yes; AAS

Key Skills & Exercises

LESSON 4.1

Key Skills

Identify corresponding parts of congruent polygons.

Given: $ABCDE \cong FGHIJ$

Identify all pairs of congruent sides and angles.

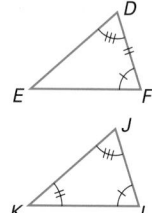

$\overline{AB} \cong \overline{FG}, \overline{BC} \cong \overline{GH}, \overline{CD} \cong \overline{HI}, \overline{DE} \cong \overline{IJ}, \overline{EA} \cong \overline{JF}$
$\angle A \cong \angle F, \angle B \cong \angle G, \angle C \cong \angle H, \angle D \cong \angle I, \angle E \cong \angle J$

Exercises

In the diagram below, $\triangle KLM \cong \triangle PQR$. Complete the following statements about congruence:

1. $\overline{LM} \cong$ _____?_____

2. $\overline{RP} \cong$ _____?_____

3. $\angle P \cong$ _____?_____

4. $\triangle QPR \cong$ _____?_____

LESSON 4.2

Key Skills

Use SSS, SAS, and ASA postulates to determine if triangles are congruent.

Which triangles below are congruent to $\triangle ABC$?

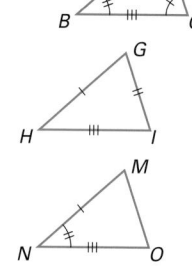

$\triangle DEF \cong \triangle ABC$ by ASA. $\triangle GHI \cong \triangle ABC$ by SSS. $\triangle JKL$ cannot be proven congruent to $\triangle ABC$ from the given information. $\triangle MNO \cong \triangle ABC$ by SAS.

Exercises

Are the triangles in each pair below congruent? State the postulate or theorem that supports your answer.

5. 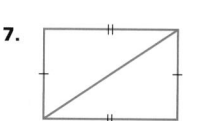   6.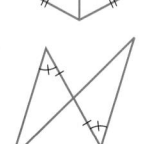

7.   8.

LESSON 4.3

Key Skills

Use AAS and HL Theorems to determine whether triangles are congruent.

In the figure at right, is $\triangle UVW \cong \triangle UST$?

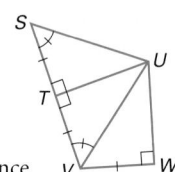

$\triangle UVW \cong \triangle UVT$ by HL and $\triangle UVT \cong \triangle UST$ by AAS, so $\triangle UVW \cong \triangle UST$ by the Transitive Property of Congruence.

Exercises

Are the triangles in each pair below congruent? State the postulate or theorem that supports your answer.

9. 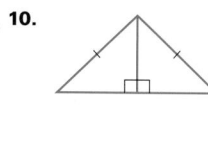   10.

11.

LESSON 4.4

Key Skills

Use triangle congruence in proofs.

In the figure below, $\overline{AB} \cong \overline{DC}$ and $\overline{BD} \cong \overline{CA}$. Prove that $\angle A \cong \angle D$.

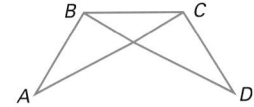

$\triangle ABC \cong \triangle DCB$ by SSS, so $\angle A \cong \angle D$ because CPCTC.

Use properties of isosceles triangles in proofs.

$QRST$ is a kite. Prove that $\overline{QS}$ bisects $\overline{TR}$.

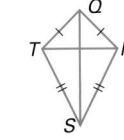

Because $\triangle QTR$ is isosceles, $\angle QTR \cong \angle QRT$. Also, $\triangle QTS \cong \triangle QRS$ by SSS, so $\angle TQS \cong \angle RQS$ because CPCTC. Then $\triangle QUT \cong \triangle QUR$ by ASA, so $\overline{TU} \cong \overline{UR}$ because CPCTC.

Exercises

Complete the proof below.

Given: quadrilateral $MNOP$ with $\angle MPN$, $\angle MNP$, $\angle OPN$, and $\angle ONP$ all congruent.

Prove: $MNOP$ is a rhombus.

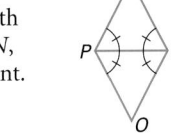

Proof:

$\triangle MNP$ and $\triangle ONP$ are isosceles by

12. _____?_____, so $\overline{MP} \cong \overline{MN}$ and $\overline{OP} \cong \overline{ON}$ by **13.** _____?_____. Also, $\triangle MPN \cong \triangle OPN$ by **14.** _____?_____, so $\overline{MP} \cong \overline{OP}$ and $\overline{MN} \cong \overline{ON}$ because **15.** _____?_____. Because all of the sides are congruent, $MNOP$ is a rhombus.

LESSON 4.5

Key Skills

Prove properties of quadrilaterals.

Quadrilateral $EFGH$ is a rhombus. Prove that $\overline{ED} \cong \overline{GD}$.

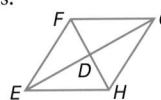

Because $EFGH$ is a rhombus, it is also a parallelogram, so the diagonals bisect each other. Therefore, $\overline{ED} \cong \overline{GD}$.

Exercises

16. In parallelogram $RSTU$, prove that $\angle R$ and $\angle S$ are supplementary.

17. In rectangle $CDEF$ with diagonals intersecting at G, prove that $\triangle CDG \cong \triangle EFG$.

18. In rhombus $NOPQ$, prove that $\angle NOQ \cong \angle POQ$.

19. In square $VWXY$, prove that $\triangle VWX \cong \triangle WXY$.

LESSON 4.6

Key Skills

Classify quadrilaterals from given information.

If quadrilateral $ABCD$ has two parallel sides and $\overline{AB} \cong \overline{CD}$, is $ABCD$ a parallelogram?

The information given is not enough to determine whether $ABCD$ is a parallelogram. For example, if the parallel sides are $\overline{AD}$ and $\overline{BC}$, $ABCD$ could be an isosceles trapezoid, as shown below.

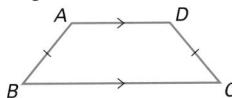

Exercises

Classify each quadrilateral according to the given information. List all terms that apply to each quadrilateral. (You may wish to draw a diagram of each quadrilateral.)

20. In $PQRS$, $\overline{PQ} \parallel \overline{RS}$ and $\overline{PQ} \cong \overline{RS}$.

21. In $KLMN$, $\overline{KL} \parallel \overline{MN}$, $\overline{KN} \parallel \overline{LM}$, and $\overline{KL} \cong \overline{LM}$.

22. In $WXYZ$, $\overline{WY}$ and $\overline{XZ}$ bisect each other, $\overline{WY} \perp \overline{XZ}$, and $\overline{WX} \perp \overline{XY}$.

23. In $EFGH$, $\overline{EF} \parallel \overline{GH}$, $\overline{EG} \cong \overline{FH}$, and $m\angle E = 90°$.

---

12. Converse of the Isosceles Triangle Theorem

13. Definition of isosceles triangle

14. ASA

15. CPCTC

16. Given: $RSTU$ is a parallelogram.
Prove: $m\angle R + m\angle S = 180°$

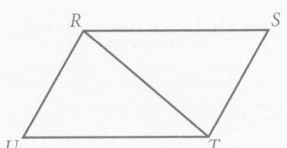

Statements	Reasons
$\overline{RU} \parallel \overline{ST}$	Definition of a parallelogram
$m\angle RTS = m\angle URT$	Alternate Interior Angles Theorem
$m\angle RTS + m\angle TRS + m\angle S = 180°$	Triangle Sum Theorem
$m\angle URT + m\angle TRS + m\angle S = 180°$	Substitution Property
$m\angle URT + m\angle TRS = m\angle R$	Angle Addition Postulate
$m\angle R + m\angle S = 180°$	Substitution Property

17. Given: $CDEF$ is a rectangle. $\overline{CE}$ and $\overline{DF}$ intersect at G.
Prove: $\triangle CDG \cong \angle EFG$

Statements	Reasons
$CDEF$ is a rectangle	Given
$CDEF$ is a parallelogram	Theorem 4.5.7
$\overline{CG} \cong \overline{GE}$ $\overline{DG} \cong \overline{GF}$	Theorem 4.5.5
$\overline{CD} \cong \overline{EF}$	Theorem 4.5.2
$\triangle CDG \cong \triangle EFG$	SSS

---

18. Given: $NOPQ$ is a rhombus.
Prove: $\angle NOQ \cong \angle POQ$

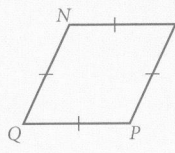

Statements	Reasons
$NOPQ$ is a rhombus	Given
$\overline{NO} \cong \overline{PO}$	Defintion of a rhombus
$\overline{OQ} \cong \overline{OQ}$	Reflexive Property of Congruence
$\overline{QN} \cong \overline{QP}$	Definition of a rhombus
$\triangle NOQ \cong \triangle POQ$	SSS
$\angle NOQ \cong \angle POQ$	CPCTC

24.

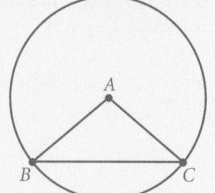

25.

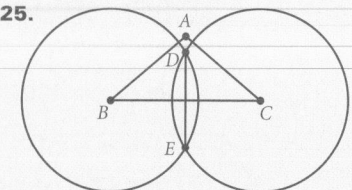

26.

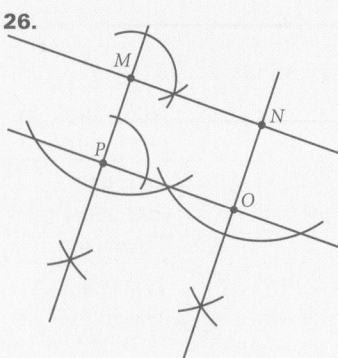

27.

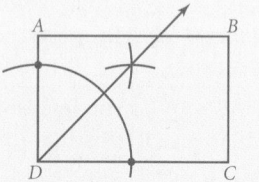

28.

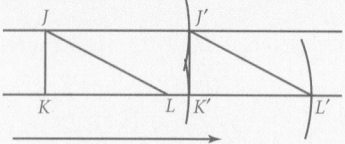

29.

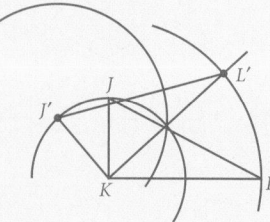

30.

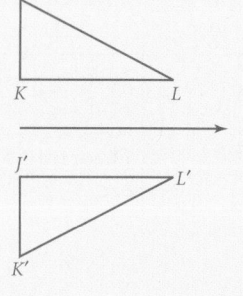

Key Skills

Construct figures by using a compass and straightedge.

Construct an angle with twice the measure of ∠A.

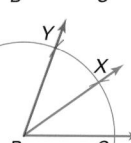

Draw $\overrightarrow{BC}$, and construct ∠CBX ≅ ∠A.

Using $\overrightarrow{BX}$, construct ∠XBY ≅ ∠A. By the Angle Addition Postulate and the Substitution Property, m∠CBY = 2 × m∠A.

Exercises

Use a compass and straightedge to complete the following constructions:

24. Construct an isosceles triangle.

25. Construct the perpendicular bisector of the base of the isosceles triangle.

26. Construct a rectangle.

27. Construct the angle bisector of any angle of a rectangle.

Key Skills

Translate, rotate, and reflect a figure by using a compass and straightedge.

Rotate △PQR about point P.

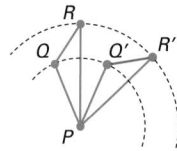

Construct arcs centered at P through points Q and R. Choose a point on the arc through Q, and label it Q′. Construct ∠Q′PR′ ≅ ∠QPR, and connect Q′ and R′ to form △PQ′R′.

Exercises

Refer to the diagram at right. Trace the figure onto paper for each exercise.

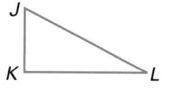

28. Translate △JKL as indicated by the translation vector.

29. Rotate △JKL about point K.

30. Reflect △JKL across the translation vector.

31. Reflect △JKL across $\overline{KL}$.

Applications

32. SURVEYING Refer to the figure at right. What measurement(s) would you need to make in order to determine the distance across the lake? (Assume that the distance cannot be measured directly.)

33. DESIGN Trace the figure shown below. Construct a congruent copy of the figure by using a compass and straightedge.

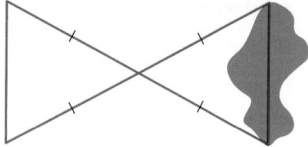

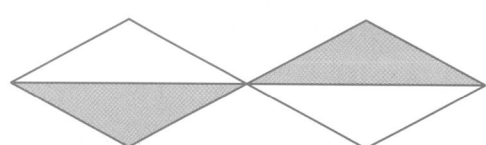

31.

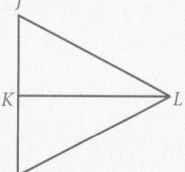

32. Since the isosceles triangles are congruent, it is only necessary to know the base of the isosceles triangle on land.

33. Check students' drawings.

Chapter Test

In the diagram below, *ABCD* ≅ *EFGH*. Complete the following statements about congruence.

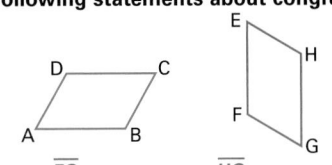

1. $\overline{BC} \cong \underline{\ ?\ }$ $\overline{FG}$
2. $\overline{DC} \cong \underline{\ ?\ }$ $\overline{HG}$
3. $\overline{FE} \cong \underline{\ ?\ }$ $\overline{BA}$
4. $\angle C \cong \underline{\ ?\ }$ $\angle G$
5. $\angle H \cong \underline{\ ?\ }$ $\angle D$
6. $CBAD \cong \underline{\ ?\ }$ $GFEH$

For Exercises 7–12, are the triangles in each pair congruent? State the postulate or theorem that supports your answer.

7. yes; SSS

8. 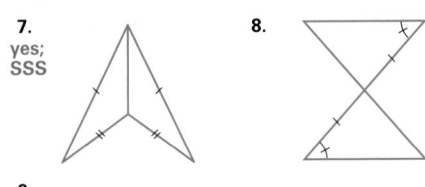 yes; ASA

9. 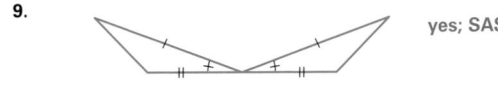 yes; SAS

10. 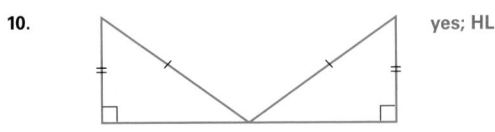 yes; HL

Find each indicated measure.

11. m∠Y 50°

12. m∠Q 60°
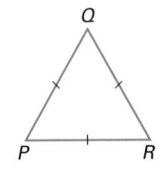

13. LN 30
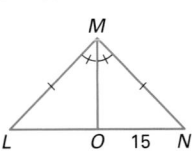

14. m∠B 100°

15. SURVEYING A surveyor needs to measure the distance across a field from point *J* to point *K*. What is the distance? Justify your answer.

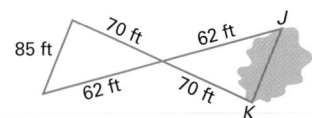

Find the indicated measure for each parallelogram.

16. *RU* 44

17. m∠XYZ 110°

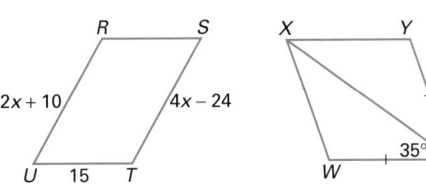

Given: *ABCD* is a parallelogram and $\overline{AB} \cong \overline{BC}$

Prove: *ABCD* is a rhombus.

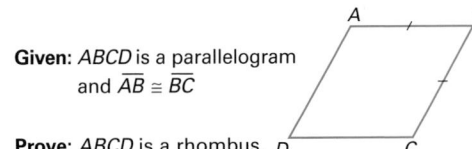

Statements	Reasons
ABCD is a parallelogram.	Given
$\overline{AB} \cong \overline{CD}$, $\overline{AD} \cong \overline{BC}$	18. ?
$\overline{AB} \cong \overline{BC}$	19. ?
$\overline{AB} \cong \overline{BC} \cong \overline{CD} \cong \overline{AD}$	20. ?
ABCD is a rhombus.	21. ?

Construct each of the following with a compass and straightedge.

22. a trapezoid

23. a rhombus

Is each of the following triangles possible? If not, why not?

24. *AB* = 17, *BC* = 20, *AC* = 22 yes

25. *DE* = 25, *EF* = 10, *DF* = 40 no; *DE* + *EF* < *DF*

26. *GH* = 9, *HI* = 14, *GI* = 9 yes

27. *XY* = 13, *YZ* = 11, *XZ* = 24 no; *XY* + *YZ* = *XZ*

15. 85 ft; The two triangles are congruent by SAS. By CPCTC, *JK* = 85 ft.

18. Opposite sides of a parallelogram are congruent.

19. Given

20. Transitive Property of Congruence

21. Definition of rhombus

22. Sample answer:

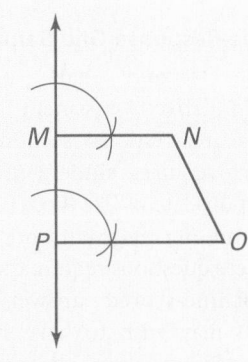

23. Sample answer:

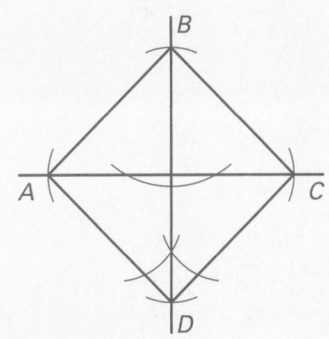

College Entrance Exam Practice

Multiple-Choice Samples

The first half of the Cumulative Assessment contains a multiple-choice section. This part of the Cumulative Assessment consists of items commonly found on standardized tests.

Free-Response Grid Samples

The second half of the Cumulative Assessment consists of a free-response section. This part requires student-produced response items like those comonly found on college entrance exams. These questions require the use of machine-scored answer grids. You may wish to have students practice answering these items in preparation for standardized tests.

1. b

2. b

3. d

4. c

5. c

6. a

7. d

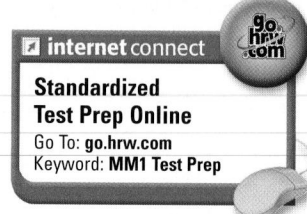
MULTIPLE-CHOICE For Questions 1–8, write the letter that indicates the best answer.

1. What is the slope of $\overline{DE}$ with endpoints $D(-7, 7)$ and $E(-3, -3)$? *(LESSON 3.8)*

   a. $-\frac{2}{5}$

   b. $-\frac{5}{2}$

   c. $\frac{5}{2}$

   d. -1

2. Refer to the diagrams below. Find the values of x and y. *(LESSON 4.1)*

   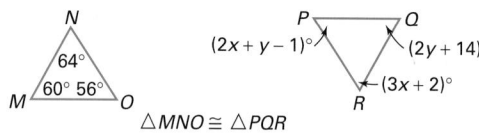

   $\triangle MNO \cong \triangle PQR$

   a. $x = 20\frac{2}{3}, y = 19\frac{2}{3}$

   b. $x = 18, y = 25$

   c. $x = 19\frac{1}{3}, y = 18\frac{1}{3}$

   d. $x = 20\frac{2}{3}, y = 23$

3. Which congruence statement indicates that the two triangles are congruent? *(LESSON 4.3)*

   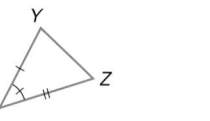 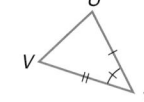

   a. $\triangle XYZ \cong \triangle TVU$, SAS
   b. $\triangle XZY \cong \triangle TVU$, HL
   c. $\triangle XZY \cong \triangle TVU$, SSA
   d. $\triangle XYZ \cong \triangle TUV$, SAS

4. Which pair of points defines a line perpendicular to $\overline{MN}$? *(LESSON 3.8)*

   a. $(0, 7), (8, -4)$
   b. $(4, -7), (-4, 4)$
   c. $(-7, 0), (4, 8)$
   d. $(7, -4), (-4, 4)$

   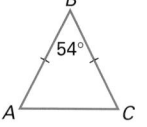

5. What is the measure of $\angle ACB$? *(LESSON 4.4)*

   a. $42°$
   b. $126°$
   c. $63°$
   d. cannot be determined

   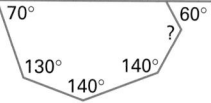

6. What is the unknown angle measure? *(LESSON 3.6)*

   a. $120°$
   b. $180°$
   c. $130°$
   d. $100°$

   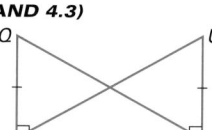

7. Which postulate or theorem justifies the congruence statement $\triangle QDA \cong \triangle UAD$? *(LESSONS 4.2 AND 4.3)*

   a. SSA
   b. SSS
   c. HL
   d. SAS

8. Which is not a feature of every rhombus?
(LESSON 4.6)
   a. parallel opposite sides
   b. congruent diagonals
   c. four congruent sides
   d. perpendicular diagonals

9. Write the converse of this statement: If trees bear cones, then the trees are conifers.
(LESSON 2.2)

10. Use the rule $T(x, y) = (x + 2, y - 2)$ to transform the figure below. What type of transformation results? *(LESSON 1.7)*

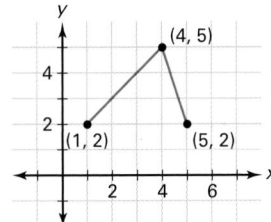

11. Lines ℓ and m are parallel. Find m∠2.
(LESSON 3.3)

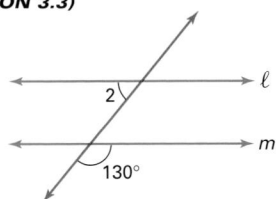

12. Are the triangles below congruent? Which postulate or theorem justifies your answer?
(LESSONS 4.2 AND 4.3)

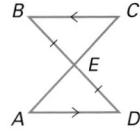

13. Which of the following measurements do not define a possible triangle? Choose all that apply. *(LESSONS 3.8 AND 4.8)*
   a. m∠A = 50°, m∠B = 85°, m∠C = 45°
   b. AB = 12, BC = 7, AC = 9
   c. m∠A = 90°, m∠B = 65°, m∠C = 15°
   d. AB = 18, BC = 6, AC = 10

FREE-RESPONSE GRID
Items 14–16 may be answered by using a free-response grid such as that commonly used by standardized-test services.

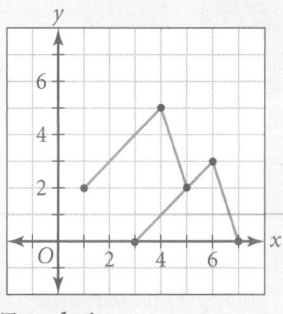

14. *QUAD* is a rectangle. Find *x*. *(LESSON 4.5)*

$QC = x$
$DC = 3x - 8$

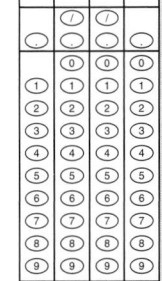

15. Find the measure of an interior angle of the regular polygon below. *(LESSON 3.6)*

16. Point *A* is the midpoint of $\overline{BF}$. Point *D* is the midpoint of $\overline{CE}$. Find *BC*. *(LESSON 3.7)*

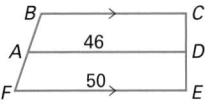

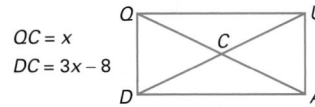

Perimeter and Area

Lesson Presentation CD-ROM
Power Point® presentations for each lesson 5.1-5.8

CHAPTER PLANNING GUIDE

Lesson	5.1	5.2	5.3	5.4	5.5	5.6	5.7	5.8	Project and Review
Pupil's Edition Pages	294–302	303–311	314–320	321–330	331–338	339–346	347–352	353–359	312–313, 360–369
Practice and Assessment									
Extra Practice (Pupil's Edition)	832	832	833	833	834	834	835	835	
Practice Workbook	29	30	31	32	33	34	35	36	
Practice Masters Levels A, B, and C	85–87	88–90	91–93	94–96	97–99	100–102	103–105	106–108	
Standardized Test Practice Masters	33	34	35	36	37	38	39	40	41
Assessment Resources	57	58	59	60	62	63	64	65	61, 66–71
Visual Resources									
Lesson Presentation Transparencies Vol. 1	110–113	114–117	118–121	122–125	126–129	130–133	134–137	138–141	
Teaching Transparencies		43	44–45			46	47–48	49	
Answer Key Transparencies	193–195	196–200	201–202	203–208	209–210	211–212	213–220	221–223	224–226
Quiz Transparencies	5.1	5.2	5.3	5.4	5.5	5.6	5.7	5.8	
Teacher's Tools									
Reteaching Masters	57–58	59–60	61–62	63–64	65–66	67–68	69–70	71–72	
Make-Up Lesson Planner for Absent Students	29	30	31	32	33	34	35	36	
Student Study Guide	29	30	31	32	33	34	35	36	
Spanish Resources	29	30	31	32	33	34	35	36	
Block Scheduling Handbook									10–11
Activities and Extensions									
Lesson Activities	76–80	81–86	87–90	91–92	93–94	95–97	98–102	103–104	
Enrichment Masters	29	30	31	32	33	34	35	36	
Cooperative-Learning Activities	29	30	31	32	33	34	35	36	
Problem-Solving/ Critical Thinking	29	30	31	32	33	34	35	36	
Student Technology Guide		28	29–30	31	32	33	34–35	36	
Long Term Projects									17–20
Writing Activities for Your Portfolio									13–15
Tech Prep Masters									21–24
Building Success in Mathematics									12–14

LESSON PACING GUIDE

Lesson	5.1	5.2	5.3	5.4	5.5	5.6	5.7	5.8	Project and Review
Traditional	2 days	2 days	2 days	2 days	2 days	2 days	2 days	2 days	2 days
Block	1 day	1 day	1 day	1 day	1 day	1 day	1 day	1 day	1 day
Two-Year	4 days	4 days	4 days	4 days	4 days	4 days	4 days	4 days	4 days

CONNECTIONS AND APPLICATIONS

Lesson	5.1	5.2	5.3	5.4	5.5	5.6	5.7	5.8	Review
Algebra	297, 299	309	317, 319	326, 327, 329	331, 332, 333	343, 344			
Geometry	294–302	303–311	314–320	321–330	331–338	339–346	347–352	353–359	360–369
Coordinate Geometry		309							
Maximum/Minimum		310					352		
Technology		311						359	
Number Theory				327					
Probability								354, 357	
Life Skills	300, 301, 302	306	318	323, 329	337			358	366
Science and Technology	297, 298, 300, 301	311	318, 319	325	337	345	347, 352	358	
Sports and Leisure	297, 298	304		329				358, 359	366
Cultural Connection: Africa		311		324					
Cultural Connection: Asia			320	322					367
Cultural Connection: Europe				322					

BLOCK-SCHEDULING GUIDE

Day	Lesson	Teacher Directed: Lesson Examples, Teaching Transparencies	Student Guided: Activity, Try This	Cooperative-Learning Activity, Lesson Activity, Student Technology Guide	Practice: Practice & Apply, Extra Practice, Practice Workbook	Assessment: Quiz, Mid-Chapter Assessment	Problem Solving, Reteaching
1	5.1	15 min	15 min	20 min	55 min	15 min	15 min
2	5.2	15 min	15 min	20 min	55 min	15 min	15 min
3	5.3	15 min	15 min	20 min	55 min	15 min	15 min
4	5.4	15 min	15 min	20 min	55 min	15 min	15 min
5	5.5	15 min	15 min	20 min	55 min	15 min	15 min
6	5.6	15 min	15 min	20 min	55 min	15 min	15 min
7	5.7	15 min	15 min	20 min	55 min	15 min	15 min
8	5.8	15 min	15 min	20 min	55 min	15 min	15 min
9	Assess.	50 min	90 min	90 min	65 min	30 min	
		PE: Chapter Review	PE: Chapter Project, Writing Activities	Tech Prep Masters	PE: Chapter Assessment, Test Generator	Chap. Assess. (A or B), Alt. Assess. (A or B), Test Generator	

PE: Pupil's Edition

Alternative Assessment

The following suggest alternative assessments for students who may benefit from a different type of assessment than the regular chapter quizzes and the mid-chapter/end-of-chapter test. Visit the HRW web site to get additional Alternative Assessment material.

☑ internet connect

Alternative Assessment
Go To: **go.hrw.com**
Keyword: **MG1 Alt Assess**

Performance Assessment

1. Draw and label three triangles with an area of 24 square centimeters. The drawings do not have to be to scale. For each figure, give the measure of the base and height.

2. Using a compass and straight edge, construct a 45-45-90 triangle and a 30-60-90 triangle. Note that a 45-45-90 triangle is half of a square, and a 30-60-90 triangle is half of an equilateral triangle. Check the angle measures of your constructions with a protractor.

3. Find the area of a regular decagon with a side length of 3 centimeters.

4. Use the Pythagorean Theorem to find the length, diagonally, from one corner of your school gym (or swimming pool) to the other. Describe how you found the lengths of the sides.

Portfolio Project

Suggest that students choose one of the following projects for inclusion in their portfolios.

1. Design a dart board in which the player has a 15% chance of hitting the bulls eye and a 30% chance of hitting another region on the board. Give the area of your dart board, as well as the area of the bulls eye and one other region.

2. There are over one hundred different proofs of the Pythagorean Theorem. Several are shown in the lesson and exercises for Lesson 5.4. Use library materials to research a proof of the Pythagorean Theorem not shown in your textbook. Present the proof on a posterboard to display in your classroom. Be sure to list the source of your information on the poster.

☑ internet connect

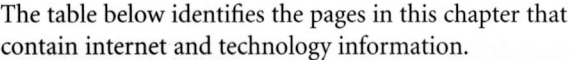

The table below identifies the pages in this chapter that contain internet and technology information.

Content Links

Activities Online	pages 298, 326, 335
Portfolio Extensions	pages 320, 359
Homework Help Online	pages 299, 308, 319, 326, 336, 343, 351, 357

Resource Links

Parents can go online and find concepts that students are learning–lesson by lesson–and questions that pertain to each lesson, which facilitate parent-student discussion.

Go To: **go.hrw.com**
Keyword: **MG1 Parent Guide**

Technical Support

The following may be used to obtain technical support for any HRW software product.

Online Help: **www.hrwtechsupport.com**

e-mail: **tschrw@hbtechsupport.com**

HRW Technical Support Center: **(800)323-9239**

7 AM to 10 PM Monday through Friday CST

Visit the HRW math web site at: **www.hrw.com/math**

Technology

Technology Objectives and Suggestions

Lesson 5.1 Perimeter and Area

In this lesson students solve problems involving fixed perimeters and fixed area. For Activities 1 and 2 on page 295, students can use spreadsheet software to complete the tables. For both activities, a graphics calculator can be used to graph equations to find the maximum area for a fixed perimeter and the minimum perimeter for a fixed area.

Lesson 5.2 Areas of Triangles, Parallelograms, and Trapezoids

In this lesson students develop area formulas and solve problems using these formulas for the areas of triangles, parallelograms, and trapezoids. Geometry software can be used to explore the formulas in this lesson. For example, the figure shows the formula developed in Part II of Activity 1. Students should construct two perpendicular lines, and then construct another line parallel to each of these. Four right angles are formed. Point J can move anywhere on the line containing points G and H. Students should drag point J and observe that the base, height, and area of $\triangle ABJ$ do not change.

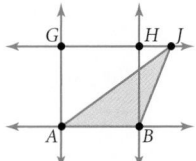

Another example is to have students use geometry software to look at the area of a parallelogram. Have them construct a rectangle out of two sets of parallel lines, and calculate its area. Have them drag one point of the figure until it is no longer a rectangle, but is still a parallelogram. Students should notice that the area stays the same because the length of the base and the height do not change, although the measures of the vertex angles do change.

Lesson 5.3 Circumferences and Areas of Circles

In this lesson students identify and apply formulas for the circumference and area of a circle, and solve problems using the formulas. A calculator with a π key will be needed for the majority of the examples and exercises in this lesson.

Geometry software can be used to create ways of approximating areas and circumferences of circles. For example, students might calculate the areas of an inner and outer square. They average these two areas and compare the result with that computed using π. Encourage students to use the software to create their own ways of approximating area and circumference.

Lesson 5.4 The Pythagorean Theorem

In this lesson students solve problems using the Pythagorean Theorem and its converse. A calculator with a square root key will be needed for the majority of examples and exercises in this lesson.

Geometry software can be used to explore the Pythagorean Theorem. Students should start with two perpendicular lines. By connecting various points on these lines they can generate many different right triangles. They should measure the three side lengths and confirm that the Pythagorean Theorem is always true. They can also use spreadsheet software to generate Pythagorean triples, using to formulas given in Exercises 31 and 32 of Lesson 5.4.

Lesson 5.5 Special Triangles and Areas of Regular Polygons

Underlying the ideas in the lesson is the fact that all 45-45-90 triangles are similar, and that all 30-60-90 triangles are similar. Have students use geometry software to create sets of similar triangles to be a visual reminder that the ratios are the same, no matter what the sizes of the triangles. Geometry software may also be used to confirm the Area of a Regular Polygon Theorem for a wide variety of cases.

Lesson 5.6 The Distance Formula and the Method of Quadrature

In this lesson students develop and apply the distance formula, and use the distance formula to develop techniques for estimating the area under a curve. Students will need a calculator with a square root key for the examples and exercises in this lesson. In addition, a graphics calculator will be helpful in graphing the curves in Exercises 36–38 of this lesson.

Lesson 5.7 Proofs Using Coordinate Geometry

In this lesson students develop coordinate proofs for the Triangle Midsegment Theorem, the diagonals of a parallelogram, and the reflection of a point about the line $y = x$. They also use the concepts of coordinate proofs to solve problems on the coordinate plane.

Ask students to use geometry software to graph the figures in Activities 1, 2, and 3 on a coordinate grid. They can use measurement options to verify their proof statements. You may also want students to use geometry software to complete or check Exercises 11–32.

Lesson 5.8 Geometric Probability

In this lesson students develop and apply the basic formula for geometric probability. You may wish for students to use calculators to convert fractions to decimals and percents. Students will need a calculator with a random number generator to complete the portfolio activity on page 351.

Background Information

This chapter builds on students' previous experiences to explore strategies for finding perimeter, circumference, and area. Students also learn key properties for right triangles, including proofs and applications of the Pythagorean Theorem.

CHAPTER RESOURCES

- Block-Scheduling Handbook
- Writing Activities for Your Portfolio
- Tech Prep Masters
- Long-Term Project
- Assessment Resources:
  Mid-Chapter Assessment
  Chapter Assessments
  Alternative Assessments
- Test and Practice Generator
- Technology Handbook

Chapter Objectives

- Identify and use the Area of a Rectangle and the Sum of Areas Postulates. [5.1]

- Solve problems involving fixed perimeters and fixed areas. [5.1]

- Develop formulas for the areas of triangles, parallelograms, and trapezoids. [5.2]

- Solve problems by using the formulas for the areas of triangles, parallelograms, and trapezoids. [5.2]

- Identify and apply formulas for the circumference and area of a circle. [5.3]

- Solve problems by using the formulas for the circumference and area of a circle. [5.3]

Perimeter and Area

THE IMPORTANCE OF THE IDEAS OF PERIMETER AND AREA is suggested by the images on these pages. How long is the oval track around the playing field? How long is the Great Wall of China? How much grain can be grown on a given field?

Sometimes, people require information about area for strange reasons. When the conceptual artist Christo devised his plan of surrounding islands by pink plastic, he needed to know how much material would be required for the project.

You may already be familiar with formulas for perimeter and area. In this chapter, you will deepen your knowledge of them by developing and proving them.

Lessons

About the Photos

The photos above, from left to right, show the Great Wall of China; an athletic field in Long Island, New York; farm fields in Prince Edward Island, Canada; and Christo's and Jeanne Claude's *Surrounded Islands*. The Great Wall of China, whose construction started in 221 B.C.E., extends approximately 4000 miles (6400 kilometers) from west to east. For the project *Surrounded Islands*, artists Christo and Jeanne Claude used a total of 603, 850 square meters (6.5 million square feet) of pink woven polypropylene fabric to surround the contour of 11 islands in Biscayne Bay, Florida. When farmers prepare land for cultivation, they must calculate the area of the farm field to be plowed.

Christo's Surrounded Islands, *Biscayne Bay, Miami, Florida, 1980-1983*

- Identify and apply the Pythagorean Theorem and its converse. [**5.4**]
- Solve problems by using the Pythagorean Theorem. [**5.4**]
- Identify and use the 45-45-90 Triangle Theorem and the 30-60-90 Triangle Theorem. [**5.5**]
- Identify and use the formula for the area of a regular polygon. [**5.5**]
- Develop and apply the distance formula. [**5.6**]
- Use the distance formula to develop techniques for estimating the area under a curve. [**5.6**]
- Develop coordinate proofs for the Triangle Midsegment Theorem, the diagonals of a parallelogram, and the reflection of a point across the line $y = x$. [**5.7**]
- Use the concepts of coordinate proofs to solve problems on the coordinate plane. [**5.7**]
- Develop and apply the basic formula for geometric probability. [**5.8**]

About the Chapter Project

In this chapter, you will study formulas for the area of common shapes such as rectangles, triangles, and circles. But there are many other shapes that cannot be included in these categories: stars, crescents, and irregular shapes that do not even have names, much less formulas for their areas.

In the Chapter Project, you will study polygons drawn on square grid paper in such a way that every vertex is at a grid point. By determining the area of the figures and searching for a pattern, you will derive the formula for the area of such figures discovered by George Pick in 1899.

After completing the Chapter Project you will be able to do the following:

- Find the areas of certain complex figures without using traditional area formulas.

About the Portfolio Activities

Throughout the chapter, you will be given opportunities to complete Portfolio Activities that are designed to support your work on the Chapter Project.

The theme of each Portfolio Activity and of the Chapter Project is the area of figures in a plane.

- In the Portfolio Activity on page 320, you will find areas of irregular polygons by dividing them into rectangles and triangles.
- In the Portfolio Activity on page 346, you will find the area of a crescent-shaped figure called a *lune*.
- In the Portfolio Activity on page 359, you will use a calculator to generate random points on a grid and use them to test the theoretical probability that a point chosen at random lies within a given area on the grid.

Portfolio Activities appear at the end of Lessons 5.3, 5.6, and 5.7. Each serves as preparation for the Chapter Project. The Portfolio Activities, as well as the Chapter Project Activities, are appropriate for inclusion in the student's portfolio. Students should be encouraged to include in their portfolios any other work in which they feel a sense of pride or a sense of accomplishment.

internet connect

Chapter Internet Features and Online Activities

Perimeter and Area

Delta of
the Nile

Nile
River

Objectives

● Identify and use the
Area of a Rectangle
and the Sum of Areas
Postulates.

● Solve problems
involving fixed
perimeters and fixed
areas.

APPLICATION
SURVEYING

In ancient Egypt, the yearly flooding of the Nile River was beneficial for crops,
but it damaged and even destroyed property boundaries. This forced the
Egyptians to devise methods for redetermining boundaries after each flood.
The later development of such methods by Greek mathematicians led to the
science of "Earth measure" for which geometry is named.

Perimeter

Perimeter

The **perimeter** of a closed plane figure is the distance around the figure.

5.1.1

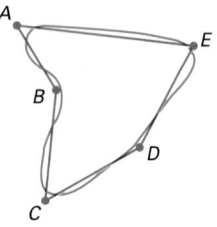

For a polygon, the perimeter is the sum of the lengths of its sides.

The perimeter of a polygon can be used to approximate the perimeter of a
closed figure in a plane. For example, the perimeter of the irregular figure at
left is approximately equal to the perimeter of polygon *ABCDE*.

CRITICAL THINKING
How could you redraw the polygon in the figure at left above to get a better
estimate of the perimeter of the irregular figure?

EXAMPLE **1** Find the perimeter of each figure.

a. pentagon *ABCDE*

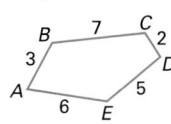

b. parallelogram *WXYZ*

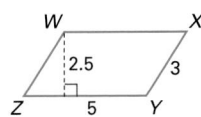

● **SOLUTION**

a. The perimeter is the sum of the sides of the pentagon.

perimeter = 3 + 7 + 2 + 5 + 6 = 23 units

b. The figure is a parallelogram, so opposite sides are of equal length.

$WX = YZ = 5$ $ZW = XY = 3$

The perimeter is the sum of the sides.

perimeter = $WX + XY + YZ + ZW$
= 5 + 3 + 5 + 3 = 16 units

Area

> Figures are **non-overlapping** if they have no points in common (except for boundary points).

Area

The **area** of a closed plane figure is the number of non-overlapping squares of a given size that will exactly cover the interior of the figure.

5.1.2

In the floor tile illustration at right, there are 3 rows of tiles with 4 tiles in each row. You can find the total number of tiles by multiplying.

$3 \times 4 = 12$

Number of rows | Number of tiles in each row

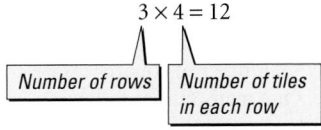

Twelve 1 × 1 tiles will exactly cover a 3 × 4 rectangle.

The following postulate is used to find the areas of nonrectangular shapes. (See Example 2 on page 296.)

Postulate: The Sum of Areas

If a figure is composed of non-overlapping regions *A* and *B*, then the area of the figure is the sum of the areas of regions *A* and *B*.   5.1.3

Interdisciplinary Connection

GEOGRAPHY When land is bought or sold, surveyors usually confirm the boundaries of the property. Have students use reference materials to find out why this is not always a simple process. They should list and discuss factors that make it difficult to establish boundary lines.

CRITICAL THINKING

Redraw the figure with more sides to approximate the shape of the curve more closely.

ADDITIONAL
EXAMPLE **1**

Find the perimeter of each figure.

a.

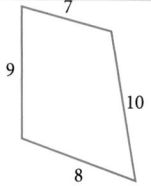

The perimeter is 34.

b.

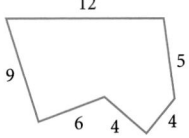

The perimeter is 40.

Teaching Tip

For a rectangle or square, the distinction between the base and the height is arbitrary. Any of the four sides can be designated as the base. The height is then one of the sides perpendicular to this base.

Find the area of each figure.

a.

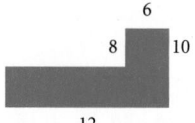

The area is 72 square units.

b.

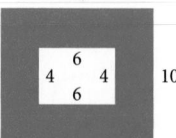

The area is 96 square units.

CRITICAL THINKING

In part **b**, the figure can be divided into two or more nonoverlapping rectangles, so the Sum of the Areas Postulate (5.1.3) is used. Since the figure in part **c** is divided into a grid of non-overlapping squares, the Sum of the Areas Postulate is used.

Area and Perimeter of Rectangles

The perimeter and area of a rectangle can be found easily by using the definitions on the previous pages. However, it is often more convenient to use the following formulas:

The Perimeter of a Rectangle

The perimeter of a rectangle with base b and height h is given by:
$P = 2b + 2h$.

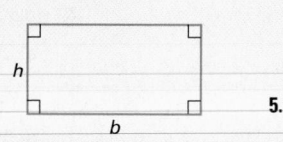

5.1.4

Postulate: Area of a Rectangle

The area of a rectangle with base b and height h is given by:
$A = bh$.

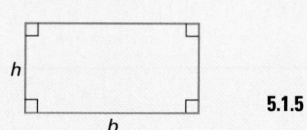

5.1.5

EXAMPLE ② 2 **Find the area of each figure below. Explain your method in each case.**

a.
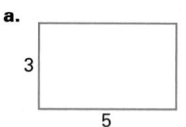

b.
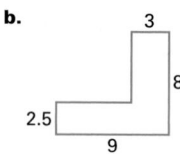

c.

● **SOLUTION**

a. Multiply the base and the height. The area is 15 square units.

b. Divide the figure into separate rectangles. One way of doing this is shown at right.

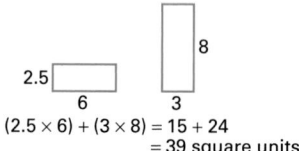
$(2.5 \times 6) + (3 \times 8) = 15 + 24$
$= 39$ square units

c. Select an appropriate grid and place it over the figure. Estimate the area by counting the number of squares inside the figure. One method is as follows: Take half the number of squares that are only partially inside the figure (these are identified with dots). Add this to the number of squares that are entirely inside the figure. There are 18 partial squares and 16 complete squares. The area is approximately $\frac{1}{2}(19) + 16 = 25\frac{1}{2}$ square units.

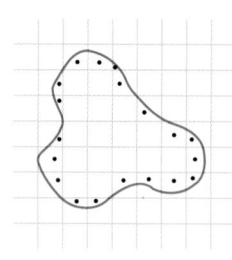

CRITICAL THINKING How is Postulate 5.1.3 (on page 295) used in Example 2?

Enrichment

The square has an area of 64 square units. The new figure has only 63 square units. What happened to the missing square? **The two pieces do not fit together exactly. The diagonals do not match exactly.**

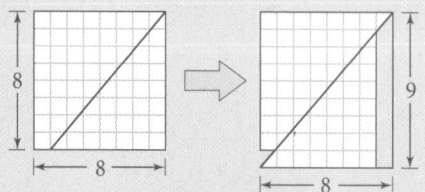

Inclusion Strategies

USING MODELS Students who have difficulty with the content of this lesson may benefit from using manipulatives. Have students draw figures made up of rectangular shapes with dimensions in centimeters. Have them fill in the areas of the rectangles with square centimeter titles. Ask them to separate each figure into two or more rectangles. Students should draw and measure each resulting rectangle.

Fixed Perimeter/Maximum Area

A gardener has 24 feet of fencing for a garden. What are the dimensions of the rectangle that will enclose the greatest area?

1. Sketch three different rectangles with a perimeter of 24 units. What is the area of each?

2. Show that $h = 12 - b$ for a rectangle with a perimeter of 24.

3. Fill in a table like the one at right. What do you observe about the area values?

b	h = 12 − b	A = bh
1	11	11
2	?	?
3	?	?
⋮	⋮	⋮

4. Plot the values for b and A from the table on a graph, with b on the horizontal axis and A on the vertical axis.

5. What are the dimensions of the rectangle that has the maximum area for the given perimeter? If you have a graphics calculator, graph $A = b(12 - b)$ by entering **Y = X(12 − X)**. Trace the graph to find the maximum.

CHECKPOINT ✔

6. What kind of rectangle is your result from Step 6? Do you think this would be the result for any fixed perimeter?

Fixed Area/Minimum Perimeter

A farmer wants to enclose a rectangular area of 3600 square feet with the minimum amount of fencing. What should be the dimensions of the rectangle?

1. Show that $h = \frac{3600}{b}$ for a rectangle with an area of 3600.

2. Fill in a table like the one at right. Let b range from 10 to 100.

b	h = 3600/b	P = 2b + 2h
10	360	740
20	?	?
30	?	?
⋮	⋮	⋮

3. Plot the values for b and P from the table on a graph, with b on the horizontal axis and P on the vertical axis.

4. What are the dimensions of the rectangle that has the minimum perimeter for the given area? If you have a graphics calculator, graph $P = 2b + \frac{7200}{b}$ by entering **Y = 2X + $\frac{7200}{X}$**. Trace the graph to find the minimum.

CHECKPOINT ✔

5. What kind of rectangle is your result from Step 5? Do you think this would be the result for any fixed area?

Reteaching the Lesson

USING MODELS Have students use graph paper to draw several rectangles with a perimeter of 32 units. They should number the rectangles in order of increasing area. Students should also draw several rectangles with an area of 64 square units. They should number these rectangles in order of decreasing perimeter.

The sample spreadsheet shown in the text is just one way to set up this problem. If the heading for the second column is difficult to understand suggest that students insert more columns in the table. For example, the first four columns could be headed b, $2b$, $24 - 2b$, and $(24 - 2b) \div 2$. The last column gives a value for h. This is then used to find the area.

For a student worksheet of this Activity and detailed Teacher Notes, see page 76 in the Lesson Activities booklet.

Math
CONNECTION

ALGEBRA In both Activities on this page, students are asked to graph equations on a graphics calculator and find the maximum or minimum points on the graphs. Explain to students that graphing allows them to find the solutions to the problems visually.

CHECKPOINT ✔

6. A square; yes, the maximum area will always be reached when the sides of the rectangle are equal.

With some spreadsheet programs, students will need more than three columns. For example, the columns might progress in this order: b, $3600 \div b = h$, $2h$, $2b$, and $2h + 2b$. Encourage students to set up the columns in a logical progression.

For a student worksheet of this Activity and detailed Teacher Notes, see page 76 in the Lesson Activities booklet.

Exercises

Communicate

APPLICATIONS

1. **RECREATION** Explain a way to estimate the perimeter of the swimming pool shown below.

2. **AVIATION** Explain a way to estimate the area of a wing of the airplane shown below.

▶ **internet** connect

Activities Online

Go To: **go.hrw.com**
Keyword:
MG1 PicksThm

3. Use the figures below to explain the following statement:

area of figure *A* + area of figure *B* ≠ area of figure *C*

4. Many applications in mathematics involve finding the area under a curve. How could you estimate the shaded area in the figure at right?

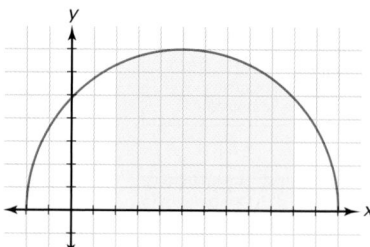

Guided Skills Practice

5. Find the perimeter of the figure at right.
   (EXAMPLE 1) 21 units

6. Find the area of the figure at right.
   (EXAMPLE 2) 16.75 units²

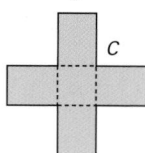

7. What is the maximum area of a rectangular garden with a perimeter of 48 feet? *(ACTIVITY 1)* **144 sq. ft**

8. What is the minimum perimeter of a rectangular region with an area of 900 square square feet? *(ACTIVITY 2)* **120 ft**

● *Practice and Apply*

For Exercises 9–18, use the figure and measurements below to find the indicated perimeters and areas:

$AD = 15$ in.	$AC = 13$ in.	$BD = 10$ in.
$DL = 11$ in.	$EI = 3$ in.	$CH = 4$ in.

9. the perimeter of rectangle *ADLI* **52 in.**

10. the area of rectangle *ADLI* **165 in^2**

11. the perimeter of rectangle *CDLK* **26 in.**

12. the area of rectangle *CDLK* **22 in^2**

13. 42 in.

13. the perimeter of hexagon *GHCDLJ*

14. 78 in^2

14. the area of hexagon *GHCDLJ*

15. 24 in.

15. the perimeter of rectangle *BCHG*

16. 32 in^2

16. the area of rectangle *BCHG*

17. 16 in^2

17. the area of $\triangle BHG$ (Hint: Use your answer to Exercise 16.)

18. 82.5 in^2

18. If points *I* and *D* were connected by a segment, what would be the area of $\triangle ADI$?

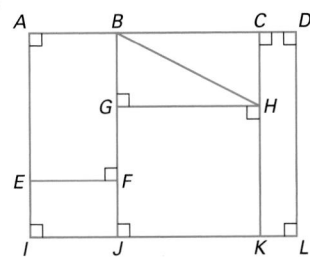

PARAGRAPH PROOF

19. Write a paragraph proof of the formula for the perimeter of a rectangle.

Algebra

In Exercises 20–23, find the area of the rectangle with vertices at the given points. You may find it helpful to sketch a graph.

20. $(0, 0), (0, 2), (5, 0), (5, 2)$ **10 units^2**

21. $(3, 1), (3, 7), (9, 1), (9, 7)$ **36 units^2**

22. $(-2, -5), (-2, 3), (4, -5), (4, 3)$ **48 units^2**

23. $(0, 0), (3, 3), (6, 0), (3, -3)$ **18 units^2**

24. $h = 9$ cm
$b = 27$ cm
$A = 243$ cm^2

24. The perimeter of a rectangle is 72 centimeters. The base is 3 times the height. What are the dimensions of the rectangle? What is the area?

25. $h = 3$ ft
$b = 9$ ft
$P = 24$ ft

25. The area of a rectangle is 27 square feet. The base is 3 more than twice the height. What are the dimensions of the rectangle? What is the perimeter?

26. $h = 5x$
$b = 35x$
$A = 175x^2$

26. The perimeter of a rectangle is $80x$. The base is 7 times the height. In terms of *x*, what are the dimensions of the rectangle? What is the area?

27. $b = 3$
$h = 6$
$A = 18$

27. The perimeter of a rectangle is equal to its area. The height is 3 more than the base. What are the dimensions of the rectangle? What is the area?

28. For a rectangle with a fixed perimeter of 100, $2b + 2h = 100$. Solve this equation for *b* or *h* and graph the resulting function. What type of function represents this relationship? What values of *b* and *h* do not make sense in the equation?

19. Let *b* be the base and *h* be the height of a rectangle. Since opposite sides of a rectangle are the same length, two sides have length *b* and two sides have length *h*. The perimeter is $b + b + h + h = 2b + 2h$.

28. $b = 50 - h$ or $h = 50 - b$

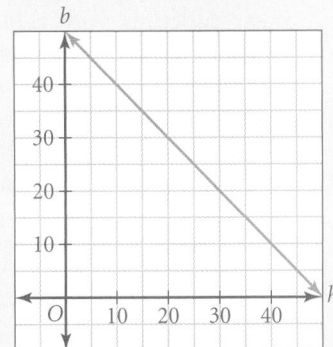

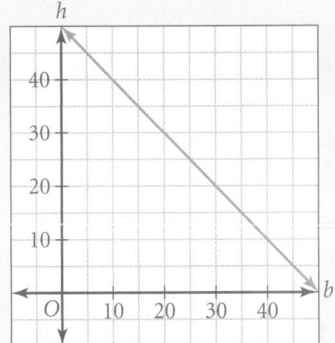

The relationship is a linear function. Any function value that produces a zero or negative value for *b* or *h* doesn't make sense for the perimeter example—specifically $b \leq 0$, $b \geq 50$ or $h \leq 0$, $h \geq 50$.

29. $b = \frac{100}{h}$ or $h = \frac{100}{b}$

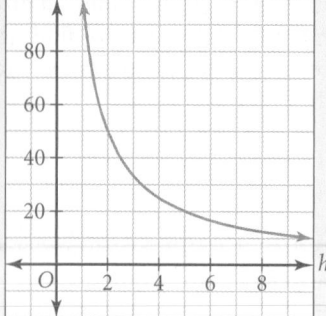

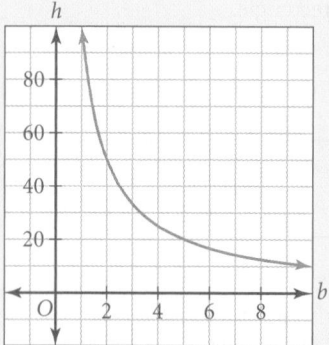

The relationship is a non-linear function. Any value which causes division by zero will make the function undefined, and a negative value for b or h doesn't make sense for a side length. That is, $b \leq 0$ or $h \leq 0$ are sets of values that don't make sense in the equation.

29. For a rectangle with a fixed area of 100, $bh = 100$. Solve this equation for b or h and graph the resulting function. What type of function represents this relationship? What values of b and h do not make sense in the equation?

30. The squares in the grid below measure 0.5 centimeters on each side. Estimate the area of the shaded figure. **3.5 sq. cm**

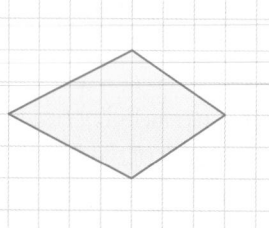

31. The squares in the grid below measure 0.5 centimeters on each side. Estimate the area of the shaded figure. **7.5 sq. cm**

CONSTRUCTION A house has a roof with the dimensions shown at right. Both rectangular halves of the roof will be covered with plywood.

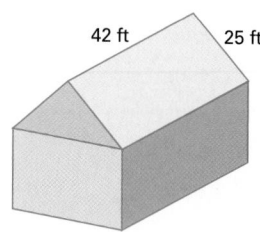

42 ft   25 ft

32. If plywood comes in pieces that measure 8 feet by 4 feet, how many pieces of plywood are needed to cover the roof? Assume that both halves of the roof are the same. **66 sheets**

33. Show a possible arrangement of the plywood pieces on one half of the roof. Try to make as few cuts as possible. How much scrap plywood is left over after covering the entire roof? **Answers will vary. 12 ft² are left over.**

34. SOLAR ENERGY Eric uses solar power to heat his home. He wants to provide 15,000 BTUs of heat from 2 solar panels. Eric wants the panels to be equal in size and 10 feet long. If 6 square feet of panels provide 1000 BTUs, what should be the dimensions of the panels that he uses? **4.5 ft wide**

The amount of energy produced by a solar panel depends on its area.

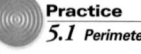

NAME _____ CLASS _____ DATE _____

Practice
5.1 Perimeter and Area

For Exercises 1–12, use the figure and measurements below to find the indicated perimeter or area. All measurements are in centimeters.

$MS = 15$	$US = 12$
$MT = 7$	$SQ = 20$
$XP = 13$	$VO = 10$

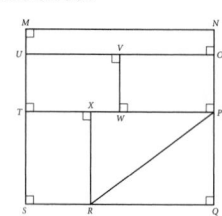

1. the area of rectangle $MNQS$
   300 square centimeters

2. the perimeter of rectangle $MNQS$
   70 centimeters

3. the area of rectangle $MNOU$
   60 square centimeters

4. the perimeter of rectangle $MNOU$
   46 centimeters

5. the area of rectangle $XPQR$
   104 square centimeters

6. the perimeter of rectangle $XPQR$
   42 centimeters

7. the perimeter of hexagon $SRXWVU$
   44 centimeters

8. the area of hexagon $SRXWVU$
   96 square centimeters

9. the area of $\triangle PQR$
   52 square centimeters

10. If points M and Q were connected by a segment, what would be the area of $\triangle MQS$?
    150 square centimeters

11. If points M and Q were connected by a segment, what would be the perimeter of $\triangle MQS$?
    60 centimeters

12. If points N and W were connected by a segment, what would be the perimeter of $\triangle NPW$?
    ≈ 29.21 centimeters

    the perimeter of trapezoid $MNWT$?
    ≈ 49.21 centimeters

---

35. AGRICULTURE You have 200 feet of fencing material to make a pen for livestock. If you make a rectangular pen, what is the maximum area you can fence in? Extend the table at right to determine the answer. **2500 ft²**

Base	Height	Perimeter	Area
1	?	200	?
2	?	200	?
5	?	200	?
20	?	200	?

36. AGRICULTURE Suppose that you need an area of 5625 square feet for grazing livestock. What is the minimum amount of fencing needed for a rectangular pen with this area? Extend the table at right to determine the answer. **300 ft**

Base	Height	Perimeter	Area
1	?	?	5625
2	?	?	5625
5	?	?	5625
20	?	?	5625

37. AGRICULTURE Repeat Exercise 35, but suppose that one side of the pen can be left out by placing the pen against the wall of the barn, as shown at left. **5000 ft²**

Base	Height	Perimeter	Area
1	?	200	?
2	?	200	?
5	?	200	?
20	?	200	?

Barn

Pen

38. AGRICULTURE Repeat Exercise 36, but suppose that one side of the pen can be left out by placing the pen against the wall of the barn, as shown at left. **212.13 ft**

Base	Height	Perimeter	Area
1	?	?	5625
2	?	?	5625
5	?	?	5625
20	?	?	5625

39. LANDSCAPING Rudy is buying sod for the lawn illustrated at right. Given the dimensions, estimate the number of square feet of sod that he needs. Assume that all angles shown are 90°. **690 ft²**

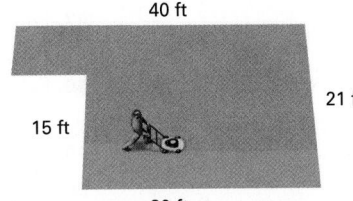

40 ft

21 ft

15 ft

30 ft

40. LANDSCAPING Dian wants to plant a rectangular vegetable garden that measures 10 feet by 16 feet. She will plant the vegetables in rows that are 10 feet long. Dian wants an equal amount of space for lettuce, tomatoes, carrots, zucchini, and peppers. In addition, she wants one 10-foot row of marigolds that will be 1 foot wide. How much space will Dian have for each type of vegetable? **30 ft² each**

Marigolds are often used by gardeners to repel insects from other plants.

Technology

In Exercises 35 and 36, a graphics calculator may be used to find the maximum or minimum area or perimeter.

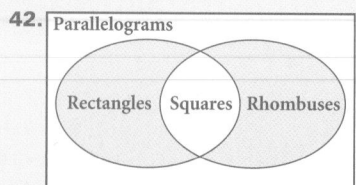
In Exercises 52 and 53, students compare the area of a circle with the area of a square. The circumference and area of a circle will be studied in Lesson 5.3.

42.

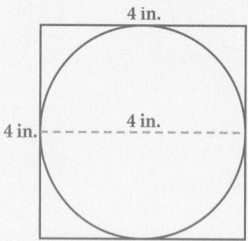

Parallelograms

Rectangles | Squares | Rhombuses

52. A circle with a diameter equal to the length of the side of the square can be inscribed inside the square as shown. Therefore, the square has the greater area.

4 in.

4 in. | 4 in.

120°

53. Divide the square into smaller squares. Then estimate the number of squares outside the circle and subtract from the total number of squares. The area of the remaining squares approximates the area of the circle.

**APPLICATION
CHALLENGE**

41. HOME IMPROVEMENT Brenda wants to paint her room, which measures 14 feet × 16 feet × 10 feet, as shown. The room has a 6 foot × 4 foot window and a 3 foot × 7 foot door that will not be painted. She will give the walls and ceiling two coats of paint: a base coat and a final coat.

One gallon of the paint for the base coat costs $10 and covers 500 square feet. One gallon of the paint for the final coat costs $20 and covers 250 square feet. The paint is sold only in 1-gallon cans. If the sales tax is 7%, how much will it cost to paint the room? **$107.00**

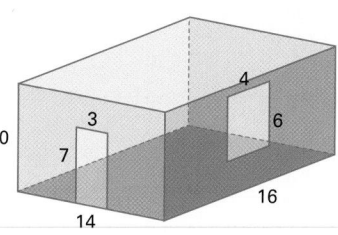

Look Back

42. Draw an Euler diagram to illustrate the relationships among parallelograms, rectangles, rhombuses, and squares. *(LESSON 3.2)*

43. Find the measure of an exterior angle of an equilateral triangle. *(LESSON 3.6)* **120°**

44. If the sum of the measures of 3 angles of a quadrilateral equals 300°, find the measure of the fourth angle. *(LESSON 3.6)* **60°**

45. Find the sum of the measures of the angles of a polygon with n sides. *(LESSON 3.6)* **$(n-2)180°$**

120° **46.** Find the measure of an interior angle of a regular hexagon. *(LESSON 3.6)*

47. Find the slope of the segment connecting the points (2, 3) and (4, −1). *(LESSON 3.8)* **−2**

48. Find the slope of a line perpendicular to the segment given in Exercise 47. *(LESSON 3.8)* **$\frac{1}{2}$**

49. What is the slope of a horizontal line? *(LESSON 3.8)* **0**

50. What is the slope of a vertical line? *(LESSON 3.8)* **undefined**

51. Find the midpoint of the segment connecting the points (−4, 6) and (6, 4). *(LESSON 3.8)* **(1, 5)**

Look Beyond

52. Which has a greater area, a square with a side length of 4 inches or a circle with a diameter of 4 inches? Explain your answer.

53. The square at right has 1-inch sides. How could you estimate the area of the circle inside the square?

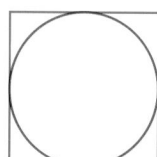

Areas of Triangles, Parallelograms, and Trapezoids

5.2

Objectives

- Develop formulas for the areas of triangles, parallelograms, and trapezoids.

- Solve problems by using the formulas for the areas of triangles, parallelograms, and trapezoids.

Why Designs on graph paper, such as this knitting pattern, suggest a method for estimating areas of geometric figures, but it is often more convenient to use exact formulas.

QUICK WARM-UP

Define each figure by describing key properties of the sides and angles.

1. right triangle 3 sides, 1 right angle

2. acute triangle 3 sides, all angles less than 90°

3. parallelogram 4 sides, opposite sides parallel and congruent

Also on Quiz Transparency 5.2

Areas of Triangles

Parts of a Triangle

Any side of a triangle can be called the **base of the triangle**. The **altitude of the triangle** is a perpendicular segment from a vertex to a line containing the base of the triangle. The **height of the triangle** is the length of the altitude.

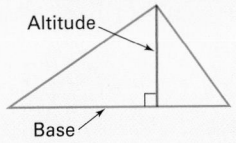

5.2.1

> *For each possible choice of the base of a triangle, there is a corresponding altitude and height.*

Activity 1

The Area Formula for Triangles

YOU WILL NEED
graph paper

Part I: Deriving a Formula for the Area of a Right Triangle

1. Draw a rectangle on graph paper. Calculate its area.

2. Draw a diagonal of your rectangle to form two right triangles. What do you know about these triangles from your study of special quadrilaterals? Based on this, what is the area of each triangle?

Teach

Why Many purchasing decisions are based on the size of the items being purchased. The size of an apartment, TV screen, tennis racket, mouse pad, back pack, etc., are all applications of the concept of area.

☞ For Activity Notes and the answer to the Checkpoint, see page 304.

Alternative Teaching Strategy

TECHNOLOGY Geometry graphics software can be used to explore the formulas in this lesson. For example, the figure at right can be used to derive the formula from Part II of Activity 1. Students should construct two sets of parallel lines whose intersections form right angles as shown at right. Point *J* can move anywhere on the line containing points *G* and *H*. Students should drag point *J* and observe that the base, height, and area of △*ABJ* do not change.

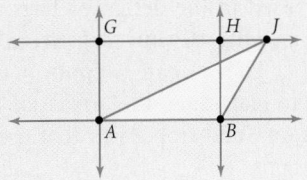

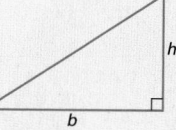

Activity 1 Notes

Students discover important relationships between the area of a right triangle and the area of a rectangle. Emphasize that the height of a triangle must be perpendicular to its base. Point out that any side of the triangle can be chosen as the base.

For a student worksheet of this Activity and detailed Teacher Notes, see page 81 in the Lesson Activities booklet.

CHECKPOINT ✔
Part I

4. $A = \frac{1}{2}bh$

CHECKPOINT✔
Part II

5. $A = \frac{1}{2}bh$

CRITICAL THINKING

Add dashed lines to complete a rectangle. The area of the rectangle is $(b + x)h = bh + xh$. The total area of the right triangles is $\frac{1}{2}(b + x)h + \frac{1}{2}xh = \frac{1}{2}bh + xh$. Subtracting, the area of the blue triangle is $\frac{1}{2}bh$.

ADDITIONAL
EXAMPLE ❶

Find the area of a triangle with a base length of 15 feet and height of 22 feet.
165 square feet

Cooperative Learning

Have students work in small groups to prove that any side of a triangle can be chosen as the base. They can use graph paper or geometry graphics software for this investigation.

---

3. If you are given a right triangle, can you always form a rectangle by fitting it together with a copy of itself? Illustrate your answer with examples.

CHECKPOINT ✔ **4.** Write a formula for the area, A, of a right triangle in terms of its base, b, and its height, h.

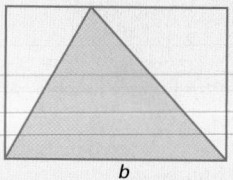

Part II: Deriving a Formula for the Area of Any Triangle

1. Copy the drawing at right onto graph paper.

2. Draw an altitude of the triangle from the top vertex to the base of the triangle. Is the altitude parallel to the sides of the rectangle? What theorem justifies your answer?

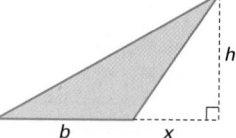

3. The altitude divides the rectangle into two smaller rectangles. Each rectangle is divided into two congruent triangles. What theorem justifies these statements?

4. What is the relationship between the area of the rectangle and the area of the shaded triangle? Explain your answer.

CHECKPOINT ✔ **5.** Write a formula for the area, A, of a triangle in terms of its base, b, and its height, h.

CRITICAL THINKING How can you use the method from Activity 1 to derive the formula for the area of an obtuse triangle with base b and height h, such as the one at right? (See Exercise 52.)

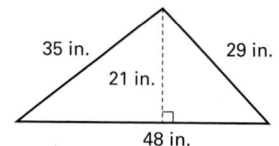

In Activity 1, you derived the following formula:

Area of a Triangle

For a triangle with base b and height h, the area, A, is given by:
$$A = \frac{1}{2}bh$$

5.2.2

EXAMPLE ❶

APPLICATION
THEATER ARTS

You are building a triangular flat, as shown at right, for a dance performance. The flat needs to be covered with cloth. What is the area you need to cover?

35 in.   29 in.
21 in.
48 in.

● **SOLUTION**
$A = \frac{1}{2}bh = \frac{1}{2}(48)(21) = 504$ in.^{2}

Interdisciplinary Connection

SOCIAL STUDIES Triangles and parallelograms are frequently used in the design of flags. Have students use an atlas to copy flags from different countries. They can apply the area formulas to calculate the areas of triangles, parallelograms, and trapezoids that are found in the flag designs.

Areas of Parallelograms

Parts of a Parallelogram

Any side of a parallelogram can be called the **base of the parallelogram**. An **altitude of a parallelogram** is a perpendicular segment from a line containing the base to a line containing the side opposite the base. The **height of the parallelogram** is the length of the altitude.

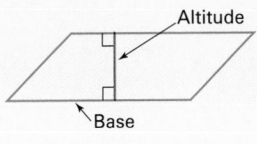

5.2.3

Activity 2 Notes

In this Activity, students derive the area formula for parallelograms. Some students may think that the area of a parallelogram must equal the product of two adjacent sides. Have students use graph paper or geometry graphics software to derive the correct area formula for parallelograms.

For a student worksheet of this Activity and detailed Teacher Notes, see page 81 in the Lesson Activities booklet.

Activity 2
The Area Formula for Parallelograms

YOU WILL NEED

graph paper and scissors

1. Copy the drawing at right onto graph paper. Draw an altitude in the parallelogram from point A to the base so that a right triangle is formed.

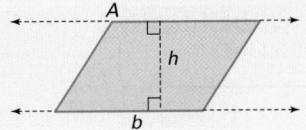

2. Cut out the parallelogram. Cut off the right triangle, translate it to the opposite side, and fit it to the figure.

   What kind of quadrilateral is formed? What is the area of the figure in terms of b and h of the original parallelogram? How does the area of the parallelogram relate to the area of the figure formed by the translation?

CHECKPOINT ✔ 3. Write a formula for the area of a parallelogram in terms of its base, b, and its height, h.

4. How do you know that the triangle will always fit, as in Step 2? To answer this question, first prove that $\triangle AEB \cong \triangle DFC$. Then prove that $ADFE$ is a rectangle.

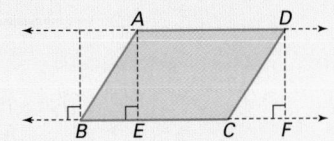

CHECKPOINT ✔
3. $A = bh$

Teaching Tip

TECHNOLOGY Have students use geometry graphics software to investigate the area of a parallelogram. Have them construct a rectangle with two sets of parallel segments and calculate its area. Then have them drag one side of the rectangle until the figure becomes a parallelogram. Students should notice that the area stays the same because the length of the base and the height do not change.

In Activity 2, you derived the following formula:

Area of a Parallelogram

For a parallelogram with base b and height h, the area, A, is given by:
$$A = bh$$

5.2.4

EXAMPLE 2 Find the area of parallelogram $ABCD$.

● **SOLUTION**

$A = bh = 5(13) = 65$ square units

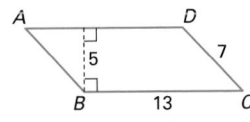

ADDITIONAL
EXAMPLE 2

Find the area of a parallelogram with base length of 27 cm and a height of 13 cm.
351 cm²

Inclusion Strategies

USING COGNITIVE STRATEGIES   Some students may attempt to memorize the area formulas without trying to understand how the formulas are developed. Explain that they are likely to forget the formulas unless they also remember a related diagram and the logical reasoning used to derive the formula. For example, to remember the formula for the area of a triangle, they should think of a triangle drawn inside a rectangle, as shown in Part II of Activity 1.

Activity 3 Notes

In this Activity, students use the formula for the area of a parallelogram to derive the formula for the area of a trapezoid.

For a student worksheet of this Activity and detailed Teacher Notes, see page 81 in the Lesson Activities booklet.

Teaching Tip

TECHNOLOGY If students use geometry graphics software for Activity 3, they should start by constructing a trapezoid. They should then reflect the trapezoid twice—first across a vertical line and then across a horizontal line. The final figure should then be fitted to the original figure to make a parallelogram.

CHECKPOINT ✔

3. $A = \frac{1}{2}(b_1 + b_2)h$

ADDITIONAL
EXAMPLE ③

Find the area of a trapezoid with a height of 12 inches and bases of 3 inches and 8 inches. **66 square inches**

Areas of Trapezoids

Parts of a Trapezoid

The two parallel sides of a trapezoid are known as the **bases of the trapezoid**. The two nonparallel sides are called the **legs of the trapezoid**. An **altitude of a trapezoid** is a perpendicular segment from a line containing one base to a line containing the other base. The **height of a trapezoid** is the length of an altitude.

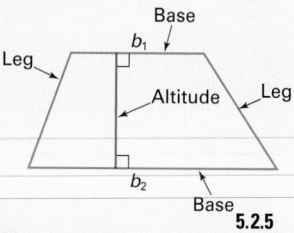

5.2.5

Activity 3
The Area Formula for Trapezoids

YOU WILL NEED
graph paper and scissors

CHECKPOINT ✔

1. Make two copies of the trapezoid at right on graph paper and cut them out. The bases of the trapezoid are b_1 and b_2, and the height is h.

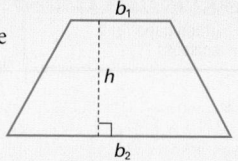

2. Find a way to fit the two copies of the trapezoid together to form a parallelogram. Sketch the parallelogram.

3. Write a formula for the area of the parallelogram involving the expression $(b_1 + b_2)$. Use the formula you wrote for the parallelogram to write a formula for the area of the original trapezoid.

In Activity 3, you derived the following formula:

Area of a Trapezoid

For a trapezoid with bases b_1 and b_2 and height h, the area, A, is given by:
$$A = \frac{1}{2}(b_1 + b_2)h$$ **5.2.6**

EXAMPLE ③

APPLICATION
LANDSCAPING

A homeowner needs to buy sod for a bare trapezoidal region of the lawn in front of his house. The dimensions of the region are shown at right. What is the area of the region?

● **SOLUTION**

$A = \frac{1}{2}(b_1 + b_2)h$

$A = \frac{1}{2}(30 + 50)(23)$

$A = 920$ ft^2

Enrichment

Heron's formula for the area of a triangle is introduced in Exercises 58–60. Have students use a spreadsheet to show that the formulas below give the area of an isosceles trapezoid.

Semiperimeter $= s = \frac{a+b+c+d}{2}$

Area $= \sqrt{(s-a)(s-b)(s-c)(s-d)}$

Area $= \frac{(a+b)h}{2}$

EXAMPLE 4 Use the diagram and measurements given below to find the areas of the indicated figures.

a. $\triangle VWZ$
c. parallelogram VWXY

b. $\triangle WXY$
d. trapezoid WXYZ

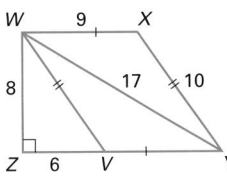

SOLUTION

a. area of $\triangle VWZ = \frac{1}{2}bh = \frac{1}{2}(6)(8) = 24$

b. area of $\triangle WXY = \frac{1}{2}bh = \frac{1}{2}(9)(8) = 36$

c. area of parallelogram $VWXY = bh = (9)(8) = 72$

d. area of trapezoid $WXYZ = \frac{1}{2}(b_1 + b_2)h = \frac{1}{2}(9 + 15)(8) = 96$

Exercises

Communicate

1. Draw a parallelogram and one of its diagonals. What can you say about the two triangles formed by the diagonal? Draw the other diagonal to form four triangles. Are any of the four triangles congruent? Which postulates and theorems can you use to prove your answer?

2. Can two triangles have the same base and height and not be congruent? Can two trapezoids have the same base and height and not be congruent? Explain your reasoning.

3. In the parallelogram shown at right, which is longer, AB or h? Use your answer to explain why the parallelogram with a given base and height that has the smallest perimeter is a rectangle.

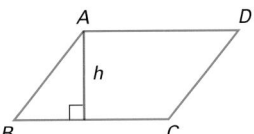

4. In a trapezoid, what is the average of the lengths of the bases? How is this quantity related to the area of the trapezoid?

5. Suppose that you are given a rectangle, a triangle, a parallelogram, and a trapezoid, each with a base of 16 and a height of 11. Which figures must have the same area? Which area cannot be determined? Explain your reasoning.

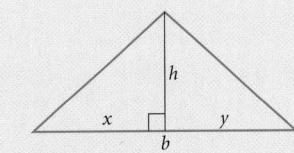
ADDITIONAL
EXAMPLE 4

Using the figure at the top of page 307, find the areas of $\triangle VWY$ and $\triangle ZWY$.

area of $\triangle VWY = 36$
area of $\triangle ZWY = 60$

Assess

Selected Answers
Exercises 6–9, 11–57 odd

ASSIGNMENT GUIDE

In Class	1–9
Core	11–47 odd
Core Plus	19–53 odd
Review	54–57
Preview	58–60

✎ Extra Practice can be found beginning on page 818.

6. Find the area of the shaded triangle.
    (ACTIVITY 1 AND EXAMPLE 1)
    300 units²

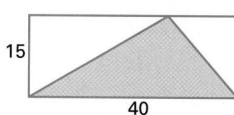

7. Find the area of the shaded parallelogram. *(ACTIVITY 2 AND EXAMPLE 2)*
    35 units²

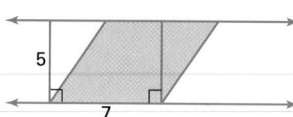

8. The two congruent trapezoids shown fit together to form a parallelogram. Find the area of each trapezoid, and find the area of the parallelogram.
    (ACTIVITY 3 AND EXAMPLE 3)
    51 units²,
    102 units²

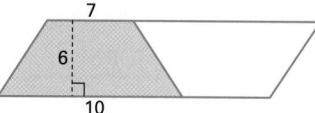

9. Use the diagram below to find the areas of the indicated figures.
    (EXAMPLE 4)

   a. 4 *ADE* 204 units²

   b. parallelogram *ABCE* 480 units²

   c. trapezoid *ABCD* 684 units²

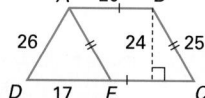

For Exercises 10–12, find the area of each triangle.

10.

14 units²

11. 12 units²

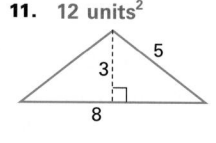

12. 42 units²

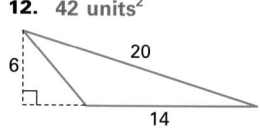

For Exercises 13–15, find the area of each parallelogram.

13. 312 units²

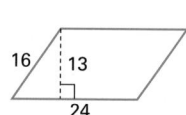

14. 22 units²

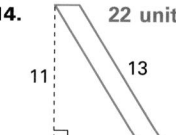

15. 48 units²

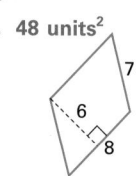

For Exercises 16–18, find the area of each trapezoid.

16. 24 units²

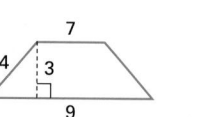

17. 6 units²

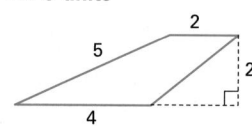

18.

18. 650 units²

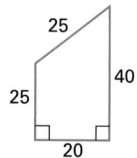

For Exercises 19–30 use the diagram and measurements below to find the area of each figure.

$\overline{AC} \parallel \overline{IL}$ $\overline{DH} \parallel \overline{IL}$

$\overline{AI} \parallel \overline{JB}$ $\overline{KC} \parallel \overline{JB}$

$BK = 14$ $KL = 14$

$IL = 32$ $DH = 25$

$FK = 7$ $EH = 17$

$JI = 8$ $DG = 18$

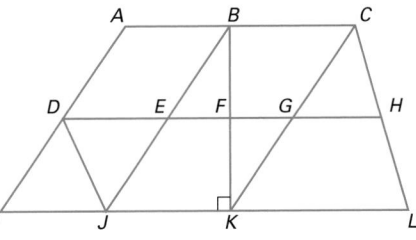

19. $\triangle KCL$ 98 units^2 **20.** $\triangle BJK$ 70 units^2 **21.** $\triangle BCK$ 70 units^2

22. $\triangle DIJ$ 28 units^2 **23.** $BCKJ$ 140 units^2 **24.** $EGKJ$ 70 units^2

25. $ABED$ 56 units^2 **26.** $ACKI$ 252 units^2 **27.** $EHLJ$ 143.5 units^2

28. $BCHE$ 94.5 units^2 **29.** $BCLJ$ 238 units^2 **30.** $ACLI$ 350 units^2

CONNECTIONS

COORDINATE GEOMETRY For Exercises 31–34, find the area of a triangle with vertices at the given points. You may find it helpful to sketch a graph.

31. $(0, 0), (0, 3), (4, 0)$ 6 units^2 **32.** $(1, 2), (5, 2), (3, 7)$ 10 units^2

33. $(1, 0), (3, 0), (0, 6)$ 6 units^2 **34.** $(-2, 1), (4, 1), (1, -3)$ 12 units^2

COORDINATE GEOMETRY For Exercises 35–38, find the area of a parallelogram with vertices at the given points. You may find it helpful to sketch a graph.

35. $(0, 0), (4, 0), (6, 2), (2, 2)$ 8 units^2

36. $(0, 1), (0, 3), (3, 5), (3, 3)$ 6 units^2

37. $(2, 3), (3, -1), (-1, -1), (-2, 3)$ 16 units^2

38. $(4, 1), (5, 1), (2, -2), (1, -2)$ 3 units^2

COORDINATE GEOMETRY For Exercises 39–42, find the area of a trapezoid with vertices at the given points. You may find it helpful to sketch a graph.

39. $(0, 2), (3, 2), (5, 0), (-1, 0)$ 9 units^2

40. $(2, 1), (4, 1), (6, 4), (-1, 4)$ 13.5 units^2

41. $(1, 0), (5, 0), (5, 5), (1, 3)$ 16 units^2

42. $(3, 1), (6, -1), (-1, -1), (-2, 1)$ 12 units^2

Algebra

43. $b = 20$ cm

43. Find the base of a triangle with a height of 10 cm and an area of 100 cm^2.

44. Find the height of a parallelogram with a base of 15 cm and an area of 123 cm^2. $h = 8.2$ cm

45. Use the diagram at right of a trapezoid and its midsegment to determine a formula for the area of a trapezoid that uses only the height of the trapezoid and the length, m, of its midsegment. $A = mh$

Math
CONNECTION

ALGEBRA In Exercise 43, students use the formula for the area of a triangle and solve for the height, h.

Error Analysis

Students who are not able to complete Exercises 48–52 may benefit from developing a plan for the proof in small groups. This may help them to complete the proof on their own.

46. The triangle with the largest area is the center one, which is an equilateral triangle. Conjecture: For triangles with a given perimeter, an equilateral triangle has the largest area.

Sample answer: This is similar to the fact that the largest area enclosed by a rectangle is enclosed by a square.

46. MAXIMUM/MINIMUM The triangles below have the same perimeter. Which has the largest area? Make a conjecture about the triangle with the largest possible area for a given perimeter. Explain your reasoning.

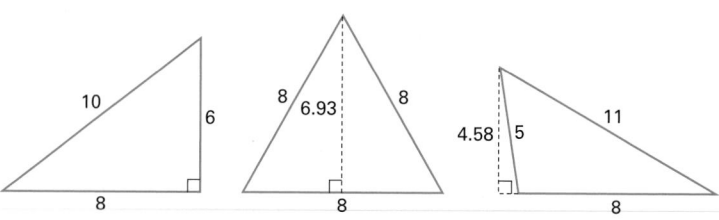

47. MAXIMUM/MINIMUM The parallelograms below have the same perimeter. Which has the largest area? Make a conjecture about the parallelogram with the largest possible area for a given perimeter. Explain your reasoning.

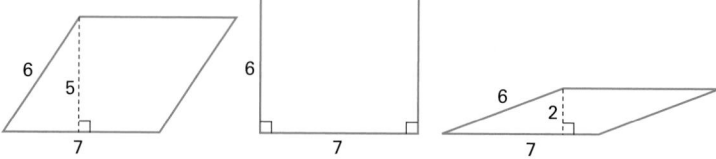

48. Write out a two-column or paragraph proof of the formula you derived in Activity 1.

49. Write out a two-column or paragraph proof of the formula you derived in Activity 2.

50. Write out a two-column or paragraph proof of the formula you derived in Activity 3.

51. A **kite** is a quadrilateral with exactly two pairs of adjacent congruent sides.

Given: The diagonals of a kite are perpendicular.
Prove: The area is equal to one-half of the product of the lengths of the diagonals.

(Hint: Use the formula for the area of a triangle.)

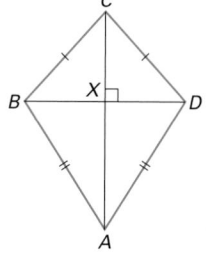

52. Use the diagram at right to prove that the formula for the area of a triangle works for an obtuse triangle.

The longest side of the shaded triangle 4 *KLM* divides the rectangle into two congruent right triangles. What is the area of each of these triangles?

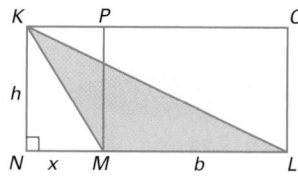

Right triangle 4 *KLN* is divided into two triangles, 4 *KLM* and 4 *KMN*. What is the area of 4 *KMN*?

Prove that the area of 4 *KLM* is $\frac{1}{2}bh$.

Practice

Practice
5.2 Areas of Triangles, Parallelograms, and Trapezoids

For Exercises 1–3, find the area of each triangle.

1.	2.	3.
24	45	32

For Exercises 4–6, find the area of each parallelogram.

4.	5.	6.
12	10	200

For Exercises 7–9, find the area of each trapezoid.

7.	8.	9.
32	190	28

Find the area of the indicated figure.

10. △*WVZ* _____ 105
11. parallelogram *WXYZ* _____ 195
12. trapezoid *WXYV* _____ 300
13. right triangle with hypotenuse $\overline{XY}$ _____ 60

47. The parallelogram with the largest area is the rectangle. Conjecture: For parallelograms with a given perimeter, the parallelogram with the largest area is a rectangle.

Sample answer: The shortest distance between a point and a line is along the perpendicular. This perpendicular distance is larger in the rectangle than in the other two parallelograms.

48. Given a right or acute triangle *ABC* with altitude *AF* of length h and base b, let *DEBC* be a rectangle with base b and height h. The altitude *AF* forms two rectangles, *DAFC* and *AEBF*. Since $\overline{AC}$ divides rectangle *DAFC* in half and $\overline{AB}$ divides rectangle *AEBF* in half, the area of the triangle is half the sum of the areas of the two smaller rectangles. The sum of the areas of the rectangles is bh, so the area of the triangle is $\frac{1}{2}bh$.

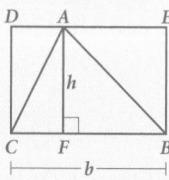

53. FARMING In order to fertilize a field, a farmer needs to estimate its area. Estimate the area of the field outlined in the photo at right. If an acre equals 43,560 square feet, how many acres is the field? If it takes 435 pounds of fertilizer to cover 1 acre, how much fertilizer is required to cover the field?

3500 ft

2800 ft

$A = 4,900,000 \text{ ft}^2 \approx 112.5 \text{ acres}$, 48,937.5 lb of fertilizer

 **Look Back**

54. Construct an Euler diagram to illustrate the relationship between scalene, isosceles, and equilateral triangles. **(LESSON 2.3)**

55. Refer to the diagram below. Prove that $\triangle ABC \cong \triangle DEF$. **(LESSON 4.4)**

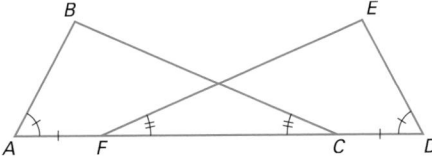

Algebra

56. $x^2 + 2xy + y^2$

56. Find the area of a square with a side length of $x + y$. **(LESSON 5.1)**

57. Find the area of a rectangle with a base of $x + 2$ and a height of $x - 3$. **(LESSON 5.1)** $x^2 - x - 6$

 **Look Beyond**

CULTURAL CONNECTION: AFRICA Heron's formula, named after a mathematician who lived in Alexandria in around 100 C.E., can be used to find the area of a triangle from the lengths of the sides. The formula is
$$A = \sqrt{s(s-a)(s-b)(s-c)},$$
where s is the *semiperimeter*—that is, half of the perimeter—and a, b, and c are the lengths of the sides.

58. Find the semiperimeter of a triangle with side lengths of 7, 8, and 9. **12 cm**

59. Find the area of the triangle described in Exercise 58. **26.83 cm²**

60. TECHNOLOGY Set up a spreadsheet to find the area of a triangle using Heron's formula when the lengths of the sides are given. Find the areas of the triangles with the sides given below.

=.5*(A2 + B2 + C2)

	A	B	C	D	E
1	a	b	c	s	A
2	7	8	9	?	?
3	3	4	5	?	?
4	2	10	10	?	?
5	x	x	x	?	?

=SQRT(D2*(D2−A2)*(D2−B2)*(D2−C2))

55.

Statements	Reasons
1. $\angle A \cong \angle D$, $\angle EFD \cong \angle BCA$, $\overline{AF} \cong \overline{CD}$	1. Given
2. $\overline{FC} \cong \overline{FC}$	2. Reflexive Property of Congruence
3. $\overline{AC} \cong \overline{FD}$	3. Overlapping Segments Theorem
4. $\triangle ABC \cong \triangle DEF$	4. ASA

60.

	A	B	C	D	E
1	a	b	c	sum(A1:C1)/2	SQRT[D1*(D1-A1)*(D1-B1)*(D1-C1)]
2	7	8	9	12	26.83
3	3	4	5	6	6
4	2	10	10	11	9.95

Technology

In Exercises 46 and 47, a graphics calculator may be used to find the maximum area for a given perimeter. Spreadsheet software may be used in Exercises 57–59.

Math
CONNECTION

ALGEBRA In Exercise 56, students will multiply binomials to find the area of a square with a side length of $x + y$.

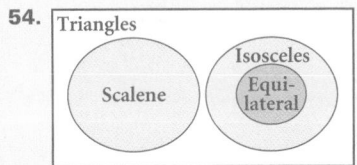 **Look Beyond**

If students have programmable calculators, they can enter and store Heron's formula. This formula, known for its tedious computations, allows students to find the area of a triangle without using its height.

54.

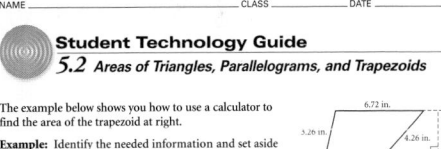

Triangles

Scalene

Isosceles
Equi-
lateral

Student Technology Guide

NAME _____ CLASS _____ DATE _____

Student Technology Guide
5.2 Areas of Triangles, Parallelograms, and Trapezoids

The example below shows you how to use a calculator to find the area of the trapezoid at right.

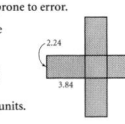

6.72 in.
3.28 in.
4.26 in.
3.28 in.
4.56 in.

Example: Identify the needed information and set aside extraneous information.
height: 3.28 in.   bases: 4.56 in. and 6.72 in.
Using the needed information, apply the formula $A = \frac{1}{2}(b_1 + b_2)h$. (Note: Use 0.5 rather than $\frac{1}{2}$.)
Press .5 [×] [(] 6.72 [+] 4.56 [)] [×] 3.28 [ENTER].
Rounded to the nearest hundredth, the area equals 18.50 in.².

Sometimes you need to write and simplify an area expression before you use a calculator. The resulting key sequence is often simpler, and you will be less prone to error.

Example: To the nearest hundredth of a unit, find the area of the shaded region at right. A square is the center of four congruent rectangles. Simply find their total area by multiplying the area of one of them by 4. Press 4 [×] 3.84 [×] 2.24 [+] 2.24 [x²] [ENTER].
To the nearest hundredth, the area equals 39.42 square units.

2.24
3.84

Using a calculator, find the area of each shaded region. Write the simplest area expression. Give answers to the nearest hundredth of a square unit.

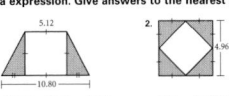 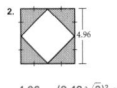

1.
5.12
10.80
$2(0.5 \times 2.84 \times 5.12) \approx$
14.54 square units

2.
4.96
$4.96 - (2.48\sqrt{2})^2 \approx$
12.30 square units

3.
5.12
10.80
.5 × 5
$(5.12 + 10.80) - (2.56\sqrt{2})$
27.65 square units

To find the length of one side of a square whose area equals 200 square units, press [2nd] [√] 200 [)] [ENTER].
√(200)
14.14213562

To the nearest tenth, find the side length of a square with given area.

4. 240 ___ **15.5 units**   5. 250 ___ **15.8 units**   6. 260 ___ **16.1 units**

EYEWITNESS MATH

AN ANCIENT WONDER

Focus

A book of ancient Chinese mathematics, ignored for centuries by Western civilization, provides examples of exploring the solution of geometric problems by cutting up figures and rearranging the pieces.

Motivate

As you review the page, help students gain an appreciation for *The Nine Chapters on the Mathematical Art*, particularly the sophistication of the mathematics in relation to what was happening in the world at that time. The book includes geometric and algebraic proofs credited to other mathematicians who did not discover them until centuries later.

The ancient Chinese book *The Nine Chapters on the Mathematical Art* is the collective effort of mathematicians over hundreds of years that may have begun long before Euclid compiled his famous work, *The Elements*. The version that exists today has been significantly edited and annotated by many people.

The Nine Chapters on the Mathematical Art has only recently received attention in the West. This ancient text contains algebraic and geometric proofs that are commonly credited to other mathematicians who didn't discover them until hundreds of years later.

On the following page, you will explore how to simplify complex problems by using a "patchwork" method of cutting up figures and rearranging the pieces.

Cooperative Learning

The following problems and methods are the work of Liú Hui , who wrote a commentary for the book around 236 C.E. The commentary, which is now considered to be a part of the book, contains explanations of solutions given in the earlier version.

1. You can begin your exploration of *The Nine Chapters on the Mathematical Art* with the following problem:

Find the side length of a square that is inscribed in a right triangle.

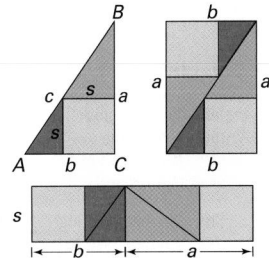

a. Draw a square inscribed in a right triangle as shown at right. Make two congruent copies of the square and triangle and fit them together to form a rectangle.

b. Cut each triangle into three pieces and reassemble them to form a long rectangle as shown at right.

c. What can you say about the area of the rectangle that you formed in part **a** compared with the area of the one that you found in part **b**?

d. Write an expression in terms of *a* and *b* for the area of the rectangle in part **a**. Write an expression in terms of *a*, *b*, and *s* for the area of the rectangle in part **b**.

e. Set the two expressions from part **d** equal to each other to form an equation in terms of *a*, *b*, and *s*. Solve the equation for *s* to find the side length of the square.

2. Now try a harder problem from *The Nine Chapters on the Mathematical Art*.

Find the radius of a circle that is inscribed in a right triangle.

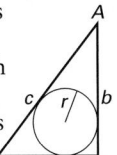
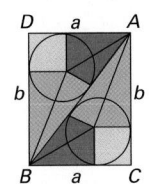

a. Refer to the diagrams at right. Two congruent copies of the triangle fit together to form a rectangle. The rectangle is divided into triangles and squares, which are reassembled to form a long rectangle. Explain why *EH* in the long rectangle is equal to *r*, the radius of the inscribed circle.

b. In rectangles *ADBC* and *EFGH*, explain why the lengths labeled *a*, *b*, and *c* are the same as the side lengths *a*, *b*, and *c* of △*ABC*.

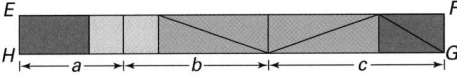

c. Write an expression in terms of *a* and *b* for the area of rectangle *ADBC*. Write an expression in terms of *a*, *b*, *c*, and *r* for the area of rectangle *EFGH*.

d. Set the two expressions from part **c** equal to each other to form an equation in terms of *a*, *b*, *c*, and *r*. Solve the equation for *r* to find the radius of the circle.

Discuss

Use the patchwork method to find the formula for the area of an isosceles trapezoid. A sample answer is given below.

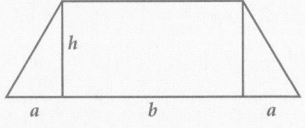

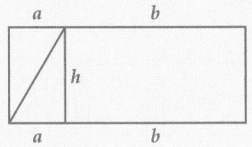

$$A = (a + b)h$$

Evaluate. Round your answers to the nearest hundredth.

1. 5^2 25

2. 6.7^2 44.89

3. $\sqrt{36}$ 6

4. $\sqrt{48}$ 6.93

5. π 3.14

6. 2π 6.28

7. $3\pi + 1.3$ 10.72

8. $3^2 \cdot \pi$ 28.27

9. $(3\pi)^2$ 88.83

10. Draw a circle. Label the center, radius, and diameter. **Check students' drawings.**

Also on Quiz Transparency 5.3

Teach

Why The ability to estimate values that cannot be calculated directly is of great importance in school, work, statistics, etc. Given the diameter of the Moon, students can estimate the diameter of the larger craters in the photograph. Have students research the diameter of Earth and use it to find the circumference of the Earth.

Circumferences and Areas of Circles

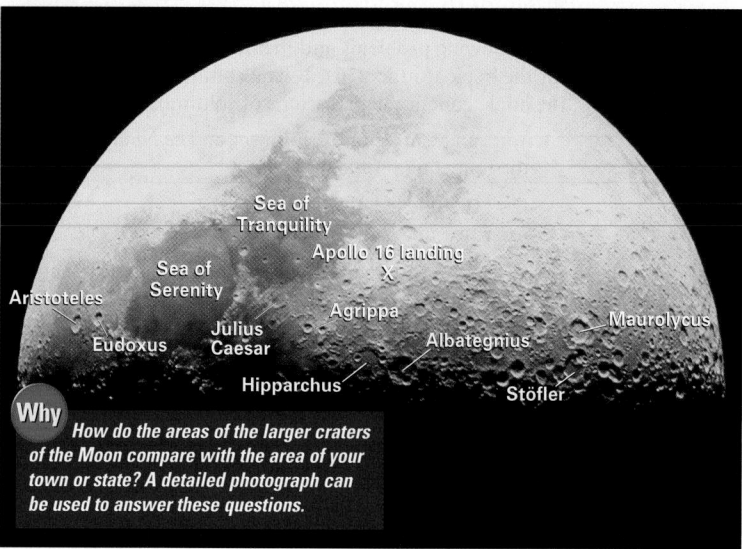

Objectives

● Identify and apply formulas for the circumference and area of a circle.

● Solve problems using the formulas for the circumference and area of a circle.

Why *How do the areas of the larger craters of the Moon compare with the area of your town or state? A detailed photograph can be used to answer these questions.*

The diameter of the Moon is about 1077 miles. You can use this information to determine the scale of this photograph. Then you can estimate the areas of the craters that are visible.

The Definition of a Circle

When you draw a circle with a compass, the distance from the point of the compass to the pencil or pen does not change. Therefore, every part of the circle you draw is the same distance from the *center* of the circle, the point where the compass is fixed. This leads to the following definition:

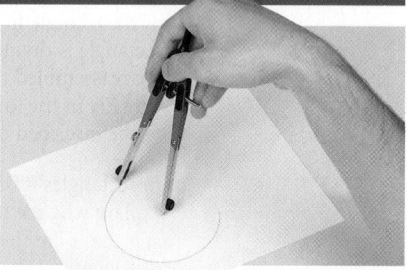

Definition: Circle

A **circle** is the set of all points in a plane that are the same distance, *r*, from a given point in the plane known as the **center** of the circle. The distance *r* is known as the **radius** of the circle. The distance $d = 2r$ is known as the **diameter** of the circle.

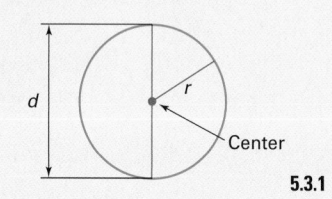

5.3.1

Alternative Teaching Strategy

TECHNOLOGY Geometry graphics software can be used to approximate the area and circumference of circles. For example, students can approximate the area of a circle by averaging the areas of its inscribed and circumscribed squares and compare the result with that calculated by using π. Encourage students to use the software to create their own ways of approximating area and circumference.

Circumferences and Areas of Circles

You are probably already familiar with the formulas for the circumferences and areas of circles from your earlier mathematics studies. The following activities will lead you to an understanding of these formulas.

Activity 1
The Circumference of a Circle

1. The distance around a circle is called its **circumference**. Measure the circumferences and diameters of several circular objects. Record the results in a table like the one at right.

Object	C	d	Ratio: $\frac{C}{d}$
1. can	31.4	10	?
2. ?	?	?	?
3. ?	?	?	?

2. The ratio $\frac{C}{d}$ is known as π, pronounced "pie." What do you notice about your values for this ratio? Find the average of the values.

3. Compare your result with the results of your classmates. How close are the results to 3.14, an approximate value of π? Press the π key on your calculator, followed by ENTER. What value does the calculator show?

CHECKPOINT ✔ 4. Write a formula for the circumference, C, of a circle. Begin by expressing π as a ratio. Then solve for C and write the formula in terms of the radius, r.

In Activity 1, you discovered the following formula:

Circumference of a Circle

The circumference, C, of a circle with diameter d and radius r is given by:
$$C = \pi d \text{ or } C = 2\pi r$$

5.3.2

EXAMPLE ① Find the circumference of each circle below. Give your answers exactly, in terms of π, and rounded to two decimal places.

a.

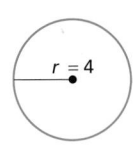
$r = 4$

b.

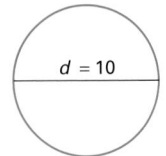
$d = 10$

● SOLUTION

a. $C = 2\pi r = 2\pi(4) = 8\pi \approx 25.13$

b. $C = \pi d = \pi(10) = 10\pi \approx 31.42$

Interdisciplinary Connection

HISTORY The history of mathematics includes stories of efforts to find approximations for the value of π. Have students use library materials to prepare a timeline that shows the increasingly precise values for π.

Enrichment

The British Museum owns a papyrus roll called the Rhind papyrus by Ahmes. Ahmes was the scribe who copied the work in about 1650 B.C.E. In problem 50, Ahmes assumed that the area of a circular field with a diameter of 9 units is the same as the area of a square with a side of 8 units. Have students find the value of π that is implied by this assumption. $\pi(4.5)^2 = 8^2$; $\pi \approx 3.1605$

Activity 2
The Area of a Circle

YOU WILL NEED

ruler, compass, protractor (optional), and scissors

1. Draw a circle. Label its radius r. Using any method you like, such as paper folding, divide the circle into eight congruent pie-shaped parts, or **sectors**.

2. Cut out the sectors and reassemble them into a single figure, as shown at right. If the curved parts of your figure were segments instead of curves, what kind of figure would you have?

3. Divide the circle into 16 congruent sectors by cutting each sector from Step 2 in half. Then reassemble the sectors as in Step 2. As the number of sectors increases, do the curved parts seem straighter?

4. What geometric figure do your sector assemblies in Steps 2 and 3 resemble? The height of your figures is approximately equal to r, the radius of the circle. What happens to this approximation as the number of sectors increases infinitely?

5. The base of your assembled figure is approximately equal to half of the circumference of your original figure. (Why?) Write an expression for the base of the figure in terms of π and the radius, r, of the circle.

CHECKPOINT ✔   6. Write an expression for the area of the figure in terms of π and r.

Area of a Circle

The area, A, of a circle with radius r is given by:
$$A = \pi r^2$$

5.3.3

CRITICAL THINKING   Why does the method you used in Activity 2 become more realistic as you increase the number of sectors?

EXAMPLE ②   Find the area of each circle below. Give your answers exactly, in terms of π, and rounded to two decimal places.

a.
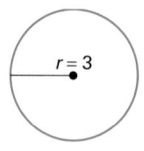
$r = 3$

b.
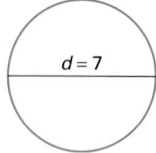
$d = 7$

● **SOLUTION**

a. $A = \pi r^2 = \pi(3^2) = 9\pi \approx 28.27$ units²

b. $r = d \div 2 = 7 \div 2 = 3.5$ units
$A = \pi r^2 = \pi(3.5^2) = 12.25\pi \approx 38.48$ units²

Inclusion Strategies

ENGLISH LANGUAGE DEVELOPMENT Have students write a paragraph about the relationship among the radius, diameter, circumference, and area of a circle. They should first try to explain their ideas without using diagrams. They should then add drawings to clarify their work.

Reteaching the Lesson

USING VISUAL MODELS Each student should draw several circles on graph paper and estimate the areas by counting the number of square units contained in each circle. Students should then compare their areas with the areas calculated with the formula.

Exercises

Communicate

1. Suppose that you have 100 feet of fence to make a play area for your dog. Does a square or a circle provide more area? What other factors might you take into consideration in designing the play area?

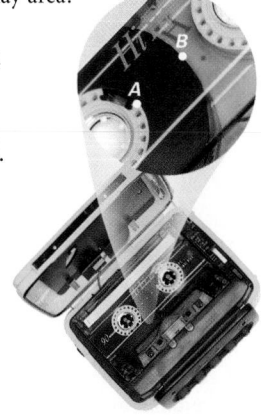

2. There are many different approximations for π. Two commonly used values are 3.14 and $\frac{22}{7}$. Compare these values with the value your calculator gives for π. Give a reason you might choose either 3.14 or $\frac{22}{7}$. Why is it necessary to estimate π when calculating the area and circumference of a circle?

3. When the cassette in the photo is rewinding, which moves faster, point A or point B? Explain your reasoning.

Guided Skills Practice

In Exercises 4–7, give your answers exactly, in terms of π, and rounded to two decimal places.

4. $6\pi \approx 18.85$

5. $25\pi \approx 78.54$

6. $25\pi \approx 78.54$

7. $196\pi \approx 615.75$

4. Find the circumference of a circle with a radius of 3. **(EXAMPLE 1)**

5. Find the circumference of a circle with a diameter of 25. **(EXAMPLE 1)**

6. Find the area of a circle with a radius of 5. **(EXAMPLE 2)**

7. Find the area of a circle with a diameter of 28. **(EXAMPLE 2)**

Practice and Apply

In Exercises 8–13, find the circumference and area of each circle.

Use 3.14 for π. Round your answers to the nearest tenth.

8. $r = 6$
   37.7; 113.0

9. $r = 10$
   62.8; 314

10. $d = 18$
    56.5; 254.3

Use $\frac{22}{7}$ for π. Leave your answers in fraction form.

11. $r = 6$
    $\frac{264}{7}, \frac{792}{7}$

12. $d = 21$
    66; $\frac{693}{2}$

13. $d = \frac{35}{8}$
    $\frac{55}{4}$; $\frac{1925}{128}$

Algebra

Find the radius of the circle with the given measurement. Give your answers exactly, in terms of π, and rounded to the nearest tenth.

14. $C = 12$ $\frac{6}{\pi} \approx 1.9$

15. $C = 62.8$ $\frac{31.4}{\pi} \approx 10.0$

16. $C = 50\pi$ 25

17. $A = 314$ $\sqrt{\frac{314}{\pi}} \approx 10.0$

18. $A = 50$ $\sqrt{\frac{50}{\pi}} \approx 4.0$

19. $A = 100\pi$ 10

Selected Answers
Exercises 4–7, 9–47 odd

ASSIGNMENT GUIDE

In Class	1–7
Core	8–19, 21–31 odd
Core Plus	15–31 odd, 33–38
Review	39–48
Preview	49–51

✎ Extra Practice can be found beginning on page 818.

Math
CONNECTION

ALGEBRA In Exercises 14–19, students use the formulas for the area and circumference of a circle to solve for the radius, r.

Technology

A calculator with a π key will be needed for the majority of the exercises in this lesson.

30. Yes; the area of the 18-inch pizza is more than three times the area of the 10-inch pizza, so the 18-inch pizza should feed at least 6 people.

20. 8π ≈ 25.13

21. 65π ≈ 204.20

22. 2.25π ≈ 7.07

23. 4π − 7.84 ≈ 4.73

24. 240 + 18π ≈ 296.55

25. 360 + 112.5π ≈ 713.43

26. 1200 − 300π ≈ 257.52

27. 56 − 8π ≈ 30.87

In Exercises 20–27, find the area of the shaded region. Give your answers exactly, in terms of π, and rounded to the nearest hundredth.

20.
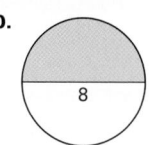

21.
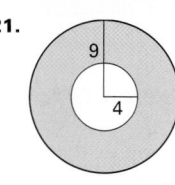

22.
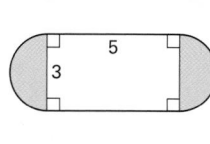

23.
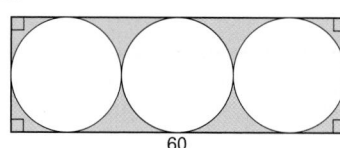

24.
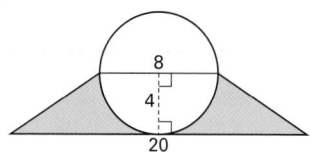

25.
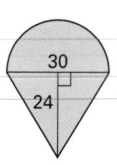

26.

27.

28. The circumference is multiplied by 2.

29. The area is multiplied by 4.

APPLICATIONS

31. The 18-inch pizza is the better deal. The 18-inch pizza gives about 17 in.² per dollar; while the 10-inch pizza gives about 16 in.² per dollar.

28. What happens to the circumference of a circle when the radius is doubled?

29. What happens to the area of a circle when the radius is doubled?

30. MEAL PLANNING If a 10-inch pizza is enough to feed 2 people, will an 18-inch pizza be enough to feed 6 people? Why or why not?

31. MEAL PLANNING If a 10-inch pizza costs $5 and an 18-inch pizza costs $15, which is the better deal? Explain your reasoning.

32. 21,895,644 ft²

32. IRRIGATION *Center pivot irrigation* is a method of agricultural irrigation using a long, wheeled arm with many nozzles that pivots about the center of a circle.

If the area inside the square is one square mile, what is the area, in square feet, of the irrigated circle? (Note: 1 mile = 5280 feet)

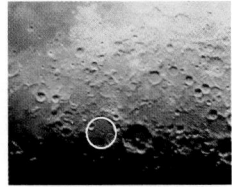

Differential gear

AUTOMOBILE ENGINEERING Tires are tested for traction by driving around a small circular track. Refer to the diagram below for Exercises 33–37.

33. What is the circumference of the circle formed by the inside tire tracks? **44 ft**

34. What is the circumference of the circle formed by the outside tire tracks? **78.5 ft**

35. Based on your answers to Exercises 33 and 34, what can you say about the speed of the inside tires compared with the speed of the outside tires?

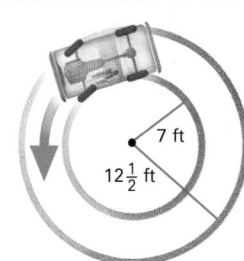

Suppose that the tires have a radius of 15 inches.

The tires of a car need to turn at different speeds on a curve. To make this possible, a device called a differential gear is used.

36. How many revolutions will the inside tires make in one lap around the circle? **5.6 revolutions**

37. How many revolutions will the outside tires make in one lap around the circle? **9.99 revolutions**

38. **ASTRONOMY** At left is a detail of the photo of the Moon from page 314. The diameter of the Moon in the photo is 12.2 cm. Use your own measurements and the fact that the actual diameter of the Moon is 1077 miles to estimate the actual diameter of the indicated crater. **44.14 mi**

Use your estimate of the diameter to find the approximate area of the crater. Is it closest in size to New York City (309 miles²), Los Angeles (469 miles²), Delaware (1955 miles²), Lake Superior (31,700 miles²), or Texas (261,914 miles²)? **approximately 1530 miles², closest in size to Delaware**

Detail of the Moon's surface, showing the Hipparchus Crater (circled)

 Look Back

Recall from algebra that a radical can be simplified by taking the square root of any factors that are perfect squares. For example:

$$2\sqrt{75} = 2\sqrt{25 \times 3} = 2 \times 5\sqrt{3} = 10\sqrt{3}$$

Simplify each expression below.

39. $3\sqrt{8}$ $6\sqrt{2}$

40. $16\sqrt{32}$ $64\sqrt{2}$

41. $3\sqrt{500}$ $30\sqrt{5}$

Find the positive solution for x.

42. $x^2 + 16 = 25$ $x = 3$

43. $x^2 + 144 = 169$ $x = 5$

44. $x^2 + 12.25 = 13.69$ $x = 1.2$

45. Given parallelogram $ABCD$ with diagonal $\overline{AC}$, prove that $\triangle ABC \cong \triangle CDA$. **(LESSON 4.5)**

46. Find the area of a triangle with a base of 9 in. and a height of 7 in. **(LESSON 5.2)** **31.5 in²**

47. Find the area of a parallelogram with a base of 5 cm and a height of 3.5 cm. **(LESSON 5.2)** **17.5 cm²**

48. Find the area of a trapezoid with a height of 3 cm and bases of 6 cm and 5 cm. **(LESSON 5.2)** **16.5 cm²**

Error Analysis

Understanding and visualizing the problems in Exercises 33–37 may be difficult for some students. You can use a bicycle wheel to demonstrate what happens to various spots on the wheel as it turns.

35. The inside tires do not have as far to go, so they are not moving as fast.

---

45. Possible proof:

Statements	Reasons
1. *ABCD* is a parallelogram	1. Given
2. $\overline{AB} \cong \overline{CD}$, $\overline{BC} \cong \overline{AD}$	2. Opposite sides of a parallelogram are $\cong$
3. $\overline{AC} \cong \overline{AC}$	3. Reflexive Property
4. $\triangle ABC \cong \triangle CDA$	4. SSS

Alternately, students can use $\angle B \cong \angle D$ (opposite angles of a parallelogram are congruent) in Step 3, and then prove that the triangles are congruent by SAS.

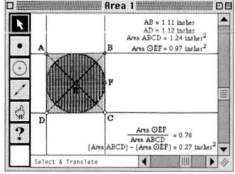

Look Beyond

In Exercises 49–51, students compare different configurations of circles with their respective geometric properties.

Portfolio Activity

The Portfolio Activity can be used as preparation for the Chapter Project or as a separate activity. In the Portfolio Activity on this page, students find the areas of irregular polygons drawn on graph paper by dividing the figures into rectangles and triangles.

PROOFS

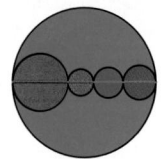

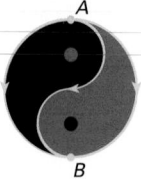

A

B

49. In the diagram at left, the centers of all the circles are collinear. Prove that the sum of the circumferences of the small circles is equal to the circumference of the large circle.

50. Show that the result you proved in Exercise 49 is true for any number of smaller circles.

51. **CULTURAL CONNECTION: ASIA** A *yin-yang* symbol is composed of circles and semicircles, as shown below. Yin and yang represent the two complementary forces, or principles, that make up all aspects and phenomena of life.

Which of the three indicated paths from point A to point B in the diagram is the longest? Explain your reasoning.

You can find the areas of some irregular polygons by dividing them into rectangles and triangles and adding the areas of these figures.

1. Find the areas of the figures below. The first one has been divided into two rectangles and a triangle to help you.

internet connect

Portfolio Extension

Go To: **go.hrw.com**
Keyword:
MG1 SquareFoot

2. Sometimes it may be easier to subtract areas than to add them. Find the areas of the figures below by subtracting areas from the area of a larger rectangle. The first rectangle has been drawn to help you.

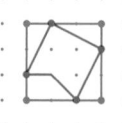 

WORKING ON THE CHAPTER PROJECT

You should now be able to complete Activity 1 of the Chapter Project.

49. Let C_L = circumference of large circle
d_L = diameter of large circle
$$C_L = \pi d_L = \pi(d_1 + d_2 + d_3 + d_4)$$
$$= \pi d_1 + \pi d_2 + \pi d_3 + \pi d_4$$
$$= C_1 + C_2 + C_3 + C_4$$

50. $C_L = \pi d_L = \pi(d_1 + d_2 + d_3 + \ldots + d_n)$
$$= \pi d_1 + \pi d_2 + \pi d_3 + \ldots + \pi d_n$$
$$= C_1 + C_2 + C_3 + \ldots + C_n$$

51. Path AB on the outer circle is $\frac{1}{2}$ the circumference of the outer circle:
$$AB_0 = \frac{1}{2}\pi d = \frac{\pi d}{2}$$
Path AB through the interior is the circumference of a circle with $\frac{1}{2}$ the diameter:
$$AB_1 = \pi\left(\frac{d}{2}\right) = \frac{\pi d}{2}$$
All three paths are the same length.

The Pythagorean Theorem

5.4

Objectives

● Identify and apply the Pythagorean Theorem and its converse.

● Solve problems by using the Pythagorean Theorem.

Why *A 4000-year-old clay tablet from ancient Babylon—in what is now Iraq—revolutionized our knowledge of ancient mathematics.*

When the tablet known as Plimpton 322 was first found, no one understood the significance of the strange columns of numbers—until a mathematician who looked at it made an exciting discovery.

Plimpton 322

CULTURAL CONNECTION: ASIA At the height of its power, according to Greek historian Herodotus (485-427 B.C.E.), Babylon was the world's most splendid city. It was surrounded by walls almost 85 feet thick with eight bronze gates. The main gate and its walls were decorated with figures composed of glazed colored brick that depicted dragons, lions, and bulls.

The tablet in the illustration above, known as Plimpton 322, comes from a much earlier period known as the Old Babylonian Empire, which included the reign of King Hammurabi, who ruled from 1792 to 1750 B.C.E. and became famous for his wise and fair code of laws.

Plimpton 322 is a piece of a larger tablet. Part of the larger tablet, including at least one column of numbers, has broken off and is lost. On the part that remains, there are columns of numbers, including the two columns shown on the next page, with four apparent errors corrected.

QUICK **WARM-UP**

Evaluate. Round your answers to the nearest hundredth.

1. $\sqrt{72}$		8.49
2. $\sqrt{10}$		3.16
3. $\sqrt{120}$		11.18
4. $\sqrt{6^2 + 4^2}$		7.21
5. $\sqrt{14^2 - 5^2}$		13.08

Also on Quiz Transparency 5.4

Why The Plimpton tablet reveals that the Babylonians knew the Pythagorean Theorem more than a thousand years before Pythagoras. Further study of the tablet in the Chapter 10 Project reveals that Babylonian mathematics was highly sophisticated.

For a student worksheet of this Activity and detailed Teacher Notes, see page 91 in the Lesson Activities booklet.

Cooperative Learning

Have students work in groups to draw the *Chou pei suan ching* diagram in varying sizes on graph paper. They should compute the areas of the five parts of each diagram in order to verify that the Pythagorean Theorem is true for a variety of right triangles.

Teaching Tip

Students may be interested in researching recent efforts to prove "Fermat's last theorem". In this theorem, Fermat stated that there are no whole numbers x, y, and z such that $x^n + y^n = z^n$ for any value of n greater than 2. For example, there are no whole numbers that satisfy $x^3 + y^3 = z^3$.

Activity
Solving the Puzzle

YOU WILL NEED

calculator

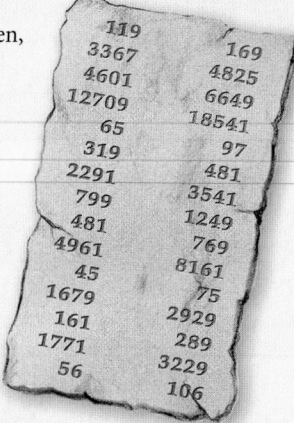

1. Work in pairs. Each person should pick two numbers at random from 50 to 5000. Use a calculator to square each number. Subtract the smaller square from the larger square. Take the square root of the difference. Is the result an integer?

2. Repeat Step 1 at least five times. How often, if ever, did you obtain an integer as the result?

3. Square the numbers in each row of the tablet at right. Subtract the smaller square from the larger square and take the square root of the difference. How often is the result an integer?

4. Do you think the Babylonians knew which sets of integers were related in this way? Explain.

Columns II and III of Plimpton 322

119	169
3367	4825
4601	6649
12709	18541
65	97
319	481
2291	3541
799	1249
481	769
4961	8161
45	75
1679	2929
161	289
1771	3229
56	106

In the Activity, you discovered that Babylonians knew about the sets of numbers now called **Pythagorean triples**—that is, sets of positive integers a, b, and c such that $a^2 + b^2 = c^2$.

It is also clear, from other evidence on the tablet, that the Babylonians knew that right triangles have the property known today as the *Pythagorean relationship*. But surprisingly, Plimpton 322 was created over a thousand years before the teacher Pythagoras (569–500 B.C.E.), the person to whom the relationship is traditionally attributed.

At present there is no direct evidence that the Babylonians could prove the Pythagorean relationship.

Proving the Relationship for Right Triangles

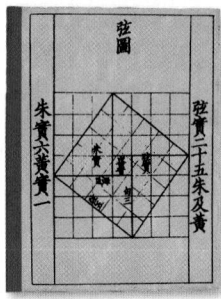

An ancient Chinese diagram.

CULTURAL CONNECTION: EUROPE The Pythagoreans were members of a secret society of followers of Pythagoras in ancient Greece. According to tradition, Pythagoras said, "Number rules the universe." We do not know how Pythagoras actually proved the theorem that now bears his name, or whether his particular proof—there are many others—originated with him.

CULTURAL CONNECTION: ASIA One very early illustration of the Pythagorean Theorem is found in an ancient Chinese source known as *Chou pei suan ching*. The source does not give details of a proof, but it does include the diagram shown at left, which suggests a proof of the theorem (see page 323). According to some scholars, the date of the diagram is at least as early as Pythagoras.

Interdisciplinary Connection

LANGUAGE ARTS In science-fiction stories and movies, people who try to prove to another life-form that humans are an intelligent species often use a diagram of the Pythagorean Theorem. Ask students to give reasons for or against using this theorem as a proof of human intelligence. What other proofs might be used?

PARAGRAPH PROOF

In outer part of the *Chou pei suan ching* diagram, four congruent right triangles form a large square with a smaller square in the center. The area of the larger square can be found by squaring the length of its sides, which are equal to $a + b$, or by adding the individual pieces that make up the figure.

By setting these two expressions equal to each other and simplifying, you obtain the famous result.

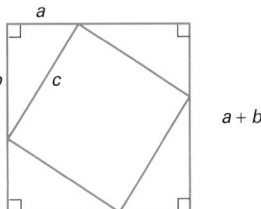

$$(a + b)^2 = 4\left(\tfrac{1}{2}ab\right) + c^2$$
$$a^2 + 2ab + b^2 = 2ab + c^2$$
$$a^2 + b^2 = c^2$$

CRITICAL THINKING   Does the shape of the right triangles matter in the proof above? How do you know that the central figure is in fact a square? (See Exercise 39 for another proof of the Pythagorean Theorem based on the *Chou pei suang ching* diagram.)

Pythagorean Theorem

For any right triangle, the sum of the squares of the lengths of the legs is equal to the square of the length of the hypotenuse.

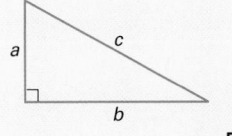

$$a^2 + b^2 = c^2$$

5.4.1

EXAMPLE ❶

APPLICATION
SAFETY

The following is a rule of thumb for safely positioning a ladder: The distance from bottom of the ladder to the wall should be one-fourth of the length of the ladder. Thus, the bottom of a 16-foot ladder should be 4 feet from the wall. How far up the wall will the ladder reach?

● **SOLUTION**
$$a^2 + b^2 = c^2$$
$$4^2 + b^2 = 16^2$$
$$16 + b^2 = 256$$
$$b^2 = 256 - 16$$
$$b^2 = 240$$
$$b = \sqrt{240} \approx 15.5 \text{ ft}$$

16 ft

4 ft

Enrichment

A text from the Old Babylonian period includes a diagram of a square and its diagonals. The number 30 is written along one side; the number 42.2535 appears along a diagonal. Explain the meanings of the numbers. **42.2535 is an approximation of the length of the diagonal and shows that the text writer had some knowledge of the Pythagorean Theorem. A calculator gives 42.4264 as an approximate value for the diagonal of a square with a side length of 30.**

CRITICAL THINKING
No, but the right triangles must be congruent. All the sides of the central figure are congruent, and the angles are right angles because they are composed of the two acute angles of the basic right triangle (see the *Chou pei suan ching* diagram on page 322.)

Teaching Tip

TECHNOLOGY Geometry graphics software may be used to explore the Critical Thinking question on page 323. Students will need to construct the figure in such a manner that they can enlarge and diminish the central square.

ADDITIONAL
EXAMPLE ❶

The legs of a right triangle are 15 cm and 34 cm. Find the length of the hypotenuse. ≈ 37.16 cm

The Converse of the Theorem

The converse of the Pythagorean Theorem is also a true theorem. It is useful for proving that two segments or lines are perpendicular.

Converse of the Pythagorean Theorem

If the square of the length of one side of a triangle equals the sum of the squares of the lengths of the other two sides, then the triangle is a right triangle.

5.4.2

CULTURAL CONNECTION: AFRICA The following proof of the Converse of the Pythagorean Theorem is taken directly from Euclid's *The Elements*, Book 1, which was translated from the original Greek by Thomas Heath. You will notice that some of Euclid's notation and use of terms are different from our own. Exercises 52–64 will help you follow the proof and put it into a modern form.

PROPOSITION 48

If in a triangle the square on one of the sides be equal to the squares on the remaining two sides of the triangle, the angle contained by the remaining two sides of the triangle is right.

For in the triangle *ABC* let the square on one side *BC* be equal to the squares on the sides *BA*, *AC*;
  I say that the angle *BAC* is right.
For let *AD* be drawn from the point *A* at right angles to the straight line *AC*, let *AD* be made equal to *BA*, and let *DC* be joined.
  Since *DA* is equal to *AB*, the square on *DA* is also equal to the square on *AB*.
  Let the square on *AC* be added to each;
  therefore the squares on *DA*, *AC* are equal to the squares on *BA*, *AC*.
  But the square on *DC* is equal to the squares on *DA*, *AC*, for the angle *DAC* is right;   [1.47]
and the square on *BC* is equal to the squares on *BA*, *AC*, for this is the hypothesis;
  therefore the square on *DC* is equal to the square on *BC*,
  so that the side *DC* is also equal to *BC*.
  And, since *DA* is equal to *AB*, and *AC* is common,
  the two sides *DA*, *AC* are equal to the two sides *BA*, *AC*;
and the base *DC* is equal to the base *BC*;
  therefore the angle *DAC* is equal to the angle *BAC*.   [1.8]
But the angle *DAC* is right;
  therefore the angle *BAC* is also right.
Therefore etc.   Q.E.D.

> The bracketed numbers in the right margin refer to propositions proven earlier in The Elements.

Inclusion Strategies

HANDS-ON STRATEGIES The squares at right can be used to form the sides of a right triangle. Have students decide which two squares to use for the legs and which one for the hypothenuse of the right triangle. By cutting and reassembling the squares for the legs, they should be able to show that the combined area of the squares used for the legs equals the area of the square used for the hypothenuse.

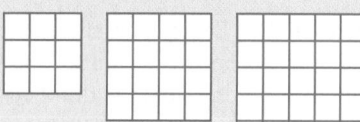

EXAMPLE ❷ A plowed field is in the shape of a triangle. If the sides have the lengths shown in the figure, is the field a right triangle?

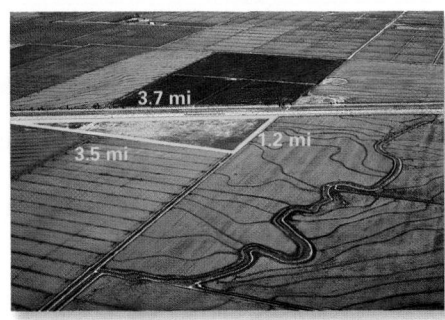

3.7 mi
1.2 mi
3.5 mi

● **SOLUTION**

If the field is a right triangle, then $a^2 + b^2 = c^2$, where a, b, and c are the lengths of the sides. The longest side of a right triangle is the hypotenuse, so if the field is a right triangle, the hypotenuse, c, must be 3.7.

$a^2 + b^2 = (1.2)^2 + (3.5)^2$
$\quad\quad\quad = 13.69$

Since $c = \sqrt{13.69} = 3.7$ miles, the field is a right triangle.

The following helpful inequalities can be derived from the Pythagorean relationship:

Pythagorean Inequalities

For $\triangle ABC$, with c as the length of the longest side:

If $c^2 = a^2 + b^2$, then $\triangle ABC$ is a right triangle.
If $c^2 > a^2 + b^2$, then $\triangle ABC$ is an obtuse triangle.
If $c^2 < a^2 + b^2$, then $\triangle ABC$ is an acute triangle.

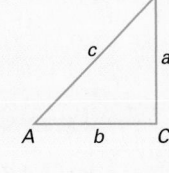

5.4.3

EXAMPLE ❸ A triangle has side lengths of 7 inches, 8 inches, and 12 inches. Is the triangle right, obtuse, or acute?

● **SOLUTION**

$12^2 \overset{?}{\underline{}} 7^2 + 8^2$
$144 > 113$

Therefore, the triangle is obtuse.

TRY THIS A triangle has side lengths of 8 inches, 8 inches, and 8 inches. Is the triangle right, obtuse, or acute?

Reteaching the Lesson

TECHNOLOGY Have students work in pairs or small groups. They should use spreadsheet software to create their own list of Pythagorean triples. Either of the formulas given in Exercises 31 and 32 may be used to generate the data. Students should use graph paper or geometry graphics software to draw triangles that match the triples.

Teaching Tip

Given two real numbers, one of the following three situations must occur: they are equal, the first is less that the second, or the first is greater than the second. This is called the Property of Trichotomy. Similarly, a triangle must be right, obtuse, or acute. There are no other possibilities.

TRY THIS
$8^2 < 8^2 + 8^2$, so the triangle is acute.

ASSIGNMENT GUIDE

In Class	1–7
Core	8–13, 15–33 odd
Core Plus	19–37 odd
Review	43–48
Preview	49–64

✐ Extra Practice can be found beginning on page 818.

Error Analysis

Students using calculators may inadvertently follow the wrong order of operations. Grouping symbols, exponents, and radicals tend to create errors during data entry. You may wish to have students record intermediate steps in pencil to prevent these errors.

8. 5

10. $\sqrt{7445}$

12. $4\sqrt{130}$

Math
CONNECTION

ALGEBRA In Exercises 8–13, students will need to solve the Pythagorean Theorem for a, b, or c. In Exercises 18–23, students will need to write an inequality involving a, b, and c to determine whether the triangles given are acute, obtuse, or right. In Exercises 43–45, students will multiply radicals to rationalize the denominator.

Exercises

● Communicate

1. State the Pythagorean Theorem in your own words.

2. What are some practical uses of the Pythagorean Theorem and its converse?

3. Explain how something called the "3-4-5 rule" could help carpenters create square corners.

4. Explain how the Greek postage stamp shown at right illustrates the Pythagorean Theorem.

Activities Online
Go To: go.hrw.com
Keyword:
MG1 Theorem

● Guided Skills Practice

5. A right triangle has one leg with a length of 48 and a hypotenuse with a length of 80. What is the length of the other leg? *(EXAMPLE 1)* 64

6. A triangle has side lengths of 7, 10, and 12. Is the triangle a right triangle? *(EXAMPLE 2)* No

7. A triangle has side lengths of 8, 15, and 18. Is the triangle right, acute, or obtuse? *(EXAMPLE 3)* obtuse

● Practice and Apply

Algebra

Homework Help Online
Go To: go.hrw.com
Keyword:
MG1 Homework Help
for Exercises 8–30

For Exercises 8–13, two side lengths of a right triangle are given. Find the missing side length. Leave your answers in radical form.

8. $a = 3, b = 4, c = \underline{\ ?\ }$ 9. $a = 10, b = 15, c = \underline{\ ?\ }$ $5\sqrt{13}$

10. $a = 46, b = 73, c = \underline{\ ?\ }$ 11. $a = \underline{\ ?\ }, b = 6, c = 8$ $2\sqrt{7}$

12. $a = 27, b = \underline{\ ?\ }, c = 53$ 13. $a = 1, b = 1, c = \underline{\ ?\ }$ $\sqrt{2}$

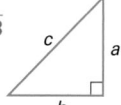

Find the perimeter of each triangle. Round your answers to the nearest tenth.

14. 12 units

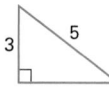

15. 129.5 units

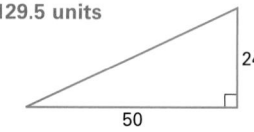

16. 48 units

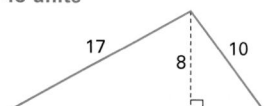

17. 63 units

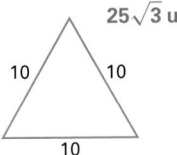
Algebra

Each of the following triples represents the side lengths of a triangle. Determine whether the triangle is right, acute, or obtuse.

18. 5, 9, 12 obtuse   **19.** 13, 15, 17 acute   **20.** 7, 24, 25 right

21. 7, 24, 26 obtuse   **22.** 3, 4, 5 right   **23.** 25, 25, 30 acute

Find the area of each figure. Leave your answers in radical form.

24. $25\sqrt{3}$ units²

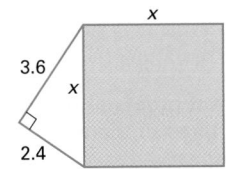

25. 72 units²

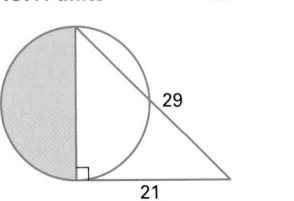

7.07 units   **26.** What is the length of a diagonal of a square with a side length of 5?

27. What is the side length of a square with a diagonal of 16? 11.31 units

Find the area of the shaded region in each figure. Round your answers to the nearest tenth.

28. 18.7 units²   **29.** 157.1 units²   **30.** 16.8 units²

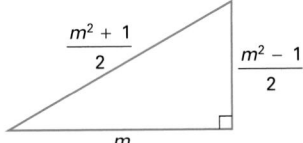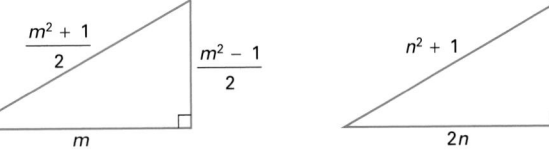

CONNECTIONS

NUMBER THEORY Mathematicians have long been fascinated with techniques for generating Pythagorean triples. For each method below, generate five sets of triples. Then use algebra to show that the method will always work.

31. Method of Pythagoreans
Let m be any odd number greater than 1.

$$\frac{m^2 + 1}{2} \qquad \frac{m^2 - 1}{2}$$
$$m$$

32. Method of Plato
Let n be any integer greater than 1.

$$n^2 + 1 \qquad n^2 - 1$$
$$2n$$

33. NUMBER THEORY Test the following conjecture for 10 different Pythagorean triples: In each Pythagorean triple, at least one of the numbers is divisible by 3, and at least one is divisible by 5. **Answers will vary.**

34. NUMBER THEORY If you multiply each number of a Pythagorean triple by the same constant, you get another Pythagorean triple. For example, (3, 4, 5) is a Pythagorean triple. Multiplying each number by 2 results in (6, 8, 10), another Pythagorean triple.

Use algebra to show that if (x, y, z) is a Pythagorean triple, then (ax, ay, az) is also a Pythagorean triple for any positive integer a.

31. Triples: 3, 4, 5   5, 12, 13
7, 24, 25   9, 40, 41
11, 60, 61

Algebraic proof:

$$m^2 + \left(\frac{m^2 - 1}{2}\right)^2$$
$$= m^2 + \left(\frac{m^4 - 2m^2 + 1}{4}\right)$$
$$= \frac{4m^2 + m^4 - 2m^2 + 1}{4}$$
$$= \frac{m^4 + 2m^2 + 1}{4}$$
$$= \frac{(m^2 + 1)^2}{4}$$
$$= \left(\frac{m^2 + 1}{2}\right)^2$$

Since the sum of the squares of the lengths of two sides equals the square of the longest side, any triple generated with this method will represent the sides of a right triangle.

32. Triples: 3, 4, 5   6, 8, 10
8, 15, 17   10, 24, 26
12, 35, 37

Algebraic proof:
$$(2n)^2 + (n^2 - 1)^2$$
$$= 4n^2 + n^4 - 2n^2 + 1$$
$$= n^4 + 2n^2 + 1$$
$$= (n^2 + 1)^2$$

Since the sum of the squares of the lengths of two sides equals the square of the longest side, any triple generated with this method will represent the sides of a right triangle.

34. Assume (x, y, z) is a Pythagorean triple. Therefore, $x^2 + y^2 = z^2$. Multiply both sides of the equation by a^2 to get: $a^2x^2 + a^2y^2 = a^2z^2$ $\Rightarrow (ax)^2 + (ay)^2 = (az)^2$. Therefore (ax, ay, az) is a Pythagorean triple.

35. Let $AB = a$, $AR = EH = b$, and $BR = c$. Then area $ABCD = a^2$, area $EFGH = b^2$, and area $BRST = c^2$. Since triangle BAR is a right triangle, $a^2 + b^2 = c^2$, showing that $BRST$ is the square whose area is equal to the sum of the areas of the two given squares.

36. Construct the new square so that the diagonal of the given square is the side of the new square, as shown.

The area of $BDEF$ is twice the area of $ABCD$.

35. CULTURAL CONNECTION: ASIA

The *Sulbasutras* were ancient Indian mathematical manuals for the design and construction of Vedic altars.

The diagram at right, from the *Sulbasutras*, demonstrates a method for constructing a square with an area equal to the sum of the areas of two given squares. Use the Pythagorean Theorem to prove that this construction works.

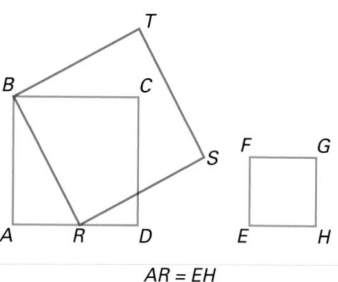

$AR = EH$

Given: $ABCD$, $EFGH$, and $BRST$ are squares.
$AR = EH$

Prove: area of $BRST$ = area of $ABCD$ + area of $EFGH$

36. Explain how to use the method from the *Sulbasutras* to construct a square with twice the area of a given square.

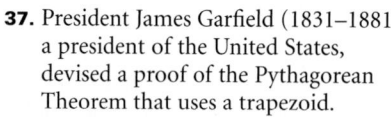
There are over 100 different proofs of the Pythagorean Theorem. Exercises 37–39 present three of them.

37. President James Garfield (1831–1881), a president of the United States, devised a proof of the Pythagorean Theorem that uses a trapezoid.

Using the figure at right, write two different expressions for the area of the trapezoid. Set the two expressions equal to each other to discover Garfield's proof.

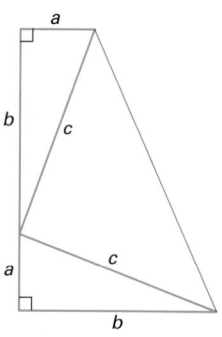

38. Explain how the diagram below is a "proof without words" of the Pythagorean Theorem. You may wish to draw the figures on a separate piece of paper and cut them into pieces to help explain the proof.

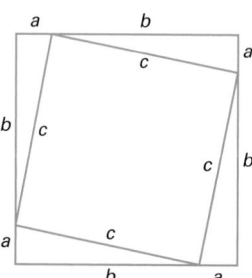
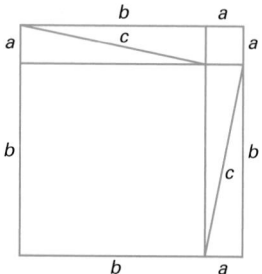

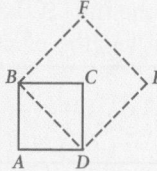
37. Use the formula for the area of a trapezoid, $A = \frac{1}{2}h(b_1 + b_2) = \frac{1}{2}(a + b)(a + b) = \frac{1}{2}(a + b)^2$. Find the sum of the area of the three right triangles, $A = \frac{1}{2}ab + \frac{1}{2}ab + \frac{1}{2}c^2 = ab + \frac{1}{2}c^2$. Then set the areas equal to one another and simplify.

$$\frac{1}{2}(a + b)^2 = ab + \frac{1}{2}c^2$$
$$(a + b)^2 = 2ab + c^2$$
$$a^2 + b^2 + 2ab = 2ab + c^2$$
$$a^2 + b^2 = c^2,$$

which is the Pythagorean Theorem.

38. Both squares have the same area, which is $(a + b)^2$. In the first square, after the triangles are removed, a square with side c and area c^2 remains. In the second square, after the triangles are removed, two squares, with sides a and b, and a total area $a^2 + b^2$, remain. Since the areas of the triangles are the same in both squares, $c^2 = a^2 + b^2$.

39. The diagram from the *Chou pei suan* suggests at least two different proofs of the Pythagorean Theorem.

a. Write an expression for c^2, the area of square *EFGH*, by adding the areas of the four inner right triangles and square *IJKL*.

$$c^2 = \underline{\ ?\ }$$

b. Write an expression for c^2, the area of square *EFGH*, by subtracting the areas of the four outer right triangles from the area of square *ABCD*.

$$c^2 = \underline{\ ?\ }$$

Simplify each equation to prove the Pythagorean Theorem.

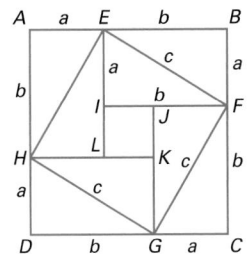

40. **PUBLIC SAFETY** If the base of the ladder in the photo at right is 8 feet off the ground, how far up the wall will the ladder reach? **43.6 ft**

41. **TRAVEL** Starting from his house, Jesse drives north 6 miles and then turns east and drives 2 miles. He turns north again and drives 4 miles and then turns east and drives 7 miles. How far is he from his house? (Hint: You may wish to draw a map.) **13.5 miles**

42. **SPORTS** A baseball diamond is a square with 90-foot sides. What is the approximate distance of a catcher's throw from home plate to second base? **127.3 ft**

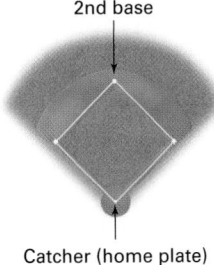

2nd base

Catcher (home plate)

37 ft

10 ft

Look Back

Algebra

Recall from algebra that a fraction with a radical in the denominator can be simplified by multiplying by a fraction equal to 1. This technique is known as *rationalizing the denominator*. For example:

$$\frac{5}{\sqrt{2}} = \frac{5}{\sqrt{2}} \times \frac{\sqrt{2}}{\sqrt{2}} = \frac{5\sqrt{2}}{\sqrt{2} \times \sqrt{2}} = \frac{5\sqrt{2}}{2}$$

Rationalize the denominator in each expression below.

43. $\frac{4}{\sqrt{2}}$ $2\sqrt{2}$

44. $\frac{13}{\sqrt{3}}$ $\frac{13\sqrt{3}}{3}$

45. $\frac{\sqrt{7}}{\sqrt{5}}$ $\frac{\sqrt{35}}{5}$

For Exercises 47–48, classify each statement as true or false. If true, explain why. If false, give a counterexample. *(LESSON 4.6)*

46. Every rhombus is a rectangle.

47. Every rhombus is a parallelogram.

48. If a diagonal divides a quadrilateral into two congruent triangles, then the quadrilateral is a parallelogram.

39. a. The area of each inner right triangle $= \frac{1}{2}ab$ and, by segment subtraction, the side of square *IJKL* is $(b - a)$. By the Sum of Areas postulate, area *EFGH* = area *IJKL* + 4 × (area of one inner right triangle). Algebraically,
$$c^2 = (b - a)^2 + 4\left(\frac{1}{2}ab\right) = (b - a)^2 + 2ab$$

b. The area of each outer right triangle $= \frac{1}{2}ab$ and the side of square *ABCD* = $(a + b)$. By the Sum of Areas postulate, area *ABCD* = 4 × (area of one outer right triangle) + area *EFGH*. Thus area *EFGH* = area *ABCD* − 4 × (area of

one outer right triangle). Algebraically,
$$c^2 = (a + b)^2 - 4\left(\frac{1}{2}ab\right) = (a + b)^2 - 2ab$$

Simplifying each:

Part a.
$$c^2 = (b - a)^2 + 2ab$$
$$= b^2 - 2ab + a^2 + 2ab$$
$$= b^2 + a^2$$
So $c^2 = a^2 + b^2$

Part b.
$$c^2 = (a + b)^2 - 2ab$$
$$= a^2 + 2ab + b^2 - 2ab$$
$$= a^2 + b^2$$
So $c^2 = a^2 + b^2$

In Exercises 49–64, students examine Euclid's proof of the converse of the Pythagorean Theorem. They are asked to draw a diagram and complete a flow-chart proof.

49. Hypothesis: In a triangle, the square of one of the sides is equal to the sum of the squares of the remaining two sides of the triangle. Conclusion: The angle contained by the remaining two sides of the triangle is right.

50.

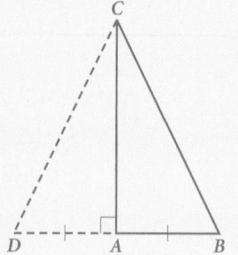

 *Look Beyond*

In Exercises 49–64, you will examine Euclid's proof of the Converse of the Pythagorean Theorem. Read through Euclid's proof on page 324.

49. What is the hypothesis of Proposition 48? What is the conclusion?

50. Draw a triangle and label the vertices A, B, and C. Use Euclid's directions to construct the triangle △DAC as in the proof.

51. In your diagram for Exercise 50, can you be sure that the figure connecting D to B is really a single segment? (Hint: Don't assume what you are trying to prove.) **No**

FLOWCHART PROOF

Complete the flowchart proof based on Euclid's proof of the Converse of the Pythagorean Theorem.

Given: $(BA)^2 + (AC)^2 = (BC)^2$

Prove: △ABC is right.

Proof:

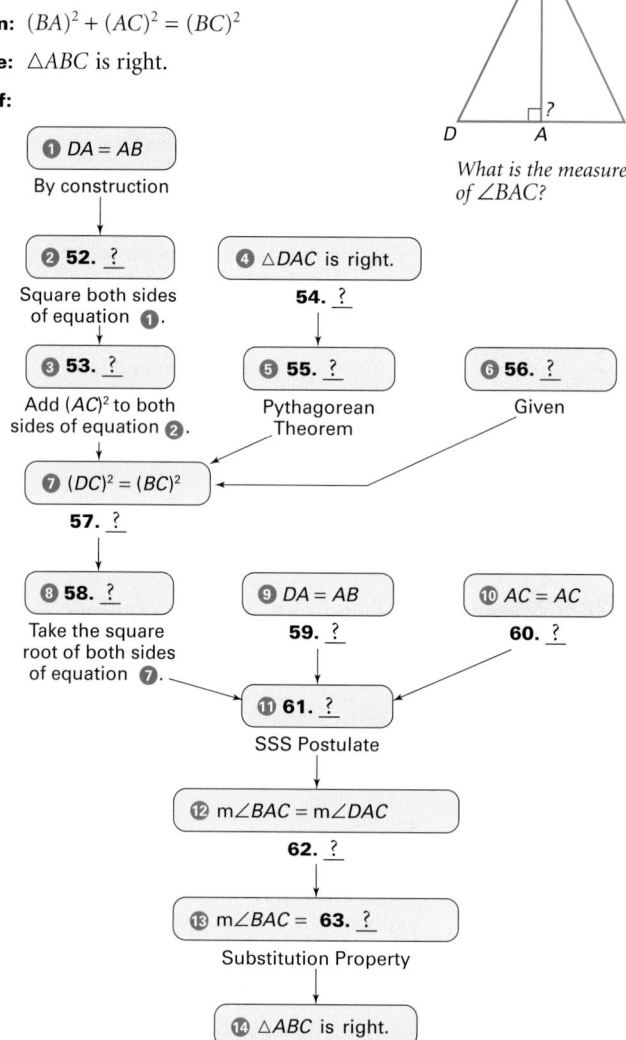

What is the measure of ∠BAC?

52. $(DA)^2 = (AB)^2$

53. $(DA)^2 + (AC)^2 = (AB)^2 + (AC)^2$

54. ∠DAC is a right angle, by construction

55. $(DA)^2 + (AC)^2 = (DC)^2$

56. $(BA)^2 + (AC)^2 = (BC)^2$

57. Substitution Property applied to equations 3, 5, and 6.

58. $DC = BC$

59. By construction

60. Reflexive Property

61. △DAC ≅ △BAC

62. CPCTC

63. 90°

64. Definition of a right triangle

Special Triangles and Areas of Regular Polygons

Why *The traditional tools of mechanical drawing include a T square and two special triangles. One of the triangles has angles measuring 30°, 60°, and 90°. The other has angles measuring 45°, 45°, and 90°. The properties of these triangles make them especially useful in geometry as well as in drawing.*

Objectives

● Identify and use the 45-45-90 Triangle Theorem and the 30-60-90 Triangle Theorem.

● Identify and use the formula for the area of a regular polygon.

This draftsman uses her T square to establish an imaginary line parallel to the base of her page. Then, using a special triangle, she draws a segment at a 30° angle to this line.

45-45-90 Triangles

If you draw a diagonal of a square, two congruent isosceles triangles are formed. Because the diagonal is the hypotenuse of a right triangle, its length can be found by using the Pythagorean Theorem.

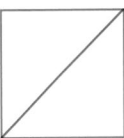

EXAMPLE **1** Use the Pythagorean Theorem to find the length of the hypotenuse of $\triangle ABC$. What is the ratio of the hypotenuse to a leg?

● **SOLUTION**

$$h^2 = 10^2 + 10^2 = 200$$
$$h = \sqrt{200} = \sqrt{100} \times \sqrt{2} = 10\sqrt{2}$$

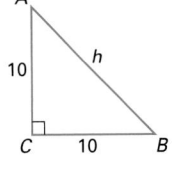

The ratio of the hypotenuse to a leg is $\frac{10\sqrt{2}}{10}$, or $\sqrt{2}$.

CRITICAL THINKING   What is the length of a diagonal of a square with a side of length *s*? What is the ratio of the diagonal to the side?

Alternative Teaching Strategy

TECHNOLOGY Explain to students that all 45-45-90 triangles are similar and all 30-60-90 triangles are similar. Have students use geometry graphics software to create sets of similar triangles. Students should discover that the ratios between corresponding parts of similar triangles remain unchanged.

Prepare

QUICK *WARM-UP*

Find the values for *a* and *b* in each radical expression.

1. $\sqrt{32} = \sqrt{a} \cdot \sqrt{2} = b\sqrt{2}$
   $a = 16, b = 4$

2. $\sqrt{54} = \sqrt{a} \cdot \sqrt{6} = b\sqrt{6}$
   $a = 9, b = 3$

3. $\sqrt{45} = \sqrt{a} \cdot \sqrt{5} = b\sqrt{5}$
   $a = 9, b = 3$

Also on Quiz Transparency 5.5

Teach

Why Ask students why 30-60-90 and 45-45-90 triangles are useful in architecture and construction. Students may point out that they are easy to construct because a 30-60-90 triangle is half of an equilateral triangle and a 45-45-90 triangle is half of a square.

CRITICAL THINKING
$\sqrt{2} \cdot s$; $\sqrt{2}$

Additional

EXAMPLE ①

A right triangle is also isosceles. The length of each leg is 5 cm. Find the length of the hypotenuse. What is the ratio of the hypotenuse to a leg? $5\sqrt{2}$; $\sqrt{2}$

Teaching Tip

Throughout the lesson, you may wish to let students write answers in the form $a\sqrt{b}$. If so, remind them that the simplest form of such expressions requires that b have no factors that are perfect squares and that the denominator, if any, does not include a radical expression.

Additional

EXAMPLE ②

Triangle ABC is a 30-60-90 triangle. ∠C is a right angle, and ∠A is 30°. If the length of $\overline{AC}$ is $6\sqrt{3}$ units, what is the length of the other leg, $\overline{CB}$, and of the hypotenuse, $\overline{AB}$?

Leg $\overline{CB}$ is 6 units; hypotenuse $\overline{AB}$ is 12 units.

Math
CONNECTION

ALGEBRA Remind students how to write $\sqrt{75}$ in simplest form. They should look for factors of 75 that are perfect squares. In this case, $75 = 25 \cdot 3$, so $\sqrt{75} = \sqrt{25 \cdot 3} = \sqrt{5^2 \cdot 3} = 5\sqrt{3}$

Notice that a diagonal of a square forms a right triangle with two 45° base angles. This triangle is known as a **45-45-90 triangle**. Because this is a right triangle, the hypotenuse can be found by applying the Pythagorean Theorem.

> ### 45-45-90 Triangle Theorem
>
> In any 45-45-90 triangle, the length of the hypotenuse is $\sqrt{2}$ times the length of a leg.   **5.5.1**

30-60-90 Triangles

If you draw an altitude of an equilateral triangle, two congruent right triangles are formed. The measures of the acute angles of each right triangle are 30° and 60°. This triangle is known as a **30-60-90 triangle**. The length of the hypotenuse of a 30-60-90 triangle is twice the length of the shorter leg.

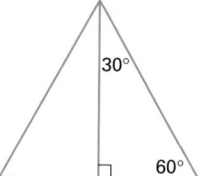

EXAMPLE ② Find the unknown lengths for the 30-60-90 triangle shown at right.

● **SOLUTION**

In a 30-60-90 triangle, the length of the hypotenuse is twice the length of the shorter leg. Thus, the length of the hypotenuse is 10.

Algebra

Use the Pythagorean Theorem to find the length of the other leg.

$$5^2 + x^2 = 10^2$$
$$25 + x^2 = 100$$
$$x^2 = 100 - 25$$
$$x^2 = 75$$
$$x = \sqrt{75} = 5\sqrt{3} \approx 8.66$$

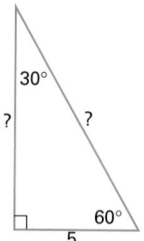

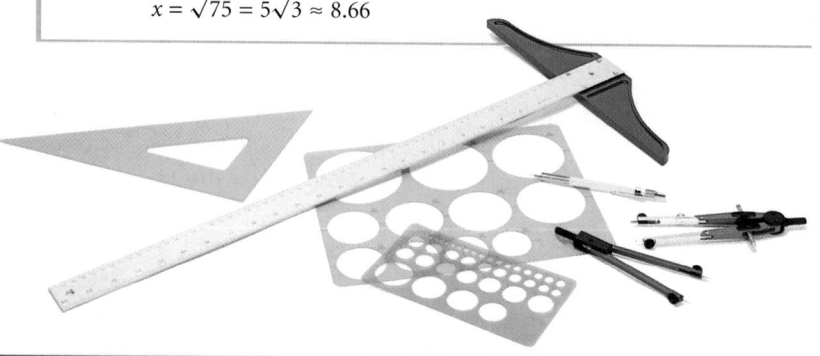

Enrichment

Patterns involving sets of special triangles can be created by drawing successive altitudes from the right angle of 45-45-90 and 30-60-90 triangles.

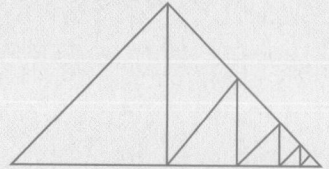

Have students work together to explore variations of these patterns. The patterns can be created for the aesthetic appeal alone or you can ask students to investigate the relationships among the dimensions of all the line segments drawn inside the triangle.

30-60-90 Triangles

1. Use the Pythagorean Theorem to fill in a table like the one below for several 30-60-90 triangles. Write your answers in simplest radical form.

Shorter leg	Hypotenuse	Longer leg
1	2	?
2	?	?
3	?	?

2. Look for a pattern in the lengths of the longer leg and make a generalization.

3. Let x be the length of the shorter leg.

   a. What is the length of the hypotenuse?

   b. What is the length of the longer leg?

CHECKPOINT ✔ 4. Use your results to label a general 30-60-90 triangle like the one at right. Then complete the theorem below.

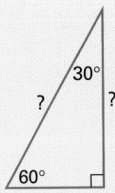

30-60-90 Triangle Theorem

In any 30-60-90 triangle, the length of the hypotenuse is ___?___ times the length of the shorter leg, and the length of the longer leg is ___?___ times the length of the shorter leg.

5.5.2

EXAMPLE 3 Jake is measuring the height of a tree that his grandfather planted as a boy. Jake uses a special instrument to find a spot where a 30° angle is formed by his line of sight and a ray parallel to the ground. Jake's eye level is 5 feet above the ground. How tall is the tree if Jake is standing 80 feet from the base?

SOLUTION

The line of sight to the tree is the hypotenuse of a right triangle. The length of the longer leg of the triangle is 80 feet. The length, x, of the shorter leg of the triangle is equal to the height of the tree above Jake's eye level.

Algebra

$$80 = x\sqrt{3}$$

$$x = \frac{80}{\sqrt{3}} = \frac{80}{\sqrt{3}} \times \frac{\sqrt{3}}{\sqrt{3}} = \frac{80\sqrt{3}}{3}, \text{ or about 46.2 feet}$$

Now find the height, h, of the tree.

$$h \approx 46.2 + 5 = 51.2, \text{ or about 51 feet}$$

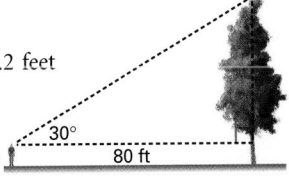
30°
5 ft   80 ft

Inclusion Strategies

HANDS-ON STRATEGIES Allow students to draw and measure examples of 45-45-90 triangles and 30-60-90 triangles to confirm that the ratios developed in the text are always true. Students can also refer to the sketches of these triangles when they need to find the measures of the sides.

In this Activity, students apply the Pythagorean Theorem to several 30-60-90 triangles. The purpose of the Activity is to draw conclusions from the data collected and to develop the 30-60-90 Triangle Theorem.

For a student worksheet of this Activity and detailed Teacher Notes, see page 93 in the Lesson Activities booklet.

CHECKPOINT ✔

In any 30-60-90 triangle, the length of the hypotenuse is 2 times the length of the shorter leg, and the length of the longer leg is $\sqrt{3}$ times the length of the shorter leg.

ADDITIONAL
EXAMPLE 3

a. Using the figure from Example 3 on page 333, what is the distance from Jake's eye to the top of the tree?   about 92.4 feet

b. Suppose that Jake stands 56 feet from a different tree and spots the top of the tree with the instrument at a 45° angle. What is the height of this tree?   61 feet

Math
CONNECTION

ALGEBRA Remind students that a fraction with a radical expression in the denominator is not in simplest form. Changing $\frac{80}{\sqrt{3}}$ to $\frac{80\sqrt{3}}{3}$ is called *rationalizing the denominator*. Give students opportunities to practice rationalizing the denominator with the following fractions:

a. $\frac{2}{\sqrt{7}}$ $\frac{2\sqrt{7}}{7}$ **b.** $\frac{10}{\sqrt{75}}$ $\frac{2\sqrt{3}}{3}$

Areas of Regular Polygons

To find the area of a regular hexagon, divide the hexagon into 6 congruent, non-overlapping equilateral triangles. Find the area of one triangle and multiply by 6 to find the area of the hexagon. Note that the altitude of the equilateral triangle is the longer leg of a 30-60-90 triangle. The altitude is $\frac{1}{2}$ of the length of the side of the hexagon multiplied by $\sqrt{3}$.

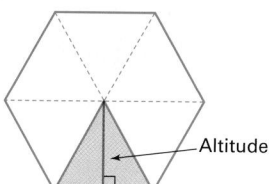

— Altitude

E X A M P L E ④ Find the area of a regular hexagon with sides of 20 centimeters.

● **SOLUTION**

Divide the hexagon into 6 equilateral triangles. Because the altitude of one of the triangles forms the longer leg of a 30-60-90 triangle and half of the side of the hexagon forms the shorter leg, the length of the altitude is $10\sqrt{3}$ centimeters.

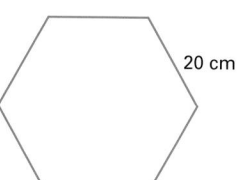

20 cm

Use the area formula to find the area of one of the triangles.

$$A = \tfrac{1}{2}(20)(10\sqrt{3})$$
$$A = 100\sqrt{3}$$

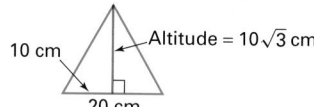

10 cm — Altitude = $10\sqrt{3}$ cm

20 cm

Because the hexagon is composed of 6 congruent triangles, the area of the hexagon is found as follows:

$$A = 6(100\sqrt{3}) = 600\sqrt{3} \approx 1039 \text{ square centimeters}$$

The Canadian dollar coin, or "loonie," is a regular 11-gon.

The method of dividing a regular hexagon into congruent triangles can be applied to find the area of any regular polygon.

An *n*-sided regular polygon can be divided into *n* non-overlapping congruent triangles. Each altitude of a triangle from the center of the polygon to a side of the polygon is called an **apothem** of the polygon.

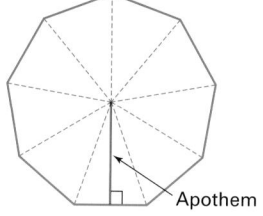

— Apothem

Area of a Regular Polygon

The area, *A*, of a regular polygon with apothem *a* and perimeter *p* is given by:

$$A = \tfrac{1}{2}ap \qquad\qquad \textbf{5.5.3}$$

EXAMPLE ⑤ Find the area of the regular pentagon shown at right.

● **SOLUTION**

The perimeter, p, is 5×10, or 50 units.

$$A = \frac{1}{2}ap$$

$$A = \frac{1}{2}(6.88)(50) = 172 \text{ square units}$$

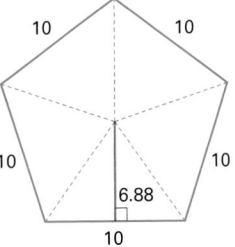

ADDITIONAL

EXAMPLE ⑤

Find the area of a regular pentagon with sides of 12 cm and an apothem of 8 cm.
240 cm²

Exercises

Selected Answers

Exercises 5–9, 11–47 odd

● *Communicate*

internet connect

Activities Online

Go To: go.hrw.com
Keyword:
MG1 Carpentry

1. Describe how two 45-45-90 triangles fit together to form a square. Describe how four 45-45-90 triangles fit together to form a square.

2. An equilateral triangle is a regular 3-sided polygon. Describe its apothem.

3. A square is a regular 4-sided polygon. Describe its apothem.

4. Can the lengths of the sides of a 45-45-90 triangle or a 30-60-90 triangle ever be a Pythagorean triple? Why or why not?

ASSIGNMENT GUIDE	
In Class	1–9
Core	11–41 odd
Core Plus	23–45 odd
Review	46–53
Preview	54–61

✐ Extra Practice can be found beginning on page 818.

Error Analyis

A right triangle with one leg twice the length of the other leg has acute angles of 63.43° and 26.57°, so it looks almost like a 30-60-90 triangle. To avoid this type of error, emphasize that the 30-60-90 triangle is one half of an equilateral triangle. The altitude is the longer leg and has a length that is a multiple of $\sqrt{3}$, not a whole number.

● *Guided Skills Practice*

5. Find the length of the hypotenuse of $\triangle RST$. *(EXAMPLE 1)*

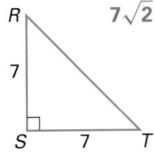

6. Find the missing lengths for $\triangle WXY$. *(EXAMPLE 2)* 14; $7\sqrt{3}$

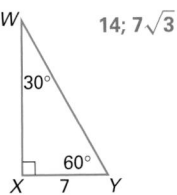

7. How tall is the tree in the diagram at left? *(EXAMPLE 3)* **34.6 ft**

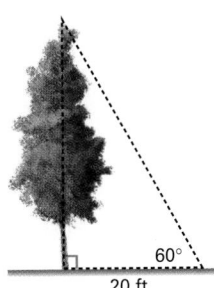

8. Find the area of a regular hexagon with sides of 12 inches. *(EXAMPLE 4)* **374.12 in²**

9. Find the area of the regular decagon at right. *(EXAMPLE 5)* **276.9 units²**

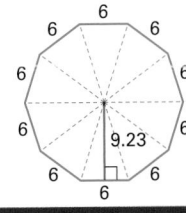

Practice and Apply

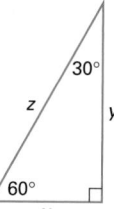

internet connect

**Homework
Help Online**

Go To: go.hrw.com
Keyword:
MG1 Homework Help
for Exercises 10-13,
18-21

**For each length given below, find the
remaining two lengths. Give your
answers in simplest radical form.**

10. $x = 6$ $y = 6\sqrt{3}$; $z = 12$

11. $y = 6$ $x = 2\sqrt{3}$; $z = 4\sqrt{3}$

12. $z = 14$ $x = 7$; $y = 7\sqrt{3}$

13. $y = 4\sqrt{3}$ $x = 4$; $z = 8$

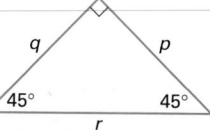

14. $p = 6$ $q = 6$; $r = 6\sqrt{2}$

15. $r = 6$ $p = 3\sqrt{2}$; $q = 3\sqrt{2}$

16. $q = 4\sqrt{2}$ $p = 4\sqrt{2}$; $r = 8$

17. $r = 10$ $p = 5\sqrt{2}$; $q = 5\sqrt{2}$

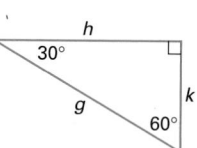

18. $k = 3.4$ $h = 3.4\sqrt{3}$; $g = 6.8$

19. $k = 6\sqrt{3}$ $h = 18$; $g = 12\sqrt{3}$

20. $g = 17$ $k = 8.5$; $h = 8.5\sqrt{3}$

21. $h = 2\sqrt{3}$ $k = 2$; $g = 4$

Find the area of each figure. Round your answers to the nearest tenth.

22. 2664.5 units²

73

23. 25.3 units²

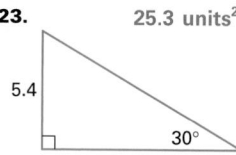

5.4

30°

24. 5.4 units²

60°

5

25. 112.5 units²

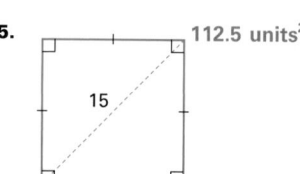

15

26. 62.4 units²

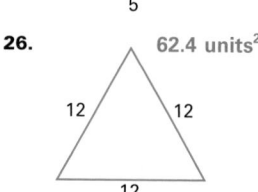

12   12

12

27. 147.8 units²

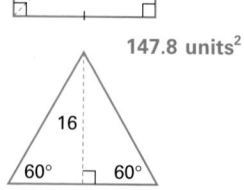

16

60°   60°

**Each of the following triples represents the side lengths of a triangle.
Determine whether the triangle is a 45-45-90 triangle, a 30-60-90
triangle, or neither. Explain your reasoning.**

28. $4, 4, 4\sqrt{2}$ 45-45-90

29. $6, 3\sqrt{2}, 3\sqrt{2}$ 45-45-90

30. $3, \sqrt{3}, 2\sqrt{3}$ 30-60-90

31. $2, 1, 2\sqrt{3}$ neither

32. $5, 5\sqrt{2}, 5\sqrt{2}$ neither

33. $\sqrt{2}, 2\sqrt{2}, \sqrt{6}$ 30-60-90

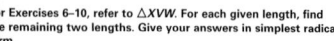
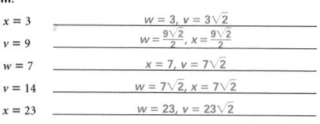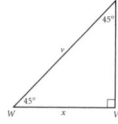 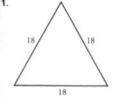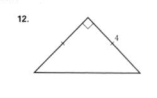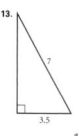

For Exercises 34–39, find the perimeter and area of each figure. Give your answers in simplest radical form.

34. a 30-60-90 triangle with a hypotenuse of 18

35. a 45-45-90 triangle with a hypotenuse of 24

36. an equilateral triangle with sides of length 8

37. a regular hexagon with sides of length 13

38. a square with a diagonal of 14

39. a regular hexagon with an apothem of 5

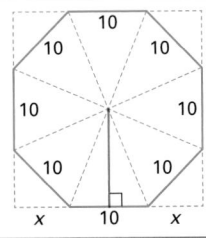

For Exercises 40–43, refer to the figure at left of a regular octagon inside a square.

40. From the diagram at left, what can you say about the triangles formed by the corners of the square, outside of the octagon? Find the value of x.

41. Find the length of the apothem of the octagon. $5\sqrt{2} + 5$ units

42. Find the area of the octagon. 482.8 units^2

43. If the sides of a regular octagon have length s, what is the area of the octagon? $A = 2s^2(\sqrt{2} + 1)$

CIVIL ENGINEERING An engineer is in charge of attaching guy wires to a tower. One wire attaches to a point on the ground 60 feet from the tower.

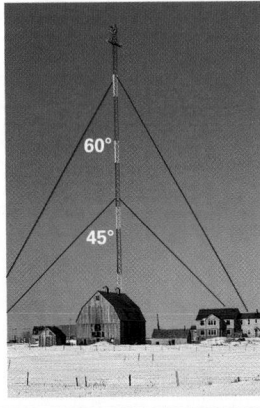

44. If the angle that the wire forms with the tower is 45°, what is the length of the wire? **84.85 ft**

45. If the angle that the wire forms with the tower is 60°, what is the length of the wire? **69.28 ft**

Look Back

46. The lines are parallel because they are perpendicular to the same line.

47. The lines are parallel by the Converse of the Corresponding Angles Postulate.

DRAFTING In drafting, a T square is used to align drawings. The T square is held against the side of the drawing board, and the straightedge stays perpendicular to the side of the drawing board. *(LESSON 3.4)*

46. An artist makes a series of horizontal lines with the straightedge of the T square by moving it up 1 inch for each new line. What is the relationship between these lines? Why?

47. With the 30-60-90 triangle positioned as shown at right, the artist uses the hypotenuse to draw a line that forms an angle of 60° with the straightedge. Without moving the T square, she slides the triangle 1 inch to the right and draws another line the same way. What is the relationship between the two lines? Why?

Technology

A calculator with a tangent feature is needed for Exercises 54–61.

34. $P = 27 + 9\sqrt{3}$ units;
$A = \frac{81}{2}\sqrt{3}$ units^2

35. $P = 24\sqrt{2} + 24$ units;
$A = 144$ units^2

36. $P = 24$ units;
$A = 16\sqrt{3}$ units^2

37. $P = 78$ units;
$A = 253.5\sqrt{3}$ units^2

38. $P = 28\sqrt{2}$ units;
$A = 98$ units^2

39. $P = 20\sqrt{3}$ units;
$A = 50\sqrt{3}$ units^2

40. They are 45-45-90 triangles;
$x = 5\sqrt{2}$

Student Technology Guide

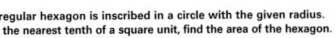

NAME _____ CLASS _____ DATE _____

Student Technology Guide
5.5 *Special Triangles and Areas of Regular Polygons*

In Lesson 5.4, you learned how to apply the Pythagorean Theorem to find side lengths in right triangles. Then in Lesson 5.5, you linked this skill to the problem of finding areas of regular polygons. This worksheet will help you use a calculator to carry out complicated calculations to solve an area problem.

Example: A regular hexagon is inscribed in a circle whose radius is 4. To the nearest tenth of a square unit, find the hexagon's area. The vertices of the hexagon and point O determine 6 congruent equilateral triangles and 12 congruent isosceles triangles. Thus, $AF = 4$ and $AX = 2$. As a result:

area of hexagon $ABCDEF = 12 \times$ area of $\triangle OAX$
$= 12(\frac{1}{2} \times 2 \times \sqrt{4^2 - 2^2})$
$= 12\sqrt{4^2 - 2^2}$

Press 12 [2nd] [x^2] 4 [x^2] [−] 2 [x^2] [)] [ENTER].
To the nearest tenth, the area is about 41.6 square units.

A regular hexagon is inscribed in a circle with the given radius. To the nearest tenth of a square unit, find the area of the hexagon.

1. $r = 10$ __259.8 square units__ 2. $r = 24$ __1496.5 square units__ 3. $r = 36$ __3367.1 square units__

When you study the diagram above of the regular hexagon inscribed in the circle, you may be able to see how to derive a formula for the area of the hexagon inscribed in a circle with a radius of r.

4. Write a formula for the area of a regular hexagon inscribed in a circle with a radius of r. area $= \frac{3r^2\sqrt{3}}{2}$

Using the formula you found in Exercise 4, find the area of a regular hexagon inscribed in a circle with the given radius.

5. $r = 10$ __259.8 square units__ 6. $r = 24$ __1496.5 square units__ 7. $r = 36$ __3367.1 square unit__

8. a. A square is inscribed in a circle with a radius of 50. Write an expression for the area of the square. area $= 2\left(\frac{50}{\sqrt{2}}\right)^2$
   b. Use a calculator to approximate the area of the square to the nearest tenth of a square unit. __5000 square units__

9. Write a formula for the area of a square inscribed in a circle with a radius of r. Use the formula to solve the problem in Exercise 8. area $= (r\sqrt{2})^2$; __5000 square units__

In Exercises 54–61, students use the tangent feature of a calculator to find the sides and angles of a triangle. Solving triangles by using trigonometric functions will be covered in a later chapter.

48. Smallest area: 1×49
$= 49$ units^2

Largest area: 25×25
$= 625$ units^2

48. Find the smallest possible area for a rectangle with a perimeter of 100 and side lengths that are whole numbers. Then find the greatest possible area. State the dimensions of each rectangle. *(LESSON 5.1)*

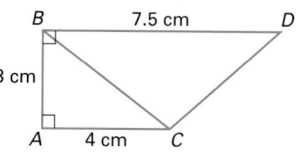

49. Find the area of the triangles in the figure at right. *(LESSON 5.2)*
area of $\triangle ABC = 6$ units^2;
area of $\triangle BCD = 11.25$ units^2

50. Find the height of a trapezoid with an area of 103.5 ft^2 and base lengths of 17.5 ft and 5.5 ft. *(LESSON 5.2)* 9 units

51. The base of an isosceles triangle is 8 m and the legs are 6 m. Find the area and perimeter. *(LESSONS 5.2 AND 5.4)* $8\sqrt{5}$ units^2; 20 units

52. In $\triangle PQR$, find PR. *(LESSON 5.4)* 218.17 units

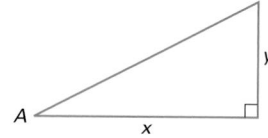

53. Find the area and perimeter of a right triangle with legs of lengths a and b. *(LESSONS 5.2 AND 5.4)* $A = \frac{1}{2}ab$; $P = a + b + \sqrt{a^2 + b^2}$

Look Beyond

Refer to the diagram at right. The ratio $\frac{y}{x}$ of the lengths of the legs of a right triangle is called the *tangent* of $\angle A$. The [TAN] button on your calculator is used to calculate tangent ratios.

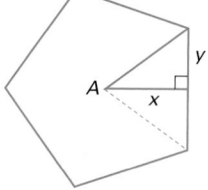

Be sure your calculator is in degree mode.

54. tangent of 45° = 1; yes; The legs of a 45-45-90 triangle are the same length.

54. Use your calculator to find the tangent of 45°. Does your answer agree with the ratio of the corresponding sides of a 45-45-90 triangle? Why or why not?

55. tangent of 30° ≈ 0.58; yes; The ratio of the shorter leg of a 30-60-90 triangle to the longer leg is $\frac{1}{\sqrt{3}} \approx 0.58$.

55. Use your calculator to find the tangent of 30°. Does your answer agree with the ratio of the corresponding sides of a 30-60-90 triangle? Why or why not?

56. In the regular pentagon at right, with its center at A, what is the measure of $\angle A$? m$\angle A = 36°$

57. Use your calculator to find the tangent of $\angle A$. tangent of $\angle A \approx 0.73$

58. If the side length of the regular pentagon is 12, what is y? $y = 6$

59. Use your results from Exercises 57 and 58, and the fact that the tangent of $\angle A$ is $\frac{y}{x}$ to find x. $x \approx 8.22$

60. $A \approx 246.6$ units^2

60. Use your results from Exercises 56–59 to find the area of a regular pentagon with a side length of 12. (Hint: What is the apothem?)

61. Use the method above to find the area of a regular 9-gon with a side length of 4. $A \approx 100.08$ units^2

5.6

The Distance Formula and the Method of Quadrature

Objectives

● Develop and apply the distance formula.

● Use the distance formula to develop techiques for estimating the area under a curve.

Why *The ability to compute the distance between two points is important in many situations. In some cases, an estimate based on measurement is sufficient. In other cases, greater precision is necessary.*

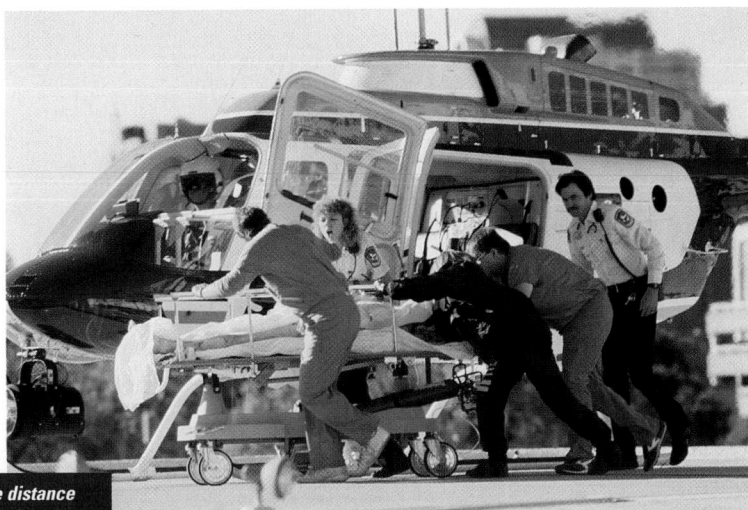

To reach his destination as quickly as possible, a helicopter pilot travels the shortest distance between two points.

Prepare

Teach

Why Planning a road trip, fencing a yard, or choosing the best theater seat requires the ability to calculate or estimate the distance between two points.

The Distance Formula

The distance between two points on the same horizontal or vertical line can be found by taking the difference of the *x*- or *y*-coordinates.

The vertical distance between points *A* and *B* is $AB = |7 - 3| = |3 - 7| = 4$.

The horizontal distance between points *B* and *C* is $BC = |5 - 2| = |2 - 5| = 3$.

Because *AC* is the hypotenuse of right triangle $\triangle ABC$, its length can be found by using the Pythagorean Theorem.

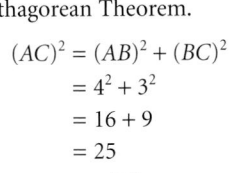

$$(AC)^2 = (AB)^2 + (BC)^2$$
$$= 4^2 + 3^2$$
$$= 16 + 9$$
$$= 25$$
$$AC = \sqrt{25} = 5$$

The Pythagorean Theorem can be used to find the distance between any two points in the coordinate plane.

Alternative Teaching Strategy

TECHNOLOGY Geometry graphics software that includes coordinate axes can be used to explore the ideas in this lesson. Students should start by replicating the figures in the text. Then they can drag points or line segments to see which relationships change and which remain constant.

Distance Formula

In a coordinate plane, the distance, d, between two points (x_1, y_1) and (x_2, y_2) is given by the following formula:

$$d = \sqrt{(x_2 - x_1)^2 + (y_2 - y_1)^2}$$ **5.6.1**

A proof of the distance formula is given below.

PROOF

Given: points $A(x_1, y_1)$ and $B(x_2, y_2)$

Prove: The distance between the two points is $\sqrt{(x_2 - x_1)^2 + (y_2 - y_1)^2}$.

Proof: Draw a right triangle with hypotenuse $\overline{AB}$ and a right angle at point $C(x_2, y_1)$. Let d be the distance between points A and B.

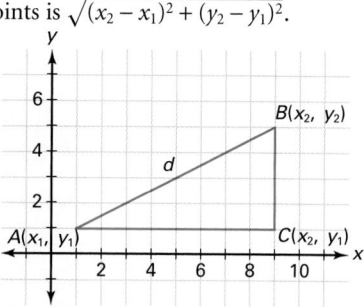

$$AC = |x_2 - x_1|$$
$$BC = |y_2 - y_1|$$

Use the Pythagorean Theorem.

$$d^2 = |x_2 - x_1|^2 + |y_2 - y_1|^2$$
$$d^2 = (x_2 - x_1)^2 + (y_2 - y_1)^2$$

You can drop the absolute-value symbols because the quantities are being squared.

Solve for d.

$$d = \sqrt{(x_2 - x_1)^2 + (y_2 - y_1)^2}$$

EXAMPLE A helicopter pilot located 1 mile east and 3 miles north of the command center must respond to an emergency located 7 miles east and 11 miles north of the the center. How far must the helicopter travel to get to the emergency site?

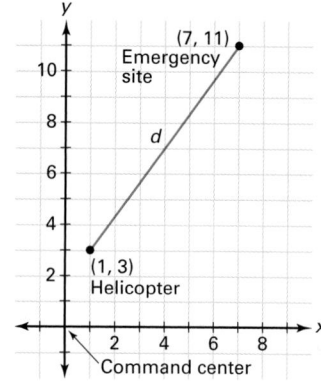

SOLUTION

Let $(0, 0)$ be the coordinates of the command center, with the positive part of the y-axis representing north. Then the helicopter's current coordinates are $(1, 3)$. The coordinates of the emergency site are $(7, 11)$. Use the distance formula.

$$d = \sqrt{(7 - 1)^2 + (11 - 3)^2}$$
$$d = \sqrt{6^2 + 8^2} = \sqrt{36 + 64} = \sqrt{100} = 10$$

The helicopter must travel 10 miles to get to the emergency site.

The Method of Quadrature

The area of an enclosed region on a plane can be approximated by the sum of the areas of a number of rectangles. This technique, called **quadrature**, is particularly important for finding the area under a curve.

Activity
Estimating the Area of a Circle

Part I: Method A (Left-Hand Rule)

1. Draw a quarter of a circle with a radius of 5 units centered at the origin, (0, 0). Draw rectangles as shown, with the upper left vertex of each rectangle touching the curve. This method is called the *left-hand rule*.

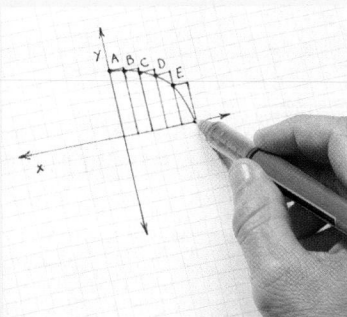

2. Find the y-coordinate of each point, A, B, C, D, and E. The segment connecting each point to the origin is the hypotenuse of a right triangle. Because the radius of the circle is 5 units, this is the length of the hypotenuse of each triangle. For point C, the y-coordinate is found as shown below.

$$y = \sqrt{5^2 - 2^2} = \sqrt{21}$$

The y-coordinate of each point is the height of a rectangle.

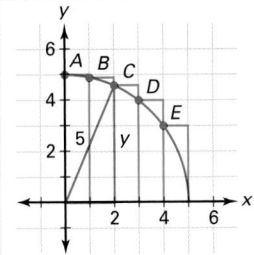

3. Find the area of each rectangle. (This is simplified by the fact that the base of each rectangle is 1.)

4. Find the sum of the areas of the rectangles by completing the pattern below.

$$\sqrt{5^2 - 0^2} + \sqrt{5^2 - 1^2} + \sqrt{5^2 - 2^2} + \cdots =$$
$$\sqrt{25} + \sqrt{24} + \sqrt{21} + \cdots \approx \underline{\ ?\ }$$

CHECKPOINT ✔

5. Multiply your sum by 4. The result is an estimate of the area of a complete circle with a radius of 5 units. Does this method overestimate or underestimate the area of the circle? Explain why.

6. Calculate the true value of the area of a complete circle to four decimal places by using $A = \pi r^2$. Find the relative error of your estimate by using the following formula:

$$E = \frac{|V_e - V_t|}{V_t} \times 100$$

where V_e = estimated value, V_t = true value, and E = percent of error

---

For a student worksheet of this Activity and detailed Teacher Notes, see page 95 in the Lesson Activities booklet.

CHECKPOINT ✔

5. $21.48 \times 4 = 85.93$ units^2; it overestimates because each rectangle covers more area than the circle in the same 1-inch "strip."

Teaching Tip

Remind students that the purpose of the Activity is not simply to find the area of a circle but to learn a technique for estimating the area of figures with irregular boundaries.

Inclusion Strategies

USING SYMBOLS The subscripted variables used in this lesson may be unfamiliar and, therefore, difficult for some students. Have students practice solving simple linear equations with subscripted variables, such as the following:

1. $4 + x_1 = 12$ $x_1 = 8$

2. $2y_2 - 1 = 9$ $y_2 = 5$

3. $3V_e = 18$ $V_e = 6$

Part II

y-coordinates:

Point A: $y = \sqrt{5^2 - 1^2} = \sqrt{24}$

Point B: $y = \sqrt{5^2 - 2^2} = \sqrt{21}$

Point C: $y = \sqrt{5^2 - 3^2} = \sqrt{16} = 4$

Point D: $y = \sqrt{5^2 - 4^2} = \sqrt{9} = 3$

Point E: $y = \sqrt{5^2 - 5^2} = \sqrt{0} = 0$

Sum of the areas:

$\sqrt{5^2 - 1^2} + \sqrt{5^2 - 2^2} + \sqrt{5^2 - 3^2} +$
$\sqrt{5^2 - 4^2} + \sqrt{5^2 - 5^2}$
$= \sqrt{24} + \sqrt{21} + \sqrt{16} + \sqrt{9} + \sqrt{0}$
≈ 16.48 units^2

$16.48 \times 4 = 65.92$ units^2

This result is an underestimate, since each rectangle covers less area than the circle in the same 1-inch "strip."

Actual area $= \pi \cdot 5^2$
≈ 78.5398 units^2

Relative Error

$E = \dfrac{|65.92 - 78.5398|}{78.5398} \times 100$

≈ 16.07 percent error

CHECKPOINT ✔
Part III

Average

$= \dfrac{85.92 + 65.92}{2} = 75.92$ units^2

Relative Error

$= \dfrac{|75.92 - 78.5398|}{78.5398} \times 100$

≈ 3.34 percent error

CRITICAL THINKING

Yes; method A will always give a result which is too big and method B will always give a result which is too small. The average of the two should give a better result.

CHECKPOINT ✔

Part II: Method B (Right-Hand Rule)

Repeat Part I with the rectangles arranged as shown at right. Here, the upper right vertex of each rectangle is touching the curve. This method is called the *right-hand rule*. Does this new method overestimate or underestimate the area of a complete circle? Explain why.

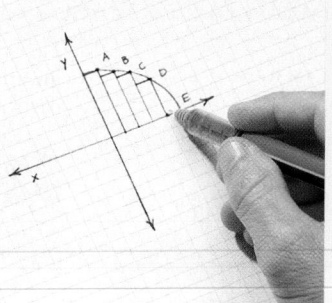

Part III: Combining Methods

CHECKPOINT ✔ Average your estimates for the area of the circle from Parts I and II. What is the relative error of your new estimate?

CRITICAL THINKING Do you think that the average of the results from methods A and B will always give more accurate results than either method by itself? Explain your answer.

Exercises

Communicate

1. Explain the relationship between the distance formula and the Pythagorean Theorem.

2. When using the distance formula, does it matter which point is (x_1, y_1) and which point is (x_2, y_2)? Why or why not?

3. Describe two methods of estimating the shaded area under the curve at right.

4. For each method you described in Exercise 3, how might you make your estimate of the area more accurate?

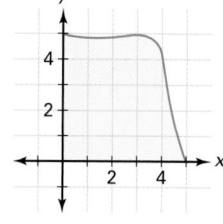

Guided Skills Practice

Find the distance between the indicated points. *(EXAMPLE)*

5.
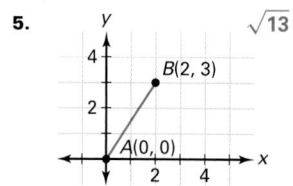

6.
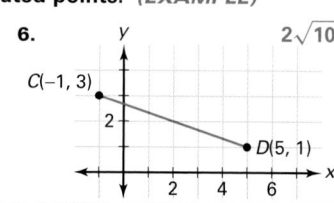

Reteaching the Lesson

COOPERATIVE LEARNING Have students work in pairs. Each student should draw five line segments on graph paper. The segments should not be horizontal or vertical. Students should then exchange papers. They should find the segment lengths by counting squares and using the Pythagorean Theorem. Then they should find the segment lengths by using the coordinates of the endpoint and the distance formula. This will help students make the connection between the Pythagorean Theorem and the distance formula.

For Exercises 7 and 8, estimate the area of the quarter-circle with a radius of 4 by finding the areas of the given rectangles. *(ACTIVITY)*

7. 13.98 units²

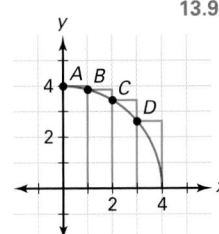

8. 9.98 units²

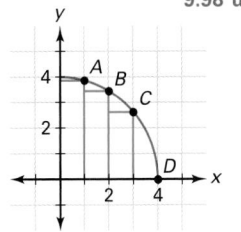

Assess

Selected Answers

Exercises 5–8, 9–47 odd

ASSIGNMENT GUIDE

In Class	1–8
Core	9–35 odd
Core Plus	15–39 odd
Review	41–47
Preview	48–51

✎ Extra Practice can be found beginning on page 818.

Error Analysis

Some students may make rounding errors in Exercises 9–20. If you suspect that this will happen, have them review place values and the criteria for rounding decimals.

● *Practice and Apply*

🖸 internet connect

Homework Help Online

Go To: go.hrw.com
Keyword:
MG1 Homework Help
for Exercises 9-20

Find the distance between each pair of points. Round your answers to the nearest hundredth.

9. $(0, 0)$ and $(5, 8)$ **9.43 units**

10. $(1, 2)$ and $(4, 6)$ **5 units**

11. $(1, 4)$ and $(3, 9)$ **5.39 units**

12. $(-3, -3)$ and $(6, 12)$ **17.49 units**

13. $(-1, 4)$ and $(-6, 16)$ **13 units**

14. $(-2, -3)$ and $(-6, -12)$ **9.85 units**

Algebra

Refer to the diagram below for Exercises 15–23.

15. $AB = $? **6.08 units**

16. $BC = $? **4.12 units**

17. $GJ = $? **3.61 units**

18. 25.36 units

19. 22.37 units

20. 10.46 units

21. The triangle is isosceles.

22. The triangle is not isosceles.

23. The triangle is not equilateral.

18. Find the perimeter of $\triangle EJH$.

19. Find the perimeter of $\triangle ABF$.

20. Find the perimeter of quadrilateral $FGHI$.

21. Is $\triangle BCD$ isosceles? Why or why not?

22. Is $\triangle AJE$ isosceles? Why or why not?

23. Is $\triangle BEH$ equilateral? Why or why not?

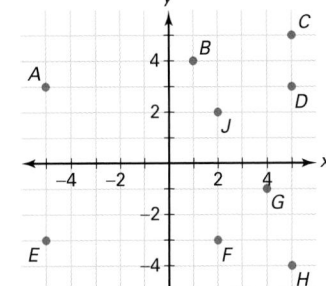

For Exercises 24–26, use the converse of the Pythagorean Theorem to determine whether the triangle with the given vertices is a right triangle. You may wish to plot the points and connect them to form the triangle.

24. $(2, 1)$, $(6, 4)$, and $(-4, 9)$ **The triangle is a right triangle.**

25. $(1, 5)$, $(6, 0)$, and $(-2, 2)$ **The triangle is a right triangle.**

26. $(1, 4)$, $(4, 2)$, and $(6, 6)$ **The triangle is not a right triangle.**

Technology

Students will need a calculator with a square-root key for most of the exercises in this lesson. A graphics calculator can be used for Exercises 36–38.

31. The midpoint of the hypotenuse is $(3, 4)$.
The distances are the same, 5 units.

32. The midpoint of the hypotenuse is $\left(2, \frac{7}{2}\right)$.
The distances are the same, $\frac{\sqrt{65}}{2}$ units.

33. Conjecture: The distances from the midpoint of the hypotenuse to any vertex of a right triangle are equal.

Practice

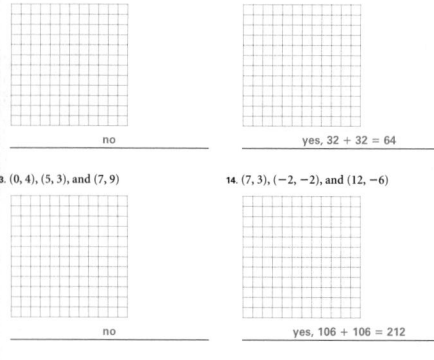

Algebra

For Exercises 27–30, give your answers in simplest radical form.

27. Suppose that the endpoints of the hypotenuse of a 45-45-90 triangle are $(3, 3)$ and $(9, 2)$. What is the length of the legs? $\frac{\sqrt{74}}{2}$ units

28. Suppose that the endpoints of one leg of a 45-45-90 triangle are $(0, 3)$ and $(4, -1)$. What is the length of the hypotenuse? **8 units**

29. $\frac{5}{2}$ units; $\frac{5\sqrt{3}}{2}$ units

29. Suppose that the endpoints of the hypotenuse of a 30-60-90 triangle are $(4, -2)$ and $(7, 2)$. What is the length of the shorter leg? of the longer leg?

30. Suppose that the endpoints of the longer leg of a 30-60-90 triangle are $(-3, 5)$ and $(2, 1)$. What is the length of the shorter leg? of the hypotenuse? $\frac{\sqrt{123}}{3}$ units; $\frac{2\sqrt{123}}{3}$ units

---

In Exercises 31–34, you will use coordinate geometry to explore a property of right triangles.

31. Find the midpoint of the hypotenuse of the right triangle with vertices at $(0, 0), (6, 0),$ and $(0, 8)$. What is the distance from this midpoint to each vertex? What do you notice?

32. Repeat Exercise 31, using a right triangle with vertices at $(0, 0), (4, 0),$ and $(0, 7)$.

33. Based on your results from Exercises 31 and 32, make a conjecture about the distance from the midpoint of the hypotenuse of a right triangle to each vertex of the triangle.

PROOF

34. Prove your conjecture from Exercise 33 by using a right triangle with vertices at $(0, 0), (x, 0)$ and $(0, y)$.

35. Use the method of quadrature to estimate the shaded area under the curve shown below. Use both the left-hand and right-hand rules, and then average your results. **Sample answer: 16.375 units²**

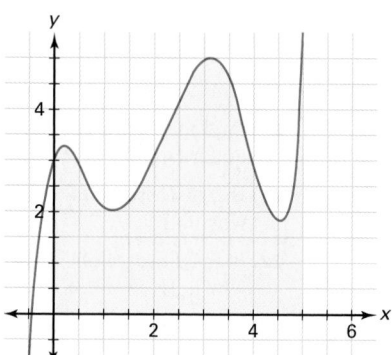

Algebra

36. Use the method of quadrature to estimate the area between $y = x^2$ and the x-axis for $0 \le x \le 4$. **Sample answer: 21.5 units²**

37. Use the method of quadrature to estimate the area between $y = x^2 + 2$ and the x-axis for $0 \le x \le 2$. **Sample answer: 6.75 units²**

38. Use the method of quadrature to estimate the area between $y = -x^2 + 4$ and the x-axis for $-2 \le x \le 2$. **Sample answer: 10.5 units²**

---

34. The hypotenuse is the segment between $(x, 0)$ and $(0, y)$. The midpoint of the hypotenuse is: $M = \left(\frac{x+0}{2}, \frac{0+y}{2}\right) = \left(\frac{x}{2}, \frac{y}{2}\right)$.

Let $A = (0, 0), B = (x, 0), C = (0, y)$.

$$AM = \sqrt{\left(\frac{x}{2} - 0\right)^2 + \left(\frac{y}{2} - 0\right)^2}$$

$$= \sqrt{\left(\frac{x}{2}\right)^2 + \left(\frac{y}{2}\right)^2} = \sqrt{\frac{x^2}{4} + \frac{y^2}{4}}$$

$$BM = \sqrt{\left(\frac{x}{2} - x\right)^2 + \left(\frac{y}{2} - 0\right)^2}$$

$$= \sqrt{\left(\frac{-x}{2}\right)^2 + \left(\frac{y}{2}\right)^2} = \sqrt{\frac{x^2}{4} + \frac{y^2}{4}}$$

$$CM = \sqrt{\left(\frac{x}{2} - 0\right)^2 + \left(\frac{y}{2} - y\right)^2}$$

$$= \sqrt{\left(\frac{x}{2}\right)^2 + \left(\frac{-y}{2}\right)^2} = \sqrt{\frac{x^2}{4} + \frac{y^2}{4}}$$

Therefore $AM = BM = CM$.

39. ENVIRONMENTAL PROTECTION
An oil pipeline runs through part of a national forest. Because of the danger to the forest, an enviromental group wants to know the length of the pipeline that lies inside the boundaries of the forest. Use the map at right to find the desired length. Each square of the grid represents 1 square mile. **8.5 mi**

40. CIVIL ENGINEERING A dam has been built on a river, and the river has begun to back up to form a lake as shown in the "map" at right. Each square on the map represents 10,000 sq ft. Use the method of quadrature to estimate the area of the lake. **75,000 ft²**

A lake formed by a dam, viewed from above, will usually have a flat side. Explain why. This makes it convenient to use the method of quadrature. What else must be true about the shape of the lake in order to use the method of quadrature as explained in this lesson?

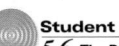 *Look Back*

41. Prove that the quadrilateral with vertices at $A(-1, 1)$, $B(7, -5)$, $C(10, -1)$, and $D(2, 5)$ is a rectangle. *(LESSON 3.8)*

42. The circular area is larger.

42. Suppose that a circle and a square both have a perimeter of 100 meters. Which has the larger area? *(LESSON 5.3)*

43. The square has the greater perimeter.

43. Suppose that a circle and a square both have an area of 225 square centimeters. Which has the larger perimeter? *(LESSON 5.3)*

44. 9.18 cm

44. The legs of a right triangle are 4.5 centimeters and 8 centimeters. What is the length of the hypotenuse? *(LESSON 5.4)*

45. 3 cm

45. The hypotenuse of a right triangle is 5 centimeters and one of the legs is 4 centimeters. What is the length of the other leg? *(LESSON 5.4)*

46. $A = 16\sqrt{3}$ in²; $P = 24$ in

46. Each side of an equilateral triangle measures 8 inches. Find the area and perimeter. *(LESSON 5.5)*

47. $A = \frac{s^2}{4}\sqrt{3}$ units²; $P = 3s$ units

47. Each side of an equilateral triangle measures s units. Find the area and perimeter. *(LESSON 5.5)*

41. Let $A = (-1, 1)$, $B = (7, -5)$, $C = (10, -1)$, $D = (2, 5)$

slope $\overline{AB} = \dfrac{-5-1}{7-(-1)} = \dfrac{-6}{8} = -\dfrac{3}{4}$

slope $\overline{BC} = \dfrac{-5-(-1)}{7-10} = \dfrac{-4}{-3} = \dfrac{4}{3}$

slope $\overline{CD} = \dfrac{-1-5}{10-2} = \dfrac{-6}{8} = -\dfrac{3}{4}$

slope $\overline{AD} = \dfrac{1-5}{-1-2} = \dfrac{-4}{-3} = \dfrac{4}{3}$

Since slope $\overline{AB} =$ slope $\overline{CD}$ and slope $\overline{BC} =$ slope $\overline{AD}$, $\overline{AB} \parallel \overline{CD}$ and $\overline{BC} \parallel \overline{AD}$ and $ABCD$ is a parallelogram. Since (slope $\overline{AB}$)(slope $\overline{BC}$) = -1, $\overline{AB} \perp \overline{BC}$ and $ABCD$ is a rectangle.

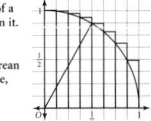
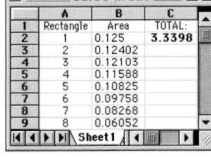

In Exercises 48–51, students estimate the area of a quarter-circle by using rectangles, trapezoids, and triangles. In Exercise 51, students are asked to write a proof of the two methods used in the Activity on pages 341 and 342.

Assessment

Portfolio Activity

The Portfolio Activity can be used as preparation for the Chapter Project or as a separate activity. In the Portfolio Activity on this page, students find the areas of semicircles and lunes.

49. The answer is an underestimate since the curve is above the top leg of each trapezoid.

50. Sample answer: An estimate for the area of the full circle is $4 \times 18.98 = 75.92$ units^2, which is the answer we obtained by averaging the Left-Hand and Right-Hand Rules.

51. The rectangle with width h and height y_1 has area $y_1 h$. The rectangle with width h and height y_2 has area $y_2 h$. The average of the two is:

$$\frac{y_1 h + y_2 h}{2} = \frac{(y_1 + y_2)h}{2}$$
$$= \frac{1}{2}(y_1 + y_2)h.$$

The area of the trapezoid is:
$$\frac{1}{2}(y_1 + y_2)h$$

The areas are the same using both methods.

The area of an enclosed region can be approximated by using shapes other than rectangles. In the figure below, the area of the quarter-circle is approximated by the areas of four trapezoids and one triangle.

48. The quarter-circle in the figure has a radius of 5 units. Use the y-values you found in the Activity on page 341 as the bases of the trapezoids and use 1 as the height to estimate the area under the curve. **18.98 units^2**

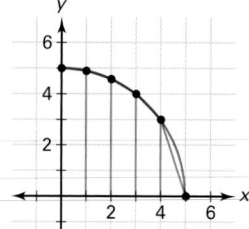

49. Is your answer an overestimate or underestimate of the actual area under the curve? Why?

50. Compare your answer to Exercise 48 to your answer to Part III of the Activity. What do you notice?

51. The area found by using trapezoids is equal to the average of the estimates found by using the left-hand and right-hand rules.

Use the diagram at right and the formulas for the area of a rectangle and the area of a trapezoid to prove this fact algebraically.

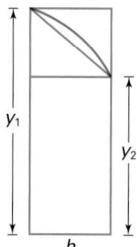

The figure below contains a 45-45-90 triangle. Three semicircles have been constructed, with their centers at the midpoint of the hypotenuse and the midpoint of each leg. The shaded crescents in the figure are called *lunes*.

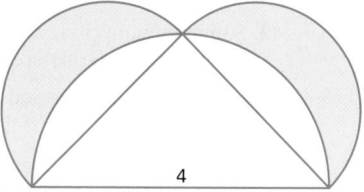

1. Find the area of the 45-90-45 triangle.
2. Find the total area of the small semicircles.
3. Find the area of the large semicircle.
4. Find the total area of the lunes.
5. What other area in the figure is equal to the total area of the lunes?

Proofs Using Coordinate Geometry

Objectives

- Develop coordinate proofs for the Triangle Midsegment Theorem, the diagonals of a parallelogram, and the reflection of a point across the line $y = x$.

- Use the concepts of coordinate proofs to solve problems on the coordinate plane.

Why The essential attributes of geometric figures can be captured in a coordinate drawing. For this reason, coordinate geometry can be used to prove geometry theorems.

APPLICATION
METEOROLOGY

Meteorologists use latitude and longitude readings to track hurricanes on a map of Earth. Near the equator, these coordinates are very similar to *xy*-coordinates.

Prepare

QUICK WARM-UP

1. Find the slope of the line containing $(-2, 4)$ and $(-1, 3)$. **−1**

2. Find the length of the segment with endpoints $(4, 7)$ and $(-2, -1)$. **10**

3. What are the the coordinates of the image of $(-2, 4)$ after a reflection across the line $y = x$? **$(4, -2)$**

Also on Quiz Transparency 5.7

Three Coordinate Proofs

Activity 1
The Triangle Midsegment Theorem

YOU WILL NEED
graph paper

CHECKPOINT ✔

1. Use the coordinates of the triangles given in the table to test the following theorem: The midsegment of a triangle is parallel to a side of the triangle and has a measure equal to half the measure of that side. (Theorem 4.6.9) Draw the first two triangles in the coordinate plane, and complete the table below.

Vertices of triangle	Coordinates of midpoint *M* and *S*		Slope		Length	
	$\overline{AB}$	$\overline{BC}$	$\overline{MS}$	$\overline{AC}$	$\overline{MS}$	$\overline{AC}$
$A(0, 0)$, $B(2, 6)$, $C(8, 0)$	$M(?, ?)$	$S(?, ?)$	?	?	?	?
$A(0, 0)$, $B(6, -8)$, $C(10, 0)$	$M(?, ?)$	$S(?, ?)$	?	?	?	?
General case (see Steps 3 & 4)	$M(?, ?)$	$S(?, ?)$	?	?	?	?

Teach

Why Studying coordinate geometry proofs helps students describe geometric shapes analytically and demonstrates the strong connection between algebra and geometry.

☞ For Activity Notes and the answer to the Checkpoint, see page 348.

Alternative Teaching Strategy

TECHNOLOGY Ask students to use geometry graphics software to draw the figures in Activities 1, 2, and 3 on a coordinate plane and use the measurement capabilities of the software to verify the corresponding proofs.

Interdisciplinary Connection

ART Computer design artists use coordinates to draw and place figures on the computer screen. Have students discuss how principles in this lesson can be applied to an artist's work.

Activity 1 Notes

In this Activity, students prove the Triangle Midsegment Theorem by using coordinates.

For a student worksheet of this Activity and detailed Teacher Notes, see page 98 in the Lesson Activities booklet.

CHECKPOINT ✔

1. See Additional Answers, beginning on page 879.

CHECKPOINT ✔

4. $\overline{MS}$ is equal to half of $\overline{AC}$, and $\overline{MS}$ and $\overline{AC}$ are parallel (they have the same slope), which proves the Midsegment Theorem.

CRITICAL THINKING

The proof in Lesson 4.6 is much longer and more complicated. Answers may vary.

Activity 2 Notes

In this Activity, students verify that the diagonals of a parallelogram bisect each other. Have students do this activity with a variety of parallelograms.

For a student worksheet of this Activity and detailed Teacher Notes, see page 98 in the Lesson Activities booklet.

CHECKPOINT ✔

2. The diagonals of a parallelogram bisect each other. The intersection of the diagonals is the midpoint of each diagonal.

---

2. In each triangle you drew in Step 1, one vertex was at the origin and another was on the *x*-axis. These vertices were chosen to keep the calculations simple. Do you think it is possible to create a triangle of any shape and size that fits these same conditions? Explain your answer.

3. The figure at right represents the general case of the Triangle Midsegment Theorem. The triangle could be of any shape and size, and it could be in any quadrant of the coordinate plane—so don't take the drawing too literally.

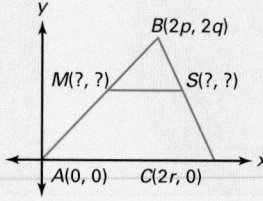

The number 2 has been used as a multiplier to keep the calculations simple. Notice that the numbers 2*p*, 2*q*, and 2*r* can represent any desired real number. Explain why this is true.

CHECKPOINT ✔

PROOF

4. Fill in the third row of the table on the previous page. Based on this information, what have you proven about the relationship of $\overline{MS}$ and $\overline{AC}$? Explain why this proves the Triangle Midsegment Theorem.

CRITICAL THINKING

Compare this proof of the Triangle Midsegment Theorem with the one that appears in Lesson 4.6. Which proof seems easier to you? Explain why.

Activity 2
The Diagonals of a Parallelogram

YOU WILL NEED

graph paper

1. Use the coordinates of the parallelograms given in the table below to test the following theorem: The diagonals of a parallelogram bisect each other. (Theorem 4.5.4) Draw the first figure in the coordinate plane. Three vertices of a parallelogram are given. Find the fourth vertex and fill in the blanks of the table.

Three vertices of a parallelogram	Fourth vertex	Midpoint of $\overline{BD}$	Midpoint of $\overline{AC}$
$A(0, 0)$, $B(2, 6)$, $D(10, 0)$	$C(?, ?)$	$(?, ?)$	$(?, ?)$
General case (see Step 2)	$C(?, ?)$	$(?, ?)$	$(?, ?)$

CHECKPOINT ✔

PROOF

2. The figure at right represents the general case of the theorem. Use the coordinates of the figure to fill in the blanks of the table. Based on this information, what have you proven about the diagonals of a parallelogram? Explain your answer.

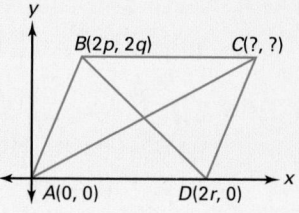

CRITICAL THINKING

How can you prove this theorem without using coordinate geometry? Which proof seems easier to you? Explain why.

Inclusion Strategies

USING COGNITIVE STRATEGIES Students may have difficulty distinguishing between traditional geometric proofs and coordinate proofs. Create a chart with the features of these two proofs. For example, the traditional approach usually refers to angle or side congruence, while the coordinate approach may refer to equivalent measures of sides or angles.

Reflection Across the Line y = x

You may recall that the effect of reversing the x- and y-coordinates of a point is to reflect the line across the line $y = x$. You can use coordinate geometry to prove this result.

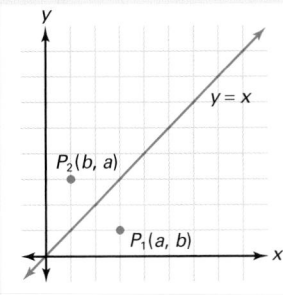

1. If you know the x-coordinate of a point on the line $y = x$, what can you conclude about the y-coordinate? (Filling in a table like the one at right will reveal the pattern.)

2. Find the midpoint between the points $P_1(a, b)$ and $P_2(b, a)$ by using the midpoint formula. Does this point lie on the line $y = x$? Explain your reasoning.

x	y
1	?
2	?
0	?
−1	?
n	?

3. Pick two points on the line $y = x$ and use them to find the slope of this line. Record your result.

4. Find the slope of the line that passes through the points $P_1(a, b)$ and $P_2(b, a)$. Record your result. Hint: You can rewrite $b - a$ as $-(a - b)$.

5. Compare your results in Steps 3 and 4. What can you conclude about the relationship between the two lines?

CHECKPOINT ✔ 6. Recall the definition of a reflection from Lesson 1.6. Explain how your results prove that the result of reversing the coordinates of a point is a reflection of the point across the line $y = x$.

Exercises

Communicate

1. In coordinate geometry proofs, why is it possible to place one vertex of the figure at the origin and one side along the x-axis?

2. In a coordinate geometry proof about triangles, could you place one vertex at the origin, one on the x-axis, and one on the y-axis? Why or why not?

3. In a coordinate geometry proof, what do you need to show to prove that two lines are parallel? that two lines are perpendicular?

4. In a coordinate geometry proof, what do you need to show to prove that two segments bisect each other?

Reteaching the Lesson

USING VISUAL MODELS Have students draw a polygon on the coordinate plane with one side along the x-axis and one vertex at $(0, 0)$. Label the vertices with coordinates that are easy to use in calculations, such as $(2a, 2b)$. Ask them to list as many statements as they can about their polygons. Have them describe how they would use coordinates to prove each statement.

ASSIGNMENT GUIDE

In Class	1–10
Core	11–15 odd, 17–26, 27–31 odd
Core Plus	17–32
Review	33–44
Preview	46–47

✐ Extra Practice can be found beginning on page 818.

Guided Skills Practice

Find the length of the segment with the given endpoints. *(ACTIVITY 1)*

5. $(0, 0)$ and (p, q) $\sqrt{p^2 + q^2}$

6. (p, q) and (r, s) $\sqrt{(r - p)^2 + (s - q)^2}$

Find the midpoint of the segment with the given endpoints.
(ACTIVITIES 1 AND 2)

7. $(0, 0)$ and $(2p, 2q)$ (p, q)

8. $(2p, 2q)$ and $(2r, 2s)$ $(p + r, q + s)$

Find the slope of the segment with the given endpoints. *(ACTIVITY 3)*

9. $(0, 0)$ and (p, q) $\dfrac{q}{p}$

10. (p, q) and (r, s) $\dfrac{s - q}{r - p}$

Practice and Apply

Determine the coordinates of the unknown vertex or vertices of each figure below. Use variables to represent any coordinates that are not completely determined by the given vertices.

11. rectangle *ABCD*
$A(0, 0)$, $B(0, p)$,
$C(\underline{\;?\;}, \underline{\;?\;})$, $D(q, 0)$ **C(q, p)**

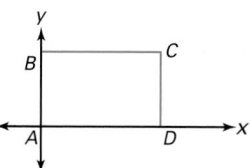

12. isosceles triangle *DEF*
$D(0, 0)$, $E(p, q)$, $F(\underline{\;?\;}, \underline{\;?\;})$
F(2p, 0)

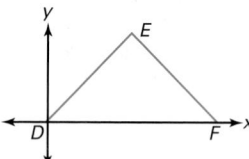

13. parallelogram *GHIJ*
$G(0, 0)$, $H(p, q)$,
$I(\underline{\;?\;}, \underline{\;?\;})$, $J(r, 0)$
I(r + p, q)

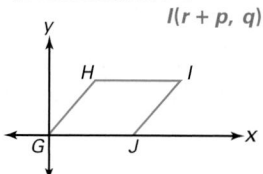

14. square *KLMN*
$K(0, 0)$, $L(\underline{\;?\;}, \underline{\;?\;})$,
$M(\underline{\;?\;}, \underline{\;?\;})$, $N(p, 0)$
M(p, p); L(0, p)

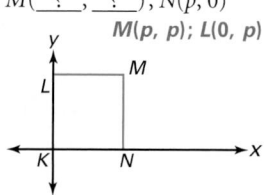

16. $V(p + \sqrt{p^2 + q^2}, q)$
$W(\sqrt{p^2 + q^2}, 0)$

15. trapezoid *PQRS*
$P(0, 0)$, $Q(p, q)$,
$R(\underline{\;?\;}, \underline{\;?\;})$, $S(\underline{\;?\;}, \underline{\;?\;})$
R(a, q); S(b, 0)

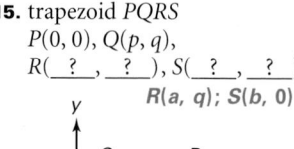

16. rhombus *TUVW*
$T(0, 0)$, $U(p, q)$,
$V(\underline{\;?\;}, \underline{\;?\;})$, $W(\underline{\;?\;}, \underline{\;?\;})$

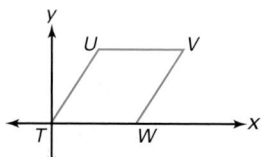

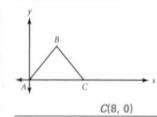
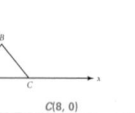
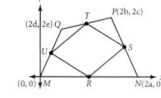

17. $C(2r, 2q)$ and $D(2s, 0)$

18. $M(p, q)$; $N(r + s, q)$

21. $M\left(\dfrac{q}{2}, \dfrac{p}{2}\right)$

In Exercises 17–20, you will prove the Trapezoid Midsegment Theorem by using coordinate geometry.

17. Trapezoid $ABCD$ at right has vertices at $A(0, 0)$, $B(2p, 2q)$, $C(2r, \underline{\ ?\ })$, and $D(2s, \underline{\ ?\ })$.

18. Find the coordinates of M, the midpoint of $\overline{AB}$, and N, the midpoint of $\overline{CD}$.

19. Find the lengths AD, BC, and MN. Prove that the length of the midsegment is the average of the lengths of the bases.

20. Find the slopes of $\overline{AD}$, $\overline{BC}$, and $\overline{MN}$. Prove that the midsegment is parallel to the bases.

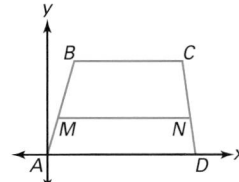

In Exercises 21–23, you will prove that the center of the circumscribed circle for a right triangle is the midpoint of the hypotenuse.

21. Find the midpoint, M, of the hypotenuse of $4\ JKL$ at right.

22. Show that the midpoint of the hypotenuse is equidistant from the three vertices.

23. Explain why point M must be the center of the circumscribed circle for $4\ JKL$.

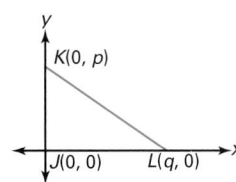

In Exercises 24–26, you will prove that the midpoints of any quadrilateral are the vertices of a parallelogram.

24. In quadrilateral $EFGH$ at right, find the coordinates of the midpoints of the sides.

25. Copy the diagram and connect the midpoints of the sides.

26. Prove that the quadrilateral formed by connecting the sides is a parallelogram.

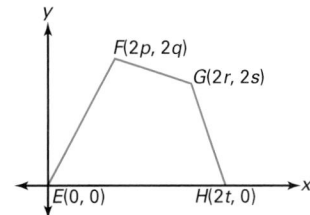

Use coordinate geometry to prove each of the following theorems:

27. The opposite sides of a parallelogram are congruent.

28. The diagonals of a square are perpendicular to each other.

29. The diagonals of a rectangle are congruent.

30. If the diagonals of a parallelogram are perpendicular, then the parallelogram is a rhombus.

31. The medians of a triangle intersect at a single point. (Hint: Find the equations of the lines containing the medians.)

internet connect

Homework Help Online
Go To: go.hrw.com
Keyword:
MG1 Homework Help
for Exercises 27-30

Technology

You may want students to use geometry graphics software for Exercises 17–26. Have students use the measurement capabilities of the software to verify the statements given in each proof.

Error Analysis

Students may not know how to begin the proofs in Exercises 27–30. It may be helpful to discuss different methods for the proofs in each of these exercises before assigning them individually.

19. $AD = 2s$; $BC = 2r - 2p$;
$MN = s + r - p$

$$\dfrac{(AD + BC)}{2} = \dfrac{(2s + 2r - 2p)}{2}$$
$$= s + r - p = MN$$

20. slope of $\overline{AD} = 0$
slope of $\overline{BC} = 0$
slope of $\overline{MN} = 0$
All three segments have the same slope. Therefore the midsegment is parallel to the bases.

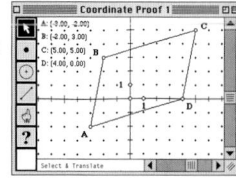
---

22. Find the lengths of JM, KM, and LM, if they are equal then the midpoint of the hypotenuse is equidistant from the three vertices.

$$JM = \sqrt{\left(\dfrac{q}{2} - 0\right)^2 + \left(\dfrac{p}{2} - 0\right)^2}$$
$$= \sqrt{\left(\dfrac{q}{2}\right)^2 + \left(\dfrac{p}{2}\right)^2}$$

$$KM = \sqrt{\left(\dfrac{q}{2} - 0\right)^2 + \left(\dfrac{p}{2} - p\right)^2}$$
$$= \sqrt{\left(\dfrac{q}{2}\right)^2 + \left(\dfrac{p}{2}\right)^2}$$

$$LM = \sqrt{\left(\dfrac{q}{2} - q\right)^2 + \left(\dfrac{p}{2} - 0\right)^2}$$
$$= \sqrt{\left(\dfrac{q}{2}\right)^2 + \left(\dfrac{p}{2}\right)^2}$$

23. M is equidistant from the three vertices. Therefore a circle can be drawn through J, K, and L and its radius is $JM = KM = LM$.

24. Midpoint of $\overline{EF} = M(p, q)$
Midpoint of $\overline{FG} = N(p + r, q + s)$
Midpoint of $\overline{GH} = O(r + t, s)$
Midpoint of $\overline{EH} = P(t, 0)$

32. a. $d_L = 100 - x$

$d_W = \sqrt{x^2 + 900}$

b. $c = 100 - x + 2\sqrt{x^2 + 900}$

$x \approx 17.3$ miles

cost $\approx \$151,962$

APPLICATION

CONNECTION

32. CIVIL ENGINEERING Points *A* and *B* are on opposite sides of a straight portion of a river 30 meters wide. Point *B* is 100 meters downstream from point *A*. An engineer is designing a road from point *A* to point *B*. To find the lowest cost road between the points, the engineer positions points *A* and *B* on a coordinate plane as shown. The road costs more to build over the river, so part of the road will be built on land, leaving the river bank at point *X*, as shown.

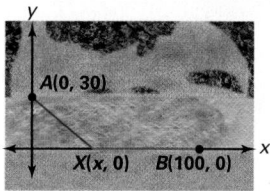

a. Use the given coordinates to find d_L, the length of the road that lies on land, and d_W, the length of the road that lies over the water.

b. MAXIMUM/MINIMUM Suppose that the cost of building the road is $1000 per meter on land and $2000 per meter over the water. Write an equation for the cost (in thousands of dollars) of the road in terms of *x*. Use a graphics calculator to graph the cost function, and trace to find the coordinates of *X* for the lowest cost road.

Look Back

Given square *ABCD* with *AB* = 5, and diagonals $\overline{AC}$ and $\overline{BD}$ intersecting at point *E*, find the indicated measures. *(LESSON 3.2, 4.5, 5.4)*

33. *BC* 5   **34.** *CD* 5   **35.** *AD* 5

36. *AC* $5\sqrt{2}$ **37.** *BD* $5\sqrt{2}$ **38.** *AE* $\frac{5\sqrt{2}}{2}$

39. *EC* $\frac{5\sqrt{2}}{2}$ **40.** *BE* $\frac{5\sqrt{2}}{2}$ **41.** *ED* $\frac{5\sqrt{2}}{2}$

42. m∠*BEA* 90°   **43.** m∠*BEC* 90°

You are given trapeziod *ABCD* with bases $\overline{AB}$ and $\overline{CD}$ and midsegment $\overline{MN}$, where *AB* = 5.7, *CD* = 8.5, and the height, *h* = 3.5. Find the indicated values. *(LESSON 3.7, 5.2)*

44. *MN* 7.1   **45.** Area of trapezoid *ABCD* 24.85

Look Beyond

PROOFS

46. Draw a scalene triangle, △*ABC*, and construct median $\overline{AM}$. Prove that the area of △*ABM* is equal to the area of △*ACM*.

47. Using △*ABC* from Exercise 46, construct the centroid (the intersection of the medians) of △*ABM* and label it *D*. Construct the centroid of △*ACM* and label it *E*. Draw $\overline{DE}$. What is the relationship between $\overline{AM}$ and $\overline{DE}$? What is the relationship between $\overline{BC}$ and $\overline{DE}$? If you are using geometry graphics software, try dragging the points to different locations to see if your results still hold. Otherwise, draw several different triangles to test your results.

46.

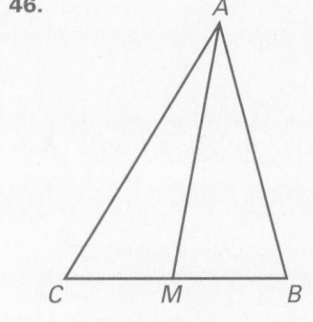

Given: scalene △*ABC* with median $\overline{AM}$

Prove: area of △*ABM* = area of △*ACM*

Proof: *M* is the midpoint of $\overline{BC}$, so *CM* = *MB* by the definition of midpoint. *A* is the vertex opposite $\overline{CM}$ in △*ACM* and also the vertex opposite $\overline{MB}$ in △*ABM*. Therefore, the two triangles have the same altitude and thus the same height. Since the bases and heights of the two triangles are the same, their areas must be equal.

47.

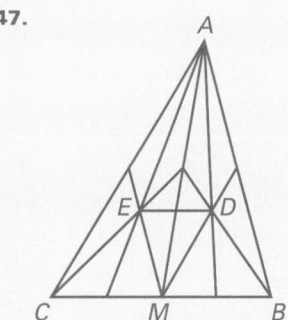

Sample answer: $\overline{AM}$ bisects $\overline{DE}$; $\overline{DE}$ is parallel to $\overline{BC}$.

Geometric Probability

Objective

● Develop and apply the basic formula for geometric probability.

Why *Some scientists believe that an asteroid or comet caused the extinction of the dinosaurs. What is the likelihood that one particular such object would have struck land? You can use geometric probability to help answer questions like this.*

YUCATÁN PENINSULA

Gulf of Mexico

MEXICO

BELIZE

Pacific Ocean

GUATEMALA

HONDURAS

EL SALVADOR

According to the fossil record, Earth experienced a sudden and drastic change about 65 million years ago that resulted in the extinction of dinosaurs like Tyrannosaurus rex. Some scientists believe that this upheaval was caused by an asteroid striking Earth near the Yucatán Peninsula.

The Probability of an Event

Consider the following question:

The surface of Earth consists of about 30 percent land and 70 percent water. Assuming that a comet or asteroid would be equally likely to strike anywhere on Earth, what is the probability that such an object would strike land instead of water?

Mathematical intuition should tell you that there is a 30 percent chance, or *probability*, that the object would strike land. **Probability** is a number from 0 to 1 (or from 0 to 100 percent) that indicates how likely an event is to occur.
•A probability of 0 (or 0 percent) indicates that the event cannot occur.
•A probability of 1 (or 100 percent) indicates that the event must occur.

For many situations, it is possible to define and calculate the *theoretical probability* of an event with mathematical precision. The theoretical probability that an event will occur is a fraction whose denominator represents all equally likely outcomes and whose numerator represents the outcomes in which the event occurs.

Alternative Teaching Strategy

HANDS-ON STRATEGIES The probability calculated in Example 1 can be converted into a simulation in which students perform experiments, record data, and find experimental probabilities. Comparing theoretical and experimental probabilities can help students better understand the difference between these two measures of chance.

Prepare

QUICK WARM-UP

Use graph paper for the following problems:

1. Draw a rectangle in which the side lengths are whole numbers.

2. Draw a circle inside the rectangle with a whole number for a radius.

3. Find the ratio of the area of the circle to the area of the rectangle.
   1. – 3. Answers will vary.

Also on Quiz Transparency 5.8

Teach

Why Weather forecasting, health and car insurance, and the lottery are all real-world applications of probability. Students need to understand the basic principles that underlie any probability situation.

Math
CONNECTION

PROBABILITY To review some probability concepts, conduct a simple demonstration in which a few counters or small objects are placed in a bag. Have students calculate the probability of drawing one object and then compare the predicted results with the observed results.

ADDITIONAL
EXAMPLE ❶

Draw three concentric circles on graph paper. Use 1, 3, and 5 units for the diameters; the middle ring is shaded to represent the target area. What is the probability that a dart will land in the target area? $\frac{8}{25}$

Teaching Tip

Ask students to discuss Additional Example 1 in order to determine whether they understand that the probability is equal to the ratio of the shaded region to the total area. You may want to simplify the Example by using a small square inside a larger square instead of three concentric circles.

ADDITIONAL
EXAMPLE ❷

A coin is tossed onto a grid composed of equilateral triangles. The diameter of the coin equals the side length of the triangles. What is the probability that the coin will touch an intersection point of the grid? The probability equals the ratio of the three shaded areas to the area of the equilateral triangle.

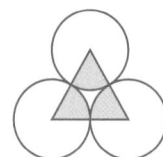

probability $= \frac{\pi\sqrt{3}}{6}$

Theoretical Probability

Consider the following two probability experiments:

a. There are 30 marbles in a bag, and 3 of them are red. A marble is drawn at random from the bag. The probability of drawing a red marble is:

$$P = \frac{3}{30} = 0.1$$

b. A point in the figure at right is selected at random. The probability that the selected point is from area A is:

$$P = \frac{\text{area } A}{\text{area } B}$$

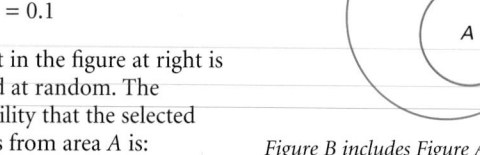

Figure B includes Figure A (the inner circle).

In cases like **a**, which you are probably already familiar with from your algebra studies, the number of possible outcomes can be counted. In cases like **b**, however, the number of possible outcomes is infinite. In these cases, the *areas* of the figures are used as measures of the sets of outcomes.

E X A M P L E ❶ If a dart lands in the blue part of the board below, a prize is given. What is the probability that a dart that hits the board at random will win?

SOLUTION

The area of the blue part of the board is 60 square units. The area of the entire board is 96 square units. The probability that the dart will land in the blue part is $\frac{60}{96}$, or $\frac{5}{8}$.

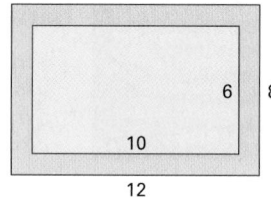

E X A M P L E ❷ In a game, pennies are tossed onto a grid of squares whose sides are equal to the diameter of a penny. To win, a penny must touch or cover an intersection of the grid. What is the probability of winning on a random toss?

SOLUTION

Imagine that circles are drawn around each intersection on the grid, as shown on the following page. If the *center* of a penny falls within one of these circles, the penny itself will touch or cover an intersection.

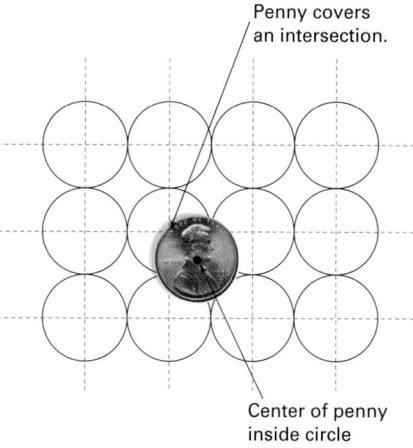

Penny covers an intersection.

Center of penny inside circle

Interdisciplinary Connection

EARTH SCIENCE Probabilities in percent form are often used in weather predictions. Have students use library materials to find out how these probabilities are calculated and why weather predictions are often wrong.

Enrichment

Have students work alone or with others to solve this problem: An infinite plane is covered with circles. Each circle touches four other circles, and the straight lines joining the centers of the circles divide the plane into congruent squares. A dart is thrown at random at the plane. What is the probability that the dart will land inside one of the circles? **Students should realize that this problem is equivalent to the penny game in Example 2. The probability is $\frac{\pi}{4}$.**

The diagram at right represents a single square of the grid. If the center of the penny lands in a shaded region, the toss will be a winning toss.

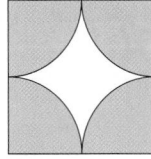

Let a unit be defined as the radius of a penny. Then the width of a square (and the diameter of a penny) will be 2 units.

Area of a shaded region: $A_c = \pi \times 1^2 = \pi$ units^2

Area of a square: $A_s = 2 \times 2 = 4$ units^2

The area of each quarter of a circle is $\frac{\pi}{4}$ units^2, so the area of the shaded regions is π units^2. The theoretical probability, P, of the penny touching or covering a vertex—and of the player winning—is found below.

$$P = \frac{\text{area of shaded regions}}{\text{area of square}} = \frac{\pi}{4}$$

Experimental Probability

In the following Activity, you will use *experimental probability* and the result from Example 2 to estimate the value of π.

Activity
A "Monte Carlo" Method for Estimating π

1. Toss a penny randomly onto the grid paper. Notice whether the coin touches or covers an intersection of two lines. Repeat for a total of 20 tosses. Record the number of tosses that touch or cover an intersection and the number that do not.

2. Share your results with the rest of your class. Find the totals for the entire class. To find the experimental proabability that the penny will touch or cover an intersection, calculate the ratio of the number of tosses that touch or cover an intersection to the total number of tosses. Call this number R.

3. For a large number of tosses, you may assume that the value of the experimental probability, R, will be close to the theoretical probability, P, that you calculated in Example 2. Using this assumption, calculate an estimate for π.

   $R \approx \frac{\pi}{4}$ (assumed to be true for large numbers)
   $\pi \approx R \times 4$

4. Compare your estimate with the known value of π. Determine the relative error of your estimate by using the formula below.

   $E = \frac{|V_t - V_e|}{V_t} \times 100$

   where E = percent of error, V_t = true value, and V_e = estimated value

Inclusion Strategies

HANDS-ON STRATEGIES   It may be difficult for some students to understand why theoretical and experimental probabilities differ. Coin-toss experiments can help explain the differences. Students are likely to know that the chance of tossing heads with a fair coin is 50%. If they work in teams to toss coins a great number of times, they will see that as the number of tosses increases, the experimental results get closer to 50%.

Reteaching the Lesson

USING MODELS   Have students work in small groups to design target games on graph paper. They should compute the probabilities of an object landing on various regions and then perform experiments to see what actually happens.

For a student worksheet of this Activity and detailed Teacher Notes, see page 103 in the Lesson Activities booklet.

Teaching Tip

Students can create their own graph paper for the Activity by using pennies and rulers to measure the grid lines.

Cooperative Learning

Have students work in small groups. Each group will need a supply of toothpicks. They should prepare a large sheet of paper with parallel lines drawn such that the distance between adjacent lines is equal to the length of a toothpick. The probability of a dropped toothpick touching one of the lines is $P = \frac{2}{\pi}$. Students should compare their experimental results with those predicted by the formula.

ASSIGNMENT GUIDE

In Class	1–8
Core	9–14, 15–33 odd
Core Plus	15–39 odd
Review	41–50
Preview	51–56

✐ Extra Practice can be found beginning on page 818.

8. Sample answer: The width of the penny is considered 1 unit, so the shaded area is π, and the square has area 4. The probability that the penny will cover a vertex is $\frac{\pi}{4}$, theoretically. Repeated tosses should produce a number close to $\frac{\pi}{4}$. The more tosses, the more accurate the estimate should be. Multiply by 4 to give an estimate for π.

7. area of unshaded region: $4 - \pi$ probability of penny not touching intersection: $\frac{4-\pi}{4} = 1 - \frac{\pi}{4} \approx 0.21$

Exercises

● *Communicate*

1. Why does a probability always have to be a number from 0 to 1 (or 0% to 100%)?

2. What does it mean for an outcome to have a probability of 0?

3. What does it mean for an outcome to have a probability of 1?

4. Assign a probability to each of the following words:

often	seldom	usually
never	maybe	frequently
sometimes	always	rarely

Rank the words in order from lowest probability to highest. Compare your list with those of your classmates. Which words do you agree on? Which words do you disagree on?

5. In the Activity, how did your estimate of π from 20 tosses compare with the estimate of π from the total tosses for the entire class? How do you think you could improve your estimate of π?

● *Guided Skills Practice*

6. Find the probability that a dart tossed at random will land in the blue area of the figure at right. *(EXAMPLE 1)*
$\frac{14}{25}$, or 0.56

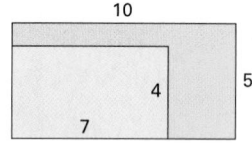

7. If the area of the shaded region at right is π square units and the area of the square is 4 square units, what is the area of the unshaded region? What is the probability that a penny tossed onto a grid with squares equal to the width of a penny will *not* touch or cover an intersection? *(EXAMPLE 2)*

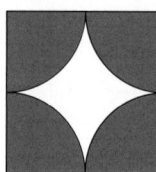

8. Explain how you could use a grid with squares equal to the width of a penny to estimate π. *(ACTIVITY)*

● *Practice and Apply*

Find the probability that a dart tossed at random onto each figure will land in the shaded area.

9. 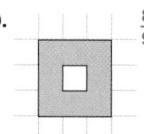 $\frac{8}{9}$

10. 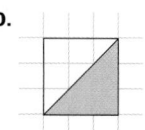 $\frac{1}{2}$

11. 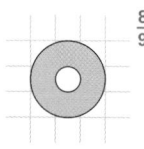 $\frac{8}{9}$

For each spinner below, find the theoretical probability that the arrow will land on red.

12.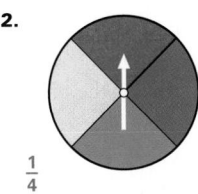

$\frac{1}{4}$

13.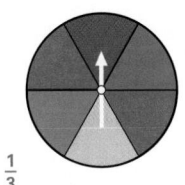

$\frac{1}{3}$

14.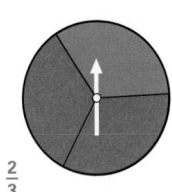

$\frac{2}{3}$

☑ internet connect

**Homework
Help Online**
Go To: go.hrw.com
Keyword:
MG1 Homework Help
for Exercises 15-23

PROBABILITY Convert each probability to a percent.

15. 0.75 75%

16. $\frac{1}{4}$ 25%

17. $\frac{2}{3}$ $66\frac{2}{3}$%

PROBABILITY Convert each percent to a decimal probability.

18. 60% 0.6

19. 50% 0.5

20. $33\frac{1}{3}$% $0.\overline{3}$

PROBABILITY Convert each percent to a fractional probability. Write your answers in lowest terms.

21. 45% $\frac{9}{20}$

22. 80% $\frac{4}{5}$

23. $66\frac{2}{3}$% $\frac{2}{3}$

For Exercises 24–27, refer to the spinner shown at left.

24. What is the probability that the arrow will land on 5? $\frac{1}{5}$

25. What is the probability that the arrow will land on an odd number? $\frac{3}{5}$

26. What is the probability that the arrow will land on an even number? $\frac{2}{5}$

27. Add your results from Exercises 25 and 26. What does this result represent in terms of probability? $\frac{3}{5} + \frac{2}{5} = 1$ One of these events must occur.

28. Design a dartboard in which the probability of a dart landing in a red circle is 0.5.

29. Design a dartboard in which the probability of a dart landing in a red triangle is $\frac{1}{3}$.

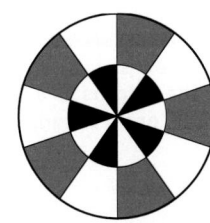

In the dartboard shown at left, the radius of the inner circle is half of the radius of the outer circle.

$\frac{3}{8}$ **30.** What is the probability that a dart thrown at random will hit a red region?

31. What is the probability that a dart thrown at random will hit a black region? $\frac{1}{8}$

32. What is the probability that a dart thrown at random will hit a white region? $\frac{1}{2}$

CHALLENGE

33. The squares in the grid at right are exactly the width of a penny. What is the probability that the center of a penny tossed at random onto the grid will land within a white square, with no part of the penny touching an intersection? $\frac{4 - \pi}{8} \approx 0.11$

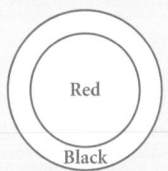
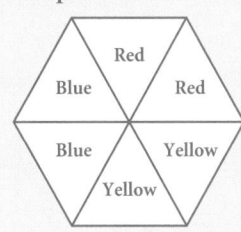

Technology

You may wish to allow students to use calculators to convert fractions to decimals and percents. Students will need a calculator with a random number generator to complete the Portfolio Activity on page 359.

34. METEOROLOGY The weather forecaster predicts an 80% chance of rain. Express this as a probability in decimal and fractional forms. $\frac{4}{5} = 0.8$

35. METEOROLOGY The area of Oklahoma is 69,956 square miles. The area of its capital, Oklahoma City, is 625 square miles. If a tornado touches down at random in Oklahoma, what is the probability it will touch down in Oklahoma City? **0.0089**

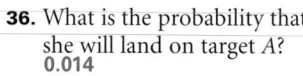

SKYDIVING A skydiver jumps from an airplane and parachutes down to the rectangular field shown below. Assume that she is equally likely to land anywhere in the field.

36. What is the probability that she will land on target *A*? **0.014**

37. What is the probability that she will land on target *B*? **0.089**

38. What is the probability that she will land on target *C*? **0.180**

39. What is the probability that she will miss all three of the targets? **0.717**

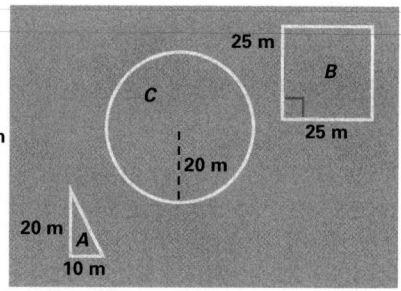

40. TRANSPORTATION At a subway stop in New York City, a train arrives every 5 minutes, waits 1 minute, and then leaves. Suppose that you go to the stop at a random time. Use the diagram below to determine the probability that you will wait 2 minutes or less for a train to arrive. $\frac{3}{5}$

5 min

1 min

Look Back

In Exercises 41–46, three lengths are given. Classify the lengths as sides of a right triangle, an acute triangle, an obtuse triangle, or no triangle. *(LESSONS 4.8 AND 5.4)*

43. not a triangle   **41.** 16, 30, 34 right   **42.** 27, 36, 50 obtuse   **43.** 16, 63, 82

46. not a triangle   **44.** 11, 60, 61 right   **45.** 10, 24, 25 acute   **46.** 8, 15, 23

Find the area of each figure described below. *(LESSONS 5.2 AND 5.3)*

47. a triangle with a base of 4 and a height of 7.5 **15 units²**

48. a parallelogram with a base of 4 and a height of 7.5 **30 units²**

49. a trapezoid with bases of 20 and 30 and a height of 12.6 **315 units²**

50. a circle with a radius of 16 (Note: Use $\frac{22}{7}$ for π.) **804.57 units²**

 **Look Beyond**

Double Ring

Triple Ring

Each sector of the board has a number from 1 to 20.

APPLICATION

RECREATION **The dimensions of a standard dartboard are given below.**

outside edge of double ring to center = 170 mm
outside edge of triple ring to center = 117 mm
outer bull's-eye diameter = 31 mm
inner bull's-eye diameter = 12.7 mm
double and triple rings inside width = 8 mm

Suppose that a thrown dart is equally likely to land anywhere on the board.

51. What is the probability that the dart will land in the double ring? **0.09**

52. What is the probability that the dart will land in the triple ring? **0.06**

53. The highest scoring section of the board is in the 20 sector of the triple ring (60 points). What is the probability of hitting this section? **0.003**

54. The inner bull's-eye (50 points) is the small red circle in the center of the board. What is the probability of hitting the inner bull's-eye? **0.001**

55. The outer bull's-eye (25 points) is the black ring around the inner bull's-eye. What is the probability of hitting the outer bull's-eye? **0.007**

56. When playing darts, why might the actual results be different from the theoretical probabilities? **When playing darts, you are aiming at a certain section instead of throwing at random.**

internet connect

Portfolio Extension
Go To: go.hrw.com
Keyword:
MG1 Buffon

In Exercises 51–56, students calculate the probabilities that a dart will hit a dartboard at various points.

ALTERNATIVE
Assessment

Portfolio Activity

The Portfolio Activity can be used as preparation for the Chapter Project or as a separate activity. In the Portfolio Activity on this page, students will use a random-number generator to generate ordered pairs. They will then calculate the probability that the points will land within a certain region on a graph.

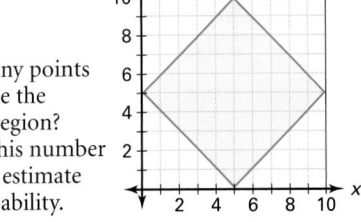

TECHNOLOGY Most scientific and graphics calculators have random-number generators, which generate a number between 0 and 1.

1. Use your calculator to generate two random numbers. Multiply each number by 10 to get a number between 0 and 10, and use the results to create an ordered pair. Round to three decimal places.

For example: 0.6984024365 and 0.3306812911 → (6.984, 3.307)

2. Use this method to generate 20 ordered pairs.

If a point chosen at random is on the 10 × 10 grid shown at right, what is the probability the point will be in the shaded region?

You will use the points generated in Steps 1 and 2 to test this probability.

3. Copy the 10 × 10 grid at right, including the shaded region. Plot the ordered pairs on your grid.

4. How many points are inside the shaded region? Divide this number by 20 to estimate the probability.

5. Was your estimate close to the probability you calculated? Compute the relative error by using the formula $E = \frac{|V_e - V_t|}{V_t} \times 100$, where V_e is the estimated (experimental) value and V_t is the true (theoretical) value.

WORKING ON THE CHAPTER PROJECT

You should now be able to complete Activity 2 of the Chapter Project.

Student Technology Guide

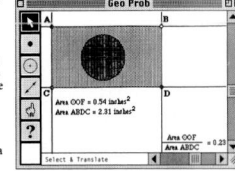

Student Technology Guide
5.8 *Geometric Probability*

To find the probability that a point chosen at random in the rectangle at right will be inside the circle, you must find the ratio of the area of the circle to the area of the rectangle. Using geometry graphics software, you can easily sketch the circle and rectangle, calculate the area of each figure, and compute the ratio of the areas.

Also, the software enables you to enlarge or shrink the circle inside the rectangle. What effect does this have on the probability that a randomly chosen point will be inside the circle?

Sketch each diagram below. Enlarge the circle to find the maximum probability that a point chosen at random will be inside the circle. (Hint: What is the largest radius the circle can have without breaking through the sides of the rectangle?)

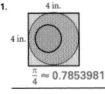
$\frac{\pi}{4} \approx 0.785398163$

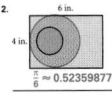
$\frac{\pi}{6} \approx 0.523598776$

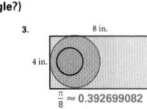
$\frac{\pi}{8} \approx 0.392699082$

4. Suppose that a rectangle has a width of 4 and length of a, where $a \geq 4$. What happens to the maximum probability as a increases? As a increases, the maximum probability decreases.

5. Using the software, construct an equilateral triangle with sides of 5 inches. Place a circle inside it. Calculate the areas of the circle and triangle. Then calculate the ratio of their areas. Find the maximum probability that a point chosen at random within the triangle is also inside the circle. Circle: 6.536 square inches; triangle: 10.825 square inches; the maximum probability is 0.602340798. The maximum probability occurs when the circle is the inscribed circle of the triangle.

6. Make a conjecture about the circle that maximizes the probability that a point chosen at random inside a regular polygon will also be inside the circle. Use the software to illustrate your conjecture for a regular pentagon. It is the inscribed circle of a regular polygon which maximizes the probability a point chosen at random inside the polygon is also inside the circle. (Check students' sketches of the regular pentagon and inscribed circle. The maximum probability is 0.8655408331.)

Focus

In this project, students generate data and derive an area formula known as *Pick's formula*. Students will find that the equation $A = \frac{Nb}{2} + Ni - 1$, or an equivalent equation, will relate the three variables in their data lists.

Motivate

Point out to students that they have learned the area formulas for circles, triangles, and a few other polygons. Ask them to describe how they would find the area of a polygon with an irregular boundary. (Possible answer: Divide the polygon into triangles, measure or calculate an altitude for each triangle, find the area of each triangle, and add. Some students might suggest using geometry graphics software, which can be used to find the area of any polygon.)

After students have read the instructions for each Activity in the project, ask if they think that they can develop an area formula that will work for any polygon. The answer is no. The formula given in the text only works if the vertices of the polygon are on the intersection points of a square grid.

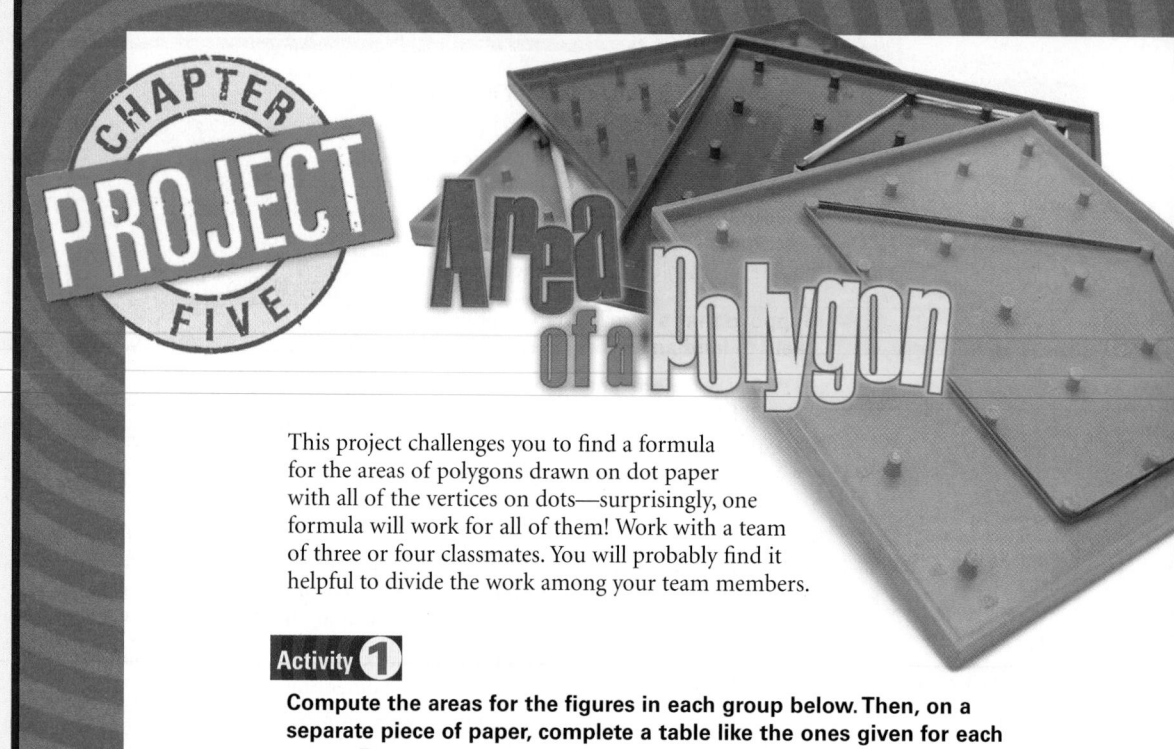

This project challenges you to find a formula for the areas of polygons drawn on dot paper with all of the vertices on dots—surprisingly, one formula will work for all of them! Work with a team of three or four classmates. You will probably find it helpful to divide the work among your team members.

Activity 1

Compute the areas for the figures in each group below. Then, on a separate piece of paper, complete a table like the ones given for each group. For each figure, N_b is the number of dots on the boundary, N_i is the number of dots in the interior, and A is the area.

GROUP A

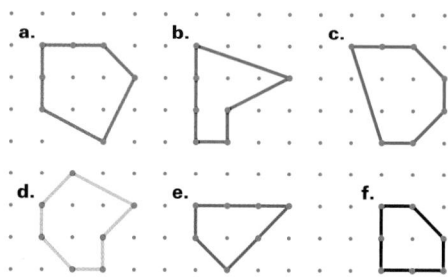

		Group A	
	N_b	N_i	A
a.	7	4	$6\frac{1}{2}$
b.	?	?	?
c.	?	?	?
d.	?	?	?
e.	?	?	?
f.	?	?	?

GROUP B

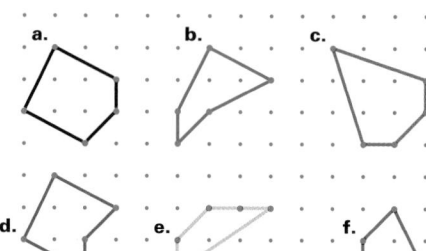

		Group B	
	N_b	N_i	A
a.	?	?	?
b.	?	?	?
c.	?	?	?
d.	?	?	?
e.	?	?	?
f.	?	?	?

Activity 1

	Part A		
	N_b	N_i	A
a.	7	4	$6\frac{1}{2}$
b.	7	2	$4\frac{1}{2}$
c.	7	4	$6\frac{1}{2}$
d.	7	3	$5\frac{1}{2}$
e.	7	1	$3\frac{1}{2}$
f.	7	1	$3\frac{1}{2}$

	Part B		
	N_b	N_i	A
a.	5	4	$5\frac{1}{2}$
b.	5	2	$3\frac{1}{2}$
c.	5	4	$5\frac{1}{2}$
d.	5	3	$4\frac{1}{2}$
e.	5	1	$2\frac{1}{2}$
f.	5	1	$2\frac{1}{2}$

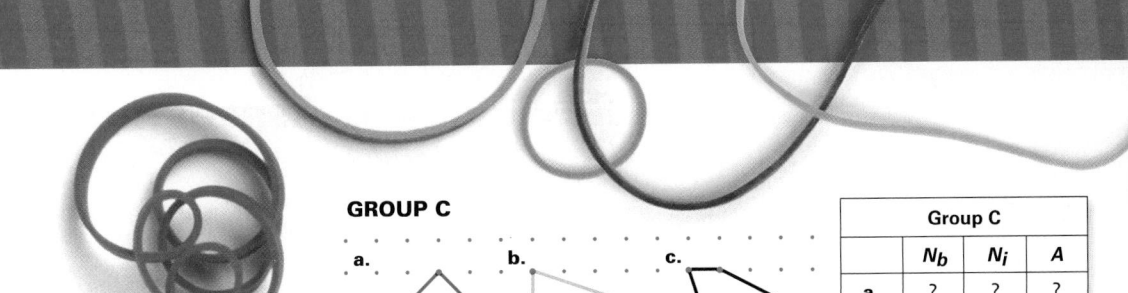

GROUP C

a. 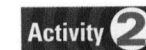   b.   c.

d.    e.   f.

Group C			
	N_b	N_i	A
a.	?	?	?
b.	?	?	?
c.	?	?	?
d.	?	?	?
e.	?	?	?
f.	?	?	?

Activity 2

To help you determine the formula, notice that all the figures in each group of Activity 1 have the same number of boundary points.

1. What is the pattern for calculating the area?

2. Write the pattern as a formula in terms of N_b and N_i.

3. Test your formula by creating new figures on dot paper and calculating their areas.

The formula you have just discovered was originally found by George Pick in 1899.

Activity 3

Refer to the figure at right.

1. Calculate the probability that a point chosen at random in the grid will be inside the shaded region.

2. Use a random-number generator to create 20 ordered pairs. Copy the figure onto graph paper and plot the ordered pairs on the figure. Divide the number of points that fall inside the shaded region by 20. Is your result close to the probability you calculated in Step 1?

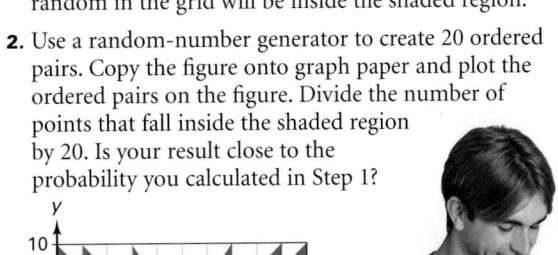

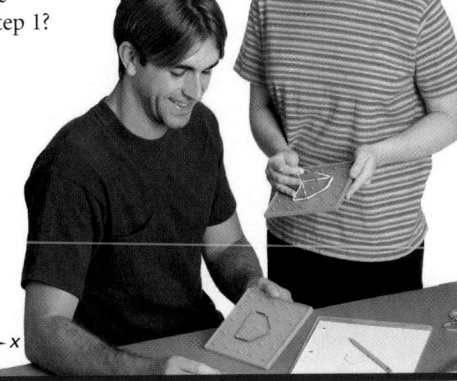

Cooperative Learning

Students doing the project with a partner or in small groups can benefit from working each problem independently and then comparing results. This allows students to correct errors that may have ocurred as they were counting grid points or unit squares.

Discuss

After all students have finished Activities 1, 2, and 3, have them share the methods they used for finding the areas and the probabilities. In Activity 3, compile the data to compute a probability for the entire class. Compare the theoretical probabilities calculated in the first part of Activity 3 with the experimental probability calculated with the class data.

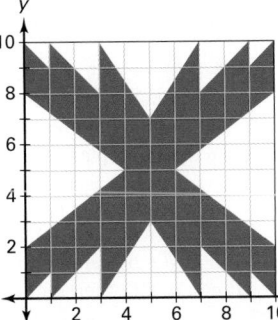

Part C			
	N_b	N_i	A
a.	6	3	5
b.	6	2	4
c.	6	4	6
d.	6	2	4
e.	6	1	3
f.	6	1	3

Activity 2

The pattern is: half the number of boundary dots plus the number of interior dots, minus one.

$$\text{Area} = \frac{1}{2}N_b + N_i - 1$$

Activity 3

1. 54%

2. Yes, the answer is close to the probability calculated in Step 1.

Chapter Review and Assessment

VOCABULARY

POSTULATES, FORMULAS, AND THEOREMS

Lesson	Number	Postulate or Theorem
5.1	5.1.3 Postulate: Sum of Areas	If a figure is composed of non-overlapping regions A and B, then the area of the figure is the sum of the areas of regions A and B.
	5.1.4 Perimeter of a Rectangle	The perimeter of a rectangle with base b and height h is $P = 2b + 2h$.
	5.1.5 Postulate: Area of a Rectangle	The area of a rectangle with base b and height h is $A = bh$.
5.2	5.2.1 Area of a Triangle	The area of a triangle with base b and height h is $A = \frac{1}{2}bh$.
	5.2.2 Area of a Parallelogram	The area of a parallelogram with base b and height h is $A = bh$.
	5.2.3 Area of a Trapezoid	The area of a trapezoid with bases b_1 and b_2 and height h is $A = \frac{1}{2}(b_1 + b_2)h$.
5.3	5.3.2 Circumference of a Circle	The circumference of a circle with diameter d and radius r is $C = \pi d$ or $C = 2\pi r$.
	5.3.3 Area of a Circle	The area of a circle with radius r is $A = \pi r^2$.
5.4	5.4.1 Pythagorean Theorem	For any right triangle, the square of the length of the hypotenuse is equal to the sum of the squares of the lengths of the legs; that is, $c^2 = a^2 + b^2$.
	5.4.2 Converse of the Pythagorean Theorem	If the square of the length of one side of a triangle equals the sum of the squares of the lengths of the other two sides, then the triangle is a right triangle.
	5.4.3 Pythagorean Inequalities	For any triangle ABC, with c as the length of the longest side: If $c^2 = a^2 + b^2$, then $\triangle ABC$ is a right triangle. If $c^2 > a^2 + b^2$, then $\triangle ABC$ is an obtuse triangle. If $c^2 < a^2 + b^2$, then $\triangle ABC$ is an acute triangle.

Chapter Test, Form A

NAME _____ CLASS _____ DATE _____

Chapter Assessment
Chapter 5, Form A, page 1

Write the letter that best answers the question or completes the statement.

d 1. For a fixed perimeter, what shape has the maximum area?
 a. parallelogram b. trapezoid c. square d. circle

c 2. The perimeter of a rectangle is 54 centimeters and the base length is twice the height. What is the area?
 a. 108 sq cm b. 154 sq cm
 c. 162 sq cm d. 324 sq cm

a 3. The perimeter of a rectangle is 40 centimeters. If the width is 8 centimeters, what is the length?
 a. 12 cm b. 16 cm c. 32 cm d. 48 cm

c 4. What is the area, in square units, of parallelogram *ABDE*?
 a. 120 b. 136
 c. 210 d. 240

a 5. What is the area, in square units, of triangle *BCD*?
 a. 60 b. 128
 c. 120 d. cannot be determined

b 6. What is the area, in square units, of trapezoid *ABCE*?
 a. 240 b. 270 c. 352 d. 420

b 7. The perimeter of *ABCE* is _____.
 a. 63 b. 69 c. 75 d. cannot be determined

d 8. What is the measure of the base of a triangle with a height of 20 centimeters and an area of 100 centimeters?
 a. 5 cm b. 6 cm c. 8 cm d. 10 cm

b 9. To the nearest tenth, the circumference of a circle with a radius of 3 centimeters is
 a. 9.4 cm b. 18.8 cm c. 28.3 cm d. 31.4 cm

b 10. A right triangle has lengths of 15 centimeters and 20 centimeters. What is the length of the hypotenuse?
 a. 18 cm b. 25 cm c. 35 cm d. 50 cm

NAME _____ CLASS _____ DATE _____

Chapter Assessment
Chapter 5, Form A, page 2

c 11. What is the area, in square units, of the shaded region?
 a. 34.4 b. 108
 c. 339.3 d. 37.7

b 12. If $AB = 10$, what is BC?
 a. 5 b. $5\sqrt{3}$
 c. $10\sqrt{3}$ d. 20

a 13. If $BC = 9$, what is AC?
 a. $3\sqrt{3}$ b. 4.5 c. $9\sqrt{3}$ d. 18

b 14. What is the area, in square units, of the figure at right?
 a. 4 b. 8
 c. 16 d. 32

b 15. The distance between $(5, -2)$ and $(-3, 4)$ is _____.
 a. 8 b. 10 c. 64 d. 100

c 16. The midpoint of the line segment from $(-3, 4)$ and $(-1, -6)$ is _____.
 a. $(-1, 5)$ b. $(-1, -2)$ c. $(-2, -1)$ d. $(-2, 1)$

b 17. The triangle formed by *H*, *I*, and *J* is _____.
 a. scalene b. isosceles
 c. right d. equilateral

c 18. What is the probability that a dart thrown at random will land in a circle at right?
 a. 52% b. 60%
 c. 79% d. 85%

c 19. What is the probability that the dart will *not* land in a circle?
 a. 48% b. 40%
 c. 21% d. 15%

Lesson	Number	Postulate or Theorem
5.5	5.5.1 45-45-90 Triangle Theorem	In any 45-45-90 triangle, the length of the hypotenuse is $\sqrt{2}$ times the length of a leg.
	5.5.2 30-60-90 Triangle Theorem	In any 30-60-90 triangle, the length of the hypotenuse is 2 times the length of the shorter leg, and the length of the longer leg is $\sqrt{3}$ times the length of the shorter leg.
	5.5.3 Area of a Regular Polygon	The area of a regular polygon with apothem a and perimeter p is $A = \frac{1}{2}ap$.
5.6	5.6.1 Distance formula	On a coordinate plane, the distance between two points (x_1, y_1) and (x_2, y_2) is $d = \sqrt{(x_2 - x_1)^2 + (y_2 - y_1)^2}$.

Key Skills & Exercises

LESSON 5.1

Key Skills

Find the perimeter of a polygon.

Find the perimeter of polygon *ABFCDE* below.

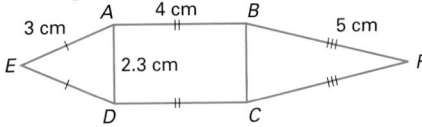

Polygon *ABCDEF* has a perimeter of
$3 + 4 + 5 + 5 + 4 + 3 = 24$ cm.

Find the area of a rectangle.

Find the area of rectangle *ABCD* above.

The base of the rectangle is 4 and the height is 2.3, so the area is $4 \times 2.3 = 9.2$ cm².

Exercises

Find the perimeter of each polygon.

1.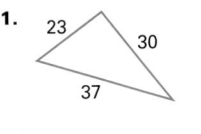

2.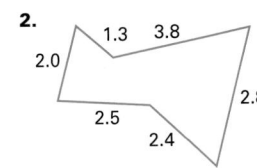

Find the area of each rectangle.

3.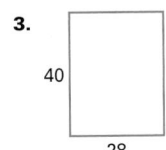

4.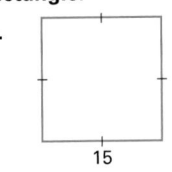

LESSON 5.2

Key Skills

Find the area of a triangle.

Find the area of a triangle with a base of 6 and a height of 8.

The area is $\frac{1}{2} \times 6 \times 8 = 24$.

Find the area of a parallelogram.

Find the area of a parallelogram with a base of 6 and a height of 8.

The area is $6 \times 8 = 48$.

Exercises

For Exercises 5–8, find the area of the given figure.

5. a triangle with a base of 4 and a height of 12

6. a triangle with a base of 7 and a height of 5

7. a parallelogram with a base of 4 and a height of 9

---

1. 90 units

2. 14.8 units

3. 1120 units²

4. 225 units²

5. 24 units²

6. 17.5 units²

7. 36 units²

Chapter Test, Form B

NAME _____ CLASS _____ DATE _____

Chapter Assessment
Chapter 5, Form B, page 1

Find the perimeter and the area of each figure.

1.
$P = 24$ cm; $A = 28$ sq cm

2.
$P = 38$ in.; $A = 42$ sq in.

3.
$P = 24$ cm; $A = 24\sqrt{3} \approx 41.6$

4. The perimeter of a rectangle is 46 cm. If its height is 8 cm, find its area. _____ 120 sq cm

5. The perimeter of a rectangle is 48 cm. If its base is three times its height, find its dimensions. _____ base = 18 cm; height = 6 cm

Given the area and the information shown in the diagram, find x.

6. $A = 28.14$ square units _____ $x = 4.2$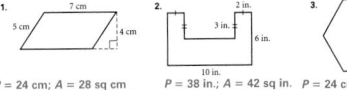

7. $A = 19.58$ square inches _____ $x = 8.9$ in.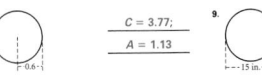

Find the circumference and area of each circle.

8.
$C = 3.77$;
$A = 1.13$

9.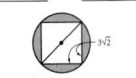
$C = 47.12$ in.;
$A = 176.71$ sq in.

Use the Triangle Inequality Theorem to determine whether the given lengths can be lengths of a triangle. If so, tell whether the triangle is acute, right, or obtuse.

10. 18, 24, 30 _____ right

11. 12, 15, 25 _____ obtuse

12. 10, 13, 24 _____ not possible

13. 15, 18, 21 _____ acute

14. Find the area of the shaded region at right. _____ ≈ 10.27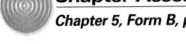

NAME _____ CLASS _____ DATE _____

Chapter Assessment
Chapter 5, Form B, page 2

Find the value of each variable.

15.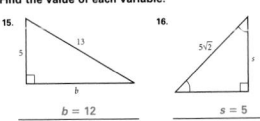
$b = 12$

16.
$s = 5$

17.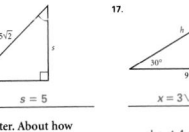
$x = 3\sqrt{3}$; $h = 9$

18. A tricycle wheel has a six-inch diameter. About how many revolutions must the wheel make to travel 6 feet? _____ about 4 revolutions

19. An isosceles trapezoid has nonparallel sides 15 inches long. One parallel base is twice as long as the other. The perimeter of the trapezoid is 60 inches. Find the area of the trapezoid. _____ $75\sqrt{2}$

For Exercises 20–22, refer to the coordinate plane at right.

20. Use the distance formula to find *AB*, *BC*, and *AC*.
$AB = 3\sqrt{10}$, $BC = 2\sqrt{10}$, $AC = \sqrt{130}$

21. Use the converse of the Pythagorean Theorem to show that $\triangle ABC$ is a right triangle. $(AB)^2 + (BC)^2 = 90 + 40 = 130 = (AC)^2$

22. Find the area of $\triangle ABC$. _____ 30 sq units

In Exercises 23 and 24, find the perimeter and area of each regular polygon.

23.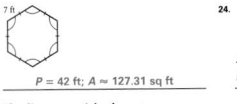
$P = 42$ ft; $A \approx 127.31$ sq ft

24.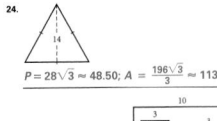
$P = 28\sqrt{3} \approx 48.50$; $A = \frac{196\sqrt{3}}{3} \approx 113.1$

25. The diagram at right shows two squares inside a rectangle. Estimate the probability that a dart randomly thrown at the rectangle will land inside one of the squares. _____ $\frac{18}{60} = 30\%$

8. 24 units^2

9. 11π units

10. 5π units

11. 100π units^2

12. $\frac{1}{4}$π units^2

13. 29

14. 20

15. acute

16. right

17. 50$\sqrt{2}$ units

18. 34$\sqrt{2}$ units

19. 9$\sqrt{3}$ units; 18$\sqrt{3}$ units

Find the area of a trapezoid.

Find the area of a trapezoid with bases of 6 and 10 and a height of 8.

The area is $\frac{1}{2}(6 + 10) \times 8 = 64$.

8. a trapezoid with bases of 3 and 5 and a height of 6

LESSON 5.3

Key Skills

Find the circumference of a circle.

Find the circumference of a circle with a diameter of 8 inches.

The circumference is $\pi \times 8 \approx 25.12$ inches.

Find the area of a circle.

Find the area of a circle with a radius of 3 meters.

The area is $\pi \times 3^2 \approx 28.26$ square meters.

Exercises

9. Find the circumference of a circle with a diameter of 11.

10. Find the circumference of a circle with a radius of 2.5.

11. Find the area of a circle with a radius of 10.

12. Find the area of a circle with a diameter of 1.

LESSON 5.4

Key Skills

Find the lengths of the sides of a right triangle by using the Pythagorean Theorem.

Find the missing leg of a right triangle with a leg of 40 and a hypotenuse of 41.

If the missing leg is a, then $a^2 + 40^2 = 41^2$, or $a^2 + 1600 = 1681$. Thus, $a^2 = 81$, and $a = 9$.

Determine whether a triangle is right, acute, or obtuse by using the Pythagorean inequalities.

A triangle has side lengths of 28, 45, and 50. Classify the triangle as right, acute, or obtuse.

$50^2 = 2500$ and $28^2 + 45^2 = 2809$
Since $2500 < 2809$, the triangle is acute.

Exercises

Two side lengths of a triangle with hypotenuse c are given. Find the third side.

13. $a = 20$ $b = 21$ $c = \underline{?}$

14. $a = \underline{?}$ $b = 99$ $c = 101$

Three side lengths of a triangle are given. Classify the triangle as right, acute, or obtuse.

15. 48, 63, and 65

16. 48, 55, and 73

LESSON 5.5

Key Skills

Find the side lengths of 45-45-90 triangles.

The hypotenuse of a 45-45-90 triangle is 18. Find the length of the legs.

The hypotenuse is $\sqrt{2}$ times the length of a leg, so the legs have lengths of
$\frac{18}{\sqrt{2}} = \frac{18\sqrt{2}}{2} = 9\sqrt{2}$.

Exercises

17. The hypotenuse of a 45-45-90 triangle is 100. Find the length of the legs.

18. The length of each leg of a 45-45-90 triangle is 34. Find the hypotenuse.

19. The longer leg of a 30-60-90 triangle is 27. Find the lengths of the shorter leg and the hypotenuse.

Find the side lengths of 30-60-90 triangles.

The longer leg of a 30-60-90 triangle is 12. Find the shorter leg and the hypotenuse.

The longer leg is $\sqrt{3}$ times the shorter leg, so the shorter leg has a length of $\frac{12}{\sqrt{3}} = \frac{12\sqrt{3}}{3} = 4\sqrt{3}$. The hypotenuse is twice the length of the shorter leg, or $8\sqrt{3}$.

20. The hypotenuse of a 30-60-90 triangle is 16. Find the lengths of the shorter leg and the longer leg.

LESSON 5.6

Key Skills

Determine the distance between two points on a coordinate plane.

Find the distance between the points $(4, 1)$ and $(2, 3)$.

$$d = \sqrt{(2-4)^2 + (3-1)^2} = \sqrt{8} = 2\sqrt{2}$$

Estimate the area under a curve by the method of quadrature.

Use the method of quadrature to estimate the area between the curve $y = x^2$ and the x-axis from $x = 1$ to $x = 4$.

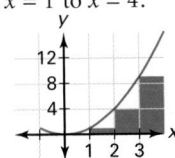 The area can be estimated by three rectangles with a base of 1 and the following heights:

$1^2 = 1 \qquad 2^2 = 4 \qquad 3^2 = 9$

Thus, the area is approximately $1 + 4 + 9 = 14$.

Exercises

Find the distance between each pair of points.

21. $(0, 0)$ and $(6, 8)$

22. $(3, 3)$ and $(7, 2)$

23. $(-2, 4)$ and $(3, -1)$

24. Use the method of quadrature to estimate the area between the curve $y = x^3$ and the x-axis from $x = 1$ to $x = 4$.

LESSON 5.7

Key Skills

Prove theorems by using coordinate geometry.

Prove that if the diagonals of a rectangle are perpendicular to each other, then the rectangle is a square.

A general rectangle can be drawn with vertices at $(0, 0)$, $(a, 0)$, (a, b), and $(0, b)$. The slopes of the diagonals are $\frac{b}{a}$ and $-\frac{b}{a}$. If the diagonals are perpendicular to each other, then $\left(\frac{b}{a}\right)\left(-\frac{b}{a}\right) = -1$, so $\frac{b^2}{a^2}$, which means that $b = a$ (or $b = -a$). Thus, the lengths of the sides of the rectangle are the same, so the rectangle is a square.

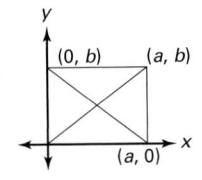

Exercises

In Exercises 25–28, you will prove that if the diagonals of a parallelogram are equal, the parallelogram is a rectangle.

25. Draw a general parallelogram in a coordinate plane with three vertices at $B(0, 0)$, $A(a, 0)$, and $C(b, c)$. Give the coordinates of the fourth vertex, D.

26. Write expressions for the lengths of the diagonals of the parallelogram you drew in Exercise 25.

27. Assume the diagonals are the same length. Set the expressions from Exercise 26 equal to each other and prove that a or b must equal 0.

28. Use your result from Exercise 27 to complete the proof.

---

20. 8 units; $8\sqrt{3}$ units

21. 10

22. $\sqrt{17}$

23. $5\sqrt{2}$

24. 67.5 units^2 Answers will vary. 63.75 units^2 is the exact answer.

25. $D(a + b, c)$

26. $BD = \sqrt{(a + b - 0)^2 + (c - 0)^2}$
$\qquad = \sqrt{(a + b)^2 + c^2}$
$AC = \sqrt{(b - a)^2 + (c - 0)^2}$
$\qquad = \sqrt{(b - a)^2 + c^2}$

27.
$$BD = AC$$
$$\sqrt{(a + b)^2 + c^2} = \sqrt{(b - a)^2 + c^2}$$
$$(a + b)^2 + c^2 = (b - a)^2 + c^2$$
$$(a + b)^2 = (b - a)^2$$
$$a^2 + 2ab + b^2 = b^2 - 2ab + a^2$$
$$4ab = 0$$
$$a = 0 \text{ or } b = 0$$

28. If $a = 0$, then A and B are the same point and C and D are the same point and $ABCD$ becomes a segment, not a parallelogram. Thus $a \neq 0$. If $b = 0$ then C is $(0, C)$. Thus C is on the y-axis and therefore $m\angle B = 90°$. Thus $ABCD$ is a rectangle.

29. 0.25

30. 0.375

31. 0.1875

32. 0.67

33. ≈ 16.52 feet; yes; ≈ 4.13 feet

34. 822.65 m, 571.33 m, 27,879.64 m²

Key Skills

Find the probability of an event.

What is the probability that a randomly generated point with $0 \leq x \leq 5$ and $0 \leq y \leq 5$ will land in a circle with radius 2 centered at (3, 3)?

The diagram below represents the points with $0 \leq x \leq 5$ and $0 \leq y \leq 5$ and the circle with a radius of 2.

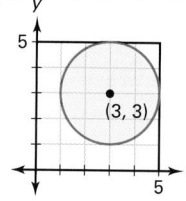

The area of the circle is 4π, and the area of the square is 25, so the probability is $\frac{4\pi}{25} \approx 0.50$.

Exercises

Find the probability that a dart tossed randomly onto each figure will land in the shaded area.

29.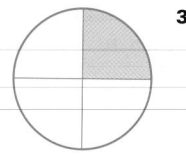

30.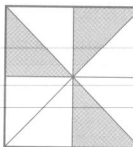

31.

32.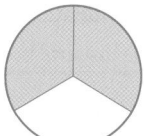

Applications

33. SAFETY You have a 20-foot ladder to climb onto a wall that is 16 feet high. Due to safety considerations, the ladder must extend at least 3 feet past the top of the wall, and the bottom of the ladder must be placed away from the wall at a distance of $\frac{1}{4}$ of the length of the ladder from the ground to where the ladder rests against the roof. Use the diagram at right to find x, the length of the ladder from the ground to the wall. Is the ladder long enough? How far will the bottom of the ladder be from the wall?

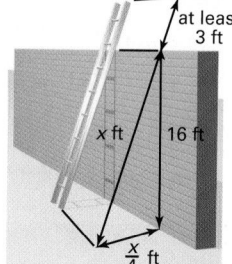

34. SPORTS A running track has straight sides and semicircular ends. The outer radius of the semicircles is 80 meters, the inner radius is 40 meters, and the straight sides are 160 meters long. What is the perimeter of the outside edge of the track? What is the perimeter of the inside edge of the track? What is the area of the track?

Chapter Test

Use the labeled figure below to find the indicated perimeters and areas.

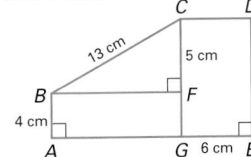

1. the perimeter of pentagon *ABCDE* **50 cm**
2. the perimeter of hexagon *ABFCDE* **54 cm**
3. the area of rectangle *CDEG* **54 cm²**
4. the area of hexagon *ABFCDE* **102 cm²**

Find the area of the given figure.

5. a triangle with a base of 9 and height of 15 **67.5 units²**
6. a parallelogram with a height of 8 and base of 12 **96 units²**
7. a trapezoid with bases of 5 and 9 and a height of 7 **49 units²**

Find the circumference and area of each circle. Use 3.14 for π.

8. *r* = 8   9. *r* = 26   10. *d* = 30   11. *d* = 42

12. **CONSUMER ECONOMICS** Which is the better deal: a 12-inch round pizza that costs $8 or a 12-inch square pizza that costs $8? Explain. **square; larger area**

Find the area of the shaded region. Use 3.14 for π.

13. **174.96 units²**

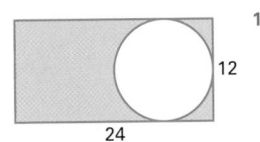

For Exercises 14–17, two side lengths of a right triangle are given. Find the missing side. Leave your answers in radical form.

14. *a* = 7, *b* = 24, *c* = _?_ **25**
15. *a* = 10, *b* = _?_, *c* = 26 **24**
16. *a* = _?_, *b* = 12, *c* = 16 **4√7**
17. *a* = 6, *b* = 6, *c* = _?_ **6√2**

18. The hypotenuse of a 45-45-90 triangle is 80. Find the lengths of the legs. **40√2**
19. The length of each leg of a 45-45-90 triangle is 28. Find the hypotenuse. **28√2**
20. The longer leg of a 30-60-90 triangle is 15. Find the lengths of the shorter leg and the hypotenuse. **5√3; 10√3**
21. The hypotenuse of a 30-60-90 triangle is 24. Find the lengths of the shorter leg and the longer leg. **12; 12√3**

Find the distance between each pair of points. Leave your answers in radical form.

22. (−3, −3) and (5, 3) **10 units**
23. (2, 3) and (5, 7) **5 units**
24. (0, 0) and (−1, −3)
25. (−6, 3) and (4, 9)

In Exercises 26–30, prove the Trapezoid Midsegment Theorem by using coordinate geometry.

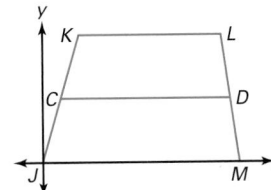

26. Trapezoid *JKLM* above has vertices at *J*(0, 0), *K*(2*p*, 2*q*), *L*(2*r*, _?_), and *M*(2*s*, _?_).
27. Find the coordinates of *C*, the midpoint of *JK*, and *D*, the midpoint of *LM*.
28. Find the lengths of *JM*, *KL*, and *CD*.
29. Prove that the length of the midsegment is the average of the lengths of the bases.
30. Find the slopes of *JM*, *KL*, and *CD*. Prove that the midsegment is parallel to the bases.

Find the probability that a dart tossed randomly onto each figure will land in the shaded area.

31. **0.625**

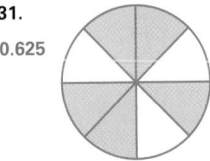

32. **0.5**

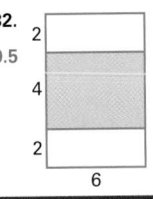

---

8. 50.24 units; 200.96 units²
9. 163.28 units; 2122.64 units²
10. 94.2 units; 706.5 units²
11. 131.88 units; 1384.74 units²
24. √10 units
25. 2√34 units
26. 2*q*; 0
27. *C*(*p*, *q*); *D*(*r* + *s*, *q*)
28. *JM* = 2*s*; *KL* = 2*r* − 2*p*; *CD* = *s* + *r* − *p*
29. $\frac{(JK + KL)}{2} = \frac{(2s + 2r - 2p)}{2}$
   $= s + r - p = CD$
30. All three slopes = 0; all three segments have the same slope. Therefore, the midsegment is parallel to the bases.

College Entrance Exam Practice

Multiple-Choice Samples

The first half of the Cumulative Assessment contains a multiple-choice section. This part of the Cumulative Assessment consists of items commonly found on standardized tests.

Free-Response Grid Samples

The second half of the Cumulative Assessment consists of a free-response section. This part requires student-produced response items like those commonly found on college entrance exams. These questions require the use of machine-scored answer grids. You may wish to have students practice answering these items in preparation for standardized tests.

1. a

2. c

3. a

4. Parallelogram

5. Hexagon

6. Trapezoid

7. Circle

8. c

MULTIPLE-CHOICE For Questions 1–3, write the letter that indicates the best answer.

1. Which of the following are possible side lengths of a triangle? *(LESSON 4.8)*

   a. 8, 18, 24
   b. 8, 16, 27
   c. 8, 12, 20
   d. 8, 8, 21

2. *RSTU* is a parallelogram. Find *ST*. *(LESSON 4.5)*

$3n - 20$ $n + 12$

   a. 16
   b. 27
   c. 28
   d. 10

3. Find the area of the circle below. *(LESSON 5.3)*

$r = 5$

   a. 25π
   b. 6.25π
   c. 10π
   d. 100π

Give the most specific name for each figure. It may help to draw a possible example.
(LESSONS 3.1, 4.6, AND 5.3)

4. quadrilateral *ABCD*, in which $AB = CD$ and $AD = BC$

5. polygon *MNOPQR*

6. quadrilateral *FGHI*, in which $FG = HI$, $\overline{FG} \nparallel \overline{HI}$, and $\overline{FI} \parallel \overline{GH}$

7. a plane figure in which all points are equidistant from point *P*

For Items 8–12, refer to the diagram below. Complete the proof that △*ABD* **is isosceles.**
(LESSON 4.4)

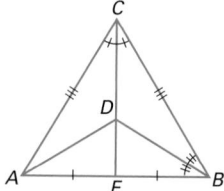

8. $\angle CAB \cong \angle CBA$ by ___?___.
   a. CPCTC
   b. ASA
   c. Isosceles Triangle Theorem
   d. Converse of the Isosceles Triangle Theorem

9. $\triangle ACD \cong \triangle BCD$ by ___?___.
 a. Isosceles Triangle Theorem
 b. SAS
 c. ASA
 d. Angle Addition Postulate

10. $\angle CAD \cong \angle CBD$ by ___?___.
 a. CPCTC
 b. ASA
 c. Isosceles Triangle Theorem
 d. definition of angle bisector

11. $\angle DAE \cong \angle DBE$ by ___?___.
 a. Isosceles Triangle Theorem
 b. Converse of the Isosceles Triangle Theorem
 c. Angle Addition Postulate
 d. SAS

12. $\triangle ABD$ is isosceles by ___?___.
 a. definition of isosceles triangle
 b. CPCTC
 c. Isosceles Triangle Theorem
 d. Converse of the Isosceles Triangle Theorem

In Items 13–16, suppose that $\triangle ABC$ has vertices at A (–1, 8), B (4, 3), and C (1, 2).

13. Find the slope of each side of $\triangle ABC$.
 (LESSON 3.8)

14. Prove that $\triangle ABC$ is a right triangle.
 (LESSON 3.8)

15. Find the midpoint of each side of $\triangle ABC$.
 (LESSON 3.8)

16. Find the length of each side of $\triangle ABC$. Round your answers to the nearest hundredth.
 (LESSON 5.6)

FREE-RESPONSE GRID

Items 17–20 may be answered by using a free-response grid such as that commonly used by standardized-test services.

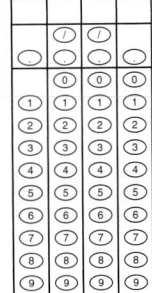

For Items 17–19, refer to the figure below.
(LESSONS 5.2 AND 5.4)

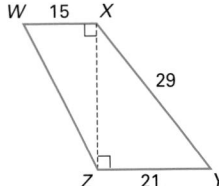

17. $XZ =$ ___?___

18. $WZ =$ ___?___

19. area of trapezoid $WXYZ =$ ___?___

20. Find the length of a side of a square with the same area as a circle with a radius of 10. Round your answer to the nearest tenth.
 (LESSONS 5.1 AND 5.3)

9. b

10. a

11. c

12. d

13. -1; $\frac{1}{3}$; -3

14. Since the slope of $\overline{BC}$ and the slope of $\overline{AC}$ are negative reciprocals, or $\frac{1}{3} \cdot (-3) = -1$, sides $\overline{BC}$ and $\overline{AC}$ are perpendicular, so $\triangle ABC$ is a right triangle.

15. $\left(\frac{3}{2}, \frac{11}{2}\right)$; $\left(\frac{5}{2}, \frac{5}{2}\right)$; $(0, 5)$

16. 7.07; 3.16; 6.32

17. $XZ = 20$ units

18. $WZ = 25$ units

19. 360 units^2

20. $s \approx 17.7$ units

Shapes in Space

CHAPTER PLANNING GUIDE

Lesson	6.1	6.2	6.3	6.4	6.5	6.6	Project and Review
Pupil's Edition Pages	372–378	379–387	388–395	396–401	402–408	409–417	418–427
Practice and Assessment							
Extra Practice (Pupil's Edition)	836	836	837	837	838	838	
Practice Workbook	37	38	39	40	41	42	
Practice Masters Levels A, B, and C	109–111	112–114	115–117	118–120	121–123	124–126	
Standardized Test Practice Masters	42	43	44	45	46	47	48
Assessment Resources	72	73	74	76	77	78	75, 79–84
Visual Resources							
Lesson Presentation Transparencies Vol. 1	142–145	146–149	150–153	154–157	158–161	162–165	
Teaching Transparencies	50–52	52–53	54–55	56–57			
Answer Key Transparencies	227–233	234–239	240–242	243–248	249–266	267–271	272–280
Quiz Transparencies	6.1	6.2	6.3	6.4	6.5	6.6	
Teacher's Tools							
Reteaching Masters	73–74	75–76	77–78	79–80	81–82	83–84	
Make-Up Lesson Planner for Absent Students	37	38	39	40	41	42	
Student Study Guide	37	38	39	40	41	42	
Spanish Resources	37	38	39	40	41	42	
Block Scheduling Handbook							12–13
Activities and Extensions							
Lesson Activities	105–109	110–116	117–119				
Enrichment Masters	37	38	39	40	41	42	
Cooperative-Learning Activities	37	38	39	40	41	42	
Problem-Solving/ Critical Thinking	37	38	39	40	41	42	
Student Technology Guide			37–38			39–40	
Long Term Projects							21–24
Writing Activities for Your Portfolio							16–18
Tech Prep Masters							25–28
Building Success in Mathematics							15–17

LESSON PACING GUIDE

Lesson	6.1	6.2	6.3	6.4	6.5	6.6	Project and Review
Traditional	2 days	2 days	2 days	2 days	2 days	2 days	2 days
Block	1 day	1 day	1 day	1 day	1 day	1 day	1 day
Two-Year	4 days	4 days	4 days	4 days	4 days	4 days	4 days

CONNECTIONS AND APPLICATIONS

Lesson	6.1	6.2	6.3	6.4	6.5	6.6	Review
Algebra	377		393	400, 401	403, 405, 406, 408		
Geometry	372–378	379–387	388–395	396–401		409–417	420–427
Coordinate Geometry			394		402–408		
Technology					407		425
Business and Economics			393				424
Life Skills	372	385		400			424
Science and Technology	377, 378	383	390, 391, 394	401	406, 407		424
Sports and Leisure	377						

BLOCK-SCHEDULING GUIDE

Day	Lesson	Teacher Directed: Lesson Examples, Teaching Transparencies	Student Guided: Activity, Try This	Cooperative-Learning Activity, Lesson Activity, Student Technology Guide	Practice: Practice & Apply, Extra Practice, Practice Workbook	Assessment: Quiz, Mid-Chapter Assessment	Problem Solving, Reteaching
1	6.1	15 min	15 min	20 min	55 min	15 min	15 min
2	6.2	15 min	15 min	20 min	55 min	15 min	15 min
3	6.3	15 min	15 min	20 min	55 min	15 min	15 min
4	6.4	15 min	15 min	20 min	55 min	15 min	15 min
5	6.5	15 min	15 min	20 min	55 min	15 min	15 min
6	6.6	10 min	20 min	20 min	55 min	15 min	15 min
7	Assess.	50 min	90 min	90 min	65 min	30 min	
		PE: Chapter Review	**PE:** Chapter Project, Writing Activities	Tech Prep Masters	**PE:** Chapter Assessment, Test Generator	Chap. Assess. (A or B), Alt. Assess. (A or B), Test Generator	

PE: Pupil's Edition

Alternative Assessment

The following suggest alternative assessments for students who may benefit from a different type of assessment than the regular chapter quizzes and the mid-chapter/end-of-chapter test. Visit the HRW web site to get additional Alternative Assessment material.

internet connect

Alternative Assessment
Go To: **go.hrw.com**
Keyword: **MG1 Alt Assess**

Performance Assessment

1. Using seven unit cubes, create a figure and draw it on isometric grid paper. Give the surface area of your figure, and draw the six different orthographic projections.

2. Using pretzel sticks and miniature marshmallows, create a figure with 12 segments and 7 parallel planes. The figure will be a rectangular prism, but not a cube. When you are finished, draw your figure and label the vertices and the parallel planes.

3. Find a box of any size. Use the formula on page 384 to find the length of the diagonal of the box. Using a piece of string and a ruler, measure the length of the diagonal. Compare this measurement with the one you computed using the formula.

Portfolio Project

Suggest that students choose one of the following projects for inclusion in their portfolios.

1. Using the perspective drawing methods on page 412, draw a poster for display in your classroom.

2. Create a grid model that allows you to plot three dimensional points. You may use any material you wish, and may get ideas from pictures in your text, such as those on page 372, 375, or 394. In your model, include the x, y, and z-axes and some kind of numerical scale. Give the coordinates of the points you plot on your grid.

3. Create polyhedra from nets such as those shown on page 387. If the month of the assignment is close to December, create Christmas ornaments for display in the classroom.

4. Using the polyhedra made in the previous portfolio activity, find the measures of several of the dihedral angles formed.

internet connect

The table below identifies the pages in this chapter that contain internet and technology information.

Content Links

Activities Online	pages 384, 413
Portfolio Extensions	pages 386, 395, 408
Homework Help Online	pages 377, 385, 393, 400, 406, 416

Resource Links

Parents can go online and find concepts that students are learning—lesson by lesson—and questions that pertain to each lesson, which facilitate parent-student discussion.

Go To: **go.hrw.com**
Keyword: **MG1 Parent Guide**

Technical Support

The following may be used to obtain technical support for any HRW software product.

Online Help: **www.hrwtechsupport.com**

e-mail: **tschrw@hbtechsupport.com**

HRW Technical Support Center: **(800)323-9239**

7 AM to 10 PM Monday through Friday CST

Visit the HRW math web site at: **www.hrw.com/math**

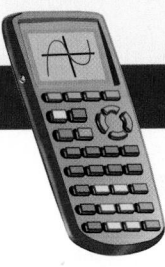

Technology

Technology Objectives and Suggestions

Lesson 6.1 Solid Shapes
In this lesson students use isometric grid paper to draw three-dimensional shapes built with cubes, and develop an understanding of orthographic projection. Students can practice drawing shapes by using the copy and paste features of geometry software. Allow students to make their own grid paper by constructing overlapping hexagons on the computer screen, and "hiding" the sides, leaving only the vertices.

Lesson 6.2 Spatial Relationships
In this lesson, students identify the relationships of points, lines, segments, planes, and angles in three-dimensional space. Geometry software can be used to reinforce how perpendicular and parallel planes can be drawn to form cubes.

Lesson 6.3 Prisms
In this lesson students examine the shapes of lateral faces of prisms, and solve problems using the diagonal measure of a right prism. Students may use geometry software to draw various prisms. Try dragging a point on a prism drawn on a computer to change it from a right prism to an oblique prism. When applying the formula for finding the length of the diagonal of a right rectangular prism, a calculator with a square root key is needed.

Lesson 6.4 Coordinates in Three Dimensions
In this lesson students identify the features of a three-dimensional coordinate system, including axes, octants, and coordinate planes. They also solve problems using the distance formula in three dimensions. If available, plot three dimensional coordinates on a graphics calculator or program that allows you to do three-dimensional graphing. To use the distance formula, students will need a calculator with a square root key.

Lesson 6.5 Lines and Planes in Space
In this lesson students define the equation of a line and the equation of a plane in space. They also solve problems using the equations of lines and planes in space. Although the exercises and problems for this lesson do not require the use of computer technology, using software or graphics calculators to graph three-dimensional images may give students a better perspective of lines and planes in space.

Lesson 6.6 Perspective Drawing
In this lesson students identify and define the basic concepts of perspective drawing, and apply perspective drawing concepts to creating perspective drawings. If available, allow students to experiment with more sophisticated graphics programs to create perspective drawings.

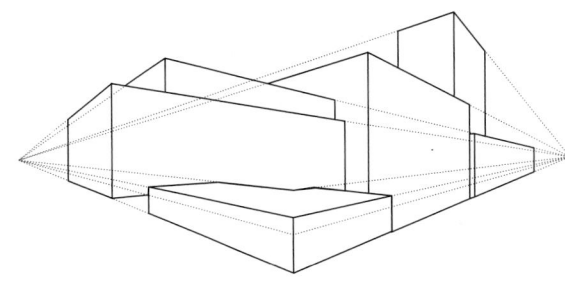

Background Information

This chapter introduces students to geometric figures in three-dimensional space. Students will extend ideas and concepts about plane figures to three-dimensional figures, or spatial figures. Volume and surface area are introduced. Formal presentations of volume and surface area are presented in Chapter 7.

CHAPTER RESOURCES

- Block-Scheduling Handbook
- Writing Activities for Your Portfolio
- Tech Prep Masters
- Long-Term Project
- Assessment Resources:
  Mid-Chapter Assessment
  Chapter Assessments
  Alternative Assessments
- Test and Practice Generator
- Technology Handbook

Chapter Objectives

- Use isometric dot paper to draw three-dimensional shapes built with cubes. **[6.1]**
- Develop an understanding of orthographic projection. **[6.1]**
- Develop a basic understanding of volume and surface area. **[6.1]**
- Define *polyhedron*. **[6.2]**
- Identify the relationships of points, lines, segments, planes, and angles in three-dimensional space. **[6.2]**
- Define *dihedral angle*. **[6.2]**

6

Shapes in Space

WHAT GIVES THE PICTURE ON THE FACING page such a dramatic effect? Even though it is printed on a flat, two-dimensional surface, the French Concorde jet has the illusion of depth. In this chapter, you will find out how to give drawings this three-dimensional quality.

As you study geometric figures in space, you will discover that three-dimensional figures have properties similar to those of two-dimensional figures. In this chapter, you will apply what you already know about points and lines in a plane to points, lines, and planes in space.

Lessons

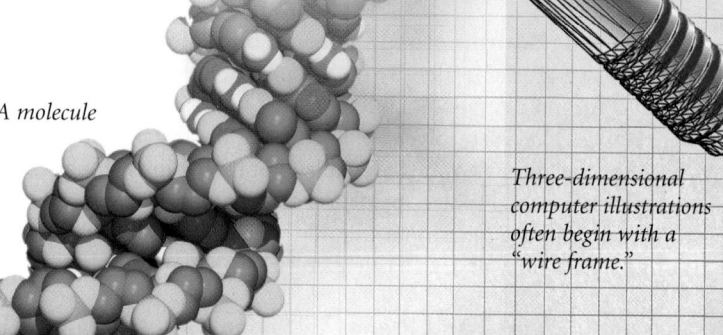

DNA molecule

Three-dimensional computer illustrations often begin with a "wire frame."

About the Photos

The photos on these pages show a computer-generated model of a DNA molecule, a computer-aided design of a spark plug, a molecular model of the local anesthetic procaine, and a Concorde jet from France. Computers make perspective drawings by converting the coordinates of three-dimensional objects into two dimensions. In this chapter, students will learn how to make a projection, the two-dimensional graph of a three-dimensional object.

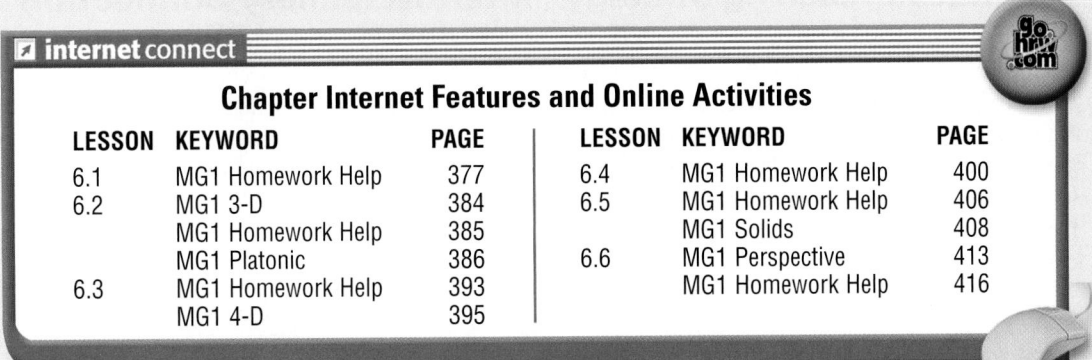

- Define *prism*, *right prism*, and *oblique prism*. **[6.3]**
- Examine the shapes of lateral faces of prisms. **[6.3]**
- Solve problems by using the diagonal measure of a right prism. **[6.3]**
- Identify the features of a three-dimensional coordinate system, including the axes, octants, and coordinate planes. **[6.4]**
- Solve problems by using the distance formula in three dimensions. **[6.4]**
- Define the equation of a line and the equation of a plane in space. **[6.5]**
- Solve problems by using the equations of lines and planes in space. **[6.5]**
- Identify and define the basic concepts of perspective drawing. **[6.6]**
- Apply these basic concepts to creating your own perspective drawings. **[6.6]**

CHAPTER PORTFOLIO ACTIVITIES PROJECT

About the Chapter Project

Solid figures that are bounded by polygons are called polyhedra. Mathematicians have long been interested in regular polyhedra, which are bounded by congruent regular polygons.

The last book of Euclid's *Elements* contains a proof that only five regular polyhedra exist. The earlier Greek philosopher Plato also knew this fact, and the five regular polyhedra are often called Platonic solids after him.

In the Chapter Project, you will explore why there are exactly five regular polyhedra.

After completing the Chapter Project you will be able to do the following:

- Visualize and understand the relationships among polygons in space.

About the Portfolio Activities

Throughout the chapter, you will be given opportunities to complete Portfolio Activities that are designed to support your work on the Chapter Project.

The theme of each Portfolio Activity and of the Chapter Project is polyhedra.

- In the Portfolio Activity on page 387, you will build the five Platonic solids from patterns called nets.
- In the Portfolio Activity on page 395, you will create two nets that "pop up" to form a dodecahedron with the help of a rubber band.
- In the Portfolio Activity on page 408, you will create semiregular polyhedra, also called Archimedean solids, from nets.

CHAPTER PORTFOLIO ACTIVITIES PROJECT

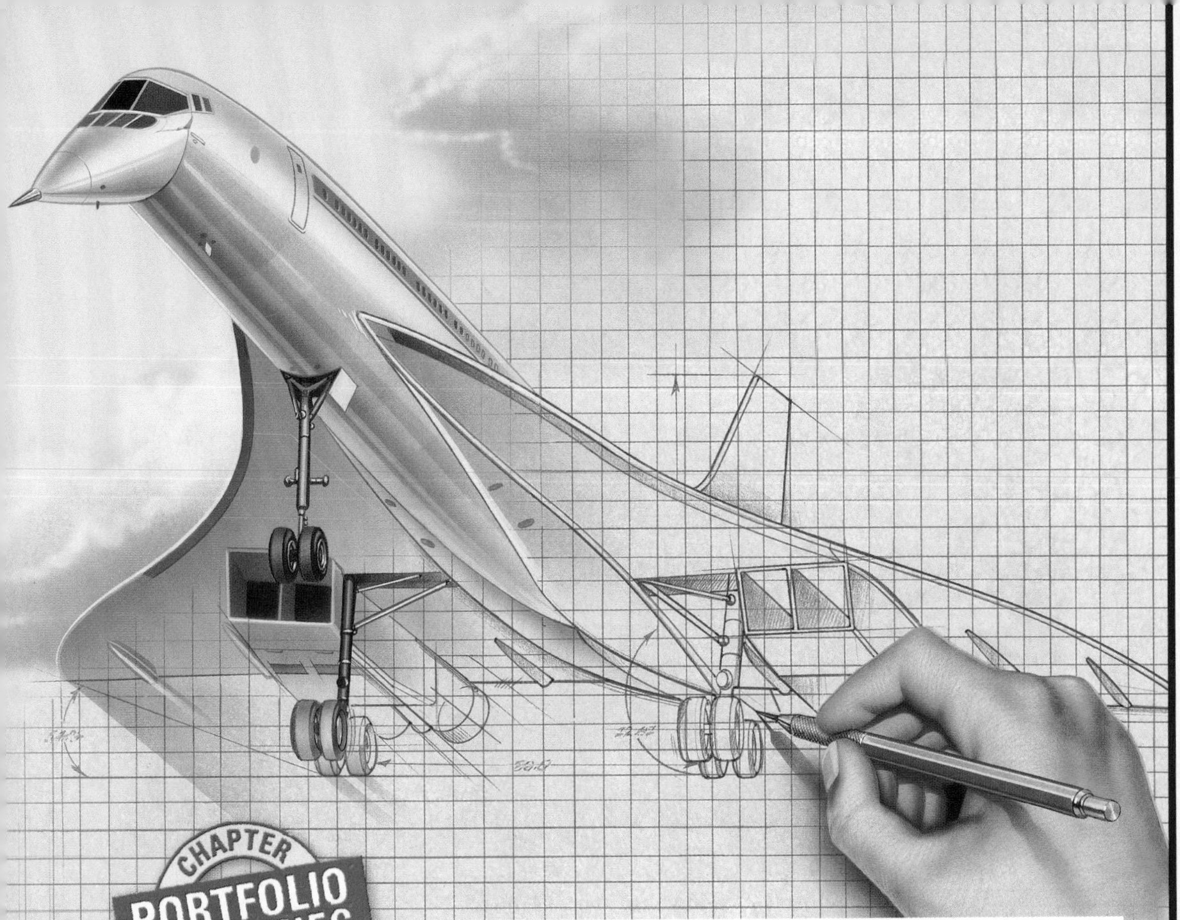

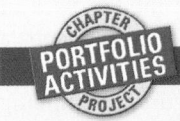

Portfolio Activities appear at the end of Lessons 6.2, 6.3, and 6.5. Each serves as preparation for the Chapter Project. The Portfolio Activities, as well as the Chapter Project Activities, are appropriate for inclusion in the student's portfolio. Students should be encouraged to include in their portfolios any other work in which they feel a sense of pride or a sense of accomplishment.

internet connect

Chapter Internet Features and Online Activities

LESSON	KEYWORD	PAGE	LESSON	KEYWORD	PAGE
6.1	MG1 Homework Help	377	6.4	MG1 Homework Help	400
6.2	MG1 3-D	384	6.5	MG1 Homework Help	406
	MG1 Homework Help	385		MG1 Solids	408
	MG1 Platonic	386	6.6	MG1 Perspective	413
6.3	MG1 Homework Help	393		MG1 Homework Help	416
	MG1 4-D	395			

1. How many faces does a cube have?   **6**

2. How many faces of a cube are ordinarily hidden from view?   **3**

3. Find the area of a square with side lengths of 1 cm.   **1 cm²**

Also on Quiz Transparency 6.1

Teach

Why Drawings on isometric dot paper and other types of drawings and models used in this lesson will give students a practice-based introduction to drawing solid figures and finding volumes and surface areas.

Solid Shapes

Objectives

- Use isometric dot paper to draw three-dimensional shapes composed of cubes.

- Develop an understanding of orthographic projection.

- Develop a basic understanding of volume and surface area.

Why Architects, engineers, and many other professionals must be skilled in the art of drawing three-dimensional shapes. In this lesson, you will learn some of the fundamental concepts of this art.

APPLICATION
MECHANICAL DRAWING

An **isometric drawing** is one in which the horizontal lines of an object are represented by lines that form 30° angles with a horizontal line in the picture.

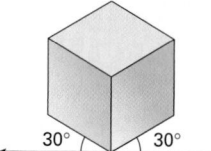

Drawing Cubes

A type of graph paper called isometric dot paper can be helpful in drawing solid shapes such as cubes. This paper has diagonal rows of dots that form a 30° angle with horizontal lines.

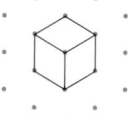

Connect the dots as shown to draw a cube. On a solid cube, at most three faces are visible at a time; however, sometimes you may wish to represent the hidden faces of a cube. This is done by using dashed lines, as shown in the second drawing.

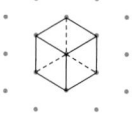

In the first and second drawings, the cube is viewed from above. In the third drawing, the cube is viewed from below.

Alternative Teaching Strategy

HANDS-ON STRATEGIES Students can explore different views of an object by taking apart various common boxes such as cereal boxes or shoe boxes. Have them construct a closed box by assembling nets made from graph paper. Have students calculate the surface area of a closed box by counting the number of squares on the graph paper used to create it.

Interdisciplinary Connection

INDUSTRIAL ARTS Some students in the class may have taken a course in mechanical drawing. Ask them to display some of their three-dimensional drawings and orthographic projections for the class. They can also talk about the various tools that they used.

Activity 1
Using Isometric Grid Paper

1. Draw each solid figure below on isometric dot paper. Then redraw the figures with the red cube or cubes removed.

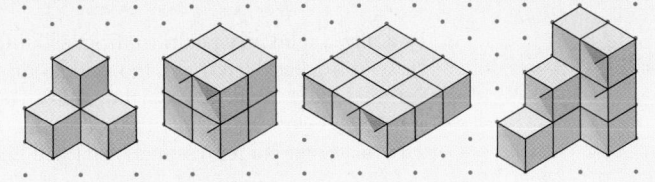

2. Draw each solid figure below on isometric dot paper. Then add a cube at each red face and draw the new figure.

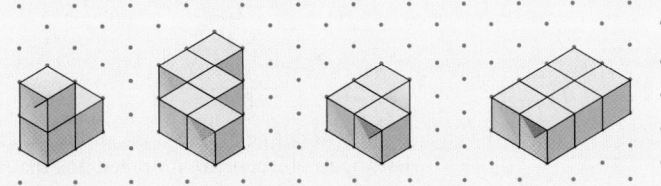

CHECKPOINT ✔ **3.** Describe how the drawings change as you add or subtract cubes.

Activity 2
Using Unit Cubes

1. Use unit cubes to build the three figures below. Are your figures the same as those of your classmates? If not, how are they different?

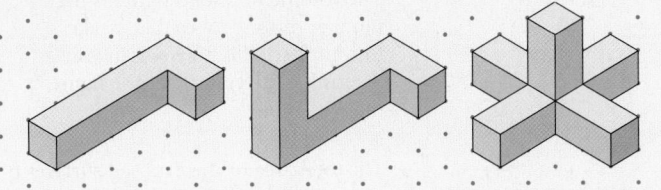

2. Build three solids of your own with at least six unit cubes in each. Draw your solids on isometric dot paper.

CHECKPOINT ✔ **3.** Could any of the drawings you made in Step 2 represent more than one solid figure? Explain your reasoning.

Inclusion Strategies

TACTILE-KINESTHETIC STRATEGIES Some students may need extra practice at drawing shapes made from unit cubes. Using actual cubes to model the figure they want to draw can help them to distinguish different views of the same object.

Activity 1 Notes

Students will practice drawing solids made from cubes. Encourage students to show hidden lines, if necessary. Make the connection to surface area by pointing out which faces of the solids are exposed.

For a student worksheet of this Activity and detailed Teacher Notes, see page 105 in the Lesson Activities booklet.

CHECKPOINT ✔
3. Sample answer: When you add cubes, segments that represent edges of cubes that get hidden must be erased. When you subtract cubes, segments must be added to represent edges of cubes that become visible.

Activity 2 Notes

In this Activity, students will draw shapes to match those in the book and then compare their drawings with those of classmates. Students will use isometric dot paper to draw the solids that they build with cubes. Use their isometric drawings to introduce students to orthographic projections.

For a student worksheet of this Activity and detailed Teacher Notes, see page 105 in the Lesson Activities booklet.

CHECKPOINT ✔
3. See Additional Answers, page 905.

Teaching Tip

In Activity 2, using blocks to create three-dimensional figures may help students distinguish different views of the figure.

Have small groups collaborate to draw different orthographic projections of various objects in the classroom, such as chairs, desks, filing cabinets, etc. Group work allows participants to share their skills and special abilities.

Activity 3 Notes

In this Activity, students use cubes to re-create an existing object and then draw all six orthographic projections of the object. Students then relate their models and drawings to surface area and volume.

For a student worksheet of this Activity and detailed Teacher Notes, see page 105 in the Lesson Activities booklet.

CHECKPOINT ✔

2. Count the number of cubes used to build the figure.

CHECKPOINT ✔

3. Add the areas of the orthographic projections.

CRITICAL THINKING

No. The figure below is a top view of two cubes whose edges are not aligned. Two faces of the top cube will appear in each of the following views: front, back, right, and left.

Front

Also, if the object has faces that form an interior wall, as in the figure below, these faces will not appear in any views.

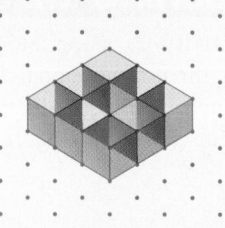

Orthographic Projections

An **orthographic projection** is a view of an object in which points of the object are "projected" onto the picture plane along lines perpendicular to the picture plane.

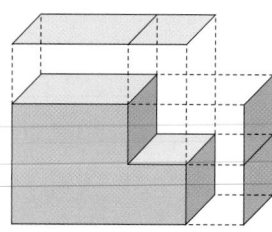

Typically, a solid may be drawn from six different views: front, back, left, right, top. and bottom. For the solid at right, all six orthographic views are shown below. Edges that cannot be seen in a particular view are represented by dashed lines.

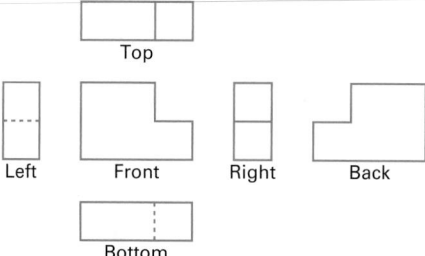

You may find it helpful to think of the different views of an object on an unfolded box that has a different view on each face.

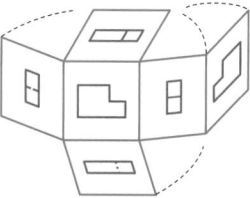

Activity 3
Volume and Surface Area

YOU WILL NEED
unit cubes and graph paper

1. Use 7 unit cubes to build a model of the solid shown at right. Then draw six orthographic projections of the object on graph paper, representing the views from the back, front, left, right, top, and bottom.

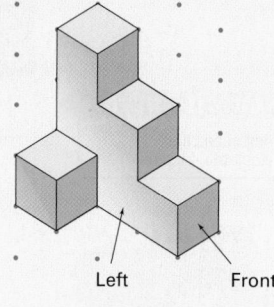

Left   Front

CHECKPOINT ✔

2. The volume of a solid figure is the number of unit cubes that it takes to completely fill it. How can you determine the volume of the figure that you built?

CHECKPOINT ✔

3. The total area of the exposed surfaces of an object is called its surface area. How can you use the orthographic projections of the object to determine its surface area?

CRITICAL THINKING

In the orthographic projections you drew in Activity 3, each exposed face of the object appeared in exactly one view. Do you think this would be true for any object you might build? Use drawings or models to explain your reasoning.

Enrichment

The diagram at left below represents the top view of a solid made from cubes. The letters represent the corners of a piece of paper, and the numbers represent the number of cubes in that stack. Draw a view of the solid from corner A.

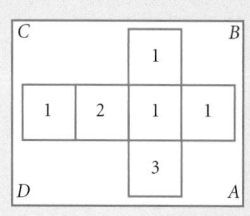

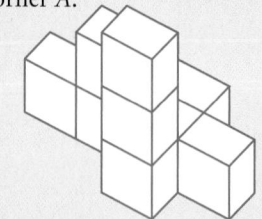

Exercises

Communicate

1. A solid built from cubes has the top view shown at right. Draw two possibilities for this solid.

2. A solid has the right view and front view shown at right. What is a possible top view for this solid?

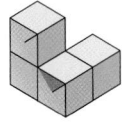
Front   Right

3. Why do you think an isometric drawing uses angles of 30° from a horizontal line? (Hint: Draw a cube on isometric grid paper and measure its edges.)

4. Describe the difference between volume and surface area. Include units in your description.

Guided Skills Practice

Use isometric dot paper for Exercises 5 and 6. *(ACTIVITY 1)*

5. Draw the solid below with the red cube removed.

6. Draw the solid below with a cube added at the red face.

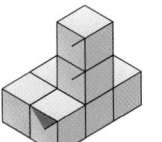

Use unit cubes for Exercises 7–10. Refer to the figure below.
(ACTIVITIES 2 AND 3)

7. Build a model of the figure out of unit cubes. Then turn the figure and draw a different view on isometric dot paper.

8. Draw six orthographic projections of the solid.

9. Find the solid's volume in cubic units. **6 units³**

10. Find the solid's surface area in square units. **26 units²**

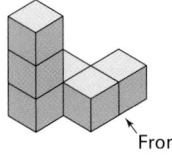
Front

Practice and Apply

For Exercises 11–16, refer to the isometric drawing of a cube shown below. Identify the letter of the indicated face.

11. front **a**   12. right **e**

13. top **c**   14. left **b**

15. bottom **f**   16. back **d**

b   d
c
a e
Front f

Reteaching the Lesson

USING VISUAL MODELS Have students draw the six different orthographic projections of the figure below. Have them find the surface area and the volume of the figure as well.

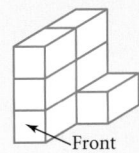
Front

surface area = 26 square units
volume = 7 cubic units

Top

Left   Front Right Back

Bottom

Selected Answers

Exercises 5–10, 11–49 odd

ASSIGNMENT GUIDE

In Class	1–10
Core	11–39 odd
Core Plus	19–39 odd, 40–43
Review	44–50
Preview	51–54

✐ Extra Practice can be found beginning on page 818.

Error Analysis

For students who have difficulty drawing the solid in Exercise 35, suggest that they use fewer than 8 cubes and build a solid with a greater surface area than the one shown.

5.
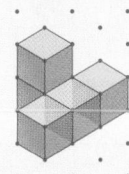

6.

7. Sample answer:
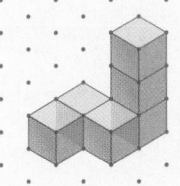

17.

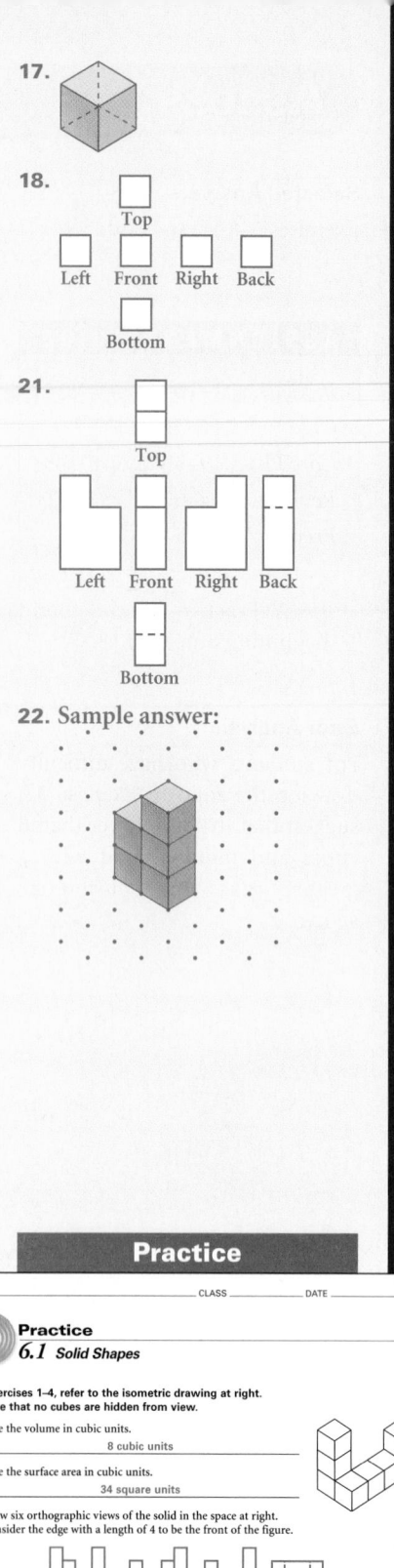

18.

Top

Left Front Right Back

Bottom

21.

Top

Left Front Right Back

Bottom

22. Sample answer:

17. What lines would you add to the drawing
at right to indicate the hidden faces?

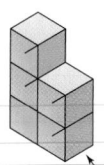

18. Draw an orthographic projection of the cube.

For Exercises 19–23, refer to the isometric drawing below.

19. Give the solid's volume in cubic units. **5 units³**

20. Give the solid's surface area in square units.
20 units²

21. Draw six orthographic views of the solid.

22. Draw the solid on isometric dot paper
from a different viewpoint.

23. Draw the solid on isometric dot paper
with a cube added at each red face.

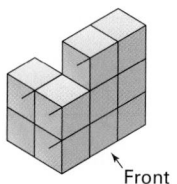

Front

**Refer to the drawing below for Exercises 24–27. Assume that there
are no hidden cubes.**

24. Each flat area of the solid is a face. How
many faces does this solid have? **10 faces**

25. Draw the front face.

26. Is any other face of the solid congruent to
the front face? **No**

27. Are any two faces of the solid congruent?
Explain your answer.

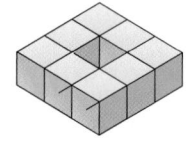

Front

The three solids below each have a volume of 4 cubic units.

a.   b.   c.

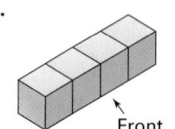  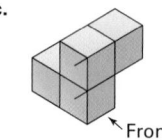

Front   Front

28. a. 18 units²
     b. 16 units²
     c. 18 units²

     **b has the least
     surface area.**

28. Find each solid's surface area in square units. Which has the smallest
surface area?

29. Draw six orthographic projections of each solid.

30. Draw two other solids with a volume of 4 cubic units that are different
solids from those above (not just different views).

Refer to the solid below for Exercises 31 and 32.

31. Draw six orthographic projections of the solid.

32. Yes; the faces which
face the hole do not
appear in any views.

32. Are there any exposed faces of the solid that do
not appear in any of the orthographic views you
drew in Exercise 31? Explain.

33. Use isometric dot paper to draw two solids that have the same volume but
different surface areas.

34. Use isometric dot paper to draw two solids that have the same surface area
but different volumes.

23.

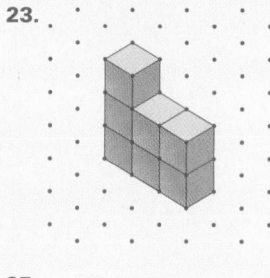

25.

27. Yes, the top view has 2 distinct faces which
are congruent because they have the same
size and shape. They are also congruent to
one face of the back view and one face of
the right view. They are rectangles that
measure 1 unit by 2 units.

35. The solid at right is made up of 8 unit cubes. Draw a solid with a smaller volume but a larger surface area.

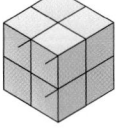

36. Draw two different solids that have the same orthographic projections from the top and bottom views.

37. Draw two different solids that have the same orthographic projections from the front and back views.

38. How many unit cubes does it take to build a $2 \times 2 \times 2$ cube? a $3 \times 3 \times 3$ cube? a $4 \times 4 \times 4$ cube? an $n \times n \times n$ cube? **8; 27; 64; n^3**

39. What is the surface area in square units of a $2 \times 2 \times 2$ cube? a $3 \times 3 \times 3$ cube? a $4 \times 4 \times 4$ cube? an $n \times n \times n$ cube?

40. The figure at right shows a $3 \times 3 \times 3$ cube built from unit cubes. Suppose that the exposed faces of the cube were painted red and then the solid was disassembled into unit cubes.

How many unit cubes would have
a. 3 red faces? **8**
b. 2 red faces? **12**
c. 1 red face? **6**
d. no red faces? **1**

41. Repeat Exercise 40 for a $4 \times 4 \times 4$ cube. **8; 24; 24; 8**

42. ARCHITECTURE Draw orthographic projections that represent the top, front, and right views of the building shown at right. Assume that the front faces are the ones on the right, in sunlight.

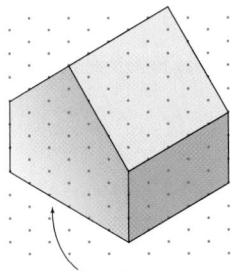

Front

43. RECREATION The isometric drawing of a tent, shown at left, can be used to show how the tent will look when it is assembled.
Draw six orthographic views of the tent. What might these views be used for?

---

35. Sample answer:

36. Sample answer:

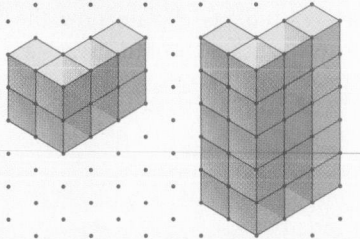

37. Sample answer:

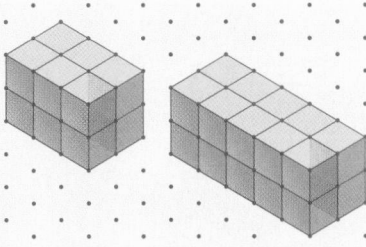

39. 24 square units; 54 square units; 96 square units; $6n^2$ square units

42.

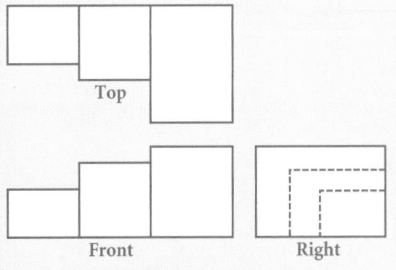

43.

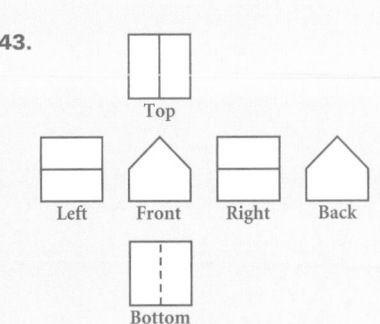

These views might be used when writing instructions on how to assemble the tent.

Look Beyond

Exercises 51–54 involve composing a three-dimensional drawing of a solid from its orthographic projections. This extends the content of the lesson and requires abstract thinking from students. Have students explain how they know which faces of their drawing are congruent.

51. Sample answer:

54. The solid has eleven congruent faces of one square and three congruent faces of two squares.

Find the area of each figure. *(LESSONS 5.2 AND 5.3)*

44. 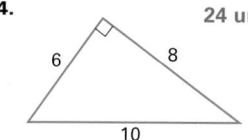   24 units²

45.   24.61 m²
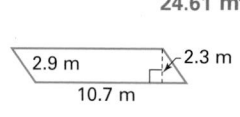

46.   24.05 ft²
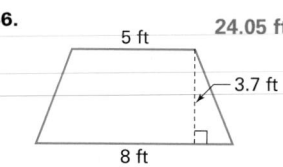

47.   ≈ 78.54 m²
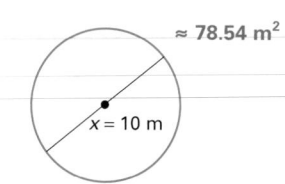

Find the unknown length in each right triangle. *(LESSON 5.4)*

48.   ≈ 6.71 cm
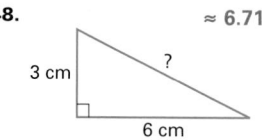

49.   ≈ 170.29 yd
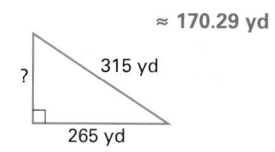

APPLICATION

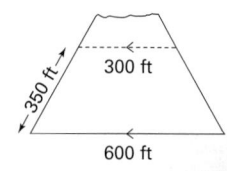

50. ARCHAEOLOGY The top of a pyramid has been destroyed. Use the diagram of a cross section at left to estimate its original height. *(LESSONS 3.7 AND 5.4)*
≈ 632 ft

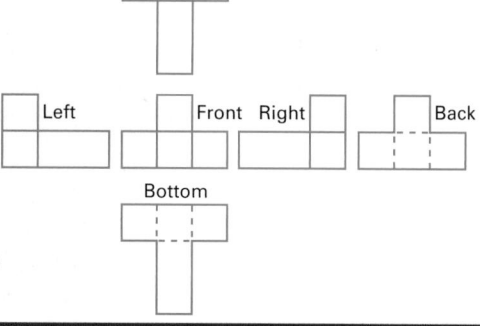

Look Beyond

Use the orthographic projections at right for Exercises 51–54.

51. Use isometric dot paper to draw two different views of the solid.

52. Are any cubes hidden in your drawings? **Yes**

53. How many faces does the solid have? **16 faces**

54. How many congruent faces does the solid have? Explain your answer.

Spatial Relationships

Objectives

- Define *polyhedron*.

- Identify the relationships among points, lines, segments, planes, and angles in three-dimensional space.

- Define *dihedral angle*.

Why The interplay of lines and planes is apparent in real-world objects such as this quartz crystal. Understanding the relationships among lines and planes in three dimensions is essential to understanding the structure of matter.

Figures in Space

A closed spatial figure made up of polygons is called a *polyhedron* (plural, *polyhedra* or *polyhedrons*). Closed spatial figures are also known as **solids**.

Definition: Polyhedron

A **polyhedron** is a closed spatial figure composed of polygons, called the **faces** of the polyhedron. The intersections of the faces are the **edges** of the polyhedron. The vertices of the faces are the **vertices** of the polyhedron.

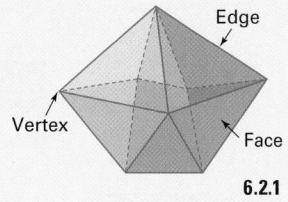

6.2.1

The polyhedron you are probably most familiar with is a cube. A cube is an example of a regular polyhedron. In a *regular polyhedron,* all of the faces are congruent regular polygons, and the same number of polygons meet at each vertex.

Alternative Teaching Strategy

USING MODELS Have students create models of polyhedrons out of index cards to help them identify the edges, vertices, and faces. Ask them to classify their models as either regular or irregular polyhedrons . If their models are not regular, ask them how the models could be changed to create regular polyhedrons.

QUICK WARM-UP

Classify each statement below as true or false.

1. Two lines in three-dimensional space that do not intersect are parallel. false

2. A cube has 6 faces. true

3. A cube has 12 edges. true

4. Parallel lines are coplanar. true

Also on Quiz Transparency 6.2

Teach

Why Relating points, lines, and planes to each other in three-dimensional space will help students understand spatial relationships among solid figures.

Activity 1 Notes

In this Activity, students identify parallel segments and planes in a cube and develop a definition of parallel planes.

For a student worksheet of this Activity and detailed Teacher Notes, see page 110 in the Lesson Activities booklet.

CHECKPOINT ✔

Part I

3. Yes. This is possible because the segments are in different planes. Examples of pairs of skew segments are $\overline{AE}$ and $\overline{GH}$, $\overline{DH}$ and $\overline{GF}$, $\overline{CG}$ and $\overline{EF}$, and $\overline{BF}$ and $\overline{EH}$.

CHECKPOINT ✔

Part II

2. Two planes are parallel if and only if they do not intersect.

Activity 2 Notes

In this Activity, students explore the characteristics of a line that is perpendicular to a plane. Have students tilt a pencil until it is perpendicular to a line on paper. Have them clarify the terminology and check each other's definition of a line perpendicular to a plane.

For a student worksheet of this Activity and detailed Teacher Notes, see page 110 in the Lesson Activities booklet.

In the Activities that follow, you will discover and develop ideas about how lines and planes relate to each other in space.

Activity 1
Parallel Lines and Planes in Space

Part I

YOU WILL NEED
no special tools

1. Sketch a cube and label its vertices as shown. Identify the segments that form the vertical edges.

   Do $\overline{AE}$ and $\overline{CG}$ seem to be coplanar? Do they seem to be parallel? Do you think that they would meet if they were extended infinitely?

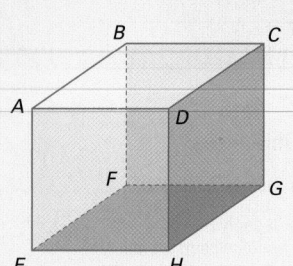

2. Which edges of the cube seem to be parallel?

CHECKPOINT ✔

3. Are there edges of the cube that are not parallel and yet would never meet if they were extended infinitely in either direction? Explain how this can be. These segments (or lines or rays) are said to be **skew**. Make a list of four pairs of skew edges in the cube.

Part II

1. How many faces does a cube have? Which faces seem to be parallel to each other?

CHECKPOINT ✔

2. Write your own definition of **parallel planes** by completing the sentence below.

Definition: Parallel Planes

Two planes are parallel if and only if ____?____.   **6.2.2**

Note: Two plane figures are parallel if and only if they are contained in parallel planes.

Activity 2
Segments and Planes

Part I

YOU WILL NEED
no special tools

1. Each edge of a cube is perpendicular to two different faces. For the cube at right, make a list of edges and the faces to which they are perpendicular.

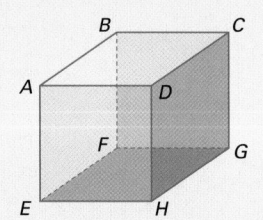

Inclusion Strategies

USING VISUAL MODELS You may help students see relationships between segments and planes by having them construct three-dimensional figures with geometry graphics software. Encourage them to use the measurement options of the software to verify the properties that appear later in this lesson.

2. What do you think it means for a segment or a line to be perpendicular to a plane? Draw a line, ℓ, on a piece of paper and label a point, P, on it. Hold your pencil so that it is perpendicular to the paper, and place the tip on point P. Is the pencil perpendicular to line ℓ?

3. Is it possible to tilt your pencil so that it is still perpendicular to line ℓ but not to the paper? Make a sketch illustrating your answer.

4. Draw a new line, m, through point P. Place your pencil on P so that it is perpendicular to ℓ and m at point P. What is the relationship between the pencil and the plane of the paper?

5. Draw several other lines through point P. When you place your pencil on P so that it is perpendicular to the paper, is it also perpendicular to these other lines?

CHECKPOINT ✔ 6. Write your own definition of a **line perpendicular to a plane** by completing the sentence below.

Definition: A Line Perpendicular to a Plane

A line is perpendicular to a plane at a point P if and only if it is perpendicular to every line in the plane that ____?____.   **6.2.3**

Part II

1. Each edge of a cube (see Step 1 of Part I) is parallel to two different faces. Make a list of edges and the faces to which they are parallel. What do you think it means for a segment or a line to be parallel to a plane?

2. Draw a line, ℓ, on a piece of paper. Hold your pencil above ℓ so that it is parallel to the line. Does the pencil seem to be parallel to the plane of the paper?

3. Turn your pencil so that it is still parallel to the paper but not to the line. Do you think that you could draw another line on the paper that is parallel to the pencil in this position?

CHECKPOINT ✔ 4. Write your own definition of a **line parallel to a plane** by completing the sentence below.

Definition: A Line Parallel to a Plane

A line that is not contained in a given plane is parallel to the plane if and only if it is parallel to ____?____.   **6.2.4**

Interdisciplinary Connection

ARCHITECTURE Parallel lines and planes in space are necessary and sometimes aesthetic components of architecture. Ask students to find pictures that emphasize parallelism in architecture.

CHECKPOINT ✔
Part I
6. intersects P

CHECKPOINT ✔
Part II
4. a line in that plane

Teaching Tip

You can help students understand the concepts in Activity 2 by providing concrete models, including noncubical rectangular prisms. Have students try to draw the models, label them, and make lists of all parallel segments and planes.

This Activity will help students understand the measures of dihedral angles.

For a student worksheet of this Activity and detailed Teacher Notes, see page 110 in the Lesson Activities booklet.

CHECKPOINT ✔
4. perpendicular

Cooperative Learning

Have small groups of students find different examples of dihedral angles in the classroom and at home. Ask them to list objects with dihedral angles equal to 90°, less than 90°, and greater than 90°.

Angles Formed by Planes

A **half-plane** is the portion of a plane that lies on one side of a line in the plane and that includes the line. The angle between two half-planes is known as a *dihedral angle*.

Definition: Dihedral Angle

A **dihedral angle** is the figure formed by two half-planes with a common edge. Each half-plane is called a **face** of the angle, and the common edge of the half-planes is called the **edge** of the angle.

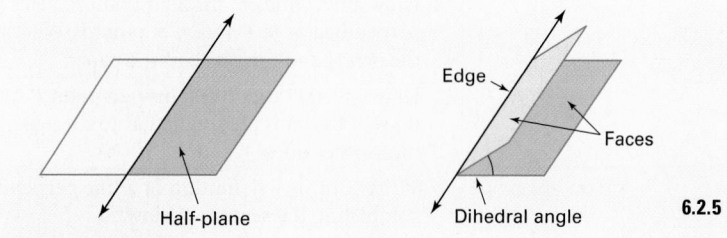

6.2.5

Activity 3
Measuring Dihedral Angles

YOU WILL NEED

scissors and either stiff folding paper or an index card

1. Some of the faces of a cube form right dihedral angles (that is, the faces are perpendicular to each other). Each face of the cube is perpendicular to how many other faces of the cube?

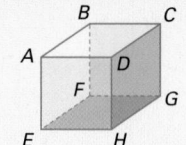

2. Draw a horizontal line, ℓ, on a piece of paper. Mark and label points A, B, and C on the line, with point B between A and C. Make a crease through point B so that line ℓ folds onto itself. What is the relationship between line ℓ and the line of the crease?

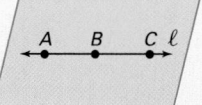

3. Open the paper slightly. The angle formed by the sides of the paper is a dihedral angle. The measure of the dihedral angle is the measure of $\angle ABC$.

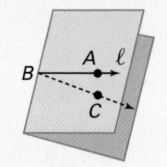

CHECKPOINT ✔
4. Write your own definition of the **measure of a dihedral angle** by completing the sentence below.

Definition: Measure of a Dihedral Angle

The measure of a dihedral angle is the measure of an angle formed by two rays that are on the faces and that are ____?____ to the edge.

6.2.6

Enrichment

Have students use the net below to predict the dihedral angles in the solid formed by the net. Have them check their answers by tracing, cutting, and folding the net.

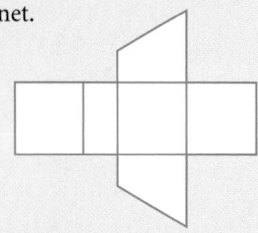

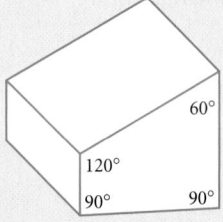

5. Open the paper and flatten it. Add points *X* and *Y* on the line of the crease. Draw two rays from each point, with one ray on each side of the crease, as shown.

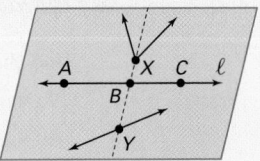

6. Fold the paper again along the same crease. Cut out pieces of paper that fit neatly into the angles formed by the different rays. Compare the shapes of the pieces of paper. Do the angles that are formed by the new rays have the same or different measures than that of ∠*ABC*?

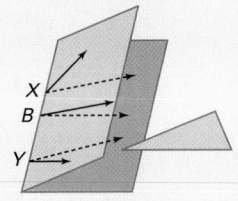

CHECKPOINT ✔ 7. By measuring the angles formed by different rays in Step 6, you can get a variety of results. What is the smallest angle that you could measure? What is the largest? Use your results to explain why a dihedral angle is measured along rays that are perpendicular to the edge.

Exercises

● *Communicate*

1. If two lines in space are perpendicular to the same line, are the two lines parallel to each other? Why or why not?

2. If a line not in a plane is perpendicular to a line in the plane, is the first line perpendicular to the plane? Why or why not?

3. If a line is perpendicular to two intersecting lines in a plane, is the first line perpendicular to the plane? Why or why not?

APPLICATION

CHEMISTRY Use the model of a sodium chloride crystal, which has a cubic structure, for Exercises 4–6.

4. Describe a pair of parallel segments, and explain why the segments are parallel.

5. Describe a pair of parallel planes, and explain why the planes are parallel.

6. Describe a pair of perpendicular planes, and explain why the planes are perpendicular.

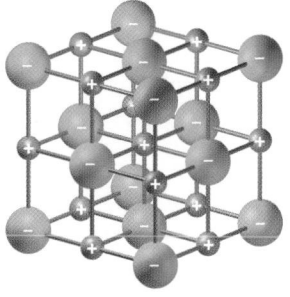

Crystal lattice of sodium chloride (table salt)

Assess

Selected Answers
Exercises 7–14, 15–39 odd

ASSIGNMENT GUIDE	
In Class	1–14
Core	15–21 odd, 23–27
Core Plus	23–30
Review	31–40
Preview	41–42

✏ Extra Practice can be found beginning on page 818.

Technology
Geometry graphics software can be used to reinforce students' understanding of how perpendicular and parallel planes intersect to form cubes.

Error Analysis
If students find Exercises 18–22 difficult to understand, use the corner of a box to illustrate the relationships between lines and planes in space.

CHECKPOINT ✔
7. Sample answer: Theoretically, the smallest angle is only slightly bigger than 0°. The largest angle is only slightly less than 180°. By using rays perpendicular to the common edge of the two faces (the crease), a nonarbitrary and standard way of measuring dihedral angles is achieved.

Reteaching the Lesson

USING VISUAL MODELS Given the cube below, have students determine whether it is a regular polyhedron. Have them name all of the parallel segments and all of the parallel planes in the cube.

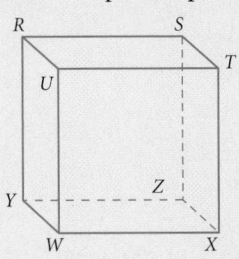

By definition, a cube is a regular polyhedron.
$\overline{RS} \parallel \overline{UT} \parallel \overline{WX} \parallel \overline{YZ}$
$\overline{RU} \parallel \overline{ST} \parallel \overline{ZX} \parallel \overline{YW}$
plane *RSTU* ∥ plane *YZXW*
plane *RUWY* ∥ plane *STXZ*
plane *RSZY* ∥ plane *UTXW*

7. Sample answer: $\overline{HI}$ and $\overline{KG}$; $\overline{LM}$ and $\overline{ON}$

8. Sample answer: $\overline{HI}$ and $\overline{MN}$; $\overline{HL}$ and $\overline{IG}$

9. Sample answer: *HIML* and *KGNO*; *HKOL* and *IGNM*

10. Sample answer: $\overline{HI}$ and *IGNM*; $\overline{HL}$ and *HIGK*

internet connect

Activities Online

Go To: **go.hrw.com**
Keyword: **MG1 3-D**

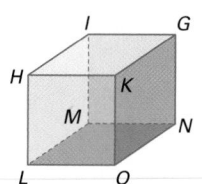

Guided Skills Practice

Use the figure of the cube at right for Exercises 7–11. *(ACTIVITY 1)*

7. Name two pairs of parallel edges.

8. Name two pairs of skew edges.

9. Name two pairs of parallel faces.

10. List two edges and the planes to which they are perpendicular.

11. List two pairs of parallel edges that do not lie on the same face of the cube.
Sample answer: $\overline{HI}$ and $\overline{ON}$; $\overline{LO}$ and $\overline{IG}$

12. In the figure at right, line *p* is perpendicular to plane $\mathcal{R}$. What is the relationship between line *p* and line *q*? line *p* and line *r*? *(ACTIVITY 2)*
perpendicular; perpendicular

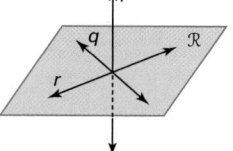

13. In the figure at right, line *m* is parallel to plane $\mathcal{S}$. What can you conclude about a certain line in plane $\mathcal{S}$? *(ACTIVITY 2)*
There exists a line in $\mathcal{S}$ that is parallel to line *m*

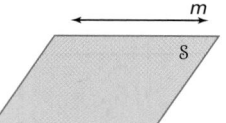

14. What is the measure of the dihedral angle in the figure at right? *(ACTIVITY 3)*
The measure of the dihedral angle is $m\angle EFG$.

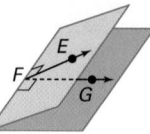

Practice and Apply

15. In the figure at right, line *p* is parallel to line *q*. What can you conclude about the relationship between line *p* and plane $\mathcal{M}$?
Line *p* and plane $\mathcal{M}$ are parallel.

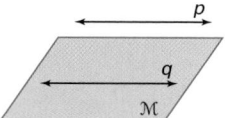

16. In the figure at right, line *m* is not parallel to line *n*. Can you draw a conclusion about the relationship between line *m* and plane $\mathcal{Q}$? Explain your answer.
No, we would only be able to draw a conclusion if we knew that *m* intersected *n*.

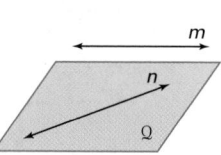

17. In the figure at right, line *p* is perpendicular to line *r* but not to line *s*. What can you conclude about the relationship between line *p* and plane $\mathcal{R}$?
Line *p* is not perpendicular to plane $\mathcal{R}$.

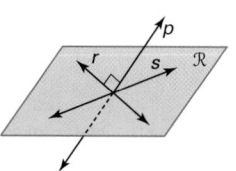

For Exercises 18–22, indicate whether each statement is true or false for a figure in space. Explain your answers by making sketches.

18. If two lines are parallel to a third line, then the two lines are parallel.

19. If two planes are parallel to a third plane, then the two planes are parallel.

20. If two planes are perpendicular to a third plane, then the two planes are parallel.

21. If two planes are perpendicular to the same line, then the planes are parallel.

22. If two lines are perpendicular to the same plane, then the lines are parallel.

Use your model from Exercise 23 to answer Exercises 24–26.

23. Check student
constructions.

23. Fold an index card in half and mark it with two segments, $\overline{AB}$ and $\overline{CD}$, so that $\overline{AB}$ is perpendicular to the folded edge and $\overline{CD}$ is not. Cut along each segment from the folded edge. Insert two more index cards to model intersecting planes, as shown at right.

24. The insert in $\overline{CD}$.

24. Which inserted card forms a plane that is not perpendicular to the edge?

25. The angle formed by the insert in $\overline{AB}$.

25. Which inserted card can be used to measure the dihedral angle formed by the folded card?

26. Which of the two angles formed by the inserted cards is larger?

APPLICATIONS

Ships navigate the surface of the ocean, which can be modeled by a plane. Airplanes navigate in space. Exercise 27–30 concern the navigation of ships and airplanes.

27. NAVIGATION Ship *A* is traveling south. Ship *B* in the same vicinity is traveling southeast. Their lines of travel _____?_____. **intersect**

28. NAVIGATION Airplane *A* is traveling south at an altitude of 23,000 feet. Nearby, airplane *B* is traveling southeast at an altitude of 18,000 feet. Their lines of travel are _____?_____. **skew**

29. NAVIGATION Two airplanes flying horizontally at the same altitude are flying in the same _____?_____. **plane**

30. NAVIGATION An airplane takes off from a runway. The airplane's nose and wingtips define the *plane of flight*. Sketch the dihedral angle formed by the airplane's plane of flight and the surface of the runway. Is the angle acute, right, or obtuse? Explain.

18. True

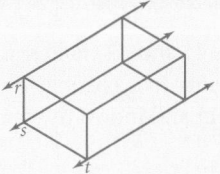

19. True

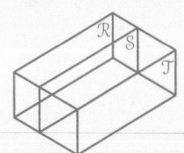

20. False

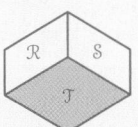

21. True

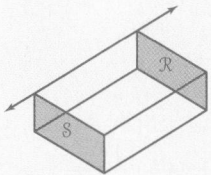

22. True

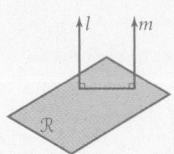

26. The dihedral angle formed by the insert in $\overline{AB}$ is larger than the angle formed by the insert in $\overline{CD}$.

30. Acute; the angle the airplane makes with the ground is less than 90°.

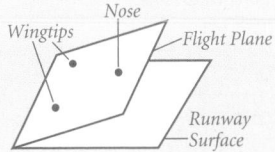

Wingtips *Nose* *Flight Plane* *Runway Surface*

Look Beyond

Exercises 41 and 42 challenge students' ability to re-arrange objects in efficient and innovative ways.

31.

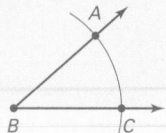

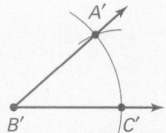

32.

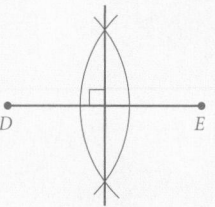

33.

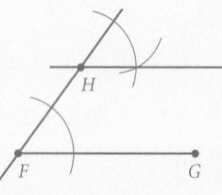

Look Back

In Exercises 31–34, use a compass and straightedge to construct each figure. *(LESSON 4.7)*

31. Draw an angle and label it ∠*ABC*. Construct a copy of ∠*ABC*.

32. Draw a segment and label it $\overline{DE}$. Construct its perpendicular bisector.

33. Draw segment and label it $\overline{FG}$. Mark a point, *H*, not on $\overline{FG}$. Construct a line through point *H* parallel to $\overline{FG}$.

34. Draw a triangle and label it △*JKL*. Construct a copy of △*JKL*.

Find the area of each polygon below. *(LESSONS 5.1, 5.2, AND 5.5)*

35. square

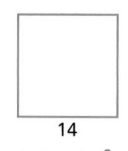

14

196 units²

36. trapezoid

22

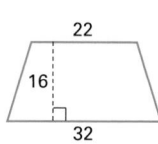

16

32

432 units²

37. parallelogram

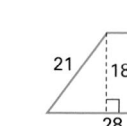

21 18

28

504 units²

38. right triangle

28 53

630 units²

39. equilateral triangle

15

≈ 97.4 units²

40. regular hexagon

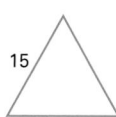

9

≈ 210.4 units²

Look Beyond

41. Arrange the figure below into exactly three triangles by moving only two toothpicks.

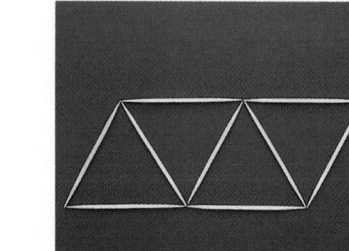

42. Arrange the figure below into exactly three squares by moving only three toothpicks.

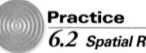
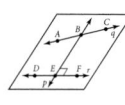
34.

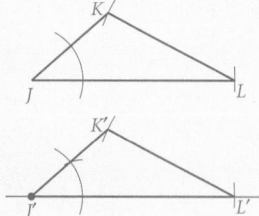

41.

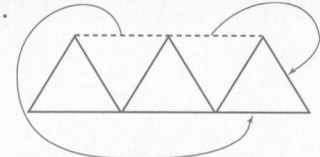

42.

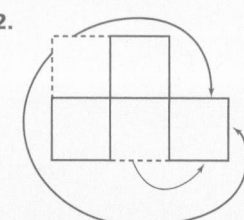

BUILDING PLATONIC SOLIDS

A **regular polyhedron** has congruent regular polygons as faces, with the same number of faces meeting at each vertex. There are exactly five different convex regular polyhedra, shown below. Because the Greek mathematician Plato proved that only these five exist, they are also known as the **Platonic solids.** Polyhedra are named for the number of faces they have.

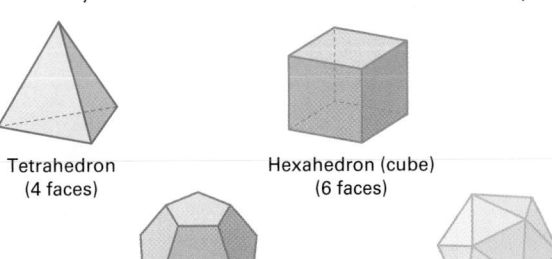

Tetrahedron
(4 faces)

Hexahedron (cube)
(6 faces)

Octahedron
(8 faces)

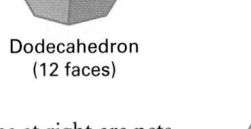

Dodecahedron
(12 faces)

Icosahedron
(20 faces)

tetrahedron

cube

octahedron

dodecahedron

icosahedron

1. The patterns at right are nets for the Platonic solids. Enlarge and copy the nets onto a sturdy material such as poster board or cardboard.

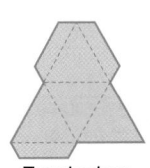

Tetrahedron

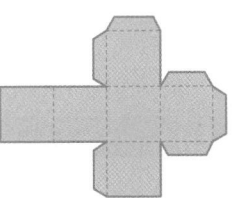

Hexahedron (cube)

2. Cut out the nets and fold along all of the dotted lines. You may wish to color the faces or decorate them in some way.

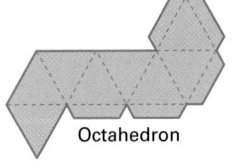

Octahedron

3. Use tape or glue to assemble each polyhedron.

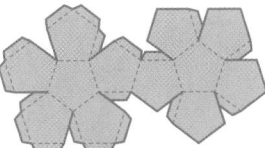

Dodecahedron

WORKING ON THE CHAPTER PROJECT

You should now be able to complete Activity 1 of the Chapter Project.

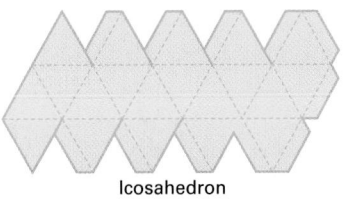

Icosahedron

ALTERNATIVE
Assessment

Portfolio Activity

The Portfolio Activity can be used as preparation for the Chapter Project or as a separate activity. In the Portfolio Activity on this page, students become familiar with the five Platonic solids, or regular convex polyhedrons, by using nets to build models of them.

Prisms

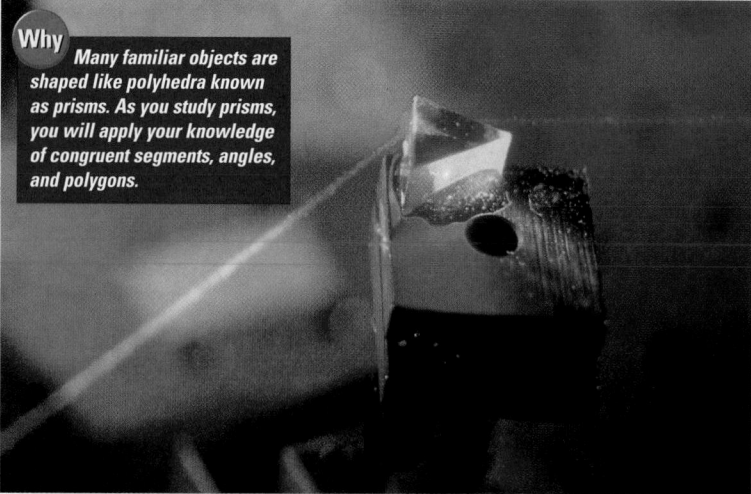

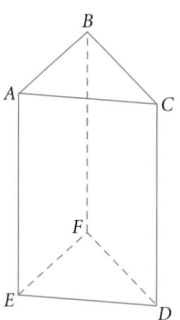
Objectives

● Define *prism*, *right prism*, and *oblique prism*.

● Examine the shapes of lateral faces of prisms.

● Solve problems by using the diagonal measure of a right prism.

A laser beam is bent as it passes through the triangular glass prism in this experiment. Laser light consists of one pure color, so it is not refracted into a spectrum as ordinary light would be.

Prisms

The figures shown below are prisms. A prism is named by the shape of its base.

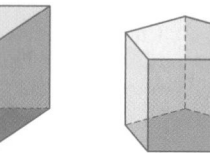

Triangular prism   Rectangular prism   Pentagonal prism   Hexagonal prism

Teach

Why This lesson will introduce students to the properties of prisms—three-dimensional figures that model many real-world objects.

A **prism** is a polyhedron that consists of a polygonal region and its translated image on a parallel plane, with quadrilateral faces connecting corresponding edges.

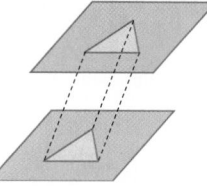

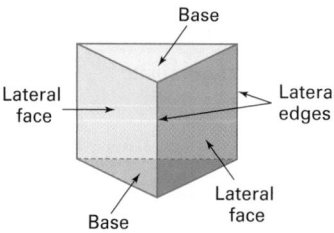

Base

Lateral face

Lateral edges

Base   Lateral face

The faces formed by the polygonal region and its image are each called a **base** of the prism. The remaining faces, which are quadrilaterals, are called **lateral faces** of the prism. The edges of the lateral faces that are not edges of either base are called **lateral edges** of the prism.

Alternative Teaching Strategy

HANDS-ON STRATEGIES Have each student use thick paper to construct a right prism and a non-right prism. Include prisms whose bases are triangles, rectangles, pentagons, hexagons, and octagons. Have them label the vertices, name the parallel planes, and name all other planes in their models.

What type of quadrilateral are the lateral faces of a prism? (Hint: Which edges are parallel? Which edges are congruent?)

Activity
The Lateral Faces of Prisms

Part I

Use uncooked spaghetti and miniature marshmallows to build a model of a rectangular prism. The lateral edges should be vertical, as shown at right.

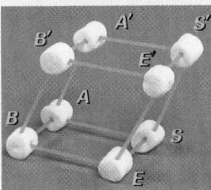

1. What type of quadrilateral is each of the following? Explain.

   a. *BASE*   **b.** *B'E'EB*   **c.** *E'S'SE*

2. List all pairs of congruent polygons in this model.

3. Translate the upper base in the direction of a slide arrow pointing from *B* to *E*, as shown at right. What type of quadrilateral is each of the following? Explain.

   a. *BASE*   **b.** *B'E'EB*   **c.** *E'S'SE*

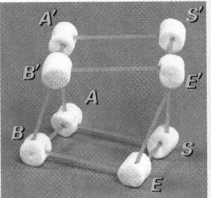

CHECKPOINT ✔
4. Which of the pairs of polygons that you listed in Step 2 are still congruent after the translation of the upper base?

Part II

Return your prism model to its original shape.

1. Rotate the upper base to the right or left, as shown at right. Is your new figure still a prism? Why or why not?

2. Are the segments that formed the corresponding edges of the bases of your original prism still coplanar?

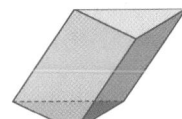

3. Why do you think prisms are defined so that one base is a translation of the other?

CHECKPOINT ✔
4. Is it possible to manipulate your prism so that it is still a prism but none of its lateral faces are rectangles? Explain.

In the Activity, you discovered that all lateral faces of a prism are rectangles or nonrectangular parallelograms. This gives an additional classification for prisms.

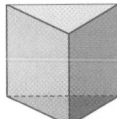

A **right prism** is a prism in which all of the lateral faces are rectangles.

An **oblique prism** has at least one nonrectangular lateral face.

Interdisciplinary Connection

CHEMISTRY   Constructing prisms may remind students of the crystal lattices of chemical compounds such as sodium chloride (salt). Ask students to do research in the library and their science books to find examples of prisms that occur in science and nature.

Activity   Notes

Students may enjoy participating in this Activity in small groups. This Activity will help students to understand that the sides of a prism must be parallelograms. Make sure that students use prime notation to identify the faces that are transformation images.

For a student worksheet of this Activity and detailed Teacher Notes, see page 117 in the Lesson Activities booklet.

CHECKPOINT ✔
Part I

4. All of them are still congruent.

CHECKPOINT ✔
Part II

4. Yes. For example, if the top base of a right rectangular prism is translated horizontally and vertically within its plane, the new prism will have lateral faces that are parallelograms.

The Diagonals of a Right Rectangular Prism

A **diagonal** of a polyhedron is a segment whose endpoints are vertices of two different faces of the polyhedron.

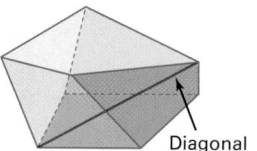

Diagonal

Diagonal of a Right Rectangular Prism

In a right rectangular prism with dimensions $\ell \times w \times h$, the length of a diagonal is given by

$$d = \sqrt{\ell^2 + w^2 + h^2}.$$ **6.3.1**

Given: rectangular prism with dimensions $\ell \times w \times h$

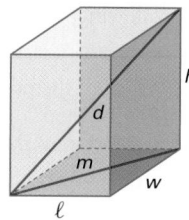

Prove: $d = \sqrt{\ell^2 + w^2 + h^2}$

Proof:

Statements	Reasons
$m^2 = \ell^2 + w^2$	Pythagorean Theorem
$d^2 = m^2 + h^2$	Pythagorean Theorem
$d^2 = \ell^2 + w^2 + h^2$	Substitution Property of Equality
$d = \sqrt{\ell^2 + w^2 + h^2}$	Take the square root of each side.

EXAMPLE ●

APPLICATION
CHEMISTRY

Iron atoms form *body-centered cubic* structures. The atoms form cubes with one atom in the center and one at each vertex, as shown at right. The cube, called a *unit cell*, is the smallest unit of a repeating pattern.

The atoms along the diagonal may be thought of as touching each other. If a unit cell measures 291 picometers (1 picometer = 1×10^{-12} meters) on each edge, what is the approximate radius of an iron atom?

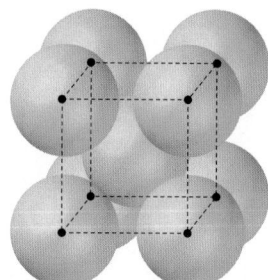

Each vertex of the dashed cube is at the center of one of the atoms.

● **SOLUTION**

First find the length of the diagonal of the unit cell.

$$d = \sqrt{\ell^2 + w^2 + h^2}$$
$$d = \sqrt{291^2 + 291^2 + 291^2} = \sqrt{3(291^2)} = 291\sqrt{3}$$
$$d \approx 504 \text{ picometers}$$

The diagonal is the length of four radii, so divide by 4.

$$r \approx 504 \div 4 \approx 126 \text{ picometers}$$

The radius of an iron atom is approximately 126 picometers, or 0.000000000126 meter.

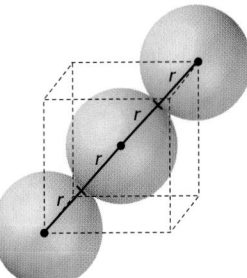

Assess

Selected Answers

Exercises 5–8, 9–44 odd

ASSIGNMENT GUIDE	
In Class	1–8
Core	9–33 odd, 34–40
Core Plus	21–33 odd, 34–44
Review	45–52
Preview	53–54

✐ Extra Practice can be found beginning on page 818.

Exercises

● Communicate

1. Why is a prism with rectangular lateral faces called a "right" prism?

2. In the right rectangular prism shown at right, which faces are congruent? Would they still be congruent if the prism were oblique? Explain your reasoning.

3. In the rectangular prism shown at right above, could more than one pair of faces be called the bases? Explain your reasoning.

4. Identify some real-world objects that are prisms, including oblique prisms and prisms that do not have rectangular bases.

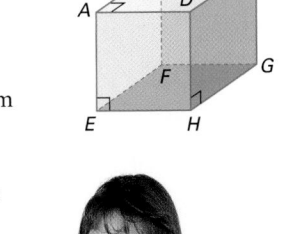

A rare example of an oblique prism

● Guided Skills Practice

5. Rectangle

6. Parallelogram

7. Rectangle

Refer to the oblique rectangular prism with base *EFGH* shown at right. Classify each quadrilateral named below. *(ACTIVITY)*

5. *EFGH* **6.** *ADHE* **7.** *ABFE*

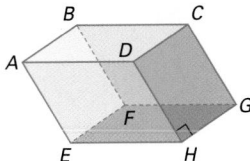

APPLICATION

8. **CHEMISTRY** Sodium has a body-centered cubic structure with a unit cell that is 430 picometers wide. Estimate the radius of a sodium atom. *(EXAMPLE)* ≈ 186 pm

Error Analysis

If students find it difficult to generalize to *n* sides in Exercise 38, have them extend the table to include more sides in the base of a prism before such a generalization is made.

Technology

Students may use geometry graphics software to draw various prisms. They may also use the software to experiment with other solids, including the Platonic solids.

Reteaching the Lesson

COOPERATIVE LEARNING Have one student draw a right rectangular prism. Have another student add information to the drawing, such as the name and length of an edge. Have students continue adding information until one student is able to name and find the length of a diagonal of the prism. Repeat the activity until all students have participated.

15. △ABC ≅ △DEF; in a prism, the bases are translated images of each other.

20. ∠ADE, ∠ADF, ∠DAC, ∠DAB, ∠BED, ∠BEF, ∠CFD, ∠CFE, ∠FCA, ∠FCB, ∠EBC, ∠EBA

24. GKLH ≅ JNMI; GKNJ ≅ HLMI; KLMN ≅ GHIJ

25. ∠KNJ, ∠JGK, ∠LMI, ∠IHL, ∠NJI, ∠IMN, ∠KGH, ∠HLK

11. Not a prism; prisms do not have curved edges

13. Not a prism; no translated bases

16. $\overline{AD}$ and $\overline{CF}$

17. Rectangle

18. $\overline{BC}$, $\overline{AB}$, and $\overline{DE}$

19. ABED and BCFE

21. $\overline{HM}$

22. Parallelogram

23. Parallelogram

● *Practice and Apply*

Which of the figures below appear to be prisms? Give the name for each prism. If the figure is not a prism, explain why not.

9. 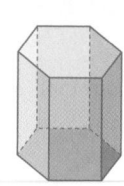 Hexagonal prism

10. 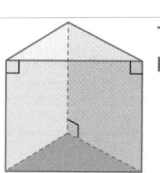 Pentagonal prism

11.

12. 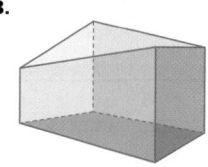 Triangular prism

13.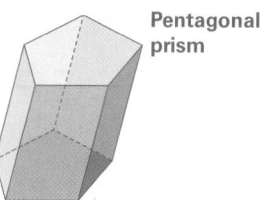

14. 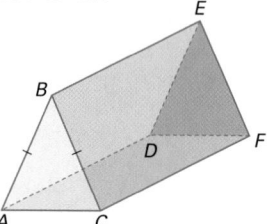 Pentagonal prism

Use the right triangular prism below for Exercises 15–20.

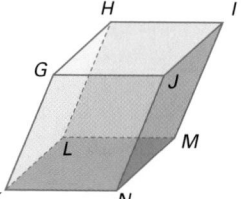

15. Which face is congruent to △ABC? Explain.

16. Name all segments congruent to $\overline{BE}$.

17. What type of quadrilateral is ACFD?

18. Name all segments congruent to $\overline{EF}$.

19. Name two congruent lateral faces.

20. List all right angles in the prism.

Use the oblique rectangular prism below for Exercises 21–25. In the prism, m∠GKN = 60° and m∠GKL = 80°.

21. Name all segments congruent to $\overline{GN}$.

22. What type of quadrilateral is GJNK?

23. What type of quadrilateral is JIMN?

24. Name all pairs of congruent faces.

25. List all obtuse angles in the prism.

Find the length of a diagonal of a right rectangular prism with the given dimensions.

26. $\ell = 4, w = 12, h = 3$ 13

27. $\ell = 10, w = 5, h = 12$ ≈16.40

28. $\ell = 7.5, w = 8, h = 8.5$ ≈13.87

29. $\ell = a, w = a, h = a$ $a\sqrt{3}$

For Exercises 30–33, refer to the right rectangular prism shown at right.

30. $\ell = \underline{\quad ? \quad}$, $w = 8$, $h = 24$, $d = 26$ **6**

31. $\ell = 7$, $w = \underline{\quad ? \quad}$, $h = 10$, $d = 15$ **≈8.72**

32. $\ell = a$, $w = a$, $h = 2a$, $d = \underline{\quad ? \quad}$ **$a\sqrt{6}$**

33. $\ell = x$, $w = x$, $h = x$, $d = 10$, $x = \underline{\quad ? \quad}$ **≈5.77**

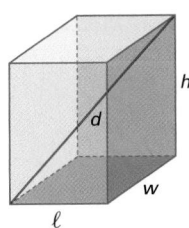

internet connect

Homework Help Online

Go To: go.hrw.com
Keyword:
MG1 Homework Help
for Exercise 34-38

For Exercises 34–38, complete the table below.

	Type of prism	Number of faces	Number of vertices	Number of edges
34.	triangular	? 5	? 6	? 9
35.	rectangular	? 6	? 8	? 12
36.	pentagonal	? 7	? 10	? 15
37.	hexagonal	? 8	? 12	? 18
38.	*n*-gonal	? $n + 2$	? $2n$	? $3n$

39. Use the pattern from the table above to determine how many faces, vertices, and edges a 20-gonal prism has.

Leonhard Euler

40. CULTURAL CONNECTION: EUROPE Swiss mathematician Leonhard Euler (1707–1783) proved a relationship between the faces, vertices, and edges of a polyhedron. Let F represent the number of faces, V represent the number of vertices, and E represent the number of edges of a prism. For each prism in the table above, calculate $V - E + F$. What do you notice?

CHALLENGE

41. The base of the right hexagonal prism shown at right is a regular hexagon with a side length of a and a height of h. Find the length of the indicated diagonal in terms of a and h. $d = \sqrt{4a^2 + h^2}$

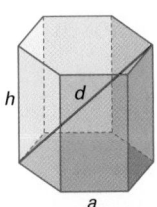

APPLICATION

42. PACKAGING A manufacturer wishes to package a candy bar in a box shaped like a right triangular prism with equilateral triangles for the bases. The unfolded pattern for the box, called a *net*, is shown at right. Explain how the net should be folded in order to produce the desired shape. Draw a net for a box shaped like a right hexagonal prism.

39. faces: $20 + 2 = 22$
vertices: $2(20) = 40$
edges: $3(20) = 60$

40. Triangular prism:
$6 - 9 + 5 = 2$
Rectangular prism:
$8 - 12 + 6 = 2$
Pentagonal prism:
$10 - 15 + 7 = 2$
Hexagonal prism:
$12 - 18 + 8 = 2$
n-gonal prism:
$(n + 2) - 3n + 2n = 2$
$V - E + F = 2$

42. Fold the top and bottom rectangle up to meet above the center rectangle. Then fold the triangles up to meet the edges of the these rectangles.

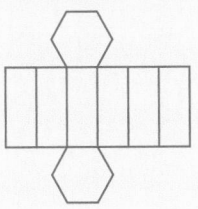

45. true, because the diagonals of a parallelogram bisect each other

46. Sample answer: false, because in a rectangle made by joining two congruent 30-60-90 triangles, the diagonal divides the vertex into 60° and 30°

47. False

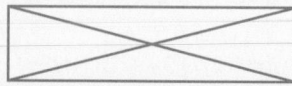

48. False; the diagonals of a rectangle divide the rectangle into two pairs of congruent triangles.

APPLICATIONS

43. $\frac{4a\sqrt{3}}{3}$

43. CHEMISTRY A given substance has a body-centered cubic structure, like that of iron and sodium. If the radius of one atom of this substance is a, find the length of a side, in terms of a, of a unit cell of the substance.

44. CHEMISTRY Another type of atomic structure is called *face-centered cubic*. This structure is a cube with one atom at each vertex and one in the center of each face, as shown at right. Calcium has a face-centered cubic structure and has an atomic radius of 197 picometers. Find the dimensions, including the diagonal, of a unit cell of calcium.

side length ≈ 557.2 pm
diagonal ≈ 965.1 pm

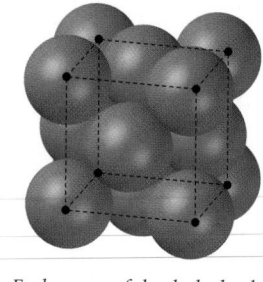

Each vertex of the dashed cube is at the center of one of the atoms.

Look Back

Classify each statement as true or false and explain your reasoning. *(LESSON 4.6)*

45. The diagonals of a rectangle that is not a square bisect each other.

46. The diagonals of a rectangle that is not a square bisect the angles of the rectangle.

47. The diagonals of a rectangle that is not a square are perpendicular to each other.

48. The diagonals of a rectangle that is not a square divide the rectangle into four congruent triangles.

CONNECTION

COORDINATE GEOMETRY **For Exercises 49–51, find the length and midpoint of the segment connecting each pair of points.** *(LESSON 5.6)*

49. $(3, 4), (3, -4)$
8; (3, 0)

50. $(4, -2), (-2, 3)$
≈ 7.81; $(1, \frac{1}{2})$

51. $(2, -2), (5, 6)$
≈ 8.54; $(\frac{7}{2}, 2)$

APPLICATION

52. OPTICS When white light passes through a glass prism, it is separated into a spectrum of different wavelengths, as shown at right. In the diagram, the light hits the prism at an angle of 36°. Use the information in the diagram to determine the angle between the red and violet light rays leaving the prism. **15°**

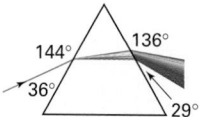

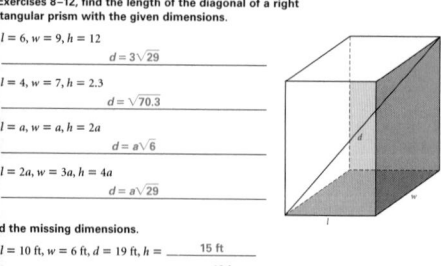

The Spider and the Fly Suppose that a spider and a fly are in a room shaped like a right rectangular prism. The fly can travel directly to any point in the room by flying, but the spider must walk on the walls, floor, or ceiling.

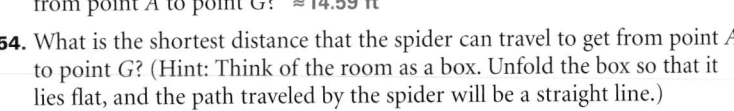

53. What is the shortest distance that the fly can travel to get from point *A* to point *G*? ≈**14.59 ft**

54. What is the shortest distance that the spider can travel to get from point *A* to point *G*? (Hint: Think of the room as a box. Unfold the box so that it lies flat, and the path traveled by the spider will be a straight line.) ≈**18.03 ft**

internet connect

Portfolio Extension

Go To: **go.hrw.com**
Keyword:
MG1 4-D

 **Look Beyond**

In Exercises 53 and 54, students compare the distance from one point to another in two-dimensional space with that in three-dimensional space.

ALTERNATIVE
Assessment

Portfolio Activity

The Portfolio Activity can be used as preparation for the Chapter Project or as a separate activity. In the Portfolio Activity on this page, students build a pop-up model of a dodecahedron.

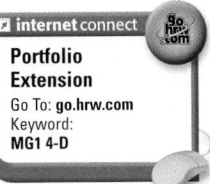 **PORTFOLIO ACTIVITY**

POP-UP DODECAHEDRON

Follow the steps below to create a model of a regular dodecahedron.

1. Make two copies of the pattern at right out of cardboard. Fold along the dotted lines.

2. Holding the two pattern pieces together, place a rubber band around them as shown.

3. Release your hold on the pieces to form the dedecahedron.

Student Technology Guide

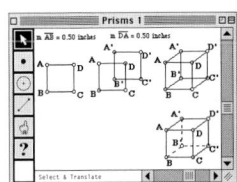

Coordinates in Three Dimensions

1. Find the midpoint of the segment joining points $(3, -5)$ and $(8, -2)$.
   $(5.5, -3.5)$

2. Name the axis containing the point $(-4, 0)$. *x*-axis

3. A square has vertices at $(5, 0)$, $(-2, 0)$, and $(-2, 7)$. What is its fourth vertex? $(5, 7)$

Also on Quiz Transparency 6.4

Teach

Why Remind students that geometry was developed to solve practical problems. The three-dimensional coordinates studied in this lesson not only are practical but also are an important tool for locating points in space.

Objectives

- Identify the features of a three-dimensional coordinate system, including the axes, octants, and coordinate planes.

- Solve problems by using the distance formula in three dimensions.

Why *By using a three-dimensional coordinate system, it is possible to locate objects anywhere in space. One method used by astronomers to identify points in space is based on this system.*

By using the *x*- and *y*-axes in a coordinate plane, you can give the location of any point on the plane. Only two numbers are required to do this, so a plane is said to be *two-dimensional*. By adding a third axis, called the *z*-axis, that intersects the *x*- and *y*-axes at right angles, you can give the location of any point in space. Three numbers are required for this, so space is said to be *three-dimensional*.

The Arrangement of the Axes

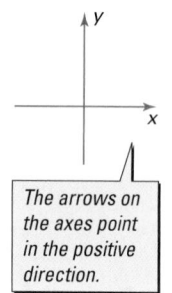

The arrows on the axes point in the positive direction.

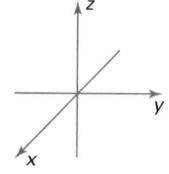

Examine the *x*- and *y*-axes at left. There are two ways to add a third axis. The positive direction of the *z*-axis can point either straight up (out of the page) from the origin or straight down (into the page). In the most common system, the *z*-axis points straight up. To represent the three axes on a page, turn the axes so that the *x*-axis points down and to the left, the *y*-axis points to the right, and the *z*-axis points straight up, as shown at right. Imagine that the *x*-axis points straight out from the page.

This arrangement of the axes is called a **right-handed system**. With your right hand, let your index finger represent the *x*-axis, your middle finger, the *y*-axis, and your thumb, the *z*-axis. Hold these fingers at right angles to each other, as shown. Each finger points in the positive direction of the axis.

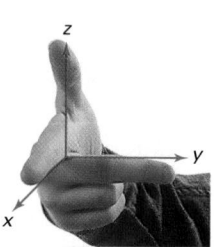

Alternative Teaching Strategy

USING DISCUSSION Students can develop an understanding of how to locate points in space by considering questions such as the following:

1. How many coordinates does a point in space have? 3

2. Which axis contains the point $(0, -2, 0)$?
   y-axis

3. Which plane contains the point $(3, -2, 0)$?
   xy-plane

4. Which axis is the intersection of the *yz*-plane and *xz*-plane? *z*-axis

5. How many planes contain a line through $(0, 0, 0)$ and $(2, 3, 4)$? infinitely many

View your right hand from different perspectives while keeping your fingers in the same positions to see different views of the right-handed system. Do the axes shown at right represent right-hand systems?

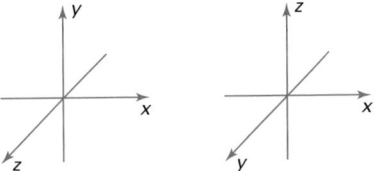

EXAMPLE ① Locate the point $P(1, 2, 3)$ in a three-dimensional coordinate system.

● **SOLUTION**

1. Starting at the origin, count 1 unit in the positive direction along the x-axis. Make a mark on the x-axis at this position.

2. From your mark on the x-axis, count 2 units in the positive direction along the y-axis, drawing a dashed line to represent the distance. Make a mark at the new position.

3. From the new position, count 3 units in the positive direction along the z-axis, drawing a dashed line to represent the distance. Label your final position as point $P(1, 2, 3)$.

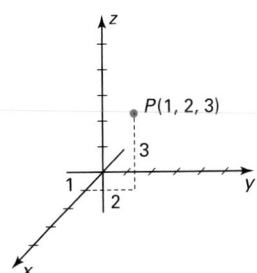

Locate the point $P(2, 3, 6)$ in a three-dimensional coordinate system.

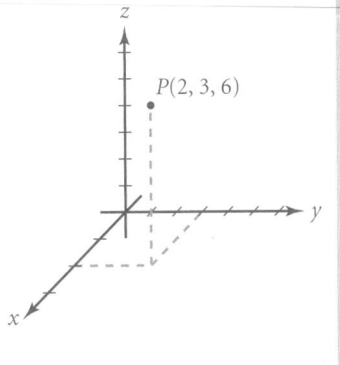

The Octants and the Coordinate Planes

Just as the x- and y-axes divide the plane into four quadrants, the x-, y- and z-axes divide space into eight **octants**. The octant in which all three coordinates of a point are positive, abbreviated $(+, +, +)$, is called the *first octant*.

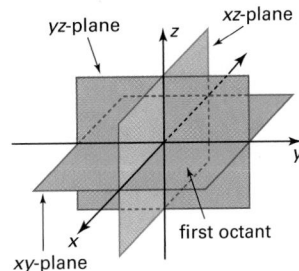

The remaining octants are described by the words *top*, *bottom*, *front*, *back*, *left*, and *right*. For example, the top-front-left octant is the octant in which each point has positive x- and z-coordinates and a negative y-coordinate, or $(+, -, +)$.

CHECKPOINT ✔ What are the signs for the coordinates of points in each of the other six octants?

Each pair of axes also determines a **coordinate plane**. There are three coordinate planes, each named by the pair of axes that determines the plane: the xy-plane, the xz-plane, and the yz-plane.

- In the xy-plane, the z-coordinate of every point is 0.
- In the xz-plane, the y-coordinate of every point is 0.
- In the yz-plane, the x-coordinate of every point is 0.

CHECKPOINT ✔ What can you say about the coordinates of a point on the x-axis? on the y-axis? on the z-axis?

Inclusion Strategies

USING MODELS Some students may have difficulty visualizing three-dimensional coordinates. You can illustrate the eight octants of a three-dimensional system with physical models of planes made of cardboard or other materials.

Draw a right rectangular prism in the first octant so that three of its faces are in the coordinate planes. Label the coordinates of the vertices.

Answers will vary. A sample answer is given below.

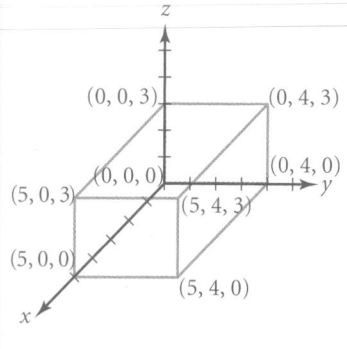

EXAMPLE 2 Draw a right rectangular prism in the first octant so that three of its faces are in the three coordinate planes. Label the coordinates of each vertex.

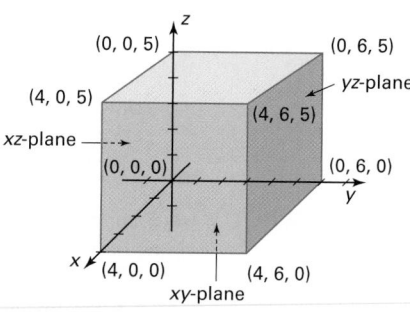

● SOLUTION

One possible solution is shown in the figure at right.

The Distance Formula in Three Dimensions

Recall from Lesson 6.3 that the length of the diagonal of a right rectangular prism with dimensions $\ell \times w \times h$ is given by $d = \sqrt{\ell^2 + w^2 + h^2}$.

To find the distance between two points in three-dimensions, construct a right rectangular prism with faces parallel to the coordinate planes and with the given points as vertices. The segment joining the two points is a diagonal of this prism. The dimensions of the prism are $\ell = |x_2 - x_1|$, $w = |y_2 - y_1|$, and $h = |z_2 - z_1|$.

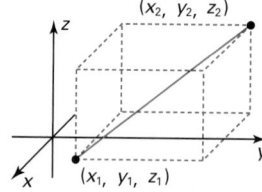

Therefore, the distance between two points in space is given by the following formula:

Distance Formula in Three Dimensions

The distance, d, between the points (x_1, y_1, z_1) and (x_2, y_2, z_2) is given by
$$d = \sqrt{(x_2 - x_1)^2 + (y_2 - y_1)^2 + (z_2 - z_1)^2}.$$ **6.4.1**

Find the distance between the points $P(2, -4, -2)$ and $Q(0, -3, 1)$.

$\sqrt{14} \approx 3.74$

EXAMPLE 3 Find the distance between the points $R(4, 6, -9)$ and $S(-3, 2, -6)$.

● SOLUTION

$$
\begin{aligned}
RS &= \sqrt{(-3 - 4)^2 + (2 - 6)^2 + [-6 - (-9)]^2} \\
&= \sqrt{(-7)^2 + (-4)^2 + (3)^2} \\
&= \sqrt{49 + 16 + 9} \\
&= \sqrt{74} \approx 8.6
\end{aligned}
$$

TRY THIS Find the distance between the points $C(3, -4, -5)$ and $D(2, 0, -1)$.

TRY THIS
$\sqrt{33} \approx 5.74$

Enrichment

Air-traffic control towers need to keep accurate information on the location of all airplanes within their radar range. Suppose that an airplane appears to be located at (8, 6) on a two-dimensional view screen with units in miles. However, the aircraft is also 2 miles above ground. Find the coordinates of the aircraft in three dimensions.
(8, 6, 2)

Exercises

Communicate

1. What can you say about the location of a point in space that has one coordinate equal to 0? two coordinates equal to 0?

Exercises 2 and 3 refer to the coordinate axes and point *P* at right.

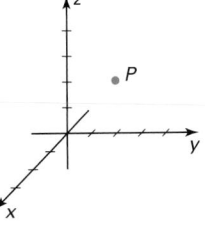

2. Is it possible to determine the coordinates of point *P* from the diagram? Explain your reasoning. If not, what information is needed?

3. Is it possible to determine the octant in which point *P* is located from the diagram? If not, is it possible to determine an octant in which *P* could not be located?

4. In the distance formula in three dimensions, why is it not necessary to use absolute-value signs on the lengths of the sides?

5. Imagine that your classroom is in a three-dimensional coordinate system. Choose a point in the room to represent the origin, and describe the coordinates of some objects in the room.

Guided Skills Practice

Locate each point in a three-dimensional coordinate system. (EXAMPLE 1)

6. $(2, 0, -1)$ **7.** $(-1, -2, -3)$ **8.** $(3, -1, 4)$

9. A right rectangular prism in the first octant of a three-dimensional coordinate system is positioned so that three of its faces are in the coordinate planes. The prism has a length of 2, width of 7, and height of 6. Find the coordinates of each vertex. *(EXAMPLE 2)*

Find the distance between each pair of points. *(EXAMPLE 3)*

10. $(0, 0, 0)$ and $(3, 6, 1)$ $\sqrt{46} \approx 6.78$ **11.** $(3, 1, 0)$ and $(5, -3, 1)$ $\sqrt{21} \approx 4.58$

12. $(-4, 7, -2)$ and $(2, 3, -5)$ $\sqrt{61} = 7.81$

17. top-front-right
18. bottom-back-right
19. *xz*-plane
20. *y*-axis
21. top-back-right
22. *xy*-plane
23. *x*-axis
24. bottom-back-left

Practice and Apply

Locate each point in a three-dimensional coordinate system.

13. $(4, 1, 5)$ **14.** $(0, -3, 0)$ **15.** $(0, -5, 2)$ **16.** $(-3, -1, 3)$

Name the octant, coordinate plane, or axis in which each point is located.

17. $(3, 1, 7)$ **18.** $(-2, 10, -4)$ **19.** $(4, 0, -2)$ **20.** $(0, -5, 0)$

21. $(-3, 1, 7)$ **22.** $(-1, 3, 0)$ **23.** $(4, 0, 0)$ **24.** $(-8, -1, -5)$

6.

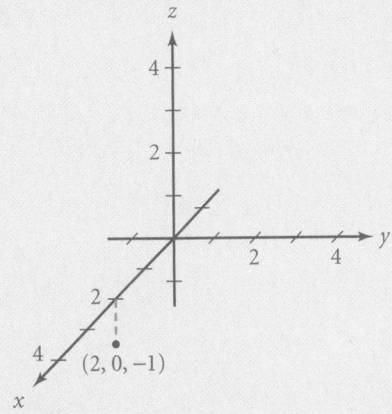

Assess

Selected Answers

Exercises 6–12, 13–59 odd

ASSIGNMENT GUIDE

In Class	1–12
Core	13–49 odd
Core Plus	25–47 odd, 49, 50
Review	51–59
Preview	60–63

✐ Extra Practice can be found beginning on page 818.

Error Analysis

In Exercises 43–48, students may not make the connection between finding the midpoint of a segment with algebra and finding the location of the midpoint in space. To help students make this connection, use a model or a drawing of a three-dimensional coordinate system to show how the midpoint in each direction $(x, y, \text{and } z)$ gives the coordinates of the midpoint in three-dimensional space.

Exercises 60–63 show applications of the distance formula in three dimensions. The units used here, astronomical units, were created so that the large numbers associated with distances in space could be managed more easily.

25. Sample answer: $(1, 2, 3)$

26. Sample answer: $(1, 2, -3)$

27. Sample answer: $(-1, -2, 3)$

28. Sample answer: $(-1, -2, -3)$

43. $\left(\dfrac{11}{2}, -\dfrac{9}{2}, \dfrac{7}{2}\right)$

44. $(2, 2, 2)$

45. $\left(\dfrac{5}{2}, -\dfrac{3}{2}, -\dfrac{5}{2}\right)$

46. $\left(1, -\dfrac{1}{2}, \dfrac{1}{2}\right)$

47. $(0, 0, 0)$

48. $(0, 0, 0)$

49. Sample answer:
$\left(\dfrac{\sqrt{3}}{3}, \dfrac{\sqrt{3}}{3}, \dfrac{\sqrt{3}}{3}\right)$

Practice

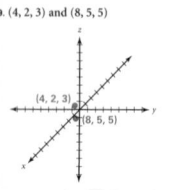

NAME _____ CLASS _____ DATE _____

Practice
6.4 *Coordinates in Three Dimensions*

Name the octant, coordinate plane, or axis for each point.

1. $(1, -8, 7)$ top-front-left octant
2. $(0, 7, 0)$ y-axis
3. $(-2, -6, -1.7)$ bottom-back-left octant
4. $(1, 8, -7)$ bottom-front-right octant
5. $(7, 0, 8)$ xz-plane
6. $(-7, 6, 2)$ top-back-right octant
7. $(0, 0, -2)$ z-axis
8. $(-2, -3, 4)$ top-back-left octant

For Exercises 9–12, locate each pair of points in a three-dimensional coordinate system. Find the distance between the points, and find the midpoint of the segment connecting them.

9. $(4, 2, 3)$ and $(8, 5, 5)$
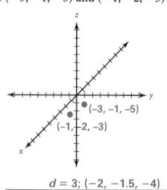
$d = \sqrt{29}$; $(6, 3.5, 4)$

10. $(8, 3, 5)$ and $(-3, -5, -8)$
$d = \sqrt{354}$; $(2.5, -1, -1.5)$

11. $(-3, -1, -5)$ and $(-1, -2, -3)$
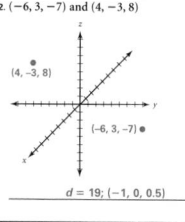
$d = 3$; $(-2, -1.5, -4)$

12. $(-6, 3, -7)$ and $(4, -3, 8)$
$d = 19$; $(-1, 0, 0.5)$

Give the coordinates of a point in the given octant.

25. first octant
26. bottom-front-right
27. top-back-left
28. bottom-back-left

Algebra

Find the distance between each pair of points.

29. $(1, 1, 1)$ and $(-1, -1, -1)$ $\sqrt{12} \approx 3.46$
30. $(2, 1, 3)$ and $(5, -2, 7)$ $\sqrt{34} \approx 5.83$
31. $(2, 0, 1)$ and $(-5, -6, -5)$ 11
32. $(7, -6, 5)$ and $(6, 4, 3)$ $\sqrt{105} \approx 10.25$

For Exercises 33–42, refer to the diagram below of a right rectangular prism.

Determine the coordinates of each point.

33. point F (0, 10, 7)
34. point H (6, 10, 7)
35. point C (6, 10, 0)
36. point E (0, 0, 7)

Find each measure.

37. AD 6
38. AG $\sqrt{85} \approx 9.22$
39. AH $\sqrt{185} \approx 13.6$
40. DF $\sqrt{185} \approx 13.6$
41. area of $EGDA$ 42 units^2
42. area of $ABCD$ 60 units^2

Algebra

The midpoint of a segment in a three dimensional coordinate system is given by the following formula:

Midpoint Formula in Three Dimensions

The midpoint of a segment with endpoints at (x_1, y_1, z_1) and (x_2, y_2, z_2) is the point $\left(\dfrac{x_1 + x_2}{2}, \dfrac{y_1 + y_2}{2}, \dfrac{z_1 + z_2}{2}\right)$

6.4.2

---

internet connect

Homework Help Online
Go To: **go.hrw.com**
Keyword:
MG1 Homework Help
for Exercises 43-48

Find the midpoint of the segment with the given endpoints.

43. $(5, -2, 3)$ and $(6, -7, 4)$
44. $(3, 2, 1)$ and $(1, 2, 3)$
45. $(-1, -4, -5)$ and $(6, 1, 0)$
46. $(2, -1, 0)$ and $(0, 0, 1)$
47. $(1, 1, 1)$ and $(-1, -1, -1)$
48. (a, b, c) and $(-a, -b, -c)$

49. Give the coordinates of a point that is not on an axis or in a coordinate plane and that is a distance of 1 unit from the origin.

50. The red light source is located above the positive y-axis and the blue light source is located above the negative y-axis.

50. **GRAPHICS** A white sphere is illuminated by two different light sources, one red and one blue. Where the red and blue light mix, magenta light is produced. Examine the illustration at right. Where are the light sources located with respect to the coordinate axes?

Look Back

Algebra

Find the slope of the segment with the given endpoints. *(LESSON 3.8)*

51. $(0, 0)$ and $(6, 4)$ $\frac{2}{3}$

52. $(1, 5)$ and $(3, 3)$ -1

53. $(3, 1)$ and $(2, -5)$ **6**

54. $(-3, 1)$ and $(6, -4)$ $-\frac{5}{9}$

Graph each line and find the *x*- and *y*-intercepts.

55. $y = 2x + 1$

56. $2x + 3y = 6$

57. $y - 3 = 4(x + 1)$

58. $3x - y = 5$

APPLICATION

59. PHYSICAL SCIENCE The relationship between Celsius and Fahrenheit temperatures can be represented by a straight line. Suppose that the *x*-coordinate represents Celsius temperature and the *y*-coordinate represents Fahrenheit temperature. Then the freezing point of water is the point $(0, 32)$ and the boiling point is $(100, 212)$.

  a. Find the slope of the line through the given points.
  b. Find the equation of the line through the given points.
  a. 1.8; b. $y = 1.8x + 32$

Look Beyond

APPLICATION

ASTRONOMY The table below gives the *x*-, *y*-, and *z*-coordinates of points in a three-dimensional coordinate system with the Sun at the origin.

The *x*-axis points in the direction of Earth's position at the first moment of the spring, or vernal, equinox. The *xy*-plane is the plane of Earth's equator at the time of the equinox.

An astronomical unit is a measure of length that is equal to the mean distance from Earth to the Sun, about 92,960,000 miles or 149,600,000 kilometers.

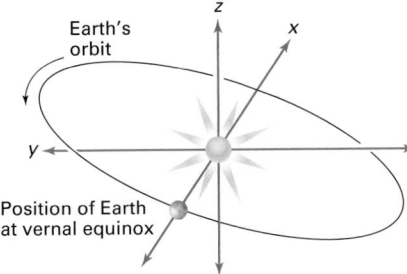

Earth's orbit

Position of Earth at vernal equinox

Sun-Centered Coordinates of Major Planets on August 17, 1990 (in astronomical units)			
Planets	**x**	**y**	**z**
Earth	0.819	−0.546	−0.237
Mars	1.389	0.145	0.029
Jupiter	−2.238	4.328	1.910
Saturn	3.935	−8.443	−3.656

Source: *The Astronomical Almanac for the year 1990*

60. ≈6.141 au,
    ≈570,867,360 mi

61. 9.686 au =
    1,449,025,600 km

62. 1.397 au; 0.127 au;
    the orbit of Mars
    is elliptical.

63. Earth

60. Calculate the distance from Earth to Jupiter on August 17, 1990. Then convert from astronomical units to miles.

61. Calculate the distance from Mars to Saturn on August 17, 1990. Then convert from astronomical units to kilometers.

62. Calculate the distance from the Sun to Mars on August 17, 1990. The mean distance from the Sun to Mars is 227,940,000 kilometers. How close to this value is your calculation? Why do you think there is a difference?

63. Of the planets in the table, which is closest to the Sun?

55.

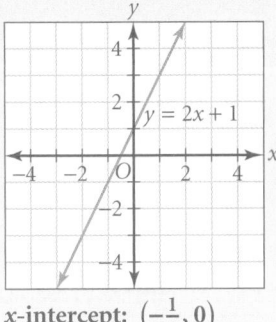

x-intercept: $\left(-\frac{1}{2}, 0\right)$
y-intercept: $(0, 1)$

56.

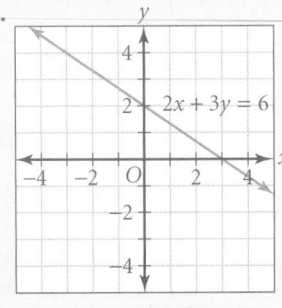

x-intercept: $(3, 0)$
y-intercept: $(0, 2)$

57.

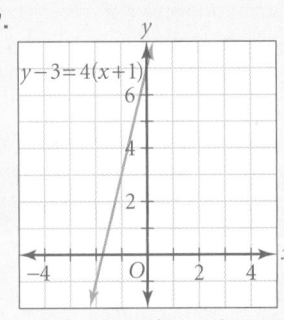

x-intercept: $\left(-\frac{7}{4}, 0\right)$
y-intercept: $(0, 7)$

58.

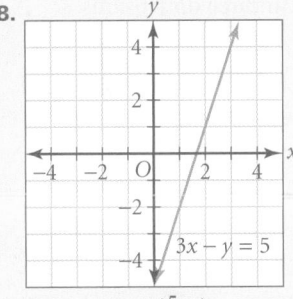

x-intercept: $\left(\frac{5}{3}, 0\right)$
y-intercept: $(0, -5)$

Prepare

1. Determine the *x*- and *y*-intercepts of the line defined by $2x + 3y = 6$.
x-intercept: 3 or $(3, 0)$
y-intercept: 2 or $(0, 2)$

2. Is the point $(-2, 4)$ on the graph of $2x + 3y = 6$?
no

3. Find the equation of the horizontal line that passes through the point $(0, -3)$.
$y = -3$

Also on Quiz Transparency 6.5

Teach

Why This lesson presents ideas from algebra to help students see the connections between graphing lines in two dimensions and in three dimensions.

Lines and Planes in Space

Objectives

● Define the equation of a line and the equation of a plane in space.

● Solve problems by using the equations of lines and planes in space.

Why *Planes in space can form three-dimensional shapes, as in these computer drawings. Each plane in the drawings can be described mathematically.*

The Equation of a Plane

Recall from algebra that the standard form of the equation of a line in a plane is

$$Ax + By = C,$$

where *A*, *B*, and *C* are real numbers, *A* is nonnegative, and *A*, *B*, and *C* are not all zero. The standard equation of a plane in space resembles the equation of a line, with an extra variable for the added dimension of space:

$$Ax + By + Cz = D,$$

where *A*, *B*, *C*, and *D* are real numbers, *A* is nonnegative, and *A*, *B*, *C*, and *D* are not all zero. For example,

$$2x + 5y - 2z = 9$$

is an equation of a plane where $A = 2$, $B = 5$, $C = -2$, and $D = 9$.

Alternative Teaching Strategy

USING VISUAL MODELS Use clear plastic squares to create a model of the octants in a right-handed three-dimensional coordinate system. Cut slits in the squares to slide them into position, and label the axes. Use straws to represent lines. Put the straws in various positions in the model, and have students identify the intercepts of each line (straw) and give the equation of the line. A sample equation would be $3x - 4y = 12$ in two dimensions and in three dimensions. Ask students to give the intercepts for both cases.

Using Intercepts in Graphing

In the coordinate plane, a line that does not lie entirely on an axis crosses the x- and y-axes at one or two points called **intercepts**. In coordinate space, a plane that does not lie entirely on an axis has one, two, or three intercepts.

EXAMPLE 1 Sketch the graph of the plane defined by the equation $2x + y + 3z = 6$.

Algebra

● **SOLUTION**

Find the intercepts where the plane crosses each axis. The x-intercept, for example, is the x-coordinate of the point where the plane crosses the x-axis. At this point, the y- and z-coordinates are 0. To find the x-intercept, set $y = 0$ and $z = 0$ and solve the equation for x.

$$2x + 0 + 3(0) = 6$$
$$2x = 6$$
$$x = 3$$

The x-intercept is $(3, 0, 0)$. Similarly, the y-intercept is $(0, 6, 0)$, and the z-intercept is $(0, 0, 2)$. Plot the intercepts as points on the axes. These three noncollinear points determine the plane. To sketch the plane, connect the points with segments and add shading. NOTE: Intercepts are often written as single numbers.

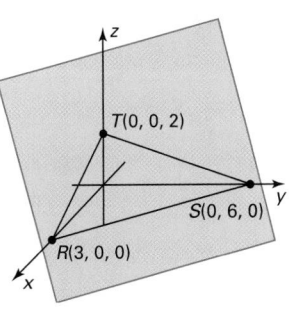

What Happens to the Equation of a Line in Space?

In a coordinate plane, $2x + 4y = 8$ is the equation of a line.

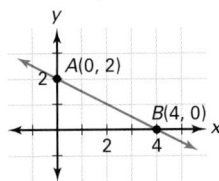

In a three dimensional coordinate space, the equation $2x + 4y = 8$ is the equation of a plane in which the coefficient of z is equal to 0.

Notice that this equation of a plane is unaffected by the values of z. Thus for any values of x and y that satisfy the equation, any value of z will also work. Thus every point directly above and below the line is in the graph of the plane.

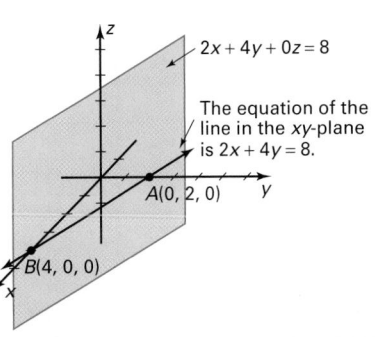

Interdisciplinary Connection

ART Artists and cartoonists who use computers frequently do their work on coordinate systems, both in two dimensions and in three dimensions.

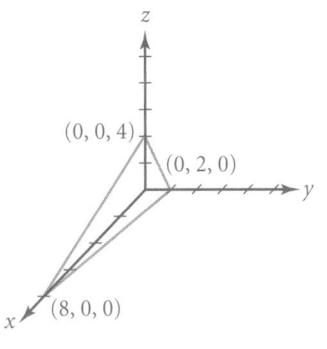

Using another variable, t, to help students plot a line or a plane in space has connections to parametric equations. These equations may be used to describe other kinds of graphs, especially graphs showing motion in space. You may want to use a graphics calculator in parametric mode to demonstrate how to graph lines defined by these kinds of equations.

Lines and Planes in Space: A Step-by-Step Procedure

Imagine that you are able to plot one point of a graph in space every minute according to a given set of instructions. Your instructions have one rule for the x-coordinates, another for the y-coordinates, and another for the z-coordinates. These rules are called **parametric equations**.

Let $t = 1, 2, 3, \ldots$ represent the time (in minutes) at which you plot each point. Your rules for each coordinate would have the following form:

$$x = [\text{an expression involving } t]$$
$$y = [\text{an expression involving } t]$$
$$z = [\text{an expression involving } t]$$

Note: t can also be 0 or negative.

ADDITIONAL
EXAMPLE ②

Use the rules $x = 2t$ and $y = -t + 3$ to plot a line in a plane, where $t = 1, 2, 3, 4, \ldots$

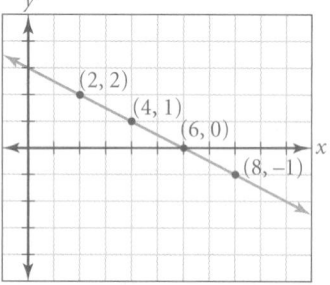

EXAMPLE ② To plot a graph in a coordinate plane, you will need just two rules. Use the rules below to plot a line in a coordinate plane for $t = 1, 2, 3, 4 \ldots$

$$x = 2t$$
$$y = 3t + 1$$

SOLUTION

Fill in a table like the one below. Then plot the points in the xy-coordinate plane.

t	x	y
1	2	4
2	4	7
3	6	10
4	8	13

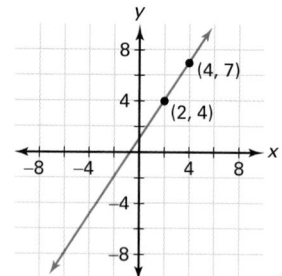

EXAMPLE ③ Use the rules below to plot a line in a coordinate system for $t = 1, 2, 3, 4 \ldots$

$$x = 2t + 1$$
$$y = 3t - 5$$
$$z = 4t$$

SOLUTION

Fill in a table like the one below. Then plot the graph in a three-dimensional coordinate system.

t	x	y	z
1	3	-2	4
2	5	1	8
3	7	4	12
4	9	7	16

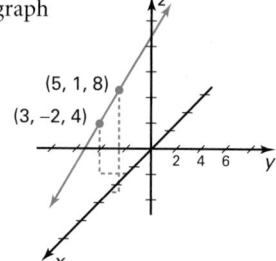

ADDITIONAL
EXAMPLE ③

Use the rules $x = t - 1$, $y = 2t + 2$, and $z = 3t$ to plot a line in space, where $t = 1, 2, 3, 4, \ldots$

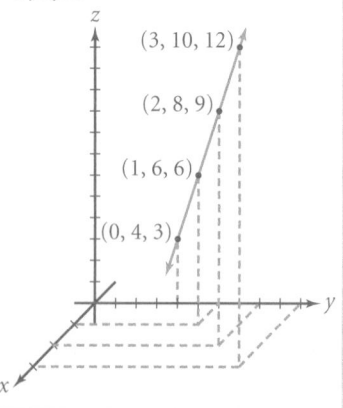

Enrichment

The radius of a sphere that is graphed in a three-dimensional coordinate system is given by $r = \sqrt{(x - i)^2 + (y - j)^2 + (z - k)^2}$, where (i, j, k) is the center of the sphere. What is the radius of a sphere with its center at $(3, -2, 4)$ and a point on the sphere at $(3, -8, 4)$? Give another point on the sphere. $r = 6$. **Answers may vary.**

Inclusion Strategies

ENGLISH LANGUAGE DEVELOPMENT Students may have difficulty reading and understanding mathematics. Suggest that students read this lesson slowly and carefully, write notes, make sketches, and develop a list of questions that they need answered. You may wish to have students form cooperative groups to discuss the questions.

Exercises

Communicate

1. What is the standard form for an equation of a plane? How is it similar to the standard equation of a line? How is it different?

2. Describe the characteristics of a plane defined by an equation in which the coefficient of x is 0.

3. Describe the characteristics of a plane defined by an equation in which the coefficient of z is 0.

4. Describe the characteristics of a plane defined by an equation in which the coefficients of y and z are 0.

5. Can a line in a three-dimensional coordinate system pass through exactly one octant? exactly two octants? exactly three octants? more than three octants? Explain your reasoning.

ASSIGNMENT GUIDE

In Class	1–11
Core	13–35 odd, 36
Core Plus	19–35 odd, 36–38
Review	39–46
Preview	47–50

Extra Practice can be found beginning on page 818.

Guided Skills Practice

Sketch the graph of each plane. *(EXAMPLE 1)*

6. $3x + 6y + 4z = 12$

7. $2x + 5y - z = 2$

Using $t = 1, 2, 3, \ldots$, create a table of x- and y-values for the equations below. Then graph each line in a coordinate plane. *(EXAMPLE 2)*

8. $x = t + 2$
   $y = 4t$

9. $x = t - 1$
   $y = 2t + 3$

Using $t = 1, 2, 3, \ldots$, create a table of x-, y-, and z-values for the equations below. Then graph each line in a coordinate system.
(EXAMPLE 3)

10. $x = t$
    $y = 2t$
    $z = t + 1$

11. $x = 3t - 2$
    $y = t + 6$
    $z = -t$

Technology

Although the exercises and problems in this lesson do not require the use of computer technology, using software to create three-dimensional images may give students a better perspective of lines and planes in space.

Error Analysis

Students may have difficulty sketching the traces for Exercises 26–29. Advise them to make a table of values for the equation of the trace.

Practice and Apply

Use intercepts to sketch the plane defined by each equation below.

12. $3x + 2y + 7z = 4$

13. $2x - 4y + z = -2$

14. $x - 2y - 2z = -4$

15. $-3x + y = 4$

16. $x - 2y = 2$

17. $x = -4$

Algebra **Plot the line for each pair of parametric equations in a coordinate plane.**

18. $x = t + 3$
    $y = 1 - t$

19. $x = 2t - 1$
    $y = -t$

20. $x = 3t$
    $y = 5$

21. $x = 2$
    $y = -4t + 6$

Reteaching the Lesson

USING DISCUSSION Review the formula for the distance between two points in the plane, and then present the formula for distance in three dimensions. Compare the formulas by discussing their similarities and differences.

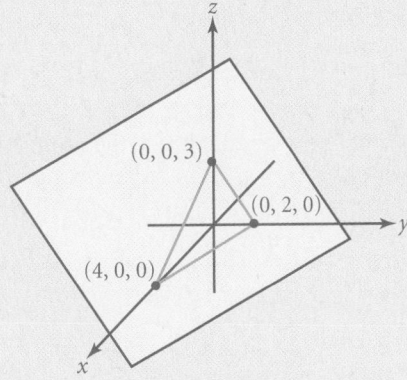

6.

22.

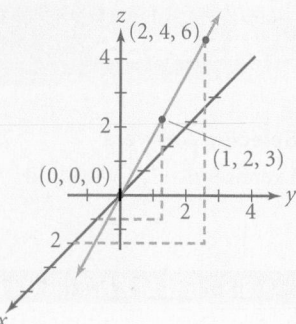

23.

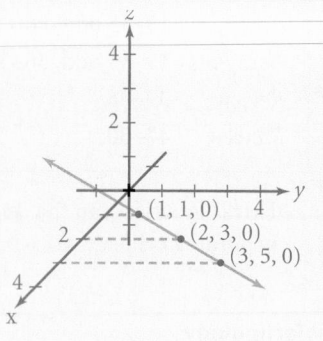

30. Sample answer:

$x = t$

$y = 2t$

$z = -\dfrac{1}{2}t$

Plot the line for each set of parametric equations in a coordinate system.

22. $x = t$
$y = 2t$
$z = 3t$

23. $x = t + 1$
$y = 2t + 1$
$z = 0$

24. $x = t$
$y = t$
$z = 1 - t$

25. $x = 3t$
$y = 4$
$z = 4t - 7$

The *trace* of a plane is its intersection with the *xy*-plane. To find the equation of the trace, set the *z*-coordinate equal to 0 (all points in the *xy*-plane have a *z*-coordinate of 0). For example:

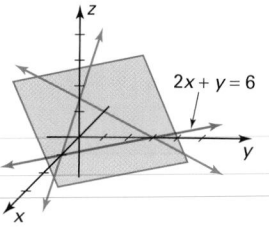

$2x + y + 3z = 6$ *Equation of the plane*

$2x + y + 3(0) = 6$ *Set z equal to 0.*

$2x + y = 6$ *Equation of the trace*

internet connect

Homework Help Online

Go To: **go.hrw.com**
Keyword:
MG1 Homework Help
for Exercises 26–29

Find the equation of the trace for each plane defined below. Sketch the plane and indicate the trace.

26. $x + 3y - z = 7$

27. $5x - 2y + z = 2$

28. $2x + 7y + 3z = 2$

29. $-4x - 2y + 2z = 1$

30. Write parametric equations to describe the line that passes through the points $(0, 0, 0)$ and $(2, 4, -1)$. (Hint: Start with the rule for the *x*-coordinate. The equation will be of the form $x = at + b$, where a and b are constants.)

31. Write parametric equations to describe the line that passes through the points $(-2, 1, 5)$ and $(6, -4, 7)$.

Sample answer: $x = t;\ y = -\dfrac{5}{8}t - \dfrac{1}{4};\ z = \dfrac{1}{4}t + \dfrac{11}{2}$

In a coordinate system, plot the pair of lines described by each set of parametric equations below. Determine whether the lines are parallel, intersecting, or skew. Explain your reasoning. (Note: Let $t = 1, 2, 3, \ldots$ and $s = 1, 2, 3, \ldots$)

32. $x = 1$ $x = 1$
$y = 2$ $y = s$
$z = t$ $z = 3$

33. $x = 1$ $x = 3$
$y = 2$ $y = 5$
$z = t$ $z = s$

34. $x = 1$ $x = 3$
$y = 2$ $y = s$
$z = t$ $z = s$

CHALLENGE

35. Yes. Set the coordinates of the two lines equal to each other and solve the system. $(0, 3, -1)$

35. Do the lines described by the parametric equations below intersect? How can you tell? If they intersect, find the point of intersection. (Hint: Set the *x*-coordinates equal to each other to form an equation. Repeat with the *y*- and *z*-coordinates to get a system of three equations in *s* and *t*, and then solve the system.)

$x = t - 1$ $x = s - 3$
$y = 2t + 1$ $y = s$
$z = -t$ $z = 2s - 7$

APPLICATION

36. HOUSE PAINTING You agree to pay a friend $5 per hour to help paint your house. The paint costs $25 per gallon, and the equipment (ladder, brushes, etc.) costs $200. An equation representing the cost, *z*, of painting for *x* hours and using *y* gallons of paint is $z = 5x + 25y + 200$. Write this equation in the standard form for an equation of a plane, and sketch a graph of the plane.

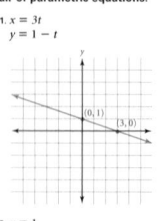 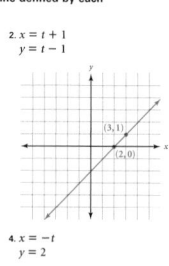
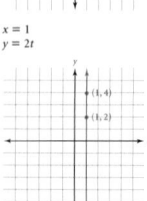 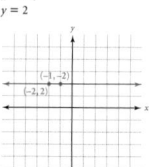

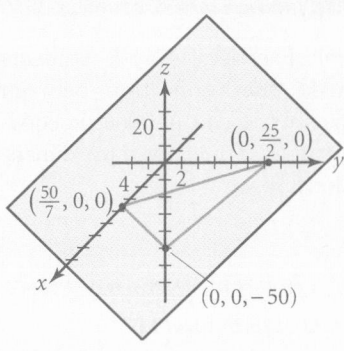

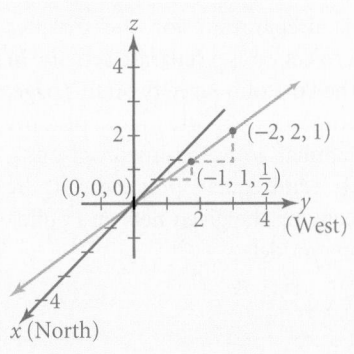

37. HOBBIES Delia rented a booth at a craft fair for $50. She is selling earrings at a profit of $7 per pair and bracelets at a profit of $4 each. Her profit, z, for selling x pairs of earrings and y bracelets is represented by the equation $z = 7x + 4y - 50$. Write this equation in the standard form for an equation of a plane, and sketch a graph of the plane. Find the trace of the plane. What is Delia's profit for points in the trace?

38. NAVIGATION Suppose that an airplane's initial path at takeoff is described by the following parametric equations:

$$x = -t$$
$$y = t$$
$$z = 0.5t$$

Plot the line described by the given equations. If the positive x-axis points north, in which direction did the airplane take off?

 Look Back

COORDINATE GEOMETRY **For each pair of lines below, state whether the lines are parallel, perpendicular, or neither.** *(LESSON 3.8)*

39. $y = 3x + 5$
$y = 3x - 7$
parallel

40. $y = x + 2$
$y = 2 - x$
perpendicular

41. $y = 2x - 1$
$y = -2x + 4$
neither

42. $2x + 3y = 6$
$3x - 2y = 6$
perpendicular

43. Given: $\angle 1 \cong \angle 2$ and $\overline{EH} \cong \overline{FG}$
Prove: $\overline{EF} \cong \overline{HG}$ *(LESSON 3.3)*

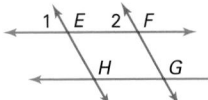

44. Given rectangle $ABCD$ with diagonals $\overline{AC}$ and $\overline{BD}$, prove $\triangle ADC \cong \triangle BCD$. *(LESSON 4.6)*

45. Given rectangle $ABCD$ with diagonals $\overline{AC}$ and $\overline{BD}$ that intersect at point E, prove $\triangle AED \cong \triangle CEB$. *(LESSON 4.6)*

46. Given quadrilateral $PQRS$ in which $\overline{PQ} \cong \overline{RS}$ and $\overline{PS} \cong \overline{QR}$, prove that $PQRS$ is a parallelogram. *(LESSON 4.6)*

43.

Statements	Reasons
1. $\angle 1 \cong \angle 2$; $\overline{EH} \cong \overline{FG}$	1. Given
2. $\overline{EH} \parallel \overline{FG}$	2. Converse of the Corresponding Angles Postulate
3. $EFGH$ is a parallelogram.	3. Theorem 4.6.2
4. $\overline{EF} \cong \overline{HG}$	4. Theorem 4.5.2

Look Beyond

For Exercises 47–50, students make the connection between symmetric and parametric equations of a line in three-dimensional space.

Assessment

Portfolio Activity

The Portfolio Activity can be used as preparation for the Chapter Project or as a separate activity. In the Portfolio Activity on this page, students explore and become familiar with Archimedean solids, or semiregular polyhedrons, by copying the given nets and building models.

49. $\dfrac{x-1}{2} = \dfrac{y-2}{-3} = \dfrac{z}{4}$

50. $x + 4 = \dfrac{y}{2} = \dfrac{z-2}{6}$

 *Algebra*

47. $x = 3t + 2$
$y = -t + 5$
$z = 4t - 1$

48. $x = t + 2$
$y = 2t - 6$
$z = -5t + 3$

internet connect

Portfolio Extension
Go To: **go.hrw.com**
Keyword:
MG1 Solids

Symmetric equations for a line in space are in the form $\dfrac{x - x_1}{a} = \dfrac{y - y_1}{b} = \dfrac{z - z_1}{c}$, where (x_1, y_1, z_1) is a point on the line and a, b, and c are constants.

The relationship between the parametric and symmetric forms is shown below.

$$\dfrac{x - x_1}{a} = \dfrac{y - y_1}{b} = \dfrac{z - z_1}{c} \longleftrightarrow t = \dfrac{x - x_1}{a} \longleftrightarrow \begin{array}{l} x = at + x_1 \\ y = bt + y_1 \\ z = ct + z_1 \end{array}$$
$$t = \dfrac{y - y_1}{b}$$
$$t = \dfrac{z - z_1}{c}$$

Write each set of symmetric equations in parametric form.

47. $\dfrac{x-2}{3} = \dfrac{y-5}{-1} = \dfrac{z+1}{4}$

48. $x - 2 = \dfrac{y+6}{2} = \dfrac{z-3}{-5}$

Write each set of parametric equations in symmetric form.

49. $x = 2t + 1$
$y = -3t + 6$
$z = 4t$

50. $x = t - 4$
$y = 2t$
$z = 6t + 2$

PORTFOLIO ACTIVITY

ARCHIMEDEAN SOLIDS

Archimedean solids, or semiregular polyhedra, are named after Greek mathematician Archimedes. All of the faces of an Archimedean solid are regular polygons, but two or more types of polygons are used. In addition, every vertex must have the exact same arrangement of faces around it. There are exactly 13 Archimedean solids.

1. Enlarge and copy the net at right. You may wish to add tabs at some of the edges to help in assembling the polyhedron. This solid is called a *rhombicuboctahedron*. It has 8 equilateral triangular faces and 18 square faces. Three squares and one triangle meet at every vertex.

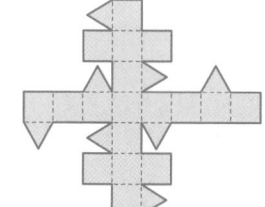

2. Enlarge and make two copies of the net at right. The net forms half of the solid known as a *truncated icosahedron*. If you truncate, or cut off, the vertices of a regular icosahedron (refer back to the Portfolio Activity on page 379), this shape is formed. A truncated icosahedron has 12 regular pentagonal faces and 20 regular hexagonal faces. You may recognize the truncated icosahedron as a common object—a soccer ball.

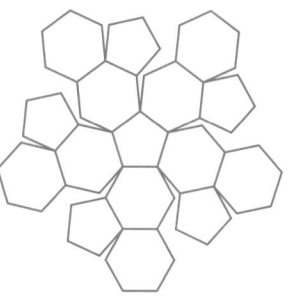

Perspective Drawing

Why You know that objects that are far away appear smaller than they would if they were close to you. In studying perspective drawing you will learn the rules for making things appear in proper relationship to each other.

Objectives

- Identify and define the basic concepts of perspective drawing.

- Apply these basic concepts to create your own perspective drawings.

European Renaissance artists in the fourteenth through sixteenth centuries rediscovered, from classical Greek and Roman art, how to create the illusion of depth in drawings and paintings. Notice the relatively flat appearance of the pre-Renaissance work on the right compared with the Renaissance painting to the left.

Perspective Drawings: Windows to Reality

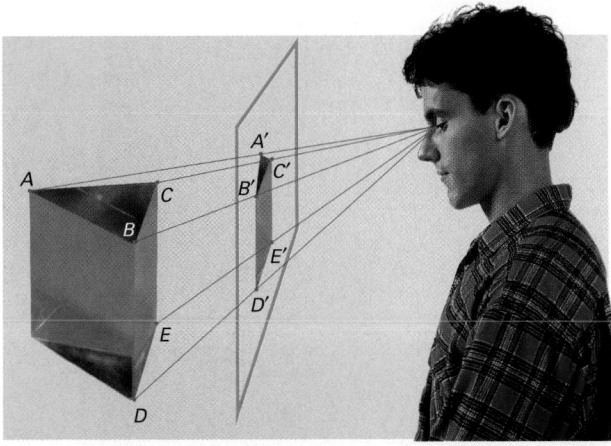

A picture-plane "window" containing a projected image

Modern perspective drawing methods, discovered by the Italian architect Fillip Brunelleschi (1377–1446), are based on the idea that a picture is like a window.

An artist creating a picture, or a person looking at the finished picture, is *thought* of as looking through the picture to the reality it portrays. (The word *perspective* comes from the Latin words meaning "looking through.")

When someone looks at an object, there is a line of sight from every point on the object to the eye. Imagine a plane, such as an empty canvas, that intersects the lines of sight. The points of intersection on the plane make up the image of the object, and the image is said to be *projected* onto the "picture plane."

Alternative Teaching Strategy

USING VISUAL MODELS To help demonstrate how a person looks at an object, use a flashlight to project the shadow of an object onto the wall. Draw imaginary lines from the center of the flashlight through the vertices of the object to the corresponding vertices of the shadow. Vary the position of the flashlight to show different shadows of the object.

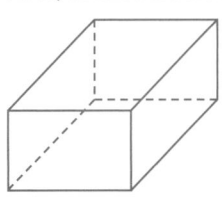

Drawing figures in perspective is an ideal activity for small group work. Students can form a circle around an object and try to draw it by applying the perspective-drawing theorems in this lesson. Each drawing should show a different perspective of the object.

CHECKPOINT ✔

A string can be used to model the line of sight from a point on the object to the eye.

Albrecht Dürer (1471–1528), a German artist, visited Italy to learn the techniques of perspective drawing. He produced a number of works that showed artists employing these techniques, as in the woodcut at right.

Two artists producing an image of a lute, one point at a time

CHECKPOINT ✔ How does the technique used by the artist in Dürer's woodcut illustrate the projection of an object onto a picture plane?

Parallel Lines and Vanishing Points

Have you ever noticed how the rails of a railroad track or the sides of a highway seem to meet as they recede into the distance? The point where parallel lines seem to meet, which is often on the horizon, is known in perspective drawing as the **vanishing point**.

$\overline{AB}$ and $\overline{CD}$ on the sidewalk in the picture below are actually the same length, but when they are projected onto the picture plane seen by the student, the image of $\overline{AB}$ is longer than the image of $\overline{CD}$.

Interdisciplinary Connection

ART Artists have used perspective drawing in their work for hundreds of years. Have students bring in books that show the works of artists they like, and have them discuss how perspectives are used by the artist.

Principles of Perspective Drawing

The two theorems on this page provide the basis for an understanding of how perspective drawings are made. They can be demonstrated by studying the way parallel lines project onto the picture plane of a perspective drawing.

Theorem: Sets of Parallel Lines

In a perspective drawing, all lines that are parallel to each other, but not to the picture plane, meet at a single point known as a vanishing point.

6.6.1

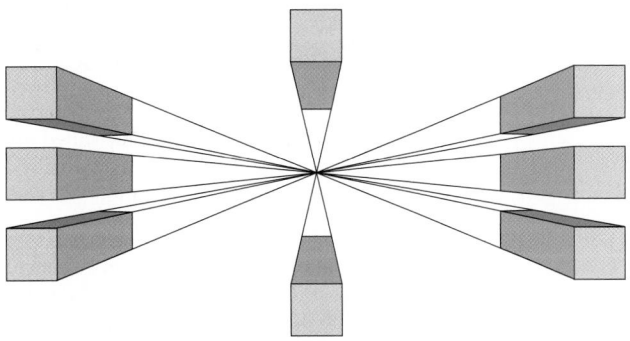

CRITICAL THINKING Do you think that the point where a set of parallel lines seem to meet has to be somewhere in the drawing? Use illustrations in your explanation.

Theorem: Lines Parallel to the Ground

In a perspective drawing, a line that is in the plane of the ground in the drawing and is not parallel to the picture plane will meet the horizon of the drawing. Any line parallel to this line will meet the horizon of the drawing at the same point.

6.6.2

CRITICAL THINKING Parallel lines that are parallel to the picture plane of a perspective drawing are usually represented with no vanishing point. In most cases this procedure causes no problems. Can you think of situations in which it would result in unrealistic drawings?

Enrichment

Computers make perspective drawings by converting the coordinates of three-dimensional objects into two dimensions. Graphing two-dimensional coordinates in place of their three-dimensional counterparts is called a *projection*. Have students draw a rectangular prism in three dimensions and label its vertices. Then have them project its vertices onto the *xy*-plane and label them with prime notation. Then have them redraw the projected image.

Inclusion Strategies

ENGLISH LANGUAGE DEVELOPMENT Students with limited English language development may experience difficulty understanding the theorems in this lesson. Demonstrate the theorems by using as many illustrations as possible.

Teaching Tip

It may be very helpful to photocopy some illustrations and then project them on an overhead screen. Use a ruler to show the lines of perspective.

CRITICAL THINKING
Sample answer: No. The vanishing point could be off the edge of the drawing.

Teaching Tip

Review Lesson 6.1 by comparing the process of drawing a cube in perspective with the process of drawing a cube on isometric dot paper.

CRITICAL THINKING
Sample answer: A very tall building would be difficult to represent in a perspective drawing. The sides of the building would appear to spread apart as they rise.

In a perspective drawing, the concept of vanishing points applies even when no parallel lines are actually shown in the drawing. In the row of telescope dishes shown below, for example, there are imaginary lines that pass through points at the tops and at the bottoms of the reflectors. These lines meet at CHECKPOINT ✔ the horizon. Explain why.

Much of the subject matter of early perspective drawings was architectural. Buildings and houses are ideally suited to the development of the theories of perspective drawing because they usually contain many lines that are parallel to each other and to the ground.

Eventually, perspective drawing began to influence architecture, as illustrated in the Church of San Lorenzo in Florence, Italy, shown below. Brunelleschi used the principles of perspective drawing in his design for this church. The vanishing point of the structural elements appears to be at the altar.

Church of San Lorenzo in Florence, Italy

Reteaching the Lesson

USING COGNITIVE STRATEGIES You may want students to restate the theorems from the lesson as they refer to some of their perspective drawings. Have them identify the vanishing point in a drawing and draw the imaginary parallel lines that connect the vanishing point with various points on the object.

Exercises

Communicate

1. Explain what is meant by a vanishing point in a perspective drawing.

2. Assume that any two lines in a perspective drawing that are parallel to each other but not to the picture plane meet at a point. How is this statement extended to apply to sets of parallel lines in Theorem 6.6.1? What theorem or postulate justifies this extension?

3. Explain why a line that is in the plane of the ground and is not parallel to the picture plane must end at the horizon. Make a sketch to explain your answer.

4. Explain why drawings of buildings and houses are ideally suited to the development of perspective drawing. Make a sketch to explain your answer.

5. In a perspective drawing, the horizon is usually represented by a horizontal line that is assumed to be at eye level. Why do you think this assumption is made?

6. What is the difference between a perspective drawing and an isometric drawing?

CRITICAL THINKING

7. The images of the two polar bears in the picture at right are actually the same size (measure them), but one appears larger than the other. Explain why.

Guided Skills Practice

8. The drawing at right is a perspective drawing of a cube. What do you know about the lines that contain $\overline{AB}$, $\overline{CD}$, and $\overline{EF}$? about the lines that contain $\overline{AC}$, $\overline{BD}$, and $\overline{GF}$? **(THEOREM 6.6.1)**

9. In the drawing of the cube, the lines that contain the vertical segments will *not* meet at a vanishing point. Use the principles of perspective drawing to explain why. **(THEOREM 6.6.1)**

10. Imagine that the drawing of the cube represents a cube-shaped building on level ground. Where will the lines that contain the nonvertical sides of the building meet? **(THEOREM 6.6.2)**

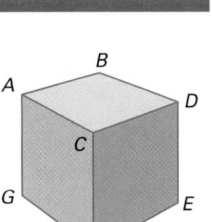

12.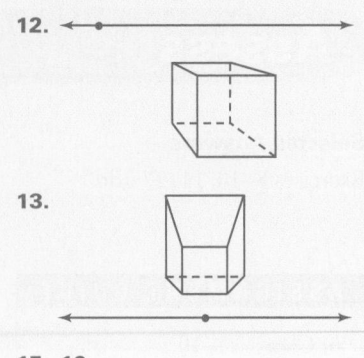

13.

15.–16.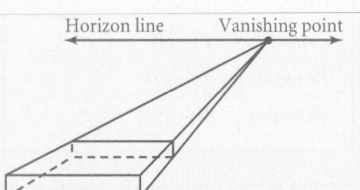

Horizon line   Vanishing point

The exercises below give the steps that are used to produce various types of perspective drawings.

11. Students follow the steps to produce a one-point perspective drawing of a cube.

11. A drawing that has just one vanishing point is said to have one-point perspective. Follow the steps below to produce a one-point perspective drawing of a cube.

   a. Draw a square. Then draw a horizontal line to represent the horizon. Mark a vanishing point on the horizon.

   b. From each corner of the square, lightly draw dashed lines to the vanishing point you marked in part **a.**

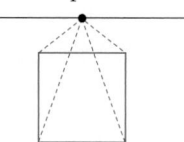

   c. Lightly draw the sides of a smaller square whose vertices touch the lines you drew in part **b.**

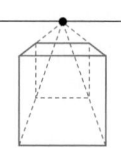

   d. Erase the perspective lines that extend "behind" the smaller cube. The dashed lines that remain indicate the edges of the cube that are hidden from view.

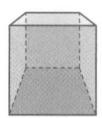

12. Repeat the steps in Exercise 11, but place the vanishing point to the left or right of the square.

13. Repeat the steps in Exercise 11, but place the horizon line and vanishing point below the square.

14. What happens if you place the vanishing point in the interior or on an edge of the square?

Only the front face will be visible. The others will be hidden.

For Exercises 15 and 16, trace the figure below onto your paper.

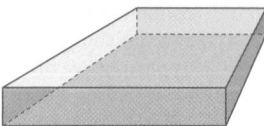

15. Locate the vanishing point for the figure.

16. Draw the horizon line.

17. Students follow the steps to produce a two-point perspective drawing of a cube.

17. A drawing that has two vanishing points is said to have **two-point perspective**. Follow the steps below to produce a two-point perspective drawing of a cube.

a. Draw a vertical segment. This will be the front edge of your cube. Draw a horizon line above the segment. Place two vanishing points on the horizon line as shown, with one on either side of the vertical segment.

b. Lightly draw lines back to each vanishing point from the endpoints of the vertical segment as shown.

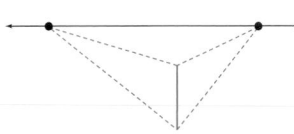

c. Draw vertical segments to complete the front sides of the cube.

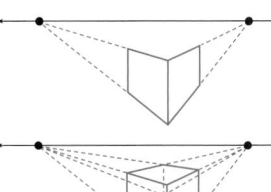

d. Lightly draw lines from the endpoints of the segments you drew in step **c** to each vanishing point. Draw a vertical dashed line between the two intersection points of the light perspective lines.

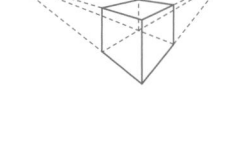

e. Erase the perspective lines that extend beyond the edges of the cube. Use dashed lines to indicate the edges of the cube that are hidden from view.

18. Repeat the steps in Exercise 17, but place the horizon line and vanishing points below the vertical line.

19. Repeat the steps in Exercise 17, but place the horizon line so that it intersects the vertical line.

20. The picture becomes narrower; the picture becomes broader.

20. What happens to a two-point perspective drawing of a cube as the vanishing points are moved closer together? farther apart?

21. What happens to a two-point perspective drawing of a cube if both vanishing points are moved to the same side of the original vertical line? if one vanishing point is directly above the vertical line?
The cube becomes distorted.

For Exercises 22 and 23, trace the figure below onto your paper.

22. Locate the vanishing points for the figure.

23. Draw the horizon line.

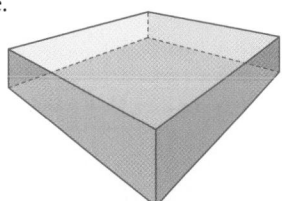

18.

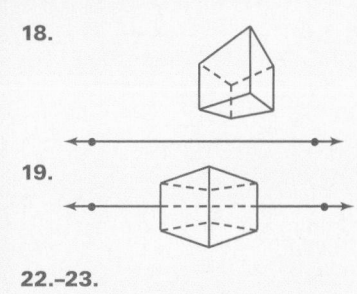

19.

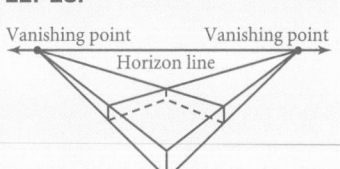

22.–23.

Vanishing point   Vanishing point
Horizon line

24. Check student drawings.

24. Follow the steps below to create a perspective drawing of your name in black letters.

   a. Draw "flat" block letters and a horizon line with a vanishing point, as shown.

   b. Draw lines from all corners and appropriate curved edges of the letters, as shown.

   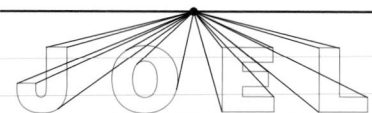

   c. Fill in the edges as shown. Erase the perspective lines to complete the drawing.

   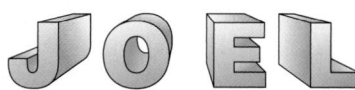

internet connect

Homework Help Online

Go To: **go.hrw.com**
Keyword:
MG1 Homework Help
for Exercise 25

25. Use one-point or two-point perspective to draw a city view. Start with boxes for the buildings and then add details.
Check student drawings.

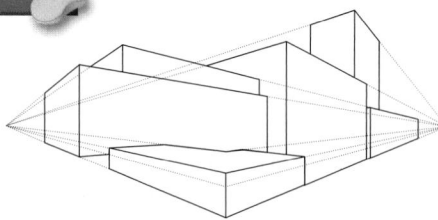

Tiling was particularly intriguing to the artists who first explored perspective studies. The pictures below suggest a technique for creating a tile pattern in a perspective drawing. Study the pictures to answer Exercises 26–28.

a.
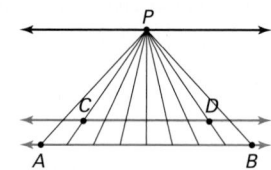

b.
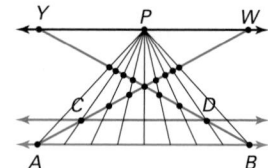

c.
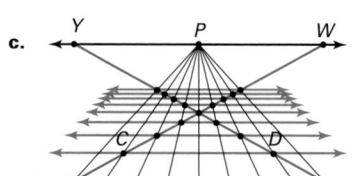

d.

Practice Master

NAME _____ CLASS _____ DATE _____

Practice
6.6 *Perspective Drawing*

In Exercises 1–4, locate the vanishing point for the figure and draw the horizon line.

1.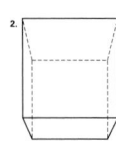

2.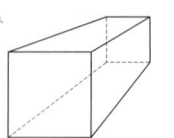

3.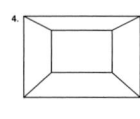

4.

5. In the space below, make a one-point perspective drawing of a rectangular solid. Place the vanishing point below the solid.
Answers may vary.
See Answer Key.

6. In the space below, make a two-point perspective drawing of a rectangular solid. Place the vanishing points below the solid.
Answers may vary.
See Answer Key.

26. A method for finding the lines parallel to $\overline{AB}$ is suggested by the diagonals. Explain how the diagonals can be used to determine these parallel lines.

27. Explain how the tile pattern could be viewed from a corner by applying two-point perspective. How would the intersecting lines be determined?

28. Create your own one- or two-point perspective drawing of a square tile pattern.

 **Look Back**

29. A solid composed of unit cubes has the top view shown at right. What are two possibilities for the solid? Make an isometric drawing of each. *(LESSON 6.1)*

30. A solid composed of unit cubes has the front and right views shown at right. What might the top view look like? *(LESSON 6.1)*

Front   Right

Locate and sketch each point in a three-dimensional coordinate system. *(LESSON 6.4)*

31. $(5, -1, -2)$ **32.** $(13, 0, 0)$ **33.** $(-2, 0, 5)$

Find the distance between each pair of points with the given coordinates in a three-dimensional coordinate system. *(LESSON 6.4)*

34. $(4, 3, 2)$ and $(-5, 2, -1)$ $\approx$ **9.54**   **35.** $(-1, 0, 1)$ and $(15, 5, -2)$ $\approx$ **17.03**

Find the midpoint of the segment with the given endpoints in a three-dimensional coordinate system. *(LESSON 6.4)*

36. $(5, 5, 5)$ and $(-3, -3, -3)$ **(1, 1, 1)**   **37.** $(0, 0, 0)$ and $(-1, 10, 9)$ $\left(-\frac{1}{2}, 5, \frac{9}{2}\right)$

 **Look Beyond**

38. The artist in the photograph at right is using an image produced by a projector to create a mural. Draw a diagram showing how an image on a slide is projected onto a screen or wall. Assume that the light source is a single point of light. (The lens arrangement of the projector makes this assumption appropriate.)

39. How is the projection process of the slide projector like the projection in perspective drawing? How is it different?

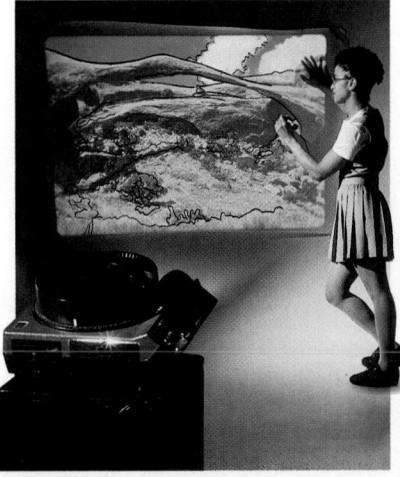

28. Check student drawings.

30. Sample answer:

From the front view we know the solid's length is 2 and there is an unseen edge. From the right view we know the solid's width is 2 and there is an unseen edge. Thus the top will show a length of 2 and a width of 2 with edges that cannot be seen from the front or right.

31.

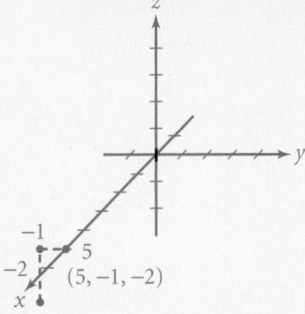

$(5, -1, -2)$

26. Segments radiating from point P divide $\overleftrightarrow{AB}$ into a number of congruent segments. $\overleftrightarrow{CD}$ is drawn parallel to $\overleftrightarrow{AB}$. Diagonals $\overline{AW}$ and $\overline{BY}$ are drawn through points C and D, respectively. The intersections of the diagonals with the segments from point P determine the vertical placement of the parallel lines.

27. Sample answer: Make a perspective drawing of a square with horizontal and vertical diagonals. Divide the horizontal diagonal equally. Pass lines from the vanishing points through the division points.

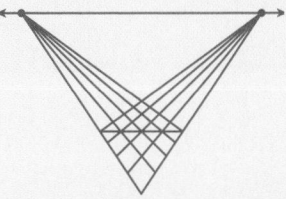

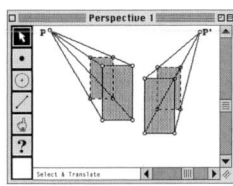

Focus

By following the steps in **Activity 1**, students can create their own stellated dodecahedrons. In **Activity 2**, students explore patterns in Platonic solids.

Motivate

Show students pictures of various stellated polyhedrons and Platonic solids. Have them search for information about these figures on the Internet.

Activity 1

Students should follow instructions to construct polyhedra.

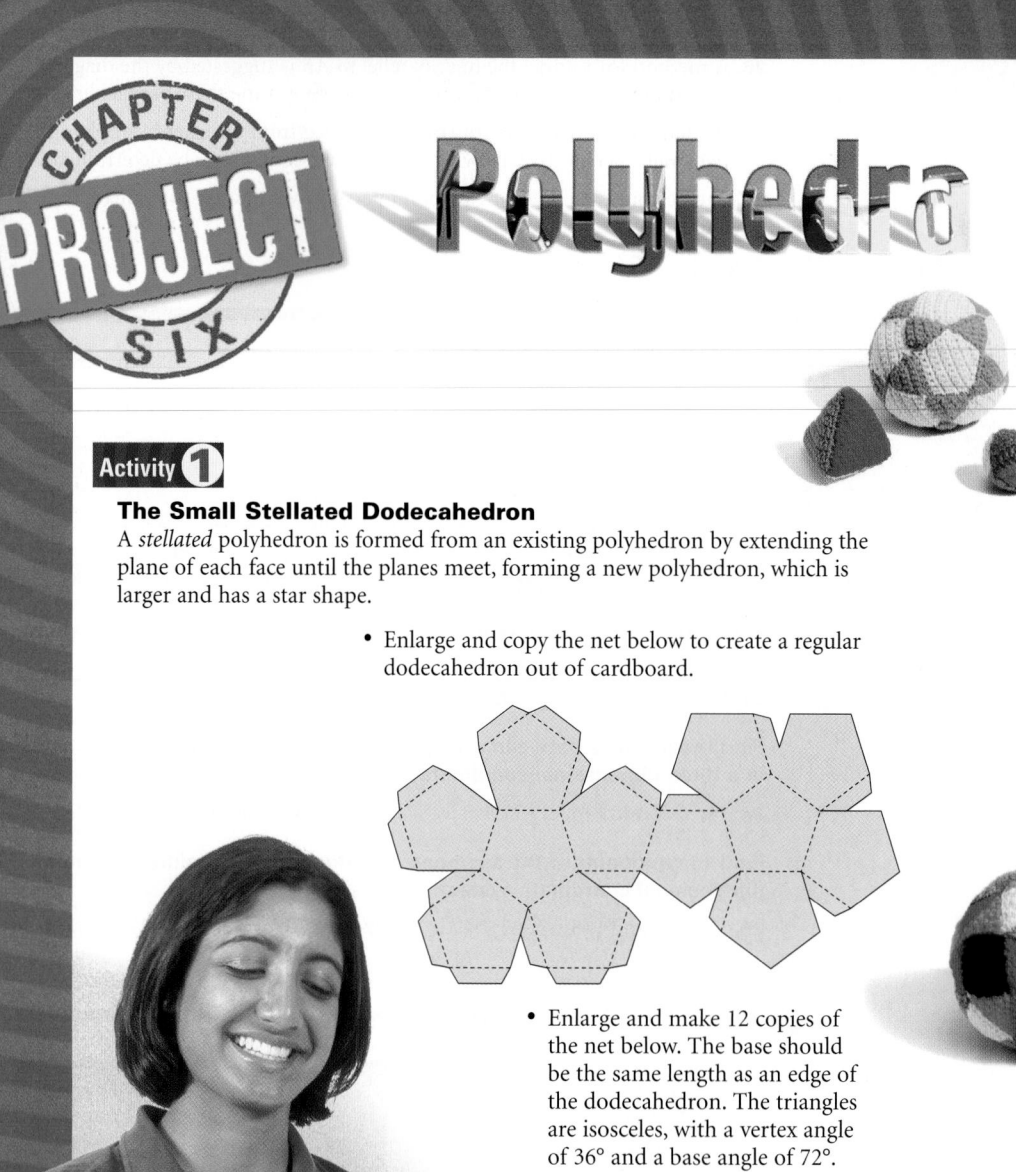

CHAPTER PROJECT SIX

Polyhedra

Activity 1

The Small Stellated Dodecahedron

A *stellated* polyhedron is formed from an existing polyhedron by extending the plane of each face until the planes meet, forming a new polyhedron, which is larger and has a star shape.

- Enlarge and copy the net below to create a regular dodecahedron out of cardboard.

- Enlarge and make 12 copies of the net below. The base should be the same length as an edge of the dodecahedron. The triangles are isosceles, with a vertex angle of 36° and a base angle of 72°. Color the triangles, if you wish.

- Fold the 12 nets above to form 12 pyramids that are open at the bottom. Use glue or tape to attach one pyramid to each face of the dodecahedron.

Activity 2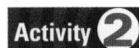

Why Are There Five?

The five convex regular polyhedra, known as the Platonic solids, have been studied extensively since the time of the ancient Greeks. To see why there are only five, consider a single vertex of each one. Since all vertices of a regular polyhedron are identical, the shape of the polyhedron is uniquely determined by one vertex.

1. The patterns below are possible arrangements of equilateral triangles at a vertex. Copy the patterns and cut them out. Glue or tape the red edges of each pattern to form a single vertex.

Which Platonic solid has 3 triangles at each vertex? Which has 4 triangles at each vertex?

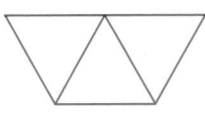
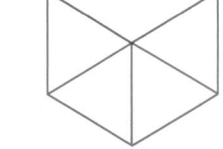

2. Try some other arrangements of equilateral triangles at a vertex. What is the minimum number possible? What is the maximum number possible? Explain your reasoning.

Johannes Kepler (1571–1630) believed that the orbits of the planets were described by nested regular polyhedra.

3. The pattern at right is a possible arrangement of squares at a vertex. Copy the pattern and cut it out. Glue or tape the red edges to form a single vertex.

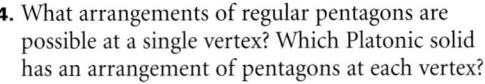

Which Platonic solid has 3 squares at each vertex? Are any other arrangements of squares possible at a single vertex?

4. What arrangements of regular pentagons are possible at a single vertex? Which Platonic solid has an arrangement of pentagons at each vertex?

5. Can any other regular polygons be arranged at a single vertex? Explain your reasoning.

Extension

1. In each vertex arrangement above, what is the sum of the angles at the vertex?

2. As the sum of the angles increases, what happens to the sharpness of the vertex? What do you think is the maximum possible sum of the angles?

3. How do you think it would be possible to prove that there are exactly 13 Archimedean solids?

Chapter Review and Assessment

VOCABULARY

Chapter Test, Form A

NAME _____ CLASS _____ DATE _____

Chapter Assessment
Chapter 6, Form A, page 1

Write the letter that best answers the question or completes the statement. Use the given figure for Exercises 1–4. Assume that no cubes are hidden from view.

b 1. Which of the following is the left side view?
a. b.
c. d.

c 2. What is the surface area if each cube edge is 1 unit in length?
a. 17 square units b. 25 square units
c. 32 square units d. 36 square units

a 3. What is the volume if each cube edge is 1 unit in length?
a. 9 cubic units b. 10 cubic units
c. 27 cubic units d. 34 cubic units

c 4. What is the measure of the dihedral angle formed by the left face and the front face?
a. 30° b. 60° c. 90° d. 120°

For Exercises 5–8, refer to the prism at right.

a 5. Which is the classification of the figure?
a. triangular prism b. oblique prism
c. rectangular prism d. pentagonal prism

b 6. What is the measure of the dihedral angle formed by the intersection of planes ADFC and BEFC?
a. 30° b. 45° c. 60° d. 90°

d 7. Which plane is parallel to plane ABC?
a. ADF b. BEF c. ADE d. DEF

c 8. Which segment is is perpendicular to $\overline{EB}$?
a. $\overline{FC}$ b. $\overline{DF}$ c. $\overline{BC}$ d. $\overline{AC}$

NAME _____ CLASS _____ DATE _____

Chapter Assessment
Chapter 6, Form A, page 2

For Exercises 9–15, use the rectangular solid at right.

b 9. The length of $\overline{GI}$ in the xy-plane is _____.
a. $3\sqrt{5}$ b. $3\sqrt{17}$
c. 13 d. $6\sqrt{30}$

d 10. The length of diagonal $\overline{GM}$ is _____.
a. $6\sqrt{5}$ b. $4\sqrt{10}$
c. $2\sqrt{13}$ d. $3\sqrt{21}$

a 11. What are the coordinates of point H?
a. (4, 12, 0) b. (12, 4, 0)
c. (4, 0, 12) d. (0, 4, 12)

c 12. What are the coordinates of point L?
a. (4, 6, 12) b. (12, 6, 4) c. (4, 12, 6) d. (6, 12, 4)

a 13. The midpoint of $\overline{LM}$ is _____.
a. (2, 12, 6) b. (2, 12, 0) c. (2, 6, 3) d. (2, $2\sqrt{3}$, 6)

d 14. Which line is skew to $\overline{GH}$?
a. $\overline{KL}$ b. $\overline{NM}$ c. $\overline{LH}$ d. $\overline{NJ}$

a 15. Name the location of point K.
a. xz-plane b. first octant c. yz-plane d. top-front-left octant

b 16. Find the distance between (−1, 5, 2) and (3, 1, −4) in space.
a. 9 b. $2\sqrt{17}$ c. $2\sqrt{22}$ d. $2\sqrt{14}$

For Exercises 17 and 18, use the figure at right.

b 17. An equation of the plane is _____.
a. $3x + 4y + 2z = 12$ b. $4x + 3y + 6z = 12$
c. $3x + 2y + 4z = 12$ d. $4x + 2y + 3z = 12$

a 18. An equation of the trace defined by $2x + 5y − 3z = 8$ is _____.
a. $2x + 5y = 8$ b. $2x − 3z = 8$
c. $5y − 3z = 8$ d. $2x + 5y = 0$

POSTULATES AND THEOREMS

Lesson	Number	Postulate or Theorem
6.3	6.3.1 Diagonal of a Right Rectangular Prism	The length of the diagonal, d, of a right rectangular prism is given by $d = \sqrt{\ell^2 + w^2 + h^2}$.
6.4	6.4.1 Distance Formula in Three Dimensions	The distance, d, between the points (x_1, y_1, z_1) and (x_2, y_2, z_2) in space is given by $d = \sqrt{(x_2 − x_1)^2 + (y_2 − y_1)^2 + (z_2 − z_1)^2}$.
	6.4.2 Midpoint Formula in Three Dimensions	The midpoint of a segment with endpoints at (x_1, y_1, z_1) and (x_2, y_2, z_2) in space is given by $\left(\dfrac{x_1 + x_2}{2}, \dfrac{y_1 + y_2}{2}, \dfrac{z_1 + z_2}{2} \right)$.
6.6	6.6.1 Theorem: Sets of Parallel Lines	In a perspective drawing, all lines that are parallel to each other, but not to the picture plane, will seem to meet at a single point known as a vanishing point.
	6.6.2 Theorem: Lines Parallel to the Ground	In a perspective drawing, a line that is on the plane of the ground and is not parallel to the picture plane will meet the horizon of the drawing. Any line parallel to this line will meet the horizon at the same point.

Key Skills & Exercises

Key Skills

Create isometric drawings of solid figures.

A solid figure is composed of seven cubes as shown.

Make a drawing of the solid on isometric dot paper.

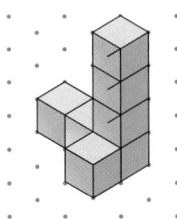

Exercises

Refer to the isometric drawing below. Assume that no cubes are hidden in the drawing.

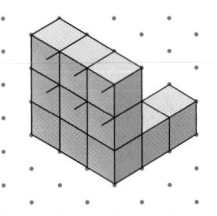

1. Make an isometric drawing of the solid from a different angle. (You may wish to create a model of the solid by using unit cubes.)

2. Draw the six orthographic views of the solid.

3. Find the volume of the solid.

4. Find the surface area of the solid.

Draw orthographic projections of solid figures.

Draw six orthographic projections of the solid above.

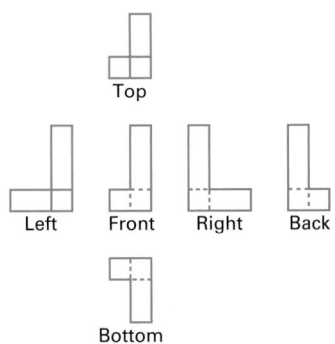

Top

Left   Front   Right   Back

Bottom

Find the volume and surface area of solids that are composed of cubes.

Find the volume and surface area of the solid above.

The solid is composed of seven cubes, so its volume is 7 cubic units. There are 30 exposed cube faces, so the surface area is 30 square units.

2.

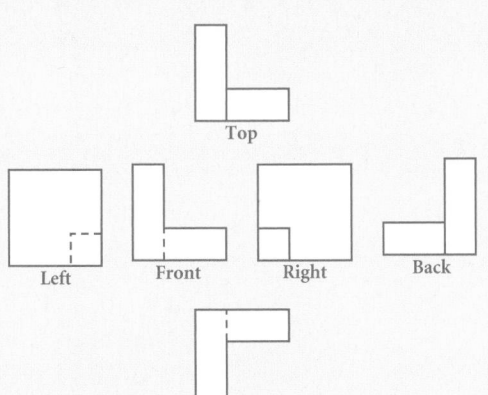

Top

Left   Front   Right   Back

Bottom

3. 11 units³

4. 38 units²

---

1. Sample answer:

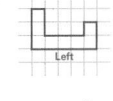
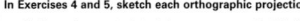
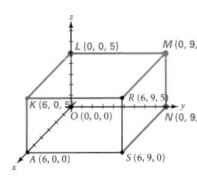
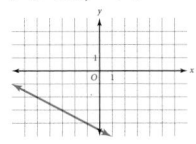
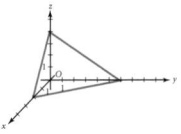
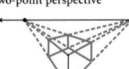
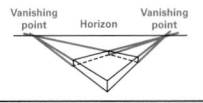

5. Sample answer: *OPQRST*
and *UVWXYZ*
PQWV and *TSYZ*

6. Sample answer: *PQWV* and
UVWXYZ

7. Sample answer: $\overline{PO}$ and $\overline{WX}$

8. Sample answer: $\overline{PV}$ and
UVWXYZ

9. Sample answer: *DEF*

10. Sample answer: *CBEF*

11. Sample answer: $\overline{CF}$

12. 26

13. top-front-right

14. bottom-back-right

Key Skills

Identify relationships among lines and planes in space.

In the cube, name a set of parallel faces, perpendicular faces, and skew edges. Then name an edge and a face that are perpendicular.

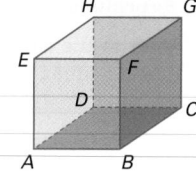

Faces *ADHE* and *BCGF* are parallel. Faces *ADHE* and *ABFE* are perpendicular. $\overline{EH}$ and $\overline{BF}$ are skew edges. $\overline{AB}$ is perpendicular to *BCGF*.

Exercises

Refer to the right hexagonal prism below. *UVWXYZ* is a regular hexagon.

5. Name two pairs of parallel faces.

6. Name a pair of perpendicular faces.

7. Name a pair of skew edges.

8. Name an edge and a face that are perpendicular.

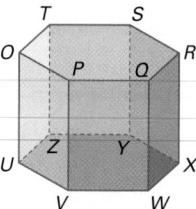

Key Skills

Identify parts of a prism.

In the pentagonal prism at right, name a base, a lateral face, and a lateral edge.

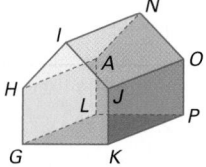

Pentagon *GHIJK* is a base, parallelogram *GLPK* is a lateral face, and $\overline{IN}$ is a lateral edge.

Find the length of a diagonal of a right rectangular prism.

A right rectangular prism has a length of 20, a width of 10, and a height of 24. What is the length of its diagonal?

The length of a diagonal is given by the formula $d = \sqrt{\ell^2 + w^2 + h^2}$.

$$d = \sqrt{20^2 + 10^2 + 24^2} = \sqrt{1076} \approx 32.8$$

Exercises

Refer to the prism below.

9. Name a base.

10. Name a lateral face.

11. Name a lateral edge.

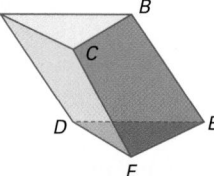

12. What is the length of a diagonal of a right rectangular prism with a length of 24, width of 8, and height of 6?

Key Skills

Locate points in a three-dimensional coordinate system.

Locate the point (3, −1, 7) in a three-dimensional coordinate system. Name the octant of the point.

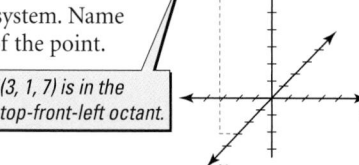

(3, 1, 7) is in the top-front-left octant.

Exercises

Locate each point in a three-dimensional coordinate system and name its octant.

13. (5, 3, 4)

14. (−6, 1, −2)

Find the distance between two points in space.

Find the distance between the points $(3, -1, 7)$ and $(2, 0, -5)$.

$$d = \sqrt{(x_2 - x_1)^2 + (y_2 - y_1)^2 + (z_2 - z_1)^2}$$
$$= \sqrt{(2 - 3)^2 + [0 - (-1)]^2 + (-5 - 7)^2}$$
$$= \sqrt{1 + 1 + 144} = \sqrt{146} \approx 12.1$$

Find the distance between each pair of points in space.

15. $(2, 4, 5)$ and $(0, 7, -1)$

16. $(-1, 6, -2)$ and $(-5, 2, 4)$

15. 7

16. ≈ 8.25

17.

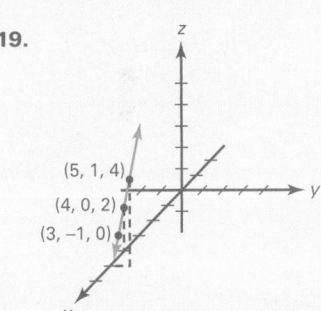

LESSON 6.5

Key Skills

Sketch planes in space.

Sketch the plane defined by the equation $x + 2y - 3z = 18$.

First, find the intercepts.

x-intercept: $x + 2(0) + 3(0) = 18$, so $x = 18$
y-intercept: $0 + 2y + 3(0) = 18$, so $y = 9$
z-intercept: $0 + 2(0) + 3z = 18$, so $z = 6$

Plot the intercepts as points and sketch the plane.

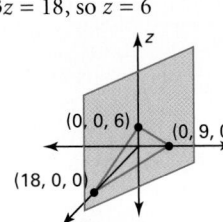

Plot lines in space.

Plot the line represented by the parametric equations below.

$$x = 2t$$
$$y = t + 1$$
$$z = -t$$

When $t = 0$, the point defined by the equations is $(0, 1, 0)$. When $t = 1$, the point defined by the equations is $(2, 2, -1)$. Locate these points and sketch the line.

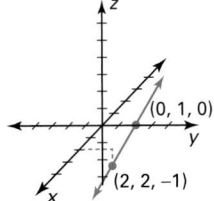

Exercises

17. Sketch the plane defined by the equation $2x + y - z = 10$.

18. Sketch the plane defined by the equation $-x + 4y + z = 8$.

19. Plot the line represented by the parametric equations below.

$$x = t + 3$$
$$y = t - 1$$
$$z = 2t$$

20. Plot the line represented by the parametric equations below.

$$x = -t + 2$$
$$y = t - 4$$
$$z = -3t + 1$$

18.

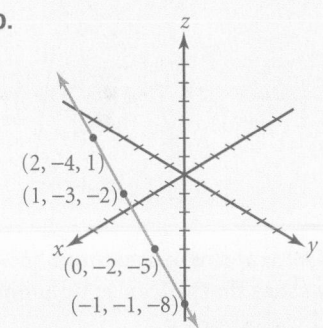

19.

20.

21.

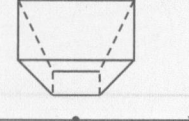

22.

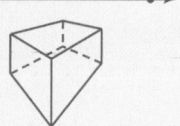

23.

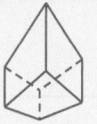

24.

25.

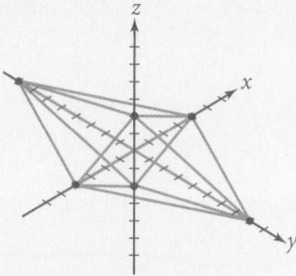

26. $3x + 5y - z = 20$

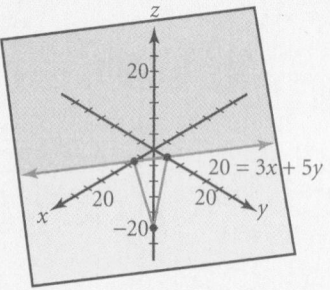

The trace represents number of washes that will raise no money (break even).

27.

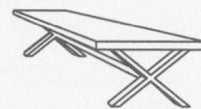

Key Skills

Use vanishing points to make perspective drawings.

Make a one-point and a two-point perspective drawing of the letter **T**.

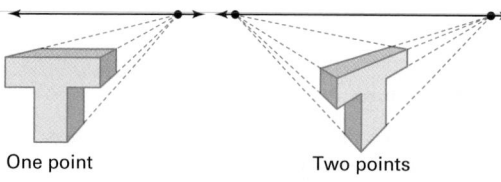

One point   Two points

Exercises

21. Make a one-point perspective drawing of a right rectangular prism with the vanishing point to the right of the prism.

22. Make a one-point perspective drawing of a right rectangular prism with the vanishing point below the prism.

23. Make a two-point perspective drawing of a right rectangular prism with the horizon above the prism.

24. Make a two-point perspective drawing of a right rectangular prism with the horizon below the prism.

Applications

25. GEOLOGY Crystals are classified by the different axes, called *crystallographic axes*, that passes through the center of the crystal. Topaz has a crystal structure called *orthorhombic*. In topaz, three axes of three different lengths all meet at right angles. Draw the three axes for topaz. Suppose that the endpoints of the axes are vertices of a crystal. Draw an outline of the crystal around your axes.

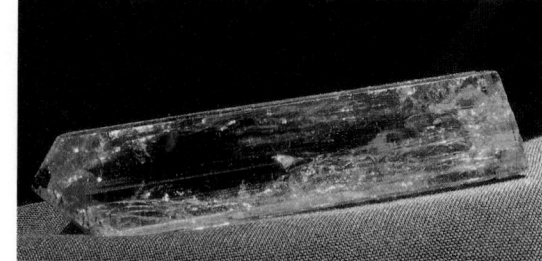

Topaz crystal

26. FUND-RAISING The school band is raising money to buy new uniforms. The band members will wash cars in a space donated by a local business. The supplies for the car wash cost $20. The charges are $3 for a regular car wash and $5 for a deluxe car wash. The amount of money, z, raised by x regular washes and y deluxe washes is given by $z = 3x + 5y - 20$.

Write this equation in the standard form and sketch the graph of the plane. Indicate the trace of the plane on your graph. What does the trace represent?

27. DESIGN The diagrams at right are three orthographic views of a table. Use the diagrams to make a perspective drawing of the table.

Top

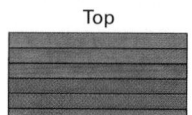

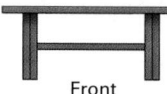

Front   Side

Chapter Test

Refer to the isometric drawing at right. Assume that no cubes are hidden in the drawing.

1. Make an isometric drawing of the solid from a different angle.

2. Draw the six orthographic views of the solid.

3. Find the volume of the solid. **12 units³**

4. Find the surface area of the solid. **40 units²**

Refer to the cube below.

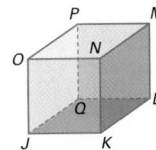

5. Name two pairs of parallel faces.

6. Name a pair of skew edges.

7. Name an edge and a face that are perpendicular.

8. Two pencils lying on the same table are lying in the same __?__. **plane**

Refer to the prism below.

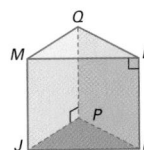

9. Which face is congruent to △QLM? **△PKJ**

10. Name a lateral face.

11. Name a lateral edge.

12. Name two right angles.

13. DESIGN Students want to hang balloons along the diagonal of a gym area that is a rectangular prism with a length of 12 feet, width of 9 feet, and height of 5 feet. How long will the balloon display be? **5√10 feet ≈ 15.8 feet**

Name the octant, coordinate plane, or axis in which each point is located.

14. $(2, 5, 9)$ 15. $(-1, 5, -3)$ 16. $(1, 0, -3)$

17. $(0, -4, 0)$ 18. $(-6, 6, 0)$ 19. $(2, 0, 0)$

For Exercises 20 and 21, find the distance between each pair of points.

20. $(1, 0, 0)$ and $(0, 1, 1)$ **√3 ≈ 1.73**

21. $(-4, 0, 1)$ and $(3, -2, 1)$ **√53 ≈ 7.28**

22. Sketch the plane defined by the equation $x + 2y + 4z = 12$.

23. Sketch the plane defined by the equation $3x + 3y + 2z = 18$.

24. Plot the line represented by the parametric equations below.
$$x = t + 1$$
$$y = 2t$$
$$z = t - 3$$

25. AVIATION Suppose that an airplane's initial path at takeoff is described by the following parametric equations:
$$x = t$$
$$y = -0.5t$$
$$z = 2t$$
Plot the line described by the given equations. If the positive x-axis points north, in which direction did the airplane take off?

26. Make a one-point perspective drawing of a cube with the vanishing point to the left of the cube.

27. Make a one-point perspective drawing of a cube with the vanishing point below the cube.

28. Make a two-point perspective drawing of a cube with the horizon below the cube.

29. Make a two-point perspective drawing of a cube with the horizon above the cube.

1. Sample answer:

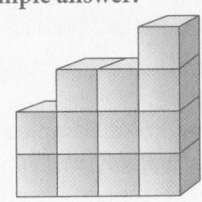

2.

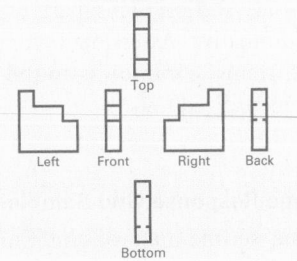

5. Sample answer: *ONKJ* and *PQLM*, *OPMN* and *JQLK*

6. Sample answer: $\overline{OP}$ and $\overline{ML}$

7. Sample answer: $\overline{JK}$ and *KLMN*

10. Sample answer: *JMQP*

11. Sample answer: $\overline{KL}$

12. Sample answer: ∠*JKL* and ∠*MLK*

14. first octant (top-front-right)

15. bottom-back-right

16. xz-plane

17. y-axis

18. xy-plane

19. x-axis

22.

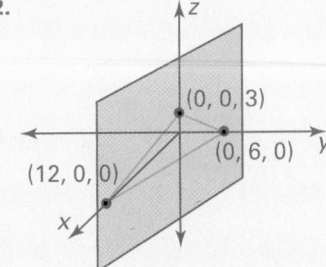

23.

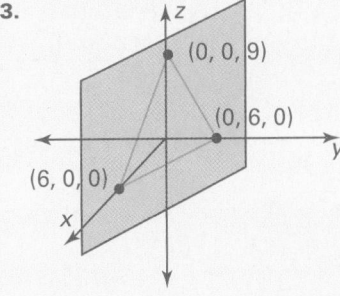

24.

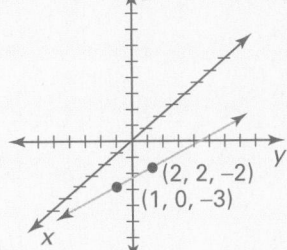

25. northeast

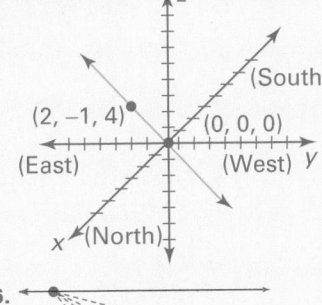

26.

27.

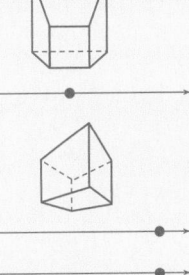

28.

29.

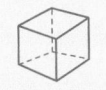

College Entrance Exam Practice

Multiple-Choice Samples

The first half of the Cumulative Assessment contains a multiple-choice section. This part of the Cumulative Assessment consists of items commonly found on standardized tests.

Free-Response Grid Samples

The second half of the Cumulative Assessment consists of a free-response section. This part requires student-produced response items like those commonly found on college entrance exams. These questions require the use of machine-scored answer grids. You may wish to have students practice answering these items in preparation for standardized tests.

1. a

2. c

3. a

4. a

5. c

6. b

7. a

MULTIPLE-CHOICE For Questions 1–10, write the letter that indicates the best answer.

1. Refer to the triangle below. Find the length of $\overline{AB}$. **(LESSON 5.5)**

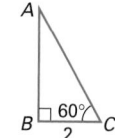

 a. $2\sqrt{3}$
 b. 2
 c. 4
 d. $2\sqrt{2}$

2. Find the distance between $(15, 0, 19)$ and $(6, 9, 10)$. **(LESSON 6.4)**

 a. $\sqrt{283}$
 b. 9
 c. $9\sqrt{3}$
 d. $9\sqrt{2}$

3. Refer to the figure below. Find the area of *EFGH*. **(LESSON 5.2)**

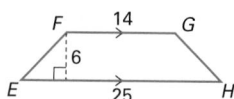

 a. 117
 b. 234
 c. 84
 d. 157

internet connect

**Standardized
Test Prep Online**

Go To: **go.hrw.com**
Keyword: **MM1 Test Prep**

4. Find the area of the figure below. **(LESSON 5.5)**

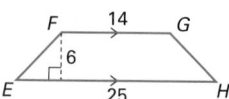

 a. 4.5
 b. 18
 c. $9\sqrt{2}$
 d. 20.25

5. What is the measure of an interior angle of a regular nonagon (9-gon)? **(LESSON 3.6)**

 a. 40°
 b. 100°
 c. 140°
 d. 160°

6. The area of a circle is 154 square inches. Approximate its circumference. **(LESSON 5.3)**

 a. 22 in.
 b. 44 in.
 c. 51 in.
 d. 77 in.

7. What is the midpoint of a segment with endpoints at $(9, 12)$ and $(-12, 9)$? **(LESSON 5.6)**

 a. $(-1.5, 10.5)$
 b. $(4.5, 6)$
 c. $(-6, -4.5)$
 d. $(10.5, -10.5)$

8. What is the midpoint of a segment with endpoints at (2, 10, 0) and (11, 10, −6)? **(LESSON 6.4)**

    a. (6.5, 0, −3)
    b. (4.5, 10, 3)
    c. (6.5, 10, −3)
    d. (5.5, 5, 3)

9. In the oblique rectangular prism shown below, find x. **(LESSON 2.2)**

    a. 60
    b. 80
    c. 100
    d. 120

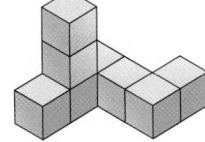

10. Find the surface area of the solid shown below. Assume that no cubes are hidden. **(LESSON 6.1)**

    a. 34 square units
    b. 45 square units
    c. 54 square units
    d. 66 square units

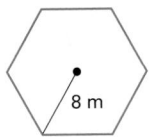

11. Is the statement below a definition? Explain your reasoning. **(LESSON 2.3)**

Skew lines are not parallel and do not intersect.

12. A circle and a square both have an area of 441 square inches. Find the circumference and perimeter, respectively. **(LESSONS 5.1 AND 5.3)**

13. A quadrilateral has vertices at (3, −1), (9, −5), (7, −8), and (1, −4). Use slopes to prove that the quadrilateral is a rectangle. **(LESSON 3.8)**

14. Find the area of the regular hexagon below. **(LESSON 5.5)**

15. Draw a rhombus and construct a reflection of it over a horizontal line. **(LESSON 4.5)**

FREE-RESPONSE GRID
Items 16–19 may be answered by using a free-response grid such as that commonly used by standardized-test services.

16. Find the area of △*PEG*. **(LESSON 5.2)**

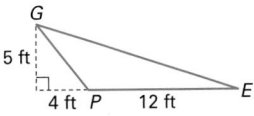

17. Find the total area of the three rectangles below. **(LESSON 5.1)**

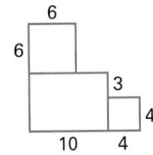

18. Refer to the diagram below. Find the value of x. **(LESSON 3.6)**

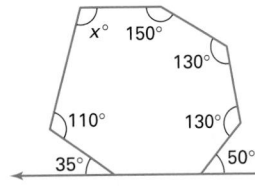

19. A point, *P*, is selected at random on the segment below. What is the probability that $2 \leq P \leq 2.5$? **(LESSON 5.7)**

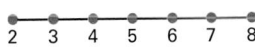

---

8. c

9. c

10. a

11. Yes, the statement and its converse are true.

12. Perimeter of square: 84 inches
Circumference of circle: ≈ 74.44 inches

13.

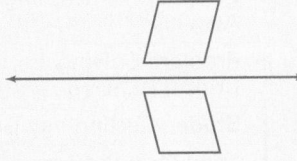

Let *A* be the vertex at (3, −1), *B* be vertex (1, −4), *C* be vertex (7, −8) and *D* be vertex (9, −5).

Slope of $\overline{AB} = \dfrac{-1-(-4)}{3-1} = \dfrac{3}{2}$

Slope of $\overline{BC} = \dfrac{-4-(-8)}{1-7} = -\dfrac{2}{3}$

Slope of $\overline{DC} = \dfrac{-8-(-5)}{7-9} = \dfrac{3}{2}$

Slope of $\overline{AD} = \dfrac{-5-(-1)}{9-3} = -\dfrac{2}{3}$

The product of the slopes of $\overline{AD}$ and $\overline{AB}$ is $\left(-\dfrac{2}{3}\right)\left(\dfrac{3}{2}\right) = -1$.

Since the slopes of $\overline{AD}$ and $\overline{BC}$ the slopes of $\overline{AB}$ and $\overline{DC}$ are equal, the opposite sides are parallel and *ABCD* is a parallelogram. Since the product of the slopes of $\overline{AD}$ and $\overline{AB}$ is −1, $\overline{AD} \perp \overline{AB}$. Thus, *ABCD* is a rectangle.

14. ≈ 166.3 square meters

15. Sample answer:

16. 30 ft²

17. 122 units²

18. 105°

19. $\dfrac{.5}{6}$ ≈ .08 or 8%

Surface Area and Volume

Lesson Presentation CD-ROM
Power Point® presentations for each lesson 7.1-7.7

CHAPTER PLANNING GUIDE

Lesson	7.1	7.2	7.3	7.4	7.5	7.6	7.7	Project and Review
Pupil's Edition Pages	430–436	437–444	445–452	453–459	460–468	469–475	478–485	476–477, 486–495
Practice and Assessment								
Extra Practice (Pupil's Edition)	839	839	840	840	841	841	842	
Practice Workbook	43	44	45	46	47	48	49	
Practice Masters Levels A, B, and C	127–129	130–132	133–135	136–138	139–141	142–144	145–147	
Standardized Test Practice Masters	49	50	51	52	53	54	55	56
Assessment Resources	85	86	87	88	90	91	92	89, 93–98
Visual Resources								
Lesson Presentation Transparencies Vol. 1	166–169	170–173	174–177	178–181	182–185	186–189	190–193	
Teaching Transparencies		58	59–60	61	62	63	64	
Answer Key Transparencies	281–284	285–287	288–290	291–293	294–295	296–297	298–300	301–303
Quiz Transparencies	7.1	7.2	7.3	7.4	7.5	7.6	7.7	
Teacher's Tools								
Reteaching Masters	85–86	87–88	89–90	91–92	93–94	95–96	97–98	
Make-Up Lesson Planner for Absent Students	43	44	45	46	47	48	49	
Student Study Guide	43	44	45	46	47	48	49	
Spanish Resources	43	44	45	46	47	48	49	
Block Scheduling Handbook								14–15
Activities and Extensions								
Lesson Activities	120–125		126–127	128–129	130–132		133–135	
Enrichment Masters	43	44	45	46	47	48	49	
Cooperative-Learning Activities	43	44	45	46	47	48	49	
Problem-Solving/ Critical Thinking	43	44	45	46	47	48	49	
Student Technology Guide			41–42	43	44	45		
Long Term Projects								25–28
Writing Activities for Your Portfolio								19–21
Tech Prep Masters								31–34
Building Success in Mathematics								18–19

LESSON PACING GUIDE

Lesson	7.1	7.2	7.3	7.4	7.5	7.6	7.7	Project and Review
Traditional	2 days	2 days	2 days	2 days	2 days	2 days	2 days	2 days
Block	1 day	1 day	1 day	1 day	1 day	1 day	1 day	1 day
Two-Year	4 days	4 days	4 days	4 days	4 days	4 days	4 days	4 days

CONNECTIONS AND APPLICATIONS

Lesson	7.1	7.2	7.3	7.4	7.5	7.6	7.7	Review
Algebra	431, 434	442, 444	450	456, 457 459		475		
Geometry	430–436	437–444	445–452	453–459	460–468	469–475	478–485	486–495
Coordinate Geometry		444						
Maximum/Minimum	432–434			457	466–468		483	
Technology					466			
Business and Economics	433, 434	443		454, 456, 458	467			492
Life Skills	435		449, 451	458, 459		475	483, 484	
Science and Technology	430, 433, 435	444	448	453, 455, 458	460, 463	471	484	492
Sports and Leisure	434	439, 443, 444			467, 468	469, 473, 474, 475	484	492
Cultural Connection: Africa			448, 449					
Cultural Connection: Asia				457				

BLOCK-SCHEDULING GUIDE

Day	Lesson	Teacher Directed: Lesson Examples, Teaching Transparencies	Student Guided: Activity, Try This	Cooperative-Learning Activity, Lesson Activity, Student Technology Guide	Practice: Practice & Apply, Extra Practice, Practice Workbook	Assessment: Quiz, Mid-Chapter Assessment	Problem Solving, Reteaching
1	7.1	15 min	15 min.	20 min	55 min	15 min	15 min
2	7.2	15 min	15 min	20 min	55 min	15 min	15 min
3	7.3	15 min	15 min	20 min	55 min	15 min	15 min
4	7.4	15 min	15 min	20 min	55 min	15 min	15 min
5	7.5	15 min	15 min	20 min	55 min	15 min	15 min
6	7.6	15 min	15 min	20 min	55 min	15 min	15 min
7	7.7	10 min	20 min	20 min	55 min	15 min	15 min
8	Assess.	50 min	90 min	90 min	65 min	30 min	
		PE: Chapter Review	PE: Chapter Project, Writing Activities	Tech Prep Masters	PE: Chapter Assessment, Test Generator	Chap. Assess. (A or B), Alt. Assess. (A or B), Test Generator	

Alternative Assessment

The following suggest alternative assessments for students who may benefit from a different type of assessment than the regular chapter quizzes and the mid-chapter/end-of-chapter test. Visit the HRW web site to get additional Alternative Assessment material.

internet connect

Alternative Assessment
Go To: **go.hrw.com**
Keyword: **MG1 Alt Assess**

Performance Assessment

1. Find the interior surface area and volume of your classroom.

2. Plot the points $(0, 7, 0)$ and $(0, 7, 9)$ on a three-dimensional grid. Rotate the segment about the z-axis. Find the volume of the resulting solid.

3. Write a paragraph explaining the similarities between the formulas for surface area and volume of cylinders and prisms and pyramids and cones. Illustrate your paragraph with diagrams.

4. Explain how the formula for the volume of a sphere is derived.

5. Explain why an oblique prism has a larger surface area than a right prism. Use the model for Cavalari's Principle on page 440 in your explanation.

Portfolio Project

Suggest that students choose one of the following projects for inclusion in their portfolios.

1. Using modeling clay or some other material, show how the formula for the volume of a cylinder is derived from the formula for the volume of a right rectangular prism $(V = \ell wh)$.

2. Draw a net for a right hexagonal prism. Find the surface area and volume for the prism.

3. Draw a net for a right square pyramid. Find the surface area and volume of the pyramid.

4. Build a hollow, open prism and a pyramid with the same height and same base area. Use these to show that the volume of the pyramid is one-third volume of the prism.

 **internet** connect

The table below identifies the pages in this chapter that contain internet and technology information.

Content Links

Activities Online	pages 441, 449, 456
Portfolio Extensions	pages 452, 485
Homework Help Online	pages 434, 442, 449, 456, 465, 474, 483

Resource Links

Parents can go online and find concepts that students are learning–lesson by lesson–and questions that pertain to each lesson, which facilitate parent-student discussion.

Go To: **go.hrw.com**
Keyword: **MG1 Parent Guide**

Technical Support

The following may be used to obtain technical support for any HRW software product.

Online Help: **www.hrwtechsupport.com**

e-mail: **tschrw@hbtechsupport.com**

HRW Technical Support Center: **(800)323-9239**

7 AM to 10 PM Monday through Friday CST

Visit the HRW math web site at: **www.hrw.com/math**

Technology

Technology Objectives and Suggestions

Lesson 7.1 Surface Area and Volume
In this lesson students explore the ratio of surface area and volume, and develop an understanding of the concept of maximizing volume while minimizing surface area.

Have each student build and measure one or more rectangular prisms and then compile the data into one spreadsheet. In Activity 2, students can use graphics calculators to enter the formula for volume as a function, and then trace the function to estimate the side length that produces the maximum volume.

Lesson 7.2 Surface Area and Volume of Prisms
In this lesson students find the surface area and volume of prisms and develop a formula for the volume of oblique prisms using Cavalari's Principle. Throughout the lesson students will need a scientific calculator to help them with computing surface area and volume.

In the lesson there are two examples that have an aquatic animal theme. In keeping with the theme, have students use the internet to research the sizes of pools that hold sea mammals and sharks at parks such as Sea World or at state and national aquariums.

Lesson 7.3 Surface Area and Volume of Pyramids
In this lesson students find the surface area of a regular pyramid and the volume of a pyramid. Throughout the lesson students will need a scientific calculator to help them with computing surface area and volume. Students with programmable calculators can program the formulas into the calculators.

Have students construct a net for a given volume. After giving students a value for the volume of a square pyramid, have them enter the formula into their graphics calculators as a function. Using the table feature, have them find a base area and height for the volume. Working backwards from the base area, they can find the side length of the square base.

Lesson 7.4 Surface Area and Volume of Cylinders
In this lesson students find the surface area and volume of a cylinder. Throughout the lesson students will need scientific calculators to aid in calculating the surface area and volume.

Students with programmable calculators may want to program the formulas into their calculators to make computations faster. The following is a graphics calculator program for the surface area and volume of a cylinder. It may need to be adapted somewhat for your individual brand of calculator.

```
:Input "BASE RADIUS" ,R
:Input "HEIGHT" , H
:2(3.1416)R•H+2(3.1416)R²→S
:3.1416R²H→V
:Disp "SURFACE AREA" ,S
:Disp "Volume" ,V
```

Lesson 7.5 Surface Area and Volume of Cones
In this lesson students find the surface area and volume of cones. Throughout the lesson students will need scientific calculators or calculators with the formulas programmed in to aid in calculations.

Have students use geometry software to aid in demonstrating the difference between pyramids and cones. Start by constructing a polygon such as a square or triangle. Demonstrate that as the number of sides of the polygon increases, the figure more closely resembles a circle. Next have students use computer graphics to draw a square pyramid, a hexagonal pyramid, and finally a pyramid for a many sided figure such as a 16-gon. They will find that the more sides the base has, the more lateral sides the pyramid has, until its surface resembles that of a cone.

Lesson 7.6 Surface Area and Volume of Spheres
In this lesson students find the surface area and volume of spheres. They will need a scientific calculator throughout the lesson, or a programmable calculator, to compute the surface area and volume.

As an aid in understanding the diagram on page 468, have students use computer graphics to recreate the diagram.

Lesson 7.7 Three-Dimensional Symmetry
In this lesson students solve problems using transformations in three-dimensional space. Students will need a scientific calculator for this lesson. Students can use software or graphics calculators with 3-D graphics capabilities to plot points on a three-dimensional coordinate plane or to create rotational solids.

CHAPTER RESOURCES

- Block-Scheduling Handbook
- Writing Activities for Your Portfolio
- Tech Prep Masters
- Long-Term Project
- Assessment Resources:
  Mid-Chapter Assessment
  Chapter Assessments
  Alternative Assessments
- Test and Practice Generator
- Technology Handbook

Chapter Objectives

- Explore ratios of surface area to volume. [7.1]
- Develop the concepts of maximizing volume and minimizing surface area. [7.1]
- Define and use the formula for finding the surface area of a right prism. [7.2]
- Define and use the formula for finding the volume of a right prism. [7.2]
- Use Cavalieri's Principle to develop the formula for the volume of a right or oblique prism. [7.2]
- Define and use the formula for the surface area of a regular pyramid. [7.3]
- Define and use the formula for the volume of a pyramid. [7.3]

Surface Area and Volume

FOR SEVEN DAYS, 120 WORKERS AND 90 ROCK climbers worked to cover the Reichstag building in Berlin with shiny silver fabric. A million square feet of fabric and over 10 miles of plastic cord were required for the feat.

Why would anyone want to wrap a huge building? This is a question that conceptual artist Christo, who conceived of the project 24 years before he was allowed to do it, has heard many times. What do you think? Is it art?

In this chapter, you will investigate the surface area of solid objects. Christo certainly needed to determine the surface area of the Reichstag to prepare for the project.

Kugel Ball, *at the Houston Museum of Natural Science*

Lessons

7.1 ● **Surface Area and Volume**

7.2 ● **Surface Area and Volume of Prisms**

7.3 ● **Surface Area and Volume of Pyramids**

7.4 ● **Surface Area and Volume of Cylinders**

7.5 ● **Surface Area and Volume of Cones**

7.6 ● **Surface Area and Volume of Spheres**

7.7 ● **Three-Dimensional Symmetry**

Chapter Project A Three-Dimensional Puzzle

Quartet *by Barry LeVa*

Pine Circles, Cone Sphere *by Chris Drury*

About the Photos

In 1995, artists Christo and Jeanne Claude wrapped the Reichstag building in Berlin with aluminized fabric. The completion of this temporary work of art required the use of 100,000 m² (1,076,000 ft²) of fabric, 15,600 m (51,181 ft) of rope, and 200 metric tons of steel. This project, which began in 1971, required calculating the surface area of the Reichstag. All of the materials used were recycled after the building was unwrapped. The sphere at the center left, created with pine cones, pine bark, pine twigs, and pine needles, is

the work of enviromental artist Chris Drury. The sphere at the top left is the *Kugel Ball*, a granite sphere with the continents etched on it that revolves over a platform in front of the Houston Museum of Natural Science. In *Quartet* (1987), sculptor Barry Le Va incorporated spherical and rectangular shapes made of mahogany, fiberboard, and concrete. In this chapter, students will learn how to calculate the surface area and volume of solids.

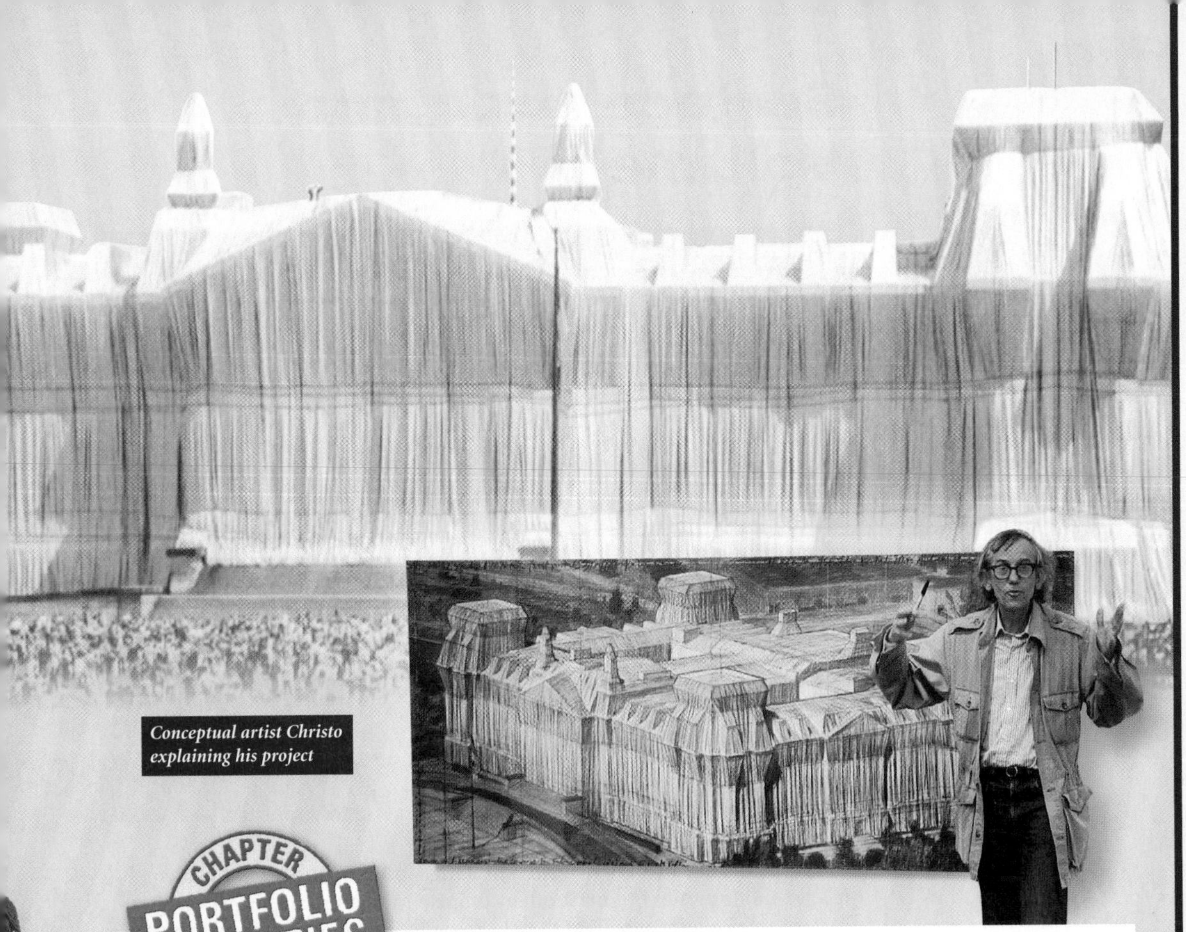

Conceptual artist Christo explaining his project

PORTFOLIO ACTIVITIES PROJECT

About the Chapter Project

Models of solid figures can be very useful in studying the properties of these figures. Throughout this chapter, you will use nets for solid figures to determine surface area.

In the Chapter Project, you will create a three-dimensional puzzle based on the tangram.

After completing the Chapter Project, you will be able to do the following:

- Assemble three-dimensional objects from nets.

- Understand the volume of a three-dimensional object in terms of the sum of its parts.

About the Portfolio Activities

Throughout the chapter, you will be given opportunities to complete Portfolio Activities that are designed to support your work on the Chapter Project.

The theme of each Portfolio Activity and of the Chapter Project is modeling solid figures.

- In the Portfolio Activity on page 452, you will build three oblique pyramids that fit together to form a cube.

- In the Portfolio Activity on page 459, you will examine the lateral surface of an oblique cylinder.

- In the Portfolio Activity on page 485, you will create a solid of revolution by using cardboard and thread.

PORTFOLIO ACTIVITIES PROJECT

Portfolio Activities appear at the end of Lessons 7.3, 7.4, and 7.7. Each serves as preparation for the Chapter Project. The Portfolio Activities, as well as the Chapter Project Activities, are appropriate for inclusion in the student's portfolio. Students should be encouraged to include in their portfolios any other work in which they feel a sense of pride or a sense of accomplishment.

🔲 internet connect

Chapter Internet Features and Online Activities

QUICK **WARM-UP**

1. Find the surface area and volume of a cube with a side length of 4 inches. surface area: 96 square inches; volume 64 cubic inches

2. Find the surface area and volume of a rectangular prism with a length of 5 inches, height of 4 inches, and width of 3 inches. surface area: 94 square inches; volume: 60 cubic inches

3. Suppose there are 20 students in a geometry class and 8 of them are boys. What is the ratio of boys to girls? $\frac{8}{12}$ or $\frac{2}{3}$

Also on Quiz Transparency 7.1

Teach

Why Maximizing volume while minimizing surface area is an important concept for students to understand because of its applications in science.

Surface Area and Volume

Objectives

● Explore ratios of surface area to volume.

● Develop the concepts of maximizing volume and minimizing surface area.

Why *The shapes and sizes of plants and animals are partly determined by the necessary ratio of surface area to volume. Sometimes a large ratio is advantageous, while other times a small ratio is better.*

APPLICATION
BIOLOGY

Desert plants must conserve water but have a lot of light. Tropical plants receive a lot of water but must often compete for light. Leaves with a large surface area collect more light than leaves with a smaller surface area, but they also lose more water through evaporation.

Surface Area and Volume

The surface area of an object is the total area of all the exposed surfaces of the object. The volume of a solid object is the number of nonoverlapping unit cubes that will exactly fill the interior of the figure.

Surface Area and Volume Formulas

The surface area, S, and volume, V, of a right rectangular prism with length ℓ, width w, and height h are

$$S = 2\ell w + 2wh + 2\ell h \quad \text{and} \quad V = \ell wh.$$

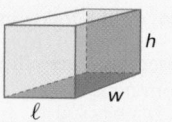

7.1.1

The surface area, S, and volume, V, of a cube with side s are

$$S = 6s^2 \quad \text{and} \quad V = s^3.$$

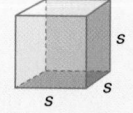

7.1.2

Alternative Teaching Strategy

COOPERATIVE LEARNING Students can explore the ratio of surface area to volume by working in small groups. Have each student build and measure one or more rectangular prisms. Data about surface area and volume should be compiled into a table for further analysis. Have each group discuss their conjectures about the ratio of surface area to volume until they agree. Have them share their conjectures with the entire class.

In Activity 1 below, you will explore the relationship between the volume and surface area of right rectangular prisms.

Activity 1
Ratio of Surface Area to Volume

Algebra

Part I

1. Draw or build several rectangular prisms with a width and height of 1 unit and a length $\ell = 1, 2, 3, \ldots$ Copy and complete the table below, or use a spreadsheet to generate a similar table.

Length, ℓ	Surface area = $2\ell w + 2\ell h + 2wh$	Volume = ℓwh	Ratio of $\frac{\text{surface area}}{\text{volume}}$
1	6	1	?
2	?	?	?
3	?	?	?
n	?	?	?

2. If $\ell = 100$, what is the ratio of surface area to volume? What happens to the ratio of surface area to volume as the value of ℓ increases? Is there a number that the ratio approaches?

Part II

Algebra

1. Draw or build several cubes with side $s = 1, 2, 3, \ldots$ Copy and complete the table below, or use a spreadsheet to generate a similar table.

Side, s	Surface area = $6s^2$	Volume = s^3	Ratio of $\frac{\text{surface area}}{\text{volume}}$
1	6	1	?
2	?	?	?
3	?	?	?
n	?	?	?

CHECKPOINT ✔

2. If $s = 100$, what is the ratio of surface area to volume? What happens to the ratio of surface area to volume as the value of s increases? What conclusions can you draw about the surface-area-to-volume ratios of smaller cubes compared to those of larger cubes?

By extending your tables from Activity 1, you can discover another important result: *For a given volume greater than 1*, the cube in Part II will have a smaller surface-area-to-volume ratio than the long rectangular prism in Part I. Compare, for example, the ratios for a cube and a long prism, each with a volume of 8. (For the cube, $s = 2$, and for the prism, $\ell = 8$.)

Interdisciplinary Connection

INDUSTRIAL TECHNOLOGY The design of a package may reflect environmental and health factors as well as the economics of maximizing volume. Have students discuss what a packaging engineer may need to know to design better packages.

Activity 1 Notes

In this Activity, students explore the ratio of surface area to volume for a rectangular prism. In Part I, students draw or build several rectangular prisms. In Part II, students draw or build cubes.

For a student worksheet of this Activity and detailed Teacher Notes, see page 120 in the Lesson Activities booklet.

Math
CONNECTION

ALGEBRA In Parts I and II of Activity 1, students find the ratio of surface area to volume for rectangular prisms and for cubes. Students can use a graphics calculator with a table or spreadsheet function to complete the Activity.

CHECKPOINT ✔
Part II

2. When $s = 100$ the ratio is $\frac{3}{50}$. As s increases, the ratio decreases. Smaller cubes have a larger ratio than larger cubes.

Cooperative Learning

Have students work in pairs to build the figures in Parts I and II of Activity 1. When finished, have one student perform the calculations in Part I while another student fills in the table. They can switch roles for Part II.

Activity 2 Notes

In this Activiy, students explore how to maximize the volume of an open box by tracing the graph of the volume function on a graphics calculator or by graphing the function on graph paper.

For a student worksheet of this Activity and detailed Teacher Notes, see page 120 in the Lesson Activities booklet.

Math
CONNECTION

MAXIMUM/MINIMUM Maximum/minimum problems are common in real-world math applications. Calculus deals extensively with these concepts.

CHECKPOINT ✔
4. a. 1.60   b. 66.15

Small animals such as hummingbirds have high surface-area-to-volume ratios. They must maintain high metabolisms to compensate for the loss of body heat.

CRITICAL THINKING

YOU WILL NEED

graphics calculator or graph paper

CONNECTION
MAXIMUM/MINIMUM

CHECKPOINT ✔

In general, a "bunched-up" object (like a cube or a sphere) will have a smaller surface-area-to-volume ratio than a long, thin object with the same volume. Also, a large object has a smaller surface-area-to-volume ratio than a small object of the same shape (compare the ratios for cubes with different side lengths).

These observations have important consequences in the sciences. In biology and physics, many important processes happen at the surface of a solid body. For example, an animal's loss of body heat to the atmosphere occurs at the surface. If its body is large and the surface area is small, the rate of heat loss will be low. If its body is small and the surface area is large, the rate of heat loss will be high.

On sunny days, a snake may stretch out in the sun to absorb heat. At night, a snake may coil its body to retain heat. Why does this strategy work?

Activity 2
Maximizing Volume

1. Create an open box from a standard 8.5 × 11 inch piece of paper by cutting squares from the corners and folding up the sides, as shown. Copy and complete the table below.

Side of square, x	Length, ℓ	Width, w	Height, h	Volume, ℓwh
1	9	6.5	1	58.5
2	?	?	?	?
3	?	?	?	?
x	?	?	?	?

2. What side length of the square will maximize the volume of the box?

3. Use a graphics calculator or graph paper to graph the volume as a function of the side length of the square. Let x be the side length of the square, and let y be the volume of the open box.

4. Use the trace function of your calculator or estimate from your graph to find the following:
   a. the side length that produces the largest volume
   b. the largest volume

A cube has a larger volume for its surface area than any other rectangular prism. However, it is not possible to make a cubic box from an 8.5 × 11 inch piece of paper by using the given method. The shape you determined in Activity 2 is the closest shape to a cube that can be created by this method.

Enrichment

Have students explore the relationship between the area and volume of similar figures. Have cooperative groups build pairs of similar rectangular prisms. Have them find the scale factor, k, between corresponding sides of the prisms and answer the following questions: What is the ratio of the surface areas for each similar pair? $\frac{1}{k^2}$ What is the ratio of the volumes for each similar pair? $\frac{1}{k^3}$

Inclusion Strategies

HANDS-ON STRATEGIES Have students use small cubes to model the surface area and volume data for Activity 1. Students can glue together sugar cubes to create their own models for their portfolios. Students should also model Activity 2 by using several sheets of 11-by-14-inch paper. You may wish to have the students create a poster to display their work on these Activities.

A cereal company is choosing between two box designs with the dimensions shown at right. Which design has the greater surface area and thus requires more material for the same volume?

Box A   Box B

10 in.

8 in.

4 in.   5 in.   2 in.   8 in.

SOLUTION

Both boxes have a volume of 160 cubic inches. The surface area of box A is $2(8)(5) + 2(4)(5) + 2(4)(8) = 184$ square inches.

The surface area of box B is $2(10)(8) + 2(2)(8) + 2(2)(10) = 232$ square inches.

Box B has the greater surface area.

CRITICAL THINKING

Most cereal boxes use a design that is more like box B than box A. What considerations other than surface area might be important in choosing a box design?

Exercises

Communicate

1. Explain why the surface area of a right rectangular prism is given by the formula $S = 2\ell w + 2\ell h + 2wh$.

2. What can you say about a solid whose surface-area-to-volume ratio is greater than 1? What can you say about a solid whose surface-area-to-volume ratio is less than 1?

3. Explain why the surface-area-to-volume ratio of a rectangular prism with dimensions of $\ell \times 1 \times 1$ approaches 4 as ℓ increases. (Hint: Which faces of the prism have the same area as ℓ increases?)

4. In Activity 2, you discovered that the surface-area-to-volume ratio of a cube approaches 0 as the side length increases. Can this ratio ever equal 0? Why or why not?

5. **BIOLOGY** Single-celled organisms are able to absorb all of their oxygen and food through their surface. Use surface-area-to-volume ratios to explain why this is not possible for larger animals.

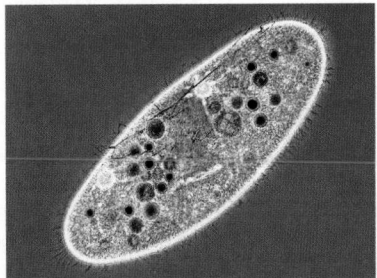

ASSIGNMENT GUIDE

In Class	1–11
Core	12–26, 27–39 odd
Core Plus	12–26, 27–45 odd
Review	46–55
Preview	56–59

✐ Extra Practice can be found beginning on page 818.

Error Analysis

Make sure that students have the correct functions to graph for Exercises 35 and 36.

Practice

NAME _____ CLASS _____ DATE _____

Practice

7.1 *Surface Area and Volume*

Determine the surface-area-to-volume ratio for a rectangular prism with the indicated dimensions. Show all of your steps.

1. $4 \times 4 \times 3$
$S = 2(4 \times 4) + 2(4 \times 3) + 2(4 \times 3);$
$V = 4 \times 4 \times 3; \frac{5}{3} \approx 1.67$

2. $80 \times 1 \times 1$
$S = 2(80 \times 1) + 2(80 \times 1) + 2(1 \times 1);$
$V = 80 \times 1 \times 1; \frac{161}{40} \approx 4.03$

3. $24 \times 24 \times 24$
$S = 2(24 \times 24) + 2(24 \times 24) + 2(24 \times 24);$
$V = 24 \times 24 \times 24; \frac{1}{4} = 0.25$

4. $7 \times 9 \times 22$
$S = 2(7 \times 9) + 2(7 \times 22) + 2(9 \times 22);$
$V = 7 \times 9 \times 22; \frac{415}{693} \approx 0.60$

5. $4 \times 16 \times 48$
$S = 2(4 \times 16) + 2(4 \times 48) + 2(16 \times 48);$
$V = 4 \times 16 \times 48; \frac{2}{3} \approx 0.67$

6. $25 \times 14 \times 33$
$S = 2(25 \times 14) + 2(25 \times 33) + 2(14 \times 33);$
$V = 25 \times 14 \times 33; \frac{1637}{5775} \approx 0.28$

Find the surface-area-to-volume ratio for each solid described below. Show all of your steps.

7. a cube with a surface area of 150 square units
$s = \sqrt{\frac{150}{6}} = \sqrt{25} = 5;$
$V = 5 \times 5 \times 5; \frac{6}{5} = 1.2$

8. a cube with a volume of 512 cubic units
$s = \sqrt[3]{512} = 8;$
$S = 6(8 \times 8); \frac{3}{4} = 0.75$

9. a rectangular prism with dimensions $4 \times 4 \times 4$
$S = 6(4 \times 4);$
$V = 4 \times 4 \times 4; \frac{3}{2} = 1.5$

10. a cube with a volume of 8000 cubic units
$s = \sqrt[3]{8000} = 20;$
$S = 6(20 \times 20); \frac{3}{10} = 0.30$

11. a rectangular prism with dimensions $4 \times 1 \times 1$
$S = 2(4 \times 1) + 2(4 \times 1) + 2(1 \times 1);$
$V = 4 \times 1 \times 1; \frac{9}{2} = 4.5$

12. a cube with a volume of 216 cubic inches
$s = \sqrt[3]{216} = 6;$
$V = 6 \times 6 \times 6; 1$

13. a rectangular prism with a diagonal length of 19 and a base of 10×15
$h = \sqrt{19^2 - (10^2 + 15^2)} = 6; S = 2(15 \times 10) + 2(15 \times 6) + 2(10 \times 6);$
$V = 15 \times 10 \times 6; \frac{2}{3} \approx .67$

---

10. 6682.72 in³
7.45 in. × 7.45 in.

APPLICATIONS

CONNECTIONS

12. 16	13. 4
14. 4	15. 48
16. 16	17. 3
18. 142	19. 105
20. 1.35	21. 7
22. 122	23. 1.45
24. 6	25. 60
26. 1.73	

Guided Skills Practice

Determine the surface-area-to-volume ratio for a rectangular prism with the given dimensions. *(ACTIVITY 1)*

6. $5 \times 1 \times 1$ **4.4**   **7.** $10 \times 1 \times 1$ **4.2**   **8.** $5 \times 5 \times 5$ **1.2**   **9.** $10 \times 10 \times 10$ **0.6**

10. HOBBIES Teresa is selling vegetables from her garden at a farmer's market. She has some rectangular pieces of cardboard that are 42×48 inches. What is the maximum volume for a box made by cutting squares out of the corners of these pieces and folding up the edges? What size squares should be cut out to make a box with this volume? *(ACTIVITY 2)*

11. MANUFACTURING A company is choosing between two box designs. The dimensions (in inches) of box A are $4 \times 2 \times 6$ and of box B are $3 \times 2 \times 8$. Which has the smaller surface area? *(EXAMPLE)* box A

Practice and Apply

Algebra

📶 internet connect

Homework Help Online

Go To: go.hrw.com
Keyword:
MG1 Homework Help
for Exercises 12–30

Copy and complete the following table for rectangular prisms:

Length	Width	Height	Surface area	Volume	Surface area / volume
2	2	1	**12.** ?	**13.** ?	**14.** ?
4	4	1	**15.** ?	**16.** ?	**17.** ?
7	3	5	**18.** ?	**19.** ?	**20.** ?
4	**21.** ?	3	**22.** ?	84	**23.** ?
2	5	**24.** ?	104	**25.** ?	**26.** ?

Find the surface-area-to-volume ratio of each of the following:

27. a cube with a volume of 64 cubic units **1.5 units**

28. a cube with a volume of 1000 cubic units **0.6 units**

29. a rectangular prism with dimensions of $n \times n \times 1$ $2 + \frac{4}{n}$

30. a prism with a square base, where the lateral edge is twice the length of a base edge $\frac{5}{s}$

MAXIMUM/MINIMUM For each situation below, determine whether you should maximize the volume or minimize the surface area. Explain your reasoning.

31. building a storage bin with a limited amount of lumber

32. designing soup cans that will hold 15 ounces of soup

33. mailing a package with dimensions whose sum is less than 72 inches

34. building a tank that must hold 100 gallons of water

35. MAXIMUM/MINIMUM Compare the surface-area-to-volume ratio of an $n \times n \times 1$ rectangular prism with that of an $n \times n \times n$ rectangular prism as n increases. $2 + \frac{4}{n}; \frac{6}{n}$

36. MAXIMUM/MINIMUM Based on your answer to Exercise 35, make a conjecture about what type of rectangular prism has the smallest surface area for a given volume. **a cube**

---

31. Sample answer: Maximize the volume; assuming you use all available lumber, the surface area is fixed. You want to create the maximum amount of storage space.

32. Sample answer: Minimize the surface area. The volume is constant, so you may want to minimize the surface area to save on packaging costs.

33. Sample answer: Minimize the surface area; assuming the item(s) to be mailed have a fixed volume, you would want to minimize the amount of packaging to reduce the weight of the package.

34. Sample answer: Minimize the surface area. The volume is constant, so you may want to minimize the surface area to save on construction materials.

CHALLENGES

37. Find the side length of a cube with a surface-area-to-volume ratio of 1. **6**

38. A prism with a square base has a lateral edge that is twice as long as the base edges. If its surface-area-to-volume ratio is 1, find the length of a base edge. **5**

APPLICATIONS

39. BIOLOGY Flatworms do not have gills or lungs. They absorb oxygen through their skin. Use surface-area-to-volume ratios to explain how the shape of a flatworm helps it absorb enough oxygen.

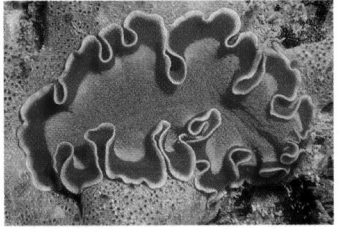

Marine flatworm

40. PHYSIOLOGY Human lungs are subdivided into thousands of air sacs. The total surface area of an average person's lungs is about 100 square yards. Use surface-area-to-volume ratios to explain why such a large surface area is necessary.

41. BOTANY Use surface-area-to-volume ratios to explain why tall trees with large, broad leaves do not usually grow in deserts.

42. COOKING A roast is finished cooking when the internal temperature is 160°F. Use surface-area-to-volume ratios to explain why a large roast takes longer to cook than a small one of the same shape.

43. CHEMISTRY Use surface-area-to-volume ratios to explain why a large block of ice will melt more quickly if it is broken up into smaller pieces.

44. PHYSIOLOGY Use surface-area-to-volume ratios to explain why chewing your food thoroughly makes digestion easier.

45. BOTANY A barrel cactus has a short, squatty form. Use surface-area-to-volume ratios to explain how this form helps the cactus conserve water in desert conditions.

Look Back

46. Show that AAA information cannot be used to prove that triangles are congruent. *(LESSON 4.3)*

Find the area of each figure below. *(LESSONS 5.1 AND 5.2)*

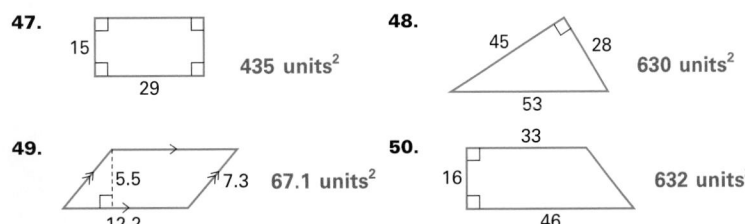

47. 15, 29 — **435 units²**

48. 45, 28, 53 — **630 units²**

49. 5.5, 7.3, 12.2 — **67.1 units²**

50. 33, 16, 46 — **632 units²**

---

43. Sample answer: When broken into smaller pieces, a block of ice will have a larger surface-area-to-volume ratio which maximizes heat exchange.

44. Sample answer: Completely chewing food "flattens" it out, thus increasing the surface area on which digestive enzymes and juices can act.

45. Sample answer: The short, squat form maximizes the volume of the cactus and therefore the amount of water it can hold. Also the surface area is kept to a minimum which minimizes water loss through transpiration and evaporation.

46. Possible counterexamples: Any two isosceles right triangles, one with legs of length a and one with legs of length b, with $a \neq b$; or any two equilateral triangles, one with a side of length a and one with a side of length b with $a \neq b$. These will satisfy AAA but will not be congruent. Any two different equilateral triangles.

Technology

Students can use the trace features of a graphics calculator to help them explain their answers to Exercises 35 and 36. Encourage students to use the table features to explore the change in y-values as the x-values increase rapidly.

Math
CONNECTION

ALGEBRA In Exercises 12–26, students use the formulas for surface area and volume to solve for length, width, and height. Exercises 31–36 are maximum/minimum problems.

39. Sample answer: The flat shape of the worm maximizes the surface-area-to-volume ratio of its body which maximizes the amount of oxygen it can take in through its skin.

40. Sample answer: The thousands of air sacs make the surface-area-to-volume ratio of the lungs larger, thereby maximizing the ability of the lungs to absorb oxygen.

41. Sample answer: Tall trees with large, broad leaves have a large surface area exposed. The large surface area is important if there is competition for light, but it also increases water loss. In desert conditions surface-area-to-volume ratios should be minimized; broad-leafed trees would soon die.

42. Sample answer: In a large roast, there is more meat to cook, so the surface-area-to-volume ratio is less than for a smaller roast. There is relatively less surface through which heat can enter to cook a larger amount of meat.

Exercises 56–59 study the relationship between the height of a person and the amount of weight the person's foot would have to bear. The question of the practical limits of the size of a human body is raised.

47. 435 units²

51. No. Sample answer:

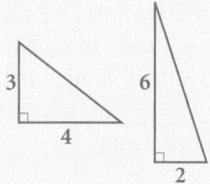

These two triangles have the same area, but are not congruent.

52. $\ell = 9$ $w = 1$ $A = 9$
$\ell = 8$ $w = 2$ $A = 16$
$\ell = 7$ $w = 3$ $A = 21$
$\ell = 6$ $w = 4$ $A = 24$
$\ell = 5$ $w = 5$ $A = 25$

---

51. If two triangles have the same area, are they necessarily congruent? If not, give a counterexample. *(LESSON 5.2)*

52. If the perimeter of a rectangle is 20 inches, give five possible areas and the dimensions for each. *(LESSON 5.2)*

53. 25π units²

53. What is the area of a circle with a circumference of 10π? *(LESSON 5.3)*

54. The legs of an isosceles triangle measure 15 feet and the base is 16 feet. Find the area. *(LESSON 5.4)* 101.51 ft²

55. The legs of an isosceles right triangle measure 6 centimeters. Find the length of the hypotenuse. *(LESSONS 5.4 AND 5.5)* 8.49 cm

 Look Beyond

Ratios of surface area to volume play a part in determining how big a person can be. As the size of a person increases, the weight increases in proportion to the volume, while certain other measurements increase in proportion to the surface area.

56. Sample answer: 40 in²

57. Sample answer: 3 lb/in²

58. 4; 8

59. Sample answer: 6 lb/in². As the size increases, the pressure on the feet increases. Because there is a limit to the pressure that the feet can withstand, the size of the human body cannot exceed certain limits.

56. Estimate the total area of the bottom of both of your feet in square inches. One way to do this is to trace your feet onto a piece of graph paper and count the squares.

57. Divide your weight in pounds by the area of the bottom of both of your feet. This gives the pressure, in pounds per square inch, on the bottom of your feet.

58. Suppose that the size of your body increased while keeping the same shape. If the length and width of your feet were doubled, the area of your feet would increase by a factor of _____?_____. Estimate the new area of the bottom of your feet. If your length, width, and height were doubled, your weight would increase by a factor of _____?_____. What would be your new weight?

59. Divide the new weight by the new area to find the new pressure on the bottoms of your feet. Explain what happens to the pressure on your feet as your size increases. How does this limit the size of the human body?

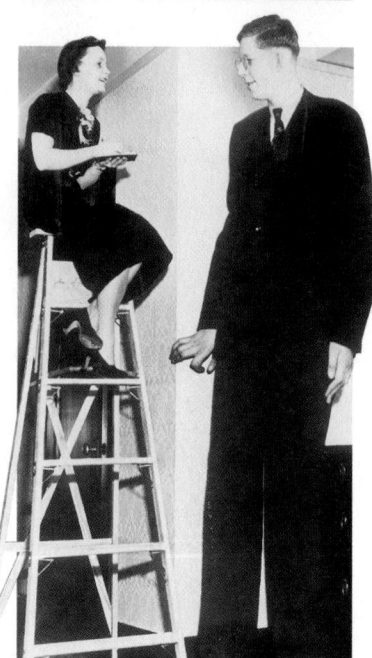

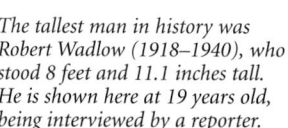

The tallest man in history was Robert Wadlow (1918–1940), who stood 8 feet and 11.1 inches tall. He is shown here at 19 years old, being interviewed by a reporter.

Surface Area and Volume of Prisms

Objectives

- Define and use a formula for finding the surface area of a right prism.

- Define and use a formula for finding the volume of a right prism.

- Use Cavalieri's Principle to develop a formula for the volume of a right or oblique prism.

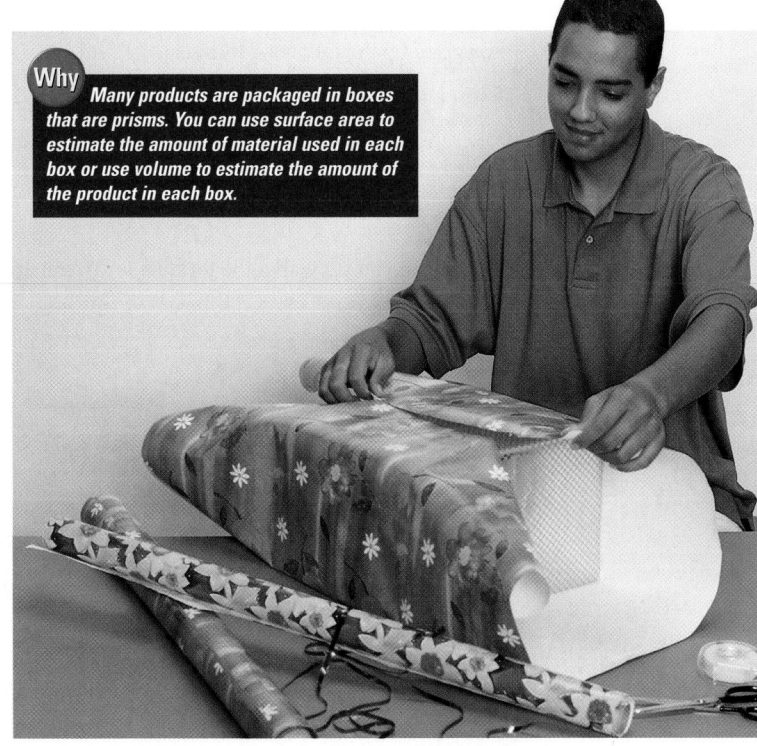

You can use the surface area of a box to estimate the amount of wrapping paper you will need to cover the box.

Recall from Lesson 6.3 that a prism has two parallel congruent faces called bases and that the remaining faces, called lateral faces, are parallelograms. In a right prism, all of the lateral faces are rectangles.

An **altitude** of a prism is a segment that has endpoints in the planes containing the bases and that is perpendicular to both planes. The **height** of a prism is the length of an altitude.

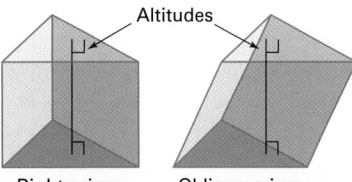

Surface Area of Right Prisms

The surface area of a prism may be broken down into two parts: the area of the bases, or base area, and the area of the lateral faces, or lateral area.

Since the bases are congruent, the base area is twice the area of one base, or $2B$, where B is the area of one base.

Prepare

QUICK WARM-UP

1. Find the area of a triangle with a height of 6 cm and base of 4 cm. $12 \, \text{cm}^2$

2. Find the distance between $(0, 3, -2)$ and $(4, -1, 3)$ in a three-dimensional coordinate system. $\sqrt{57}$

3. Find the surface area and volume of a cube with a side length 8 inches. surface area: 384 square inches; volume: 512 cubic inches

Also on Quiz Transparency 7.2

Teach

Why This lesson helps students extend their ideas about surface area and volume by introducing the idea that the volume of a prism is equal to the area of its base times its height. Ask students to compare this with the formula for volume: $V = \ell wh$. Point out that length times width is also the area of the base of a rectangular prism or box.

Alternative Teaching Strategy

HANDS-ON STRATEGIES Have students use modeling materials such as clay to build prisms. Using the same amount of clay, have them construct a prism and find its volume and surface area. Then have them construct another prism of a different shape and find the volume and surface area. If the same amount of clay was used, both prisms should have the same volume, but probably different surface areas. This activity can be used to link this lesson to Lesson 7.1 because students can find the minimum surface area for a given volume.

CRITICAL THINKING

You may want to use models of right prisms with different bases to emphasize the applicability of the formula given in the text. By building a rectangular prism, trapezoidal prism, etc., from a net and then unfolding the net, you can see that the perimeter of the base of the prism is equal to the length of the rectangle that is formed by the lateral surface.

Cooperative Learning

Have students work in pairs and use nets of prisms to make models in class. Then have them use the models to find lateral area and surface area.

Teaching Tip

Emphasize that a prism is named by the shape of its base. The faces that are not bases are called lateral faces and are always parallelograms. Point out that the bases of a prism are always congruent.

To find the lateral area of a right prism, you may find it helpful to use a net. Because the net folds up to form the prism, the area of the net is equal to the surface area of the prism.

If the sides of the base are s_1, s_2, and s_3 and the height is h, then the lateral area is given by the following formula:

$$L = s_1h + s_2h + s_3h = h(s_1 + s_2 + s_3)$$

Because $s_1 + s_2 + s_3$ is the perimeter of the base, we can write the lateral area as $L = hp$, where p is the perimeter of the base.

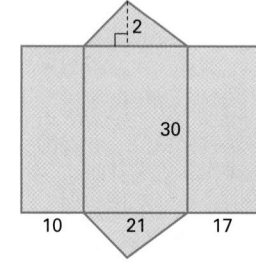

CRITICAL THINKING How can you show that the formula for lateral area, $L = hp$, is true for right prisms with bases that are not triangles?

The surface area of a prism is the sum of the base area and the lateral area.

Surface Area of a Right Prism

The surface area, S, of a right prism with lateral area L, base area B, perimeter p, and height h is
$$S = L + 2B \quad \text{or} \quad S = hp + 2B.$$

7.2.1

EXAMPLE ①

The net for a right triangular prism is below. What is its surface area?

SOLUTION

The area of each base is
$B = \frac{1}{2}(2)(21) = 21$.

The perimeter of each base is $p = 10 + 21 + 17 = 48$, so the lateral area is
$L = hp = 30(48) = 1440$.

Thus, the surface area is
$S = L + 2B = 1440 + 2(21) = 1440 + 42 = 1482$.

Volumes of Right Prisms

Recall from Lesson 7.1 that the volume of a right rectangular prism with length ℓ, width w, and height h is given by $V = \ell wh$. Because the base area, B, of this type of prism is equal to ℓw, you can also write the formula for the volume as $V = Bh$.

Does this formula work for right prisms with bases that are not rectangles? The discussion that follows begins by considering a triangular prism with a height of 1.

Interdisciplinary Connection

MANUFACTURING Cardboard boxes are often manufactured as nets with extra flaps attached to help keep the boxes closed when assembled. Have students create examples of nets by taking boxes apart. Find the surface area of each net and compare it with the surface area of a rectangular prism with the same dimensions.

The prism at right has a height of 1. The number of unit cubes needed to fill the prism is equal to the number of unit squares needed to cover the base, which is equal to the area of the base.

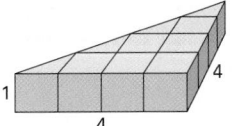

A prism with a height of h can be made by stacking h prisms with a height of 1 on top of each other. The volume of this prism is h times the volume of the prism with a height of 1, or h times the area of the base. Thus, the volume of any right prism with a base area of B and height of h is $V = Bh$.

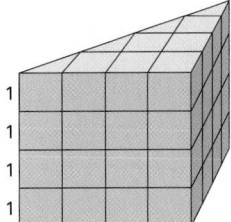

EXAMPLE ② An aquarium in the shape of a right rectangular prism has dimensions of $110 \times 50 \times 7$ feet. Given that 1 gallon ≈ 0.134 cubic feet, how many gallons of water will the aquarium hold? Given 1 gallon of water ≈ 8.33 pounds, how much will the water weigh?

APPLICATION
AQUARIUMS

● **SOLUTION**

The volume of the aquarium is found by using the volume formula.
$V = Bh = \ell wh = (110)(50)(7) = 38,500$ cubic feet

To approximate the volume in gallons, divide by 0.134.
$V = 38,500 \div 0.134 \approx 287,313$ gallons

To approximate the weight, multiply by 8.33.
weight $\approx (287,313)(8.33) \approx 2,393,317$ pounds

EXAMPLE ③ An aquarium has the shape of a right regular hexagonal prism with the dimensions shown at right. Find the volume of the aquarium.

● **SOLUTION**

The base of the aquarium has a perimeter of $(14)(6)$, or 84, inches and an apothem of $7\sqrt{3}$ inches, so the base area is found as follows:

$B = \frac{1}{2}ap = \frac{1}{2}(84)(7\sqrt{3}) = 294\sqrt{3} \approx 509.22$ square inches

The volume is $V = Bh = (294\sqrt{3})(48) = 14112\sqrt{3} \approx 24,443$ cubic inches.

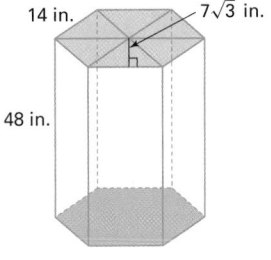

14 in. $7\sqrt{3}$ in.

48 in.

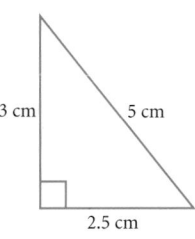

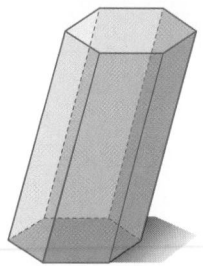

In an oblique prism, the lateral edges are not perpendicular to the bases, and there is no simple general formula for surface area. However, the formula for the volume is the same as that for a right prism. To understand why this is true, consider the explanation below.

Stack a set of index cards in the shape of a right rectangular prism. If you push the stack into the shape of an oblique prism, the volume of the solid does not change because the number of cards does not change.

Both stacks have the same number of cards, and each prism is the same height. Also, because every card has the same size and shape, they all have the same area. Any card in either stack represents a cross section of each prism.

The prisms above illustrate an important geometry concept.

Cavalieri's Principle

If two solids have equal heights and the cross sections formed by every plane parallel to the bases of both solids have equal areas, then the two solids have equal volumes.

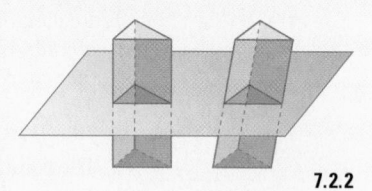

7.2.2

The solids will have equal volumes if they are the same height and all cross-sectional areas are equal.

Every oblique prism can be compared with a right prism with the same base and height. As the stacks of cards suggest, every cross section of a prism is congruent to its bases, so all of the cross sections of the oblique prism are equal to the cross sections of the right prism. Thus, by Cavalieri's Principle, they have equal volumes.

The formula for the volume of any prism is given below.

Volume of a Prism

The volume, V, of a prism with height h and base area B is
$$V = Bh.$$

7.2.3

Exercises

Communicate

1. Explain how to find the surface area of a right prism.

2. Explain the formula for the volume of a right prism.

3. Explain Cavalieri's Principle and how it can be used to find the volume of an oblique prism.

4. Can Cavalieri's Principle be used for two prisms of the same height if the base of one is a triangle and the base of the other is a hexagon? Why or why not?

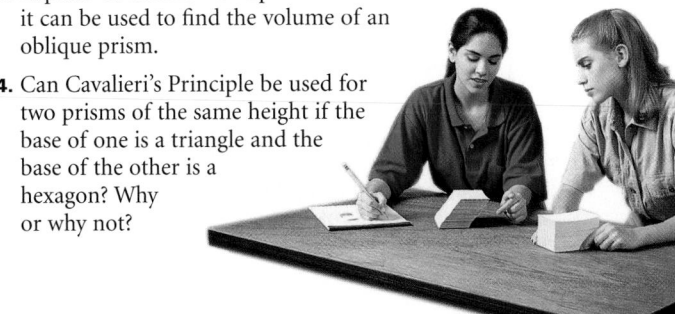

internet connect

Activities Online
Go To: **go.hrw.com**
Keyword: **MG1 Aquariums**

Selected Answers
Exercises 5–8, 9–47 odd

ASSIGNMENT GUIDE

In Class	1–8
Core	9–35 odd
Core Plus	21–37
Review	40–47
Preview	48–50

✐ Extra Practice can be found beginning on page 818.

Error Analysis

Students will alternate between finding volumes and surface areas. Encourage students to be very careful with their calculations and to double-check all answers. Having a list of the formulas for area, surface area, and volume handy will help them as they complete their assignment.

Guided Skills Practice

Determine the surface area of the prism formed by each net.
(EXAMPLE 1)

5. 888 units^2

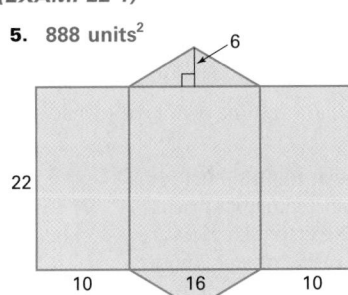

6. 176 units^2

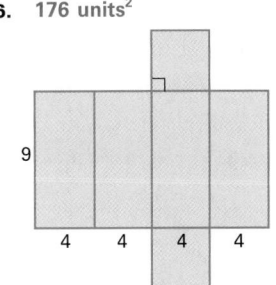

Find the volume of each prism. (EXAMPLES 2 AND 3)

7. right rectangular prism
840 units^3

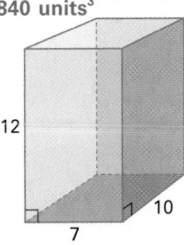

8. right regular octagonal prism
27,811.74 units^3

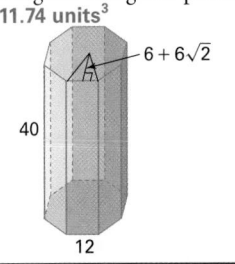

9. Sample answer:

10. Sample answer:

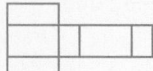

11. Sample answer:

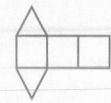

12. Sample answer:

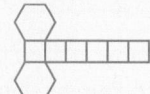

internet connect

Homework Help Online

Go To: **go.hrw.com**
Keyword:
MG1 Homework Help
for Exercises 13-20

● *Practice and Apply*

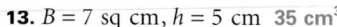

Draw a net for each figure named below.

9. cube

10. right rectangular prism

11. right equilateral triangular prism

12. right regular hexagonal prism

Find the volume of a prism with the given dimensions.

13. $B = 7$ sq cm, $h = 5$ cm **35 cm^3**

14. $B = 9$ sq m, $h = 6$ m **54 m^3**

15. $B = 17$ sq in., $h = 23$ in. **391 in^3**

16. $B = 32$ sq ft, $h = 17$ ft **544 ft^3**

Find the surface area and volume of a right rectangular prism with the given dimensions.

17. $\ell = 5, w = 7, h = 2$ **118 units^2; 70 units^3**

18. $\ell = 16, w = 9, h = 10$ **788 units^2; 1440 units^3**

19. $\ell = \frac{1}{2}, w = \frac{2}{3}, h = 1$ **3 units^2; $\frac{1}{3}$ units^3**

20. $\ell = 1.3, w = 4, h = 0.5$ **15.7 units^2; 2.6 units^3**

Algebra

21. Find the height of a rectangular prism with a surface area of 286 in.^{2} and a base measuring 7×9 in. **5 in.**

22. The height of a right regular hexagonal prism is 25 cm, and the sides of its base measure 18 cm. Find its surface area. **4383.55 cm^2**

23. The height of a right regular hexagonal prism is 20 cm, and the apothem of its base is $4\sqrt{3}$ cm. Find its surface area. **1292.55 cm^2**

24. If a cube has a volume of 343 yd^3, what is its surface area? **294 yd^2**

25. The figure at right is a net for an oblique rectangular prism. The 3×2 rectangular faces are the bases of the prism. Find its volume and surface area. It may help to copy the net and fold it to form the prism.

16.8 units^3; 41.2 units^2

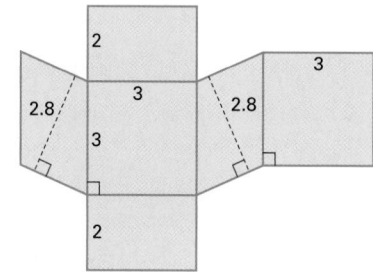

26. The figure at right is a net for an oblique triangular prism. Find its volume and surface area. It may help to copy the net and fold it to form the prism.

8.55 units^3; 31.31 units^2

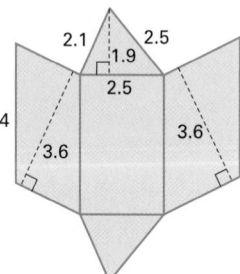

---

27. Find the volume of a right trapezoidal prism if the bases of the trapezoid measure 6 m and 8 m, the height of the trapezoid is 7 m, and the height of the prism is 18 m. **882 m³**

575 cm³ **28.** Find the volume of a right triangular prism whose base is an isosceles right triangle with a hypotenuse of 10 cm and whose height is 23 cm.

29. Find the surface area and volume of a right triangular prism whose base is an isosceles triangle with legs measuring 4 in. and a base of 6 in. and whose height is 13 in. **197.87 in²; 103.18 in³**

30. If the height of a prism is doubled and the bases are unchanged, what happens to the volume of the prism? **The volume is doubled.**

31. If the edges of a cube are doubled, what happens to the surface area? What happens to the volume?

32. If the edges of a cube are tripled, what happens to the surface area? What happens to the volume?

33. The height of a right regular hexagonal prism is equal to the side length of its base. If these side lengths are doubled, what happens to the volume? What happens to the surface area?

CHALLENGE

34. What happens to the side length of a cube if its volume is doubled? What happens to the surface area?

APPLICATIONS

35. MANUFACTURING Find the surface area and volume of each box of cat food shown below.

35. Feline Feast:
surface area = 198 in²
volume = 162 in³
Kitty Krunchies:
surface area = 190 in²
volume = 126 in³

2160 ft³ ≈ 16,119.4 gal **36. RECREATION** The swimming pool shown below is a right prism with concave hexagonal bases. Use the given dimensions to find the volume in cubic feet and in gallons (1 gallon ≈ 0.134 cubic foot).

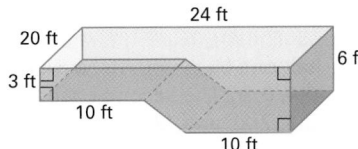

896 ft²; 2.24 gal **37. RECREATION** Find the surface area of the sides and bottom of the swimming pool shown above. If 1 gallon of paint covers 400 square feet, how many gallons of paint will be needed to paint the inside of the pool?

Technology

Students can use a scientific calculator for most of the exercises. Establish rules for rounding. The provided answers are rounded to the nearest hundredth.

Math

CONNECTION

ALGEBRA In Exercise 21, students need to set up an equation with the formula for surface area and solve for the height, h.

31. The surface area is multiplied by $2^2 = 4$, and the volume is multiplied by $2^3 = 8$.

32. The surface area is multiplied by $3^2 = 9$, and the volume is multiplied by $3^3 = 27$.

33. The volume is multiplied by $2^3 = 8$, and the surface area is multiplied by $2^2 = 4$.

34. The side is multiplied by $\sqrt[3]{2}$, and the surface area is multiplied by $2^{\frac{2}{3}}$.

In Exercises 48–50, students plot the vertices of a prism on a three-dimensional coordinate system and find the volume and surface area of the prism.

48.

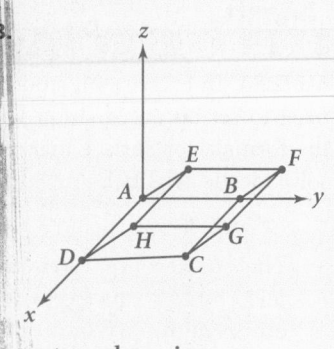

rectangular prism

38. RECREATION The tent shown at right is a right triangular prism. The bases of the prism are isosceles triangles. Find the surface area of the tent, including the floor. **105.5 ft²**

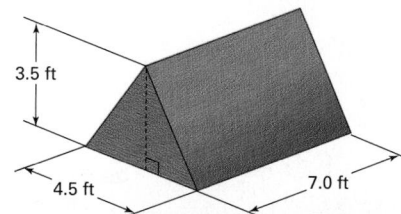

3.5 ft
4.5 ft
7.0 ft

39. ARCHITECTURE Use the lateral area of the building shown at right to estimate how much glass was used to cover the outside walls of the building. Each story is 12 feet high, and the base of the building is a square with sides of 48 feet. **23,040 ft²**

Look Back

Algebra

Simplify each radical expression. Give answers in simplified radical form. *(ALGEBRA REVIEW)*

40. $\sqrt{20}$ $2\sqrt{5}$

41. $(\sqrt{18})(\sqrt{2})$ 6

42. $(5\sqrt{7})^2$ 175

43. $(2\sqrt{2})(3\sqrt{8})$ 24

44. $\dfrac{2}{\sqrt{2}}$ $\sqrt{2}$

45. $\sqrt{27} + \sqrt{12}$ $5\sqrt{3}$

Exercises 46 and 47 refer to a 30-60-90 triangle with a hypotenuse of 10 inches. *(LESSON 5.5)*

46. Find the length of the shorter leg. 5 in.

47. Find the length of the longer leg. 8.66 in.

Look Beyond

COORDINATE GEOMETRY A prism in a three-dimensional coordinate system has one base with its vertices at $A(0, 0, 0)$, $B(0, 5, 0)$, $C(4, 5, 0)$, and $D(4, 0, 0)$. The second base has its vertices at $E(2, 2, 2)$, $F(2, 7, 2)$, $G(6, 7, 2)$, and $H(6, 2, 2)$.

48. Draw the prism on a set of coordinate axes. Name the type of prism.

49. Find the volume of the prism. **40 units³**

50. Find the surface area of the prism. $40 + 36\sqrt{2} \approx 90.9$ **units²**

Surface Area and Volume of Pyramids

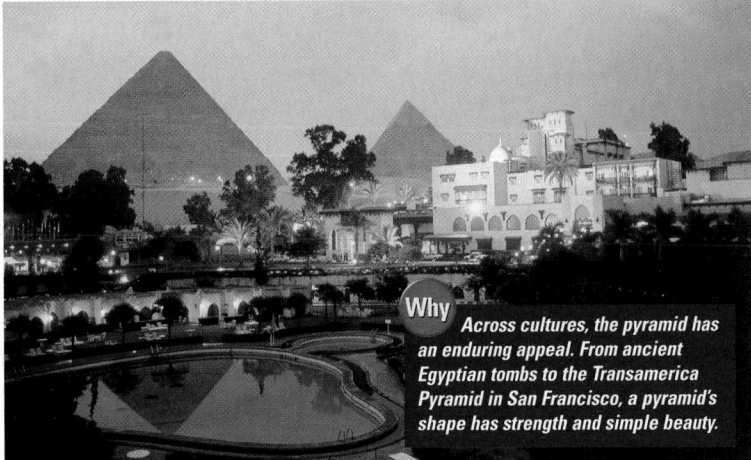

Objectives

- Define and use a formula for the surface area of a regular pyramid.

- Define and use a formula for the volume of a pyramid.

Why Across cultures, the pyramid has an enduring appeal. From ancient Egyptian tombs to the Transamerica Pyramid in San Francisco, a pyramid's shape has strength and simple beauty.

Pyramids

A **pyramid** is a polyhedron consisting of a **base**, which is a polygon, and three or more **lateral faces**. The lateral faces are triangles that share a single vertex, called the **vertex of the pyramid**. Each lateral face has one edge in common with the base, called a **base edge**. The intersection of two lateral faces is a **lateral edge**.

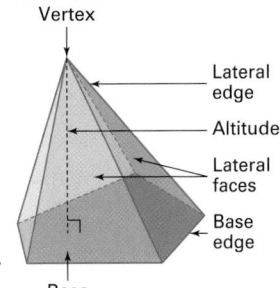

Vertex

Lateral edge

Altitude

Lateral faces

Base edge

Base

The **altitude** of a pyramid is the perpendicular segment from the vertex to the plane of the base. The **height** of a pyramid is the length of its altitude.

A **regular pyramid** is a pyramid whose base is a regular polygon and whose lateral faces are congruent isosceles triangles. In a regular pyramid, all of the lateral edges are congruent, and the altitude intersects the base at its center. The length of an altitude of a lateral face of a regular pyramid is called the **slant height** of the pyramid.

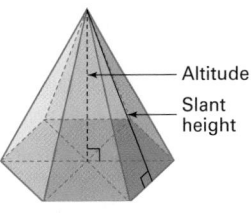

Altitude

Slant height

Pyramids, like prisms, are named by the shape of their base.

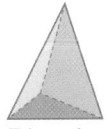
Triangular pyramid

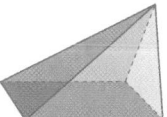
Rectangular pyramid

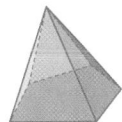
Pentagonal pyramid

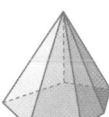
Hexagonal pyramid

Prepare

QUICK WARM-UP

1. Find the volume of a right rectangular prism with a length of 12, width of 8, and height of 4.
   384 cubic units

2. Find the height of an isosceles triangle with sides of 8, 8, and 6.
   $\sqrt{55}$, or ≈ 7.42

3. Find the lateral area of a right triangular prism with an equilateral base length of 5 cm and a height of 10 cm. **150 cm²**

Also on Quiz Transparency 7.3

Teach

Why This lesson gives students formulas for finding the surface area and volume of pyramids. Encourage students to bring in pictures of pyramids to display in the classroom.

Alternative Teaching Strategy

USING MODELS Use modeling clay to help students visualize how prisms can be cut into pyramids as in the Activity on page 447. Begin by forming the clay into a triangular prism. Slice through one upper corner and the diagonal of the lower base to divide the shape into a pyramid and a six-sided polyhedron. Slice through the six-sided polyhedron to create two more pyramids.

The Surface Area of a Pyramid

To analyze the surface area of a pyramid, it is helpful to use a net. The area of the net is the same as the surface area of the pyramid. For pyramids that are not regular, the area of each face must be calculated separately, and then the areas must be added together. For regular pyramids, however, there is an easier way.

EXAMPLE ① Find the surface area of a regular square pyramid whose slant height is ℓ and whose base edge length is s.

● **SOLUTION**

The surface area is the sum of the lateral areas and the base area.

$$S = L + B$$
$$S = 4\left(\tfrac{1}{2}s\ell\right) + s^2$$

Area of each triangle = $\tfrac{1}{2}s\ell$

This can be rewritten as follows:

$$S = \tfrac{1}{2}\ell(4s) + s^2$$

Because $4s$ is the perimeter of the base,

$$S = \tfrac{1}{2}\ell p + s^2.$$

The lateral area of any regular pyramid is equal to $\tfrac{1}{2}\ell p$. The surface area is found by adding the base area, B, to this value.

Surface Area of a Regular Pyramid

The surface area, S, of a regular pyramid with lateral area L, base area B, perimeter of the base p, and slant height ℓ is

$$S = L + B \quad \text{or} \quad S = \tfrac{1}{2}\ell p + B. \qquad \textbf{7.3.1}$$

EXAMPLE ② The roof of a gazebo is a regular octagonal pyramid with a base edge of 4 feet and a slant height of 6 feet. Find the area of the roof. If roofing material costs $3.50 per square foot, find the cost of covering the roof with this material.

● **SOLUTION**

The area of the roof is the lateral area of the pyramid.

$$L = \tfrac{1}{2}\ell p = \tfrac{1}{2}(6)(8 \times 4) = 96 \text{ square feet}$$

96 square feet × $3.50 per square foot = $336.00

Interdisciplinary Connection

CHEMISTRY Tetrahedrons, which are pyramids with all triangular faces, model the structure of carbon-based molecules like methane and carbon tetrachloride. Ask students to use straws or connecting sticks (sold in toy stores) to construct models of molecules based on tetrahedrons that they find in chemistry books. Ask students if they can think of a reason for this shape. (The arrangement places the atoms attached to the carbon atom as far apart as possible and places each atom the same distance from the carbon atom in the center.)

The Volume of a Pyramid

The volume of a pyramid has an interesting relationship to the volume of a prism with the same base and height as the pyramid.

Activity
Pyramids and Prisms

1. Using construction paper, make a right square prism and a regular square pyramid with the same base and the same height. Seal the lateral edges with tape. (Do not seal the base edges.)

2. Fill the pyramid with dry cereal or packing material. Pour the contents into the prism. Repeat as necessary to fill the prism completely. How many times did you have to fill the pyramid in order to fill the prism?

CHECKPOINT ✔ 3. Make a conjecture about the relationship of the volume of a pyramid to the volume of a prism with the same base and height. Express your conjecture as a formula for the volume of a pyramid.

This relationship between triangular prisms and pyramids can be verified by dividing a triangular prism into three triangular pyramids, as shown below. The pyramids are not congruent to each other, but they each have the same volume.

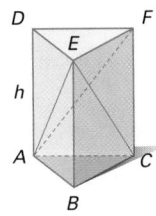 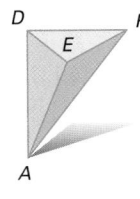 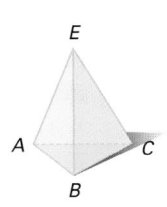 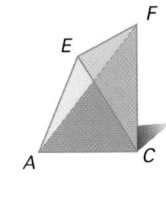

Triangular prism   Pyramid I   Pyramid II   Pyramid III

Consider the pyramids in pairs, as follows:

Pair A: Pyramid I has base *DEF* and height *h*.
Pyramid II has base *ABC* and height *h*.

Notice that the bases are congruent and the heights are equal. It can be shown, using Cavelieri's principle, that pyramids with congruent bases and equal heights have equal volumes.

Pair B: Pyramid II has base *EBC*.
Pyramid III has base *EFC*.

The bases of these pyramids are congruent because $\overline{EC}$ is a diagonal of *BEFC*. The heights are also equal because both pyramids include vertex *A*. Therefore, these two pyramids have equal volumes.

Enrichment

Have students use straws, connecting sticks (sold in toy stores), or nets printed on cardboard to construct various polyhedrons from pyramids. Have them describe how to find the area and volume of the figures they construct.

Inclusion Strategies

HANDS-ON STRATEGIES Some students may be interested in exploring various polyhedrons by examining cross sections of a prism, including those that result in pyramids. Have them use modeling clay and a knife to create these models. Ask them to make a diagram of their results and name the polygon that describes the cross section.

Divide the pyramid into a number of triangular pyramids which each contain the vertex of the original pyramid. Each triangular pyramid is one-third the volume of a prism which contains it, as in the illustrations on page 447. The sum of the volume of all the prisms is equal to the volume of the prism which contains the original pyramid.

ADDITIONAL EXAMPLE 3

Find the volume of a regular square pyramid with a height of 112 feet and side lengths of 99 feet. 365,904 cubic feet

---

By the Transitive Property, the three pyramids all have equal volume, so each must be one-third of the original prism. This suggests the following formula:

Volume of a Pyramid

The volume, V, of a pyramid with height h and base area B is

$$V = \frac{1}{3}Bh.$$

7.3.2

CRITICAL THINKING Once you know that the above formula works for all triangular pyramids, how would you show that it works for pyramids with other bases?

EXAMPLE 3

APPLICATION
ARCHAEOLOGY

CULTURAL CONNECTION: AFRICA The pyramid of Khufu is a regular square pyramid with a base edge of approximately 776 feet and an original height of 481 feet. The limestone used to construct the pyramid weighs approximately 167 pounds per cubic foot. Estimate the weight of the pyramid of Khufu. (Assume the pyramid is solid.)

SOLUTION

The volume of the pyramid is is found as follows:

$V = \frac{1}{3}Bh$

$\approx \frac{1}{3}(776^2)(481)$

$\approx 96{,}548{,}885$ cubic feet

The weight in pounds is 96,548,885 cubic feet × 167 pounds per cubic foot ≈ 16,123,663,850 pounds, or 8,061,831 tons.

Exercises

Communicate

1. Define *pyramid*. Is there any type of pyramid in which more than one face could be considered the base?

2. Explain how to find the surface area of a regular pyramid.

3. Explain how to find the volume of a pyramid.

Reteaching the Lesson

COOPERATIVE LEARNING Have cooperative groups construct models of pyramids from straws or connecting sticks and use the models to summarize the lesson for each other. Ask them to write sample questions for a quiz. Collect the questions and use them for review.

4. In a regular pyramid, which is larger, the height or the slant height? Explain your reasoning.

5. Explain the relationship between the volume of a pyramid and the volume of a prism with the same base and height.

These strings of lights form the lateral edges of a 39-gonal pyramid.

Assess

Selected Answers
Exercises 6–9, 11–55 odd

ASSIGNMENT GUIDE	
In Class	1–9
Core	11–33 odd, 35–43, 45
Core Plus	25–33 odd, 35–47
Review	48–55
Preview	56–60

✎ Extra Practice can be found beginning on page 818.

Error Analysis

Students may confuse the formulas for surface area and volume. Encourage students to double-check all answers and be sure that their solutions provide answers to the questions in the exercises.

Guided Skills Practice

Determine the surface area of each regular pyramid. *(EXAMPLE 1)*

6. 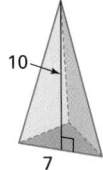   126.22 units²

7. 132 units²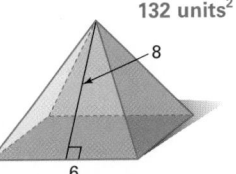

APPLICATION

8. **CONSTRUCTION** The roof of a gazebo is a regular decagonal pyramid with a base edge of 4 ft and a slant height of 7 ft. Find the surface area of the roof. *(EXAMPLE 2)* 140 ft²

9. **CULTURAL CONNECTION: AFRICA** The pyramid of Khafre is a regular square pyramid with a base edge of 708 ft and a height of 471 ft. It is constructed of limestone, which weighs approximately 167 pounds per cubic foot. Estimate the weight of the pyramid. *(EXAMPLE 3)* 13,142,640,826 pounds

Practice and Apply

Draw a net for each regular pyramid named below.

10. a square pyramid

11. a triangular pyramid

12. a pentagonal pyramid

13. a hexagonal pyramid

Find the surface area of each regular pyramid with side length *s* and slant height ℓ given below. The number of sides of the base is given by *n*.

14. $s = 8$ 135.71 units²
    $\ell = 9$
    $n = 3$

15. $s = 6$ 120 units²
    $\ell = 7$
    $n = 4$

16. $s = 10$ 619.81 units²
    $\ell = 12$
    $n = 6$

Find the volume of each rectangular pyramid.

17.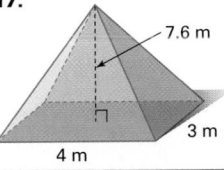

18.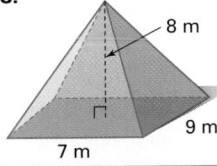

19.

10.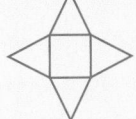

11.

17. 30.4 m³

18. 168 m³

19. 468 in³

12.

13.

ALGEBRA In Exercises 31–34, students set up equations with the formulas for surface area and volume and solve for the height, *h*.

20. $\frac{385}{3}$ units^3

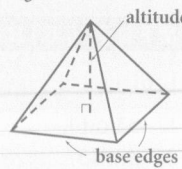

altitude

base edges

21. $\frac{224}{3}$ units^3

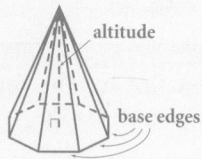

altitude

base edges

22. 100 units^3

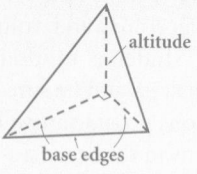

altitude

base edges

23. $\frac{64\sqrt{2}}{3}$ units^3

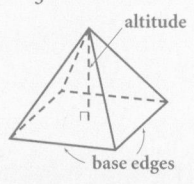

altitude

base edges

In Exercises 20–23, draw a diagram of each pyramid and label the base edges and the altitude. Find the volume of each pyramid. Give exact answers.

20. a rectangular pyramid with a 5×7 base and a height of 11

21. an octagonal pyramid with a base area of 16 and a height of 14

22. a right triangular pyramid with base edges of 5, 12, and 13, and a height of 10

23. a square pyramid with a base edge of 4 and a height equal to the diagonal of the base

Use the diagram of the pyramid below for Exercises 24–30.

24. 49 units^2

24. Find the area of the base, *BCDE*.

Find the area of each lateral face.

25. $\triangle ABC$ 17.5 units^2

26. $\triangle ACD$ $\approx$ 20.41 units^2

27. $\triangle ADE$ $\approx$ 14.85 units^2

28. $\triangle AEB$ $\approx$ 12.62 units^2

29. Find the surface area of the pyramid.   114.38 units^2

30. Find the volume of the pyramid.   49 units^3

Algebra

Find the height of each pyramid described below.

31. a pentagonal pyramid with a base area of 24 square units and a volume of 104 cubic units   13 units

32. a regular square pyramid with a base edge of 10 units and a volume of 500 cubic units   15 units

33. a regular triangular pyramid with a base perimeter of 12 units and a volume of 8 cubic units $\approx$ 3.46 units

34. a regular hexagonal pyramid with a base edge of 2 units and a slant height of 2 units   1 unit

35. 5
36. 8
37. 5
38. 6
39. 10
40. 6
41. $n + 1$
42. $2n$
43. $n + 1$

Copy and complete the table below for the vertices, edges, and faces of pyramids. For each entry in the last row, explain your reasoning.

Number of sides of base, *n*	Number of vertices, *V*	Number of edges, *E*	Number of faces, *F*
3	4	6	4
4	**35.** ?	**36.** ?	**37.** ?
5	**38.** ?	**39.** ?	**40.** ?
n	**41.** ?	**42.** ?	**43.** ?

TABLE PROOF

44. Use the last row of the table above to prove that $V - E + F = 2$ for all pyramids. $V - E + F = (n + 1) - 2n + (n + 1) = n - 2n + n + 1 + 1 = 2$

45. Find the volume and surface area of the rectangular pyramid below.

46. CONSTRUCTION The entrance of the Louvre museum in France is a square pyramid with a base area of 225 m² and a height of 15 m. What is the volume of the pyramid? **1125 m³**

47. CONSTRUCTION How much glass would it take to cover the pyramidal entrance of the Louvre? (Ignore the trapezoidal doorway and do not include the floor.) **503.1 m²**

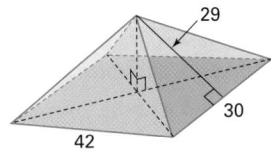

Look Back

48. An angle measures 51°. What is the measure of its complement? What is the measure of its supplement? *(LESSON 1.3)* **39°; 129°**

Quadrilateral ABCD has vertices at A(0, 0), B(5, 0), C(7, 6), and D(2, 6). *(LESSONS 3.8 AND 5.6)*

49. What type of special quadrilateral is *ABCD*? Prove your answer.

50. Find the perimeter of the quadrilateral.

51. Find the area of the quadrilateral. **30 units²**

52. Give the vertices of a quadrilateral with the same area as *ABCD* but a different perimeter. **Sample answer: A(0, 0), B(5, 0), C(5, 6), D(0, 6)**

Find the area of the shaded region in each figure. *(LESSONS 5.1 AND 5.3)*

53.

54.

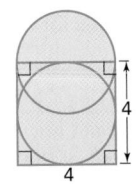

55.

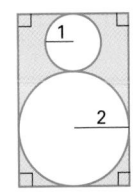

Technology

Students can use a scientific calculator for most of the exercises. Establish rules for rounding. The provided answers are rounded to the nearest hundredth. Some students may wish to create calculator programs for finding surface area and volume.

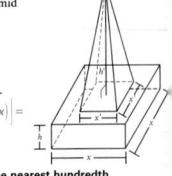

Student Technology Guide

NAME _____ CLASS _____ DATE _____

Student Technology Guide

7.3 Surface Area and Volume of Pyramids, page 1

Suppose that you and some classmates have been asked to participate in designing a new park monument–a tall pyramid sitting on top of a solid rectangular base. The following relationships exist among the dimensions:

$$x' = \tfrac{1}{2}x \qquad h' = 10h \qquad h = \tfrac{1}{4}x$$

Your job is to use the diagram at the right to carry out the calculations needed for the construction of the monument. Recall the formulas below.

1.b. $V_p = \tfrac{1}{3}\left|(10h)\left(\tfrac{1}{2}x\right)\left(\tfrac{1}{2}x\right)\right| =$

Volume of a prism = height × area of base

Volume of a pyramid = $\tfrac{1}{3}$ × height × area of base

Use a calculator as needed. Round your answers to the nearest hundredth.

1. a. Show that V_b, the volume of the base of the monument, can be written as $\tfrac{1}{4}x^3$. $V_b = x \times x \times \tfrac{1}{4}x = \tfrac{1}{4}x^3$

   b. Show that V_p, the volume of the pyramidal top of the monument, can be written as $\tfrac{10}{48}x^3$.   See above. $\tfrac{1}{3}\left(\left(\tfrac{10}{4}x\right)\left(\tfrac{1}{2}x\right)\left(\tfrac{1}{2}x\right)\right) = \tfrac{10}{48}x^3$

   c. Show that V, the volume of the entire monument, can be written as $\tfrac{22}{48}x^3$. $V = V_b + V_p = \tfrac{1}{4}x^3 + \tfrac{10}{48}x^3 = \tfrac{22}{48}x^3$

To evaluate $\tfrac{22}{48}x^3$ on a calculator for a value of x, such as 5, press 22 [÷] 48 [×] 5 [^] 3 [ENTER].

2. Find the volume of the monument for each value of x below.
   a. 10 feet   b. 12 feet   c. 18 feet   d. 20 feet
   458.33 cubic feet   792 cubic feet   2673 cubic feet   3666.67 cubic feet

3. The monument is to be made of concrete. If x is given in feet, then $\tfrac{22}{48}x^3$ gives the volume in cubic feet. Modify your expression so that the volume is in cubic yards. (cubes with an edge length of one yard)   In cubic yards, $V = \tfrac{1}{27}\left(\tfrac{22}{48}x^3\right) = \tfrac{22}{48 \times 27}x^3$

   Find the volume in cubic yards if x is 15.5 feet.   63.21 cubic yards

The visible surfaces of the monument will be covered with copper sheathing. Using the relationships among height and length, you can express the monument's visible surface area as $\left(\tfrac{7 + \sqrt{101}}{4}\right)x^2$. Recall that [2nd] [$x^2$] accesses the square root function.

4. Find the visible surface area of the monument for each value of x.
   a. 10 feet   b. 12 feet   c. 18 feet   d. 20 feet
   426.25 square feet   613.80 square feet   1381.04 square feet   1704.99 square fee

In Exercises 56–60, students use the illustration and the series of questions to show that pyramids with equal heights and bases have the same volume.

Portfolio Activity

The Portfolio Activity can be used as preparation for the Chapter Project or as a separate activity. In the Portfolio Activity on this page, students construct three pyramids from nets and use them to form a cube. They use the model to illustrate the formula for the volume of a pyramid.

 *Look Beyond*

In Exercises 56–60, write a conjecture for each question. If a theorem you know supports your reasoning, state the theorem.

In the proof of the formula for the volume of a pyramid, it was stated that if two pyramids have the same height and bases of the same area, then they have equal volumes. Examine the illustration below.

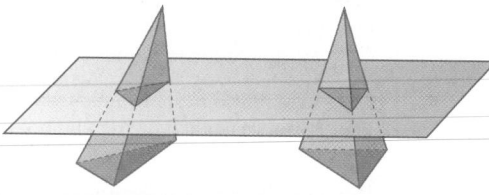

56. In the illustration, the intersecting plane is parallel to the plane of the bases of the pyramid, and it intersects each altitude at its midpoint. How does the intersecting plane seem to divide the lateral edges of the pyramids? **The intersecting plane divides the lateral edges in half.**

57. How do you think the lengths of the sides of the red triangles compare with the lengths of the sides of the bases in each pyramid?

58. How do you think the areas of the red triangles compare with the areas of the bases? How do you think they compare with each other?

59. Do you think the results from Exercise 58 would be the same if the intersecting plane intersected the altitudes at points other than their midpoints?

60. How do your results show that the pyramids have equal volumes? **This is another application of Cavalieri's Principle.**

NETS FOR OBLIQUE PYRAMIDS

The net at right is for an oblique square pyramid.

1. Find BE. Which sides are congruent to $\overline{BE}$?

2. Find FC. Which side is congruent to $\overline{FC}$?

3. Make three copies of the net and fold them into three congruent square pyramids. Fit the three pyramids together to form a cube. How does this illustrate the formula for the volume of a pyramid?

Adjacent edges of the pyramid must be equal in length.

WORKING ON THE CHAPTER PROJECT

You should now be able to complete Activity 1 of the Chapter Project.

internet connect

Portfolio Extension

Go To: go.hrw.com
Keyword:
MG1 IsoTet

57. The lengths of the sides of the red triangles are half the lengths of the sides of the bases.

58. The areas of the red triangles are $\frac{1}{4}$ the areas of the bases. The red triangles have the same area.

59. Yes and no. The areas of the red triangles will be the same no matter where the plane intersects the pyramids as long as it intersects the pyramids in a way that is parallel to the plane of the bases of the pyramid. However the comparison between the bases and the red triangles depends on exactly where the plane intersects the pyramids.

Surface Area and Volume of Cylinders

Objectives

- Define and use a formula for the surface area of a right cylinder.

- Define and use a formula for the volume of a cylinder.

Why *Many everyday objects are cylindrical in shape. You can use the volume of these objects to find how much liquid they will hold.*

APPLICATION
ENGINEERING

The gasoline you buy at a pump is stored in underground tanks. How could you use the dimensions of an underground gasoline tank to estimate the number of car tanks that could be filled from it? (See Example 2, page 455.)

Cylinders

A **cylinder** is a solid that consists of a circular region and its translated image on a parallel plane, with a **lateral surface** connecting the circles.

The faces formed by the circular region and its translated image are called the **bases** of the cylinder.

Bases

An **altitude** of a cylinder is a segment that has endpoints in the planes containing the bases and is perpendicular to both planes. The **height** of a cylinder is the length of an altitude.

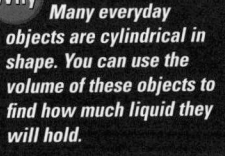

The **axis** of a cylinder is the segment joining the centers of the two bases.

If the axis of a cylinder is perpendicular to the bases, then the cylinder is a **right cylinder**. If not, it is an **oblique cylinder**.

Alternative Teaching Strategy

USING MODELS Students can develop an understanding of the lateral area and surface area of a cylinder by cutting out models that can be assembled into cylinders. Have them cut out a rectangle and two circles whose circumference equals a side of the rectangle. Then tape together the three pieces. For an oblique cylinder, use a nonright parallelogram instead of a rectangle.

The lateral area equals the area of the rectangle (or parallelogram) and the surface area equals the sum of the areas of the three pieces. Ask students to think about whether the choice of a rectangle or parallelogram affects the volume.

Teaching Tip

Emphasize the similarity in the formulas for the lateral areas of a prism and a cylinder. Since the circumference of a circle is its perimeter, the formula for the lateral area of a cylinder is the perimeter of the base times the height.

Cylinders and Prisms

As the number of sides of a regular polygon increases, the figure becomes more and more like a circle.

Similarly, as the number of lateral faces of a regular polygonal prism increases, the figure becomes more and more like a cylinder.

This fact suggests that the formulas for surface areas and volumes of prisms and cylinders are similar.

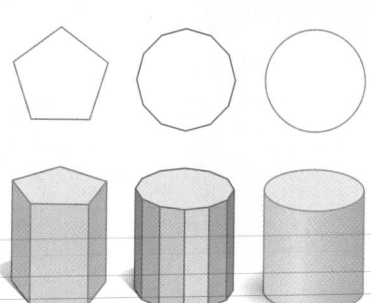

The Surface Area of a Right Cylinder

The surface area of a right cylinder with a radius of r and a height of h can be found by using a net. The net for a right cylinder, shown at right, includes the two circular bases and the lateral surface, which becomes a rectangle. The length of this rectangle is the circumference of the base of the cylinder, or $2\pi r$. The height of the rectangle is the height of the cylinder, or h.

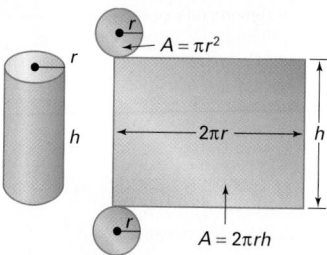

Thus, the lateral area of the cylinder is $2\pi rh$, and the area of each base is πr^2. The surface area of a cylinder is the sum of the lateral area and the base areas.

Surface Area of a Right Cylinder

The surface area, S, of a right cylinder with lateral area L, base area B, radius r, and height h is

$$S = L + 2B \quad \text{or} \quad S = 2\pi rh + 2\pi r^2.$$

7.4.1

EXAMPLE 1

APPLICATION
COINS

A penny is a right cylinder with a diameter of 19.05 millimeters and a thickness of 1.55 millimeters. Ignoring the raised design, estimate the surface area of a penny.

SOLUTION

The radius of a penny is half of the diameter, or 9.525 millimeters. Use the formula for the surface area of a right cylinder.

$$S = 2\pi rh + 2\pi r^2$$
$$S = 2\pi(9.525)(1.55) + 2\pi(9.525)^2 \approx 663.46 \text{ square millimeters}$$

Pennies, at one time made of pure copper, are now made of copper-plated zinc.

Interdisciplinary Connection

SCIENCE Scientists use test tubes, beakers, and graduated cylinders to find the volume of liquids. Have students find the surface area and volume of graduated cylinders. Students can compare the calculated volumes with those indicated by the markings on the cylinders.

Enrichment

For delivery, cylindrical food containers are often packed tightly in boxes that are rectangular prisms. Ask students to find the ratio of the volume of six regular food cans to the volume of the box containing them. $\frac{\pi}{4}$

Volumes of Cylinders

For the following Activity, recall the method you used in Lesson 5.3 to find the area of a circle.

Activity
Analyzing the Volume of a Cylinder

The formula for the area of a circle was found by dividing the circle into sectors and fitting them together to form a shape that was close to that of a rectangle. The same idea can be used to find the volume of a cylinder.

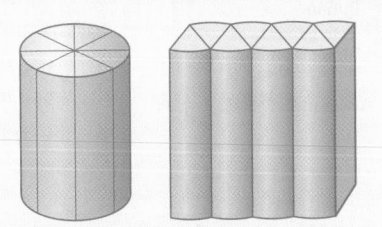

1. Refer to the figure above. What geometric solid does the cylinder approximate when the sections are arranged as they are at right above?

CHECKPOINT ✔ 2. Use the length, width, and height of this arrangement to write a formula for the volume of a cylinder in terms of its radius and height.

CRITICAL THINKING   How would you show that the formula for the volume of an oblique cylinder is the same as the formula for the volume of a right cylinder? Use sketches to illustrate your answer.

Volume of a Cylinder

The volume, V, of a cylinder with radius r, height h, and base area B is
$$V = Bh \quad \text{or} \quad V = \pi r^2 h. \qquad \textbf{7.4.2}$$

EXAMPLE ❷   The tank in the illustration on page 453 has a length of 31 feet $6\frac{1}{2}$ inches and an outer diameter of 8 feet 0 inches. Assuming a wall thickness of about 2 inches, what is the volume of the tank? At 15 gallons per car, how many car tanks could be filled from the storage tank if it starts out completely full of gasoline?

APPLICATION
ENGINEERING

● **SOLUTION**

The tank is not perfectly cylindrical, because of its hemispherical heads, but you can approximate its volume by a slightly shorter cylindrical tank, say, 29 feet long. Subtracting the wall thickness from the dimensions of the tank,

$$V = \pi r^2 h \approx \pi (3.833)^2 (28.667) \approx 1323 \text{ cubic feet}$$

Convert from cubic feet to gallons.

Note: 1 cubic foot ≈ 7.48 gallons.

1323 cubic feet × 7.48 gallons per cubic foot ≈ 9896 gallons

Thus the tank could deliver about $\frac{9896}{15}$, or ≈ 660, 15-gallon fill-ups.

Inclusion Strategies

ENGLISH LANGUAGE DEVELOPMENT   Some students may have difficulty understanding the terminology in this lesson. Have them draw and label several cylinders with each of the following parts: bases, radius, altitude, axis, and lateral surface. Include oblique cylinders. Extend the exercise by having them draw and label a net for each cylinder.

Reteaching the Lesson

COOPERATIVE LEARNING   Bring examples of cylinders to class and have the students work in groups to measure the radius and height of each. Ask each group to calculate the lateral area, surface area, and volume of each cylinder.

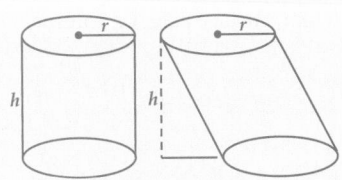

Assess

Selected Answers

Exercises 6–11, 13–47 odd

ASSIGNMENT GUIDE

In Class	1–11
Core	13–35 odd
Core Plus	21–33 odd, 34–39
Review	40–47
Preview	48–51

✐ Extra Practice can be found beginning on page 818.

Error Analysis

Students may confuse the formulas for surface area and volume. Encourage students to double-check all answers and be sure that their solutions provide answers to the questions in the exercises.

Technology

Students can use a scientific calculator for most of the exercises. Some students may wish to create calculator programs for finding surface area and volume.

Math
CONNECTION

ALGEBRA In Exercises 14–25, students set up equations with the formulas for surface area and volume and solve for the radius, height, surface area, or volume.

Exercises

internet connect
Activities Online
Go To: go.hrw.com
Keyword: **MG1 Recording**

Communicate

1. Explain the difference between an altitude and the axis of a cylinder.

2. Is a cylinder a polyhedron? Why or why not?

3. Explain how to find the surface area of a right cylinder.

4. Write the formula for the surface area of a right cylinder in factored form. Which form do you prefer, and why?

5. How are cylinders and prisms alike? How are they different?

Guided Skills Practice

APPLICATION

COINS The dimensions of various coins are given in the table below.

Coin	Diameter	Thickness
nickel	21.21 mm	1.95 mm
dime	17.91 mm	1.35 mm
quarter	24.26 mm	1.75 mm

Find the surface area of each coin. *(EXAMPLE 1)*

6. nickel **7.** dime **8.** quarter

Find the volume of each coin. *(EXAMPLE 2)*

9. nickel **10.** dime **11.** quarter

6. 836.58 mm²
7. 579.82 mm²
8. 1057.86 mm²
9. 688.98 mm³
10. 340.11 mm³
11. 808.93 mm³

Practice and Apply

Draw a net for the right cylinders shown below. Label the dimensions of the net.

12.
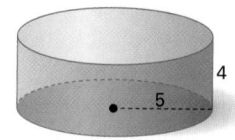
4
5

13.
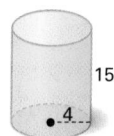
15
4

internet connect
Homework Help Online
Go To: go.hrw.com
Keyword: **MG1 Homework Help** for Exercises 14–28

Find the unknown value for a right cylinder with radius *r*, height *h*, and surface area *S*. Round your answers to the nearest tenth.

14. $r = 5, h = 4, S = $ ___?___

15. $r = 4, h = 15, S = $ ___?___

16. $r = \frac{1}{2}, h = 1, S = $ ___?___

17. $r = 3, h = $ ___?___ $, S = 72\pi$

18. $r = 7, h = $ ___?___ $, S = 550$

19. $r = $ ___?___ $, h = 2, S = 70\pi$

12.

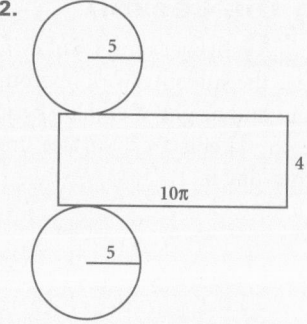

5

10π

4

5

13.

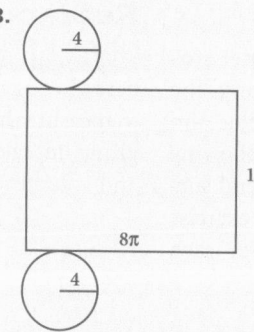

4

15

8π

4

14. 282.7 units²

15. 477.5 units²

16. 4.7 units²

17. 9 units

18. 5.5 units

19. 5 units

Find the unknown value for a right cylinder with radius *r*, height *h*, and volume *V*. Give your answers in exact form.

20. 100π units³

21. 240π units³

22. 3 units

23. $\frac{24}{\pi}$ units

24. 3 units

25. $\frac{4\sqrt{5}}{3}$ units

20. $r = 5, h = 4, V = \underline{\ ?\ }$

21. $r = 4, h = 15, V = \underline{\ ?\ }$

22. $r = 2, h = \underline{\ ?\ }, V = 12\pi$

23. $r = 8, h = \underline{\ ?\ }, V = 1536$

24. $r = \underline{\ ?\ }, h = 6, V = 54\pi$

25. $r = \underline{\ ?\ }, h = 9, V = 80\pi$

26. The surface area of a cylinder is 200 cm². The diameter is equal to the height. Find the radius. **3.26 cm**

27. The volume of a cylinder is 360π mm³ and the height is 10 mm. Find the circumference of the base. **37.7 mm**

28. A semicircular cylinder is formed by cutting a solid cylinder with a radius of 8 ft and height of 10 ft in half along a diameter. Find the volume and surface area of the semicircular cylinder. **1005.31 ft³; 612.39 ft²**

29. How does doubling the height of a cylinder affect the volume?

30. How does doubling the radius of a cylinder affect the volume?

31. How does doubling the height and radius of a cylinder affect the volume?

32. How does doubling the height and radius of a cylinder affect the surface area? **It multiplies the surface area by 4.**

CHALLENGE

33. The volume of a cylinder is equal to its surface area. Prove that the radius and height must both be greater than 2.

CONNECTION

34. **MAXIMUM/MINIMUM** A right cylinder has a volume of 16π cubic units. To find the minimum possible surface area, first solve the volume equation $\pi r^2 h = 16\pi$ for *h* and substitute the expression for *h* in the surface area formula. **r = 2 units; s ≈ 75.40 units²; h = 4 units**

Graph this formula for surface area on a graphics calculator, using *x* for the radius. (Hint: Use a viewing window with $0 \le x \le 10$ and $0 \le y \le 200$.)

Use the trace function of the calculator to estimate the radius for a cylinder with the minimum surface area. What is the height of the cylinder? What is the surface area?

35. **CULTURAL CONNECTION: ASIA** In ancient Mesopotamia, *cylinder seals* were used to make impressions in clay tablets. These seals, which were shaped like cylinders, had carved designs and were rolled on soft clay to create a repeating pattern. Suppose that an archaeologist discovers an impression from a cylinder seal that is 4.7 cm wide. What was the radius of the cylinder seal? **0.75 cm**

29. It doubles the volume of a cylinder.

30. It multiplies the volume by 4.

31. It multiplies the volume by 8.

33.
$$S = V$$
$$2\pi r^2 + 2\pi rh = \pi r^2 h$$
$$2r + 2h = rh$$
$$\frac{2r + 2h}{rh} = 1$$
$$\frac{2}{h} + \frac{2}{r} = 1$$

If either *h* or *r* is less than or equal to 2, then $\frac{2}{h} > 1$ or $\frac{2}{r} > 1$ and $\frac{2}{h} + \frac{2}{r} > 1$, so both *h* and *r* must be greater than 2.

40.

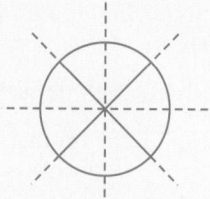

The symbol has 90°, 180° and 270° rotational symmetry.

41.

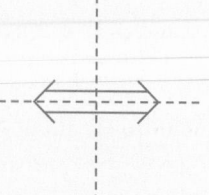

The symbol has 180° rotational symmetry.

42.

The symbol has only trivial rotational symmetry.

43.

The symbol has 180° rotational symmetry.

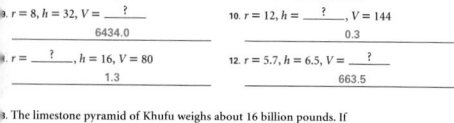
---

36. $\approx$ 38.52

37. 60 times more

36. MARINE BIOLOGY The Giant Ocean Tank in Boston's New England Aquarium, shown at right, is a cylinder that is 23 ft high and has a volume of 200,000 gal. Find the diameter of the tank. (Note: 1 gal $\approx$ 0.134 ft^3)

37. PUBLIC HEALTH A scientist researching the health risks of cigars and filtered cigarettes wishes to compare the amount of tobacco in each. A cigarette is a cylinder with a diameter of 0.16 in. and a length (without the filter) of 2.56 in. A cigar is a cylinder with a diameter of 0.75 in. and a length of 7 in. How many times more tobacco is contained in the cigar than in the filtered cigarette?

38. MANUFACTURING A processing plant needs storage tanks to hold at least half a million gallons of waste water. How many cylindrical tanks must be built to hold the water if each tank has a diameter of 50 ft and a height of 25 ft? (Note: 1 ft^3 = 7.48 gal) **2 tanks**

39. PRODUCT PACKAGING A manufacturer is designing a can that will hold 64 fluid ounces, or 115.5 in.^{3}, and that uses the least amount of materials. Use the method described in Exercise 34 to find the can with the minimum surface area. $r \approx 2.64$ in., $s \approx 131.29$ in^2, $h \approx 5.28$ in.

Look Back

Copy each symbol. Draw all of the lines of symmetry and describe all rotational symmetries of each symbol. *(LESSON 3.1)*

40.   41. ⬌   42. ↖   43. $

Find the surface area and volume of each solid. *(LESSONS 7.2 AND 7.3)*

44. 276 units^2; 280 units^3

45. 337.5 units^2; 421.88 units^3

46. 156 units^2; 72 units^3

47. 360 units^2; 400 units^3

44. right rectangular prism

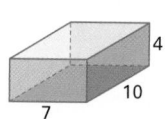

45. cube

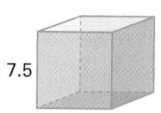

46. right triangular prism

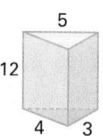

47. regular square pyramid

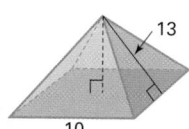

Look Beyond

APPLICATION

CARPENTRY **The cross section of the strongest beam that can be cut from a cylindrical log is shown at right.**

The diameter is divided into three equal segments, and perpendicular segments are drawn as shown. The points where these segments intersect the circle are the vertices of the beam.

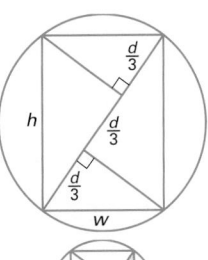

48. Use the Pythagorean Theorem to write an equation relating the quantities h, w, and d for the shaded triangle at right. $d^2 = w^2 + h^2$

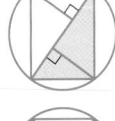

49. Use the Pythagorean Theorem to write an equation relating the quantities w, d, and x for the shaded triangle at right. $w^2 = \frac{d^2}{9} + x^2$

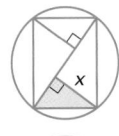

50. Use the Pythagorean Theorem to write an equation relating the quantities h, d, and x for the shaded triangle at right. $h^2 = x^2 + \frac{4d^2}{9}$

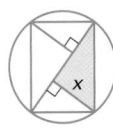

Algebra

51. Use the equations you found in Exercises 48–50 to find the height, h, and width, w, of the strongest beam that can be cut from a log with a diameter of 18 inches. Round your answers to the nearest tenth. **14.7 in.; 10.4 in.**

PORTFOLIO ACTIVITY

NETS FOR OBLIQUE CYLINDERS

A net for the lateral surface of an oblique cylinder has an interesting shape. To examine this, construct an oblique cylinder from a right cylinder.

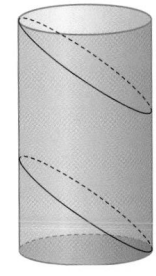

- Create a right cylinder by rolling up a rectangle of heavy paper. Tape the edges together. Your diameter should be at least 2 inches.

- Use a pencil to mark two parallel circles at an angle to the bases. Cut along the circles to form an oblique.

- Unroll the paper to see the net. What do you notice about the sides of the net?

Exercises 48–51 show applications of a cylinder and a right rectangular prism. The Pythagorean Theorem is used to find the area of the base of the prism.

ALTERNATIVE
Assessment

Portfolio Activity

The Portfolio Activity can be used as preparation for the Chapter Project or as a separate activity. In the Portfolio Activity on this page, students create a net for an oblique cylinder.

Student Technology Guide

NAME _____ CLASS _____ DATE _____

Student Technology Guide
7.4 *Surface Area and Volume of Cylinders*

You can make a pipe by constructing an *annulus*, which is a pair of distinct coplanar circles with the same center, and then *extruding* the annulus straight up into three-dimensional space.

The surface area S of a pipe depends on its length, h. It also depends on both the inside radius, s, and the outside radius, r, of the annulus. The volume V of the material used to make the pipe also depends on the thickness, t, of the pipe.

Using a spreadsheet, you can efficiently calculate the surface area and volume of pipes of varying thickness.

Refer to the spreadsheet below. It has data for a pipe with an outside radius of 3 inches and a length of 10 feet (120 inches). The thickness of the pipe varies.

1. Set up columns A, B, C, and D in a spreadsheet like the one shown. Check students' spreadsheets.

2. a. Write a formula in terms of r, t, and h for the surface area of the pipe and the volume of the material used to make the pipe.

   $S =$ ___ $2\pi(2hr - ht + 2rt - t^2)$ ___ $V =$ ___ $\pi h(2rt - t^2)$ ___

   b. Write spreadsheet formulas for S and V, and enter them into cells E2 and F2. Then **FILL DOWN** columns E and F.

   E2: ___ 2*3.14159*(2*D2*A2-2*D2*B2-2*A2*B2-B2^2) ___   F2: ___ 3.14159*D2*(2*A2*B2-B2^2) ___

3. Both S and V are functions of the thickness t. Describe how S and V change as t decreases.
   From the spreadsheet, surface area increases and volume decreases as thickness decreases.

An *open can* is a pipe whose bottom is plugged by a thin cylinder with a radius of s and height of h'.

4. Write formulas for the surface area and volume of an open can.

   $S =$ ___ $2\pi(r^2 + 2hr - ht)$, or $\pi(2r^2 - s^2) + 2\pi(rh - sh + sh')$ ___

   $V =$ ___ $\pi h(2rt - t^2) + \pi h'(r^2 - 2rt + t^2)$, or $\pi(r^2h - s^2h + s^2h')$ ___

5. Use a spreadsheet to explore how S and V change given the information above and $h' = t$.

   Again, surface area increases and volume decreases as thickness decreases.

Prepare

Teach

Why This lesson brings together ideas about surface area and volume to help students understand the connection between pyramids, cylinders, and cones. Motivate students by having them list real-world examples of cones.

Surface Area and Volume of Cones

Why *The properties of cones can be used to model the physical properties of real-world objects that have approximately conic shapes.*

Objectives

- Define and use the formula for the surface area of a cone.

- Define and use the formula for the volume of a cone.

As a volcano erupts and deposits lava and ash over a period of time, it forms a cone. Volcanic cones may be different shapes and sizes, depending on factors such as the rate at which the lava and ash are deposited, how fast the lava cools, etc.

Cones

A **cone** is a three-dimensional figure that consists of a circular **base** and a curved **lateral surface** that connects the base to a single point not in the plane of the base, called the **vertex**.

The **altitude** of a cone is the perpendicular segment from the vertex to the plane of the base. The **height** of the cone is the length of the altitude.

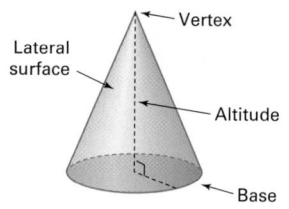

If the altitude of a cone intersects the base of the cone at its center, the cone is a **right cone**. If not, it is an **oblique cone**.

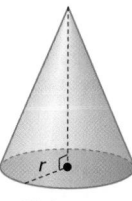

Right cone

Oblique cone

Alternative Teaching Strategy

HANDS-ON STRATEGIES Some students will benefit from building cones from nets. Use nets of right circular cones to emphasize that the radius of a sector that forms a cone equals its slant height. The same net can be used to show that the area of the sector equals the lateral area of the cone. Ask students to find a net for an oblique cone and give its dimensions.

Just as a cylinder resembles a prism, a cone resembles a pyramid. As the number of sides of the base of a regular pyramid increases, the figure becomes more and more like a right cone. The illustrations below suggest that the formulas for the surface areas and volumes of prisms and cylinders are similar.

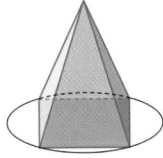 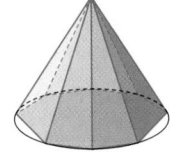 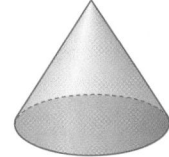

The Surface Area of a Right Cone

YOU WILL NEED

no special tools

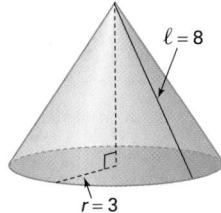

ℓ = 8

r = 3

The surface area of a right cone is found by using a method similar to the one used for a right pyramid. The net for a right cone includes the circular base and the flattened lateral surface, which becomes a portion of a circle known as a *sector*. The radius, ℓ, of the sector is the **slant height** of the cone.

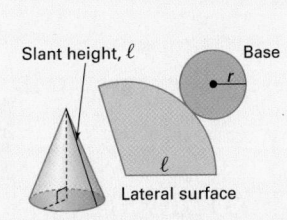

Slant height, ℓ   Base

r

ℓ

Lateral surface

The surface area of a right cone can be found by adding the area of the lateral surface and the area of the base. Use the steps below to find the surface area of the cone at left.

1. The lateral surface of a right cone occupies a part of a circle. The arc from A to B matches the circumference of the base of the cone. The length, c, of this arc is equal to the circumference of the base. Find c for the given cone by using the radius of the base ($r = 3$).

2. Find C, the circumference of the larger circle, for the given cone by using the slant height ($ℓ = 8$) as the radius.

3. Divide c by C. This number tells you what fractional part the lateral surface occupies in the larger circle.

4. Find the area of the larger circle for the given cone by using the slant height ($ℓ = 8$). Multiply this number by the fraction from Step 3. The result is the lateral area, L, of the cone.

CHECKPOINT ✔

5. Find B, the area of the base, for the given cone by using the radius ($r = 3$). Add this to the lateral area, L, of the cone from Step 4. What does your answer represent?

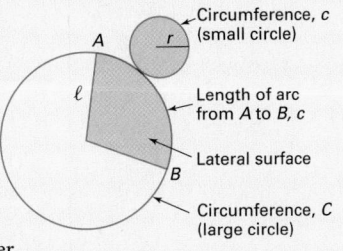

Circumference, c (small circle)

A   r

ℓ

Length of arc from A to B, c

Lateral surface

B

Circumference, C (large circle)

Interdisciplinary Connection

PHYSICAL SCIENCE   Open pits formed during mining operations often have the shape of an inverted cone. Ask students to draw diagrams of why this may be so. Have them observe that sand passing through an hour-glass forms a cone much like the piles of salt, sand, coal, or asphalt piled up by construction trucks.

Activity 2 **Notes**

In this Activity, students find the formula for the surface area of a right cone. Emphasize the connections between the steps in the Activity and the use of variables in the formula for the surface area of a cone. Draw examples of cones on the board or overhead, and use the formula to find their surface area.

For a student worksheet of this Activity and detailed Teacher Notes, see page 130 in the Lesson Activities booklet.

CHECKPOINT ✔
3. The results are the same.

EXAMPLE ❶ Find the surface area of a right cone with the indicated measurements.

● **SOLUTION**

The circumference of the base is $c = 2\pi r = 14\pi$.

The lateral area is a sector of a circular region with circumference $C = 2\pi\ell = 30\pi$.

The portion of the circular region occupied by the sector is $\frac{c}{C} = \frac{14\pi}{30\pi} = \frac{7}{15}$.

Calculate the area of the sector (lateral area).

$$\pi\ell^2 = 225\pi$$

$$L = \frac{7}{15} \cdot 225\pi = 105\pi$$

Calculate the base area and add the lateral area.

$$B = \pi r^2 = 49\pi$$

$$B + L = 49\pi + 105\pi = 154\pi \approx 483.8$$

$\ell = 15$
$r = 7$

YOU WILL NEED

no special tools

CHECKPOINT ✔

Activity 2
The Surface Area Formula for a Right Cone

1. Follow the steps in Example 1 above and in Activity 1 on the previous page to find the surface area of a general right cone. That is, use the variables r, h, and ℓ instead of numerical measures. Your answer will be a formula for the surface area of a right cone.

2. Use algebra to simplify your formula. Show your steps in an organized way and save your work in your notebook.

3. Compare your results with the formula given below. Are they the same? Explain.

Surface Area of a Right Cone

The surface area, S, of a right cone with lateral area L, base of area B, radius r, and slant height ℓ is

$$S = L + B \quad \text{or} \quad S = \pi r\ell + \pi r^2. \qquad \textbf{7.5.1}$$

Many families celebrate special occasions by placing lateral surfaces of cones on their heads.

Enrichment

Write the converse of the following true statement and prove or disprove it: If the altitudes of a cone and cylinder are equal and the diameters of their bases are equal, then the volume of the cone is equal to one-third of the volume of the cylinder.

Inclusion Strategies

ENGLISH LANGUAGE DEVELOPMENT Students may have difficulty reading and understanding mathematics. Suggest that students read this lesson slowly and carefully, compare it with the lesson on pyramids, and write a list of questions they need answered. You may wish to have students discuss the questions in cooperative groups.

The Volume of a Cone

In Lesson 7.3, you found that a pyramid had one-third of the area of a prism with a congruent base and equal height. Now imagine performing a similar experiment with a cone and a cylinder. The result of the experiment should be the same—you can try it—because cones and cylinders are like many-sided pyramids and prisms.

This similarity leads to the formula for the volume of a cone.

Volume of a Cone

The volume, V, of a cone with radius r, height h, and base area B is

$$V = \frac{1}{3}Bh \quad \text{or} \quad V = \frac{1}{3}\pi r^2 h.$$

7.5.2

CRITICAL THINKING

Is the formula for the volume of an oblique cone the same as the formula for the volume of a right cone? What principle justifies your answer? Illustrate your answer with sketches.

EXAMPLE ② A volcanologist is studying a violent eruption of a cone-shaped volcano. The original volcanic cone had a radius of 5 miles and a height of 2 miles. The eruption removed a cone-shaped area from the top of the volcano. This cone had a radius of 1 mile and a height of $\frac{1}{2}$ mile. What percent of the total volume of the original volcano was removed by the eruption?

APPLICATION
GEOLOGY

● **SOLUTION**

Find the volume of the original volcano.

$$V = \frac{1}{3}\pi r^2 h = \frac{1}{3}(5^2)(2) \approx 52.4 \text{ cubic miles}$$

July 22nd, 1980 Eruption of Mount Saint Helens.

Find the volume of the destroyed cone.

$$V = \frac{1}{3}\pi r^2 h = \frac{1}{3}(1^2)(0.5) \approx 0.52 \text{ cubic miles}$$

Find the percent of the original volcano removed by the eruption.

$$\left(\frac{0.52}{52.4}\right)100 \approx 1\%$$

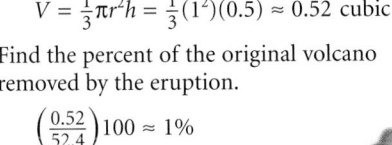

Reteaching the Lesson

COOPERATIVE LEARNING Have students work in cooperative groups and repeat the Activity from Lesson 7.3 (page 447), using cylinders and cones instead of prisms and pyramids. Have the groups express their results as a conjecture for the volume of a cone. Have the students make a list of similarities between prisms, cylinders, pyramids, and cones.

Teaching Tip

Many suppliers of educational materials sell sets of hollow plastic cones, cylinders, pyramids, prisms, etc., with congruent bases and heights. These solids have removable bases so that liquids can be poured into them. If these are available in your classroom, perform the experiment at the top of page 459, using water or sand.

CRITICAL THINKING

The formulas for the volumes of right and oblique cones are the same. This follows from Cavalieri's Principle.

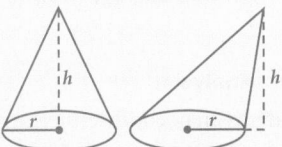

ADDITIONAL
EXAMPLE ②

A sugar cone for ice cream is a cone with a radius of 3 cm and a height of 15 cm. A cake cone looks like a sugar cone with the tip removed. If the volume of a cake cone is 24π cm³, how much larger is the sugar cone than the cake cone? Express your answer as a percent. 187.5%

Selected Answers

Exercises 6–9, 11–53 odd

ASSIGNMENT GUIDE

In Class	1–9
Core	11–35 odd, 36–45
Core Plus	25–35 odd, 36–49
Review	50–54
Preview	55

✐ Extra Practice can be found beginning on page 818.

Error Analysis

Students may confuse the formulas for surface area and volume. Encourage students to double-check all answers and be sure that their solutions provide answers to the questions in the exercises.

Exercises

● Communicate

1. Explain how to find the surface area of a right cone.

2. Explain how to find the volume of a right cone.

3. In a right cone, which is longer, the altitude or the slant height? Explain your reasoning.

4. Slant height is not defined for an oblique cone. Explain why.

5. What happens to the volume of a cone if the radius is doubled? if the height is doubled? if both are doubled?

● Guided Skills Practice

6. The diagram below represents a net for a right cone. Find the surface area of the cone. *(ACTIVITY 1)*

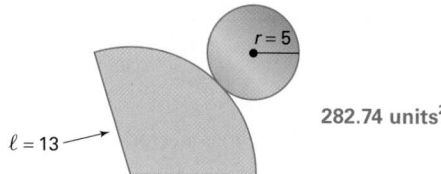

282.74 units²

7. Find the surface area of a right cone with the measures shown below. *(EXAMPLE 1)*

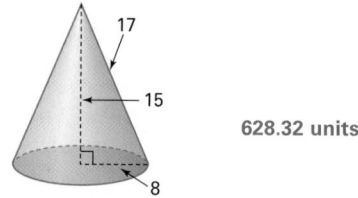

628.32 units²

8. A formula for the surface area of a right cone is
   S (surface area) = L (lateral area) + B (base area).

   Explain how to rewrite the formula by substituting values for L and B in terms of π, r, and ℓ, where r is the radius of the base of the cone and ℓ is the slant height of the cone. *(ACTIVITY 2)* $S = \pi r \ell + \pi r^2$

9. Find the volume of the oblique cone shown below. *(EXAMPLE 2)*

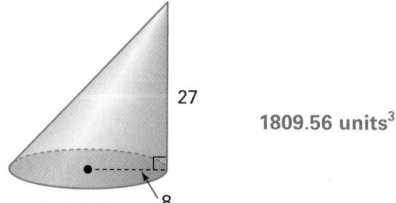

1809.56 units³

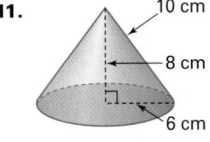

● *Practice and Apply*

Find the surface area of each right cone.

10.
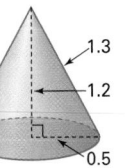
5 cm
4 cm
3 cm

11.
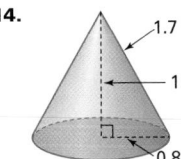
10 cm
8 cm
6 cm

12.
2 in.
1.6 in.
1.2 in.

13.
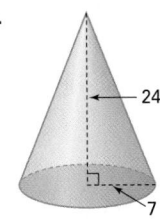
1.3
1.2
0.5

14.
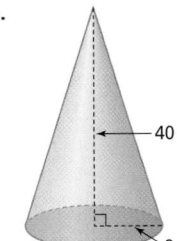
1.7
1.5
0.8

15.
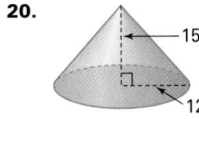
5.3
2.8
4.5

16. A right cone has a radius of 3 in. and a height of 4 in. What is the slant height of the cone? **5 in.**

17. A right cone has a radius of 5 in. and a height of 6 in. What is the slant height of the cone? **7.81 in.**

Find the surface area of each right cone.

18. 703.72 units²

19. 1413.72 units²

20. 1176.57 units²

21. 3171.72 units²

22. 24.88 units²

23. 40.72 units²

18.
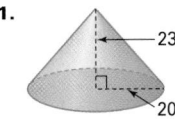
24
7

19.
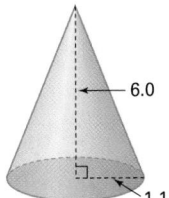
40
9

20.
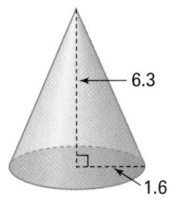
15
12

21.
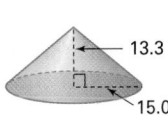
23
20

22.
6.0
1.1

23.
6.3
1.6

Find the volume of each cone.

24. 3133.74 units³

25. 2309.07 units³

26. 1251.61 units³

24.
13.3
15.0

25.
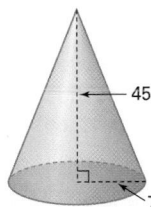
45
7

26.
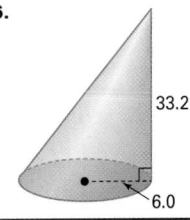
33.2
6.0

Technology

Students can use a scientific calculator for most of the exercises. Some students may wish to create calculator programs for finding surface area and volume.

10. 75.40 cm²

11. 301.59 cm²

12. 12.06 in²

13. 2.83 units²

14. 6.28 units²

15. 138.54 units²

30. A cone with $r = 20$ and slant height $= \sqrt{30^2 + 20^2} = \sqrt{1300}$ is formed; 3522.07 units^2

27. 19,396.19 units^3

28. 5875.62 units^3

29. 5598.32 units^3

27.

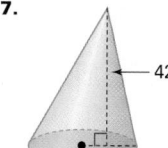

28.

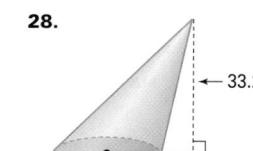

29.

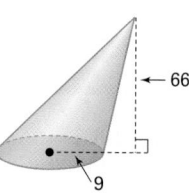

30. If trianguar region ABC below is rotated in three dimensions about $\overleftrightarrow{AB}$, what solid figure is formed? Find its surface area.

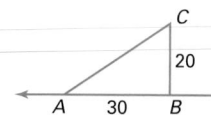

31. 6031.86 units^3; 2513.27 units^3

32. $\frac{2}{\pi}$

31. A right triangle has sides of 10, 24, and 26. Rotate the triangle about each leg, and find the volume of the figures formed.

32. A regular square pyramid is inscribed in a right cone of the same height. The radius of the base and the height of the cone are each 10 cm. What is the ratio of the volume of the pyramid to the volume of the cone?

33. A right cone has a radius of 5 in. and a surface area of 180 in.^{2} What is the slant height of the cone? 6.46 in.

34. Find the surface area of a right cone whose base area is 25π cm^2 and whose height is 13 cm. 297.33 cm^2

35. An oblique cone has a volume of 1000 cm^3 and a height of 10 cm. What is the radius of the cone? 9.77 cm

CONNECTIONS

MAXIMUM/MINIMUM A right cone has a slant height of 10. What are the radius and height of a cone with the maximum volume? Copy and complete the chart for Exercises 36–42.

36. 41.04

37. $\sqrt{91}$

38. 89.91

39. $\sqrt{84}$

40. 153.56

41. $\sqrt{75}$

42. 226.72

Radius	Height	Volume
1	$\sqrt{99}$	$\frac{1}{3}\pi \times 1^2 \times \sqrt{99} \approx 10.4$
2	$\sqrt{96}$	**36.** ?
3	**37.** ?	**38.** ?
4	**39.** ?	**40.** ?
5	**41.** ?	**42.** ?

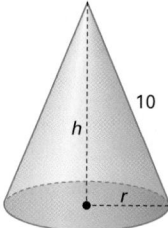

TECHNOLOGY Use a graphics calculator and the information from Exercises 36–42 to graph the values of the radius on the *x*-axis and the values of the volume on the *y*-axis. Set the viewing window so that Ymax is at least 410. Trace the graph to find the following:

43. the radius (x-value) that produces the largest volume 8.2 units

44. the largest volume 403.1 units^3

Practice

NAME ________ CLASS ________ DATE ________

Practice
7.5 Surface Area and Volume of Cones

n Exercises 1 and 2, find the surface area and volume of each ight cone. Express your answers in terms of π.

1.

$S = 216\pi$; $V = 324\pi$

2.

$S = 450\pi$; $V = 1080\pi$

Find the surface area of each right cone with radius r, height h, and slant height ℓ. Express your answers in terms of π.

3. $r = 5.1, h = 2, \ell = 5.5$ — $S = 54.06\pi$

4. $r = 13, h = 17, \ell = 21.4$ — $S = 447.2\pi$

5. $r = 4.2, h = 3.8, \ell = 5.7$ — $S = 41.58\pi$

6. $r = 1.1, h = 3, \ell = 3.2$ — $S = 4.73\pi$

Find the volume of each right cone with radius r and height h. Express your answers in terms of π.

7. $r = 0.5, h = 4$ — $V = \frac{1}{3}\pi$

8. $r = 15, h = 20$ — $V = 1500\pi$

9. $r = 24, h = 30$ — $V = 5760\pi$

10. $r = 8.2, h = 9$ — $V = 201.72\pi$

11. The volume of a right cone is 1680π cm^3. The radius of its base is 12 cm. What is the height of the cone? 35 cm

12. The surface area of a right cone is 300π cm^2. Its slant height and the diameter of its base are equal. Find its radius. 10 cm

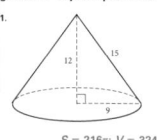

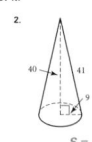

45. RECREATION An oblique conical tower is to be built at an amusement park, as shown in the architect's sketch below. The vertex is to be directly above the edge of the circular base. The tower is 150 ft tall, and the diameter of the base is 60 ft. What is the volume of the cone? ≈ **141,372 ft³**

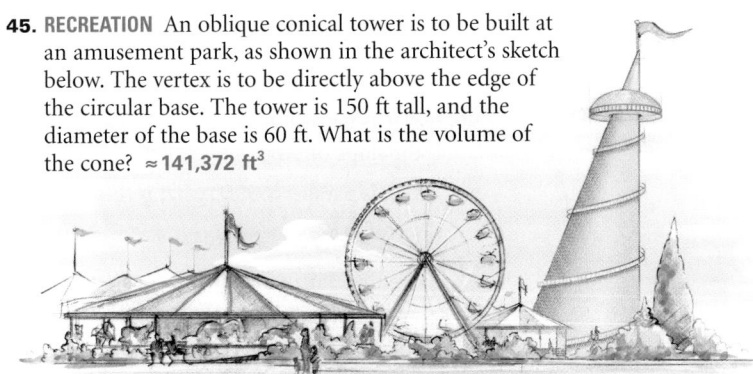

46. SMALL BUSINESS A businesswoman is selecting glasses for her new ice-cream parlor. Estimate the volumes of the two glasses shown at right.

cylindrical glass: **594 cm³**;
conical glass: **271 cm³**

7.5 cm

7.2 cm

18.4 cm

14.6 cm

MANUFACTURING Cone-shaped paper cups can be manufactured from patterns shaped like sectors of a circle. The figures below show two patterns. The first has a straight angle (180°). The second has an angle of 120°. Use the information provided in the diagrams for Exercises 47–49.

47. 56.55 cm²;
   37.70 cm²

47. If each sector has a radius of 6 cm, what will be the area of each sector? (Hint: What fraction of the circle is included in the sector?)

48. Find the volume of the cup formed by the sector with the straight angle. (Hint: The length of the arc of the sector is equal to the circumference of the base of the cone.) **48.97 cm³**

49. Find the volume of the cup formed by the sector with a 120° angle.

23.70 cm³

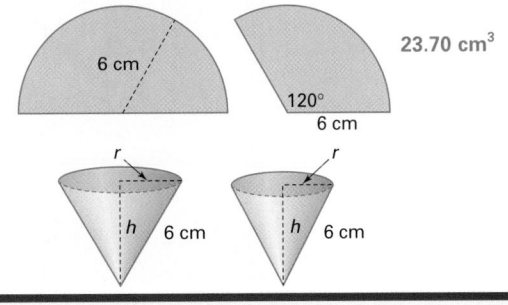

6 cm

120°

6 cm

r

h   6 cm

r

h   6 cm

Student Technology Guide

NAME _____ CLASS _____ DATE _____

Student Technology Guide
7.5 Surface Area and Volume of Cones

The following example shows how to use a calculator to find the volume of a cone:

Example: A cone has a height of 18 inches and a base radius of 8 inches. What is the volume of the cone?

Evaluate $V = \frac{1}{3}\pi r^2 h$, or $V = \frac{\pi}{3} r^2 h$.

Press [2nd] [^] [÷] 3 [×] 8 [x²] [×] 18 [ENTER].

The volume of the cone is about 1206.37 cubic inches.

`π/3*8²*18`
`        1206.371579`

To the nearest tenth of a cubic unit, find the volume of each cone described below.

1. $r = 10$ and $h = 20$ _____ 2094.4   2. $r = 10.5$ and $h = 200$ _____ 23,090.7

3. $r = 24$ inches
   and $h = 24$ inches _____ 14,476.5   4. $r = 10$ feet
   and $h = 1$ foot _____ 104.7

A *frustum* of a cone is the section that is left when the top of the cone is cut off by a plane parallel to the plane containing the cone's base. In the figure at right, the shaded frustum results when a cone whose height is h' units is removed. The radius and height of the removed cone are proportional to the radius and height of the full cone.

Example: A cone has a height of 18 inches and a base radius of 8 inches. The top quarter of the cone is removed. Find the volume of the frustum.

Evaluate $V = \frac{1}{3}\pi r^2 h - \frac{1}{3}\pi(r')^2 h'$, given $r' = \frac{1}{4}r$ and $h' = \frac{1}{4}h$.

• Evaluate $V = \frac{1}{3}\pi(8^2)(18) - \frac{1}{3}\pi\left(\frac{8}{4}\right)^2\left(\frac{18}{4}\right)$, or $\frac{\pi}{3}\left[18(8^2) - \left(\frac{8}{4}\right)^2\left(\frac{18}{4}\right)\right]$.

Press [2nd] [^] [÷] 3 [×] [(] 18 [×] 8 [x²] [−] [(] 8
[÷] 4 [)] [x²] [×] [(] 18 [÷] 4 [)] [)] [ENTER].

The volume of the frustum is about 1187.5 cubic inches.

A cone has a height of 18 inches and a base radius of 8 inches. Approximate the volume of each frustum if the following parts of the cone are removed:

5. top half _____ 1055.6 in.³   6. top tenth _____ 1205.2 in.³   7. top two-thirds _____ 848.9 in.³

8. top 5% _____ 1206.2 in.³   9. top 10% _____ 1205.2 in.³   10. top 90% _____ 326.9 in.³

Exercise 55 applies concepts learned in the lesson to the scale factor of a model train engine. This exercise looks ahead to the study of similar figures.

 **Look Back**

Refer to the graph below for Exercises 50 and 51. *(LESSON 3.8)*

50. If $\overleftrightarrow{AB}$ is perpendicular to $\overleftrightarrow{CD}$, what is the slope of $\overleftrightarrow{AB}$? $\frac{5}{3}$

51. If $\overleftrightarrow{AB}$ is parallel to $\overleftrightarrow{CD}$, what is the slope of $\overleftrightarrow{AB}$? $-\frac{3}{5}$

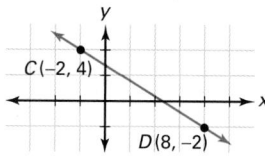

52. Prove that a diagonal divides a parallelogram into two congruent triangles. *(LESSON 4.6)*

53. Find the distance between $(2, 4)$ and $(-5, 8)$ in a coordinate plane. *(LESSON 5.6)* 8.06 units

CONNECTION

54. ≈ 2.35 × 2.35 in.
 ≈ 2.35 × 13.30 × 7.30 in.
 ≈ 228.16 in.³

54. MAXIMUM/MINIMUM A manufacturing company makes cardboard boxes of varying sizes by cutting out square corners from rectangular sheets of cardboard that are 12 in. wide and 18 in. long. The cardboard is then folded to form a box. If a customer wants a box with the greatest possible volume for the given sheet of cardboard, what should be the dimensions of the squares that are cut from the corners? What will be the dimensions of the box? What will be the volume of the box? *(LESSON 7.1)*

 Look Beyond

APPLICATION

55. SCALE MODELS An "HO gauge" model train engine is $\frac{1}{87}$ of the size of a real train engine. If the model's width is 1.5 in. and its length is 5.5 in., what is the length of the real engine? 478.5 in., or 39.875 ft

52. Sample answer:
Given: Parallelogram $ABCD$ with diagonal $\overline{AC}$
Prove $\triangle ADC \cong \triangle CBA$

1. $ABCD$ is a parallelogram with diagonal $\overline{AC}$
2. $\overline{AD} \cong \overline{CB}$, $\overline{DC} \cong \overline{AB}$

3. $\overline{AC} \cong \overline{AC}$
4. $\triangle ADC \cong \triangle CBA$

1. Given
2. Opposite sides of a parallelogram are congruent
3. Reflexive Property of Congruence
4. SSS Postulate

Surface Area and Volume of Spheres

The balloon is being inflated by fanning hot air into the cloth "envelope."

Objectives

- Define and use the formula for the surface area of a sphere.

- Define and use the formula for the volume of a sphere.

Why *How much cloth do you think is needed to make a hot-air balloon? You can use the formula for a sphere to estimate the balloon's surface area.*

The Volume of a Sphere

A **sphere** is the set of all points in space that are the same distance, r, from a given point known as the center of the sphere. To find the formula for the volume of a sphere, we first show that a sphere has the same volume as a cylinder with a double cone cut out of it. Then, by using the formulas for cones and cylinders, we derive the formula for the volume of each figure. The discussion that follows begins with a numerical calculation before moving on to the general case.

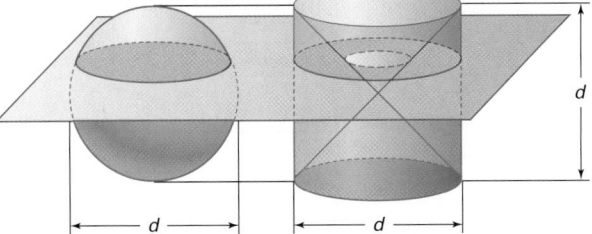

The height and diameter of the cylinder are the same as the sphere's diameter.

Suppose that the sphere has a diameter of 50 units and that the cylinder has a height and diameter of 50 units. Assume that a blue plane cuts through the sphere and the cylinder 10 units above their centers. You can prove that the two red cross sections—the circular region in the sphere and the **annulus** (the ring-shaped figure) in the cylinder—are both equal to 525π square units.

Teach

Why Spherical objects are so common that students should be able to understand their geometry thoroughly. Emphasize that spheres are common objects by having each student in the room name an example of a sphere.

Alternative Teaching Strategy

USING VISUAL MODELS Bring in various examples of spheres to class. Have students measure the circumference of each sphere and use the formula for circumference to find the radius. Students can also find the volume of each sphere.

Teaching Tip

Review the Pythagorean Theorem and the conditions of Cavalieri's Principle before deriving the formula for the volume of a sphere.

There are three red right triangles in the figure below: one in the sphere and two in the cylinder with the cones removed. The triangle in the sphere can be solved for b, the radius of the disc. The two triangles in the cylinder are isosceles and have known side lengths, as you can convince yourself by studying the figure. (The blue cutting plane is parallel to the bases of the hollow cones in the cylinder.) The information from the triangles is used to calculate the areas of the cross sections.

In the cylinder, the acute angles of the large red triangle are congruent. You can show that the acute angles of the small triangle are congruent, as well.

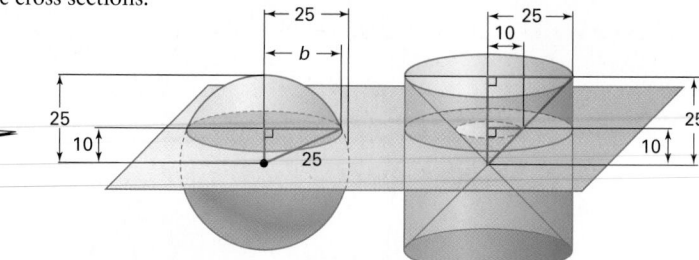

Area of the circle in the sphere

$b^2 + 10^2 = 25^2$

$b^2 + 100 = 625$

$b^2 = 525$

$b = \sqrt{525}$

$\text{area} = \pi b^2 = \pi \left(\sqrt{525}\right)^2$

$= 525\pi$ square units

Area of the annulus

area of large circle − area of small circle

$= \pi 25^2 - \pi 10^2$

$= \pi(25^2 - 10^2)$

$= \pi(625 - 100)$

$= \pi(525)$

area $= 525\pi$ square units

PROOF

Thus, the areas are equal. Next, *generalize* the procedure to get the formulas for a sphere and a cylinder of radius r. Let r be the radius of the sphere and y be the distance from the centers of the figures to the blue cutting plane.

Area of the circle in the sphere

$b^2 + y^2 = r^2$

$b^2 = r^2 - y^2$

$b^2 = \sqrt{r^2 - y^2}$

$\text{area} = \pi\left(\sqrt{r^2 - y^2}\right)^2 = \pi(r^2 - y^2)$

Area of the annulus

area of large circle − area of small circle

$= \pi r^2 - \pi y^2$

area $= \pi(r^2 - y^2)$

This shows that the formulas are true for all planes parallel to the bases of each figure and for all values of y.

The corresponding cross sections have equal areas.

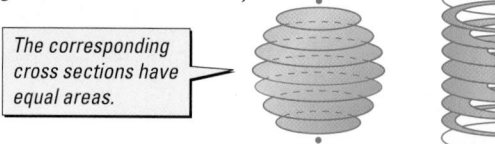

Therefore, the conditions of Cavalieri's Principle are satisfied. The volume of the sphere equals the volume of the cylinder with the double cones removed.

$V(\text{sphere}) = V(\text{cylinder}) - V(\text{cones})$

$V(\text{sphere}) = \pi r^2(2r) - 2\left(\frac{1}{3}\pi^2\right)(r)$

$= 2\pi r^3 - \frac{2}{3}\pi r^3$

$= \frac{4}{3}\pi r^3$

Interdisciplinary Connection

PHYSICAL SCIENCE Spheres occur often in the natural world. Air bubbles trapped in water are spherical and so are soap bubbles. If students have ever seen footage of astronauts in zero gravity, they may have noticed that uncontained liquids form floating spheres. Discuss with students what physical properties make these objects assume the shape of spheres as opposed to some other shape.

CRITICAL THINKING When the cutting plane in the proof of the volume formula cuts through the center of the sphere and the center of the cylinder, what happens to the annulus? How does this affect the calculations? Explain.

EXAMPLE ❶ The envelope of a hot-air balloon has a radius of 27 feet when fully inflated. Approximately how many cubic feet of hot air can it hold?

SOLUTION

$V = \frac{4}{3}\pi r^3$

$\quad = \frac{4}{3}\pi (27)^3$

$\quad = \frac{4}{3}(19{,}683)\pi$

$\quad = 26{,}244\pi$ cubic feet $\approx 82{,}488$ cubic feet

The Surface Area of a Sphere

APPLICATION
CARTOGRAPHY

In previous lessons, you analyzed the surface areas of three-dimensional figures by unfolding them to form a net on a flat surface. But, as mapmakers know, a sphere cannot be unfolded smoothly onto a flat surface.

The most common map of the world uses a Mercator projection of Earth's surface onto a flat plane. On this kind of map, the landmasses near the North and South Poles, such as Greenland and Antartica, appear to be much larger in relation to other landmasses than they actually are.

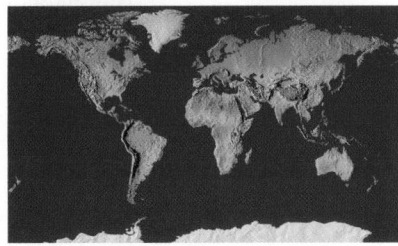

This map is a Mercator projection. You will learn more about projecting a sphere onto a flat surface in Lesson 11.7.

The formula for the surface area of a sphere can nevertheless be derived by using some clever techniques.

Enrichment

From physics, students learn that a soap bubble has tension at its surface that holds in its volume of air. If it does not burst, the bubble changes from a sphere to a hemisphere when it lands on a table. Have students determine how much larger the diameter of the hemisphere is.
$2^{\frac{1}{3}}$ or $\sqrt[3]{2}$ **times as larger**

Inclusion Strategies

GUIDED RESEARCH Some students may be interested in the chemical connection between the volume of a metal sphere and its density. The density of an object is the ratio of its mass to its volume, and so mass = density × volume. Have students find the mass of some common metal balls, given their density and volume.

Teaching Tip

Cut a piece of paper into triangles and glue them onto a large ball to completely cover it. Verify the area formula for a sphere by comparing the area of the wrapping paper with 4π times the square of the radius of the ball.

Cooperative Learning

Divide the students into five cooperative groups and assign one of the Communicate exercises to each group. Have the groups present their answers to the whole class.

To derive the formula for the surface area of a sphere, begin with an approximation. Imagine the surface of a sphere as a large number of polygons, as in the geodesic dome in the photo at right. The smaller these polygons are, the more closely they approximate the surface of a sphere.

Doppler radar tower

Consider each polygon to be the base of a pyramid with its vertex at the center of the sphere. The volumes of these pyramids added together will approximate the volume of the sphere. The height of each pyramid is the radius of the sphere. Therefore, the volume of each pyramid is $\frac{1}{3}Br$, where B is the area of the base of the pyramid and r is the radius of the sphere.

$$\text{Volume of a sphere} \approx \frac{1}{3}B_1r + \frac{1}{3}B_2r + \cdots + \frac{1}{3}B_nr$$
$$\approx \frac{1}{3}r(B_1 + B_2 + \cdots + B_n)$$

If the total area of the bases of the pyramids is assumed to equal the surface area of the sphere, S, then the volume, V, of the sphere can be written as

> The larger the number of pyramids, the closer the approximation will be to the actual volume of the sphere. The approximation is said to be exact "in the limit."

$$V = \frac{1}{3}r(S).$$

Solve for S to get

$$S = \frac{3V}{r}.$$

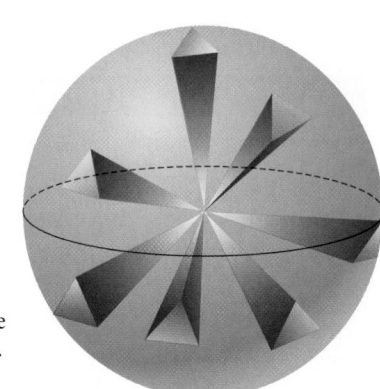

You now have a formula for the surface area of a sphere in terms of its volume. Now substitute in the formula for volume from the previous page.

$$S = \frac{3V}{r}$$
$$= \frac{3\left(\frac{4}{3}\pi r^3\right)}{r}$$
$$= 4\pi r^2$$

Surface Area of a Sphere

The surface area, S, of a sphere with radius r is
$$S = 4\pi r^2.$$

7.6.2

Reteaching the Lesson

INVITING PARTICIPATION Have one student draw a sphere on the board or overhead. Have another student give the measure of its radius or diameter. Ask a third student to calculate its surface area and a fourth student to calculate its volume. Repeat the process until all students have participated.

| EXAMPLE | 2 |

APPLICATION
HOT-AIR BALLOONING

The envelope of a hot-air balloon is 54 feet in diameter when inflated. The cost of the fabric used to make the envelope is $1.31 per square foot. Estimate the total cost of the fabric for the balloon envelope.

SOLUTION

First estimate the surface area of the inflated balloon envelope. The balloon is approximately a sphere with a diameter of 54 feet, so the radius is 27 feet.

$$S = 4\pi r^2$$
$$= 4\pi(27)^2$$
$$= 4(729)\pi$$
$$= 2916\pi \approx 9160.9 \text{ square feet}$$

Now multiply the surface area of the fabric by the cost per square foot to find the approximate cost of the fabric.

9160.9 square feet × $1.31 per square foot ≈ $12,000

Assess

Exercises

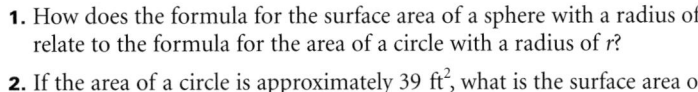

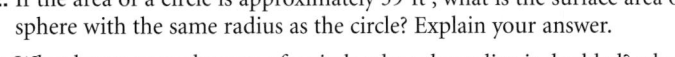

Communicate

1. How does the formula for the surface area of a sphere with a radius of *r* relate to the formula for the area of a circle with a radius of *r*?

2. If the area of a circle is approximately 39 ft², what is the surface area of a sphere with the same radius as the circle? Explain your answer.

3. What happens to the area of a circle when the radius is doubled? when it is tripled?

4. What happens to the surface area of a sphere when the radius is doubled? when it is tripled?

5. What happens to the volume of a sphere when the radius is doubled? when it is tripled?

Guided Skills Practice

6. Find the approximate volume of a sphere that has a radius of 40 ft. Round your answer to the nearest whole number. *(EXAMPLE 1)* 268,083 ft³

7. Find the surface area of a sphere that has a diameter of 48 ft. Find the cost of constructing the sphere if the cost for materials and labor is $1.50 per square foot of surface area. Round your answer to the nearest whole dollar. *(EXAMPLE 2)* 7238.23 ft²; $10,857

8. a. 64π units^2; $\frac{256}{3}\pi$ units^3
   b. 201.06 units^2;
       268.08 units^3

9. a. 256π units^2; $\frac{2048}{3}\pi$ units^3
   b. 804.25 units^2;
       2144.66 units^3

10. a. 6724π units^2;
       $\frac{275,684}{3}\pi$ units^3
   b. 21,124.07 units^2;
       288,695.61 units^3

11. a. 4356π units^2;
       $47,916\pi$ units^3
   b. 13,684.78 units^2;
       150,532.55 units^3

12. a. 719.3124π units^2;
       3215.33π units^3
   b. 2259.79 units^2;
       10,101.25 units^3

13. a. 597.3136π units^2;
       2433.06π units^3
   b. 1876.52 units^2;
       7643.68 units^3

14. a. 324π units^2; 972π units^3
   b. 1017.88 units^2;
       3053.63 units^3

15. a. 256π units^2; $\frac{2048}{3}\pi$ units^3
   b. 804.25 units^2;
       2144.66 units^3

Practice

internet connect

Homework Help Online
Go To: **go.hrw.com**
Keyword:
MG1 Homework Help
for Exercises 8-31

Practice and Apply

In Exercises 8–19, find the surface area and volume of each sphere, where *r* is the radius of the sphere and *d* is the diameter. Express your answers in two ways: (a) as an exact answer in terms of π, and (b) as an approximate answer rounded to the nearest hundredth.

8. $r = 4$	**9.** $r = 8$	**10.** $r = 41$
11. $r = 33$	**12.** $r = 13.41$	**13.** $r = 12.22$
14. $d = 18$	**15.** $d = 16$	**16.** $d = 22.34$
17. $d = 11.48$	**18.** $r = 12.33$	**19.** $r = 99.98$

In Exercises 20–25, find the area and volume of each sphere with the given radius or diameter. Give exact answers in terms of π and a variable.

20. $r = x$	**21.** $r = 2y$	**22.** $d = 12x$
23. $d = 4y$	**24.** $r = \frac{x}{2}$	**25.** $r = \frac{y}{4}$

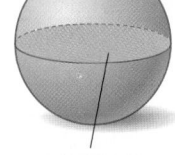

Area of cross section, *A*

In Exercises 26–31, find the surface area of the sphere at left based on the area, *A*, of a cross section through its center. (Hint: See Exercise 1.)

26. $A = 225$ 900 units^2 **27.** $A = 125$ 500 units^2 **28.** $A = 32.30$ 129.2 units^2

29. $A = 11.22$ 44.88 units^2 **30.** $A = 16\pi$ 201.06 units^2 **31.** $A = 225\pi$

31. 2827.43 units^2

32. 904.78 in^3

33. 600 units^2;
    483.60 units^2

34. 1728 units^3
    2388.06 units^3

32. What is the volume of the largest ball that will fit into a cubical box with edges of 12 in.?

33. A cube and a sphere both have a volume of 1000 cubic units. What are their surface areas?

34. A cube and a sphere both have a surface area of 864 square units. What are their volumes?

CHALLENGE

35. The figures shown at right are a hemisphere, a right circular cone, and a right circular cylinder. Each has the same volume, and each has a base with a radius of 10 in. Find the altitude of each. 10 in.; 20 in.; 6.67 in.

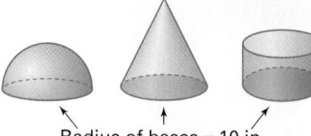

Radius of bases = 10 in.

APPLICATIONS

36. 38.48 in^2; 22.45 in^3

37. 26.42 in^2; 12.77 in^3

In Exercises 36 and 37, round your answers to the nearest tenth.

36. SPORTS Find the surface area and volume of the softball at right.

37. SPORTS Find the surface area and volume of the baseball at right.

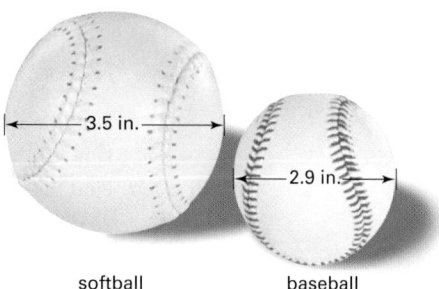

3.5 in.

2.9 in.

softball   baseball

16. a. 499.0756π units^2; 1858.22π units^3
   b. 1567.89 units^2; 5837.79 units^3

17. a. 131.7904π units^2; 252.16π units^3
   b. 414.03 units^2; 792.18 units^3

18. a. 608.1156π units^2; 2499.36π units^3
   b. 1910.45 units^2; 7851.96 units^3

19. a. $39,984.0016\pi$ units^2;
       $1,332,533.49\pi$ units^3
   b. 125,613.45 units^2; 4,186,277.43 units^3

20. $4x^2\pi$; $\frac{4}{3}x^3\pi$

21. $16y^2\pi$; $\frac{32}{3}y^3\pi$

22. $144x^2\pi$; $288x^3\pi$

23. $16y^2\pi$; $\frac{32}{3}y^3\pi$

24. $x^2\pi$; $\frac{x^3}{6}\pi$

25. $\frac{y^2}{4}\pi$; $\frac{y^3}{48}\pi$

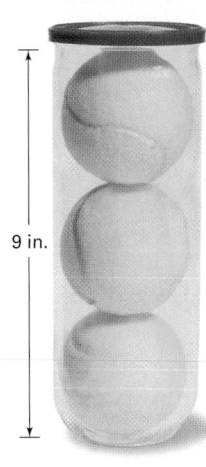

9 in.

38. SPORTS A can of tennis balls has 3 balls stacked tightly in it. The approximate height of the stacked balls and of the interior of the can is 9 in. How much space to do the tennis balls occupy? Approximately what percent of the space inside the can do the balls occupy? Round your answers to the nearest tenth. **42.4 in³; ≈66.7%**

39. GEOGRAPHY Earth's radius is approximately 4000 mi. If two-thirds of Earth's surface is covered by water and one-third is land, estimate the land area on Earth. **67,020,643 mi²**

FOOD Hosea buys an ice-cream cone. The ice cream is a sphere with a radius of 1.25 in. The cone has a height of 8 in. and a diameter of 2.5 in. Round your answers for Exercises 40–42 to the nearest hundredth.

40. What is the volume of the ice cream?

41. What is the volume of the cone?

42. If the ice cream sits in the cone so that it forms a hemisphere (a half of a sphere) above the rim of the cone, what is the total surface area of the hemisphere of ice cream and the cone?

40. 8.18 in³

41. 13.09 in³

42. 41.61 in²

43. 2.60 in.

43. METALWORK A metal sculptor has a solid bronze sphere with a radius of 10 in. She melts the sphere and casts a hollow sphere with an inner radius of 10 in. What is the thickness of the shell of the hollow sphere?

Pouring molten metal into a mold

Look Back

Find the surface area of each figure. *(LESSONS 7.2, 7.4, AND 7.5)*

44. right rectangular prism: ℓ = 3 in., w = 10 in., h = 5 in. **190 in²**

45. right cone: r = 15 cm, slant height = 45 cm **2827.43 cm²**

46. right cylinder: r = 9 m, h = 10 m **1074.42 m²**

Find the volume of each figure. *(LESSONS 7.3, 7.4, AND 7.5)*

47. 420 ft³

47. rectangular pyramid: base length = 15 ft, base width = 7 ft, altitude = 12 ft

48. right cone: r = 5 in., h = 10 in. **261.80 in³**

49. oblique cylinder: r = 7.5 m, h = 20 m **3534.29 m³**

Look Beyond

Algebra

50. $\frac{r}{3}$

51. $V = \frac{S\sqrt{\pi S}}{6\pi}$

52. 1.5 units

53. $\frac{6}{\pi}$

50. What is the ratio of the volume of a sphere to the surface area of the same sphere? Express your answer in terms of the radius, r, of the sphere.

51. Write a formula for the volume of a sphere in terms of its surface area, S.

52. If the radius of a sphere is 4.5 units, by what number can you multiply the surface area of the sphere to find the volume?

53. A sphere with a diameter of $2r$ is contained in a cube with edges of $2r$. What is the ratio of the volume of the cube to the volume of the sphere?

Technology

Students can use a scientific calculator for most of the exercises. Some students may wish to create calculator programs for finding surface area and volume.

Math

C O N N E C T I O N

ALGEBRA In Exercises 50–53, students find the ratio of the surface of a sphere to its volume.

Look Beyond

In Exercises 50–53, students analyze the relationship between the surface area of a sphere and its volume by finding the corresponding ratio.

Student Technology Guide

NAME _____ CLASS _____ DATE _____

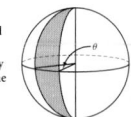

Student Technology Guide
7.6 *Surface Area and Volume of Spheres*

To find the surface area S and volume V of a sphere with radius r, apply the formulas $S = 4\pi r^2$ and $V = \frac{4}{3}\pi r^3$. You can use them to find the surface area and volume of a spherical lune such as the one shown at right. The lune is a two-dimensional surface determined by the angle marked . To find the surface area of a lune and the volume of its associated spherical wedge, see the calculator example below.

Example: Find the surface area of the lune and the volume of a wedge formed by a 60° angle in a sphere whose radius is 12 inches.

Since 60° is $\frac{1}{6}$ of a complete revolution, find one-sixth of the surface area and one-sixth of the volume of the sphere.

• Surface area: $\frac{4\pi r^2}{6} = \frac{2\pi r^2}{3}$; Volume: $\frac{\frac{4}{3}\pi r^3}{6} = \frac{2\pi r^3}{9}$

• Press 2 [2nd] [^] [×] 12 [x²] [÷] 3 [ENTER].
Press [2nd] [ENTER] to edit the expression.
Use [◄] to replace the exponent 2, and press [^] [2nd] [DEL] 3. Use [►] to replace the denominator, 3. Press 9 [ENTER].

• The lune covers about 301.6 in.² The wedge holds about 1206.4 in.³

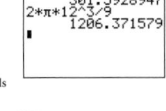

The figures and formulas at right are for a spherical cap and a spherical zone.

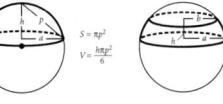

$S = \pi p^2$
$V = \frac{h\pi p^2}{6}$

$S = 2\pi r h$
$V = \frac{3a^2 + 3b^2 + h^2}{6}\pi h$

A sphere has a radius of 12 inches. Find the surface area and volume of each shape to the nearest tenth, according to the given information.

1. = 45°
226.2 in.²; 904.8 in.³

2. $h = 3, a = 5, p = \sqrt{34}$
106.8 in.²; 53.4 in.³

3. $h = 3, a = 5, b = 2$
226.2 in.²; 150.8 in.³

4. $h = \sqrt{10}, a = \sqrt{20}, b = \sqrt{8}$
238.4 in.²; 155.6 in.³

5. = 120°
603.2 in.²; 2412.7 in.³

6. $h = 12, a = 12, p = 12\sqrt{2}$
904.8 in.²; 1809.6 in.³

Focus

The actual reconstruction of a robe found in the tomb of Tutankhamun provides a case study of geometric patterns in archaeology.

Motivate

After students read the news excerpt, have them discuss why the discovery of Tutankhamun's tomb was so significant. Note the following:

- Despite some early plundering, the tomb and its contents were left virtually intact, especially compared to the ravaged state of other discovered tombs of ancient Egyptian rulers.

- The mummy of Tutankhamun was still there.

- Many archaeologists believed that the search for the tomb of Tutankhamun would be fruitless.

To help students understand the background of the reconstruction, ask "What would happen if Carter had picked up the robe when he first opened the box?" (It would have crumbled and the bead pattern might have been lost.)

THE EYEWITNESS MATH

TREASURE of King Tut's Tomb

GEM-STUDDED RELICS AMAZE EXPLORERS
Special Cable to the New York Times

LONDON, Nov. 30, 1922—The Cairo correspondent of the London Times, in a dispatch to his paper, describes how Lord Carnarvon and Howard Carter unearthed below the tomb of Ramses VI, near Luxor, two rooms containing the funeral paraphernalia of King Tutankhamen, who reigned about 1350 b.c.

A sealed outer door was carefully opened, then a way was cleared down some sixteen steps and along a passage of about twenty-five feet. A door to the chambers was found to be sealed as the outer door had been and as on the outer door, there were traces of re-closing.

The excitement of unearthing the tomb was followed by three years of hard work. Unpacking a box of priceless relics took three painstaking weeks. The items, which had been stored for thousands of years, could have been easily damaged if not handled with great care.

A crumpled robe in Box 21 presented a dilemma for Howard Carter. Should he preserve the garment as is and lose the chance to learn its full design? Or should he handle it and sacrifice the cloth in order to examine the complete robe?

Excerpt from Howard Carter's notes

. . . by sacrificing the cloth, picking it carefully away piece by piece, we could recover, as a rule, the whole scheme of decoration. Later, in the museum, it will be possible to make a new garment of the exact size, to which the original ornamentation—bead-work, gold sequins, or whatever it may be—can be applied. Restorations of this kind will be far more useful, and have much greater archaeological value, than a few irregularly shaped pieces of cloth and a collection of loose beads and sequins.

Men build a single track railway through the Valley of Kings for transporting cases of relics from Tutankhamen's tomb in Luxor, 1922.

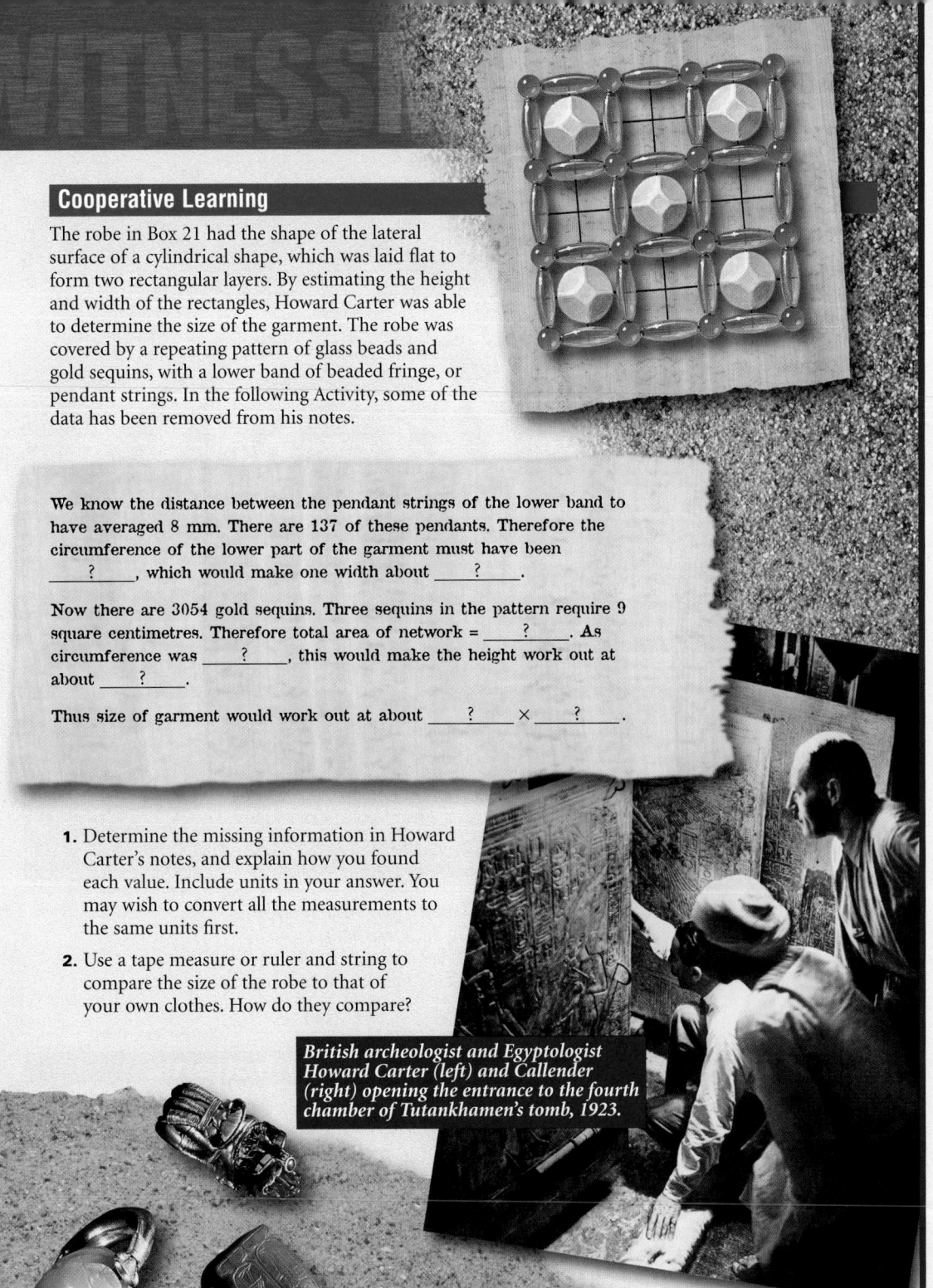

Cooperative Learning

The robe in Box 21 had the shape of the lateral surface of a cylindrical shape, which was laid flat to form two rectangular layers. By estimating the height and width of the rectangles, Howard Carter was able to determine the size of the garment. The robe was covered by a repeating pattern of glass beads and gold sequins, with a lower band of beaded fringe, or pendant strings. In the following Activity, some of the data has been removed from his notes.

We know the distance between the pendant strings of the lower band to have averaged 8 mm. There are 137 of these pendants. Therefore the circumference of the lower part of the garment must have been _____?_____, which would make one width about _____?_____.

Now there are 3054 gold sequins. Three sequins in the pattern require 9 square centimetres. Therefore total area of network = _____?_____. As circumference was _____?_____, this would make the height work out at about _____?_____.

Thus size of garment would work out at about _____?_____ × _____?_____.

1. Determine the missing information in Howard Carter's notes, and explain how you found each value. Include units in your answer. You may wish to convert all the measurements to the same units first.

2. Use a tape measure or ruler and string to compare the size of the robe to that of your own clothes. How do they compare?

British archeologist and Egyptologist Howard Carter (left) and Callender (right) opening the entrance to the fourth chamber of Tutankhamen's tomb, 1923.

Cooperative Learning

Have students work in pairs or small groups on the Activity. Students may find it helpful to use calculators. In his notes, Carter's use of the terms *circumference* and *width* might be confusing. Students should use a diagram to see that the "width" of the garment is half the "circumference." Remind students that one square meter is 10,000 square centimeters.

Discuss

Have the class make up their own reconstruction problem about an ancient garment, piece of jewelry, or other relic. Be sure the students include enough information about the contents and patterns of the discovered item so that the size or appearance of the original article can be determined.

Three-Dimensional Symmetry

1. Find the image of (–2, 4) reflected across the *y*-axis. (2, 4)

2. Find the image of (3, –4) reflected across the *x*-axis. (3, 4)

3. In which plane of a three-dimensional coordinate system is point *P*(0, –2, 4) located? *yz*-plane

Also on Quiz Transparency 7.7

Teach

Why This lesson extends the ideas of symmetry to include reflectional and rotational symmetry in space. Students need to be able to describe a three-dimensional object as well as find its area and volume. This lesson also presents many examples of objects that have symmetry.

Objectives

● Define various transformations in three-dimensional space.

● Solve problems by using transformations in three-dimensional space.

Why So far, the definitions of symmetry have been limited to a plane. But three-dimensional figures, like the tiger in the photo, may also have symmetry.

Tiger! Tiger! burning bright
In the forests of the night,
What immortal hand or eye
Could frame thy fearful symmetry?
—William Blake

Three-Dimensional Reflections

A three-dimensional figure may be reflected across a plane, just as a two-dimensional figure can be reflected across a line. What happens to each point in the preimage of a figure as a result of a reflection across a plane? You will investigate this question in Activity 1.

Alternative Teaching Strategy

USING VISUAL MODELS Orient a rectangular box in the first octant of a three-dimensional coordinate system with one vertex at (0, 0, 0) and one face of the box in the *xy*-plane. Measure the box and determine the coordinates of its other vertices. Find the reflection of each vertex across the *xy*-plane, the *xz*-plane, and the *yz*-plane. Then rotate the original box so that it is in a different octant with one vertex still at (0, 0, 0) and one face in a plane. Repeat the reflections. Are the image points the same? no

Activity 1
Reflections in Coordinate Space

Part I

1. Graph the point $A(1, 1, 1)$ in a three-dimensional coordinate space. Use dashed lines to make the location of the point in space evident.

2. Multiply the x-coordinate of point A by -1 and graph the resulting point. Label the new point A'.

3. Point A' is the image of point A reflected across a plane. Name this plane. If you connect points A and A' to form $\overline{AA'}$, what is the relationship between this segment and the plane of reflection?

4. Write your definition for the reflection of a point across a plane in a three-dimensional coordinate space.

5. Experiment with other reflections of point A. What happens if you multiply the x-coordinates by -1? the z-coordinates?

6. Now study the reflection of an entire segment, such as $\overline{AB}$, across a plane. (For example, multiply the x-coordinates of points A and B by -1 and connect the resulting points.) Experiment with the reflections of other figures, such as cubes.

CHECKPOINT ✔
7. Write your own definition for the reflection of a figure in a three-dimensional coordinate space across a plane.

Part II

1. Fill in the table below. Use the terms *front, back, left, right, top,* and *bottom* to describe the octant of a point.

		Octant of image	Coordinates of image
$A(2, 3, 4)$	Reflection across the xy-plane	front-right-bottom	$(2, 3, -4)$
	Reflection across the xz-plane	?	?
	Reflection across the yz-plane	?	?
$B(-4, 5, 6)$	Reflection across the xy-plane	?	?
	Reflection across the xz-plane	?	?
	Reflection across the yz-plane	?	?

CHECKPOINT ✔
2. Generalize your findings for point $P(x, y, z)$.
   a. What are the coordinates of the reflection of P across the xy-plane?
   b. What are the coordinates of the reflection of P across the xz-plane?
   c. What are the coordinates of the reflection of P across the yz-plane?

YOU WILL NEED

no special tools

Interdisciplinary Connection

ART Computer artists work with the symmetry of figures by using a coordinate grid. Have students use computer drawing programs to practice drawing solid figures.

Each figure is symmetrical across a plane, or mirror, dividing it into two values. A figure has a reflectional symmetry in space if and only if its reflected image across a plane coincides with its preimage. The plane is called a plane of symmetry.

Teaching Tip

It may be very helpful to photocopy some illustrations and have students draw the lines of symmetry on them.

Activity 2 Notes

This Activity will help students extend their ideas about symmetry in space by having them rotate coordinates in space. This generates a three-dimensional figure. Have students compare their definitions in Step 5 with those of their classmates.

For a student worksheet of this Activity and detailed Teacher Notes, see page 133 in the Lesson Activities booklet.

CHECKPOINT ✔

5. Sample answer: When a figure is rotated around an axis in space, each point in the figure moves around the axis of rotation in a circle contained in a plane that is perpendicular to the axis of rotation. The radius of the circle is the distance from the point to the axis of rotation and the center of the circle is on the axis of rotation.

CHECKPOINT ✔ The three-dimensional figures below have reflectional symmetry. Explain why. Where is the reflection "mirror" in each case? Create your own definition of reflectional symmetry in space. (Use the earlier definition of reflectional symmetry in a plane as a model.)

Activity 2

Rotations in Coordinate Space

YOU WILL NEED
no special tools

1. Graph $\overline{AB}$ in a three-dimensional coordinate space. Use dashed lines to make the location of the segment evident.

2. Multiply the x- and y-coordinates of points A and B by -1 and graph the resulting points, A' and B'. Connect the new points to form $\overline{A'B'}$.

3. $\overline{A'B'}$ is a rotation image of $\overline{AB}$ about the z-axis. About what point on the z-axis has point A been rotated?

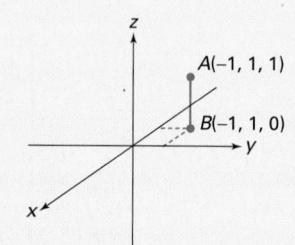

4. Imagine the segment rotating about the z-axis, as suggested by the picture. Does it seem to you that each of the rotation images of point A is in the same plane? of point B? What is the relationship between these planes and the z-axis?

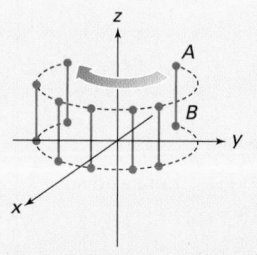

CHECKPOINT ✔

5. Write your own definition of the rotation of a figure about an axis in coordinate space.

Enrichment

Have students do a collage of figures with symmetry. Ask them to select a specific theme and use magazines to find examples of the themes. Possible themes include items in nature with symmetry, buildings, or food.

Inclusion Strategies

HANDS-ON STRATEGIES Some students may have trouble finding image points under reflectional or rotational symmetry. Have students work with model prisms and experiment with the effects of a rotation on a model of a three-dimensional coordinate system made with drinking straws.

The three-dimensional figures below have rotational symmetry. Explain why. What is the axis of rotation in each case? Create your own definition of rotational symmetry in space. (Use the earlier definition of rotational symmetry in the plane as a model.)

Each figure can be rotated about a line and have its image coincide with its preimage. The axis of rotation in each case is the line passing through the center of the figure. The beach ball and the polyhedron have multiple axes of rotation; the iris and the starfish have only one. A figure in space has rotational symmetry if and only if it has at least one rotation image, not counting rotations of 0° or multiples of 360°, that coincides with the original image.

Revolutions in Coordinate Space

If you rotate a figure about an axis, a spatial figure is formed. The spatial figure is the set of points through which the original figure passes in one complete revolution.

E X A M P L E ❶ You are given $\overline{AB}$ with endpoints $A(0, 5, 0)$ and $B(0, 5, 5)$. Sketch, describe, and give the dimensions of the figure that results when

a. $\overline{AB}$ is rotated about the z-axis.

b. $\overline{AB}$ is rotated about the y-axis.

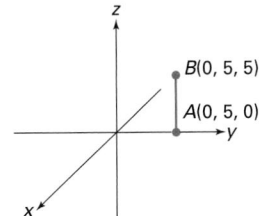

● **SOLUTION**

a.

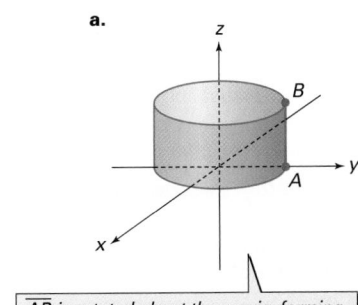

$\overline{AB}$ is rotated about the z-axis, forming the lateral surface of a cylinder with a radius of 5 and height of 5.

b.

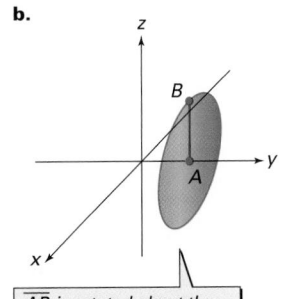

$\overline{AB}$ is rotated about the y-axis, forming a circular region with a radius of 5.

TRY THIS Describe or sketch the spatial figure that would be formed by rotating each of these plane figures about the red line.

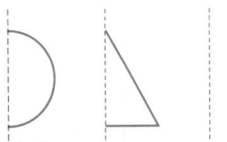

E X A M P L E ❶

You are given segment $\overline{XY}$ with endpoints $X(0, 6, 0)$ and $Y(0, 6, 4)$. Sketch the figure that is formed if segment $\overline{XY}$ is rotated about the z-axis, and find its volume.

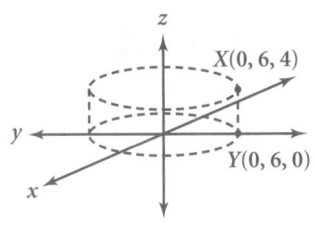

Volume = $144\pi \approx 452.4$

TRY THIS
The figures formed would be, from left to right, a sphere, a cone, and a torus or doughnut.

Reteaching the Lesson

USING DISCUSSION You may want to have students restate the definitions they wrote in this lesson and include some examples of their drawings. Have students identify the preimage and image points.

ASSIGNMENT GUIDE

In Class	1–9
Core	11–33 odd, 34
Core Plus	23–31 odd, 32–40
Review	41–45
Preview	46–47

✏ Extra Practice can be found beginning on page 818.

9.

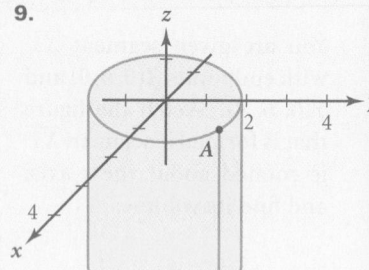

10.

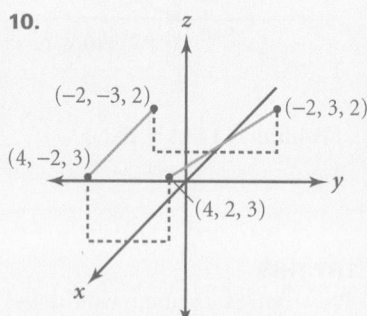

11.

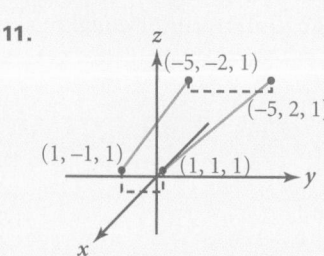

12.

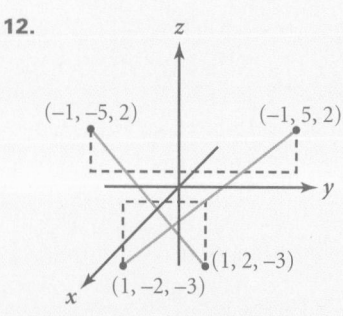

Exercises

Communicate

1. Describe the similarities and differences betweeen three-dimensional reflectional symmetry and two-dimensional reflectional symmetry.

2. What spatial figure is formed by rotating rectangle *ABCD* about $\overline{AB}$?

3. What spatial figure is formed by rotating △*EFG* about $\overline{EG}$?

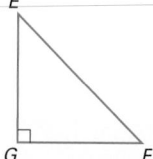

4. List some objects in your classroom that have three-dimensional rotational symmetry.

Guided Skills Practice

Graph the point *A*(2, 1, −1) in a three-dimensional coordinate space. Give the coordinates of the image if point *A* is reflected across each coordinate plane listed below. *(ACTIVITY 1)*

5. *xy*-plane (2, 1, 1)   6. *xz*-plane (2, −1, −1)   7. *yz*-plane (−2, 1, −1)

8. Graph $\overline{AB}$ with endpoints *A*(1, 3, −2) and *B*(1, 3, 0) in a three-dimensional coordinate space. Multiply the *y*- and *z*-coordinates of points *A* and *B* by −1 and graph the resulting points, *A'* and *B'*. About what axis is $\overline{AB}$ rotated to get $\overline{A'B'}$? *(ACTIVITY 2)* **about the *x*-axis**

9. Sketch the spatial figure that results when $\overline{AB}$, with endpoints *A*(1, 2, 0) and *B*(1, 2, −4), is rotated about the *z*-axis. *(EXAMPLE)*

Practice and Apply

Draw three-dimensional coordinate systems and graph segments with the given endpoints. Reflect each segment by multiplying the *y*-coordinates by −1.

10. (4, −2, 3) and (−2, −3, 2)   11. (−5, 2, 1) and (1, 1, 1)

12. (1, −2, −3) and (−1, 5, 2)   13. (3, 2, −3) and (−4, 3, −2)

14. (5, 3, 2) and (−1, −1, −1)   15. (1, 2, 3) and (1, 3, 3)

What are the coordinates of the image if each point below is reflected across the *xy*-plane in a three-dimensional coordinate system?

16. (6, 5, 8) (6, 5, −8)   17. (−2, 3, −1) (−2, 3, 1)   18. (1, 1, 1) (1, 1, −1)

19. (4, −2, 3) (4, −2, −3)   20. (−5, 2, 1) (−5, 2, −1)   21. (1, −2, −3) (1, −2, 3)

13.

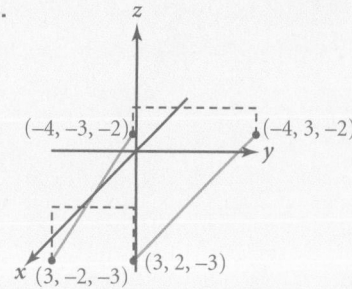

14.

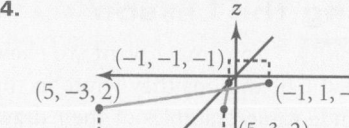

15.

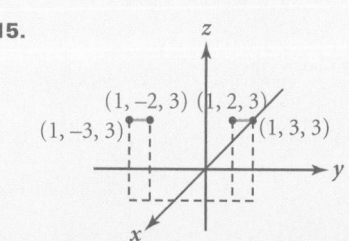

What are the octants and coordinates of the image if each point below is reflected across the *xz*-plane in a three-dimensional coordinate system?

22. $(6, -2, 8)$ **23.** $(-4, -4, -1)$ **24.** $(1, 0, 1)$

25. $(1, 1, 0)$ **26.** $(4, -4, 1)$ **27.** $(2, 2, -8)$

Homework Help Online

Go To: **go.hrw.com**
Keyword:
MG1 Homework Help
for Exercises 28-34

In Exercises 28–31, $\overline{AB}$ has endpoints $A(5, 0, 10)$ and $B(5, 0, 0)$.

28. What spatial figure is formed by rotating $\overline{AB}$ about the *x*-axis?

29. What is the area of the figure formed by rotating $\overline{AB}$ about the *x*-axis?

30. What spatial figure is formed by rotating $\overline{AB}$ about the *z*-axis?

31. What is the volume of the figure formed by rotating $\overline{AB}$ about the *z*-axis?

In Exercises 32 and 33, $\overline{CD}$ has endpoints $C(4, 0, 0)$ and $D(0, 0, 4)$.

33. 67.02 units^3

32. What spatial figure is formed by rotating $\overline{CD}$ about the *z*-axis?

33. What is the volume of the figure formed by rotating $\overline{CD}$ about the *z*-axis?

CONNECTION
CHALLENGE

34. **MAXIMUM/MINIMUM** The area of a right triangle with a fixed perimeter is maximized when the triangle is a 45-45-90 triangle. Suppose that you rotate a 45-45-90 triangle about one leg to create a cone. Does the cone have the maximum volume for the given slant height? Why or why not? (See Exercises 43–44 in Lesson 7.5, page 466.) **No**

APPLICATION

POTTERY A potter is making pots according to certain patterns. For each half of a pattern given below, sketch what the complete pot will look like. (Rotate about the dashed red line.)

35. 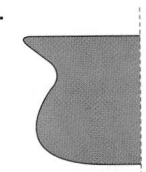   **36.** 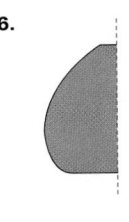   **37.**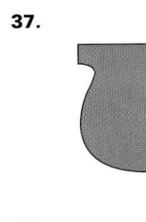

38.    **39.**    **40.**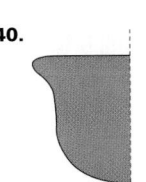

 Look Back

Find the volume and surface area of each prism. *(LESSON 7.2)*

41. 445.48 units^3;
    456.68 units^2

42. 1122.37 units^3;
    619.06 units^2

41.

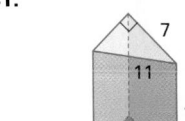

Right-triangular prism

42.

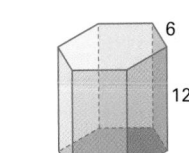

Regular hexagonal prism

37.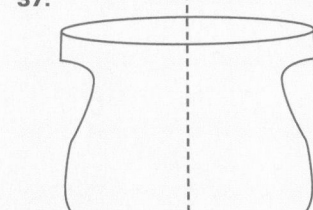

38.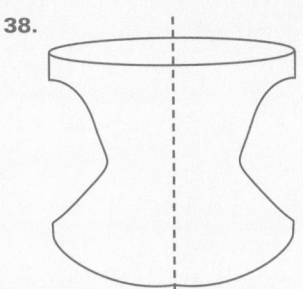

Error Analysis

Students need to be sure that they are creating the correct solid when performing the rotations in Exercises 28–34. Encourage students to double-check their work.

Technology

Students can use software or a graphics calculator with three-dimensional graphing capabilities to graph the coordinates of three-dimensional objects in the exercises.

22. front-right-top, $(6, 2, 8)$

23. back-right-bottom, $(-4, 4, -1)$

24. in the *xz*-plane, $(1, 0, 1)$

25. in the *xy*-plane, $(1, -1, 0)$

26. front-right-top, $(4, 4, 1)$

27. front-left-bottom, $(2, -2, -8)$

28. The resulting figure is a circle and its interior. Its center is at $(5, 0, 0)$ with radius = 10.

29. The area of this circle
$= \pi(10)^2$
$= 100\pi$
≈ 314.16 units^2

30. A cylinder with $h = 10$ and $r = 5$, centered on the *z*-axis.

31. $V = \pi(5)^2(10) = 250\pi$
≈ 785.40 units^3

32. A cone with $r = 4$, $h = 4$ and $l = 4\sqrt{2}$

35.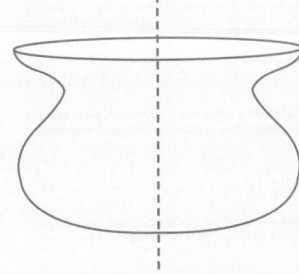

36.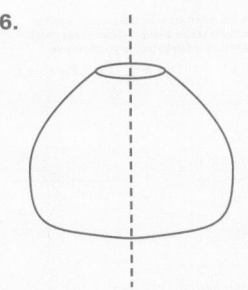

43. TRANSPORTATION A truck's storage space is shaped like a rectangular prism that measures 7 ft 10 in. × 8 ft 2 in. × 24 ft. What is the volume of the storage space? *(LESSON 7.2)*

44. Find the surface area and volume of a regular square pyramid with a base area, B, of 36 cm² and a height, h, of 5 cm. *(LESSON 7.3)*

APPLICATION

45. HOBBIES Seeing barbecue smokers displayed next to bags of potting soil, Dolores decides to change her old smoker at home into a planter. The center of her smoker comes up to her waist and is thus half of her height of 5 ft. Use this information to estimate other dimensions in the photo, and then determine how much soil it will take to fill the smoker to the centerline of the barrel from which it is made. *(LESSON 7.4)*

Look Beyond

APPLICATION

OPTICS A spinning fan creates the illusion of being solid. A strobe light, which emits a flash at certain time intervals, can "freeze" the motion of the fan.

46. Suppose that a fan with four equally spaced, identical blades is turning at 12 revolutions per second. How often will the strobe need to flash to make the fan appear to be frozen?

47. If the strobe flashes 36 times per second, at what speeds (in revolutions per second) can the fan move and still appear to be frozen?

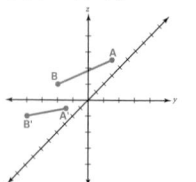
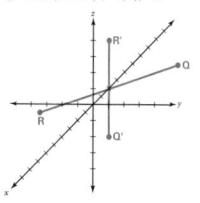

CREATING SOLIDS OF REVOLUTION

If a region in a plane is rotated about an axis that lies in the same plane, the resulting figure is called a *solid of revolution*. Follow the directions below to create your own solids of revolution with cardboard, foam core board, a needle and thread.

1. On a piece of graph paper, draw a shape that has at least one vertical edge, such as the one shown at right. This edge will be the axis of revolution of your solid. Your shape should be at least 10 units high.

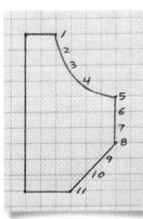

2. Number the intersections of the grid lines with the edge of your figure, as shown. For each intersection, set your compass to the width of your figure, and draw a circle on a piece of cardboard. Number the circle according to the number of the intersection.

internet connect

Portfolio Extension

Go To: go.hrw.com
Keyword:
MG1 SolRev

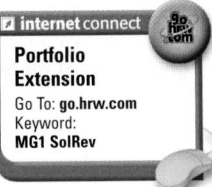

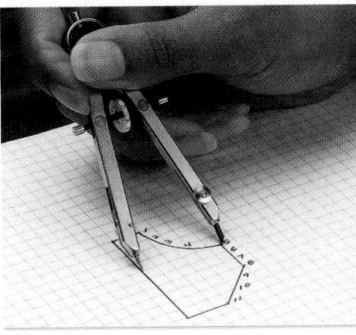

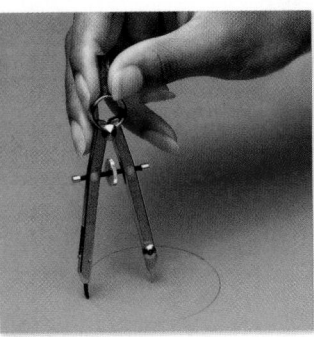

3. Cut the circles out of the cardboard. These circles represent cross-sections of your solid of revolution. Cut 10 small pieces of foam core board that are approximately the thickness of the squares on your graph paper to use as spacers.

4. Using a small needle, thread the circles of cardboard in order, alternating with the pieces of foam core. Secure the string at the bottom with a small piece of tape. Tie a loop in the string at the top for a hanger.

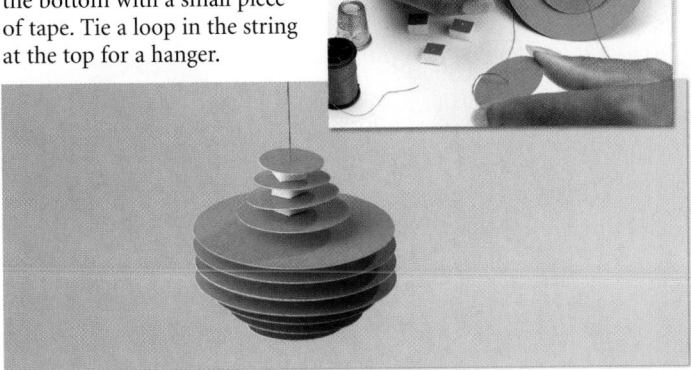

ALTERNATIVE

Assessment

Portfolio Activity

The Portfolio Activity can be used as preparation for the Chapter Project or as a separate activity. In the Portfolio Activity on this page, students create a solid of revolution by gluing together tissue paper.

Focus

Students assemble the nets of six solids and fit them together to form a cube. Have students identify each solid as it is assembled, such as right triangular prism, oblique trapezoidal pyramid, etc.

Motivate

When students finish the activities, have them try to create their own three-dimensional puzzle from heavy paper or modeling materials. Students who are knowledgeable in woodworking may want to create such a puzzle as a woodworking project.

CHAPTER PROJECT SEVEN

A THREE-DIMENSIONAL PUZZLE

This puzzle is a three-dimensional version of the ancient Chinese puzzle known as the tangram (see page 359).

Activity 1

THE PUZZLE Copy the nets shown below and on the next page onto heavy paper and assemble them to form six solids. Then see if you can fit them together to make a cube. The measurements are given in centimeters.

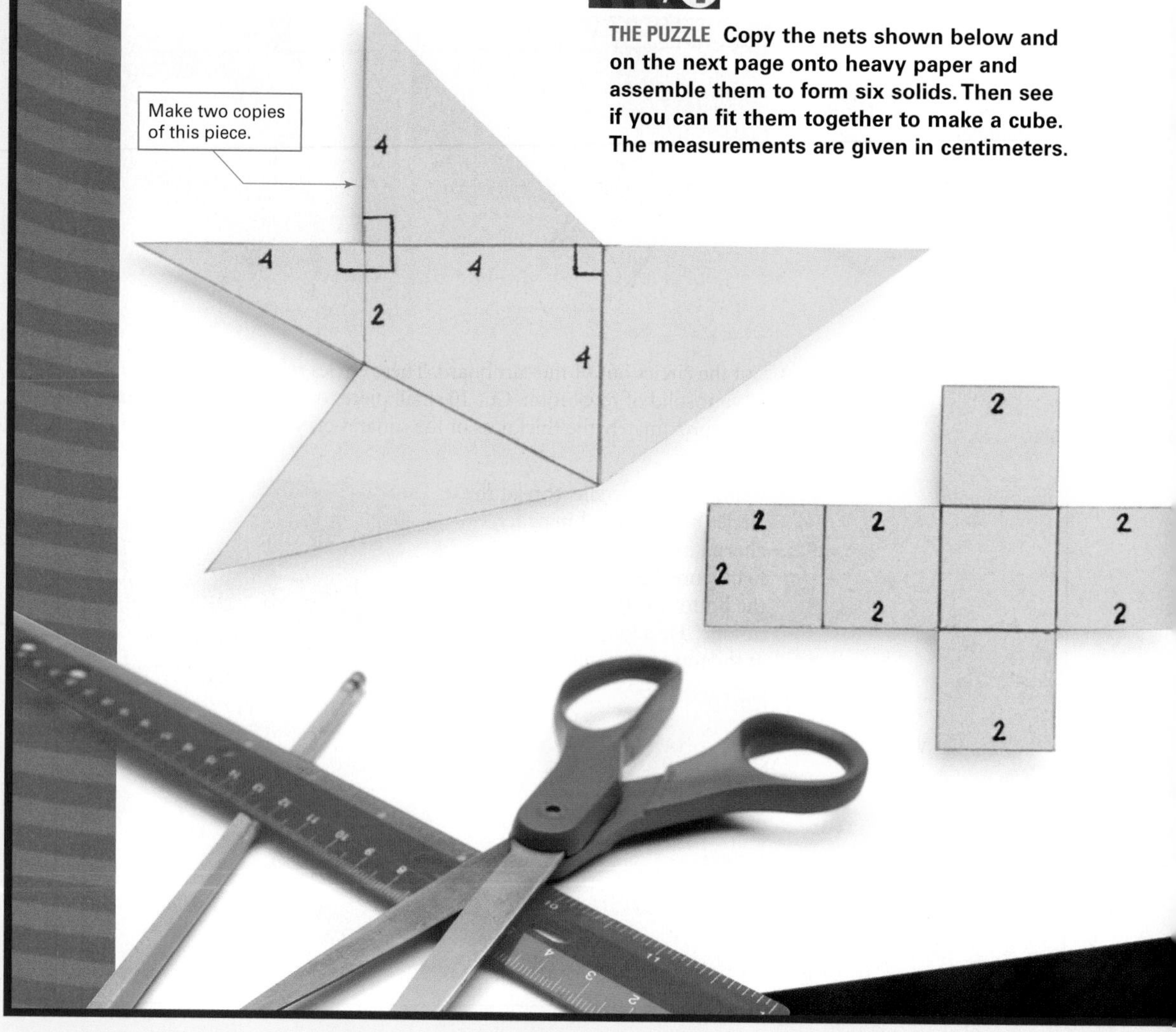

Make two copies of this piece.

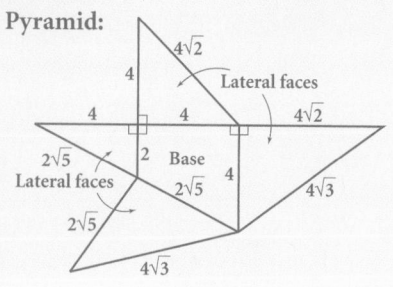

Activity 1

Check students' models.

Activity 2

The pieces are a non-regular pyramid with a trapezoidal base, a cube, a prism with trapezoidal base, and a long and a short triangular prism.

Pyramid:

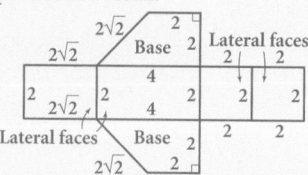

Volume = 16 cm³
Surface area ≈ 45.11 cm²

Cube: All edges have a length of 2 cm.
Volume = 8 cm³
Surface area = 24 cm²

Trapezoidal Prism:

Volume = 12 cm³
Surface area ≈ 33.66 cm²

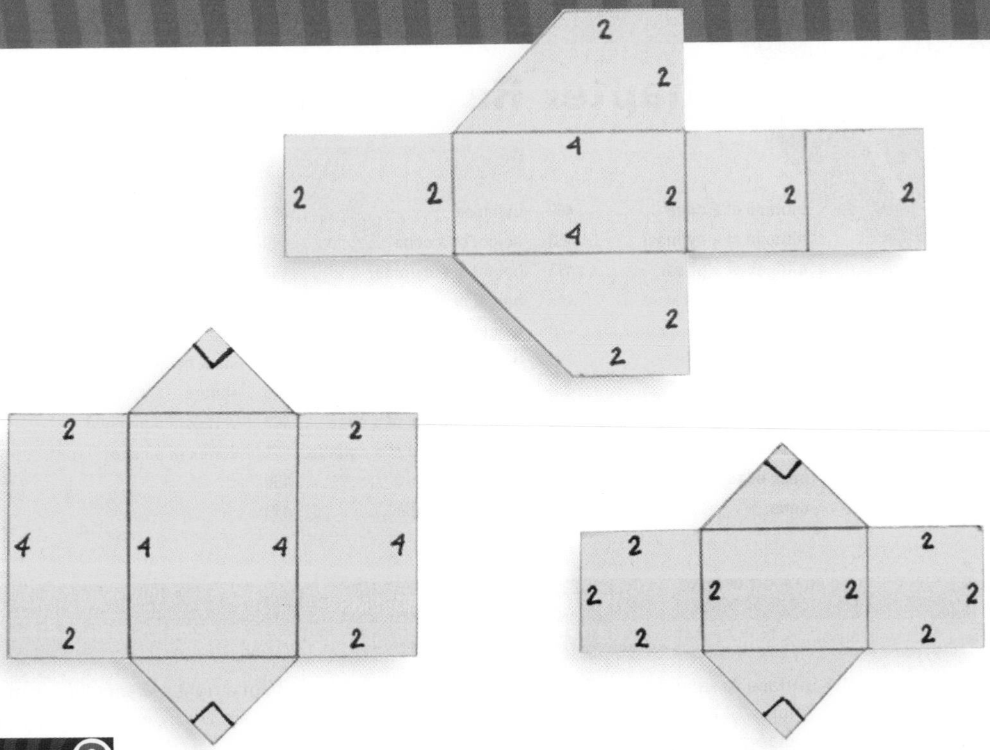

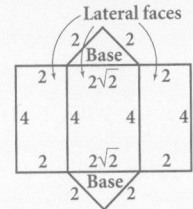

Cooperative Learning

Cooperative learning groups are very appropriate for this hands-on activity. Have students construct the polyhedrons and offer each other ideas about original patterns.

Discuss

After all students have finished making their models and solving the puzzle, discuss surface area and volume. Which sides of each solid contribute to the surface area of the cube? Can you easily find the volume of each individual solid? How can you use the volume of the cube, which is the solution to the puzzle, to find the volumes of the individual solids?

Activity ②

Describe each piece in as much detail as possible. What shape is it? What are the lateral faces and bases? Find the lengths of all edges that are not given. Then find the volume of each piece.

Extension

1. What are the dimensions of the cube that is formed when the pieces are assembled? Does the sum of the volumes of the individual pieces equal the volume of the cube?

2. Can you assemble the pieces to form any other interesting solids?

Triangular Prisms:

Lateral faces
Base
$2\sqrt{2}$
$2\sqrt{2}$
Base

Volume = 8 cm³
Surface area ≈ 31.31 cm²

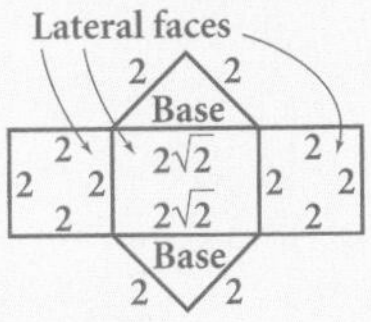

Lateral faces
Base
$2\sqrt{2}$
$2\sqrt{2}$
Base

Volume = 4 cm³
Surface area ≈ 17.66 cm²

Chapter Review and Assessment

VOCABULARY

Chapter Test, Form A

NAME _____ CLASS _____ DATE _____

Chapter Assessment

Chapter 7, Form A, page 1

Write the letter that best answers the question or completes the statement.

c 1. What is the ratio of surface area to volume for a cube with a volume of 27 cm³?

   a. 1:2   b. 1:3   c. 2:1   d. 3:1

d 2. What is the volume of a cube with a surface area of 600 cm²?

   a. 100 cm³   b. 216 cm³   c. 750 cm³   d. 1000 cm³

c 3. The height of an oblique rectangular prism is 4 cm. The length of the base is 3 cm. If the volume of the prism is 96 cm³, the width of the base is _____.

   a. 4 cm   b. 5 cm   c. 8 cm   d. 12 cm

b 4. The base of a triangular prism is a right triangle. The lengths of the legs of the base are 3 cm and 4 cm. The height of the prism is 8 cm. What is the surface area of the prism?

   a. 96 cm²   b. 108 cm²   c. 148 cm²   d. 480 cm²

a 5. The base of a triangular prism is a right triangle. The lengths of the legs of the base are 3 cm and 4 cm. The height of the prism is 8 cm. What is the volume of the prism?

   a. 48 cm³   b. 96 cm³   c. 108 cm³   d. 148 cm³

b 6. A right isosceles trapezoidal prism has a height of 6 m. The parallel sides and altitude of the trapezoid measure 8 m, 10 m, and 5 m, respectively. What is the volume of the prism?

   a. 138 m³   b. 270 m³   c. 540 m³   d. 2400 cm³

b 7. The surface area of the pyramid is _____.

   a. 48 cm²   b. 96 cm²
   c. 144 cm²   d. 156 cm²

a 8. The volume of the pyramid is _____.

   a. 48 cm³   b. 72 cm³
   c. 96 cm³   d. 144 cm³

a 9. A square pyramid has a height of 9 cm and a volume of 75 cm³. What is the length of a side of the base?

   a. 5 cm   b. $8\frac{1}{3}$ cm   c. 9 cm   d. 25 cm

NAME _____ CLASS _____ DATE _____

Chapter Assessment

Chapter 7, Form A, page 2

d 10. A cylinder has a radius of 6 ft and a height of 10 ft. The surface area of the cylinder is about _____.

   a. 200.96 ft²   b. 602.8 ft²   c. 640 ft²   d. 603.19 ft²

c 11. A cylindrical tank is being constructed to hold 20,000 ft³ of water. The diameter of the tank will be 40 ft. What is the minimum height needed for the tank?

   a. 14 ft   b. 15 ft   c. 16 ft   d. 17 ft

c 12. A semicircular solid is formed by cutting a cylinder in half lengthwise. If the diameter of the cylinder is 6 ft and its height is 10 ft, the surface area of the semicircular solid is about _____.

   a. 182.5 ft²   b. 276.7 ft²   c. 122.52 ft²   d. 465.1 ft²

c 13. The surface area of the cone is _____.

   a. ≈ 78.5 cm²   b. ≈ 204.1 cm²
   c. ≈ 282.7 cm²   d. ≈ 267.0 cm²

b 14. The volume of the cone is about _____.

   a. 260 cm³   b. 314 cm³
   c. 360 cm³   d. 780 cm³

d 15. The surface area of a sphere with a diameter of 12 in. is about _____.

   a. 113.0 in.²   b. 150.7 in.²   c. 266.1 in.²   d. 452.4 in.²

c 16. The diameter of a golf ball is about 4.2 cm. The volume of the golf ball is about _____.

   a. 8.8 cm³   b. 18.5 cm³   c. 38.8 cm³   d. 55.4 cm³

d 17. The surface area of a sphere is 144π m². What is the volume of the sphere?

   a. 6π m³   b. 36π m³   c. 144π m³   d. 288π m³

a 18. The point A(2, 7, −3) is reflected across the xy-plane. What are the coordinates of the image?

   a. (2, 7, 3)   b. (−2, 7, 3)   c. (2, −7, 3)   d. (−2, 7, −3)

b 19. Segment $\overline{AB}$ with endpoints A(3, 0, 5) and B(3, 0, 0) is rotated about the z-axis. The volume of the figure formed is about _____.

   a. 94.2 cubic units   b. 141.4 cubic units
   c. 235.5 cubic units   d. 423.9 cubic units

POSTULATES AND THEOREMS

Lesson	Number	Postulate or Theorem
7.1	7.1.1 Surface Area and Volume of a Right Rectangular Prism	The surface area, S, and volume, V, of a right rectangular prism with length ℓ, width w, and height h, are $S = 2\ell w + 2wh + 2\ell h$ and $V = \ell wh$.
	7.1.2 Surface Area and Volume of a Cube	The surface area, S, and volume, V, of a cube with side length s are $S = 6s^2$ and $V = s^3$.
7.2	7.2.1 Surface Area of a Right Prism	The surface area, S, of a right prism with lateral area L, base area B, perimeter p, and height h is $S = L + 2B$ or $S = hp + 2B$.
	7.2.2 Cavalieri's Principle	If two solids have equal heights and the cross sections formed by every plane parallel to the bases of both have equal areas, then the two solids have equal volumes.
	7.2.3 Volume of a Prism	The volume, V, of a prism with height h and base area B is $V = Bh$.
7.3	7.3.1 Surface Area of a Regular Pyramid	The surface area, S, of a regular pyramid with lateral area L, base area B, perimeter of the base p, and slant height ℓ is $S = L + B$ or $S = \frac{1}{2}\ell p + B$.
	7.3.2 Volume of a Pyramid	The volume, V, of a pyramid with height h and base area B is $V = \frac{1}{3}Bh$.
7.4	7.4.1 Surface Area of a Right Cylinder	The surface area, S, of a right cylinder with lateral area L, base area B, radius r, and height h is $S = L + 2B$ or $S = 2\pi rh + 2\pi r^2$.
	7.4.3 Volume of a Cylinder	The volume, V, of a cylinder with radius r, height h, and base area B is $V = Bh$ or $V = \pi r^2 h$.

Lesson	Number	Postulate or Theorem
7.5	7.5.1 Surface Area of a Right Cone	The surface area, S, of a right cone with lateral area L, base area B, radius r, and slant height ℓ is $S = L + B$ or $S = \pi r \ell + \pi r^2$.
	7.5.2 Volume of a Cone	The volume, V, of a cone with radius r, height h, and base area B is $V = \frac{1}{3}Bh$ or $V = \frac{1}{3}\pi r^2 h$.
7.6	7.6.1 Volume of a Sphere	The volume, V, of a sphere with radius r is $V = \frac{4}{3}\pi r^3$.
	7.6.2 Surface Area of a Sphere	The surface area, S, of a sphere with radius r is $S = 4\pi r^2$.

Key Skills & Exercises

LESSON 7.1

Key Skills

Solve problems by using the ratio of surface area to volume.

A cube has a volume of 27,000 cubic millimeters. What is its ratio of surface area to volume?

The length of each edge of the cube is $\sqrt[3]{27,000} = 30$ millimeters, so the area of each face is $30 \times 30 = 900$ square millimeters. There are 6 faces, so the surface area is $6(900) = 5400$ square millimeters. Thus, the ratio of surface area to volume is $\frac{5400}{27,000} = 0.2$.

Exercises

1. Find the ratio of surface area to volume for a cube with a volume of 64 cubic inches.

2. Find the ratio of surface area to volume for a cube with a volume of 100 cubic inches.

3. Compare the ratios of surface area to volume for a cube with an edge length of 4 and for a right rectangular prism with the dimensions $4 \times 7 \times 3$.

4. A right rectangular prism has a square base, and its height is triple the base edge. Find the ratio of its surface area to volume.

LESSON 7.2

Key Skills

Find the surface area of a right prism.

Find the surface area of the right triangular prism below.

The lateral area is hp, or $100(36 + 77 + 85) = 19,800$.

The area of each base is $\frac{1}{2}(36)(77) = 1386$.

The surface area is $S = L + 2B$, or $19,800 + 2(1386) = 22,572$.

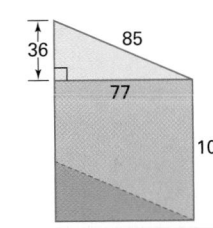

Exercises

Refer to the right hexagonal prism with a regular hexagonal base below.

5. Find the lateral area of the prism.

6. Find the surface area of the prism.

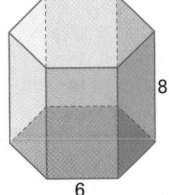

1. $\frac{3}{2}$

2. 1.29

3. cube: $\frac{3}{2}$

   prism: $\frac{61}{42}$

   The ratio is larger for the cube.

4. $\frac{14}{3s}$

5. 288 units^2

6. 475.06 units^2

 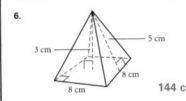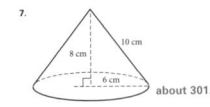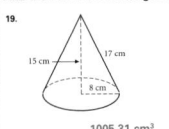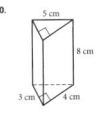

7. 748.25 units^3

8. 5986 units^3

9. 18 units^2

10. 33.59 units^2

11. 5.20 units^3

12. 10.39 units^3

13. 72π units^2 $\approx$ 226.19 units^2

14. 104π $\approx$ 326.73 units^2

15. 144π $\approx$ 452.39 units^3

16. Doubling the height will double the volume of the cylinder, but doubling the radius will multiply the volume by 4. Thus, doubling the radius increases the volume more than doubling the height.

Find the volume of a prism.

Find the volume of the prism on the previous page.

The area of the base is 1386, so the volume is $V = Bh$, or $1386(100) = 138{,}600$.

7. Find the volume of the prism on the previous page.

8. Find the volume of the prism on the previous page if the dimensions are doubled.

LESSON 7.3
Key Skills

Find the surface area of a regular pyramid.

Find the surface area of the regular square pyramid below.

The lateral area is $\frac{1}{2}\ell p$, or $\frac{1}{2}(25)(56) = 700$.

The area of the base is $14^2 = 196$.

The surface area is $S = L + B$, or $700 + 196 = 896$.

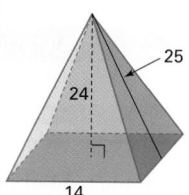

Find the volume of a pyramid.

Find the volume of the pyramid above.

The area of the base is 700, so the volume is $V = \frac{1}{3}Bh$, or $\frac{1}{3}(700)(24) = 5600$.

Exercises

Refer to the regular pyramid below.

9. Find the lateral area of the pyramid.

10. Find the surface area of the pyramid.

11. Find the volume of the pyramid.

12. Find the volume of the pyramid if the height is doubled.

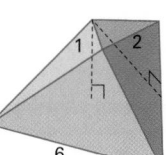

LESSON 7.4
Key Skills

Find the surface area of a right cylinder.

Find the surface area of the right cylinder below.

The lateral area is $2\pi rh$, or $2\pi(12)(7) = 168\pi$.

The area of each base is πr^2, or $\pi(12^2) = 144\pi$.

The surface area is $S = L + 2B$, or $168\pi + 2(144\pi) = 456\pi$.

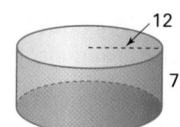

Find the volume of a cylinder.

Find the volume of the right cylinder above.

The area of the base is 144π, so the volume is $V = Bh$, or $144\pi(7) = 1008\pi$.

Exercises

A right cylinder has a radius of 4 and a height of 9.

13. Find the lateral area of the cylinder.

14. Find the surface area of the cylinder.

15. Find the volume of the cylinder.

16. Which increases the volume of a cylinder more, doubling the height or doubling the radius? Explain your reasoning.

17. $580\pi \approx 1822.12$ units^2

18. $980\pi \approx 3078.76$ units^2

19. $2800\pi \approx 8796.46$ units^3

20. $5600\pi \approx 17{,}592.92$ units^3

21. $100\pi \approx 314.16$ units^2

22. $\frac{500}{3}\pi \approx 523.60$ units^3

23. 3 units

24. 1.61 units

25.

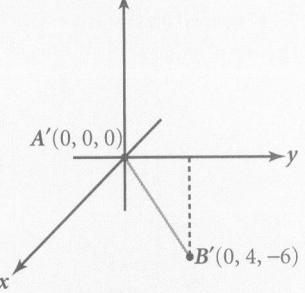

26.

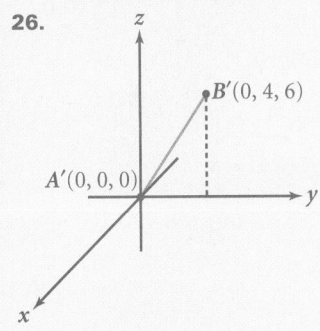

Key Skills

Find the surface area of a right cone.

Find the surface area of the right cone below.

The lateral area, L, is $\pi r \ell$, or
$\pi(6)(10.9) = 65.4\pi$.

The area of the base, B, is πr^2, or $\pi(6^2) = 36\pi$.

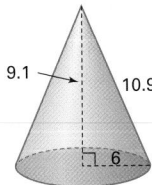

The surface area is
$S = L + B$, or
$65.4\pi + 36\pi = 101.4\pi$.

Find the volume of a cone.

Find the volume of the cone above.

The area of the base is 36π, so the volume is
$V = \frac{1}{3}Bh$, or $\frac{1}{3}36\pi(9.1) = 109.2\pi$.

Exercises

Refer to the right cone below.

17. Find the lateral area of the cone.

18. Find the surface area of the cone.

19. Find the volume of the cone.

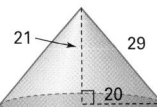

20. Find the volume of the cone if the height is doubled. Round your answer to the nearest hundredth.

Key Skills

Find the surface area of a sphere.

Find the surface area of a sphere with a radius of 9.

The surface area is $S = 4\pi r^2$, or $4\pi 9^2 = 324\pi$.

Find the volume of a sphere.

Find the volume of a sphere with a radius of 9.

The volume is $V = \frac{4}{3}\pi r^3$, or $\frac{4}{3}\pi 9^3 = 972\pi$.

Exercises

21. Find the surface area of a sphere with a radius of 5.

22. Find the volume of a sphere with a radius of 5.

23. Find the radius of a sphere with a volume of 36π.

24. What is the side length of a cube that has the same volume as a sphere with a radius of 1?

Key Skills

Reflect a figure in a three-dimensional coordinate system.

Reflect the segment with endpoints at $(5, 5, 2)$ and $(5, 5, -2)$ across the xz-plane.

The image has endpoints at $(5, -5, 2)$ and $(5, -5, -2)$.

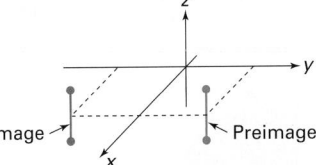

Exercises

A segment has endpoints at $A(0, 0, 0)$ and $B(0, 4, 6)$.

25. Draw a reflection of the segment across the xy-plane. Give the coordinates of the endpoints of the image.

26. Draw a reflection of the segment across the yz-plane. Give the coordinates of the endpoints of the image.

27.

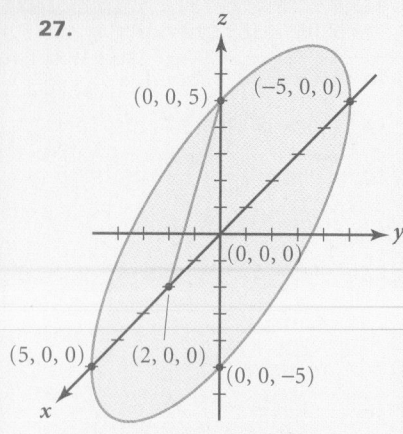

The figure is a flat disk on the xz-plane.

28.

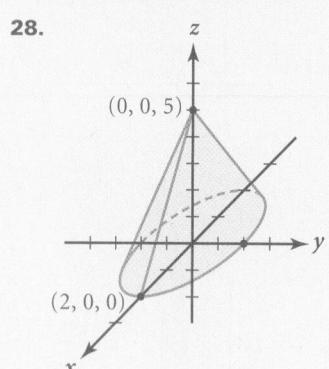

29. ≈ 55,850 ft

30. 283.53 in²; 448.92 in³; 48%

31. 6.30 cm

Key Skills, continued

Sketch the spatial figure formed by rotating a figure about an axis.

Sketch the spatial figure formed by revolving the segment with endpoints at (5, 5, 2) and (5, 5, −2) about the z-axis.

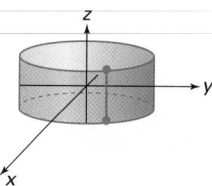

A triangle has vertices at (0, 0, 0), (2, 0, 0), and (0, 0, 5).

27. Sketch the spatial figure formed by rotating the triangle about the y-axis.

28. Sketch the spatial figure formed by rotating the triangle about the z-axis.

Applications

29. PRINTING A roll of paper used in printing has a diameter of 3 feet, a hollow core of 4 inches, and a width of 5 feet. If the thickness of the paper is 0.0015 inch, what is the length of the paper on the roll?

30. SPORTS A basketball has a radius of approximately 4.75 inches when filled. How much material is needed to make one? How much air will it hold? If the basketball is stored in a cubic box whose edges are 9.5 inches long, what percent of the box is not filled by the basketball? (Round your answers to the nearest hundredth.)

31. PHYSICS A spherical soap bubble with a radius of 5 centimeters lands on a flat surface and becomes a hemisphere. What is the radius of the hemisphere?

7 Chapter Test

Find the surface-area-to-volume ratio for each of the following:

1. a cube with a volume of 125 cubic units **1.2**

2. a rectangular prism with dimensions $n \times 1 \times 1$ $\dfrac{4 + \frac{2}{n}}{}$

3. a right rectangular prism with a square base and a height that is quadruple the base edge $\dfrac{9}{2s}$

4. **PHYSICS** Use surface-area-to-volume ratios to explain why solar panels are shaped like large thin rectangular prisms.

Find the surface area and volume of a right rectangular prism with the given dimensions.

5. $l = 5, w = 8, h = 7$ 6. $l = 6, w = 4, h = 0.25$

7. Find the volume of a right triangular prism whose base is an isosceles right triangle with a leg of 12 centimeters and whose height is 15 centimeters. **1080 cm³**

8. **CAMPING** The tent shown is a right triangular prism. The bases of the prism are isosceles triangles. Find the surface area of the tent, including the floor. **115.30 ft²**

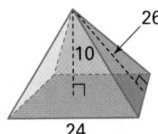

5.0 ft   6.0 ft   4.5 ft

For Exercises 9–12, refer to the regular square pyramid below.

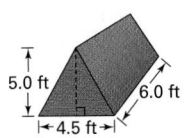

26   10   24

9. Find the lateral area of the pyramid. **1248 units²**

10. Find the surface area of the pyramid. **1824 units²**

11. Find the volume of the pyramid. **1920 units³**

12. Find the volume of the pyramid if the height is doubled. **3840 units³**

A right cylinder has a radius of 6 inches and a height of 8 inches.

13. Find the lateral area of the cylinder. **96π ≈ 301.59 in²**

14. Find the surface area of the cylinder. **168π ≈ 527.79 in²**

15. Find the volume of the cylinder. **288π ≈ 904.79 in³**

16. The volume of a cylinder is 100π cm³ and the height is 4 cm. Find the circumference of the base. **10π ≈ 31.4 cm**

Find the surface area and volume of each right cone.

17.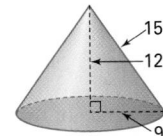
15   12   9

18.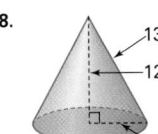
13   12   5

19. Find the surface area of a sphere with a radius of 12. **576π ≈ 1809.56 units²**

20. Find the volume of a sphere with a radius of 12.

21. Find the surface area and volume of a sphere with a diameter of 6x. Give an exact answer in terms of π and the variable.

22. Find the surface area and volume of a kickball with a diameter of 9 inches. Round your answers to the nearest hundredth.

What are the coordinates of the image if each point below is reflected across the xy-plane in a three-dimensional coordinate system?

23. $(1, 5, 9)$
$(1, 5, -9)$

24. $(-3, 2, -5)$
$(-3, 2, 5)$

25. $(6, -1, -4)$
$(6, -1, 4)$

$\overline{JK}$ **has endpoints J(3, 0, 6) and K(3, 0, 0). What spatial figure is formed in each case?**

26. rotating $\overline{JK}$ about the x-axis

27. rotating $\overline{JK}$ about the z-axis

---

4. Sample answer: The large rectangular form maximizes the surface area to absorb solar power while the thinness of the panel minimizes the volume of the panel.

5. 262 units²;
280 units³

6. 53 units²;
6 units³

17. 216π ≈ 678.24 units²;
324π ≈ 1017.88 units³

18. 90π ≈ 282.74 units³;
100π ≈ 314.16 units³

20. 2304π ≈ 7238.23 units³

21. 36πx² units²;
36πx³ units³

22. 81π ≈ 254.47 in.²;
121.5π ≈ 381.70 in.³

26. circle and its interior with center at (3, 0, 0) and r = 6

27. a cylinder centered on the z-axis with h = 6 and r = 3

Multiple-Choice Samples

The first half of the Cumulative Assessment contains a multiple-choice section. This part of the Cumulative Assessment consists of items commonly found on standardized tests.

Free-Response Grid Samples

The second half of the Cumulative Assessment consists of a free-response section. This part requires student-produced response items like those commonly found on college entrance exams. These questions require the use of machine-scored answer grids. You may wish to have students practice answering these items in preparation for standardized tests.

1. b

2. c

3. b

4. d

5. c

6. b

7. b

MULTIPLE-CHOICE For Questions 1–10, write the letter that indicates the best answer.

1. Find the volume of the figure below. **(LESSON 7.5)**

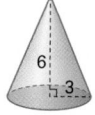

   a. 36π **b.** 18π
   c. 54π **d.** 72π

2. Refer to the triangle below. Which of the following statements is true? **(LESSON 3.5)**

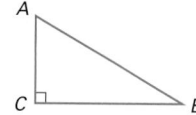

   a. $\angle A$ and $\angle B$ are adjacent to $\angle C$.
   b. $m\angle A + m\angle B = 180$
   c. $m\angle A + m\angle B = m\angle C$
   d. $\angle A$ and $\angle B$ are complements of $\angle C$.

3. Refer to the prism below. Find the area of *RSTU*. **(LESSON 6.3)**

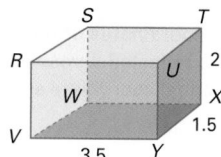

   a. 7 units²   **b.** 5.25 units²
   c. 3 units²   **d.** 10.5 units²

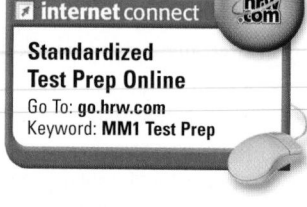

📶 internet connect

Standardized Test Prep Online
Go To: go.hrw.com
Keyword: **MM1 Test Prep**

4. Refer to the figure below. What can be concluded about the measures of $\angle 1$ and $\angle 2$? **(LESSON 3.3)**

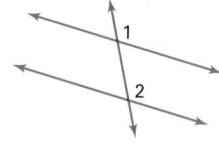

   a. $m\angle 1 = m\angle 2$
   b. $m\angle 1 + m\angle 2 = 180°$
   c. $m\angle 1 + m\angle 2 = 90°$
   d. No conclusion can be drawn.

5. Vertical angles are _____?_____. **(LESSON 2.5)**

   a. complementary
   b. supplementary
   c. congruent
   d. obtuse

6. If lines are parallel, then same-side interior angles are _____?_____. **(LESSON 3.3)**

   a. complementary
   b. supplementary
   c. congruent
   d. obtuse

7. Which of the following is valid for proving that triangles are congruent?
   (LESSONS 4.2 AND 4.3)

   a. AAA
   b. ASA
   c. SSA
   d. none of the above

8. The diagonals of a rhombus are _____?_____.
(LESSON 4.5)
   a. congruent
   b. parallel
   c. perpendicular
   d. convex

9. The diagonals of a rectangle are _____?_____.
(LESSON 4.5)
   a. congruent
   b. parallel
   c. perpendicular
   d. convex

10. Find x in the diagram below. *(LESSON 3.7)*

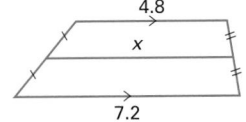

   a. 12
   b. 3.6
   c. 6
   d. 5.4

11. Write the conclusion, if any, that follows from the given statements. *(LESSON 2.3)*

If a person exercises regularly, then that person is in shape.

Vanessa is in shape.

12. Complete the Euler diagram below.
(LESSON 3.2)

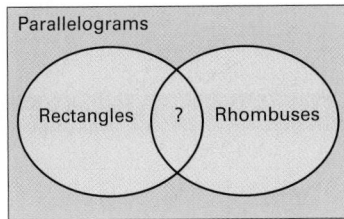

13. What is the area of a circle with a radius of 7? Give an exact answer. *(LESSON 5.3)*

14. What is the radius of a circle with an area of 7? Give an exact answer. *(LESSON 5.3)*

15. Draw six orthographic views of the solid below. *(LESSON 6.1)*

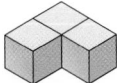

FREE-RESPONSE GRID
Items 16–18 may be answered by using a free-response grid such as that commonly used by standardized-test services.

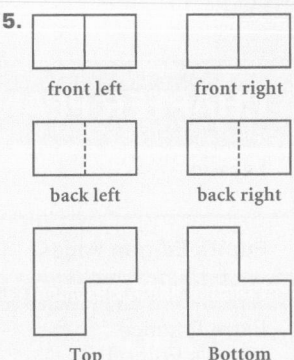

16. Find the length of the diagonal of the right rectangular prism below. Round your answer to the nearest tenth. *(LESSON 6.3)*

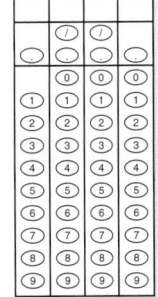

17. Find the surface area of the right rectangular prism above. *(LESSON 7.1)*

18. Find the volume of the regular hexagonal pyramid below. Round your answer to the nearest unit. *(LESSON 7.3)*

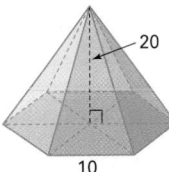

8. c

9. a

10. c

11. No conclusion

12. Squares

13. 49π units^2

14. $\sqrt{\dfrac{7}{\pi}}$ units

15.

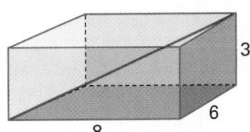

front left   front right

back left   back right

Top   Bottom

16. 10.4 units

17. 180 units^2

18. 1732 units^3

Similar Shapes

CHAPTER PLANNING GUIDE

Lesson	8.1	8.2	8.3	8.4	8.5	8.6	Project and Review
Pupil's Edition Pages	498–506	507–516	517–524	525–532	533–542	543–551	552–561
Practice and Assessment							
Extra Practice (Pupil's Edition)	842	843	843	844	844	845	
Practice Workbook	43	44	45	46	47	48	
Practice Masters Levels A, B, and C	148–150	151–153	154–156	157–159	160–162	163–165	
Standardized Test Practice Masters	57	58	59	60	61	62	63
Assessment Resources	99	100	101	103	104	105	102, 106–111
Visual Resources							
Lesson Presentation Transparencies Vol. 2	1–4	5–8	9–12	13–16	17–20	21–24	
Teaching Transparencies	65–66		67	68–69	70–73		
Answer Key Transparencies	304–310	311–313	314–317	318–321	322–324	325–327	328–330
Quiz Transparencies	8.1	8.2	8.3	8.4	8.5	8.6	
Teacher's Tools							
Reteaching Masters	99–100	101–102	103–104	105–106	107–108	109–110	
Make-Up Lesson Planner for Absent Students	50	51	52	53	54	55	
Student Study Guide	50	51	52	53	54	55	
Spanish Resources	50	51	52	53	54	55	
Block Scheduling Handbook							16–17
Activities and Extensions							
Lesson Activities	136–140		141–145			146–151	
Enrichment Masters	50	51	52	53	54	55	
Cooperative-Learning Activities	50	51	52	53	54	55	
Problem-Solving/ Critical Thinking	50	51	52	53	54	55	
Student Technology Guide	46–47	48		49	50–51	52	
Long Term Projects							29–32
Writing Activities for Your Portfolio							22–24
Tech Prep Masters							35–38
Building Success in Mathematics							20–22

LESSON PACING GUIDE

Lesson	8.1	8.2	8.3	8.4	8.5	8.6	Project and Review
Traditional	2 days	2 days	2 days	2 days	2 days	2 days	2 days
Block	1 day	1 day	1 day	1 day	1 day	1 day	1 day
Two-Year	4 days	4 days	4 days	4 days	4 days	4 days	4 days

CONNECTIONS AND APPLICATIONS

Lesson	8.1	8.2	8.3	8.4	8.5	8.6	Review
Algebra	499, 504	513, 516	524	529		551	
Geometry	498–506	507–516	517–524	525–532	533–542	543–551	552–561
Coordinate Geometry	506			530, 532			
Business and Economics						550	
Life Skills	498	515	523			550	
Science and Technology	501, 505	510, 511, 514		531, 532	533, 540, 541	546, 549	
Sports and Leisure	504	515, 516		531		550	
Other							
Cultural Connection: Africa				527	542		

BLOCK-SCHEDULING GUIDE

Day	Lesson	Teacher Directed: Lesson Examples, Teaching Transparencies	Student Guided: Activity, Try This	Cooperative-Learning Activity, Lesson Activity, Student Technology Guide	Practice: Practice & Apply, Extra Practice, Practice Workbook	Assessment: Quiz, Mid-Chapter Assessment	Problem Solving, Reteaching
1	8.1	15 min	15 min	20 min	55 min	15 min	15 min
2	8.2	15 min	15 min	20 min	55 min	15 min	15 min
3	8.3	15 min	15 min	20 min	55 min	15 min	15 min
4	8.4	15 min	15 min	20 min	55 min	15 min	15 min
5	8.5	15 min	15 min	20 min	55 min	15 min	15 min
6	8.6	10 min	20 min	20 min	55 min	15 min	15 min
7	Assess.	50 min	90 min	90 min	65 min	30 min	
		PE: Chapter Review	PE: Chapter Project, Writing Activities	Tech Prep Masters	PE: Chapter Assessment, Test Generator	Chap. Assess. (A or B), Alt. Assess. (A or B), Test Generator	

PE: Pupil's Edition

Alternative Assessment

The following suggest alternative assessments for students who may benefit from a different type of assessment than the regular chapter quizzes and the mid-chapter/end-of-chapter test. Visit the HRW web site to get additional Alternative Assessment material.

internet connect

Alternative Assessment
Go To: **go.hrw.com**
Keyword: **MG1 Alt Assess**

Performance Assessment

1. Trace the figure at right on your own paper. Choose a center for dilation and use a scale factor of 3.

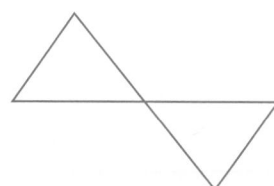

2. Write a paragraph proof explaining why any two regular polygons (such as equilateral triangles, squares, regular hexagons, etc.) are similar.

3. Find the distance across the river. Explain what theorem or theorems allowed you to find the answer.

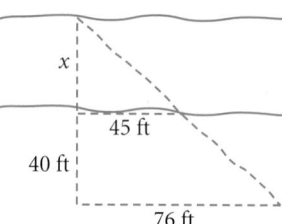

Portfolio Project

Suggest that students choose one of the following projects for inclusion in their portfolios.

1. Draw a floor plan of a house using a scale factor of 1 cm to 10 feet. List the proportions relating to the measurements on the floor plan to the actual dimensions of the house.

2. Find a picture from a magazine or newspaper. Draw centimeter grid lines over the picture. On a large posterboard, draw a 2 or 3 centimeter grid. Recreate the original picture by drawing exactly what is in each grid square onto the larger grid square. Give the scale factor of the drawing you create.

3. Research the sculpting of Mount Rushmore. Most sculptures are copied from a smaller original sculpture, which is the case in Mount Rushmore. What is the scale factor of the original sculpture to the finished monument? What methods did the artist use to create the giant sculpture?

internet connect

The table below identifies the pages in this chapter that contain internet and technology information.

Content Links

Activities Online	**pages 502, 511, 520**
Portfolio Extensions	**pages 506, 542**
Homework Help Online	**pages 503, 512, 521, 528, 537, 548**

Resource Links

Parents can go online and find concepts that students are learning–lesson by lesson–and questions that pertain to each lesson, which facilitate parent-student discussion.

Go To: **go.hrw.com**
Keyword: **MG1 Parent Guide**

Technical Support

The following may be used to obtain technical support for any HRW software product.

Online Help: **www.hrwtechsupport.com**

e-mail: **tschrw@hbtechsupport.com**

HRW Technical Support Center: **(800)323-9239**
7 AM to 10 PM Monday through Friday CST

Visit the HRW math web site at: **www.hrw.com/math**

Technology

Technology Objectives and Suggestions

Lesson 8.1 Dilations and Scale Factors

In this lesson students construct a dilation of a closed plane figure using a scale factor. This lesson can be introduced using geometry software. In most programs, dilations can be done by highlighting a figure, accessing dilations as a menu option, and giving the computer a scale factor. The transformation is done automatically. Students can print out the similar figures created in this way and measure them, giving ratios for linear measurements and for area.

In addition, the construction in Activity 2 can be done using geometry software. It might be interesting for students to transform the figure using the lines and the center of dilation and comparing that to the dilation that can be done automatically using a menu command.

Lesson 8.2 Similar Polygons

In this lesson students define similar polygons and use properties of proportions and scale factors to solve problems involving similar polygons. Students will find using a calculator helpful for finding solutions to problems in the exercise set.

Lesson 8.3 Triangle Similarity Postulates

In this lesson students develop the AA Triangle Similarity Postulate and the SSS and SAS Triangle Similarity Theorems. Geometry software can be used with the three activities in this lesson to draw the figures as described in Steps 1 and 2 of each activity.

Geometry software may also be helpful in answering the Graphic Design questions in Exercises 35–38. Have students recreate the designs mentioned in each exercise.

Lesson 8.4 The Side-Splitting Theorem

In this lesson students develop and prove the Side-Splitting Theorem and use the theorem to solve problems. Students will need calculators for many of the exercises in this lesson.

To reteach the Side-Splitting Theorem, have students construct several triangles and trapezoids with their midsegments using geometry software. Then have the computer measure the corresponding segments and calculate the ratios of the segments. Students can then drag the figures to change lengths, but see that the segments are still proportional. While not a proof, this can be a convincing demonstration of the theorem.

Lesson 8.5 Indirect Measurement and Additional Similarity Theorems

In this lesson students use triangle similarity to measure distances indirectly. They also develop and use similarity theorems for altitudes and medians of triangles. Students will need calculators for the exercises in this lesson. You can have them draw figures for their proofs using geometry software, such as those for Exercises 26–41.

Have students research latest technology. Surveyors use special instruments to measure distances that cannot be measured directly. Have students research the type of tools used by surveyors, including modern laser instrument.

Lesson 8.6 Area and Volume Ratios

In this lesson students develop and use ratios for areas and volumes of similar figures and solids. They also explore relationships between cross-sectional area, weight, and height. Calculators are needed for the exercises in this lesson. For Activity 3, students can compile the data using a spreadsheet.

Background Information

In this chapter, students explore similar figures through dilations in the plane, constructions, and formal proofs. Students use similarity to measure distances indirectly and explore the area and volume of similar figures.

CHAPTER RESOURCES

- Block-Scheduling Handbook
- Writing Activities for Your Portfolio
- Tech Prep Masters
- Long-Term Project
- Assessment Resources:
  Mid-Chapter Assessment
  Chapter Assessments
  Alternative Assessments
- Test and Practice Generator
- Technology Handbook

Chapter Objectives

- Construct a dilation of a segment and a point by using a scale factor. [8.1]
- Construct a dilation of a closed plane figure. [8.1]
- Define *similar polygons*. [8.2]
- Use Properties of Proportions and scale factors to solve problems involving similar polygons. [8.2]
- Develop the AA Triangle Similarity Postulate and the SSS and SAS Triangle Similarity Theorems. [8.3]
- Develop and prove the Side-Splitting Theorem. [8.4]
- Use the Side-Splitting Theorem to solve problems. [8.4]

Similar Shapes

IN 1948 KORCZAK ZIOLKOWSKI (1908–1982) began work on the world's largest sculpture, the Crazy Horse Memorial, in the Black Hills of South Dakota. The sculpture is 563 feet high, more than 9 times the height of the sculptures on Mount Rushmore. The completed face of Crazy Horse is nine stories tall.

In the photo on the page facing, a scale model of the finished statue stands in front of the actual work, which is 1 mile behind it. The model is $\frac{1}{34}$ the height of the mountain carving in progress. Such figures are known as *similar* figures. You will be studying similar figures throughout this chapter.

Korczak Ziolkowski with scale model

Lessons

8.1 ● **Dilations and Scale Factors**

8.2 ● **Similar Polygons**

8.3 ● **Triangle Similarity Postulates**

8.4 ● **The Side-Splitting Theorem**

8.5 ● **Indirect Measurement and Additional Similarity Theorems**

8.6 ● **Area and Volume Ratios**

Chapter Project Indirect Measurement

Precision explosives were used to create the outline of the form

About the Photos

Located in the Black Hills of South Dakota, the Crazy Horse Memorial is a cultural and humanitarian project dedicated to the Native Americans of North America. In 1964, sculptor Korczak Ziolkowski (1908–1982) created a 16-ton plaster model that is $\frac{1}{34}$ the size of the 563-foot high sculpture. In 1994, a fiberglass cast of the scale model was created to facilitate computer-imaging programs that were used to measure the mountain. This monumental mountain-carving project and its scale models illustrate an important use of similar figures. In this chapter, students will learn about properties and applications of similar figures.

Crazy Horse 1/34ᵗʰ Scale Model
© KORCZAK, Sculptor

- Use triangle similarity to measure distances indirectly. [8.5]
- Develop and use similarity theorems for altitudes and medians of triangles. [8.5]
- Develop and use ratios for areas of similar figures. [8.6]
- Develop and use ratios for volumes of similar solids. [8.6]
- Explore relationships between cross-sectional area, weight, and height. [8.6]

Portfolio Activities appear at the end of Lessons 8.1 and 8.5. Each serves as preparation for the Chapter Project. The Portfolio Activities, as well as the Chapter Project Activities, are appropriate for inclusion in the student's portfolio. Students should be encouraged to include in their portfolios any other work in which they feel a sense of pride or a sense of accomplishment.

About the Chapter Project

Similarity is a term used to describe objects that are the same shape, but not necessarily the same size. One application of similarity is a scale model, which is an object built in the exact shape of another object, but usually smaller. Scale models are often used to get an overall view of a large object or area.

In the Chapter Project, you and your classmates build a scale model of your or school, starting with a map.

After completing the Chapter Project you will be able to do the following:

- Work as a team to apply the techniques of measurement to real-world objects.
- Create maps and models of large structues that cannot be measured directly.

About the Portfolio Activities

Throughout the chapter, you will be given opportunities to complete Portfolio Activities that are designed to support your work on the Chapter Project.

The theme of each Portfolio Activity and of the Chapter Project is scale models.

- In the Portfolio Activity on page 506, you will enlarge a design by using grids.
- In the Portfolio Activity on page 542, you will use indirect measurement to find the dimensions of a building.

🖳 internet connect

Chapter Internet Features and Online Activities

LESSON	KEYWORD	PAGE	LESSON	KEYWORD	PAGE
8.1	MG1 Dilations	502	8.3	MG1 Similarity	520
	MG1 Homework Help	503		MG1 Homework Help	521
	MG1 Scale	506	8.4	MG1 Homework Help	528
8.2	MG1 Maps	511	8.5	MG1 Homework Help	537
	MG1 Homework Help	512		MG1 Indirect	542
			8.6	MG1 Homework Help	548

Dilations and Scale Factors

QUICK **WARM-UP**

1. Find the distance between the points $(-2, 1)$ and $(4, 3)$.
$\sqrt{40}$, or $2\sqrt{10}$

2. What are the slopes of the lines given by $y = -3x + 2$? $y = -3x - 3$? -3

3. The points $(-2, 4)$, $(3, 4)$, and $(3, -1)$ are vertices of a square. Find the fourth vertex of this square. $(-2, -1)$

4. Write the equation, in slope-intercept form, of the line containing the points $(-1, 5)$ and $(1, 3)$.
$y = -x + 4$

Also on Quiz Transparency 8.1

Objectives

- Construct a dilation of a segment and a point by using a scale factor.
- Construct a dilation of a closed plane figure.

Why A camera obscura projects an inverted image through a small opening into a dark room. Dilations can be used to explain the process.

The principle of the camera obscura, the forerunner of the modern camera, was discovered by physicist Ibn al-Haitham (965–1039 C.E.).

APPLICATION
PHOTOGRAPHY

The camera obscura was used to draw accurate images of an object or a scene. The artist traced the image that was projected through a pinhole onto a drawing surface. In a modern camera, the pinhole has been replaced by a lens and the drawing surface has been replaced by light-sensitive photographic film.

Teach

Why Students need to understand the properties of dilations because they are the analytical basis for constructing similar figures and drawing scale models.

Dilations

So far you have studied three types of transformations: translations, rotations, and reflections. These transformations are called rigid because they preserve size and shape.

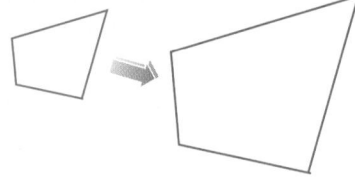

A **dilation** is an example of a transformation that is not rigid. Dilations preserve the shape of an object, but they may change its size.

The size of the figure is changed by a dilation.

A dilation of a point in a coordinate plane can be found by multiplying the x- and y-coordinates of a point by the same number, n.

$$D(x, y) = (nx, ny)$$

The number n is called the **scale factor** of the transformation.

Alternative Teaching Strategy

TECHNOLOGY Students can explore dilations and other transformations with geometry graphics software, which can also be used to measure the lengths of image segments and to explore the collinearity of transformed points. Ask students to verify their conjectures with computer-drawn images.

EXAMPLE ① What is the image of the point (2, 3) transformed by the dilation $D(x, y) = (4x, 4y)$? What is the scale factor?

● **SOLUTION**

The image is the point $D(2, 3) = (4 \cdot 2, 4 \cdot 3) = (8, 12)$. The scale factor is the multiplier, 4.

Activity 1
Dilations in the Coordinate Plane

YOU WILL NEED

graph paper, ruler, and calculator
OR
geometry graphics software

1. Plot $A(3, 4)$ on a coordinate plane. Use the distance formula or a ruler to find the distance, OA, from $O(0, 0)$ to A. Use dilations with the scale factors given below to transform point A, and then copy and complete the table.

A	OA	Scale factor	Image, A′	OA′	Ratio $\frac{OA'}{OA}$
(3, 4)	?	2	(?, ?)	?	?
(3, 4)	?	0.5	(?, ?)	?	?
(3, 4)	?	−1	(?, ?)	?	?
(3, 4)	?	n	(?, ?)	?	?

2. Plot point A and its image, A', for each dilation above. What is the simplest geometric figure that contains all of these points? Add this figure to your graph.

3. Complete the following conjecture:

CHECKPOINT ✔

Conjecture

The distance from the origin to the image of a point transformed by a dilation with scale factor *n* is ____?____ the distance from the origin to the preimage.

4. Plot point A again and a new point, $B(5, 6)$, on a new coordinate plane. Use the distance formula or a ruler to find the length of $\overline{AB}$. Use dilations with the scale factors given below to transform points A and B, and then copy and complete the table.

B	AB	Scale factor	Image, B′	A′B′	Ratio $\frac{A'B'}{AB}$
(5, 6)	?	2	(?, ?)	?	?
(5, 6)	?	0.5	(?, ?)	?	?
(5, 6)	?	−1	(?, ?)	?	?
(5, 6)	?	n	(?, ?)	?	?

Algebra

5. Find the slope of each dilated segment, $\overline{A'B'}$.

6. Complete the following conjecture:

CHECKPOINT ✔

Conjecture

The image of a segment transformed by a dilation with a scale factor of *n* is ____?____ the length of the preimage, and the slopes of the image and preimage are ____?____.

Interdisciplinary Connection

PHOTOGRAPHY Photographers use dilations when they reduce or enlarge photographs. Have students research the techniques and types of equipment used by photographers to do enlargements. Have them calculate the area of a 3-inch-by-5-inch photograph and compare it with the area of an enlargement by a scale factor of 2. **The enlargement will have 4 times the area, or 60 square inches.**

ADDITIONAL
EXAMPLE ①

What is the image of the point (−1, −3) after a transformation by the dilation $D(x, y) = \left(\frac{1}{3}x, \frac{1}{3}y\right)$? What is the scale factor? $\left(-\frac{1}{3}, -1\right); \frac{1}{3}$

Activity 1 **Notes**

In this Activity, students will perform dilations on points and segments by using different scale factors. This will help students understand that the ratio of the image-point distance to the preimage-point distance (from the origin) is equal to the scale factor.

For a student worksheet of this Activity and detailed Teacher Notes, see page 136 in the Lesson Activities booklet.

Teaching Tip

The prime notation that is used to label the image points may confuse some students. Point out that it is used to make the correspondence between preimage points and image points more obvious.

CHECKPOINT ✔
3. *n* times

CHECKPOINT ✔
6. *n* times, equal

In a dilation, each point and its image lie on a straight line that passes through a point known as the **center of dilation**. In Activity 1, that point was the origin. Every dilation has a center of dilation.

APPLICATION
OPTICS

The colored portion of the eye, known as the *iris*, changes shape to let in more or less light. When the iris is relatively open it is said to be *dilated*. A similar device in a camera, also known as the *iris*, is placed behind the lens to let in more or less light to accommodate different lighting conditions.

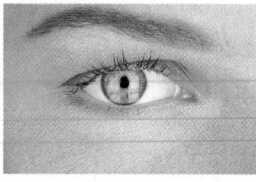 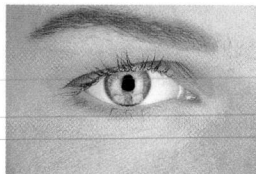

In the photo on the left, the eye is contracted for relatively bright light. In the photo on the right, it is dilated for less bright conditions.

You may have noticed that the size of images of segments that have been dilated varies according to the scale factor. If the size of a figure is reduced by a dilation, the dilation is called a **contraction**. If the size of a figure is enlarged by a dilation, the dilation is called an **expansion**. For a dilation with a scale factor of n:

If $|n| < 1$, the dilation is a contraction.

If $|n| > 1$, the dilation is an expansion.

CRITICAL THINKING

How is the image of a point or segment affected by a negative scale factor?

Activity 2
Drawing a Dilation

YOU WILL NEED

ruler and calculator
OR
geometry graphics software

1. Draw a polygon, such as a triangle, and a center of dilation. Draw lines that pass through each vertex of the figure and the center of dilation.

2. Decide on a scale factor n for your dilation. Choose one of the vertices and measure the distance, x, from the center of dilation to the vertex. Multiply this distance by n to obtain a new value, x'. On the line containing the chosen vertex, plot a point that is a distance of x' from the center of dilation. This new point is the image of the chosen vertex.

3. Repeat for each of the other vertices of your figure.

CHECKPOINT ✔

4. Connect the images of the vertices to form the dilated image of the figure.

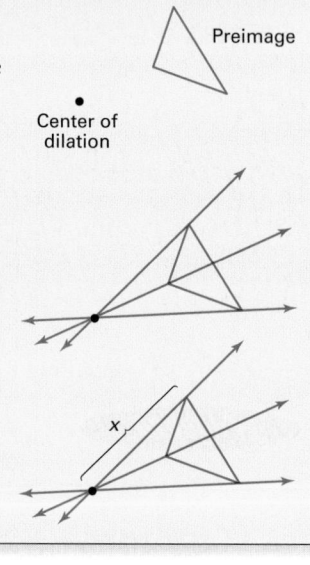

CRITICAL THINKING

How could you draw an approximate dilation of a curved plane figure?

Enrichment

A pantograph is an instrument that is used to reproduce maps and other drawings. Have students research how it works and write a report. Extend the assignment by having students build a pantograph.

Inclusion Strategies

USING COGNITIVE STRATEGIES Help students make generalizations by having them keep track of the scale factor, preimage distance, and image distance for each dilation they do. Encourage them to make conjectures about the properties of dilations as they do each one and to verify the conjectures as they go along.

EXAMPLE 2

The students shown below are using a pinhole to observe a solar eclipse. The process may be understood as a dilation, with the pinhole as the center of dilation. The diameter of the Sun is approximately 870,000 mi. If the image is 0.25 inch in diameter, what is the scale factor of the dilation?

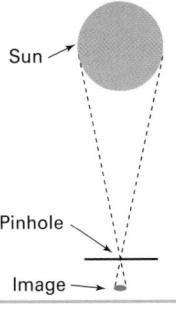

● **SOLUTION**

The diagram at right (not to scale) shows the path of rays of light from the edges of the Sun as seen from Earth. The scale factor is negative because the image is on the opposite side of the center of dilation (the pinhole). The ratio of the image to the preimage is

$\frac{0.25 \text{ inch}}{55,123,200,000 \text{ inch}} \approx 4.5 \times 10^{-12}$, so the

scale factor is approximately -4.5×10^{-12}.

Sun →
Pinhole →
Image →

ADDITIONAL
EXAMPLE 2

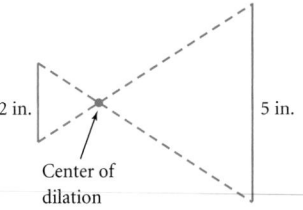

The figure below shows a segment and its image under a dilation. What is the scale factor of the dilation? 2.5

2 in.   5 in.

Center of dilation

Teaching Tip

Summarize dilations by pointing out that if k is the scale factor with $|n| > 1$, then the dilation will be an expansion. If $0 < |n| < 1$, then the dilation will be a contraction.

Reteaching the Lesson

INVITING PARTICIPATION Have a student draw a regular polygon on the board or overhead and choose a center of dilation and a scale factor. Have the rest of the class perform the dilation and compare their results. Have another student draw another polygon and repeat the activity. Continue until a variety of polygons, scale factors, and centers of dilation have been used.

Selected Answers

Exercises 7–13, 15–49 odd

ASSIGNMENT GUIDE

In Class	1–13
Core	15–33 odd, 35–39
Core Plus	19–33 odd, 35–44
Review	45–50
Preview	51–54

✐ Extra Practice can be found beginning on page 818.

7. (3, 15)

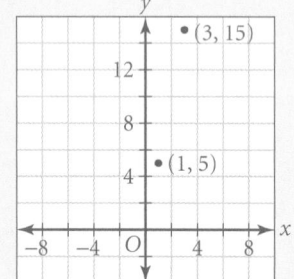

8. (−2, 8)

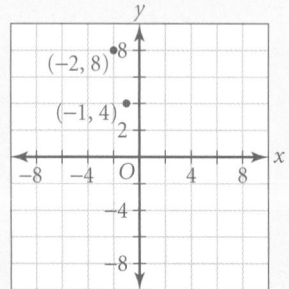

9. (1.5, −0.5)

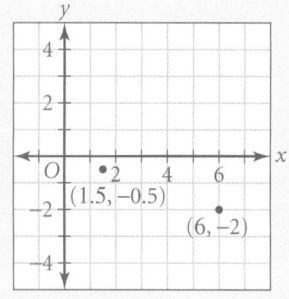

Exercises

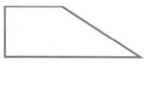

internet connect

Activities Online

Go To: **go.hrw.com**
Keyword:
MG1 Dilations

Communicate

1. What is a dilation? How is a dilation different from other transformations you have studied?

2. What is a scale factor? How can you determine the scale factor of a dilation by looking at a segment and its image?

Explain how the image of a figure transformed by a dilation would be affected by the following scale factors:

3. 2   **4.** 0.5   **5.** −1   **6.** 1

Guided Skills Practice

Find the image of each point transformed by the given dilation. Plot the point and its image on a coordinate plane. *(EXAMPLE 1 AND ACTIVITY 1)*

7. (1, 5); $D(x, y) = (3x, 3y)$ **8.** (−1, 4); $D(x, y) = (2x, 2y)$

9. (6, −2); $D(x, y) = (0.25x, 0.25y)$ **10.** (2, 3); $D(x, y) = (−2x, −2y)$

Copy each figure below and draw a dilation with the given scale factor. *(ACTIVITY 2)*

11. scale factor of 2   **12.** scale factor of −1

13. The figure at right shows a segment and its image under a dilation. What is the scale factor of the dilation? *(EXAMPLE 2)* −0.5

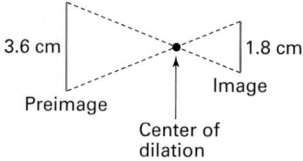

Practice and Apply

For Exercises 14–17, the vertices of a figure and a scale factor, *n*, are given. Use the dilation $D(x, y) = (nx, ny)$ to transform each figure, and plot the preimage and image on a coordinate plane.

14. (1, 3), (2, 5), (4, 3)
$n = 2$

15. (−3, 5), (8, 9), (2, −6)
$n = \frac{1}{3}$

16. (0, 0), (6, 0), (4, 4), (2, 3)
$n = -\frac{1}{2}$

17. (1,1), (3, −1), (−2, −3)
$n = 1.6$

10. (−4, −6)

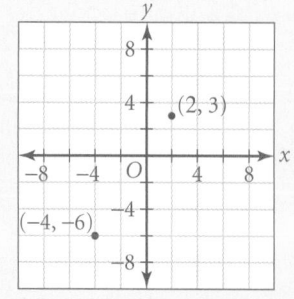

11.

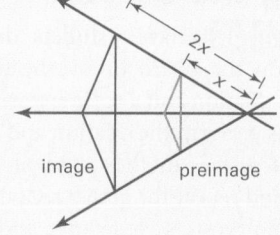

The answers to Exercises 12 and 14–17 can be found in Additional Answers beginning on page 879.

In Exercises 18–21, the blue figures represent the preimages of dilations, and the red figures represent the images. Find the scale factor of each dilation.

Homework Help Online
Go To: go.hrw.com
Keyword:
MG1 Homework Help
for Exercises 18-21

18. 3

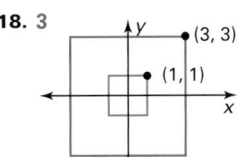

19. $\frac{1}{2}$

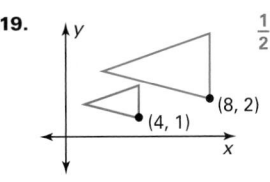

20. −1

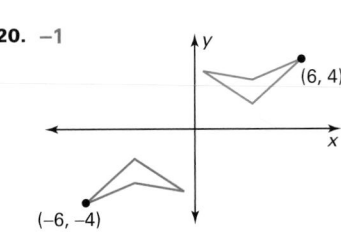

21. $-\frac{1}{3}$

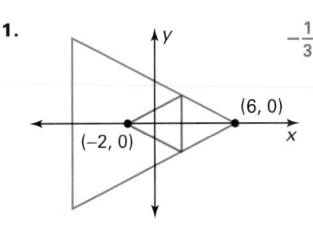

Copy each figure below and draw a dilation with the given scale factor, *n*.

22. $n = 3$ **23.** $n = \frac{3}{4}$ **24.** $n = -2$

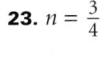 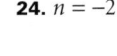 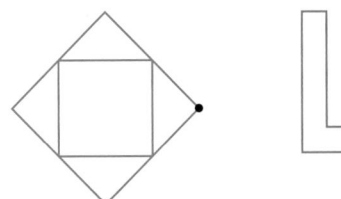

25. Copy the figure below. Draw three dilations of the triangle with a scale factor of 2, using the three given points as centers of dilation. How are the images alike? How are they different?

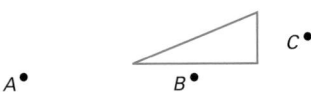

26. Copy the figure below. Locate the center of dilation and find the scale factor.

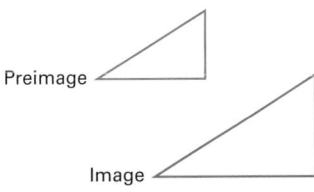

25. b.

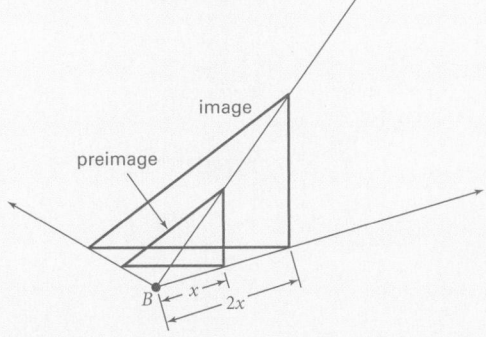

25. c.

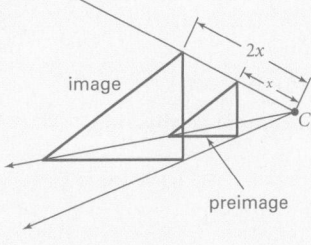

Technology
Geometry graphics software can be used in Exercises 22–24. Encourage students to experiment with other scale factors.

22.

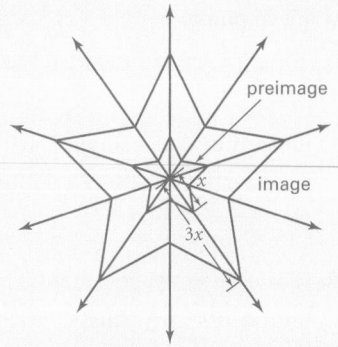

23.

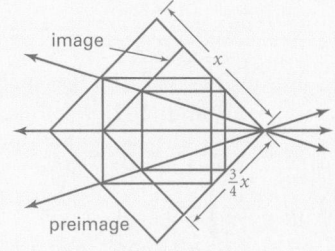

24.

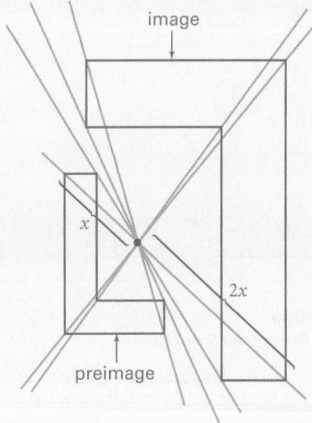

25. a.

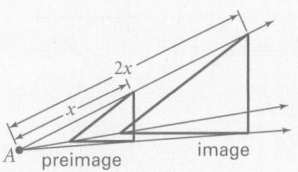

The answer to Exercise 26 can be found in Additional Answers beginning on page 879.

Error Analysis

Students may have difficulty with Exercises 27–34. For Exercises 27–30, a review of the slope formula from algebra may be helpful. For Exercises 31–34, it may be necessary to review how to find an equation of a line that contains two given points.

27. $m = \frac{3-0}{5-1} = \frac{3}{4}$

Endpoints of image are (2, 0) and (10, 6). So, $m = \frac{6-0}{10-2} = \frac{6}{8} = \frac{3}{4}$.

28. $m = \frac{1-3}{3-(-2)} = -\frac{2}{5}$

Endpoints of image are (−10, 15) and (15, 5). So, $m = \frac{5-15}{15-(-10)} = -\frac{10}{25} = -\frac{2}{5}$.

29. $m = \frac{8-4}{4-(-2)} = \frac{4}{6} = \frac{2}{3}$

Endpoints of image are (1, −2) and (−2, −4). So, $m = \frac{-4-(-2)}{-2-1} = \frac{-2}{-3} = \frac{2}{3}$.

30. $m = \frac{4-1}{2-1} = \frac{3}{1} = 3$

Endpoints of image are (1.7, 1.7) and (3.4, 6.8). So, $m = \frac{6.8-1.7}{3.4-1.7} = \frac{5.1}{1.7} = 3$.

31. $y = \frac{1}{5}x$.

$0 = \frac{1}{5}(0) \Rightarrow 0 = 0$

So, the origin is on this line.

Practice

NAME _____ CLASS _____ DATE _____

(((•))) **Practice**

8.1 *Dilations and Scale Factors*

In Exercises 1–4, the dashed figures represent the preimages of dilations, and the solid figures represent the images. Find the scale factor of each dilation.

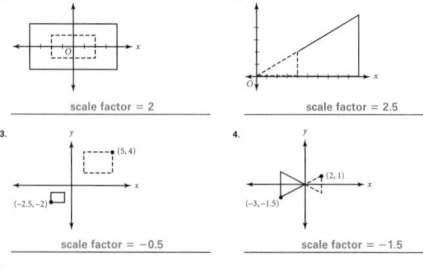

1.   scale factor = 2   2.   scale factor = 2.5

3. (5, 4) (−2.5, −2) scale factor = −0.5   4. (2, 1) (−3, −1.5) scale factor = −1.5

For Exercises 5–10, given a point and a scale factor, find the line that passes through the preimage and image, and show that it contains the origin.

5. (2, 3); $n = 3$ $y = \frac{3}{2}x$; substituting $O(0, 0)$ gives $0 = 0$, which is true. Thus, the origin is on this line.
6. (1, 4); $n = -2$ $y = 4x$; substituting $O(0, 0)$ gives $0 = 0$, which is true. Thus, the origin is on this line.
7. (−1, 2); $n = \frac{1}{3}$ $y = -2x$; substituting $O(0, 0)$ gives $0 = 0$, which is true. Thus, the origin is on this line.
8. (3, 4); $n = -1$ $y = \frac{4}{3}x$; substituting $O(0, 0)$ gives $0 = 0$, which is true. Thus, the origin is on this line.
9. (−4, 3); $n = 2$ $y = -\frac{3}{4}x$; substituting $O(0, 0)$ gives $0 = 0$, which is true. Thus, the origin is on this line.
10. (−3, −3); $n = -1$ $y = x$; substituting $O(0, 0)$ gives $0 = 0$, which is true. Thus, the origin is on this line.

For Exercises 27–30, the endpoints of a segment and a scale factor n are given. Show that the dilation image of the segment has the same slope as the preimage.

27. (1, 0) and (5, 3); $n = 2$ **28.** (−2, 3) and (3, 1); $n = 5$

29. (−2, 4) and (4, 8); $n = -\frac{1}{2}$ **30.** (1, 1) and (2, 4); $n = 1.7$

For Exercises 31–34, a point and a scale factor are given. Find the line that passes through the preimage and image, and show that this line contains the origin.

31. (5, 1); $n = 4$ **32.** (−2, 3); $n = \frac{5}{6}$

33. (3, −5); $n = -3$ **34.** (4, 7); $n = 2.5$

The figure below shows a right rectangular prism in a three-dimensional coordinate system.

35. Draw the image of the prism transformed by the dilation $D(x, y, z) = (2x, 2y, 2z)$. Label the coordinates of the image's vertices.

36. What is the ratio of the lengths of the edges in the image to the lengths of the edges in the preimage? **2**

37. What is the ratio of the surface area of the image to the surface area of the preimage? **4**

38. What is the ratio of the volume of the image to the volume of the preimage? **8**

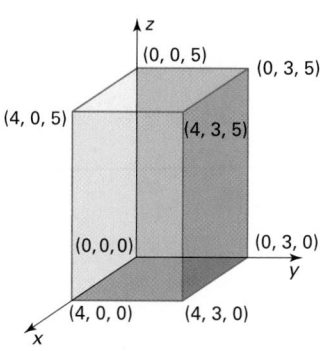

CHALLENGE

39. An example of a dilation that is not centered at the origin of a coordinate plane is $D(x, y) = (2x - 4, 2y - 3)$. Use this dilation to transform the segment with endpoints (2, 3) and (5, 5). Plot the preimage and the image on a coordinate plane. Determine the scale factor and locate the center of dilation. Write the rule for a dilation with a center of (2, 1) and a scale factor of 4.

APPLICATION

Cubed lattice

40. HOBBIES A quilter has a pattern for a 4-in. square quilt block and wishes to enlarge it to a 12-in. block. What is the scale factor of the enlargement? The pattern is traced onto a grid of 1-in. squares, as shown at left. One of the shapes in the pattern has vertices at (0, 1), (1, 1), (2, 2), and (1, 2). What are the coordinates of the image of this shape in the 12-in. block?

scale factor: **3**
coordinates of image: **(0, 3), (3, 3), (6, 6), (3, 6)**

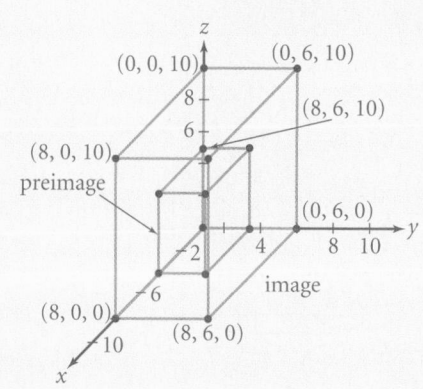

32. $y = -\frac{3}{2}x$.

$0 = \frac{-3}{2}(0) \Rightarrow 0 = 0$

So, the origin is on this line.

33. $y = \frac{-5}{3}x$.

$0 = -\frac{5}{3}(0) \Rightarrow 0 = 0$

So, the origin is on this line.

34. $y = \frac{7}{4}x$.

$0 = \frac{7}{4}(0) \Rightarrow 0 = 0$

So, the origin is on this line.

35.

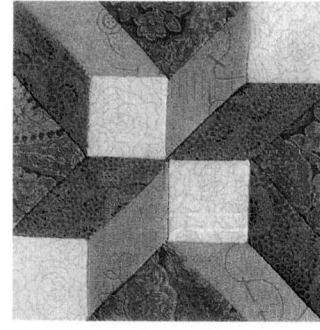

OPTICS The diagram below shows a part of a camera obscura. The image is projected through a small hole into a dark room or chamber. This projection is an example of a dilation.

41. What part of the camera obscura acts as the center of dilation?

42. Is the scale factor positive or negative? Explain your answer.

43. Explain why the projected image is inverted.

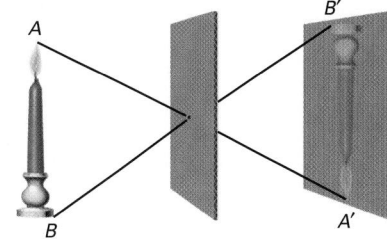

44. **GRAPHIC ARTS** An artist is using a photocopier to reduce a design. The original design is 5 in. wide. The copy should be 2 in. wide. What is the scale factor of the dilation? $\frac{2}{5}$

41. The small opening in the center plate.

42. The scale factor is negative because the image is inverted.

43. The image is inverted because the light from point A travels in a straight line to A'; the light from point B travels in a straight light to point B'; and so on for each point on the preimage.

Look Back

45. The base of an isosceles triangle is 6 m and the legs are 8 m each. Find the perimeter and area of the triangle. *(LESSONS 5.2 AND 5.4)* P = 22 m
A ≈ 22.25 m²

46. A leg of a 45-45-90 triangle is 7 cm long. What is the length of the hypotenuse? *(LESSON 5.5)* 9.9 cm

47. **ENGINEERING** A spherical gas tank has an outer diameter of 40 ft. The tank is made with 1-in. thick steel. Find the difference between the surface area of the outside and of inside of the tank. *(LESSON 7.6)* 6019.29 in²

48. **ENGINEERING** Suppose that 1 gal of paint covers 400 ft². How many gallons of paint are needed to paint the inside and outside of the tank described in Exercise 47? *(LESSON 7.6)* 25 gal

49. **EARTH SCIENCE** The circumference of a great circle of Earth is about 40,000 km. What is the radius of Earth? *(LESSON 7.6)* 6366.2 km

50. **EARTH SCIENCE** The height of Earth's atmosphere is about 550 km. Use this information and your answer to Exercise 49 to find the volume of Earth and its atmosphere. *(LESSON 7.6)* 1.39×10^{12} km³

Earth's atmosphere, as photographed by a Russian cosmonaut

39.

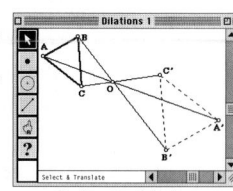

scale factor: 2
center of dilation: (4, 3)
$D(x, y) = (4x - 2, 4y - 1)$

Student Technology Guide

NAME _____ CLASS _____ DATE _____

Student Technology Guide
8.1 *Dilations and Scale Factors, page 1*

One day Debra sketched △ABC, located and labeled point O outside of △ABC, chose O as a center of dilation, and then dilated △ABC. Without thinking, she chose a negative number, −1.5, as the scale factor. Shown here is the image △A'B'C' that she got.

It appears that △A'B'C' is related to △ABC in another way. Whether the scale factor is positive or negative is important.

Use geometry graphics software as directed.

Check students' sketches.

1. Sketch △ABC with heavy lines and point O as shown. Using O as center of dilation and scale factor −1.5, construct the image of △ABC with dashed lines. Show its vertex labels.

2. Based only on observations, describe how A'B'C' is obtained from △ABC.
   Rotate △ABC about O clockwise 180°. Then using point O as a center of dilation, enlarge the image by a scale factor of 1.5.

3. Modify △ABC or drag point O to make different dynamic drawings. Do your observations from Exercise 2 change or remain the same? Explain your response.
   As long as the scale factor is −1.5, the observations from Exercise 2 stays the same.

4. Use experimentation, reasoning, and the software to complete the summary table below.

Scale factor, s	Description of transformation
s > 1	The figure is enlarged and there is no rotation.
s = 1	The dilation has no effect on the original figure.
0 < s < 1	The figure is shrunk or contracted and there is no rotation.
s = 0	The image is a point, the center of the dilation.
−1 < s < 0	The figure is shrunk and rotated 180°.
s = −1	The figure is rotated 180° and there is no enlargement or contraction.
s < −1	The figure is rotated 180° and is enlarged.

Exercises 51–54 extend the idea of dilations to nonrigid dilations in which all *x*-coordinates are multiplied by one scale factor and all *y*-coordinates are multiplied by a different scale factor.

Portfolio Activity

The Portfolio Activity can be used as preparation for the Chapter Project or as a separate activity. In the Portfolio Activity on this page, students use a grid to enlarge their choice of a design, cartoon, or photo.

51.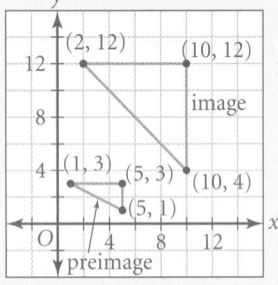

52. The preimage and the image are both right triangles with one vertical leg and one horizontal leg. The image is twice as wide and 4 times as high as the preimage, and the slopes of the hypotenuses are different. The form of the rule, $T(x, y) = (mx, ny)$ is similar to the rule for a dilation, but the scale factors are different for the *x*- and *y*-coordinates. In a dilation, the scale factors for the coordinates must be the same.

53. $\frac{1}{2}$

 **Look Beyond**

Another type of transformation that is not rigid can be described in a coordinate plane by $T(x, y) = (mx, ny)$, where the *x*- and *y*-values are multiplied by two different scale factors.

51. Draw a triangle with vertices at $(1, 3)$, $(5, 1)$, and $(5, 3)$ in a coordinate plane. Transform the figure by using the transformation $T(x, y) = (2x, 4y)$.

52. How are the preimage and image alike? How are they different? How is this type of transformation like a dilation? How is it different?

53. Using the segment with endpoints $(1, 3)$ and $(5, 1)$, what is the ratio of its slope to the slope of its image?

54. Use the transformation $T(x, y) = (3x, -y)$ to transform the figure at left.

Calvin and Hobbes   by Bill Watterson

USING GRIDS TO ENLARGE A DESIGN

Choose a design, cartoon, or photo that you wish to enlarge. You may wish to photocopy the design so that you do not damage the original.

1. Using a ruler, draw a grid of $\frac{1}{2}$-cm squares on the design.

2. Draw a grid with the same number of squares on a larger piece of paper. These squares should be at least twice as big.

3. Copy the pattern that appears in each square of the first grid onto the corresponding square in the larger grid. It may help to cover all but one square of the design as you copy it.

Be sure to copy one square at a time—do not try to draw larger parts of the design.

Once all the squares are copied, you will have an accurate enlargement of the design.

WORKING ON THE CHAPTER PROJECT

You should now be able to complete Activity 1 of the Chapter Project.

▣ internet connect

Portfolio Extension
Go To: **go.hrw.com**
Keyword:
MG1 Scale

54.

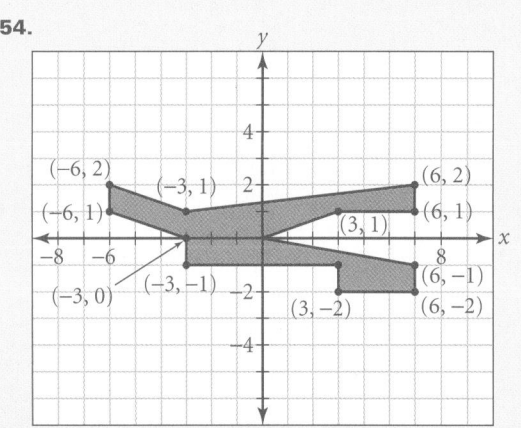

Similar Polygons

Objectives

● Define *similar polygons*.

● Use Properties of Proportions and scale factors to solve problems involving similar polygons.

Why *Scale models are used by architects, city planners, movie set designers, and hobbyists. To create a scale model, properties of similar figures are needed.*

QUICK **WARM-UP**

1. Given $\triangle ABC \cong \triangle DEF$, name the congruent angles and sides.
$\angle ABC \cong \angle DEF$;
$\angle ACB \cong \angle DFE$;
$\angle BAC \cong \angle EDF$;
$\overline{AB} \cong \overline{DE}$; $\overline{AC} \cong \overline{DF}$;
$\overline{BC} \cong \overline{EF}$

2. The dilation image of a segment with a length of 4 has a length of 6. What is the scale factor? **1.5**

3. Solve the equation $6x = 45$. $x = 7.5$

Also on Quiz Transparency 8.2

Similar Polygons

When a figure undergoes a dilation (Lesson 8.1), the preimage and image have the same shape but are not necessarily the same size. They are said to be *similar*.

Definition: Similar Figures

Two figures are **similar** if and only if one is congruent to the image of the other by a dilation.
8.2.1

Teach

Why Creating similar polygons is an important skill for students because of its application to photography, maps, scale drawings, and scale models. This lesson will help students make connections to algebra and can be extended to include three-dimensional figures.

In the dilation at right, the corresponding angles of the triangles are congruent, and the ratios of the lengths of the corresponding sides are all equal to the absolute value of the scale factor of the dilation—which is 2 in this case.

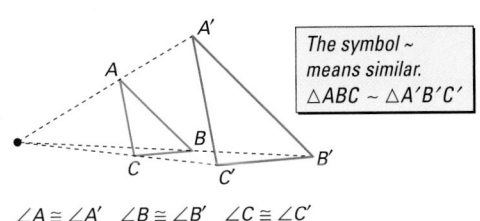

The symbol ~ means similar.
$\triangle ABC \sim \triangle A'B'C'$

$\angle A \cong \angle A' \quad \angle B \cong \angle B' \quad \angle C \cong \angle C'$

$\dfrac{A'B'}{AB} = \dfrac{B'C'}{BC} = \dfrac{A'C'}{AC} = 2 \quad$ Scale factor

Alternative Teaching Strategy

USING VISUAL MODELS Use a photocopier to enlarge and reduce a cartoon or drawing. Transfer the copies to transparency sheets and superimpose the similar figures on the overhead. Measure and compare the similar sides and angles in the figures.

Cooperative Learning

Have several students write their definition of similar polygons on the board. Have the class compare them and agree on a definition. Discuss the importance of including "if and only if" in the definition.

CRITICAL THINKING

No; in the square and the rectangle, all pairs of corresponding angles are congruent since each angle is a right angle, but all pairs of corresponding sides are not proportional. In the square and the rhombus, all pairs of corresponding sides are proportional, but all pairs of corresponding angles are not congruent since each angle in the square is a right angle, and the rhombus contains no right angles.

ADDITIONAL EXAMPLE ❶

Are the two parallelograms shown below similar?

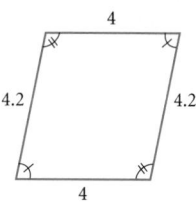

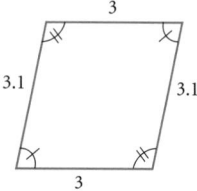

The corresponding angles are congruent, but the corresponding sides are not proportional. Therefore, the parallelograms are not similar.

When the ratios of corresponding sides of two polygons are equal (as in the illustration on the previous page), the sides are said to be **proportional**. A statement of the equality of two ratios is called a **proportion**. The concept of proportion is important in the Polygon Similarity Postulate.

> ### Polygon Similarity Postulate
>
> Two polygons are similar if and only if there is a way of setting up a correspondence between their sides and angles such that the following conditions are met:
>
> • Each pair of corresponding angles is congruent.
> • Each pair of corresponding sides is proportional.   **8.2.2**

In a similarity statement, as in a congruence statement, the letters of the vertices must be written in corresponding order.

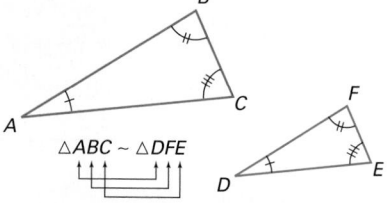

$\triangle ABC \sim \triangle DFE$

CRITICAL THINKING

Is either condition in the Polygon Similarity Postulate, taken separately, enough to guarantee that two polygons are similar? Use the figures at right to explain your answer.

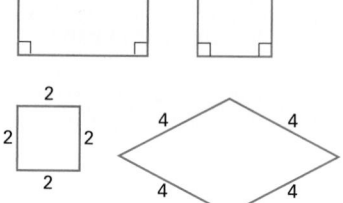

E X A M P L E ❶ Are the triangles at right similar?

SOLUTION

It is given that $\angle P \cong \angle S$, $\angle Q \cong \angle T$, and $\angle R \cong \angle U$, so the first condition of the Polygon Similarity Postulate is satisfied. To see whether the second condition is satisfied, check each ratio of corresponding sides.

$$\frac{ST}{PQ} = \frac{57}{38} = \frac{3}{2} \qquad \frac{TU}{QR} = \frac{30}{20} = \frac{3}{2} \qquad \frac{SU}{PR} = \frac{39}{26} = \frac{3}{2}$$

Because the ratios are equal, the corresponding sides are proportional, and the second condition of the Polygon Similarity Postulate is satisfied. Therefore, $\triangle PQR \sim \triangle STU$.

Interdisciplinary Connection

BIOLOGY Microscopes are used to study the characteristics of plant and animal cells that cannot be seen with the naked eye. Borrow several microscopes and some prepared slides from your school's science department, and have students draw what they see. Then have them find the scale factor between the specimens and the drawings.

Properties of Proportions

When working with similar figures, it is often helpful to know the following Properties of Proportions:

Properties of Proportions

Let a, b, c, and d be any real numbers.

Cross-Multiplication Property

If $\frac{a}{b} = \frac{c}{d}$ and b and $d \neq 0$, then $ad = bc$. **8.2.3**

Reciprocal Property

If $\frac{a}{b} = \frac{c}{d}$ and a, b, c, and $d \neq 0$, then $\frac{b}{a} = \frac{d}{c}$. **8.2.4**

Exchange Property

If $\frac{a}{b} = \frac{c}{d}$ and a, b, c, and $d \neq 0$, then $\frac{a}{c} = \frac{b}{d}$. **8.2.5**

"Add-One" Property

If $\frac{a}{b} = \frac{c}{d}$ and b and $d \neq 0$, then $\frac{a+b}{b} = \frac{c+d}{d}$. **8.2.6**

TRY THIS   Verify each property for the proportion $\frac{1}{3} = \frac{2}{6}$.

CRITICAL THINKING   Why do you think the last property in the list above is called the "Add-One" Property? (Hint: Separate each side of the equation into two separate fractions.)

E X A M P L E  In the figure at right, pentagon $ABCDE \sim$ pentagon $FGHIJ$. Find AB.

● **SOLUTION**

Because the pentagons are similar, the Polygon Similarity Postulate states that the sides are proportional; thus,
$$\frac{AB}{FG} = \frac{BC}{GH} = \frac{CD}{HI} = \frac{DE}{IJ} = \frac{EA}{JF}.$$

Because all the ratios of the sides are equal, any two ratios are equal. For example:

$$\frac{AB}{FG} = \frac{CD}{HI}$$

$$\frac{AB}{16} = \frac{15}{20} \qquad \text{\textit{Substitute the known lengths into the proportion.}}$$

$$16 \cdot \frac{AB}{16} = \frac{15}{20} \cdot 16 \qquad \text{\textit{To solve, multiply both sides by 16.}}$$

$$AB = \frac{240}{20} = 12$$

Enrichment

Have students draw a simple floor plan of a house with a scale of 1 inch to 16 feet. Ask them to list proportions that relate the measurements on the floor plan to the actual dimensions of the house.

Inclusion Strategies

 Emphasize that the properties of proportions are used to simplify the work associated with similar figures. Point out that in everyday language, the word *proportion* is sometimes used incorrectly to mean a single fraction rather than the equality of two fractions. For example, the statement, "the proportion of students getting an A is 1 out of 5" is not mathematically correct.

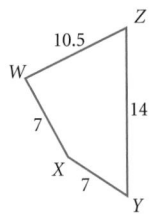
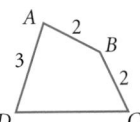

EXAMPLE ③

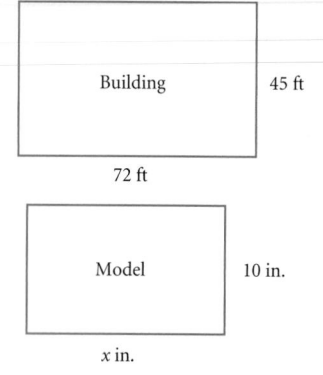

Cooperative Learning

Have groups of five students review the Properties of Proportions. Have one student write a proportion and each of the other four students restate the proportions in terms of the properties. Key concepts related to proportions are more likely to be understood if they are stated in more than one way.

Proportionality Within a Figure

Consider the two similar rectangles shown at right. By the Polygon Similarity Postulate,

$$\frac{KL}{WX} = \frac{LM}{XY} = \frac{MN}{YZ} = \frac{NK}{ZW}.$$

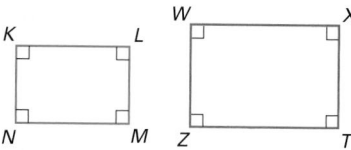

You can form a proportion with any two of these ratios, such as $\frac{MN}{YZ} = \frac{NK}{ZW}$.

Notice that each ratio in the proportion contains a side length from each rectangle. By using the Exchange Property, you get $\frac{MN}{NK} = \frac{YZ}{ZW}$.

Thus, the ratio of the long side to the short side is the same in each rectangle. This provides another way of thinking about similarity: the ratio of any two sides in one polygon is the same as the ratio of the corresponding sides in a similar polygon.

EXAMPLE ③

APPLICATION
ARCHITECTURE

Amber and Adrianne are making a scale model of a building with a rectangular foundation, as shown below. If the long sides of the model are 24 inches, how long are the short sides?

SOLUTION

Because the scale model is similar to the original building, the ratio of the shorter side to the longer side is the same in each rectangle.

$$\frac{x}{24} = \frac{18}{32}$$

$$24 \cdot \frac{x}{24} = \frac{18}{32} \cdot 24$$

$$x = \frac{432}{32}, \text{ or } 13.5 \text{ inches.}$$

Reteaching the Lesson

USING VISUAL MODELS Have each student draw and label a polygon and measure each side and angle. Ask them to construct a similar polygon with a scale factor of 3 and another polygon with a scale factor of $\frac{1}{3}$. Verify the Properties of Proportions for the corresponding sides of the similar polygons.

Exercises

Communicate

Classify each statement as true or false and explain your reasoning.

1. If $\triangle ABC \sim \triangle DEF$, then $\triangle DEF \sim \triangle ABC$.

2. If $\triangle ABC \sim \triangle DEF$, then $\triangle ABC \sim \triangle EFD$.

3. If two figures are congruent, then they are similar.

4. If two figures are similar, then they are congruent.

5. Any two regular polygons with the same number of sides are similar.

Guided Skills Practice

Determine whether each pair of figures is similar. Explain your reasoning. *(EXAMPLE 1)*

6.
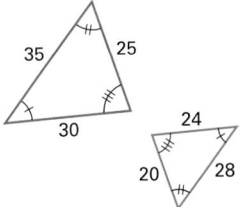

7.
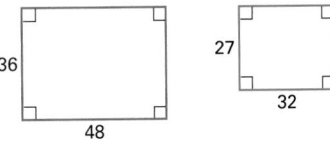

In Exercises 8 and 9, the polygons in each pair are similar. Find the missing length. *(EXAMPLE 2)*

8. *EH* **19.8**
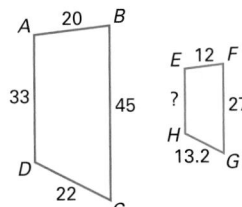

9. *JK* **15**
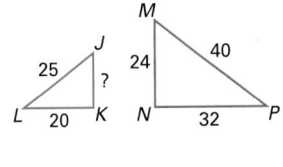

APPLICATION

10. ARCHITECTURE A scale model of a building has the dimensions shown below. Find the length of the actual building. **90 ft**

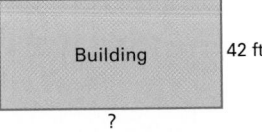

Selected Answers
Exercises 6–10, 11–55 odd

ASSIGNMENT GUIDE

In Class	1–10
Core	11–43 odd
Core Plus	17–47 odd
Review	49–56
Preview	57–60

✐ Extra Practice can be found beginning on page 818.

6. Yes, corresponding angles are congruent, and corresponding sides are proportional:
$\frac{20}{25} = \frac{24}{30} = \frac{28}{35}$.

7. No; $\frac{27}{36} \neq \frac{32}{48}$, corresponding sides are not proportional.

16. Since every angle is right, then corresponding angles are congruent. $\frac{3.6}{2.4} = \frac{5.4}{3.6}$, so corresponding sides are proportional. Thus quadrilateral $ABCD \sim$ quadrilateral $EFGH$.

17. $\angle J \cong \angle P$, $\angle K \cong \angle M$, and $\angle L \cong \angle N$, so corresponding angles are congruent. $\frac{30}{22.5} = \frac{20}{15} = \frac{27}{20.25}$, so corresponding sides are proportional. Thus, $\triangle JKL \sim \triangle PMN$.

18. $\angle S \cong \angle U$, $\angle Q \cong \angle V$, and $\angle R \cong \angle T$, so corresponding angles are congruent. $\frac{30}{19.5} = \frac{16}{10.4} = \frac{34}{22.1}$, so corresponding sides are proportional. Thus, $\triangle RSQ \sim \triangle TUV$.

19. $\frac{25}{18} \neq \frac{30}{20}$, so corresponding sides are not proportional. Thus, pentagon $KLMNJ \not\sim$ pentagon $PQRST$.

20. $(3)(20) = (15)(4) \Rightarrow 60 = 60$

21. $\frac{2}{5} = \frac{4}{10} \Rightarrow \frac{2}{5} = \frac{2}{5}$

22. $\frac{6}{2} = \frac{9}{3} \Rightarrow 3 = 3$

24. Ratio of sides: $\frac{1}{2}$

Ratio of areas: $\frac{1}{4}$

Observe that $\left(\frac{1}{2}\right)^2 = \frac{1}{4}$.

25. Ratio of sides: 3
Ratio of areas: 9
Observe that $(3)^2 = 9$.

Practice and Apply

11. Given the proportionality statement $\frac{SG}{MW} = \frac{GT}{WR} = \frac{TS}{RM}$ for two similar triangles, write a similarity statement that shows the correct correspondence. $\triangle SGT \sim \triangle MWR$

12. Given the proportionality statement $\frac{JD}{LE} = \frac{DC}{EB} = \frac{CP}{BH} = \frac{PJ}{HL}$ for two similar rectangles, write a similarity statement that shows the correct correspondence. rectangle $JDCP \sim$ rectangle $LEBH$

13. Given $\triangle ABC \sim \triangle XYZ$, write a proportionality statement for the ratios between the sides. $\frac{AB}{XY} = \frac{BC}{YZ} = \frac{CA}{ZX}$

14. Given quadrilateral $EFGH \sim$ quadrilateral $VWXY$, write a proportionality statement for the ratios between the sides. $\frac{EF}{VW} = \frac{FG}{WX} = \frac{GH}{XY} = \frac{HE}{YV}$

15. Given pentagon $JKLMN \sim$ pentagon $PQRST$, write a proportionality statement for the ratios between the sides. $\frac{JK}{PQ} = \frac{KL}{QR} = \frac{LM}{RS} = \frac{MN}{ST} = \frac{NJ}{TP}$

internet connect
Homework Help Online
Go To: go.hrw.com
Keyword:
MG1 Homework Help
for Exercises 16-19

For Exercises 16–19, determine whether the polygons are similar. Explain your reasoning.

16.

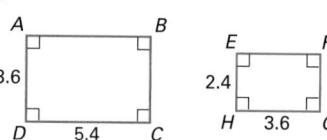

17.

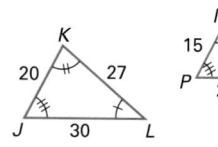
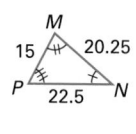

18.

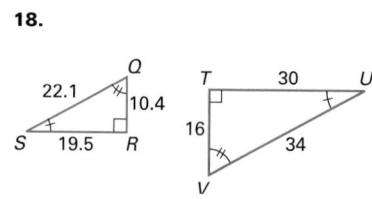

19.

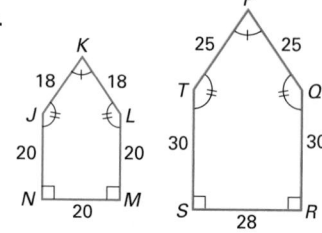

20. Verify the Cross-Multiplication Property for the proportion $\frac{3}{4} = \frac{15}{20}$.

21. Verify the Reciprocal Property for the proportion $\frac{5}{2} = \frac{10}{4}$.

22. Verify the Exchange Property for the proportion $\frac{6}{9} = \frac{2}{3}$.

23. Given $\frac{x}{4} = \frac{y}{8}$, find $\frac{x}{y}$. $\frac{1}{2}$

For each pair of similar figures below, compare the ratio of the sides of the figures with the ratio of the areas of the figures.

24.

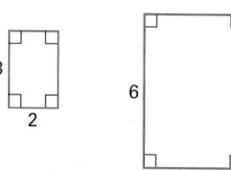

25.

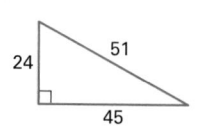

Algebra

For Exercises 26–29, the given polygons are similar. Find x.

26.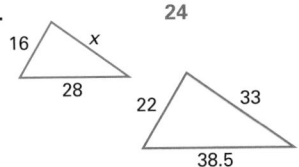

27.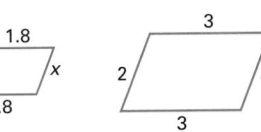

28.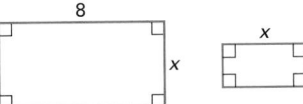

29.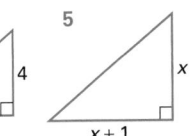

30. Use the diagram below to determine whether quadrilateral *GHIJ* ~ quadrilateral *KLMN*. Explain your answer.

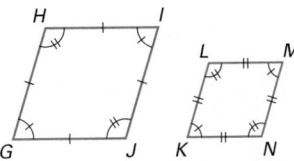

Algebra

Solve each proportion for x.

31. $\frac{6x}{24} = \frac{27}{9}$ x = 12

32. $\frac{4.8}{x} = \frac{6}{8.4}$ x = 6.72

33. $\frac{\frac{2}{5}}{8} = \frac{\frac{7}{10}}{x}$ x = 14

34. $\frac{6}{x} = \frac{x}{150}$ x = 30

35. $\frac{3}{x-4} = \frac{7}{x+4}$ x = 10

36. $\frac{5-2x}{8} = \frac{3x+1}{4}$ x = $\frac{3}{8}$

37. In general, what do you think is true about the ratio of the sides and the ratio of the areas of two similar polygons?
    If the ratio of the sides is *n*, then the ratio of the areas is *n*².

For real numbers *a*, *b*, *c*, and *d*, where $b \neq 0$ and $d \neq 0$, $\frac{a}{b} = \frac{c}{d}$.
Determine whether each proportion is true for all values of the variables for which the proportion is defined. If a proportion is not true, give a numerical counterexample.

38. $\frac{a+x}{b} = \frac{c+x}{d}$

39. $\frac{a+x}{b+x} = \frac{c+x}{d+x}$

40. $\frac{a+b}{b} = \frac{c+d}{d}$

41. $\frac{a}{a+b} = \frac{c}{c+d}$

CHALLENGE

42. Verify that the following proportion is true for all real numbers *a*, *b*, *c*, *d*, *e*, and *f*, where *b*, *d*, and *f* $\neq 0$: If $\frac{a}{b} = \frac{c}{d} = \frac{e}{f}$, then $\frac{a}{b} = \frac{a+c+e}{b+d+f}$.

Error Analysis

Some students may find Exercises 38–41 difficult to understand without concrete examples. Encourage students to restate the questions with numbers in place of the variables.

30. $\angle G \cong \angle K$, $\angle H \cong \angle L$, $\angle I \cong \angle M$, and $\angle J \cong \angle N$, so corresponding angles are congruent. $\overline{GH}$, $\overline{HI}$, $\overline{IJ}$. and $\overline{JG}$ are proportional to $\overline{KL}$, $\overline{LM}$, $\overline{MN}$, and $\overline{NK}$, respectively, so corresponding sides are proportional. Thus, quadrilateral *GHIJ* ~ quadrilateral *KLMN*.

38. False; it is true that $\frac{3}{4} = \frac{6}{8}$. But, $\frac{3+1}{4} = \frac{4}{4} = 1$, and $\frac{6+1}{8} = \frac{7}{8}$. So, $\frac{3+1}{4} \neq \frac{6+1}{8}$.

39. False; it is true that $\frac{3}{4} = \frac{6}{8}$. But, $\frac{3+1}{4+1} = \frac{4}{5}$ and $\frac{6+1}{8+1} = \frac{7}{9}$.

40. True; it is true by the "Add-One" Property.

41. True; if $\frac{a}{b} = \frac{c}{d}$ then
    $ad = bc$
    $\Rightarrow ac + ad = ac + bc$
    $\Rightarrow a(c+d) = c(a+b)$
    $\Rightarrow \frac{a}{a+b} = \frac{c}{c+d}$.

42. $\frac{a}{b} = \frac{c}{d} = \frac{e}{f}$, then
    $ad = bc$
    $af = be$
    $ab = ba$
    Thus,
    $ab + ad + af = ba + bc + be$
    $a(b+d+f) = b(a+c+e)$
    $\frac{a}{b} = \frac{a+c+e}{b+d+f}$

Technology

Use geometry graphics software to demonstrate how to enlarge or reduce figures to scale, as required in Exercises 46 and 47.

APPLICATIONS

43. WILDLIFE MANAGEMENT A method used to estimate the size of wild-animal populations uses proportions. Suppose that a scientist catches 300 fish from a lake and then tags and releases them. After a short time, the scientist comes back, catches 100 fish, and finds that 8 of them are already tagged. Assuming that the proportion of tagged fish in this catch is equal to the proportion of tagged fish in the entire population, estimate the number of fish in the lake. **3750 fish**

44. Estimates should vary from 110 mi² to 120 mi² and from 70,000 to 75,000 acres.

44. WILDLIFE MANAGEMENT On the map of the Aransas National Wildlife Refuge shown below, the scale is 1 cm = 3.5 mi. Estimate the area of the refuge in square centimeters, and then use that value to estimate the area of the refuge in square miles. How many acres are in the refuge? (Note: 1 square mile = 640 acres)

The Aransas National Wildlife Refuge on the Texas coast contains important nesting sites for endangered whooping cranes. The delicate salt-marsh environment must be carefully maintained in order to help this and other species of wildlife survive.

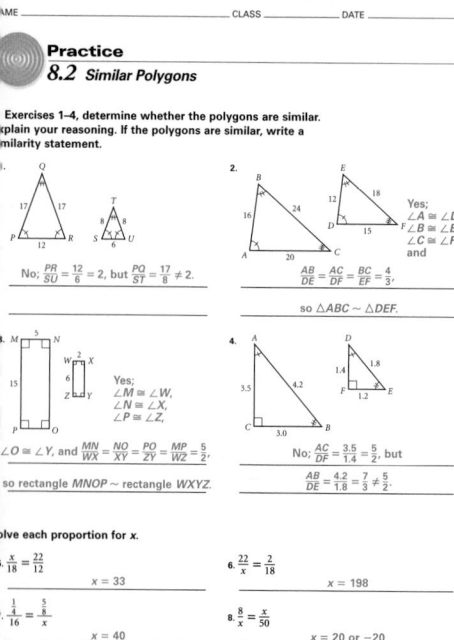

45. LANDSCAPING In the landscaping diagram below, the scale is 1 cm = 15 ft. Measure the diagram and use the given scale to determine how far apart the trees should be planted. **25.5 ft**

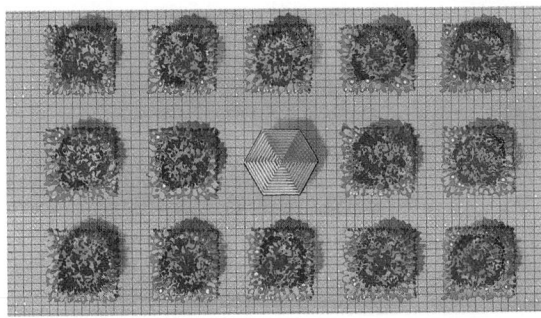

46. FINE ART Brenda is attempting to paint a reproduction of the *Mona Lisa* from a print that is 16 in. by 24 in. If her canvas is 15 in. wide, how tall should it be in order for the reproduction to be proportional to the print? **22.5 ft**

47. FINE ART In the print of the *Mona Lisa* described in Exercise 46, the face is 6 in. tall and 4 in. wide. What should be the dimensions of the face in Brenda's reproduction? $5\frac{5}{8}$ in. $\times 3\frac{3}{4}$in.

Mona Lisa (1503-6) by Leonardo da Vinci (1452-1519). Louvre, Paris, France

48. INTERIOR DECORATING Fernando is drawing a floor plan of his house to help him in arranging furniture. His dining room is a 12 ft × 15 ft rectangle, and his table is a regular octagon with 2 ft 3 in. sides. If the room on the floor plan is 8 in. × 10 in., how long should the sides of the table in the floor plan be? **1.5 in.**

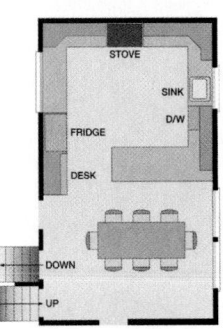

Designers use floor plans to help in determining the layout of a room.

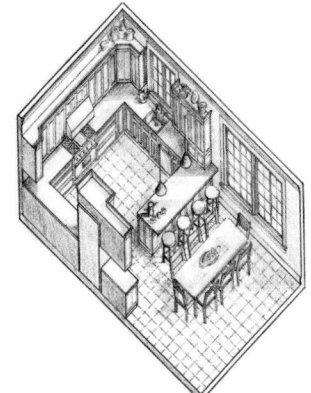
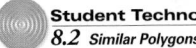

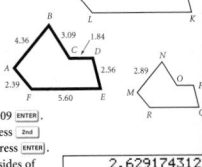
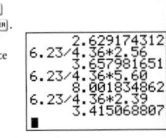
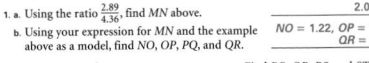

Exercises 57–60 introduce students to indirect measurement, which will be studied in greater depth in Lesson 8.5.

57. Anthony can note where point A is, and then pace off or measure the distance to it from where he is standing. This distance should be equivalent to the width of the river.

59. $\triangle SNR \cong \triangle SNA$ by ASA, because $\angle ASN \cong \angle RSN$, $\angle SNA \cong \angle SNR$ (because both are right angles), and $\overline{SN} \cong \overline{SN}$.

60. Answers may vary. Sample answer: It is difficult to keep your head at the same angle, and it's difficult to know when you are keeping your eyes level to pick out appropriate points.

Look Back

Algebra

49. The angles of a triangle measure $(x + 5)°$, $(5x + 12)°$, and $(2x + 3)°$. Find the measures of the angles. *(LESSON 3.5)* **25°, 112°, and 43°**

Which of the following can be used to prove triangle congruence? *(LESSONS 4.2 AND 4.3)*

50. ASA **valid**   **51.** AAA **invalid**   **52.** SAA **valid**   **53.** SAS **valid**

54. Find the measure of a base angle of an isosceles triangle whose vertex angle is 92°. *(LESSON 4.4)* **44°**

55. A right triangle has legs of 5 cm and 7 cm. Find the length of the hypotenuse. **8.6 cm**

56. A right triangle has a leg of 5 cm and a hypotenuse of 7 cm. Find the length of the other leg. $2\sqrt{6}$ **cm ≈ 4.9 cm**

Look Beyond

INDIRECT MEASUREMENT **Anthony uses the following method to estimate the width of a river:**

Anthony stands at point N and adjusts the visor of his cap until it is in his line of sight to point R on the opposite shore. Without changing the position of his cap, he turns and sights along the visor to point A on his side of the river.

57. Explain how Anthony can find the width of the river.

58. Which segment in the figure has the same length as $\overline{NR}$? $\overline{NA}$

59. Which postulate can be used to prove that $\triangle SNR \cong \triangle SNA$? Explain your answer.

60. What are some possible problems with using this method of indirect measurement?

8.3

Triangle Similarity

Objective

- Develop the AA Triangle Similarity Postulate and the SSS and SAS Triangle Similarity Theorems.

Why *Similar triangles have interesting mathematical properties.*

Prepare

QUICK WARM-UP

1. List the Triangle Congruence Postulates.
   SSS, SAS, ASA

2. What is the ratio of the corresponding side lengths for two congruent triangles? **1 to 1**

Also on Quiz Transparency 8.3

Triangle Similarity

Activities 1, 2, and 3 suggest some shortcuts for determining triangle similarity.

Activity 1

AA Triangle Similarity Postulate

YOU WILL NEED

ruler and protractor
OR
geometry graphics software

1. Draw △ABC with m∠A = 45° and m∠B = 65°. What is m∠C? Measure the sides of △ABC.

2. Draw △DEF with m∠D = 45° and m∠E = 65° such that DE is longer than AB. What is m∠F? Measure the sides of △DEF.

3. Use your measurements to complete the table below.

	Sides			Angles		
△ABC	AB = ?	BC = ?	AC = ?	m∠A = 45°	m∠B = 65°	m∠C = ?
△DEF	DE = ?	EF = ?	DF = ?	m∠D = 45°	m∠E = 65°	m∠F = ?
Ratio	$\frac{DE}{AB}$ = ?	$\frac{EF}{BC}$ = ?	$\frac{DF}{AC}$ = ?			

4. What is the relationship between corresponding sides? between corresponding angles? Are the triangles similar? Explain your reasoning.

CHECKPOINT ✔

5. Based on your results, complete the postulate below.

AA (Angle-Angle) Similarity Postulate

If two ____?____ of one triangle are congruent to two ____?____ of another triangle, then the triangles are ____?____. **8.3.1**

Alternative Teaching Strategy

HANDS-ON STRATEGIES Some students may benefit from the kinesthetic experience of doing hands-on measurements. Have those students use a ruler and protractor to measure the sides and angles of pairs of similar triangles. Make a chart showing the measurements and the ratio of the corresponding sides and angles. Include triangles that are acute, obtuse, and right. Have students do as many examples as necessary to formulate a postulate for similarity between two triangles.

Teach

Why Developing the similarity postulates through explorations and activities should help students better understand the relationships among similar figures.

Activity 1 Notes

Have students compare the AA Triangle Similarity Postulate developed in this Activity with the ASA Congruence Postulate for triangles. They should be able to associate similarity with same shape and congruency with same size.

For a student worksheet of this Activity and detailed Teacher Notes, see page 141 in the Lesson Activities booklet.

CHECKPOINT ✔
5. angles, angles, similar

EXAMPLE ❶

Are the triangles shown below similar?

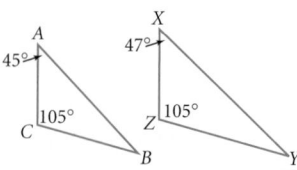

Two corresponding angles are not congruent, so the triangles are not similar.

Activity 2 Notes

This Activity will help students understand that when all pairs of corresponding sides in two triangles are in the same proportion, the triangles are similar. Ask students to compare this theorem with the SSS Congruence Postulate. They should notice the importance of same size for triangle congruence.

For a student worksheet of this Activity and detailed Teacher Notes, see page 141 in the Lesson Activities booklet.

CHECKPOINT ✔
5. sides, sides, similar

EXAMPLE ❷

Are the triangles shown below similar?

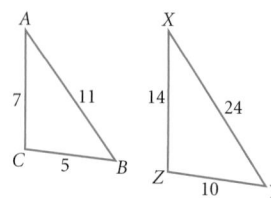

The three sides of △ABC are not proportional to the three sides of △XYZ, so the triangles are not similar.

EXAMPLE ❶ Are the triangles at right similar?

● SOLUTION

By the Triangle Sum Theorem, $m\angle J = 180° - 90° - 35° = 55°$, so $m\angle J = m\angle P$ ($\angle J \cong \angle P$) and $m\angle K = m\angle M$ ($\angle K \cong \angle M$).

Thus, by the AA Similarity Postulate, $\triangle JKL \sim \triangle PMN$.

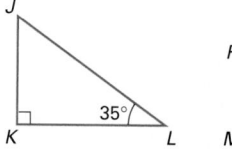

Activity 2
SSS Triangle Similarity Theorem

YOU WILL NEED

ruler, protractor, and compass

OR

geometry graphics software

1. Draw △ABC with $AB = 2$ cm, $BC = 3$ cm, and $AC = 4$ cm. Measure the angles of △ABC to the nearest degree.

2. Draw △DEF with $DE = 6$ cm, $EF = 9$ cm, and $DF = 12$ cm. Measure the angles of △DEF to the nearest degree.

3. Use your measurements to complete the table below.

	Sides			Angles
△**ABC**	$AB = 2$	$BC = 3$	$AC = 4$	$m\angle A = ?$ $m\angle B = ?$ $m\angle C = ?$
△**DEF**	$DE = 6$	$EF = 9$	$DF = 12$	$m\angle D = ?$ $m\angle E = ?$ $m\angle F = ?$
Ratio	$\frac{DE}{AB} = ?$	$\frac{EF}{BC} = ?$	$\frac{DF}{AC} = ?$	

4. What is the relationship between corresponding sides? between corresponding angles? Are the triangles similar? Explain your reasoning.

CHECKPOINT ✔
5. Based on your results, complete the theorem below, which you will be asked to prove in Exercises 25–27.

SSS (Side-Side-Side) Similarity Theorem

If the three _____?_____ of one triangle are proportional to the three _____?_____ of another triangle, then the triangles are _____?_____.

8.3.2

EXAMPLE ❷ Are the triangles at right similar?

● SOLUTION

The ratios of the three sides are as follows:

$$\frac{QR}{UT} = \frac{4}{2.4} = \frac{5}{3} \qquad \frac{RS}{TV} = \frac{7}{4.2} = \frac{5}{3} \qquad \frac{QS}{UV} = \frac{7}{4.2} = \frac{5}{3}$$

Thus, the sides of the triangle are proportional and, by the SSS Similarity Theorem, $\triangle QRS \sim \triangle UTV$.

Interdisciplinary Connection

ARTS Many graphic designs can be done with computers. Ask students to find examples of computer-drawn graphic designs in books or magazines. Ask them to describe how designs and photographs are enlarged or reduced to fit a particular space on a page.

Enrichment

Many books on fractal geometry show the Sierpinski triangle, which contain an infinite number of self-similar triangles. Have students use the library to find books on fractals. Use a computer to generate a color drawing of the Sierpinski triangle.

Activity 3

SAS Triangle Similarity Theorem

1. Draw △ABC with AB = 3 cm, BC = 4 cm, and m∠B = 60°. Measure the sides and angles of △ABC.

2. Draw △DEF with DE = 6 cm, EF = 8 cm, and m∠E = 60°. Measure the sides and angles of △DEF.

3. Use your measurements to complete the table below.

	Sides			Angles		
△**ABC**	AB = 3	BC = 4	AC = ?	m∠A = ?	m∠B = 60°	m∠C = ?
△**DEF**	DE = 6	EF = 8	DF = ?	m∠D = ?	m∠E = 60°	m∠F = ?
Ratio	$\frac{DE}{AB}$ = ?	$\frac{EF}{BC}$ = ?	$\frac{DF}{AC}$ = ?			

4. What is the relationship between corresponding sides? between corresponding angles? Are the triangles similar? Explain your reasoning.

CHECKPOINT ✔

5. Based on your results, complete the theorem below, which you will be asked to prove in Exercises 28–30.

SAS (Side-Angle-Side) Similarity Theorem

If two ___?___ of one triangle are proportional to two ___?___ of another triangle and their ___?___ ___?___ are congruent, then the triangles are ___?___.   **8.3.3**

E X A M P L E ③ Are the triangles shown below similar?

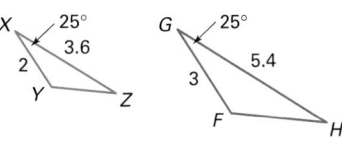

SOLUTION

The ratios of the given sides are as follows:

$$\frac{GF}{XY} = \frac{3}{2} \qquad \frac{GH}{XZ} = \frac{5.4}{3.6} = \frac{3}{2}$$

Thus, the sides of the triangle are proportional, and the included angles of these sides are congruent. By the SAS Similarity Theorem, △QRS ~ △UTV.

CRITICAL THINKING Why are ASA and AAS not included in a list of triangle similarity theorems?

Inclusion Strategies

INVITING PARTICIPATION Some students may be comfortable using a more analytical approach to the triangle similarity postulates and theorems. Give these students the answers to the Checkpoint question in each activity and have them use technology or a ruler and protractors to verify the postulates and theorems and to give examples.

Reteaching the Lesson

COOPERATIVE LEARNING Divide the class into small groups. Have each group construct similar triangles from straws and use them to summarize the lesson. Ask them to create sample questions to be used as a review for the whole class.

Are the triangles shown below similar?

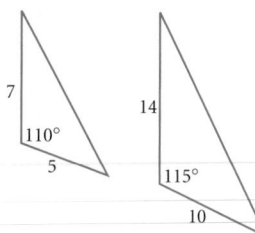

The two given angles are not congruent, so the triangles are not congruent.

Assess

Selected Answers

Exercises 5–7, 9–47 odd

ASSIGNMENT GUIDE

In Class	1–7
Core	9–21 odd, 22–30
Core Plus	22–30, 31–37 odd
Review	39–47
Preview	48–51

✐ Extra Practice can be found beginning on page 818.

Exercises

● *Communicate*

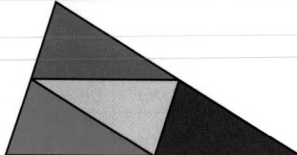

1. In the figure below, the midsegments divide the large triangle into four smaller triangles. How would you show that each of the smaller triangles is similar to the large triangle?

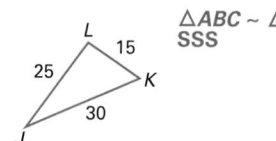

2. Recall the HL Congruence Theorem from Lesson 4.3. Could there be an HL Similarity Theorem? Why or why not?

3. Use a counterexample to explain why there is no AAA Similarity Postulate for quadrilaterals. That is, if three angles of one quadrilateral are congruent to three corresponding angles of another quadrilateral, explain why the quadrilaterals are not necessarily similar.

4. Use a counterexample to explain why there is no SSSS Similarity Postulate for quadrilaterals. That is, if four sides of one quadrilateral are congruent to four corresponding sides of another quadrilateral, explain why the quadrilaterals are not necessarily similar.

● *Guided Skills Practice*

Each pair of triangles below can be proven similar by using AA, SSS, or SAS information. Write a similarity statement for each pair, and identify the postulate or theorem used. *(EXAMPLES 1, 2, AND 3)*

5.

△ABC ~ △LKJ
SSS

6.
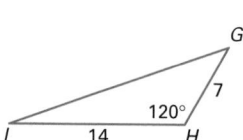

△EFD ~ △QPR
AA

7.
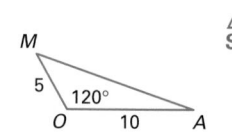

△GHI ~ △MOA
SAS

Practice and Apply

Determine whether each pair of triangles can be proven similar by using AA, SSS, or SAS similarity. If so, write a similarity statement, and identify the postulate or theorem used.

8.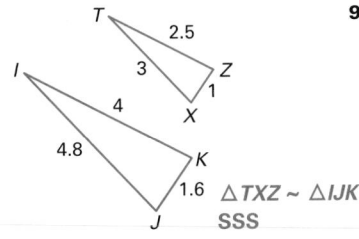

$\triangle TXZ \sim \triangle IJK$
SSS

9.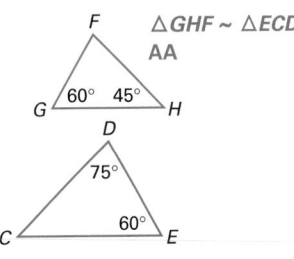

$\triangle GHF \sim \triangle ECD$
AA

10. Cannot be proven similar

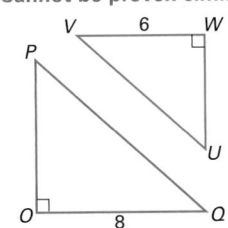

11.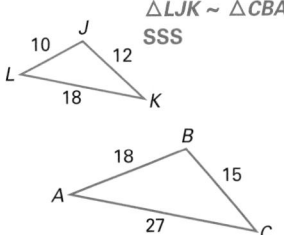

$\triangle LJK \sim \triangle CBA$
SSS

12. Cannot be proven similar

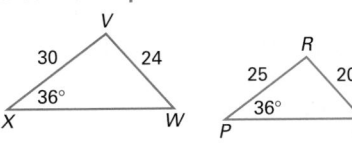

13.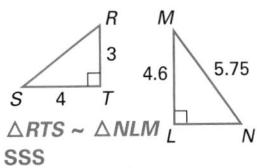

$\triangle RTS \sim \triangle NLM$
SSS

In Exercises 14 and 15, can the pairs of triangles be proven similar? Why or why not?

14.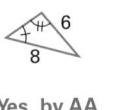

Yes, by AA

15.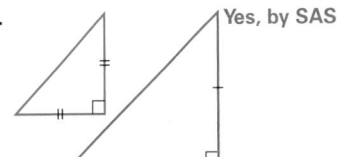

Yes, by SAS

In Exercises 16 and 17, indicate which figures are similar. Explain your reasoning.

16.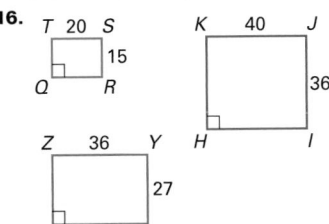

17.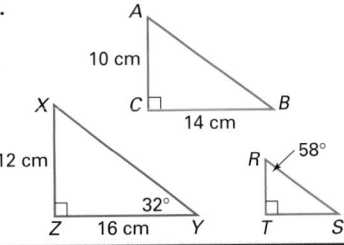

16. rectangle $QRST \sim$ rectangle $WXYZ$

17. $\triangle XYZ \sim \triangle RST$; by the Angle-Sum Theorem, $m\angle S = 32°$. Also, $\angle Z \cong \angle T$ since both are right angles. Then, apply the AA Similarity Postulate.

18. Yes; corresponding angles are congruent, so since $\overline{BC} \parallel \overline{DE}$, then $\angle D \cong \angle B$. Also, $\angle A$ is shared by both $\triangle ABC$ and $\triangle ADE$, so by AA, $\triangle ABC \sim \triangle ADE$.

19. Yes; $\overline{AB}$ and $\overline{AC}$ are proportional to $\overline{AD}$ and $\overline{AE}$, respectively. Also, the included angle, $\angle A$, is shared by both $\triangle ABC$ and $\triangle ADE$, so by SAS, $\triangle ABC \sim \triangle ADE$.

20. Yes; since $AD = DB$ and $AE = EC$, then $AB = 2AD$ and $AC = 2AE$. Thus, $\overline{AB}$ and $\overline{AC}$ are proportional to $\overline{AD}$ and $\overline{AE}$, respectively. Also, the included angle, $\angle A$, is shared by both $\triangle ABC$ and $\triangle ADE$, so by SAS, $\triangle ABC \sim \triangle ADE$.

21. Not enough; need $\angle D \cong \angle B$ to ensure $\triangle ABC \sim \triangle ADE$.

22. Yes, by SSS

23. No; $\triangle QRS$ is not a right triangle so $\triangle QRS$ is not a dilation of $\triangle KLM$.

25. Corresponding angles are congruent, so since $\overline{GH} \parallel \overline{BC}$, then $\angle AGH \cong \angle ABC$. Also, $\angle A$ is shared by both triangles, so by AA, $\triangle AGH \cong \triangle ABC$.

Practice

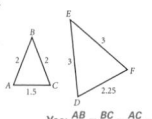
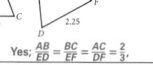
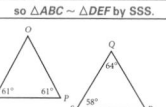
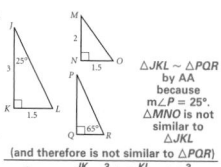

NAME _____ CLASS _____ DATE _____

Practice

8.3 Triangle Similarity Postulates

Determine whether each pair of triangles can be proven similar by using AA, SSS, or SAS. If so, write a similarity statement, and identify the postulate or theorem used.

1. Yes; $\frac{AB}{ED} = \frac{BC}{EF} = \frac{AC}{DF} = \frac{2}{3}$, so $\triangle ABC \sim \triangle DEF$ by SSS.

2. Yes; m$\angle G = 60°$, so $\triangle GHI \sim \triangle LKM$ by AA.

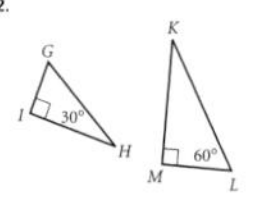

3. No; no sides are given, so you would have to use the AA Similarity Postulate. Although m$\angle O = 58°$ and m$\angle R = 58°$, AA does not apply here.

4. Yes; $\frac{TU}{WX} = \frac{UV}{XY} = 2$ and m$\angle U = m\angle X = 90°$, so $\triangle TUV \sim \triangle WXY$ by SAS.

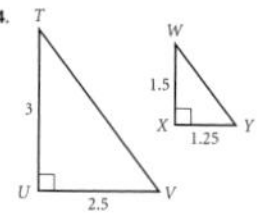

Exercises 5 and 6, indicate which figures are similar. Explain your reasoning.

5. $\triangle ABC \sim \triangle GHI$ by SSS because $\frac{AB}{GH} = \frac{BC}{DF} = \frac{AC}{GI} = 1.2$. $\triangle DEF$ is not similar to $\triangle ABC$ (and therefore is not similar to $\triangle GHI$) because $\frac{AB}{DE} = 3$ but $\frac{AC}{DF} = \frac{12}{5} \neq 3$.

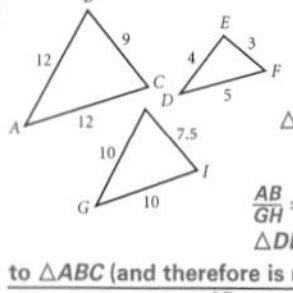

6. $\triangle JKL \sim \triangle PQR$ by AA because m$\angle P = 25°$. $\triangle MNO$ is not similar to $\triangle JKL$ (and therefore is not similar to $\triangle PQR$) because $\frac{JK}{MN} = \frac{3}{2}$ but $\frac{KL}{NO} = 1 \neq \frac{3}{2}$.

For Exercises 18–21, refer to the diagram below. Is the given information enough to prove that $\triangle ABC \sim \triangle ADE$? Explain your reasoning.

18. $\overline{BC} \parallel \overline{DE}$

19. $\frac{AB}{AD} = \frac{AC}{AE}$

20. $AD = DB$ and $AE = EC$

21. $\frac{AB}{AD} = \frac{BC}{DE}$

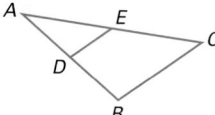

For Exercises 22–24, refer to the diagram below.

22. Draw $\triangle NOP$ in which each side is twice as long as each side of $\triangle KLM$. Is $\triangle NOP \sim \triangle KLM$? Why or why not?

23. Draw $\triangle QRS$ in which each side is 1 cm longer than the corresponding side of $\triangle KLM$. Is $\triangle QRS \sim \triangle KLM$? Why or why not?

24. Draw $\triangle TUV \sim \triangle KLM$ such that $\frac{TU}{KL} = \frac{UV}{LM} = \frac{VT}{MK} = \frac{3}{2}$. What is the ratio of the perimeter of $\triangle TUV$ to the perimeter of $\triangle KLM$? $\frac{3}{2}$

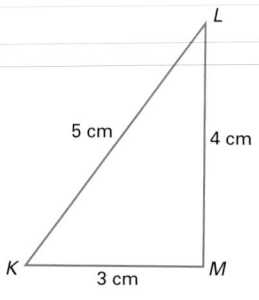

PROOF

In Exercises 25–27 you will prove the SSS Similarity Theorem. In the diagram at right, the sides of $\triangle ABC$ are proportional to the sides of $\triangle DEF$, and $\frac{AB}{DE} = \frac{BC}{EF} = \frac{CA}{FD}$. Also, $\overline{GH}$ has been added such that $AG = DE$ and $\overline{GH} \parallel \overline{BC}$.

25. Use the AA Similarity Postulate to prove that $\triangle AGH \sim \triangle ABC$.

26. Use the result of Exercise 25 to prove that $\triangle AGH \cong \triangle DEF$.

27. Use the results of Exercises 25 and 26 to prove that $\triangle ABC \sim \triangle DEF$.

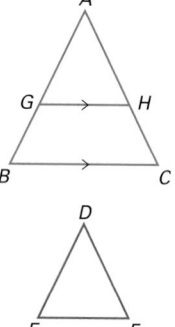

PROOF

In Exercises 28–30 you will prove the SAS Similarity Theorem. In the diagram at right, two sides of $\triangle UVW$ are proportional to two sides of $\triangle XYZ$, $\frac{UV}{XY} = \frac{UW}{XZ}$, and $\angle U \cong \angle X$. Also $\overline{ST}$ has been added such that $US = XY$ and $\overline{ST} \parallel \overline{VW}$.

28. Use the AA Similarity Postulate to prove that $\triangle UST \sim \triangle UVW$.

29. Use the result of Exercise 28 to prove that $\triangle UST \cong \triangle XYZ$.

30. Use the results of Exercises 28 and 29 to prove that $\triangle UVW \sim \triangle XYZ$.

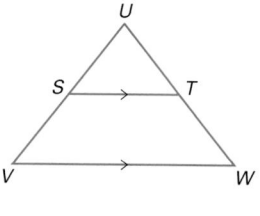
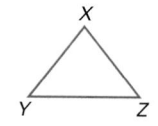

---

26. $\triangle AGH \sim \triangle ABC$, so the sides of $\triangle AGH$ are proportional to the sides of $\triangle ABC$. But, it is given that the sides of $\triangle ABC$ are proportional to the sides of $\triangle DEF$. Thus, the sides of $\triangle AGH$ are proportional to the sides of $\triangle DEF$, or $\frac{AG}{DE} = \frac{GH}{EF} = \frac{HA}{FD}$. But, $EF = GH$ and $HA = FD$. So, it follows that $\triangle AGH \cong \triangle DEF$ by SSS.

27. Since $\triangle AGH \cong \triangle DEF$, then $\triangle AGH \sim \triangle DEF$. But $\triangle AGH \sim \triangle ABC$, so $\triangle ABC \sim \triangle DEF$.

28. Corresponding angles are congruent, so since $\overline{ST} \parallel \overline{VW}$, then $\angle UST \cong \angle UVW$. Also, U is shared by both triangles, so by AA, $\triangle UST \sim \triangle UVW$.

29. $\triangle UST \sim \triangle UVW$ so, $\overline{US}$ and $\overline{UT}$ are proportional to $\overline{UV}$ and $\overline{UW}$ respectively. That is, $\frac{US}{UV} = \frac{UT}{UW}$. But, it is given that $\frac{UV}{XY} = \frac{UW}{XZ}$. So, $\frac{US}{XY} = \frac{UT}{XZ}$. But, $US = XY$ which means that $UT = XZ$. Also, $\angle U \cong \angle X$, so $\triangle UST \cong \triangle XYZ$ by SAS.

PROOF

Recall from Lesson 2.4 that a relationship is called an *equivalence relation* if it satisfies the Reflexive, Symmetric, and Transitive Properties. Exercises 31–33 establish the fact that similarity is an equivalence relation.

31. Prove that △ABC ~ △ABC (Reflexive Property).

32. Prove that if △ABC ~ △DEF, then △DEF ~ △ABC (Symmetric Property).

33. Prove that if △ABC ~ △DEF and △DEF ~ △GHI, then △ABC ~ △GHI (Transitive Property).

CHALLENGE

34. Refer to the diagram at right. Suppose that there is a quadrilateral STUV with ∠T ≅ ∠X and ∠V ≅ ∠Z and that all four sides of STUV are proportional to all four sides of WXYZ. Show that the two quadrilaterals are similar. Generalize your results from this case to make a conjecture about a set of conditions that can be used to prove similarity in quadrilaterals.

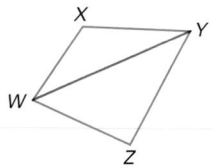

Artists creating a mural by enlarging smaller images.

In Exercises 35–38, refer to the following situation: Four artists are copying a design to a larger scale. The design contains a triangle, and the artists need to make a triangle that is similar to it.

APPLICATIONS

35. GRAPHIC DESIGN Tony first draws one side of his triangle and then copies the angles from the original triangle at the endpoints of that side. He then extends the angles until they meet. Will Tony's triangle be similar to the original? Explain your reasoning.

36. GRAPHIC DESIGN Biata measures the sides of the original triangle and then multiplies each length by 5 to get the sides of the triangle in her design. Will Biata's triangle be similar to the original? Explain your reasoning.

37. GRAPHIC DESIGN Miki copies one angle from the original triangle, then measures the two sides adjacent to that angle, and multiplies the lengths by 5. She then draws sides with these lengths adjacent to the copied angle and connects the endpoints to form a triangle. Will Miki's triangle be similar to the original? Explain your reasoning.

38. GRAPHIC DESIGN George measures two sides of the triangle and multiplies the lengths by 5. He draws one of the sides and copies the nonincluded angle at one of the endpoints. He extends the angle and then draws the remaining side from the other endpoint so that it intersects the side of the copied angle. Will George's triangle be similar to the original? Explain your reasoning.

30. Since △UST ≅ △XYZ, then △UST ~ △XYZ. But, △UST ~ △UVW so that △UVW ~ △XYZ.

31. ∠A ≅ ∠A and ∠B ≅ ∠B, so by AA, △ABC ~ △ABC.

32. If △ABC ~ △DEF then ∠A ≅ ∠D and ∠B ≅ ∠E. But, this means that ∠D ≅ ∠A and ∠E ≅ ∠B. So by AA, △DEF ~ △ABC.

33. If △ABC ~ △DEF, then ∠A ≅ ∠D and ∠B ≅ ∠E, and if △DEF ~ △GHI, then ∠D ≅ ∠G and ∠E ≅ ∠H. So, it follows that ∠A ≅ ∠G and ∠B ≅ ∠H. Thus by AA, △ABC ~ △GHI.

34. Since all 4 sides of STUV are proportional to all 4 sides of WXYZ, then $\overline{WX}$ and $\overline{XY}$ are proportional to $\overline{ST}$ and $\overline{TU}$, respectively. ∠T and ∠X are each included angles, so by SAS, △WXY ~ △STU. Similarly, it follows that △WYZ ~ △SUV. Appending △WXY and △STU to △WYZ and △SUV, respectively, two similar quadrilaterals are formed. Conjecture: If the sides of one quadrilateral are proportional to the sides of a second quadrilateral, and one pair of opposite angles are congruent to their corresponding angles, then the quadrilaterals are similar.

35. Yes; since Tony copied two angles from the original triangle, then by AA, the two triangles are similar.

36. Yes; since Biata multiplied each length by 5 to get the sides of her triangle, then by SSS, the two triangles are similar.

37. Yes; since Miki copied an angle which is adjacent to two sides of her triangle, of which the lengths are 5 times the lengths of two sides of the original triangle, then by the SAS Similarity Theorem, the two triangles are similar.

38. No; even though the lengths of two sides of George's triangle are 5 times the lengths of two sides of the original triangle, and he copied an angle from the original triangle, the angle is not adjacent to those two sides. So, the SAS Similarity Theorem cannot be applied.

In Exercises 48–51, students copy the given similar triangles and then draw lines connecting the three corresponding angles of each pair of similar triangles. Students' discoveries should extend the properties of similar triangles that were presented in the lesson.

48.

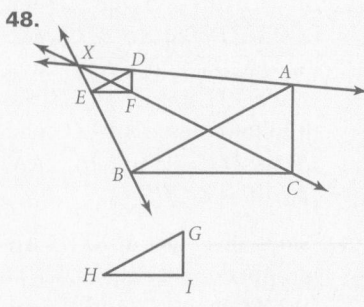

49.

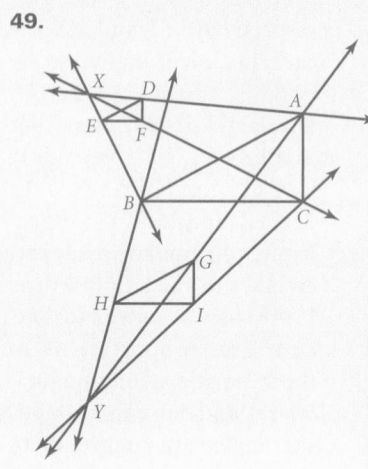

50.

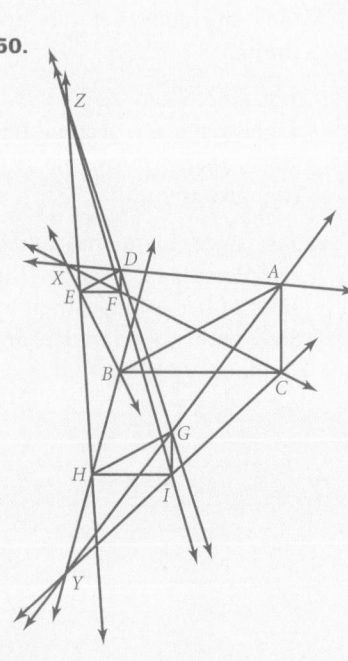

 **Look Back**

Refer to the diagram at right.
$\overline{AB} \cong \overline{BC}$ and line *j* ∥ line *k*.
Find each angle listed below.
(LESSON 3.3)

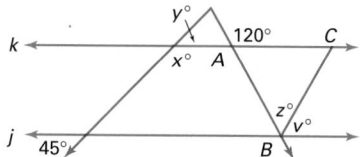

39. *x* 135   **40.** *y* 45

41. *z* 60   **42.** *v* 60

 Algebra

Plot point *A*(3, 5) on a coordinate plane and draw a line through this point and the origin.

43. What is the slope of this line? *(LESSON 3.8)* $\frac{5}{3}$

44. What is the distance from the origin to point *A*? *(LESSON 5.6)* $\sqrt{34} \approx 5.83$

45. Draw a right triangle with point *A* as one vertex, the origin as one vertex, and one side on the *x*-axis. Find the ratio of the length of the longer leg to the length of the the shorter leg. *(LESSON 8.2)* $\frac{5}{3}$

46. Choose another point anywhere on the original line and label it *B*. Draw a right triangle with point *B* as one vertex, the origin as one vertex, and one side on the *x*-axis. Find the ratio of the length of the longer leg to the length of the shorter leg. *(LESSON 8.2)* $\frac{5}{3}$

47. Are the triangles in Exercises 45 and 46 similar? Explain your reasoning. *(LESSON 8.2)*
Yes; both have a right angle and share an angle with its vertex at the origin. Thus, by AA similarity, they are similar.

Look Beyond

In the diagram below, the three triangles are similar, and $\overline{AC}$, $\overline{DF}$, and $\overline{GI}$ are parallel.

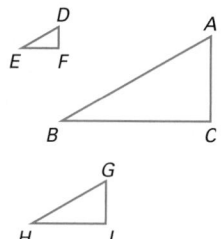

48. Copy the diagram and draw lines $\overleftrightarrow{AD}$, $\overleftrightarrow{BE}$, and $\overleftrightarrow{CF}$. Label the point of intersection *X*.

49. Draw lines $\overleftrightarrow{AG}$, $\overleftrightarrow{BH}$, and $\overleftrightarrow{CI}$. Label the point of intersection *Y*.

50. Draw lines $\overleftrightarrow{DG}$, $\overleftrightarrow{EH}$, and $\overleftrightarrow{FI}$. Label the point of intersection *Z*.

51. What do you notice about points *X*, *Y*, and *Z*? **They are collinear.**

The Side-Splitting Theorem

Objectives

- Develop and prove the Side-Splitting Theorem.

- Use the Side-Splitting Theorem to solve problems.

Why *When you hear the phrase "side splitting," you may think of laughter. In geometry, it refers to a useful theorem.*

The capital letter A may be embellished in endless ways for reasons of beauty and style. But in its most basic form, it suggests a geometry theorem.

The Side-Splitting Theorem

As you will see in Example 2, people have been solving problems about proportions in triangles since ancient times. One useful result is the Side-Splitting Theorem.

Recall from the Triangle Midsegment Theorem in Lesson 4.6 that the midsegment of a triangle is parallel to one side of the triangle. The following theorem applies to any segment that is parallel to one side of a triangle.

Side-Splitting Theorem

A line parallel to one side of the triangle divides the other two sides proportionally.   8.4.1

Alternative Teaching Strategy

HANDS-ON STRATEGIES Students can develop an understanding of the Side-Splitting Theorem by cutting out paper triangles. Ask students to draw a triangle on a piece of paper and make a photocopy using the Enlarge/Reduce feature of the copier. After cutting out both triangles, students should superimpose the corresponding angles to show that they are congruent. They should also measure the corresponding sides to show that they are proportional.

Cooperative Learning

Have students work in pairs. Each student should draw and label an example to illustrate the Side-Splitting Theorem. Ask them to measure the angles and sides and to verify each step of the proof of the theorem in terms of their example. Have students exchange papers and check each other's work.

CRITICAL THINKING

$\frac{AD}{DB} = \frac{AE}{EC}$, $\frac{AD}{AB} = \frac{AE}{AC}$, $\frac{DB}{AB} = \frac{EC}{AC}$,

$\frac{AD}{AE} = \frac{DB}{EC}$

More answers are possible.

EXAMPLE ❶

Use the Side-Splitting Theorem to find x in the triangle below.

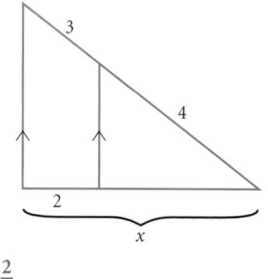

$4\frac{2}{3}$

Given: $\overline{DE} \parallel \overline{BC}$

Prove: $\frac{DB}{AD} = \frac{EC}{AE}$

Proof:

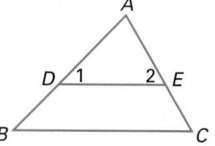

Statements	Reasons
1. $\overline{DE} \parallel \overline{BC}$	Given
2. $m\angle B = m\angle 1$ $m\angle C = m\angle 2$	If $\parallel$ lines are cut by a transversal, corresponding angles are $\cong$.
3. $\triangle ABC \sim \triangle ADE$	AA Similarity Postulate
4. $\frac{AB}{AD} = \frac{AC}{AE}$	Polygon Similarity Postulate
5. $AD + DB = AB$ $AE + EC = AC$	Segment Addition Postulate
6. $\frac{AD + DB}{AD} = \frac{AE + EC}{AE}$	Substitution Property
7. $\frac{AD}{AD} + \frac{DB}{AD} = \frac{AE}{AE} + \frac{EC}{AE}$	Addition of fractions
8. $1 + \frac{DB}{AD} = 1 + \frac{EC}{AE}$	Simplify.
9. $\frac{DB}{AD} = \frac{EC}{AE}$	Subtraction Property of Equality

CRITICAL THINKING What other proportions can you find in $\triangle ABC$ by using the Side-Splitting Theorem?

The diagram at right may help you remember the proportions in a triangle with a segment parallel to one side. The following are some of the proportions resulting from the Side-Splitting Theorem:

$\frac{\text{upper left}}{\text{lower left}} = \frac{\text{upper right}}{\text{lower right}}$ 　　 $\frac{\text{upper left}}{\text{upper right}} = \frac{\text{lower left}}{\text{lower right}}$

$\frac{\text{upper left}}{\text{whole left}} = \frac{\text{upper right}}{\text{whole right}}$ 　　 $\frac{\text{lower left}}{\text{whole left}} = \frac{\text{lower right}}{\text{whole right}}$

EXAMPLE ❶ Use the Side-Splitting Theorem to find x in the triangle below.

● **SOLUTION**

Choose a proportion that includes x as a single term.

$\frac{\text{upper left}}{\text{lower left}} = \frac{\text{upper right}}{\text{lower right}}$

$\frac{12}{16} = \frac{15}{x}$

$12x = 240$

$x = 20$

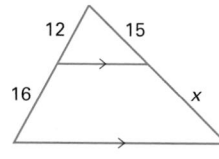

Inclusion Strategies

ENGLISH LANGUAGE DEVELOPMENT Some students may have difficulty understanding the terminology in this lesson. Have them list all concepts that are confusing and then verify the postulates and theorems in terms of those concepts.

Enrichment

The Eiffel Tower offers many examples of the Side-Splitting Theorem and of triangular bracing. Ask students to write a report about the Eiffel Tower in which they identify specific triangular shapes that relate to the concepts presented in this lesson.

The following is a corollary of the Side-Splitting Theorem:

Two-Transversal Proportionality Corollary

Three or more parallel lines divide two intersecting transversals proportionally.

8.4.2

In the diagram at right, lines ℓ, m, and n are parallel, with transversals s and t.

One proportion that results from Corollary 8.4.2 is $\frac{AB}{BC} = \frac{XY}{YZ}$.

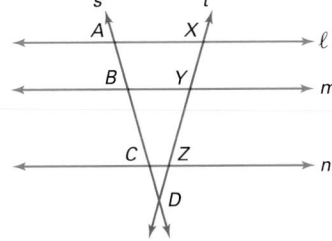

CRITICAL THINKING How can you tell that $\triangle ADX \sim \triangle BDY$?

EXAMPLE ② **CULTURAL CONNECTION: AFRICA** Students in ancient Egypt studied geometry to solve practical problems involving the pyramids. This problem is based on a problem in a papyrus copied in 1650 B.C.E. by the scribe Ahmes from a source that may date back to 2000 B.C.E.

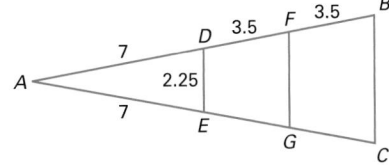

In the diagram above, $\overline{DE}$, $\overline{FG}$, and $\overline{BC}$ are parallel, $AD = AE = 7$ cubits, $DF = FB = 3.5$ cubits, and $DE = 2.25$ cubits. Find the remaining lengths.

● **SOLUTION**

Use the Two-Transversal Proportionality Corollary:

$$\frac{AD}{DF} = \frac{AE}{EG} \quad \Rightarrow \quad \frac{7}{3.5} = \frac{7}{EG} \quad \Rightarrow \quad EG = 3.5$$

and

$$\frac{DF}{FB} = \frac{EG}{GC} \quad \Rightarrow \quad \frac{3.5}{3.5} = \frac{3.5}{GC} \quad \Rightarrow \quad GC = 3.5$$

Use the definition of similar triangles:

$$\frac{AD}{AF} = \frac{DE}{FG} \quad \Rightarrow \quad \frac{7}{10.5} = \frac{2.25}{GF} \quad \Rightarrow \quad GF = 3.375$$

$$\frac{AD}{AB} = \frac{DE}{BC} \quad \Rightarrow \quad \frac{7}{14} = \frac{2.25}{BC} \quad \Rightarrow \quad BC = 4.5$$

Reteaching the Lesson

COOPERATIVE LEARNING Arrange the class into groups to review the terms used in this lesson, including *parallel*, *transversal*, and *proportional*. Ask each group to create examples of the Side-Splitting Theorem and its corollary, and use the examples as review for the entire class. Having students work together may help them make connections between the algebra and geometry used in this lesson.

CRITICAL THINKING

Since $\ell \parallel m$ then $\angle XAD$ and $\angle YBD$ are corresponding angles so that $\angle XAD \cong \angle YBD$. Also, $\angle D$ is shared by both $\triangle ADX$ and $\triangle BDY$. So, by the AA Similarity Postulate, $\triangle ADX \sim \triangle BDY$.

ADDITIONAL
EXAMPLE ②

Land plots between two streets are laid out according to the plan shown below. The horizontal lot boundaries are parallel to each other. Find the missing lengths.

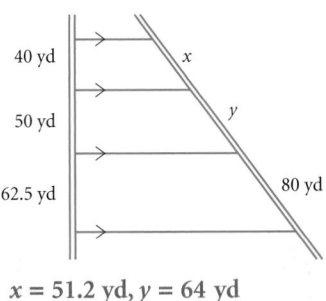

$x = 51.2$ yd, $y = 64$ yd

ASSIGNMENT GUIDE	
In Class	1–12
Core	13–25 odd, 26–30, 31–35 odd
Core Plus	21–25 odd, 26–30, 31–39 odd
Review	40–47
Preview	48–50

✐ Extra Practice can be found beginning on page 818.

Exercises

Communicate

1. Use the examples on page 526 to make a list of proportions in the figure at right.

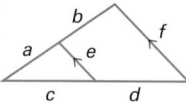

2. Are all isosceles triangles similar? Explain or give a counterexample.

3. Are all equilateral triangles similar? Explain or give a counterexample.

4. Are all right triangles similar? Explain or give a counterexample.

5. How does the capital letter A relate to the Side-Splitting Theorem? If the cross bar is horizontal, what is true of the places where it intersects the sides of the letter? Would this be true of an italic (slanted) A, as well?

Guided Skills Practice

Use the Side-Splitting Theorem to find *x*. *(EXAMPLE 1)*

6. 15

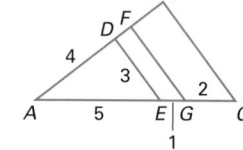

7. 6

8.

Find the indicated measurements. *(EXAMPLE 2)*

9. DF $\frac{4}{5} = 0.8$

10. FB $\frac{8}{5} = 1.6$

11. FG $\frac{18}{5} = 3.6$

12. BC $\frac{24}{5} = 4.8$

Practice and Apply

🖸 internet connect

Homework Help Online

Go To: go.hrw.com
Keyword:
MG1 Homework Help
for Exercises 13-20

In Exercises 13–20, use the Side-Splitting Theorem to find *x*. In some exercises, there may be more than one possible value for *x*.

13. 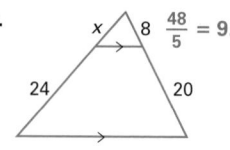 $\frac{48}{5} = 9.6$

14. 8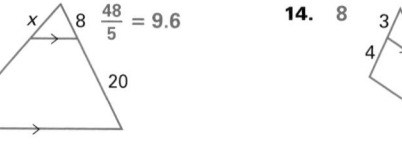

15. 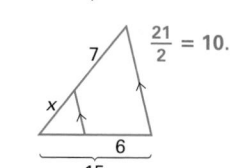 $\frac{21}{2} = 10.5$

16. $\frac{33}{4} = 8.25$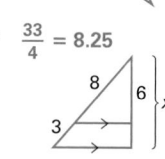

17. 6

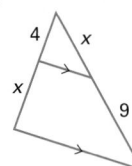

18. 5

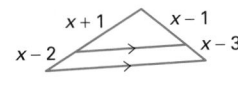

19. 1 or 2

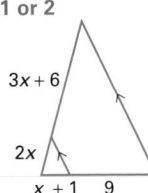

20. $\frac{3}{2}$ or 3

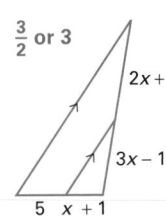

Name all similar triangles in each figure. State the postulate or theorem that justifies each similarity.

21.

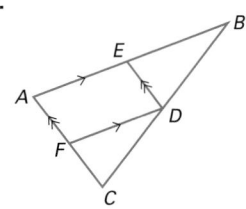

22.

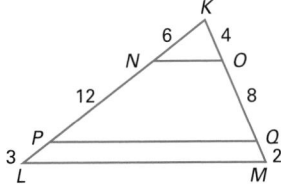

Algebra Find *x* and *y* in each figure below.

23. *QRST* is a parallelogram.

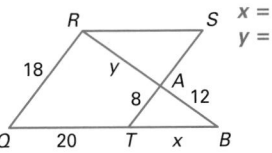

$x = 16$
$y = 15$

24.

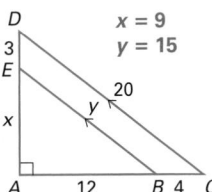

$x = 9$
$y = 15$

25.

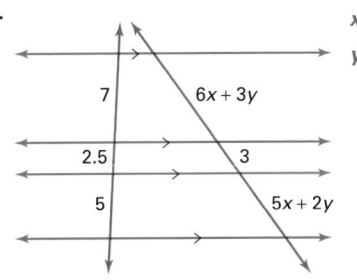

$x = \frac{2}{5}$
$y = 2$

---

21. By the AA Similarity Theorem, $\triangle BED \sim \triangle BAC$. Similarly, $\triangle BAC \sim \triangle DFC$. Then, by the Transitive Property, $\triangle BED \sim \triangle DFC$.

22. $\triangle NKO \sim \triangle PKQ$, $\triangle NKO \sim \triangle LKM$, and $\triangle PKQ \sim \triangle LKM$ by the SAS Similarity Theorem.

26. Sample answer:

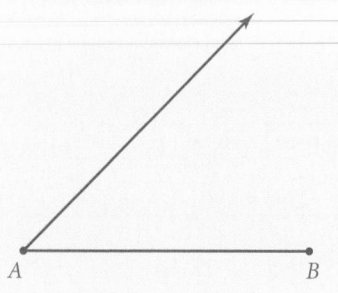

27. Sample answer:

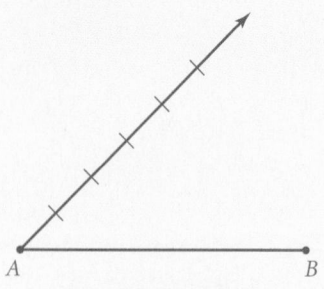

The Side-Splitting Theorem can be used to divide a segment into any number of congruent parts using a compass and straightedge.

26. Draw a segment and label its endpoints A and B. Using your straightedge, draw a ray extending from point A to form an acute angle with $\overline{AB}$.

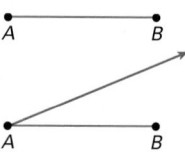

27. Set your compass to some small length (such as 1 cm) and mark off several lengths along the ray, as many as the number of parts you wish to divide the segment into.

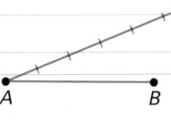

28. Connect the last mark to point B with a segment. Construct lines parallel to this segment through each mark of the compass.

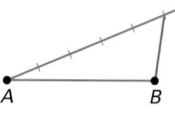

29. Explain why the parallel lines in the last figure divide $\overline{AB}$ into five congruent segments.

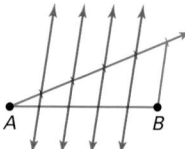

30. Draw a segment that is 15 cm long. Use the method described above to divide the segment into seven congruent segments.

CONNECTION

COORDINATE GEOMETRY **The distance between two parallel lines is measured along a line perpendicular to both. In Exercises 31–34, you will explore the distances between parallel lines in a coordinate plane.**

31. Graph the lines $y = x$, $y = x + 2$, and $y = x - 3$ in a coordinate plane. How can you verify that these lines are are parallel? **They all have a slope of 1.**

32. Graph the line $y = -x$ in the same plane. How can you verify that this line is perpendicular to the lines $y = x$, $y = x + 2$, and $y = x - 3$?

33. Give the coordinates of the points where the line $y = -x$ intersects the lines $y = x$ and $y = x + 2$. Find the distance between these parallel lines.

34. Graph the horizontal line $y = 3$ in the same plane. Use the Two-Transversal Proportionality Corollary to write a proportion, and solve it to find the distance between the lines $y = x$ and $y = x - 3$.

PROOF

35. Prove the converse of the Side-Splitting Theorem: If a segment divides two sides of a triangle proportionally, then the segment is parallel to the third side.

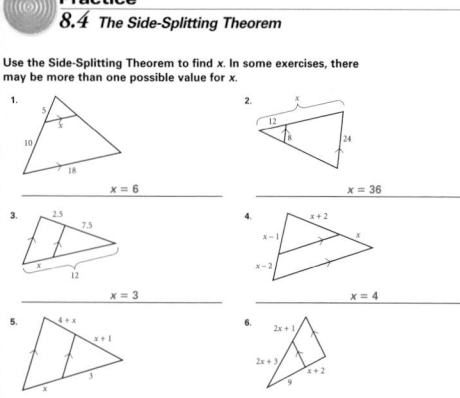
28. Sample answer:

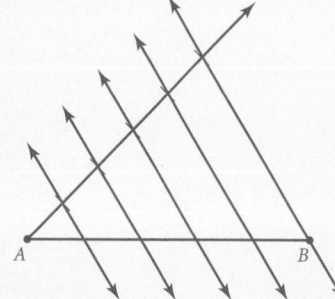

29. By the Two-Transversal Proportionality Corollary, the parallel lines divide $\overline{AB}$ and the ray proportionally. But the ray was divided into congruent parts, so $\overline{AB}$ is divided into congruent parts.

30.

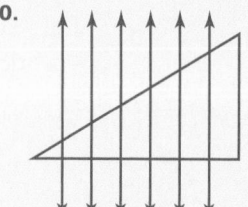

36. In the diagram at right, $\frac{AE}{AG} = 1.3$, $\frac{AB}{AE} = 2.2$, and $AD = 10$ cm. Find AB. **13 cm**

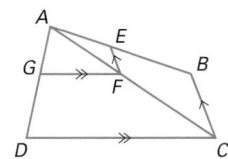

37. MUSIC A simple stringed instrument can be created with parallel strings on an isosceles trapezoidal frame, as shown. If the note for the first (lowest) string, with a length of 40 cm, is an F, then a string with a length of 20 cm would be one *octave* higher, also an F.

If a string for the note C is $\frac{2}{3}$ of the length of the string for the lower F, where would you place the string on the frame?

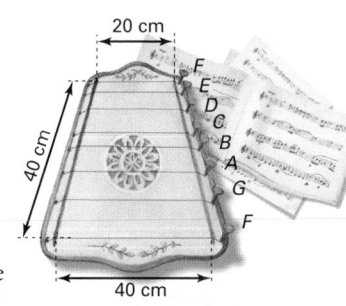

38. MUSIC For the instrument above, the string for the note G is $\frac{8}{9}$ of the length of the string for the lower F. Where would you place the string on the frame?

39. BIOLOGY An orb spider web consists of a spiral of straight segments attached to a series of radial segments that meet at the center. Each sector of the web can be modelled by nested similar triangles. Suppose the outside of a spider web is approximately a regular 16-gon with a side length of 19.5 cm and the 9 inner segments are spaced at equal intervals. Use the diagram to estimate the length of silk used to build the web. **2516 cm**

The diagram represents one sector of the web.

50 cm

19.5 cm

Notice the spiral of straight segments in this orb spider web. Though spider webs in nature are not perfectly symmetrical, or the spacing of their spirals even, you can model a web mathematically and get a good estimate of the length of the silk in its spiral.

31. They all have a slope of 1.

32. The slope of $y = -x$ is -1, which is the negative reciprocal of the slope of the other lines, which is 1.

33. $(0, 0)$ and $(-1, 1)$; $\sqrt{2} \approx 1.41$

34. $\frac{\sqrt{2}}{d} = \frac{2}{3}$; $d = \frac{3\sqrt{2}}{2} \approx 2.12$

37. The string for the note C should be placed $13\frac{1}{3}$ cm from the string for the higher F.

38. The string for the note G should be placed $31\frac{1}{9}$ cm from the string for the higher F.

The answer to Exercise 35 can be found in Additional Answers beginning on page 879.

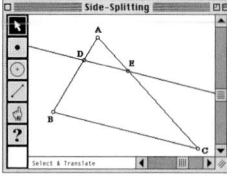

Look Beyond

In Exercises 48–50, students extend the properties of similar triangles to include the special case of the 30-60-90 right triangle.

41.

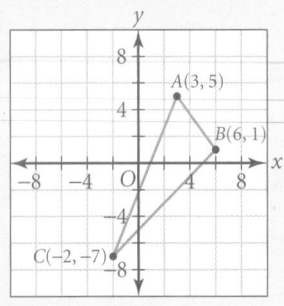

43. $D\left(\dfrac{9}{2}, 3\right)$; $E(2, -3)$;

$F\left(\dfrac{1}{2}, -1\right)$

45. Check students' drawings, $\sqrt{65} \approx 8.06$

46. Check students' drawings, $\dfrac{\sqrt{137}}{2} \approx 5.85$

47. Check students' drawings, $\dfrac{\sqrt{569}}{2} \approx 11.93$

48. Yes; each triangle has a right angle and a 30° angle. Thus, by the AA Similarity Postulate, they are similar.

49. $\dfrac{1}{2}$; The ratio is the same for both triangles.

50. $\dfrac{1}{2}$

Look Back

40. ENGINEERING Find the height of the stack of pipes shown at right. The diameter of each pipe is 8 ft. (Hint: Connect the centers of the circles to form an equilateral triangle.)
$8 + 4\sqrt{3} \approx 14.9$ ft

COORDINATE GEOMETRY For Exercises 41–47, refer to △ABC with vertices A(3, 5), B(6, 1), and C(−2, −7) in a coordinate plane. *(LESSON 5.6)*

41. Draw △ABC in a coordinate plane.

42. Find the perimeter of △ABC. $18 + 8\sqrt{2} \approx 29.31$

43. Find the midpoint of each side. Label them D, E, and F.

44. Find the perimeter of △DEF. $9 + 4\sqrt{2} \approx 14.66$

Recall that the median of a triangle is a segment that connects a vertex to the midpoint of the opposite side. *(LESSON 5.6)*

45. Draw the median of △ABC from A, and find its length.

46. Draw the median of △ABC from B, and find its length.

47. Draw the median of △ABC from C, and find its length.

Look Beyond

Refer to the figures below for Exercises 48–49.

48. Are the triangles similar? Why or why not?

49. Find the ratio of the length of the shorter leg to the length of the hypotenuse for each triangle. What do you notice?

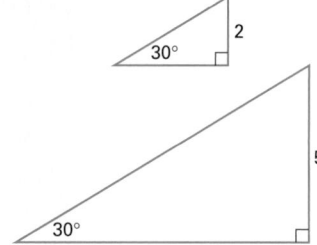

50. The shorter leg of another right triangle with a 30° angle has a length of 17. What is the ratio of the length of the shorter leg to the length of the hypotenuse for this triangle?

Indirect Measurement and Additional Similarity Theorems

Objectives

- Use triangle similarity to measure distances indirectly.

- Develop and use similarity theorems for altitudes and medians of triangles.

Why *Directly measuring the distance across a lake or the height of a mountain can be difficult or even impossible. A more practical method uses similar triangles to estimate the desired measurement.*

Monument Valley in Arizona

Using Similar Triangles to Measure Distances

Inaccessible distances can often be measured by using similar triangles.

EXAMPLE 1 Suppose that a military engineer needs to know the distance across a river in order to build a temporary bridge. Using a point on the opposite side of the river as a reference point, the engineer sets up right triangles along the bank of the river. Use the diagram at right to find the distance across the river.

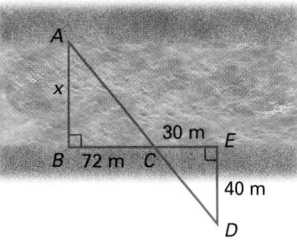

SOLUTION

Using the right angles and vertical angles, $\triangle ABC \sim \triangle DEC$ by AA Similarity, so $\frac{AB}{DE} = \frac{BC}{EC}$. Therefore:

$$\frac{x}{40} = \frac{72}{30}, \text{ or } \frac{x}{40} = \frac{12}{5}$$

$$x = \frac{12}{5} \cdot 40 = 96 \text{ meters}$$

Teach

Why Over the centuries, similarity has been important to surveyors and mapmakers. Students need to learn how to use similarity in order to measure distances that would not otherwise be measurable.

ADDITIONAL EXAMPLE 1

Find x in the diagram below.

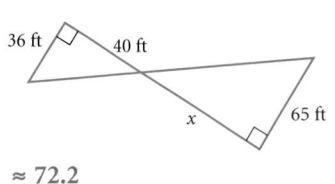

≈ 72.2

Alternative Teaching Strategy

COOPERATIVE LEARNING Divide the class into cooperative groups and have each group devise an experiment for estimating the size of something in or near the school building that would otherwise be difficult to measure. Examples include the height of the classroom, the height of a stairwell, or the height of a flagpole. Ask each group to describe their experiment and their results to the class.

Find *x* in the diagram below.

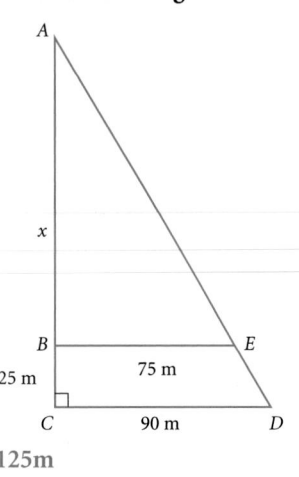

125m

90° angles are easy to measure, but other angles would also lead to similar triangles.

Teaching Tip

In Example 3, students solve a proportion that they learned in algebra. At this time, you may want to review the method of solving proportions by cross multiplying.

Mao wants to know the height of a light pole in his neighborhood. Mao is 170 cm tall. If the shadow of the light pole is 840 cm long and his shadow is 310 cm long, how tall is the light pole?

≈ 460.65 cm

EXAMPLE 2 The diagram at right illustrates another method of setting up right triangles along the river bank. Use this diagram to find the distance across the river.

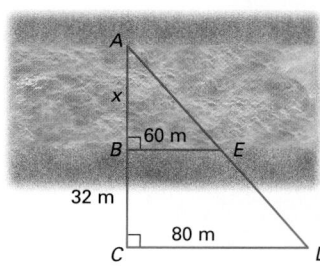

● **SOLUTION**

Because $\angle ABE \cong \angle ACD$ and $\angle A \cong \angle A$, $\triangle ABE \sim \triangle ACD$ by AA Similarity, and so $\frac{AB}{AC} = \frac{BE}{CD}$. Therefore:

$$\frac{x}{x+32} = \frac{60}{80}, \text{ or } \frac{x}{x+32} = \frac{3}{4}$$
$$4x = 3x + 96$$
$$x = 96 \text{ meters}$$

CRITICAL THINKING   Why do you think the engineer used 90° angles? Could other angles be used?

EXAMPLE 3 Kim wants to know the height of a basketball hoop. The regulation height is 10 feet. Use Kim's height and the length of the shadows in the diagram at right to find the height of the basketball hoop. Is the hoop at regulation height?

● **SOLUTION**

Because the Sun is so far away, the sun's rays are essentially parallel, and $\angle 1 = \angle 2$. Both Kim and the post of the basketball hoop are perpendicular to the ground, so the two triangles are similar by the AA Similarity Postulate.

$$\frac{x}{70} = \frac{85}{50}$$
$$70 \cdot \frac{x}{70} = \frac{85}{50} \cdot 70$$
$$x = 119 \text{ inches, or 9 feet}$$
$$\text{and 11 inches.}$$

The hoop is 1 inch below regulation height.

Interdisciplinary Connection

SURVEYING   Surveyors use special instruments to measure distances that cannot be measured directly. Have students research the types of tools that are used by surveyors, including modern laser instruments and standard transits. Ask them to explain how similar triangles are used to make measurements with these tools.

Additional Similarity Theorems

As you know, the sides of similar triangles are proportional. As you will see below, other parts of similar triangles are also proportional.

Proportional Altitudes Theorem

If two triangles are similar, then their corresponding altitudes have the same ratio as their corresponding sides.

8.5.1

Recall that an altitude of a triangle may be inside the triangle, outside the triangle, or a side of the triangle.

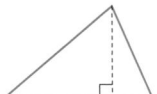  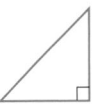

The following proof shows the case of an altitude inside the triangle. You will be asked to prove the case of an altitude outside the triangle in Exercise 26.

PARAGRAPH PROOF

Given: $\triangle ABC \sim \triangle XYZ$; $\overline{AD}$ is an altitude of $\triangle ABC$ and $\overline{XW}$ is an altitude of $\triangle XYZ$.

Prove: $\dfrac{AD}{XW} = \dfrac{AB}{XY}$

Proof:

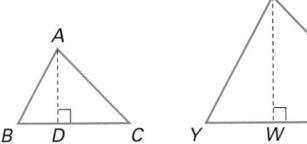

Because $\triangle ABC \sim \triangle XYZ$, $\angle B \cong \angle Y$ by the Polygon Similarity Postulate, and $\angle BDA$ and $\angle YWX$ are right angles by definition of altitude. Thus, $\triangle ABD \sim \triangle XYW$ by the AA Similarity Postulate, and $\dfrac{AD}{XW} = \dfrac{AB}{XY}$ by the Polygon Similarity Postulate.

CRITICAL THINKING   How would you prove the case of an altitude that is a side of the triangle?

EXAMPLE 4   The triangles in the diagram are similar. Solve for *x*.

● **SOLUTION**

$$\frac{x}{12} = \frac{16}{20}$$

$$12 \cdot \frac{x}{12} = \frac{16}{20} \cdot 12$$

$$x = 9.6$$

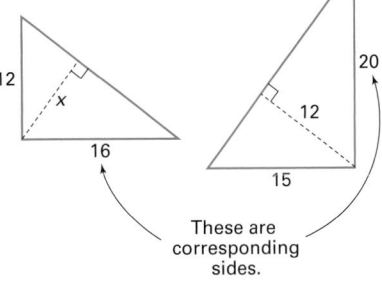

These are corresponding sides.

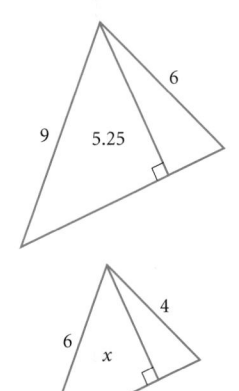

The triangles in the diagram below are similar. Solve for *x*.   5.6

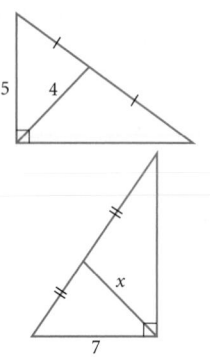

Recall that a *median* of a triangle is a segment that joins a vertex to the midpoint of the opposite side.

Proportional Medians Theorem

If two triangles are similar, then their corresponding medians have the same ratio as their corresponding sides.   **8.5.2**

You will be asked to prove this theorem in Exercises 35–41.

E X A M P L E ⑤ The triangles in the diagram are similar. Solve for *x*.

● **SOLUTION**

$$\frac{4}{6} = \frac{3.2}{x}$$

$$4x = 19.2$$

$$x = 4.8$$

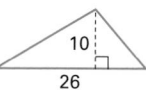
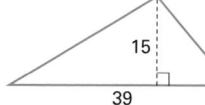

Exercises

Teaching Tip

Remind students that although different geometric segments are used in different problems, solving the proportions involves the same algebraic method each time.

● *Communicate*

1. Explain how triangle similarity is used in indirect measurement.

2. Are the triangles in the diagram at right necessarily similar? Why or why not?

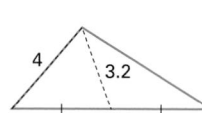
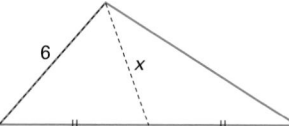

3. How could you find the height of an object from the length of its shadow?

4. Suppose that the shadow of the object is not visible. Describe another method of finding the height of the object with indirect measurement.

Assess

Selected Answers
Exercises 5–9, 11–47 odd

ASSIGNMENT GUIDE

In Class	1–9
Core	11–25 odd, 27–33, 35–41
Core Plus	27–41, 43–47
Review	48–53
Preview	54–56

✎ Extra Practice can be found beginning on page 818.

Reteaching the Lesson

USING COGNITIVE STRATEGIES Have students list the ratios for the similar triangles below that equal the scale factor between the triangles. Include the ratios for altitudes $\overline{BE}$ and $\overline{SV}$ and medians $\overline{BD}$ and $\overline{SU}$.

$$\frac{AB}{RS} = \frac{BC}{ST} = \frac{AC}{RT} = \frac{BE}{SV} = \frac{AE}{RV} = \frac{BD}{SU} = \frac{ED}{VU}$$

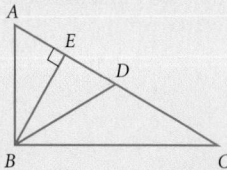
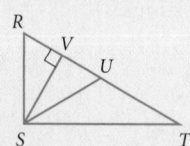

Guided Skills Practice

5. Use the diagram at right to estimate the width of the lake. Justify your answer.
(EXAMPLE 1) $\frac{x}{10} = \frac{5.4}{3}$; $x = 18$ km

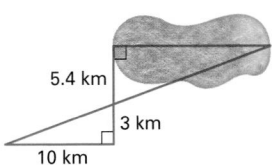

5.4 km
3 km
10 km

6. Use the diagram at right to estimate the width of the lake. Justify your answer.
(EXAMPLE 2) $\frac{x}{} = \frac{9}{}$; $x = 18$ km

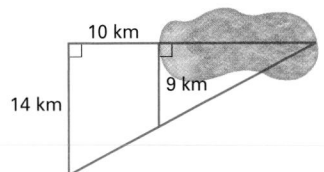

10 km
9 km
14 km

7. Tranh wants to know the height of a streetlight in his neighborhood. He measures his shadow and the pole's shadow at the same time of day. Tranh is 170 cm tall. Find the height of the streetlight. *(EXAMPLE 3)*
≈ 460.65 cm

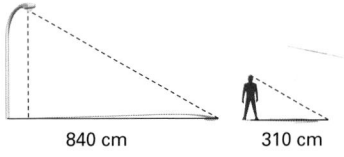

840 cm
310 cm

In Exercises 8 and 9, the triangles are similar. Find x. *(EXAMPLES 4 AND 5)*

8.

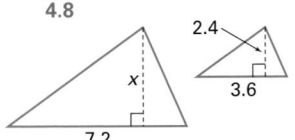

4.8
2.4
x
3.6
7.2

9.

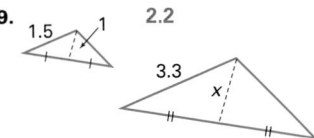

1.5
1
2.2
3.3
x

Practice and Apply

internet connect

Homework Help Online
Go To: go.hrw.com
Keyword:
MG1 Homework Help
for Exercises 10-20

For Exercises 10–13, use the diagrams to find the height of each building.

10.

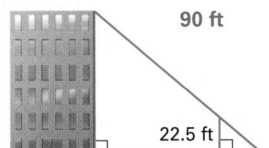

90 ft
22.5 ft
81 ft   27 ft

11.

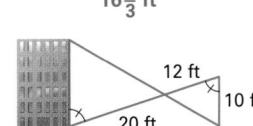

$16\frac{2}{3}$ ft
12 ft
10 ft
20 ft

12.

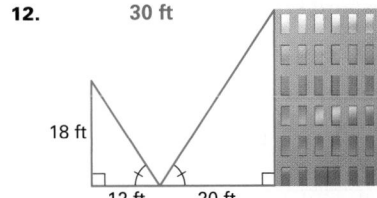

30 ft
18 ft
12 ft   20 ft

13.

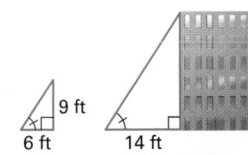

21 ft
9 ft
6 ft   14 ft

21. Yes, $\frac{MK}{JK} = \frac{KJ}{KL} = \frac{4}{3}$ and $\angle MKJ \cong \angle JKL$ since both are right angles, so the triangles are similar by the SAS Similarity Theorem.

22. $\triangle JKL$: 15; $\triangle MKJ$: 20

23. $\triangle JKL$: 54; $\triangle MKJ$: 96

24. $\triangle JKL$: 7.2; $\triangle MJK$: 9.6

25. Yes, $\frac{9.6}{7.2} = \frac{MK}{JK} = \frac{4}{3}$.

26. Since $\triangle ABC \sim \triangle XYZ$, $\angle ABC \cong \angle XYZ$ by the Polygon Similarity Postulate. $\angle ABD$ and $\angle XYW$ are supplementary to $\angle ABC$ and $\angle XYZ$, respectively, so $\angle ABD \cong \angle XYW$. Thus, $\triangle ABD \sim \triangle XYW$ by the AA Similarity Postulate. So, $\frac{AD}{XW} = \frac{AB}{XY}$ by the Polygon Similarity Postulate.

For Exercises 14–17, the triangles are similar. Find x.

14. 75

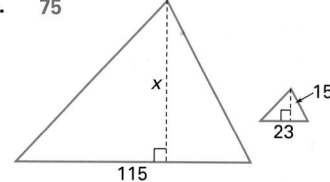

15. 10

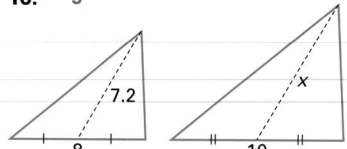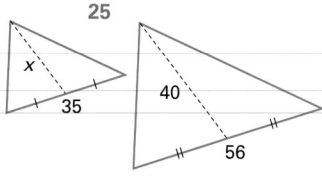

16. 9

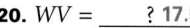

17. 25

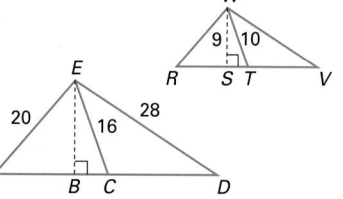

In the diagram at right, $\triangle ADE \sim \triangle RVW$, $AC = CD$, and $RT = TV$.

18. $RW = $ ___?___ 12.5   **19.** $EB = $ ___?___ 14.4

20. $WV = $ ___?___ 17.5

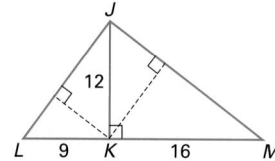

For Exercises 21–25, refer to the triangle below.

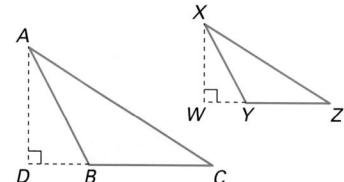

21. Are $\triangle JKL$ and $\triangle MKJ$ similar? Explain.

22. Find the length of the hypotenuse of $\triangle JKL$ and of $\triangle MKJ$.

23. Find the area of $\triangle JKL$ and of $\triangle MKJ$.

24. Find the lengths of the dashed altitudes of $\triangle JKL$ and of $\triangle MKJ$.

25. Is the ratio of the altitudes of $\triangle JKL$ and $\triangle MKJ$ the same as that of the corresponding sides of $\triangle JKL$ and $\triangle MKJ$? Explain your reasoning.

PROOF

26. Use the diagram below to prove the Proportional Altitudes Theorem for the case when the altitude is outside the triangle.

Complete the proof of the following theorem:

Proportional Angle Bisectors Theorem

If two triangles are similar, then their corresponding angle bisectors have the same ratio as the corresponding sides.   8.5.3

Given: $\triangle ABC \sim \triangle XYZ$, $\overline{AD}$ bisects $\angle BAC$, and $\overline{XW}$ bisects $\angle YXZ$.

Prove: $\dfrac{AD}{XW} = \dfrac{AB}{XY}$

Proof:

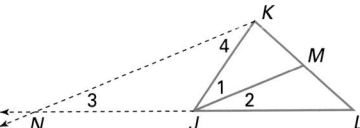

Statements	Reasons
$\triangle ABC \sim \triangle XYZ$	**27.** ___?___ Given
$\angle B \cong \angle Y$	**28.** ___?___ Polygon Similarity Postulate
$m\angle BAC = m\angle YXZ$	**29.** ___?___ Polygon Similarity Postulate
$\overline{AD}$ bisects $\angle BAC$; $\overline{XW}$ bisects $\angle YXZ$.	Given
30. ___?___ $= 2 \cdot m\angle BAD$ ___?___ $= 2 \cdot m\angle YXW$	Definition of angle bisector
31. ___?___	Substitution Property and Division Property
32. ___?___	AA Similarity Postulate
33. ___?___	Polygon Similarity Postulate

30. $m\angle BAC$
    $m\angle YXZ$

31. $m\angle BAD = m\angle YXW$

32. $\triangle ABD \sim \triangle XYW$

33. $\dfrac{AD}{XW} = \dfrac{AB}{XY}$

---

34. Prove the following theorem:

Proportional Segments Theorem

An angle bisector of a triangle divides the opposite side into two segments that have the same ratio as the other two sides.   8.5.4

Given: In $\triangle JKL$, $\overline{JM}$ bisects $\angle KJL$.

Prove: $\dfrac{KM}{LM} = \dfrac{JK}{JL}$

Hint: First construct a line parallel to $\overline{JM}$ through K. Extend $\overline{JL}$ to intersect this line at point N.

34.

Statements	Reasons
1. $\dfrac{KM}{LM} = \dfrac{NJ}{JL}$	1. Side-Splitting Theorem
2. $\angle 1 \cong \angle 4$	2. Alternate Interior Angles Theorem
3. $\angle 2 \cong \angle 3$	3. Corresponding Angles Postulate
4. $\angle 1 \cong \angle 2$	4. Definition of angle bisector
5. $\angle 3 \cong \angle 4$	5. Substitution Property
6. $NJ = JK$	6. Converse of the Isosceles Triangle Theorem
7. $\dfrac{KM}{LM} = \dfrac{JK}{JL}$	7. Substitution Property

For Exercises 35–41, complete the proof of the Proportional Medians Theorem.

Given: $\triangle EFG \sim \triangle STU$, $\overline{EH}$ is the median of $\triangle EFG$, and $\overline{SV}$ is the median of $\triangle STU$.

Prove: $\dfrac{EH}{SV} = \dfrac{EF}{ST}$

Proof:

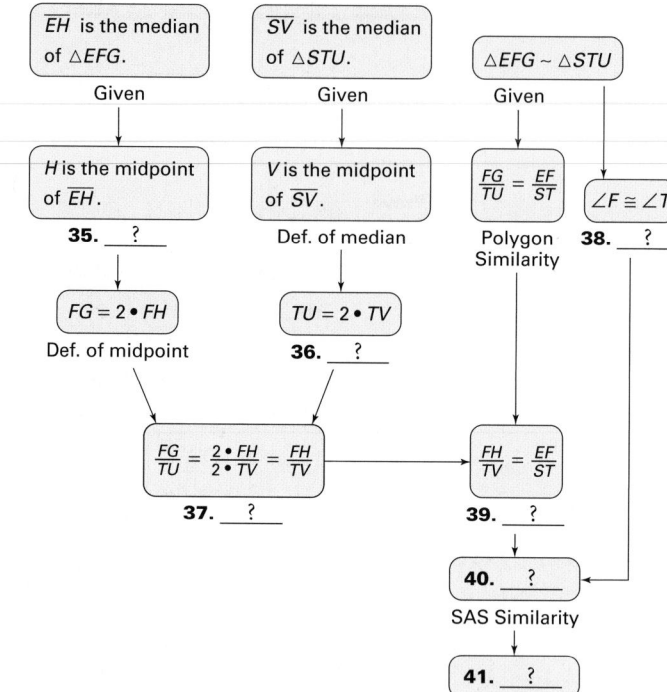

$\overline{EH}$ is the median of $\triangle EFG$.	$\overline{SV}$ is the median of $\triangle STU$.	$\triangle EFG \sim \triangle STU$
Given	Given	Given

H is the midpoint of $\overline{EH}$.   **35.** ___?___

V is the midpoint of $\overline{SV}$.   Def. of median

$\dfrac{FG}{TU} = \dfrac{EF}{ST}$ Polygon Similarity

$\angle F \cong \angle T$ **38.** ___?___

$FG = 2 \cdot FH$ Def. of midpoint

$TU = 2 \cdot TV$ **36.** ___?___

$\dfrac{FG}{TU} = \dfrac{2 \cdot FH}{2 \cdot TV} = \dfrac{FH}{TV}$ **37.** ___?___

$\dfrac{FH}{TV} = \dfrac{EF}{ST}$ **39.** ___?___

40. ___?___   SAS Similarity

41. ___?___   Polygon Similarity

35. Def. of median

36. Def. of midpoint

37. Substitution and Division Property

38. Polygon Simiarity Postulate

39. Substitution Property

40. $\triangle EFH \sim \triangle STV$

41. $\dfrac{EH}{SV} = \dfrac{EF}{ST}$

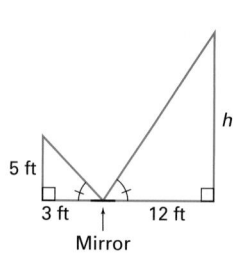

The laws of optics guarantee that the indicated angles are congruent.

42. PALEONTOLOGY You can use a mirror to estimate the height of an object, as shown in the photo. According to the laws of optics, the light reflects off a mirror at the same angle from which it strikes the mirror. Use the diagram to estimate the height of the dinosaur skeleton. **20 ft**

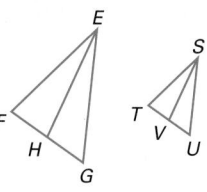

Practice

Practice

8.5 *Indirect Measurement and Additional Similarity Theorems*

In Exercises 1–4, use the diagrams to find the height of each building.

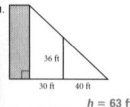 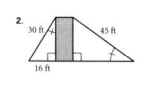

1. $h = 63$ ft

2. $h = 24$ ft

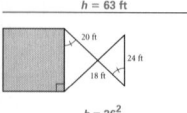 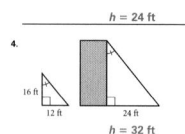

3. $h = 26\frac{2}{3}$

4. $h = 32$ ft

In Exercises 5–8, the triangles are similar. Find x.

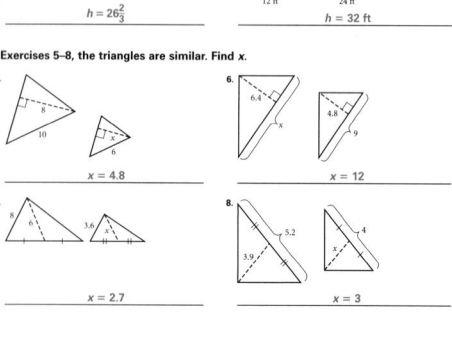

5. $x = 4.8$

6. $x = 12$

7. $x = 2.7$

8. $x = 3$

OPTICS The diagram at right shows a *convex lens*, which is thicker in the middle than at the edges. A convex lens bends light rays to form an image of an object on the opposite side of the lens. Light rays through the center of the lens are not bent, so $\overline{BB'}$ and $\overline{CC'}$ are straight segments. Thus, $\triangle ABC$ and $\triangle AB'C'$ are similar and $\dfrac{BC}{B'C'} = \dfrac{AC}{AC'}$, or $\dfrac{\text{object size}}{\text{image size}} = \dfrac{\text{object distance}}{\text{image distance}}$.

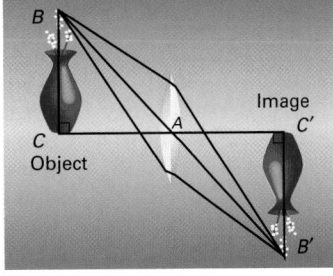

43. Prove that $\triangle ABC \sim \triangle AB'C'$.

44. If an object 6 cm from a convex lens forms an image 15 cm from the lens on the opposite side, what is the ratio of the size of the object to the size of the image? $\dfrac{2}{5}$

45. Rosa placed a lens 25 cm from an object 10 cm tall. An image was formed 5 cm from the lens on the opposite side. How tall was the image? **2 cm**

46. How could you arrange an object, a lens, and an image so that the object and the image are the same size?

47. How could you arrange an object, a lens, and an image so that the image is 20 times taller than the object?

43. Since vertical angles are congruent, then $\angle BAC \cong \angle B'AC'$. Since $\overline{BC} \parallel \overline{B'C'}$, then $\angle B$ and $\angle B'$ are alternate interior angles so $\angle B \cong \angle B'$. Thus, by the AA Similarity Postulate, $\triangle ABC \sim \triangle AB'C'$.

46. Arrange them so that the distance from the object to the lens is the same as the distance from the image to the lens.

47. Arrange them so that the distance from the image to the lens is 20 times greater than the distance from the object to the lens.

![Look Back icon] **Look Back**

Find the shaded area of each figure. *(LESSONS 5.2, 5.3, 5.4, AND 8.4)*

48. $16 + 8\pi \approx 41.13$

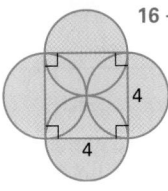

49. 4

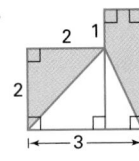

50. 4.32

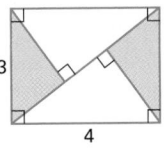

51. $\dfrac{13}{2}$

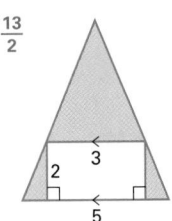

Find the volume of each solid. *(LESSONS 7.3 AND 7.6)*

52. $\dfrac{1372}{3}\pi \approx 1436.76$

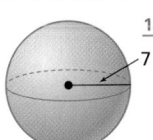

53. 10

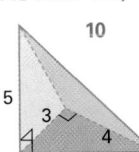

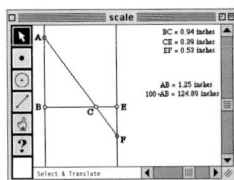

Look Beyond

The Cultural Connection to Africa in Exercises 54–56 extends the ideas of indirect measurement and proportions and connects these problem-solving methods with some history of astronomy.

Assessment

Portfolio Activity

The Portfolio Activity can be used as preparation for the Chapter Project or as a separate activity. In the Portfolio Activity on this page, students are asked to use methods of indirect measurement that have been presented in this lesson in order to measure some distances in their school or neighborhood that may be difficult to measure directly.

CULTURAL CONNECTION: AFRICA Eratosthenes, a Greek astronomer who lived in northern Africa, used indirect measurement to estimate Earth's circumference in approximately 200 B.C.E. Eratosthenes knew that the Sun was directly above the city of Aswan at noon on the summer solstice, so a vertical rod would cast no shadow. He measured the angle that the Sun's rays formed with a vertical rod in Alexandria at the same time and found it to be 7.5°.

54. Explain why $\angle 1 \cong \angle 2$ in the diagram below. Remember that the Sun's rays are considered to be parallel.

55. The distance from Aswan to Alexandria is 5000 *stades*, or about 575 mi. Complete the proportion $\frac{7.5°}{360°} = \frac{?}{?}$ and solve it to find Eratosthenes' estimate of Earth's circumference. $\frac{575 \text{ mi}}{\text{circumference of Earth}}$; **27,600 mi**

56. Given that Earth's circumference is approximately 24,900 mi, find the relative error in Eratosthenes' estimate. **0.11, or 11%**

☑ internet connect Go hrw com
Portfolio Extension
Go To: **go.hrw.com**
Keyword: **MG1 Indirect**

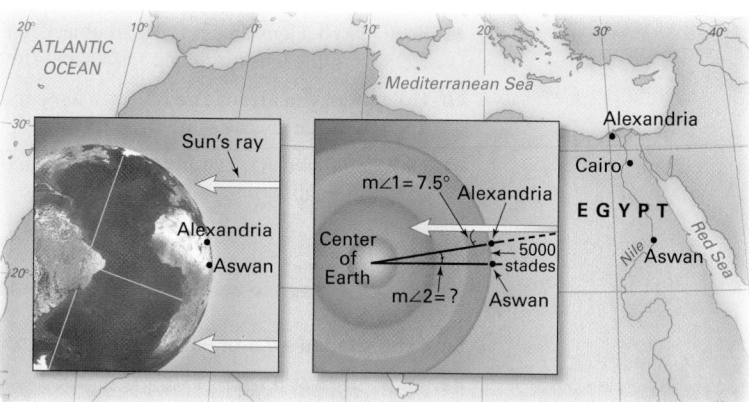

TECHNIQUES FOR INDIRECT MEASUREMENT

Use proportions and at least two of the methods listed below to find the dimensions of a building or other structure at your school or in your neighborhood.

• Measure the shadow of the building and the shadow of a person or object with a known height.

• Use a mirror to create similar triangles, as shown in Exercise 43.

• Take a photograph of the building with a person or object of known height standing in front of it. Measure the building and the person in the photograph.

WORKING ON THE CHAPTER PROJECT

You should now be able to complete Activity 2 of the Chapter Project.

54. The angles are alternate interior angles where the parallel lines are the sun's rays and the transversal is the radius from the center of Earth to Alexandria.

Area and Volume Ratios

Why If one spherical tank has twice the diameter of another, how much more will the larger tank hold? This lesson will show you how to answer questions like these.

Spherical tanks are best for storing contents under pressure, such as liquid butane and related chemicals like butylene and butadiene.

Objectives

● Develop and use ratios for areas of similar figures.

● Develop and use ratios for volumes of similar solids.

● Explore relationships between cross-sectional area, weight, and height.

Changing Dimensions of Figures

If the diameter of the spherical container in the photo is doubled, what would happen to the volume? In the Activities that follow, you will answer this and related questions.

Activity 1
Ratios of Areas of Similar Figures

YOU WILL NEED

calculator

For each pair of similar figures below, the ratio of a pair of corresponding measures is given. Find the ratio of the areas for each pair.

1. Two squares

$$\frac{\text{side of square A}}{\text{side of square B}} = \frac{3}{1}$$

$$\frac{\text{area of square A}}{\text{area of square B}} = \frac{?}{?}$$

2. Two triangles

$$\frac{\text{side of triangle A}}{\text{side of triangle B}} = \frac{2}{1}$$

$$\frac{\text{area of triangle A}}{\text{area of triangle B}} = \frac{?}{?}$$

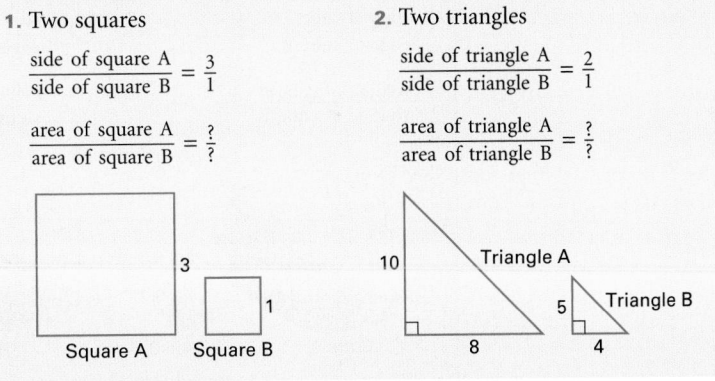

Alternative Teaching Strategy

HANDS-ON STRATEGIES Have students use graph paper to explore the ratio of the areas of similar figures. Ask them to cut two similar figures from graph paper. Then they should count the squares that make up the area and find the ratio of the areas. They can likewise use cubes to explore the ratio of the volumes of two similar figures. Create two similar shapes with cubes, count the cubes that make up the volumes, and find the ratio of the volumes.

Prepare

QUICK WARM-UP

1. Find the area of a circle with a radius of 4 cm. 50.24 cm^2

2. Find the volume of a rectangular prism with dimensions 4 cm by 6 cm by 7 cm. 168 cm^3

3. Find the volume of a right cylinder with a radius of 5 cm and a height of 6 cm. 471 cm^3

Also on Quiz Transparency 8.6

Teach

Why In this lesson, students compare the areas and volumes of similar figures. Students must know this if they are to understand the effect that a change in linear dimensions has on area and volume. Emphasize that the area is directly proportional to the square of the change in dimension and that the volume is directly proportional to the cube of the change in dimension.

Activity 1 Notes

This Activity will help students understand that similar figures with a scale factor of $\frac{a}{b}$ have areas with a ratio of $\frac{a^2}{b^2}$. Use similar figures cut from transparent graph paper to create visual models for the overhead projector. Students can count the squares in each figure to verify the ratio of the areas.

For a student worksheet of this Activity and detailed Teacher Notes, see page 146 in the Lesson Activities booklet.

Teaching Tip

Try to include oblique figures in the discussion about similar figures. Emphasize that any two similar figures share the same shape, that their area is proportional to the square of the scale factor, and that their volume is porportional to the cube of the scale factor.

Activity 2 Notes

This Activity will help students understand that similar figures with a scale factor of $\frac{a}{b}$ have volumes with a ratio of $\frac{a^3}{b^3}$.

For a student worksheet of this Activity and detailed Teacher Notes, see page 146 in the Lesson Activities booklet.

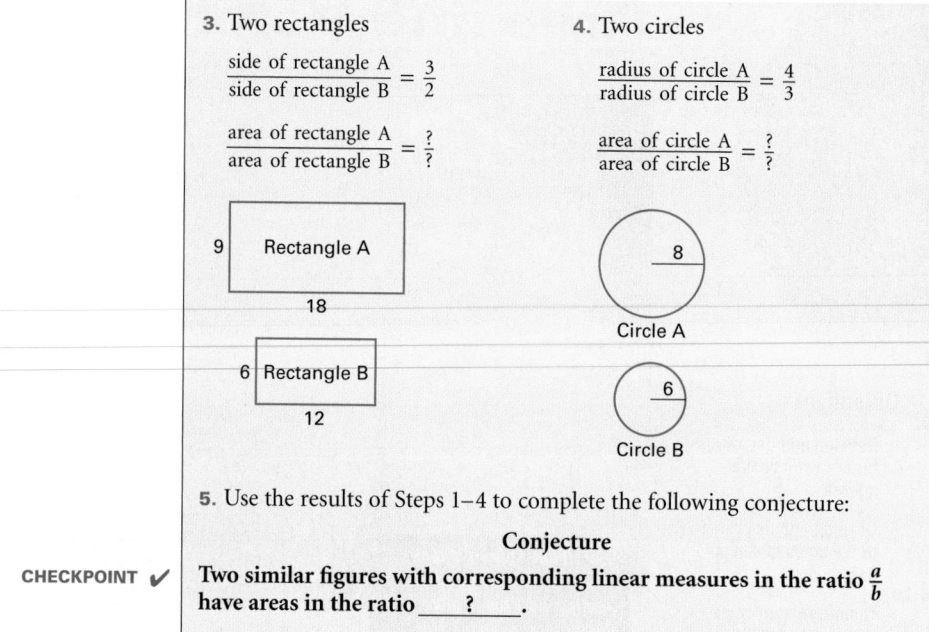

3. Two rectangles

$\dfrac{\text{side of rectangle A}}{\text{side of rectangle B}} = \dfrac{3}{2}$

$\dfrac{\text{area of rectangle A}}{\text{area of rectangle B}} = \dfrac{?}{?}$

9 | Rectangle A
18

6 | Rectangle B
12

4. Two circles

$\dfrac{\text{radius of circle A}}{\text{radius of circle B}} = \dfrac{4}{3}$

$\dfrac{\text{area of circle A}}{\text{area of circle B}} = \dfrac{?}{?}$

8
Circle A

6
Circle B

5. Use the results of Steps 1–4 to complete the following conjecture:

Conjecture

CHECKPOINT ✔   Two similar figures with corresponding linear measures in the ratio $\frac{a}{b}$ have areas in the ratio _____?_____.

Solids are similar if they are the same shape and all corresponding linear dimensions are proportional. For example, two rectangular prisms are similar if the lengths and widths of the corresponding faces and bases are proportional. In Activity 2, you will explore the ratios of the volumes of similar solids.

Activity 2
Ratios of Volumes of Similar Figures

YOU WILL NEED

calculator

For each pair of similar solids below, the ratio of a pair of corresponding measures is given. Find the ratio of the volumes for each pair.

1. Two cubes

$\dfrac{\text{edge of cube A}}{\text{edge of cube B}} = \dfrac{3}{1}$

$\dfrac{\text{volume of cube A}}{\text{volume of cube B}} = \dfrac{?}{?}$

2. Two rectangular prisms

$\dfrac{\text{edge of prism A}}{\text{edge of prism B}} = \dfrac{3}{2}$

$\dfrac{\text{volume of prism A}}{\text{volume of prism B}} = \dfrac{?}{?}$

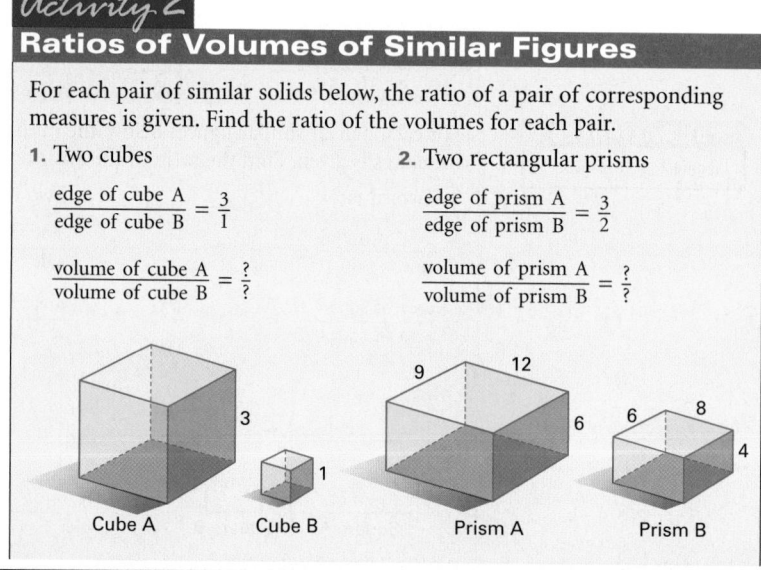

Cube A   Cube B   Prism A   Prism B

Interdisciplinary Connection

ART   Sculptors often use clay or other heavy materials when they make figures. Have students discuss how the volume, and therefore the weight, of a figure changes as the dimensions change. Ask them to estimate the weight of a statue that is 5 times taller than a similar statue weighing 2 pounds.

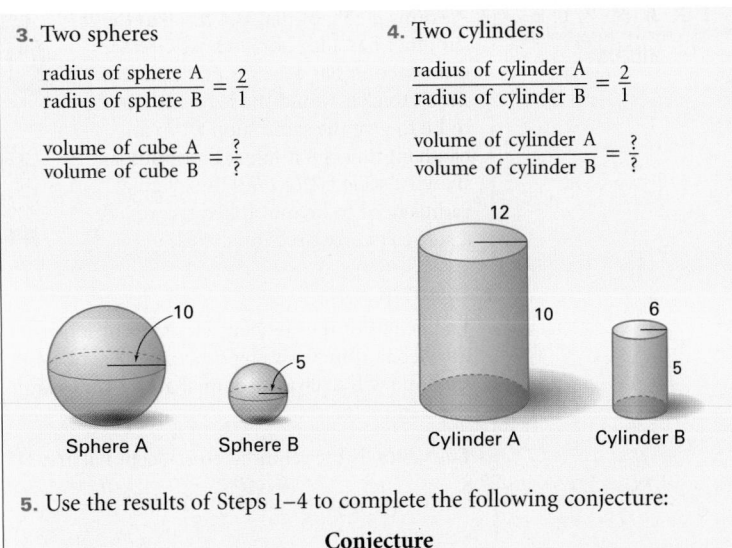

3. Two spheres

$$\frac{\text{radius of sphere A}}{\text{radius of sphere B}} = \frac{2}{1}$$

$$\frac{\text{volume of cube A}}{\text{volume of cube B}} = \frac{?}{?}$$

4. Two cylinders

$$\frac{\text{radius of cylinder A}}{\text{radius of cylinder B}} = \frac{2}{1}$$

$$\frac{\text{volume of cylinder A}}{\text{volume of cylinder B}} = \frac{?}{?}$$

Sphere A   Sphere B   Cylinder A   Cylinder B

5. Use the results of Steps 1–4 to complete the following conjecture:

Conjecture

CHECKPOINT ✔ **Two similar solids with corresponding linear measures in the ratio $\frac{a}{b}$ have volumes in the ratio ____?____.**

CRITICAL THINKING   Are all cubes similar? Why or why not? Area all spheres similar? Why or why not? Are all cylinders similar? Why or why not?

TRY THIS

a. The corresponding sides of two similar triangles are in the ratio $\frac{5}{9}$. What is the ratio of their areas?

b. The surface areas of two similar rectangular prisms are in the ratio $\frac{9}{4}$. What is the ratio of their corresponding sides?

c. One sphere has a radius of 5 meters. Another sphere has a radius of 15 meters. What is the ratio of their volumes?

Cross-Sectional Areas, Weight, and Height

The amount of weight that a structure can support is proportional to its cross-sectional area. For example, column A, whose radius is 3 times the radius of column B, can support 9 times more weight than column B. This is because the cross-sectional area of column A is 9 times that of column B.

In Example 1 and Activity 3, you will investigate the mathematical consequences of some animals' support requirements.

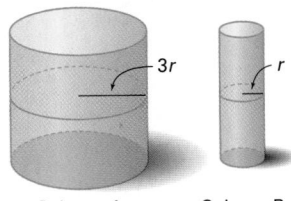

Column A   Column B

The load-bearing capacities of the columns depend on their cross-sectional areas.

In Example 1, you may want to review the algebra needed to find the radius of a circle when the area is known. Point out that the radius is the scale factor that is used to find the cross-sectional area of the leg.

A young elephant that is 3 feet tall at the shoulder has a leg bone whose circular cross section has a radius of about 1 inch. How much greater would cross-sectional area of the leg bone need to be in order to give the same support to an elephant that is 9 feet tall at the shoulder? By what scale factor does the original radius need to be multiplied to provide a sufficient cross-sectional area? 27 times greater; scale factor of about 5.2

Activity 3 Notes

Use the chart created in this Activity to generalize the ratios of the cross-sectional areas and volumes of similar solids. Students will discover that if the dimensions of similar solids have a ratio of 1 to n^2, then their cross-sectional areas have a ratio of 1 to n^2 and their volumes have a ratio of 1 to n^3. Have students compare the measurements in the charts with these ratios.

For a student worksheet of this Activity and detailed Teacher Notes, see page 146 in the Lesson Activities booklet.

CHECKPOINT ✔

3. The cross-sectional radius scale factor increases faster than the height scale factor, so the legs would be proportionally thicker.

EXAMPLE ❶
BIOLOGY

A young elephant that is 4 ft tall at the shoulder has a leg bone whose circular cross section has a radius of 4 cm. How much thicker would the leg bone need to be to give the same support to an elephant that is 8 ft tall at the shoulder? By what scale factor does the original radius need to be multiplied to provide a sufficient cross-sectional area?

● **SOLUTION**

The height of the elephant increases by a factor of 2. However, the volume of the elephant increases by a factor of 8. Thus, the larger elephant would need a leg bone with a cross-sectional area 8 times larger than that of the original elephant.

Thus, $8\pi r^2$ is the required cross-sectional area.

Let R represent the radius of the leg bone of the larger elephant.

$$\pi R^2 = 8\pi r^2$$
$$R^2 = 8r^2$$
$$R = \sqrt{8}r$$

Thus, the radius of the larger bone must be $\sqrt{8}$, or about 2.8, times the radius of the smaller bone.

Have you read stories of giants hundreds of feet tall? Do you think giants like this could be real? The elephant example shows how increased linear dimensions affect the structures that must support the increased bulk. Think about the question of giants as you do the next Activity.

Activity 3
Increasing Height and Volume

YOU WILL NEED
calculator (optional)

1. Complete the chart to determine the necessary scale factors for the cross-sectional radius of leg bones of animals.

Height scale factor	Volume scale factor	Cross-sectional radius scale factor
2	8	$\sqrt{8}$
3	27	$\sqrt{27}$
4	64	?
5	?	?
100	?	?

CHECKPOINT ✔

2. If a normal-sized horse requires a leg bone with a cross-sectional radius of 2.5 cm, what must the cross-sectional radius of a leg bone be for a horse that is 20 times taller? 50 times taller? 100 times taller?

3. What effect does increasing the size of the horse have on the relative proportions of the legs? Would the legs of a larger horse be proportionally thicker or thinner than those of a normal-sized horse?

Reteaching the Lesson

USING MODELS Ask students to do the following activity:

1. Draw a pair of similar rectangular prisms.

2. Measure the length, width, and height of each.

3. Find the surface area of each.

4. Predict the ratio of the surface areas and then verify the prediction.

5. Find the volume of each.

6. Predict the ratio of the volumes and then verify the prediction.

Exercises

Communicate

1. What is the relationship between the ratio of the edges of two cubes and the ratio of their volumes?

2. Why are all spheres similar? Why are all cubes similar? Is this true for other three-dimensional figures? If so, name or describe them.

3. If you know the surface areas of two similar prisms, explain how you can find the ratio of their volumes.

4. What happens to the volume of a cylinder if the radius is doubled but the height stays the same?

5. How does the cross-sectional area of a bone relate to an animal's weight and height?

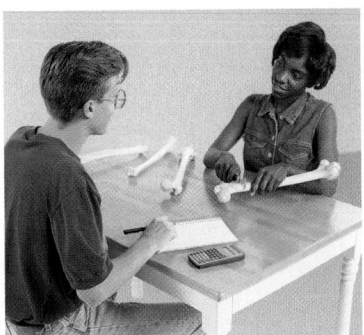

Assess

Selected Answers

Exercises 6–8, 9–57 odd

ASSIGNMENT GUIDE	
In Class	1–8
Core	9–37 odd, 41, 43
Core Plus	15–37 odd, 41–47 odd
Review	48–57
Preview	58

✐ Extra Practice can be found beginning on page 818.

Guided Skills Practice

6. Of two similar parallelograms, one has a base of 8 and a height of 3. The linear dimensions of the other parallelogram are one-half the linear dimensions of the first. Find the ratio of the areas. *(ACTIVITY 1)* $\frac{4}{1}$

7. Of two similar pyramids, one has a square base with each side equal to 6 and a height of 4. The linear dimensions of the other pyramid are one-half the linear dimensions of the first. Find the ratio of the volumes. *(ACTIVITY 2)* $\frac{8}{1}$

8. The leg bone of an animal has a circular cross section with an area of about 4π cm². If the animal's height were tripled without changing the proportions of its measurements, how would the volume of the animal be affected? How would the cross-sectional area of the leg bone be affected? *(EXAMPLE 1 AND ACTIVITY 3)*

8. The volume would be multiplied by 27, while the cross-sectional area would be multiplied by only 9.

● *Practice and Apply*

In the figure at right, $\overline{EC} \parallel \overline{AB}$.

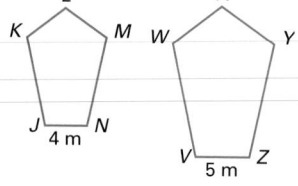

9. Find the ratio of the perimeters of $\triangle ABD$ and $\triangle ECD$. $\frac{10}{3}$

10. Find the ratio of the areas of $\triangle ABD$ and $\triangle ECD$. $\frac{100}{9}$

In the figure at right, JKLMN ~ VWXYZ. JKLMN has a perimeter of 24 m and an area of 50 m².

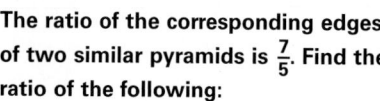

11. Find the perimeter of *VWXYZ*. **30 m**

12. Find the area of *VWXYZ*. ≈ **78.13 m²**

13. In the figure at right, *ABCDEF ~ PQRSTU*. *ABCDEF* has a perimeter of 42 m and an area of 96 m². Find the perimeter and area of *PQRSTU*.
   $P = 28$ m $A ≈ 42.67$ m²

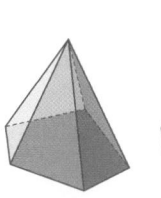 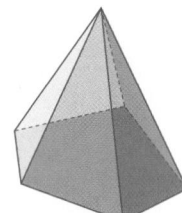

The ratio of the corresponding edges of two similar pyramids is $\frac{7}{5}$. Find the ratio of the following:

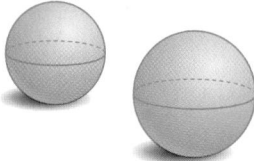

14. the perimeters of their bases $\frac{7}{5}$

15. the areas of their bases $\frac{49}{25}$

16. their volumes $\frac{343}{125}$

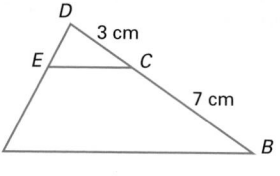
Two spheres have radii of 5 cm and 7 cm. Find the ratio of the following:

 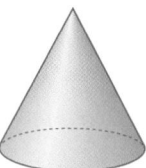

17. the circumferences of their great circles $\frac{5}{7}$

18. their surface areas $\frac{25}{49}$

19. their volumes $\frac{125}{343}$

The ratio of the heights of two similar cones is $\frac{7}{9}$. Find the ratio of the following:

20. their radii $\frac{7}{9}$

21. their volumes $\frac{343}{729}$

22. the areas of their bases $\frac{49}{81}$

The ratio of the base areas of two similar cylinders is $\frac{16}{25}$. Find the ratio of the following:

23. their heights $\frac{4}{5}$

24. the circumferences of their bases $\frac{4}{5}$

25. their volumes $\frac{64}{125}$

The ratio of the surface areas of two spheres is $\frac{144}{169}$. Find the ratio of the following:

26. their radii $\frac{12}{13}$

27. their volumes $\frac{1728}{2197}$

28. the circumferences of their great circles $\frac{12}{13}$

29. 12 cm

29. Two similar cylinders have base areas of 16 cm² and 49 cm². If the larger cylinder has a height of 21 cm, find the height of the smaller cylinder.

Two similar cones have surface areas of 225 cm² and 441 cm².

30. $8\frac{4}{7}$ cm

30. If the height of the larger cone is 12 cm, find the height of the smaller cone.

31. If the volume of the smaller cone is 250 cm³, find the volume of the larger cone. **686 cm³**

The ratio of the volumes of two prisms is $\frac{64}{125}$. Find the ratio of the following:

32. their surface areas $\frac{16}{25}$ 33. the perimeters of their bases $\frac{4}{5}$

34. corresponding base diagonals $\frac{4}{5}$ 35. the areas of their bases $\frac{16}{25}$

The small cone is formed by cutting off the lower part of the larger cone. Use the figure at right for Exercises 36 and 37.

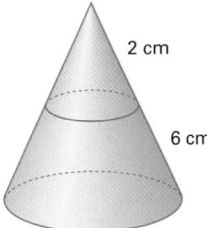

2 cm

6 cm

36. If the larger cone has a volume of 288 cm³ find the volume of the smaller cone. **4.5 cm³**

37. If the base of the smaller cone has an area of 3.6 cm², find the area of the base of the larger cone. **57.6 cm²**

38. A cross section of the leg bone of a horse is a circle with an area of 19 cm². What is the diameter of the leg bone? **≈ 5 cm**

39. What should the diameter of the cross section in Exercise 38 be to support a horse twice as tall? **14.14 cm**

40. A 100-ft tall cylindrical tower has a cross-sectional radius of 26 ft. What would be the radius of a 350-ft tall tower that is similar to the 100-ft tall tower? **91 ft**

41. **ASTRONOMY** The circumferences of Earth and the Moon are about 40,200 km and 10,000 km, respectively. Find the ratio of their diameters and the ratio of their volumes. $\frac{201}{50}$; $\frac{8,120,601}{125,000}$

Earthrise, viewed from the moon

42. The area of the 8-inch pizza is 16π, so the ratio of pizza to its price is $\frac{16\pi}{4} = 4\pi$. The area of the 16-inch pizza is 64π, so the ratio of this pizza to its price is $\frac{64\pi}{8} = 8\pi$. Thus, since the 16-inch pizza has the bigger ratio, then this one is a better deal.

43. No; doubling all the dimensions would increase the volume of the box by a factor of $2^3 = 8$. To double the volume, all the dimensions would be increased by a factor of $\sqrt[3]{2} \approx 1.26$.

44. People will use $\frac{9}{4}$, or 2.25, times the amount of toothpaste as before.

42. FOOD A new pizzeria sells an 8-in. diameter pizza for $4.00 and a 16-in. diameter pizza for $8.00. Which pizza is the better deal? Explain.

43. PACKAGING A 12-oz box of noodles is a rectangular prism that measures 15 cm × 20 cm × 4 cm. To make 24-oz box, should the company double all of the dimensions of the 12-oz box? Explain.

44. PACKAGING A toothpaste company packages its product in a tube that has a circular opening with a radius of 2 mm. The company increases the radius of the opening to 3 mm. Predict what will happen to the amount of toothpaste used if people continue to use the same length of toothpaste on their toothbrushes.

45. BUSINESS The area of the parking lot at Jerome's Restaurant is 400 m². Jerome buys some adjoining land and expands the parking lot to 1.5 times as wide and 1.75 times as long as the original lot. Find the area of the expanded lot. **1050 m²**

46. SPORTS The diameter of a standard basketball is about 9.5 in. A company that makes basketballs is contracted to make promotional basketballs with a diameter of 5 in. The materials for a standard-sized basketball cost $1.40. How much will the materials cost for a promotional basketball made of the same materials? **≈39¢**

47. STORAGE A city stores rock salt for winter road maintenance in a dome-shaped building that is 82 ft in diameter at the base. The building holds 3366 tons of salt. Because the city is growing, the city planners decide to build a second, smaller dome. The linear dimensions of the new dome will be three-fourths the linear dimensions of the old dome. Estimate the storage capacity of the new dome. **≈1420 tons**

Practice

NAME _____ CLASS _____ DATE _____

Practice

8.6 Area and Volume Ratios

1. In $\triangle ABC$, D and E are midpoints. What fraction of the area of $\triangle ABC$ is $\triangle ADE$? $\frac{\text{area of } \triangle ADE}{\text{area of } \triangle ABC} = \frac{1}{4}$

The ratio of the corresponding sides of two similar triangles is $\frac{3}{5}$. Find the ratio of the following:

2. their altitudes $\frac{3}{5}$ **3.** their perimeters $\frac{3}{5}$ **4.** their areas $\frac{9}{25}$

The side lengths of two squares are 4 cm and 9 cm. Find the ratio of the following:

5. their diagonals $\frac{4}{9}$ **6.** their perimeters $\frac{4}{9}$ **7.** their areas $\frac{16}{81}$

Two spheres have radii of 6 cm and 8 cm. Find the ratio of the following:

8. the circumferences of their great circles $\frac{3}{4}$ **9.** their surface areas $\frac{9}{16}$ **10.** their volumes $\frac{27}{64}$

The ratio of the base areas of two similar cones is $\frac{16}{25}$. Find the ratio of the following:

11. the circumference of their bases $\frac{4}{5}$ **12.** their heights $\frac{4}{5}$ **13.** their volumes $\frac{64}{125}$

14. Two cubes have volumes of 3375 and 1331. What is the ratio of their heights? $\frac{15}{11}$

15. Suppose that the triangle from Exercise 1 is extruded in space to form a triangular prism. What is the ratio of volumes of the prism with $\triangle ADE$ as a base and the prism with $\triangle ABC$ as a base? $\frac{1}{8}$

48. The measure of each interior angle of a regular polygon is 165°. How many sides does the polygon have? *(LESSON 3.6)* **24**

49. The exterior angles of a regular polygon each measure 40°. How many sides does the polygon have? *(LESSON 3.6)* **9**

50. Which of the lines below are parallel? *(LESSON 3.8)* **b and c**
 a. $5x + 4y = 18$
 b. $-10x + 8y = 21$
 c. $30x - 24y = 45$

51. Write an equation for the line that is parallel to $5x + 4y = 18$ and passes through the point $(3, 8)$. *(LESSON 3.8)* $y = -\frac{5}{4}x + \frac{47}{4}$

52. Write an equation of the line that is perpendicular to $5x + 4y = 18$ and passes through the point $(3, 8)$. *(LESSON 3.8)* $y = \frac{4}{5}x + \frac{28}{5}$

Solve each proportion. Give all possible values of x. *(LESSON 8.2)*

53. $\frac{x+1}{4} = \frac{3x}{8}$ **x = 2**

54. $\frac{2}{x+2} = \frac{x-1}{2}$ **x = -3 or 2**

55. $\frac{x-1}{x+1} = \frac{x+2}{2x+2}$ **x = 4**

56. In a rectangle, the ratio of the length of the long side to the length of the short side is $\frac{5}{3}$. If the short side has a length of 9, what is the length of the long side? *(LESSON 8.2)* **15**

57. In the triangle below, $\overline{DE} \| \overline{BC}$. Solve for x. *(LESSON 8.4)* **x = 17**

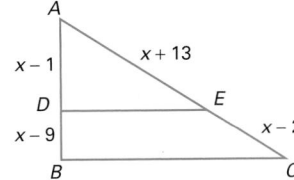

Exercise 58 introduces the concept of the convergence of a geometric series. Encourage students to try to find the limit of the sum. As a hint, have students refer to the results from Exercises 22–28 of Lesson 2.1 on pages 84 and 85.

 Look Beyond

CHALLENGE

 Algebra

ABCD is a square with side lengths of 1. The squares inside ABCD are formed by connecting the midpoints of the larger square.

58. What happens to the perimeter of each successive square? Using a calculator, add the perimeters of the first eight squares. Keep track of the sums as each perimeter is added. Do you think the sum of the perimeters of all the squares generated this way will ever reach 14? Explain.

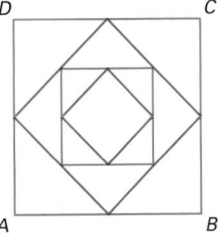

58. The perimeter of each successive square is $\frac{1}{\sqrt{2}}$ times the perimeter of its larger square. The sum of the first eight perimeters is about 12.8. No, the sum of the perimeters would be:

$$4 + \frac{4}{\sqrt{2}} + 2 + \frac{2}{\sqrt{2}} + 1 + \frac{1}{\sqrt{2}} + \frac{1}{2} + \frac{1}{2\sqrt{2}} + \dots$$

$$= \left(4 + 2 + 1 + \frac{1}{2} + \dots\right) + \left(\frac{4}{\sqrt{2}} + \frac{2}{\sqrt{2}} + \frac{1}{\sqrt{2}} + \frac{1}{2\sqrt{2}} + \dots\right)$$

$$= 4\left(1 + \frac{1}{2} + \frac{1}{4} + \dots\right) + \frac{4}{\sqrt{2}}\left(1 + \frac{1}{2} + \frac{1}{4} + \dots\right)$$

$$= 4(2) + \frac{4}{\sqrt{2}}(2)$$

$$\approx 13.66$$

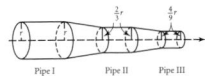
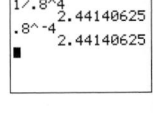

In this project, students will apply the properties of similarity and use indirect measurement to create scale maps and models of familiar neighborhoods.

Motivate

Look with students at a map of the neighborhood around the school. Discuss the locations of stores, trees, or street signs that are in the area.

Activity 1

Check students' drawings.

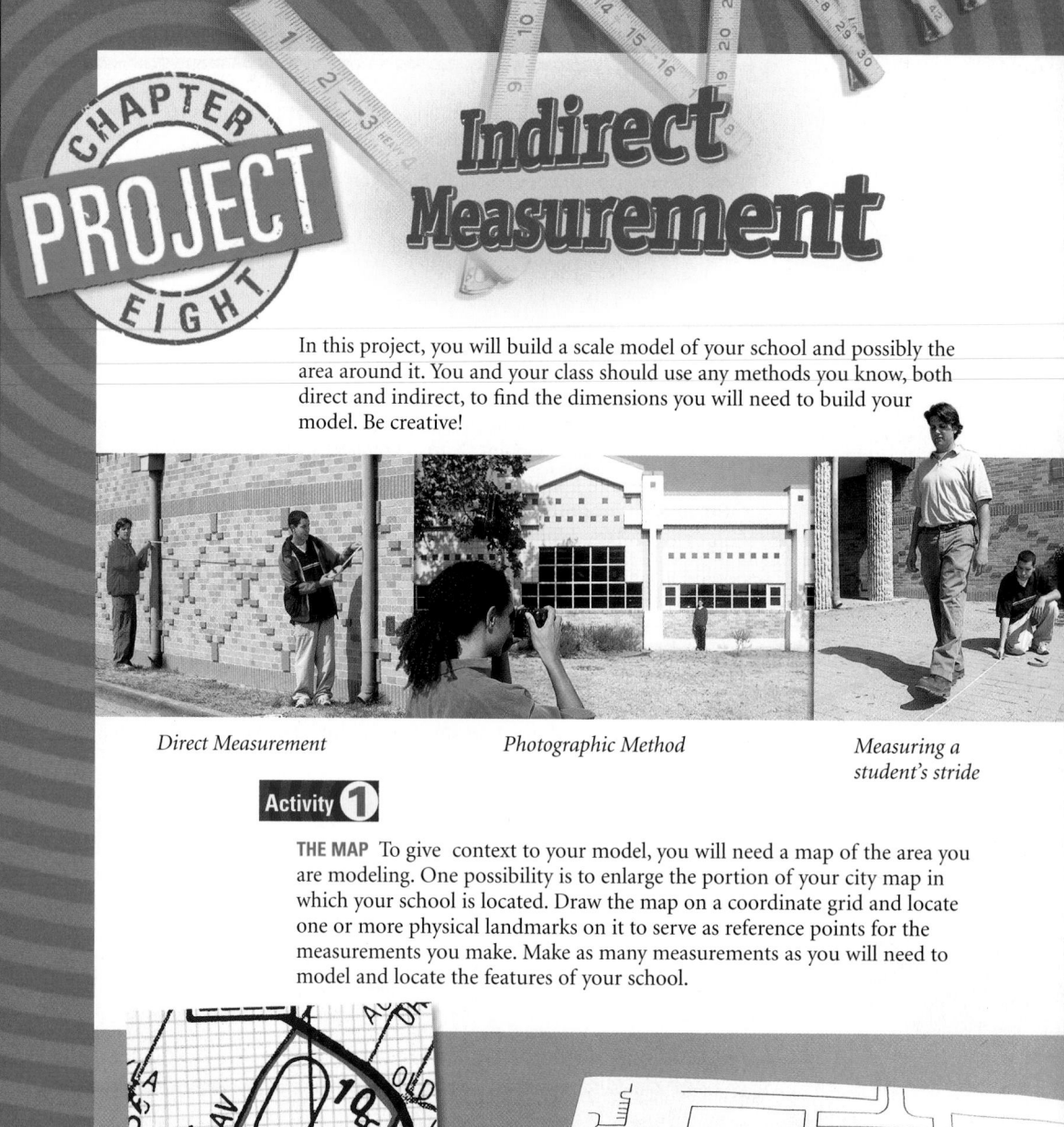

Indirect Measurement

CHAPTER PROJECT EIGHT

In this project, you will build a scale model of your school and possibly the area around it. You and your class should use any methods you know, both direct and indirect, to find the dimensions you will need to build your model. Be creative!

Direct Measurement

Photographic Method

Measuring a student's stride

Activity 1

THE MAP To give context to your model, you will need a map of the area you are modeling. One possibility is to enlarge the portion of your city map in which your school is located. Draw the map on a coordinate grid and locate one or more physical landmarks on it to serve as reference points for the measurements you make. Make as many measurements as you will need to model and locate the features of your school.

Students can work in small groups to create the maps and models in Activities 1 and 2. When creating the map, students can help each other remember details of the area that should be included. When creating the model, students should choose individual buildings to construct, as well as the trees and other objects that surround the buildings.

Activity ②

Lay out a coordinate grid on the area where your model will be built. Use the grid to give organization and structure to the measurements you took in Activity 1. Create physical models of the buildings and other structures on your campus and place them on the grid.

Discuss

Have each student or group present their map and model to the class. Ask students to discuss how they made sure that their map and model were made to scale.

Activity 2

Check students' models.

8 Chapter Review and Assessment

VOCABULARY

POSTULATES AND THEOREMS

Lesson	Number	Postulate or Theorem
8.2	8.2.2 Polygon Similarity Postulate	Two polygons are similar if and only if there is a way of setting up a correspondence between their sides and angles such that: • Each pair of corresponding angles is congruent. • Each pair of corresponding sides is proportional.
	8.2.3 Cross-Multiplication Property	For real numbers a, b, c, and d such that b, $d \neq 0$, if $\frac{a}{b} = \frac{c}{d}$, then $ad = bc$.
	8.2.4 Reciprocal Property	For real numbers a, b, c, and d such that a, b, c, $d \neq 0$, if $\frac{a}{b} = \frac{c}{d}$, then $\frac{b}{a} = \frac{d}{c}$.
	8.2.5 Exchange Property	For real numbers a, b, c, and d such that a, b, c, $d \neq 0$, if $\frac{a}{b} = \frac{c}{d}$, then $\frac{a}{c} = \frac{b}{d}$.
	8.2.6 "Add-One" Property	For real numbers a, b, c, and d such that b, $d \neq 0$, if $\frac{a}{b} = \frac{c}{d}$, then $\frac{a+b}{b} = \frac{c+d}{d}$.
8.3	8.3.1 AA (Angle-Angle) Similarity Postulate	If two angles of one triangle are congruent to two angles of another triangle, then the triangles are similar.
	8.3.2 SSS (Side-Side-Side) Similarity Theorem	If the three sides of one triangle are proportional to the three sides of another triangle, then the triangles are similar.
	8.3.3 SAS (Side-Angle-Side) Similarity Theorem	If two sides of one triangle are proportional to two sides of another triangle and their included angles are congruent, then the triangles are similar.
8.4	8.4.1 Side-Splitting Theorem	A line parallel to one side of a triangle divides the other two sides proportionally.
	8.4.2 Two-Transversal Proportionality Corollary	Three or more parallel lines divide two intersecting transversals proportionally.
8.5	8.5.1 Proportional Altitudes Theorem	If two triangles are similar, then their corresponding altitudes have the same ratio as their corresponding sides.
	8.5.2 Proportional Medians Theorem	If two triangles are similar, then their corresponding medians have the same ratio as their corresponding sides.
	8.5.3 Proportional Angle Bisectors Theorem	If two triangles are similar, then their corresponding angle bisectors have the same ratio as their corresponding sides.
	8.5.4 Proportional Segments Theorem	An angle bisector of a triangle divides the opposite side into two segments that have the same ratio as the other two sides.

Chapter Test, Form A

NAME _____ CLASS _____ DATE _____

Chapter Assessment

Chapter 8, Form A, page 1

Write the letter that best answers the question or completes the statement.

___d___ 1. Find the image of $P(6, 12)$ after it is transformed by the dilation $D(x, y) = \left(-\frac{2}{3}x, -\frac{2}{3}y\right)$.
   a. $(9, 18)$ b. $(-9, -18)$ c. $(4, 8)$ d. $(-4, -8)$

___b___ 2. The scale factor of a dilation is $\frac{3}{2}$. What is the equation of the line through $P(2, -4)$ and its image?
   a. $y = 2x$ b. $y = -2x$ c. $x = 2y$ d. $x = -2y$

___c___ 3. The dashed-line preimage has been transformed to form the solid-line image. What is the scale factor of the dilation?
   a. $\frac{2}{5}$ b. 2
   c. $\frac{5}{2}$ d. 5

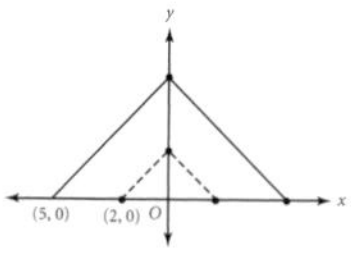

___c___ 4. The two triangles below are similar. If $YZ = 6$, then $AB =$ ___.
   a. 2.5 b. 6 c. 10 d. 18

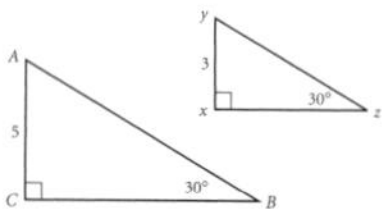

___c___ 5. The perimeter of $\triangle GHI$ is 18, the perimeter of $\triangle PQR$ is 30, and $\triangle GHI \sim \triangle PQR$. If GH is 10, what is PQ?
   a. 3.60 b. 6.00 c. 16.67 d. 27.78

___d___ 6. Which proportion illustrates the Side-Splitting Theorem?
   a. $\frac{AD}{AB} = \frac{AL}{AE}$ b. $\frac{AD}{EL} = \frac{AB}{AL}$
   c. $\frac{AD}{AE} = \frac{AL}{AB}$ d. $\frac{AD}{DB} = \frac{AE}{EL}$

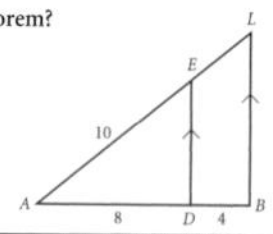

NAME _____ CLASS _____ DATE _____

Chapter Assessment

Chapter 8, Form A, page 2

___c___ 7. In $\triangle ADE$ on the previous page, if $ED = 6$, what is BL?
   a. 3 b. 5 c. 9 d. 12

___b___ 8. What is the value of x in the figure at the right?
   a. 8 b. 12
   c. 14 d. 16

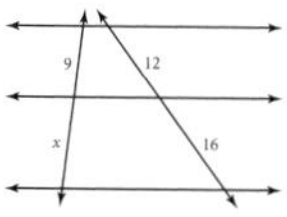

___b___ 9. The shadow of a man 6 feet tall is 30 inches long. At the same time of day, a building casts a shadow 125 inches long. How tall is the building?
   a. 15 feet   b. 25 feet   c. 30 feet   d. 50 feet

___a___ 10. Dan placed a lens 16 cm away from an 8-cm object. An image formed 10 cm away from the lens. How tall was the image?
   a. 5 cm   b. 10 cm   c. 15 cm   d. 20 cm

___a___ 11. In the figure at right, $\triangle MNO \sim \triangle HIJ$. What is the length of $\overline{NQ}$?
   a. 6 b. 10
   c. 12 d. 15

___b___ 12. What is the length of $\overline{IK}$?
   a. 9 b. 10
   c. 12 d. 15

___b___ 13. $MN =$
   a. 10 b. 11
   c. 12 d. 15

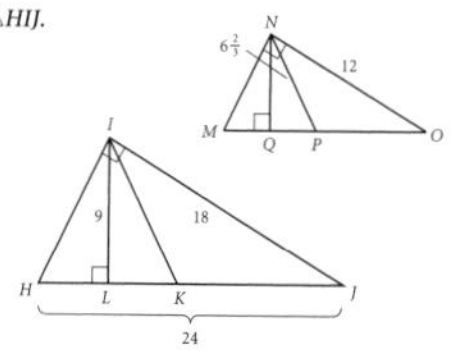

___c___ 14. Two spheres have radii of 3 cm and 5 cm. What is the ratio between the areas of their great circles?
   a. $3:5$ b. $6:10$ c. $9:25$ d. $27:125$

___c___ 15. It cost $144 to refinish a floor that is 9 feet by 12 feet. At the same rate, how much will it cost to refinish a floor that is 12 feet by 16 feet?
   a. $81   b. $108   c. $256   d. $576

___d___ 16. The area of one side of a cube is 36 ft^2. If the edges of the cube are tripled, what is the volume of the new cube?
   a. 36 ft^3 b. 196 ft^3 c. 324 ft^3 d. 5832 ft^3

Key Skills & Exercises

The answers to Exercises 5 and 6 can be found on page 556.

LESSON 8.1
Key Skills

Draw a dilation on a coordinate plane.

Find the image of a segment with endpoints at (1, 3) and (4, 0) that is transformed by the dilation $D(x, y) = (2x, 2y)$.

The endpoints of the image are (2, 6) and (8, 0).

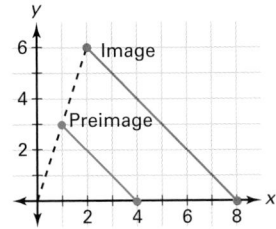

Draw a dilation in a plane.

Draw an equilateral triangle and dilate it about one vertex by a scale factor of 2.

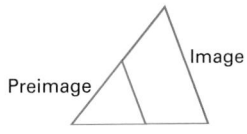

Exercises

Find the image of a segment with endpoints at (–2, 1) and (3, 4) that is transformed by the given dilation.

1. $D(x, y) = (3x, 3y)$

2. $D(x, y) = (-x, -y)$

Copy each figure below and dilate the figure about the given point by the given scale factor.

3. $n = 3$

4. $n = \dfrac{1}{2}$

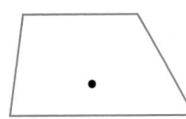

1. The endpoints of the image are (–6, 3) and (9, 12).

2. The endpoints of the image are (2, –1) and (–3, –4).

LESSON 8.2
Key Skills

Determine whether polygons are similar.

Are the rectangles in the diagram below similar? Why or why not?

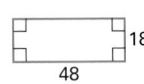 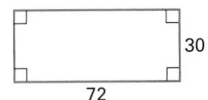

If the rectangles were similar, the proportion $\dfrac{18}{30} = \dfrac{48}{72}$ would be true. Cross-multiplying gives $18 \cdot 72 = 30 \cdot 48$, or $1296 = 1440$, so the rectangles are not similar.

Exercises

Determine whether the polygons in each pair are similar. Explain your reasoning.

5.

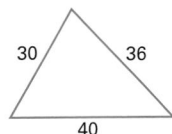 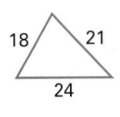

6.

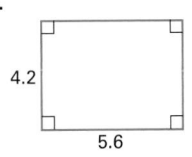 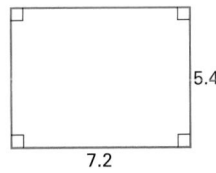

3.

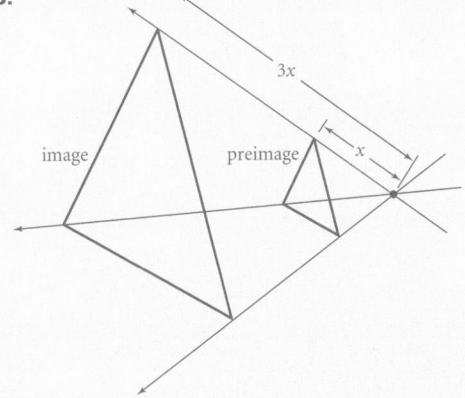

4.

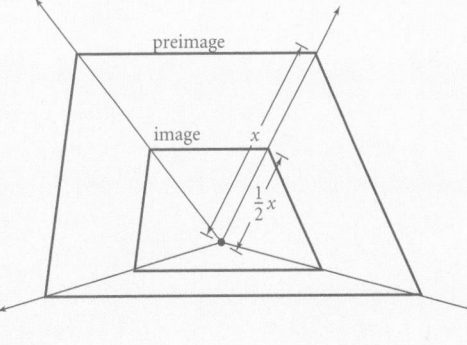

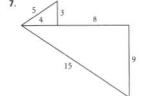 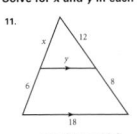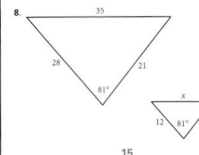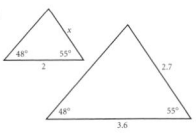
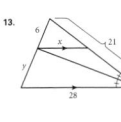
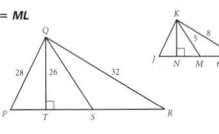

5. If the triangles were similar, the proportion $\frac{30}{18} = \frac{36}{21}$ would be true. By cross-multiplying, we obtain $30 \cdot 21 = 36 \cdot 18$ or $630 = 648$, so the triangles are not similar.

6. If the rectangles were similar, the proportion $\frac{4.2}{5.4} = \frac{5.6}{7.2}$ would be true. By cross-multiplying, we obtain $4.2 \cdot 7.2 = 5.6 \cdot 5.4$ or $30.24 = 30.24$, so the rectangles are similar.

7. 5.4

8. 36

9. SAS Similarity Theorem

10. SSS Similarity Theorem

11. AA Similarity Postulate

12. SAS Similarity Theorem

13. 12

14. 39

15. 9.6

16. 2.94

Use proportions to find the side lengths of similar figures.

The triangles in the diagram below are similar. Find x.

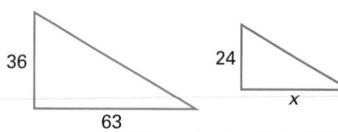

$$\frac{36}{24} = \frac{63}{x} \rightarrow 36x = 24 \cdot 63$$
$$x = \frac{1512}{36}, \text{ or } 42$$

The figures in each pair below are similar. Find x.

7.

8.

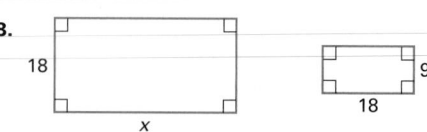

LESSON 8.3

Key Skills

Use the AA Similarity Postulate and SSS and SAS Similarity Theorems to determine whether triangles are similar.

Name the postulate or theorem that can be used to prove that the triangles are similar.

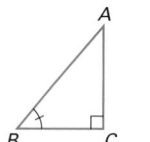 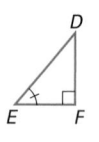

$\angle B \cong \angle E$ and $\angle C \cong \angle F$, so the triangles are similar by the AA Similarity Postulate.

Exercises

For each pair of triangles, name the postulate or theorem that can be used to prove that the triangles are similar.

9.

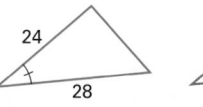 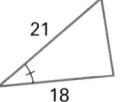

10.

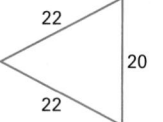

11.

12.

 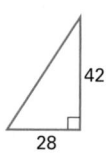

LESSON 8.4

Key Skills

Use the Side-Splitting Theorem to solve problems involving triangles.

In the triangle below, $\overline{DE} \parallel \overline{BC}$. Find x.

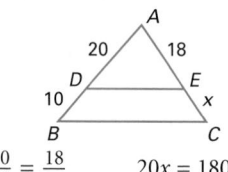

$$\frac{20}{10} = \frac{18}{x} \qquad 20x = 180 \qquad x = 9$$

Exercises

Use the Side-Splitting Theorem to find x.

13.

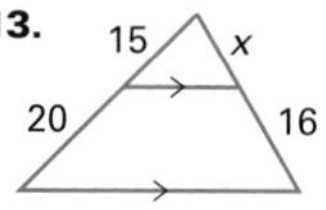

15, x, 20, 16

14.

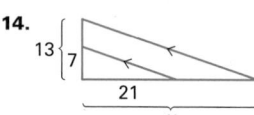

15.

12, 15, 12, x

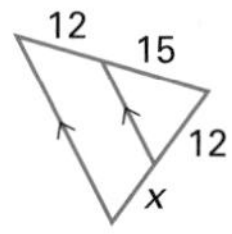

16.

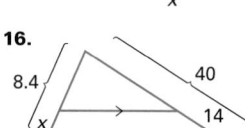

17. 8 m

18. $2916\frac{2}{3}$ m

19. 24

20. 15

21. $\frac{25}{1}$

22. $\frac{1}{4}$

LESSON 8.5

Key Skills

Use similar triangles to measure distance indirectly.

A right square pyramid has a base edge of 40 ft. If the pyramid casts a shadow 60 ft long at the same time that a yardstick casts a shadow 6 ft long, what is the height of the pyramid?

Because the sun's rays are considered parallel, the triangles formed by the pyramid and the yardstick are similar. The horizontal side of the large triangle is 60 ft plus 20 ft, or 80 ft.

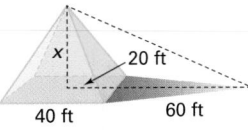

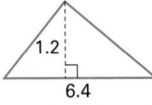

$$\frac{6}{80} = \frac{3}{x} \qquad 6x = 240 \qquad x = 40$$

The height of the pyramid is 40 ft.

Use similarity theorems to solve problems involving altitudes and medians of triangles.

The triangles below are similar. Find x.

Because the triangles are similar, their altitudes are proportional.

$$\frac{x}{1.2} = \frac{5.6}{6.4}$$

$$1.2 \cdot \frac{x}{1.2} = \frac{5.6}{6.4} \cdot 1.2$$

$$x = 1.05$$

Exercises

17. Catherine is 1.6 m tall and casts a shadow of 3.5 m. At the same time, a house casts a shadow of 17.5 m. Find the height of the house.

Refer to the diagram below.

18. A surveyor made the measurements shown below. $\overline{ZY}$ is parallel to $\overline{WX}$. Find the distance across the base of the hill.

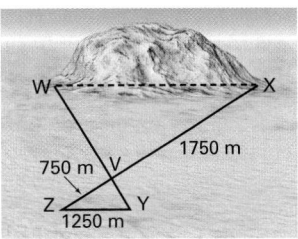

The triangles in each pair below are similar. Find x.

19.

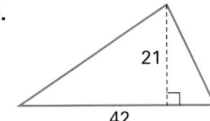 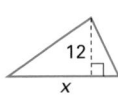

20.

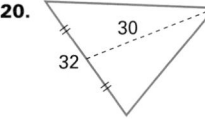

LESSON 8.6

Key Skills

Find the ratio of the areas of similar figures.

The ratio of the sides of two similar quadrilaterals is $\frac{2}{3}$. Find the ratio of their areas. The ratio of their areas is $\frac{2^2}{3^2} = \frac{4}{9}$.

Exercises

21. The ratio of the sides of two similar triangles is $\frac{5}{1}$. What is the ratio of their areas?

22. The ratio of the sides of two similar parallelograms is $\frac{1}{2}$. What is the ratio of their areas?

23. $\frac{1}{1000}$

24. $\frac{1}{8}$

25. 14.4 in^2

26. No

27. $\left(\frac{1209}{35}\right)^3$, or $\frac{1,767,172,329}{42,875}$

Find the ratio of the volumes of similar solids.

The ratio of the edges of two similar pyramids is $\frac{2}{3}$. Find the ratio of their volumes.

The ratio of their volumes is $\frac{2^3}{3^3} = \frac{8}{27}$.

23. The ratio of the edges of two similar prisms is $\frac{1}{10}$. What is the ratio of their volumes?

24. The ratio of the radii of two similar spheres is $\frac{1}{2}$. What is the ratio of their volumes?

Applications

25. **GRAPHIC DESIGN** An artist uses a photocopier to enlarge a design by 120%. If the area of the original design is 10 in.^{2}, what will be the area of the enlarged design?

26. **LANDSCAPING** Brian needs to cut down a tree that is 30 m away from his house. To estimate the height of the tree, he places a mirror on the ground 15 m from the base of the tree and stands 1 m away from the mirror, in which he can see the reflection of the top of the tree. Brian's eyes are 1.8 m above the ground. If the tree falls toward Brian's house, will it hit the house?

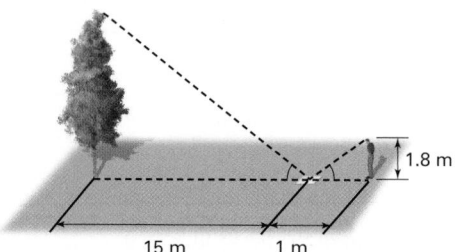

27. **FINE ARTS** The sculpture *Batcolumn* by Claes Oldenburg and Coosje van Bruggen is a giant replica of a baseball bat and is 100 ft and 9 in. tall. What is the ratio of the volume of this sculpture to the volume of a real baseball bat that is 35 in. long?

Chapter Test

Find the image of a segment with endpoints at (6, −1) and (−4, 2) that is transformed by the given dilation.

1. $D(x, y) = (-2x, -2y)$ **2.** $D(x, y) = (0.5x, 0.5y)$

3. Copy the figure at right and dilate the figure about the given point by a scale factor of 2.

Determine whether the polygons in each pair are similar. Explain your reasoning.

4.

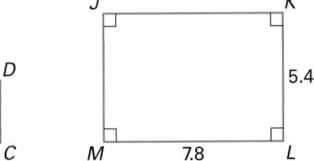

5.

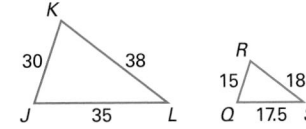

The figures in the pair below are similar. Find x.

6.   5.8

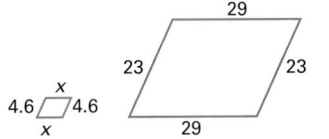

7. Clair is enlarging a rectangular flag from a diagram that is 3 inches by 5 inches. If she wants the flag to be 18 inches long, how wide should the flag be? **10.8 in.**

For each pair of triangles, name the postulate or theorem that can be used to prove that the triangles are similar.

8.   AA

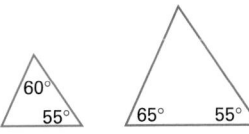

9.

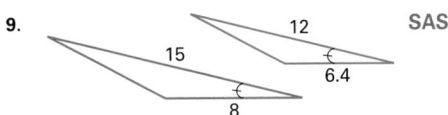

SAS

10.

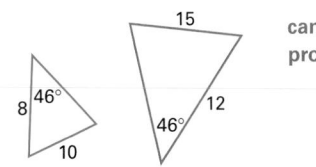

cannot be proven similar

For Exercises 11–12, use the Side-Splitting Theorem to find x.

11.

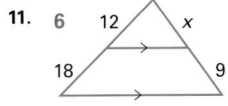

12.

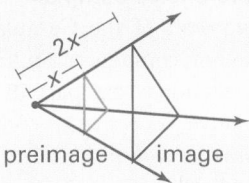

13. SCALE DRAWING Andy is 1.2 meters tall and casts a shadow of 1.5 meters. At the same time, a lamppost casts a shadow of 25 meters. Find the height of the lamppost. **20 m**

In Exercises 14 and 15, the triangles in each pair are similar. Find x.

14.

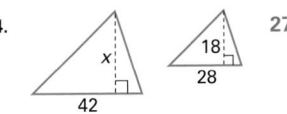

15.

16. The ratio of the sides of two similar triangles is $\frac{8}{1}$. What is the ratio of their areas? $\frac{64}{1}$

17. The ratio of the edges of two similar pyramids is $\frac{2}{5}$. What is the ratio of their volumes? $\frac{8}{125}$

18. The ratio of the heights of two similar cylinders is $\frac{4}{3}$. What is the ratio of their volumes? $\frac{64}{27}$

1. The endpoints of the image are (−12, 2) and (8, −4).

2. The endpoints of the image are (3, −0.5) and (−2, 1).

3.

preimage   image

4. Since every angle is right, corresponding angles are congruent. $\frac{2.6}{7.8} = \frac{1.8}{5.4}$, so corresponding sides are proportional. Quadrilateral *ABCD* is similar to quadrilateral *JKLM*.

5. $\frac{38}{18} \neq \frac{30}{15}$, so corresponding sides are not proportional. The triangles are not similar.

College Entrance Exam Practice

1-8

College Entrance Exam Practice

Multiple-Choice Samples
The first half of the Cumulative Assessment contains a multiple-choice section. This part of the Cumulative Assessment consists of items commonly found on standardized tests.

Free-Response Grid Samples
The second half of the Cumulative Assessment consists of a free-response section. This part requires student-produced response items like those commonly found on college entrance exams. These questions require the use of machine-scored answer grids. You may wish to have students practice answering these items in preparation for standardized tests.

1. c

2. a

3. b

4. a

5. d

6. b

MULTIPLE-CHOICE For Questions 1–10, write the letter that indicates the best answer.

1. Find the sum of *AC* and *BD*. **(LESSON 1.2)**

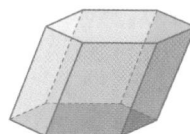

   a. 8
   b. 9
   c. 11
   d. 5

2. What is the sum of the first 15 positive odd numbers? **(LESSON 2.1)**

   a. 225
   b. 240
   c. 120
   d. 255

3. What is the sum of the number of lateral edges and the number of faces of a hexagonal prism? **(LESSON 6.3)**

   a. 12
   b. 14
   c. 26
   d. 24

4. Find the lateral area of the cone below to the nearest hundredth. **(LESSON 7.5)**

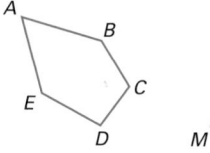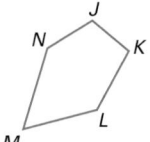

   a. 4.44
   b. 7.58
   c. 1.05
   d. 4.19

5. If line ℓ has a slope of 5 and line *m* has a slope of 0.2, then the lines are ___?___.
   (LESSON 3.8)
   a. parallel
   b. perpendicular
   c. vertical
   d. none of the above

6. If pentagon *ABCDE* ≅ pentagon *MNJKL*, which of the following is true? **(LESSON 4.1)**

   a. $\angle C \cong \angle K$
   b. $\angle A \cong \angle M$
   c. $\overline{AE} \cong \overline{MN}$
   d. cannot be determined from the given information

7. Which of the following quadrilaterals has congruent diagonals? *(LESSON 4.5)*
  a. trapezoid
  b. rectangle
  c. parallelogram
  d. rhombus

8. Which of the following quadrilaterals can have exactly two right angles? *(LESSON 4.5)*
  a. trapezoid
  b. rectangle
  c. parallelogram
  d. rhombus

9. Find the midpoint of a segment with endpoints at $(5, 1)$ and $(1, -3)$. *(LESSON 5.6)*
  a. $(6, -2)$
  b. $(2, 2)$
  c. $(3, -1)$
  d. $(4, -2)$

10. The point $(3, -1, 2)$ is in which octant of a three-dimensional coordinate system?
(LESSON 6.4)
  a. first octant
  b. top-right-back
  c. top-front-left
  d. bottom-front-right

11. Find the y-intercept of the plane defined by the equation $5x + 2y - z = 6$. *(LESSON 6.5)*

12. Solve the proportion for x. *(LESSON 8.2)*
$$\frac{2 + x}{3} = \frac{x}{2}$$

13. Which of the following is not valid for proving triangles similar? *(LESSON 8.3)*
  a. SSS
  b. AA
  c. SSA
  d. SAS

14. Find the ratio of the volumes of a sphere with a radius of 3 and a sphere with a radius of 6. *(LESSON 8.6)*

FREE-RESPONSE GRID

Items 15–18 may be answered by using a free-response grid such as that commonly used by standardized-test services.

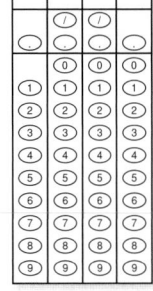

15. What is the sum of the interior angles of the figure below? *(LESSON 3.6)*

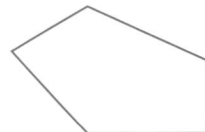

16. What is the sum of the exterior angles of the figure above? *(LESSON 3.6)*

17. Find the value of x in the figure below. Round your answer to the nearest tenth.
(LESSON 5.4)

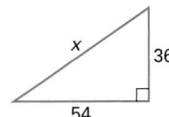

18. In the figure below, $\overline{BC} \parallel \overline{DE}$. Find BD.
(LESSON 8.4)

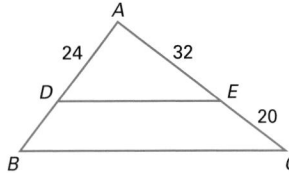

7. b

8. a

9. c

10. c

11. $(0, 3, 0)$

12. 4

13. c

14. $\frac{1}{8}$

15. 540°

16. 360°

17. 64.9

18. 15

Circles

Lesson Presentation CD-ROM
Power Point® presentations for each lesson 9.1-9.6

CHAPTER PLANNING GUIDE

Lesson	9.1	9.2	9.3	9.4	9.5	9.6	Project and Review
Pupil's Edition Pages	564–572	573–579	580–587	588–597	600–609	610–617	598–599, 618–627
Practice and Assessment							
Extra Practice (Pupil's Edition)	845	846	846	847	847	848	
Practice Workbook	56	57	58	59	60	61	
Practice Masters Levels A, B, and C	166–168	169–171	172–174	175–177	178–180	181–183	
Standardized Test Practice Masters	64	65	66	67	68	69	70
Assessment Resources	112	113	114	116	117	118	115, 119–124
Visual Resources							
Lesson Presentation Transparencies Vol. 2	25–28	29–32	33–36	37–40	41–44	45–48	
Teaching Transparencies	75–76	77–78	79–80	81–83		84–85	
Answer Key Transparencies	331–333	334–336	337–339	340–342	343–344	345–350	351–354
Quiz Transparencies	9.1	9.2	9.3	9.4	9.5	9.6	
Teacher's Tools							
Reteaching Masters	111–112	113–114	115–116	117–118	119–120	121–122	
Make-Up Lesson Planner for Absent Students	56	57	58	59	60	61	
Student Study Guide	56	57	58	59	60	61	
Spanish Resources	56	57	58	59	60	61	
Block Scheduling Handbook							18–19
Activities and Extensions							
Lesson Activities	152–155	156–159	160–164	165–170	171–177		
Enrichment Masters	56	57	58	59	60	61	
Cooperative-Learning Activities	56	57	58	59	60	61	
Problem-Solving/ Critical Thinking	56	57	58	59	60	61	
Student Technology Guide	53–54	55–56	57	58	59	60–61	
Long Term Projects							33–36
Writing Activities for Your Portfolio							25–27
Tech Prep Masters							41–44
Building Success in Mathematics							23–25

LESSON PACING GUIDE

Lesson	9.1	9.2	9.3	9.4	9.5	9.6	Project and Review
Traditional	2 days	2 days	2 days	2 days	4 days	2 days	2 days
Block	1 day	1 day	1 day	1 day	2 days	1 day	1 day
Two-Year	4 days	4 days	4 days	4 days	8 days	4 days	4 days

CONNECTIONS AND APPLICATIONS

Lesson	9.1	9.2	9.3	9.4	9.5	9.6	Review
Algebra	570	575, 578, 579		594, 595, 597	601, 602, 604, 606, 607, 609	610, 611, 614, 615, 617	
Geometry	564–572	573–579	580–587	588–597	600–609	610–617	620–627
Coordinate Geometry						612	
Technology						611–613	
Trigonometry	572						
Life Skills	571	578	580, 584, 585, 586	588, 591, 593, 597	603	616, 617	624
Science and Technology	571	578, 579	583, 587	597	600, 603, 604, 608	616	624
Sports and Leisure	571, 567						
Other	571				608		

BLOCK-SCHEDULING GUIDE

Day	Lesson	Teacher Directed: Lesson Examples, Teaching Transparencies	Student Guided: Activity, Try This	Cooperative-Learning Activity, Lesson Activity, Student Technology Guide	Practice: Practice & Apply, Extra Practice, Practice Workbook	Assessment: Quiz, Mid-Chapter Assessment	Problem Solving, Reteaching
1	9.1	15 min	15 min	20 min	55 min	15 min	15 min
2	9.2	15 min	15 min	20 min	55 min	15 min	15 min
3	9.3	15 min	15 min	20 min	55 min	15 min	15 min
4	9.4	15 min	15 min	20 min	55 min	15 min	15 min
5	9.5	15 min	15 min	20 min	55 min	15 min	15 min
6	9.6	15 min	15 min	20 min	55 min	15 min	15 min
7	Assess.	50 min	90 min	90 min	65 min	30 min	
		PE: Chapter Review	**PE:** Chapter Project, Writing Activities	Tech Prep Masters	**PE:** Chapter Assessment, Test Generator	Chap. Assess. (A or B), Alt. Assess. (A or B), Test Generator	

PE: Pupil's Edition

Alternative Assessment

The following suggest alternative assessments for students who may benefit from a different type of assessment than the regular chapter quizzes and the mid-chapter/end-of-chapter test. Visit the HRW web site to get additional Alternative Assessment material.

internet connect

Alternative Assessment
Go To: **go.hrw.com**
Keyword: **MG1 Alt Assess**

Performance Assessment

1. List all the theorems in your text from Lessons 9.3–9.5 and illustrate each one with a diagram.

2. Using inscribed angles, find the center of a circular object such as a coffee can lid or a paper plate. Give step-by-step instructions on how you found the center point.

3. Write and graph an equation for a circle with an 8 unit diameter whose center is at (2, 4). Translate the circle down 5 units and write the new equation.

4. Draw two squares on grid paper, both with sides of 6 units. In one square inscribe a circle, and circumscribe a circle about the other square. For each figure label any radii, tangents, chords, secants, and special angles.

5. There are six figures drawn on page 588. Draw these figures on your own paper. For each figure, describe the relationship between the measures of the special angles and their intercepted arcs.

Portfolio Project

Suggest that students choose one of the following projects for inclusion in their portfolios.

1. Using a compass and straightedge, construct a figure like the one shown below.

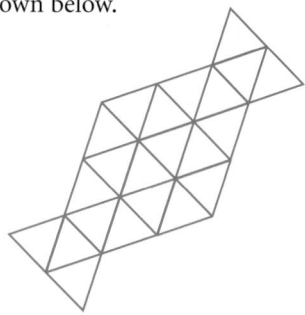

2. Using geometry software, create an electronic page to demonstrate the Tangent Theorem on page 574. Include instructions on how to drag the figure and display measurements.

---

internet connect

The table below identifies the pages in this chapter that contain internet and technology information.

Content Links

Activities Online	**pages 569, 576, 585**
Portfolio Extensions	**pages 572, 579**
Homework Help Online	**pages 570, 578, 585, 593, 606, 614**

Resource Links

Parents can go online and find concepts that students are learning–lesson by lesson–and questions that pertain to each lesson, which facilitate parent-student discussion.

Go To: **go.hrw.com**
Keyword: **MG1 Parent Guide**

Technical Support

The following may be used to obtain technical support for any HRW software product.

Online Help: **www.hrwtechsupport.com**

e-mail: **tschrw@hbtechsupport.com**

HRW Technical Support Center: **(800)323-9239**

7 AM to 10 PM Monday through Friday CST

Visit the HRW math web site at: **www.hrw.com/math**

Technology

Technology Objectives and Suggestions

Lesson 9.1 Chords and Arcs
In this lesson students define a circle and its associated parts, and use them in constructions. Geometry graphics software may be used for the constructions in this lesson and for measuring central angles and arcs in degrees. Some programs will also give arc length.

Lesson 9.2 Tangents to Circles
In this lesson students define tangents and secants of circles, and understand the relationship between tangents and certain radii of circles. The discoveries in the activities can be made using geometry software. For instance, in Activity 1 students can drag the chord until it becomes a tangent, while at the same time observing the angle measure of $\angle PQR$. For exercises involving the Pythagorean Theorem, students may want to use a calculator with a square root key to make computations easier.

Lesson 9.3 Inscribed Angles and Arcs
In this lesson students define inscribed angle and intercepted arc. They also develop and use the Inscribed Angle Theorem and Corollaries. Using geometry software, students can draw a circle with an inscribed angle and display the measures of the angle and its arc. By dragging the angle to change its measure, students can discover that the measure of the inscribed angle is half the measure of its intercepted arc. Similar demonstrations can be done with the Right-Angle Corollary and the Arc-Intercept Corollary.

Geometry software can be used to recreate the figures in the exercise set. Students can use the software to check angle measures they find through computation or logical reasoning.

Lesson 9.4 Angles Formed by Secants and Tangents
In this lesson students define angles formed by secants and tangents of circles, and develop and use theorems about measures of arcs intercepted by these angles. Students can use geometry software for the activities in this lesson. You may need to model for students how the find the measure of an arc in degrees. For example, in *The Geometer's Sketchpad*, students must select the two endpoints of the arc and the circle that contains the arc. Then they use *Arc Angle* from the Measure menu to find the number of degrees in the arc.

Lesson 9.5 Segments of Tangents, Secants, and Chords
In this lesson students define special cases of segments related to circles, and develop and use theorems about measures of these segments. The theorems developed in the activities can be demonstrated using geometry software. For instance, for Activity 3, construct a circle with two intersecting chords, with chord-segments of lengths a, b, c, and d. Then create two rectangles, one with sides a and b, and the other with sides c and d, corresponding to the four chord-segments. Confirm the cross-product relationship by showing that the areas of the rectangles are equal.

In the exercise set students will find calculators helpful for computations.

Lesson 9.6 Circles in the Coordinate Plane
In this lesson students develop and use the equation of a circle. They also adjust a circle equation to move the center on a coordinate plane. Graphics calculators will be very helpful to students in this lesson. When graphing circles, a square window setting is needed to keep the circles from looking like ellipses. The following settings and format for the TI-83 will work well for most of the activities in the lesson:

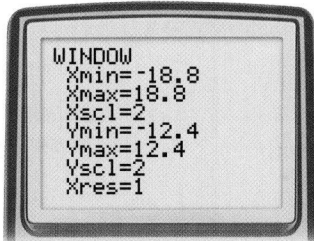

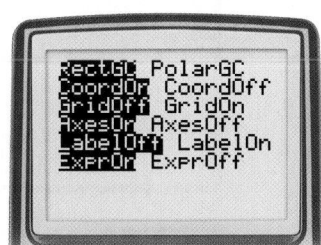

This chapter focuses on the relationships among parts of circles and between circles and various angles, segments, and arcs. Students develop theorems and rules through the activities and then apply those rules in problem-solving situations.

CHAPTER RESOURCES

- Block-Scheduling Handbook
- Writing Activities for Your Portfolio
- Tech Prep Masters
- Long-Term Project
- Assessment Resources:
  Mid-Chapter Assessment
  Chapter Assessments
  Alternative Assessments
- Test and Practice Generator
- Technology Handbook

Chapter Objectives

- Define a circle and its associated parts, and use them in constructions. [9.1]
- Define and use the degree measure of arcs. [9.1]
- Define and use the length measure of arcs. [9.1]
- Prove a theorem about chords and their intercepted arcs. [9.1]
- Define *tangents* and *secants* of circles. [9.2]
- Understand the relationship between tangents and certain radii of circles. [9.2]
- Understand the geometry of a radius perpendicular to a chord of a circle. [9.2]
- Define *inscribed angle* and *intercepted arc*. [9.3]

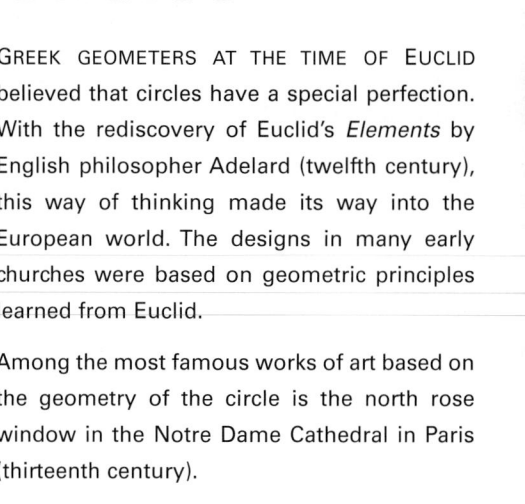

Circles

GREEK GEOMETERS AT THE TIME OF EUCLID believed that circles have a special perfection. With the rediscovery of Euclid's *Elements* by English philosopher Adelard (twelfth century), this way of thinking made its way into the European world. The designs in many early churches were based on geometric principles learned from Euclid.

Among the most famous works of art based on the geometry of the circle is the north rose window in the Notre Dame Cathedral in Paris (thirteenth century).

The circular structures in the aerial photo of the Pueblo ruins at Chaco Canyon are known as *kivas*. The circular design of these ceremonial structures reflects the Pueblo belief, suggested by the recurrence of celestial phenomena, that time is circular in nature.

Lessons

9.1 ● **Chords and Arcs**

9.2 ● **Tangents to Circles**

9.3 ● **Inscribed Angles and Arcs**

9.4 ● **Angles Formed by Secants and Tangents**

9.5 ● **Segments of Tangents, Secants, and Chords**

9.6 ● **Circles in the Coordinate Plane**

Project Tangent Curves

Rose window, Notre Dame Cathedral

Chaco Canyon

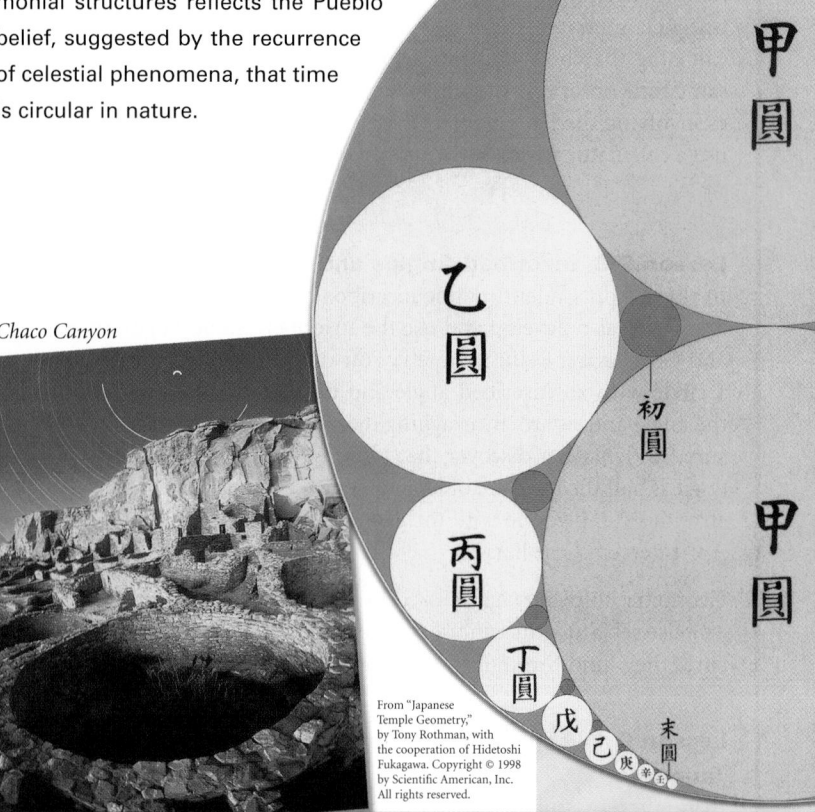

甲圓

乙圓

甲圓

丙圓

初圓

丁圓

末圓

戊

乙 庚 辛

From "Japanese Temple Geometry," by Tony Rothman, with the cooperation of Hidetoshi Fukagawa. Copyright © 1998 by Scientific American, Inc. All rights reserved.

About the Photos

Circles have fascinated people for centuries. In many countries of the world, circles have cultural significance. The photos above show the north rose window of the Notre Dame Cathedral in Paris; the Pueblo ruins at Chaco Canyon with its circular structures known as *kivas*; a water-filled sinkhole, or cenote, commonly found in Yucatán, Mexico; and the construction of a circle on paper. In all of these structures, circles hold special meaning.

Natural well or cenote at Chichén Itzá

- Develop and use the Inscribed Angle Theorem and its corollaries. [**9.3**]
- Define angles formed by secants and tangents of circles. [**9.4**]
- Develop and use theorems about measures of arcs intercepted by these angles. [**9.4**]
- Define special cases of segments related to circles, including secant-secant, secant-tangent, and chord-chord. [**9.5**]
- Develop and use theorems about measures of the segments. [**9.5**]
- Develop and use the equation of a circle. [**9.6**]
- Adjust the equation for a circle to move the center in a coordinate plane. [**9.6**]

About the Chapter Project

The art of constructing elegant egg shapes from curves of different radii is an ancient art that goes back to prehistoric times. Certain principles of geometry in this chapter will enable you to construct objects like the egg shape shown at right.

After completing the Chapter Project, you will be able to do the following:

- Construct artistic eggs from given examples.
- Create your own designs by using reverse curves.

About the Portfolio Activities

Throughout this chapter, you will be given opportunities to complete Portfolio Activities that are designed to support your work on the Chapter Project.

- The basic construction for circle flowers is the topic of the Portfolio Activity on page 572.

- A method for creating a single smooth curve from two curves of different radii is shown in the Portfolio Activity on page 579. With this method, you can construct the egg shapes in the Chapter Project.

- The method of creating a reverse curve, or "S-curve," by connecting two arcs smoothly is given in the Portfolio Activity on page 609. This principle, which is important in art, engineering, and architecture, will enable you to construct the figures in the Chapter Project, as well as figures of your own design.

Portfolio Activities appear at the end of Lessons 9.1, 9.2, and 9.5. Each serves as preparation for the Chapter Project. The Portfolio Activities, as well as the Chapter Project Activities, are appropriate for inclusion in the student's portfolio. Students should be encouraged to include in their portfolios any other work in which they feel a sense of pride or a sense of accomplishment.

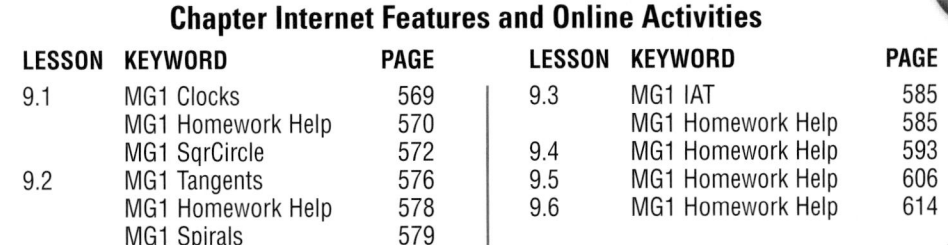

◢ internet connect

Chapter Internet Features and Online Activities

Define the following terms:

1. circle
   A set of points in a plane that are equidistant from a given point.

2. radius
   A line segment that connects the center of a circle with any point on the circle.

3. diameter
   A line segment that connects two points on a circle and passes through the center.

Also on Quiz Transparency 9.1

Teach

Why After students have examined the circle flowers in the text, point out that these designs are created with overlapping circles. Colors can be added, parts can be erased, etc. Many artistic designs are often the result of overlapping circles or parts of circles.

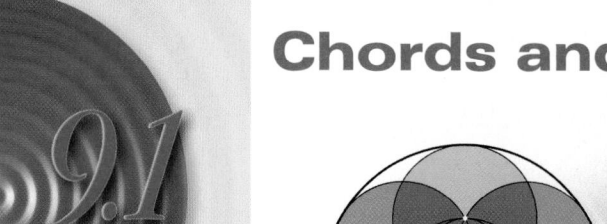

Chords and Arcs

Objectives

- Define a circle and its associated parts, and use them in constructions.

- Define and use the degree measure of arcs.

- Define and use the length measure of arcs.

- Prove a theorem about chords and their intercepted arcs.

Why To study the geometry of circles, you will need to know some basic definitions and ideas. With these, you can quickly learn to make interesting constructions.

The mathematical features of these attractive designs, which were created with geometry graphics software, can easily be understood from the ideas in this lesson.

Circles: A Formal Definition

A circle can be named by using the symbol $\odot$ and the center of the circle. The circle in the illustration below is $\odot P$, or circle P.

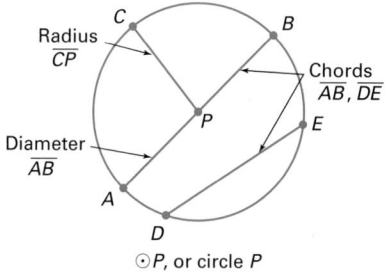

$\odot P$, or circle P

Definition: Circle

A circle is the set of all points in a plane that are equidistant from a given point in the plane known as the center of the circle. A **radius** (plural, *radii*) is a segment from the center of the circle to a point on the circle. A **chord** is a segment whose endpoints line on a circle. A **diameter** is a chord that contains the center of a circle.

9.1.1

Alternative Teaching Strategy

TECHNOLOGY Geometry graphics software may be used for the constructions in this lesson and for measuring central angles. Some software programs will give arc measures in linear units as well as in degrees.

In Activity 1 below, you will use concepts from the definition of a circle to perform and analyze a construction.

Activity 1
Constructing a Hexagon in a Circle

1. Draw a circle with a compass. Label the center *P*. Choose a point on the circle and label it *A* (figure **a**).

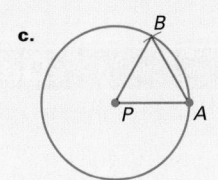

a.

2. Without changing your compass setting, place the point of your compass on point *A*. Draw an arc that intersects the circle at a new point. Label the new point *B* (figure **b**).

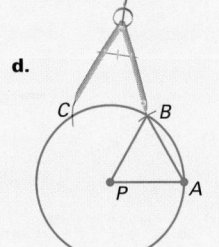

b.

3. Draw chord $\overline{AB}$. Draw radii $\overline{PA}$ and $\overline{PB}$ (figure **c**).

CHECKPOINT ✔ 4. What kind of triangle is △*ABP*? What are the measures of its angles? Explain your reasoning.

c.

5. Without changing your compass setting, place the point of your compass on point *B*. Draw an arc that intersects the circle at a new point. Label the new point *C* (figure **d**).

6. Draw chord $\overline{BC}$ and the new radius $\overline{PC}$.

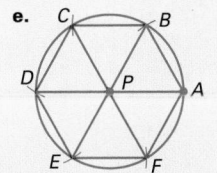

d.

7. Continue drawing new points, chords, and radii until you have completed a figure like the one shown in figure **e**.

8. Is polygon *ABCDEF* a regular hexagon? Explain your answer.

CHECKPOINT ✔ 9. An angle such as ∠*APB* is known as a *central angle* of a circle. (See page 566.) Are all the central angles in the figure congruent? Does the sum of their measures equal 360°? Explain your reasoning.

e.

Interdisciplinary Connection

FINE ARTS Designs based on congruent circles, either overlapping or separate, are found in textile and rug patterns from many different cultures. Have students collect samples of such designs for discussion. Portions of the original circles that are hidden or erased to make the final designs may be of particular interest.

Activity 1 Notes

In this Activity, students use a compass and straightedge (or geometry graphics software) to construct a hexagon. Have students note that the hexagon they construct is made up of equilateral triangles.

For a student worksheet of this Activity and detailed Teacher Notes, see page 152 in the Lesson Activities booklet.

Cooperative Learning

In pairs, have students experiment with the construction in Activity 1. Encourage them to develop alternate ways to accomplish this task. Then have students experiment with the constructions of other regular polygons.

CHECKPOINT ✔
4. △*ABP* is an equilateral triangle with 60° angles. You know this because all of the sides of the triangle are congruent.

CHECKPOINT ✔
9. Yes, they are all 60° angles. Yes, in the construction, a rotation was divided into 6 equal parts of 60° each.

Major and Minor Arcs

An **arc** is an unbroken part of a circle. Any two distinct points on a circle divide the circle into two arcs. The points are called the **endpoints** of the arcs.

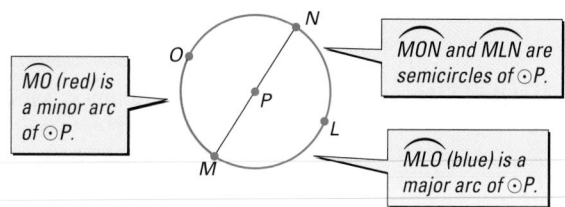

$\overset{\frown}{MO}$ (red) is a minor arc of ⊙P.

$\overset{\frown}{MON}$ and $\overset{\frown}{MLN}$ are semicircles of ⊙P.

$\overset{\frown}{MLO}$ (blue) is a major arc of ⊙P.

A **semicircle** is an arc whose endpoints are endpoints of a diameter. A semicircle is informally called a half-circle. A semicircle is named by its endpoints and another point that lies on the arc.

A **minor arc** of a circle is an arc that is shorter than a semicircle of that circle. A minor arc is named by its endpoints.

A **major arc** of a circle is an arc that is longer than a semicircle of that circle. A major arc is named by its endpoints and another point that lies on the arc.

Degree Measures of Arcs

Central angles of circles are used to find the measures of arcs.

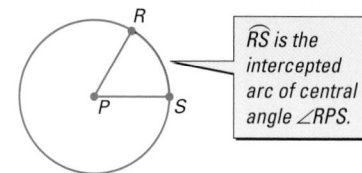

$\overset{\frown}{RS}$ is the intercepted arc of central angle ∠RPS.

Definitions: Central Angle and Intercepted Arc

A **central angle** of a circle is an angle in the plane of a circle whose vertex is the center of the circle. An arc whose endpoints lie on the sides of the angle and whose other points lie in the interior of the angle is the **intercepted arc** of the central angle.

9.1.2

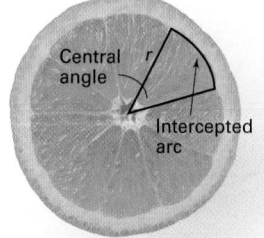

Central angle *r*

Intercepted arc

An orange may consist of nine wedges, seen in cross section here. Thus, an average wedge would form a central angle of about one-ninth of the full circle, or 40°. (When you look at a typical orange wedge, does it seem to be about 40°?)

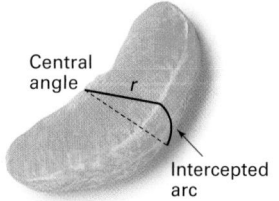

Central angle *r*

Intercepted arc

Arcs can be measured in terms of degrees.

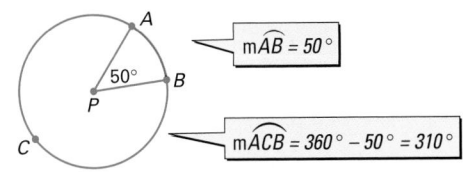

$m\overarc{AB} = 50°$

$m\overarc{ACB} = 360° - 50° = 310°$

Definition: Degree Measure of Arcs

The degree measure of a minor arc is the measure of its central angle. The degree measure of a major arc is 360° minus the degree measure of its minor arc. The degree measure of a semicircle is 180°.

9.1.3

EXAMPLE ① Find the measures of $\overarc{RT}$, $\overarc{TS}$, and $\overarc{RTS}$.

● **SOLUTION**

The measures of $\overarc{RT}$ and $\overarc{TS}$ are found from their central angles.

$m\overarc{RT} = 100°$ $m\overarc{TS} = 90°$

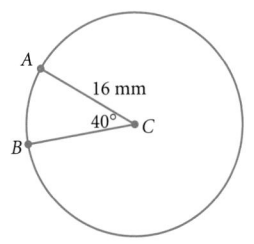

$\overarc{RT}$ and $\overarc{TS}$, which have just one endpoint in common, are called adjacent arcs. Add their measures to find the measure of $\overarc{RTS}$.

$m\overarc{RTS} = m\overarc{RT} + m\overarc{TS} = 100° + 90° = 190°$

Arc Length

A second way to measure an arc is in terms of its length. To find the length of an arc, you need to know the radius of the circle of which the arc is a part.

EXAMPLE ② Find the length of the indicated arc. Express your answer to the nearest millimeter. (There are 20 equal sectors on a dartboard.)

APPLICATION
GAMES

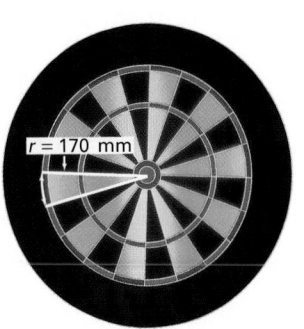

r = 170 mm

● **SOLUTION**

The length of the arc is $\frac{1}{20}$ of the circumference of the circle. Remember that $C = 2\pi r$.

$$\text{Length of arc} = \frac{1}{20}(2\pi \times 170)$$
$$= 17\pi \approx 53.4 \approx 53 \text{ mm}$$

Inclusion Strategies

HANDS-ON STRATEGIES If students have difficulty with the construction in Activity 1, have them cut six congruent circles from tracing paper and draw a triangle such as △ABP on each circle (follow Steps 1–3 of Activity 1). Have students use a black pen and a straightedge to construct the hexagon over the six superimposed circles.

Yes, it is possible if the arcs are on two different circles with different radii. It is not possible if the arcs are on the same circle or on congruent circles.

In general, the length of an arc can be found by using the formula below.

Arc Length

If r is the radius of a circle and M is the degree measure of an arc of the circle, then the length, L, of the arc is given by the following:

$$L = \frac{M}{360°}(2\pi r)$$

9.1.4

Activity 2 **Notes**

In this Activity, students construct a circle with congruent arcs and discover that in the same circle or in congruent circles, arcs of congruent chords are congruent.

For a student worksheet of this Activity and detailed Teacher Notes, see page 152 in the Lesson Activities booklet.

TRY THIS Find the length of $\overset{\frown}{RS}$. Round your answer to the nearest hundredth.

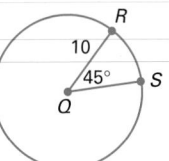

CRITICAL THINKING Is it possible for two arcs to have the same degree measure but different lengths? Explain why or why not.

CHECKPOINT ✔
3. The arcs are congruent.

Activity 2
Chords and Arcs Theorem

YOU WILL NEED
compass and straightedge

1. In the figure, chords $\overline{AB}$ and $\overline{CD}$ are congruent. The minor arcs $\overset{\frown}{AB}$ and $\overset{\frown}{CD}$ are called the arcs of the chords. Do you think that $\overset{\frown}{AB}$ and $\overset{\frown}{CD}$ have equal measures? To find the answer, make your own drawing of the figures and construct central angles $\angle APB$ and $\angle CPD$.

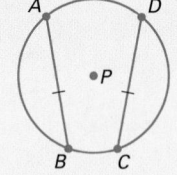

2. Prove that $\triangle APB$ and $\triangle CPD$ are congruent. What can you conclude about central angles $\angle APB$ and $\angle CPD$? about $\overset{\frown}{AB}$ and $\overset{\frown}{CD}$?

The diagram does not contain any specific features other than the given information; therefore, it represents the general case.

CHECKPOINT ✔
3. If two chords of a circle are congruent, what can you conclude about their arcs?

4. How can you extend this Activity to congruent chords of two or more different congruent circles? Explain and include appropriate diagrams.

PARAGRAPH PROOF
5. Present your discovery as a paragraph proof. Complete the theorem below.

In the process of discovering your result, you have proved a theorem.

CHECKPOINT ✔

CHECKPOINT ✔
5. congruent

Chords and Arcs Theorem

In a circle, or in congruent circles, the arcs of congruent chords are ___?___.

9.1.5

Reteaching the Lesson

USING REVIEW Have students list all the mathematical terms used in this lesson. For each term, they should do the following:

1. Write a definition of the term.
2. Make several drawings to illustrate the term.
3. Write a sentence using the term's mathematical meaning.

Exercises

Communicate

In Exercises 1–5, classify each statement as true or false and explain your reasoning.

1. Every diameter of a circle is also a chord of the circle.

2. Every radius of a circle is also a chord of the circle.

3. Every chord of a circle contains exactly two points of the circle.

4. If two chords of a circle are congruent, then their arcs are also congruent.

5. If two arcs of a circle are congruent, then their chords are also congruent.

6. How can you show that two arcs of a circle are congruent? Is there more than one way? Discuss.

7. Are all semicircles of a circle congruent? Explain.

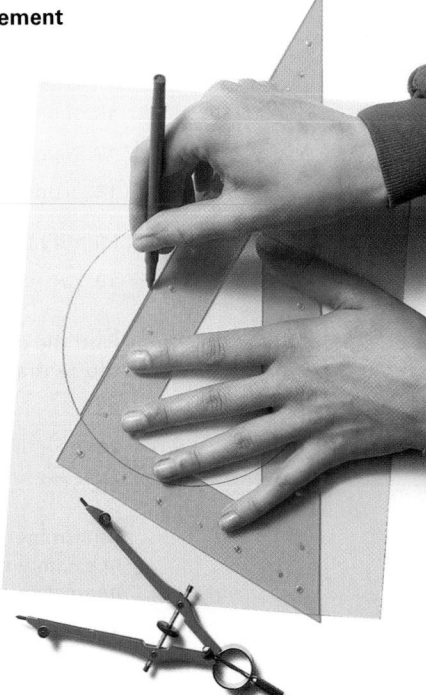

internet connect

Activities Online
Go To: go.hrw.com
Keyword:
MG1 Clocks

Assess

Selected Answers
Exercises 8–10, 11–51 odd

ASSIGNMENT GUIDE

In Class	1–10
Core	11–22, 23–43 odd
Core Plus	19–43 odd, 45–47
Review	48–52
Preview	53–57

✐ Extra Practice can be found beginning on page 818.

Guided Skills Practice

Use the figure below for Exercises 8 and 9.

8. Find the degree measures of $\overarc{AB}$, $\overarc{BC}$, and $\overarc{CA}$. *(EXAMPLE 1)* 70°; 160°; 130°

9. Find the length of $\overarc{AB}$. Round your answer to the nearest hundredth. *(EXAMPLE 2)* 3.67

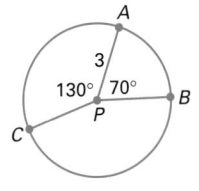

PARAGRAPH PROOF

It is given that ⊙P and ⊙M are congruent, and that $\overline{QR}$ and $\overline{NO}$ are congruent. Form two triangles by adding the radii $\overline{PR}$, $\overline{PQ}$, $\overline{MN}$, and $\overline{MO}$. Then △PQR and △MNO are congruent by SSS. Thus, ∠QPR and ∠NMO are congruent and hence the arcs $\overarc{QR}$ and $\overarc{NO}$ are congruent.

10. Write a paragraph proof of the Chords and Arcs Theorem for the case in which the congruent chords are in different congruent circles. Use the diagram below to begin your proof. *(ACTIVITY 2 AND THEOREM 9.1.5)*

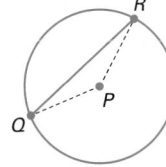
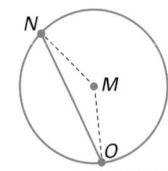

Error Analysis

In Exercises 31–38, it may be helpful to students to make a sketch of
the circle and arc before finding
the answer.

Math
CONNECTION

ALGEBRA In Exercises 34–37, students set up and solve an equation
in order to find the degree measure of each arc.

34. $\frac{36°}{\pi} \approx 11.46°$

35. $\frac{36°}{\pi} \approx 11.46°$

36. $\frac{36°}{\pi} \approx 11.46°$

37. $\frac{36°}{\pi} \approx 11.46°$

38. $\overline{XY} = 5$, since $\overline{XB}$ and $\overline{AY}$ are
diameters and $\angle ARB$ and
$\angle YRX$ are congruent central
angles. Hence $\triangle BRA$ is congruent to $\triangle XRY$.

Practice

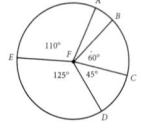
● *Practice and Apply*

Use the figure of ⊙P below for Exercises 11–22.

11. Name the center of the circle. **point P**

12. Name a radius of the circle. **Sample answer: $\overline{PB}$**

13. Name a chord of the circle. **$\overline{ED}$ or $\overline{AC}$**

14. Name a diameter of the circle. **$\overline{AC}$**

15. Name a central angle of the circle. **Sample answer: $\angle BPC$**

16. Name a semicircle of the circle. **Sample answer: $\widehat{ABC}$**

17. Name two minor arcs of the circle. **Sample answer: $\widehat{AB}$ and $\widehat{BC}$**

18. Name two major arcs of the circle. **Sample answer: $\widehat{DAB}$ and $\widehat{CEB}$**

Identify the given part of ⊙P.

19. $\overline{AP}$ **radius**   **20.** $\overline{AC}$ **diameter**   **21.** $\overline{ED}$ **chord**   **22.** $\angle APB$ **central angle**

Find the degree measure of each arc by using the central angle measures given in ⊙Q at right.

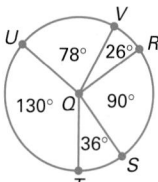

23. $\widehat{TU}$ **130°**   **24.** $\widehat{TSU}$ **230°**   **25.** $\widehat{RT}$ **126°**

26. $\widehat{UR}$ **104°**   **27.** $\widehat{VS}$ **116°**   **28.** $\widehat{US}$ **166°**

29. $\widehat{SUV}$ **244°**   **30.** $\widehat{VTR}$ **334°**

Determine the length of an arc with the given central angle measure, m∠P, in a circle with the given radius, r. Round your answer to the nearest hundredth.

31. m∠P = 90°; r = 10   **15.71**   **32.** m∠P = 60°; r = 3   **3.14**   **33.** m∠P = 30°; r = 120   **62.83**

Determine the degree measure of an arc with the given length, L, in a circle with the given radius, r.

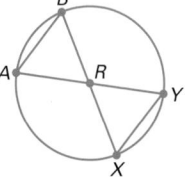

34. L = 14; r = 70   **35.** L = 20; r = 100

36. L = 3; r = 15   **37.** L = 5; r = 25

38. In ⊙R at right, if m∠ARB = 43° and AB = 5,
find XY. Explain your reasoning.

Suppose that *ABCDE* is a regular pentagon inscribed in ⊙Q and that AQ = 2. Find the following:

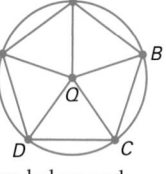

39. m∠AQB **72°**   **40.** m$\widehat{AE}$ **72°**   **41.** m$\widehat{ACE}$ **288°**

42. length of $\widehat{AE}$ **$\frac{4\pi}{5} \approx 2.51$**   **43.** length of $\widehat{ACE}$ **$\frac{16\pi}{5} \approx 10.05$**

PROOF

44. Complete the converse of the Chords and Arcs Theorem below and
prove your result.

The Converse of the Chords and Arcs Theorem

In a circle or in congruent circles, the chords of congruent arcs
are   ?   .

9.1.6

---

45. LANDSCAPING Sixty tulips are planted around the base of a circular fountain. If the fountain is 20 ft in diameter and the tulips are placed 1 ft away from the base, what is the length of the arc between consecutive tulips? $\frac{11\pi}{30} \approx 1.15$ ft

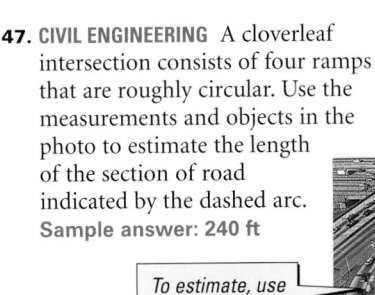

46. DEMOGRAPHICS Smith High School has 450 freshmen, 375 sophomores, 400 juniors, and 325 seniors. Create a pie chart that shows the distribution of students. First find the percent of the total student body for each class. Then multiply each percent by 360° to find the measure of the central angle for each section of the chart.

47. CIVIL ENGINEERING A cloverleaf intersection consists of four ramps that are roughly circular. Use the measurements and objects in the photo to estimate the length of the section of road indicated by the dashed arc.
Sample answer: 240 ft

To estimate, use 1 car ≈ 16 feet.

 Look Back

48. What effect does doubling the radius of a sphere have on its volume?
(LESSONS 7.1 AND 7.6) It increases the volume by a factor of 8.

SPORTS Find the volume of each object. *(LESSONS 7.2 AND 7.6)*

49.

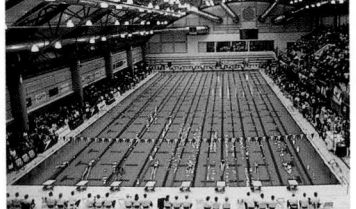

1.45 in.

12.77 in³

50.

$\ell = 59$ m
$w = 25$ m
$h = 2$ m

2950 m³

SPORTS A bicycle tire has a diameter of 26 in. *(LESSON 5.3)*

51. Find the circumference of the tire. $26\pi \approx 81.68$ in.

52. If the bicycle travels so that the wheel makes 100 complete turns, how far does the bicycle travel? ≈ 680.7 ft, or ≈ 680 ft 8 in.

Technology

Geometry graphics software can be used to perform the constructions in this exercise set. Students may need a scientific calculator with a π key to help with computations.

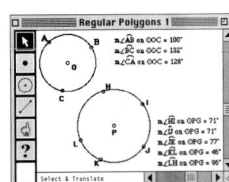

In Exercises 53–57, students explore the properties of tangent lines and of lines perpendicular to tangents.

ALTERNATIVE
Assessment

Portfolio Activity

The Portfolio Activity can be used as preparation for the Chapter Project or as a separate activity. The Portfolio Activity on this page features the construction of a circle flower. Have students make flowers with more than and less than 6 petals.

CONNECTION

A(1, 0), *B*(0, 1),
C(–1, 0), *D*(0, –1)

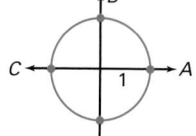
📶 internet connect

**Portfolio
Extension**

Go To: go.hrw.com
Keyword:
MG1 SqrCircle

TRIGONOMETRY The circle with its center at the origin and a radius of 1 is called the **unit circle.** You will study functions defined on the unit circle in trigonometry.

53. What are the coordinates of *A*, *B*, *C*, and *D* on the unit circle shown at right?

54. What is the circumference of the unit circle? 2π

Use the graph of the unit circle to find the following:

55. m$\widehat{AB}$ and 90°
length of $\widehat{AB}$ $\frac{\pi}{2}$

56. m$\widehat{ABC}$ and 180°
length of $\widehat{ABC}$ π

57. m$\widehat{ABD}$ and 270°
length of $\widehat{ABD}$ $\frac{3\pi}{2}$

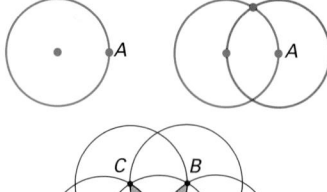

PORTFOLIO ACTIVITY

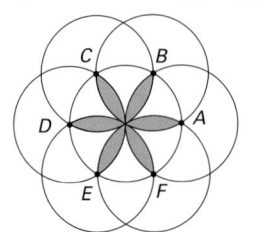

CONSTRUCTING A CIRCLE FLOWER A circle flower is formed when three or more circles meet at the center of another circle, forming "petals." A six-petal circle flower has a construction that is similar to that of the regular hexagon in Activity 1.

1. First draw a circle with a compass or geometry graphics software. Choose a point on the circle and label it *A*. Draw another circle congruent to the first with point *A* as its center.

2. Choose one of the points where circle *A* intersects the original circle and label it *B*. Using this point as a center, draw another circle congruent to the first. Repeat, going around the circle in one direction, until you have completed the flower.

3. How is the flower related to the hexagon in Activity 1? Think about why the petals meet at the center of the flower. You may wish to consider the measures of the arcs and central angles in the figure.

WORKING ON THE CHAPTER PROJECT

You should now be able to complete Activity 1 of the Chapter Project.

Many other interesting circle patterns are possible. This "quilt" was assembled by a computer from a pattern on page 223 of this book.

Tangents to Circles

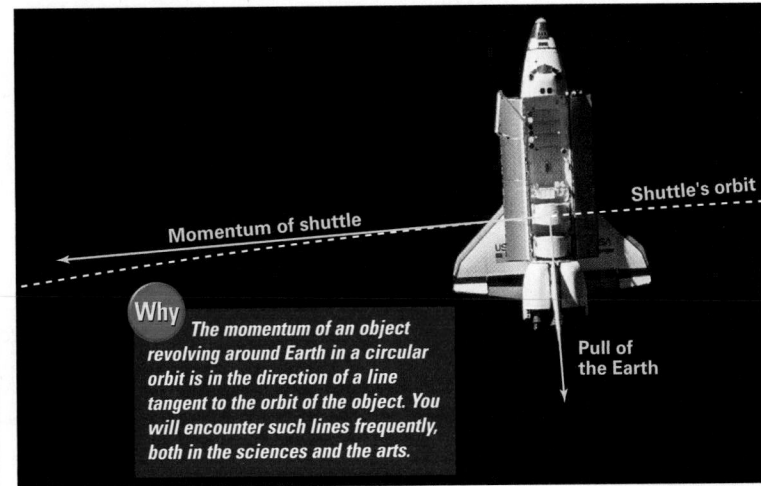

Momentum of shuttle

Shuttle's orbit

Why The momentum of an object revolving around Earth in a circular orbit is in the direction of a line tangent to the orbit of the object. You will encounter such lines frequently, both in the sciences and the arts.

Pull of the Earth

This photograph was taken by an astronaut during a space walk. The space shuttle is orbiting Earth from right to left. If it were not for the gravitational pull of the Earth, the shuttle would continue in a straight line in the direction of its momentum. It would quite literally "go off on a tangent."

Objectives

- Define *tangents* and *secants* of circles.
- Understand the relationship between tangents and certain radii of circles.
- Understand the geometry of a radius perpendicular to a chord of a circle.

Secants and Tangents

A line in the plane of a circle may or may not intersect the circle. There are three possibilities.

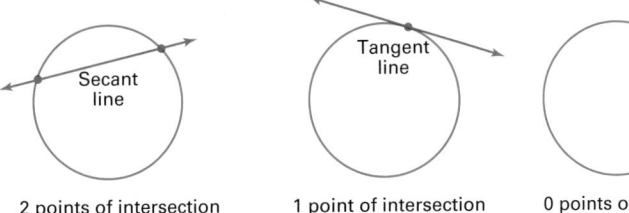

Secant line

Tangent line

2 points of intersection

1 point of intersection

0 points of intersection

Secants and Tangents

A **secant** to a circle is a line that intersects the circle at two points. A **tangent** is a line in the plane of the circle that intersects the circle at exactly one point, which is known as the **point of tangency**. **9.2.1**

CRITICAL THINKING

The word *tangent* comes from the Latin word meaning "to touch." The word *secant* comes from the Latin word meaning "to cut." Why are these words appropriate names for the lines they describe?

Alternative Teaching Strategy

HANDS-ON STRATEGIES Circular geoboards can be used for modeling problems and measuring angles in the Activities and exercises in this lesson.

Interdisciplinary Connection

PHYSICS The inertial tendency of an object to travel in a straight line is demonstrated by a stone in a sling. The sling is swung in a circular motion and then released. The stone travels in a line tangent to the point of release.

Activity 1 Notes

In this Activity, students discover that a tangent to a circle is perpendicular to the radius at the point of tangency.

For a student worksheet of this Activity and detailed Teacher Notes, see page 156 in the Lesson Activities booklet.

CHECKPOINT ✔

4. If a line is tangent to a circle, then the line is perpendicular to a radius of the circle drawn to the point of tangency.

Teaching Tip

TECHNOLOGY The discovery in Activity 1 can be made by using geometry graphics software. Students can observe the angle measure of ∠PQR as they drag the chord until it becomes a tangent.

Cooperative Learning

Students should work in small groups so that several different cases can be constructed, analyzed, and discussed.

Activity 2 Notes

In this Activity, students discover a relationship between radii of circles and chords to which they are perpendicular. Ask students to compare this theorem with the one they wrote in Activity 1.

For a student worksheet of this Activity and detailed Teacher Notes, see page 156 in the Lesson Activities booklet.

CHECKPOINT ✔

4. A radius that is perpendicular to a chord of a circle bisects the chord.

Activity 1
Radii and Tangents

YOU WILL NEED

compass, ruler, and protractor

OR

geometry graphics software

1. Draw ⊙P with radius $\overline{PQ}$.

2. Locate a point R on the circle and draw $\overleftrightarrow{QR}$. Measure ∠PQR.

3. Repeat Step 2 with point R closer to point Q, but still on the circle. If you are using geometry graphics software, drag point R toward point Q. What do you observe about m∠PQR as point R moves closer to Q? What happens when R coincides with Q?

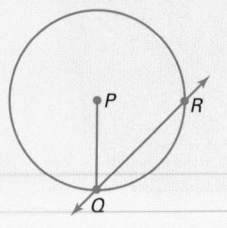

CHECKPOINT ✔

4. Make a conjecture about the relationship between a tangent to a circle and a radius drawn to the point of tangency. Based on your observations, complete the theorem below, which you will be asked to prove in Exercises 31–34.

Tangent Theorem

If a line is tangent to a circle, then the line is ____?____ to a radius of the circle drawn to the point of tangency.

9.2.2

Activity 2
Radii Perpendicular to Chords, Part 1

YOU WILL NEED

compass, ruler, and protractor

OR

geometry graphics software

1. Draw ⊙P with chord $\overline{AB}$.

2. Construct a radius perpendicular to chord $\overline{AB}$. Label the point of intersection X.

3. Measure $\overline{AX}$ and $\overline{BX}$. What do you observe?

4. Repeat Steps 1–3 with different circles and chords. If you are using geometry graphics software, experiment by changing the size of the circle and by dragging the chords to different locations. Make a conjecture about radii that are perpendicular to chords in circles. Based on your conjecture, complete the theorem below.

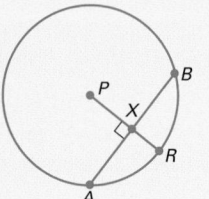

Radius and Chord Theorem

A radius that is perpendicular to a chord of a circle ____?____ the chord.

9.2.3

CHECKPOINT ✔

TWO-COLUMN PROOF

5. Draw $\overline{PA}$ and $\overline{PB}$ in one of your circles. Using this diagram, write a two-column proof of the Radius and Chord Theorem.

Enrichment

Line designs, sometimes called string art because they can be made from string, apply the concept of tangents to a circle or to other curve. A set of tangent lines can be drawn so that they suggest the shape of a curve.

Inclusion Strategies

ENGLISH LANGUAGE DEVELOPMENT Some students may need extra help with the new vocabulary introduced in this chapter. Draw a circle on the chalkboard and have students demonstrate their understanding of various terms by identifying the parts of the circle.

EXAMPLE ● ⊙P has a radius of 5 in. and PX is 3 in. $\overline{PR}$ is perpendicular to $\overline{AB}$ at point X. Find AB.

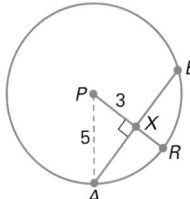

Algebra ● **SOLUTION**

By the Pythagorean Theorem:

$$(AX)^2 + 3^2 = 5^2$$
$$(AX)^2 = 5^2 - 3^2$$
$$(AX)^2 = 16$$
$$AX = 4$$

By the Radius and Chord Theorem, $\overline{PR}$ bisects $\overline{AB}$, so $BX = AX = 4$. Therefore, $AB = AX + BX = 4 + 4 = 8$.

ADDITIONAL

E X A M P L E ❶

In the Example on page 575, Example 1, suppose that $PX = 6$ and $AB = 10$. Find PA.   ≈ 7.81

Math
C O N N E C T I O N

ALGEBRA Students use the Pythagorean Theorem to find AX. Review methods for solving an equation that contains squares and square roots.

Activity 3
Radii Perpendicular to Chords, Part 2

YOU WILL NEED

compass, ruler, and protractor
OR
geometry graphics software

1. Explain how this diagram is like the one in Activity 2.

2. Imagine moving point X to point R. What happens to intersection points A and B as X gets closer to R? What happens when X touches R? What conjecture does this suggest? Use your answer to complete the theorem below, which is proved on the following page.

CHECKPOINT ✔

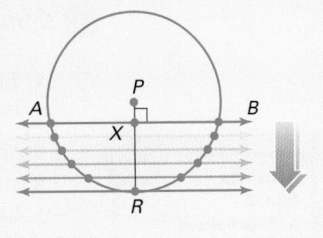

Converse of the Tangent Theorem

If a line is perpendicular to a radius of a circle at its endpoint on the circle, then the line is ___?___ to the circle.   **9.2.4**

CRITICAL THINKING   The proof on the following page uses the fact that the hypotenuse is the longest side of a right triangle. How could you argue that this must be true?

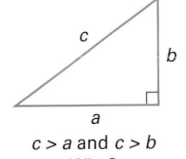

$c > a$ and $c > b$
Why?

Activity 3 **Notes**

Students have already discovered that a radius perpendicular to a chord bisects the chord. In this Activity, they consider the case of a perpendicular chord as the line containing it becomes a tangent.

For a student worksheet of this Activity and detailed Teacher Notes, see page 156 in the Lesson Activities booklet.

CHECKPOINT ✔

2. If a line is perpendicular to a radius of a circle at its endpoint on the circle, then the line is tangent to the circle.

CRITICAL THINKING

In a triangle, the side opposite the largest angle is the longest side. In a right triangle, the largest angle is the right angle, so the hypotenuse is the longest side.

Reteaching the Lesson

COOPERATIVE LEARNING In pairs, have students copy the Tangent Theorem given in their text and then create five different ways of restating the theorem. The new statements may read "If a line passing through the endpoint of a radius. . ." or "A radius of a circle is perpendicular to a line . . ." Students should be prepared to defend, orally or in writing, the fact that their new statements have exactly the same meaning as the original statement.

Selected Answers

Exercises 6–9, 11–29 odd

ASSIGNMENT GUIDE

In Class	1–9
Core	10–23
Core Plus	11–23 odd, 24–26
Review	27–30
Preview	31–34

✐ Extra Practice can be found beginning on page 818.

The Converse of the Tangent Theorem

The following proof uses the definition of a circle in an interesting way. When you understand the proof, you should be able to summarize it quickly in your own words (see Exercise 18).

PARAGRAPH PROOF

Given: Point P is on $\odot O$, and $\overline{OP}$ is perpendicular to $\overleftrightarrow{AB}$.

Prove: $\overline{AB}$ is tangent to $\odot O$ at point P.

Proof: Choose any point on $\overleftrightarrow{AB}$ other than point P and label it Q. Draw right triangle OPQ. Since $\overline{OQ}$ is the hypotenuse of a right triangle, it is longer than $\overline{PO}$, which is a radius of the circle. Therefore, point Q does not lie on the circle. This is true for all points on $\overleftrightarrow{AB}$ except point P, so $\overleftrightarrow{AB}$ touches the circle at just one point. By definition, $\overleftrightarrow{AB}$ is tangent to $\odot O$ at point P.

Exercises

● Communicate

1. Explain the three possible relationships between a line and a circle in a plane.

2. Explain how a secant intersects a circle.

3. How many lines are tangent to a circle? Explain.

4. How many lines are tangent to a circle at a given point? Explain.

5. Describe a point of tangency in the photo at right.

☑ internet connect

Activities Online

Go To: go.hrw.com
Keyword:
MG1 Tangents

● Guided Skills Practice

6. $\overleftrightarrow{KL}$ is tangent to $\odot M$ at K. If $KM = 1$ and $LM = 2$, find KL. $\sqrt{3}$
(ACTIVITY 2 AND THEOREM 9.2.2)

7. In $\odot P$, $QR = 3$. Find RS. 3
(ACTIVITY 2 AND THEOREM 9.2.3)

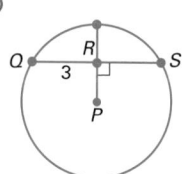

8. $\odot X$ has a radius of 13, $XW = 5$, and $\overline{XV} \perp \overline{YZ}$. Find YZ. *(EXAMPLE AND THEOREM 9.2.3)* 24

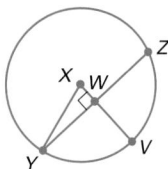

9. Verify that $\overleftrightarrow{AB}$ is tangent to $\odot C$ at B. *(ACTIVITY 3 AND THEOREM 9.2.4)*

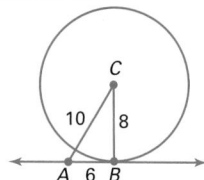

● *Practice and Apply*

For Exercises 10–12, refer to $\odot R$, in which $\overline{RY} \perp \overline{XZ}$ at W.

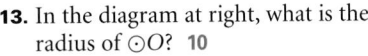

10. $\overline{XW} \cong \underline{}$ WZ

11. If $RY = 7$ and $RW = 2$, what is XW? What is WZ? $\sqrt{45} = 3\sqrt{5}, \sqrt{45} = 3\sqrt{5}$

12. If $RY = 3$ and $RW = 2$, what is XW? What is WZ? $\sqrt{5}, \sqrt{5}$

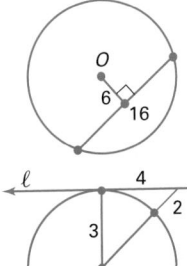

13. In the diagram at right, what is the radius of $\odot O$?   10

14. In $\odot N$, verify that line ℓ is a tangent by using the Converse of the Tangent Theorem.

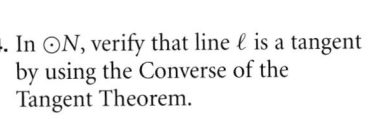

PARAGRAPH PROOF

15. In Activity 2 you proved a theorem about a radius that is perpendicular to a chord. Write a paragraph proof of the following related theorem:

Theorem

The perpendicular bisector of a chord passes through the center of the circle.

9.2.5

CONSTRUCTION

16. Use Theorem 9.2.5 above to construct a circle through any three noncollinear points. First draw three points not on a straight line. Label them A, B, and C. Draw $\overline{AB}$ and $\overline{BC}$. Construct the perpendicular bisector of each segment. Where is the center of the circle that contains A, B, and C? Complete the construction. How does this construction relate to Activity 2 in Lesson 1.5?

17. Use the Converse of the Tangent Theorem to construct a tangent to a circle at a given point. First draw a circle and label the center P. Choose any point on the circle and label it A. Draw $\overline{AP}$. How is the tangent line at A related to $\overline{AP}$? Complete the construction.

Technology

Geometry graphics software can be used to perform the constructions in this exercise set. Students may need a scientific calculator with a square root key for exercises in which they must use the Pythagorean Theorem.

14. The triangle formed has sides of length 3, 4, and 5 units. Since $3^2 + 4^2 = 9 + 16 = 25 = 5^2$, the triangle is a right triangle and line l is perpendicular to the radius as its endpoint on the circle, so it is tangent to the circle.

15. Let $\overline{PQ}$ be a chord in $\odot M$ and consider the line l which is the perpendicular bisector of $\overline{PQ}$, intersecting $\overline{PQ}$ at point X. Let $\overline{MN}$ be the radius of $\odot M$ which is perpendicular to $\overline{PQ}$. By the Radius and Chord Theorem, $\overline{MN}$ bisects $\overline{PQ}$ so $\overline{MN}$ also passes through the point X. Thus both $\overline{MN} \perp \overline{PQ}$ and $l \perp \overline{PQ}$. Since both l and $\overline{MN}$ pass through the point X and are both $\perp \overline{PQ}$, they must coincide. Hence, the perpendicular bisector of $\overline{PQ}$, l, passes through the center of the circle, M.

16. Sample answer:

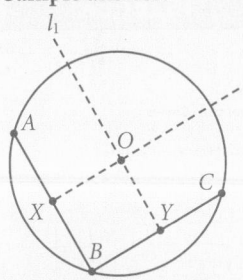

The center of the circle is the intersection of the two perpendicular bisectors. This is the same as the construction of a circumscribed circle.

17. Sample answer: The tangent line is $\perp \overline{AP}$.

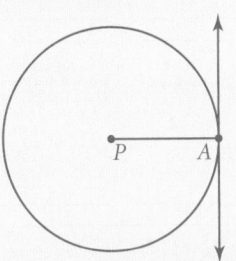

ALGEBRA In Exercises 19–23, students will need to apply the Pythagorean Theorem to find some of the solutions.

Error Analysis

Be sure students know that they need to convert between miles and feet in Exercise 24.

18. Sample answer: Any point on the line $\overleftrightarrow{AB}$ which is not P is not on the circle because of the Pythagorean Theorem. Thus, line $\overleftrightarrow{AB}$ touches the circle at exactly one point.

PARAGRAPH PROOF

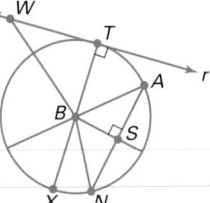
Algebra

18. Summarize the proof of the Converse of the Tangent Theorem from page 576 in your own words.

Use the diagram of $\odot B$ to find the indicated lengths. Line r is tangent to $\odot B$ at T, $BT = 2$, $BS = 1$, and $WT = 5$. Round your answers to the nearest hundredth.

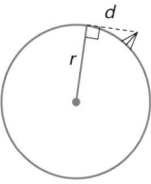

19. $BA = \underline{\quad ? \quad}$ 2

20. $SA = \underline{\quad ? \quad}$ $\sqrt{3} \approx 1.73$

21. $SN = \underline{\quad ? \quad}$ $\sqrt{3} \approx 1.73$

22. $BW = \underline{\quad ? \quad}$ $\sqrt{29} \approx 5.39$

23. $XT = \underline{\quad ? \quad}$ 4

24. COMMUNICATIONS A radio station installs a VHF radio tower that stands 1500 ft tall. What is the maximum effective signal range of the tower? The diagram suggests a way to use tangents to solve the problem. Use the Pythagorean Theorem to find d. (The diameter of Earth is approximately 8000 mi.) $d \approx 47.67$ mi

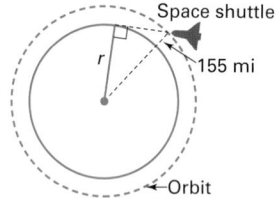

25. SPACE FLIGHT The space shuttle orbits at 155 mi above Earth. How far is it from the shuttle to the horizon? (The diameter of Earth is approximately 8000 mi.) $d \approx 1124.29$ mi

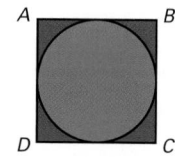

26. DESIGN An artist wants to draw the largest circle that will fit into a square. She uses the following method:

Draw square $ABCD$. Connect the midpoints of sides $\overline{AB}$ and $\overline{CD}$ with a segment. Connect the midpoints of sides $\overline{AD}$ and $\overline{BC}$ with another segment.

Construct the desired circle. How can you prove that this is the largest circle that will fit in the square? How can you prove that no part of the circle lies outside the square?

Look Back

Algebra

27. A triangle has a perimeter of 24 cm and an area of 24 cm². What are the perimeter and area of a larger similar triangle if the scale factor is $\frac{2}{1}$? **(LESSONS 8.1 AND 8.6)** $P = 48$ cm $A = 96$ cm²

$P = 58\frac{2}{3}$ ft $A = 156\frac{4}{9}$ ft² **28.** A rectangle has a perimeter of 22 ft and an area of 22 ft². What are the perimeter and area of a larger similar rectangle if the scale factor is $\frac{8}{3}$? **(LESSONS 8.1 AND 8.6)**

26. Sample answer: The center of the circle is the intersection of the segments that join the midpoints. The segments that join the midpoints are diameters of the circle. The circle is tangent to the sides of the square at the midpoints. As in the proof of the Converse of the Tangent Theorem, every point on a side of the square, except the midpoint, is outside of the circle, so no part of the circle lies outside of the square. Any larger circle would have diameters longer than the sides of the square, so the ends of the diameters perpendicular to the sides of the square would have to lie outside the square.

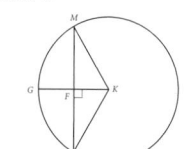

29. A rectangular prism has dimensions $\ell = 12$ in., $w = 8$ in., and $h = 15$ in. What is the volume of a larger similar rectangular prism if the scale factor is $\frac{5}{3}$? *(LESSONS 8.1 AND 8.6)* $6666\frac{2}{3}$ in^3

APPLICATION

30. ENGINEERING A cylindrical water tower has a radius of 30 ft and a height of 100 ft. What is the volume of a larger similar water tower if the scale factor is $\frac{7}{5}$? *(LESSONS 8.1 AND 8.6)* $246,960\pi \approx 775,848$ ft^3

Look Beyond

CHALLENGE

PARAGRAPH PROOF

Answer the questions below to prove the Tangent Theorem.

31. Suppose that the Tangent Theorem is *false*. That is, suppose that line m is tangent to $\odot O$ at point A, but that line m is *not* perpendicular to $\overline{OA}$. If this is true, then there *is* some segment with endpoint O, different from $\overline{OA}$, that is perpendicular to line m. Call that segment $\overline{OB}$. Then $\triangle OBA$ is a right triangle. What is the hypotenuse of $\triangle OBA$? Which segment is longer, $\overline{OA}$ or $\overline{OB}$?

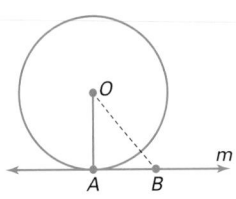

32. Point B must be in the exterior of the circle because m is a tangent line. What does this imply about the relative lengths of $\overline{OA}$ and $\overline{OB}$? Explain.

33. Compare your answers to Exercises 31 and 32. What do you observe?

34. If an assumption leads to a contradiction, it must be rejected. This is the basis for a type of proof known as an *indirect proof* or a *proof by contradiction*. Explain how the argument above leads to the desired conclusion.

internet connect

Portfolio Extension
Go To: go.hrw.com
Keyword:
MG1 Spirals

CONSTRUCTING SMOOTH CURVES A curve may be made up of arcs of more than one circle. In order for arcs from two different circles to join smoothly at a point, they must have the same tangent at that point.

1. Using a compass and straightedge or geometry graphics software, draw $\odot P$ with radius $\overline{PR}$. Construct line ℓ tangent to $\odot P$ at R; that is, construct a line perpendicular to $\overline{PR}$ at R (refer to Exercise 17). For another circle to have the same tangent, ℓ, its center must be on the line $\overleftrightarrow{PR}$. Why?

2. Choose a point on $\overline{PR}$ that is not P or R and label it Q. Construct a circle centered at Q with radius $\overline{QR}$. $\odot P$ and $\odot Q$ will have the same tangent at R.

3. Try tracing part of your construction in a different color. Starting on $\odot P$ near R, trace until you get to R, and then continue tracing on $\odot Q$. The curves should join smoothly at R.

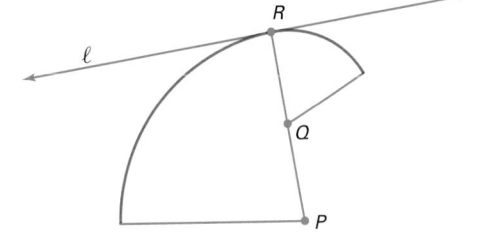

WORKING ON THE CHAPTER PROJECT

You should now be able to complete Activity 2 of the Chapter Project.

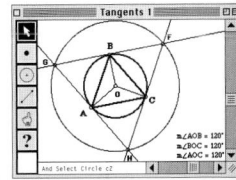
31. $\overline{OA}$ is the hypotenuse of $\triangle OBA$. Since the hypotenuse is the longest side of a right triangle, $OA > OB$.

32. Since B is outside the circle, $OB > OA$.

33. Contradictory answers

34. Since a contradiction is obtained, we reject the assumption that m is not perpendicular to $\overline{OA}$. We conclude that $m \perp \overline{OA}$.

Prepare

QUICK WARM-UP

A circle and an angle are drawn in the same plane. The vertex of the angle is on the circle.

1. Find and sketch all the possible ways that the two figures can be arranged. Neither, both, or one of the rays of the angle passes through the circle.

2. For each arrangement, give the number of intersection points.
   neither: 1; both: 3; one: 2

Also on Quiz Transparency 9.3

Teach

Why Ask students to devise various ways of finding the center of a circle. They can refer to the construction in the application problem on page 584 for ideas.

CRITICAL THINKING
Circle *P*. Yes; a similar result is true for all inscribed angles because an inscribed angle divides a circle into a major arc and a minor arc.

Inscribed Angles and Arcs

Objectives

- Define *inscribed angle* and *intercepted arc*.
- Develop and use the Inscribed Angle Theorem and its corollaries.

Why You can find the center of a circular object, such as the tabletop shown here, by using a carpenter's square. The process involves inscribed angles.

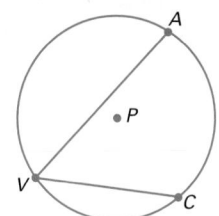

APPLICATION
CARPENTRY

A carpenter's square is used both for creating right angles in structures and for testing whether angles in structures are right angles. It can also be used to find the center of a circle. The method for finding the center of a circle relies on an important geometry corollary (see the application on page 584).

The Inscribed Angle Theorem

An **inscribed angle** is an angle whose vertex lies on a circle and whose sides are chords of the circle.

$\angle AVC$ is inscribed in $\odot P$.
$\angle AVC$ intercepts $\overarc{AC}$.

If you know the measure of the inscribed angle, you can determine the measure of its intercepted arc. How do you think this can be done? In the activities on the following pages, you will discover the answer.

CRITICAL THINKING In the illustration above, what is $m\overarc{AVC} + m\overarc{AC}$? Would a similar result be true for all possible inscribed angles in a circle? Explain.

Alternative Teaching Strategy

HANDS-ON STRATEGIES Students can use a compass and straightedge, geometry graphics software, or paper folding to construct the figures in this lesson. Encourage students to create different cases for each figure.

Teaching Tip

Contrast the definition of *inscribed angle* with that of *central angle*, which was introduced in Lesson 9.1. Provide examples and have students classify each example as either inscribed or central.

Activity 1
Angles and Intercepted Arcs: A Conjecture

1. Draw three different figures in which inscribed angle ∠AVC intercepts an arc of the circle. Include one minor arc, one major arc, and one semicircle.

2. Measure the inscribed angle and the intercepted arc in each figure. (You will need to draw central angles to determine the measures of the arcs.)

3. Compare the measures of the inscribed angle and its intercepted arc in each case.

CHECKPOINT ✔

4. Make a conjecture about the relationship between the measure of an inscribed angle and the measure of its intercepted arc.

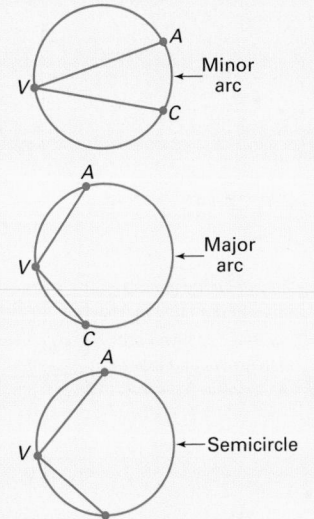

Activity 2
Proving the Conjecture

To prove your conjecture from Activity 1, you will need to consider three separate cases.

Part I

1. In the figure at right, one side of the inscribed angle contains the center of the circle. What is the relationship between m∠1 and m∠2?

2. Notice that ∠3 is an exterior angle of △AVP. What is the relationship among m∠3, m∠1, and m∠2?

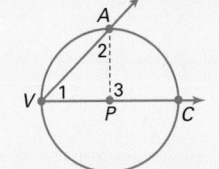

The center of the circle is on one side of the angle.

TABLE PROOF

3. Complete the table below. For each entry in the last row, give a reason.

m∠1	m∠2	m∠3	m$\widehat{AC}$
20°	?	?	?
30°	?	?	?
40°	?	?	?
x°	?	?	?

> As you justify each entry of the table for the general case, you are also proving a theorem.

CHECKPOINT ✔

4. What does your table show about the relationship between m∠1 and m$\widehat{AC}$?

Activity 1 Notes

In this Activity, students discover that the measure of an inscribed angle is equal to one-half the measure of its intercepted arc. Before they begin the Activity, ask them to define these terms: *major arc*, *minor arc*, *semicircle*, and *intercepted arc*. If necessary, review the meanings of these terms with the class.

For a student worksheet of this Activity and detailed Teacher Notes, see page 160 in the Lesson Activities booklet.

CHECKPOINT ✔
4. The measure of an inscribed angle is half the measure of its intercepted arc.

Activity 2 Notes

In this Activity, students prove the relationship that they discovered in Activity 1. The figure in Part III can be created with geometry graphics software. Students can drag points *A* or *C* on the circle to change the measure of ∠AVC.

For a student worksheet of this Activity and detailed Teacher Notes, see page 160 in the Lesson Activities booklet.

CHECKPOINT ✔
4. m∠1 = $\frac{1}{2}$m$\widehat{AC}$

Interdisciplinary Connection

PHYSICS The inside of a cylinder has a reflective surface. A beam of light starts on the rim of a circular cross section of the cylinder and moves toward any other point on the circle, following the path of a secant line. When the beam hits the circle, it is reflected so that the angle of incidence (formed by the secant and a tangent line) equals the angle of reflection (formed by another secant and the same tangent line). Have students make a sketch to model the path of the beam of light. Students should discover that the resulting secant lines intercept two arcs that are the same length.

Have students work on the Activities in groups of three. Students should display their group's work on large sheets of paper and share their conjectures with the whole class.

3. $m\angle AVC = \frac{1}{2}m\widehat{AXC}$

The measure of an inscribed angle of a circle is equal to one-half the measure of the intercepted arc.

ADDITIONAL

EXAMPLE ❶

In Example 1, suppose that the measure of $\angle XVY$ is 26°. What is the measure of the intercepted arc?   52°

Part II

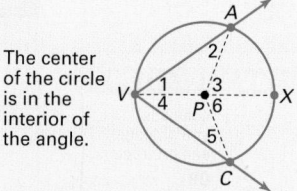

1. In the figure at right, the center of the circle is in the interior of the inscribed angle. What is the relationship between $m\angle 1$ and $m\widehat{AX}$? between $m\angle 4$ and $m\widehat{XC}$?

The center of the circle is in the interior of the angle.

2. Complete the table below. For each entry in the last row of the table, give a reason.

TABLE PROOF

> As you justify each entry of the table for the general case, you are also proving a theorem.

m∠1	m$\widehat{AX}$	m∠4	m$\widehat{CX}$	m∠AVC	m$\widehat{AXC}$
20°	?	20°	?	?	?
30°	?	20°	?	?	?
40°	?	50°	?	?	?
x°	?	y°	?	?	?

3. What does your table show about the relationship between $m\angle AVC$ and $m\widehat{AXC}$?

Part III

One more case remains to be proved. In this case, the center of the circle is in the *exterior* of the inscribed angle. Draw your own figure for this case and prove your conjecture for it. (You may want to make a table.)

The center of the circle is in the exterior of the angle.

Create your own figure.

Conclusion

Complete the theorem below, which your work in this Activity has proven.

Inscribed Angle Theorem

The measure of an angle inscribed in a circle is equal to ____?____ the measure of the intercepted arc.

9.3.1

EXAMPLE ❶ Find the measure of $\angle XVY$.

SOLUTION

$\angle XVY$ is inscribed in $\odot P$ and intercepts $\widehat{XY}$. By the Inscribed Angle Theorem:

$m\angle XVY = \frac{1}{2}m\widehat{XY} = \frac{1}{2}(45°) = 22\frac{1}{2}°$

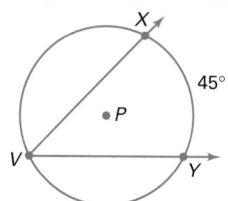

Inclusion Strategies

HANDS-ON STRATEGIES Many of the figures in this lesson can be created on circular geoboards. The experience of creating the figures with rubber bands may help some students understand the relationships between various angles and the circle.

Two Results of the Inscribed Angle Theorem

In the figure at right, ∠A is inscribed in a semicircle. Therefore, the arc it intercepts is also a semicircle. Thus, the measure of the intercepted arc is 180° (why?), and the measure of ∠A is $\frac{1}{2} \times 180°$, or 90°.

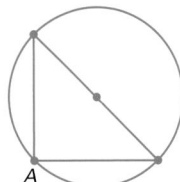

This calculation illustrates the following important corollary to the Inscribed Angle Theorem:

Right-Angle Corollary

If an inscribed angle intercepts a semicircle, then the angle is a right angle.

9.3.2

In the figure at right, the measure of $\overset{\frown}{AC}$ is twice the measure of ∠D, or $2 \times 50° = 100°$. The measure of ∠B is one-half the measure of $\overset{\frown}{AC}$, or $\frac{1}{2} \times 100° = 50°$.

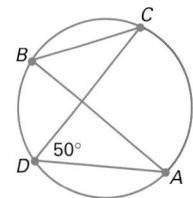

These calculations illustrate an important principle. Notice that ∠D and ∠B intercept the same arc. As you found, they have the same angle measure. This leads to another corollary of the Inscribed Angle Theorem.

Arc-Intercept Corollary

If two inscribed angles intercept the same arc, then they have the same measure.

9.3.3

EXAMPLE 2 A person's effective field of vision is about 30°. In the diagram of the amphitheater, a person sitting at point A can see the entire stage. What is the measure of ∠B? Can the person sitting at point B view the entire stage?

APPLICATION
OPTICS

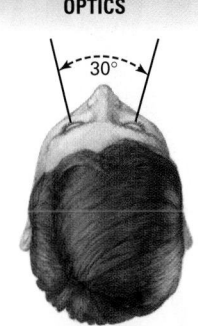

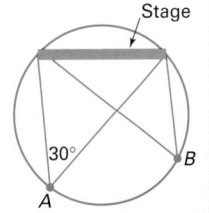

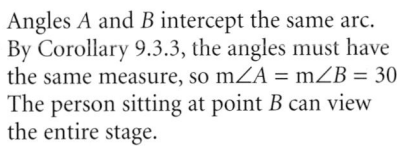

● **SOLUTION**

Angles A and B intercept the same arc. By Corollary 9.3.3, the angles must have the same measure, so m∠A = m∠B = 30°. The person sitting at point B can view the entire stage.

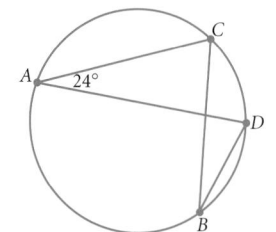
Enrichment

Extend the Interdisciplinary Connection exercise on page 581 by asking students the following questions: If the beam returns to its starting point, what types of paths could it have followed? What initial condition is necessary in order for the beam to return to its starting point? Remind students that each time the beam of light hits the rim of the circle, the angle of incidence equals the angle of reflection.

$\overline{AB}$ is parallel to $\overline{CD}$. What is the measure of $\overset{\frown}{AC}$? Explain.

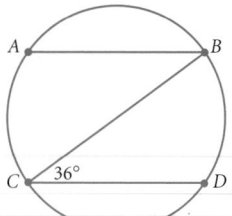

$\angle ABC$ is congruent to $\angle DCB$ because they are alternate interior angles. Since $\angle ABC$ has a measure of 36°, $\overset{\frown}{AC}$ has a measure of 72°.

E X A M P L E ③ In ⊙P with diameter $\overline{AB}$, $m\overset{\frown}{CB} = 110°$, and $m\overset{\frown}{BD} = 130°$. Find the measures of $\angle 1$, $\angle 2$, $\angle 3$, $\angle 4$, $\angle APC$, $\angle ADB$, and $\angle CAD$.

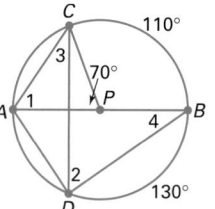

● **SOLUTION**

Copy the figure and add information as you work through the solution.

Begin by labeling as many arc measures as you can. Since $\overline{AB}$ is a diameter, $\overset{\frown}{ACB}$ and $\overset{\frown}{ADB}$ have a measure of 180°. Therefore, $m\overset{\frown}{AC} = 70°$ and $m\overset{\frown}{AD} = 50°$.

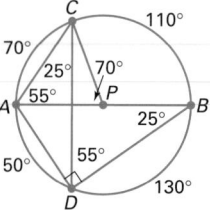

$\angle 1$ and $\angle 2$ intercept $\overset{\frown}{CB}$. Thus, $m\angle 1 = m\angle 2 = \frac{1}{2}(110) = 55°$.

$\angle 3$ and $\angle 4$ intercept $\overset{\frown}{AD}$. Thus, $m\angle 3 = m\angle 4 = \frac{1}{2}(50) = 25°$.

$\angle APC$ is a central angle. Thus, $m\angle APC = m\overset{\frown}{AC} = 70°$.

$\angle ADB$ is inscribed in a semicircle. Thus, $m\angle ADB = 90°$.

$\angle CAD$ intercepts $\overset{\frown}{CBD}$. Thus, $m\angle CAD = \frac{1}{2}(110 + 130) = 120°$.

APPLICATION
CARPENTRY

A carpenter needs to find the center of a small tabletop. How can she do this by using a carpenter's square?

The carpenter inscribes two right angles, $\angle ABC$ and $\angle DEF$, in the circle. She then draws the hypotenuses of the triangles by connecting points A and C and points D and F. The intersection of $\overline{AC}$ and $\overline{DF}$ is the center of the table.

You will be asked to perform a similar construction in Exercise 49.

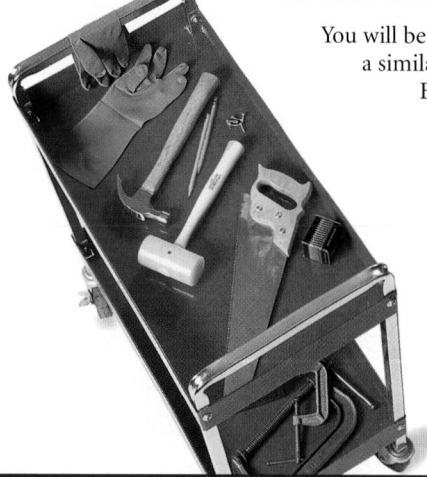

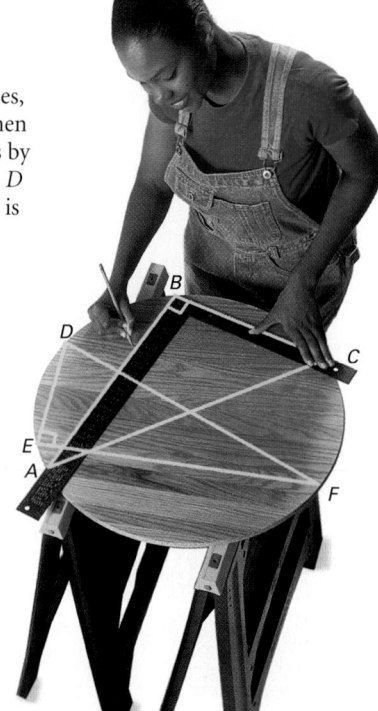

Reteaching the Lesson

COOPERATIVE LEARNING Have students work in pairs to review the lesson. Each student in a pair should draw a figure similar to the ones used in the Activities and Examples. For example, the figure could illustrate an inscribed angle and the related central angle. Have students exchange papers. Each should measure just one angle and then use the lesson concepts to determine the measures of all other angles and arcs in the figure.

Exercises

Selected Answers

Exercises 7–10, 11–61 odd

Communicate

Refer to ⊙O in the photo for Exercises 1–5.

1. Name an inscribed angle in ⊙O.

2. Explain how to find m$\overset{\frown}{BD}$.

3. Explain how to find m∠1.

4. Explain how to find m$\overset{\frown}{AC}$.

5. Explain how to find m∠2.

6. Explain why two inscribed angles that intercept the same arc have the same measure.

internet connect
Activities
Online
Go To: **go.hrw.com**
Keyword:
MG1 IAT

Error Analysis

If students use geometry graphics software, remind them that they must determine the measures of angles and arcs in the figures based on the information given in the problem and then use the software only to confirm their answers.

Guided Skills Practice

APPLICATION

PHOTOGRAPHY A person standing at a point *M* on the edge of a circular gallery takes a photograph. Suppose that $\overset{\frown}{OP}$ represents everything that is included in the photograph and that m$\overset{\frown}{OP}$ = 70°.

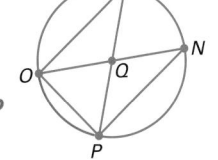

35° **7.** What is m∠OMP, known as the *picture angle* of the camera? *(ACTIVITIES 1 AND 2, THEOREM 9.3.1, AND EXAMPLE 1)*

35° **8.** What is m∠ONP? *(COROLLARY 9.3.3 AND EXAMPLE 2)*

In ⊙F, $\overline{HK}$ is a diameter, m$\overset{\frown}{GH}$ = 50°, and m$\overset{\frown}{JK}$ = 90°. *(THEOREM 9.3.1 AND EXAMPLE 3)*

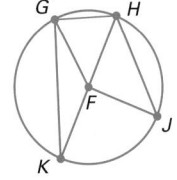

9. Find the measures of ∠GFH, ∠GKH, ∠JFK, and ∠KHJ.

10. Find the measure of ∠KGH.

m∠GFH = 50°
m∠GKH = 25°
m∠JFK = 90°
m∠KHJ = 45°
m∠KGH = 90°

Practice and Apply

internet connect
Homework
Help Online
Go To: **go.hrw.com**
Keyword:
MG1 Homework Help
for Exercises 11-28

In ⊙W, m$\overset{\frown}{XZ}$ = 60°, m∠VYZ = 40°, and $\overleftrightarrow{YZ}$ is a diameter. Find the following:

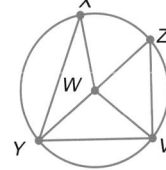

11. m∠XYW 30°

12. m∠WXY 30°

13. m∠XWY 120°

14. m∠XWZ 60°

15. m$\overset{\frown}{YXZ}$ 180°

16. m∠YVZ 90°

17. m$\overset{\frown}{XY}$ 120°

18. m$\overset{\frown}{VZ}$ 80°

19. m$\overset{\frown}{VY}$ 100°

20. m∠VZY 50°

Technology

Geometry graphics software can be used to re-create the figures in the exercise set. Students can use the software to check the accuracy of angle measures that they found through computation or logical reasoning.

For Exercises 21–28, refer to ⊙P with diameter $\overline{AC}$. Find the following:

21. m∠A 29°   22. m∠B 90°

23. m∠BCA 61°   24. m$\widehat{AB}$ 122°

25. m∠PCD 50°   26. m∠CPD 80°

27. m$\widehat{DC}$ 80°   28. m$\widehat{AD}$ 100°

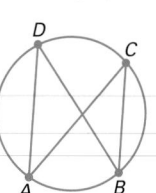

For Exercises 29–32, refer to the circle below.

29. m$\widehat{AB}$ = 68°   m∠C = ___?___ 34° m∠D = ___?___ 34°

30. m∠D = 30°   m$\widehat{AB}$ = ___?___ 60° m∠C = ___?___ 30°

31. m$\widehat{CD}$ = 87°   m∠B = ___?___ 43.5° m∠A = ___?___ 43.5°

32. m∠B = a°   m$\widehat{CD}$ = ___?___ 2a° m∠A = ___?___ a°

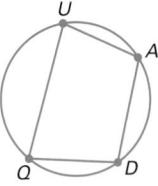

TABLE PROOF

> As you justify each entry of the table for the general case, you are also proving a theorem.

Quadrilateral *QUAD* is inscribed in a circle, as shown below. Copy and complete the following table. For each entry in the last row, give a reason.

m∠U	m$\widehat{QDA}$	m$\widehat{QUA}$	m∠D	m∠U + m∠D	m∠Q + m∠A
100°	33. ?	34. ?	35. ?	36. ?	37. ?
38. ?	160°	39. ?	40. ?	41. ?	42. ?
x°	43. ?	44. ?	45. ?	46. ?	47. ?

48. Based on the information in the table, state a theorem about the angles of a quadrilateral that is inscribed in a circle.
If a quadrilateral is inscribed in a circle, then the opposite angles are supplementary.

APPLICATIONS

In Exercise 49, you will use a piece of paper with right-angle corners to model the use of a carpenter's square to find the center of a circular object, as described on page 584.

49. **CARPENTRY** Draw a large circle on a sheet of paper. Place one right-angle corner of a sheet of paper on the circle so that both sides of the angle cross the circle. Mark the three points where the right angle at the corner of the paper touches the circle, and connect them to form a triangle. Repeat these steps to form another triangle in the circle. Identify the point where the longest sides of your triangles intersect. Explain why this point is the center of the circle.

m∠A = 40°
m∠B = 80°

50. **STAINED GLASS** An artist is creating a circular stained-glass window with the design shown at right. The artist wants the arc intercepted by ∠A and ∠B to measure 80°. What should the measures be of ∠A and ∠B?

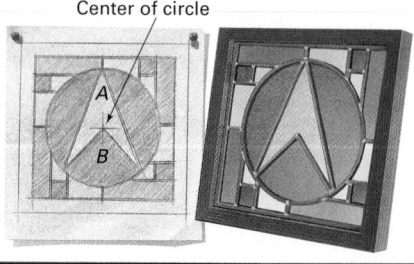

Center of circle

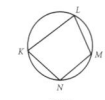
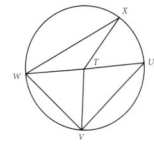

FLOWCHART PROOF

For Exercises 51–61, complete the flowchart proof below.

Given: Line ℓ is the angle bisector of $\angle BXC$, and line m is the angle bisector of $\angle CXD$.

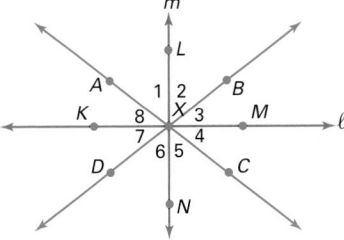

Prove: Line ℓ and m are perpendicular.

51. m is the angle bisector of $\angle CXD$.

52. Vertical Angles Theorem

53. $m\angle 2 = m\angle 5$

54. Addition Property

55. Angle Addition Postulate

56. Substitution Property

57. Linear Pair Property

58. $2 \cdot m\angle MXL = 180°$
$m\angle MXL = 90°$

59. Substitution Property and Division Property

60. $\ell \perp m$

61. Definition of perpendicular

Proof:

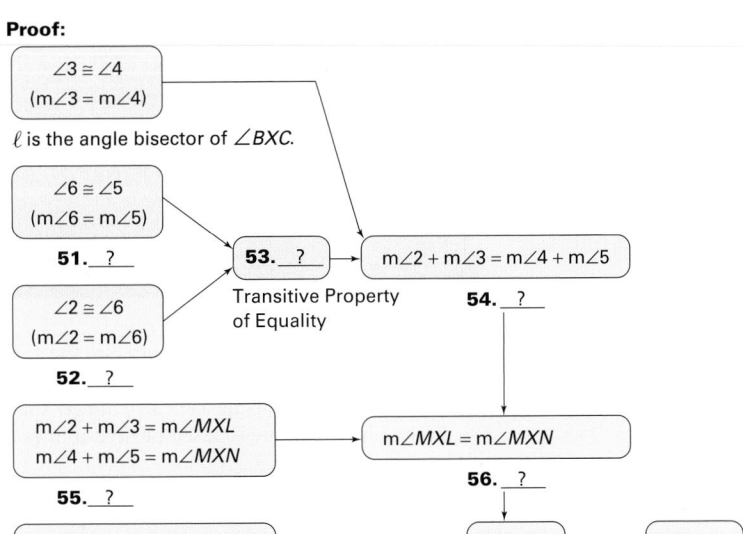

Look Beyond

APPLICATION

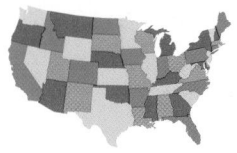

This map of the United States uses only four colors.

CARTOGRAPHY A famous theorem in mathematics is related to mapmaking. The *four-color theorem* states that any map in a plane can be colored with a maximum of four colors so that no two adjacent areas are the same color. The first "proof" of this theorem, by Appel and Haken in 1976, was controversial because parts of the proof used a computer and could not be verified by hand.

62. Explain why the "map" below cannot be colored with fewer than four colors.

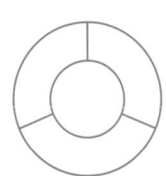

63. Make up your own map and try to color it with only four colors.

Look Beyond

Exercises 62 and 63 are applications of the four-color theorem in cartography. Concepts such as this will be studied further in Chapter 11.

Student Technology Guide

NAME _____ CLASS _____ DATE _____

Student Technology Guide

9.3 Inscribed Angles and Arcs

Since geometry graphics software allows you to measure angles and arcs, you will find it helpful in studying problems that involve inscribed angles.

To find the measure of $\widehat{AD}$, select A, the circle, and then D. From the main menu, select **Measure** **Arc Angle**.

Use geometry graphics software in the exercises below.

1. Create a diagram something like the one at right. Find and display m$\angle ABD$, m$\angle ACD$, and m$\widehat{AD}$. Do your measurements confirm a theorem that you learned in Lesson 9.3? What theorem?

2. Display AB, DC, AE, DE, BE, and CE. Calculate the ratios AB:DC, AE:DE, and BE:CE. What can you conclude about $\triangle ABE$ and $\triangle DCE$? Explain.

3. Devise and carry out a method different from the method in Exercise 2 to find out how $\triangle ABE$ and $\triangle DCE$ are related. What relationship did you discover?

4. Can you draw the same conclusion about $\triangle AED$ and $\triangle BEC$ as you did in Exercise 2? Explain. _____

5. Move points A, B, C, and D around the circle until point E coincides with point O, the center of the circle. What can you say about $\angle ABC$, $\angle BCD$, $\angle CDA$, and $\angle DAB$? What figure is formed by points A, B, C, and D?

6. Find and display m$\widehat{BCD}$ + m$\widehat{CDA}$ + m$\widehat{DAB}$ + m$\widehat{ABC}$. How does this calculation help confirm that the sum of the measures of the interior angles of a convex quadrilateral is 360°? _____

A circle and an angle are drawn in the same plane. Find all the possible ways in which the circle and angle intersect at two points. Possibilities are as follows: The circle intersects the angle at the vertex and on one ray. The circle is inside the angle with intersection points at tangents to the circle. One ray intersects the circle at two at points other than the vertex. The vertex is in the interior of the circle and the circle intersects each ray at one point.

Also on Quiz Transparency 9.4

Teach

Why The application on page 591 focuses on the concept of the *circle of danger*. Ask students to think about other navigation situations in which angle measures might be important.

Angles Formed by Secants and Tangents

Why The principles of circle geometry are used by navigators at sea. For example, the "horizontal angle of danger" enables a navigator to stay a safe distance from a dangerous region.

Objectives

- Define angles formed by secants and tangents of circles.
- Develop and use theorems about measures of arcs intercepted by these angles.

APPLICATION
NAVIGATION

A navigator on board a ship can measure the angle between two lines of sight to lighthouses on the coastline. This measurement tells the navigator whether the ship is inside or outside a *circle of danger* that contains hazardous rocks or shallows. The technique is based on secant lines that intersect the circle of danger.

Classification of Angles With Circles

Angles formed by pairs of lines that intersect a circle in two or more places can be studied systematically. There are three cases to consider, according to the placement of the vertex of the angles.

Case 1: Vertex is on the circle.   **Case 2: Vertex is inside the circle.**

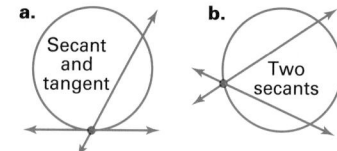

Case 3: Vertex is outside the circle.

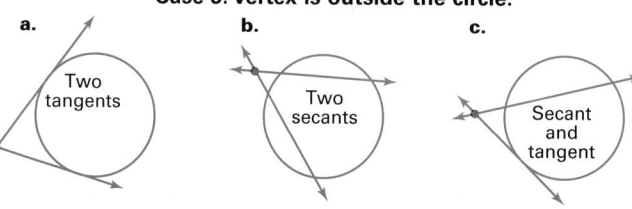

Alternative Teaching Strategy

TECHNOLOGY Students can use geometry graphics software for the Activities in this lesson. You may need to show students how to find the measure of an arc in degrees. For example, some basic steps are as follows:

1. Select the two endpoints of the arc to measure and the circle that contains the arc.

2. Select the appropriate command from the measure menu to find the number of degrees in the arc.

Activity 1
Vertex on Circle—Secant and Tangent (Case 1a)

In this Activity, you will examine three configurations of secant-tangent angles.

1. The secant-tangent angle is a right angle. (The secant contains the center of the circle.)

$$m\angle AVC = \underline{\quad?\quad} \qquad m\widehat{AV} = \underline{\quad?\quad}$$

CHECKPOINT ✔

How does this relationship compare with the one between an inscribed angle and its intercepted arc?

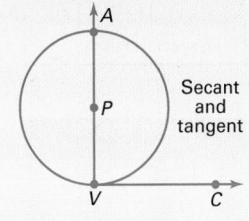

∠AVC is a right angle.

TABLE PROOFS

2. The secant-tangent angle is acute. Copy and complete the following table:

As you justify each entry of the table for the general case, you are also proving a theorem.

m$\widehat{AV}$	m∠1	m∠2	m∠PVC	m∠AVC
120°	120°	30°	?	60°
100°	?	?	?	?
80°	?	?	?	?
x°	?	?	?	?

CHECKPOINT ✔

Complete the following statement:

The measure of an acute secant-tangent angle with its vertex on a circle is __?__ the measure of its intercepted arc.

3. The secant-tangent angle is obtuse. Copy and complete the following table:

m$\widehat{AXV}$	m∠1	m∠2	m∠PVC	m∠AVC
200°	160°	10°	?	100°
220°	?	?	?	?
240°	?	?	?	?
x°	?	?	?	?

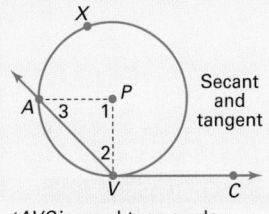

∠AVC is an obtuse angle.

CHECKPOINT ✔

Complete the following statement:

The measure of an obtuse secant-tangent angle with its vertex on a circle is __?__ the measure of its intercepted arc.

CHECKPOINT ✔

4. Based on your results, complete the following theorem:

Theorem

If a tangent and a secant (or a chord) intersect on a circle at the point of tangency, then the measure of the angle formed is __?__ the measure of its intercepted arc.

9.4.1

CRITICAL THINKING

Case 1b (vertex on circle, two secants) has already been studied in this book. Where? What theorem corresponds to this case?

Interdisciplinary Connection

LANGUAGE ARTS Have students collect examples of poetry or song lyrics that mention circles. Examples may be accompanied by student-made illustrations or cartoons to make an interesting bulletin-board display.

Activity 1 Notes

In this Activity, students discover that the measure of a secant-tangent angle with its vertex on a circle is one-half the measure of the intercepted arc. Before students begin, explain that the angle in this Activity is called a secant-tangent angle because one ray is a secant and one ray is a tangent. Have students identify each ray as either a secant or tangent.

For a student worksheet of this Activity and detailed Teacher Notes, see page 165 in the Lesson Activities booklet.

CHECKPOINT ✔

1. The relationship is the same as that between an inscribed angle and its intercepted arc.
$$m\angle AVC = \tfrac{1}{2}m\widehat{AV}$$

CHECKPOINT ✔

2. The measure of an acute secant-tangent angle with its vertex on a circle is one-half the measure of its intercepted arc.

CHECKPOINT ✔

3. The measure of an obtuse secant-tangent angle with its vertex on a circle is one-half the measure of its intercepted arc.

CHECKPOINT ✔

4. If a tangent and a secant (or a chord) intersect on a circle at the point of tangency, then the measure of the angle formed is one-half the measure of its intercepted arc.

CRITICAL THINKING

Section 9.3; Inscribed Angle Theorem

All three Activities in this lesson lend themselves to cooperative-learning groups. Students can divide the work to complete the tables. By discussing the relationships shown in the figures, students can develop a better understanding of relevant concepts.

Activity 2 Notes

This Activity focuses on the chord-chord case of secants that intersect inside a circle. Students discover that the measure of an angle formed by them equals one-half the sum of the measures of the intercepted arcs.

For a student worksheet of this Activity and detailed Teacher Notes, see page 165 in the Lesson Activities booklet.

CHECKPOINT ✔

3. The measure of an angle formed by two secants or by two chords intersecting in the interior of a circle is one-half the sum of the measures of the arcs intercepted by the angle and its vertical angle.

Activity 3 Notes

The angle in this Activity has its vertex outside the circle and is formed by two secants. Students discover that the measure of the angle equals one-half the difference of the measures of the intercepted arcs. Two other cases in which the vertex of the angle is outside the circle are covered in the Exercises: secant-tangent and tangent-tangent.

For a student worksheet of this Activity and detailed Teacher Notes, see page 165 in the Lesson Activities booklet.

Activity 2
Vertex Inside Circle—Two Secants (Case 2)

YOU WILL NEED
no special tools

TABLE PROOF

1. ∠AVC is an exterior angle of △ADV. What is the relationship between the measure of ∠AVC and the measures of ∠1 and ∠2?

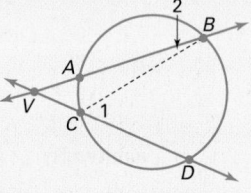
Two secants

2. Copy and complete the following table:

> As you justify each entry of the table for the general case, you are also proving a theorem.

m$\widehat{AC}$	m$\widehat{BD}$	m∠1	m∠2	m∠AVC	m∠DVB
160°	40°	80°	20°	100°	100°
180°	70°	?	?	?	?
200°	60°	?	?	?	?
$x_1°$	$x_2°$	?	?	?	?

3. Based on your results, complete the theorem below.

CHECKPOINT ✔

> ### Theorem
> The measure of an angle formed by two secants or chords that intersect in the interior of a circle is __?__ the __?__ of the measures of the arcs intercepted by the angle and its vertical angle. **9.4.2**

Activity 3
Vertex Outside Circle—Two Secants (Case 3b)

1. ∠1 is an exterior angle of △BVC. What is the relationship between the measure of ∠1 and the measures of ∠2 and ∠AVC?

Two secants

TABLE PROOF

2. Copy and complete the following table:

> As you justify each entry of the table for the general case, you are also proving a theorem.

m$\widehat{BD}$	m$\widehat{AC}$	m∠1	m∠2	m∠AVC
200°	40°	100°	20°	80°
250°	60°	?	?	?
100°	50°	?	?	?
$x_1°$	$x_2°$	?	?	?

3. Based on your results, complete the theorem below.

CHECKPOINT ✔

> ### Theorem
> The measure of an angle formed by two secants that intersect in the exterior of a circle is __?__ the __?__ of the measures of the intercepted arcs. **9.4.3**

You will explore cases 3a and 3c in Exercises 27–37.

Enrichment

Have students construct a circle that passes through three noncollinear points. Possible method is as follows: Choose any three points A, B, and C such that the points are not on the same line. Draw segments $\overline{AB}$ and $\overline{BC}$. Construct the perpendicular bisectors of the segments. Label the intersection of the bisectors at point O, which is the center of the desired circle.

The illustration below shows a ship at point *F* and two lighthouses at points *A* and *B*. The circle encloses a region of dangerous rocks. The measure of an angle inscribed in $\overset{\frown}{AGB}$, such as ∠*C*, is known as the horizontal angle of danger. Navigation charts contain information about horizontal angles of danger for different regions.

The ship's navigator measures ∠*F*, the angle formed by his lines of sight to the two lighthouses. He knows that he is outside the circle of danger because m∠*F* is less than the horizontal angle of danger for the area. Why does this method work?

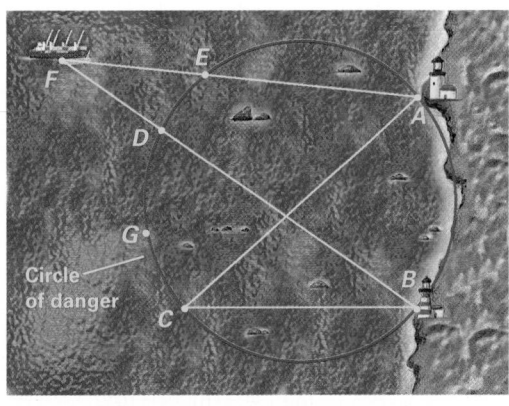

Notice that ∠*C* is an inscribed angle and that it intercepts $\overset{\frown}{AB}$. Thus, if the ship were on the circle, ∠*F* would equal ∠*C*. Therefore, it should be obvious that if m∠*F* is less than m∠*C*, the ship is outside the circle.

CRITICAL THINKING   How can you use theorems from this lesson to prove the "obvious" fact mentioned above? How can you prove that if m∠*F* were greater m∠*C*, then the ship would be *inside* the circle?

EXAMPLE ① Find m∠*AVC* in each figure.

a. 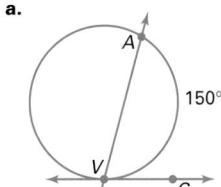   b. 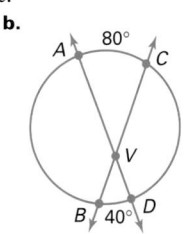   c.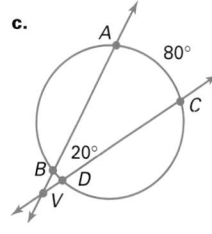

● **SOLUTION**

a. ∠*AVC* is formed by a secant and a tangent that intersect on the circle. By Theorem 9.3.1, m∠*AVC* = $\frac{1}{2}$m$\overset{\frown}{AV}$ = $\frac{1}{2}$(150°) = 75°.

b. ∠*AVC* is formed by two secants that intersect inside the circle. By Theorem 9.3.2, m∠*AVC* = $\frac{1}{2}$(m$\overset{\frown}{AC}$ + m$\overset{\frown}{BD}$) = $\frac{1}{2}$(80° + 40°) = 60°.

c. ∠*AVC* is formed by two secants that intersect outside the circle. By Theorem 9.3.3, m∠*AVC* = $\frac{1}{2}$(m$\overset{\frown}{AC}$ − m$\overset{\frown}{BD}$) = $\frac{1}{2}$(80° − 20°) = 30°.

CHECKPOINT ✔
3. The measure of an angle formed by two secants intersecting in the exterior of a circle is one-half the difference of the measures of the intercepted arcs.

CRITICAL THINKING
When the ship is on the circle, m∠*F* = m∠*C*. Theorem 9.4.2 states that if the vertex of ∠*F* is inside the circle, m∠*F* > m∠*C*. Theorem 9.4.3 states that if the vertex of ∠*F* is outside the circle then m∠*F* < m∠*C*.

ADDITIONAL
E X A M P L E ①

Given   m∠*AVC* = 60° and m$\overset{\frown}{AC}$ = 130°, find m$\overset{\frown}{BD}$. 10°

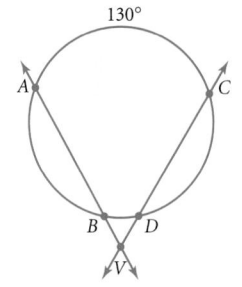

Inclusion Strategies

USING VISUAL MODELS   Students may understand the concepts in the lessson but become confused by the names of the various lines and angles. Suggest that they color code different types of lines. For example, they could use red for tangents and blue for secants.

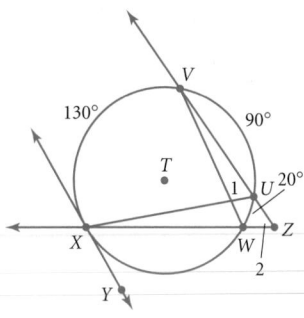

Given: $\overleftrightarrow{XY}$ is tangent to $\odot T$
at point X.
$m\widehat{UV} = 90°$;
$m\widehat{VX} = 130°$;
$m\widehat{UW} = 20°$

Find: **a.** m∠WXY
b. m∠1
c. m∠2

a. 60°; **b.** 105; **c.** 55°

EXAMPLE 2 **Given:** $\overleftrightarrow{TU}$ is tangent to $\odot P$ at point T.
$m\widehat{QR} = 90°$
$m\widehat{RT} = 150°$
$m\widehat{QS} = 50°$
Find the following:
a. m∠STU **b.** m∠1 **c.** m∠2

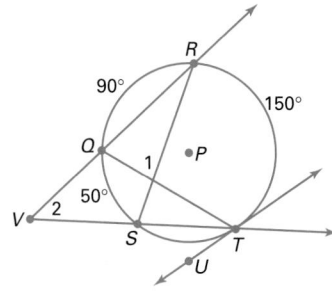

● **SOLUTION**

Make your own sketch of the figure. As you obtain new information, add it to
the figure.

a. Because the vertex of ∠STU is on
$\odot P$ and ∠STU is formed by a tangent
and a secant line, $m\angle STU = \frac{1}{2}m\widehat{ST}$.
To find $m\widehat{ST}$, note that
$50° + 90° + 150° + m\widehat{ST} = 360°$.
Thus, $m\widehat{ST} = 70°$ and
$m\angle STU = \frac{1}{2}(70°) = 35°$.

b. Because the vertex of ∠1 is in
the interior of $\odot P$ and is formed
by two intersecting chords,
$m\angle 1 = \frac{1}{2}(m\widehat{QR} + m\widehat{ST}) = \frac{1}{2}(90° + 70°) = 80°$.

c. Because the vertex of ∠2 is outside the circle and is formed by two
secants, $m\angle 2 = \frac{1}{2}(m\widehat{RT} - m\widehat{QS}) = \frac{1}{2}(150° - 50°) = 50°$.

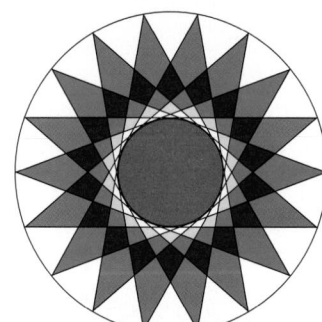

Exercises

● *Communicate*

1. The measure of an acute secant-
tangent angle with its vertex on
a circle is ___?___ .

2. The measure of an obtuse secant-
tangent angle with its vertex on
a circle is ___?___ .

3. The measure of a secant-secant
(or chord-chord) angle with its
vertex inside a circle is ___?___ .

4. The measure of a secant-secant
angle with its vertex outside a
circle is ___?___ .

*In this figure, every chord of the outer
circle is tangent to the inner circle.*

Reteaching the Lesson

COOPERATIVE LEARNING Have students work in
groups to copy the figures given at the beginning
of this lesson. Ask students to look for relationships between various angles and their intercepted
arcs. Then they should describe in writing the
relationships found in each of the figures.

Guided Skills Practice

5. In ⊙Z, m$\widehat{WY}$ = 138°. If $\overleftrightarrow{XY}$ is tangent to ⊙Z at Y, find m∠WYX. *(THEOREM 9.4.1 AND EXAMPLE 1)*

69°

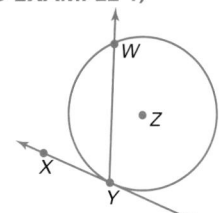

6. In ⊙Q, m$\widehat{MN}$ = 60° and m$\widehat{OP}$ = 150°. Find m∠1. *(THEOREM 9.4.2 AND EXAMPLE 1)*

105°

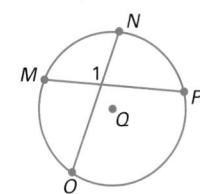

7. In ⊙A, m$\widehat{BC}$ = 84° and m$\widehat{DE}$ = 40°. Find m∠BFC. *(THEOREM 9.4.3 AND EXAMPLE 1)*

22°

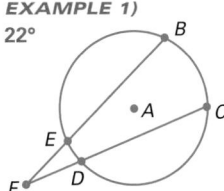

8. $\overleftrightarrow{AB}$ is tangent to ⊙G at B, m$\widehat{BC}$ = 70°, m$\widehat{BF}$ = 100°, and m$\widehat{EF}$ = 130°. Find the following. *(EXAMPLE 2)*

35° **a.** m∠ABC
100° **b.** m∠1
80° **c.** m∠2
20° **d.** m∠D

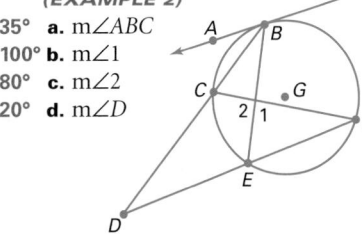

APPLICATION

9. NAVIGATION Lighthouses are located at points *A* and *B* on the circle of danger. If a ship is located at point *X* (not shown) and m∠BXA = 27°, is the ship inside or outside the circle of danger? Explain. *(APPLICATION)*
The ship is outside the circle of danger, since m∠BXA < m∠BCA.

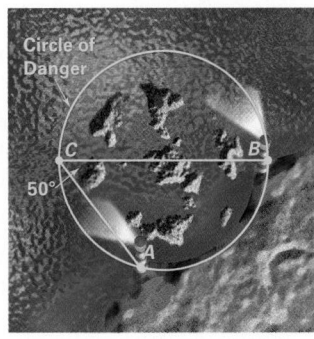

Practice and Apply

🔲 **internet**connect

Homework Help Online
Go To: go.hrw.com
Keyword:
MG1 Homework Help
for Exercises 10-26

In ⊙X, m$\widehat{WT}$ = 36°, m$\widehat{TU}$ = 148°, and m$\widehat{UY}$ = 70°. $\overleftrightarrow{VY}$ is tangent to ⊙X at Y. Find each of the following:

10. m∠SYV 127°

11. m∠VSY 17°

12. m∠SVY 36°

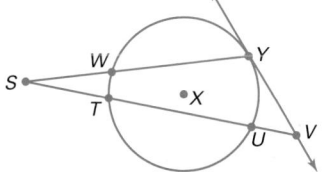

Assess

Selected Answers
Exercises 5–9, 11–63 odd

ASSIGNMENT GUIDE

In Class	1–9
Core	10–16, 17–25 odd, 27–30
Core Plus	31–39 odd, 40–56
Review	57–63
Preview	64–69

✎ Extra Practice can be found beginning on page 818.

ALGEBRA In Exercises 17–20, students set up and solve an equation to find the degree measure of each arc or angle. In Exercise 38, students must justify the statement algebraically. Exercises 64–69 involve conic sections.

28. a. 125°
  b. 30°
  c. 95°
  d. 100°
  e. 20°
  f. 80°
  g. 65°
  h. 20°
  i. 45°
  j. 35°
  k. 15°
  l. 20°
  m. $\frac{x_1{}^\circ}{2}$; m∠1 = $\frac{1}{2}$m$\overset{\frown}{BC}$
  n. $\frac{x_2{}^\circ}{2}$; m∠2 = $\frac{1}{2}$m$\overset{\frown}{AC}$
  o. $\frac{1}{2}(x_1{}^\circ - x_2{}^\circ)$;
  m∠AVC = m∠1 − m∠2

29. m∠AVC = $\frac{1}{2}\left(\text{m}\overset{\frown}{BC} - \text{m}\overset{\frown}{AC}\right)$

In the figure, $\overline{AB}$ and $\overline{CD}$ are chords, m$\overset{\frown}{CB}$ = 60°, and m$\overset{\frown}{AD}$ = 110°. Find each of the following:

13. m∠1 85° **14.** m∠2 95°

15. m∠3 85° **16.** m∠4 95°

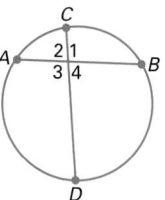

Algebra

In the figure, $\overset{\leftrightarrow}{VA}$ is tangent to ⊙P at V.

17. If m$\overset{\frown}{VC}$ = 150°, find m∠AVC. 75°

18. If m∠AVC = 80°, find m$\overset{\frown}{VC}$. 160°

19. If m$\overset{\frown}{VC}$ = 2x + 4, find m∠AVC. (x + 2)°

20. If m∠AVC = 3x − 1, find m$\overset{\frown}{VC}$. (6x − 2)°

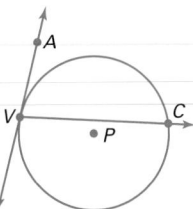

Refer to ⊙O for Exercises 21–26. $\overset{\leftrightarrow}{AF}$ is tangent to ⊙O at A, m$\overset{\frown}{CD}$ = 105°, m$\overset{\frown}{BC}$ = 47°, and m$\overset{\frown}{AB}$ = m$\overset{\frown}{DC}$. Find the following:

21. m$\overset{\frown}{AB}$ 105° **22.** m$\overset{\frown}{AD}$ 103°

23. m∠AED 28° **24.** m∠CAF 104°

25. m∠CQD 105° **26.** m∠BQC 75°

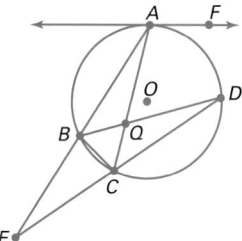

In Exercises 27–37, you will explore cases 3a and 3c from page 588.

Case 3c: A secant and a tangent intersect outside the circle.

Refer to ⊙P for Exercises 27–30. $\overset{\leftrightarrow}{VC}$ is tangent to ⊙P at C.

27. What is the relationship between the measure of ∠1 and the measures of ∠2 and ∠AVC?
m∠1 = m∠2 + m∠AVC

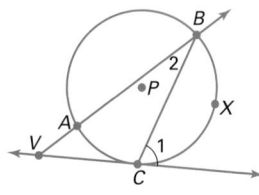

TABLE PROOF

28. Copy and complete the table. For each entry in the last row, give a reason.

m$\overset{\frown}{BXC}$	m$\overset{\frown}{AC}$	m∠1	m∠2	m∠AVC
250°	60°	**a.** ?	**b.** ?	**c.** ?
200°	40°	**d.** ?	**e.** ?	**f.** ?
130°	40°	**g.** ?	**h.** ?	**i.** ?
70°	30°	**j.** ?	**k.** ?	**l.** ?
$x_1{}^\circ$	$x_2{}^\circ$	**m.** ?	**n.** ?	**o.** ?

29. Write an equation that describes m∠AVC in terms of m$\overset{\frown}{BC}$ and m$\overset{\frown}{AC}$.

30. Complete the following theorem:

Theorem

The measure of a secant-tangent angle with its vertex outside the circle is _____?_____.

9.4.4

Case 3a: Two tangents intersect outside the circle.

Refer to ⊙*M* for Exercises 31–37. $\overleftrightarrow{VA}$ and $\overleftrightarrow{VC}$ are tangent to ⊙*M* at *A* and *C*, respectively.

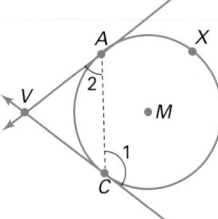

31. What is the relationship between the measure of ∠1 and the measures of ∠2 and ∠*AVC*? m∠1 = m∠*AVC* + m∠2

32. The measure of ∠1 is half the measure of its intercepted arc. Name the arc. $\overparen{AXC}$

33. The measure of ∠2 is half the measure of its intercepted arc. Name the arc. $\overparen{AC}$

34. Given m$\overparen{AXC}$ = 260°, find m$\overparen{AC}$. 100°

> **TABLE PROOF**

35. Copy and complete the table. For each entry in the last row, give a reason.

m$\overparen{AXC}$	m$\overparen{AC}$	m∠1	m∠2	m∠*AVC*
300°	a. ?	b. ?	c. ?	d. ?
250°	e. ?	f. ?	g. ?	h. ?
220°	i. ?	j. ?	k. ?	l. ?
200°	m. ?	n. ?	o. ?	p. ?
x°	q. ?	r. ?	s. ?	t. ?

> *As you justify each entry of the table for the general case, you are also proving a theorem.*

36. Write an equation that describes m∠*AVC* in terms of m$\overparen{AXC}$ and m$\overparen{AC}$.

37. Complete the following theorem:

Theorem

The measure of a tangent-tangent angle with its vertex outside the circle is _____?_____.

9.4.5

> one-half the difference of the measures of the intercepted arcs.

38. Justify the following statement algebraically by using the diagram of ⊙*M* above:

$$\frac{m\overparen{AXC} - m\overparen{AC}}{2} = m\overparen{AXC} - 180°$$

(Hint: m$\overparen{AXC}$ + m$\overparen{AC}$ = ___?___)

> **CHALLENGE**

39. Use the statement in Exercise 38 to write an alternative version of Theorem 9.4.5.

38.
$$\frac{m\overparen{AXC} - m\overparen{AC}}{2}$$

$$= \frac{m\overparen{AXC} + m\overparen{AXC} - m\overparen{AXC} - m\overparen{AC}}{2}$$

$$= \frac{2 \cdot m\overparen{AXC} - \left(m\overparen{AXC} + m\overparen{AC}\right)}{2}$$

$$= \frac{2 \cdot m\overparen{AXC} - 360°}{2}$$

$$= m\overparen{AXC} - 180°$$

Technology

The theorems in Exercises 28–37 can be explored by using geometry graphics software.

Theorem 9.4.4
One-half the difference of the measures of the intercepted arcs.

Error Analysis

In Exercises 30 and 37, students must complete two theorems. Check the students' statements for form and content, or allow groups of students to check each other's statements.

35. a. 60°
  b. 150°
  c. 30°
  d. 120°
  e. 110°
  f. 125°
  g. 55°
  h. 70°
  i. 140°
  j. 110°
  k. 70°
  l. 40°
  m. 160°
  n. 100°
  o. 80°
  p. 20°
  q. 360° − *x*°;
    m$\overparen{AC}$ = 360 − m$\overparen{AXC}$
  r. $\frac{x°}{2}$; m∠1 = $\frac{1}{2}$m$\overparen{AXC}$
  s. 180° − $\frac{x°}{2}$;
    m∠2 = $\frac{1}{2}$m$\overparen{AC}$
  t. *x*° − 180°;
    m∠*AVC* = m∠1 − m∠2

36. m∠*AVC* = $\frac{1}{2}\left(m\overparen{AXC} - m\overparen{AC}\right)$

39. The measure of a tangent-tangent angle with its vertex outside the circle is the measure of the major arc minus 180°.

In Exercises 27–37, you completed the investigations of angle-arc relationships in circles. In Exercises 40–55, you will summarize the angles studied. Complete the table below.

SUMMARY OF ANGLE-ARC RELATIONSHIPS

Diagram	Location of vertex	Sides of angle	Formula for m∠AVC
V is the center of the circle.	center of circle	central angle: sides formed by 2 radii	m∠AVC = m⌢AC
	40. ___?___	41. ___?___	42. ___?___
	43. ___?___	inscribed angle: sides formed by 2 secants or chords	44. ___?___
$\overleftrightarrow{VC}$ tangent at V	45. ___?___	46. ___?___	47. ___?___
	48. ___?___	exterior of circle	sides formed by 2 tangents	49. ___?___
	50. ___?___	51. ___?___	52. ___?___
$\overleftrightarrow{VC}$ tangent at C	53. ___?___	54. ___?___	55. ___?___

40. interior of circle

41. two secants or chords

42. $\frac{1}{2}\left(m\widehat{AC} + m\widehat{DB}\right)$

43. on the circle

44. $\frac{1}{2}m\widehat{AC}$

45. on the circle

46. secant and tangent

47. $\frac{1}{2}m\widehat{AV}$

48.

49. $\frac{1}{2}\left(m\widehat{AXC} - m\widehat{AC}\right)$
$= m\widehat{AXC} - 180°$

50. exterior of circle

51. two secants

52. $\frac{1}{2}\left(m\widehat{BD} - m\widehat{AC}\right)$

53. exterior of circle

54. secant and tangent

55. $\frac{1}{2}\left(m\widehat{BC} - m\widehat{AC}\right)$

NAME _____ CLASS _____ DATE _____

Practice
9.4 Angles Formed by Secants and Tangents

In ⊙L, m⌢IK = 15°, m⌢KM = 180°, and m⌢HM = 50°. $\overleftrightarrow{NH}$ is tangent to ⊙L at H. Find each of the following:

1. m∠JHN _____ 122.5° _____

2. m∠NJH _____ 17.5° _____

3. m∠JNH _____ 40° _____

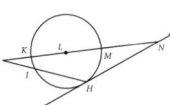

In the figure, $\overrightarrow{QR}$ is tangent to ⊙S at Q.

4. If m⌢QP = 105°, find m∠RQP.
_____ 127.5° _____

5. If m∠RQP = 110°, find m⌢QP.
_____ 140° _____

6. If m∠RQP = 90°, find m⌢QP.
_____ 180° _____

7. If m⌢QP = x + 35, find m∠RQP.
_____ 180° − 0.5(x + 35)° _____

8. If m⌢QP = 2x − 17, find m∠RQP.
_____ 180° − 0.5(2x − 17)° _____

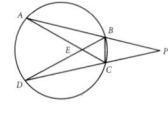

In the figure at right, $\overline{PA}$ and $\overline{PD}$ are secants to the circle, chords $\overline{AC}$ and $\overline{BD}$ intersect at E, $\overline{BA} \cong \overline{CD}$, m⌢BC = 40°, and m∠ABD = 60°. Find the following:

9. m⌢AB _____ 100° _____

10. m∠ACD _____ 60° _____

11. m∠AEB _____ 100° _____

12. m∠BDP _____ 20° _____

13. m∠P _____ 40° _____

56. COMMUNICATIONS The maximum distance that a radio signal can reach directly is the length of the segment tangent to the curve of Earth's surface. If the angles formed by the tangent radio signals and the tower are 89.5°, what is the measure of the intercepted arc on Earth? Given that the radius of Earth is approximately 4000 mi, estimate how far the signal can reach. **1°, approx. 69.8 mi**

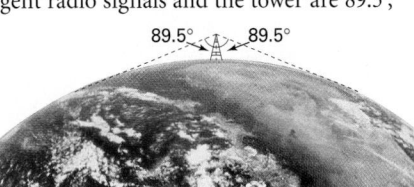

89.5°   89.5°

Look Back

Find the indicated angle measures. *(LESSON 3.6)*

57.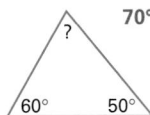
70°
?
60°   50°

58.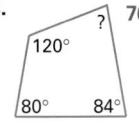
76°
?
120°
80°   84°

59.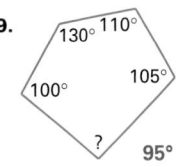
130°   110°
100°   105°
?   95°

≈ 96,566,924 ft³

60. ARCHAEOLOGY The Great Pyramid of Giza in Egypt, built by Khufu (*Cheops* in Greek), is considered by some to be the greatest structure ever built. Its roughly square base measures an average of $775\frac{3}{4}$ ft on a side, and its original height, before erosion, was $481\frac{2}{5}$ ft. Find the volume of the original structure. [*Source: Encyclopedia Britannica*] *(LESSON 7.3)*

DESIGN A photocopier was used to reduce △ABC by a factor of 0.75.

Yes; everything was reduced by a common factor.

61. Are △ABC and △A′B′C′ similar? Explain your reasoning. *(LESSON 8.2)*

62. Find the following ratios: $\frac{A'B'}{AB}$, $\frac{A'C'}{AC}$, and $\frac{C'B'}{CB}$. *(LESSON 8.6)* $\frac{3}{4}, \frac{3}{4}, \frac{3}{4}$

63. Find A′C′. *(LESSON 8.6)* **9**

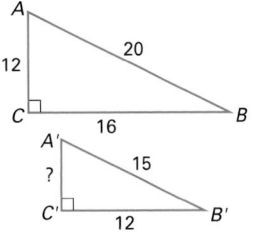

A
12   20
C   16   B

A′
?   15
C′   12   B′

Look Beyond

A *conic section* is the intersection of a right double cone and a cutting plane. In the diagram below, the intersection of the plane and the cone forms a circle. Other geometric figures are formed when the angle at which the plane cuts the double cone is adjusted.

Describe how the cutting plane should be adjusted in order to produce each conic section listed below. You may use a sketch or a verbal description.

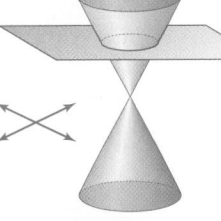

64. ellipse

65. parabola

66. one point ·

67. two intersecting lines

68. hyperbola

69. one line

Look Beyond

In Exercises 64–69, students consider how a cutting plane could be adjusted to produce six different conic sections.

64. Tilt the plane so that it is no longer perpendicular to the axis of the cone, but so that it still intersects both 'sides' of the top or the bottom of the double cone.

65. Tilt the plane so that it only intersects one side of either the top or the bottom of the double cone, but not through the vertex.

66. The plane should be perpendicular to the axis and intersect the double cone at the vertex.

67. The plane should contain the axis of the double cone.

68. The plane should be parallel to the axis of the double cone, but not contain the axis.

69. Tilt the plane as in Exercise 65, then move it up or down until it intersects the vertex of the double cone.

Student Technology Guide

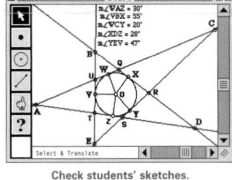

Focus

The 1994 earthquake that struck Los Angeles serves as a case study for developing an equation that relates seismic wave data to the distance from the epicenter. Students use the geometry of circles to plot the location of the epicenter.

Motivate

Discuss the Los Angeles earthquake. Students may have read or heard about it, or some students may have experienced an earthquake themselves. Discuss with students the following information about seismic waves:

- They are vibrations of the earth.
- The two types referred to in this feature are called S waves and P waves.
- S waves are transverse; that is, the vibrations are perpendicular to the path of the wave, like ocean waves.
- P waves are longitudinal; that is, the vibrations are along the path of the wave, like sound waves.
- Seismographs are instruments that record seismic waves; the display that is produced is called a seismogram.

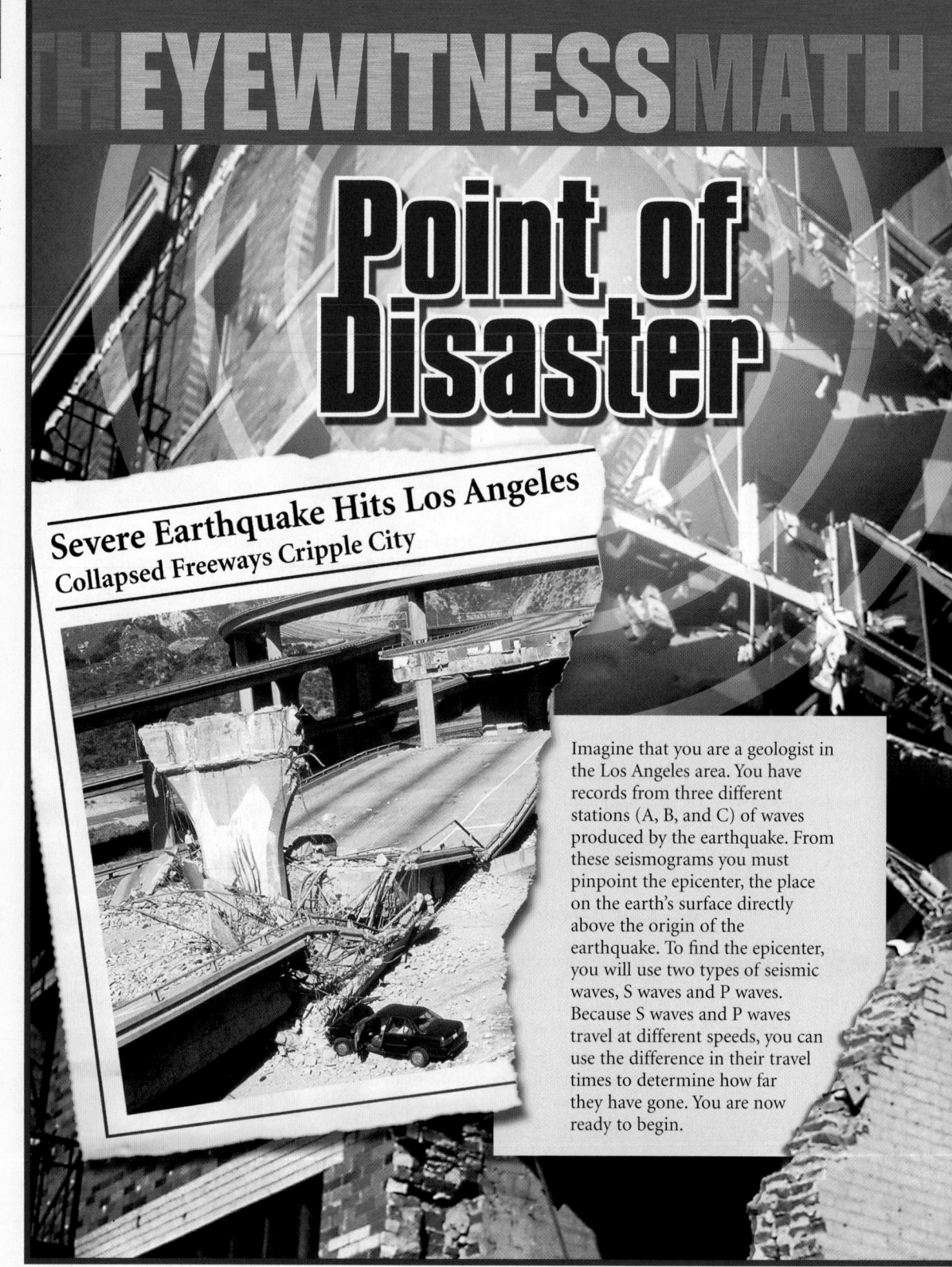

EYEWITNESSMATH

Point of Disaster

Severe Earthquake Hits Los Angeles
Collapsed Freeways Cripple City

Imagine that you are a geologist in the Los Angeles area. You have records from three different stations (A, B, and C) of waves produced by the earthquake. From these seismograms you must pinpoint the epicenter, the place on the earth's surface directly above the origin of the earthquake. To find the epicenter, you will use two types of seismic waves, S waves and P waves. Because S waves and P waves travel at different speeds, you can use the difference in their travel times to determine how far they have gone. You are now ready to begin.

WAVE SPEED VALUES

Speed of P wave
$v_p = 6$ km/s

Speed of S wave
$v_s = 3.5$ km/s

STATION DATA
D_t **(seconds)**

Station A	4.4
Station B	5.5
Station C	6.3

This seismogram shows S waves and P waves.

Cooperative Learning

1. First you will examine how the time difference between the arrival of the S and P waves depends on the distance from the station to the epicenter of the earthquake. Let d represent the distance from the station to the epicenter.

a. S waves travel at 3.5 kilometers per second (km/s). Write an equation in terms of d for t_S, the time (in seconds) for the S waves to travel to the station.

$$t_S = \underline{\quad ? \quad}$$

b. P waves travel at 6 km/s. Write an equation in terms of d for t_P, the time (in seconds) for the P waves to travel to the station.

$$t_P = \underline{\quad ? \quad}$$

c. Use the equations from parts **a** and **b** above to write an equation in terms of d for D, the difference in time between the arrival of the S waves and P waves.

$$D = t_S - t_P = \underline{\quad ? \quad}$$

d. Solve your equation from part **c** for d to find an equation in terms of D for the distance from the station to the epicenter.

$$d = \underline{\quad ? \quad}$$

2. Use your equation from part **d** above to find the distances in kilometers from each station to the epicenter.

Station	D	d
A	4.4	?
B	5.5	?
C	6.3	?

3. Make a scale drawing of the locations of the three stations, with 1 km = 1 cm. Draw three circles centered at the three stations, using the distances you calculated above as the radii of the circles. The epicenter is in the region where the three circles overlap.

For more precision, draw chords connecting the intersections of each pair of circles. The epicenter is the intersection of these chords. Label the epicenter on your drawing.

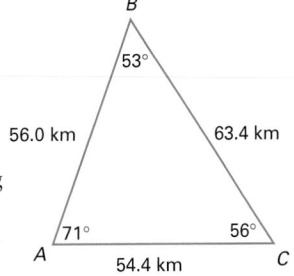

4. All of the seismograph readings used to locate the epicenter of the 1994 Los Angeles earthquake came from stations that were southeast of the epicenter. How would that make finding the location of the epicenter more difficult?

Discuss

What are the advantages of having a global network of seismograph stations? What is the minimum number of stations needed to locate the epicenter of a quake? Would more stations make locating earthquake epicenters more precise? Why or why not?

Segments of Tangents, Secants, and Chords

United States

The concentration of cenotes is greatest along a semicircular arc.

Chichén Itzá cenote

Chichén Itzá

YUCATAN

Mexico

Belize

Guatemala

Honduras

El Salvador

Nicaragua

Temple of Kukulkan at Chichén Itzá

Objectives

- Define special cases of segments related to circles, including secant-secant, secant-tangent, and chord-chord segments.

- Develop and use theorems about measures of the segments.

Why A theorem in this lesson will extend your list of techniques for finding the center of circular objects.

APPLICATION
GEOLOGY

A semicircular arc of natural wells (cenotes) like the one at Chichén Itzá stretches across the Yucatán peninsula. Some scientists believe that the cenote arc is evidence of an impact by an asteroid or comet, perhaps explaining the extinction of the dinosaurs (see page 353). The center of the arc would be the point where the asteroid or comet actually struck Earth.

Exploring Segment Relationships in Circles

In the previous lesson, you investigated special angles and arcs formed by secants and tangents of circles. As you will see, segments formed by secants and tangents of circles also have interesting relationships. The terms in the list below are used to classify such segments.

In the illustration, $\overleftrightarrow{XA}$ is a tangent line and $\overleftrightarrow{XB}$ is a secant line.

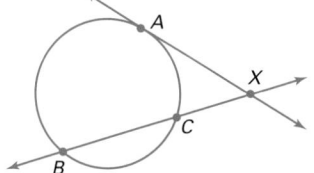

$\overline{XA}$ is a **tangent segment**.

$\overline{XB}$ is a **secant segment**.

$\overline{XC}$ is an **external secant segment**.

$\overline{BC}$ is a **chord**.

Segments Formed by Tangents

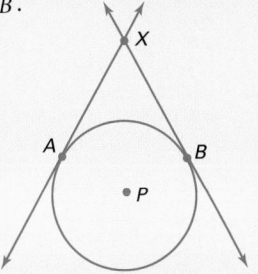

1. Construct ⊙P with tangent lines $\overleftrightarrow{XA}$ and $\overleftrightarrow{XB}$.

2. Measure the lengths of $\overline{XA}$ and $\overline{XB}$.

3. Make a conjecture about the lengths of two segments that are tangent to a circle from the same external point.

4. Add segments $\overline{AP}$, $\overline{BP}$, and $\overline{XP}$ to your figure. Write a paragraph proof of your conjecture.

5. Complete the theorem below.

Theorem

If two segments are tangent to a circle from the same external point, then the segments _____?_____.

9.5.1

Segments Formed by Secants

Part I

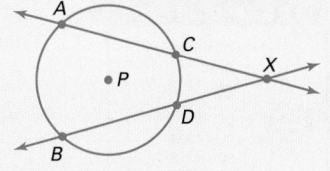

1. Construct ⊙P with secant lines $\overleftrightarrow{XA}$ and $\overleftrightarrow{XB}$.

2. Construct $\overline{AD}$ and $\overline{BC}$ and label their intersection point O.

3. Name the two large triangles in your figure that have a vertex at X. What can you conclude about them? (Hint: Use Theorem 9.3.3.) Complete the following similarity statement:

$$\triangle AXD \sim \underline{\quad ? \quad}$$

4. Complete the proportion below. Cross multiply and state your result.

$$\frac{AX}{?} = \frac{XD}{?}$$

> *In the process of discovering your result, you are proving a theorem.*

5. Present your discovery as a paragraph proof. Complete the theorem below.

Theorem

If two secants intersect outside a circle, the product of the lengths of one secant segment and its external segment equals _____?_____.

(Whole × Outside = Whole × Outside)

9.5.2

Interdisciplinary Connection

EARTH SCIENCE Craters such as the one shown at the beginning of this lesson are formed by waves spreading out from a center point. Have students share their knowledge of wave motions from science courses.

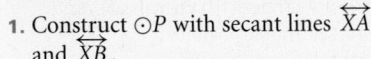 **Notes**

Make students aware that geometry graphics software can be used to measure and compare the lengths of two segments. However, this does not constitute a formal proof that the two segments are equal.

For a student worksheet of this Activity and detailed Teacher Notes, see page 171 in the Lesson Activities booklet.

CHECKPOINT ✔

5. If two segments are tangent to a circle from the same external point, then the segments are congruent.

Activity 2 **Notes**

In this Activity, students look for four segments so that they can write a relationship involving the products of the lengths of the segments. That is, if the lengths of the segments are a, b, c, and d, students should be able to establish equations such as $ab = cd$ or $ac = bd$.

For a student worksheet of this Activity and detailed Teacher Notes, see page 171 in the Lesson Activities booklet.

☞ For the answer to the Checkpoint, see page 602.

Math
CONNECTION

ALGEBRA You may want to review solving proportions for Step 4 of this Activity.

CHECKPOINT ✔

5. If two secants intersect outside a circle, the product of the lengths of one secant segment and its external segment equals the product of the lengths of the other secant segment and its external segment.

CHECKPOINT ✔

3. If a secant and a tangent intersect outside a circle, then the product of the lengths of one secant segment and its external segment equals the length of the tangent segment squared.

Activity 3 **Notes**

As in Activity 2, students look for four segments so that they can write a relationship involving products of lengths of segments. Students using geometry graphics software can drag point *X* to change the shapes of the triangles in the diagram.

For a student worksheet of this Activity and detailed Teacher Notes, see page 171 in the Lesson Activities booklet.

Math
C O N N E C T I O N

ALGEBRA In Step 5 of Activity 3, students should complete the proportion by using the segments in the circle.

CHECKPOINT ✔

6. If two chords intersect inside a circle, then the product of the lengths of the segments of one chord equals the product of the lengths of the segments of the other chord.

Part II

1. Imagine moving point *B* along the circle so that $\overleftrightarrow{XB}$ becomes a tangent line. Points *B* and *D* will coincide at the point of tangency. What is the relationship between *XB* and *XD*?

2. Substitute *XB* for *XD* in your result from Step 5 of Part I.

3. Based on your result, complete the theorem below, which you will be asked to prove in Exercise 38.

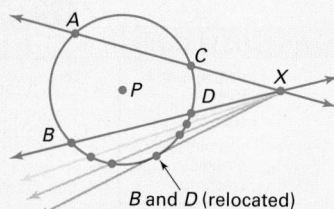

B and *D* (relocated)

CHECKPOINT ✔

Theorem

If a secant and a tangent intersect outside a circle, then the product of the lengths of the secant segment and its external segment equals _____?_____ .

(Whole × Outside = Tangent Squared)   **9.5.3**

Activity 3

Segments Formed by Interesting Chords

> **YOU WILL NEED**
> compass and ruler
> **OR**
> geometry graphics software

Algebra

PARAGRAPH PROOF

1. Construct a circle with chords $\overline{AC}$ and $\overline{DB}$ intersecting at point *X*.

2. Draw $\overline{AD}$ and $\overline{BC}$. Name the two triangles formed.

3. Name two angles of the triangles that intercept the same arc of the circle. What can you conclude about these angles?

4. What other angles of the triangles can you show to be congruent? What can you conclude about the triangles?

5. Complete the proportion below by relating two sides of one triangle to two sides of the other triangle. Cross multiply and state your result.
$$\frac{DX}{XA} = \frac{?}{?}$$

> *In the process of discovering your result, you are proving a theorem.*

6. Present your discovery as a paragraph proof. Complete the theorem below.

CHECKPOINT ✔

Theorem

If two chords intersect inside a circle, then the product of the lengths of the segments of one chord equals _____?_____ .   **9.5.4**

Enrichment

Have students construct a tangent to a circle from a point outside the circle. The resulting figure will resemble the diagram for Activity 1. Possible method: Given circle *O* and point *A* outside the circle, draw $\overline{AO}$. Bisect $\overline{AO}$ to get its midpoint, *M*. Draw circle *M* with radius $\overline{MO}$. This circle intersects circle *O* at points *B* and *C*. $\overleftrightarrow{AB}$ and $\overleftrightarrow{AC}$ are both tangent to circle *O*.

Applying the Segment Theorems

EXAMPLE ① Global positioning satellites are used in navigation. If the range of the satellite, *AX*, is 16,000 miles, what is *BX*?

APPLICATION
NAVIGATION

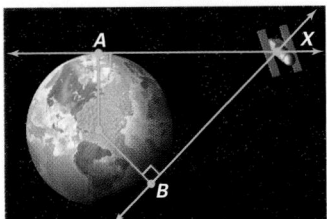

● **SOLUTION**

$\overleftrightarrow{AX}$ and $\overleftrightarrow{BX}$ are tangents to a circle from the same external point. By Theorem 9.5.1, they are equal.

$$AX = BX = 16,000 \text{ miles}$$

EXAMPLE ② In the figure, *EX* = 1.31, *GX* = 0.45, and *FX* = 1.46. Find *HX*. Round your answer to the nearest hundredth.

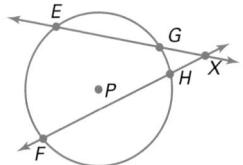

● **SOLUTION**

$\overleftrightarrow{EX}$ and $\overleftrightarrow{FX}$ are secants that intersect outside the circle.

By Theorem 9.5.2, Whole × Outside = Whole × Outside.

$$EX \cdot GX = FX \cdot HX$$
$$1.31 \cdot 0.45 = 1.46 \cdot HX$$
$$1.46 \cdot HX = 0.5895$$
$$HX \approx 0.40$$

EXAMPLE ③ An eagle is released from captivity at a ranger station near an approximately circular lake, and it builds a nest in a tree on the opposite shore. The distance from the station to the lake is 300 yd along a road tangent to the lake, and 50 yd along a line straight to the nest. How far is the nest from the ranger station?

APPLICATION
WILDLIFE MANAGEMENT

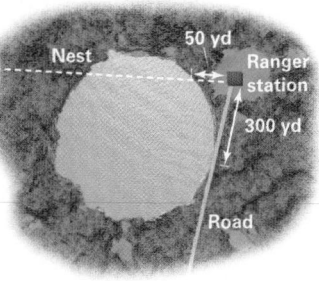

● **SOLUTION**

The road is a tangent and the line to the nest is a secant. They intersect outside the circle. Let *d* be the distance from the ranger station to the nest.

By Theorem 9.5.3, Whole × Outside = Tangent Squared.

$$d \times 50 = 300^2$$
$$d \times 50 = 90,000$$
$$d = 1800 \text{ yd} \approx 1 \text{ mi} \quad (1 \text{ mi} = 1760 \text{ yd})$$

Inclusion Strategies

ENGLISH LANGUAGE DEVELOPMENT It is particularly important that students understand the vocabulary given on the first page of the lesson. If the vocabulary continues to present difficulties, have students use different colors to sketch and name particular segments in the diagrams.

ADDITIONAL
EXAMPLE ①

Find *BX*. 15 cm

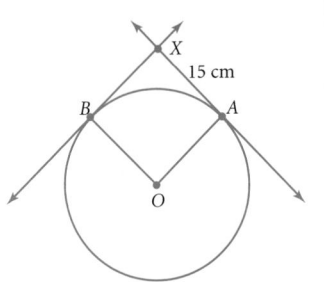

ADDITIONAL
EXAMPLE ②

In circle *O*, *AB* = 13, *BX* = 6, and *CD* = 11. Find *DX*. *DX* ≈ 6.51

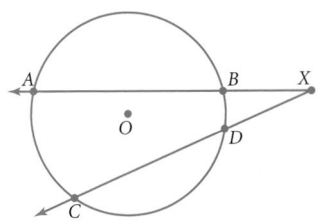

ADDITIONAL
EXAMPLE ③

In circle *O*, *BX* = 24 and *CX* = 30. Find *AB*.
AB = 37.5

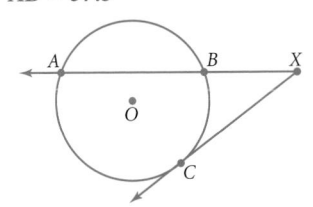

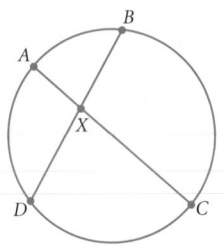
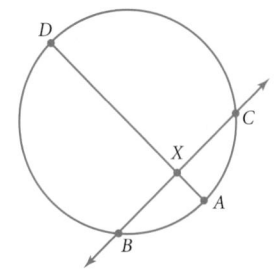
EXAMPLE ④

In the figure, *AX* = 0.26, *XC* = 0.91, and *DX* = 0.27. Find *XB*.

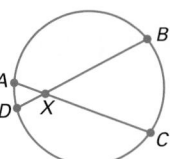

SOLUTION

$\overline{AX}$, $\overline{XC}$, $\overline{DX}$, and $\overline{XB}$ are chords that intersect inside a circle. By Theorem 9.5.4:

$$AX \cdot XC = DX \cdot XB$$
$$0.26 \cdot 0.91 = 0.27 \cdot XB$$
$$0.27 \cdot XB = 0.2366$$
$$XB \approx 0.88$$

EXAMPLE ⑤

APPLICATION
GEOLOGY

The map shows the Yucatán peninsula, where *A* and *B* are points on the cenote ring. Assuming that the cenote ring defines the outer edge of an impact crater of an asteroid or comet, where did the object hit the Earth? Use a ruler and the map with the given scale to find the point of impact.

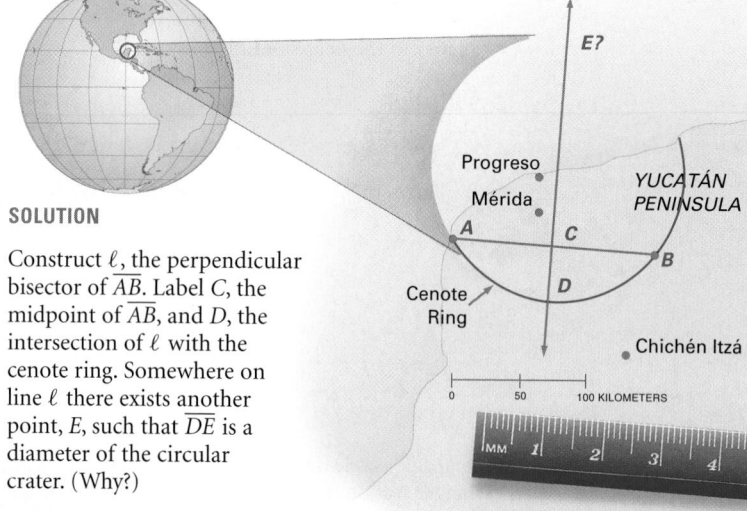

SOLUTION

Construct ℓ, the perpendicular bisector of $\overline{AB}$. Label *C*, the midpoint of $\overline{AB}$, and *D*, the intersection of ℓ with the cenote ring. Somewhere on line ℓ there exists another point, *E*, such that $\overline{DE}$ is a diameter of the circular crater. (Why?)

The diagram below shows the position of points *A*, *B*, and *D* on the circle. By measuring the map, *AC* = *BC* = 1.7 cm, and *CD* = 0.9 cm.

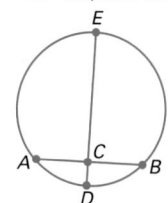

By Theorem 9.5.4:
$$CD \cdot CE = AC \cdot BC$$
$$0.9 \cdot CE = 1.7 \cdot 1.7$$
$$CE \approx 3.2 \text{ cm}$$

$\overline{DE}$ is a diameter of the circle, and $\overline{DE} = 0.9 + 3.2 = 4.1$ cm.

For the given map scale, the actual diameter is about 180 km. Using a radius of 2.05 cm for the cenote ring, you can locate the center point on ℓ. This point is 0.35 cm southeast of Progreso on the map. Using the given map scale, the center of the possible impact crater is a little over 15 km southeast of Progreso. This would be where the comet or asteroid actually hit the Earth.

Reteaching the Lesson

TECHNOLOGY The key theorem developed in Activity 3 can be demonstrated by using geometry graphics software. You can either present this demonstration to the class or assign the construction to small groups of students. Construct a circle and two intersecting chords with chord segment lengths of *a*, *b*, *c*, and *d*. Then create two rectangles, one with sides *a* and *b*, and the other with sides *c* and *d*, corresponding to the lengths of the four chord segments. Confirm the cross-product relationship by showing that the areas of the rectangles are equal.

Exercises

Communicate

For Exercises 1 and 2, explain the meaning of the statement and state which theorem is related to it.

1. Whole × Outside = Whole × Outside

2. Whole × Outside = Tangent Squared

3. Suppose that two secants or chords intersect on a circle. Can you determine anything about the lengths of the segments formed? Why or why not?

4. Suppose that a tangent and a secant intersect on a circle. Can you determine anything about the length of the segment formed? Why or why not?

5. List as many methods as you can for finding the center of a circle. How much of the circle must be known for each method?

Guided Skills Practice

6. $\overleftrightarrow{AX}$ and $\overleftrightarrow{BX}$ are tangent to $\odot P$ and $AX = 7$. Find BX.
(THEOREM 9.5.1 AND EXAMPLE 1)
7

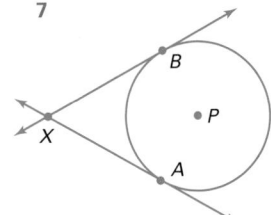

7. In $\odot Q$, $AX = 16$, $CX = 9$, and $BX = 18$. Find DX.
(THEOREM 9.5.2 AND EXAMPLE 2)
8

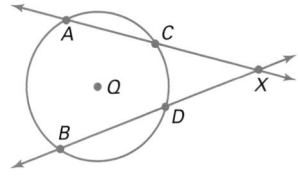

8. $\overleftrightarrow{BX}$ is tangent to $\odot R$ at B, $AX = 8$, and $CX = 2$. Find BX.
(THEOREM 9.5.3 AND EXAMPLE 3)
4

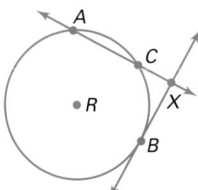

9. In the diagram, $AX = 6$, $BX = 3$, and $DX = 8$. Find CX.
(THEOREM 9.5.4 AND EXAMPLE 4)
4

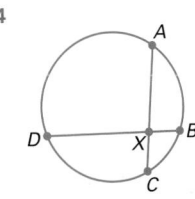

Assess

Selected Answers

Exercises 6–9, 11–49 odd

ASSIGNMENT GUIDE

In Class	1–10
Core	11–28
Core Plus	17–29 odd, 30–40
Review	41–49
Preview	50

✐ Extra Practice can be found beginning on page 818.

Error Analysis

Students may benefit from sketching the figures and color coding the segments.

ALGEBRA In Exercises 16–29, students set up and solve proportions to find segment lengths.

10. In the diagram at right, $\overleftrightarrow{FG}$ is the perpendicular bisector of $\overline{DE}$, $FG = 12$, and $DE = 48$. Find the diameter of the circle that contains points D, E, and G. *(EXAMPLE 5)* **60**

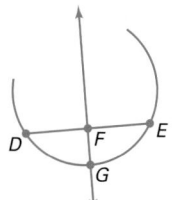

Practice and Apply

$\overleftrightarrow{AB}$ is tangent to $\odot R$ at C. Identify each of the following:

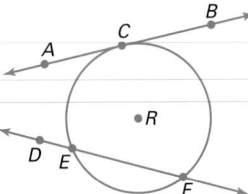

11. a tangent segment $\overline{AC}$ or $\overline{BC}$

12. a secant segment $\overline{EF}$

13. an external secant segment $\overline{DE}$

Identify each of the following in the figure at right:

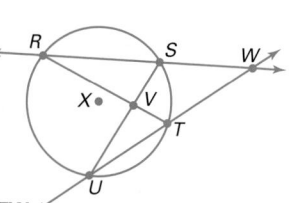

14. two pairs of congruent angles

15. a pair of similar triangles

14. $\angle SRT$ and $\angle SUT$; $\angle RVS$ and $\angle TVU$

15. $\triangle SVR$ and $\triangle TVU$ or $\triangle USW$ and $\triangle RTW$

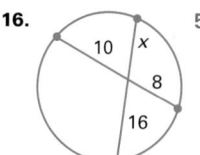

Find x in each circle below.

16.

10 x 5
8
16

17.

6 4 6
9 x

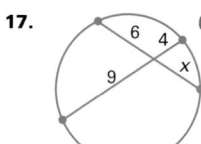

internet connect

Homework Help Online
Go To: **go.hrw.com**
Keyword:
MG1 Homework Help
for Exercises 18-21

In the figure below, $\overleftrightarrow{VA}$ and $\overleftrightarrow{VB}$ are tangent to $\odot P$, the radius of $\odot P$ is 3 cm, and $VA = 6$ cm. Find the following:

18. VB **6 cm**   **19.** AP **3 cm**

20. $PV \approx$ **6.71 cm** **21.** $XV \approx$ **3.71 cm**

22. Name an angle congruent to $\angle AVP$.

23. Name an angle congruent to $\angle APV$.

24. Name an arc congruent to $\overparen{AX}$.

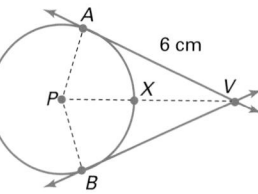
6 cm

22. $\angle BVP$

23. $\angle BPV$

24. $\overparen{BX}$

Algebra

Use the figure below for Exercises 25–27. **(Hint: You may need to use the quadratic formula.)**

25. Given $WA = 4$, $WB = 10$, and $WC = 5$, find CD. **3**

26. Given $WB = x$, $WA = 6$, $WD = x + 3$, and $WC = 5$, find x. **15**

27. Given $WB = x$, $WA = x - 16$, $WD = 8$, and $WC = 5$, find x. $\approx$ **18.198**

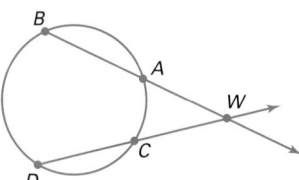

Use the figure at right for Exercises 28 and 29.

28. Given $AE = x$, $BE = x - 2$, $CE = 3$, and $DE = 8$, find x. **6**

29. Given $AB = 10$, $CE = 2$, and $CD = 12$, find AE.
$\dfrac{10 \pm \sqrt{20}}{2} \approx 2.76$ or 7.24

In Exercises 30–37, you will summarize the properties of segments formed by secants, tangents, and chords.

30. exterior of circle
31. 2 secant segments
32. $AV \cdot BV = CV \cdot DV$

33.

34. $VC \cdot VA = (VB)^2$

35. interior of circle
36. 2 chords
37. $AV \cdot VB = CV \cdot VD$

SUMMARY: SECANTS, TANGENTS, AND CHORDS			
Diagram	Location of vertex	Types of segments	Theorem
$\vec{VA}$ tangent at A / $\vec{VC}$ tangent at C	exterior of circle	2 tangent segments	$AV = CV$
	30. ?	31. ?	32. ?
	33. ?	exterior of circle	1 secant segment and 1 tangent segment	34. ?
	35. ?	36. ?	37. ?

38. Prove Theorem 9.5.3 by using the diagram at right. (Hint: Draw segments $\overline{AB}$ and $\overline{BC}$. Then set up a proportion with the quantities in the theorem.)

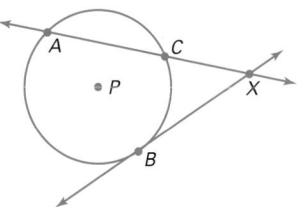

Technology

Geometry graphics software can be used to construct the figures in this exercise set. Students will find calculators helpful for the computations.

38. By the Inscribed Angle Theorem, $m\angle BAC = \frac{1}{2}m\widehat{BC}$. By Theorem 9.4.1, $m\angle CBX = \frac{1}{2}m\widehat{BC}$. Thus by the Transitive Property, $\angle BAC \cong \angle CBX$.

$\angle AXB \cong \angle BXC$ by the Reflexive Property of Congruence. Thus, by the AA Similarity Postulate, $\triangle AXB \sim \triangle BXC$ and so $\frac{AX}{BX} = \frac{BX}{CX}$ or $AX \cdot CX = (BX)^2$.

41. $210\pi \approx 659.73$ cm^2;
$400\pi \approx 1256.64$ cm^3

42. 840 sq cm; 1600 cm^3

43. $5\pi\sqrt{281} + 25\pi \approx 341.85$ cm^2;
$\frac{400\pi}{3} \approx 418.88$ cm^3

44. Opposite sides of a parallelogram are congruent.

45. Definition of a parallelogram.

46. Alternate Interior Angles Theorem

47. Alternate Interior Angles Theorem

48. ASA Postulate

49. CPCTC

39. LUNAR EXPLORATION The diameter of a lunar crater, from measurements made on Earth, is known to be 32 km. How far is it from the lunar lander, located at L, to point X on the far rim of the crater, where a rock sample is to be taken? **45.22 km**

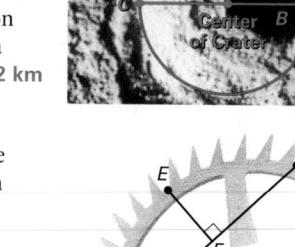
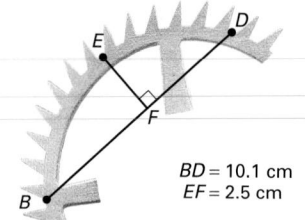

40. ENGINEERING Jeff is restoring a clock and needs a new gear drive to replace the broken one shown at right. To make a new gear, he must determine the diameter of the original drive. In the picture, F is the midpoint of $\overline{BD}$. Use the product of the chord segments to find the diameter of the drive.

$BD = 10.1$ cm $EF = 2.5$ cm
12.7 cm

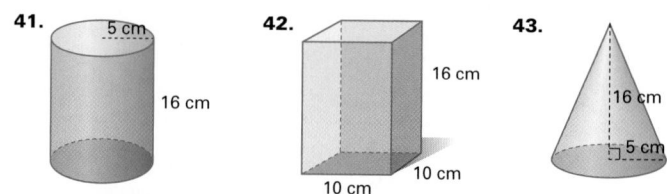

BD = 10.1 cm
EF = 2.5 cm

Look Back

PACKAGE DESIGN A manufacturer is considering the three potential package designs shown below. Draw a net for each one. Then calculate the surface area and volume for each.
(LESSONS 6.1, 7.1, 7.4, AND 7.5)

41. 5 cm 16 cm

42. 16 cm 10 cm 10 cm

43. 16 cm 5 cm

A flowchart proof of Theorem 4.5.2 is shown below. In Exercises 44–49, state the reason for each step of the proof. *(LESSON 4.5)*

Given: parallelogram $ABCD$ with diagonals intersecting at point X

Prove: $AX = CX$ and $DX = BX$

Proof:

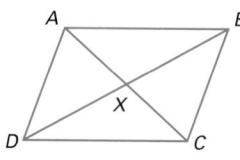

$AD = BC$

44. ?

$\overline{AD} \parallel \overline{BC}$ → $\angle CAD \cong \angle ACB$

45. ? **46.** ?

→ $\triangle AXD \cong \triangle CXB$

48. ?

$\angle ADB \cong \angle CBD$

47. ?

$AX = CX$
$DX = BX$

49. ?

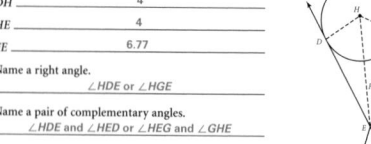
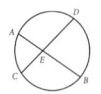
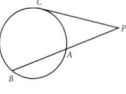

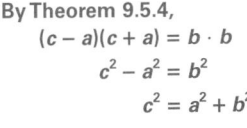 *Algebra*

50. The drawing below suggests a visual proof of the Pythagorean Theorem. Use chord-segment products to explain how the proof works.

By Theorem 9.5.4,

$$(c - a)(c + a) = b \cdot b$$
$$c^2 - a^2 = b^2$$
$$c^2 = a^2 + b^2$$

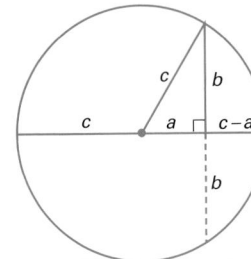

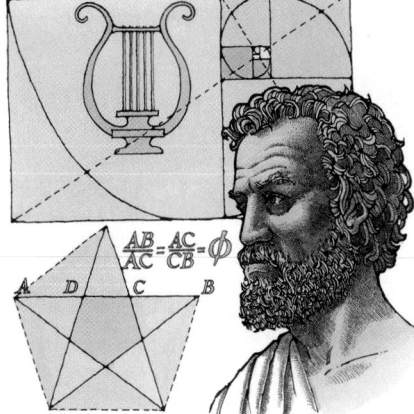

Artist's conception of Pythagoras, with related images. See Lesson 11.1.

 **Look Beyond**

In Exercise 50, students are asked to confirm a visual proof of the Pythagorean Theorem by using products of chord segments.

ALTERNATIVE
Assessment

Portfolio Activity

The Portfolio Activity can be used as preparation for the Chapter Project or as a separate activity. The Portfolio Activity on this page has students construct S-shaped curves.

REVERSE CURVES In the Portfolio Activity for Lesson 9.2, you constructed smooth curves by using arcs of different circles with the same tangent. If the center of one circle is outside the other circle, the direction of the curve is reversed, creating an **S** shape.

1. Using a compass and straightedge or geometry graphics software, draw ⊙P with radius $\overline{PR}$. Construct line ℓ tangent to P at R (that is, a line perpendicular to $\overline{PR}$ at R).

2. Extend $\overline{PR}$ past R, choose a point on $\overline{PR}$ in the exterior of ⊙P, and label it Q. Construct a circle centered at Q with radius $\overline{QR}$. ⊙P and ⊙Q have the same tangent at R.

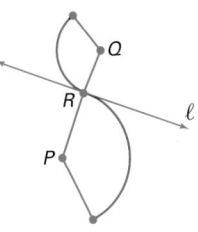

3. Try constructing a "snake" of reverse curves like the one shown at right. The lines of construction have been left in the picture to help you.

WORKING ON THE CHAPTER PROJECT

You should now be able to complete Activity 3 of the Chapter Project.

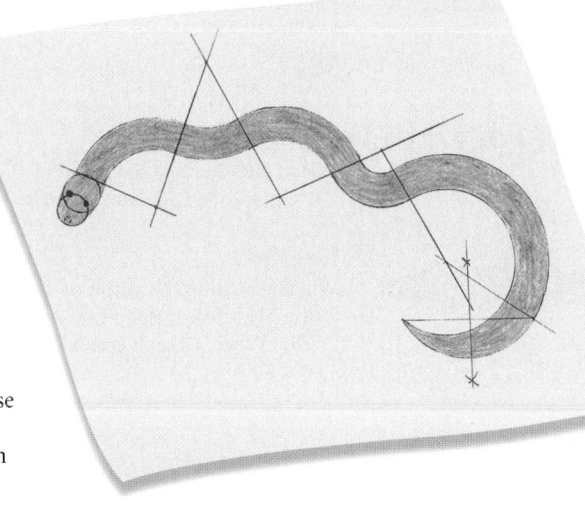

Circles in the Coordinate Plane

Solve for x.

1. $x^2 + 4^2 = 5^2$
$x = \pm 3$

2. $x^2 + 12^2 = 13^2$
$x = \pm 5$

3. $3^2 + x^2 = 6^2$
$x = \pm 3\sqrt{3}$ or $x = \pm 5.2$

4. $y^2 + x^2 = 4^2$
$x = \pm \sqrt{16 - y^2}$, where $16 - y^2 \geq 0$

Also on Quiz Transparency 9.6

Teach

Why Investigating circle relationships with a circle that is graphed on a coordinate plane can help students integrate ideas from geometry with ideas from algebra.

Objectives

● Develop and use the equation of a circle.

● Adjust the equation for a circle to move the center in a coordinate plane.

Why *Computer graphics software can create a variety of geometric objects, such as points, lines, and circles. To do this, the software has subroutines that use algebraic representations of these objects—that is, equations.*

Graphing a Circle From an Equation

In your work in algebra, you may have investigated graphs of equations such as $y = 2x - 3$ (a line), $y = x^2 - 3$ (a parabola), and $y = 3 \cdot 2^x$ (an exponential curve). In this lesson, you will investigate equations in which both x and y are squared.

EXAMPLE ① Given: $x^2 + y^2 = 25$
Sketch and describe the graph by finding ordered pairs that satisfy the equation. Use a graphics calculator to verify your sketch.

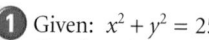

● **SOLUTION**

When sketching the graph of a new type of equation, it is often helpful to locate the intercepts. To find the x-intercept(s), find the value(s) of x when $y = 0$. (When a graph crosses the x-axis, $y = 0$.)

$$x^2 + 0^2 = 25$$
$$x^2 = 25$$
$$x = \pm 5$$

Thus, the graph has two x-intercepts, $(5, 0)$ and $(-5, 0)$.

Alternative Teaching Strategy

TECHNOLOGY Geometry graphics software that includes a coordinate system can help students investigate and understand the equations for circles in the coordinate plane. By dragging the center of the circle to a different position, students can see how the equation changes as the position of the circle changes.

To find the y-intercept(s), find the value(s) of y when $x = 0$.

$$0^2 + y^2 = 25$$
$$y^2 = 25$$
$$y = \pm 5$$

Thus, the graph has two y-intercepts, $(0, 5)$ and $(0, -5)$.

Now set x to some other value, such as 3.

$$3^2 + y^2 = 25$$
$$y^2 = 16$$
$$y = \pm 4$$

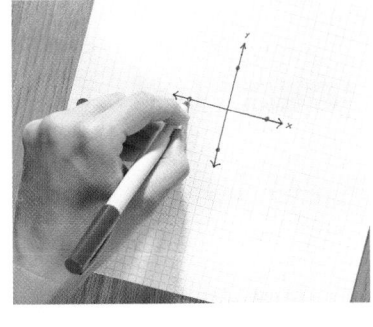

Thus, there are two points with an x-value of 3: $(3, 4)$ and $(3, -4)$.

Similarly, by choosing other convenient values for x, a table like the one below is obtained.

x	y	Points on graph
3	±4	$(3, 4), (3, -4)$
-3	±4	$(-3, 4), (-3, -4)$
4	±3	$(4, 3), (4, -3)$
-4	±3	$(-4, 3), (-4, -3)$

Add these new points to the graph.
The graph of a circle with a radius of 5 and its center at the origin, $(0, 0)$, begins to appear. Sketch the curve.

CRITICAL THINKING How does the graph change if 25 in the equation is changed to 49? to 81? to 51?

Using Graphing Technology

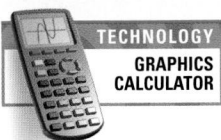

TECHNOLOGY
GRAPHICS CALCULATOR

You can also graph the curve on a graphics calculator or a computer with graphing software. You can use the trace function of the calculator or software to find the coordinates of individual points on the graph.

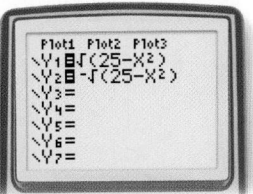

Graphing technology normally requires an equation to be in the form $y = $ ___. You will need to solve the circle equation for y. For example:

Algebra

$$x^2 + y^2 = 25$$
$$y^2 = 25 - x^2$$
$$y = \pm\sqrt{25 - x^2}$$

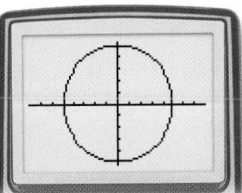

Thus, you will need to graph two separate curves:
$$y = \sqrt{25 - x^2} \qquad y = -\sqrt{25 - x^2}$$

Interdisciplinary Connection

SOCIAL STUDIES The latitude and longitude lines shown on a globe or map form a type of coordinate system. Have students research how the grid on a globe works. Suppose that a circle with a radius of 69 miles is centered at $(0, 0)$, which is on the equator directly south of Greenwich, England. What places would this circle include? What latitude and longitude lines would it intersect or touch? Have students experiment with circles centered at other places on the globe.

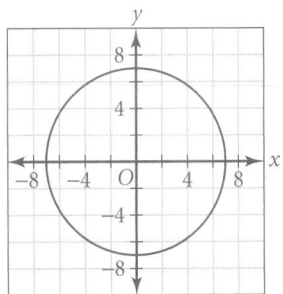

Before students read the text on
this page, have them work in small
groups to graph several circles
centered at the origin. Each graph
should be labeled with its equation. After students have finished,
ask them to suggest a general
equation for the graph of any circle with its center at the origin.

Teaching Tip

Derive the formula for a circle by
showing the right triangles whose
legs have lengths of $|x|$ and $|y|$. Ask
students how they can use the
Pythagorean Theorem to relate
the sides of the triangles to each
other.

CHECKPOINT ✔

The point (x_1, y_1) is some distance r_1 from the origin, where
$r_1 \neq r$, and so $x_1^2 + y_1^2 = r_1^2 \neq r$.
That is, $x_1^2 + y_1^2 \neq r$.

CRITICAL THINKING

With the absolute values as
shown, the signs of h and k do
not affect the relationship
shown.

Deriving the Equation of a Circle

CONNECTION
COORDINATE GEOMETRY

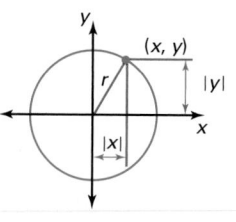

In a circle, all of the points are a
certain distance, r, from a fixed point.
In the simplest case, that fixed point
is the origin, as shown.

For any point (x, y) on the circle that
is not on the x- or y-axis, you can
draw a right triangle whose legs
have lengths of $|x|$ and $|y|$. The
length of the hypotenuse is the
distance, r, from the point to the
origin. For any such point:

$$x^2 + y^2 = r^2 \quad \textit{Equation 1}$$

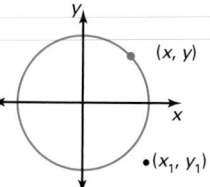

By substituting 0 for x or y in the
equation, you can see that it is also
true for any point (x, y) on the circle
that is on the x- or y-axis.

If a point (x_1, y_1) is *not* on the circle, then its distance from the orgin is some
value not equal to r, and the equation will not be true for it. That is, for any
point *not* on the circle:

$$x_1^2 + y_1^2 \neq r^2$$

CHECKPOINT ✔ Explain why this is true.

Notice that equation 1 satisfies the following two conditions:

• It is true of all points (x, y) that are on the circle.
• It is not true of any point (x_1, y_1) that is not on the circle.

Thus, equation 1 is the equation of a circle.

Moving the Center of the Circle

To find the *standard form of the
equation* of a circle centered at a
point (h, k) that is not at the origin,
study the diagram at right. For such
a circle:

$$(x - h)^2 + (y - k)^2 = r^2 \quad \textit{Equation 2}$$

It should be clear that the equation
is *not* true for a point that does not
lie on the circle, so equation 2 is
the general equation of the circle.

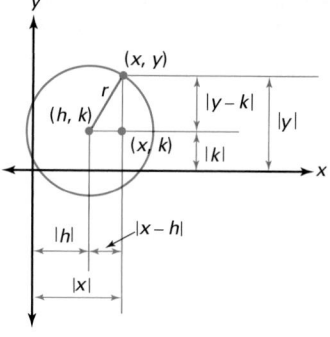

CRITICAL THINKING   How can you show that the relationships shown in the diagram are the same
for points (h, k) in Quadrants II, III, and IV?

Enrichment

Students can investigate equations of the form
$(x - h)^2 - (y - k)^2 = r^2$. Ask them to describe the
resulting graphs in writing and explain how different values of h, k, and r change the graphs.

Inclusion Strategies

HANDS-ON STRATEGIES   Have students make a
coordinate system with four pieces of standard-
size paper. Then have students cut circles from
transparent paper. The diameters of the circles
that match the units of the grid should be whole
numbers. Students can move their transparent
circles around on the grid to center them at
various intersection points. Have students find the
radius, center and equation for each placement.

● **SOLUTION**

Comparing the given equation with the standard form of the equation of the circle, you find the following correspondences:

The standard form of the equation	The given equation
$(x - h)^2$	$(x - 7)^2$
$(y - k)^2$	$(y + 3)^2$ or $[y - (-3)]^2$
r^2	36

From this you can conclude that

$$h = 7, k = -3, \text{ and } r = 6.$$

That is, the center of the circle is $(7, -3)$, and the radius is 6 units.

TRY THIS For each equation below, find the radius and the center of the circle represented. Graph the equation and compare the graph with your values for the radius and the center of the circle.

a. $(x + 3)^2 + (y - 3)^2 = 49$ **b.** $(x - 3)^2 + (y + 3)^2 = 49$
c. $(x - 4)^2 + (y - 5)^2 = 30$ **d.** $(x + 2)^2 + (y - 5)^2 = 50$

Exercises

● *Communicate*

1. Explain how to find the x- and y-intercepts of $x^2 + y^2 = 4$.

2. Explain how to find the x- and y-intercepts of $(x - 2)^2 + (y + 2)^2 = 4$.

3. Is it possible for a circle to have no x- or y-intercepts? What would the graph look like?

4. **TECHNOLOGY** Most graphics calculators will graph only equations of the form $y = $ ___. How would you solve the standard equation of a circle for y?

5. Use the distance formula (Lesson 5.6) to write an expression for the distance from point (h, k) to point (x, y). How does this expression relate to the standard equation of a circle?

Reteaching the Lesson

TECHNOLOGY Using a graphics calculator, have students enter and graph the following equations:

$$y_1 = \sqrt{36 - (x - 7)^2} - 3$$
$$y_2 = -\sqrt{36 - (x - 7)^2} - 3$$

These equations give the graph of the circle in Example 2. Once the equations are entered,

students should experiment by entering different values for values of 7 and 3 as well as changing the corresponding signs. Students will get circles with a radius of 6 but with different centers. To change the radius, students should enter different values for 36.

ADDITIONAL
E X A M P L E ②

Find the center and radius of the circle with the equation $(x + 4)^2 + y^2 = 20$.
center: $(-4, 0)$; radius: $\sqrt{20}$

TRY THIS
a. center: $(-3, 3)$; radius: 7

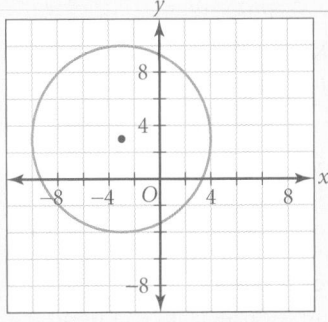

b. center: $(3, -3)$; radius: 7

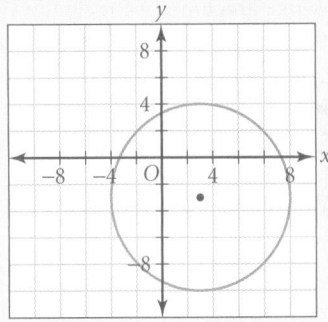

c. center: $(4, 5)$;
radius: $\sqrt{30} \approx 5.48$

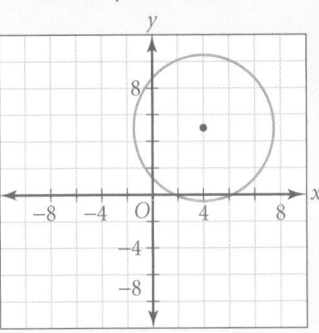

d. center: $(-2, 5)$;
radius: $\sqrt{50} \approx 7.07$

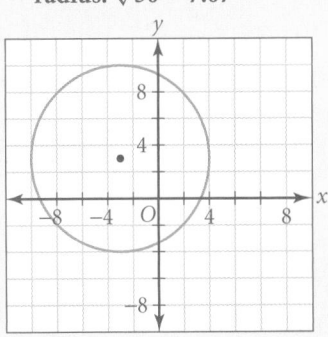

Selected Answers

Exercises 6–11, 13–59 odd

ASSIGNMENT GUIDE

In Class	1–11
Core	13–51 odd
Core Plus	17–51 odd, 52–54
Review	55–60
Preview	61–63

✐ Extra Practice can be found beginning on page 818.

Error Analysis

Point out that the graph of a circle on a graphics calculator will look like an ellipse if the window setting is not a square window.

● *Guided Skills Practice*

For Exercises 6–8, refer to the equation $x^2 + y^2 = 100$. *(EXAMPLE 1)*

6. Find the *x*- and *y*-intercepts. *x*-intercepts: (10, 0) and (–10, 0)

7. Complete the table below. *y*-intercepts: (0, 10) and (0, –10)

x	y	Points on graph	
0	?±10	?	(0, 10) and (0, –10)
?±10	0	?	(10, 0) and (–10, 0)
6	?±8	?	(6, 8) and (6, –8)
–6	?±8	?	(–6, 8) and (–6, –8)
8	?±6	?	(8, 6) and (8, –6)
–8	?±6	?	(–8, 6) and (–8, –6)

8. Plot the points from the table in a coordinate plane and sketch the circle.

For Exercises 9–11, refer to the equation $(x - 4)^2 + (y - 3)^2 = 25$. *(EXAMPLE 2)*

9. Find the *x*- and *y*-intercepts. *x*-intercepts: (8, 0) and (0, 0)

10. Complete the table below. *y*-intercepts: (0, 6) and (0, 0)

x	y	Points on graph	
0	?6, 0	?	(0, 6) and (0, 0)
?8, 0	0	?	(8, 0) and (0, 0)
1	?7, –1	?	(1, 7) and (1, –1)
–1	?3	?	(–1, 3)
4	?8, –2	?	(4, 8) and (4, –2)
7	?7, –1	?	(7, 7) and (7, –1)
8	?6, 0	?	(8, 6) and (8, 0)
9	?3	?	(9, 3)

11. Plot the points from the table in a coordinate plane and sketch the circle.

● *Practice and Apply*

☑ internet connect

Homework Help Online

Go To: go.hrw.com
Keyword:
MG1 Homework Help
for Exercises 12-32

Find the *x*- and *y*-intercepts for the graph of each circle.

12. $x^2 + y^2 = 64$ *x*-int. (8, 0), (–8, 0); *y*-int. (0, 8), (0, –8)

13. $x^2 + y^2 = 50$ *x*-int. (5√2, 0), (–5√2, 0); *y*-int. (0, 5√2), (0, –5√2)

14. $x^2 + (y - 4)^2 = 25$ *x*-int. (3, 0), (–3, 0); *y*-int. (0, 9), (0, –1)

15. $(x - 2)^2 + y^2 = 9$ *x*-int. (5, 0), (–1, 0); *y*-int. (0, √5), (0, –√5)

16. $(x - 6)^2 + (y - 8)^2 = 100$ *x*-int. (12, 0), (0, 0); *y*-int. (0, 16), (0, 0)

8.

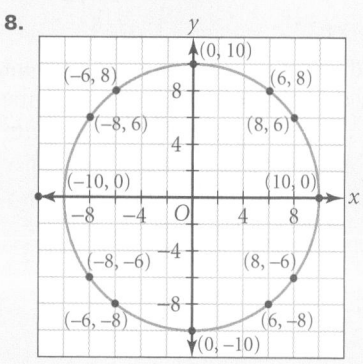

11.

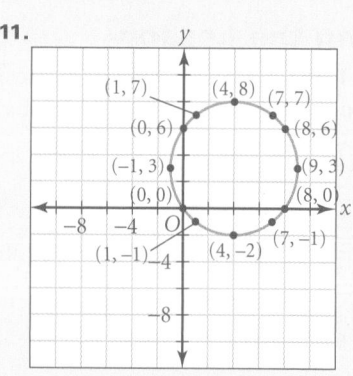

Write an equation for the circle with the given center and radius.

17. center: $(0, 0)$; radius $= 6$

18. center: $(0, 0)$; radius $= 2.5$

19. center: $(0, 0)$; radius $= \sqrt{13}$

20. center: $(2, 3)$; radius $= 4$

21. center: $(0, 6)$; radius $= 5$

22. center: $(4, -5)$; radius $= 7$

23. center: $(1, -7)$; radius $= 10$

24. center: $(4, -3)$; radius $= \sqrt{7}$

Algebra

Find the center and radius of each circle.

25. $x^2 + y^2 = 100$ **(0, 0); 10**

26. $x^2 + y^2 = 36$ **(0, 0); 6**

27. $x^2 + y^2 = 101$ **(0, 0); $\sqrt{101}$**

28. $(x - 6)^2 + y^2 = 9$ **(6, 0); 3**

29. $x^2 + (y - 3)^2 = 4$ **(0, 3); 2**

30. $(x + 5)^2 + (y - 2)^2 = 16$ **(–5, 2); 4**

31. $y^2 + (x + 3)^2 = 49$ **(–3, 0); 7**

32. $(x + 1)^2 + (y + 3)^2 = 19$ **(–1, –3); $\sqrt{19}$**

Write an equation for each circle.

33. $x^2 + y^2 = 16$

34. $(x - 2)^2 + (y - 1)^2 = 9$

35. $(x + 2)^2 + (y - 3)^2 = 4$

33.
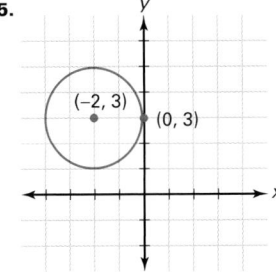

34.
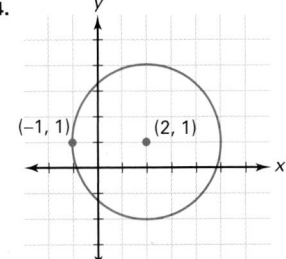

35.

Algebra

Draw a circle on graph paper with the given intercepts and find its equation. Some exercises may have more than one possible answer.

	x-intercept(s)	y-intercept(s)
36.	3, –3	3, –3
37.	2, 6	none
38.	0	0, 8
39.	none	5
40.	none	none

39. Sample answer: $(x - 1)^2 + (y - 5)^2 = 1$
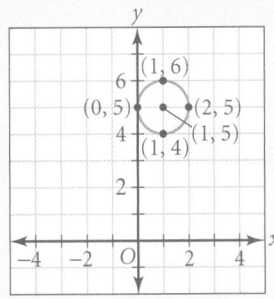

40. Sample answer: $(x - 2)^2 + (y - 2)^2 = 1$
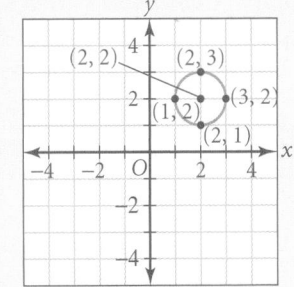

ALGEBRA In Exercises 12–16 and 25–32, students solve equations for x and y. Students write equations for Exercises 36–40. You may want to review how to solve equations that involve squares and square roots.

17. $x^2 + y^2 = 36$

18. $x^2 + y^2 = 6.25$

19. $x^2 + y^2 = 13$

20. $(x - 2)^2 + (y - 3)^2 = 16$

21. $x^2 + (y - 6)^2 = 25$

22. $(x - 4)^2 + (y + 5)^2 = 49$

23. $(x - 1)^2 + (y + 7)^2 = 100$

24. $(x - 4)^2 + (y + 3)^2 = 7$

36. $x^2 + y^2 = 9$
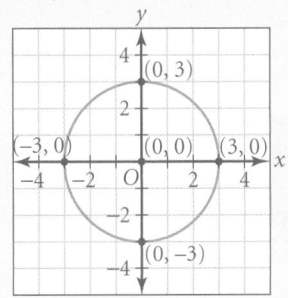

37. Sample answer: $(x - 4)^2 + y^2 = 4$
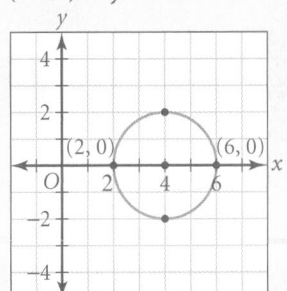

38. Sample answer: $x^2 + (y - 4)^2 = 16$
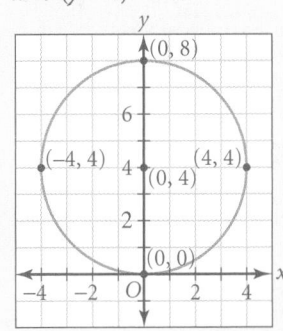

Technology

A graphics calculator or geometry graphics software is suggested for Exercises 47–53.

47. $(x - 3)^2 + (y + 5)^2 = 4$

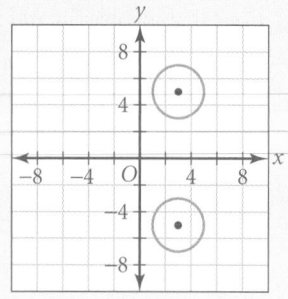

48. $(x + 4)^2 + (y - 2)^2 = 1$

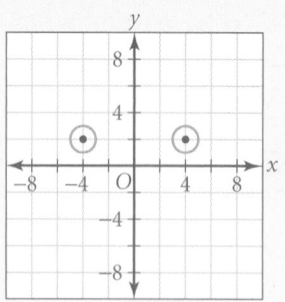

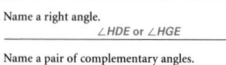
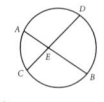
 **Write an equation for the circle with the given characteristics. It may be helpful to sketch a graph.**

41. center: (2, 3); tangent to the x-axis $(x - 2)^2 + (y - 3)^2 = 9$

42. center: (2, 3); tangent to the y-axis $(x - 2)^2 + (y - 3)^2 = 4$

43. center: (0, 1); contains the point (4, 4) $x^2 + (y - 1)^2 = 25$

44. center: (2, 3); contains the point (8, 3) $(x - 2)^2 + (y - 3)^2 = 36$

45. center: (2, 3); contains the point (8, 11) $(x - 2)^2 + (y - 3)^2 = 100$

46. has (1, 3) and (5, 3) as endpoints of a diameter $(x - 3)^2 + (y - 3)^2 = 4$

TECHNOLOGY Use geometry graphics software or graph paper for Exercises 47–53.

47. Sketch the graph of $(x - 3)^2 + (y - 5)^2 = 4$. Reflect the graph across the x-axis, and sketch the image. Write an equation for the image.

48. Sketch the graph of $(x - 4)^2 + (y - 2)^2 = 1$. Reflect the graph across the y-axis, and sketch the image. Write an equation for the image.

49. Sketch the graph of $(x - 2)^2 + y^2 = 9$. Translate the graph 6 units to the right, and sketch the image. Write an equation for the image.

50. Sketch the graph of $(x - 6)^2 + (y - 4)^2 = 9$. Translate the graph 2 units to the right and 1 unit down, and sketch the image. Write an equation for the image.

51. Sketch the graph of $(x - 5)^2 + (y - 4)^2 = 9$. Rotate the graph 180° about the origin, and sketch the image. Write an equation for the image.

52. Find the equation of the line tangent to the circle $x^2 + y^2 = 100$ at the point $(-6, 8)$.

CHALLENGE **53.** Draw a triangle with vertices at (0, 0), (0, 6), and (8, 0). Find the equation of the circle that circumscribes the triangle.

APPLICATION

$(x - 320)^2 + (y + 240)^2 = 240^2$

54. COMPUTER GRAPHICS A computer screen is a coordinate plane in which each pixel is one unit. A typical computer screen measures 640 pixels horizontally and 480 pixels vertically. The origin of the plane is the upper left corner of the screen, so all y-coordinates are negative.

A computer programmer wants to create a circle that is as large as possible for a typical screen. What is the equation of the circle? (More than one answer is possible, depending on your choice of the center.)

Look Back

55. Suppose that a dart is tossed at random onto the graph of $x^2 + y^2 = 100$. What is the probability that it will land within the graph of the circle $x^2 + y^2 = 25$? *(LESSON 5.7)* $\frac{1}{4}$

APPLICATION **56. ASTRONOMY** You can view an eclipse by using a pinhole camera. If the image of

$2x; \frac{1}{2}x; 0.9x$

the Sun measures x mm in diameter when the distance from the image to the pinhole is 50 cm, how large will the image be when the distance is 100 cm? 25 cm? 45 cm? *(LESSON 8.1)*

49. $(x - 8)^2 + y^2 = 9$

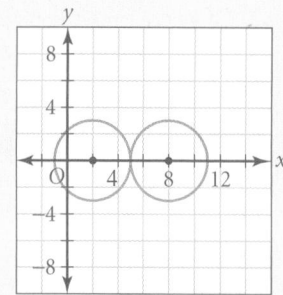

50. $(x - 8)^2 + (y - 3)^2 = 9$

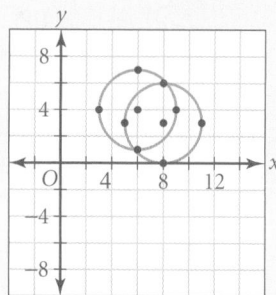

APPLICATION

STRUCTURAL DESIGN A manufacturer is designing a wheelchair to use in wheelchair basketball. The wheel diameter is 24.5 in. *(LESSONS 5.3 AND 9.3)*

3.06π in. ≈ 9.6 in.

$\frac{1128}{24.5\pi} \approx 14.7$ rotations

57. If the wheels rotate through 45°, how far will the wheelchair travel?

58. A basketball court is 94 ft long. How many rotations of the wheels will it take to get from one end of the court to the other?

Wheelchair basketball is a highly competitive organized sport. The players' chairs are often custom-made.

59. Find *DE*. *(LESSON 9.5)*
9

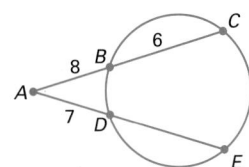

60. Find *DE*. *(LESSON 9.5)*
$\frac{48}{7} \approx 6.86$

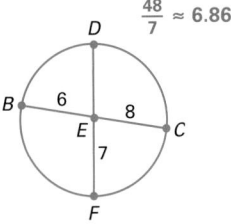

Look Beyond

Algebra

In Exercises 61–63, you will prove that an angle inscribed in a semicircle measures 90°. Refer to the diagram below.

61. Use the equation for a circle with radius r centered at $(0, 0)$ to explain why the coordinates of a point on the circle are $(p, \sqrt{r^2 - p^2})$.

62. Find the slopes of the segments that form the inscribed angle in the diagram.

63. Prove that the inscribed angle measures 90°. (Hint: If two lines are perpendicular, the product of their slopes is __?__.)

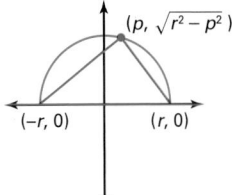

62.
$$m_1 = \frac{\sqrt{r^2 - p^2}}{p - r}; \quad m_2 = \frac{\sqrt{r^2 - p^2}}{p + r}$$

63. $m_1 \cdot m_2 = \dfrac{\sqrt{r^2 - p^2}}{p - r} \cdot \dfrac{\sqrt{r^2 - p^2}}{p + r}$

$$= \frac{\left(\sqrt{r^2 - p^2}\right)^2}{(p - r)(p + r)} = \frac{r^2 - p^2}{p^2 - r^2} = -1$$

Since the product of the slopes is –1, the line segments are perpendicular. That is, the inscribed angle measures 90°.

Look Beyond

In Exercises 61–63, students preview a basic trigonometric concept known as the unit circle.

61. The equation of the circle is $x^2 + y^2 = r^2$. When $x = p$, then substitute p for x and solve for y to get the y-coordinate.

$$p^2 + y^2 = r^2$$
$$y^2 = r^2 - p^2$$
$$y = \pm\sqrt{r^2 - p^2}$$

The coordinates are $\left(p, \sqrt{r^2 - p^2}\right)$. Note that this is the top half of the circle, so the positive value is used for y.

Student Technology Guide

To graph a function on a graphics calculator, you enter the expression that defines the function. The equation $x^2 + y^2 = r^2$ defines a circle, but y is not a function of x. To graph a circle given its equation, you must write two equations that define y.

$x^2 + y^2 = 25$
$y^2 = 25 - x^2$
$y = \pm\sqrt{25 -}$

Example: To graph $x^2 + y^2 = 25$, you need to carry out three steps.
- Write and enter two expressions, $\sqrt{25 - x^2}$ and $-\sqrt{25 - x^2}$.
  [Y=] [2nd] [x^2] 25 [−] [X,T,θ,n] [x^2] [)] [ENTER] [(−)] [2nd] [x^2] 25 [−] [X,T,θ,n] [x^2] [)] [ENTER]
- Select window settings that give a true circle. An x-range that is about 1.5 times as long as the y-range will accomplish this. On the display below, $-9 \le x \le 9$ and $-6 \le y \le 6$.
- Press [GRAPH].

The required displays are shown below.

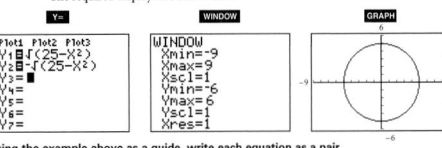

Using the example above as a guide, write each equation as a pair of equations for y in terms of x, and then graph the equations.

1. $x^2 + y^2 = 16$ $y = \pm\sqrt{16 - x^2}$ 2. $x^2 + y^2 = 36$ $y = \pm\sqrt{36 - x^2}$
3. $x^2 + y^2 = 49$ $y = \pm\sqrt{49 - x^2}$ 4. $x^2 + y^2 = 64$ $y = \pm\sqrt{64 - x^2}$
5. $x^2 + y^2 = 81$ $y = \pm\sqrt{81 - x^2}$ 6. $x^2 + y^2 = 100$ $y = \pm\sqrt{100 - x^2}$
7. $(x - 1)^2 + (y - 2)^2 = 3^2$
8. $(x + 1)^2 + (y + 2)^2 = 3^2$ $y = -2 \pm \sqrt{9 - (x + 1)^2}$
9. $(x - 1)^2 + (y + 2)^2 = 3^2$ $y = -2 \pm \sqrt{9 - (x - 1)^2}$
10. a. Graph $(x - 2)^2 + (y - 1)^2 = 4$ and $(x - 4)^2 + (y - 1)^2 = 1$ on the same calculator display.
   b. At how many points do the graphs intersect?   two points
   c. Use [Trace] to approximate the coordinates of the points of intersection.   approximate coordinates: (3.74, 1.98) and (3.74, 0.02)

Focus

In these activities, students will analyze and construct circles, arcs, smooth curves, and reverse curves to create circle flowers, eggs, and various other designs.

Motivate

Have students turn at least one of the activities into a complete work of art that is suitable for display. Bring art books to class, particularly books by abstract artists, to motivate students with some examples.

Many students will be eager to create and display their artwork. Allow students to use any medium of their choice. Many will use pencils, pens, and poster boards, but some may wish to use oil, acrylic, or watercolor paints, pastels, or even computer graphics.

Activity 1

Sample answer: Draw a circle with the same center as the original circle, but with radius twice as large as the original set of circles. Draw new circles, congruent to the one just drawn, with their centers at the points where the original circles touch the larger circle (they should be tangent). Finally draw a circle with the same center as the original circle, but with radius twice as large as the set of circles just drawn (four times as large as the radius of the original circle).

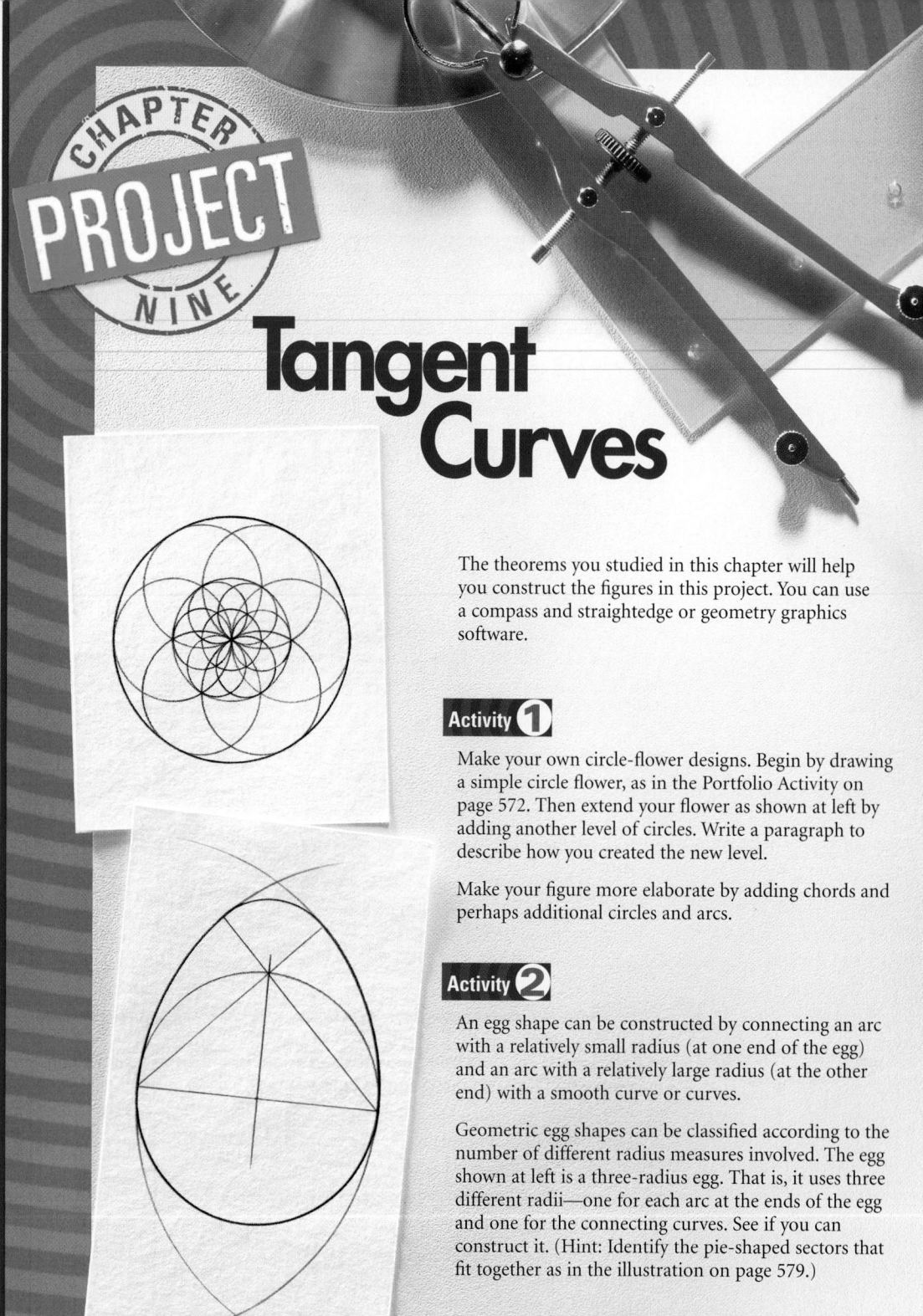

CHAPTER PROJECT NINE

Tangent Curves

The theorems you studied in this chapter will help you construct the figures in this project. You can use a compass and straightedge or geometry graphics software.

Activity ❶

Make your own circle-flower designs. Begin by drawing a simple circle flower, as in the Portfolio Activity on page 572. Then extend your flower as shown at left by adding another level of circles. Write a paragraph to describe how you created the new level.

Make your figure more elaborate by adding chords and perhaps additional circles and arcs.

Activity ❷

An egg shape can be constructed by connecting an arc with a relatively small radius (at one end of the egg) and an arc with a relatively large radius (at the other end) with a smooth curve or curves.

Geometric egg shapes can be classified according to the number of different radius measures involved. The egg shown at left is a three-radius egg. That is, it uses three different radii—one for each arc at the ends of the egg and one for the connecting curves. See if you can construct it. (Hint: Identify the pie-shaped sectors that fit together as in the illustration on page 579.)

Activity 2

Check student's drawings. For the three-radius egg, the circle forms the bottom of the egg. Work from the bottom of the egg to the top. For the four-radius egg, start with a small circle and a pair of perpendicular diameters extended beyond the circle. The larger circle that forms the bottom of the egg has radius equal to the diameter of the small circle. Work from the bottom of the egg to the top, noting that each side involves arcs with two different radii.

For the five-radius egg, start with four congruent circles with centers on the same line. The circles line up vertically in the center of the egg in the diagram. Add circles on each side of the central circles to form the circle flowers. Work from the bottom of the egg to the top. The centers of the arcs that form the sides of the egg are shown with larger dots. Note that the radius of the arc forming the egg changes where a line intersects the outside of the egg.

The egg below is a four-radius egg. See if you can construct it. It may help to work from an enlarged photocopy of the illustration.

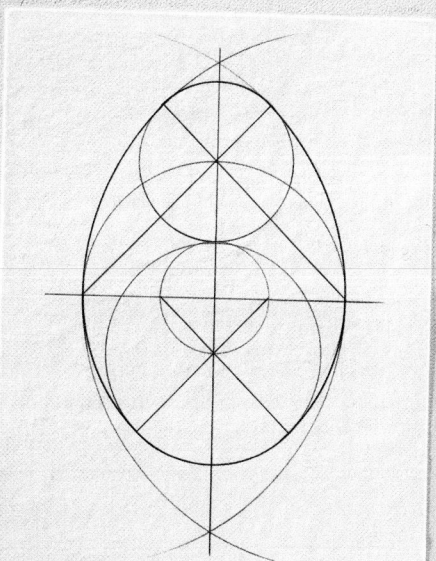

The egg below is a five-radius egg. See if you can construct it. The principle of the circle flower was used to construct the interior of the egg.

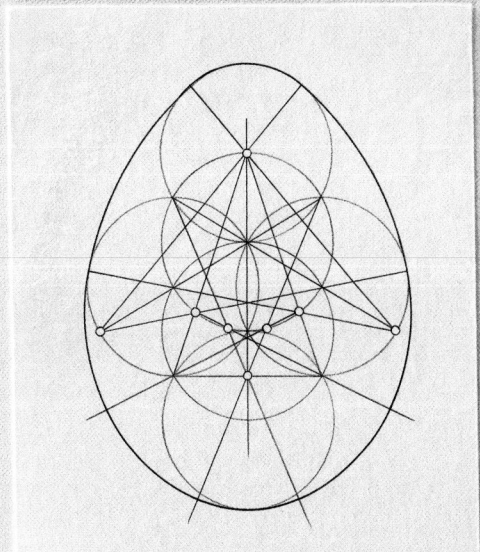

Activity 3

There are a number of reverse curves in the drawings below (see page 609). How many can you identify? Try constructing figures of your own that use reverse curves.

Cooperative Learning

Working in groups may be particularly helpful for Activity 2. The constructions in this activity are relatively complex, and group work can facilitate their completion.

Discuss

After completing Activities 1–3, have students discuss how they completed each construction. For Activity 2, ask groups or individuals to give step-by-step instructions on how to construct each egg. Students can use the board or overhead to demonstrate.

Activity 3

The bird and the fish each involve at least six reverse curves.

Chapter Review and Assessment

POSTULATES AND THEOREMS

Lesson	Number	Postulate or Theorem
9.1	9.1.5 Chords and Arcs Theorem	In a circle, or in congruent circles, the arcs of congruent chords are congruent.
	9.1.6 Converse of the Chords and Arcs Theorem	In a circle, or in congruent circles, the chords of congruent arcs are congruent.
9.2	9.2.2 Tangent Theorem	If a line is tangent to a circle, then the line is perpendicular to a radius of the circle drawn to the point of tangency.
	9.2.3 Radius and Chord Theorem	A radius that is perpendicular to a chord of a circle bisects the chord.
	9.2.4 Converse of the Tangent Theorem	If a line is perpendicular to a radius of a circle at its endpoint on the circle, then the line is tangent to the circle.
	9.2.5 Theorem	The perpendicular bisector of a chord passes through the center of the circle.
9.3	9.3.1 Inscribed Angle Theorem	The measure of an angle inscribed in a circle is equal to one-half the measure of the intercepted arc.
	9.3.2 Right Angle Corollary	If an inscribed angle intercepts a semicircle, then the angle is a right angle.
	9.3.3 Arc-Intercept Corollary	If two inscribed angles intercept the same arc, then they have the same measure.
9.4	9.4.1 Theorem	If a tangent and a secant (or a chord) intersect on a circle at the point of tangency, then the measure of the angle formed is one-half the measure of its intercepted arc.
	9.4.2 Theorem	The measure of an angle formed by two secants or chords that intersect in the interior of a circle is one-half the sum of the measures of the arcs intercepted by the angle and its vertical angle.
	9.4.3 Theorem	The measure of an angle formed by two secants that intersect in the exterior of a circle is one-half the difference of the measures of the intercepted arcs.

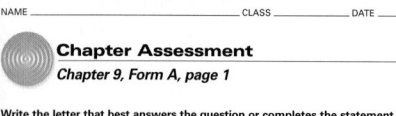

Chapter Test, Form A

NAME _____ CLASS _____ DATE _____

Chapter Assessment
Chapter 9, Form A, page 1

Write the letter that best answers the question or completes the statement.

In $\odot B$, $\overline{CD}$ is tangent to B at C and $m\widehat{AC} = 70°$.

__c__ 1. $m\angle ABC =$
 a. 35°   b. 55°
 c. 70°   d. 140°

__a__ 2. $m\angle ACD =$
 a. 35°   b. 55°
 c. 70°   d. 140°

__b__ 3. $m\angle BCA =$
 a. 35°   b. 55°
 c. 70°   d. 140°

__b__ 4. In a circle, the length of an arc intercepted by a central angle is 3π and the radius of the circle is 9 inches. What is the measure of the central angle?
 a. 30°   b. 60°   c. 90°   d. 120°

__c__ 5. A regular octagon is inscribed in a circle with a radius of 12 inches. What is the length of one arc of the circle intercepted by one side of the octagon?
 a. 45°   b. 8 inches   c. 3π inches   d. 12π inches

In $\odot Q$, the radius is 15, $DG = 6$, and $JK = 24$.

__d__ 6. $FH =$
 a. 9   b. 12
 c. 15   d. 24

__d__ 7. $FK =$
 a. 24   b. 36
 c. 39   d. 54

__a__ 8. If $m\widehat{HJ} = 70°$, then $m\angle K =$ _____.
 a. 20°   b. 30°   c. 40°   d. 90°

__c__ 9. $QD =$
 a. 3   b. 6   c. 9   d. 12

NAME _____ CLASS _____ DATE _____

Chapter Assessment
Chapter 9, Form A, page 2

For Exercises 10–11, refer to $\odot P$.

__d__ 10. $m\angle RPU =$
 a. 30°   b. 50°
 c. 80°   d. 140°

__a__ 11. $m\angle TVU =$
 a. 30°   b. 50°
 c. 80°   d. 100°

For Exercises 12–14, use the given circle.

__b__ 12. $AF =$
 a. 6   b. 8
 c. 12   d. 15

__d__ 13. $AC =$
 a. 6   b. 8
 c. 10   d. 14

__b__ 14. $DC =$
 a. 3   b. 7   c. 12   d. 16

__c__ 15. In a circle, chord $\overline{CD}$ intersects chord $\overline{AB}$ at point X and divides $\overline{AB}$ into segments with a ratio of 9 to 2. If $CX = 9$ and $DX = 8$, what is the length of $\overline{AB}$?
 a. 2   b. 4   c. 22   d. 72

__a__ 16. What is the radius of the circle defined by $x^2 + y^2 = 10$?
 a. $\sqrt{10}$ b. 10   c. $\sqrt{20}$ d. 100

__b__ 17. The center of the circle defined by $(x - 5)^2 + y^2 = 25$ is _____.
 a. $(-5, 0)$ b. $(5, 0)$ c. $(0, -5)$ d. $(0, 5)$

__d__ 18. What is the equation of a circle centered at $(3, -2)$ with a radius of 4?
 a. $(x + 3)^2 + (y - 2)^2 = 4$ b. $(x + 3)^2 + (y - 2)^2 = 16$
 c. $(x - 3)^2 + (y + 2)^2 = 4$ d. $(x - 3)^2 + (y + 2)^2 = 16$

__d__ 19. What are the x-intercepts of the circle defined by $(x - 5)^2 + y^2 = 25$?
 a. $x = 20$ and $x = 3$ b. $x = 25$ c. $x = 5$ d. $x = 0$ and $x = 10$

Lesson	Number	Postulate or Theorem
	9.4.4 Theorem	The measure of a secant-tangent angle with its vertex outside the circle is one-half the difference of the measures of the intercepted arcs.
	9.4.5 Theorem	The measure of a tangent-tangent angle with its vertex outside the circle is one-half the difference of the measures of the intercepted arcs, or the measure of the major arc minus 180°.
9.5	9.5.1 Theorem	If two segments are tangent to a circle from the same external point, then the segments are of equal length.
	9.5.2 Theorem	If two secants intersect outside a circle, then the product of the lengths of one secant segment and its external segment equals the product of the lengths of the other secant segment and its external segment. (Whole × Outside = Whole × Outside)
	9.5.3 Theorem	If a secant and a tangent intersect outside a circle, then the product of the lengths of the secant segment and its external segment equals the length of the tangent segment squared. (Whole × Outside = Tangent Squared)
	9.5.4 Theorem	If two chords intersect inside a circle, then the product of the lengths of the segments of one chord equals the product of the lengths of the segments of the other chord.

Key Skills & Exercises

LESSON 9.1

Key Skills

Identify parts of a circle.

In ⊙M, $\overline{AB}$ and $\overline{CD}$ are chords. $\overline{CD}$ is also a diameter. $\overline{MC}$, $\overline{ME}$, and $\overline{MD}$ are radii. Central angle ∠CME intercepts minor arc $\overset{\frown}{CE}$.

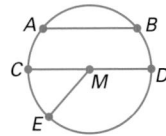

Find central angle measures.

In ⊙N, find m∠ONP.

$$90° + 60° + 90° + m\angle ONP = 360°$$
$$m\angle ONP = 120°$$

Find arc measures and lengths.

In ⊙N, find m$\overset{\frown}{OP}$ and the length of $\overset{\frown}{OP}$.

$$m\overset{\frown}{OP} = m\angle ONP = 120°$$
$$\text{length of } \overset{\frown}{OP} = \frac{120°}{360°} \times 2\pi(21 \text{ cm}) \approx 44 \text{ cm}$$

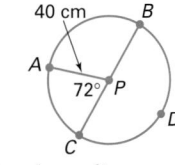

Exercises

Refer to ⊙P below.

1. Name a chord, a radius, a central angle, and a major arc.

2. Find m∠APB.

3. Find m$\overset{\frown}{AC}$.

4. Find the length of $\overset{\frown}{AC}$. Round your answer to the nearest centimeter.

Chapter Test, Form B

NAME _____ CLASS _____ DATE _____

Chapter Assessment
Chapter 9, Form B, page 1

In ⊙O, $\overline{EH}$ is a diameter, m∠HOI = 42°, m$\overset{\frown}{FG}$ = 42°, and HI = 9.

1. Find m∠GOH. **96°**   2. Find m$\overset{\frown}{HI}$. **42°**
3. Find FG. **9**   4. Find m$\overset{\frown}{EI}$. **138°**
5. Find m∠EHI. **69°**   6. Find m∠FIH. **69°**
7. If the radius of the circle is 30, find the length of $\overset{\frown}{HI}$. **7π**
8. If the length of $\overset{\frown}{GH}$ is 8π, find the circumference of the circle. **30π**

In ⊙O, KM⊥JL.

9. If the radius is 13 and ON = 5, find JN. **12**
10. If ON = 8 and JL = 30, find the radius. **17**
11. If the radius is 15 and JL = 24, find NM. **6**
12. If the radius is 8 and ON = 6, find JL to the nearest tenth. **≈10.6**
13. If ON = 6 and JN = 1, find the radius to the nearest tenth. **6.1**

In the given circle, m$\overset{\frown}{UR}$ = 140°, m$\overset{\frown}{RS}$ = 100°, and m$\overset{\frown}{ST}$ = 30°.

14. Find m∠RSU. **70°**   15. Find m∠RVU. **55°**
16. Find m∠USV. **110°**   17. Find m∠RWS. **95°**
18. If m$\overset{\frown}{RU}$ = 120° and m∠RVU = 35°, find m$\overset{\frown}{ST}$. **50°**

Find m$\overset{\frown}{AB}$ in each circle.

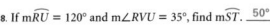
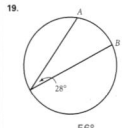
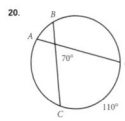
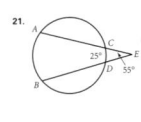

19. **56°**   20. **30°**   21. **135°**

NAME _____ CLASS _____ DATE _____

Chapter Assessment
Chapter 9, Form B, page 2

In the given circle, $\overline{KH}$ is tangent at point H.

22. JG = **9**
23. HK = **15**

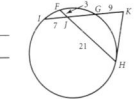

24. A regular hexagon is inscribed in a circle. If one side of the hexagon is 10 cm, what is the circumference of the circle? **20π, or 62.8 cm**

25. The angle formed by two tangents from a communications satellite measures 20°. What is the measure of the arc on the Earth that is visible from the satellite? **160°**

26. In a circle, two chords intersect at a 90° angle. What is the sum of the measures of two of the intercepted arcs? Explain why. **180°; the measure of an angle formed by two chords inside a circle is half the sum of the measures of the intercepted arcs.**

Find the x- and y-intercepts for each equation.

27. $x^2 + y^2 = 16$ **(−4, 0), (4, 0); (0, −4), (0, 4)**   28. $x^2 + y^2 = 20$ **(−√20, 0), (√20, 0); (0, −√20), (0, √20)**

Find the center and radius of each circle.

29. $x^2 + y^2 = 36$ **(0, 0); 6**   30. $x^2 + (y − 2)^2 = 49$ **(0, 2); 7**   31. $(x − 1)^2 + (y + 3)^2 = 10$ **(1, −3); √10**

Write an equation of a circle with the given center and radius.

32. center at (4, 0); radius = 5 **$(x − 4)^2 + y^2 = 25$**

33. center at (−1, 5); radius = 2 **$(x + 1)^2 + (y − 5)^2 = 4$**

34. Write an equation for the circle graphed at right. **$(x + 1)^2 + (y + 2)^2 = 4$**

35. Find an equation of a circle centered at (3, −2) and containing the point (0, −2). **$(x − 3)^2 + (y + 2)^2 = 9$**

5. 13

6. 33.8

7. True; $\triangle QPX \cong \triangle RPX$.

8. False; if $\overline{ST}$ were tangent to $\odot P$ at S, then $\triangle PST$ would be a right triangle. However, $(PS)^2 + (ST)^2 = 7^2 + 10^2 = 149$ and $(PT)^2 = 12^2 = 144$

9. 20°

10. 50°

11. 25°

12. 90°

13. 47°

14. 47°

15. 94°

16. 88°

Key Skills

Use properties of secants and tangents to solve problems.

In the figure below, $\overleftrightarrow{EF}$ is tangent to $\odot G$ at F, $BF = 10$, $CG = 3$, and $EF = 12$. Find AD and EG.

By the Pythagorean Theorem:

$(CG)^2 + (AC)^2 = (AG)^2$

$3^2 + (AC)^2 = 5^2$

$AC = 4$

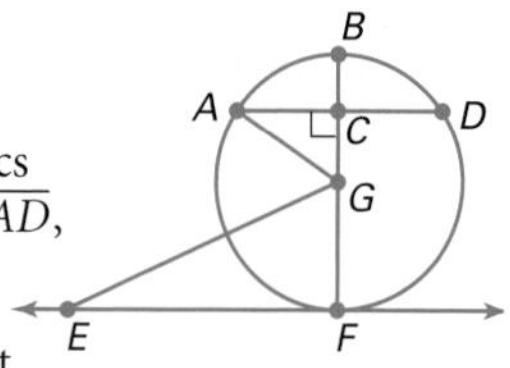

By the Chords and Arcs Theorem, $\overline{BG}$ bisects $\overline{AD}$, so $AD = 2 \cdot AC = 8$.

Because $\overleftrightarrow{EF}$ is tangent to $\odot G$, $\overline{GF} \perp \overleftrightarrow{EF}$.

$(GF)^2 + (EF)^2 = (EG)^2$

$12^2 + 5^2 = (EG)^2$

$EG = 13$

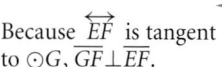

Exercises

In $\odot M$ below, $\overleftrightarrow{NL}$ is tangent to $\odot M$ at N, $NP = 24$, $MO = 5$, and $NL = 31.2$.

5. Find the radius of $\odot M$.

6. Find ML.

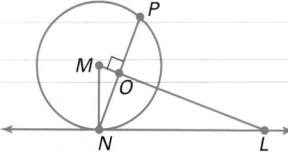

In $\odot P$, $PS = 7$, $ST = 10$, $PT = 12$, $QX = 6$, and $RX = 6$. Classify each statement as true or false and explain your reasoning.

7. $\overline{PS} \perp \overline{QR}$

8. $\overline{ST}$ is tangent to $\odot P$ at S.

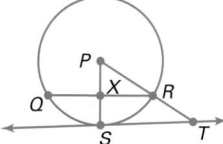

Key Skills

Find the measure of an inscribed angle and its intercepted arc.

Find $m\overset{\frown}{AD}$ and $m\angle C$.

$m\overset{\frown}{AD} = 2 \times m\angle B = 100°$

$\angle B$ and $\angle C$ both intercept $\overset{\frown}{AD}$, so $m\angle C = m\angle B = 50°$.

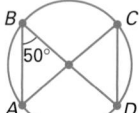

Exercises

In $\odot D$ below, $m\overset{\frown}{HE} = 40°$ and $m\angle FHG = 25°$. Find the following:

9. $m\angle F$

10. $m\overset{\frown}{FG}$

11. $m\angle E$

12. $m\angle I$

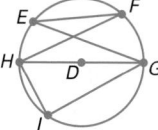

Key Skills

Use the angles formed by secants and tangents to solve problems.

In $\odot X$, $m\angle YUZ = 66°$, $m\angle VZU = 35°$, and $m\overset{\frown}{VY} = 80°$. Find $m\overset{\frown}{YZ}$, $m\overset{\frown}{UV}$, $m\angle YTZ$, and $m\angle VYW$.

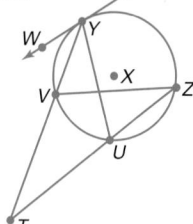

Exercises

In $\odot G$ below, $\overleftrightarrow{HJ}$ is tangent to $\odot G$ at H, $\overleftrightarrow{HJ} \parallel \overleftrightarrow{NP}$, $m\overset{\frown}{HP} = 136°$, and $m\overset{\frown}{KN} = 42°$. Find the following:

13. $m\angle KMN$

14. $m\angle KHJ$

15. $m\overset{\frown}{KH}$

16. $m\overset{\frown}{NP}$

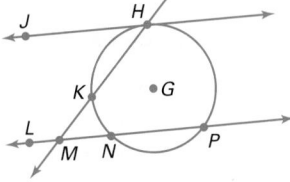

LESSON 9.4
Key Skills, continued

$\text{m}\widehat{YZ} = 2 \times \text{m}\angle YUZ = 2 \times 66° = 132°$

$\text{m}\widehat{UV} = 2 \times \text{m}\angle VZU = 2 \times 35° = 70°$

$\text{m}\angle YTZ = \dfrac{\text{m}\widehat{YZ} - \text{m}\widehat{UV}}{2} = \dfrac{132° - 70°}{2} = 31°$

$\text{m}\angle VYW = \dfrac{1}{2} \times \text{m}\widehat{VY} = \dfrac{1}{2} \times 80° = 40°$

LESSON 9.5
Key Skills

Use segments formed by tangents, secants, and chords to solve problems.

In $\odot A$, $BF = 24$, $CF = 32$, and $DF = 48$. Find EF.

By Theorem 9.5.4:

$BF \times DF = CF \times EF$

$24 \times 48 = 32 \times EF$

$EF = 36$

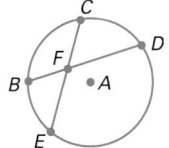

Exercises

$\odot P$ has a diameter of 75, $\overleftrightarrow{SQ}$ and $\overleftrightarrow{ST}$ are tangent to $\odot P$ at Q and T, and $RS = 60$. Find the following:

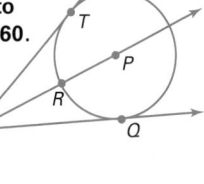

17. SQ

18. ST

In $\odot X$, $SY = 12$, $SZ = 18$, $SV = 4$, $VW = 8$, $WZ = 6$, and $WT = 5$. Find the following:

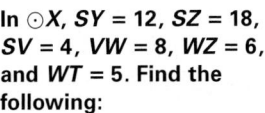

19. SU

20. WY

LESSON 9.6
Key Skills

Sketch a circle from its equation.

Sketch the circle $(x + 3)^2 + (y - 4)^2 = 16$.

center: $(-3, 4)$
radius $= 4$

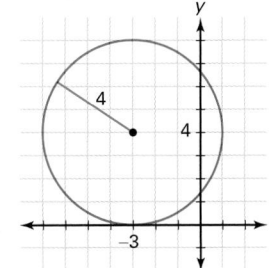

Exercises

Sketch the following circles:

21. $x^2 + y^2 = 49$

22. $(x - 1)^2 + (y + 2)^2 = 25$

Write the equation of the circle with the given center and radius.

23. center: $(0, 0)$; radius $= 1$

24. center: $(6, -2)$; radius $= 8$

Write the equation of a given circle.

Write the equation of a circle with center $(5, 0)$ and radius 6.

$(x - h)^2 + (y - k)^2 = r^2$ $h = 5$, $k = 0$, $r = 6$

$(x - 5)^2 + y^2 = 36$

21.

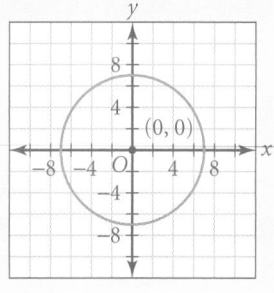

22.

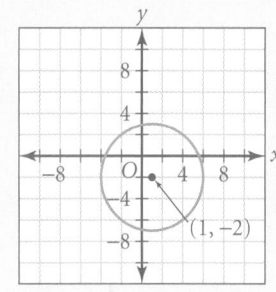

23. $x^2 + y^2 = 1$

24. $(x - 6)^2 + (y + 2)^2 = 64$

25. $340\pi \approx 1068$ in.

26. Take the perpendicular bisectors of two chords; their intersection is the center of the circle.

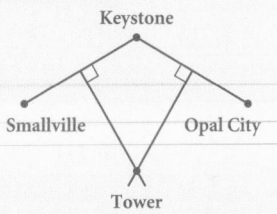

27. 11.35 mi

28. ≈ 0.72 mi^2; $\approx 72\%$

25. SURVEYING A simple way to get a rough estimate of property boundaries is to use a measuring wheel. Attach an 8-in. diameter wheel to a long handle, put a marker on the edge of the wheel, and count the number of rotations the wheel makes as you walk along the property line, rolling the wheel. If the wheel makes 42.5 rotations, how long is the property line?

26. COMMUNICATIONS A radio station wishes to locate its broadcasting tower an equal distance from three small towns. Trace the figure shown and locate the center of the circle that passes through the three points in order to find the location of the tower.

27. NAVIGATION A kayaker is rowing toward a lighthouse. If the light is 85 ft above sea level, how far away is the kayaker when he first sees the light? Assume that the water is calm and visibility is good. (1 mi = 5280 ft, radius of Earth = 4000 mi)

28. AGRICULTURE A field is irrigated by a pipe that extends from the center of the field to its outer edge and sweeps around in an arc. The corner of the field is blocked by the farmer's house and yard. If the area of the square field is 1 mi^2, what is the area of the irrigated sector? To find the area of the sector, use the following formula:

$$\text{area of sector} = \frac{C}{360°} \times \text{area of circle},$$

where C is the degree measure of the arc that bounds the sector

What percent of the entire field is the irrigated sector?

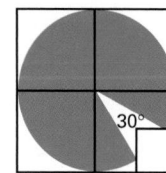

Chapter Test

For Exercises 1–4, refer to ⊙C.

1. Name a chord, a radius, a central angle, and a major arc.

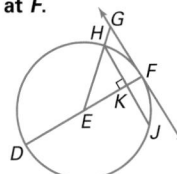

2. Find m∠QCR **95°**

3. Find m$\overarc{RS}$ **85°**

4. Find the length of $\overarc{RS}$. Round your answer to the nearest centimeter. **107 cm**

$\overleftrightarrow{FG}$ **is tangent to ⊙E below at F.**

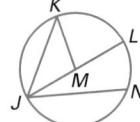

5. If EF = 6 and EG = 10, find FG. **8**

6. If KJ = 6, find HJ. **12**

In ⊙C at right, $\overline{CE} \perp \overline{AD}$ **at B.**

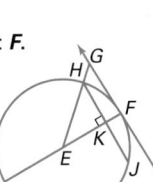

7. $\overline{AB} \cong$? $\overline{BD}$

8. If CE = 6 and CB = 4, what is AB? $2\sqrt{5}$

9. If CB = 7 and AD = 48, what is the radius? **25**

In ⊙M, m$\overarc{KL}$ = 80°, m∠MJN = 25°, and $\overline{JL}$ **is a diameter. Find the following:**

10. m∠KJL **40°**
11. m$\overarc{JK}$ **100°**
12. m$\overarc{LN}$ **50°**
13. m$\overarc{JN}$ **130°**

14. **DESIGN** A jeweler is making a circular pin with the design shown below. She wants the arc intercepted by ∠C and ∠D to measure 120°. If C is the center of the circle, what should the measure of each angle be? **120°; 60°**

Use the drawing below for Exercises 15–18.
$\overleftrightarrow{AG}$ **is tangent to ⊙O at A, m∠ADF = 52°, m∠BFD = 36°, and m$\overarc{AB}$ = 130°. Find each measure.**

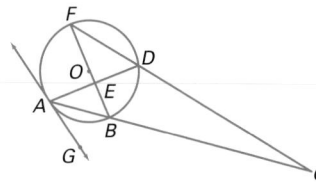

15. m$\overarc{AF}$ **104°**
16. m$\overarc{BD}$ **72°**
17. m∠AEF **88°**
18. m∠BAG **65°**

Find x in each circle below.

19. **6**

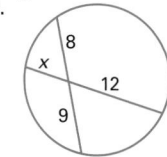

20. **20.125**

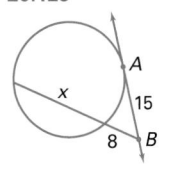

21. $\overleftrightarrow{CE}$ and $\overleftrightarrow{CF}$ are tangent to ⊙H, ⊙H has a radius of 9 centimeters, and CE = 12 centimeters. Find HC. **15 cm**

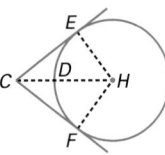

Sketch each circle.

22. $x^2 + y^2 = 36$
23. $x^2 + y^2 = 64$
24. $(x + 2)^2 + (y - 3)^2 = 9$

Write the equation of the circle with the given center and radius.

25. center: (0, 0); radius = 2 $x^2 + y^2 = 4$
26. center: (1, −5); radius = 4 $(x - 1)^2 + (y + 5)^2 = 16$

1. Sample answer: $\overline{QS}$, $\overline{CR}$, ∠QCR, $\overarc{RSQ}$

22.

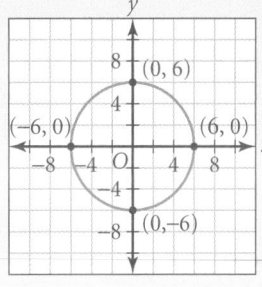

23.

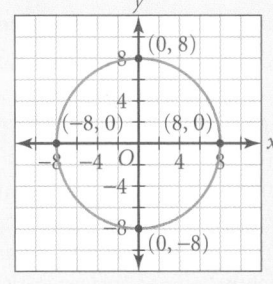

24.

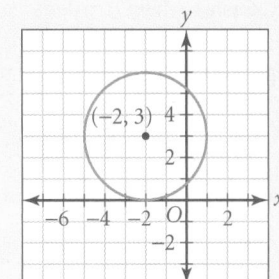

College Entrance Exam Practice

Multiple-Choice Samples
The first half of the Cumulative Assessment contains a multiple-choice section. This part of the Cumulative Assessment consists of items commonly found on standardized tests.

Free-Response Grid Samples
The second half of the Cumulative Assessment consists of a free-response section. This part requires student-produced response items like those commonly found on college entrance exams. These questions require the use of machine-scored answer grids. You may wish to have students practice answering these items in preparation for standardized tests.

1. c

2. b

3. b

4. d

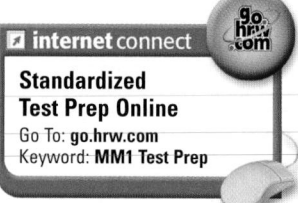
MULTIPLE-CHOICE For Questions 1–6, write the letter that indicates the best answer.

1. Find the sum of the surface areas of figure *A* and figure *B*. *(LESSON 6.1)*

Assume that there are no hidden cubes.

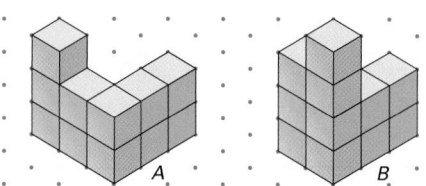

    a. 38 units²
    b. 78 units²
    c. 76 units²
    d. 86 units²

2. What is the ratio of the area of circle *A* to the area of circle *B*? *(LESSON 5.3)*

    a. $\frac{36}{25}$

    b. $\frac{25}{36}$

    c. $\frac{5}{6}$

    d. $\frac{125}{216}$

3. Refer to the figure below. What can be concluded about the slopes of lines *l* and *m*? *(LESSON 3.8)*

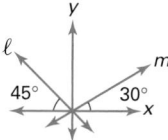

    a. The slope of line *l* is greater than the slope of line *m*.
    b. The slope of line *l* is less than the slope of line *m*.
    c. The slope of line *l* is 15° greater than the slope of line *m*.
    d. The slope of line *l* is $1\frac{1}{2}$ times the slope of line *m*.

4. Refer to the diagram below. Find the area of triangle *ABC*. *(LESSON 5.2)*

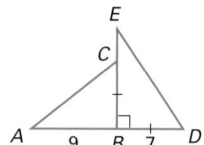

    a. 25 units²
    b. 63 units²
    c. 11.4 units²
    d. 31.5 units²

5. Choose the most complete and accurate description of the two polygons below. *(LESSON 4.6)*

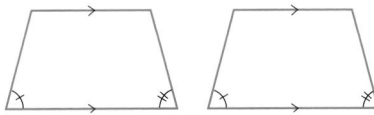

   a. quadrilaterals
   b. trapezoids
   c. similar trapezoids
   d. congruent trapezoids

6. In ⊙*O*, which angle or arc measures 60°? *(LESSON 9.3)*

   a. ∠*ABD*
   b. ∠*BDC*
   c. $\overarc{AD}$
   d. $\overarc{BC}$

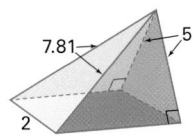

7. The oblique pyramid has a rectangular base. Find its volume. *(LESSON 7.3)*

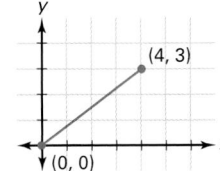

8. The ratio of the volumes of two spheres is 27:1. If the smaller sphere has a radius of 15 in., what is the radius of the larger sphere? *(LESSONS 7.1 AND 7.6)*

For Exercises 9–10, refer to the figure below.

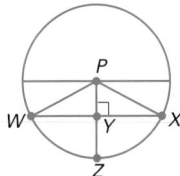

9. Construct a rotation of the segment. Rotate the segment 180° counterclockwise about the endpoint (0, 0). *(LESSON 4.9)*

10. Give the coordinates of the endpoints of the rotated segment from Item 9. *(LESSON 1.7)*

11. Write a paragraph proof that △*PYW* and △*PYX* in ⊙*P* below are congruent. *(LESSONS 4.2 AND 4.3)*

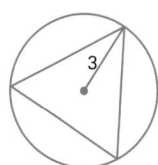

12. An equilateral triangle is inscribed in a circle with a radius of 3 units. If a point is picked at random anywhere inside the circle, what is the probability that the point will *not* be inside the triangle? *(LESSON 5.7)*

FREE-RESPONSE GRID
Items 13–15 may be answered by using a free-response grid such as that commonly used by standardized-test services.

13. Find the volume of a cylinder with a radius of 2 and a height of 7. Round your answer to the nearest tenth. *(LESSON 7.4)*

14. Find the slope of a line that passes through the points (2, 6) and (9, 12). *(LESSON 3.8)*

15. Find the area of the parallelogram below. Round your answer to the nearest hundredth. *(LESSON 5.5)*

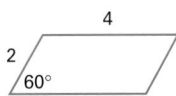

---

5. b.

6. d.

7. $8\sqrt{6} \approx 19.6$ units³

8. 45 in.

9.

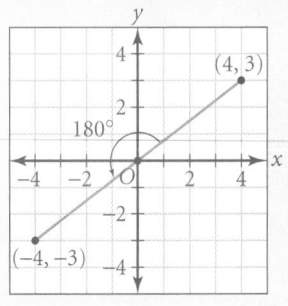

10. (0, 0) and (−4, −3)

11. $\overline{PX} \cong \overline{PW}$ since both are radii of ⊙*P*. ∠*PYX* ≅ ∠*PYW* since both are right angles. $\overline{PY} \cong \overline{PY}$ by the Reflexivity Property. Thus, △*PYX* ≅ △*PYW* by HL.

12. 0.59

13. 88.0 units³

14. $\frac{6}{7}$

15. 6.93 units²

10 Trigonometry

Lesson Presentation CD-ROM
Power Point® presentations for each lesson 10.1-10.7

CHAPTER PLANNING GUIDE

Lesson	10.1	10.2	10.3	10.4	10.5	10.6	10.7	Project and Review
Pupil's Edition Pages	630–638	639–646	647–653	654–662	663–669	672–679	680–685	670–671, 686–695
Practice and Assessment								
Extra Practice (Pupil's Edition)	848	849	849	850	850	851	851	
Practice Workbook	62	63	64	65	66	67	68	
Practice Masters Levels A, B, and C	184–186	187–189	190–192	193–195	196–198	199–201	202–204	
Standardized Test Practice Masters	71	72	73	74	75	76	77	78
Assessment Resources	125	126	127	129	130	131	132	128, 133–138
Visual Resources								
Lesson Presentation Transparencies Vol. 2	49–52	53–56	57–60	61–64	65–68	69–72	73–76	
Teaching Transparencies	86–87	88–89	90–92	93–95		96		
Answer Key Transparencies	355–356	357–358	359–361	362–366	367–369	370–372	373–379	380–382
Quiz Transparencies	10.1	10.2	10.3	10.4	10.5	10.6	10.7	
Teacher's Tools								
Reteaching Masters	123–124	125–126	127–128	129–130	131–132	133–134	135–136	
Make-Up Lesson Planner for Absent Students	62	63	64	65	66	67	68	
Student Study Guide	62	63	64	65	66	67	68	
Spanish Resources	62	63	64	65	66	67	68	
Block Scheduling Handbook								20–21
Activities and Extensions								
Lesson Activities	178–182	183–188	189–191	192–193			194–195	
Enrichment Masters	62	63	64	65	66	67	68	
Cooperative-Learning Activities	62	63	64	65	66	67	68	
Problem-Solving/ Critical Thinking	62	63	64	65	66	67	68	
Student Technology Guide	62	63		64	65	66–67	68	
Long Term Projects								37–40
Writing Activities for Your Portfolio								28–30
Tech Prep Masters								45–48
Building Success in Mathematics								26–27

LESSON PACING GUIDE

Lesson	10.1	10.2	10.3	10.4	10.5	10.6	10.7	Project and Review
Traditional	2 days	2 days	2 days	2 days	2 days	2 days	2 days	2 days
Block	1 day	1 day	1 day	1 day	1 day	1 day	1 day	1 day
Two-Year	4 days	4 days	4 days	4 days	4 days	4 days	4 days	4 days

CONNECTIONS AND APPLICATIONS

Lesson	10.1	10.2	10.3	10.4	10.5	10.6	10.7	Review
Algebra	635	645			667	679	684	
Geometry	630–638	639–646	647–653	654–662	663–669	672–679	680–685	688–695
Coordinate Geometry					669	679	685	
Polar Coordinates				662				
Technology		644						
Trigonometry					669		685	
Life Skills	636	645		657, 661	668	679	685	692
Science and Technology	636	641, 645	649, 652, 653	660, 661	668	673, 678	684	692
Sports and Leisure		640, 645			668	673, 675, 676, 678		
Cultural Connection: Africa	637							

BLOCK-SCHEDULING GUIDE

Day	Lesson	Teacher Directed: Lesson Examples, Teaching Transparencies	Student Guided: Activity, Try This	Cooperative-Learning Activity, Lesson Activity, Student Technology Guide	Practice: Practice & Apply, Extra Practice, Practice Workbook	Assessment: Quiz, Mid-Chapter Assessment	Problem Solving, Reteaching
1	10.1	15 min	15 min.	20 min	55 min	15 min	15 min
2	10.2	15 min	15 min	20 min	55 min	15 min	15 min
3	10.3	15 min	15 min	20 min	55 min	15 min	15 min
4	10.4	15 min	15 min	20 min	55 min	15 min	15 min
5	10.5	15 min	15 min	20 min	55 min	15 min	15 min
6	10.6	15 min	15 min	20 min	55 min	15 min	15 min
7	10.7	15 min	15 min	20 min	55 min	15 min	15 min
8	Assess.	50 min	90 min	90 min	65 min	30 min	
		PE: Chapter Review	**PE:** Chapter Project, Writing Activities	Tech Prep Masters	**PE:** Chapter Assessment, Test Generator	Chap. Assess. (A or B), Alt. Assess. (A or B), Test Generator	

PE: Pupil's Edition

Alternative Assessment

The following suggest alternative assessments for students who may benefit from a different type of assessment than the regular chapter quizzes and the mid-chapter/end-of-chapter test. Visit the HRW web site to get additional Alternative Assessment material.

☑ internet connect

Alternative Assessment
Go To: **go.hrw.com**
Keyword: **MG1 Alt Assess**

Performance Assessment

1. Write a proof of the Law of Sines for right triangles. Use the diagram below.

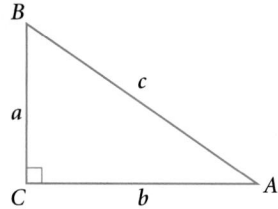

2. Use the diagram below to write the proof of the obtuse case of The Law of Cosines.

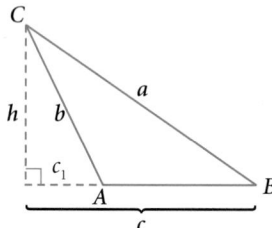

Portfolio Project

Suggest that students choose one of the following projects for inclusion in their portfolios.

1. Using geometry software, design an electronic page that will display trigonometric ratios. Draw a right triangle whose size can be changed by dragging a point on the triangle. As the size changes, have the computer program calculate the trigonometric ratios for the acute angles of the triangle.

2. Describe how you could measure your school flag pole, or some other very tall structure using trigonometric ratios. Then find the actual height of the structure. (There are several ways of finding an angle of elevation. One way is to draw and cut out a right triangle from cardboard, and tape a straw to the hypotenuse side. When looking through the straw, the angle next to the eye is the angle of elevation. Make sure the bottom side of the triangle is parallel to the ground.)

☑ internet connect

The table below identifies the pages in this chapter that contain internet and technology information.

Content Links

Activities Online	**pages 657, 666, 676**
Portfolio Extensions	**pages 638, 646**
Homework Help Online	**pages 635, 644, 652, 659, 667, 677, 683**

Resource Links

Parents can go online and find concepts that students are learning–lesson by lesson–and questions that pertain to each lesson, which facilitate parent-student discussion.

Go To: **go.hrw.com**
Keyword: **MG1 Parent Guide**

Technical Support

The following may be used to obtain technical support for any HRW software product.

Online Help: **www.hrwtechsupport.com**

e-mail: **tschrw@hbtechsupport.com**

HRW Technical Support Center: **(800)323-9239**

7 AM to 10 PM Monday through Friday CST

Visit the HRW math web site at: **www.hrw.com/math**

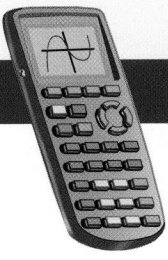

Technology

Technology Objectives and Suggestions

Lesson 10.1 Tangent Ratios

In this lesson students develop the tangent ratio and solve problems using tangent ratios. A scientific calculator will be necessary for this lesson.

The first two activities are appropriate for use with geometry software and spreadsheet software. (Students should discuss how to set up the columns in a spreadsheet.) The class data can be compiled and graphed in the second activity. In groups students can use geometry software to add more data points to the graphs in Activity 2. They can also change the acute angle by 5° or 1° increments to better see the curved shape of the resulting graph.

Lesson 10.2 Sines and Cosines

In this lesson students explore the relationship between the measure of an angle and its sine and cosine. They also solve problems using sine and cosine ratios. A scientific calculator will be necessary for this lesson.

The activities may be done using geometry software such as The Geometer's Sketchpad or Cabrii. Students should construct any right triangle and then measure one acute angle and all three sides. They can have the computer calculate the sine and cosine, and other needed relationships. By dragging on the vertices of the triangle to change the size of the acute angle, students can make and test conjectures.

Lesson 10.3 Extending the Trigonometric Ratios

In this lesson students use a rotating ray on a coordinate plane to define angles measuring greater than 90° and less than 0°. They also define sine, cosine, and tangent for angles of any size.

Geometry software that includes a coordinate system may be used for activities with the unit circle. If students are using Cabrii Geometry II, they choose the Show Axes and the Define Grid tools from the Draw toolbox to set up a coordinate system. They should construct a circle of any size centered at the origin, and a segment from the origin to a point on the circle.

Students measure the angle between this segment and a segment on the x-axis and then use Calculate from the Measure toolbox to find the sine and cosine of this angle. By moving the segment around the circle, students can see the signs of the sine and cosine change in the different quadrants.

Lesson 10.4 The Law of Sines

In this lesson students develop and use the Law of Sines to solve triangles. The example and activity in this lesson, as well as may of the exercises, can be explored or confirmed using geometry software. If the software includes trigonometric functions in the menu or toolbox, computations can be checked directly. If the trigonometric functions are not available, students can construct triangles with the given measures and confirm results by measuring sides or angles.

Lesson 10.5 The Law of Cosines

In this lesson students use the Law of Cosines to solve triangles. Students can use algebraic manipulation software such as $f(g)$ Scholar to follow along with the examples and apply the Law of Cosines. Geometry graphics software that includes the trigonometric functions can be used to show that the Law of Cosines is true for any triangle.

Lesson 10.6 Vectors in Geometry

In this lesson students add vectors and use vector addition to solve problems. Geometry software can be used to construct vector diagrams and find vector sums. Have students copy the figures used in the text, examples, and exercises. By dragging on initial and final points of vectors, students can see what happens to the magnitude and direction as the vectors change.

Lesson 10.7 Rotations in the Coordinate Plane

In this lesson students use transformation equations to rotate points. Geometry software can be used to explore rotations of points and figures on a coordinate plane. Although students do not need to use trigonometry to manipulate the computer software, an understanding of how the computer performs the rotations will be useful to students in other types of problem-solving situations.

Background Information

In this chapter, students develop the tangent, sine, and cosine ratios from both right triangles and unit circles. They explore some trigonometric identities, apply trigonometric rotation equations, and use the law of sines and the law of cosines to solve problems. The chapter concludes with an introduction to vectors, vector addition, and rotations of objects by using transformation equations.

Chapter Objectives

- Develop the tangent ratio by using right triangles. [10.1]

- Use a chart or graph to find the tangent of an angle or the angle for a given tangent. [10.1]

- Solve problems by using tangent ratios. [10.1]

- Explore the relationship between the measure of an angle and its sine and cosine. [10.2]

- Solve problems by using sine and cosine ratios. [10.2]

- Develop two trigonometric identities. [10.2]

Trigonometry

HAVE YOU EVER WONDERED HOW HIGHWAY engineers are able to make sure that a section of a freeway or overpass will correctly match up with a section that is under construction a considerable distance away? Accurate measurements and calculations are necessary to ensure success. In this kind of work, trigonometry is an indispensable tool.

Trigonometry, like much of geometry, depends on triangles. The simple study of the ratios of the sides of right triangles quickly leads to more sophisticated calculation techniques that are widely used in surveying, navigation, and the sciences.

About the Photos

The photos above, from left to right, show the constuction of the M4 bridge across the Neath River, in Baglan Bay, Wales; the construction of the Dartford bridge over the Thames River, in London, England; and the Confederation bridge in Prince Edward Island, Canada. Engineers and surveyors apply trigonometry in the design, construction, and supervision of projects.

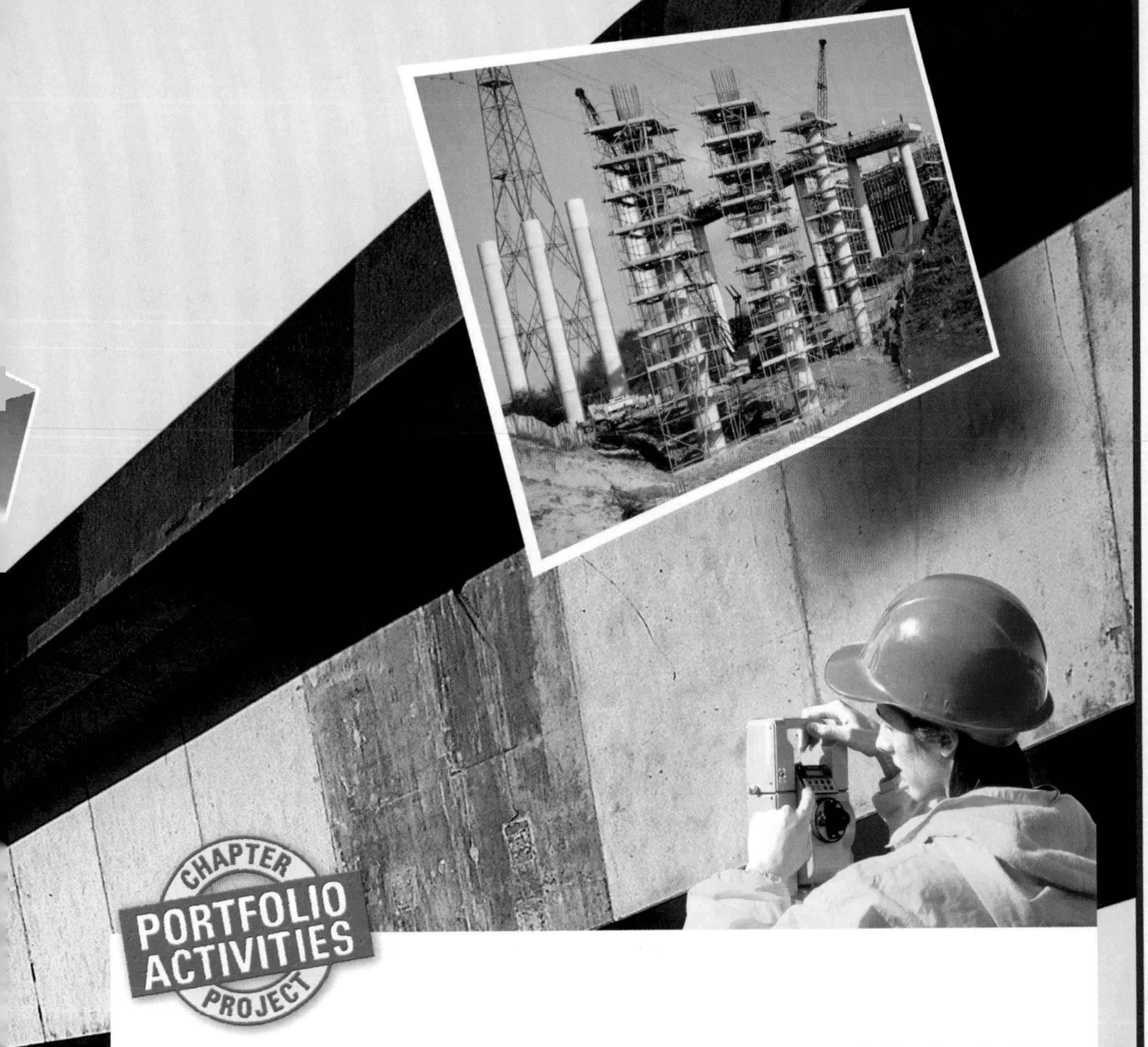

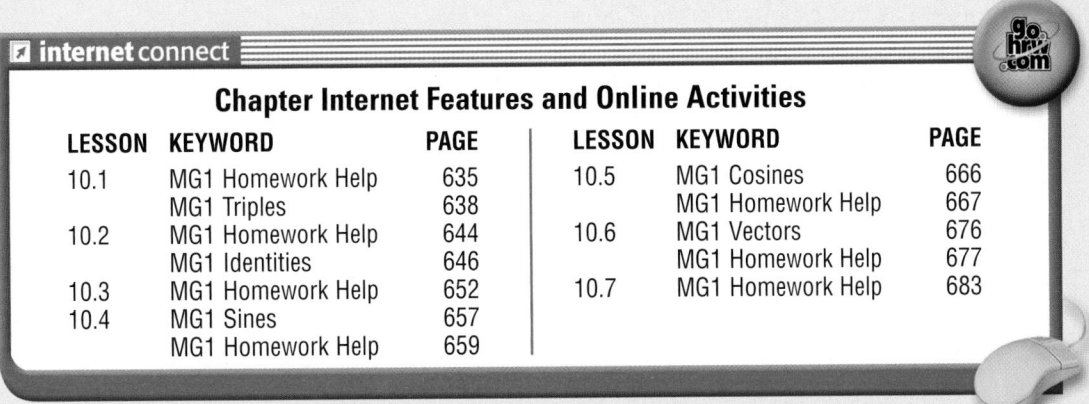

- Use a rotating ray on a coordinate plane to define angles measuring greater than 90° and less than 0°. **[10.3]**
- Define *sine*, *cosine*, and *tangent* for angles of any size. **[10.3]**
- Develop the law of sines. **[10.4]**
- Use the law of sines to solve triangles. **[10.4]**
- Use the law of cosines together with the law of sines to solve triangles. **[10.5]**
- Prove the acute case of the law of cosines. **[10.5]**
- Define *vector*. **[10.6]**
- Add two vectors. **[10.6]**
- Use vectors and vector addition to solve problems. **[10.6]**
- Use transformation equations to rotate points. **[10.7]**
- Use a rotation matrix to rotate points or polygons. **[10.7]**

Portfolio Activities appear at the end of Lessons 10.1 and 10.2. Each serves as preparation for the Chapter Project. The Portfolio Activities, as well as the Chapter Project Activities, are appropriate for inclusion in the student's portfolio. Students should be encouraged to include in their portfolios any other work in which they feel a sense of pride or a sense of accomplishment.

About the Chapter Project

Plimpton 322, the Babylonian clay tablet that you studied in Lesson 5.4, contains some very sophisticated trigonometry.

In the Chapter Project, *Plimpton 322 Revisited*, you will study the arrangement of the numbers in the table and the meaning of the values in the remaining column.

After completing the Chapter Project, you will be able to do the following:

- Read and write numbers in *cuneiform*, a system used in ancient Babylon.
- Use the Babylonian number system to analyze a cuneiform tablet.

About the Portfolio Activities

Throughout the chapter, you will be given opportunities to complete Portfolio Activities that are designed to support your work on the Chapter Project.

The theme of each Portfolio Activity and of the Chapter Project is the Plimpton 322 tablet.

- In the Portfolio Activity on page 638, you will examine the reason for the order of the values in the table.
- In the Portfolio Activity on page 646, you will discover a trigonometric identity that you will use to in examine the tablet.

internet connect

Chapter Internet Features and Online Activities

Construct any right triangle. Label the right angle *C* and the other two angles *A* and *B*. Name each part of the triangle indicated below.

1. the hypotenuse
$\overline{AB}$

2. the leg opposite $\angle A$
$\overline{BC}$

3. the leg adjacent to $\angle A$
$\overline{AC}$

4. the leg opposite $\angle B$
$\overline{AC}$

5. the leg adjacent to $\angle B$
$\overline{BC}$

Also on Quiz Transparency 10.1

Teach

Why Trigonometry relates the sides and angles of a single right triangle. Right triangles appear in numerous applications, such as architecture, topography, and art. The study of trigonometry can help students understand the functional and aesthetic uses of right triangles.

Tangent Ratios

Objectives

- Develop the tangent ratio by using right triangles.
- Use a chart or graph to find the tangent of an angle or the angle for a given tangent.
- Solve problems by using tangent ratios.

Why *Trigonometry is an essential tool of surveyors. One famous survey was the Great Trigonometric Survey of India, which began in 1802.*

Mount Everest is named for Sir George Everest, who was the superintendent of the Great Trigonometric Survey of India from 1823 to 1843. The mountain's height, about 5.5 miles, was computed by using trigonometry.

Tangent Ratios

In the Activities that follow, you will examine one of the three important ratios of trigonometry, the *tangent* ratio.

Activity 1
A Familiar Ratio

YOU WILL NEED

ruler, protractor, and calculator
OR
geometry graphics software

CHECKPOINT ✓

1. Draw an angle between 30° and 50° such that one side of the angle is horizontal. Label the vertex *A*. Draw a vertical segment to create a right triangle.

2. Measure the leg of the triangle opposite *A*.

3. Measure the leg of the triangle adjacent to *A*.

4. Divide the length of the opposite leg by the length of the adjacent leg.

5. Repeat Steps 1–4, using the same angle but different side lengths. What do you notice?

Alternative Teaching Strategy

TECHNOLOGY The first two Activities in this lesson are appropriate for use with spreadsheet software. Students should discuss how to set up the columns. Then class data can be compiled and, for the second Activity, graphed.

CRITICAL THINKING Recall that the slope of a line is its rise divided by its run. How is the ratio you calculated in Activity 1 related to the concept of slope?

Examine the triangles at right. $\angle A$ is congruent to $\angle M$, and $\angle C$ and $\angle O$ are right angles. Thus, by AA Similarity, $\triangle ABC \sim \triangle MNO$.

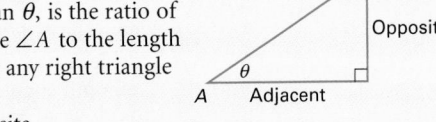

By the Polygon Similarity Postulate, $\frac{BC}{AC} = \frac{NO}{MO}$. This leads to the following definition:

Tangent Ratio

> *It is common in trigonometry to use the Greek letter θ (theta) to represent the measure of an angle.*

For a given acute angle $\angle A$ with a measure of $\theta°$, the **tangent** of $\angle A$, or tan θ, is the ratio of the length of the leg opposite $\angle A$ to the length of the leg adjacent to $\angle A$ in any right triangle having A as one vertex, or

$$\tan \theta = \frac{\text{opposite}}{\text{adjacent}}.$$

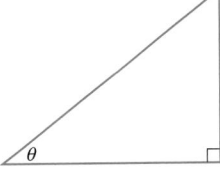

10.1.1

Note: In trigonometry, the letter of the vertex of an angle is often used to indicate the measure of the angle. Thus, the tangent of $\angle A$ can also be written as tan A.

EXAMPLE 1 Measure the legs of the triangle below to find tan θ.

SOLUTION

The leg opposite the angle is 2.9 centimeters, and the leg adjacent to the angle is 3.8 centimeters.

$$\tan \theta = \frac{2.9}{3.8} \approx 0.76$$

TRY THIS Use the given measurements to find the tangent of the other acute angle in the triangle from Example 1.

CRITICAL THINKING What is the measure of an angle with a tangent of 1? Explain your reasoning.

Inclusion Strategies

ENGLISH LANGUAGE DEVELOPMENT Students can easily become confused by the word *tangent*. (The same term is used for a line that touches a circle at one point.) In this lesson, a tangent is both a ratio of two leg lengths and a single number, the decimal equivalent of that ratio. Students may need to review the idea of a ratio, specifically that a ratio can be converted to a fraction or a decimal.

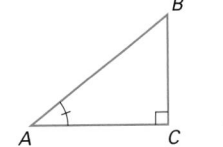 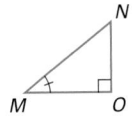

In this Activity, students discover that the tangent ratio depends only on the size of an acute angle. After they have finished the Activity, discuss why their result will be true for all right triangles with one 30° angle. Students should mention the properties of similar triangles.

For a student worksheet of this Activity and detailed Teacher Notes, see page 178 in the Lesson Activities booklet.

CHECKPOINT ✔
5. Using the same angle but different side lengths produces approximately the same ratio.

CRITICAL THINKING
The ratio is the slope of the hypotenuse of the right triangle.

ADDITIONAL EXAMPLE 1

Draw a right triangle and label one of the acute angles as θ. Measure the legs of the triangle and find tan θ.
Answers will vary. Check students' work.

TRY THIS
Let β be the other angle. Then $\tan \beta = \frac{3.8}{2.9} \approx 1.31$.

CRITICAL THINKING
If $\tan \theta = 1$, then $\frac{\text{length of opposite side}}{\text{length of adjacent side}} = 1$. Since the ratio of the two sides is 1, the sides must be of equal length. The triangle must be an isosceles right triangle, so the measure of the angle must be 45°.

Activity 2 Notes

In this Activity, students describe the behaviour of the tangent function by using their own measurements to complete a table of values for tan θ and graphing the function. Students may not be familiar with this type of coordinate system. It may be helpful to display the axes on which students are to graph the ordered pairs (θ, tan θ).

For a student worksheet of this Activity and detailed Teacher Notes, see page 178 in the Lesson Activities booklet.

CHECKPOINT ✔

4. The graph increases at an increasing rate. The graph gets steeper as you go to the right.

Teaching Tip

In Step 4 of Activity 2 (the Checkpoint question), students may not be sure what is meant by a graph that *increases* or *decreases*. Give examples of graphs of increasing and decreasing functions.

Activity 3 Notes

Have students read through the entire Activity before they begin. Explain that they are to find the actual width of a canyon by using scale drawing and the tangent function.

For a student worksheet of this Activity and detailed Teacher Notes, see page 178 in the Lesson Activities booklet.

CHECKPOINT ✔

4. Sample answer: $\tan Z = \frac{XY}{YZ}$ so $0.70 = \frac{XY}{YZ}$ and $0.70 = \frac{XY}{1.2}$.
So, $XY = (0.70)(1.2) \approx 0.84$.

Activity 2
Graphing the Tangent

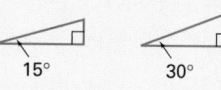

YOU WILL NEED

ruler, protractor, and graph paper
OR
geometry graphics software

1. Draw five triangles as shown, with angles of 15°, 30°, 45°, 60°, and 75°.

2. For each triangle, measure the opposite and adjacent legs to find the tangent of the angle. Copy and complete the table below.

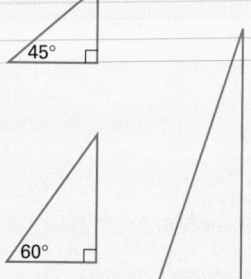

θ	tan θ
15°	?
30°	?
45°	?
60°	?
75°	?

3. Plot the ordered pairs (θ, tan θ) from the table above. Connect the points with a smooth curve.

CHECKPOINT ✔

4. Does your graph increase or decrease? Describe its behavior.

Activity 3
Using the Tangent

YOU WILL NEED

ruler, protractor, and your tangent graph from Activity 2

PROBLEM SOLVING

In this Activity you will model the calculations of a surveying crew measuring the distance across a canyon. Make a scale drawing, and let 1 centimeter = 10 meters.

1. **Make a diagram** showing the sides of a canyon, and draw a line across the canyon representing a line of sight from point X to point Y.

2. Draw a line through Y perpendicular to $\overleftrightarrow{XY}$. Choose a point on this line and label it Z. Draw $\overline{XZ}$.

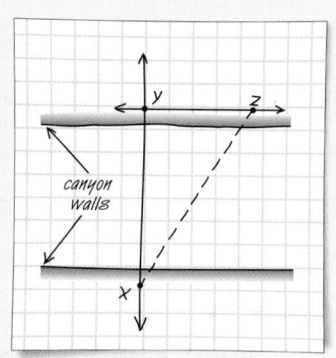

3. Measure $\angle Z$ and $\overline{YZ}$. (Remember, $\overline{XY}$ and $\overline{XZ}$ span the canyon, so they cannot be measured directly.) Use your graph from Activity 2 to estimate the tangent of $\angle Z$.

CHECKPOINT ✔

4. Substitute the values for tan Z and YZ into the equation below.
$$\tan Z = \frac{XY}{YZ}$$
Solve to find XY, the distance across the canyon.

5. Compare the value you calculated for XY with the actual distance on your drawing. How accurate is your answer? What could be some possible sources of error in your estimate?

Interdisciplinary Connection

LANGUAGE ARTS   In "The Musgrave Ritual," a Sherlock Holmes story by Arthur Conan Doyle, Holmes uses similar triangles to find the height of a tree. Have students rewrite the episode so that Holmes uses the tangent ratio to find the tree's height.

Enrichment

Have students work in pairs or small groups to construct a scale model of the pyramid illustrated in the Look Beyond section of the exercises at the end of the lesson. Students may choose any convenient scale factor, such as 1 cm = 18 cubits. Determining the altitudes for the triangles that form the lateral sides is a challenging and interesting problem.

Calculating Tangent Ratios

So far, you have measured the sides of triangles to find tangent ratios. However, since measurements can often be inaccurate, you will usually use a scientific or graphics calculator, or a table such as the one in the infobank in the back of this book.

Use the [TAN] function key on your scientific or graphics calculator to find the tangent of an angle. (Be sure your calculator is in degree mode.)

E X A M P L E ② Use a calculator to find tan 45°. Verify your answer by using a right triangle.

● **SOLUTION**

Using the [TAN] function key on a calculator, tan 45° = 1.

In the triangle at right, m∠A = 45°. By the Triangle Sum Theorem and the Converse of the Isosceles Triangle Theorem, the triangle is isosceles, so the opposite and adjacent legs are equal. Thus, the ratio of the opposite leg to the adjacent leg must equal 1.

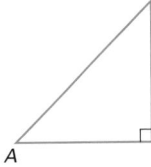

Sometimes it is necessary to find the angle measure for a given tangent ratio. To find the angle with a tangent of $\frac{a}{b}$, you can draw a right triangle with legs of length a and b, and measure the angle.

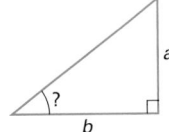

To find the angle using a calculator, use the [TAN⁻¹] key. For any positive number r, the [TAN⁻¹] key gives the measure of an angle between 0° and 90° whose tangent is r. The tan⁻¹ function is called the **inverse tangent** function. Notice that the input and output for the tangent and inverse tangent functions are reversed.

Function and key	Given (input)	Want to find (output)
tangent [TAN]	angle measure	tangent ratio
inverse tangent [TAN⁻¹]	tangent ratio	angle measure

E X A M P L E ③ Use your calculator to find an angle that has a tangent of $\frac{2}{3}$. Round your answer to the nearest degree. Verify your answer using a right triangle.

● **SOLUTION**

Using a calculator, $\tan^{-1} \frac{2}{3} = 33.69 \approx 34°$.

The triangle at right has a leg of 2 cm and a leg of 3 cm. The measure of the angle is approximately 34°.

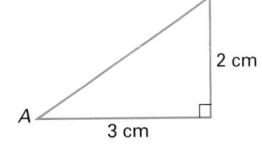

Reteaching the Lesson

COOPERATIVE LEARNING Have students work in small groups to draw a variety of right triangles on graph paper. The legs of the triangles should be whole numbers so that the tangent ratios are easy to calculate. For each triangle, students should measure the appropriate acute angle to the nearest degree. Then they should compare their results with those given by a scientific calculator.

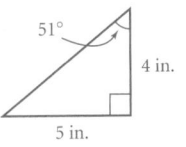

ASSIGNMENT GUIDE

In Class	1–11
Core	12–17, 19–35 odd
Core Plus	19–29 odd, 30–39
Review	40–46
Preview	47–49

✐ Extra Practice can be found beginning on page 818.

Technology

Most scientific calculators can be used to find the tangent of a given angle or the angle for a given tangent.

Exercises

● Communicate

1. Where do you think the word *trigonometry* comes from? (Hint: What could be another name for a tri-gon?)

2. Does the tangent ratio increase or decrease as an angle gets larger? Explain your answer.

3. What happens to the tangent ratio as an angle approaches 0°? Use your calculator the find the tangent of 0°. Does your answer make sense? Why or why not?

4. What happens to the tangent ratio as an angle approaches 90°? Try to find the tangent of 90° by using your calculator. What happens? Explain why in terms of right triangles.

5. When measuring the sides of a triangle to find the tangent ratio, does it matter what units you use? What happens to the units in your answer?

● Guided Skills Practice

Measure the sides of the triangles below to find tan A. *(EXAMPLE 1)*

6. 0.62; Answers may vary due to inaccuracies in measurement.

7. 1.42; Answers may vary due to inaccuracies in measurement.

6.

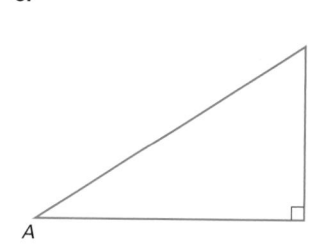

7.

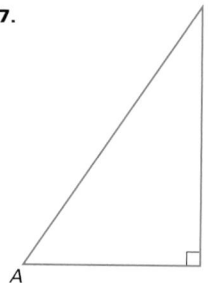

Use the graph you created in Activity 2 to estimate the tangent of each angle. Then find the tangent by using your calculator, and compare your answers. *(ACTIVITY 2 AND EXAMPLE 2)*

8. 20° 0.36   **9.** 40° 0.84   **10.** 70° 2.75

11. Use your calculator to find an angle that has a tangent of $\frac{3}{4}$. Round to the 37° nearest degree. Verify your answer by using a right triangle. *(EXAMPLE 3)*

Practice and Apply

Find tan *A* for each triangle below.

12. 0.5714

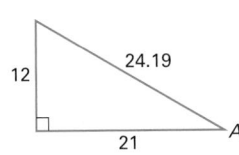
12   24.19

21

A

13. 2.5

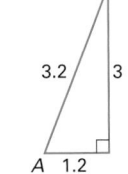
3.2   3

A 1.2

14. 0.6667

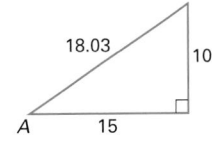
18.03   10

A   15

15. 2.4

26   24

A 10

16. 0.75

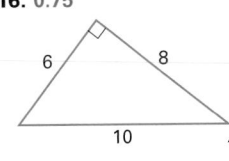
6   8

10   *A*

17. tan ∠*A* = 1

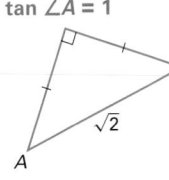
√2

A

Use a scientific or graphics calculator to find the tangent of each angle below. Round to the nearest hundredth.

18. 25° 0.47

19. 67° 2.36

20. 19° 0.34

21. 53° 1.33

22. 75° 3.73

23. 89° 57.29

Use a scientific or graphics calculator to find the inverse tangent of each ratio below. Round to the nearest degree.

24. $\frac{3}{8}$ 21°

25. $\frac{7}{5}$ 54°

26. 3 72°

27. 9.5 84°

28. 1 45°

29. 0 0°

Algebra

For Exercises 30–34, use the definition of tangent ratio to write an equation involving *x*. Find the tangent of the given angle by using a calculator, and solve the equation to find the unknown side of the triangle. Round your answer to the nearest hundredth.

30. 13.56

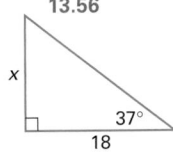
x

37°

18

31. 2

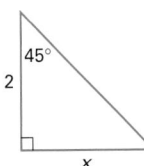
45°

2

x

32. 37.25

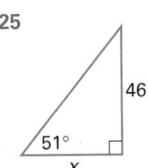
46

51°

x

33. 22.57

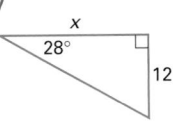
x

28°

12

CHALLENGE

34. Use the tangent ratio and the Pythagorean Theorem to find *x* and *y* in the triangle at right. Round to the nearest tenth.

x ≈ 11.5; *y* ≈ 9.7

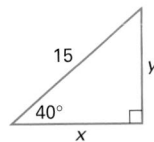
15   *y*

40°

x

35. SURVEYING Use the diagram at right to determine $\overline{AB}$, the distance across the lake. **494.2 m**

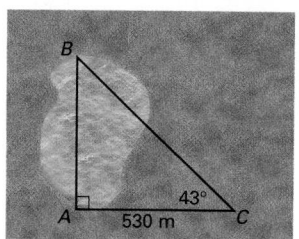

36. ENGINEERING The steepness, or grade, of a highway or railroad is expressed as a percent. In the photo of the cog railway at Pike's Peak, in Colorado, the grade is 18 percent. Thus, for every 100 ft of horizontal run, the train rises 18 ft. Find the angle of inclination of the railway. **10.2°**

37. ENGINEERING The maximum grade of the railway at Pike's Peak is 25 percent. Find the angle of inclination of the railway at this point. **14.0°**

38. INDIRECT MEASUREMENT Use the figure at left to estimate the height of the flagpole. Round to the nearest meter. **18.0 m**

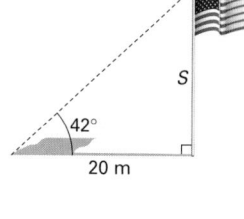

39. SURVEYING In the photo below of Glen Canyon, Utah, point P is on the north rim, point Q is on the south rim, $PR = 300$ ft, $\overline{PQ} \perp \overline{PR}$, and $m\angle R = 75°$. Find PQ, the width of the canyon. **1119.6 ft**

NAME _____ CLASS _____ DATE _____

Practice
10.1 *Tangent Ratios*

Find tan *A* for each triangle below.

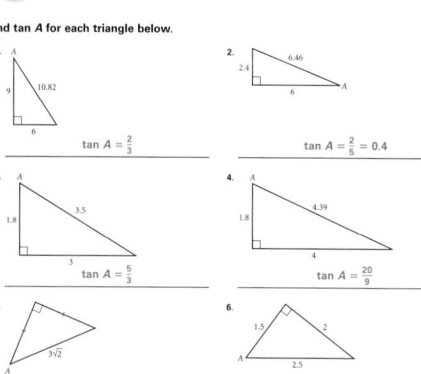

1. $\tan A = \dfrac{2}{3}$

2. $\tan A = \dfrac{2}{5} = 0.4$

3. $\tan A = \dfrac{5}{3}$

4. $\tan A = \dfrac{20}{9}$

5. $\tan A = 1$

6. $\tan A = \dfrac{4}{3}$

For Exercises 7–9, use the definition of tangent ratio to write an equation involving *x*. Find the tangent of the given angle with a calculator, and solve the equation to find the unknown side of the triangle. Round your answers to the nearest hundredth.

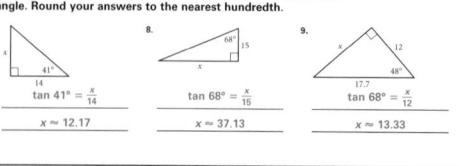

7. $\tan 41° = \dfrac{x}{14}$

$x \approx 12.17$

8. $\tan 68° = \dfrac{x}{15}$

$x \approx 37.13$

9. $\tan 68° = \dfrac{x}{12}$

$x \approx 13.33$

Look Back

Find the volume and surface area of each solid.

40. $V = 240$ units^3
$S = 248$ units^2

41. $V \approx 1099.56$ units^3
$S \approx 596.90$ units^2

42. $V \approx 728.64$ units^3
$S \approx 502.09$ units^2

43. $V \approx 65,449.85$ units^3
$S \approx 7853.98$ units^2

40. right prism
(LESSON 7.2)

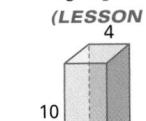

41. cylinder
(LESSON 7.4)

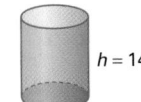

42. cone
(LESSON 7.5)

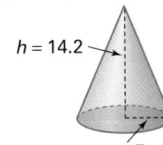

43. sphere
(LESSON 7.6)

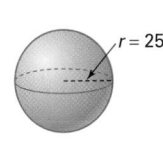

Use ⊙P, with $\overline{MN} \perp \overline{PR}$ for Exercises 44–46. *(LESSONS 9.4 AND 9.5)*

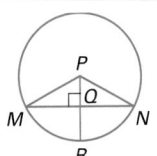

44. $MQ = $ ___**? NQ**

45. $PR = 8, PQ = 3$
$MQ = $ ___**? 7.42** units $\quad QN = $ ___**? 7.42** units

46. $PR = 12, PQ = 4$
$MQ = $ ___**? 11.31** units $\quad QN = $ ___**? 11.31** units

Look Beyond

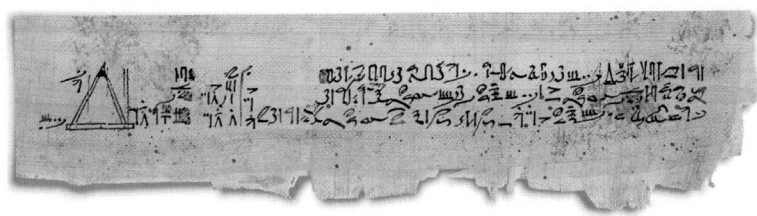

CULTURAL CONNECTION: AFRICA **Trigonometry has been used by many cultures for over 4000 years. The following problem is taken from an ancient Egyptian papyrus:**

> *If a pyramid is 250 cubits high and the side of its base is 360 cubits long, what is its seked?* —Problem 56 of the Rhind papyrus

A *cubit* is a measure of length equal to about 21 inches, and the *seked* is a measure of steepness.

47. In the diagram at right, what is the ratio of the run to the rise of the pyramid? (Notice this is not the same as the slope.) **0.72**

48. How is the ratio of run to rise related to the concept of the tangent ratio? **reciprocal of the tangent ratio**

49. The seked is the ratio you computed in Exercise 47, converted into palms per cubit. Since 1 cubit = 7 palms, this is found by multiplying the ratio by 7. Find the seked of the pyramid. **5.04**

rise = 250 cubits

run = 180 cubits

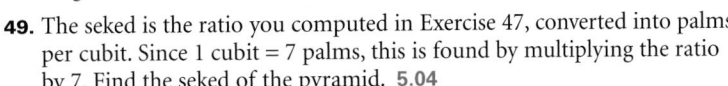
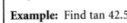

Exercises 47–49 extend the concept of the tangent ratio to a related ratio, the *seked*, from ancient Egypt.

Portfolio Activity

The Portfolio Activity can be used as preparation for the Chapter Project or as a separate activity. In the Portfolio Activity on this page, students explore and discover the reason for the order of the values on the Plimton 322 tablet.

TRIGONOMETRY IN BABYLONIAN MATHEMATICS

Recall from the Plimpton 322 tablet (see Lesson 5.4) that two columns represented a leg and the hypotenuse of Pythagorean triples. The arrangement of the numbers may seem random, but the reason for the arrangement becomes clear when you consider the angles formed in the right triangles.

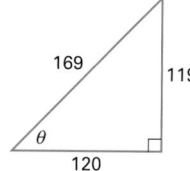

1. Complete the table below. What do you notice about the arrangement of the values of the angles? Why do you think this table might have been useful?

Column II	Column III			
Opposite leg	Hypotenuse	Adjacent leg	tan θ	θ
119	169	120	?	?
3367	4825	3456	?	?
4601	6649	4800	?	?
12,709	18,541	13,500	?	?
65	97	72	?	?
319	481	360	?	?
2291	3541	2700	?	?
799	1249	960	?	?
481	769	600	?	?
4961	8161	6480	?	?
45	75	60	?	?
1679	2929	2400	?	?
161	289	240	?	?
1771	3229	2700	?	?
56	106	90	?	?

internet connect

Portfolio Extension

Go To: **go.hrw.com**
Keyword:
MG1 Triples

2. The first angle is almost equal to 45°. Is it possible to find a Pythagorean triple that corresponds to a 45° angle? Why or why not?

WORKING ON THE CHAPTER PROJECT

You should now be able to complete Activity 1 of the Chapter Project.

Sines and Cosines

Objectives

- Explore the relationship between the measure of an angle and its sine and cosine.

- Solve problems by using sine and cosine ratios.

- Develop two trigonometric identities.

Why Sine and cosine ratios can be used to solve many types of problems. For example, if you know the length and the approximate angle of the rope, you can estimate the height of a parasailer above the water.

Prepare

QUICK WARM-UP

One acute angle in a right triangle increases by 5° increments. What happens to the other acute angle? The other acute angle decreases by 5° increments.

Also on Quiz Transparency 10.2

Teach

Why In Lesson 10.1, students learned to use the tangent of an angle to find missing measures of a right triangle. In this lesson, they will see how the sine and cosine of an angle can be used for similar purposes.

Trigonometric Ratios

In Lesson 10.1, you learned that the tangent of an angle is the ratio of two sides of a right triangle containing the given angle. However, there are other trigonometric ratios that can be formed by using different sides of the same triangle. The three most important ratios are the tangent, the sine, and the cosine.

Sine and Cosine Ratios

For a given angle $\angle A$ with a measure of $\theta°$, the **sine** of $\angle A$, or $\sin \theta$, is the ratio of the length of the leg opposite A to the length of the hypotenuse in a right triangle with A as one vertex, or

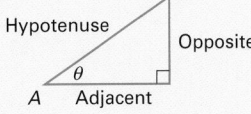

$$\sin \theta = \frac{\text{opposite}}{\text{hypotenuse}}.$$ **10.2.1**

The **cosine** of $\angle A$, or $\cos \theta$, is the ratio of the length of the leg adjacent to A to the length of the hypotenuse, or

$$\cos \theta = \frac{\text{adjacent}}{\text{hypotenuse}}.$$ **10.2.2**

Alternative Teaching Strategy

USING TECHNOLOGY The Activities in this lesson may be done with geometry graphics software. Students should construct a right triangle and measure one acute angle and all three sides. Then they can use the features of the software to find the sine, cosine, and other relationships. By dragging the vertices of the triangle to change the size of the acute angle, students should be able to make and test conjectures about the relationships between the measures.

Activity 1 Notes

In this Activity, students investigate the relationship between the size of an angle and its sine or cosine. Ask students to predict the shapes of the curves before they make the sine and cosine graphs. Students should be able to tell from their data that they will not get straight lines for the graphs; that is, the functions are not linear.

For a student worksheet of this Activity and detailed Teacher Notes, see page 183 in the Lesson Activities booklet.

CHECKPOINT ✔

3. Sample answer: The values of sin θ increase as θ increases from 0° to 90°. The values of cos θ decrease as θ increases from 0° to 90°. The values of both sin θ and cos θ are between 0 and 1 for values of θ between 0° and 90°.

CRITICAL THINKING

In a right triangle, the hypotenuse is always the longest side, so the sine, $\frac{\text{opposite}}{\text{hypotenuse}}$, or cosine, $\frac{\text{adjacent}}{\text{hypotenuse}}$, of an angle can never be greater than 1.

Teaching Tip

To facilitate students' comprehension, have them draw a figure similar to the one in Example 1 as they listen to you read the problem.

ADDITIONAL
EXAMPLE ❶

A 12-foot guy wire is to be attached to a tree at a 30° angle to the ground. At what height on the tree should the guy wire be attached? How far from the tree should the wire be attached to the ground? **6 ft; ≈10.4 ft**

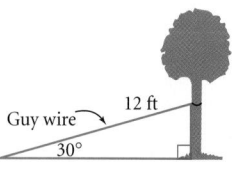

Activity 1
Sines and Cosines

YOU WILL NEED

scientific or graphics calculator and graph paper

1. Examine the triangles in the diagram below. As θ increases, what happens to the value of sin θ? Does it increase or decrease? What happens as θ gets close to 0°? to 90°? Write a conjecture about the sine of 0° and of 90°.

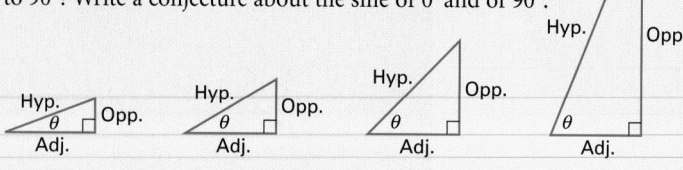

In this sequence of triangles, the adjacent legs stay the same. What happens to the other sides?

2. Repeat Step 1 for cos θ.

CHECKPOINT ✔

3. Compare the behavior of the sine and cosine ratios as the measure of an angle increases from 0° to 90°.

4. Copy and complete the table below. (Use the ⎡SIN⎤ and ⎡COS⎤ keys on your calculator.) Round your answers to the nearest hundredth.

Be sure your calculator is in degree mode.

θ	0°	10°	20°	30°	40°	50°	60°	70°	80°	90°
sin θ	?	?	?	?	?	?	?	?	?	?
cos θ	?	?	?	?	?	?	?	?	?	?

5. Plot the pairs (θ, sin θ) for the angles in the table. Draw a smooth curve through the points. Repeat for the pairs (θ, cos θ). Do your graphs verify the conclusions you made in Step 3?

CRITICAL THINKING Could the sine or cosine of an angle ever be bigger than 1? Explain your answer in terms of right triangles.

EXAMPLE ❶

APPLICATION
RECREATION

A paraglider is towed behind a boat by 400-ft ropes attached to the boat at a point 15 ft above the water. The spotter in the boat estimates the angle of the ropes to be 35° above the horizontal. Estimate the paraglider's height above the water.

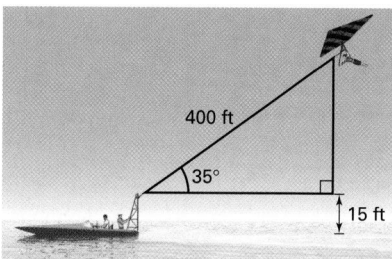

SOLUTION

Label the hypotenuse and the opposite and adjacent legs for the given angle. The hypotenuse is 400 ft. Since the height is the opposite side, use the sine ratio.

$$\sin 35° = \frac{\text{opposite}}{\text{hypotenuse}} = \frac{x}{400} \qquad \textit{Using a scientific calculator, sin 35° = 0.5736.}$$

$$0.5736 = \frac{x}{400}$$

$$x = 400(0.5736) \approx 229 \text{ ft}$$

Thus, the height of the paraglider is approximately 229 + 15 = 244 ft.

Interdisciplinary Connection

PHYSICS Snell's law describes how a ray of light is refracted as it passes from air or a vacuum into any denser material such as water, oil, or glass. One expression of this principle is $\sin \theta_1 \div \sin \theta_2 = \frac{n_2}{n_1}$. The ratio on the right side is a ratio of the two indices of refraction. Have students use their science textbooks or library resources to research Snell's law. Their written reports should include diagrams like the one shown at right.

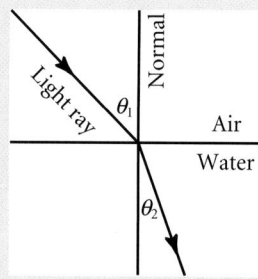

EXAMPLE 2

APPLICATION
ASTRONOMY

Charles, an amateur astronomer, has been keeping records of the position of the sunrise each day as viewed from a fixed point, A. He marks the position of the sunrise on the autumnal equinox and the winter solstice. (On the autumnal equinox, there are equal hours of daylight and darkness, and on the winter solstice, there are the fewest hours of daylight.) What is the measure of ∠A, the angle between the positions of sunrise on these two days?

● SOLUTION

Charles uses ropes to mark the lines of sight from point A in the direction of the sunrise on the two days. He uses another rope to make a line perpendicular to the equinox line as shown, and measures the distances from point A to this rope.

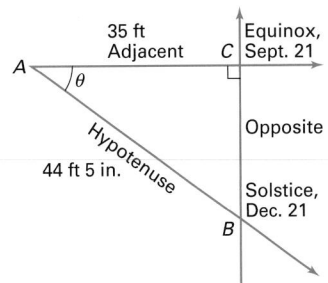

$$\cos \theta = \frac{AC}{AB} = \frac{35 \text{ ft}}{44.4 \text{ ft}} \approx 0.788$$

To find the angle whose cosine is 0.788, use the inverse cosine, or $\cos^{-1}$ key on your scientific or graphics calculator.

$$\cos^{-1} 0.788 \approx 38°$$

An **identity** is an equation that is true for all values of the variables in the equation for which all terms in the equation are defined. For example, the equation $(a + b)(a - b) = a^2 - b^2$ is an identity, because it is true for all values of a and b. In Activity 2, you will discover two trigonometric identities.

Activity 2
Two Trigonometric Identities

YOU WILL NEED
scientific or graphics calculator

Part I

1. Copy and complete the table below by using a scientific or graphics calculator. Leave at least three digits after the decimal point for each value.

θ	sin θ	cos θ	$\frac{\sin \theta}{\cos \theta}$	tan θ
20°	?	?	?	?
40°	?	?	?	?
60°	?	?	?	?

CHECKPOINT ✔

2. What do you notice about the values in the tangent column? Write a trigonometric identity involving the sine, cosine, and tangent ratios.

3. What happens when you simplify the right side of the equation below?

$$\frac{\sin \theta}{\cos \theta} = \frac{\dfrac{\text{opposite}}{\text{hypotenuse}}}{\dfrac{\text{adjacent}}{\text{hypotenuse}}}$$

Does the equation prove your identity from Step 2? Explain.

Inclusion Strategies

USING COGNITIVE STRATEGIES   Some students may have difficulty remembering which side lengths to use in the sine, cosine, and tangent ratios. Here is one mnemoic device: **S**ome **O**ld **H**en **C**aught **A**nother **H**en **T**aking **O**ne **A**way. The initial letters of the words (SOHCAHTOA) indicate the ratios: sine = opposite ÷ hypotenuse, cosine = adjacent ÷ hypotenuse, and tangent = opposite ÷ adjacent.

Enrichment

Have students prove the following statement: The ratio of two legs of any right triangle is equal to the ratio of the sines of the angles opposite those legs. Sketches will help students determine the answer.

$$\sin A = \frac{a}{h} \qquad\qquad \sin B = \frac{b}{h}$$
$$h = \frac{a}{\sin A} \qquad\qquad h = \frac{b}{\sin B}$$

$$\frac{a}{\sin A} = \frac{b}{\sin B}$$
$$\frac{a}{b} = \frac{\sin A}{\sin B}$$

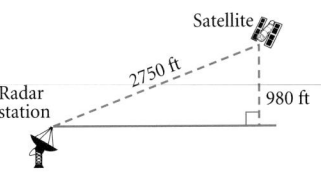

2. All the values are very near 1.
$(\sin \theta)^2 + (\cos \theta)^2 = 1$

Teaching Tip

In Step 3 of Part II in Activity 2, make sure that students are careful with the placement of the parentheses in the expressions.

Part II

1. Copy and complete the table below by using a scientific or graphics calculator. Leave at least three digits after the decimal point for each value.

This table will help you discover a Pythagorean identity (see page 646).

θ	$\sin \theta$	$\cos \theta$	$(\sin \theta)^2 + (\cos \theta)^2$
20°	?	?	?
40°	?	?	?
60°	?	?	?

CHECKPOINT ✔

2. What do you notice about the entries in the last row of the table? Write a trigonometric identity involving sine and cosine ratios.

PROOF

3. The following is a partial proof of the identity from Step 2. For each step in the proof, give a reason.

$$(\sin \theta)^2 + (\cos \theta)^2 = \left(\frac{a}{c}\right)^2 + \left(\frac{b}{c}\right)^2$$

$$= \frac{a^2}{c^2} + \frac{b^2}{c^2}$$

$$= \frac{a^2 + b^2}{c^2}$$

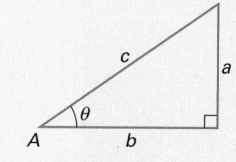

Compare the numerator and denominator of the fraction in the last step. How could you complete the proof of the identity?

Summary

As you saw in Activity 1, the sine of an angle increases from 0 to 1 as the measure of the angle increases from 0° to 90°, while the cosine of an angle decreases from 1 to 0 as the angle increases from 0° to 90°. The graphs are shown at right.

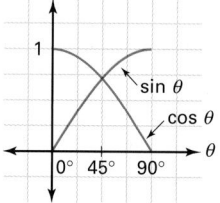

In Activity 2, you discovered and proved the following two important trigonometric identities:

Two Trigonometric Identities

$\tan \theta = \dfrac{\sin \theta}{\cos \theta}$ $(\sin \theta)^2 + (\cos \theta)^2 = 1$

10.2.3   10.2.4

Mathematical ideas often have intriguing connections like these, which is part of the fun of mathematics. In your later studies, you will see these identities often—they seem to pop up everywhere!

Reteaching the Lesson

COOPERATIVE LEARNING Have students work in pairs to draw four similar right triangles of different sizes. The triangles should not be isosceles; that is, they should not have two 45° angles. Have students use subscripted variables for the vertices: $C_1, C_2, C_3,$ and C_4 for the four right angles; $A_1, A_2, A_3,$ and A_4 for one set of congruent acute angles; and $B_1, B_2, B_3,$ and B_4 for the other set of congruent acute angles.

Students should measure the side lengths to the neareast tenth of a centimeter. Then they should write the sine, cosine, and tangent ratios for all the acute angles. Using four triangles of different sizes will help emphasize that the trigonometric ratios are related only to the size of the acute angle and are independent of the side lengths.

Exercises

Communicate

1. Think of a mnemonic (memory aid) to help you remember the parts of the tangent, sine, and cosine ratios.

$$\tan \theta = \frac{\text{opposite}}{\text{adjacent}} \qquad \sin \theta = \frac{\text{opposite}}{\text{hypotenuse}} \qquad \cos \theta = \frac{\text{adjacent}}{\text{hypotenuse}}$$

2. Describe three ways you could find θ in the triangle at right. Do your answers agree for the three ways?

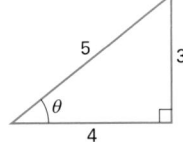

3. In Activity 2, you found that $\tan \theta = \frac{\sin \theta}{\cos \theta}$. Use the behavior of $\sin \theta$ and $\cos \theta$ to explain the behavior of $\tan \theta$ for values of θ close to $0°$ and close to $90°$.

4. Which of the following equations are identities? Explain your reasoning in each case.
a. $2x + 5 = 7$
b. $n + 2n = 3n$
c. $a^2 + b^2 = c^2$
d. $\sin \theta = \tan \theta \cdot \cos \theta$

Guided Skills Practice

Determine the height of each triangle. Round to the nearest foot.
(EXAMPLE 1)

5. 14 feet

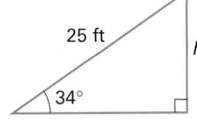

6. 7 feet

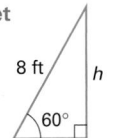

7. 13 feet

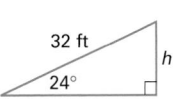

Find θ in each triangle. Round to the nearest degree. *(EXAMPLE 2)*

8. 34°

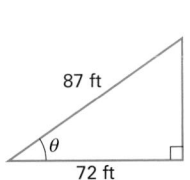

9. 59°

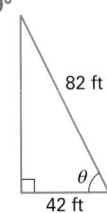

Selected Answers
Exercises 5–9, 11–59 odd

ASSIGNMENT GUIDE

In Class	1–9
Core	11–37, 38–47 odd
Core Plus	38–51
Review	52–60
Preview	61–62

✐ Extra Practice can be found beginning on page 818.

Technology

A scientific calculator is needed for most of the exercises in this lesson, although students could use a table of trigonometric function values. Geometry graphics software can be helpful for the co-function investigations in Exercises 41–43.

Error Analysis

The notation used in Exercises 30–35 may confuse some students. Explain that an expression such as $\sin^{-1} 0.3$ means the angle with a sine equal to 0.3. Since expressions of this type look so different from the typical angle notation $\angle ABC$, some students may need extra practice to become comfortable with the notation.

● **Practice and Apply**

For Exercises 10–17, refer to △CDE.
Find each of the following:

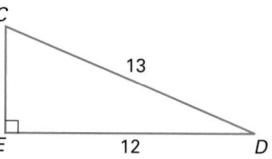

10. $\sin C$ 0.9231   11. $\sin D$ 0.3846

12. $\cos C$ 0.3846   13. $\cos D$ 0.9231

14. $\tan C$ 2.4   15. $\tan D$ 0.4167

16. $m\angle C$ 67.38° 17. $m\angle D$ 22.62°

For Exercises 18–23, refer to △XYZ.
Find each of the following:

18. $\sin \underset{\angle X}{\underline{\ ?\ }} = \frac{15}{17}$ 19. $\cos \underline{\ ?\ } = \frac{15}{17}$ $\angle Y$

20. $\sin \underset{\angle Y}{\underline{\ ?\ }} = \frac{8}{17}$ 21. $\cos \underline{\ ?\ } = \frac{8}{17}$ $\angle X$

22. $\tan \underset{\angle X}{\underline{\ ?\ }} = \frac{15}{8}$ 23. $\tan \underline{\ ?\ } = \frac{8}{15}$ $\angle Y$

TECHNOLOGY Use a scientific or graphics calculator to find the following. Round your answers to the nearest hundredth.

24. $\sin 35°$ 0.57   25. $\cos 72°$ 0.31   26. $\sin 57°$ 0.84

27. $\cos 52°$ 0.62   28. $\sin 45°$ 0.71   29. $\cos 45°$ 0.71

Round your answers to the nearest degree.

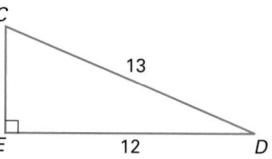

Homework Help Online
Go To: **go.hrw.com**
Keyword:
MG1 Homework Help
for Exercises 30-35

30. $\sin^{-1} 0.3$ 17°   31. $\sin^{-1} 0.875$ 61°   32. $\cos^{-1} 0.56$ 56°

33. $\cos^{-1} 0.125$ 83°   34. $\sin^{-1} 0.5$ 30°   35. $\cos^{-1} 0.95$ 18°

Use trigonometric ratios to find the area of each figure.

36. triangle ABC **17.2 units²**   37. parallelogram $KLMN$ **67.5 units²**

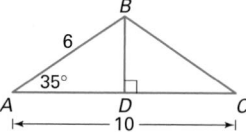 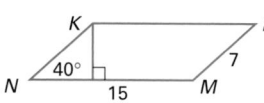

TECHNOLOGY Use a scientific or graphics calculator to answer Exercises 38–40.

38. Choose an angle between 0° and 90°. Find the sine of your angle, then find the inverse sine of your answer. Repeat for four more angles. Based on your observation, complete the following identity:
$$\sin^{-1}(\sin \theta) = \underline{\quad ? \quad} \qquad \sin^{-1}(\sin \theta) = \theta$$

39. Repeat Exercise 38 for the cosine and inverse cosine. Complete the following identity:
$$\cos^{-1}(\cos \theta) = \underline{\quad ? \quad} \qquad \cos^{-1}(\cos \theta) = \theta$$

40. Repeat Exercise 38 for the tangent and inverse tangent. Complete the following identity:
$$\tan^{-1}(\tan \theta) = \underline{\quad ? \quad} \qquad \tan^{-1}(\tan \theta) = \theta$$

Practice

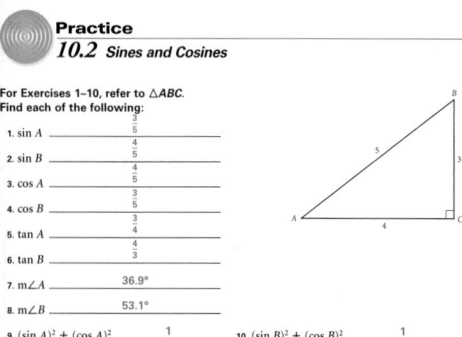

NAME _____ CLASS _____ DATE _____

Practice
10.2 *Sines and Cosines*

For Exercises 1–10, refer to △ABC.
Find each of the following:

1. sin A _____ $\frac{3}{5}$
2. sin B _____ $\frac{4}{5}$
3. cos A _____ $\frac{4}{5}$
4. cos B _____ $\frac{3}{5}$
5. tan A _____ $\frac{3}{4}$
6. tan B _____ $\frac{4}{3}$
7. m∠A _____ 36.9°
8. m∠B _____ 53.1°
9. (sin A)² + (cos A)² _____ 1   10. (sin B)² + (cos B)² _____ 1

For Exercises 11–16, refer to △DEF.
Find each of the following:

11. sin _D_ = $\frac{1}{\sqrt{5}}$
12. cos _E_ = $\frac{1}{\sqrt{5}}$
13. cos _D_ = $\frac{2}{\sqrt{5}}$ 14. sin _E_ = $\frac{2}{\sqrt{5}}$
15. tan _D_ = $\frac{1}{2}$ 16. tan _E_ = 2

Use a scientific or graphics calculator for Exercises 17–25. Round your answers to the nearest hundredth.

17. sin 22° 0.37   18. cos 78° 0.21   19. tan 12° 0.21
20. cos 33° 0.84   21. sin 18° 0.31   22. tan 2° 0.03
23. cos 54° 0.59   24. sin 82° 0.99   25. tan 76° 4.01

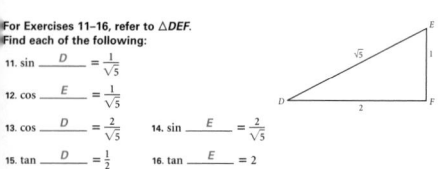

644 LESSON 10.2

The prefix *co-* in cosine indicates a certain relationship of the cosine to the sine of an angle. The exercises below develop this relationship.

41. In △*ABC* below, find sin *A* and cos *B*. Using the right triangle, explain why these ratios are the same.

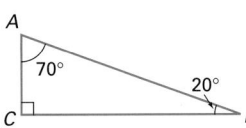

42. Complete the following statements. You may wish to draw right triangles to help determine your answers.

sin 30° = cos ___?__ **60°**   sin 65° = cos ___?__ **25°**
sin ___?__ **50°** = cos 40°   sin ___?__ **45°** = cos 45°

43. What is the relationship between the pairs of angles in Exercises 41 and 42? Use your answer to complete the following identities:

sin θ = cos ___?__ **(90 − θ)**   cos θ = sin ___?__ **(90 − θ)**

Algebra

Use the identities tan θ = $\frac{\sin θ}{\cos θ}$ and (sin θ)² + (cos θ)² = 1 to simplify the following expressions:

44. tan θ · cos θ **sin θ**

45. $\frac{\sin θ}{\tan θ}$ **cos θ**

46. 1 − (sin θ)² **(cos θ)²**

47. $\frac{1}{(\cos θ)^2}$ − 1 **(tan θ)²**

CHALLENGE

48. Use the identity (sin θ)² + (cos θ)² = 1, together with factoring and substitution, to prove the following identity:
(sin θ)⁴ − (cos θ)⁴ = 2(sin θ)² − 1

APPLICATIONS

49. FORESTRY A spruce tree is approximately the shape of a cone with a slant height of 20 ft. The angle formed by the tree with the ground measures 72°. Estimate the height of the tree. Round to the nearest foot. **19 feet**

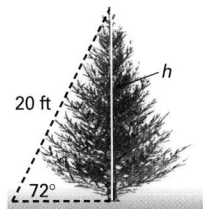

50. RECREATION A water slide is a straight ramp 25 m long that starts from the top of a tower 21 m high. Find the angle the slide forms with the tower. **33°**

51. CONSTRUCTION According to the Americans With Disabilities Act, a ramp can rise no more than 1 ft for every 12 ft of horizontal distance. What is the maximum angle that the ramp can form with the ground? **4.8°**

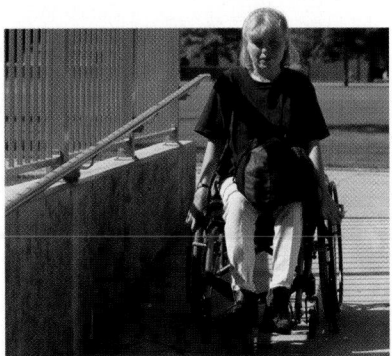

41. sin 70° = 0.9397; cos 20° = 0.9397; sin 70° and cos 20° are the same because the side opposite the 70° angle is the side adjacent to the 20° angle.

48. Since (sin θ)² + (cos θ)² = 1, then
(cos θ)² = 1 − (sin θ)² and
[(cos θ)²]² = [1 − (sin θ)²]² or
(cos θ)⁴ = 1 − 2 · (sin θ)² + (sin θ)⁴. So,
(sin θ)⁴ − (cos θ)⁴
= (sin θ)⁴ − [1 − 2 · (sin θ)² + (sin θ)⁴]
        by substitution
= (sin θ)⁴ − 1 + 2 · (sin θ)² − (sin θ)⁴,
        by the Distributive Property
= −1 + 2 · (sin θ)²
= 2(sin θ)² − 1

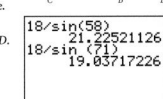

Exercises 61 and 62 extend the concepts of the sine, cosine, and tangent functions to their reciprocal functions, which are the cosecant, secant, and cotangent functions, respectively.

Assessment

Portfolio Activity

The Portfolio Activity can be used as preparation for the Chapter Project or as a separate activity. In the Portfolio Activity on this page, students will derive a trigonometric identity that is used to examine the Plimton 322 tablet in the Chapter Project.

61. cot θ is the reciprocal of tan θ, sec θ is the reciprocal of cos θ, and csc θ is the reciprocal of sin θ.

 **Look Back**

For each length given, find the remaining two lengths.
(LESSON 5.5)

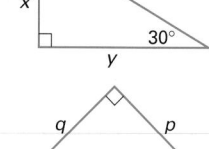

52. $x = 7$ $y = \underline{\ ?\ }$ $7\sqrt{3}$ $z = \underline{\ ?\ }$ 14

53. $x = \underline{\ ?\ }$ $\frac{14\sqrt{3}}{3}$ $y = 14$ $z = \underline{\ ?\ }$ $\frac{28\sqrt{3}}{3}$

54. $x = \underline{\ ?\ }$ 6.5 $y = \underline{\ ?\ }$ $6.5\sqrt{3}$ $z = 13$

55. $p = 1$ $q = \underline{\ ?\ }$ 1 $r = \underline{\ ?\ }$ $\sqrt{2}$

56. $p = \underline{\ ?\ }$ 3 $q = 3$ $r = \underline{\ ?\ }$ $3\sqrt{2}$

57. $p = \underline{\ ?\ }$ $8\sqrt{2}$ $q = \underline{\ ?\ }$ $8\sqrt{2}$ $r = 16$

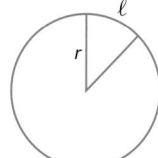

For each arc length, ℓ, and radius, r, given below, find the measure of the central angle. *(LESSON 9.1)*

58. $\ell = 12\pi, r = 20$ **108°**   **59.** $\ell = 10\pi, r = 100$ **18°**

60. $\ell = 3\pi, r = 25$ **21.6°**

 **Look Beyond**

The three ratios in a right triangle that we have explored are tangent, sine, and cosine. There are three other ratios to be considered.

cotangent	secant	cosecant
$\cot \theta = \dfrac{\text{adjacent}}{\text{opposite}}$	$\sec \theta = \dfrac{\text{hypotenuse}}{\text{adjacent}}$	$\csc \theta = \dfrac{\text{hypotenuse}}{\text{opposite}}$

61. How are the ratios above related to the tangent, sine, and cosine ratios?

62. In Exercises 41–43, you discovered a relationship between the sine and cosine in the form of two identities. Is there a similar relationship between tangent and cotangent? between secant and cosecant? Explore, using right triangles or your calculator, and then complete the following identities:

a. $\tan \theta = \cot \underline{\ ?\ }$ $(90 - \theta)$ **b.** $\cot \theta = \tan \underline{\ ?\ }$ $(90 - \theta)$

c. $\sec \theta = \csc \underline{\ ?\ }$ $(90 - \theta)$ **d.** $\csc \theta = \sec \underline{\ ?\ }$ $(90 - \theta)$

📶 **internet** connect

Portfolio Extension
Go To: **go.hrw.com**
Keyword:
MG1 Identities

 PORTFOLIO ACTIVITY

PYTHAGOREAN IDENTITIES

The identity $(\sin \theta)^2 + (\cos \theta)^2 = 1$ is often called a Pythagorean identity because it is derived from the Pythagorean Theorem. There are two other Pythagorean identities that can easily be derived from this identity.

1. Divide each term in the identity $(\sin \theta)^2 + (\cos \theta)^2 = 1$ by $(\sin \theta)^2$, and simplify as much as possible. Express your identity using one of the ratios in Exercises 61 and 62.

2. Divide each term in the identity $(\sin \theta)^2 + (\cos \theta)^2 = 1$ by $(\cos \theta)^2$, and

simplify as much as possible. Express your identity using one of the ratios in Exercises 61 and 62.

WORKING ON THE CHAPTER PROJECT

You should now be able to complete Activity 2 of the Chapter Project.

10.3

Extending the Trigonometric Ratios

Objectives

- Use a rotating ray in a coordinate plane to define angles measuring greater than 90° and less than 0°.

- Define *sine, cosine,* and *tangent* for angles of any size.

Why So far, trigonometric ratios have been defined in terms of acute angles of right triangles. But it is important to define the trigonometric ratios for other angles. In this lesson, you will learn how this can be done.

The curve shown on the oscilloscope is a graph of the sound wave generated by the synthesizer. Such a curve is known as a sine curve.

Prepare

QUICK WARM-UP

Use a protractor to draw an angle with each measure indicated below.

1. 90°
2. 180°
3. 270°
4. 360°
5. 0°

Check students' drawings.

Also on Quiz Transparency 10.3

Extending Angle Measure

Imagine a ray with its endpoint at the origin of a coordinate plane and extending along the positive *x*-axis. Then imagine the ray rotating a certain number of degrees, say *θ*, counterclockwise about the origin. As the illustration shows, *θ* can be any number of degrees, including numbers greater than 360°. A figure formed by a rotating ray and a stationary reference ray, such as the positive *x*-axis, is called an **angle of rotation**.

Angles of Rotation

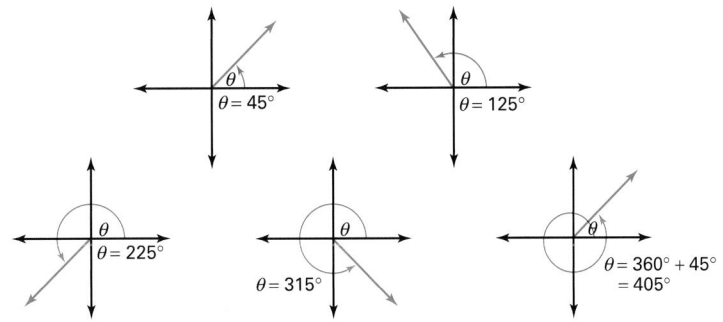

CRITICAL THINKING How do you think an angle of rotation could have a negative measure?

Teach

Why There are many physical phenomena that are periodic, which means recurring or repeating at regular intervals. Ask students to give examples of objects whose movement is periodic. The motions of yo-yos or pendulums are two such examples.

CRITICAL THINKING
An angle of rotation might have a negative measure if it is measured clockwise.

Alternative Teaching Strategy

TECHNOLOGY Geometry graphics software that includes a coordinate system may be used for activities with the unit circle. Some graphics tools may require you to activate the coordinate system in order to see it displayed. Have students construct a circle of any size centered at the origin and a segment connecting the origin with any point on the circle.

Students can measure the angle between this segment and a segment on the *x*-axis and find the sine and cosine of this angle. By moving the endpoint of the segment around the circle, students can see the signs of the sine values and cosine values change in the different quadrants.

Activity 1 **Notes**

In Example 1, students learn how to find the coordinates of a point on the unit circle. In this Activity, students extend that knowledge to discover the sine values and cosine values of four rotated angles.

For a student worksheet of this Activity and detailed Teacher Notes, see page 189 in the Lesson Activities booklet.

CHECKPOINT ✔

4. See Additional Answers beginning on page 879.

CHECKPOINT ✔

6. $y; x$

Teaching Tip

Sometimes students have difficulty remembering which coordinate represents the cosine value and which coordinate represents the sine value. Have students think of the right triangle created by any angle in the first quadrant. Then remind them that cosine is the ratio that includes the horizontal leg (x-coordinate) and the sine is the ratio that includes the vertical leg (y-coordinate).

The Unit Circle

To define the trigonometric ratios for all possible rotation angles you can use the *unit circle*. The **unit circle** is a circle with its center at the origin and a radius of 1. In the language of transformations, it consists of all the rotation images of the point $P(1, 0)$ about the origin.

EXAMPLE ① Find the coordinates of P', the image of point $P(1, 0)$ rotated 150° about the origin.

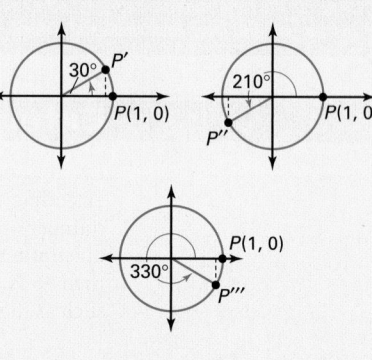

● **SOLUTION**

PROBLEM SOLVING

Use a graph. Draw a perpendicular segment from P' to the x-axis as shown. Label the intersection Q. Since $\angle P'OQ$ is supplementary to $\angle P'OP$, its measure is 30°. Thus, $\triangle P'OQ$ is a 30-60-90 triangle with its hypotenuse equal to 1.

$QO = \frac{\sqrt{3}}{2}$ and $P'Q = \frac{1}{2}$. The x-coordinate of P' is $\frac{-\sqrt{3}}{2}$ or ≈ -0.866.

The y-coordinate of P' is $\frac{1}{2} = 0.5$. Thus, the coordinates of P' are $(-0.866, 0.5)$.

Activity 1
Redefining the Trigonometric Ratios

YOU WILL NEED

scientific or graphics calculator

1. Let P' be the 30° rotation image of $P(1, 0)$ about the origin. Use the rules from the 30-60-90 Right Triangle Theorem to find the coordinates of P'.

2. Let P'' be the 210° rotation image of $P(1, 0)$ about the origin. Find the coordinates of P''.

3. Let P''' is the 330° rotation image of $P(1, 0)$ about the origin. Find the coordinates of P'''.

CHECKPOINT ✔

4. **a.** Use the results of Steps 1–3 to complete columns 2 and 3 of the table below.

   b. Use the ⎡SIN⎤ and ⎡COS⎤ keys of a scientific calculator to complete columns 4 and 5 of the table.

See Example 1.

Rotation angle, θ	x-coordinate of image point	y-coordinate of image point	$\cos \theta$	$\sin \theta$
30°	?	?	?	?
150°	$\frac{-\sqrt{3}}{2} \approx -0.866$	$\frac{1}{2} = 0.5$	?	?
210°	?	?	?	?
330°	?	?	?	?

Interdisciplinary Connection

MUSIC When a string or wire with fixed endpoints is plucked, a standing wave is created. For the simplest vibration of the string, the graph of the vertical position of the string's center point versus time is a sine curve. However, the string may vibrate in several different ways at the same time. This type of vibration is related to the sounds made by many musical instruments. Students may enjoy carrying out simple experiments with a stretched string or wire to see what kinds of waves they can create.

5. What relationships do you see in the table?

6. Complete the following definition for the sine and cosine of an angle.

CHECKPOINT ✔

Unit Circle Definition of Sine and Cosine

Let θ be a rotation angle. Then **sin θ** is the __?__-coordinate of the image of point $P(1, 0)$ rotated $\theta°$ about the origin, and **cos θ** is the __?__-coordinate.

CRITICAL THINKING

Negative rotations are represented by *clockwise* rotations of the point $P(1, 0)$. Test the unit circle definition of sine and cosine as follows:

a. Sketch the image point P' of $P(1, 0)$ as a result of a $-30°$ rotation (clockwise) about the origin.

b. Find the coordinates of the image point P' and use this information to give the sine and cosine of $-30°$.

Check your answers by using a scientific calculator.

EXAMPLE ②

APPLICATION
ROTARY MOTION

A wheel with a 1-ft radius is turning slowly at a constant velocity of 1° per second and has a light mounted on its rim. A distant observer watching the wheel from the edge sees the light moving up and down in a vertical line. Write an equation for the vertical position, h, of the light starting from the horizontal position at time $t = 0$. What will be the vertical position of the light after 1 min? after 5 min? after 24 hr?

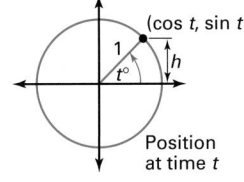

Position at time t

● **SOLUTION**

Imagine a coordinate system with the origin at the center of the wheel as shown. After t seconds have elapsed, the value of θ is $t°$. The coordinates of the light are $(\cos t, \sin t)$. Notice that $\sin t$, the second coordinate, is the vertical position of the light. Thus, at time t the vertical position, h, of the light is given by the equation

$$h = \sin t.$$

At $t = 1$ min, or 60 sec, $h = \sin 60° \approx 0.867$ units.

At $t = 5$ min, or 300 sec, $h = \sin 300° \approx -0.867$ units.

At $t = 24$ hr, or 86,400 sec, $h = \sin 86,400° = 0$ units.

Enrichment

Explain that angles can also be measured in radians. Have students work together to find out what a radian is. Then have students redraw the graphs from Activity 2 with radians rather than degrees on the horizontal axis. Explain that a radian is a linear measure that is often used in applications. Have students discuss the advantages and disadvantages of using radians versus degrees.

Inclusion Strategies

ENGLISH LANGUAGE DEVELOPMENT The fact that the words *sign* and *sine* are pronounced the same may cause some confusion during this lesson. One way to avoid this problem is to not use the term *sign* in spoken discussions. Instead, talk about whether the sine value of an angle is positive or negative.

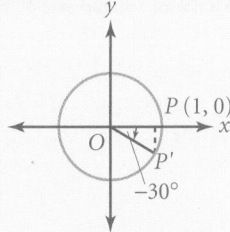
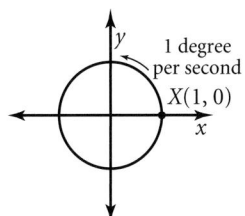

Activity 2
Graphing the Trigonometric Ratios

YOU WILL NEED

scientific or graphics calculator and graph paper

1. Extend the graphs of each of the trigonometric ratios below for values of θ from 0° to 360° by plotting points at intervals of 30°.

2. Use your graphs to determine the intervals in which the sine and cosine are positive and those in which they are negative. Complete the table below.

CHECKPOINT ✔

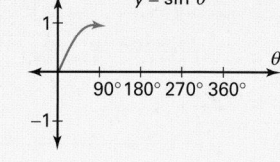

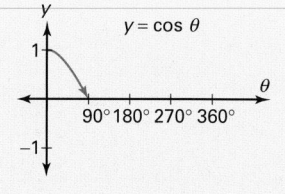

Quadrant of image point	Sign of sin θ	Sign of cos θ
I (0° to 90°)	+	?
II (90° to 180°)	+	?
III (180° to 270°)	−	?
IV (270° to 360°)	?	?

Using a Calculator to Find an Angle

In Activity 1, you saw that sin 30° and sin 150° both equal 0.5. There are, in fact, infinitely many angles for which sin θ has a given value from 0 to 1, but in working with triangles, you will want to find angles between 0° and 180°. The example which follows illustrates how to use your calculator along with a graph to find the desired angles.

EXAMPLE ③ Find two values of θ between 0° and 180° such that sin $\theta \approx 0.9397$.

● **SOLUTION**

Use the [SIN⁻¹] (inverse sine) key to find the measure of an angle whose sin is 0.9397. The calculator will give you approximately 70°. Note: For a given value from 0 to 1, the $\sin^{-1}$ function of a calculator will always return an angle measure from 0° to 90°.

PROBLEM SOLVING | **Draw a graph** of the sine function. You can see that there is another angle that has the same sine value. To find that angle, subtract the first angle from 180°.

$$180° - 70° = 110°$$

Thus, the values are 70° and 110°.

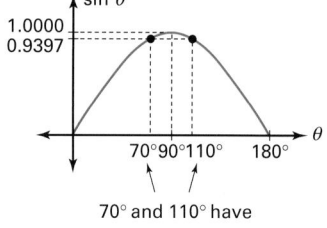

70° and 110° have the same sine value.

TRY THIS Find two values of θ between 0° and 180° such that sin $\theta \approx 0.5736$.

Exercises

Communicate

1. Explain how a rotating ray is used to extend the definition of an angle.

2. How is the unit circle used to extend the trigonometric ratios beyond 90°?

Explain how to use the unit circle definition of sine and cosine to find each of the following:

3. sin 90° and cos 90°

4. sin 180° and cos 180°

Explain how to use the unit circle definition of sine and cosine to find the sign of each of the following:

5. cosine in Quadrant III

6. sine in Quadrant II

APPLICATION

7. **ENGINEERING** The drive mechanism of the oil pump converts the circular motion of point *X* to vertical motion in a straight line at point *Y*. Explain how a trigonometric function can represent the vertical component of the motion of point *X*.

Guided Skills Practice

8. Find the coordinates of *P′*, the image of point *P*(1, 0) rotated 210° about the origin. **(EXAMPLE 1)** $\left(-\frac{\sqrt{3}}{2}, -\frac{1}{2}\right)$

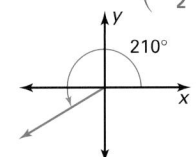

9. Use the unit circle to find the sine and cosine of *θ* from the coordinates of the given point. **(ACTIVITY 1)** sin *θ* = 0.6428 cos *θ* = 0.7660

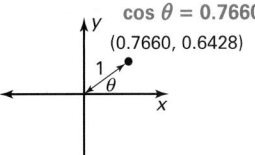

10. A wheel with a radius of 1 unit turns at a rate of 1° per second. Write an equation for the vertical position of point *X* starting from the horizontal position shown at *t* = 0. What is the vertical position of point *X* after 10 min? 20 min? 1 hr? 12 hr? **(EXAMPLE 2)**

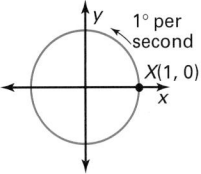

11. In Quadrants I and IV, cosine is positive. In Quadrants II and III, cosine is negative.

11. Use the graph of the cosine ratio to determine the sign of the cosine in Quadrants I through IV of the coordinate plane. **(ACTIVITY 2)**

12. Find two values of *θ* between 0° and 180° such that sin *θ* ≈ 0.7071. **(EXAMPLE 3)** 45°; 135°

Assess

Selected Answers

Exercises 8–12, 13–55 odd

ASSIGNMENT GUIDE

In Class	1–12
Core	13–45 odd
Core Plus	21–47 odd
Review	48–55
Preview	56–58

✐ Extra Practice can be found beginning on page 818.

10. The vertical position of point *X* is sin *t*, where *t* represents the number of seconds. After 10 minutes (600 sec), the vertical position is −0.866. After 20 minutes, the vertical position is 0.866. After 1 hour, the vertical position is 0. After 12 hours, the vertical position is 0.

29. $x = \cos 30° \approx 0.8660$;
$y = \sin 30° = 0.5$

30. $x = \cos 60° = 0.5$;
$y = \sin 60° \approx 0.8660$

31. $x = \cos 90° = 0$;
$y = \sin 90° = 1$

32. $x = \cos 120° = -0.5$;
$y = \sin 120° \approx 0.8660$

33. $x = \cos 180° = -1$;
$y = \sin 180° = 0$

34. $x = \cos 210° \approx -0.8660$;
$y = \sin 210° = -0.5$

35. $x = \cos 300° = 0.5$;
$y = \sin 300° \approx -0.8660$

36. $x = \cos 360° = 1$;
$y = \sin 360° = 0$

13. 0.9063

14. 0.9063

15. −0.9063

16. −0.9063

17. 0.4226

18. −0.4226

19. −0.4226

20. 0.4226

21. $\sin 45° \approx 0.7071$;
$\cos 45° \approx 0.7071$

22. $\sin 135° \approx 0.7071$;
$\cos 135° \approx 0.7071$

23. $\sin 225° \approx -0.7071$;
$\cos 225° \approx -0.7071$

24. $\sin 315° \approx -0.7071$;
$\cos 315° \approx 0.7071$

25. $\sin 30° = 0.5$;
$\cos 30° \approx 0.8660$

26. $\sin 150° = 0.5$;
$\cos 150° \approx -0.8660$

27. $\sin 210° = -0.5$;
$\cos 210° \approx -0.8660$

28. $\sin 330° = -0.5$;
$\cos 330° \approx 0.8660$

Homework Help Online
Go To: go.hrw.com
Keyword:
MG1 Homework Help
for Exercises 37-45

Practice and Apply

TECHNOLOGY **In Exercises 13–20, use a calculator to find each of the following, rounded to four decimal places:**

13. $\sin 65°$ **14.** $\sin 115°$ **15.** $\sin 245°$ **16.** $\sin 295°$

17. $\cos 65°$ **18.** $\cos 115°$ **19.** $\cos 245°$ **20.** $\cos 295°$

In Exercises 21–28, use graph paper and a protractor to sketch a unit circle and a ray with the given angle θ with the positive x-axis. Find the coordinates of the point on the ray at a distance of 1 from the origin, rounded to four decimal places. Use these values and the unit circle definition of sine and cosine to give the sine and cosine of each angle, rounded to four decimal places.

21. 45°   **22.** 135°   **23.** 225°   **24.** 315°

25. 30°   **26.** 150°   **27.** 210°   **28.** 330°

TECHNOLOGY **In Exercises 29–36, use a calculator to find the sine and cosine of each angle. Use these values to give the x- and y-coordinates of a point at the given angle on the unit circle. Round your answers to four decimal places.**

29. 30°   **30.** 60°   **31.** 90°   **32.** 120°

33. 180°   **34.** 210°   **35.** 300°   **36.** 360°

In Exercises 37–44, give two values of θ between 0° and 180° for the given value of $\sin \theta$. Express your answers to the nearest degree.

37. 0.7071   **38.** 0.8660   **39.** 0.5000   **40.** 0.9659

41. 0.3217   **42.** 0.9900   **43.** 0.9990   **44.** 0.9999

45. If $\sin \theta = 0.4756$, what are all the possible values of θ between 0° and 360°, rounded to the nearest degree? **28°; 152°**

46. If $\cos \theta = -0.7500$, what are all the possible values of θ between 0° and 360°, rounded to the nearest degree? **139°; 221°**

47. ASTRONOMY An astronomer observes a satellite that is moving around a planet at the rate of 1° per hour. Assuming that the radius of the satellite's orbit is 1 unit, what is the horizontal position of the satellite after 2 days? after 5 days?

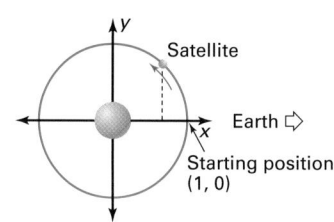

View of planet from above

37. 45°; 135°

38. 60°; 120°

39. 30°; 150°

40. 75°; 105°

41. 19°; 161°

42. 82°; 98°

43. 87°; 93°

44. 89°; 91°

47. The horizontal position is given by cos t, where t represents the number of hours. After 2 days (48 hrs), the position is given by point (0.67, 0.74). After 5 days, the position is given by point (−0.5, 0.87).

 **Look Back**

Use △ABC for Exercises 48–55. *(LESSONS 10.1 AND 10.2)*

52. $(\cos A)^2$

53. $\tan A$

54. $\cos (90° - m\angle A)$
    or $\cos B$

55. $\sin (90° - m\angle A)$
    or $\sin B$

48. $\sin A = \frac{?}{c}$ a

49. $\cos B = \frac{?}{c}$ a

50. $\tan A = \frac{?}{b}$ a

51. $\cos A = \frac{?}{c}$ b

52. $(\sin A)^2 + \underline{?} = 1$

53. $\frac{\sin A}{\cos A} = \underline{?}$

54. $\sin A = \cos \underline{?}$

55. $\cos A = \sin \underline{?}$

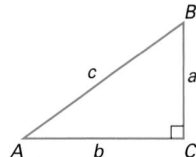

 **Look Beyond**

APPLICATIONS

CHALLENGES

56. a. III ○ II IV
       Moon I is
       behind Jupiter.

   b. ○ II I III IV

   c. IV III II I ○

57. I, Io: 1.75 days

   II, Europa: 3.5 days

   III, Ganymede: 7 days

   IV, Callisto: 16 days

ASTRONOMY The graph shows the positions of the four "Galilean" moons of Jupiter as seen from Earth at midnight on April 1–16, 1993. The parallel lines in the center of the graph represent the visible width of Jupiter.

56. Galileo was able to observe Jupiter's four largest moons with his small telescope. When viewed from Earth, they appear to move back and forth in an approximately straight line through the center of the planet. The names of the four moons are listed below.

   I.   Io
   II. Europa
   III. Ganymede
   IV. Callisto

   Use the graph to sketch the positions of the planet Jupiter and its four Galilean moons as they would appear to a person with a small telescope or a pair of binoculars on (a) April 4, (b) April 8, and (c) April 12. Use dots for the moons and a circle for Jupiter.

57. Use the graph to estimate the orbital periods of each of the four Galilean moons of Jupiter.

58. What kind of curve do the lines for the moons' orbits appear to be? Explain why they have this shape. (Note: The orbits of the four moons are nearly circular.)

Configuration of satellites I-IV for April 1–16, 1993 at midnight Greenwich Mean Time (GMT)

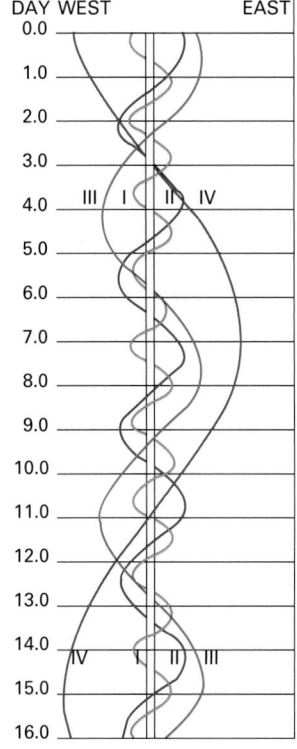

Source: *The Astronomical Almanac for the Year 1993*

 **Look Beyond**

Exercises 56–58 refer to the motions of the moons of Jupiter as an extension of the concept of sine curves.

58. Sine waves. They are repeatedly traveling around Jupiter in a circular motion that can be modeled by a point rotating around the unit circle. This creates the sine wave pattern.

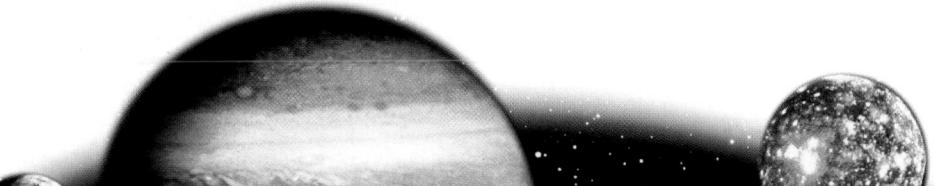

The Law of Sines

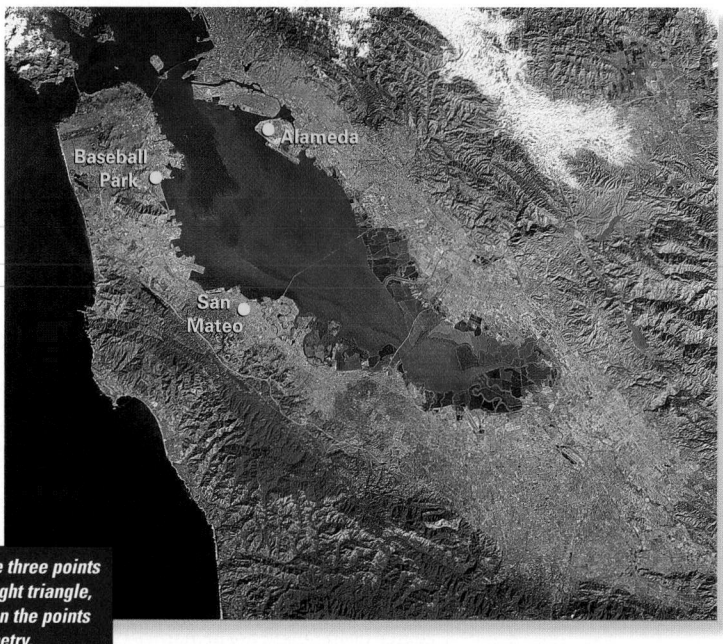

1. Draw any non-right triangle. Label the vertices A, B, and C. Use a, b, and c to label the sides, with side a opposite $\angle A$, side b opposite $\angle B$, and side c opposite $\angle C$. Check students' drawings.

2. List all possible ways to name two angles and one side of the triangle you drew above.
$\angle A, \angle B, a$; $\angle A, \angle B, b$;
$\angle A, \angle B, c$; $\angle A, \angle C, a$;
$\angle A, \angle C, b$; $\angle A, \angle C, c$;
$\angle C, \angle B, a$; $\angle C, \angle B, b$;
$\angle C, \angle B, c$

Also on Quiz Transparency 10.4

Objectives

- Develop the law of sines.

- Use the law of sines to solve triangles.

Why *The triangle formed by the three points on the satellite photo is not a right triangle, however, the distances between the points can be found by using trigonometry.*

Satellite photo of San Francisco Bay area

The Law of Sines

The trigonometric ratios you have studied all relate to right triangles. In the following Activity, you will explore the law of sines, a theorem involving sine ratios that applies to all triangles.

Teach

Why If students know at least one side and any two other measures of a triangle, they can find the missing three measures by using the law of sines or the law of cosines. In this lesson, students derive and apply the law of sines.

Activity
Law of Sines

YOU WILL NEED
geometry graphics software or ruler and protractor, and scientific or graphics calculator

1. Draw an acute triangle, a right triangle, and an obtuse triangle. Label the sides and edges as shown. Measure the sides and angles of each triangle.

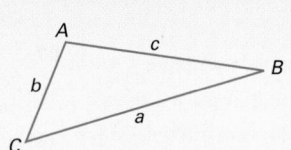

2. Copy and complete the table below, using your measurements from Step 1.

	m∠A	m∠B	m∠C	a	b	c	$\frac{\sin A}{a}$	$\frac{\sin B}{b}$	$\frac{\sin C}{c}$
acute	?	?	?	?	?	?	?	?	?
right	?	?	?	?	?	?	?	?	?
obtuse	?	?	?	?	?	?	?	?	?

CHECKPOINT ✓

3. Write a conjecture using the data in the last three columns of the table.

Alternative Teaching Strategy

TECHNOLOGY The examples in this lesson, as well as many of the exercises, can be explored or confirmed with geometry graphics software. If the software includes trigonometric functions, computations can be checked directly. If the trigonometric functions are not available, students can construct triangles with the given measures and confirm computations by measuring sides or angles.

Your conjecture from Activity 1 can be stated as the following theorem:

The Law of Sines

For any triangle $\triangle ABC$ with sides a, b, and c:

$$\frac{\sin A}{a} = \frac{\sin B}{b} = \frac{\sin C}{c}$$

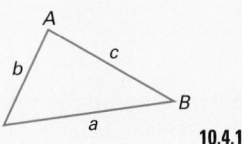

10.4.1

The proof of the law of sines is divided into three cases: acute, right, and obtuse triangles. The following is a proof of the acute case; you will prove the remaining cases in Exercises 33–49.

PROOF

Given: acute $\triangle ABC$ with sides a, b, and c

Prove: $\dfrac{\sin A}{a} = \dfrac{\sin B}{b} = \dfrac{\sin C}{c}$

Proof:

Draw the altitudes from B and C. Label the intersections of the altitudes with $\overline{AB}$ and $\overline{AC}$ as D_1 and D_2, with lengths h_1 and h_2, respectively. Then $\sin A = \dfrac{h_1}{b}$ and $\sin B = \dfrac{h_1}{a}$. Solving each equation for h_1, $h_1 = b \sin A$ and $h_1 = a \sin B$. By substitution, $b \sin A = a \sin B$. Dividing both sides of the equation by ab gives:

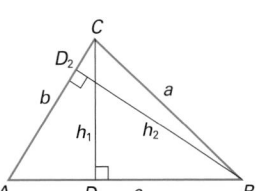

$$\frac{b \sin A}{ab} = \frac{a \sin B}{ab}$$
$$\frac{\sin A}{a} = \frac{\sin B}{b}$$

Similarly, $\sin A = \dfrac{h_2}{c}$ and $\sin C = \dfrac{h_2}{a}$, so $c \sin A = a \sin C$, which gives:

$$\frac{\sin A}{a} = \frac{\sin C}{c}$$

Therefore, $\dfrac{\sin A}{a} = \dfrac{\sin B}{b} = \dfrac{\sin C}{c}$

Solving Triangles

The law of sines can be used to find the measures of sides and angles in a triangle, if the measures of some sides and angles are given. You can use the law of sines to solve a triangle if you are given the measures of

1. two angles and one side

or

2. two sides and an angle that is opposite one of the sides.

Interdisciplinary Connection

PHYSICS Trigonometry applications in astronomy often involve an extremely long "skinny" triangle whose base is a segment between two points on the surface of Earth. The opposite vertex is at some very distant point, and the angle at this vertex is extremely small. The two long sides of the triangle are almost parallel. Have students sketch this type of triangle and discuss application problems in which it might occur. If students are interested, suggest that they investigate how the parallax method is used in situations of this type.

Cooperative Learning

Have students work in groups of 3 to complete the Activity. Each student in a group draws and measures a different type of triangle. After completing the table, students can work together to write a conjecture.

Activity **Notes**

In this Activity, students experiment with various triangles to discover the law of sines. Point out that Step 1 asks for one right, one acute, and one obtuse triangle. Ask students why all three are needed.

For a student worksheet of this Activity and detailed Teacher Notes, see page 192 in the Lesson Activities booklet.

CHECKPOINT ✔

3. Conjecture: For any triangle with sides a, b and c and with angles A, B, and C opposite the given sides, respectively, $\dfrac{\sin A}{a} = \dfrac{\sin B}{b} = \dfrac{\sin C}{c}$.

Teaching Tip

Point out that the law of sines is actually three separate "laws" or formulas. Have students take the ratios in pairs and write out the three equations.

Find r and s in △RST below.
r ≈ 7.1 and s ≈ 6.7

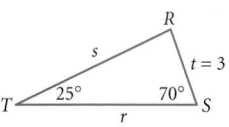

TRY THIS
b = 13.8, c = 12.2

Find m∠X in △XYZ below.
m∠X ≈ 36°

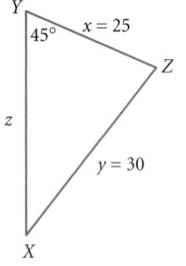

TRY THIS
≈ 42°

CRITICAL THINKING
No, because the given side opposite the given angle is longer than the other given side (see the SSA Conjecture, page 233).

EXAMPLE 1 Find b and c in the triangle at right.

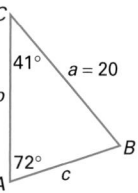

● **SOLUTION**

Set up a proportion using two ratios from the law of sines. Start with the known angle whose opposite side is also known, in this case, ∠A and a.

Then by the law of sines:

$$\frac{\sin A}{a} = \frac{\sin C}{c}$$

$$\frac{\sin 72°}{20} = \frac{\sin 41°}{c}$$

$$c \sin 72° = 20 \sin 41°$$

$$c = \frac{20 \sin 41°}{\sin 72°} \approx 13.8$$

$$\frac{\sin A}{a} = \frac{\sin B}{b}$$

$$\frac{\sin 72°}{20} = \frac{\sin 67°}{b}$$

$$b \sin 72° = 20 \sin 67°$$

$$b = \frac{20 \sin 67°}{\sin 72°} \approx 19.4$$

m∠B = 180° − (72° + 41°) = 67°

TRY THIS Given △ABC from Example 1 with a = 15, m∠A = 70°, and m∠C = 50°, find b and c.

EXAMPLE 2 Find m∠Q in the triangle at right.

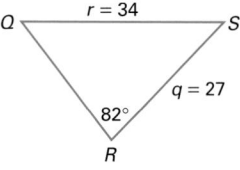

● **SOLUTION**

$$\frac{\sin R}{r} = \frac{\sin Q}{q}$$

$$\frac{\sin 82°}{34} = \frac{\sin Q}{27}$$

$$34 \sin Q = 27 \sin 82°$$

$$\sin Q = \frac{27 \sin 82°}{34} \approx 0.786$$

$$Q \approx \sin^{-1} 0.786 \approx 52°$$

TRY THIS Given △QRS from Example 2 with q = 18, r = 26, and m∠R = 75°, find m∠Q.

Note: When solving triangles by using the law of sines where two sides and a nonincluded angle are given, it may be possible to have two different answers. This is known as the *ambiguous case* of the law of sines.

In the example, sin⁻¹ was used to find the angle in question. However, the inverse sine function will only give values of angles less than 90°. To find the possible obtuse angle, subtract the value from 180°.

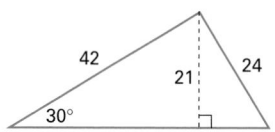
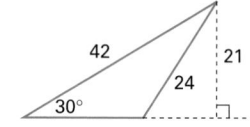

Ambiguous case of the law of sines

CRITICAL THINKING Is the situation in Example 2 ambiguous? Why or why not?

Enrichment

The diagrams above illustrate an ambiguous case of the law of sines. In another type of ambiguous case, there is no triangle at all. Have students try to find m∠B for △ABC with ∠C = 55°, c = 9 cm, and b = 25 cm. Students will get sin B ≈ 2.2754, which is impossible because the sine of an angle cannot be greater than 1. Trying to construct the triangle will also show that the dimensions will not work.

Inclusion Strategies

TECHNOLOGY Students will most likely be using graphics or scientific calculators for the exercises and problems in this lesson. They may find that it is easier to apply the law of sines if they convert it to one of the following forms:

$a = \sin A \cdot b \div \sin B$ $\sin A = a \cdot \sin B \div b$
$b = \sin B \cdot c \div \sin C$ $\sin B = b \cdot \sin C \div c$
$c = \sin A \cdot a \div \sin A$ $\sin C = c \cdot \sin A \div a$

EXAMPLE ③ A boater travels regularly from San Mateo to Alameda across San Francisco Bay, a distance of 12.8 mi. Suppose that one day he needs to stop at the baseball park. If the angle formed at the baseball park is 95° and the angle formed at Alameda is 53°, how far does the boater have to travel to get to Alameda via the baseball park?

● **SOLUTION**

Use the law of sines

$$\frac{\sin 95°}{12.8} = \frac{\sin 53°}{b}$$

$$b \sin 95° = 12.8 \sin 53°$$

$$c = \frac{12.8 \sin 53°}{\sin 95°} \approx 10.3$$

$$\frac{\sin 95°}{12.8} = \frac{\sin 32°}{b}$$

$$b \sin 95° = 12.8 \sin 32°$$

$$c = \frac{12.8 \sin 32°}{\sin 95°} \approx 6.8$$

The total distance to Alameda is approximately 10.3 + 6.8 = 17.1 mi.

ADDITIONAL

EXAMPLE ③

A nonstop flight from Trenton to Austin is 340 miles long. The airline also has a connecting flight to Austin through Dixie. If the angle formed at Austin is 28° and the angle formed at Dixie is 120°, what is the total distance for the connecting flight?

≈ 392.4 mi

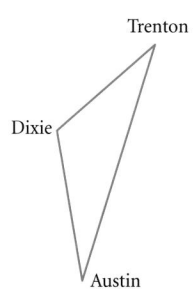

Exercises

● *Communicate*

1. Explain why the law of sines cannot be used to solve △ABC.

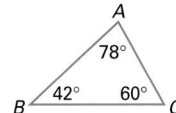

🌐 **internet** connect

Activities Online

Go To: **go.hrw.com**
Keyword:
MG1 Sines

2. Explain why the law of sines cannot be used to solve △DEF.

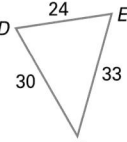

3. Explain why the law of sines cannot be used to solve △GHI.

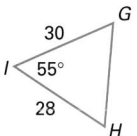

4. Explain how the ambiguous case of the law of sines relates to the SSA combination and the "swinging door" effect (see Lesson 4.3).

Reteaching the Lesson

COOPERATIVE LEARNING Have students work in small groups to verify the law of sines for an arbitrary triangle. Students should measure the three angles and the three sides of the triangle. Then they can compute the three ratios to verify the law of sines.

ASSIGNMENT GUIDE	
In Class	1–7
Core	9–21 odd, 22–28, 29, 31, 33–49
Core Plus	22–28, 33–49, 51–56
Review	57–63
Preview	64–66

✐ Extra Practice can be found beginning on page 818.

Technology

Geometry graphics software can be used to check results obtained for the law of sines. Have students construct triangles with the given measures. Then they can measure the needed angle or side to check their answers.

16. m∠R = 110°; r ≈ 14.62; p ≈ 7.78

17. m∠P = 40°; q ≈ 6.16; p ≈ 4.57

18. m∠Q = 72°; p = 12; r ≈ 7.42

19. m∠Q ≈ 42.34°; m∠R ≈ 77.66°; r ≈ 10.15

20. m∠R ≈ 35.26°; m∠P ≈ 24.74°; p ≈ 5.80

21. *Case 1:* m∠P ≈ 76.48; m∠Q ≈ 58.52°; q ≈ 9.65

    Case 2: m∠P ≈ 103.52°; m∠Q ≈ 31.48°; q ≈ 5.91

Guided Skills Practice

5. Find b and c in $\triangle ABC$.
   (EXAMPLE 1) $b \approx 10.1$ $c \approx 11.9$

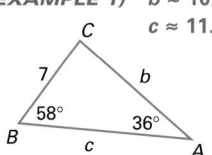

6. Find m∠Q in $\triangle QRS$.
   (EXAMPLE 2) 111.74°

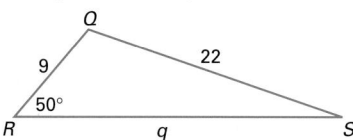

7. How much greater is the distance from point X to point Y to point Z than the distance from point X to point Z? *(EXAMPLE 3)* 2.59 ft

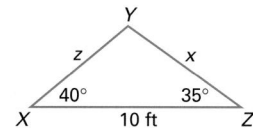

Practice and Apply

In Exercises 8–15, find the indicated measures. Assume that all angles are acute. It may be helpful to sketch the triangle roughly to scale.

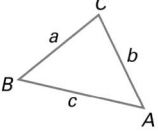

8.	m∠A = 56°	m∠B = 24°	b = 1.22 cm	c = __?__ 2.954 cm
9.	m∠B = 29°	a = 3.12 cm	b = 3.28 cm	m∠A = __?__ 27.46°
10.	m∠B = 29°	a = 3.12 cm	b = 3.28 cm	c = __?__ 5.64 cm
11.	m∠B = 67°	b = 7.36 cm	c = 2.13 cm	m∠C = __?__ 15.45°
12.	m∠B = 67°	b = 7.36 cm	c = 2.13 cm	a = __?__ 7.93 cm
13.	m∠B = 73°	m∠C = 85°	a = 3.14 cm	b = __?__ 8.02 cm
14.	m∠A = 35°	m∠B = 44°	c = 2.4 cm	a = __?__ 1.40 cm
15.	m∠A = 53°	m∠C = 72°	c = 2.34 cm	b = __?__ 2.02 cm

Find all unknown sides and angles in each triangle. If the triangle is ambiguous, give both possible angles. It may be helpful to sketch the triangle roughly to scale.

16. m∠P = 30°, m∠Q = 40°, q = 10

17. m∠Q = 60°, m∠R = 80°, r = 7

18. m∠P = 72°, m∠R = 36°, q = 12

19. m∠P = 60°, p = 9, q = 7

20. m∠Q = 120°, q = 12, r = 8

21. m∠R = 45°, p = 11, r = 8

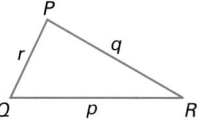

In Exercises 22–28, you will investigate the ambiguous case of the law of sines. Suppose that two sides of a triangle, *a* and *b*, and an angle opposite one side, ∠*A*, are given.

22. Case I: m∠*A* < 90°

Draw a line and choose point *A* on the line. Choose measurements for ∠*A* and *b*, and use a ruler and protractor to draw side *b*.

Use a calculator to find *b* sin *A*. Explain why there is exactly one possible triangle if *a* = *b* sin *A*.

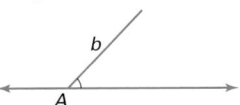   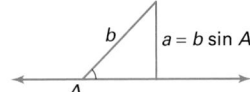

23. If *a* < *b* sin *A*, how many triangles are possible? Explain your reasoning.

24. If *a* ≥ *b*, how many triangles are possible? Explain your reasoning.

25. If *b* sin *A* < *a* < *b*, how many triangles are possible? Explain your reasoning.

26. Case II: m∠*A* ≥ 90°

Draw a line and choose point *A* on the line. Choose measurements for ∠*A* and *b*, and use a ruler and protractor to draw side *b*.

Explain why there is exactly one possible triangle if *a* > *b*.

   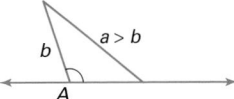

27. If *a* ≤ *b*, how many triangles are possible? Explain your reasoning.

28. Summarize your results from Exercises 22–27 above by writing a rule in your own words for deciding whether a given set of SSA measurements determine one triangle, two possible triangles, or no triangle.

In Exercises 29–32, two sides of a triangle, *a* and *b*, and an angle opposite one side, ∠*A*, are given. Use your results from Exercises 22–28 to explain whether the given measurements determine one triangle, two possible triangles, or no triangle.

29. 1 triangle possible

30. 1 triangle possible

31. 2 triangles possible

32. no triangle is possible

29. m∠*A* = 70°, *a* = 4, *b* = 3

30. m∠*A* = 110°, *a* = 7, *b* = 5

31. m∠*A* = 30°, *a* = 4, *b* = 6

32. m∠*A* = 145°, *a* = 3, *b* = 6

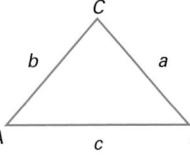

PROOF

For Exercises 33–35, use the diagram to prove the law of sines for right triangles.

33. $\sin A = \frac{a}{c}$; $\sin B = \frac{b}{c}$

34. $\frac{\sin A}{a} = \frac{1}{c}$;

$\frac{\sin B}{b} = \frac{1}{c}$

$\frac{\sin C}{c} = \frac{1}{c}$

33. Write the sine of angles *A* and *B* in terms of *a*, *b*, and *c*.

34. Write the following ratios in terms of *a*, *b*, and *c*, and simplify:

$\frac{\sin A}{a}$ $\frac{\sin B}{b}$ $\frac{\sin C}{c}$

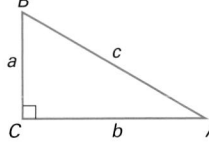

35. Complete the proof of the law of sines for right angles.

28. Sample answer: Let *A* be the angle, *b* the side nearest *A* and *a* the side across from *A*.

Case I: If m∠*A* < 90° and

(i)   *a* < *b* sin *A*, then no triangle is possible.

(ii) *a* ≥ *b*, then there is 1 triangle possible.

(iii) *b* sin *A* < *a* < *b*, then there are 2 triangles possible.

Case II: If m∠*A* ≥ 90° and

(i)   *a* > *b*, there is 1 triangle possible.

(ii) *a* ≤ *b*, then no triangle is possible.

35. $\frac{\sin A}{a} = \frac{\sin B}{b} = \frac{\sin C}{c}$, since they all are equal to the same quantity.

Error Analysis

Students may have difficulty explaining their reasoning in Exercises 22–28. Encourage them to make a sketch for each of these exercises before making any conclusions. Students may then be able to compare the different cases more easily.

22. Sample answer: If *a* = *b* sin *A*, the triangle is a right triangle. Since ∠*A* is fixed and another angle is a right angle, there is only one possibility for the third angle.

23. Sample answer: If *a* < *b* sin *A*, then side *a* is too short to form any triangle, so no triangle is possible.

24. Sample answer: If *a* ≥ *b*, then there is only 1 triangle possible because side *a* must "swing" to the right.

25. Sample answer: If *b* sin *A* < *a* < *b*, there are 2 triangles possible, one by "swinging" side *a* to the left and the other by "swinging" side *a* to the right.

26. Sample answer: The only way that side *a* can "swing" to touch the line in a different point is to the left side of ∠*A*, in which case, the triangle formed does not include ∠*A*, so only one triangle is possible.

27. Sample answer: If *a* ≤ *b* then no triangle is possible since for an obtuse or a right triangle, the side across from the obtuse or right angle must be the longest side of the triangle.

36. $b \sin A$

37. $a \sin B$

38. Multiplication Property of Equality

39. $a \sin B = b \sin A$

40. Division Property of Equality (both sides divided by ab)

41. $\dfrac{h_2}{a}$

42. Definition of sine

43. $c \sin A$

44. $a \sin C$

45. Multiplication Property of Equality

46. Substitution Property of Equality

47. $\dfrac{\sin A}{a} = \dfrac{\sin C}{c}$

48. $\dfrac{\sin A}{a} = \dfrac{\sin B}{b} = \dfrac{\sin C}{c}$

49. Transitive or Substitution Property of Equality

50. Since the left side and the right side both equal $\dfrac{a}{\sin A}$, the two sides are equal by the Transitive Property.

Practice

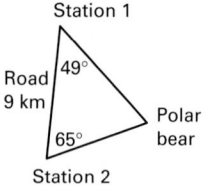

Practice

10.4 The Law of Sines

In Exercises 1–9, find the indicated measures. Assume that all angles are acute. It may be helpful to sketch the triangle roughly to scale. Round your answers to the nearest tenth.

1. $m\angle A = 48°$ $m\angle B = 73°$ $b = 1.7$ cm $a = $?   1.3 cm

2. $m\angle A = 37°$ $m\angle B = 80°$ $a = 3.4$ cm $c = $?   5.0 cm

3. $m\angle B = 78°$ $a = 1.45$ cm $b = 2.63$ cm $m\angle A = $?   32.6°

4. $m\angle B = 78°$ $a = 1.45$ cm $b = 2.63$ cm $c = $?   2.5 cm

5. $m\angle B = 25°$ $m\angle C = 80°$ $a = 5.2$ cm $b = $?   2.3 cm

6. $m\angle A = 40°$ $m\angle B = 64°$ $c = 3.62$ cm $a = $?   2.4 cm

7. $m\angle A = 41°$ $m\angle B = 58°$ $a = 14$ cm $b = $?   18.1 cm

8. $m\angle A = 72°$ $m\angle B = 40°$ $c = 15$ cm $a = $?   15.4 cm

9. $m\angle A = 35°$ $a = 8$ cm $b = 12$ cm $m\angle B = $?   59.4°

Find all unknown sides and angles for each triangle described below. If the triangle is ambiguous, give both possible angles. It may be helpful to sketch the triangle roughly to scale.

10. $m\angle P = 25°$, $m\angle Q = 55°$, $q = 10$
   $m\angle R = 100°$,
   $r \approx 12$, $p \approx 5.2$

11. $m\angle Q = 30°$, $m\angle R = 70°$, $r = 8$
   $m\angle P = 80°$,
   $p \approx 8.4$, $q \approx 4.3$

12. $m\angle P = 42°$, $m\angle R = 34°$, $q = 9$
   $m\angle Q = 104°$,
   $p \approx 6.2$, $r \approx 5.2$

13. $m\angle R = 48°$, $p = 3$, $r = 2.5$
   $m\angle P \approx 63.1°$, $m\angle Q \approx 68.9°$, $q \approx 3.1$ or
   $m\angle P \approx 116.9°$, $m\angle Q \approx 15.1°$, $q \approx 0.9$

14. $m\angle P = 35°$, $m\angle R = 41°$, $q = 23$
   $m\angle Q \approx 104°$,
   $p \approx 13.6$, $r \approx 15.6$

15. $m\angle Q = 53°$, $m\angle R = 72°$, $p = 26$
   $m\angle P \approx 55°$,
   $q \approx 25.3$, $r \approx 30.2$

TWO-COLUMN PROOF

For Exercises 36–49, use the diagram at right to prove the law of sines for obtuse triangles.

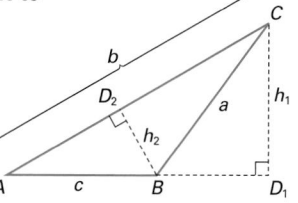

Given: obtuse $\triangle ABC$ with sides a, b, and c

Prove: $\dfrac{\sin A}{a} = \dfrac{\sin B}{b} = \dfrac{\sin C}{c}$

Proof:

Draw altitudes from B and C. Label the intersections of the altitudes with $\overleftrightarrow{AB}$ and $\overline{AC}$ as D_1 and D_2, with lengths h_1 and h_2, respectively.

Statements	Reasons
$\sin A = \dfrac{h_1}{b}$ $\quad\sin B = \dfrac{h_1}{a}$	Definition of sine
$h_1 = $ **36.** ? $\quad h_1 = $ **37.** ?	**38.** ?
39. ?	Substitution Property
$\dfrac{\sin A}{a} = \dfrac{\sin B}{b}$	**40.** ?
$\sin A = \dfrac{h_2}{c}$ $\quad\sin C = $ **41.** ?	**42.** ?
$h_2 = $ **43.** ? $\quad h_2 = $ **44.** ?	**45.** ?
$c \sin A = a \sin C$	**46.** ?
47. ?	Division Property
48. ?	**49.** ?

CHALLENGE

50. Use the law of sines to prove that the identity below is true for any triangle $\triangle ABC$.

$$\frac{a - b}{\sin A - \sin B} = \frac{a + b}{\sin A + \sin B}$$

APPLICATION

51. **WILDLIFE MANAGEMENT** Scientists are tracking polar bears that have been fitted with radio collars. The scientists have two stations that are 9 km apart along a straight road. At station 1, the signal from one of the collars comes from a direction 49° from the road. At station 2, the signal from the same collar comes from a direction 65° from the road. How far is the polar bear from station 2? What is the shortest distance from the road to the bear? **7.44 km; 6.74 km**

52. ARCHITECTURE A real-estate developer wants to build an office building on a triangular lot between Oak Street and 3rd Avenue. The dimensions of the lot are shown at right. Use the diagram to find the measures of the angles formed by the sides of the lot. **55.82°; 50.19°**

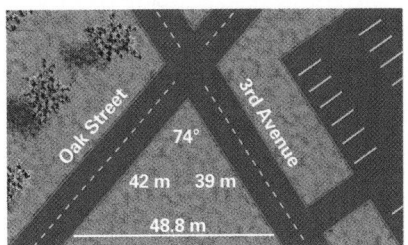

53. FORESTRY A plume of smoke is spotted from two different fire towers that are 5 mi apart. From tower A, the angle between the smoke plume and tower B is 80°. From tower B, the angle between the smoke plume and tower A is 70°. Which tower is closer to the smoke plume? How far is the smoke from the tower? **Tower A is closer to the smoke plume. The smoke is 9.40 miles from the tower.**

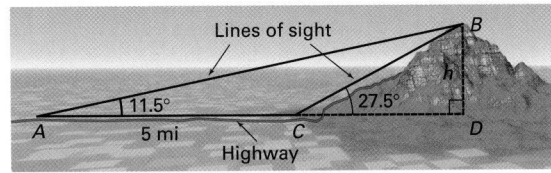

SURVEYING Suppose that you are driving toward a mountain. You stop and measure the angle of elevation from your eye to the top of the mountain, which is 11.5°. You then drive 5 mi on level ground. You stop again and measure the new angle of elevation, which is 27.5°.

54. Determine the measure of ∠ABC. **16°**

55. Find BC. **3.62 mi**

56. Use your result from Exercise 55 to determine the height of the mountain. **h = 1.67 mi**

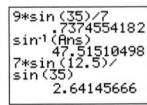

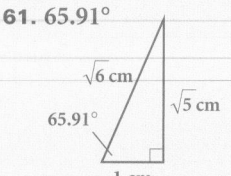
In Exercises 64–66, students are asked to use their knowledge of trigonometry to convert between polar coordinates and rectangular coordinates.

61. 65.91°

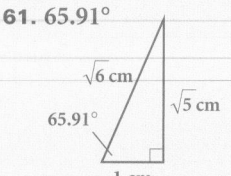

√6 cm
65.91°
√5 cm
1 cm

62. 60°

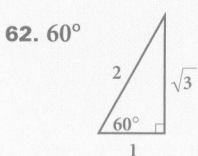

2
√3
60°
1

63. 45°

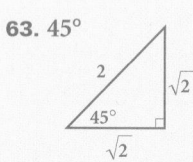

2
√2
45°
√2

64.

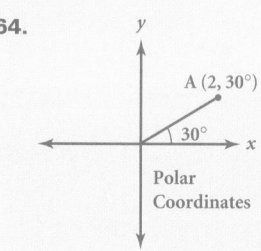

y
A (2, 30°)
30°
x
Polar
Coordinates

Rectangular Coordinates:
(√3, 1)

65.

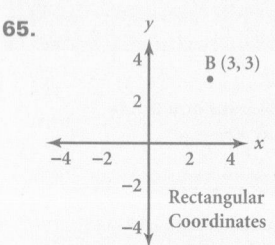

y
4
B (3, 3)
2
-4 -2   2   4 x
-2
Rectangular
-4 Coordinates

Polar Coordinates:
(3√2, 45°)

Look Back

57. Find the surface area and volume of a sphere with a radius of 15 meters. **(LESSON 7.6.)** $S = 2827.4 m^2$; $V = 14{,}137.2 m^3$

58. If the radius were tripled, the surface area would be multiplied by $3^2 = 9$. The volume would be multiplied by $3^3 = 27$.

58. What happens to the volume and surface area of a sphere if the radius is tripled? **(LESSONS 7.6 AND 8.6)**

Find *x* in each diagram below. (LESSONS 8.3 AND 8.4)

59.

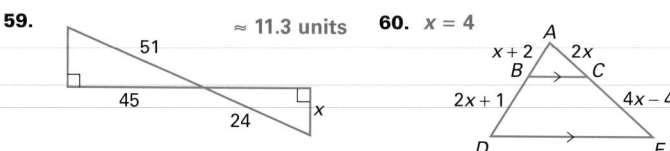

51
45
24
x

≈ 11.3 units

60. $x = 4$

A
x + 2   2x
B   C
2x + 1   4x − 4
D   E

Use a scientific or graphics calculator to find the following. Draw a triangle to illustrate each angle. (LESSONS 10.1 AND 10.2)

61. $\tan^{-1} \sqrt{5}$ **62.** $\sin^{-1}\left(\dfrac{\sqrt{3}}{2}\right)$ **63.** $\cos^{-1}\left(\dfrac{\sqrt{2}}{2}\right)$

Look Beyond

CONNECTION

POLAR COORDINATES In a rectangular coordinate system, a point is represented by *x* and *y*, the horizontal and vertical distance from the origin. In a *polar coordinate* system, a point is represented by a distance from the origin, *r*, and an angle from the positive *x*-axis, θ.

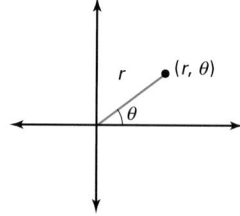

r • (r, θ)
θ

64. Point *A* has coordinates (2, 30°) in a polar coordinate system. Plot point *A* on a graph and find its rectangular coordinates.

65. Point *B* has coordinates (3, 3) in a rectangular coordinate system. Plot point *B* on a graph and find its polar coordinates.

A good window to use:
$0 \le \theta \le 360$
with θ step = 1
$-1.5 \le x \le 1.5$
$-1 \le y \le 1$

66. Use a graphics calculator with a polar graphing mode to graph some of the following functions. Draw or describe the graphs.

$r = \sin \theta$ $r = \sin(2\theta)$ $r = \sin(3\theta)$
$r = \cos \theta$ $r = \cos(2\theta)$ $r = \cos(3\theta)$

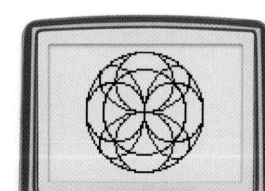

In polar coordinates, the simplest graph is a circle. Many graphs in polar coordinates have the shapes of flowers, stars, or spirals.

66. $r = \sin \theta$ is a circle of diameter 1, with the bottom edge touching the origin.
$r = \sin(2\theta)$ is a four-petaled flower shape, centered at the origin, with its petals between the coordinate axes.
$r = \sin(3\theta)$ is a three-petaled flower shaped, centered at the origin, with one petal on the negative *y*-axis.
$r = \cos \theta$ is a circle of diameter 1, with the left edge touching the origin.
$r = \cos(2\theta)$ is a four-petaled flower shape, centered at the origin, with its petals on the coordinate axes.

$r = \cos(3\theta)$ is a three-petaled flower shape, centered at the origin, with one petal on the positive *x*-axis.

The Law of Cosines

Objectives

- Use the law of cosines, together with the law of sines, to solve triangles.

- Prove the acute case of the law of cosines.

Why *The law of cosines can be used to determine whether three groups of campers can communicate from their campsites by using two-way radios.*

At a campground there are three groups of campers at three different campsites. The campers carry two-way radios with a range of about 1 mile. What information is needed to determine whether the three groups can all communicate directly with each other?

The Law of Cosines

In the previous lesson, you used the law of sines to solve triangles given two angles and a side or two sides and an angle opposite a given side.

Suppose that you have a triangle with two known sides in which the included angle is known. By SAS congruence, the triangle is uniquely determined. However, the triangle cannot be solved using the law of sines because every possible proportion in the law of sines will have two unknowns.

You can use the law of cosines to solve a triangle if you are given the measures of
  1. two sides and the included angle *or*
  2. three sides.

The Law of Cosines

For any triangle $\triangle ABC$ with sides a, b, and c:

$$a^2 = b^2 + c^2 - 2bc \cos A$$
$$b^2 = a^2 + c^2 - 2ac \cos B$$
$$c^2 = a^2 + b^2 - 2ab \cos C$$

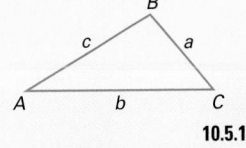

10.5.1

Alternative Teaching Strategy

USING TECHNOLOGY Students can use geometry graphics software that includes trigonometric functions to show that the law of cosines is true for any triangle.

Prepare

QUICK WARM-UP

1. Draw any non-right triangle. Label the vertices A, B, and C. Use a, b, and c to label the sides, with side a opposite $\angle A$, side b opposite $\angle B$, and side c opposite $\angle C$.
Check students' drawings.

2. List all possible ways to name two sides and the included angle for the triangle you drew above.
$a, b, \angle C$; $a, c, \angle B$; $b, c, \angle A$

Also on Quiz Transparency 10.5

Teach

Why Solving problems requires making decisions about when and how to use available information. In some problem-solving situations, the law of sines cannot be used because not enough appropriate information is given. In this lesson, students develop the law of cosines and decide when to use it, according to the information given.

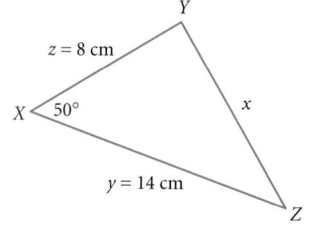
TRY THIS
$b \approx 20.65$

CRITICAL THINKING
The information given yields these ratios:

$$\frac{\sin 43°}{b} = \frac{\sin A}{28} = \frac{\sin C}{32}$$

Since any two of the three ratios has two missing variables, it is not possible to solve for any of the missing variables.

TRY THIS
$m\angle D \approx 29.4°$;
$m\angle E \approx 117.9°$;
$m\angle F = 32.7°$

664 LESSON 10.5

EXAMPLE 1 Find b in the triangle at right.

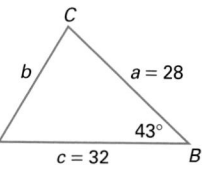

SOLUTION
Using the second equation in the law of cosines,

$$b^2 = a^2 + c^2 - 2ac \cos B$$
$$b^2 = 28^2 + 32^2 - 2(28)(32) \cos 43°$$
$$b^2 \approx 497.4$$
$$b \approx 22.3$$

TRY THIS Given $\triangle ABC$ from Example 1 with $a = 17$, $c = 25$, and $m\angle B = 55°$, find b.

CRITICAL THINKING Explain why you cannot use the law of sines to solve the triangle in Example 1.

EXAMPLE 2 Find the measures of $\angle D$, $\angle E$, and $\angle F$ in the triangle at right.

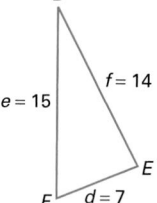

SOLUTION

Step 1
Use the law of cosines to find the first angle. It is convenient to start with the largest angle, which is opposite the longest side, e.

$$e^2 = d^2 + f^2 - 2df \cos E$$
$$15^2 = 7^2 + 14^2 - 2(7)(14) \cos E$$
$$225 = 245 - 196 \cos E$$
$$-20 = -196 \cos E$$
$$E = \cos^{-1}\left(\frac{-20}{-196}\right) \approx 84°$$

Step 2
Use the law of sines to find the second angle. Since the remaining angles are acute (why?), you do not have to worry about the ambiguous case.

$$\frac{\sin F}{14} = \frac{\sin 84°}{15}$$
$$\sin F = \frac{14 \sin 84°}{15} \approx 0.93$$
$$F = 68°$$

Step 3
Use the Triangle Sum Theorem to find the third angle.

$$m\angle D \approx 180° - (84° + 68°) \approx 28°$$

TRY THIS Given $\triangle DEF$ from Example 2 with $d = 10$, $e = 18$, and $f = 11$, find the measures of $\angle D$, $\angle E$, and $\angle F$.

CRITICAL THINKING Could the law of cosines have been used in Step 2 of Example 2? Why do you think the law of sines was used instead?

Interdisciplinary Connection

PHYSICS When designing highways, railroads, or buildings, engineers and architects sometimes use formulas involving the trigonometric functions to calculate various stresses on the foundation and other supporting structures. Invite an engineer or architect to visit the class and show examples of formulas involving the sine and cosine functions.

EXAMPLE ③ Suppose three campsites are positioned as shown. The campers communicate by using two-way radios with a range of about 1 mi, or 5280 ft. Will the campers at sites 1 and 3 be able to communicate directly with each other?

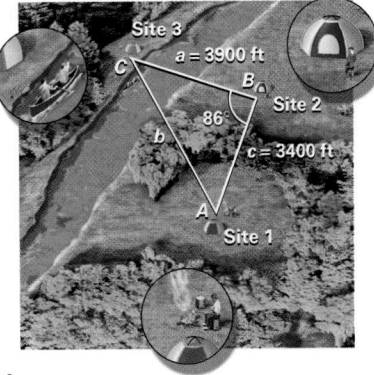
Site 3
$a = 3900$ ft
B
Site 2
86
$c = 3400$ ft
b
A
Site 1

● **SOLUTION**

Use the law of cosines.

$b^2 = a^2 + c^2 - 2ac \cos B$

$b^2 = 3900^2 + 3400^2 - 2(3900)(3400)\cos 86°$

$b^2 \approx 24{,}920{,}058$

$b \approx 4992$ ft

The campers at sites 1 and 3 will be able to communicate directly by using the two-way radios.

TRY THIS If the distance from site 1 to site 2 is 3600 ft, the distance from site 2 to site 3 is 4000 ft, and m∠B = 84°, will the campers at sites 1 and 3 be able to communicate directly with each other? Why or why not?

Proof of the Law of Cosines: Acute Case

The following is a proof of the acute case of the law of cosines. You will be asked to prove the obtuse case in Exercises 26–31.

PROOF

Given: acute △ABC with sides a, b, and c

Prove: $a^2 = b^2 + c^2 - 2bc \cos A$

Proof:

Draw the altitude from C, with height h, which divides $\overline{AB}$ into two segments, c_1 and c_2. Then, by definition of cosine,

$$\cos A = \frac{c_1}{b}$$

$$(2bc) \cos A = \frac{c_1}{b}(2bc) \qquad \text{Multiply both sides by 2bc.}$$

$$2bc \cos A = 2cc_1 \qquad \text{Simplify.}$$

From the Pythagorean Theorem:

$a^2 = h^2 + c_2{}^2$

$a^2 = h^2 + (c - c_1)^2 \qquad c_2 = c - c_1$

$a^2 = h^2 + c^2 - 2cc_1 + c_1{}^2 \qquad \text{Distribute.}$

$a^2 = (h^2 + c_1{}^2) + c^2 - 2cc_1 \qquad \text{Rearrange terms.}$

$a^2 = b^2 + c^2 - 2cc_1 \qquad b^2 = h^2 + c_1{}^2$

$a^2 = b^2 + c^2 - 2bc \cos A \qquad \text{Substitution}$

C
b / h \ a
c_1 c_2
A   c   B

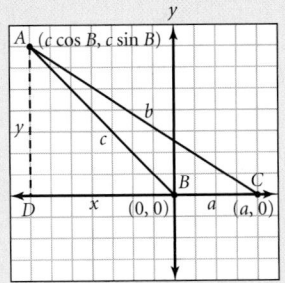

ASSIGNMENT GUIDE

In Class	1–7
Core	9–19 odd, 20–32
Core Plus	20–37
Review	38–45
Preview	46–50

✐ Extra Practice can be found beginning on page 818.

Error Analysis

Many students will make errors if they try to solve equations in just one step. Have them write out solutions with several steps. Then they should measure the needed sides or angles to check their answers.

Exercises

Communicate

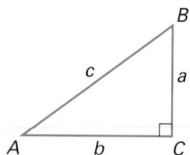

1. Suppose that △*ABC* is a right triangle. Use the law of cosines to find c^2, and simplify as much as possible. What do you notice?

i internet connect

Activities Online

Go To: go.hrw.com
Keyword:
MG1 Cosines

Which rule should you use, the law of sines or the law of cosines, to find each indicated measurement below? Explain your reasoning.

2.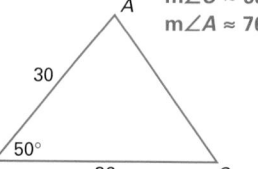

3.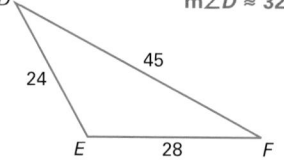

4. Suppose that three angles of a triangle are given. Could you use the law of sines or the law of cosines to solve the triangle? Explain your reasoning.

Guided Skills Practice

Use the law of cosines and the law of sines to solve each triangle.

5. *(EXAMPLE 1)* $b ≈ 28.42$
    $m\angle C ≈ 53.96°$
    $m\angle A ≈ 76.03°$

6. *(EXAMPLE 2)* $m\angle E ≈ 119.66°$
    $m\angle F ≈ 27.61°$
    $m\angle D ≈ 32.73°$

7. Campers at sites 2 and 3 cannot communicate directly since their two-way radios have a range of only 7920 feet, and the distance between them is about 8192 feet.

7. Suppose that three campers have two-way radios with a range of 7920 ft. The distance between sites 1 and 2 is 5750 ft, the distance between sites 1 and 3 is 6690 ft, and the angle formed with site 1 at the vertex is 82°. Can the campers at sites 2 and 3 commumicate directly? Why or why not? *(EXAMPLE 3)*

Reteaching the Lesson

COOPERATIVE LEARNING Have students work in pairs to find all three angles in a triangle with sides measuring 10, 16, and 20 units. They should use the law of cosines to find the largest angle first. This angle is opposite the 20-unit side and has a cosine of –0.1375, so it is obtuse.

Make sure that students know how to interpret the negative number they get for the cosine. A second angle can be found by using the law of sines, and the third angle can be found by subtracting the sum of the first two angles from 180°. **The three angles measure 97.9°, 52.4°, and 29.7°.**

Practice and Apply

For Exercises 8–13, find the indicated measures. It may help to sketch the triangle roughly to scale.

8. $a = 12$ $b = \underline{\ ?\ }$ $c = 17$ $m\angle B = 33°$

9. $a = 2.2$ $b = 4.3$ $c = \underline{\ ?\ }$ $m\angle C = 52°$

10. $a = \underline{\ ?\ }$ $b = 68.2$ $c = 23.6$ $m\angle A = 87°$

11. $a = 10$ $b = 7$ $c = 8$ $m\angle A = \underline{\ ?\ }$

12. $a = 3.6$ $b = 3.6$ $c = 2.5$ $m\angle C = \underline{\ ?\ }$

13. $a = 27$ $b = 41$ $c = 15$ $m\angle B = \underline{\ ?\ }$

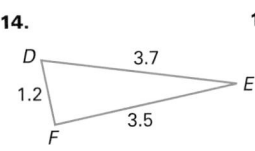

Solve each triangle.

14. 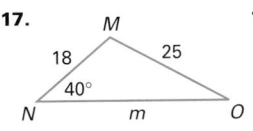   **15.** 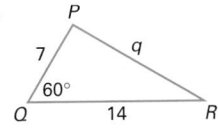   **16.**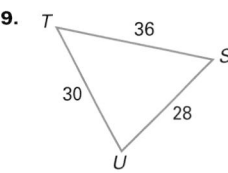

17.   **18.**   **19.**

Exercises 20–25 summarize the methods of solving triangles by using the law of sines and the law of cosines.

20. law of sines

21. law of cosines

22. two sides and an angle that is opposite one of the given sides

23. Sample answer:

24. Sample answer:

25. law of cosines

SUMMARY OF TRIANGLE-SOLVING PROCEDURES		
Given	**Example**	**Theorem(s)**
three angles	105° 35° 40°	cannot be solved
two angles and one side	80° 10 60°	**20.** __?__
two sides and included angle	24 50° 16	**21.** __?__
22. __?__	**23.** __?__	law of sines (ambiguous case)
three sides	**24.** __?__	**25.** __?__

Technology

As in the previous lesson, geometry graphics software can be used to check students results. Students should construct triangles with the given measures and then measure the needed sides or angles to check their answers.

8. $b \approx 9.53$ units

9. $c \approx 3.42$ units

10. $a \approx 70.99$ units

11. $m\angle A \approx 83.33°$

12. $m\angle C \approx 40.64°$

13. $m\angle B \approx 153.84°$

14. $m\angle F = 90°$; $m\angle E \approx 18.92°$; $m\angle D \approx 71.08°$

15. $m\angle G = 72°$; $g = 8.8$; $h \approx 5.44$

16. $k \approx 3.63$; $m\angle J \approx 24.96°$; $m\angle L \approx 25.04°$

17. $m\angle O \approx 27.57°$; $m\angle M \approx 112.43°$; $m \approx 35.95$

18. $q \approx 12.12$; $m\angle R \approx 30.01°$; $m\angle P \approx 89.99°$

19. $m\angle U \approx 76.65°$; $m\angle S \approx 54.18°$; $m\angle T \approx 49.17°$

31.
$$a^2 = h^2 + (c_1 + c)^2$$
$$a^2 = h^2 + c_1{}^2 + 2c_1c + c^2$$
$$a^2 = b^2 + 2(-b \cos A) \cdot c + c^2$$
$$a^2 = b^2 - 2bc \cos A + c^2$$
$$a^2 = b^2 + c^2 - 2bc \cos A$$

In Exercises 26–31, you will prove the obtuse case of the law of cosines. Refer to the diagram at right.

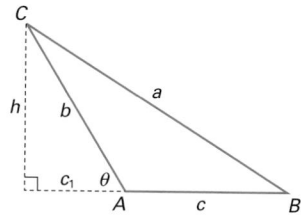

26. $\dfrac{c_1}{b}$

27. $b \cos \theta = c_1$

28. $-b \cos A = c_1$

29. $a^2 = h^2 + (c_1 + c)^2$

30. $b^2 = h^2 + c_1{}^2$

26. By definition of cosine, $\cos \theta = $? .

27. Multiply both sides of the equation from Exercise 26 by b, and simplify.

28. Use the identity $\cos(180 - A) = -\cos A$ to rewrite your equation from Exercise 27 in terms of $\cos A$.

29. By the Pythagorean Theorem, $a^2 = $? + ? .

30. By the Pythagorean Theorem, $b^2 = $? + ? .

31. Expand the square of the binomial in the equation from Exercise 29. Substitute the results of Exercises 28 and 30 into this equation and simplify to complete the proof of the law of cosines.

CHALLENGE

32. Using the law of cosines, prove the Triangle Inequality Theorem: For $\triangle ABC$ with sides a, b, and c, $a + b > c$. Also prove that $c > a - b$ for $a > b$.

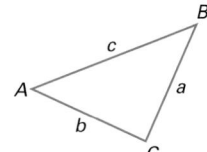

APPLICATIONS

33. SPORTS On a standard baseball field, the bases form a square with 90-ft sides. The pitcher's mound is 60.5 ft from home plate on a diagonal from home plate to second base. Find the distance from the pitcher's mound to first base. **63.72 feet**

34. NAVIGATION Mark and Stephen walk into the woods along lines that form a 72° angle. If Mark walks at 2.8 mph and Stephen walks at 4.2 mph, how far apart will they be after 3 hr? **12.80 miles**

35. The road will be 24.88 km long, and the angle formed is 42.24°.

35. NAVIGATION The distance from Greenfield to Brownsville is 37 km, and the distance from Greenfield to Red River is 25 km. The angle formed by the roads at Greenfield is 42°. The state highway department plans to build a straight road from Brownsville to Red River. How long will the road be, and what angle will the new road form with the road from Greenfield to Brownsville?

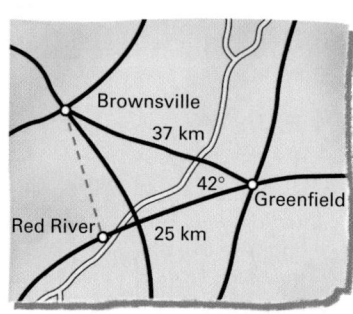

36. INDIRECT MEASUREMENT From point A, the distance to one end of a small lake is 300 yd, the distance to the other end is 500 yd, and the angle formed by the lines of sight is 105°. Estimate the width of the lake. $c \approx 646.26$ yards

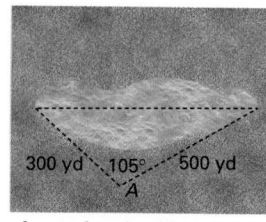

37. $m\angle A \approx 41.65°$;
$m\angle B \approx 52.89°$;
$m\angle C \approx 85.46°$

37. LANDSCAPING Nina wants to use three long boards as a border for a triangular garden. The boards have lengths of 10 ft, 12 ft, and 15 ft. Find the angles at the corners of the garden.

Practice

NAME _____ CLASS _____ DATE _____

((( ))) **Practice**
10.5 The Law of Cosines

Exercises 1–5, find the indicated measures. [It] may be helpful to sketch the triangle roughly to [sc]ale. Round your answers to the nearest tenth.

. $a = 19$	$b = 20$	$c =$?	$m\angle C = 50°$	16.5
. $a = 8$	$b = 9$	$c = 10$	$m\angle C =$?	71.8°
. $a = 6$	$b = 6$	$c = 9$	$m\angle B =$?	41.4°
. $a = 5$	$b = 6$	$c = 8$	$m\angle A =$?	38.6°
$a = 3$	$b = 4$	$c = \sqrt{33}$	$m\angle C =$?	109.5°

[So]lve each triangle.

6.
$m\angle D = 84.3°$;
$m\angle E = 32.4°$; $m\angle F = 63.3°$

7.
$m\angle G = 100°$; $g \approx 3.8$; $i \approx 2.4$

8.
$j \approx 2.8$; $m\angle L = m\angle K = 62.5°$

9.
$m\angle M \approx 59.4°$; $m\angle N \approx 85.6°$; $n \approx 3.5$
(There are two possible triangles with the dimensions given; but from the diagram, $\angle M$ must be acute, so only one of the triangles is possible.)

10.
$m\angle C = 103°$; $b \approx 1.4$; $a \approx 1.8$

11.
$m\angle Z = 133°$; $x \approx 4.4$; $z \approx 8.7$

32. Proof of the Triangle Inequality Theorem:
Given the law of cosines
$c^2 = a^2 + b^2 - 2ab \cos C$, then
$c^2 + 2ab \cos C = a^2 + b^2$.
Also $(a + b)^2 = a^2 + 2ab + b^2 > a^2 + b^2$,
since a and b are both positive.
So, by using substitution,
$(a + b)^2 > c^2 + 2ab \cos C$ and
$(a + b)^2 > c^2 + 2ab \cos C > c^2$, since $\cos C$
is positive for all angles, C, between 0 and
180°. Therefore, $(a + b)^2 > c^2$ and by taking
the square root of both sides, $a + b > c$.

Proof that for $a > b$, $c > a - b$:
Given $c^2 = a^2 + b^2 - 2ab \cos C$, and $a > b$.
Also $(a - b)^2 = a^2 - 2ab + b^2$
$< a^2 + b^2 - 2ab \cos C$, since $\cos C < 1$ for all
angles, C, between 0 and 180°.
Therefore, $(a - b)^2 < a^2 + b^2 - 2ab \cos C = c^2$,
so $(a - b)^2 < c^2$ and by taking the square
root of both sides, $a - b < c$.

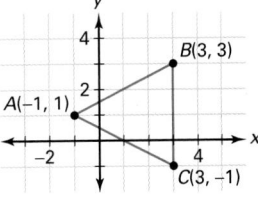

CONNECTION

COORDINATE GEOMETRY For Exercises 38–41, draw the image of △ABC under the given transformation. *(LESSONS 1.7 AND 8.1)*

38. $T(x, y) = (3x, 3y)$

39. $T(x, y) = (x - 2, y - 2)$

40. $T(x, y) = \left(\frac{3}{4}x, \frac{3}{4}y\right)$

41. $T(x, y) = (2x + 1, 2y - 3)$

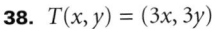

42. width = $6\frac{2}{3}$ cm
    length = $33\frac{1}{3}$ cm

42. The perimeter of a rectangle is 80 cm. The ratio of its width to length is $\frac{1}{5}$. Find the length and width of the rectangle. *(LESSON 8.2)*

43. Yes, SAS-similarity

44. No, the congruent angles are not between the corresponding sides in the triangles.

Determine whether the triangles in each pair can be proven similar. Explain your reasoning. *(LESSON 8.3)*

43.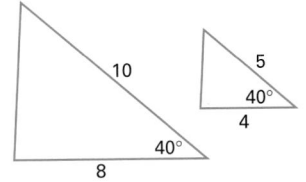

44.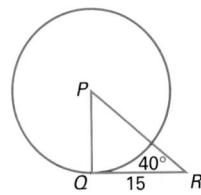

45. In the diagram at right, $\overline{QR}$ is tangent to ⊙P. Find radius PQ. *(LESSONS 9.2 AND 10.1)* PQ ≈ 12.59

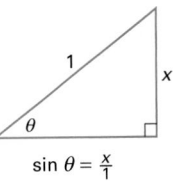

 Look Beyond

Exercises 46–50 extend the concepts learned in this lesson to show students how to evaluate composite trigonometric functions by using a right triangle.

38.

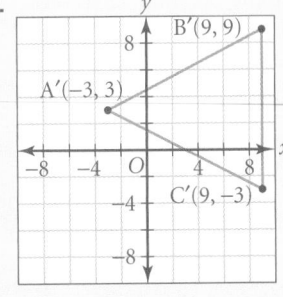

39.

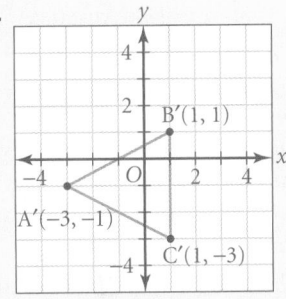

CONNECTION

TRIGONOMETRY You can use right triangles to simplify expressions involving inverse trigonometric functions. For example, consider the expression $\cos(\sin^{-1} x)$.

46. $b = \sqrt{1 - x^2}$

46. $\sin^{-1} x$ is an angle that has a sine ratio of $\frac{x}{1}$. This angle can be represented by represented by drawing a right triangle with a hypotenuse of 1 and a leg of length x. Use the Pythagorean Theorem to find an expression for the length of the adjacent leg.

$\sin \theta = \frac{x}{1}$

$\theta = \sin^{-1} x$

47. $\cos \theta = \sqrt{1 - x^2}$

47. Use the definition of cosine and your result from Exercise 47 to write an expression for $\cos \theta$, which is equal to $\cos(\sin^{-1} x)$.

Simplify the following expressions:

48. $\tan(\sin^{-1} x)$ $\dfrac{x}{\sqrt{1 - x^2}}$ **49.** $\sin(\cos^{-1} x)$ $\sqrt{1 - x^2}$ **50.** $\sin(\tan^{-1} x)$ $\dfrac{x}{\sqrt{x^2 + 1}}$

40.

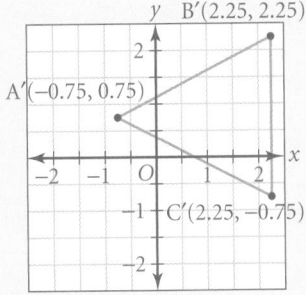

41.

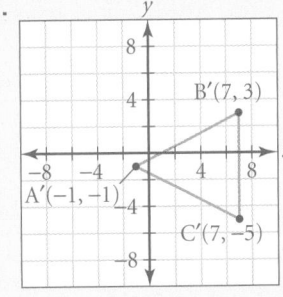

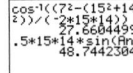

Focus

An application for the law of cosines is found in astronomer Jim Herrnstein's method of measuring the universe. Herrnstein claims that his method of direct measurement is more accurate than the method currently used by NASA astronomers.

Motivate

The article on this page includes terms that may be unfamiliar to students. Discuss the following points with students:

- A *light-year* is the distance light travels in a year. A light-year is approximately 9.46×10^{12} kilometers.

- *Radio astronomy* is the study of the universe through observations of the natural radio waves emitted by cosmic objects.

- *The National Aeronautics and Space Administration*, or NASA, is a civilian agency of the U.S. government responsible for the development of advanced technology in aviation and space exploration.

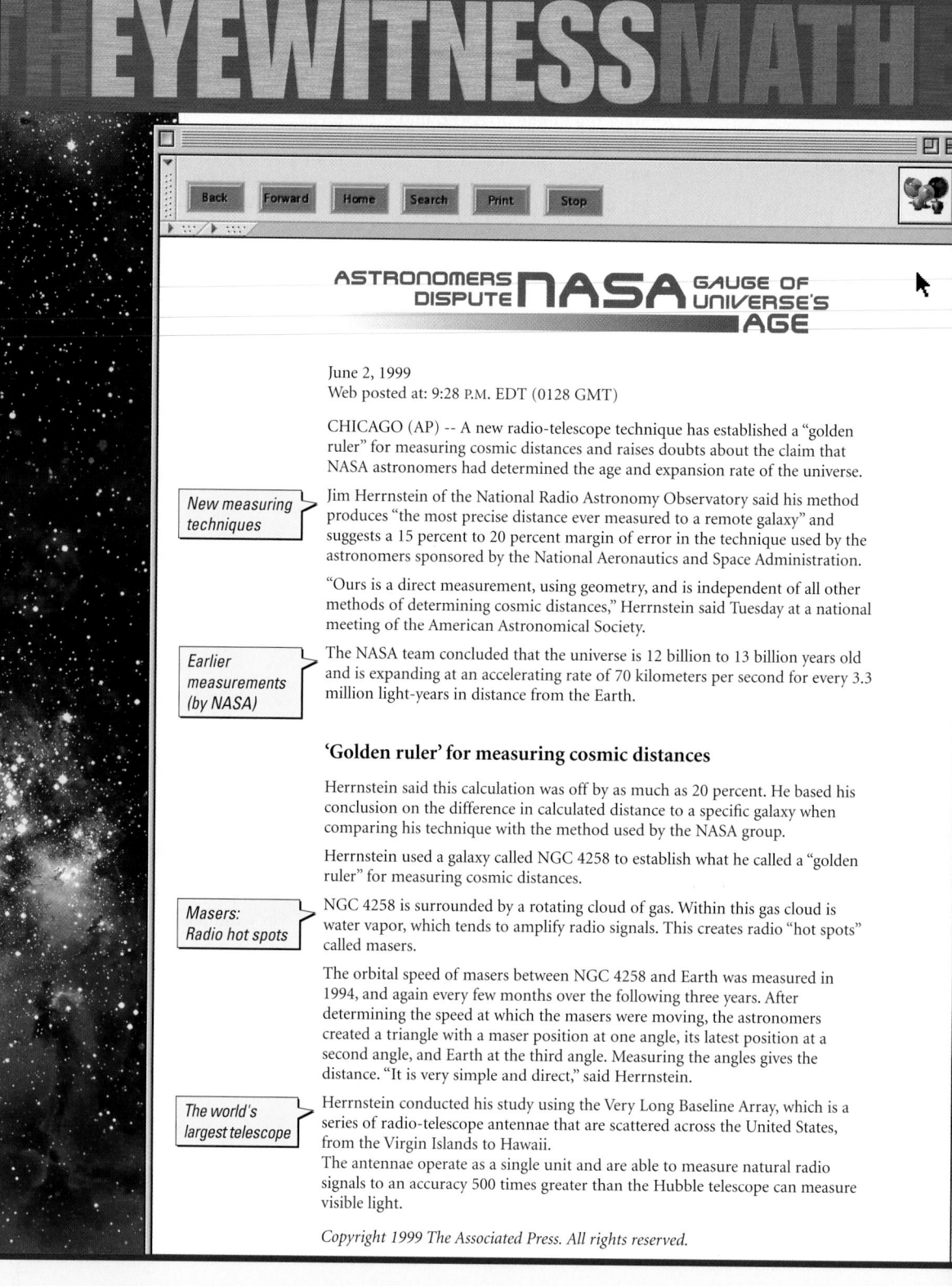

THE EYEWITNESS MATH

ASTRONOMERS DISPUTE **NASA** GAUGE OF UNIVERSE'S AGE

June 2, 1999
Web posted at: 9:28 P.M. EDT (0128 GMT)

CHICAGO (AP) -- A new radio-telescope technique has established a "golden ruler" for measuring cosmic distances and raises doubts about the claim that NASA astronomers had determined the age and expansion rate of the universe.

New measuring techniques

Jim Herrnstein of the National Radio Astronomy Observatory said his method produces "the most precise distance ever measured to a remote galaxy" and suggests a 15 percent to 20 percent margin of error in the technique used by the astronomers sponsored by the National Aeronautics and Space Administration.

"Ours is a direct measurement, using geometry, and is independent of all other methods of determining cosmic distances," Herrnstein said Tuesday at a national meeting of the American Astronomical Society.

Earlier measurements (by NASA)

The NASA team concluded that the universe is 12 billion to 13 billion years old and is expanding at an accelerating rate of 70 kilometers per second for every 3.3 million light-years in distance from the Earth.

'Golden ruler' for measuring cosmic distances

Herrnstein said this calculation was off by as much as 20 percent. He based his conclusion on the difference in calculated distance to a specific galaxy when comparing his technique with the method used by the NASA group.

Herrnstein used a galaxy called NGC 4258 to establish what he called a "golden ruler" for measuring cosmic distances.

Masers: Radio hot spots

NGC 4258 is surrounded by a rotating cloud of gas. Within this gas cloud is water vapor, which tends to amplify radio signals. This creates radio "hot spots" called masers.

The orbital speed of masers between NGC 4258 and Earth was measured in 1994, and again every few months over the following three years. After determining the speed at which the masers were moving, the astronomers created a triangle with a maser position at one angle, its latest position at a second angle, and Earth at the third angle. Measuring the angles gives the distance. "It is very simple and direct," said Herrnstein.

The world's largest telescope

Herrnstein conducted his study using the Very Long Baseline Array, which is a series of radio-telescope antennae that are scattered across the United States, from the Virgin Islands to Hawaii.
The antennae operate as a single unit and are able to measure natural radio signals to an accuracy 500 times greater than the Hubble telescope can measure visible light.

Copyright 1999 The Associated Press. All rights reserved.

Cooperative Learning

Refer to the diagram at right. The angle, θ, between two positions of a given maser in NGC 4258 was measured by the Very Long Baseline Array, or VLBA. The two positions of the maser form two vertices of an isosceles triangle, and the position of Earth forms the third vertex.

1. The maser in galaxy NGC 4258 rotates at about 1100 km per second. Use this to determine the distance that the maser traveled in 3 years between the two indicated positions in the triangle. Use 1 yr = 31536000 sec.

2. For extremely small vertex angles, such as θ, the altitude of an isosceles triangle is very nearly equal to the lengths of its legs. Thus the altitude can be used to approximate the lengths, d, of the legs of the triangle in the diagram. Find the altitude as an estimate of d.

3. According to Herrnstein's measurements, NGC 4258 is about 23.5 million light-years away. A light-year is approximately 9.46×10^{12} km. How close is your answer to Herrnstein's?

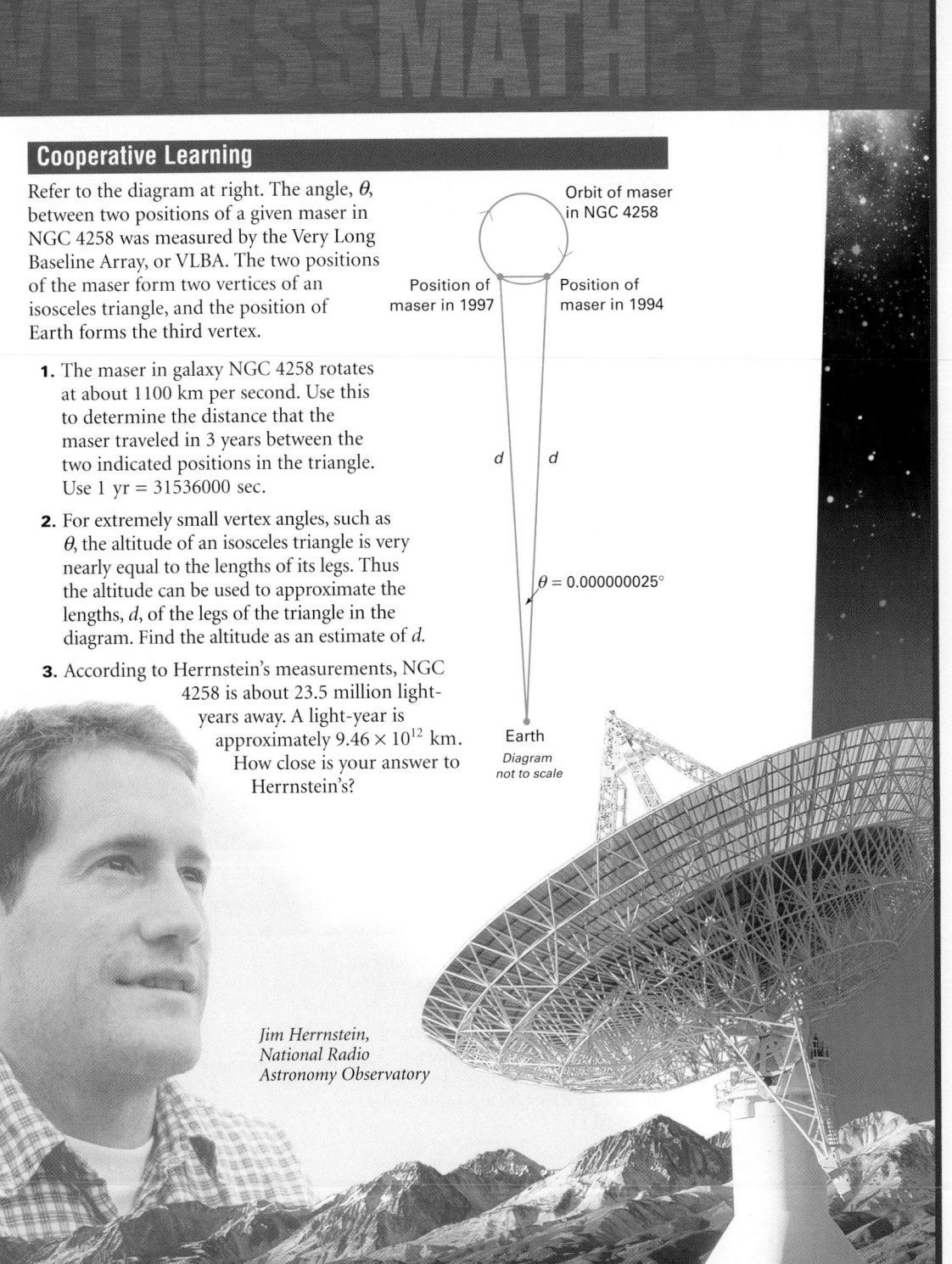

Orbit of maser in NGC 4258

Position of maser in 1997   Position of maser in 1994

d d

$\theta = 0.000000025°$

Earth
Diagram not to scale

Jim Herrnstein, National Radio Astronomy Observatory

Cooperative Learning

Working in groups of 2 or 3, have students discuss how to use trigonometry to solve for the altitude of an isosceles triangle, given the vertex angle and the measure of the base. They should understand why this method can be used to approximate the measures of the legs of the triangle when the vertex angle is very small. (NOTE: The vertex angle is too small to use the law of cosines to solve the triangle on most student calculators.)

Discuss

Discuss the following questions:

- Herrnstein claims that the calculations of the NASA astronomers for the age and rate of expansion of the universe were incorrect by as much as 20 percent. How significant is this, given the very large numbers used in the calculations?

- All objects give off natural radio signals. These signals are what radio telescopes measure. What exactly did the author mean by *radio "hot spots"* in the description of *masers*?

Vectors in Geometry

Objectives

- Define *vector*.
- Add two vectors.
- Use vectors and vector addition to solve problems.

Why *The path of a ball in a sports event is affected by the wind. For this reason, stadiums for certain sports have flags that give a visual indication of the wind's direction and its magnitude. Because wind has these two properties it can be represented by a vector.*

Teach

Why Vectors are frequently found in problem-solving situations, particularly in the physical sciences and engineering. Any quantity that has a magnitude and a direction can be represented by a vector. Examples of vectors include speed and force. Students can find vector-related situations in science textbooks for discussion and display.

What Is a Vector?

A **vector** is a mathematical object that has both a **magnitude** (a numerical measure) and a **direction**. Arrows are used to represent vectors, because arrows have both magnitude and direction. The length of a vector arrow represents the magnitude of the vector.

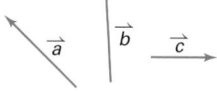

$\vec{a}$, $\vec{b}$, and $\vec{c}$ are vectors.

A lowercase letter with a vector arrow over it may be used to indicate a vector. The symbol $|\vec{v}|$ denotes the magnitude of $\vec{v}$.

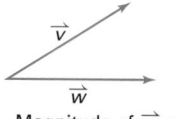

Magnitude of $\vec{v} = |\vec{v}|$
Magnitude of $\vec{w} = |\vec{w}|$

Alternative Teaching Strategy

USING TECHNOLOGY Geometry graphics software can be used to construct vectors and find vector sums. Have students re-create the figures used in the text, examples, and exercises. By dragging on initial and final points of vectors, students can see what happens to the magnitudes and directions as the vectors change.

Anything that has both magnitude and direction can be modeled by a vector. In the examples below, vectors are used to represent a velocity, a force, and a displacement (a relocation).

CHECKPOINT ✔ Describe the magnitude and direction of the vector in each.

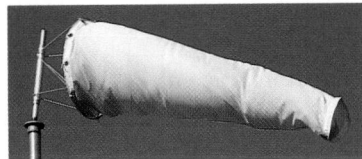

Velocity *The wind sock indicates the strength and direction of the wind.*

Force *The tugboat applies a force on the barge in the direction the barge is pointed.*

Displacement *The car speeds down the quarter-mile track.*

To describe the direction of a vector, you need a direction to use as a reference. On a coordinate grid this is usually the positive *x*-axis direction.

CHECKPOINT ✔ What are the reference directions in the three pictorial examples?

Vectors with the same direction, such as vectors $\vec{a}$ and $\vec{b}$ in the illustration, are said to be *parallel*. Vectors at right angles are *perpendicular*.

Vector Sums

In many situations it is appropriate to combine vectors to get a new vector called a *resultant*. The process of combining vectors represented in the example below is called *vector addition*.

A hiker travels 3 miles northeast, then 1 mile east. The resultant vector (red) is the total distance traveled in a specific direction from the starting point.

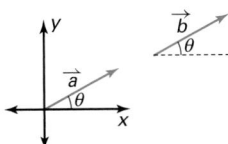

Two tractors pull on a tree stump with the forces and directions shown. The combined force exerted by the tractors in the direction shown is the resultant vector.

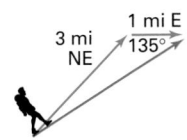
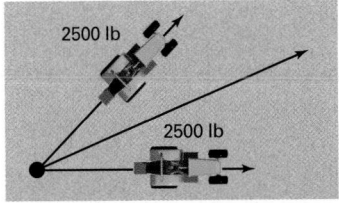

CHECKPOINT ✔
Windsock: the magnitude is the speed of the wind, and the direction of the wind.

Tugboat: the magnitude is the speed of the tugboat, and the direction is the direction of the motion of the tugboat.

Car: the magnitude is the speed of the car, and the direction is the direction of travel along the track.

CHECKPOINT ✔
In each picture, a possible reference direction is North. Other answers are possible, such as the direction of the current could be used for a reference direction of the tugboat, and the orientation of the racetrack could be used as a reference direction of the car.

Cooperative Learning

Have students work in pairs to draw a triangle and a translation vector. Ask them to translate the triangle on an *xy*-coordinate system, by using the information represented by the vector. Students should write an explanation about how this use of vectors is similar to or different from the examples given in this lesson.

Interdisciplinary Connection

GEOGRAPHY The north-south-east-west symbol used on maps is called a *compass rose*. Have students find examples of this symbol on road maps or maps in social studies books. Working in pairs, students should choose a departure point and a destination point for an imaginary trip. Then they should make a sketch that includes a vector to describe their trip.

ADDITIONAL
EXAMPLE ❶

Find the sum of $\vec{a}$ and $\vec{b}$ by using the head-to-tail method.

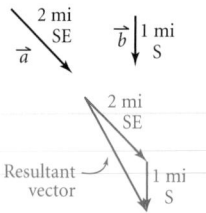

TRY THIS

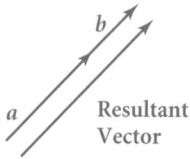

ADDITIONAL
EXAMPLE ❷

Find the sum of vectors $\vec{a}$ and $\vec{b}$ by using the parallelogram method.

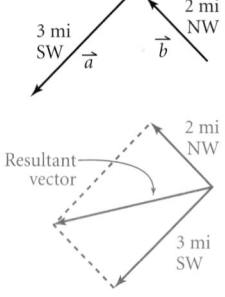

TRY THIS

If two equal forces are applied in opposite directions, the resultant vector would have length 0. If the forces are unequal, then the resultant vector would have a length which is the difference of the lengths of the two vectors.

CRITICAL THINKING

The diagram shows that vector $\vec{w}$ is parallel and congruent to the opposite side of the parallelogram, so that either placement of the vector would produce the same resultant vector.

The Head-to-Tail Method of Vector Addition

In the hiker example, each vector represents a displacement—that is, a relocation of the hiker from a given point. In such cases, the *head-to-tail method* is a natural way to do vector addition.

EXAMPLE ❶ Find the sum of vectors $\vec{a}$ and $\vec{b}$.

● **SOLUTION**

To use the head-to-tail method of vector addition, place the tail of one vector at the head of the other. The vector sum is a new displacement vector from the tail of the first vector to the head of the second vector.

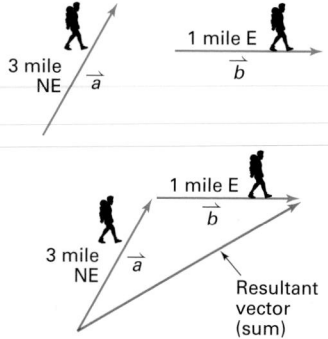

TRY THIS Use the head-to-tail method to find the sum of two parallel vectors.

The Parallelogram Method of Vector Addition

In many cases, the *parallelogram method* is a natural way to think of combining vectors. This is especially true when two forces act on the same point, such as in the tractor example on the previous page.

EXAMPLE ❷ Find the sum of vectors $\vec{m}$ and $\vec{n}$.

● **SOLUTION**

To find the sum of the vectors, complete a parallelogram by adding two segments to the figure. The vector sum is a vector along the diagonal of the parallelogram, starting at the common point.

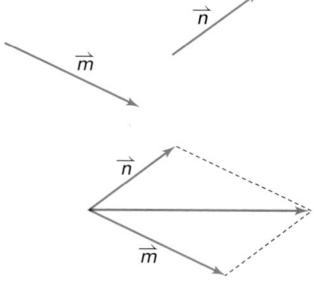

TRY THIS What is the vector sum if two equal forces are applied to a point in opposite directions? What happens to the sum if two such forces are unequal?

CRITICAL THINKING The parallelogram method and the head-to-tail method are equivalent since they produce the same resultant. How does the diagram at right show that the methods are equivalent?

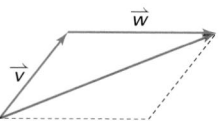

Enrichment

How steep does a plane have to be before an object resting on it will slide? In the diagram, w is the weight of an object, and $w \sin \beta$ is the force acting on the object in a direction parallel to the surface of the plane. If $w \sin \beta$ is greater than f, the frictional force that tends to keep the object from sliding, then the object will move. A related idea is the *angle of repose* for a material. This is the greatest base angle that a conical pile of the material can have.

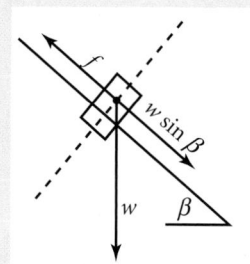

Teaching Tip

Before students study Examples 3 and 4, ask them to suggest ways of adding vectors that do not involve constructing a diagram. Possible ideas might include using the Pythagorean Theorem if the two vectors make right angles or using trigonometry for vector pairs that form angles other than 90°.

E X A M P L E ❸ A swimmer heads perpendicular to a 3-mph current in a river. Her speed in still water is 2.5 mph. Find the actual speed and direction of the swimmer.

APPLICATION

SPORTS

SOLUTION

Because the vectors are at right angles, the parallelogram method for addition gives a rectangle. Thus, you can solve for x by using the Pythagorean Theorem.

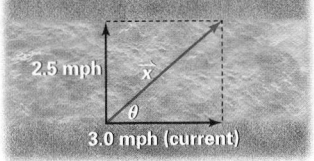

$$x^2 = 3^2 + 2.5^2$$
$$= 15.25$$
$$x = \sqrt{15.25} \approx 3.9 \text{ mph (the actual speed of the swimmer)}$$

There is more than one way to find θ. For example:

$$\tan \theta = \frac{2.5}{3.0} \approx 0.833$$
$$\theta \approx \tan^{-1}(0.833) \approx 40°$$

The swimmer swims at an angle of 40° with the current of the river.

ADDITIONAL

E X A M P L E ❸

The current in a lake is moving due east. A boat heads due north, but is pushed off course by the current so that it is moving along a line 40° north of east at a speed of 2.5 mph. How fast is the current moving?
2.5 cos 40° ≈ 1.9 mph

E X A M P L E ❹ The swimmer in Example 3 changes the direction of her effort so that she is heading at an angle of 40° with the direction of the current. What is her actual speed and direction?

SOLUTION

PROBLEM SOLVING

Draw a diagram. One way to solve such problems is to draw an accurate vector diagram and measure the resultant vector and its angle θ.

If you wish to solve the problem using trigonometry, begin by labeling the sides of a triangle as shown below. Then use known properties of parallelograms to find m∠C, which is 140°. Then use the law of cosines.

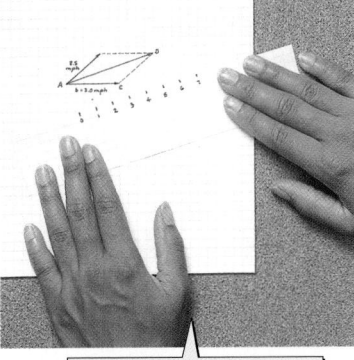

$$c^2 = a^2 + b^2 - 2ab \cos C$$
$$= 2.5^2 + 3^2 - 2(2.5)(3.0) \cos 140°$$
$$\approx 6.25 + 9 - 15(-0.766) = 26.74$$
$$c \approx \sqrt{26.74} \approx 5.17$$

If you are using graph paper, you can make a ruler from a strip of the same paper.

Teaching Tip

Students may need help identifying the three segments used for the Pythagorean Theorem in Example 3.

Use the law of sines to find m∠BAC, which is the same as m∠θ.

$$\frac{\sin A}{a} = \frac{\sin C}{c}$$
$$\frac{\sin A}{2.5} \approx \frac{.643}{5.17}$$
$$\sin A \approx 0.3109$$
$$\sin^{-1}(0.3109) \approx 18°$$

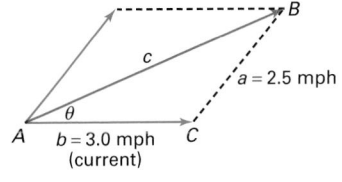

The swimmer's actual speed is 5.17 mph at angle of 18° with the current.

ADDITIONAL

E X A M P L E ❹

Two ropes are attached to a heavy object. Two people pull on the ropes, exerting forces of 15 and 22 newtons. The angle between the ropes is 100°. Find the magnitude of the force exerted on the object. A parallelogram diagram shows that the acute angles of the parallelogram are 80°. Substituting the known information into the equation for the law of cosines gives $x^2 = 15^2 + 22^2 - 2(15)(22) \cos 80°$, so $x \approx 24.4$ newtons.

Inclusion Strategies

USING MODELS Have students draw a compass rose and a vector that begins at its center point. Have them use two different angles to describe the direction of the vector, such as 30° north of east or 60° east of north. This activity will help to illustrate how the measure of the reference angle for a vector can vary.

Selected Answers

Exercises 6–9, 11–29 odd

ASSIGNMENT GUIDE

In Class	1–9
Core	10–17, 19–23 odd
Core Plus	11–23 odd, 24–26
Review	27–33
Preview	34–39

✐ Extra Practice can be found beginning on page 818.

6.

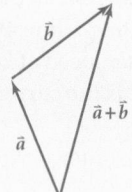

7.

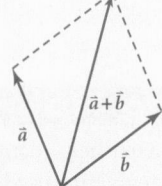

Exercises

● Communicate

📶 internet connect

Activities Online
Go To: **go.hrw.com**
Keyword:
MG1 Vectors

1. What is the magnitude of a vector? Explain.

2. What is the direction of a vector? Explain.

Describe the magnitude and direction of the vector or vectors in each of the following:

3. an airplane flying northwest at 175 knots

4. a boat going 15 knots upstream against a current of 3 knots

5. two equal tug-of-war teams pulling in opposite directions on a rope

● Guided Skills Practice

6. Find the sum of vectors $\vec{a}$ and $\vec{b}$ by using the head-to-tail method. *(EXAMPLE 1)*

7. Find the sum of vectors $\vec{a}$ and $\vec{b}$ by using the parallelogram method. *(EXAMPLE 2)*

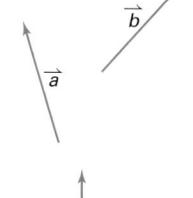

APPLICATIONS

8. **SPORTS** A swimmer heads perpendicular to the direction of a current whose speed is 1.5 mph. The speed of the swimmer in still water is 2 mph. Find the swimmer's actual speed and her direction angle, θ, with respect to the direction of the current. *(EXAMPLE 3)* **2.5 mph; 53.13°**

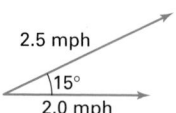

9. **SPORTS** A swimmer heads in a downstream direction at an angle of 15° with the direction of a 2-mph current. The speed of the swimmer in still water is 2.5 mph. Find the swimmer's actual speed and her direction angle, θ, with respect to the direction of the current. *(EXAMPLE 4)* **4.46 mph; 8.34°**

Reteaching the Lesson

USING MODELS To help students understand why the study of vectors relates to trigonometry, explain that it is often convenient to resolve a vector into its horizontal and vertical components. For example, vector F in the diagram at right represents the force exerted as the person pulls on the wagon. To resolve vector F into its horizontal and vertical vectors, F_x and F_y, students should use the equations $F_x = F \cos \theta$ and $F_y = F \sin \theta$.

Have students work in pairs to draw five different vectors and resolve them into their horizontal and vertical components.

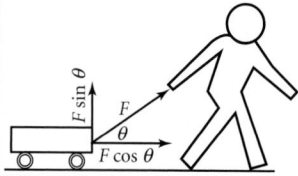

Practice and Apply

Copy each pair of vectors and draw the sum $\vec{a} + \vec{b}$ by using the head-to-tail method. You may need to translate one of the vectors.

10. 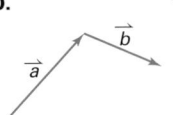   **11.** 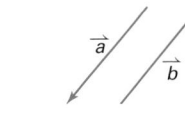   **12.** 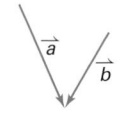   **13.**

Copy each pair of vectors and draw the sum $\vec{a} + \vec{b}$ by using the parallelogram method. You may need to translate one of the vectors.

14. 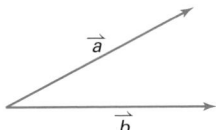   **15.**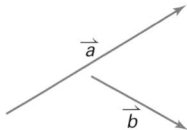

16. 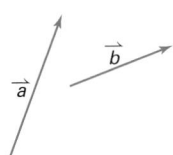   **17.**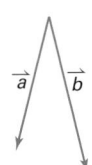

For each given pair of vectors $\vec{a}$ and $\vec{b}$ in Exercises 18–23, use the parallelogram method to find the vector sum $\vec{c}$. Use the law of cosines to find $|\vec{c}|$, the magnitude of $\vec{c}$. Use the law of sines to find the angle that $\vec{c}$ makes with $\vec{b}$.

18. $|\vec{c}|$ 13.17; 17.02°

19. $|\vec{c}|$ 10.71; 22.39°

20. $|\vec{c}|$ 9.92; 19.11°

21. $|\vec{c}|$ 13.76; 24.19°

22. $|\vec{c}|$ 6.87; 49.11°

23. $|\vec{c}|$ 6.78; 13.71°

18.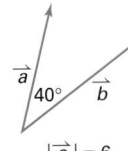
$|\vec{a}| = 6$
$|\vec{b}| = 8$

19.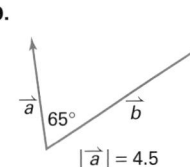
$|\vec{a}| = 4.5$
$|\vec{b}| = 8$

20.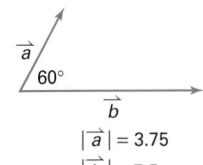
$|\vec{a}| = 3.75$
$|\vec{b}| = 7.5$

21.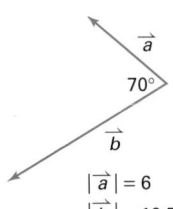
$|\vec{a}| = 6$
$|\vec{b}| = 10.5$

22.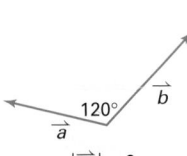
$|\vec{a}| = 6$
$|\vec{b}| = 7.5$

23.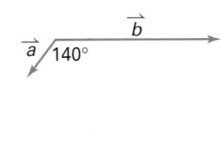
$|\vec{a}| = 2.5$
$|\vec{b}| = 8.5$

16. 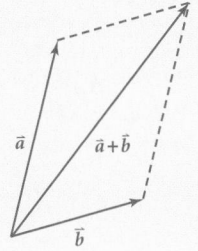   **17.**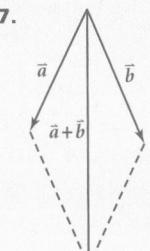

Error Analysis

For Exercises 6–13, students will need tracing paper or graph paper to make exact copies of the vectors.

Technology

Students may use geometry graphics software to check their answers to Exercises 14–19.

10.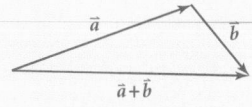

11.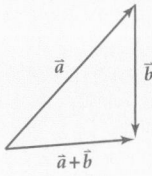

12.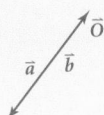

13.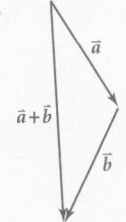

14.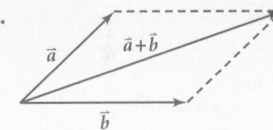

15.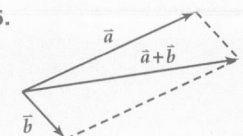

24. a. Using trigonometry, the wind correction angle is $-5.18°$, so the pilot should set out on a heading of 297.82 to achieve a true course of 303. Her ground speed will be 108.33 knots.

b. approximately 7.8 minutes.

24. AVIATION Mina, a student pilot, is making her first solo flight from Austin to Llano, Texas. Using a plotter and an aeronautical chart, she determines that her true course direction should be 303. According to the weather report, the wind at her cruising altitude of 4500 ft has a speed of 10 knots and a heading of 220 (the direction *from* which it is blowing). The true air speed of her plane, at cruising speed, is 110 knots.

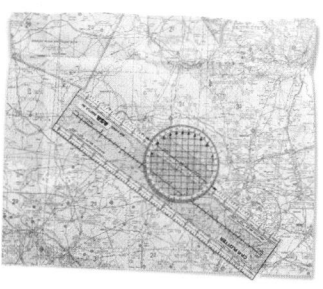

Using a mechanical computer, Mina determines that she must hold a heading of 298 to achieve her desired true course. According to the computer, her ground speed will be a little over 108 knots.

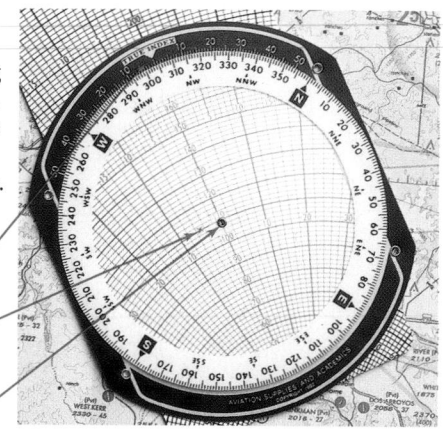

True course is set under the index.

Wind correction angle ($-5°$) is read under the *x*.

Ground speed is read under the center of the wheel.

a. Use trigonometry to check the computer's values for Mina's ground speed and heading. Round your answers to two decimal places.

In "dead reckoning," pilots use visible landmarks to check their course.

b. Mina has established a visible landmark at the bend of a river, 14 nautical miles from her point of departure. After how many minutes into her flight should she be over this landmark? (1 knot = 1 nautical mile per hour)

25.

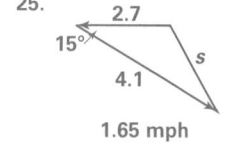

2.7
15°
4.1
s
1.65 mph

25. SCUBA DIVING Dan is investigating a boat sunk at the mouth of a river. To reach the wreck, Dan must swim against the current of 2.7 mph. Suppose Dan dives and starts swimming downward at 4.1 mph (still water speed) at a 15° angle with the water's surface. Find the actual speed Dan will swim as a result of the current.

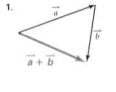
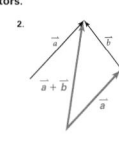
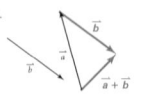
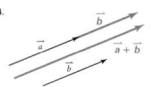
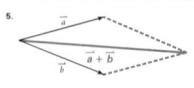
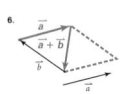
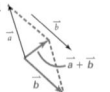

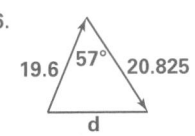
26.

19.6 /57°\ 20.825

d

a. 3.22 hours
b. 58.32°

26. NAVIGATION A boat leaves port and sails in still water at 5.6 mph for 3.5 hr. The boat then turns at a 57° angle and sails at 4.9 mph for 4.25 hr.

   a. If the boat turns and heads directly for port at 6 mph, how long will it take it to reach port?

   b. At what angle will the boat have to turn to head directly back to port in part a?

In Exercises 34–39, students are required to use their knowledge of vector addition to discover properties of vectors and vector addition in the coordinate plane.

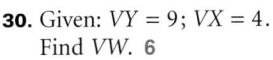
Look Back

Use the figures at right for Exercises 27–33. *(LESSON 9.5)*

27. Given: $VA = 5$; $VB = 8$; $VC = 6$. Find CD. **$12\frac{2}{3}$**

Algebra

28. Given: $VB = x$; $VA = 7$; $VD = x - 1$; $VC = 9$. Find x. **$x = 4.5$**

29. Given: $VB = x$; $VA = x - 6$; $VD = 9$; $VC = 3$. Find x. **$x = 9$**

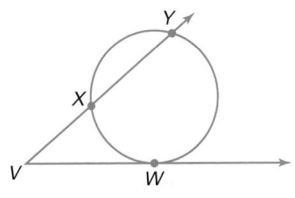

30. Given: $VY = 9$; $VX = 4$. Find VW. **6**

31. Given: $VW = 8$; $VX = 4$. Find VY. **16**

Algebra

32. Given: $AY = x$; $YB = x - 2$; $YC = 4$, $YD = 6$. Find x. **$x = 6$**

33. Given: $AB = 9$; $CD = 4$; $DY = 12$. Find AY. **$AY \approx 15.28$**

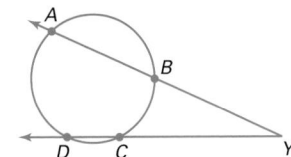

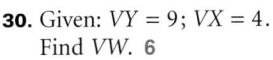
Look Beyond

COORDINATE GEOMETRY Vectors $\vec{m}$ and $\vec{n}$ have their tails at the origin, $(0, 0)$, on a coordinate plane. The head of vector $\vec{m}$ is at point $M(4, 5)$, and the head of vector $\vec{n}$ is at point $N(-3, 8)$.

34. Find the magnitude of the vector sum $\vec{p}$. $\approx$ **13.04**

35. Find the angle θ that $\vec{p}$ makes with the positive x-axis. $\approx$ **85.60°**

36. Find the coordinates of the head of $\vec{p}$. **(1, 13)**

37. What do you notice about the coordinates of the head of $\vec{p}$? Given the coordinates of the heads of two vectors that have their tails at the origin, write a rule for finding the coordinates of the head of the vector sum of the two given vectors.

38. Write a paragraph proof for your rule from Exercise 37. Use diagrams to illustrate. (Hint: Use the head-to-tail method of vector addition.)

39. Given the coordinates of the head of a vector that has its tail at the origin, write a rule for finding the angle θ that it makes with the positive x-axis.

37. Rule: To find the coordinates of the head of the vector sum of two given vectors with tails at the origin, add the x-coordinates and add the y-coordinates.

38.
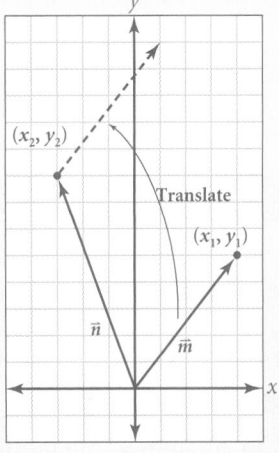

38. Let (x_1, y_1) be the coordinates of the head of vector $\vec{m}$ and let (x_2, y_2) be the coordinates of the head of vector $\vec{n}$. Vector $\vec{m}$ has slope $\frac{y_1 - 0}{x_1 - 0} = \frac{y_1}{x_1}$. Translate vector $\vec{m}$ so that the tail of $\vec{m}$ touches the head of $\vec{n}$. Since $\vec{m}$ has a slope $\frac{y_1}{x_1}$, the coordinates of the head after the translation can be found by "rising" y_1 and "running" x_1 from the point (x_2, y_2). This gives coordinates $(x_2 + x_1, y_2 + y_1)$.

39. Rule: If a vector with tail at the origin has coordinates (x_1, y_1), then the angle it makes with the positive x-axis is $\theta = \tan^{-1}\left(\frac{y_1}{x_1}\right)$. This is true if the angle θ is between $-90°$ and $90°$.

1. Draw an *xy*-coordinate system. Draw △*DEF* in the first quadrant with the vertices at intersection points. **Check students' drawings.**

2. Rotate the triangle you just drew 180° counterclockwise about the origin. **Check students' drawings.**

Also on Quiz Transparency 10.7

Teach

Why Rotations have many innovative applications. Computer screen savers, virtual reality games, and animated billboards are only a few examples. Students who have used geometry graphics software or other computer graphics programs should be familiar with the commands for rotating objects. Discuss the need to choose both a center of rotation and the magnitude of the rotation in degrees.

CRITICAL THINKING
A rotation of the point (0, 0) leaves the point unchanged. According to the transformation equations, $R(0, 0) = (0, 0)$

Rotations in the Coordinate Plane

Why *Computers use geometric transformations to show moving objects. In this lesson, you will explore rotations in a coordinate plane.*

Astronauts prepared for the Hubble Space Telescope repair mission by studying computer simulations of the planned event.

Objectives

● Use transformation equations to rotate points.

● Use a rotation matrix to rotate points or polygons.

Transformation Equations

In earlier lessons, you studied translations, reflections, and dilations in a coordinate plane. You also studied the special case of a rotation by 180° about the origin. Using trigonometry, it is possible to rotate a figure in a coordinate plane by any angle measure.

The transformation $R(x, y) = (x', y')$, where

$x' = x \cos \theta - y \sin \theta$ and
$y' = x \sin \theta + y \cos \theta,$

is a rotation by $\theta°$ about the origin. The above equations are known as **transformation equations** above.

CRITICAL THINKING What happens to the point (0, 0) under a rotation? Explain, using the transformation equations above.

Rotating a Point

YOU WILL NEED
scientific or graphics calculator, graph paper, ruler, and protractor

Part I

1. Draw coordinate axes on graph paper and choose a point *P* on the graph. Label the *x*- and *y*-coordinates of point *P*.

2. Choose a value of θ between 0° and 180°.

Alternative Teaching Strategy

TECHNOLOGY Geometry graphics software can be used to explore rotations of points and figures in a coordinate plane. Although students do not need to use trigonometry to use the computer software, an understanding of how the computer performs the rotations will be useful to students in other types of problem-solving situations.

3. Use your values for x, y, and θ in the transformation equations to find P'. Plot P' on your graph. Draw segments connecting P and P' to the origin and measure the angle of rotation. How does the angle of rotation relate to the value of θ?

Part II

1. Copy and complete the table below.

θ	0°	90°	180°	270°	360°
$\sin\theta$	?	?	?	?	?
$\cos\theta$	?	?	?	?	?

CHECKPOINT ✔

2. Substitute the values of $\sin\theta$ and $\cos\theta$ into the transformation equations for each value of θ, and simplify. Use the resulting equations to complete the following transformations:

   0° rotation: $R(x, y) = (\underline{\ ?\ }, \underline{\ ?\ })$
   90° rotation: $R(x, y) = (\underline{\ ?\ }, \underline{\ ?\ })$
   180° rotation: $R(x, y) = (\underline{\ ?\ }, \underline{\ ?\ })$
   270° rotation: $R(x, y) = (\underline{\ ?\ }, \underline{\ ?\ })$
   360° rotation: $R(x, y) = (\underline{\ ?\ }, \underline{\ ?\ })$

3. Experiment with negative values for θ, and values greater than 360°. Describe your results.

EXAMPLE ❶ According to the treasure map below, some jewels are buried at a point 60 paces east and 35 paces north of a well. Suppose that the direction of north on the map indicates true north and that you are are using a magnetic compass, which points in a direction 5.5° west of true north. How would you find the location of the jewels?

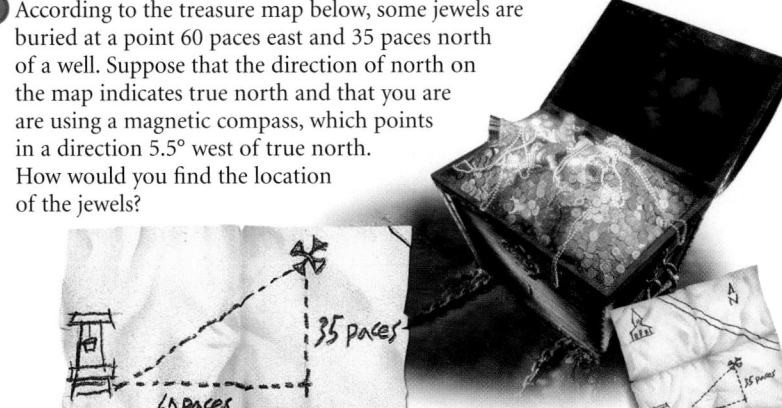

Detail of map

● **SOLUTION**

Assume the map is on a coordinate plane, with the well at the origin. If true north is represented by the y-axis, then point X is at coordinates $(60, 35)$. Because true north is 5.5° east of the value of north on your compass, the point must be rotated about the origin clockwise by 5.5°. (Recall that clockwise rotations are negative.) Use the rotation equations.

$x' = x\cos\theta - y\sin\theta = 60\cos(-5.5°) - 35\sin(-5.5°) \approx 63$ paces
$y' = x\sin\theta + y\cos\theta = 60\sin(-5.5°) + 35\cos(-5.5°) \approx 29$ paces

Using your compass, walk 63 paces east and 29 paces north to find the jewels.

Enrichment

The Activity and examples in this lesson deal with rotating objects about the origin. Have students work in small groups, and challenge them to find a way to rotate a triangle about a point other than the origin. Their solutions should be algebraic in nature; that is, they should not use technology such as geometry graphics software.

Inclusion Strategies

USING MODELS Some students may better understand this lesson if they think in terms of rotating the entire coordinate system, not just the point or polygon. Have students draw an xy-coordinate system and plot any triangle. Then have them trace the axes and the triangle on it. By rotating the tracing paper about the origin, they can see what happens to the axes and the triangle.

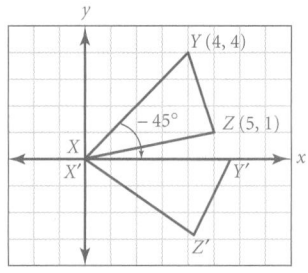
Rotation Matrices

Recall from algebra that a **matrix** is a set of numbers arranged in rows and columns and enclosed in brackets. Matrices are a convenient way to manipulate and display data.

A rotation in a coordinate plane can be represented in terms of the 2×2 matrix $\begin{bmatrix} \cos\theta & -\sin\theta \\ \sin\theta & \cos\theta \end{bmatrix}$, called a **rotation matrix**. When using matrices to transform points in a coordinate plane, each point is written as a column matrix $\begin{bmatrix} x \\ y \end{bmatrix}$. The image is the product of these two matrices.

To multiply two matrices, each row in the first matrix is multiplied by each column in the second matrix. The product of a row and a column is found by multiplying the terms in the row by the terms in the column and then adding.

For example, $\begin{bmatrix} 4 & 1 \end{bmatrix} \times \begin{bmatrix} 2 \\ 3 \end{bmatrix} = 4(2) + 1(3) = 11$.

For the image of a point transformed by a rotation matrix, the first row multiplied by the column is the entry in the first row of the product matrix, and the second row multiplied by the column is the entry in the second row of the product matrix.

$$\begin{bmatrix} \cos\theta & -\sin\theta \\ \sin\theta & \cos\theta \end{bmatrix} \times \begin{bmatrix} x \\ y \end{bmatrix} = \begin{bmatrix} x\cos\theta - y\sin\theta \\ x\sin\theta + y\cos\theta \end{bmatrix}$$

> The expressions in the product matrix agree with the transformation equations at the beginning of the lesson.

EXAMPLE ② Rotate △ABC with vertices $A(0, 0)$, $B(5, 0)$, and $C(3, 4)$ by 60° (counterclockwise) about the origin.

● **SOLUTION**

When rotating more than one point, it is convenient to represent the coordinates in a single matrix with several columns. The matrix representing the vertices of △ABC can be written as $\begin{bmatrix} 0 & 5 & 3 \\ 0 & 0 & 4 \end{bmatrix}$.

The rotation matrix is $\begin{bmatrix} \cos 60° & -\sin 60° \\ \sin 60° & \cos 60° \end{bmatrix} = \begin{bmatrix} 0.5 & -0.866 \\ 0.866 & 0.5 \end{bmatrix}$.

Multiply this matrix by the matrix representing △ABC.

$$\begin{bmatrix} 0.5 & -0.866 \\ 0.866 & 0.5 \end{bmatrix} \times \begin{bmatrix} 0 & 5 & 3 \\ 0 & 0 & 4 \end{bmatrix} = \begin{bmatrix} 0 & 2.5 & -1.964 \\ 0 & 4.33 & 4.598 \end{bmatrix}$$

$$\quad\quad A \;\; B \;\; C \quad\quad A' \;\; B' \;\; C'$$

Plot points A', B', and C', and connect them to form a triangle.

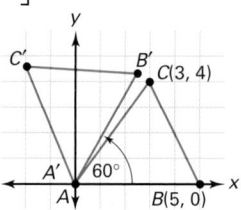

TRY THIS Rotate △DEF with vertices $D(0, 1)$, $E(2, 0)$, and $F(1, 3)$ by 30° (counterclockwise) about the origin.

Exercises

Communicate

1. Give examples of two rotations, one counterclockwise and one clockwise, that produce the same result. Use either transformation equations or rotation matrices to show that the results are the same.

2. Choose two supplementary angles and find their rotation matrices. How are they alike? How are they different?

3. Choose two complementary angles and find their rotation matrices. How are they alike? How are they different?

4. Find the rotation matrices for $\theta = 0°, 90°, 180°, 270°$, and $360°$ by finding the sine and cosine values. How do the matrices compare with the rotations you found in the Activity?

ASSIGNMENT GUIDE
In Class   1–10
Core   11–33 odd, 34, 35
Core Plus   17–33 odd, 34–37
Review   38–49
Preview   50–53

✐ Extra Practice can be found beginning on page 818.

Guided Skills Practice

Use transformation equations to rotate the point (5, 4) by the given angle. *(EXAMPLE 1)*

5. 10°
$x' \approx 4.23; y' \approx 4.81$

6. 90°
$x' = -4; y' = 5$

7. −30°
$x' \approx 6.33; y' \approx 0.96$

Find the rotation matrix for the given angle, and use it to rotate $\triangle ABC$**, with vertices** $A(1, 1)$**,** $B(3, 0)$**, and** $C(2, 2)$**.** *(EXAMPLE 2)*

8. 45° $A' = (0, 1.41)$
$B' = (2.12, 2.12)$
$C' = (0, 2.83)$

9. 72° $A' = (-0.64, 1.26)$
$B' = (0.93, 2.85)$
$C' = (-1.28, 2.52)$

10. −60° $A' = (1.37, -0.37)$
$B' = (1.5, -2.60)$
$C' = (2.73, -0.73)$

Practice and Apply

For Exercises 11–16, a point, P**, and an angle of rotation,** θ**, are given. Determine the coordinates of the image** P'**.**

11. $P(0, 5), \theta = 45°$ (−3.54, 3.54)

12. $P(-1, 1), \theta = 80°$ (−1.16, −0.81)

13. $P(-3, -2), \theta = 30°$ (−1.60, −3.23)

14. $P(2, 0), \theta = 140°$ (−1.53, 1.29)

15. $P(5, 5), \theta = 20°$ (2.99, 6.41)

16. $P(2, 6), \theta = 400°$ (−2.32, 5.88)

For Exercises 17–22, the coordinates of a point P **and an image** P' **are given. Determine of the angle of rotation,** θ**.**

17. $P(4, 2), P'(-2, 4)$ **90°**

18. $P(3, 7), P'(-3, -7)$ **180°**

19. $P(-10, -9), P'(-9, 10)$ **270° or −30°**

20. $P(3, -2), P'(3, -2)$ **0° or 360°**

21. $P(\sqrt{2}, \sqrt{2}), P'(0, 2)$ **45°**

22. $P\left(\frac{1}{2}, \frac{\sqrt{3}}{2}\right), P'\left(\frac{\sqrt{3}}{2}, \frac{1}{2}\right)$ **330° or −30°**

Find the rotation matrix for each angle below by finding the sine and cosine values. Round your answers to the nearest hundredth.

23. 45°

24. 30°

25. 120°

26. 320°

Error Analysis

Students may have trouble learning the algebraic techniques in this lesson. For Exercises 11–22, have them do the first few exercises in each set with graph paper and a protractor. Plotting points and measuring angles may help make the ideas more concrete.

23. $\begin{bmatrix} 0.71 & -0.71 \\ 0.71 & 0.71 \end{bmatrix}$

24. $\begin{bmatrix} 0.87 & -0.5 \\ 0.5 & 0.87 \end{bmatrix}$

25. $\begin{bmatrix} -0.5 & -0.87 \\ 0.87 & -0.5 \end{bmatrix}$

26. $\begin{bmatrix} 0.77 & 0.64 \\ -0.64 & 0.77 \end{bmatrix}$

Find the image of the polygon with vertices represented by the given matrix, rotated by the angle θ. Draw the polygon and its image on the same coordinate axes. You may wish to use a graphics calculator for the matrix multiplication.

27. $\begin{bmatrix} 0 & 3 & 2 \\ 1 & 2 & 5 \end{bmatrix}$, $\theta = 90°$

28. $\begin{bmatrix} -2 & 1 & 3 \\ 1 & 2 & -4 \end{bmatrix}$, $\theta = 40°$

29. $\begin{bmatrix} 2 & 1 & 2 & 4 \\ 5 & 3 & 1 & 2 \end{bmatrix}$, $\theta = 225°$

30. $\begin{bmatrix} 0 & 2 & 4 & 6 \\ 0 & 2 & 1 & 0 \end{bmatrix}$, $\theta = 70°$

31. Find the rotation matrix for a 30° angle and the rotation matrix for a 40° angle. Show that their product is the same as the rotation matrix for a 70° angle. Explain why in terms of the rotation of a point.

32. One vertex of a regular pentagon centered at the origin is (4, 0). Describe a procedure for finding the coordinates of the other four vertices. Carry out your procedure and plot all five vertices on a coordinate plane.

Algebra

33. A point that is rotated by 0° stays in the same place. Find $[R_0]$, the rotation matrix for a 0° angle. This is called the *identity matrix*. Choose any rotation matrix and multiply it by the identity matrix. What do you notice?

34. Let $[R_{60}]$ denote a 60° rotation matrix. Use a graphics calculator to multiply $[R_{60}] \times [R_{60}] \times [R_{60}] \times [R_{60}] \times [R_{60}] \times [R_{60}]$. Show that the product is the same as $[R_0]$ (see Exercise 33). Explain why in terms of the rotation of a point.

35. A point that is rotated by $\theta°$, then rotated by $-\theta°$ will end up at its original position. Find the rotation matrix for a 35° angle and the rotation matrix for a −35° angle. These matrices are called *inverse matrices* of each other. Multiply the two matrices together. What do you notice?

CHALLENGE

36. Use the transformation equations on page 680 to prove that for any point A and any rotation about the origin, the distance OA is equal to OA'. (Hint: Label the coordinates of $A(x, y)$. Then $(OA)^2 = $ ___?___ and $(OA')^2 = $ ___?___ . Simplify the expression for $(OA')^2$ to show that the distances are equal.)

APPLICATION

37. **ARCHITECTURE** A restaurant atop a tall building revolves 360° each hour to give patrons a panoramic view. If a certain table has a location of (30′, −42′) relative to the center, find its coordinates:
   a. after 20 minutes
   b. after t minutes

27.

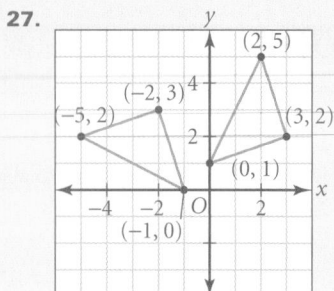

28.

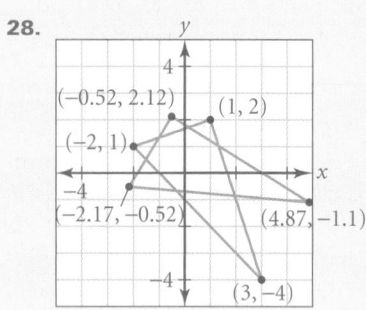

29.

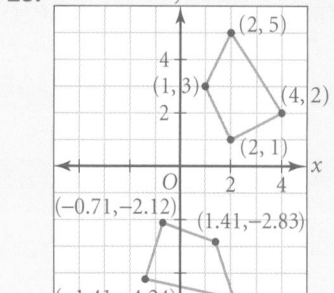

30.

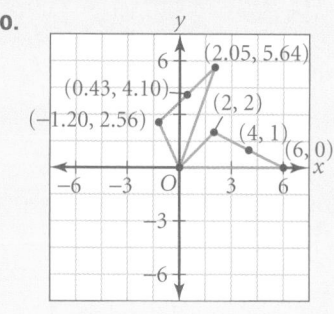

Look Back

CONNECTION

COORDINATE GEOMETRY Graph each line and use the inverse tangent to determine the angle formed by the line and the *x*-axis. *(LESSON 10.1)*

38. $y = 2x$ **39.** $y = 0.5x$ **40.** $y = 3x + 1$

Find the sine, cosine, and tangent of each angle below. Round to the nearest hundredth. *(LESSONS 10.1 AND 10.2)*

41. $37°$ **42.** $90°$ **43.** $250°$ **44.** $-10°$

Find the angle   between 0° and 90° for each trigonometric ratio given. Round to the nearest degree. *(LESSONS 10.1 AND 10.2)*

45. $\tan \theta = 2.75$ **46.** $\cos \theta = 0.36$ **47.** $\cos \theta = 0.81$ **48.** $\sin \theta = 0.77$
       70°   69°   36°   50°

APPLICATION

49. PUBLIC SAFETY A slide in a playground must have a maximum average angle of elevation of 30°. The dimensions of a slide are shown in the diagram at right. Does the slide meet the safety requirements? *(LESSON 10.1)*
The slope is too steep. $\theta \approx 30.96°$

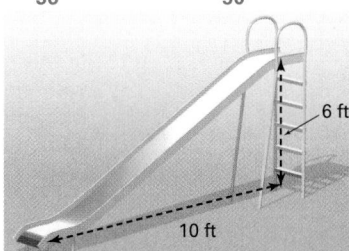

Look Beyond

CONNECTION

TRIGONOMETRY So far you have used degrees to measure angles. Another common unit of angle measure is called a *radian*. A radian is the central angle measure of an arc in a circle with a length equal to the radius of the circle.

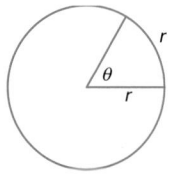

m∠θ = 1 radian

50. How many radians are in 360°? (Hint: Find the circumference of the circle.) 2π

Find the radian measure for each angle below.

51. $180°$ π radians **52.** $90°$ $\dfrac{\pi}{2}$ radians **53.** $60°$ $\dfrac{\pi}{3}$ radians

40.

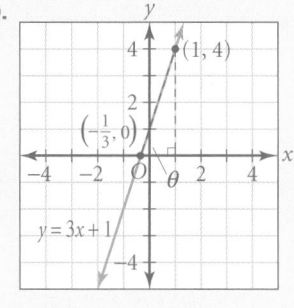

71.57°

41. $\sin 37° \approx 0.60$; $\cos 37° \approx 0.80$; $\tan 37° \approx 0.75$

42. $\sin 90° = 1$; $\cos 90° = 0$; $\tan 90°$ is undefined.

43. $\sin 250° \approx -0.94$; $\cos 250° \approx -0.34$; $\tan 250° \approx 2.75$

44. $\sin(-10°) \approx -0.17$; $\cos(-10°) \approx 0.98$; $\tan(-10°) \approx -0.18$

Look Beyond

Exercises 50–53 introduce students to radian measure and require students to convert some common degree measures to radian measures.

38.

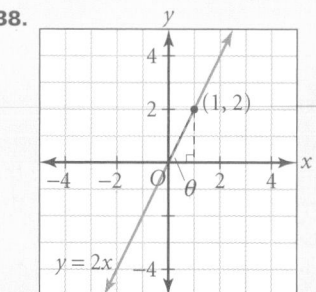

63.43°

39.

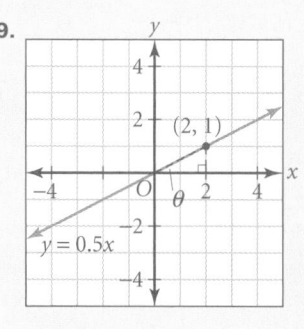

26.57°

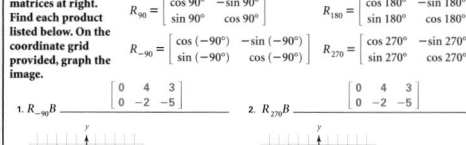
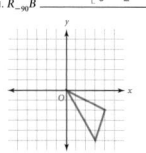 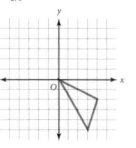

Focus

In the project, students learn to read and interpret the Babylonian number system. Using this knowledge, they examine the arrangement of the numbers in the columns of the Plimpton 322 tablet.

Motivate

Have students find other information about the Babylonians on the Internet and share it with the class.

CHAPTER PROJECT TEN

PLIMPTON 322 *Revisited*

In this project, you will re-examine the Babylonian tablet known as Plimpton 322.

Activity 1

THE BABYLONIAN NUMBER SYSTEM The Babylonians used a base 60 number system. Some remnants of this system that still survive are the 360° circle and our units of time, 60 minutes per hour and 60 seconds per minute. The following are the symbols for some of the Babylonian numerals written in the Babylonian alphabet, called *cuneiform*.

1	6	11	16	30
2	7	12	17	40
3	8	13	18	50
4	9	14	19	60
5	10	15	20	

Numbers greater than 60 were expressed by using these same numerals in different positions. All digits but the last position are multiplied by a power of 60. For example, the Babylonian number ⟨symbols⟩ or 1, 16, 41, is expressed in our number system as $1 \times 60^2 + 16 \times 60 + 41 \times 1 = 4601$.

Activity 1

In Column II the ninth and thirteenth entries are errors. In Column III the second and fifteenth entries are errors.

Activity 2

Column I

1.983402778	1.785192901	1.5625
1.949158552	1.719983676	1.48941684
1.918802127	1.692773438	1.450017361
1.886247907	1.642669444	1.43023882
1.815007716	1.586122566	1.387160494

The values in Column I are one more than the square of the values in the tan θ column on page 638. This is $\sec^2 \theta$.

In order to make the numbers easier to read, commas are used to separate the digits and a semicolon to represent the equivalent of the decimal point. However, the Babylonians did not use gaps or punctuation marks, but relied on context to determine the positions of the digits.

A number with a fractional part, such as 1; 38, 33, 36, 36, is expressed in our number system as

$$1 + \frac{38}{60} + \frac{33}{60^2} + \frac{36}{60^3} + \frac{36}{60^4} = 1.64266944444\ldots$$

The table below contains the transcription of the numbers on Plimpton 322. Some of the damaged numbers have been supplied by researchers. Compare the values in the transcription with columns II and III on page 638. Four entries in columns II and III below are errors. Can you find them?

Column I	Column II	Column III	Column IV
1; 59, 0, 15	1, 59	2, 49	1
1; 56, 56, 58, 14, 50, 6, 15	56, 7	3, 12, 1	2
1; 55, 7, 41, 15, 33, 45	1, 16, 41	1, 50, 49	3
1; 53, 10, 29, 32, 52, 16	3, 31, 49	5, 9, 1	4
1; 48, 54, 1, 40	1, 5	1, 37	5
1; 47, 6, 41, 40	5, 19	8, 1	6
1;43, 11, 56, 28, 26, 40	38, 11	59, 1	7
1; 41, 33, 59, 3, 45	13, 19	20, 49	8
1; 38, 33, 36, 36	9, 1	12, 49	9
1; 35, 10, 2, 28, 27, 24, 26, 40	1, 22, 41	2, 16, 1	10
1; 33, 45	45	1, 15	11
1; 29, 21, 54, 2, 15	27, 59	48, 49	12
1; 27, 0, 3, 45	7, 12, 1	4, 49	13
1; 25, 48, 51, 35, 6, 40	29, 31	53, 49	14
1; 23, 13, 46, 40	56	53	15

 **Activity 2**

THE FIRST COLUMN Use the table above to find the values of the entries in the first column of Plimpton 322. Use as many digits after the decimal point as you can find.

Square the entries in the tan θ column on page 638. Compare them to the values in column I of Plimpton 322. What do you notice? Use the second Pythagorean identity that you found on page 646 to identify the ratio described in column I.

Cooperative Learning

For Activity 1, students can work in small groups to convert the Babylonian numbers in the tablet. Then the groups can compare answers and work together to correct any discrepancies. For Activity 2, students may work individually and then compare and discuss their answers in small groups.

Discuss

Ask students to discuss the reasons why they think that the errors in columns II and III were made. Have students compare these errors with errors that they made and corrected during the project.

1. $\frac{3}{7}$

2. $\frac{7}{3}$

3. $23.20°$

4. $66.80°$

Chapter Review and Assessment

POSTULATES AND THEOREMS

Lesson	Number		Postulate or Theorem
10.2	10.2.3	Identity	$\tan \theta = \frac{\sin \theta}{\cos \theta}$
	10.2.4	Identity	$(\sin \theta)^2 + (\cos \theta)^2 = 1$
10.4	10.4.1	The Law of Sines	For any triangle $\triangle ABC$ with sides a, b, and c: $\frac{\sin A}{a} = \frac{\sin B}{b} = \frac{\sin C}{c}$
10.5	10.5.1	The Law of Cosines	For any triangle $\triangle ABC$ with sides a, b, and c: $a^2 = b^2 + c^2 - 2bc \cos A$ $\quad$ $b^2 = a^2 + c^2 - 2ac \cos B$ $c^2 = a^2 + b^2 - 2ab \cos C$

Key Skills & Exercises

LESSON 10.1
Key Skills

Use right triangles to find tangent ratios.

Find the tangent of 45°.

The triangle at right is a
45-45-90 right triangle,
so the opposite leg and
adjacent leg are equal.
Thus,
$\tan 45° = \frac{\text{opposite}}{\text{adjacent}} = 1$.

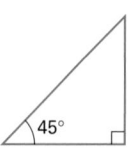

Find an angle that has a given tangent ratio.

Find an angle with a tangent of $\frac{1}{2}$.

Using a calculator, $\tan^{-1} \frac{1}{2} = 26.565 \approx 27°$.

Exercises

Use the given right triangle to find the tangent of each angle.

1. $\tan A = $ _____?_____

2. $\tan B = $ _____?_____

Use a scientific or graphics calculator to find the measure of each angle.

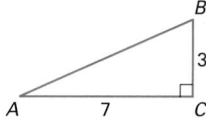

3. $m\angle A = $ _____?_____

4. $m\angle B = $ _____?_____

Chapter Test, Form A

NAME _____ CLASS _____ DATE _____

Chapter Assessment
Chapter 10, Form A, page 1

Write the letter that best answers the question or completes the statement. For Exercises 1–3, refer to $\triangle ACB$.

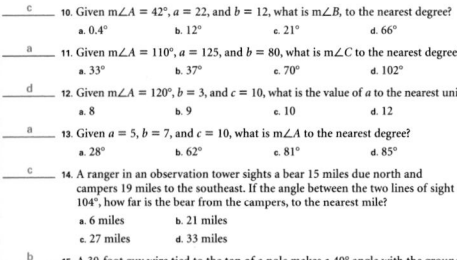

1. Which equation can be used to find $m\angle A$?
   a. $\sin A = \frac{10}{16}$
   b. $\cos A = \frac{10}{16}$
   c. $\tan A = \frac{10}{16}$
   d. $\cos A = \frac{10}{16}$

2. What is $m\angle B$ to the nearest degree?
   a. 32°   b. 39°   c. 51°   d. 90°

3. What is AC to the nearest unit?
   a. 10   b. 12   c. 16   d. 20

4. To the nearest degree, the inverse tangent of 0.667 is _____.
   a. 30°   b. 33°   c. 34°   d. 60°

5. To the nearest thousandth, $\sin 240°$ is _____.
   a. −0.500   b. 0.500   c. 0.866   d. −0.866

6. A kite string is 60 feet long and makes an angle of 42° with the ground. To the nearest foot, how high above the ground is the kite?
   a. 40 feet   b. 44 feet
   c. 54 feet   d. 90 feet

7. A 12-foot ladder leans against a building and makes an angle of 65° with the ground. To the nearest foot, how far from the building is the base of the ladder?
   a. 3 feet   b. 4 feet
   c. 5 feet   d. 6 feet

8. A 6-foot person walks 75 feet from a tree. The angle formed by the person's line of sight and the horizontal is 25°. About how tall is the tree?
   a. 34 feet   b. 35 feet
   c. 41 feet   d. 50 feet

For Exercises 9–13, find the indicated measure.

9. Given $m\angle A = 110°$, $m\angle B = 45°$, and $a = 10$, what is the value of b to the nearest unit?
   a. 7   b. 8   c. 12   d. 14

NAME _____ CLASS _____ DATE _____

Chapter Assessment
Chapter 10, Form A, page 2

10. Given $m\angle A = 42°$, $a = 22$, and $b = 12$, what is $m\angle B$, to the nearest degree?
    a. 0.4°   b. 12°   c. 21°   d. 66°

11. Given $m\angle A = 110°$, $a = 125$, and $b = 80$, what is $m\angle C$ to the nearest degree?
    a. 33°   b. 37°   c. 70°   d. 102°

12. Given $m\angle A = 120°$, $b = 3$, and $c = 10$, what is the value of a to the nearest unit?
    a. 8   b. 9   c. 10   d. 12

13. Given $a = 5$, $b = 7$, and $c = 10$, what is $m\angle A$ to the nearest degree?
    a. 28°   b. 62°   c. 81°   d. 85°

14. A ranger in an observation tower sights a bear 15 miles due north and campers 19 miles to the southeast. If the angle between the two lines of sight is 104°, how far is the bear from the campers, to the nearest mile?
    a. 6 miles   b. 21 miles
    c. 27 miles   d. 33 miles

15. A 30-foot guy wire tied to the top of a pole makes a 40° angle with the ground. If the pole is tilted away from the guy wire and makes a 75° angle with the ground, what is the length of the pole, to the nearest foot?
    a. 19 feet   b. 20 feet
    c. 45 feet   d. 89 feet

16. What is the magnitude of the resultant vector?
    a. 11.7 miles   b. 16.1 miles
    c. 24.5 miles   d. 93.2 miles

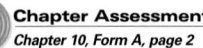

17. To the nearest degree, the angle that the resultant vector makes with the horizontal is _____.
    a. 2°   b. 28°   c. 72°   d. 108°

18. If point $P(2, 0)$ is rotated by 45° (counterclockwise) about the origin, its image is _____.
    a. (1.414, 1.414)   b. (−1.414, −1.414)
    c. (1.414, −1.414)   d. (−1.414, 1.414)

19. The coordinates of $P(−2, 1)$ are rotated to $P'(−1, −2)$. What is the angle of rotation?
    a. 360°   b. 270°   c. 180°   d. 90°

LESSON 10.2

Key Skills

Use right triangles to find sine and cosine ratios.

Find the sine and cosine of 30°.

The triangle at right is a 30-60-90 right triangle with a hypotenuse of length 2, so the opposite leg has length 1 and adjacent leg has length $\sqrt{3}$.

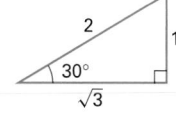

$\sin 30° = \dfrac{\text{opposite}}{\text{hypotenuse}} = \dfrac{1}{2}$

$\cos 30° = \dfrac{\text{adjacent}}{\text{hypotenuse}} = \dfrac{\sqrt{3}}{2}$

Find an angle that has a given sine or cosine ratio.

Give an angle with a cosine of $\dfrac{1}{2}$.

Using a calculator, $\cos^{-1}\dfrac{1}{2} = 60°$.

Exercises

Use the given right triangle to find the sine and cosine of each angle.

5. $\sin A =$ ___?___ $\cos A =$ ___?___

6. $\sin B =$ ___?___ $\cos B =$ ___?___

Use a scientific or graphics calculator to find the measure of each angle.

7. $m\angle A =$ ___?___

8. $m\angle B =$ ___?___

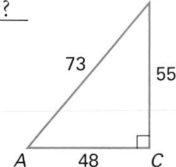

LESSON 10.3

Key Skills

Find the coordinates of a point on the unit circle corresponding to a given angle.

Find the coordinates (x, y) of point P on the unit circle at right.

Point P has coordinates $(\cos 40°, \sin 40°) \approx (0.77, 0.64)$.

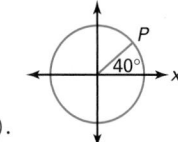

Find the measures of angles with a given sine or cosine.

Find all angles between 0° and 360° with a sine of 0.7.

Using a calculator, $\sin^{-1} 0.7 = 44.43° \approx 44°$. From the graph of $\sin \theta$, you can see that there is another angle with the same sine value, namely $180° - 44° = 136°$.

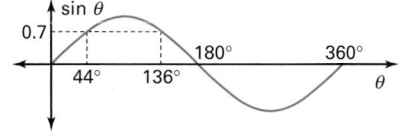

Exercises

For Exercises 9–12, use a sketch to illustrate your answer.

9. Find the coordinates of a point on the unit circle corresponding to a 70° angle of rotation.

10. Find the coordinates of a point on the unit circle corresponding to a 130° angle of rotation.

11. Find all angles between 0° and 360° with a sine of 0.2.

12. Find all angles between 0° and 360° with a cosine of 0.8.

---

5. $\sin A = \dfrac{55}{73}; \cos A = \dfrac{48}{73}$

6. $\sin B = \dfrac{48}{73}; \cos B = \dfrac{55}{73}$

7. 48.89°

8. 41.11°

9. (0.34, 0.94)

10. (−0.64, 0.77)

11. 11.54°; 168.46°

12. 36.87°; 323.13°

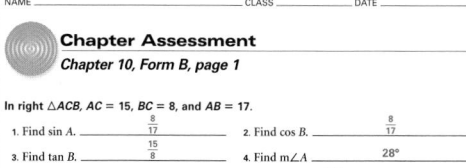
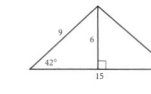
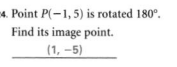

13. $m\angle R = 21.26°$;
    $m\angle Q = 43.74°$

14. 30.52

15. 107°

16. $t \approx 13.99$; $s \approx 7.95$

17. 18.06

18. $m\angle E \approx 94.62°$
    $m\angle F \approx 45.38°$

19. $m\angle I \approx 107.24°$

20. $m\angle H \approx 42.39°$;
    $m\angle G \approx 30.37°$

Key Skills

Solve triangles by using the law of sines.

Use the law of sines to solve the triangle below.

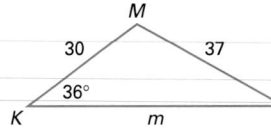

$\frac{\sin 36°}{37} = \frac{\sin L}{30}$, so $\sin L = 30\left(\frac{\sin 36°}{37}\right) \approx 0.477$.
Thus, $m\angle L = \sin^{-1} 0.477 \approx 28°$, so by the
Triangle Sum Theorem,
$m\angle M = 180° - (36° + 28°) = 116°$.

(Note: Because $\angle M$ is obtuse, $\angle L$ must be acute,
so the triangle is not ambiguous.)

Then $\frac{\sin 36°}{37} = \frac{\sin 116°}{m}$, so
$m = \frac{37 \sin 116°}{\sin 36°} \approx 56.6$.

Exercises

Refer to the diagram below.

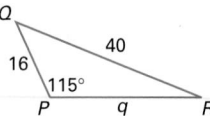

13. Find the unknown angles in $\triangle PQR$.

14. Find the unknown side of $\triangle PQR$.

Refer to the diagram below.

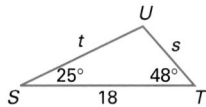

15. Find the unknown angle in $\triangle STU$.

16. Find the unknown sides in $\triangle STU$.

Key Skills

Solve triangles by using the law of cosines.

Use the law of cosines to solve the triangle below.

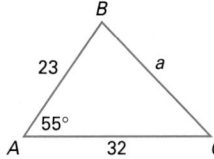

$a^2 = b^2 + c^2 - 2bc \cos A \approx 709$, so $a \approx 27$.
Use the law of sines to find $m\angle B$.
$\frac{\sin 55°}{27} = \frac{\sin B}{32}$, so $m\angle B \approx 76°$
By the Triangle Sum Theorem,
$m\angle C = 180° - (55° + 76°) = 49°$.

Exercises

Refer to the diagram below.

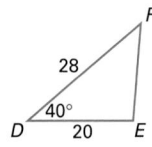

17. Find the unknown side in $\triangle DEF$.

18. Find the unknown angles in $\triangle DEF$.

Refer to the diagram below.

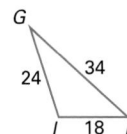

19. Find the largest unknown angle in $\triangle GHI$.

20. Find the remaining unknown angles in $\triangle GHI$.

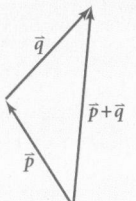

LESSON 10.6

Key Skills

Find the sum of two vectors by using the head-to-tail method and the parallelogram method.

Find the sum of vectors $\vec{a}$ and $\vec{b}$ below.

Head-to-tail method:

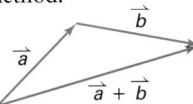

Parallelogram method:

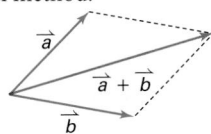

Exercises

Copy vectors $\vec{p}$, $\vec{q}$, and $\vec{r}$ below.

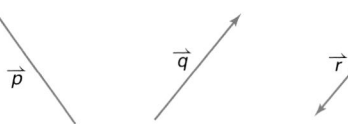

21. Find the sum of $\vec{p}$ and $\vec{q}$ by using the head-to-tail method.

22. Find the sum of $\vec{p}$ and $\vec{r}$ by using the head-to-tail method.

23. Find the sum of $\vec{p}$ and $\vec{q}$ by using the parallelogram method.

24. Find the sum of $\vec{p}$ and $\vec{r}$ by using the parallelogram method.

22.

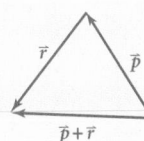

23.

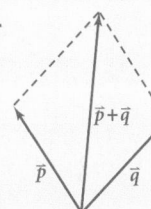

24.

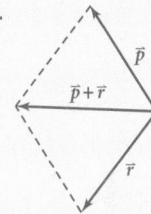

LESSON 10.7

Key Skills

Rotate a point in a coordinate plane by a given angle.

What is the image of the point (3, 2) under a 90° rotation?

The transformation equations are
$x' = x \cos 90° - y \sin 90° = x(0) - y(1) = -y$
$y' = x \sin 90° + y \cos 90° = x(1) + y(0) = x$
so the image is the point $(-2, 3)$.

Rotate a polygon in a coordinate plane by a given angle.

What is the image of $\triangle ABC$ with vertices $A(0, 0)$, $B(1, 2)$, and $C(1, 0)$ under a 60° rotation?

Multiply the rotation matrix by the triangle matrix.

$$\begin{bmatrix} 0.5 & -0.866 \\ 0.866 & 0.5 \end{bmatrix} \begin{bmatrix} 0 & 1 & 1 \\ 0 & 2 & 0 \end{bmatrix} = \begin{bmatrix} 0 & -1.232 & 0.5 \\ 0 & 1.866 & 0.866 \end{bmatrix}$$

The image $\triangle A'B'C'$ has vertices $A'(0, 0)$, $B'(-1.232, 1.866)$, and $C'(0.5, 0.866)$.

Exercises

25. What is the image of the point (1, 5) under a 270° rotation?

26. What is the image of the point (4, −3) under a 45° rotation?

27. Find the image of $\triangle DEF$ with vertices $D(0, 0)$, $E(3, 1)$, and $F(3, 2)$ under a 30° rotation.

28. Find the image of $\triangle GHI$ with vertices $G(1, 1)$, $H(-1, 1)$, and $I(1, -1)$ under a 135° rotation.

25. $(5, -1)$

26. $(4.95, 0.71)$

27. $D'(0, 0)$, $E'(2.10, 2.37)$ and $F'(1.60, 3.23)$

28. The image $\triangle G'H'I'$ has vertices $G'(-1.41, 0)$, $H'(0, -1.41)$, and $I'(0, 1.41)$.

29. 39.20 m

30. 113.89 mi

31. A force of 1041.89 lb is required to push the truck uphill.

32. 2.52 $\times$ 10^{13} ft

Applications

29. INDIRECT MEASUREMENT Estimate the height of the building in the diagram at right.

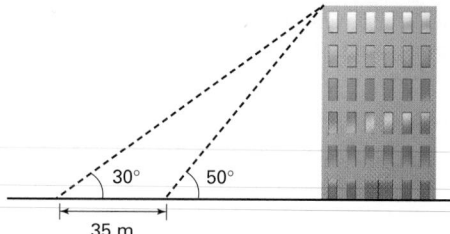

30° 50°

35 m

30. NAVIGATION Two planes set off from an airport, one with a heading of 045 at 100 mph, the other with a heading of 150 at 115 mph. How far apart will the planes be after 40 min?

31. ENGINEERING A truck weighing 6000 lb is parked on a ramp that forms a 10° angle with the ground. To find the force required to move the truck uphill, the vector representing the truck's weight (pointing straight down) is broken down into two perpendicular vectors, one parallel to the ramp, labeled $\vec{a}$, and one perpendicular to the ramp, labeled $\vec{b}$. The sum $\vec{a} + \vec{b}$ equals the vector representing the weight. Find the magnitude of $\vec{a}$, the force required to push the truck uphill.

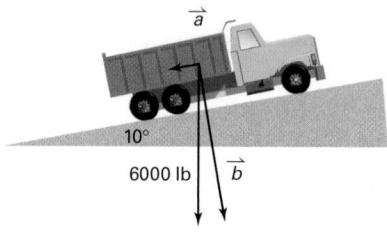

$\vec{a}$

10°

6000 lb $\vec{b}$

32. ASTRONOMY The diagram at right (not to scale) shows Alpha Centauri, the closest star to the Sun, and two positions of Earth, A and B, at six months apart in its orbit around the Sun. Use the law of cosines and the information in the diagram to estimate d, the distance from Earth at position A to Alpha Centauri.

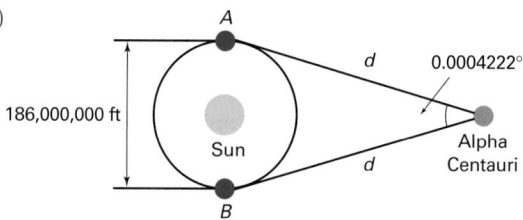

A

d   0.0004222°

186,000,000 ft

Sun

Alpha Centauri

d

B

10 Chapter Test

Use the diagram below for Exercises 1–4.

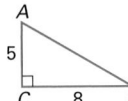

Find the tangent of each angle.

1. $\tan A = \underline{\ ?\ }$ $\dfrac{8}{5}$

2. $\tan B = \underline{\ ?\ }$ $\dfrac{5}{8}$

Use a scientific or graphics calculator to find the measure of each angle.

3. $m\angle A = \underline{\ ?\ } \approx 57.99°$

4. $m\angle B = \underline{\ ?\ } \approx 32.01°$

Use the diagram at right for Exercises 5–8. Find each of the following.

5. $\sin A = \underline{\ ?\ }$ $\dfrac{15}{17}$ $\cos A = \underline{\ ?\ }$ $\dfrac{8}{17}$

6. $\sin B = \underline{\ ?\ }$ $\dfrac{8}{17}$ $\cos B = \underline{\ ?\ }$ $\dfrac{15}{17}$

Use a scientific or graphics calculator to find the measure of each angle.

7. $m\angle A = \underline{\ ?\ } \approx 61.93°$

8. $m\angle B = \underline{\ ?\ } \approx 28.07°$

9. Find the coordinates of a point on the unit circle corresponding to a 36° angle of rotation. **(0.81, 0.59)**

10. Find the coordinates of a point on the unit circle corresponding to a 115° angle of rotation. **(−0.42, 0.91)**

11. Find all angles between 0° and 360° with a sine of 0.6. **36.87° and 143.13°**

12. Find all angles between 0° and 360° with a cosine of 0.3. **72.54° and 287.46°**

For Exercises 13–14, refer to the diagram at right.

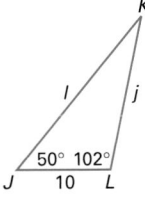

13. Find the unknown angle in △JKL. $m\angle K = 28°$

14. Find the unknown sides in △JKL. $j \approx 16.32$ units
$l \approx 20.84$ units

For Exercises 15–16, refer to the diagram below.

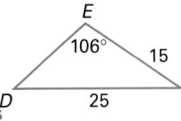

15. Find the unknown angles in △DEF. $m\angle D \approx 35.22°$
$m\angle F \approx 38.78°$

16. Find the unknown side of △DEF. $f \approx 16.29$ units

For Exercises 17–18, refer to the diagram below.

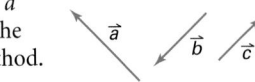

17. Find the unknown side in △QRS. $q \approx 22.18$ units

18. Find the unknown angles in △QRS. $m\angle R \approx 30.54°$
$m\angle S \approx 39.46°$

19. NAVIGATION Nathan and Julia are walking along lines that form a 38° angle. If Nathan walks at 3.6 miles per hour and Julia walks at 4.1 miles per hour, how far apart will they be after 2 hours? **5.10 miles**

For Exercises 20–21, copy vectors $\vec{a}$, $\vec{b}$ and $\vec{c}$.

20. Find the sum of $\vec{a}$ and $\vec{b}$ by using the head-to-tail method.

21. Find the sum of $\vec{a}$ and $\vec{c}$ by using the parallelogram method.

22. SPORTS A swimmer heads perpendicular to a 2-miles-per-hour current. His speed in still water is 1.8 miles per hour. Find the actual speed of the swimmer and his direction angle, θ, with respect to the direction of the current.

23. What is the image of the point $(3, -1)$ under a 90° rotation? **(1, 3)**

24. Find the rotation matrix for a 150° angle.

20.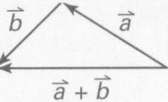

21.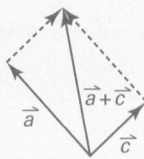

22. $\sqrt{7.24} \approx 2.7$ mph; 41.99°

24. $\begin{bmatrix} -\dfrac{\sqrt{3}}{2} & -\dfrac{1}{2} \\ \dfrac{1}{2} & -\dfrac{\sqrt{3}}{2} \end{bmatrix}$

CHAPTERS CUMULATIVE ASSESSMENT

1–10 **College Entrance Exam Practice**

Multiple-Choice Samples

The first half of the Cumulative Assessment contains a multiple-choice section. This part of the Cumulative Assessment consists of items commonly found on standardized tests.

Free-Response Grid Samples

The second half of the Cumulative Assessment consists of a free-response section. This part requires student-produced response items like those commonly found on college entrance exams. These questions require the use of machine-scored answer grids. You may wish to have students practice answering these items in preparation for standardized tests.

1. c

2. a

3. c

4. d

5. b

MULTIPLE-CHOICE For Questions 1–10, write the letter that indicates the best answer.

1. Refer to the figure below. Suppose that a point is chosen randomly in the large circle. What is the probability that the point is inside the shaded area? Round your answer to the nearest hundredth. **(LESSON 5.8)**

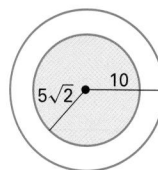

5√2 10

   a. 0.71
   b. 1.41
   c. 0.50
   d. 4.93

2. Find the surface area of the prism below. **(LESSON 7.2)**

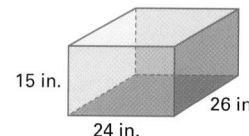

15 in. 26 in.
24 in.

   a. 2748 in²
   b. 1374 in²
   c. 9360 in²
   d. 130 in²

3. Refer to the figure below. Find the tan E to the nearest hundredth. **(LESSON 10.1)**

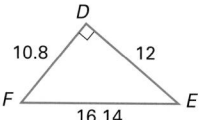

D
10.8   12
F   E
16.14

   a. 1.11
   b. 0.74
   c. 0.90
   d. 0.67

4. The value of sin θ is given by which ratio? **(LESSON 10.2)**

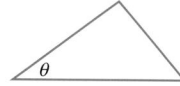

θ

   a. $\dfrac{\text{adjacent}}{\text{hypotenuse}}$
   b. $\dfrac{\text{adjacent}}{\text{opposite}}$
   c. $\dfrac{\text{opposite}}{\text{adjacent}}$
   d. $\dfrac{\text{opposite}}{\text{hypotenuse}}$

5. The transformation $T(x, y) = (-x, y)$ is a ___?___ . **(LESSONS 1.7, 8.1, AND 10.7)**
   a. translation
   b. reflection
   c. rotation
   d. dilation

6. The transformation $T(x, y) = (-x, -y)$ is a
_____?_____. *(LESSONS 1.7, 8.1, AND 10.7)*
 a. translation
 b. reflection
 c. rotation
 d. dilation

7. Which theorem can be used to find m∠A in
△ABC? *(LESSONS 10.5 AND 10.6)*

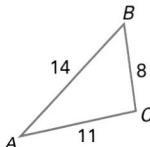

 a. law of sines
 b. law of cosines
 c. Pythagorean Theorem
 d. none of the above

8. Suppose that ∠A ≅ ∠B and ∠B ≅ ∠C. Then
∠A ≅ ∠C by _____?_____. *(LESSON 2.4)*
 a. Reflexive Property
 b. Symmetric Property
 c. Transitive Property
 d. Addition Property

9. m∠4 = _____?_____ *(LESSON 3.5)*

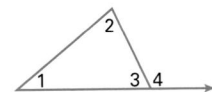

 a. m∠1 + m∠2
 b. m∠1 + m∠3
 c. m∠2 + m∠3
 d. m∠1 + m∠2 + m∠3

10. The lines $y = 2x + 3$ and $y = -\frac{1}{2}x - 1$ are
_____?_____. *(LESSON 3.8)*
 a. parallel
 b. skew
 c. vertical
 d. perpendicular

11. A point with coordinates (3, 5) is rotated by
45° about the origin. Give the coordinates of
the image point. *(LESSON 10.7)*

12. The magnitude of $\vec{a}$ is 6 and the magnitude
of $\vec{b}$ is 10. If $\vec{a}$ and $\vec{b}$ are parallel, what is the
magnitude of $\vec{a} + \vec{b}$? *(LESSON 10.6)*

13. Give the number of vertices, edges, and faces of
an oblique pentagonal prism. *(LESSON 6.3)*

14. The rectangles at right
are similar. Find x.
(LESSON 8.2)

24   36   36   x

15. Find k in △KLM at right.
(LESSON 10.4)

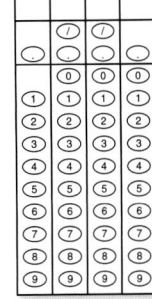

FREE-RESPONSE GRID
**Items 16–18 may be answered
by using a free-response grid
such as that commonly used
by standardized-test services.**

16. Find the radius of a circle that has the same
area as a square with a side length of 5.
(LESSON 5.3)

17. Find the radius of a sphere that has the same
volume as a cube with an edge length of 5.
(LESSON 7.6)

18. Find an angle 0° < θ < 180° such that
sin θ = −cos θ. *(LESSON 10.3)*

6. c

7. b

8. c

9. a

10. d

11. (−1.41, 5.66)

12. 16

13. Vertices = 10, faces = 7,
edges = 15

14. 54

15. $k \approx 7.07$

16. $r \approx 2.82$

17. $r \approx 3.10$

18. $\theta = 135°$

Taxicabs, Fractals, and More

Lesson Presentation CD-ROM
Power Point® presentations for each lesson 11.1-11.7

CHAPTER PLANNING GUIDE

Lesson	11.1	11.2	11.3	11.4	11.5	11.6	11.7	Project and Review
Pupil's Edition Pages	698–705	706–711	712–720	721–728	729–737	738–746	747–755	756–765
Practice and Assessment								
Extra Practice (Pupil's Edition)	852	852	853	853	854	854	855	
Practice Workbook	69	70	71	72	73	74	75	
Practice Masters Levels A, B, and C	205–207	208–210	211–213	214–216	217–219	220–222	223–225	
Standardized Test Practice Masters	79	80	81	82	83	84	85	86
Assessment Resources	139	140	141	142	144	145	146	143, 147–152
Visual Resources								
Lesson Presentation Transparencies Vol. 2	77–80	81–84	85–88	89–92	93–96	97–100	101–104	
Teaching Transparencies	97–99	100	101–102	103	104	105–108	109	
Answer Key Transparencies	383–385	386–391	392–396	397–401	402–410	411–415	416–419	420–425
Quiz Transparencies	11.1	11.2	11.3	11.4	11.5	11.6	11.7	
Teacher's Tools								
Reteaching Masters	137–138	139–140	141–142	143–144	145–146	147–148	149–150	
Make-Up Lesson Planner for Absent Students	69	70	71	72	73	74	75	
Student Study Guide	69	70	71	72	73	74	75	
Spanish Resources	69	70	71	72	73	74	75	
Block Scheduling Handbook								22–23
Activities and Extensions								
Lesson Activities	196–201	202–205	206–207			208–211	212–216	
Enrichment Masters	69	70	71	72	73	74	75	
Cooperative-Learning Activities	69	70	71	72	73	74	75	
Problem-Solving/ Critical Thinking	69	70	71	72	73	74	75	
Student Technology Guide			69	70–71		72–73	74	
Long Term Projects								41–44
Writing Activities for Your Portfolio								31–33
Tech Prep Masters								51–54
Building Success in Mathematics								28–30

LESSON PACING GUIDE

Lesson	11.1	11.2	11.3	11.4	11.5	11.6	11.7	Project and Review
Traditional	1 day	1 day	1 day	1 day	2 days	1 day	1 day	2 days
Block	$\frac{1}{2}$ day	$\frac{1}{2}$ day	$\frac{1}{2}$ day	$\frac{1}{2}$ day	1 day	$\frac{1}{2}$ day	$\frac{1}{2}$ day	1 day
Two-Year	2 days	2 days	2 days	2 days	4 days	2 days	2 days	4 days

CONNECTIONS AND APPLICATIONS

Lesson	11.1	11.2	11.3	11.4	11.5	11.6	11.7	Review
Algebra	700, 704	710		727				
Geometry	698–705	706–711	712–720	721–728	729–737	738–746	747–755	756–765
Coordinate Geometry	705							
Trigonometry			719					
Knot Theory				728				
Business and Economics	705		719	727				
Life Skills	705	710	718	728				
Science and Technology			719	721		741		762
Social Studies						739		762
Sports and Leisure	705					744		
Cultural Connection: Africa			718					
Cultural Connection: Asia						745		

BLOCK-SCHEDULING GUIDE

Day	Lesson	Teacher Directed: Lesson Examples, Teaching Transparencies	Student Guided: Activity, Try This	Cooperative-Learning Activity, Lesson Activity, Student Technology Guide	Practice: Practice & Apply, Extra Practice, Practice Workbook	Assessment: Quiz, Mid-Chapter Assessment	Problem Solving, Reteaching
1	11.1	8 min	8 min.	10 min	55 min	8 min	8 min
	11.2	7 min	7 min	10 min	55 min	7 min	7 min
2	11.3	8 min	8 min	10 min	25 min	8 min	8 min
	11.4	7 min	7 min	10 min	20 min	7 min	7 min
3	11.5	15 min	15 min	20 min	55 min	15 min	15 min
4	11.6	8 min	8 min	10 min	25 min	8 min	8 min
	11.7	7 min	7 min	10 min	20 min	7 min	7 min
5	Assess.	50 min	90 min	90 min	65 min	30 min	
		PE: Chapter Review	PE: Chapter Project, Writing Activities	Tech Prep Masters	PE: Chapter Assessment, Test Generator	Chap. Assess. (A or B), Alt. Assess. (A or B), Test Generator	

PE: Pupil's Edition

Alternative Assessment

The following suggest alternative assessments for students who may benefit from a different type of assessment than the regular chapter quizzes and the mid-chapter/end-of-chapter test. Visit the HRW web site to get additional Alternative Assessment material.

internet connect

Alternative Assessment
Go To: **go.hrw.com**
Keyword: **MG1 Alt Assess**

Performance Assessment

1. Using a city map with square grid intersections, find the taxidistance between two points of interest in miles or kilometers. Then give the actual distance between the points.

2. Looking at a floor plan of your home or school, draw a network that represents the circulation of the building. Determine whether the network represents an Euler circuit.

3. Using clay or some other modeling material, perform a demonstration of how two objects can be topologically equivalent, as in the case of a doughnut and a coffee cup.

Portfolio Project

Suggest that students choose one of the following projects for inclusion in their portfolios.

1. Looking in books or magazines, collect pictures that contain golden rectangles and/or golden spirals.

2. Using library materials, research DNA molecules. What could be the purpose of its double-helix design?

3. On a posterboard, draw several different types of fractals, such as a Koch Snowflake or a Sierpinski Gasket.

internet connect

The table below identifies the pages in this chapter that contain internet and technology information.

Content Links

Activities Online	pages 702, 724, 733
Portfolio Extensions	pages 720, 746
Homework Help Online	pages 704, 709, 718, 735

Resource Links

Parents can go online and find concepts that students are learning–lesson by lesson–and questions that pertain to each lesson, which facilitate parent-student discussion.

Go To: **go.hrw.com**
Keyword: **MG1 Parent Guide**

Technical Support

The following may be used to obtain technical support for any HRW software product.

Online Help: **www.hrwtechsupport.com**
e-mail: **tschrw@hbtechsupport.com**
HRW Technical Support Center: **(800)323-9239**
7 AM to 10 PM Monday through Friday CST

Visit the HRW math web site at: **www.hrw.com/math**

Technology

Technology Objectives and Suggestions

Lesson 11.1 Golden Connections
In this lesson students discover the relationship known as the golden ratio, and solve problems using the golden ratio. Have students create a spreadsheet that gives the length of a golden rectangle for a given width. Geometry software can be used for the constructions on pages 703, 705, and 707.

Lesson 11.2 Taxicab Geometry
In this lesson students develop a non-Euclidean geometry based on taxi movements on a street grid known as taxicab geometry. They also solve problems within a taxicab geometry system. Geometry software is appropriate for most of the exercises. The coordinate geometry mode is necessary for completing taxicab problems. Be sure that students use the taxicab geometry properties for finding distances.

Lesson 11.3 Graph Theory
In this lesson students determine whether a given graph has an Euler path. They also use Euler paths to solve problems involving graphs. For extra practice use software that creates networks in order to demonstrate some of the concepts in this lesson. Encourage students to experiment with different paths as they try to traverse a network on the computer.

Students may use computer-aided design programs to draw floor plans of houses for Exercises 16 and 17 or for other floor plans.

Lesson 11.4 Topology: Twisted Geometry
In this lesson students explore and develop concepts of topology including knots, Mobius strips, and toruses. They also use theorems of topology to solve problems. In this exercise set, use geometry software to draw figures that are topologically equivalent. Various graphics options can be used to distort shapes.

Lesson 11.5 Euclid Unparalleled
In this lesson students prove quadrilateral conjectures by using triangle congruence postulates and theorems. Have students use geometry software to make the parallelograms for this lesson. Students should be sure to construct figures that retain their properties, such as parallel segments, when vertices or segments are dragged.

Lesson 11.6 Fractal Geometry
In this lesson students discover the basic properties of fractals, including self-similarity and iterative processes. They also build fractal designs using iterative steps. Have students use fractal programs to create images. Ask them to use the computer to explore the Mandelbrot Set and to create the Sierpinski Triangle.

You may want students to use fractal software to draw the Koch Snowflake in Exercises 19–22. Have them do many iterations. Students may also use the script features of geometry software programs to create fractal images.

Lesson 11.7 Other Transformations: Projective Geometry
In this lesson students develop the concepts of affine transformations and geometric projection. They also solve problems and make conjectures using the Theorem of Pappus and the Theorem of Desargues. Use geometry software to transform plane figures. Ask students to draw a figure on the coordinate grid and use the affine transformation tool to transform it. While in the dialog box, have them define the transformation of (x, y) to (x', y').

You may want students to use geometry software to do Exercises 17 and 18. Have students label points as they draw them to avoid errors.

In this chapter, students explore different topics in mathematics, including taxicab geometry, networks, topology, two non-Euclidean geometries, projective geometry, and fractals.

CHAPTER RESOURCES

- Block-Scheduling Handbook
- Writing Activities for Your Portfolio
- Tech Prep Masters
- Long-Term Project
- Assessment Resources:
  Mid-Chapter Assessment
  Chapter Assessments
  Alternative Assessments
- Test and Practice Generator
- Technology Handbook

Chapter Objectives

- Discover the relationship known as the golden ratio. **[11.1]**

- Solve problems by using the golden ratio. **[11.1]**

- Develop a non-Euclidean geometry, known as *taxicab geometry*, based on a taxicab's movements on a street grid. **[11.2]**

- Solve problems within a taxicab geometry system. **[11.2]**

- Determine whether a given graph has an Euler path. **[11.3]**

- Use Euler paths to solve problems involving graphs. **[11.3]**

Taxicabs, Fractals, and More

CAN YOU IMAGINE A SYSTEM OF MATHEMATICS IN which a coffee cup is the same as a doughnut? Or one in which there are no such things as parallel lines? You will explore many strange ideas like these in this chapter.

By questioning commonly held assumptions or taking imaginative leaps, mathematicians create entirely new areas of mathematics that often turn out to be rewarding fields of study.

Before branching out to more recent discoveries in mathematics, you will first study an idea that goes back to classical times—the *golden ratio*. This ratio appears again and again in mathematics and nature. As you will see, the sunflower on the facing page has an interesting connection to the golden ratio.

Lessons

About the Photos

The images on these pages relate to some of the themes of the chapter. The computer-generated spiral pattern is an example of a *fractal*, that is, an image that contains self-similar images of itself within its structure. Magnification of portions of the image reveal endless levels of smaller and smaller images that resemble the larger whole. The doughnut and coffee cup suggest the subject of topology. The two objects are said to be topologically equivalent because one can be transformed into the other by stretching, bending, twisting, etc., but without cutting, tearing, or compressing any part of the object to a single point. The number of spirals in the sunflower (there are two sets of them) are Fibonacci numbers, which bear a special relationship to the golden ratio. Finally, the taxicab suggests a special geometry known as *taxicab geometry*, which is based on travel along segments in a rectangular grid like the streets in a city.

- Explore and develop concepts of topology, including knots, Möbius strips, and tori. [**11.4**]
- Use theorems of topology to solve problems. [**11.4**]
- Explore and develop general notions for spherical and hyperbolic geometries. [**11.5**]
- Develop informal proofs and solve problems by using concepts of non-Euclidean geometries. [**11.5**]
- Discover the basic properties of fractals, including self-similarity and iteration. [**11.6**]
- Build fractal designs by using iterative steps. [**11.6**]
- Understand the concepts of affine transformations and geometric projection. [**11.7**]
- Solve problems and make conjectures by using the Theorem of Pappus and the Theorem of Desargues. [**11.7**]

About the Chapter Project

You may be familiar with the puzzle known as the tower of Hanoi, invented by Edouard Lucas in 1883. In the puzzle, a stack of disks in order from smallest to largest rests on one of three pegs. The object of the puzzle is to move the disks, one at a time, to another peg without ever placing any disk onto a smaller disk.

In the Chapter Project, you will study both the number of moves required for a given number of disks and the puzzle's relationship to the fractal known as the *Sierpinski gasket*.

After completing the Chapter Project, you will be able to do the following:

- Determine the fewest number of moves required to solve the tower of Hanoi puzzle with a given number of disks.
- Use a graph to represent the possible states of the tower of Hanoi puzzle and to find a solution.

About the Portfolio Activities

Throughout the chapter, you will be given opportunities to complete Portfolio Activities that are designed to support your work on the Chapter Project.

The theme of each Portfolio Activity and of the Chapter Project is the tower of Hanoi.

- In the Portfolio Activity on page 720, you will study how puzzles can be represented graphically and how solutions can be found by using graphs.
- In the Portfolio Activity on page 746, you will generate a fractal shape by using random processes, such as rolling a number cube or generating random points on a graphics calculator.

Portfolio Activities appear at the end of Lessons 11.3 and 11.6. Each serves as preparation for the Chapter Project. The Portfolio Activities, as well as the Chapter Project Activities, are appropriate for inclusion in the student's portfolio. Students should be encouraged to include in their portfolios any other work in which they feel a sense of pride or a sense of accomplishment.

---

internet connect

Chapter Internet Features and Online Activities

Golden Connections

11.1

Prepare

QUICK WARM-UP

1. Solve the proportion $\frac{6}{x} = \frac{12}{8}$. $x = 4$

2. Use the quadratic formula to solve the equation $x^2 - 2x - 4 = 0$.
$x = \pm\sqrt{5} + 1$

3. Use a calculator to find $\frac{1+\sqrt{5}}{2}$. $x \approx 1.618$

Also on Quiz Transparency 11.1

Teach

Why Studying the golden ratio should help students appreciate how seemingly unrelated objects may have a mathematical connection. The connection to the golden ratio is especially important because of its use in art and architecture and its appearance in nature.

Objectives

● Discover the relationship known as the golden ratio.

● Solve problems by using the golden ratio.

Why *The golden ratio can be used to find the dimensions of the golden rectangle, which has been considered for centuries by artists and architects to be the "ideal" rectangle.*

The Parthenon is an ancient Greek temple dedicated to the goddess Athena. Viewed from the front, the proportions of the temple suggest the golden ratio, a mathematical idea which captured the imagination of classical Greek thinkers.

Golden Rectangles

A **golden rectangle** is a rectangle with the following property: If a square is cut from one end of the rectangle, the remaining piece is similar to the original rectangle.

If the long side of the large rectangle has a length of ℓ and the short side has a length of s, then the sides of the smaller rectangle are s and $\ell - s$.

The ratio of the long side to the short side of a golden rectangle is called the **golden ratio**. Since the large and small rectangles are similar, the relationship between their sides can be expressed by the proportion below.

$$\frac{\ell}{s} = \frac{s}{\ell - s}$$

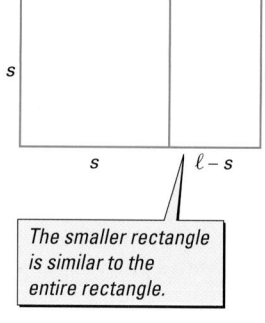

The smaller rectangle is similar to the entire rectangle.

Alternative Teaching Strategy

USING DISCUSSION Begin by reviewing the quadratic formula. Use Activity 1 as an outline for a class presentation. An overhead graphics calculator will help students see the pattern in the ratio. Conclude the discussion by calculating the golden ratio. Students should then be able to complete Activities 2 and 3 on their own.

Teaching Tip

Motivate students by having them find the ratio of their full height to their navel height, a ratio that may be golden. Point out that the golden ratio is a ratio between lengths of line segments.

Activity 1
The Dimensions of a Golden Rectangle

YOU WILL NEED

graph paper and calculator

PROBLEM SOLVING

1. Orient your graph paper horizontally and draw a vertical segment 10 units long at the left edge. This will be the short side of your golden rectangle.

2. **Make a table.** Your rectangle must satisfy the proportion $\frac{\ell}{s} = \frac{s}{\ell - s}$.

   Copy the table below and experiment with different possible values for ℓ until the values in the third and fourth columns match up to two decimal places.

ℓ	s	$\frac{\ell}{s}$	$\frac{s}{\ell - s}$
20	10	2	1
?	10	?	?
?	10	?	?
?	10	?	?

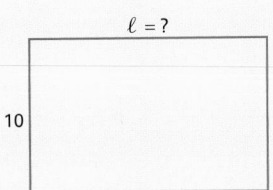

$\ell = ?$

10

3. Complete your drawing of the golden rectangle with the correct value for ℓ.

CHECKPOINT ✔

4. Complete the following statement: The golden ratio, $\frac{\ell}{s}$, is ___?___.

Activity 2
Seashells

YOU WILL NEED

ruler, compass, and your golden rectangle from Activity 1

1. Label your golden rectangle *ABCD* and draw square *EBCF*.

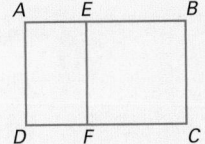

2. Draw square *AEGH*. Use your compass and the squares on the graph paper to make your squares as accurate as possible.

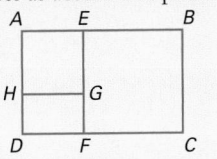

CHECKPOINT ✔

3. Repeat this process three more times. Do the rectangles that are formed seem to be golden rectangles? Find the ratio $\frac{\ell}{s}$ for each rectangle by measuring the sides.

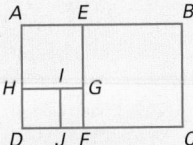

4. Use a compass to make a quarter-circle in each square, as shown below. The resulting curve approximates a *logarithmic spiral*, which models the growth pattern of seashells such as the chambered nautilus.

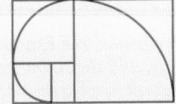

Interdisciplinary Connection

ARCHITECTURE Have students find examples of the golden ratio in architecture, especially skyscrapers. Ask interested students to make a simple model of a building using the golden ratio.

Activity 1 Notes

In this Activity, students will construct a golden rectangle from a vertical segment given as the width of the rectangle. The Activity emphasizes that a rectangle is golden if the ratio of its length to its width is a certain proportion. Students discover this proportion by substituting different values for the length of the rectangle.

For a student worksheet of this Activity and detailed Teacher Notes, see page 196 in the Lesson Activities booklet.

CHECKPOINT ✔
4. approximately 1.62

Activity 2 Notes

Students will use a compass to divide the golden rectangle from Activity 1 into similar nested golden rectangles. The Activity will help students see how nested golden rectangles create a spiral that models the growth of certain natural formations.

For a student worksheet of this Activity and detailed Teacher Notes, see page 196 in the Lesson Activities booklet.

CHECKPOINT ✔
3. Yes. Answers will vary. The ratio of the sides of the rectangle, $\frac{\ell}{s}$, should be approximately 1.62.

Math
CONNECTION

ALGEBRA To compute the golden ratio algebraically, students need to use the quadratic formula to find the length of a segment. Review the quadratic formula at this time, including how to simplify radicals.

CRITICAL THINKING ✔
The ratio $\frac{\ell}{s}$ is the same for any value of s.

Computing the Golden Ratio

Algebra

The proportion for the golden ratio can be solved algebraically to find the exact value for $\frac{\ell}{s}$. You must first notice that in a golden rectangle, the proportion must be true for any value of s. For convenience, let the value of s be 1, and find a value of ℓ that satisfies the proportion.

$$\frac{\ell}{s} = \frac{s}{\ell - s}$$

$$\frac{\ell}{1} = \frac{1}{\ell - 1}$$

$\ell(\ell - 1) = 1$ *Cross multiply.*

$\ell^2 - \ell = 1$ *Distribute.*

$\ell^2 - \ell - 1 = 0$

Notice that the resulting equation is a quadratic equation of the form $ax^2 + bx + c = 0$, where $a = 1$, $b = -1$, and $c = -1$. To solve, substitute these values into the quadratic formula, $x = \frac{-b \pm \sqrt{b^2 - 4ac}}{2a}$. Because ℓ is a length, consider only the positive value.

$$\ell = \frac{-(-1) + \sqrt{(-1)^2 - 4(1)(-1)}}{2(1)} = \frac{1 + \sqrt{5}}{2} \approx 1.618$$

Next, substitute the values of ℓ and s into the ratio.

$$\frac{\ell}{s} = \frac{\ell}{1} \approx \frac{1.618}{1}$$

This is the value of the golden ratio. It is often represented by the Greek letter ϕ (phi):

$$\phi = \frac{1 + \sqrt{5}}{2} \approx 1.618033989\ldots$$

CRITICAL THINKING Do you think you would get the same value for $\frac{\ell}{s}$ if some value other than 1 is used for s? Try several other values for s. What do you discover?

The Parthenon, the Pentagon Building, and the chambered nautilus all involve the golden ratio.

Enrichment

Construct a pyramid from straws so that the pyramid has the same proportions as the Great Pyramid. Find the ratio of the slant height to one-half the base length. It should be very close to the golden ratio.

Inclusion Strategies

ENGLISH-LANGUAGE DEVELOPMENT Encourage students to make a list of all new vocabulary words. Explain specific meanings of words and how each word applies to the golden ratio. Make sure students understand the steps in the construction of a golden rectangle with a compass and straightedge.

Constructing a Golden Rectangle

CONSTRUCTION
COMPASS and
STRAIGHTEDGE

1. Construct a square and label the vertices *ABCD*. Let the length of $\overline{AB}$ be 2 units. Extend side $\overline{AB}$ as shown.

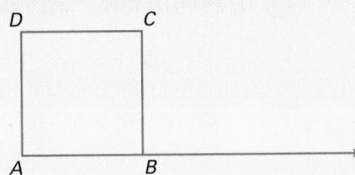

2. Construct the midpoint of $\overline{AB}$ and label it *M*. Draw $\overline{MC}$. Since *ABCD* is a square, *BC* = __?__ units. Since *M* is the midpoint of $\overline{AB}$, *MB* = __?__ units. By the Pythagorean Theorem, *MC* = __?__ units.

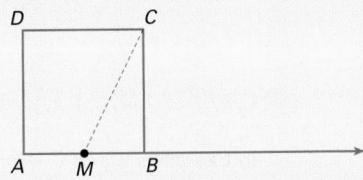

3. Set the point of your compass at *M*, and draw an arc through *C* that intersects $\overrightarrow{AB}$ at *E*.

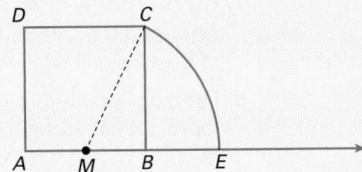

CHECKPOINT ✔ **4.** Extend $\overline{DC}$, and construct a perpendicular to $\overrightarrow{AB}$ at *E*. Label the intersection *F*. How do you know that rectangle *AEFD* is a golden rectangle? Complete the following statements:

$AD =$ __?__ $AE =$ __?__ $\dfrac{AE}{AD} =$ __?__

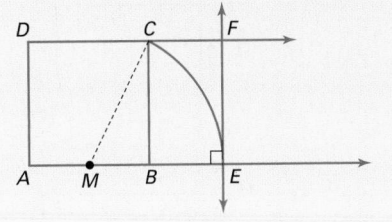

CRITICAL THINKING   In the diagram for Step 4 of Activity 3, *AEFD* is a golden rectangle. Name another golden rectangle in this diagram. How do you know it is golden?

Reteaching the Lesson

HANDS-ON STRATEGIES Have each student draw a vertical line of any length on a sheet of paper. Use the line to construct a golden rectangle with a compass and straightedge. Ask them to measure the rectangle precisely and explain in writing how they know they constructed the golden rectangle.

Activity 3 Notes

In this Activity, students learn how to construct a golden rectangle with a compass and straightedge. Point out the use of the Pythagorean Theorem to construct a segment whose length is irrational. Have students discuss how to construct a golden rectangle.

For a student worksheet of this Activity and detailed Teacher Notes, see page 196 in the Lesson Activities booklet.

Cooperative Learning

Have groups of three students review the lesson. Each student should describe one of the Activities in detail to the other members of their group.

CHECKPOINT ✔

4. You know that *AEFD* is a golden rectangle because the ratio of $\frac{\ell}{w}$ is $\frac{1+\sqrt{5}}{2}$.
$AD = 2$, $AE = 1 + \sqrt{5}$,   and $\frac{AE}{AD} = \frac{1+\sqrt{5}}{2}$

CRITICAL THINKING

CFEB is a golden rectangle. The ratio $\frac{\ell}{w}$ is equal to $\frac{1+\sqrt{5}}{2}$.

ASSIGNMENT GUIDE

In Class	1–8
Core	9–32
Core Plus	13–35
Review	36–43
Preview	44–45

✐ Extra Practice can be found beginning on page 818.

Exercises

☑ internet connect

Activities Online
Go To: **go.hrw.com**
Keyword:
MG1 Golden

● Communicate

1. The golden rectangle is said to be *self-replicating*. Use your results from Activity 2 to explain what this means.

2. Explain the construction of the golden rectangle. How is the Pythagorean Theorem used in this construction?

3. Measure the sides of several rectangles around you, such as a sheet of paper or a poster. Find the ratios of the long side to the short side. Name some that are close to the golden ratio.

4. Find the ratio of your total height to your height up to your navel. Is your answer close to the golden ratio?

● Guided Skills Practice

In Exercises 5 and 6, determine the unknown side length of the golden rectangle with the given side length. *(ACTIVITY 1)*

5. $s = 10$ $\ell = \underline{\ ?\ }$ 16.18

6. $s = \underline{\ ?\ }$ $\ell = 9$ 5.56

7. Sample answer:
$\overline{AC} \perp \overline{DE}$

The intersection seems to be the center of the spiral.

7. Use the rectangle and spiral that you created in Activity 2, or construct a new one. Draw $\overline{AC}$ and $\overline{DE}$. What seems to be true about the intersection of these segments? *(ACTIVITY 2)*

8. Use the method from Activity 3 to construct a golden rectangle whose short side has a length of 1 in. *(ACTIVITY 3)* The long side of the rectangle should be ≈ 1.6 in.

● Practice and Apply

Determine the indicated side length of each golden rectangle. Round your answers to the nearest hundredth.

9.
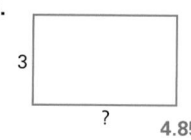
3
? 4.85

10.
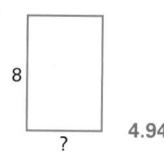
8
? 4.94

11.
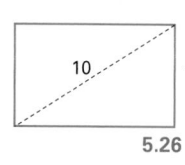
10
? 5.26

12.
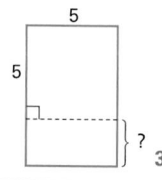
5
5
? 3.09

Error Analysis

You may wish to review the law of sines with students for Exercises 21 and 24.

Technology

Students can use geometry graphics software for the constructions in Exercises 16, 17, and 22.

The value of the golden ratio, ϕ, has some interesting properties. Refer to the diagram below for Exercises 13–15.

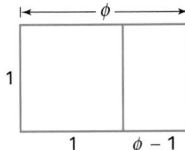

13. $\dfrac{\phi}{1} = \dfrac{1}{\phi - 1}$

14. $\phi^2 \approx 2.618033989$
$\phi^2 - \phi = 1$
$\phi^2 = \phi + 1$

15. $\phi^{-1} \approx 0.6180339887$
From Exercise 13,
$\dfrac{\phi}{1} = \dfrac{1}{\phi - 1} \Rightarrow \dfrac{1}{\phi} = \phi - 1$
$\phi^{-1} = \phi - 1$

13. Set up a proportion involving the side lengths of the two golden rectangles in the figure.

14. Enter the value of ϕ in your calculator, using as many decimal places as your calculator will hold, and find ϕ^2. Compare this value with ϕ. What do you notice? Using your proportion from Exercise 13, explain why this is true.

15. Enter the value of ϕ in your calculator, using as many decimal places as your calculator will hold, and find ϕ^{-1}. Compare this value with ϕ. What do you notice? Using your proportion from Exercise 13, explain why this is true.

CONSTRUCTION

Regular Pentagon

a. Draw $\odot A$ and diameter $\overline{BC}$. Construct M, the midpoint of $\overline{AB}$.

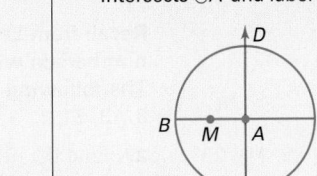

b. Construct line $\ell \perp \overline{BC}$ through A. Choose either point where ℓ intersects $\odot A$ and label it D.

c. Place your compass point at M and draw an arc through D that intersects $\overline{BC}$. Label this point P.

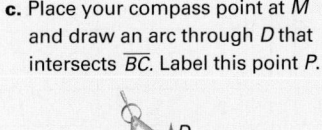

d. Set your compass to the length DP. Mark off length DP around $\odot A$. Draw segments to connect the arcs, forming a regular pentagon.

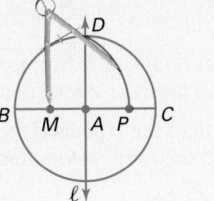

16. Check students' drawings.

17. Sample answer: In a circle of radius 3 cm, $DE \approx 3.5$, and $EH \approx 5.7$.

18. Sample answer: $\dfrac{EH}{DE} \approx 1.605$ The difference is about 0.01.

16. Use the directions above to construct regular pentagon $DEFGH$ in a circle with a radius of at least 3 cm.

17. Draw $\overline{EH}$. Measure $\overline{DE}$ and $\overline{EH}$ to the nearest millimeter.

18. Find $\dfrac{EH}{DE}$. How close to ϕ is your answer?

For Exercises 19–21, refer to the diagram of the regular pentagon at right. You will use trigonometry to verify that the ratio $\frac{EH}{DE}$ is equal to ϕ.

19. Use the fact that $DEFGH$ is a regular pentagon to find m$\angle D$. **108°**

20. $\triangle DEH$ is isosceles, so m$\angle DEH$ = m$\angle DHE$ = __?__. **36°**

21. Use the law of sines and properties of proportions to write an expression for the ratio $\frac{EH}{DE}$. Evaluate this expression on your calculator. What do you get? $\frac{EH}{DE} = \frac{\sin 108°}{\sin 36°} \approx 1.618033989 \approx \phi$

If you extend the sides of a regular pentagon, you get a five-pointed star known as the star of Pythagoras. This star was the symbol of the followers of Pythagoras.

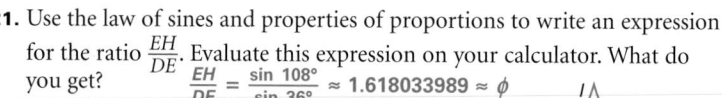

22. Construct a regular pentagon and extend the sides to form a star of Pythagoras.

23. Compute the measures of the angles in $\triangle DEI$.

24. Use the law of sines to find $\frac{EI}{DE}$. What do you notice? **1.618033989 $\approx \phi$**

25. Measure $\overline{IL}$ and $\overline{IF}$ to the nearest millimeter. Find $\frac{IL}{IF}$. $\frac{IL}{IF} \approx 1.6 \approx \phi$ What do you notice?

26. Measure $\overline{IF}$ and $\overline{IE}$ to the nearest millimeter. Find $\frac{IF}{IE}$. $\frac{IF}{IE} \approx 1.6 \approx \phi$ What do you notice?

Recall from Lesson 2.2 that the Fibonacci sequence is a sequence of numbers in which each term is the sum of the two terms preceding it. The following is the beginning of the Fibonacci sequence: 1, 1, 2, 3, 5, 8, 13, 21,…

27. Find the next five terms of the Fibonacci sequence above.

28. Copy and complete the table below by dividing each term in the sequence by the preceding term.

Term	1	1	2	3	5	8	13	21	?	?	?	?	?
Ratio	—	$\frac{1}{1}$	$\frac{2}{1}$	?	?	?	?	?	?	?	?	?	?
Value	—	1	2	?	?	?	?	?	?	?	?	?	?

29. What number do the ratios approach? Write a conjecture about the ratios of consecutive terms of the Fibonacci sequence.

30. The 28th term of the Fibonacci sequence is 317,811, and the 29th term is 514,229. Find their ratio. Does this agree with your conjecture?

31. Let $\phi' = \frac{1 - \sqrt{5}}{2}$, the negative solution to the quadratic equation on page 700. Then the nth term of the Fibonacci sequence is given by $\frac{\phi^n - (\phi')^n}{\sqrt{5}}$. Use this expression to find the 20th term of the Fibonacci sequence.

32. Use the expression for the nth term of the Fibonacci sequence from Exercise 31 to explain why the ratios in the table for Exercise 28 approach ϕ. (Hint: What happens to $(\phi')^n$ as n increases?)

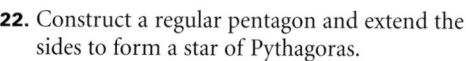
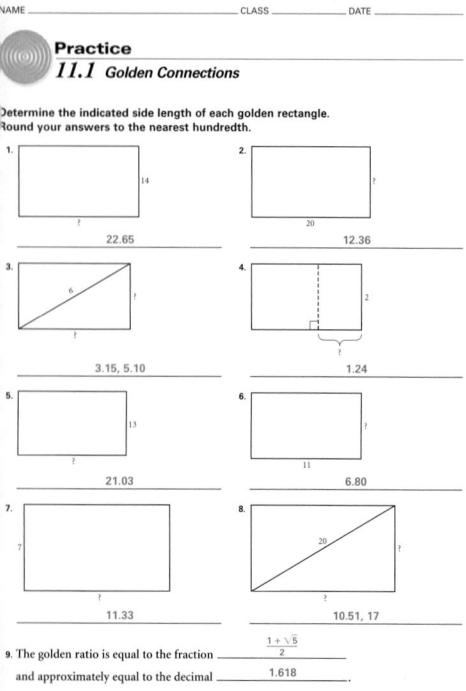

APPLICATIONS

33. $\frac{5}{3} \approx 1.67$

$\frac{7}{5} = 1.4$

$\frac{10}{8} = 1.25$

$\frac{14}{11} \approx 1.27$

$\frac{20}{16} = 1.25$

3×5 is closest to the golden ratio.

33. PHOTOGRAPHY A photography studio offers prints in the following sizes: $3 \times 5, 5 \times 7, 8 \times 10, 11 \times 14$, and 16×20. Find the ratio of the long side to the short side for each of these rectangles. Which is closest to the golden ratio?

34. MARKETING A manufacturer wishes to make a cereal box in the shape of a golden rectangle, based on the theory that this shape is the most pleasing to the average customer. If the front of the box has an area of 104 in.^{2}, what should the dimensions be? Round to the nearest inch. **8 in. × 13 in.**

35. FINE ARTS An artist wishes to make a canvas in the shape of a golden rectangle. He only has enough framing material for a perimeter of 36 in. What should the dimensions of the canvas be? **7 in. × 11 in.**

Look Back

CONNECTION

COORDINATE GEOMETRY Find the distance between each pair of points below. Round your answers to the nearest tenth. *(LESSON 5.6)*

36. $(0, 0)$ and $(4, 7)$ **8.1**

37. $(1, 3)$ and $(4, 2)$ **3.2**

38. $(-6, 0)$ and $(5, -2)$ **11.2**

39. $(-1, -3)$ and $(3, -5)$ **4.5**

40. $(x - 2)^2 + (y - 1)^2 = 9$

41. $(0.6428, 0.7660)$

42. $130°$

43.

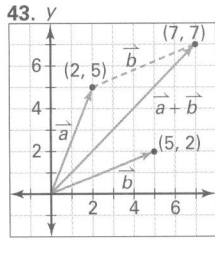

40. Write an equation for the set of points in a coordinate plane that are a distance of 3 from the point $(2, 1)$. *(LESSON 9.6)*

41. Find the coordinates of a point on the unit circle that is the image of the point $(1, 0)$ under a 50° rotation about the origin. *(LESSON 10.3)*

42. The point $(-0.643, 0.766)$ is the image of the point $(1, 0)$ under a rotation about the origin. What is the angle of rotation? *(LESSON 10.3)*

43. Vector $\vec{a}$ in a coordinate plane points from the origin to $(2, 5)$ and vector $\vec{b}$ points from the origin to $(5, 2)$. What is the sum of these two vectors? Sketch a diagram to illustrate your answer. *(LESSON 10.6)*

Look Beyond

Fibonacci numbers, which are numbers in the Fibonacci sequence, occur frequently in nature. For example, the numbers of spirals in pinecones, artichokes, sunflowers, and other objects are often Fibonacci numbers.

44. 34 and 55. They are in the Fibonacci Sequence.

44. Examine the photo of the sunflower at right. Notice that the seeds form two sets of spirals in opposite directions. Try to count the number of spirals to the left and to the right. Both numbers should be Fibonacci numbers. What are they?

45. Answers will vary.

45. The numbers of petals of many types of flowers are Fibonacci numbers. Find some pictures of flowers and count their petals. Which ones are Fibonacci numbers?

Look Beyond

In Exercises 44 and 45, students explore the Fibonacci sequence and the golden ratio in nature.

Taxicab Geometry

1. The points (–4, 4), (1, 4), (1, –1), and (–4, –1) are the vertices of a quadrilateral. Name the type of quadrilateral. **square**

2. Find the shortest distance between points (1, 1) and (5, 7). $2\sqrt{13}$, or ≈ 7.21

3. Find the distance between the points (1, 1) and (5, 7) along grid lines of graph paper. **10 units**

Also on Quiz Transparency 11.2

Objectives

● Develop a non-Euclidean geometry, known as *taxicab geometry*, based on a taxicab's movements on a street grid.

● Solve problems within a taxicab geometry system.

Why A geometry student would probably use the Pythagorean Theorem to find the distance between two points on a city map. However, a taxicab driver might have a very different idea. In this lesson, you will study a geometry based on a taxicab driver's definition of the distance between two points.

Teach

Why Taxicab geometry, a non-Euclidean geometry, has various real-world applications. This lesson helps students understand that various geometric systems can coexist in ways that are logical, consistent, and applicable to the real world.

Taxidistance

In **taxicab geometry**, points are located on a special kind of map or coordinate grid. The horizontal and vertical lines of the grid represent streets. Unlike points in a traditional coordinate plane, points in a taxicab grid can be only at intersections of two "streets." Thus, the coordinates are always integers.

In taxicab geometry, the distance between two points, known as the **taxidistance**, is the smallest number of grid units, called **blocks**, that a taxi must travel to get from one point to the other. On the map shown, the taxidistance between the two points is 5.

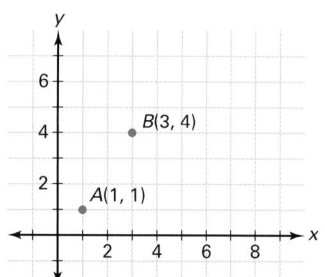

In this taxicab grid, the taxidistance from point A to point B is 5.

Alternative Teaching Strategy

USING MODELS Photocopy a portion of a local map onto a transparent grid. On the overhead, demonstrate taxicab pathways that are familiar to students. Find the shortest taxicab distance between two intersection points. Compare different paths between the two points that have the same taxicab distance.

Activity 1
Exploring Taxidistances

Part I: The taxidistance from a central terminal

Assume that all taxis leave for their destinations from a central terminal at point $O(0, 0)$.

YOU WILL NEED

graph paper

1. Draw the six destination points in a taxicab grid, as shown in the diagram at right. Label the points A through F and give their coordinates.

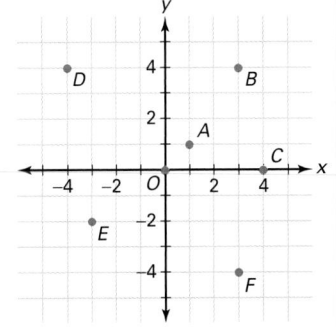

PROBLEM SOLVING

2. **Make a table.** Find the taxidistance from O to each of the six destination points. (Make sure that you have found the smallest number of blocks in each case.) Copy and complete the table below.

Point	Coordinates	Taxidistance from O
A	(?, ?)	?
B	(?, ?)	?
C	(?, ?)	?
D	(?, ?)	?
E	(?, ?)	?
F	(?, ?)	?

3. Based on your results, complete the following conjecture:

CHECKPOINT ✔

The taxidistance between the points $(0, 0)$ and (x, y) is ____?____.

Part II: The taxidistance between any two points

PROBLEM SOLVING

1. **Make a table.** Use the diagram from Part I to find the taxidistance between the pairs of points given in the table below. Copy and complete the table below.

(x_1, y_1)	(x_2, y_2)	x_1	x_2	y_1	y_2	Taxidistance
A	B	1	3	1	4	5
A	D	?	?	?	?	?
A	E	?	?	?	?	?
C	D	?	?	?	?	?
D	E	?	?	?	?	?
E	F	?	?	?	?	?
B	F	?	?	?	?	?

2. Based on your results, complete the following conjecture:

CHECKPOINT ✔

The taxidistance between the points (x_1, y_1) and (x_2, y_2) is ____?____.

Interdisciplinary Connection

GEOGRAPHY Traveling by roads on a road map is like taking a taxicab in a city in that one is constrained to taking actual physically existing roads rather than going strictly "as the crow flies." Could a road-distance geometry be created? What would be similarities and differences between such a geometry and taxicab geometry?

Enrichment

Extend Activity 2 by having students develop or verify the area formulas for taxicab triangles, quadrilaterals (including special cases), and hexagons.

Teaching Tip

Stress the importance of using integers for the coordinates in a taxicab geometry system. Ask students to find instances when negative integers, as well as positive integers, would be applicable.

Activity 1 Notes

In this Activity, students measure the taxidistance between points in a coordinate system. This Activity will help students understand that there is a minimum taxidistance between two points of a grid.

For a student worksheet of this Activity and detailed Teacher Notes, see page 202 in the Lesson Activities booklet.

Cooperative Learning

Divide students into groups of four to explore the number of minimum pathways that are possible as the taxidistance between two points increases. Have each student draw a small grid that corresponds to a map and make sketches that correspond to the minimum pathways between two points on the map.

PART I
CHECKPOINT ✔
3. The taxidistance between the points $(0, 0)$ and (x, y) is $|x| + |y|$.

PART II
CHECKPOINT ✔
2. The taxi distance between the points (x_1, y_1) and (x_2, y_2) is $|x_2 - x_1| + |y_2 - y_1|$.

Two Points Determine...?

For any two given points in Euclidean geometry, there is just one minimum-distance pathway between them. Is this true in taxicab geometry? Consider points *A* and *B* below. In both arrangements, *A* and *B* are 3 taxicab units apart. In the second arrangement, there is more than one minimum-distance pathway between them.

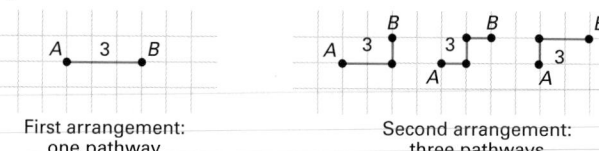

First arrangement: one pathway

Second arrangement: three pathways

CRITICAL THINKING Under what conditions is there just one minimum-distance path between two points in taxicab geometry? Under what conditions is there more than one?

Activity 2
Taxicab Circles

YOU WILL NEED

graph paper

In Euclidean geometry, a circle is the set of points that are a fixed distance from a given point. What happens when this definition is applied to taxicab geometry?

1. Plot a point *P* on graph paper. Then plot all the points that are located 1 block from point *P*. The very uncircular-looking result is a **taxicab circle** with a radius of 1.

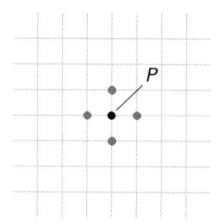

A taxicab circle with a radius of 1

2. Draw additional taxicab circles with radii of 2, 3, 4, and 5. Count the number of points on each circle. Find the circumference of each circle by adding the taxidistances between the points on the circle.

PROBLEM SOLVING

3. **Make a table.** Copy and complete the table below to find expressions for the number of points on a taxicab circle and the circumference in terms of its radius.

Radius	Number of points on circle	Circumference
1	4	8
2	?	?
3	?	?
4	?	?
5	?	?
r	?	?

CHECKPOINT ✔

4. Use the information in the chart to determine a taxicab equivalent for π.

$\left(\text{Hint: } \pi = \dfrac{\text{circumference}}{\text{diameter}}\right)$

Inclusion Strategies

USING MODELS Students may benefit from developing their own taxicab maps. Have them use graph paper to develop a map of the area between the school building and their home. Ask them to use the map to show the pathway that corresponds to the minimum distance between their home and the school.

Reteaching the Lesson

USING VISUAL MODELS Have each student draw and label a taxicab grid. Using the origin as the center point, locate all points that are a taxidistance of 3 units from the center. Describe the resulting shape. **a taxicab circle with a radius of 3**

Exercises

Communicate

1. Why is the geometry studied in this lesson called taxicab geometry?

2. Explain how to determine the distance between two points in taxicab geometry.

3. Why is π not used in taxicab geometry? What is the taxicab equivalent for π?

4. What are some practical applications of taxicab geometry?

5. What are some factors other than distance that a taxicab driver might take into consideration when planning a route?

Guided Skills Practice

For points A(0, 1), B(2, 3), and C(–2, 1), find the taxidistances below.
(ACTIVITY 1)

6. *AB* 4   7. *BC* 6   8. *AC* 2

Plot the taxicab circle with the given center and radius. *(ACTIVITY 2)*

9. center: (3, 2); radius: 2   10. center: (–1, 4); radius: 4

Practice and Apply

Find the taxidistance between each pair of points.

11. (0, 0) and (7, 3) 10   12. (5, –3) and (–2, 4) 14

13. (1, 5) and (–2, –3) 11   14. (–9, –3) and (–3, –1) 8

15. (–11, 4) and (–3, 9) 13   16. (–129, 43) and (152, 236) 474

Find the number of points on the taxicab circle with the given radius.

17. $r = 2$ 8   18. $r = 4$ 16   19. $r = 12$ 48

Find the circumference of the taxicab circle with the given radius.

20. $r = 1$ 8   21. $r = 5$ 40   22. $r = 10$ 80

Plot the taxicab circle with center P and radius r.

23. $P(0, 0); r = 3$ 24. $P(5, –2); r = 5$ 25. $P(–1, 3); r = 2$

26. Identify two pairs of points in a taxicab grid that have a taxidistance of 4. One of these pairs should have only one minimum-distance pathway between them, and the other should have several minimum-distance pathways.

27. Sketch all of the different minimum-distance pathways between the points (1, 2) and (4, 1) in a taxicab grid. How many pathways are there?

28. Sketch all of the different minimum-distance pathways between the points (3, 0) and (5, 2) in a taxicab grid. How many pathways are there?

internet connect

Homework Help Online
Go To: go.hrw.com
Keyword:
MG1 Homework Help
for Exercises 20-22

24.
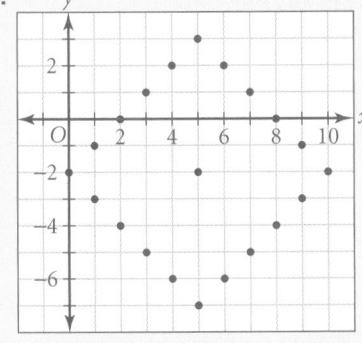

25.
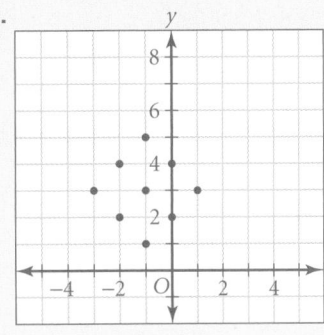

26. (0, 0) and (0, 4) have one path between them.
(0, 0) and (1, 3) have several paths.

Assess

Selected Answers
Exercises 6–10, 11–43 odd

ASSIGNMENT GUIDE

In Class	1–10
Core	11–25 odd, 26–35
Core Plus	23–25 odd, 29–37
Review	38–44
Preview	45–47

✐ Extra Practice can be found beginning on page 818

9.
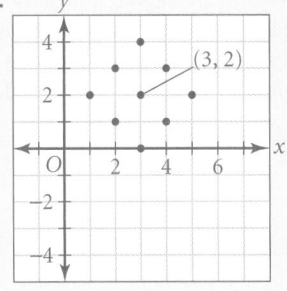

10.
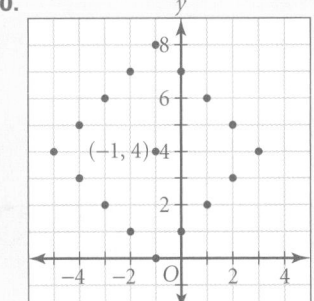

23.
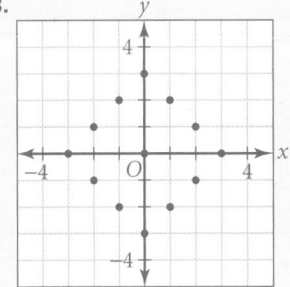

Recall from Chapter 4 that a point lies on a segment's perpendicular bisector if and only if it is equidistant from the segment's endpoints. Use this theorem in Exercises 29–33 to discover the taxicab equivalent of a perpendicular bisector.

29. **a.** On a taxicab grid, plot points $A(0, 0)$ and $B(4, 2)$. Locate all points that are a taxidistance of 3 from both A and B.

   b. On the same diagram, locate all points that are a taxidistance of 4 from both A and B.

   c. Continue locating points that are the same taxidistance from both A and B until you understand the form of a taxicab perpendicular bisector.

30. How is the perpendicular bisector you constructed in Exercise 29 similar to a perpendicular bisector in Euclidean geometry? How is it different?

31. Use the method from Exercise 29 to construct a taxicab perpendicular bisector for the points $(3, 3)$ and $(4, 0)$. How is it like the taxicab perpendicular bisector you constructed in Exercise 29? How is it different?

32. Use the method from Exercise 29 to construct a taxicab perpendicular bisector for the points $(1, 2)$ and $(5, 2)$. What do you notice?

33. Use the method from Exercise 29 to construct a taxicab perpendicular bisector for the points $(2, 0)$ and $(3, 2)$. What do you notice?

Algebra

CHALLENGE

APPLICATIONS

34. Use your formula from Activity 1 for the taxidistance between two points to derive an equation for a taxicab circle with a radius of r.

35. For the points $A(0, 0)$ and $B(2, 2)$, find the set of all points P such that the sum of the taxidistances AP and BP is equal to 6. Describe your results.

36. **PUBLIC SAFETY** The mayor of a city wants to install police call boxes at intersections around the city so that from every point in the city there will be a maximum distance of two blocks to a call box. Describe a way this can be done. Try to find a way that uses the minimum number of boxes. (Hint: Draw a pattern of taxicab circles with a radius of 2.)

37. **COMMUTING** Jenny works at an office during the day and then works in a retail shop in the evenings. Because she plans to ride her bicycle to work, she would like to live within 10 blocks of her day job and within 8 blocks of her night job so that she does not have to ride as far in the dark. Use the taxicab grid at left to determine all of the possible locations where Jenny should look for an apartment.

Shop (5, 4)

(0, 0) Office

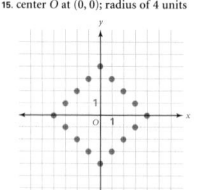
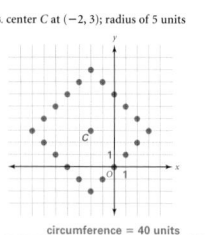
29. **a.** $(1, 2), (2, 1), (3, 0)$
    b. $(3, -1), (1, 3)$
    c.

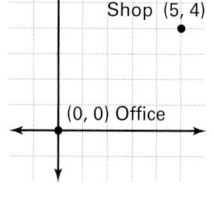

30. The perpendicular bisector is defined in taxicab geometry in the same way as it was defined in Euclidian geometry. Graphically, there appears to be a bend in the bisector.

Find y for Exercises 38 and 39. *(LESSON 9.5)*

38.

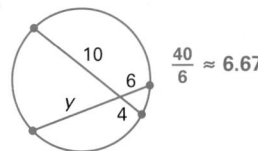

$\frac{40}{6} \approx 6.67$

39.

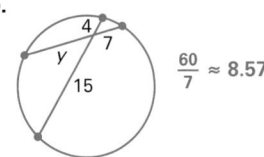

$\frac{60}{7} \approx 8.57$

40. Let $BC = 5$, $AC = 22$, and $CD = 6$. Find DE.
(LESSON 9.5)

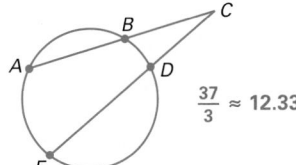

$\frac{37}{3} \approx 12.33$

41. Let $MP = 18$ and $MN = 7$. Find MR.
(LESSON 9.5)

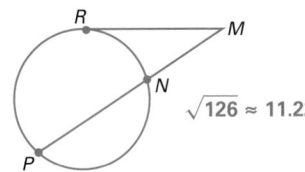

$\sqrt{126} \approx 11.22$

Refer to the triangle below. Find the value of each expression. *(LESSONS 10.1 AND 10.2)*

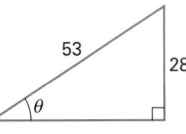

42. $\tan \theta$

$\frac{28}{45} \approx 0.62$

43. $\cos \theta$

$\frac{45}{53} \approx 0.85$

44. θ

$\approx 31.89°$

 Look Beyond

The following is a geometric "proof" that 0 = 1. Trace the steps and try to find the error in the proof.

45. a. On graph paper, draw an 8×8 square. Divide it into three regions, A, B, and C, as shown.

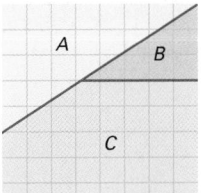

c. Sample answer: The pieces do not quite fit together to form the rectangle. Note that in the rectangle, the piece labeled B has a base of 5 units, while in the original square, the base is shorter than 5 units.

b. Cut out the pieces and fit them together to form a 5×13 rectangle, as shown.

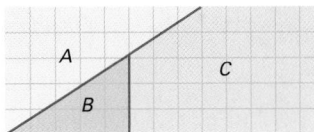

c. The area of the square is 8×8, or 64. The area of the rectangle is 5×13, or 65. However, since both areas are the sum of the areas of A, B, and C, the areas must be equal, so $64 = 65$. Subtracting 64 from both sides of the equation gives $0 = 1$. What is wrong with this proof?

In Exercise 45, students look at a geometric "proof" that $0 = 1$. Students are instructed to find the error in this proof.

QUICK WARM-UP

1. How is the term *network* commonly used?
   Answers will vary.

2. How is the number of vertices of a polygon related to the number of sides?
   They are equal.

3. Define the term *traverse*.
   Answers will vary; one meaning is "to pass through."

Also on Quiz Transparency 11.3

Teach

Why Studying networks will help students recognize how points and lines can be used to create mathematical models. Learning whether a network is traversable will show students how to design pathways through a network.

Graph Theory

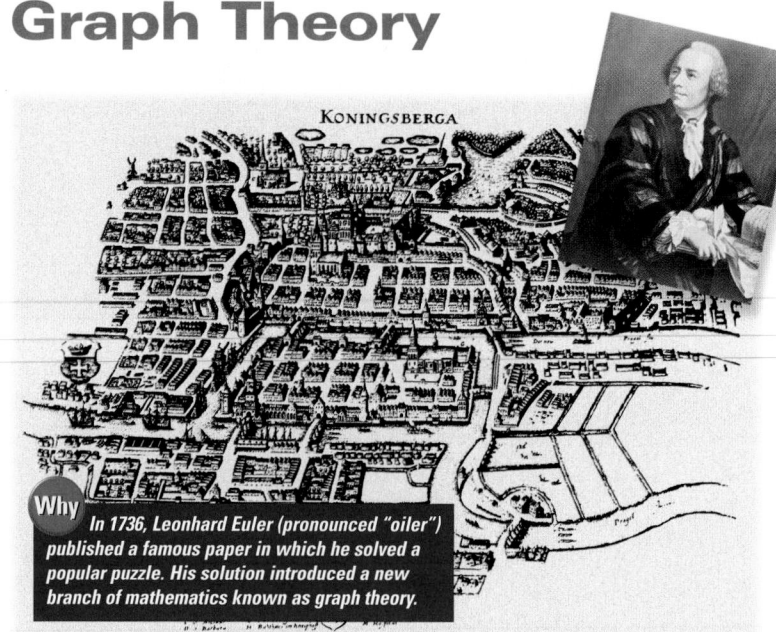

Why In 1736, Leonhard Euler (pronounced "oiler") published a famous paper in which he solved a popular puzzle. His solution introduced a new branch of mathematics known as graph theory.

Objectives

- Determine whether a given graph has an Euler path.

- Use Euler paths to solve problems involving graphs.

A local pastime in the city of Königsberg, Prussia (now Kaliningrad, Russia) was to try to walk across each of the seven bridges over the river without crossing any bridge twice.

The Bridge Problem

The following is an excerpt from Euler's paper, "The Solution to a Problem Relating to the Geometry of Position":

The problem, which I am told is widely known, is as follows: In Königsberg in Prussia, there is an island A, called the Kneiphoff; the river which surrounds it is divided into two branches, and these branches are crossed by seven bridges, a, b, c, d, e, f, and g. Concerning these bridges, it was asked whether anyone could arrange a route in such a way

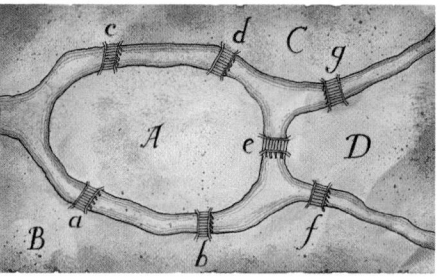

that he would cross each bridge once and only once. I was told that some people asserted that this was impossible, while others were in doubt; but nobody would actually assert that it could be done.

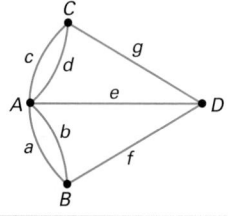

Euler's analysis of the problem can be visualized by using a diagram called a **graph**. A graph is composed of points, called **vertices**, and segments or curves linking the vertices, called **edges**. In the graph at left, each land area (refer to the map above) is represented by a vertex, and each bridge is represented by an edge. The bridge problem now becomes that of finding an **Euler path**—a continuous path that travels along each edge exactly once.

Alternative Teaching Strategy

TECHNOLOGY For extra practice, use software that creates networks to demonstrate some of the concepts in this lesson. Encourage students to experiment with different paths as they try to traverse a network on the computer.

Odd and Even Vertices

You may have tried to find an Euler path in the graph of the seven bridges and decided that it was impossible. The people of Königsberg thought so, but no one had proved it mathematically until Euler presented his proof.

Euler's proof was based on the number of times each land area would have to be visited in a solution of the problem, which is related to the number of bridges to the land area. A graph makes his reasoning easier to understand.

In a graph, the number of edges at each vertex is called the **degree** of the vertex. In the graph of the bridge problem, the degree represents the number of bridges connected to each land area. **Even vertices** have an even degree and **odd vertices** have an odd degree.

In the graph at right, Q has degree 4, so it is an even vertex. R and S both have degree 3, so they are odd vertices. T has degree 2, so it is an even vertex. Does the graph contain an Euler path?

In the following Activity, you will explore the relationship between the degree of the vertices of a graph and whether the graph contains an Euler path.

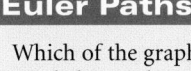

Activity
Euler Paths

PROBLEM SOLVING

Which of the graphs below contain an Euler path? **Make a table** like the one below and see if you can discover the relationship.

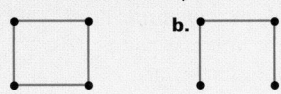
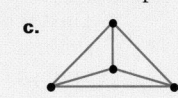
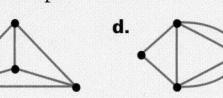

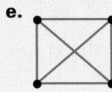
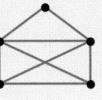
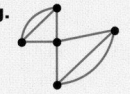
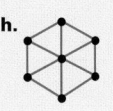

> *In some graphs, edges may intersect at a point that is not a vertex.*

	Number of vertices	Number of odd vertices	Number of even vertices	Is there an Euler path?
a.	?	?	?	?
b.	?	?	?	?
c.	?	?	?	?
	⋮	⋮	⋮	⋮

CHECKPOINT ✔

Make a conjecture based on your completed table. Use your conjecture to complete the following theorem:

Theorem

A graph contains an Euler path if and only if there are at most __?__ odd vertices.   **11.3.1**

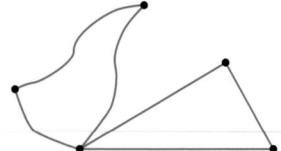

The graph does not contain any odd vertices (all vertices are even), so it contains an Euler path.

Refer to the house floor plan below. Is it possible to walk through the house by going through each door exactly once?

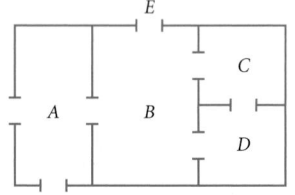

Yes; since the graph has only 2 odd vertices, A and E, it contains an Euler path.

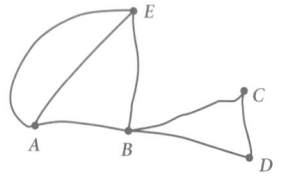

CRITICAL THINKING
Start at one of the odd vertices and end at the other odd vertex.

E X A M P L E 1 Does the graph of the bridges of Königsberg contain an Euler path?

SOLUTION

All four vertices in the graph are odd, so there are more than two odd vertices. Therefore, by Theorem 11.3.1, the graph does not contain an Euler path. That is, there is no route that crosses each bridge (each edge of the graph) exactly once.

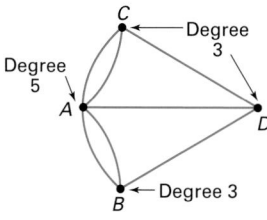

E X A M P L E 2 Refer to the floor plan of a house given at right. Is it possible to walk through all the doors in the house by going through each door exactly once?

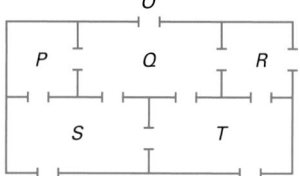

SOLUTION

PROBLEM SOLVING

Make a model. A graph can be used to model the floor plan, with a vertex for each room and for the area outside the house and an edge for each doorway.

The graph has only two odd vertices (Q and R), so it contains a Euler path. Thus, it is possible to walk through all the doors in the house by going through each door exactly once.

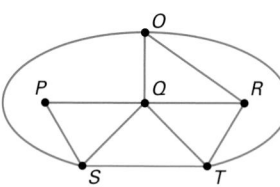

CRITICAL THINKING

How could you find an Euler path in the graph above?

Starting and Stopping

The following problem may provide some insight into Theorem 11.3.1:

While hiking on a snowy day, you startle a rabbit. It runs from bush to bush, hiding under each one. It covers all paths between the bushes (see the diagram) without following the same path more than once. Then it stops. Which bush is the rabbit hiding under?

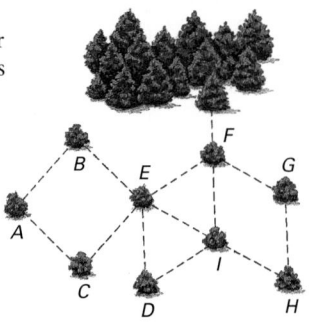

After trying a few routes, you should see that the rabbit always ends up under bush E, an odd vertex. The only other odd vertex is the woods, where the rabbit began its journey.

Inclusion Strategies

USING COGNITIVE STRATEGIES   Some students may feel more comfortable using the following analytical approach to analyzing a network: (1) A traversable network has either 0 or 2 odd vertices. Networks that have more that 2 odd vertices cannot be traversed. (2) If the network has 2 odd vertices, start at one of them and end at the other. (3) If the network has 0 odd vertices, start anywhere.

Suppose that there is an even vertex in a graph and you do not start from it. You will never get stuck at this even vertex for the following reason: *If there is an unused edge leading into the even vertex, there must be an unused edge leading out of it.*

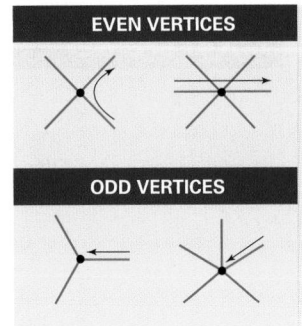

EVEN VERTICES

ODD VERTICES

On the other hand, suppose that there is an odd vertex in the graph and that you do not start from it. If you are able to get to this vertex and keep coming back to it, then you will eventually get stuck at it. *There will be no unused paths leading away from it.*

Notice that if you start at an odd vertex, it becomes, in effect, an even vertex once you have left it; therefore, you will never get stuck at it.

If there are exactly two odd vertices in a graph, you can start at one and try to end at the other. But if there are more than two odd vertices in a graph, there is no chance of finding an Euler path through the graph.

Thus, the "only if" part of Theorem 11.3.1, which states that there is an Euler path through a graph *only if* there are at most two odd vertices, has been proven informally. And as you can see, the two odd vertices must be the starting and ending points of the graph. In the rabbit puzzle, one odd vertex (the woods), is the starting point, and the other odd vertex (bush *E*), is the ending point.

The "if" part of the theorem, which states that there actually *is* such a path when there are at most two odd vertices, has not been proven. However, in Exercises 49–51 of this lesson, you will learn a method, known as *Fleury's algorithm*, for finding such a path when the conditions are right.

CRITICAL THINKING Do you think it is possible for a graph to have just one odd vertex?

Euler Circuits

A graph with only even vertices contains an Euler path that can start at any vertex. There is another important feature of such a graph: The vertex that is the starting point of the Euler path is also the ending point of the path. An Euler path that starts and ends at the same vertex is called an **Euler circuit.**

EXAMPLE ③ A street sweeper is cleaning the area shown at right. Is it possible to sweep both sides of every street exactly once and end at the starting point?

SOLUTION

PROBLEM SOLVING

Make a model. Model the problem with a graph, using an edge to represent each side of each street. The vertices represent intersections.

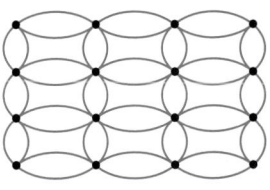

Because every vertex is even, the graph must contain an Euler path. The street sweeper will be able to plan the desired route.

Reteaching the Lesson

COOPERATIVE LEARNING Have students draw a network and use it to summarize the lesson to other students in a group. Ask each group to draw their network on the board, and have the class decide whether it is traversable.

CRITICAL THINKING
No; if all of the other vertices were even, then one path out of the odd vertex would be untraveled.

ADDITIONAL
EXAMPLE ③

Trace a path in the following graph so that you pass through each vertex only once and you start and end at the same vertex.

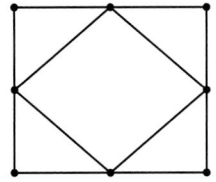

Answers will vary. One possible path is shown below, with each line numbered in succession.

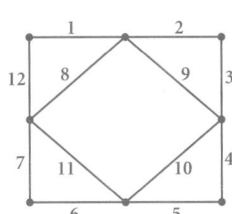

Selected Answers

Exercises 5–9, 11–47 odd

ASSIGNMENT GUIDE

In Class	1–9
Core	11–17 odd, 18–32
Core Plus	18–37
Review	38–48
Preview	49–51

 Extra Practice can be found beginning on page 818.

8.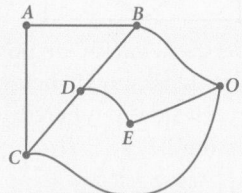

The graph does not contain an Euler path since vertices *B*, *C*, *D*, and *O* are odd.

9. Yes

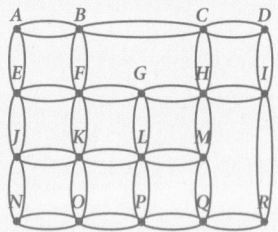

10. Euler path; no Euler circuit

11. No Euler path or circuits

12. No Euler path or circuits

Exercises

Communicate

1. Explain the Königsberg bridge problem in your own words. Why is there no solution?

2. Suppose that the residents of Königsberg decided to build more bridges until there was a way to cross each bridge exactly once. Discuss several ways this could be done.

3. Explain the difference between an Euler circuit and an Euler path.

4. Why do you think Euler called the Königsberg bridge problem "a problem relating to the geometry of position"?

Guided Skills Practice

Determine whether each graph below contains an Euler path.
(EXAMPLE 1)

5. 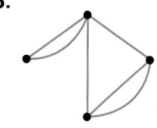 Yes

6.  No

7.  Yes

8. Draw a graph to represent the floor plan at right, with a vertex for each room and the area outside the house and an edge for each doorway. Does the graph of the floor plan contain an Euler path?
(EXAMPLE 2)

9. A truck is collecting materials for recycling. The driver's route must cover both sides of every street exactly once and end at the starting point. In the map area shown at right, is such a route possible? Use a graph to illustrate your answer. **(EXAMPLE 3)**

Practice and Apply

In Exercises 10–15, determine whether each graph contains an Euler path, an Euler circuit, or neither.

10.

11.

12.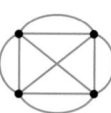

13.  **14.**  **15.**

For each floor plan below, draw a graph with a vertex for each room and the area outside the house and with an edge for each doorway. Determine whether the graph contains an Euler circuit.

16.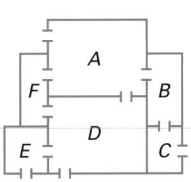

17.

For Exercises 18–20, refer to the diagram below, taken from Euler's paper, "The Solution of a Problem Relating to the Geometry of Position."

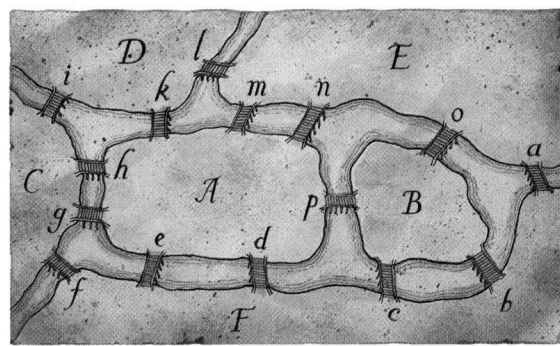

18. Draw a graph representing the land areas and bridges in the diagram.

19. Does the graph contain an Euler path? an Euler circuit?

20. If you wanted to walk across each bridge in the diagram exactly once, where would you start? Where would you finish?

In Exercises 21–29, you will explore some simple concepts of graph theory. Copy and complete the table below.

Graph	Number of edges	Sum of degrees of vertices
△	**21.** ? 5	**22.** ? 10
⧖	**23.** ? 6	**24.** ? 12
⬠	**25.** ? 8	**26.** ? 16

Technology

Students may use computer design programs to draw house floor plans in Exercises 16 and 17.

13. Euler path and circuit

14. Euler path; no Euler circuit

15. Euler path and circuit

16.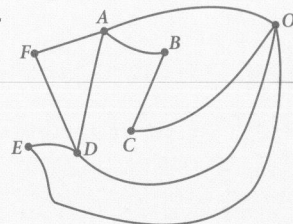

Yes; there is an Euler circuit.

17.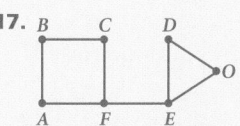

There is no Euler circuit.

18.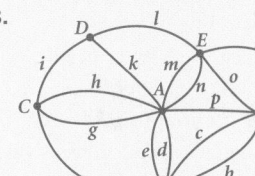

19. Euler path; no Euler circuit

20. Answers may vary. Sample answer: Note that the path must start at vertex D or E. (E)a-b-c-d-e-f-g-h-i-k-p-o-n-m-l(D)

27. The sum of the degrees of the vertices of a graph is twice the number of edges.

28. Each edge has two endpoints; therefore, it is counted twice when summing the degrees of the vertices.

29. Let S_e be the sum of the degrees of the even vertices and let S_o be the sum of the degrees of the odd vertices and E the number of edges. Then $S_e + S_o = 2E$. Note that S_e is even since the sum of even numbers is even. Then $S_o = 2E - S_e$ is even, since the difference of two even numbers is even, by transitivity of equality. Therefore, S_o is an even number, that is, the sum of degrees of odd vertices is an even number. But the only way a sum of odd numbers is even is if there were an even number of those odd numbers. Therefore the number of odd vertices must be even.

Practice

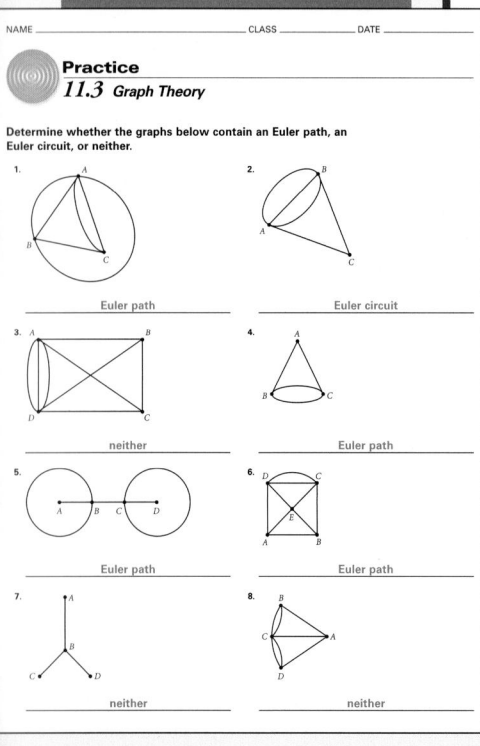

27. Make a conjecture based on your results from Exercises 21–26. Use your conjecture to complete the following theorem:

The sum of the degrees of the vertices of a graph is __?__ the number of edges.

28. Use the fact that each edge of a graph has two endpoints to explain why the theorem in Exercise 27 is true.

29. An immediate corollary of the theorem in Exercise 27 is that the sum of the degrees of the vertices of a graph must be even. Use this fact to prove that the number of odd vertices in a graph must be even.

Determine whether each figure below can be traced completely without lifting the pencil from the page or retracing any part of the figure.

30.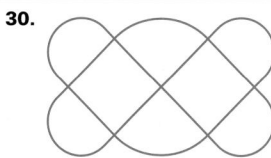

Yes

31.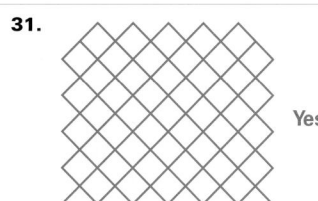

Yes

32. CULTURAL CONNECTION: AFRICA A popular game of children in the Shongo tribe in Zaire is to trace *networks*, like those shown at right, in the sand with a finger. The networks must be traced in a single stroke, without retracing. Can the network at right be traced in this way? If so, how could it be done?

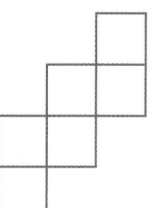

33. In the diagram at right, is it possible to draw a continuous curve that intersects every segment exactly once? (The attempt shown intersects all but two segments.) If it is possible, draw the curve. If it is not possible, explain why.

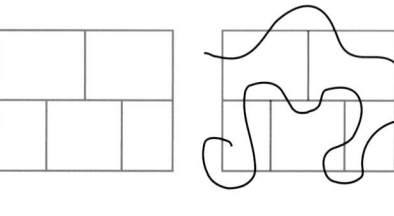

LAW ENFORCEMENT Refer to the diagram at right. A police officer is checking parking meters (represented by the red dots) along the streets. The officer wishes to go down each row of meters exactly once and return to the starting point.

34. Draw a graph that represents this situation.

35. Determine whether the officer can complete the desired route.

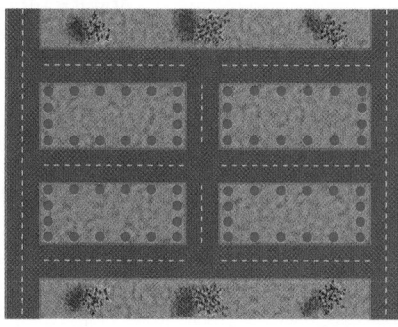

32. Yes, one solution is shown.

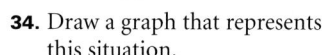
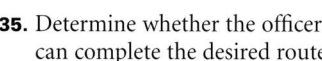
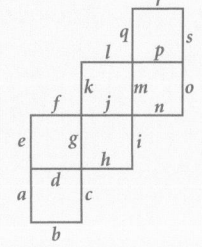

Travel the edges in alphabetical order.

33. Answers may vary. Sample answer: It is not possible. Consider each box and the outside to be a vertex and each segment between two boxes to be an edge. The vertices representing the three larger boxes will have degree 5, so there can be no Euler circuit.

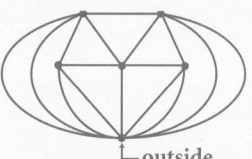

└outside

36. WILDLIFE MANAGEMENT A naturalist observing a pack of wolves comes across a set of trails on the ground, shown at right. Could the trails be traveled by a wolf in one trip, starting at the den and not retracing any of the trails? Why or why not?

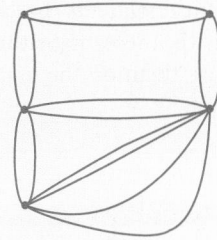

37. FUND-RAISING Jo and Tamara are trying to raise money for a school trip by selling magazines door to door. Jo decides to cover each side of every street separately, while Tamara wants to alternate from one side of the street to the other, covering it in one trip. Refer to the map at right. Draw a graph for Jo's route. Is there an Euler path for Jo's route? Draw a graph for Tamara's route. Is there an Euler path for Tamara's route?

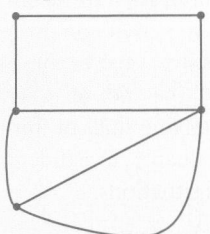

 **Look Back**

Copy each figure and reflect it with respect to the given line.
(LESSON 1.6)

38.

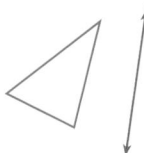

39.

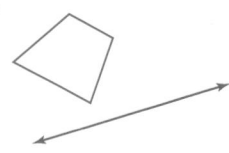

In Exercises 40–44, write each statement as a conditional. Then write its converse. *(LESSON 2.2)*

40. All people who live in California live in the United States.

41. Every square is a rectangle.

42. A square is a parallelogram with four congruent sides and four congruent angles.

43. A dodecagon is a polygon with 12 sides.

44. TRIGONOMETRY For any angle $\theta < 45°$, $\sin \theta < \cos \theta$.

Each solid below has a volume of 100 cubic units. Find *h* or *r* for each figure. Round your answers to the nearest tenth.

45.

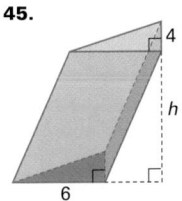

46.

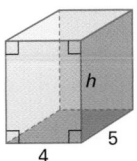

47.

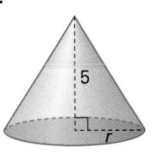

48.

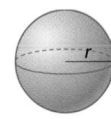

45. 8.3

46. 5

47. 4.4

48. 2.9

38.

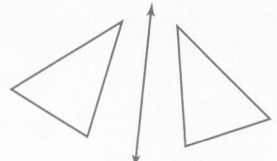

39.

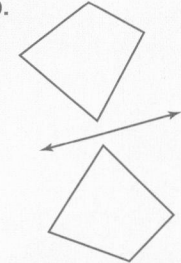

40. If people live in California, then they live in the United States. Converse: If people live in the United States, then they live in California.

---

36. No. The associated graph has 4 odd vertices, so there is no Euler path or circuit.

37. Graph for Jo

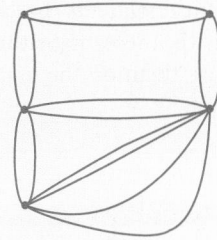

In Jo's graph, every vertex is even, so there is an Euler path.

Graph for Tamara

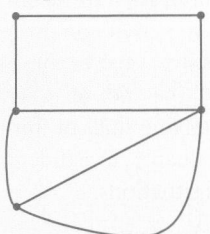

In Tamara's graph, there are two odd vertices, so there is an Euler path.

Student Technology Guide

NAME _____ CLASS _____ DATE _____

Student Technology Guide
11.3 **Graph Theory**

The diagram at the left below represents a group of five people in a network who communicate with one another in some way. For example, *P* and *Q* can communicate directly with each other. Person *S* can communicate directly with *R*, but *R* cannot communicate directly with *S*.

Matrix *A* represents communication lines among the members. An entry of 1 indicates direct communication from one member to another. An 0 entry indicates no direct line of communication between them.

A power of matrix *A* reveals which members can communicate with one another via intermediary members.

To find A^2, follow the example below.

Example: · Enter the entries in matrix *A*. Press MATRIX EDIT
1:[A] ENTER 5 ENTER 5 ENTER 0 ENTER 1 ENTER 1 ENTER 1 ...
0 ENTER 0 ENTER
· Then press MATRIX NAMES 1:[A] x^2 ENTER.
The display at the right shows the result.

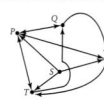

Use a graphics calculator as needed.

1. a. How many communication paths are there from member *Q* to member *T* via one group member? Compare this number with the entry in row 2 and column 5 of A^2. __2; the entry in row 2 and column 5 of A^2 is also 2.__
   b. What do you think A^2 tells you about the network? __A^2 indicates how many communication paths there are from one member to another via one intermediary.__

2. a. Find A^3. (Use ^ 3 rather than x^2 in the key sequence above.) __See above.__
   b. How many communication paths are there from member *R* to member *P* via two group members? (You may enter and exit *P* along the way if you wish.) Compare this number with the entry in row 3 and column 1 of A^3. __4; The entry in row 3 and column 1 of A^3 is also 4.__
   c. What do you think A^3 tells you about the network? __A^3 indicates how many communication paths there are from one member to another via two intermediaries.__

Exercises 49–51 present two algorithms for finding Euler circuits in graphs. Once students become familiar with these algorithms, they can use them on the graphs in the lesson.

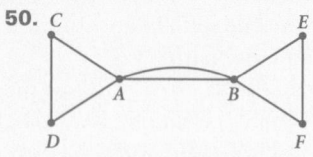
Assessment

Portfolio Activity

The Portfolio Activity can be used as preparation for the Chapter Project or as a separate activity. In the Portfolio Activity on this page, students solve puzzles or problems with the aid of a graph. Problem-solving skills of this type will be valuable to students in numerous situations.

49. Edge $\overline{CE}$ is a bridge.

50.

No edge is a bridge.

51. Answers may vary. Sample answer: *ABCDEFECBGHCH DEHGBFGAFA*

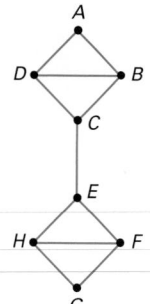

In Exercises 49 and 50, you will learn about *bridges*, which are another feature of graphs. In Exercise 51, you will apply *Fleury's algorithm*, a step-by-step procedure that uses bridges to find an Euler circuit in a graph.

49. An edge in a graph is called a *bridge* if removing it would split the graph into two separate pieces. Identify a bridge in the graph at left.

50. Draw a graph that contains an Euler circuit. Is any edge a bridge? Can a graph that contains an Euler circuit have a bridge? Why or why not?

51. *Fleury's algorithm*, given below, is a procedure that uses bridges to find an Euler circuit in a graph that has no odd vertices.

Fleury's Algorithm

Step a: Verify that the graph contains an Euler circuit. Choose a starting vertex.

Step b: Travel any edge that is not a bridge in the untraveled part of the graph.

Step c: Repeat Step **b** until you have completed an Euler circuit.

Note: In Step **b**, consider only the untraveled part of the graph. If a bridge is removed from this part of the graph, there will be edges that you are unable to get back to.

Use Fleury's algorithm to find an Euler ciruit in the graph at right. List the vertices in the order they are traveled.

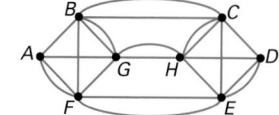

USING GRAPHS TO SOLVE PUZZLES

Suppose that you have three bottles. One holds 8 quarts, one holds 5 quarts, and one holds 3 quarts. The 8-quart bottle is full of water. How can you get exactly 4 quarts of water in both the 5-quart and 8-quart bottles?

Problems like this can be solved with the aid of a graph. The amount of water in the bottles at any given time can be written as an ordered triple. For example, the initial state in the puzzle can be written as (8, 0, 0), meaning 8 quarts in the 8-quart bottle, 0 quarts in the 5-quart bottle, and 0 quarts in the 3-quart bottle. Each state is represented by a vertex in the graph, and two vertices are connected by an edge if it is possible to move from one vertex to the other in one step. The "legal moves" consist of the following: pouring all of the water from one bottle into another bottle, or pouring as much water as possible from one bottle to another bottle.

1. The first level of the graph for this problem is shown at right. Explain how to get from the initial state to each of the states represented by vertices in the graph.

2. What moves are possible from the position (3, 5, 0)? What moves are possible from the position (5, 0, 3)? Complete the next level of the graph.

3. Extend the graph until you find the position (4, 4, 0). Describe how to pour the water from the bottles to get exactly 4 quarts in both the 8-quart and 5-quart bottles.

WORKING ON THE CHAPTER PROJECT

You should now be able to complete Activity 2 of the Chapter Project.

Topology: Twisted Geometry

Why Topology is an exciting topic and serves as the basis for many riddles and puzzles. It also has many important applications, such as robotics.

Transforming a coffee cup into a doughnut

Objectives

- Explore and develop concepts of topology, including knots, Möbius strips, and tori.

- Use theorems of topology to solve problems.

Prepare

QUICK WARM-UP

1. Imagine a line drawn from the North Pole to a point on the equator. Is the line straight? Why or why not? Is it the shortest distance? Explain.

   It is on a "line" because it is on the shortest path between the two points, but it is not straight.

2. Imagine a balloon with a printed message on it. Describe how the message would look if the balloon were deflated.

   Answers will vary. Sample: The message would be smaller and not as "stretched out."

Also on Quiz Transparency 11.4

Knots, Pretzels, Molecules, and DNA

Topology is a branch of mathematics that studies the most basic properties of figures. In topology, it is important to know whether a point is inside, outside, or on the boundary of a figure, but the distance between two points or whether they are on a straight line is not significant.

In topology, figures are deformed by stretching and squashing, as in the coffee cup example above, but cutting or tearing is not allowed. Do you think it is possible to unlink the rings of the "pretzel" at right by deforming it without cutting or tearing it?

APPLICATION
BIOCHEMISTRY

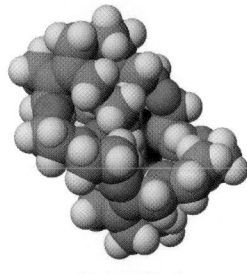

Möbius molecule

The shape at right is known as a **Möbius strip**, a classic topological figure with intriguing properties. (You will explore Möbius strips in Exercises 25–30.) The mathematical analysis of such objects has proven useful in other fields, such as chemistry and biology.

David Walba and his co-workers at the University of Colorado have synthesized a molecule in the shape of a Möbius strip. Such structures can help researchers understand why certain drugs with the same molecular structure can have vastly different effects.

Researchers have also used a branch of topology called *knot theory* to study the structures of complex, tangled structures such as DNA molecules.

Teach

Why Topology is a good introduction to non-Euclidean geometry because students may already be familiar with some of the applications to chemistry, biology, and graphic design.

Alternative Teaching Strategy

HANDS-ON STRATEGIES Help students understand topological equivalence by having them compare several ordinary objects. Using modeling clay, have them do the demonstration on page 727. They can form a coffee cup and then mold the cup into a doughnut similar to the way it was done in the illustration. Have them do similar demonstrations with other shapes, such as a disk or a ball.

Topological Equivalence

Two figures are **topologically equivalent** if one of them can be stretched, shrunk, or otherwise distorted into the other without cutting, tearing, or intersecting itself or compressing a segment or curve to a point. In a plane, all of the shapes below are topologically equivalent.

A shape that is topologically equivalent to a circle is called a **simple closed curve**. A simple closed curve does not intersect itself. The curves below are not topologically equivalent to a circle; thus, they are not simple closed curves.

CRITICAL THINKING Are any of the shapes above topologically equivalent to each other? Explain.

Jordan Curve Theorem

A fundamental theorem in topology was first stated by French mathematician Camille Jordan (1838–1922) in the nineteenth century.

> **Jordan Curve Theorem**
>
> Every simple closed curve in a plane divides the plane into two distinct regions, the inside and the outside. Every curve that connects a point on the inside to a point on the outside must intersect the curve.
>
>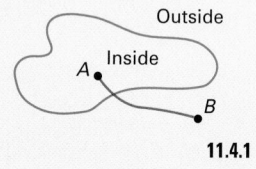
>
> 11.4.1

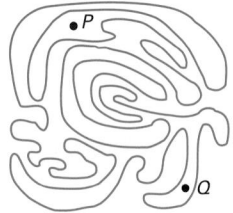

The theorem, which is very difficult to prove mathematically, seems obvious for the curve above. But for some curves it is not so simple. Are points *P* and *Q* on the inside or the outside of the simple closed curve at left? For curves like the snowflake at right, which are called *fractals* (see Lesson 11.7), it can be very difficult to determine the region that contains points near the curve.

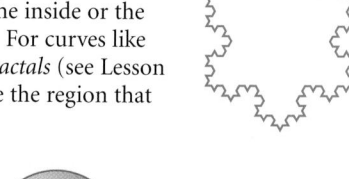

The Jordan Curve Theorem is also true for simple closed curves on a sphere, but not for those on a doughnut-shaped surface called a **torus** (plural, *tori*). Thus, a torus is not topologically equivalent to a sphere.

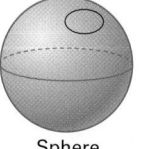 Sphere   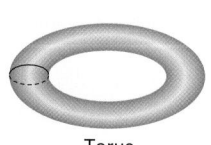 Torus

Invariants

Properties that stay the same no matter how a figure is deformed without breaking are called **invariants**. For example, if the pentagon at right is distorted into some other curve, as shown, the order of the points stays the same. Thus, the order of the points is an invariant.

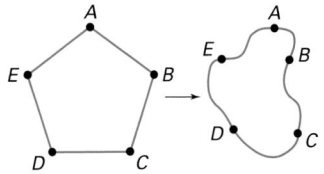

Teaching Tip

String models may help students identify the invariant properties of plane figures.

One important invariant in topology comes from Euler's formula for polyhedra, which you may recall from Lessons 6.3 and 7.3.

Euler's Formula

For any polyhedron with V vertices, E edges, and F faces,

$$V - E + F = 2.$$

11.4.2

The number $V - E + F$ is an invariant of a surface and is called the **Euler characteristic** (or Euler number). Thus, if a polyhedron is deformed into a surface that is topologically equivalent to the polyhedron, the Euler characteristic remains the same.

For example, imagine that a tetrahedron is deformed into a sphere, as shown at right. The number of vertices, edges, and faces are the same, although the edges are no longer straight and the faces are no longer flat. Thus, the Euler characteristic of a sphere is the same as that of the tetrahedron, 2.

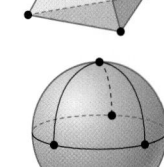

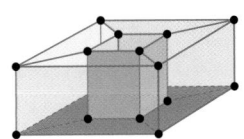

However, consider the figure at left, which is topologically equivalent to a torus. It has 16 vertices, 32 edges, and 16 faces, so its Euler characteristic is $V - E + F = 16 - 32 + 16 = 0$. Thus, the Euler characteristic of a torus is also 0. A simpler figure for finding the Euler characteristic of a torus is shown below, with 1 vertex, 2 edges, and 1 face. Again, $V - E + F = 1 - 2 + 1 = 0$. The edges and vertices can be drawn in any way, as long as each face is topologically equivalent to a disk; that is, it can be flattened out and deformed to a filled-in circle without holes in it.

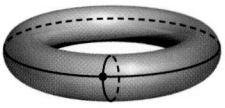  

Finally, considering everything you have learned about topology, it may not surprise you to learn that the "pretzel" shown at the beginning of this lesson can indeed be unlinked.

ASSIGNMENT GUIDE

In Class	1–9
Core	10–30
Core Plus	17–34
Review	35–41
Preview	42–44

✎ Extra Practice can be found beginning on page 818.

Exercises

● Communicate

1. Explain what it means for two shapes to be topologically equivalent.

Are the shapes below topologically equivalent? Explain your reasoning.

2.

3.

☑ internet connect

Activities Online

Go To: **go.hrw.com**
Keyword:
MG1 Topology

4. Explain two methods of proving that a sphere and a torus are not topologically equivalent.

5. Why do you think it might be easier to prove that figures are not topologically equivalent than to prove that figures are topologically equivalent?

● Guided Skills Practice

6. Which of the figures below are simple closed curves? *(TOPOLOGICAL EQUIVALENCE)* a, b, d

a.  **b.**  **c.**  **d.**

7. Which of the figures below are topologically equivalent to ∞? *(TOPOLOGICAL EQUIVALENCE)* b, c

a. 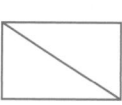 **b.**  **c.**  **d.**

Verify Euler's formula for the polyhedra below. *(EULER'S FORMULA)*

8. cube $8 - 12 + 6 = 2$ **9.** octahedron $6 - 12 + 8 = 2$

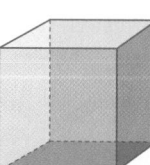

Practice and Apply

For Exercises 10–15, refer to the simple closed curve at right.

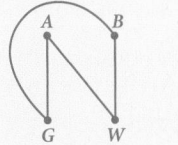

10. Is point P on the inside or the outside of the curve? **inside**

11. Is point Q on the inside or the outside of the curve? **outside**

12. Draw a ray from point P in any direction. How many times does the ray intersect the curve? (Points where the ray is tangent to the curve do not count as intersections.) **Answers may vary from 1 to 5.**

13. Repeat Exercise 12 for several rays in different directions. What can you say about the number of times each ray intersects the curve?

14. Draw several rays from point Q in several different directions. What can you say about the number of times each ray intersects the curve?

15. Complete the following conjecture: A ray from a point on the inside of a curve intersects the curve an __?__ number of times, and a ray from a point on the outside of a curve intersects the curve an __?__ number of times.

odd
even

16. Use your conjecture from Exercise 15 to determine whether the point is on the inside or the outside of the simple closed curve at right.
inside

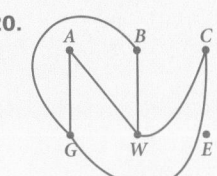

The following is a famous graph theory problem: Can three houses be connected to three utilities (gas, water, and electricity) without any intersecting lines?

17. Draw vertices A, B, and C to represent the three houses and vertices G, W, and E to represent the three utilities.

18. Connect both A and B to both G and W with edges that do not intersect. (Remember that edges do not have to be straight.) The edges should form a simple closed curve.

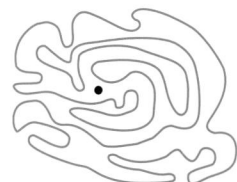

Can the houses be connected to the utilities without any intersecting lines?

19. Use your graph from Exercise 18 to explain why C and E must be both inside or both outside the simple closed curve.

20. If necessary, move vertices C and E outside the simple closed curve. Connect C to G and W with edges that do not intersect any existing edges. Use the Jordan Curve Theorem to show that either A or B cannot be connected to E without any intersecting edges.

21. Repeat Exercise 20 with C and E inside the simple closed curve. Explain why all the houses cannot be connected to all the utilities without any intersecting lines.

Technology

In this exercise set, geometry graphics software can be used to draw figures that are topologically equivalent. Various features of the software can be used to distort shapes.

13. The number of times the ray from P crosses the curve is always odd.

14. The number of times the ray from Q crosses the curve is always even.

17.–18. Sample answer:

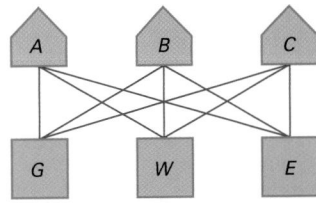

19. If one were inside and one were outside, there would be no way to connect C to E without intersecting the simple closed curve.

20.

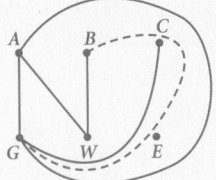

Sample answer: note that the curve $BGCW$ is a simple closed curve with A on the inside and E on the outside. Every curve connecting A to E intersects an already connected utility line. (The other case follows similarly.)

21.

Sample answer: note that B can be connected to G. But then $BGAW$ forms a simple closed curve with C on the inside and E on the outside. Every curve connecting C to E intersects an already connected utility line. (The other cases follow for similar reasons.)

24. Sample answer:

$$6 - 12 + 8 = 2$$

25. The path covers both sides of the Möbius strip.

$20 - 30 + 12 = 2$
The polyhedron has 12 faces; so it is a dodecahedron.

You must traverse the strip twice in order to get back to the starting point.

The result is a loop of paper with four half-twists.

The result is two linked loops of paper. The small loop is a Möbius strip, and the large loop has four half-twists.

The result is one loop with four half-twists.

Suppose that a polyhedron is deformed into a sphere and then all the edges and vertices are moved to the top half of the sphere. The vertices and edges form a graph, which can be drawn in a plane. Each region in the graph is a face, and *the region outside the graph is also a face*.

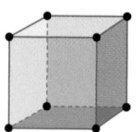 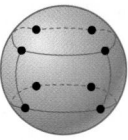 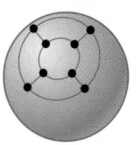 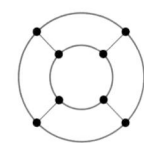

22. For the graph above, count the number of vertices, edges, and faces. Verify that $V - E + F = 2$. **8 – 12 + 6 = 2**

23. Verify that $V - E + F = 2$ for the graph at right. Remember to count the region outside the graph as a face. Based on the number of faces, what polyhedron do you think the graph represents?

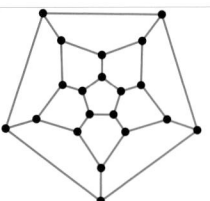

24. Draw a graph of your own and use it to verify Euler's formula. Make your graph so that no two edges intersect except at a vertex.

In Exercises 25–30, you will explore the properties of the Möbius strip. You will need paper, tape, scissors, and a pencil. Cut a strip of paper about 1 in. wide and 11 in. long. Bring the ends of the strip together to form a loop, but flip one end over to form a half-twist before joining the ends. Tape both sides of the seam securely.

25. Starting at the tape, use your pencil to draw the path of an imaginary bug crawling along the center of your strip. Continue until you reach the starting point. What do you notice?

26. Create another Möbius strip, and draw a path about one-third of the way in from the edge of the strip, continuing until you reach the starting point. Describe your results.

27. Use your scissors to cut along the path that you drew on the Möbius strip in Exercise 25. Describe your results.

28. Use your scissors to cut along the path that you drew on the Möbius strip in Exercise 26. Describe your results.

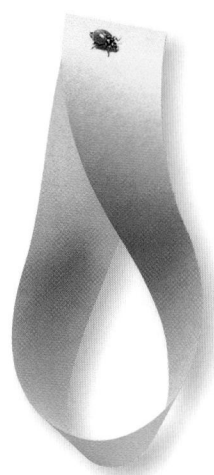

Construct a double Möbius strip as follows: Hold two strips of paper together and put a half-twist in one end. Tape the ends on both sides to form two nested Möbius strips.

29. Run your finger along the inside of the two strips, keeping them nested. What do you notice? Do the strips seem to be separate? **yes**

30. Pull the strips apart. What do you notice? As a challenge, try to nest the strips back together again.

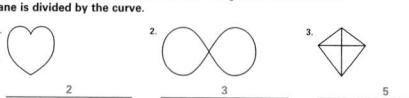
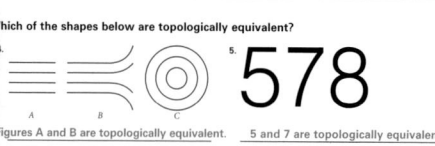
578

A Möbius strip is a surface with only one side. Another one-sided surface, which is theoretically a four-dimensional object, is called a _Klein Bottle_. A three-dimensional representation of a Klein Bottle is shown at right.

Construct a paper model of a Klein bottle as follows: Cut a 4-in. × 11-in. rectangle of paper and fold it lengthwise down the center. Tape the long edges together to form a flattened tube.

Hold the tube vertically, and cut a horizontal slit through the side of the tube nearest to you, about one-fourth of the way from the top.

Insert the bottom end of the tube into the slit, and align the two ends of the tube. Tape the ends together to produce one "hole" from above.

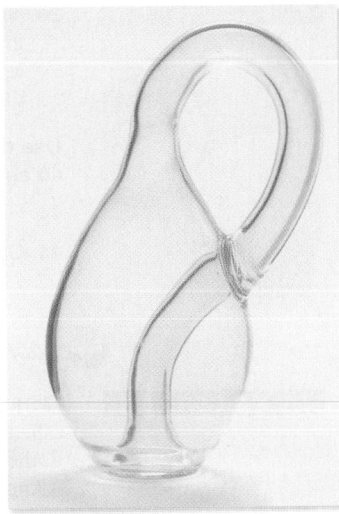

A Klein bottle in three-dimensional space must intersect itself.

31. Describe your paper model. Can any two points on the surface be connected by a path along the surface? **yes**

32. Lay your model flat and cut it in half lengthwise. Without cutting or tearing your paper, unfold the two halves of your bottle as much as possible. Describe the resulting shapes. **two Möbius strips**

33. It is possible to cut the paper model into one Möbius strip. Can you discover how it is done?

34. MANUFACTURING Sometimes long conveyor belts are shaped like Möbius strips. What do you think is the advantage of this shape?

Sample answer: The belt's life will be longer, since both sides of the belt get used.

 Look Back

Algebra

35. Find *x* in the diagram at right. **3**
(LESSON 2.6)

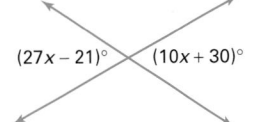
$(27x - 21)°$ $(10x + 30)°$

For Exercises 36–38, refer to the diagram at right. Assume that line $\ell \parallel$ line *m*.
(LESSONS 3.3 AND 3.5)

36. Name two pairs of congruent angles.

37. Name three pairs of supplementary angles.

38. Prove that $m\angle 2 + m\angle 5 + m\angle 6 = 180°$ without using the Triangle Sum Theorem.

38.

Statements	Reasons
1. $m\angle 1 + m\angle 2 + m\angle 3 = 180°$	1. Angle Addition Postulate
2. $m\angle 5 = m\angle 1$ $m\angle 6 = m\angle 3$	2. Alternate Interior Angles Theorem
3. $m\angle 5 + m\angle 2 + m\angle 6 = 180°$	3. Substitution

---

Math
CONNECTION

ALGEBRA In Exercise 35, students use the Vertical Angles Theorem from Lesson 2.6 to set up an equation and solve for *x*.

33. Cut the bottle "diagonally." Start at any point on the surface and return to it without crossing itself. By avoiding the joined edges entirely, this produces a surface without dividing it into more than one piece. Then cut around the self-intersection.

36. Sample answer:
$\angle 1 \cong \angle 5$; $\angle 3 \cong \angle 6$

37. Sample answer:
$\angle 1$ and $\angle 4$, $\angle 3$ and $\angle 7$, $\angle 6$ and $\angle 7$

Student Technology Guide

NAME _____ CLASS _____ DATE _____

Student Technology Guide
11.4 Topology: Twisted Geometry, page 1

The diagram at right shows a horizontal line segment. Vertical arrows of differing lengths show points that determine a curved, or bent, line segment. Using a function, you can deform a line segment to get a topologically equivalent bent line segment.

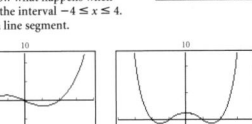

The graphics calculator displays below show what happens when some simple functions of *x* are applied to the interval $-4 \le x \le 4$. Each curve is topologically equivalent to a line segment.

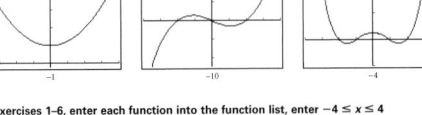

In Exercises 1–6, enter each function into the function list, enter $-4 \le x \le 4$ into the window settings, and then graph the function. Describe how the function bends the line segment defined by $-4 \le x \le 4$.

1. $y = 2x + 1$
   The line segment stays the same.
   It does not bend.

2. $y = |x| + 1$ (Press **MATH**, select **NUM 1:abs(**)
   The line segment becomes
   V shaped.

3. $y = 2x^2 - 1$
   The line segment becomes U shaped.

4. $y = 0.25(x + 2)(x)(x + 2)$
   The line segment becomes S shaped.

5. $y = \sqrt{16 - x^2}$
   The line segment becomes a semicircle.

6. $y = 0.01(x + 4)(x + 2)(x - 2)(x - 4)$
   The line segment becomes W shaped.

When you choose a function whose domain is an interval and whose graph consists of more than one piece, the new figure is not topologically equivalent to the original.

7. Let $-4 \le x \le 4$ and $y = \frac{1}{(x + 2)(x - 2)}$. Graph this function. What effect does the function have on the interval?   The interval splits into three branches.

Exercises 42–44 are an introduction to knot theory. Students are asked to identify which knots are equivalent. Knot theory has numerous applications in various fields, including biology and chemistry.

43. Answer may vary. Sample answer:

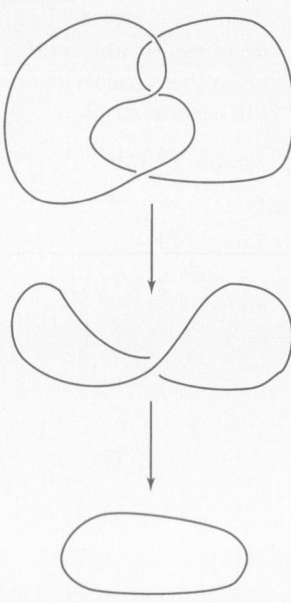

APPLICATION

39. FOOD Suppose that you have a recipe for fruit salad that calls for one large grapefruit and that you wish to substitute an equal amount of oranges. If a large grapefruit is about 6 in. in diameter and an orange is about 4 in. in diameter, about how many oranges will you need? $3\frac{3}{8}$ **oranges**

Use the law of cosines and △*ABC* for Exercises 40 and 41. *(LESSON 10.5)*

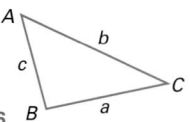

40. Given $b = 25$, $c = 20$, and m∠$A = 55°$, find a. **21.2**

41. Given $a = 104$, $c = 47$, and m∠$B = 92°$, find b. **115.6**

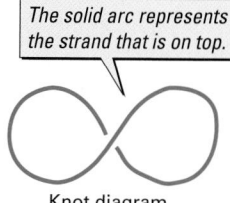

Look Beyond

CONNECTION

KNOT THEORY In the branch of mathematics called *knot theory*, a *knot* is a continuous loop; that is, the ends are joined so that it cannot be untied. A knot can be represented by a knot diagram, which indicates crossings with a solid curve and a broken curve. Two knots are *equivalent* if one can be deformed into the other in three-dimensional space without cutting.

> The solid arc represents the strand that is on top.

Knot diagram

One way of determining whether two knots are equivalent is by using the *Reidemeister moves.* Kurt Reidemeister (1893–1971) proved that if two knots are equivalent, the diagram of one can be transformed into the diagram of the other by a combination of the three moves shown below.

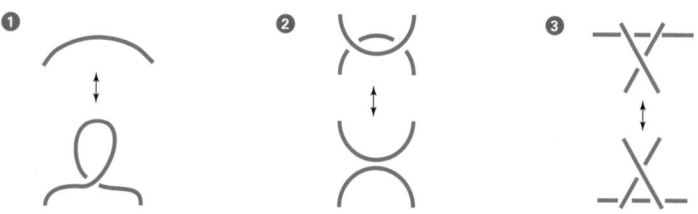

42. The following is a sequence of knot diagrams. Each diagram results from performing a Reidemeister move on the preceding diagram, showing that the knots are equivalent. For each diagram, identify which of the Reidemeister moves was used.

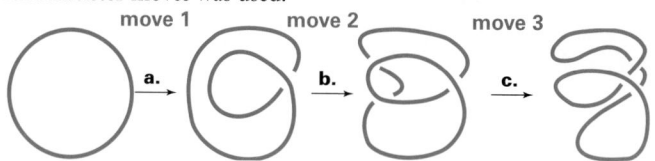

43. Draw a knot diagram and transform it by a sequence of at least two different Reidemeister moves.

44. Use a strip of paper about 1 in. wide and 17 in. long to form a loop. Before joining the ends, put three half-twists in the strip. Then tape the ends securely. If you cut the strip down the middle, you should get a knotted loop. Flatten your shape down as much as possible, and arrange it as simply as you can. Draw a knot diagram for your shape, ignoring twists in the paper. This knot is called a *trefoil knot.*

44.

Euclid Unparalleled

Objectives

- Explore and develop general notions for spherical and hyperbolic geometries.

- Develop informal proofs and solve problems by using concepts of non-Euclidean geometries.

Why *Many revolutionary concepts in math begin by questioning traditional assumptions. For example, a number of non-Euclidean geometries reject Euclid's assumption about parallel lines, with interesting and useful results.*

Imagine that you live on this two-dimensional surface. As you move outward from the center, everything gets smaller—including yourself. Would you ever reach the edge of your "universe"? This woodcut by M. C. Escher, known as Circle Limit 3, *is the artist's conception of a non-Euclidean geometry.*

Non-Euclidean Geometries

Euclid's geometry is based on five assumptions or postulates:

1. A line may be drawn between any two points.

2. Any segment may be extended indefinitely.

3. A circle may be drawn with any given point as the center and with any given radius.

4. All right angles are equal.

5. If two lines are met by another line and if the sum of the internal angles on one side is less than the sum of two right angles, then the two lines will meet.

If $m\angle 1 + m\angle 2 < 180°$, the lines will intersect.

Alternative Teaching Strategy

COOPERATIVE LEARNING Have cooperative groups measure the angles of a triangle drawn on a globe. Have groups compare their results and formulate a statement about a "spherical triangle." Ask each group to report their results to the class.

Prepare

QUICK WARM-UP

1. Consider the measure of the sum of the angles of a triangle drawn on the surface of a sphere. Is the measure greater than or less than 180°?
   greater than

2. Can you distort a triangle so that the sum of its angles is less than 180°?
   Yes; on a saddle, the sides of a triangle are concave and the sum of its angles is less than 180°.

Also on Quiz Transparency 11.5

Teach

Why Postulates and theorems in geometry are principles whose value hold only for the branch of geometry to which they belong. Raising students' awareness that questioning or denying a postulate can lead to an entirely different branch of geometry is essential to understanding how different geometries are structured.

Teaching Tip

Have students list theorems and postulates that may need to be changed if the Parallel Postulate is not true.

Because the fifth postulate in the list seemed less obvious than the others, many mathematicians wished to prove it in terms of the other four. No one ever succeeded in doing so, but along the way, a number of discoveries were made. In particular, several statements were found to be logically equivalent to the fifth postulate. Two statements are said to be **logically equivalent** if each can be derived from the other. Some examples of statements that are logically equivalent to the fifth postulate are as follows:

- If a line intersects one of two parallel lines, it will intersect the other.
- Lines that are parallel to the same line are parallel to each other.
- Two lines that intersect one another cannot be parallel to the same line.
- In a plane, if line ℓ and point P not on ℓ are given, then there exists one and only one line through P that is parallel to ℓ.

The last statement is perhaps the most useful of the four and is the version that many mathematicians refer to as the Parallel Postulate. (See Lesson 3.5, page 170.)

For years, mathematicians tried to prove the Parallel Postulate, until it was finally shown that it was impossible. Many even wondered whether the Parallel Postulate is in fact true in the real world.

If the Parallel Postulate is rejected, then any theorems that depend on it must be questioned. One such theorem is the Triangle Sum Theorem. (Recall that in the proof of the theorem, you must construct a line through one vertex that is parallel to the opposite base.)

The great mathematician Karl Friedrich Gauss (1777–1855) went so far as to measure the angles of a triangle formed by the points of three different mountain tops about 50 miles apart to see if the measures added up to 180°. (Within the limits of the accuracy of his measurements, they did.)

Does $m\angle 1 + m\angle 2 + m\angle 3 = 180°$?

Some mathematicians adopted a different attitude. They found that they could develop entirely new systems of geometry without using the Parallel Postulate. Systems in which the Parallel Postulate does not hold are examples of **non-Euclidean geometries**.

Euclidean geometry is based on figures in a plane. The figures in the non-Euclidean geometries that you will study here are in curved surfaces. Thus, the concept of a straight line no longer applies. In these geometries, a line will be defined as the shortest path between two points along the surface.

Interdisciplinary Connection

PHYSICS Outer space has often been described as non-Euclidean because if may fit a hyperbolic geometry model. Have students research the paths that may be followed by a probe traveling through the solar system.

Spherical Geometry

Geometry on a sphere such as Earth's surface, or **spherical geometry**, is an example of a type of non-Euclidean geometry known as Riemannian geometry, after its discoverer, G.F.B. Riemann (1826–1866).

In spherical geometry, a line is defined as a **great circle** of the sphere; that is, a circle that divides the sphere into two equal halves. The shortest distance between two points in a sphere is always a path along a great circle. In this geometry, as in any Riemannian geometry, there are no parallel lines at all because all great circles intersect.

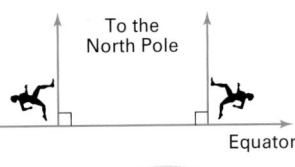

All great circles intersect.

Imagine two superhuman runners who start running at the equator. Their paths form right angles with the equator. What happens to their paths as they approach the North Pole? How does this differ from the result you would expect in a plane?

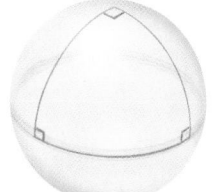

Theorems that depend on the Parallel Postulate for their proof may actually be false in spherical geometry. On a sphere, for example, the sum of the measures of the angles of a triangle is always greater than 180°, as in the spherical triangle at right.

CRITICAL THINKING Do Euclid's first four postulates seem to be true on the surface of a sphere? If so, should all the theorems that follow from them be true on a sphere?

In this triangle, the sum of the measures of the angles is 270°.

Hyperbolic Geometry on a "Saddle"

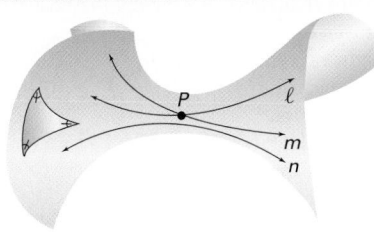

Lines ℓ and m are parallel to line n but not to each other.

Imagine that you are standing on the surface of a sphere. In every direction, the surface curves away from you. On the surface of a saddle, the surface curves away from you in some directions and toward you in other directions. On such a surface, there is more than one line through a point that is parallel to a given line. In fact, there are infinitely many.

The geometry of a saddle is an example of **hyperbolic geometry**, or Lobachevskian geometry. Nikolai Lobachevsky (1773–1856) was one of two mathematicians who discovered this type of geometry independently, the other being János Bolyai (1802–1860).

In hyperbolic geometry, just as in the other geometries you have studied, a line is defined as the shortest path between two points. In the illustration, lines ℓ and m pass through point P, and both are parallel to n (because they will never intersect line n).

Once again, the Triangle Sum Theorem does not hold. On a saddle, the sum of the angles in a triangle is always less than 180°.

Enrichment

M. C. Escher used hyperbolic models in some of his graphic work. Have students find examples of Escher's hyperbolic graphics.

CRITICAL THINKING
Yes; theorems that depend on only the first four postulates would also be true for a sphere.

Cooperative Learning

Introduce the idea of denying the Parallel Postulate by asking students to consider lines on a sphere. Then divide the class into small groups and assign one of the three types of geometries to each. Have the groups read the text and develop a class presentation about their particular geometry.

Poincaré's Model of Hyperbolic Geometry

The "universe" in the Escher woodcut on page 729 is a representation of a three-dimensional universe imagined by the French mathematician Henri Poincaré (1854–1912). Poincaré, whose many interests included physics and thermodynamics, imagined physical reasons for the geometric properties of his imagined universe.

In the model of Poincaré's universe below, the surface is not curved, but the measurement of size and distance is defined in a different way from Euclidean geometry so that it represents a curved surface.

Poincaré's universe can be represented by a circle. The temperature is greatest at the center of the circle and drops to absolute zero at the edges. According to Poincaré's rules for his universe, no one would be aware of temperature changes.

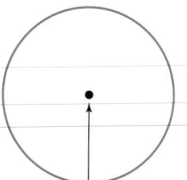

Temperature and size are greatest at the center.

In Poincaré's universe, the size of an object is proportional to its temperature. An object would grow smaller as it moved away from the center.

Because everything, including rulers, would shrink in size, there would be no way to detect the change. In fact, distance measures would keep shrinking, so you would never get any closer to the edge of the universe, no matter how long you traveled.

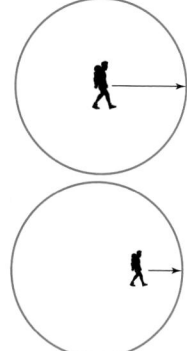

Everything shrinks as you move toward the edge of Poincaré's universe.

The arcs in the circle represent the paths that rays of light would travel in Poincaré's universe. These are defined as lines in this system. Light rays that stay close to the edge of the universe appear more curved than those that pass close to the center. Lines through the center are diameters of the circle. Another feature of these lines is that they are **orthogonal** to the circle; that is, they form right angles with the circle at the intersections.

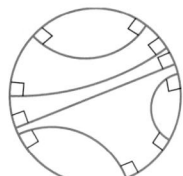

Orthogonal arcs

In Poincaré's system, a line segment connecting two points is a part of the arc that is orthogonal to the circle. This is the shortest distance between the two points. If you wanted to walk from *A* to *B*, the curved line would be shorter than the straight line connecting them because your steps would get larger as you approached the center of the circle. You would cover more distance with each step as you moved toward the center than you would by walking directly to the point.

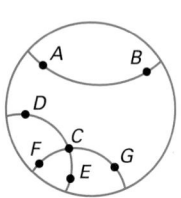

Inclusion Strategies

ENGLISH LANGUAGE DEVELOPMENT Students who may have difficulty interpreting mathematics from textbooks should read this lesson slowly and carefully and redraw the diagrams if necessary. Students should be encouraged to explain, both verbally and in writing, each diagram in terms of the model it represents.

Notice also on the previous page that more than one line can be drawn through a point that is parallel to a given line. For example, $\overleftrightarrow{DE}$ and $\overleftrightarrow{FG}$ pass through the point C, and both are parallel to $\overleftrightarrow{AB}$. Thus, the geometry of Poincaré's universe is an example of a hyperbolic, or Lobachevskian, geometry.

Finally, notice that the sum of the measures of the angles in a triangle in Poincaré's universe is always less than 180°, just as it is on the surface of a saddle.

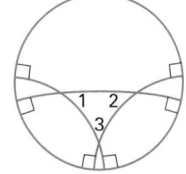

$m\angle 1 + m\angle 2 + m\angle 3 < 180°$

Years Later—Applications

In Einstein's general theory of relativity, space is non-Euclidean. In fact, due to the influence of gravity or—equivalently—acceleration, space is curved. The non-Euclidean geometries discovered years earlier proved to be both an inspiration and a useful tool to Einstein in formulating his fundamental laws of the universe.

Albert Einstein, 1879–1955

Exercises

Communicate

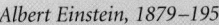

1. Which postulate of Euclid was questioned by mathematicians? State it in your own words.

2. What are the alternatives to the Parallel Postulate that have yielded two non-Euclidean geometries? Complete the following statements and explain your reasoning:

   a. In a sphere, if line ℓ and point P not on ℓ are given, then there exist(s) _____?_____ line(s) through P parallel to ℓ.

   b. In a hyperbolic surface, if line ℓ and point P not on ℓ are given, then there exist(s) _____?_____ line(s) through P parallel to ℓ.

3. In spherical geometry, how does a line extend indefinitely?

4. In Poincaré's universe, how does a line extend indefinitely?

Assess

Selected Answers
Exercises 5–12, 15–47 odd

ASSIGNMENT GUIDE

In Class	1–13
Core	14–27
Core Plus	28–40
Review	41–47
Preview	48–50

✐ Extra Practice can be found beginning on page 818.

Technology

You may want students to draw spheres with three-dimensional geometry graphics software. Have them reproduce the spheres in Exercises 14–18.

15. Sample answer:

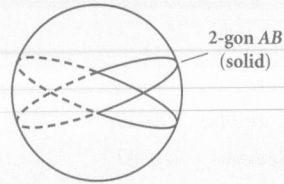

2-gon *AB*
(solid)

The vertices must be opposite each other on the sphere.

16. The sum of the measures of the angles of a 2-gon is greater than 0° and less than 360° since each angle will measure 0° and 180°.

17. Sample answer:

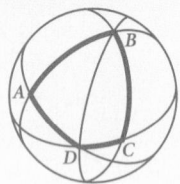

The diagonal is $\overline{BD}$.

18. The sum of the angles of the quadrilateral is greater than 360° because the 4-gon can be divided into two triangles by the diagonal. In spherical geometry the sum of the angles in a triangle is greater than 180°, so the sum of the angles of two triangles would be greater than 360°.

If *A*, *B*, and *C* are angles of a triangle, $m\angle A + m\angle B + m\angle C > 180°$ in spherical geometry and $m\angle A + m\angle B + m\angle C < 180°$ in hyperbolic geometry.

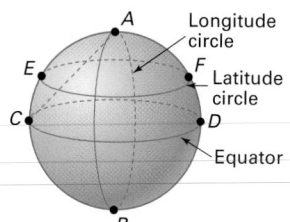

A   Longitude circle
E   F   Latitude circle
C   D
Equator
B

Guided Skills Practice

5. What can you say about the sum of the measures of the angles of a triangle in spherical geometry? in hyperbolic geometry? Express your answers as inequalities. *(SPHERICAL GEOMETRY AND HYPERBOLIC GEOMETRY)*

Refer to the diagram of a sphere at right. Determine whether each figure below is a line in spherical geometry. *(SPHERICAL GEOMETRY)*

6. $\overleftrightarrow{AB}$ Yes   **7.** $\overleftrightarrow{AC}$ No

8. $\overleftrightarrow{EF}$ No   **9.** $\overleftrightarrow{CD}$ Yes

Refer to the diagram of Poincaré's model at right. Determine whether each figure below is a line in hyperbolic geometry. *(HYPERBOLIC GEOMETRY)*

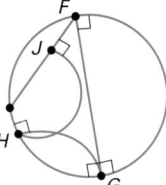

10. $\overleftrightarrow{FG}$ Yes   **11.** $\overleftrightarrow{FJ}$ No

12. $\overleftrightarrow{GH}$ Yes   **13.** $\overleftrightarrow{HJ}$ No

Practice and Apply

14. Suppose that you cut an orange in half around its "equator" and then cut each half twice at right angles through the poles. You would have divided the orange peel into 8 congruent triangles. If you were to add the measures of the angles of one of these triangles, what would be the result?
270°

15. Draw a 2-gon, or a polygon with two edges, on a sphere. (Note: The edges of a polygon must be segments, as they are in a plane.) What can you say about the vertices of the 2-gon?

16. In spherical geometry, what can you say about the sum of the measures of the angles of a 2-gon? Express your answer as an inequality.

17. Draw a quadrilateral on a sphere. Include one diagonal in your drawing.

18. What can you say about the sum of the measures of the angles of the quadrilateral from Exercise 17? How do you know? Express your answer as an inequality.

In Exercises 19–27, you will explore constructions in Poincaré's system.

Part I: Constructing a line through a given point

19. Construct ⊙P and choose a point, A, on the circle. To construct a line through A in Poincaré's system, first construct line ℓ tangent to ⊙P at A. Choose a point, C, on line ℓ. Place your compass point at C and draw a circle through A. The arc of the circle that lies inside ⊙P is a line through A in Poincaré's system.

20. Repeat Exercise 19, placing C at several different locations on line ℓ. What happens as C gets farther and farther away from A? What would happen if C were infinitely far from A?

21. Explain why the center of an arc that is orthogonal to a circle at a point must lie on the line tangent to the circle at that point.

22. Construct ⊙Q and choose a point, B, inside the circle. To construct a line through B in Poincaré's system, first choose a point, E, on ⊙Q. Construct line m tangent to ⊙Q at E, and construct line n, the perpendicular bisector of $\overline{BE}$. Place your compass point at the intersection of lines m and n and draw a circle through B. The arc of the circle that lies inside ⊙Q is a line through B in Poincaré's system.

23. Repeat Exercise 22, placing E at several different locations on ⊙Q. Explain how the line varies as the location of E changes.

Part II: Constructing a line through two given points

24. Construct ⊙P and choose two points, A and B, on the circle that are not endpoints of a diameter. To construct a line through A and B in Poincaré's system, construct line ℓ tangent to ⊙P at A and line m tangent to ⊙P at B. Place your compass point at the intersection of lines ℓ and m, and draw a circle through A and B. The arc of the circle that lies inside ⊙P is a line through A and B in Poincaré's system.

25. Why would the construction in Exercise 24 fail if A and B were the endpoints of a diameter? What would be the line through A and B in this case?

26. Construct ⊙Q and choose two points, C and D, inside the circle. To construct a line through C and D in Poincaré's system, first draw $\overleftrightarrow{CQ}$. Construct line n perpendicular to $\overleftrightarrow{CQ}$ at C. Line n intersects the circle at two points. Construct lines p and q tangent to ⊙Q at these points, and label as C' the intersection of p and q. Construct a circle that passes through points C, D, and C'. The arc of the circle that lies inside ⊙Q is a line through C and D in Poincaré's system.

Henri Poincaré (1854–1912)

27. Draw a line ℓ and a point A not on line ℓ in Poincaré's system. How many lines can you draw through point A that do not intersect line ℓ? How is this illustration related to the Parallel Postulate?

23. As E gets closer to the radial line passing through B, the radii of the arcs get very large.

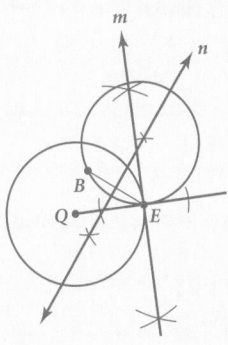

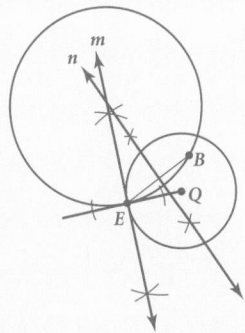

19. Sample answer:

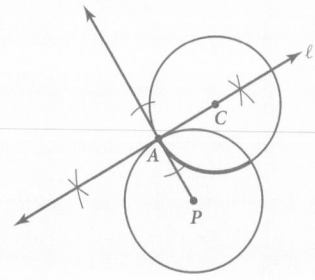

20. As the distance from A increases, the radius of the arc increases. The other endpoint of the arc on the circle moves farther from point A. If C were infinitely far from A we would obtain the diameter $\overline{AB}$.

21. If an arc is orthogonal to a circle at a point, then the tangent of the arc is perpendicular to the tangent of the circle at that point. By the Tangent Theorem, the tangent of the arc is also perpendicular to a radius of the circle drawn to the point of tangency. The tangent of the circle at the given point must contain the radius of the arc, and so it must contain the center of the arc.

22.

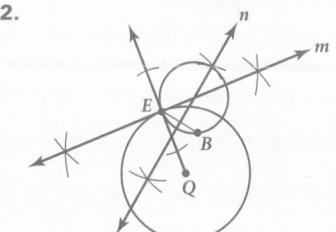

28.

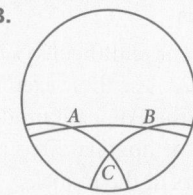

29. The sum of the measures of the angles in the triangle is less than 180°. Look at the drawing or measure the angles with a protractor to get a fairly accurate reading. In Euclidean geometry, the sum of the angle measures in a triangle is 180°, while in spherical geometry it would be greater than 180°.

30.

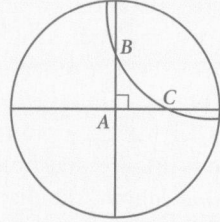

The sum of the acute angles in Poincaré's model is less than 90°, since the angle sum of the triangle is less than 180°.

In Exercises 28–33, you will explore polygons in Poincaré's system.

28. Draw three lines that intersect to form a triangle in Poincaré's system. Measure the angles of the triangle.

29. What is the sum of the angles of the triangle from Exercise 28? How does this compare with the sum of the angles of a triangle in Euclidean geometry and in spherical geometry?

30. Draw a right triangle in Poincaré's system. (Hint: The diameter of the circle is a line in Poincaré's system.) What can you say about the acute angles of a right triangle in this geometry? Express your answer as an inequality.

31. Draw four lines that intersect to form a quadrilateral in Poincaré's system. Measure the angles of the quadrilateral.

32. What is the sum of the angles of the quadrilateral from Exercise 31?

33. Write a conjecture about the sum of the angles in a polygon with n sides in Poincaré's system. (Your conjecture should involve an inequality.)

You can use equilateral triangles to build models of surfaces in spherical, Euclidean, and hyperbolic geometries. Construct an equilateral triangle with a side length of about 3 cm. Copy the triangle carefully onto heavy paper or cardboard and cut it out. Make at least 20 copies of the triangle.

34. Tape 5 triangles together at a single vertex, as shown. This is a model of part of a surface in spherical geometry. Describe your model.

35. Tape 6 triangles together at a single vertex, as shown. This is a model of part of a surface in Euclidean geometry. Describe your model.

36. Tape 7 triangles together at a single vertex, as shown. This is a model of part of a surface in hyperbolic geometry. Describe your model.

37. How are the models you made in Exercises 34–36 alike? How are they different? Explain how the surfaces model the different geometries.

38. What happens if you extend the model in Exercise 34 to create a surface with 5 equilateral triangles at every vertex? (If you are not sure, make more triangles and experiment with them.) Could the model for spherical geometry extend to infinitely many triangles? Why or why not?

39. What happens if you extend the model in Exercise 35 to create a surface with 6 equilateral triangles at every vertex? Could the model for Euclidean geometry extend to infinitely many triangles? Why or why not?

40. What happens if you extend the model in Exercise 36 to create a surface with 7 equilateral triangles at every vertex? Model a surface in hyberbolic geometry with at least 20 triangles. Could this model extend to infinitely many triangles? Why or why not?

---

31. Sample answer:

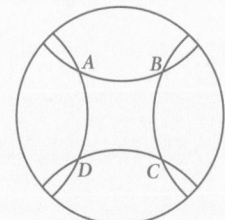

Quadrilateral *ABCD*

$$m\angle A = m\angle B = m\angle C = m\angle D \approx 40°$$

32. The sum of the measures of the angles in the quadrilateral is less than 360° because there are two triangles in a quadrilateral (divided by a diagonal). In Exercise 29, it was shown that the sum of the angle measures of a triangle is less than 180°, so if you added the sum of the angles of the two triangles, it would be less than 360°.

33. Euclidean geometry: Sum of the angles of an n-gon $= (n-2)180°$
Hyperbolic geometry: Sum of the angles of an n-gon $< (n-2)180°$

In Exercises 41–43, you will construct a regular octagon. *(LESSONS 4.7 AND 5.4)*

41. Construct square *ABCD*. Draw diagonals $\overline{AC}$ and $\overline{BD}$ and label their intersection *E*. Place your compass point at *A* and draw an arc through *E* intersecting $\overline{AB}$ and $\overline{AD}$.

42. Place your compass point at *B* and draw an arc through *E* in a similar fashion. Repeat at *C* and *D*. Connect the eight points where the arcs intersect the square to form an octagon.

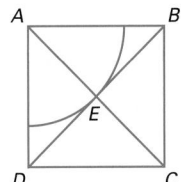

43. Prove that the octagon you constructed in Exercises 41 and 42 is regular.

44. Draw a floor plan for your house or apartment. Draw a graph that represents the circulation of traffic in it. Remember that the rooms are the vertices and the doors are the edges of the graph. *(LESSON 11.3)*

45. Does your graph from Exercise 44 contain an Euler path? an Euler circuit? Why or why not? *(LESSON 11.3)*

In Exercises 46 and 47, you will explore topological properties of letters of the alphabet.

A B C D E F G H I J K L M N O P Q R S T U V W X Y Z

C G I J L M N S U V W Z   **46.** Identify all letters of the alphabet (drawn as shown above) that are topologically equivalent to the letter Z. *(LESSON 11.4)*

47. Are the words SIDE and CLOT topologically equivalent? Are LAST and COZY? Explain your reasoning. *(LESSON 11.4)*
SIDE and CLOT are topologically equivalent. LAST and COZY are not topologically equivalent since A is not equivalent to O.

 Look Beyond

The area of triangle *ABC* on a sphere with a radius of *r* is given by the formula

$$A = \pi r^2 \left(\frac{m\angle A + m\angle B + m\angle C}{180°} - 1 \right).$$

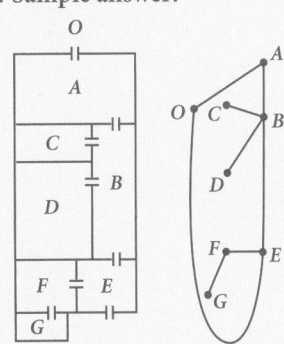

$\frac{1}{8}$ of the surface area
$$A = \pi r^2 \left(\frac{90° + 90° + 90°}{180°} - 1 \right)$$
$$= \pi r^2 \left(\frac{1}{2} \right)$$
$$= \frac{1}{8}(4\pi r^2)$$
which is $\frac{1}{8}$ of the surface area of the sphere.

48. Verify the formula for the triangle on the sphere shown at right. (Hint: What fraction of the surface area of the sphere is covered by the triangle?)

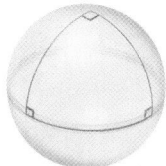

49. In spherical geometry, the sum of the angles of a triangle is always greater than 180°. What can you say about the area of a triangle in which the sum of the angles is very close to 180°? **The area is close to 0.**

50. The area of a triangle on a sphere can not be greater than the surface area of the sphere. Use this fact to complete the following inequality involving the sum of the angles of a triangle on a sphere:
$$180° < m\angle A + m\angle B + m\angle C < \underline{\ ?\ } \quad 900°$$

44. Sample answer:

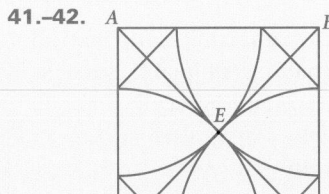

45. Answers may vary. Sample answer: the graph has four odd vertices, so no Euler path or circuit exists.

In Exercises 48–50, students find the area of a triangle on a sphere. Have students find and compare the area of a triangle drawn on a sphere with the area of a triangle drawn on a plane.

41.–42.

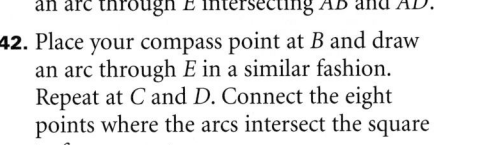

43. Consider the point where the arc made by placing the compass point at *B* and drawing the arc through *E* intersects $\overline{AB}$. Label this point *P*. Similarly, label the point where the arc through *E* drawn from *A* intersects $\overline{AB}$ as *Q*. Also label the point where the arc through *E* drawn from *D* intersects $\overline{AD}$ as *R*. To show the figure is equilateral, it is enough to prove $\overline{RP} \cong \overline{PQ}$, since all other sides can be proven congruent similarly.

Since the diagonals of square *ABCD* are perpendicular bisectors, △*AEB* is an isosceles right triangle. For simplicity, assume *AE* = *EB* = 1. Then *AB* = $\sqrt{2}$. Now *AP* = *QB* = $\sqrt{2} - 1$ since $\overline{AQ}$ and $\overline{PB}$ are congruent to $\overline{AE}$ and $\overline{BE}$ by construction. So
PQ
$= \sqrt{2} - (AP + QB)$
$= \sqrt{2} - (\sqrt{2} - 1) - (\sqrt{2} - 1)$
$= \sqrt{2} - \sqrt{2} + 1 - \sqrt{2} + 1$
$= 2 - \sqrt{2}$

Since *AR* = *AP* = $\sqrt{2} - 1$ and △*ARP* is an isosceles right triangle,
$RP = (\sqrt{2} - 1)\sqrt{2}$
$\quad = (\sqrt{2})^2 - \sqrt{2}$
$\quad = 2 - \sqrt{2}$

Thus $\overline{RP} \cong \overline{PQ}$, and so the octagon is equilateral. A simple argument shows that the interior angles measure 135°.

QUICK *WARM-UP*

> Describe the mathematical qualities of plants such as ferns and broccoli. **Answers will vary. Students with previous experience with fractals will be more likely to mention fractal qualities.**

Also on Quiz Transparency 11.6

Teach

Why The study of fractals introduces students to a world of mathematics that is enhanced by computers. Point out that the photographic quality of fractal images is the product of a computer, not of a camera.

Fractal Geometry

Objectives

● Discover the basic properties of fractals, including self-similarity and iteration.

● Build fractal designs by using iterative steps.

Why *The self-similarity of fractals enables a computer programmer to write a relatively short program for drawing such structures.*

A fractal, such as the computer-generated fern above, is a self-similar structure. Notice that each subdivision of the leaves of the fern has basically the same structure as the leaves themselves—all the way down to the curving tips.

Self-Similarity in Fractals

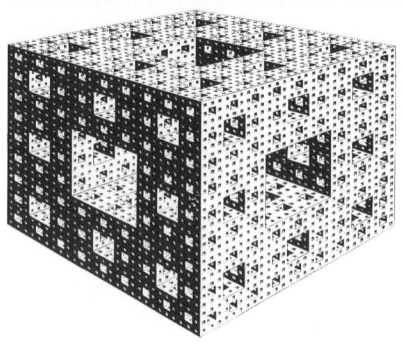

A **fractal**, like the **Menger sponge** at left, is a geometric object that exhibits some type of **self-similarity**. This means that the structure of the object always looks the same, whether seen in a highly magnified view, from a moderate distance, or from far away. If you were to cut off a small cube-shaped portion of the Menger sponge and examine it, you would find it to be a miniature copy of the entire sponge.

In mathematically created fractals like the Menger sponge, this process can theoretically be continued forever, and the self-similarity will always be evident.

Fractals can be created by repeating a simple procedure over and over. The Menger sponge, for example, is created by starting with a certain shape (a cube) and changing it according to a certain rule (removing a part of the cube). This same rule is applied to the newly changed shape. The process is then continued. This repetitive application of the same rule is called **iteration**.

Alternative Teaching Strategy

TECHNOLOGY Have students use computer programs to create fractal images. Ask them to use the computer to explore the Mandelbrot set and to create a Sierpinski gasket.

How Long Is a Coastline?

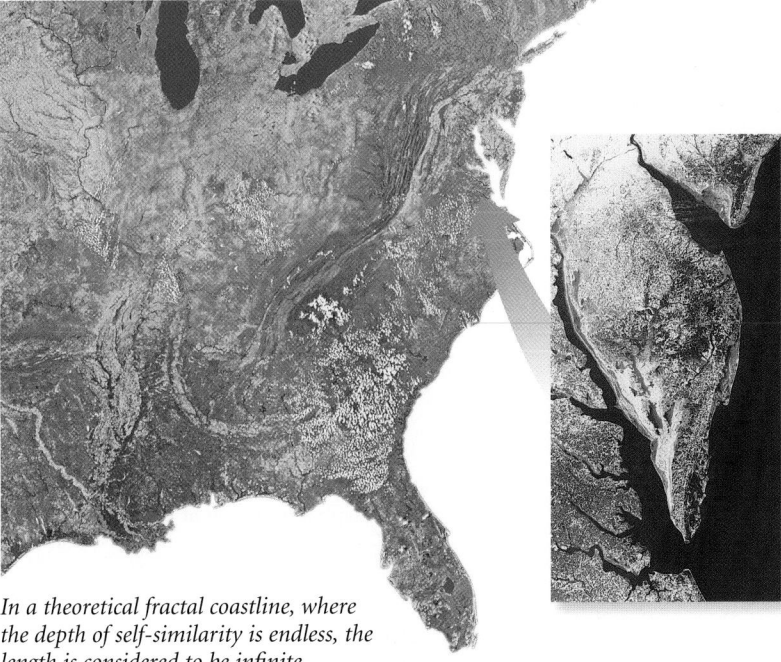

In a theoretical fractal coastline, where the depth of self-similarity is endless, the length is considered to be infinite.

A coastline is a good example of a self-similar structure in nature. Jagged irregularities such as bays, capes, and inlets can be seen from an orbiting space shuttle. The same basic structures are evident over a smaller section of the coastline viewed from an airplane and over even smaller sections viewed during a drive or a walk.

How long is a coastline? Unlike ordinary geometric segments or curves, which can often be readily measured, the "length" of a fractal coastline depends on how closely you move in to measure it.

During the American Revolution, the British Royal Navy attempted to blockade the American coastline. Although fairly successful, the blockade was not able to prevent shipping in and out of many harbors. What does a close examination of the coastline reveal about its "length" that explains the difficulty faced by the British to completely blockade it?

Many objects found in nature have some of the properties of fractals. Notice, for example, the self-similarity of the broccoli at right.

Interdisciplinary Connection

ART Graphics artists often use fractal images in the creation of their artwork. Have students find examples of fractal images in art. Explain to students that the actual computer commands to create the drawings may be simple. However, each command must be executed many times.

Activity 1
Creating the "Cantor Dust"

YOU WILL NEED
a ruler and pencil with an eraser

One of the simplest fractals was discovered by Georg Cantor (1845–1918) years before fractals were defined and studied. As you will see, Cantor's fractal is a one-dimensional version of the Menger sponge.

1. Draw a line 27 cm long.

2. Erase the middle third of the segment. You should now have two segments that are 9 cm long, with a 9-cm gap between them.

3. Erase the middle third of each of the two 9-cm segments. You will now have four segments, each 3 cm long.

4. Continue erasing the middle third of each of these segments until you are left with a scattering of point-like segments, known as the **Cantor dust**.

	27 cm
1 iteration	9 cm
2 iterations	3 cm 3 cm
3 iterations	— — — —
4 iterations	- - - - - - - -

PROBLEM SOLVING

5. Calculate the number of segments and their combined length after each iteration. **Make a table** like the one below.

Iteration	0	1	2	3	4	5	n
Number of segments	1	?	?	?	?	?	?
Combined length (cm)	27	?	?	?	?	?	?

CHECKPOINT ✔

6. As the number of iterations increases, what happens to the number of segments? What happens to the combined lengths of the segments?

CRITICAL THINKING

Describe the result if the process were repeated infinitely many times. How many segments would there be? What would be their combined length?

APPLICATION
TELECOMMUNICATIONS

When an electric current transmits data over a wire, a certain amount of "noise" occurs, which can cause errors in transmission. The noise seems to occur in random bursts, with "clean" spaces in between the bursts. Benoit Mandelbrot, the discoverer of fractal geometry, showed that the noise patterns closely matched the pattern of the Cantor dust. This geometric representation allowed the development of new strategies for reducing the transmission noise to a minimum.

The Sierpinski Gasket

The **Sierpinski gasket**, like the Menger sponge and the Cantor dust, is created by repeatedly applying a single rule to an initial shape. The rules are as follows:

1. Start with a solid triangle. Find the midpoint of each side.

2. Connect the midpoints of the sides to form four congruent isosceles triangles in the interior of the original triangle. Remove the center triangle. This is one iteration.

3. Each new iteration is performed on the remaining triangles.

Mathematical ideas often turn out to have surprising connections to seemingly unrelated fields. In Activity 2, you will learn about a connection between the Sierpinski gasket and Pascal's triangle.

Activity 2
Pascal and Sierpinski

YOU WILL NEED

graph paper

Recall that Pascal's triangle is a triangular array beginning with 1 at the top. Each number is the sum of the two numbers directly above it. The pattern is repeated endlessly.

1. Build a Pascal's triangle with at least 24 rows of squares, as indicated in the diagram. As the sums get large, you may wish to add only the final digits of the numbers. The more rows you can complete, the more impressive the result will be.

CHECKPOINT ✔

2. Shade each square that contains an odd number, and leave the other squares unshaded. Describe your results.

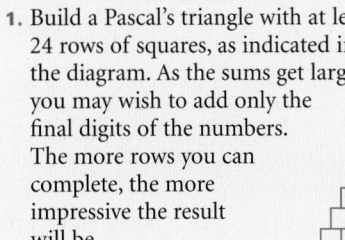

Reteaching the Lesson

USING DISCUSSION Show students examples of fractal images on the overhead. Have students identify self-similar images in the fractals. Ask them to think of analogous objects in nature.

Activity 2 **Notes**

In this Activity, students will create fractal-like images from Pascal's triangle. They are expected to discover and describe the resulting patterns in their fractals. Encourage students to compare their results with those of their classmates.

For a student worksheet of this Activity and detailed Teacher Notes, see page 208 in the Lesson Activities booklet.

Teaching Tip

In algebra, students learned how to expand a binomial by using Pascal's triangle. It would be helpful to review this idea, including how Pascal's triangle is generated. Point out that the triangle can be extended infinitely.

CHECKPOINT ✔

2. The figure looks like a Sierpinski gasket.

Cooperative Learning

Use cooperative groups to discuss the different types of fractal images in this lesson. Students should explain why an object or image is self-similar.

Selected Answers

Exercises 5–7, 9–35 odd

ASSIGNMENT GUIDE

In Class	1–7
Core	8–27
Core Plus	19–29
Review	30–36
Preview	37–38

✏ Extra Practice can be found beginning on page 818.

6. If the two boxes above a box are shaded or unshaded, then the box is left unshaded. However, if one of the boxes is shaded and one is unshaded, the box beneath them is shaded.

7. The shaded areas are inverted pyramids with increasingly longer bases arranged in self-similar patterns, as in Activity 2.

Exercises

Communicate

1. What is meant by self-similarity? You may wish to use an object to aid in your explanation.

2. Name an object from nature that has some type of self-similarity and describe how it is self-similar.

3. Mimi, Arnold, and Roberto measure a section of coastline by counting paces. Arnold's stride is 3 ft long, and he counts 24 paces. Roberto's stride is 2.5 ft long, and he counts 36 paces. Mimi's stride is 2 ft long, and she counts 47 paces. How long is the coastline for each person's set of measurements?

4. Explain how a fractal coastline could be considered to have an infinite length.

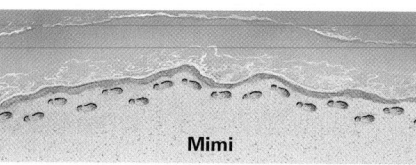

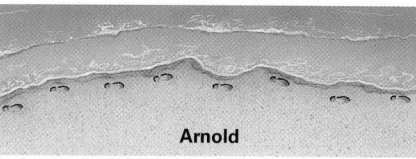

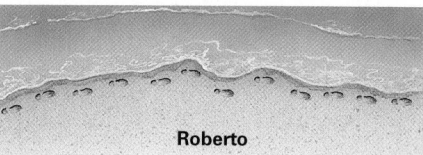

Three different measurements of a coastline

Guided Skills Practice

The points where segments connect to the base form the **Cantor dust**.

5. Draw an equilateral triangle and divide the sides into thirds. Draw segments connecting the sides to the base, forming two smaller equilateral triangles. Repeat as shown. How is the resulting fractal related to the Cantor dust? *(ACTIVITY 1)*

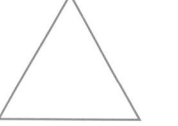

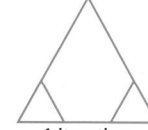 1 iteration

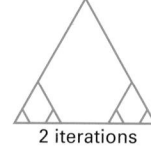 2 iterations

6. Use the following rules to explain the shading pattern in Pascal's triangle from Activity 2:

even + even = even   odd + odd = even   even + odd = odd

Based on these rules, devise a method for shading the squares without calculating the values in Pascal's triangle. *(ACTIVITY 2)*

7. Build a Pascal's triangle with at least 12 rows of squares. Shade each square that contains a number divisible by 3, and leave the other squares unshaded. How is the resulting pattern like the pattern from Activity 2? How is it different? *(ACTIVITY 2)*

Practice and Apply

In Exercises 8–18, you will explore the area and perimeter of the Sierpinski gasket. Copy and complete the table below.

	Iterations	0	1	2	3
11. 10.125	Shaded area	**8.** ? 24	**9.** ? 18	**10.** ? 13.5	**11.** ?
	Perimeter	**12.** ? 24	**13.** ? 36	**14.** ?54	**15.** ?81

16. Use the terms in the table to complete the following statement: At each iteration, the area is multiplied by ___?___ . $\frac{3}{4}$

The area decreases, and its limit is 0.

17. As the number of iterations increases, what happens to the area? Does it seem to have a limit as the number of iterations approaches infinity?

The perimeter increases without bound.

18. As the number of iterations increases, what happens to the perimeter? Does it seem to have a limit as the number of iterations approaches infinity?

In Exercises 19–23, you will construct a fractal called the Koch snowflake and explore its properties.

19. Construct the first iteration of the Koch snowflake by following the directions below.

1st iteration

a. **b.** **c.**
 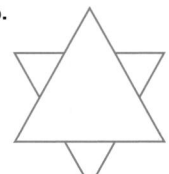 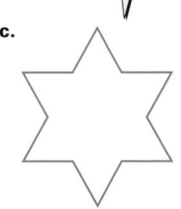

  a. Construct an equilateral triangle with a side length of 18 cm.

  b. Divide each side in thirds, and construct an equilateral triangle on each side of the triangle with the middle third as the base.

  c. Erase the base of each new triangle. You should now have a six-pointed star. This is the first iteration.

20. Continue the construction of the Koch snowflake. Repeat the steps above to complete at least three iterations.

Step 0: $P = 3 \times 18 = 54$
Step 1: $P = 12 \times 6 = 72$
Step 2: $P = 48 \times 2 = 96$

21. Find the perimeter of the Koch snowflake after the first two iterations. Remember that the snowflake starts with a perimeter of 3×18 cm, or 54 cm.

22. Does the perimeter increase or decrease? Does it increase or decrease by a greater amount or a smaller amount from each iteration to the next? What does this tell you about the perimeter?

23. Look at your snowflake and consider how it would change if you completed many more iterations. Would the area of the snowflake increase or decrease as the number of iterations increase? Would the area ever become infinite? (Hint: Can the snowflake always be enclosed by a circle?)

23. The area of the snowflake is always increasing. The area never becomes infinite because the snowflake can always be enclosed in a circle of radius 10.4 cm.

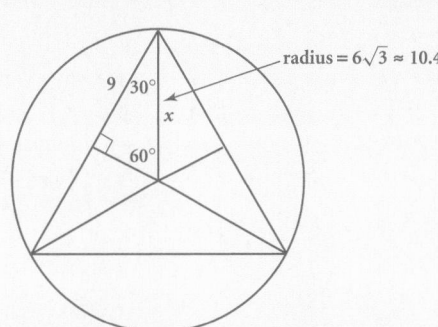

Error Analysis

Students may be discouraged by exercises in which they are asked to draw fractals because of the time required to draw them. Use these exercises to emphasize the importance of computers in drawing fractal images.

Technology

You may want students to use fractal software to draw the Koch snowflake in Exercises 19–23. Students may also use the script feature of geometry graphics software.

19. Check student constructions. Make sure directions are followed.

20. Step 2:

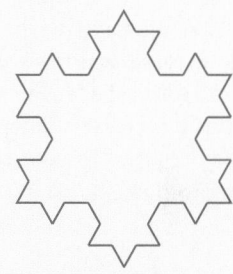

Step 3:

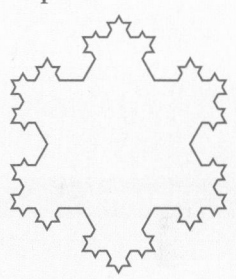

22. The perimeter is increasing. It increases by $\frac{4}{3}$ each time. The perimeter increases by a larger amount each time, so that means the perimeter will continue to increase without bound.

24. Sample answer:

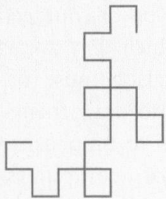

25. Sample answer:

4 iterations

5 iterations

27. The area increases as iterations increase, but not without bound, because the amount it increases is always less than half of the increase of the previous iteration.

Practice

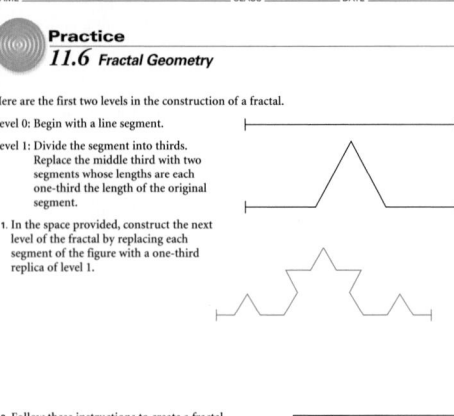

In Exercises 24–27 you will explore a fractal known as the dragon curve.

24. Take a long strip of paper and fold it in half and then in half again (in the same direction), repeating as many times as possible. Open each fold to 90°. The result, when viewed from the edge of the paper, is a dragon curve. Draw your dragon curve on paper.

25. The following is an algorithm for drawing a dragon curve: Start with two segments that intersect at a 90° angle. For each iteration, replace every $\vee$ with $\sqcup\!\!\sqcap$, as shown below. Draw a dragon curve with at least five iterations.

1 iteration   2 iterations   3 iterations

26. Suppose that the dragon curve in Exercise 25 starts with a length of 2 units. What is the length after the first iteration? after two iterations? after three iterations? Divide your results from successive iterations to complete the following statement: $2\sqrt{2}$; 4; $4\sqrt{2}$

The length of a dragon curve after $n + 1$ iterations is __?__ times the length after n iterations. $\sqrt{2}$

27. How does the area covered by the dragon curve change after each iteration? Do you think the area covered by the curve approaches infinity as the number of iterations increases? Explain your reasoning.

CHALLENGE

28. Refer to the drawing of the Menger sponge at the beginning of this lesson. Describe the construction of the sponge. If the volume before the first iteration is 1 cubic unit, what is the volume after one iteration? after two iterations? 1 iteration, $\frac{20}{27}$ units^3; 2 iterations, $\frac{400}{729}$ units^3

APPLICATION

29. HOBBIES A kite may be made in the shape of a tetrahedron, with paper or fabric covering two of the faces. Four of these kites may be joined at the corners to form a larger kite, also in the shape of a tetrahedron (the first iteration). Four of these units may be joined at the corners to form a larger tetrahedral kite (the second iteration). As the units are joined, the kite will begin to resemble a three-dimensional Sierpinksi gasket. How many of the original kites are required for the second iteration? How many are required for the third iteration? **16, 64**

Look Back

Find the distance between each pair of points. *(LESSON 5.6)*

30. $(4, -2), (2, -1)$ **2.2**   **31.** $(5, -10), (-2, 3)$ **14.8**   **32.** $(15, 2), (-6, 5)$ **21.2**

Find the distance between each pair of points in a three-dimensional coordinate system. *(LESSON 6.4)*

33. $(4, 3, 2), (2, -3, 5)$ **34.** $(18, 1, 0), (0, -1, 5)$ **35.** $(5, 1, -5), (2, -12, 0)$

36. Find the measure of the diagonal of a cube in the top-front-right octant, with a vertex at $(0, 0, 0)$, and with a side length of 8. *(LESSON 6.4)*

33. 7

34. 18.8

35. 14.2

36. 13.9

37. CULTURAL CONNECTION: ASIA
For centuries, women of India have used fascinating curve patterns known as Kolams for ritual and decorative purposes. A fractal known as the Hilbert curve may remind you of the Kolams of India. Steps for creating a Hilbert curve are given below.

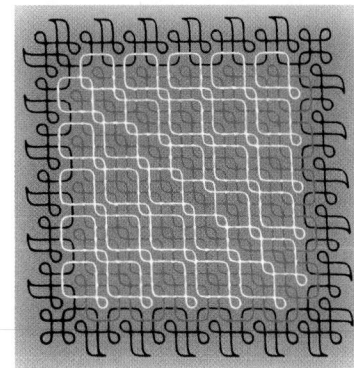

a. Start with a three-sided "square" with no bottom edge.

b. Draw a new (smaller) three-sided figure at each vertex. Be sure your new figures are oriented as shown.

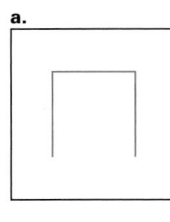 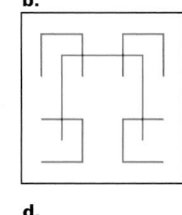

c. Connect the new figures as shown to form a single continuous line. Erase the original figure. This is the first iteration.

d. For each new three-sided figure that you just created in part **b**, create a new (smaller) figure at each vertex. Connect the new figures and erase the original figures, as in part **c**. This is a new iteration.

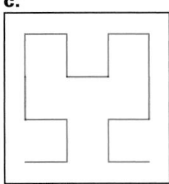 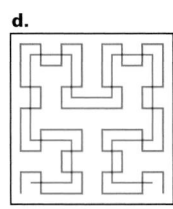

38. The resulting curve is known as a space-filling curve or a "monster curve." If you were to continue this process indefinitely, what would the result look like? **The entire paper would appear shaded.**

  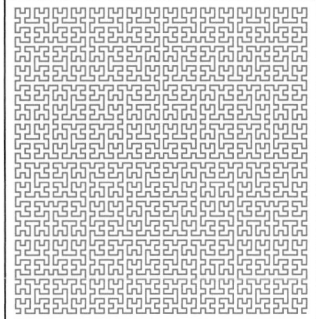

37. See figure in text.

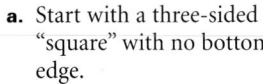
In Exercises 37 and 38, students draw the Hilbert curve, or "monster curve." Consider having students draw this with fractal software.

Student Technology Guide

NAME _____ CLASS _____ DATE _____

Student Technology Guide
11.6 Fractal Geometry, page 1

The four diagrams below illustrate the first four stages in the construction of the Sierpinski triangle, also called the Sierpinski gasket, based on an initial equilateral triangle. As the construction process continues, the higher the stage number, the more "holes" in the triangle and the more boundary lines around the holes.

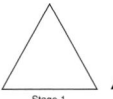  

Stage 1   Stage 2   Stage 3   Stage 4

Suppose that at stage 1 you have an equilateral triangle with sides of 1 unit. Let n represent the stage number, s represent the number of shaded triangles at stage n, L represent the length of each shaded triangle at stage n, and P represent the perimeter of all the shaded triangles at stage n.

In Exercises 1–3, write a formula for the specified variable in terms of n.

1. s
$$s = 3^{n-1}$$

2. L
$$L = \left(\frac{1}{2}\right)^{n-1}$$

3. P
$$P = 3Ls = 3\left(\frac{3}{2}\right)^{n-1}$$

4. Using your formulas from Exercises 1–3, create a spreadsheet like the one shown at right.
Check students' spreadsheets.

5. What can you conclude about P as n increases?
P increases without bound as n increases.

6. a. The area A of an equilateral triangle with sides of L units is approximated by $A = 0.4330L^2$. Modify your spreadsheet to find the total area of all shaded triangles at stage n.
Check students' spreadsheets.
b. What can you say about the total area of the shaded triangles as n increases?
As n increases, the total area of the shaded triangles decreases and approaches 0.
$$A = (0.43301)(3^{n-1})\left(\frac{1}{2}\right)^{n-1 \, 2}$$
Check students' spreadsheets.

7. a. Modify your spreadsheet to find the total area of all unshaded triangles at stage n.
b. What can you say about the total area of the unshaded triangles as n increases?
As n increases without bound, the total area of the unshaded triangles approaches the area of the original triangle, or 0.43301 square units.

ALTERNATIVE
Assessment

Portfolio Activity

The Portfolio Activity can be used as preparation for the Chapter Project or as a separate activity. In the Portfolio Activity on this page, students play the "chaos game." They should discover that the points plotted in the triangle generate an image of a Sierpinski gasket.

THE CHAOS GAME

Can there be order in a random process? The following "chaos game" may lead you to ask some deep questions.

internet connect

Portfolio Extension

Go To: **go.hrw.com**
Keyword:
MG1 Fractals

1. Draw equilateral triangle △*ABC* with a side length of 10 centimeters. Roll a number cube or use some other method to randomly select vertex *A*, *B*, or *C*. You can let 1 and 2 represent *A*, let 3 and 4 represent *B*, and let 5 and 6 represent *C*. Mark the selected vertex, which is called the *seed point*.

2. Roll the number cube again to select another vertex. This time, mark the point halfway between the seed point and the selected vertex. This point becomes the new seed point. Repeat the process several times, using the new seed point each time.

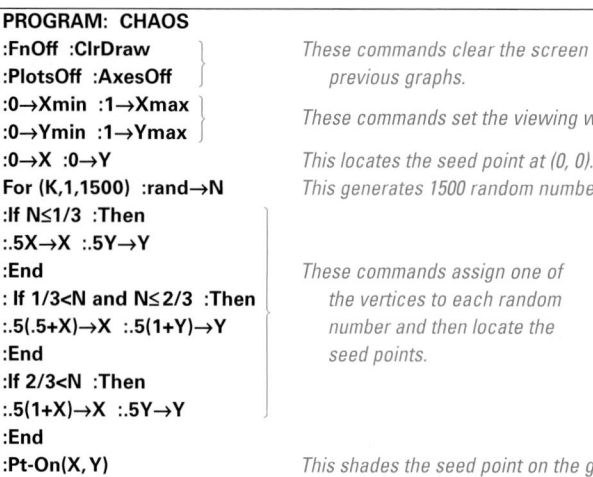

First vertex selected: *A*
Second vertex selected: *B*
Third vertex selected: *C*

3. If you were to repeat this process several hundred times, a pattern would start to emerge. Combine your results with your classmates' results and see if you can guess the pattern. The program below, which will work with a graphics calculator, simulates the chaos game for a triangle with vertices at $(0, 0)$, $(0.5, 1)$, and $(1, 0)$ and an initial seed point at $(0, 0)$ for 1500 repetitions. Study the program to see how the simulation works. Then run the program and describe the result.

```
PROGRAM: CHAOS
:FnOff :ClrDraw
:PlotsOff :AxesOff
:0→Xmin :1→Xmax
:0→Ymin :1→Ymax
:0→X :0→Y
For (K,1,1500) :rand→N
:If N≤1/3 :Then
:.5X→X :.5Y→Y
:End
: If 1/3<N and N≤2/3 :Then
:.5(.5+X)→X :.5(1+Y)→Y
:End
:If 2/3<N :Then
:.5(1+X)→X :.5Y→Y
:End
:Pt-On(X,Y)
:End
```

These commands clear the screen of any previous graphs.

These commands set the viewing window.

This locates the seed point at (0, 0).
This generates 1500 random numbers.

These commands assign one of the vertices to each random number and then locate the seed points.

This shades the seed point on the graph.

Other Transformations and Projective Geometry

Objectives

- Understand the concepts of affine transformations and geometric projection.

- Solve problems and make conjectures by using the Theorem of Pappus and the Theorem of Desargues.

Why *When you stretch or otherwise deform a figure, will you still be able to recognize it? In what ways will it be the same? These are questions that are considered in the study of projective geometry.*

Prepare

QUICK WARM-UP

1. List the transformations that preserve the size and shape of an object.
   reflections, rotations, and translations
2. Which is preserved by dilations: size or shape?
   shape
3. List the transformations that preserve the property that parallel lines are transformed into parallel lines.
   translations, dilations, rotations, and reflections

Also on Quiz Transparency 11.7

Nonrigid Transformations

The three rigid transformations—reflections, rotations, and translations—preserve the shapes and sizes of objects. Dilations, on the other hand, preserve shape but not size. There are types of transformations that shrink, expand, or stretch an object in different directions so that neither shape nor size are necessarily preserved. One of these types of nonrigid transformations is called an *affine transformation* and is defined as follows:

Definition: Affine Transformation

An **affine transformation** transforms each preimage point P in a plane to an image point P' in such a way that

1. collinear points are transformed into collinear points,

2. straight lines are transformed into straight lines,

3. intersecting lines are transformed into intersecting lines, and

4. parallel lines are transformed into parallel lines.   **11.7.1**

Teach

Why Projective geometry shows students how to draw in perspective and how to represent objects as projections, such as the Earth projected onto a flat piece of paper. Point out that projective geometry studies the invariant properties of an object.

Alternative Teaching Strategy

TECHNOLOGY Use geometry graphics software to transform plane figures. Ask students to draw a figure on the coordinate plane and use the affine transformation feature to transform it. In the dialog box, have them define the transformation from (x, y) to (x', y').

In this Activity, students use a compass to construct an affine transformation in the coordinate plane. The purpose of this Activity is to confirm the four conditions in the definition of an affine transformation. Have students repeat the Activity with a variety of shapes.

For a student worksheet of this Activity and detailed Teacher Notes, see page 212 in the Lesson Activities booklet.

Teaching Tip

When discussing affine transformations, ask students which properties of a figure are preserved. Use Activity 1 to emphasize that the property of being a line is invariant under an affine transformation, while the property of being a circle or a square is not.

CHECKPOINT ✔
Part I

6. The images of the sides of the square are the sides of the parallelogram, so collinear points project to collinear points, straight lines project to straight lines, intersecting lines project to intersecting lines, and parallel lines project to parallel lines.

CHECKPOINT ✔
Part II

4. The image of each side of the square is a side of a rectangle, so collinear points project to collinear points, straight lines project to straight lines, intersecting lines project to intersecting lines, and parallel lines project to parallel lines.

An Affine Transformation

Part I

1. Draw a 4 × 4 square and a 4 × 4 parallelogram on graph paper as shown below. (Use more than one grid square for each unit square in order to make the figures large enough to work with.)

2. Inscribe a circle in the square. Draw a square and its diagonals inside the circle.

3. Mark the points of intersection of the circle with the grid lines. Mark the vertices of the inner square.

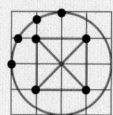

4. Mark the same points in the parallelogram grid. The first few are marked in the diagram below.

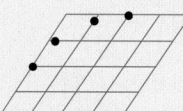

5. To draw the image of the figure under an affine transformation, connect the image points of the square with segments (including the diagonals) and draw a smooth curve through the image points of the circle.

CHECKPOINT ✔

6. How does the resulting figure illustrate each of the four conditions in the definition of an affine transformation?

Part II

Affine transformations can be represented by using coordinates. For example, multiplying the x- and y-coordinates of the points of a figure by two different scale factors is an example of an affine transformation.

Preimage point	Multiply x by 2 and y by 3.	Image point
(x, y)	$\longrightarrow$	$(2x, 3y)$

1. Draw a square and its diagonals in the first quadrant of a coordinate plane.

2. Multiply the x-coordinates of the vertices of the square by 2 and the y-coordinates by 3. Plot the resulting points and connect them with segments (including the diagonals). Describe the resulting figure.

3. Multiply the x-coordinates of the vertices of the square by 3 and the y-coordinates by 2. Plot the resulting points and connect them with segments (including the diagonals). Describe the resulting figure.

CHECKPOINT ✔

4. How do the resulting figures illustrate each of the four conditions in the definition of an affine transformation?

Interdisciplinary Connection

ART Artists can increase their understanding of perspective drawing through a study of projective geometry. Have students revisit the projection diagrams (pages 409–410) in Lesson 6.6 (*Perspective Drawing*) as they study the concepts in this lesson.

Projections and Projective Geometry

In the photograph at the beginning of this lesson, the image of the apple is distorted by a transformation. Is it an affine transformation? Study the two images below.

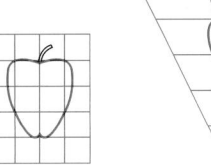

As you can see, the sides of the grid are parallel in the original image but not in the transformed image. Therefore, parallel lines do not transform to parallel lines, as required by the definition of an affine transformation. Thus, it is not an affine transformation.

The transformation of the image of the apple is an example of a class of transformations known as **central projections**. In such transformations, there is a central point known as the **center of projection**, and the **projected points** lie on rays containing the center of projection and the original points.

The diagram at right shows a projection between two lines in the same plane. The points on line ℓ are projected onto line m from the center of projection, point O. The rays drawn from the center of the projection are called the **projective rays**. These rays intersect line ℓ at points A, B, and C. The intersections of the rays with line m determine the projected points, A', B', and C'.

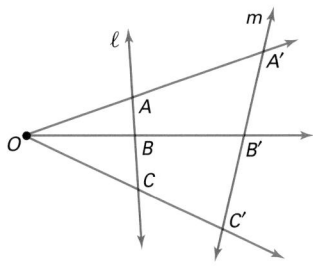

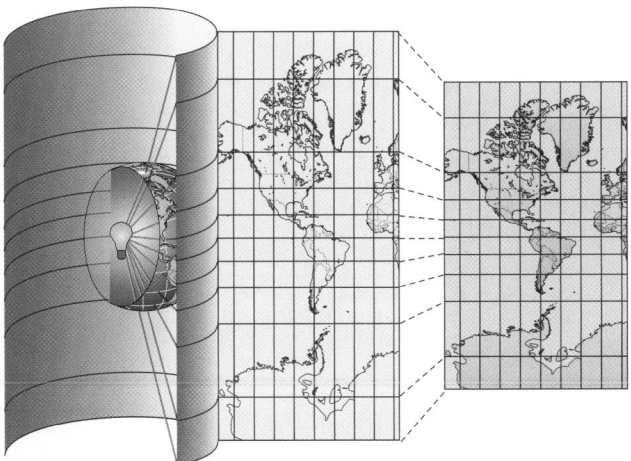

To make the map a Mercator projection, its vertical dimensions must be adjusted. (See page 176.)

The surface of a sphere can be projected onto a plane by means of a *cylindrical projection*. Notice that it is impossible to project the points at the North and South Poles onto the map, and that it is impractical to project points near the poles.

Geography teachers have long pointed out that Greenland is really much smaller in relation to other countries than it appears to be on cylindrical projection maps. Deformities like these arise because it is impossible to flatten Earth's spherical surface into a rectangle without some distortion of size, shape, distance, and direction.

Cooperative Learning

Have cooperative groups discuss how to preform an affine transformation in the coordinate plane. Have each student in the group draw a figure in the plane and transform it according to a rule developed by the group (for example $T(x, y) = (3x, 4y)$). Ask each student to describe how to apply each part of the definition of affine transformation.

Activity 2 Notes

This Activity will help students understand two of the basic theorems of projective geometry: the Theorem of Pappus and the Theorem of Desargues. Encourage students to compare their results with those of their classmates. Repeat the Activity with a different set of points.

For a student worksheet of this Activity and detailed Teacher Notes, see page 212 in the Lesson Activities booklet.

CHECKPOINT ✔

4. X, Y, and Z are collinear. If A_1, B_1, and C_1 are three distinct points on one line and A_2, B_2, and C_2 are three distinct points on a second line, then the intersections of $\overline{A_1B_2}$ and $\overline{A_2B_1}$, of $\overline{A_1C_2}$ and $\overline{A_2C_1}$, and of $\overline{B_1C_2}$ and $\overline{B_2C_1}$ are collinear.

For years, artists used the concept of projection to given their works realism and depth (see Lesson 6.6). On the other hand, mathematicians used projections to develop an entire system of geometry. The chart below summarizes its features.

Main Features of Projective Geometry

1. Projective geometry is the study of the properties of figures that do not change under a projection.
2. There is no concept of size, measurement, or congruence.
3. Its theorems state facts about such things as the positions of points and the intersections of lines.
4. An unmarked straightedge is the only tool allowed for drawing figures. **11.7.2**

YOU WILL NEED

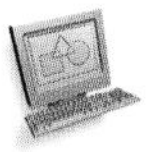

straightedge
OR
geometry graphics software

Activity 2
Two Projective Geometry Theorems

Part I: A Theorem of Pappus

CULTURAL CONNECTION: AFRICA Pappus was a mathematician who lived in Alexandria, Egypt, in the fourth century C.E. His work was very important in the development of projective geometry many centuries later. In this Activity, you will discover one of his theorems.

PROBLEM SOLVING

1. **Make a diagram.** Orient your paper horizontally. Mark a point, O, toward the left edge of the paper. Draw two rays from point O.
2. Mark A_1, B_1, and C_1 on one ray. In the same order, mark A_2, B_2, and C_2 on the other ray.

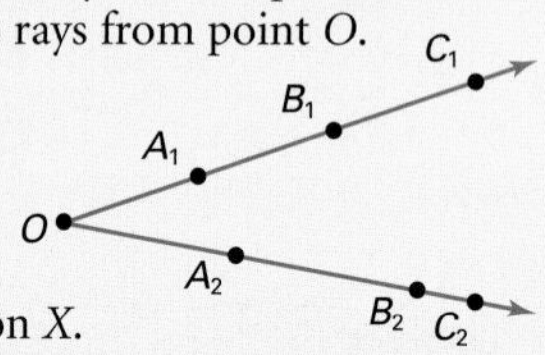

3. Draw $\overline{A_1B_2}$ and $\overline{A_2B_1}$. Label their intersection X.
Draw $\overline{A_1C_2}$ and $\overline{A_2C_1}$. Label their intersection Y.
Draw $\overline{B_1C_2}$ and $\overline{B_2C_1}$. Label their intersection Z.

CHECKPOINT ✔

4. What appears to be true about X, Y, and Z? Make a conjecture, and compare it with those of your classmates. (If you are using geometry graphics software, drag the rays in various ways to see whether your conjecture still holds.) Based on your conjecture, complete the following theorem:

The Theorem of Pappus

If A_1, B_1, and C_1 are three distinct points on one line and A_2, B_2, and C_2 are three distinct points on a second line, then the intersections of $\overline{A_1B_2}$ and $\overline{A_2B_1}$, of $\overline{A_1C_2}$ and $\overline{A_2C_1}$, and of $\overline{B_1C_2}$ and $\overline{B_2C_1}$ are ___?___. **11.7.3**

Inclusion Strategies

HANDS-ON STRATEGIES Ask students to draw several geometric shapes in the first quadrant and project them onto the x-axis. Which shapes are projected as a line segment? Conversely, suppose that the projection image of a shape onto the x-axis is a line segment. What could the pre-image shape look like?

Part II: A Theorem of Desargues

Girard Desargues (1593–1662) was a French mathematician whose ideas are among the most basic in projective geometry. Follow the steps below to discover one of his most important theorems. (Use as large a piece of paper as is practical.)

PROBLEM SOLVING

1. **Make a diagram.** Draw $\triangle ABC$ near the center of the paper. This can be any type of triangle, but it should be small so that the resulting construction will fit on your paper.

2. Mark a point outside $\triangle ABC$ and label it O. This will be the center of projection. Draw $\overrightarrow{OA}$, $\overrightarrow{OB}$, and $\overrightarrow{OC}$.

3. Mark a random point, A', as the projection of A on $\overrightarrow{OA}$. (For clarity of construction, it should be farther from point O than from point A.)

   Repeat for B', a random point on $\overrightarrow{OB}$, and C', a random point on $\overrightarrow{OC}$. Draw $\triangle A'B'C'$.

4. Extend $\overline{AB}$ and $\overline{A'B'}$ until they intersect, and label the point of intersection X. (If the extended segments are parallel or intersect at a point past the edge of your paper, reposition the image points.)

5. Extend $\overline{AC}$ and $\overline{A'C'}$ until they intersect, and label the point of intersection Y.

6. Extend $\overline{BC}$ and $\overline{B'C'}$ until they intersect, and label the point of intersection Z.

CHECKPOINT ✔

7. What appears to be true about points X, Y, and Z? Make a conjecture, and compare it with those of your classmates. (If you are using geometry graphics software, drag the rays and points in various ways to see whether your conjecture still holds.) Based on your conjecture, complete the following theorem:

The Theorem of Desargues

If one triangle is a projection of another triangle, then the intersections of the lines containing the corresponding sides of the two triangles are _____?_____.   **11.7.4**

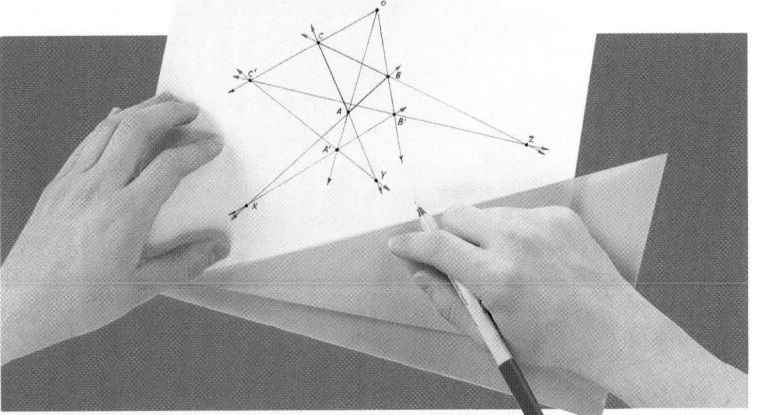

Teaching Tip

Allow students to adjust their choices for the original triangle so that all of the intersection points will fit on the page. They should compare their results with those of their classmates.

CHECKPOINT ✔

7. X, Y, and Z are collinear. If one triangle is a projection of another triangle, then the intersections of the lines containing the corresponding sides of the two triangles are collinear.

Reteaching the Lesson

USING VISUAL MODELS Have students draw a geometric shape in the coordinate plane. Label the vertices as preimage points. Ask them to find the image of the shape under an affine transformation such as $T(x, y) = (2x, 4y)$. They should use small numbers for the coordinates of the points of the triangle so that the image will fit on the page.

ASSIGNMENT GUIDE

In Class	1–8
Core	9–13 odd, 14–29
Core Plus	13–29
Review	30–39
Preview	40

✐ Extra Practice can be found beginning on page 818.

5.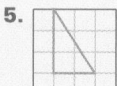

7. yes

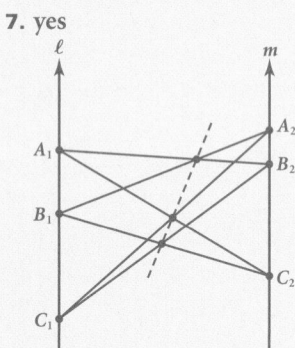

A rectangle with vertices (0, 0), (0, 3), (4, 3) and (4, 0)

Exercises

● Communicate

1. Describe what an affine transformation is. Which of the following are special cases of affine transformations? Explain your reasoning.
 a. dilations
 b. translations
 c. reflections
 d. rotations

2. Describe what a central projection is. Which of the following are special cases of central projections? Explain your reasoning.
 a. dilations
 b. translations
 c. reflections
 d. rotations

3. Explain how projective geometry is different from the other types of geometry you have studied.

4. In a Mercator projection, why do countries near the equator seem less distorted than countries near the North and South Poles?

● Guided Skills Practice

5. Copy the grids at right. Draw a right triangle on the square grid, and use the parallelogram grid to transform the triangle. *(ACTIVITY 1)*

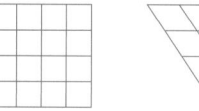 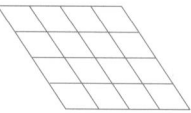

6. Draw a square in a coordinate plane with vertices at (0, 0), (0, 1), (1, 1), and (1, 0). Use the affine transformation $T(x, y) = (4x, 3y)$ to transform the square. What is the resulting figure? *(ACTIVITY 1)*

7. Draw two parallel lines ℓ and m. On line ℓ, mark points A_1, B_1, and C_1. On line m, mark points A_2, B_2, and C_2. Find the intersections of $\overline{A_1B_2}$ and $\overline{A_2B_1}$, of $\overline{A_1C_2}$ and $\overline{A_2C_1}$, and of $\overline{B_1C_2}$ and $\overline{B_2C_1}$. Are the three points of intersection collinear? *(ACTIVITY 2)*

8. Copy △ABC and point O below and draw the image, △A′B′C′, transformed by a projection centered at O. Extend the corresponding sides of the triangles until they intersect, and draw the line through the three points of intersection. *(ACTIVITY 2)*

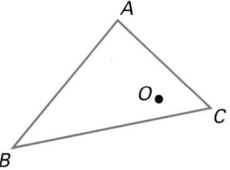

8. The points of intersection should be collinear.

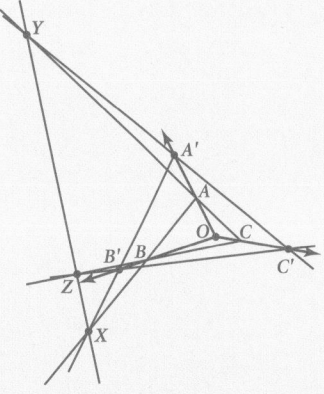

Practice and Apply

Sketch the preimage and image for each affine transformation below.

9. preimage: square $(0, 0)$, $(4, 0)$, $(4, 4)$, $(0, 4)$
transformation: $T(x, y) = (3x, -2y)$

10. preimage: rectangle $(0, 0)$, $(5, 0)$, $(5, 8)$, $(0, 8)$
transformation: $S(x, y) = (2x, 0.5y)$

11. preimage: triangle $(4, 7)$, $(-1, -1)$, $(0, 8)$
transformation: $R(x, y) = \left(\frac{5}{2}x, 5y\right)$

On a coordinate plane, draw a circle with a radius of 5 centered at the origin. Use the parallelogram grids below to transform the circle.

12.

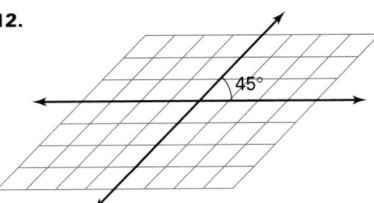

13.

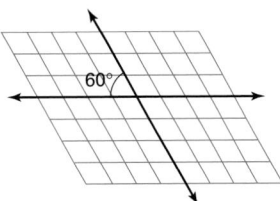

For Exercises 14–16, refer to the figures below. For each projection, identify the following:
a. the center of projection
b. the projective rays

15. a. N
    b. $\overrightarrow{NJ}$, $\overrightarrow{NK}$, $\overrightarrow{NL}$

16. a. M
    b. $\overrightarrow{MG}$, $\overrightarrow{MH}$, $\overrightarrow{MI}$

14. projection of points on ℓ_1 onto ℓ_2
    a. P
    b. $\overrightarrow{PR}$, $\overrightarrow{PS}$, $\overrightarrow{PT}$

15. projection of points on m_2 onto m_1

16. projection of points on m_2 onto m_3

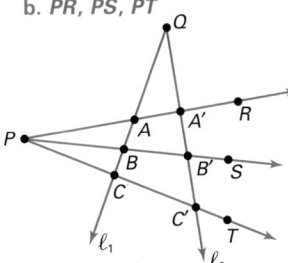

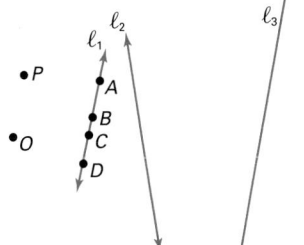

Copy the diagram below and draw each projection.

17. Using O as the center of projection, project points A, B, C, and D onto line ℓ_2. Label the projected points A', B', C', and D'.

18. Using P as the center of projection, project points A', B', C', and D' onto line ℓ_3. Label the projected points A'', B'', C'', and D''.

9.

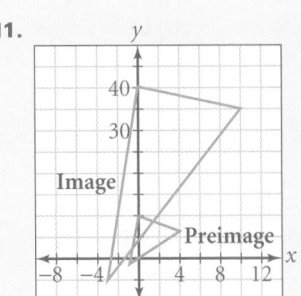

10.

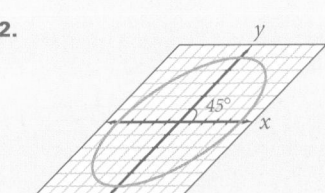

11.

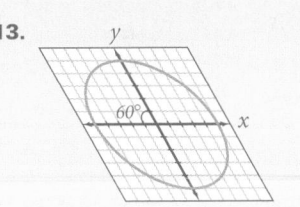

12.

13.

17.

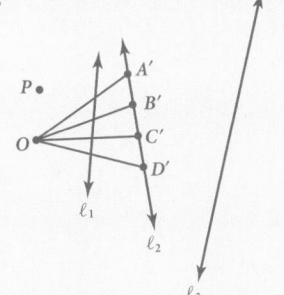

18.

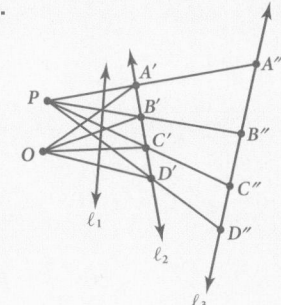

LESSON 11.7   **753**

For Exercises 19–21, use the figure at right.

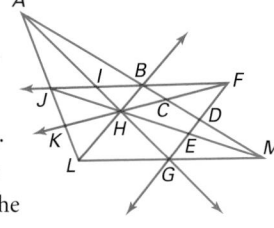

19. If point A is the center of projection, then the projection of B onto $\overline{FK}$ is ____?__ **C** , the projection of I onto $\overline{FK}$ is ____?__ **H** , and the projection of J onto $\overline{FK}$ is ____?__ **K** .

20. If point L is the center of projection, then the projection of H onto $\overline{MA}$ is ____?__ **B** , and the projection of J onto $\overline{MA}$ is ____?__ **A** .

21. If point F is the center of projection, then the projection of ____?__ **B** onto $\overline{AG}$ is I, and the projection of ____?____ onto $\overline{AG}$ is G. **D or E**

In Exercises 22–29, you will explore the converse of the Theorem of Desargues. The following construction begins where Activity 2 ended. By working backward, you will locate the original center of projection.

22. Draw ℓ and label X and Y at arbitrary locations on it.

23. Choose a point, B, not on ℓ. Draw $\overrightarrow{XB}$ and mark a random point, A, on it.

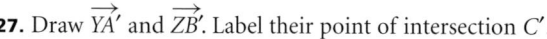

24. Choose a point, B', not on ℓ. Draw $\overrightarrow{XB'}$ and mark a random point, A', on it.

25. Draw $\overrightarrow{YA}$ and mark a random point, C, on it.

26. Draw $\overleftrightarrow{CB}$. Label the intersection of $\overleftrightarrow{CB}$ and ℓ Z.

27. Draw $\overrightarrow{YA'}$ and $\overrightarrow{ZB'}$. Label their point of intersection C'.

28. Assuming the converse of Desargue's Theorem, what can you conclude about $\triangle ABC$ and $\triangle A'B'C'$. **One triangle is a projection of the other.**

29. Locate the center of projection, point O, by drawing $\overleftrightarrow{AA'}$, $\overleftrightarrow{BB'}$, and $\overleftrightarrow{CC'}$.

 Look Back

Logical chain:

If the wind blows, then the trees shake.

If the trees shake, then the apple falls.

If the apple falls, then the worm squirms.

Conditional:

If the wind blows, then the worm squirms.

30. Arrange the statements below to form a logical chain, and write the conditional that follows from the logical chain. (*LESSON 2.2*)

If the wind blows, then the trees shake.
If the apple falls, then the worm squirms.
If the trees shake, then the apple falls.

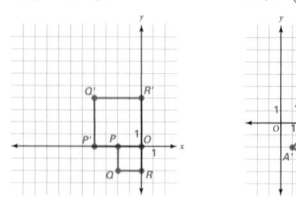
22.–27., 29.

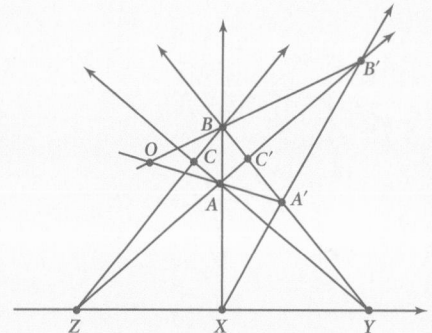

For Exercises 31–33, tell whether each argument is an example of inductive or deductive reasoning. *(LESSON 2.5)*

deductive   **31.** All squares are rectangles. *ABCD* is a square. Therefore, *ABCD* is a rectangle.

inductive   **32.** There has never been a freeze recorded in Florida in August. Today is August 12th. Therefore, it will not freeze in Florida today.

deductive   **33.** Debra likes every type of fruit. Mangoes are a type of fruit. Therefore, Debra likes mangoes.

Look Beyond

Exercise 40 is a puzzle that can be solved by using the Theorem of Pappus. Encourage students to use models to solve the problem.

34. Describe what is meant by the golden ratio. *(LESSON 11.1)*

Find the taxidistance between each pair of points. *(LESSON 11.2)*

35. $(4, 3), (2, 1)$ 4   **36.** $(-3, 2), (1, 1)$ 5   **37.** $(1, 3), (5, 5)$ 6

38. Does the figure at right contain an Euler path? an Euler circuit? Why or why not? *(LESSON 11.3)*
yes, yes; The vertices are all even.

39. In spherical geometry, how many lines are parallel to a given line through a point not on the line? *(LESSON 11.5)* 0

Look Beyond

40. The Nine-Coin Puzzle In the picture of the nine coins below, you can identify eight rows with three coins each. Can you rearrange the coins to form ten rows with three coins each? (Hint: You can use the Theorem of Pappus to solve this puzzle.)

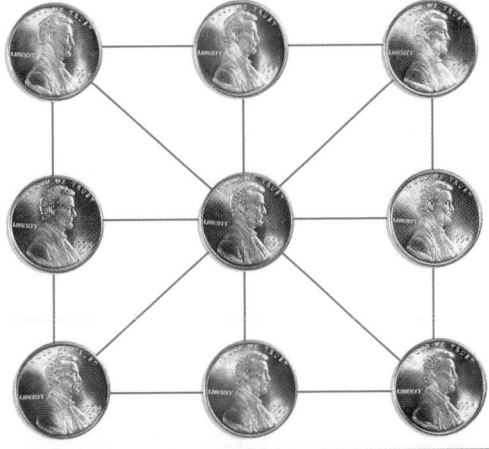

34. The golden ratio is the ratio of the sides of a golden rectangle. A golden rectangle is a rectangle with short side of length s and long side of length l where $\frac{l}{s} = \frac{s}{l-s}$.

40. Sample answer:

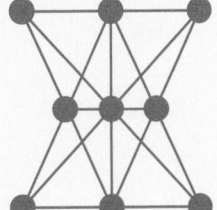

Focus

Students will analyze the game called the tower of Hanoi. They will discover an interesting link between the strategy for playing this game and the pattern of a Sierpinski gasket.

Motivate

Bring several tower of Hanoi games to class. Allow students to play the games and become familiar with the strategies needed to solve the puzzle. If games are not available, have students make their own games out of pegs and washers or similar materials. Computer simulations of the game are also available as educational software.

CHAPTER PROJECT ELEVEN

The Tower of Hanoi

In this project, you will examine a puzzle, called the tower of Hanoi, that was invented in 1883 by French mathematician Edouard Lucas.

Activity 1

THE PUZZLE The puzzle begins with three pegs, one of which has a stack of disks that increase in size from the top of the peg to the bottom. The object is to move all of the disks to another peg, following these rules:
- Only one disk may be moved at a time
- A disk must be placed on a larger disk or on an empty peg.

1. What is the least possible number of moves needed to solve the puzzle for 1 disk? for 2 disks? Complete the table at right.

2. Can you find a formula for the number of moves in terms of the number of disks? Extend the table if necessary.

Number of disks	Fewest number of moves
1	?
2	?
3	?
4	?

3. According to a popular story, 64 gold disks on 3 diamond pegs are attended by priests who move the disks according to the rules above. When they have completed the puzzle, the world will end. How many moves are required? If the priests can move one disk per second, how long will it take them to complete the puzzle?

Activity 1

1.

Number of disks	Fewest possible moves
1	1
2	3
3	7
4	15

2. The number of moves for n disks is $2^n - 1$

3. For 64 disks, the number of moves is $2^{64} - 1 = 18,446,744,073,709,551,615$. It will take over 584 million years to move the 64 disks.

Activity 2

STRANGE CONNECTIONS A graph may be used to represent the states of the tower of Hanoi. When the vertices of the graph are arranged in a certain way, the graph forms the same pattern as the Sierpinski gasket.

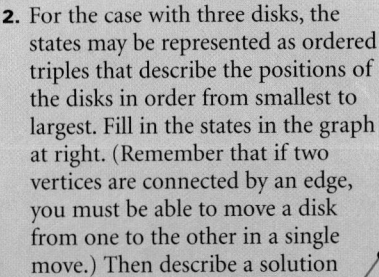

Start with the simplest case, a tower of Hanoi with only one disk. The pegs in the puzzle are labeled *A*, *B*, and *C*. Thus, the initial state can be represented by a vertex labeled *A*. It is possible to move the disk to either *B* or *C* in one move, so the graph for the puzzle can be drawn as shown. Notice that it is also possible to move the disk from *B* to *C* (or *C* to *B*), so *B* and *C* are connected by an edge.

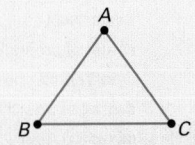

For the case with two disks, the states can be represented by ordered pairs. For example, if the smaller disk is on peg *B* and the larger disk is on peg *A*, the position can be written as (*B*, *A*). The graph for this puzzle is shown at right.

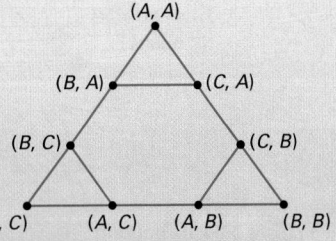

1. Which states in the graph for two disks represent solutions to the puzzle? Describe a possible solution to the puzzle.

2. For the case with three disks, the states may be represented as ordered triples that describe the positions of the disks in order from smallest to largest. Fill in the states in the graph at right. (Remember that if two vertices are connected by an edge, you must be able to move a disk from one to the other in a single move.) Then describe a solution to the puzzle.

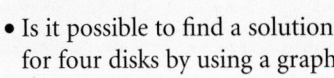

Cooperative Learning

Assign one game to each group. Students can divide the responsibilities, with one person moving the disks on the game and another person recording the number of moves. Ask students to compare and discuss results with the rest of the class before writing a formula.

Discuss

Fractals are a recent discovery in mathematics. This is due partially to the fact that fractals require massive amounts of computing before a clear pattern emerges. Ask students how increased computing capacity has changed mathematics. Have students answer the following questions:

- How can you use a graph to find the solution to the puzzle when using two disks? three disks?

- Is it possible to find a solution for four disks by using a graph?

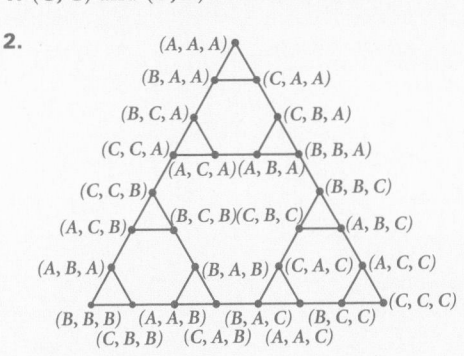

Activity 2

1. (*C*, *C*) and (*B*, *B*) are solution states

2.

(A, A, A)
(B, A, A) (C, A, A)
(B, C, A) (C, B, A)
(C, C, A) (B, B, A)
(A, C, A) (A, B, A)
(C, C, B) (B, B, C)
(A, C, B) (B, C, B)(C, B, C) (A, B, C)
(A, B, A) (B, A, B) (C, A, C) (A, C, C)
(B, B, B) (A, A, B) (B, A, C) (B, C, C)
(C, B, B) (C, A, B) (A, A, C)   (C, C, C)

Here is a particular solution, reading down the right leg:
Move the smallest disk to peg *C*.
Move the medium disk to peg *B*.
Move the smallest disk to peg *B*.
Move the largest disk to peg *C*.
Move the smallest disk to peg *A*.
Move the medium disk to peg *C*.
Move the smallest disk to peg *C*.

11 Chapter Review and Assessment

POSTULATES AND THEOREMS

Lesson	Section	Postulate or Theorem
11.3	11.3.1 Theorem	A graph contains an Euler path if and only if there are at most two odd vertices.
11.4	11.4.1 Jordan Curve Theorem	Every simple closed curve divides the plane into two distinct regions, the inside and the outside. Every curve that connects a point on the inside to a point on the outside must intersect the curve.
	11.4.2 Euler's Formula	For any polyhedron with V vertices, E edges, and F faces, $V - E + F = 2$.
11.7	11.7.3 The Theorem of Pappus	If A_1, B_1, and C_1 are three distinct points on one line and A_2, B_2, and C_2 are three distinct points on a second line, then the intersections of $\overline{A_1B_2}$ and $\overline{A_2B_1}$, of $\overline{A_1C_2}$ and $\overline{A_2C_1}$, and of $\overline{B_1C_2}$ and $\overline{B_2C_1}$ are collinear.
	11.7.4 The Theorem of Desargues	If one triangle is a projection of another triangle, then the intersections of the lines containing the corresponding sides of the two triangles are collinear.

Key Skills & Exercises

LESSON 11.1

Key Skills

Determine side lengths of golden rectangles.

A golden rectangle has a short side length of 5 units. Find the length of the long side.

The ratio of the long side to the short side of a golden rectangle is the golden ratio, $\phi \approx 1.618$.

$$\frac{\ell}{5} = \phi$$
$$\ell = \phi(5) \approx 8.09$$

Exercises

1. A golden rectangle has a short side length of 3 units. Find the length of the long side.

2. A golden rectangle has a long side length of 9 units. Find the length of the short side.

Chapter Test, Form A

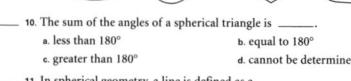

NAME _____ CLASS _____ DATE _____

Chapter Assessment
Chapter 11, Form A, page 1

Write the letter that best answers the question or completes the statement.

__a__ 1. Approximately what value of x would make the figure at right a golden rectangle?
 a. 10 b. 8
 c. 15 d. 16

__b__ 2. Approximately what value of x would make the figure at right a golden rectangle?
 a. 3 b. 7
 c. 12 d. 19

__d__ 3. What is the taxidistance between points A and B?
 a. 3 b. 4
 c. 5 d. 7

__b__ 4. If the taxidistance between two points is 3 units, what is the minimum number of pathways between the two points?
 a. 3 b. 4 c. 6 d. 9

__d__ 5. What is the circumference of a taxicab circle with a radius of 4?
 a. 4 b. 8 c. 16 d. 32

__c__ 6. A graph contains an Euler path if and only if there are at most how many odd vertices?
 a. 0 b. 1
 c. 2 d. 3

__b__ 7. How many odd vertices are shown in the graph at right?
 a. 0 b. 2
 c. 4 d. 6

__b__ 8. Properties that stay the same no matter how a figure is transformed are called _____.
 a. vertices b. invariants
 c. fractals d. iterations

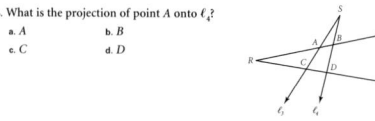

NAME _____ CLASS _____ DATE _____

Chapter Assessment
Chapter 11, Form A, page 2

__c__ 9. Which pair of figures is *not* topologically equivalent?

__c__ 10. The sum of the angles of a spherical triangle is _____.
 a. less than 180° b. equal to 180°
 c. greater than 180° d. cannot be determined

__d__ 11. In spherical geometry, a line is defined as a _____.
 a. radius b. chord c. semicircle d. great circle

__c__ 12. Which of the following shows a triangle in Poincaré's model of hyperbolic geometry?

__d__ 13. Which of the following is *not* an example of a fractal?
 a. Cantor dust b. Sierpenski gasket
 c. Koch snowflake d. Klein bottle

__b__ 14. What is the projection of point A onto ℓ_4?
 a. A b. B
 c. C d. D

__d__ 15. Which of the following is a nonrigid, affine transformation?
 a. $T(x, y) = (2x, 2y)$ b. $T(x, y) = (x + 1, y - 3)$
 c. $T(x, y) = (-x, -y)$ d. $T(x, y) = (3x, 2y)$

1. 4.85

2. 5.56

Construct a golden rectangle.

Given square $ABCD$, construct a golden rectangle whose short side is $\overline{AB}$.

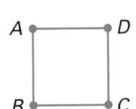

Extend side $\overline{BC}$. Locate M, the midpoint of $\overline{BC}$, and draw an arc centered at M that intersects $\overrightarrow{BC}$ at E. Construct a line perpendicular to $\overrightarrow{BC}$ at E and extend $\overline{AD}$ to intersect the perpendicular at F. $ABEF$ is a golden rectangle.

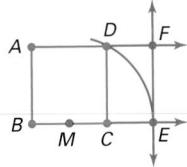

3. Draw square $WXYZ$ with a side length of 3 centimeters, and construct a golden rectangle whose short side is $\overline{XY}$.

4. Identify another golden rectangle in the diagram you constructed for Exercise 3.

LESSON 11.2

Key Skills

Find the taxidistance between two points.

Find the taxidistance between $(3, 1)$ and $(5, 2)$.

The taxidistance between (x_1, y_1) and (x_2, y_2) is

$$|x_2 - x_1| + |y_2 - y_1| =$$
$$|5 - 3| + |2 - 1| = 3.$$

Draw a taxicab circle with a given center and radius.

Draw the taxicab circle centered at $(1, -1)$ with a radius of 3.

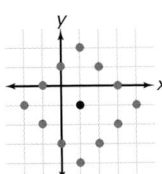

Exercises

Find taxidistance between the following pairs of points:

5. $(6, 1)$ and $(4, 5)$

6. $(7, 2)$ and $(-1, 0)$

Draw the taxicab circle with the given center and radius.

7. center: $(0, 0)$; radius: 2

8. center: $(2, 3)$; radius: 4

LESSON 11.3

Key Skills

Determine whether a graph contains an Euler path, an Euler circuit, or neither.

Does the graph below contain an Euler path, an Euler circuit, or neither?

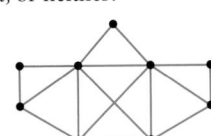

The graph contains 7 even vertices and 2 odd vertices. Since there are odd vertices, the graph does not contain an Euler circuit. However, there are not more than 2 odd vertices, so the graph contains an Euler path.

Exercises

For each graph below, determine whether the graph contains an Euler path, an Euler circuit, or neither.

9. **10.**

11. **12.**

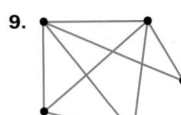

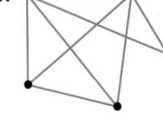

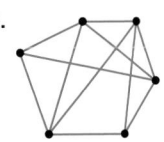

3.
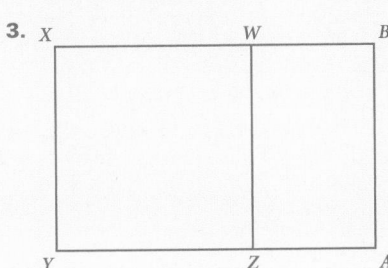

4. $AZWB$ is also a golden rectangle.

5. 6

6. 10

7.
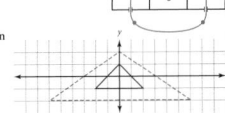

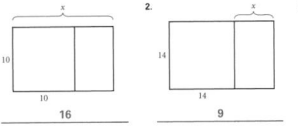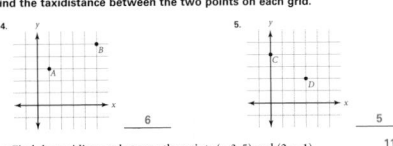
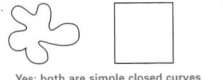

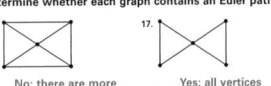

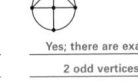
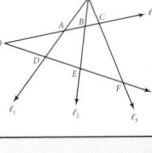

13. Yes; both are simple closed curves.

14. Note that the figure on the left intersects itself once and the figure on the right intersects itself twice. They are not topologically equivalent.

15. −4

16. A sphere has Euler characteristic 2. This is a topological invariant, so any surface with different Euler characteristic is not topologically equivalent.

17. The two curves that intersect at the endpoints of a diameter are both lines in spherical geometry.

18. The curve that wraps around the ball, intersecting each of the other curves in two places, is not a line in spherical geometry.

LESSON 11.4
Key Skills

Determine whether two figures are topologically equivalent.

Are the figures below topologically equivalent?

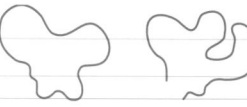

The figure on the left is a simple closed curve. It encloses a region of the plane. The figure on the right does not enclose any area. There is no way to deform one into the other without cutting the figure on the left. Thus, the figures are not topologically equivalent.

Find the Euler characteristic of a figure.

Find the Euler characteristic of the figure below.

The figure is topologically equivalent to the figure below, which has 28 vertices, 56 edges, and 26 faces.

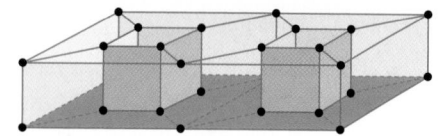

The Euler characteristic is
$V − E + F = 28 − 56 + 26 = −2$.

Exercises

Determine whether the figures in each pair below are topologically equivalent. Explain your reasoning.

13.

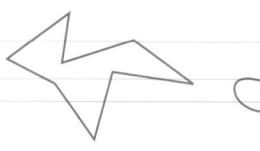

14.

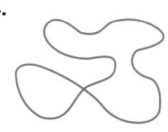 

15. Find the Euler characteristic of the figure below.

16. Use the Euler characteristic to prove that the figure above is not topologically equivalent to a sphere.

LESSON 11.5
Key Skills

Identify lines in spherical geometry.

Which of the following is a line in spherical geometry?

By definition, a line in spherical geometry is a great circle—a circle that divides the sphere into two hemispheres. Thus, figure CD is a line, but figure AB is not a line because it does not divide the circle into two hemispheres.

Exercises

Refer to the basketball below.

17. Describe a curve on the basketball that is a line in spherical geometry.

18. Describe a curve on the basketball that is not a line in spherical geometry.

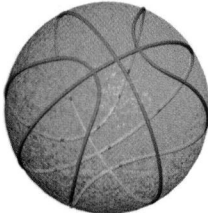

Identify lines in hyperbolic geometry.

Which figures in the diagram below are lines in Poincaré's model of hyperbolic geometry?

According to Poincaré's model, lines are defined as diameters of the outer circle or arcs that are orthogonal to the outer circle. Thus, figures *XY*, *YZ*, and *YU* are lines, but figure *VW* is not a line because it is not orthogonal to the outer circle.

Refer to the diagram of Poincaré's model of hyperbolic geometry below.

19. Name three figures in the diagram that are lines in hyperbolic geometry.

20. Draw a curve in the diagram that is not a line in hyperbolic geometry, and explain why it is not a line.

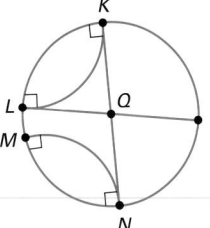

20.

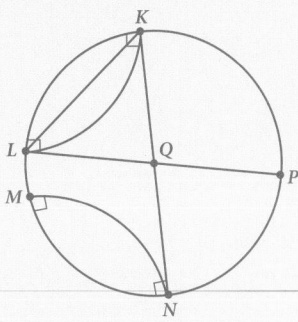

Drawing the chord from *K* to *L* on the circle gives us a curve which is not a line, since it fails to intersect the outer boundary of the circles at right angles.

21.

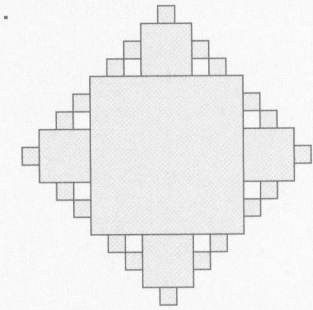

LESSON 11.6
Key Skills

Create fractals by using iterations.

To form the Sierpinski carpet, start with a solid square. Divide the square into a grid of 9 squares, and remove the center square. Repeat on the remaining squares. Draw the first three iterations of the Sierpinski carpet.

1 iteration

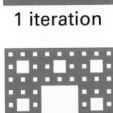

2 iterations   3 iterations

Exercises

The original figure and first iteration of a fractal are shown below.

21. Draw the second iteration of the fractal.

22. Draw the third iteration of the fractal.

23. Does the area of the fractal above increase or decrease with successive iterations? If it decreases, does it approach 0? If it increases, does it approach infinity?

24. Does the perimeter of the fractal above increase or decrease with successive iterations? If it decreases, does it approach 0? If it increases, does it approach infinity?

22.

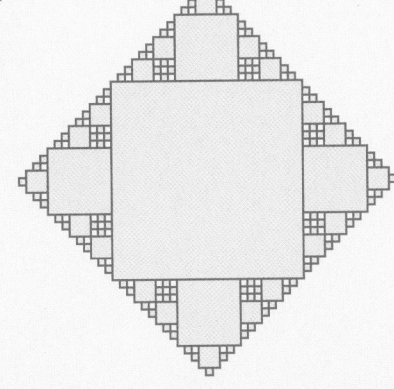

23. The area of the *n*th iteration is $\left(\frac{4}{9}\right)^n s^2$ more than the area of the $(n-1)$st iteration. The area of the fractal increases, but is bounded.

24. The perimeter of the fractal increases by $\frac{2}{3}s$ for each square added in the first iteration. Thus it increases by $\frac{8}{3}s$ in the first iteration. Therefore the perimeter increases without bound as *n* increases.

25.

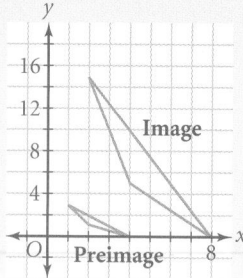

26.

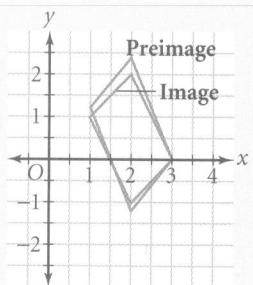

27.

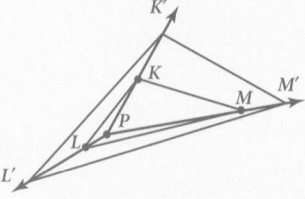

28.

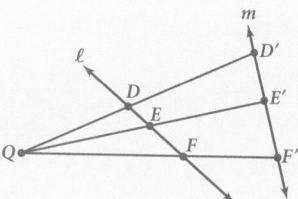

29. 73 ft or 28 ft

30. He should be flying in a great circle around the planet.

31. The photos were taken from different heights. Alex's photos have more detail, so his measurements may include small inlets or peninsulas that are not visible in the satellite photo.

Key Skills

Transform figures by using affine transformations.

Draw the preimage and image of the rectangle with vertices at $(0, 0)$, $(0, 6)$, $(3, 6)$, and $(3, 0)$ transformed by $T(x, y) = (3x, 4y)$.

$T(0, 0) = (0, 0)$
$T(0, 6) = (0, 24)$
$T(3, 6) = (9, 24)$
$T(3, 0) = (9, 0)$

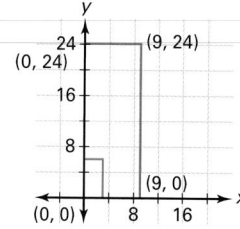

Transform figures by using central projections.

Draw a central projection of △ABC centered at point O.

 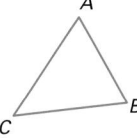

Draw projective rays $\overrightarrow{OA}$, $\overrightarrow{OB}$, and $\overrightarrow{OC}$, and choose points A', B', and C' on the rays. Draw △$A'B'C'$.

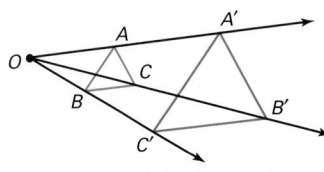

Exercises

Draw the preimage and image of the figure with the given vertices under the given transformation.

25. $(1, 3)$, $(2, 1)$, $(4, 0)$; $R(x, y) = (2x, 5y)$

26. $(2, -1)$, $(3, 0)$, $(2, 2)$, $(1, 1)$; $S(x, y) = (x, 1.2y)$

27. Draw a central projection of △KLM centered at P.

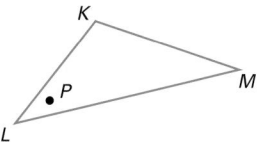

28. Draw a central projection of points D, E, and F onto line m centered at Q.

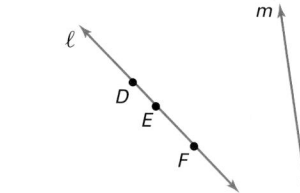

Applications

29. ARCHITECTURE An architect is designing an office building whose front face is to be in the shape of a golden rectangle. If the building will be 45 ft tall, how wide should its front face be?

30. GEOGRAPHY A pilot is flying from Tokyo to London. Use spherical geometry to describe the path the pilot should take in order to fly the shortest distance.

31. GEOGRAPHY Tani and Alex are trying to measure the coast of Oregon. Tani measures the coast from a satellite photo and finds that it is 490 km long. Alex uses a set of aerial photos and finds that it is 560 km long. Explain the difference in their results.

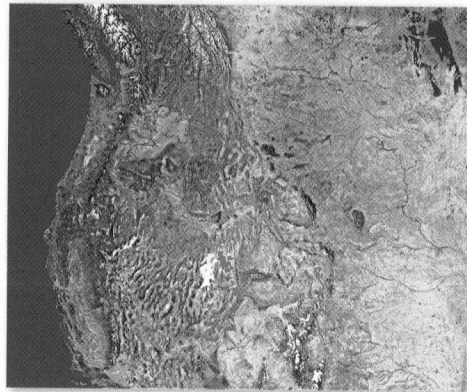

Chapter Test

Given one side length of a golden rectangle, find the unknown side length to the nearest hundredth.

1. $s = 15$ $l = \underline{?}$ **24.27** **2.** $s = \underline{?}$ $l = 8$ **4.94**

3. Draw square $ABCD$ with a side length of 2 inches, and construct a golden rectangle whose short side is $\overline{BC}$.

4. DESIGN A designer wants to make a sign in the shape of a golden rectangle. If the width of the sign is 21 inches, what should the length be? Round to the nearest inch. **34 in.**

Find the taxidistance between the following pairs of points:

5. $(4, 6)$ and $(1, 5)$ **4** **6.** $(8, 2)$ and $(-3, -1)$ **14**

Draw the taxicab circle with the given center and radius.

7. center $(0, 0)$; radius 5

8. center $(1, 4)$; radius 3

For each graph below, determine whether the graph contains an Euler path, an Euler circuit, or neither.

9. 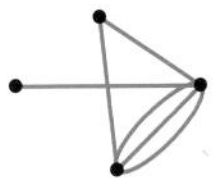  **10.**

11. WILDLIFE MANAGEMENT A park ranger comes across a set of valley bear trails, shown below. Could a bear travel all of the trails in one trip, starting at the river and not retracing any of the trails? Why or why not?

12. Verify Euler's formula for a rectangular prism.
$8 - 12 + 6 = 2$

13. Find the Euler characteristic of the figure at right.
$16 - 32 + 16 = 0$

For Exercises 14 and 15, determine whether the figures in each pair are topologically equivalent. Explain your reasoning.

14.

15.

For Exercises 16 and 17, refer to the diagram of a sphere below. Determine whether each figure is a line in spherical geometry.

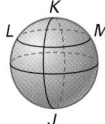

16. $\overleftrightarrow{LM}$ **no**

17. $\overrightarrow{JK}$ **yes**

18. Can a triangle in spherical geometry have two right angles? Explain.

19. Draw three lines that intersect to form a triangle in Poincare's system.

The original figure and first iteration of a fractal are shown at right.

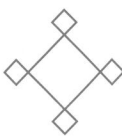

20. Draw the second iteration of the fractal.

21. Draw the third iteration of the fractal.

22. Does the area of the fractal increase or decrease with successive iterations? **increases**

23. Does the perimeter of the fractal increase or decrease with successive iterations? **increases**

Draw the preimage and image of the figure with the given vertices under the given transformation.

24. $(0, 0), (1, 4), (5, 2)$; $S(x, y) = (2x, 0.5y)$

25. $(1, 0), (1, 3), (5, 0), (5, 3)$; $R(x, y) = (-2x, 3y)$

3.
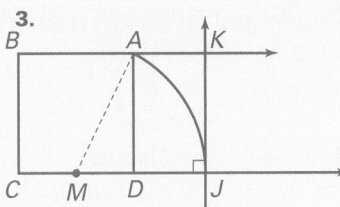

7.
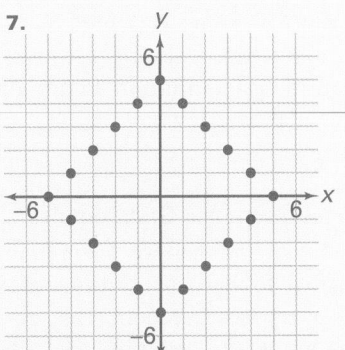

8.
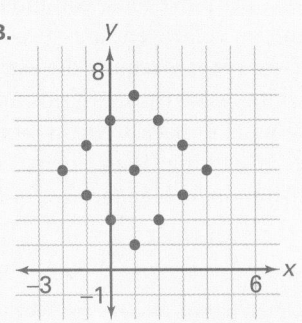

9. Euler path; no Euler circuit

10. Euler path and circuit

11. Yes; the associated graph has only two odd vertices, so the path is an Euler path.

14. Yes; just smooth out the corners in the figure on the right.

15. No; the figure on the left intersects itself while the figure on the right has a hole.

18. Yes, because the sum of the measures of the angles of a triangle is greater than 180°.

19. Sample answer:

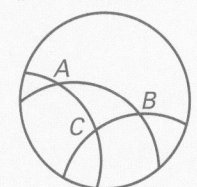

20.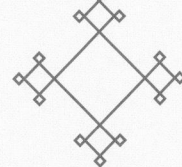

21.

24.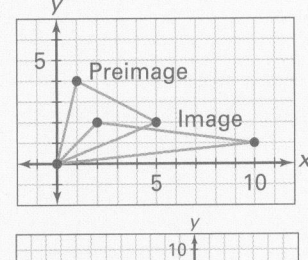

25.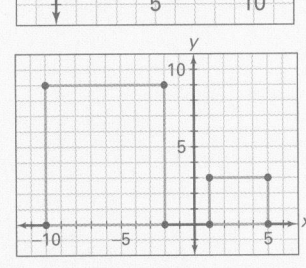

CHAPTERS **CUMULATIVE ASSESSMENT**

1-11

College Entrance Exam Practice

Multiple-Choice Samples

The first half of the Cumulative Assessment contains a multiple-choice section. This part of the Cumulative Assessment consists of items commonly found on standardized tests.

Free-Response Grid Samples

The second half of the Cumulative Assessment consists of a free-response section. This part requires student-produced response items like those commonly found on college entrance exams. These questions require the use of machine-scored answer grids. You may wish to have students practice answering these items in preparation for standardized tests.

1. c

2. a

3. b

4. a

MULTIPLE-CHOICE For Questions 1–8, write the letter that indicates the best answer.

1. Refer to the figure below. If $m\angle 1 = m\angle 2$, which statement is true? *(LESSON 3.4)*

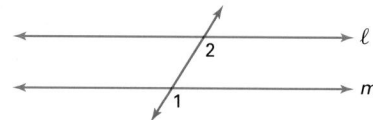

   a. $\angle 1$ and $\angle 2$ are alternate interior angles.
   b. $\angle 1$ and $\angle 2$ are vertical angles.
   c. Line l is parallel to line m.
   d. Line l is perpendicular to line m.

2. Refer to the figure below. Which statement is true? *(LESSON 5.4)*

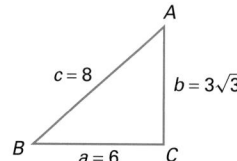

   a. $c^2 > a^2 + b^2$
   b. $c^2 = a^2 + b^2$
   c. $c^2 < a^2 + b^2$
   d. $c > a + b$

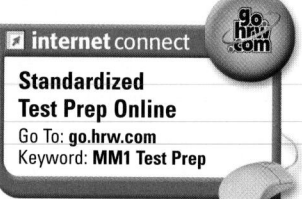

3. If the length of a side of a regular hexagon is 2, what is the area of the hexagon to the nearest tenth? *(LESSON 5.5)*

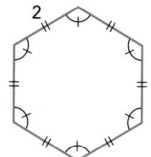

   a. 12.0
   b. 10.4
   c. 3.5
   d. 20.8

4. The figure below is a right rectangular prism. Find AG. *(LESSON 6.3)*

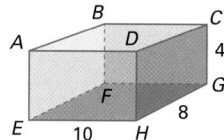

   a. $6\sqrt{5}$
   b. $4\sqrt{5}$
   c. $2\sqrt{29}$
   d. $4\sqrt{41}$

5. Find the *x*-intercept of the line given by the parametric equations below. *(LESSON 6.5)*

$$x = 4t - 8$$
$$y = 2t - 3$$
$$z = t + 4$$

a. $(0, 2, 6)$
b. $(0, 1, 6)$
c. $(0, 0, 0)$
d. none of the above

6. As the side length of a cube is increased, the ratio of surface area to volume __?__. *(LESSON 7.1)*

a. increases
b. remains constant
c. is equal to 1
d. decreases

7. If the radius of a sphere is doubled, the volume is increased by a factor of __?__. *(LESSON 8.6)*

a. 2
b. 4
c. 6
d. 8

8. The value of $\sin \theta$ is given by the ratio __?__. *(LESSON 10.2)*

a. $\dfrac{\text{opposite}}{\text{adjacent}}$

b. $\dfrac{\text{adjacent}}{\text{hypotenuse}}$

c. $\dfrac{\text{opposite}}{\text{hypotenuse}}$

d. $\dfrac{\text{hypotenuse}}{\text{opposite}}$

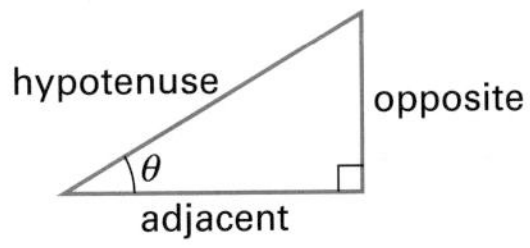

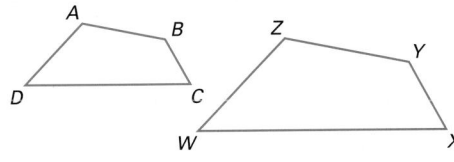

9. For the two similar quadrilaterals below, write a proportionality statement by using the ratios between the sides. *(LESSON 8.2)*

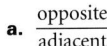
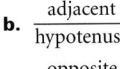
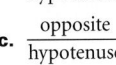

10. Determine whether the graph at right contains an Euler path, an Euler circuit, or neither. *(LESSON 11.3)*

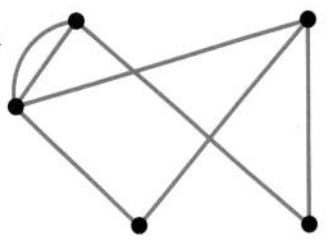

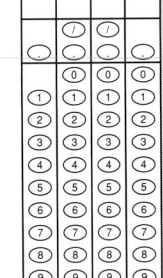

FREE-RESPONSE GRID
Items 11–15 may be answered by using a free-response grid such as that commonly used by standardized-test services.

11. How many axes of symmetry does the figure below have? *(LESSON 3.1)*

12. Find the slope of the segment with endpoints at $(2, 4)$ and $(3, -1)$. *(LESSON 3.8)*

13. Find *x* in the triangle below. *(LESSON 4.4)*

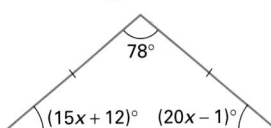
$78°$
$(15x + 12)°$ $(20x - 1)°$

14. Find *s* in the golden rectangle below. *(LESSON 11.1)*

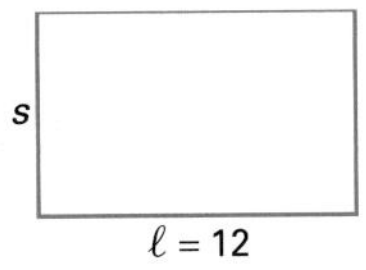

s
$\ell = 12$

15. Find the taxidistance between the points $(1, 4)$ and $(7, 10)$. *(LESSON 11.2)*

5. d

6. d

7. d

8. c

9. Answers may vary. Sample answer: $\dfrac{AB}{ZY} = \dfrac{DC}{WX}$

10. Euler path; no Euler circuit

11. 5

12. −5

13. $x = \dfrac{13}{5}$

14. 7.42

15. 12

A Closer Look at Proof and Logic

Lesson Presentation CD-ROM
Power Point® presentations for each lesson 12.1-12.5

CHAPTER PLANNING GUIDE

Lesson	12.1	12.2	12.3	12.4	12.5	Project and Review
Pupil's Edition Pages	768–775	776–781	784–790	791–797	798–804	782–783, 805–815
Practice and Assessment						
Extra Practice (Pupil's Edition)	855	856	856	857	857	
Practice Workbook	76	77	78	79	80	
Practice Masters Levels A, B, and C	226–228	229–231	232–234	235–237	238–240	
Standardized Test Practice Masters	87	88	89	90	91	92
Assessment Resources	153	154	155	157	158	156, 159–164
Visual Resources						
Lesson Presentation Transparencies Vol. 2	105–108	109–112	113–116	117–120	121–124	
Teaching Transparencies	110	111			112–113	
Answer Key Transparencies	426–429	430–434	435–442	443–446	447–450	451–453
Quiz Transparencies	12.1	12.2	12.3	12.4	12.5	
Teacher's Tools						
Reteaching Masters	151–152	153–154	155–156	157–158	159–160	
Make-Up Lesson Planner for Absent Students	76	77	78	79	80	
Student Study Guide	76	77	78	79	80	
Spanish Resources	76	77	78	79	80	
Block Scheduling Handbook						24–25
Activities and Extensions						
Lesson Activities		217–218			219–220	
Enrichment Masters	76	77	78	79	80	
Cooperative-Learning Activities	76	77	78	79	80	
Problem-Solving/ Critical Thinking	76	77	78	79	80	
Student Technology Guide		75–76		77		
Long Term Projects						45–48
Writing Activities for Your Portfolio						34–36
Tech Prep Masters						55–58
Building Success in Mathematics						31–33

LESSON PACING GUIDE

Lesson	12.1	12.2	12.3	12.4	12.5	Project and Review
Traditional	1 day	1 day	1 day	1 day	2 days	2 days
Block	$\frac{1}{2}$ day	$\frac{1}{2}$ day	$\frac{1}{2}$ day	$\frac{1}{2}$ day	1 day	1 day
Two-Year	2 days	2 days	2 days	2 days	4 days	4 days

CONNECTIONS AND APPLICATIONS

Lesson	12.1	12.2	12.3	12.4	12.5	Review
Algebra				796	804	
Geometry	768–775	776–781	784–790	791–797	798–804	808–815
Number Theory				795		
Business and Economics		780	784	795		
Life Skills	774		790			
Science and Technology		781	788	796	798, 803, 804	
Other	774					

BLOCK-SCHEDULING GUIDE

Day	Lesson	Teacher Directed: Lesson Examples, Teaching Transparencies	Student Guided: Activity, Try This	Cooperative-Learning Activity, Lesson Activity, Student Technology Guide	Practice: Practice & Apply, Extra Practice, Practice Workbook	Assessment: Quiz, Mid-Chapter Assessment	Problem Solving, Reteaching
1	12.1	8 min	8 min	10 min	25 min	8 min	8 min
	12.2	7 min	7 min	10 min	20 min	7 min	7 min
2	12.3	8 min	8 min	10 min	25 min	8 min	8 min
	12.4	7 min	7 min	10 min	20 min	7 min	7 min
3	12.5	15 min	15 min	20 min	55 min	15 min	15 min
4	Assess.	50 min	90 min	90 min	65 min	30 min	
		PE: Chapter Review	PE: Chapter Project, Writing Activities	Tech Prep Masters	PE: Chapter Assessment, Test Generator	Chap. Assess. (A or B), Alt. Assess. (A or B), Test Generator	

PE: Pupil's Edition

Alternative Assessment

The following suggest alternative assessments for students who may benefit from a different type of assessment than the regular chapter quizzes and the mid-chapter/end-of-chapter test. Visit the HRW web site to get additional Alternative Assessment material.

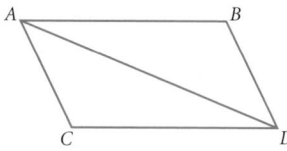

internet connect

Alternative Assessment

Go To: **go.hrw.com**
Keyword: **MG1 Alt Assess**

Performance Assessment

1. Define *modus ponens* and *modus tollens*, and give an example of each type of argument.

2. Draw a truth table like the one on page 786. Copy and complete the table for ~(*p* AND *q*) OR *q*.

3. Write the converse, inverse, and contrapositive of each conditional. Determine whether each new statement is true or false.
a. If two parallel lines are cut by a transversal, then alternate interior angles are parallel.
b. If the figure is a triangle, then the sum of its angle measures is 180°.

Portfolio Project

Suggest that students choose one of the following projects for inclusion in their portfolios.

1. Use coordinate geometry and the diagram at right to prove the following: If $\triangle ABD \cong \triangle DCA$, then *ABDC* is a parallelogram. Draw the diagram on a coordinate grid.

2. Complete the proof from Exercise 1 without coordinate geometry. Which proof is easier? Explain why.

3. Draw logic gates for the following statements:
a. ~*p* OR *q*
b. (*p* AND *q*) OR ~*p*
c. ~(*p* AND ~*q*) OR *q*

internet connect

The table below identifies the pages in this chapter that contain internet and technology information.

Resource Links

Parents can go online and find concepts that students are learning–lesson by lesson–and questions that pertain to each lesson, which facilitate parent-student discussion.

Go To: **go.hrw.com**
Keyword: **MG1 Parent Guide**

Technical Support

The following may be used to obtain technical support for any HRW software product.

Online Help: **www.hrwtechsupport.com**

e-mail: **tschrw@hbtechsupport.com**

HRW Technical Support Center: **(800)323-9239**

7 AM to 10 PM Monday through Friday CST

Visit the HRW math web site at: **www.hrw.com/math**

Technology

Technology Objectives and Suggestions

Lesson 12.1 Truth and Validity in Logical Arguments

In this lesson students define and use the valid argument forms Modus Ponens and Modus Tollens. They also define and illustrate the invalid argument forms of Affirming the Consequent and Denying the Antecedent.

Have students prove algebraic arguments. Use a graphics calculator to find all values of x for which the final conclusion is true. Then write the second sentence of the argument. For example:

Sentence 1: If x is a real number, then $x^2 + x - 3 \leq 0$.

Sentence 2: a is a value of x such that ...

Sentence 3: Therefore $a^2 + a - 3 \leq 0$.

Lesson 12.2 *And, Or,* and *Not* in Logic

In this lesson students solve logic problems using conjunction, disjunction, and negation. Use the **TEST LOGIC** menu of a graphics calculator to help create a truth table for Exercises 20 and 29–31. First store values for p, q, and r in memory. Use parentheses and the logic menu to find the truth value of individual expressions.

Lesson 12.3 A Closer Look at If-Then Statements

In this lesson students create truth tables for conditionals and converses, inverses, and contrapositives of conditionals. They also use if-then statements and forms of valid argument for problems involving logical reasoning.

Use the following program to demonstrate how if-then logic may be used in a graphics calculator program. This program uses the converse of the Pythagorean Theorem to determine whether a triangle is a right triangle.

```
PROGRAM:PYTHAG
:Input "INPUT SIDE A", A
:Input "INPUT SIDE B", B
:Input "INPUT SIDE C", C
:If A^2 + B^2 = C^2
:Then :Disp "RIGHT"
:Else :Disp "NOT RIGHT"
:End
```

Lesson 12.4 Indirect Proof

In this lesson students develop the concept of indirect proof and use indirect proof with problems involving logical reasoning. For Exercises 18–33 use geometry software to redraw the figures for each proof. For Exercises 41–44 use geometry software to try to draw each triangle described, if possible.

Lesson 12.5 Computer Logic

In this lesson students explore on-off tables, logic gates, and computer logic networks. They also solve problems by using computer logic. Students can use the **TEST LOGIC** menu of a graphics calculator to give truth values for compound sentences in Exercises 12–27. Store the values of p, q and r as 0 or 1 and use the menu to calculate the truth value of the compound sentence.

Background Information

In this chapter, students explore the validity of arguments and the different rules of formal logic, including if-then statements. They also explore computer logic and establish proofs by indirect reasoning and by using coordinate geometry.

CHAPTER RESOURCES

- Block-Scheduling Handbook
- Writing Activities for Your Portfolio
- Tech Prep Masters
- Long-Term Project
- Assessment Resources:
  Mid-Chapter Assessment
  Chapter Assessments
  Alternative Assessments
- Test and Practice Generator
- Technology Handbook

Chapter Objectives

- Define and use the valid argument forms *modus ponens* and *modus tollens*. [**12.1**]

- Define and illustrate the invalid argument forms of affirming the consequent and denying the antecedent. [**12.1**]

- Define *conjunction, disjunction,* and *negation*. [**12.2**]

- Solve logic problems by using conjunction, disjunction, and negation. [**12.2**]

- Create truth tables for conditionals and for converses, inverses, and contrapositives of conditionals. [**12.3**]

- Use if-then statements and forms of valid argument for problems involving logical reasoning. [**12.3**]

A Closer Look at Proof and Logic

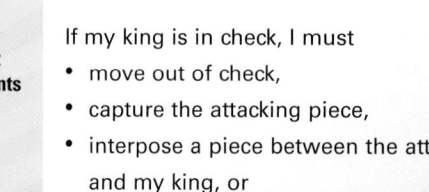

CAN A MACHINE THINK? AT TIMES IT CERTAINLY seems as if they can—as when the computer Deep Blue beat the world chess champion Gary Kasparov in a chess match. Whether they can actually think or not, machines can clearly follow rules and make "decisions" based on given conditions. The decisions made by chess-playing computers and other "smart" machines are made possible by formal logic.

Formal logic reduces logical procedures to their essential elements, which can be implemented by computers or other machines. For example, the following rule in chess can be programmed into a chess computer.

If my king is in check, I must

- move out of check,
- capture the attacking piece,
- interpose a piece between the attacker and my king, or
- resign.

Lessons

About the Photos

When world chess champion Garry Kasparov played against IBM's Deep Blue, the world's most powerful chess playing machine, both players put their logic to work. Kasparov's defeat can be explained by the fact that Deep Blue's unique software engineering and massive parallel processing allow the machine to generate up to 200,000,000 positions per second when searching for the optimal move. Technological innovations of this kind exemplify the importance of logic in decision making, strategic planning, and problem solving.

- Develop the concept of indirect proof (*reductio ad absurdum*) or proof by contradiction. [**12.4**]

- Use indirect proof with problems involving logical reasoning. [**12.4**]

- Explore on-off tables, logic gates, and computer logic networks. [**12.5**]

- Solve problems by using computer logic. [**12.5**]

PORTFOLIO ACTIVITIES PROJECT

Portfolio Activities appear at the end of Lessons 12.1 and 12.4. Each serves as preparation for the Chapter Project. The Portfolio Activities, as well as the Chapter Project Activities, are appropriate for inclusion in the student's portfolio. Students should be encouraged to include in their portfolios any other work in which they feel a sense of pride or a sense of accomplishment.

About the Chapter Project

Number theory is a branch of mathematics that is at least as old as Euclid. In fact, three books of Euclid's *Elements* are devoted to number theory. In the *Elements*, Euclid uses segments of different lengths to represent positive whole numbers and proves theorems about numbers geometrically.

In the Chapter Project, you will study two famous theorems from number theory. The first is that there are infinitely many prime numbers, and the second is that the square root of 2 is irrational.

After completing the Chapter Project you will be able to do the following:

- Have a deeper understanding of the techniques of proof.

- Understand the need for a class of numbers that are not rational.

About the Portfolio Activities

Throughout the chapter, you will be given opportunities to complete Portfolio Activities that are designed to support your work on the Chapter Project.

The theme of each Portfolio Activity and of the Chapter Project is number theory.

- In the Portfolio Activity on page 775, you will explore the Euclidean algorithm for finding the greatest common divisor (gcd) of two numbers

- In the Portfolio Activity on page 797, you will learn how to prove the Euclidean algorithm.

Truth and Validity in Logical Arguments

1. Convert the following statement to if-then form: A frog is an amphibian. **If an animal is a frog, then it is an amphibian.**

2. Write the converse of the following statement: If Jamie scores 95% on the test, then she will get an A in the course. **If Jamie gets an A in the course, then she scored 95% on the test.**

Also on Quiz Transparency 12.1

Objectives

- Define and use the valid argument forms *modus ponens* and *modus tollens*.

- Define and illustrate the invalid argument forms of *affirming the consequent* and *denying the antecedent.*

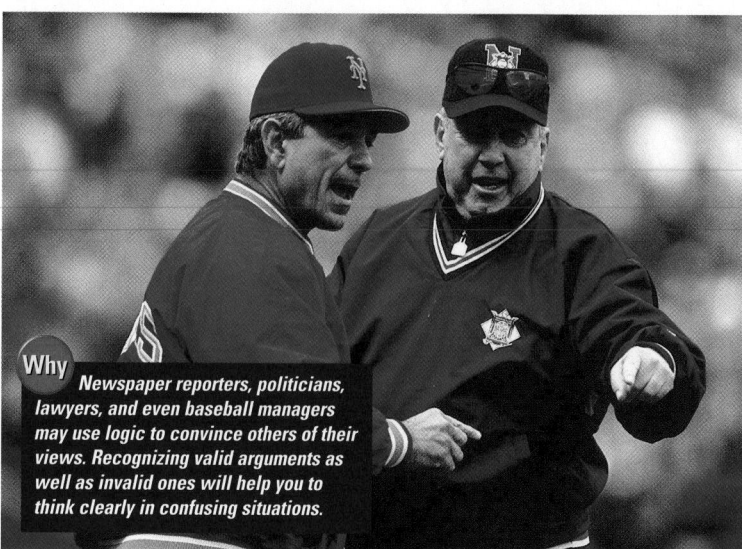

Why *Newspaper reporters, politicians, lawyers, and even baseball managers may use logic to convince others of their views. Recognizing valid arguments as well as invalid ones will help you to think clearly in confusing situations.*

Teach

Why Studying the rules of formal logic should help students recognize the importance of rules in determining whether arguments are valid. The applicability of valid and convincing arguments ranges from basic human interactions to the practice of law.

Valid Arguments

In logic, an **argument** consists of a sequence of statements. The final statement of the argument is called the **conclusion**, and the statements that come before it are known as **premises**. The following is an example of a logical argument:

If an animal is an amphibian, then it is a vertebrate.
Frogs are amphibians.   *Premises*

Therefore, frogs are vertebrates.   *Conclusion*

In this argument, the conclusion is said to follow logically from the premises. The premises force the conclusion. An argument of this kind is known as a **valid argument**, and the conclusion of such an argument is said to be a **valid conclusion**.

A valid argument makes the following "guarantee": *If the premises are all true, then the conclusion is true.* In the valid argument above, both premises are true. Therefore, the conclusion must be true.

Alternative Teaching Strategy

USING VISUAL MODELS Students may be able to visualize the hypothesis and conclusion of an argument with a Venn diagram. To do so, it may be helpful to rephrase a conditional as a declarative statement. For example, "If a car is a Corvette, then it is a Chevrolet" can be rephrased as "All Corvettes are Chevrolets." The declarative statement may be easier for the students to diagram.

Invalid Arguments

Now consider a different argument:

Invalid argument (Don't use one like this!)

Some vertebrates are warm-blooded.
Frogs are vertebrates.   *Premises*

Therefore, frogs are warm-blooded.   *Conclusion(?)*

This new argument is invalid. Both of the premises are true, but the conclusion is false. *This is never the case in a valid argument.*

CRITICAL THINKING Consider the second argument. Suppose you did not know that a frog is not a warm-blooded animal. Would you have questioned the conclusion anyway? Is there something basically wrong with the argument? If you think there is, try to describe what it is.

TRY THIS Write your own examples of valid and invalid arguments.

A Valid Argument Form: *Modus Ponens*

Logicians can tell whether an argument is valid or invalid without knowing anything about the truth of its premises or conclusion. They are able to do this by analyzing the form of the argument. The valid argument on the previous page, for example, has the following form:

> Note: *p* and *q* represent **statements**: that is, sentences that can be true or false—not questions, commands, etc.

Argument Form: *Modus Ponens*

If p then q
p *Premises*

Therefore, q *Conclusion*   **12.1.1**

This argument form is sometimes referred to by its Latin name, **modus ponens**, or the "proposing mode." Any argument that has this form is valid, regardless of the statements that are substituted for p and q. The following nonsense argument, for example, is valid because it has a valid form:

If flivvers twiddle, then bokes malk.
Flivvers twiddle.   *Premises*

Therefore, bokes malk.   *Conclusion*

This argument's form guarantees that if the first two statements should somehow turn out to be true, then the third statement (the conclusion) would be true as well.

Interdisciplinary Connection

LAW Lawyers are trained to recognize and use logical reasoning. This includes using information that is known and using deductive reasoning to reach a logical conclusion. Ask students to write a court scene involving a brief argument that a lawyer might use.

Teaching Tip

Emphasize the difference between a valid argument and an invalid argument. A valid argument guarantees that if the premises are all true, then the conclusion is also true. An argument in which the premises are true but the conclusion is false is not valid. Discuss why the argument at the top of page 769 is invalid. Point out that the form of an argument rather than its content determines its validity. Refer to the valid argument on page 770 in which the content is not true.

CRITICAL THINKING
Yes; the conclusion should be questioned even if it is not known that a frog is not a warm-blooded animal. The problem with the argument is that the first premise does not state that every vertebrate is warm-blooded.

TRY THIS
Sample answer: Valid argument: If an animal is an amphibian, then it is a vertebrate. Frogs are amphibians. Therefore, frogs are vertebrates. Invalid argument: Some reptiles are venomous. Chameleons are reptiles. Therefore, chameleons are venomous.

Cooperative Learning

Have groups of students practice composing arguments that use the *modus ponens* or *modus tollens* pattern. Have group members critique each other's arguments.

Another Valid Argument Form: *Modus Tollens*

Consider the following argument:

If a shirt is a De Morgan, then it has a blackbird logo.
This shirt does not have a blackbird logo.   *Premises*

Therefore, this shirt is not a De Morgan.   *Conclusion*

Does this argument seem valid to you? If you knew that the premises were true, would you be certain that the conclusion was true? The argument above has the following form:

> *The symbol ~ means "not."*

Argument Form: *Modus Tollens*

If p then q
$\sim q$ *Premises*

Therefore, $\sim p$ *Conclusion*   **12.1.2**

This is a valid argument form. It is sometimes referred to by its Latin name, ***modus tollens***, or the "removing mode." In more recent times, it has come to be known as the law of indirect reasoning.

False Premises

If the premises of an argument are false, then there is no guarantee that the conclusion is true, even though the argument might be valid. The following *modus ponens* argument is valid, but its conclusion is false.

If an animal is an amphibian, then it can fly.
A frog is an amphibian.   *Premises*

Therefore, frogs can fly.   *Conclusion*

The conclusion, though false, is a valid conclusion of the argument because the form of the argument is valid. Remember, a valid argument guarantees that its conclusion is true *only if its premises are true*. There is no guarantee if one or more of the premises are false.

CRITICAL THINKING

CRITICAL THINKING
Yes. Sample answer: If something is yellow, then it is a fruit. A banana is yellow. Therefore, a banana is a fruit.

The valid argument above has a false premise and a false conclusion. Do you think it is possible for a valid argument to have a false premise and a true conclusion? If so, give an example. If not, explain why not.

Enrichment

Research Lewis Carroll's use of logical arguments in *Symbolic Logic*. Explain how the use of nonsensical statements supports the importance of form in a logical argument.

Inclusion Strategies

USING SYMBOLS Encourage students to use symbols to represent premises and conclusions. This will emphasize the importance of form in determining the validity of an argument. Make sure students understand how the form of an argument relates to the validity of the conclusion.

Invalid Forms

You should be careful not to confuse the *modus ponens* form with the following *invalid* form:

Invalid Form: Affirming the Consequent

If *p* then *q*

q   *Premises*

Therefore, *p*   *Not a valid conclusion*   **12.1.3**

> The "consequent" is the conclusion, *q*, of the conditional.

Form 12.1.3 is the form of a common logical mistake, or fallacy, known as "affirming the consequent." The conclusion does not follow logically from the premises even if it is a true statement. Be sure that you understand the difference between this form and the *modus ponens* form, which it closely resembles. An example of affirming the consequent is as follows:

If Sancho is having plum pudding for dessert,
    then he is happy.   *Premises*
Sancho is happy.

Therefore, Sancho is having plum pudding for dessert.   *Conclusion (?)*

Clearly, the argument doesn't work, because Sancho may be happy for some other reason.

Picasso, Pablo. *Don Quixote. 1955.* Musee d'Art et d'Histoire, St. Denis, France. © 2001 Estate of Pablo Picasso/Artists Rights Society (ARS), New York. Scala/Art Resource.

Don Quixote's squire, Sancho Panza, loved good food.

Another important invalid form that you will need to recognize closely resembles the *modus tollens* form.

Invalid Form: Denying the Antecedent

If *p* then *q*

~*p*   *Premises*

Therefore, ~*q*   *Not a valid conclusion*   **12.1.4**

> The "antecedent" is the hypothesis, *p*, of the conditional.

An example of denying the antecedent is as follows:

If Susan overslept, then she is running late.
Susan did not oversleep.   *Premises*

Therefore, Susan is not running late.   *Conclusion (?)*

Again, the argument isn't valid because Susan might be running late for any number of reasons.

ASSIGNMENT GUIDE	
In Class	1–9
Core	11–27
	29–35 odd
Core Plus	11–27 odd, 28–36
Review	37–45
Preview	46, 47

✎ Extra Practice can be found beginning on page 818.

Exercises

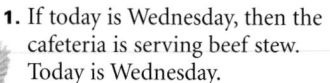

Communicate

In Exercises 1–4, determine whether the argument is valid. Explain why or why not.

1. If today is Wednesday, then the cafeteria is serving beef stew.
   Today is Wednesday. *Premises*

   Therefore, the cafeteria is serving beef stew. *Conclusion (?)*

2. If pigs fly, then today is February 30.
   Today is February 30. *Premises*

   Therefore, pigs fly. *Conclusion (?)*

3. If Jan is a man, then Jan is mortal.
   Jan is not a man. *Premises*

   Therefore, Jan is not mortal. *Conclusion (?)*

4. If $y = x$, then $a = b$.
   $a \neq b$ *Premises*

   Therefore, $y \neq x$. *Conclusion (?)*

5. Is it possible for a valid argument to have a false conclusion? Explain your reasoning.

Guided Skills Practice

In Exercises 6–9, analyze the form of each argument and give its traditional name. Then state whether it is valid or invalid.

6. If the weather takes a turn for the worse, then the local farmers will suffer a loss of income.
   The weather takes a turn for the worse. *Modus Ponens;* **valid**

   Therefore, the local farmers will suffer a loss of income.
   (ARGUMENT FORM 12.1.1)

7. If Stokes was in top form, then he won the competition.
   Stokes did not win the competition.

   Therefore, Stokes was not in top form. *Modus Tollens;* **valid**
   (ARGUMENT FORM 12.1.2)

Dusenberg

8. If the car is a Dusenberg, then it is a classic.
   The car is a classic.

   Therefore, the car is a Dusenberg. **Affirming the consquent; invalid**
   (ARGUMENT FORM 12.1.3)

9. If Sean stuffed himself at lunch, then he is feeling sleepy now.
   Sean did not stuff himself at lunch.

   Therefore, Sean is not feeling sleepy now. **Denying the antecedent; invalid**
   (ARGUMENT FORM 12.1.4)

In Exercises 10–11, write a valid conclusion from the given premises, and identify the form of the argument.

10. If the team won on Saturday, then the team is in the playoffs.
The team is not in the playoffs.

11. If Sabrina finished her assignment on time, then she did a stupendous amount of work at the last minute.
Sabrina finished her work on time.

In Exercises 12–15, arrange the sentences to form an argument. Identify the argument form, and state whether the argument is valid or invalid.

12. Therefore, Samantha is ill.
Samantha is absent.
If Samantha is ill, then she is absent.

13. Sims is not a man of good moral character.
If Sims is a man of good moral character, then he is innocent.
Therefore, Sims is not innocent.

14. Hedgehogs are tone deaf.
Therefore, hedgehogs will seldom be seen at symphony concerts.
If hedgehogs are tone deaf, they will seldom be seen at symphony concerts.

15. There were unpleasant surprises.
If the plan was foolproof, then there were no unpleasant surprises.
Therefore, the plan was not foolproof.

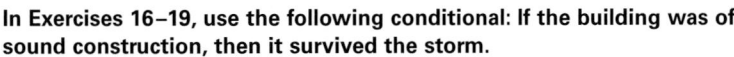

In Exercises 16–19, use the following conditional: If the building was of sound construction, then it survived the storm.

16. Write a *modus ponens* argument, using the given conditional as one of its premises.

17. Write a *modus tollens* argument, using the given conditional as one of its premises.

18. Write an invalid argument in the form of affirming the consequent, using the given conditional as one of its premises. Label the argument as invalid.

19. Write an invalid argument in the form of denying the antcedent, using the given conditional as one of its premises. Label the argument as invalid.

You are given the following premises:

If a student studies, the student will succeed.
Eleanor studies.
Tamara does not study.
Jose will succeed.
Mary will not succeed.

Which of the following conclusions are valid?

20. Eleanor will succeed. valid

21. Eleanor will not succeed. invalid

22. Tamara will succeed. invalid

23. Tamara will not succeed. invalid

24. Jose studies. invalid

25. Jose does not study. invalid

26. Mary studies. invalid

27. Mary does not study. valid

(See Exercise 14.)

18. If the building was of sound construction, then it survived the storm. The building survived the storm. Therefore, it was of sound construction. Invalid.

19. If the building was of sound construction, then it survived the storm. The building was not of sound construction. Therefore, it did not survive the storm. Invalid.

---

10. By the *modus tollens* argument form: Therefore, the team did not win on Saturday.

11. By the *modus ponens* argument form: Therefore, she did a stupendous amount of work at the last minute.

12. If Samantha is ill, then she is absent. Samantha is absent. Therefore, Samantha is ill.
Argument form: affirming the consequent; invalid

13. If Sims is a man of good moral character, then he is innocent. Sims is not a man of good moral character. Therefore, Sims is not innocent.
Argument form: denying the antecedent; invalid

14. If hedgehogs are tone deaf, they will seldom be seen at symphony concerts. Hedgehogs are tone deaf. Therefore, hedgehogs will seldom be seen at symphony concerts.
Argument form: *modus ponens;* valid

15. If the plan was foolproof there were no unpleasant surprises. There were unpleasant surprises. Therefore, the plan was not foolproof.
Argument form: *modus tollens;* valid

16. If the building was of sound construction, then it survived the storm. The building was of sound construction. Therefore, it survived the storm.

17. If the building was of sound construction, then it survived the storm. The building did not survive the storm. Therefore, it was not of sound construction.

Technology

Students can use geometry graphics software to construct the quadrilaterals in Exercises 28–31. Have them use the measurement feature of the software to verify the conclusion of each argument.

Error Analysis

Students may have difficulty with the symbols used to represent premises in Exercises 32–34. Encourage them to replace the symbols with verbal statements if necessary.

28. Valid by *modus ponens*

29. False. The diagonals of a parallelogram are congruent only if it is a rectangle.

30. The conclusion is a true statement without regard to the rest of the argument, based on the definitions of a rectangle and a square.

31. False. The diagonals of a parallelogram are congruent only if it is a rectangle.

32. Not necessarily. If the premises of an argument are false, then there is no guarantee that the conclusion is true, even though the argument might be valid.

In Exercises 28–31, consider the following argument:
> If a quadrilateral is a parallelogram, then its diagonals are congruent.
> Quadrilateral *PQRS* is a parallelogram.
> Therefore, the diagonals of quadrilateral *PQRS* are congruent.

28. Is the argument valid? Explain your reasoning.

29. Is the first premise true or false? Explain your reasoning.

30. If quadrilateral *PQRS* is a rectangle or a square, is the conclusion of the argument true or false? Explain your reasoning.

31. If quadrilateral *PQRS* is not a rectangle or a square, is the conclusion of the argument true or false? Explain your reasoning.

☑ internet connect

Homework Help Online
Go To: go.hrw.com
Keyword:
MG1 Homework Help
for Exercises 32-34

A valid argument has premises *a*, *b*, *c*, and *d* and conclusion *r*.

32. Does the validity of the argument guarantee that the premises and conclusion must be true? Explain your reasoning.

33. If the four premises are true, does the validity of the argument guarantee that *r* is true? Explain your reasoning.

34. Under what circumstances might *r* be false?

In Exercises 35 and 36, use two of the given premises to write a valid conclusion. Identify the form of the argument that you used.

35. SPORTS In football, if a team does not move the ball 10 yd in 4 downs, then they lose possession of the ball.
The Mammoths did not lose possession of the ball.
The Voyageurs moved the ball 10 yd in 4 downs.
The Cheetahs did not move the ball 10 yd in 4 downs.

CHALLENGE

36. LANDSCAPING If tulips are not planted in the fall, then they will not flower in the spring.
Adrianne's tulips did not flower in the spring.
Nuna's tulips flowered in the spring.
Geraldo planted tulips in the fall.

 *Look Back*

Can each pair of triangles below be proven congruent? Why or why not?
(LESSONS 4.2 AND 4.3)

37. Yes, by using the SSS postulate.

38. No, since none of the congruence postulates apply.

39. Yes, by using the SAS postulate.

37. 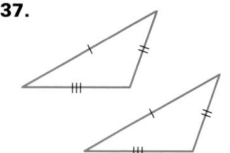 38. 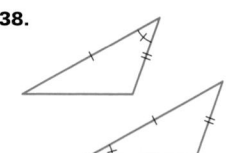 39.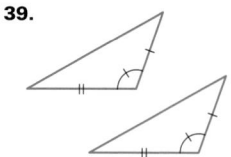

For Exercises 40–42, refer to the figure below, in which m∠1 = 20°, m∠2 = 35°, and m$\widehat{SR}$ = 80°.
(LESSONS 9.3 AND 9.4)

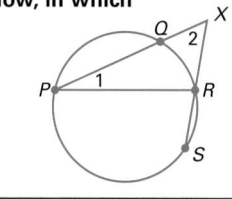

40. m$\widehat{QR}$ = _____?_____ 40°

41. m$\widehat{PS}$ = _____?_____ 110°

42. m$\widehat{PQ}$ = _____?_____ 130°

Practice

NAME _____ CLASS _____ DATE _____

Practice
12.1 Truth and Validity in Logical Arguments

In Exercises 1–6, write a valid conclusion from the given premises. Identify the form of the argument.

1. If it is a weekend, then José is not at work. It is a weekend.
   Therefore, José is not at work. (*modus ponens*)

2. If it is a weekday, then José is at work. José is not at work.
   Therefore, it is not a weekday. (*modus tollens*)

3. If José is not at work, then he is with Anna. José is not with Anna.
   Therefore, José is at work. (*modus tollens*)

4. If José is at work, then he is not with Anna. José is with Anna.
   Therefore, José is not at work. (*modus tollens*)

5. If it is a weekday, then José is at work. If José is at work, then he is wearing a tie. It is a weekday.
   Therefore, José is wearing a tie. (*modus ponens*)

6. If it is a weekend, then José is not at work. If José is not at work, then he is wearing jeans. It is a weekend.
   Therefore, José is wearing jeans. (*modus ponens*)

You are given the following premises:

If you exercise, then you are energized.   Jean was energized.
Ian exercised.   Johan was not energized.
Jon did not exercise.

Which of the following conclusions are valid? Explain.

7. Ian was energized. _____ Valid; modus ponens
8. Jon was energized. _____ Invalid; the statement "If you exercise, then you are energized" says nothing about what happens when you do not exercise.
9. Jean had exercised. _____ Invalid; affirming the consequent
10. Johan had exercised. _____ Invalid; modus tollens
11. Ian was not energized. _____ Invalid; modus ponens
12. Jon was not energized. _____ Invalid; denying the antecedent
13. Jean had not exercised. _____ Invalid; Jean could have exercised.
14. Johan had not exercised. _____ Valid; modus tollens

33. Yes, since a valid argument guarantees that its conclusion is true if its premises are true.

34. *r* might be false if one or more of the premises is false.

35. Sample answer: In football, if a team does not move the ball 10 yards in 4 downs, then they lose possession of the ball. The Mammoths did not lose possession of the ball. Therefore, the Mammoths moved the ball at least 10 yards in 4 downs. (*modus tollens*)

36. Sample answer: If tulips are not planted in the fall, then they will not flower in the spring. Nina's tulips flowered in the spring. Therefore, she must have planted them in the fall. (*modus tollens*)

44. An angle in the second quadrant cannot have the same tangent because the tangent is negative in the second quadrant.

θ is an angle in the first quadrant with a tangent of $\sqrt{3}$. *(LESSON 10.3)*

43. Find θ, $\sin\theta$, and $\cos\theta$. $\theta = 60°$; $\sin\theta = \frac{\sqrt{3}}{2} \approx 0.866$; $\cos\theta = \frac{1}{2} = 0.5$

44. Can an angle in the second quadrant have the same tangent as θ? Why or why not?

45. Give the measure of an angle in the third quadrant with the same tangent as θ. **240°**

 *Look Beyond*

Recall the If-Then Transitive Property from Lesson 2.2:
   If *p* then *q*
   If *q* then *r*
   Therefore, if *p* then *r*

List all conclusions that can be drawn from the premises given below.

46. If *x* then *y*
   If *y* then *k*
   ~*k*

   ~y and ~x

47. If *n* then *m*
   If *q* then *r*
   If *m* then *q*
   n
   m, q, and r

THE EUCLIDEAN ALGORITHM

An algorithm for finding the greatest common divisor (gcd) of two numbers is found in Book VII of Euclid's *Elements*. Trace the steps in the following example of finding the gcd of 1320 and 546:

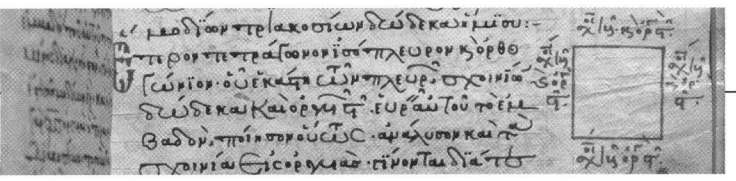

$$1320 = 546(2) + 228 \quad\longleftarrow\quad \text{\textit{To find the numbers in this row, note that } } 1320 \div 546 = 2 \text{ remainder } 228.$$

$$546 = 228(2) + 90$$

$$228 = 90(2) + 48$$

$$90 = 48(1) + 42$$

$$48 = 42(1) + \boxed{6} \quad\longleftarrow\quad \text{\textit{The last nonzero entry in this column is the gcd.}}$$

$$42 = 6(7) + 0$$

1. Follow the steps used in the example to find the gcd of 630 and 165, of 280 and 150, and of 462 and 120.

2. Two numbers are *relatively prime* if their gcd is 1. Use the Euclidean algorithm to show that 330 and 91 are relatively prime and that 560 and 429 are relatively prime.

1. Rewrite the following two sentences as one sentence connected by the word *and*: It is raining today. School is in session. **It is raining today *and* school is in session.**

2. Rewrite the two sentences in Exercise 1 as one sentence connected by the word *or*: **It is raining today *or* school is in session.**

3. Add the word *not* in order to negate the following sentence: Albert did get permission to see a movie. **Albert did *not* get permission to see a movie.**

Also on Quiz Transparency 12.2

Teach

Why Logical arguments make use of conjunctions, disjunctions, and negations. The connectors *AND*, *OR*, and *NOT* must be used appropriately in order to create logical arguments that are valid.

And, Or, and Not in Logic

12.2

Objectives

- Define *conjunction*, *disjunction*, and *negation*.

- Solve logic problems by using conjunction, disjunction, and negation.

Why *The words* and, or, *and* not *are used frequently in everyday situations. These words have precise mathematical meanings in logical arguments.*

In logic, a statement is a sentence that is either true or false. The sentence "Belinda ordered pepperoni on her pizza" is a statement because it must be either true or false. A **compound statement** is formed when two or more statements are connected. The sentence "John had a soda, and Belinda had tea" is a compound statement. A compound statement, like a simple statement, is either true or false.

Conjunctions

A compound statement that uses the word *and* to connect simple statements is called a **conjunction**.

Sentence p:	Today is Tuesday.
Sentence q:	Tonight is the first varsity track meet.
Conjunction p AND q:	Today is Tuesday, and tonight is the first varsity track meet.

A conjunction is true *if and only if* both of its statements are true. If one or both of its statements are false, the conjunction is false. The four possibilities for a conjunction can be illustrated in a **truth table.**

All possible combinations of truth values for the two statements that form the conjunction are placed in the first two columns. The last column indicates the truth values for the conjunction. In the first combination, for example, both of the statements that make up the conjunction are true. In this case, the conjunction is true.

p	q	p AND q
T	T	T
T	F	F
F	T	F
F	F	F

Alternative Teaching Strategy

INVITING PARTICIPATION Have groups of students form the conjunction and disjunction of mathematical statements. Ask them to determine as a group whether the compound sentences they form are true or false. Have them compare their sentences with those of other groups.

Interdisciplinary Connection

BUSINESS Joint checking accounts are set up so that only one of the two people on the account has to endorse a check before it is valid. Have students explain how this exemplifies a disjunction. Ask them to explain how to set up a checking account so that it is like a conjuction.

EXAMPLE 1 Determine whether the following conjunctions are true:

a. George Washington was the first president of the United States, and John Adams was the second.

b. The sum of the measures of the angles of a triangle is 200°, and blue is a color.

SOLUTION

a. The conjunction is true because both of its statements are true.

b. The conjunction is false because one of its statements is false.

Disjunctions

Two statements may also be combined into a single compound statement by the word *or*. This type of compound statement is known as a **disjunction**.

When used in everyday language, the word *or* often means "one or the other, but not both." For example, if a waitress says, "You may have soup or salad with your dinner," she means that you may choose just one of the two. This use of *or* is known as the **exclusive** *or*.

In mathematics and logic, *or* means "one or the other, or both." This use of *or* is known as the **inclusive** *or*. If someone asks how John will spend his Saturday afternoons, the answer might be, "He will go swimming or bowling." This sentence can be written in logical form as *p* OR *q*, where *p* and *q* are identified as shown below.

$$\underbrace{\text{He will go swimming}}_{p} \text{ or } \underbrace{\text{[he will go] bowling.}}_{q}$$

The statement is false only if John does neither one. There is nothing about the statement which implies that John won't do both. Notice that in the truth table, only the fourth combination gives a value of false for the disjunction. If John goes swimming or bowling or both, the disjunction is true, as the values for the first three combinations show.

p	*q*	*p* OR *q*
T	T	T
T	F	T
F	T	T
F	F	F

EXAMPLE 2 Determine whether each disjunction below is true.

a. A square is a rectangle, or a pentagon has five sides.

b. Dogs can play golf, or 5 − 3 = 2.

SOLUTION

a. The disjunction is true because both of the statements are true.

b. The disjunction is true because one of the statements is true.

Cooperative Learning

In groups, students will use truth tables to determine the truth value of the conjunctions and disjunctions in Examples 1 and 2. Have students check their tables with other group members.

Activity **Notes**

In this Activity, students learn how to write the negation of a conjunction and of a disjunction. Point out that the negation of a conjunction of two statements is logically equivalent to the disjunction of the negation of each statement. Make sure students understand that symbols are used for statements so that truth can be determined easily by form rather than by content.

For a student worksheet of this Activity and detailed Teacher Notes, see page 212 in the Lesson Activities booklet.

CHECKPOINT ✔

3. The last column from Step 1 and the last column from Step 2 are equivalent. ~ (*p* AND *q*) is equivalent to ~ *p* OR ~ *q*.

CRITICAL THINKING

The combinations of T and F should be listed in exactly the same way in the first two columns of each truth table to make the comparison possible. For the three statements, *p*, *q*, and *r*, there are eight different combinations of T and F.

p	*q*	*r*
T	T	T
T	T	F
T	F	T
T	F	F
F	T	T
F	T	F
F	F	T
F	F	F

Negations

Scene from the film Singing in the Rain

Consider each of the following statements:

It is raining outside.   It is not raining outside.

The second statement is the *negation* of the first one. If *p* is a statement, then NOT *p* is its **negation**. The negation of *p* can also be written as ~*p*.

It is raining outside.   It is not raining outside.
          p   ~*p*

Examine the truth table for the negation. Notice the following: When a statement, *p*, is true, its negation, ~*p*, is false. When a statement, *p*, is false, its negation, ~*p*, is true.

p	~*p*
T	F
F	T

In the following Activity, you will explore one of De Morgan's laws, which are named after British mathematician Augustus De Morgan (1806–1871).

Activity
The Negation of a Conjunction

YOU WILL NEED
no special tools

1. Copy and complete the truth table for the negation of a conjunction. Notice that the fourth column represents the negation of the third column.

p	*q*	*p* AND *q*	~(*p* AND *q*)
T	T	?	?
T	F	?	?
F	T	?	?
F	F	?	?

2. Copy and complete the truth table for a disjunction of two negations. Notice that the values for ~*p* and ~*q* are used to determine the truth values for ~*p* OR ~*q*.

p	*q*	~*p*	~*q*	~*p* OR ~*q*
T	T	?	?	?
T	F	?	?	?
F	T	?	?	?
F	F	?	?	?

CHECKPOINT ✔

3. Compare the last column in the first table with the last column in the second table. Explain what you observe.

4. When two logic statements have the same truth values, they are said to be **truth functionally equivalent**. Complete the statement below, which is one of De Morgan's laws.

~(*p* AND *q*) is truth functionally equivalent to ____?____.

CRITICAL THINKING

In the two truth tables in the Activity, why should the combinations of T and F be listed in exactly the same order in the first two columns of each table? If you list T and F values for the three statements *p*, *q*, and *r*, how many different combinations of T and F will there be? What is a good order for listing these combinations?

Reteaching the Lesson

USING REVIEW Have each student write the conjunction and disjunction of the following statements:

p: △*ABC* is a right triangle.
q: △*ABC* has two complementary angles.

Conjunction: △*ABC* is a right triangle AND △*ABC* has two complementary angles.
Disjunction: △*ABC* is a right triangle OR △*ABC* has two complementary angles.

Exercises

 Communicate

1. Explain the conditions necessary for a conjunction to be true.

2. Explain the conditions necessary for a disjunction to be true.

3. Explain the difference between an inclusive *or* and an exclusive *or*.

4. Give three different ways for the statement "Kimba likes pizza and Lin likes spaghetti" to be false.

 Guided Skills Practice

Indicate whether each compound statement is true or false. Explain your reasoning. *(EXAMPLES 1 AND 2)*

5. $4 + 5 = 9$ and $4 \cdot 5 = 9$ **False, because only one of the statements is true.**

6. All triangles have three angles, or all triangles have four angles. **True, because one of the statements is true.**

7. Three noncollinear points determine a unique plane, and a segment has two endpoints. **True, because both of the statements are true.**

8. All squares are hexagons, or all triangles are squares. **False, because both of the statements are false.**

Write a statement that is truth functionally equivalent to each statement below. *(ACTIVITY)*

9. ~r OR ~s
   ~(r AND s)

10. ~(t AND u)
    ~t OR ~u

 Practice and Apply

Write a conjunction for each pair of statements. State whether the conjunction is true or false.

11. A carrot is a vegetable. Florida is a state.

12. A ray has only one endpoint. Kangaroos can sing.

13. The sum of the measures of the angles of a triangle is 180°. Two points determine a line.

Write a disjunction for each pair of statements. State whether the disjunction is true or false.

14. Triangles are circles.
    Squares are parallelograms.

15. Points in a plane equidistant from a given point form a circle. The sides of an equilateral triangle are congruent.

16. An orange is a fruit.
    Cows have kittens.

Selected Answers
Exercises 5–10, 11–39 odd

ASSIGNMENT GUIDE	
In Class	1–10
Core	11–31 odd, 32, 33
Core Plus	22–35
Review	35–41
Preview	42, 43

✐ Extra Practice can be found beginning on page 818.

11. A carrot is a vegetable, and Florida is a state. The conjunction is true because both statements are true.

12. A ray has only one endpoint, and kangaroos can sing. The conjunction is false because one statement (kangaroos can sing) is false.

13. The sum of the measures of the angles of a triangle is 180°, and two points determine a line. The conjunction is true because both statements are true.

14. Triangles are circles, or squares are parallelograms. The disjunction is true because one of the statements (squares are parallelograms) is true.

15. Points in a plane equidistant from a given point form a circle, or the sides of an equilateral triangle are congruent. The disjunction is true because both of the statements are true.

16. An orange is a fruit, or cows have kittens. The disjunction is true because one of the statements (an orange is a fruit) is true.

17. The figure is not a rectangle.

18. My client is guilty.

19. Rain does not make the road slippery.

20. Triangles do not have six sides.

21. a.

p	$\sim p$	$\sim(\sim p)$
T	F	T
F	T	F

b. $\sim(\sim p)$ is logically equivalent to p because they both have the same truth values.

---

22. $\triangle ABC$ is not isosceles.

23. $\triangle ABC$ has two equal angles, or $\triangle ABC$ is isosceles.

24. $\triangle ABC$ is isosceles, and $\triangle ABC$ has two equal angles.

25. $\triangle ABC$ does not have two equal angles.

26. $\angle 1$ and $\angle 2$ are not (both) acute angles.

27. $\angle 1$ and $\angle 2$ are adjacent, or $\angle 1$ and $\angle 2$ are acute angles.

28. $\angle 1$ and $\angle 2$ are acute angles, and $\angle 1$ and $\angle 2$ are not adjacent.

✓ internet connect

Homework Help Online
Go To: go.hrw.com
Keyword:
MG1 Homework Help
for Exercises 30-32

Write the negation of each statement.

17. The figure is a rectangle.

18. My client is not guilty.

19. Rain makes the road slippery.

20. Triangles have six sides.

21. a. Copy and complete the truth table for $\sim(\sim p)$.

p	$\sim p$	$\sim(\sim p)$
T	?	?
F	?	?

b. What statement is equivalent to the statement $\sim(\sim p)$? Explain your reasoning.

For Exercises 22–29, write the statement expressed by the symbols, where p, q, r, and s represent the statements below.

p: $\triangle ABC$ is isosceles.
q: $\triangle ABC$ has two equal angles.
r: $\angle 1$ and $\angle 2$ are adjacent.
s: $\angle 1$ and $\angle 2$ are acute angles.

22. $\sim p$
23. q OR p
24. p AND q
25. $\sim q$
26. $\sim s$
27. r OR s
28. s AND $\sim r$
29. q OR $\sim s$

For Exercises 30–32, construct a truth table for the given compound statement. When is the compound statement false?

30. $(p$ AND $q)$ AND r

31. $(p$ OR $q)$ OR r

32. $(p$ AND $q)$ OR $(r$ AND $s)$

> *Use this table for Exercise 30.*

p	q	r	p AND q	$(p$ AND $q)$ AND r
T	T	T	? T	? T
T	T	F	? T	? F
T	F	T	? F	? F
T	F	F	? F	? F
F	T	T	? F	? F
F	T	F	? F	? F
F	F	T	? F	? F
F	F	F	? F	? F

33. Explain all of the logical possibilities that would make this sentence true: Flora will cook or wash the dishes, and Vernon will vacuum or wash the windows.

34. ADVERTISING An advertisement for a set of holiday lights contains the following statement: "Not all bulbs go out when one bulb burns out." Is this the same as "When one bulb burns out, all of the remaining bulbs continue burning"?

29. $\triangle ABC$ has two equal angles, or $\angle 1$ and $\angle 2$ are not (both) acute angles.

33. The sentence is true if any of the following combinations are true:
Flora will cook; Vernon will vacuum.
Flora will cook; Vernon will wash the windows.
Flora will cook; Vernon will vacuum and wash the windows.
Flora will wash the dishes; Vernon will vacuum.

Flora will wash the dishes; Vernon will wash the windows.
Flora will wash the dishes; Vernon will vacuum and wash the windows.
Flora will cook and wash the dishes; Vernon will vacuum.
Flora will cook and wash the dishes; Vernon will wash the windows.
Flora will cook and wash the dishes; Vernon will vacuum and wash the windows.

34. No

APPLICATION

35. COMPUTER DATABASES Computer database software can perform AND and OR operations on stored data. Study the sample database printout below, which contains 10 records with 5 fields each.

Last name	First name	State	Year of birth	Annual income
Craighead	Alicia	TX	1955	$ 25,000
Nicar	Bill	MN	1942	$ 45,000
Tuggle	Lawrence	LA	1972	$ 20,000
Mallo	Elizabeth	TX	1956	$ 50,000
Torres	Ernest	AK	1940	$ 38,000
Tong	Jun	TX	1952	$ 18,000
Jurek	Chandra	AZ	1944	$ 31,000
Brookshier	Mary	OH	1960	$ 62,000
Lamb	Charles	TX	1951	$ 41,000
Raemsch	Martin	OK	1965	$ 32,000

a. List all the records in which the individual was born after 1950 AND the annual income is greater then $30,000.

b. List all the records in which the individual lives in Texas OR the annual income is less than $30,000.

Look Back

Can each pair of triangles below be proven similar? Explain why or why not. *(LESSONS 8.3 AND 8.4)*

36.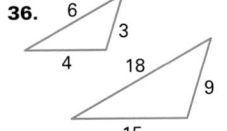

36. No, since $\frac{18}{6} = \frac{9}{3} \neq \frac{15}{4}$

37.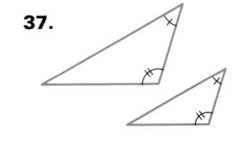

37. Yes, by AA Similarity

38.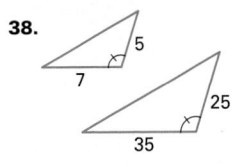

38. Yes, by SAS Similarity

Find the indicated ratios for the triangle below. *(LESSONS 10.1 AND 10.2)*

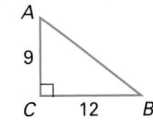

39. sin A $\frac{4}{5}$ **40.** cos B $\frac{4}{5}$

41. tan B $\frac{3}{4}$ **42.** cos A $\frac{3}{5}$

Look Beyond

43. Can you determine whether the sentence below is true or false? Explain your reasoning.

> This sentence is false.

CHALLENGE

44. A woman is on an island and is trying to determine whether she should go east or west in order to get back to the mainland. Two different groups of people live on the island. One group always tells the truth. The other group always lies. The groups dress differently, but the woman does not know which is which. She approaches two islanders who are dressed differently to ask the direction to the mainland. What one question can she ask one of the islanders to determine the correct direction?

Look Beyond

In Exercises 43 and 44, students use the skills they learned in this chapter to solve nonroutine problems or puzzles.

43. The sentence cannot be true, because that would contradict the sentence's statement. But the sentence cannot be false because then the sentence's statement would be true. Therefore it is neither true nor false.

44. The woman should ask one of the people what the other one would say if he or she were asked the correct direction. Then take the opposite of the answer. If the woman asked the truthful person, the person would give the incorrect answer, because the person would truthfully say what the liar would say. If the woman asked the liar, he/she would give the incorrect answer because he/she always lies.

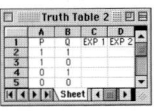
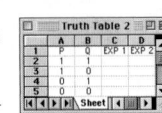

EYEWITNESS MATH

Focus

A news article about the use of computers that can translate Mayan symbols leads into a study of the Mayan calender. Students informally explore modular arithmetic and greatest common factors to understand a time-keeping system based on two years of different lengths.

Motivate

Before students read the article, discuss some background information about the Mayan civilization:

- The Mayan civilization reached its peak around 800 C.E.

- In what is now Mexico, Guatemala, and Honduras, the Mayans built great centers of religion and politics.

- The Mayans had advanced knowledge of agriculture and astronomy.

- Hundreds of years before the Spanish explorers reached the Americas, the Mayans abandoned their cities and monuments. The reason for their abandonment is still unclear.

After students read the article, discuss how a double calendar would allow you to identify many more dates without specifying the year.

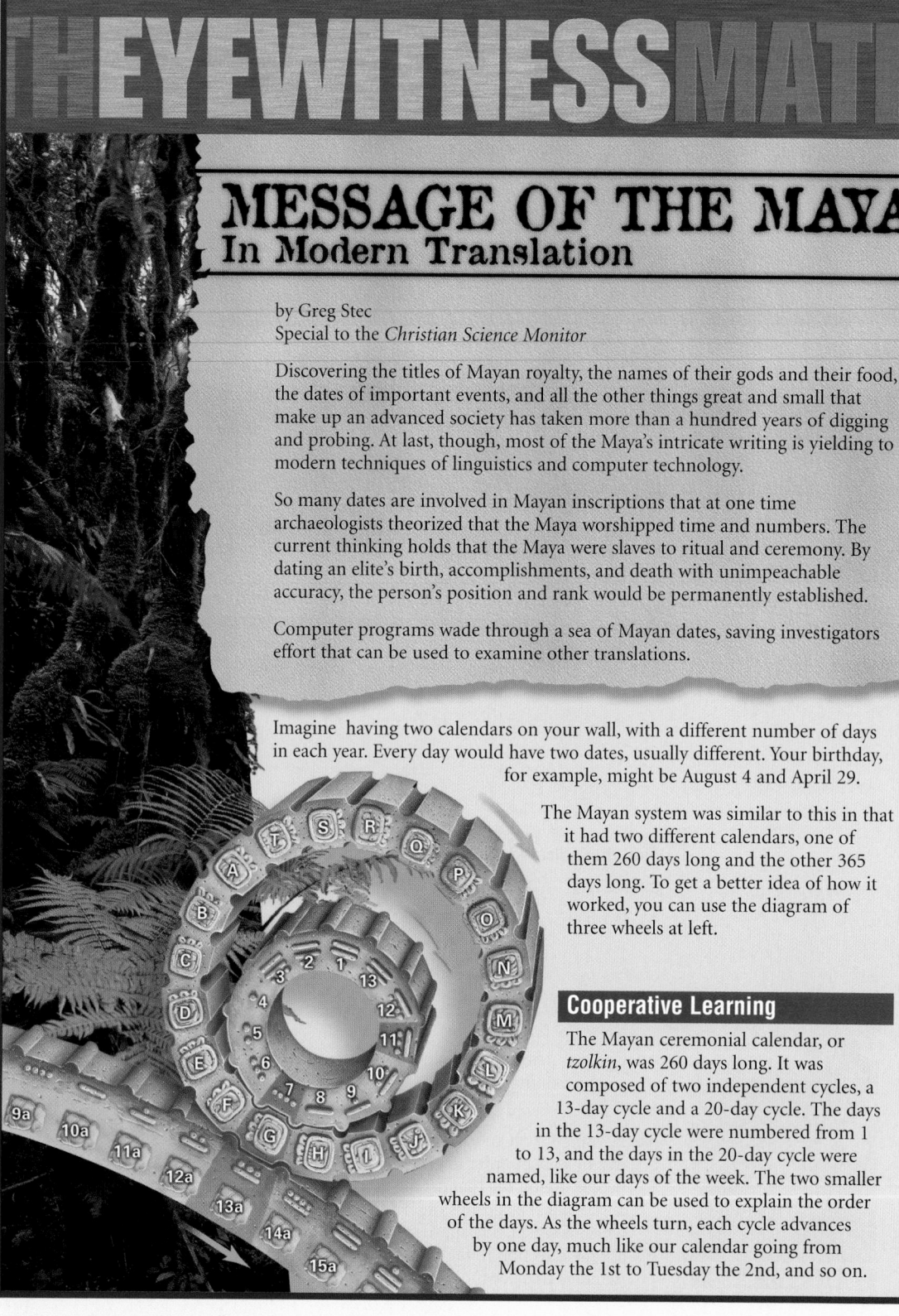

MESSAGE OF THE MAYA
In Modern Translation

by Greg Stec
Special to the *Christian Science Monitor*

Discovering the titles of Mayan royalty, the names of their gods and their food, the dates of important events, and all the other things great and small that make up an advanced society has taken more than a hundred years of digging and probing. At last, though, most of the Maya's intricate writing is yielding to modern techniques of linguistics and computer technology.

So many dates are involved in Mayan inscriptions that at one time archaeologists theorized that the Maya worshipped time and numbers. The current thinking holds that the Maya were slaves to ritual and ceremony. By dating an elite's birth, accomplishments, and death with unimpeachable accuracy, the person's position and rank would be permanently established.

Computer programs wade through a sea of Mayan dates, saving investigators effort that can be used to examine other translations.

Imagine having two calendars on your wall, with a different number of days in each year. Every day would have two dates, usually different. Your birthday, for example, might be August 4 and April 29.

The Mayan system was similar to this in that it had two different calendars, one of them 260 days long and the other 365 days long. To get a better idea of how it worked, you can use the diagram of three wheels at left.

Cooperative Learning

The Mayan ceremonial calendar, or *tzolkin*, was 260 days long. It was composed of two independent cycles, a 13-day cycle and a 20-day cycle. The days in the 13-day cycle were numbered from 1 to 13, and the days in the 20-day cycle were named, like our days of the week. The two smaller wheels in the diagram can be used to explain the order of the days. As the wheels turn, each cycle advances by one day, much like our calendar going from Monday the 1st to Tuesday the 2nd, and so on.

For convenience, let the days in the 20-day cycle be represented by letters instead of the Mayan names. Then each day has a date composed of a number from 1 to 13 followed by a letter from A to T. For example, the first day of the year would be 1A.

1. How many possible dates are there? Explain your reasoning.

2. The date shown in the diagram below is 7G. What would be the date one day later? 13 days later? 20 days later?

3. To help you understand the arrangements of the dates, consider some "calendars" with fewer days.

   a. Suppose that a calendar has cycles of 8 days and 5 days. The days in the 5-day cycle will be represented by letters from A to E. How many dates are possible? To find the order of the dates, repeat the numbers from 1 to 8 in one row and the letters from A to E in another, as follows:

   1 2 3 4 5 6 7 8 1 2 3 4 5 6 7 8 1 2...
   A B C D E A B C D E A B C D E A B C...

   Continue until the date 1A repeats. Does every possible date occur?

   b. Suppose that a calendar has cycles of 6 days and 4 days. The days in the 4-day cycle will be represented by letters from A to D. How many dates are possible? To find the order of the dates, repeat the numbers from 1 to 6 in one row and the letters from A to D in another, as follows:

   1 2 3 4 5 6 1 2 3 4 5...
   A B C D A B C D A B C...

   Continue until the date 1A repeats. Does every possible date occur?

   c. For every possible date to occur, what must be true of the number of days in the cycles? (Hint: Look at the factors of the number of days in each cycle.)

The Mayan solar calendar, or *haab*, was 365 days long. It was composed of 18 months of 20 days each (numbered from 0 to 19) and one month of 5 days; the days of the 5-day month were considered unlucky. The order of dates in the solar calendar was similar to our calendar. Using lowercase letters to represent the names of the months, the dates would start at 0a, then 1a, 2a, and so on. The solar calendar is represented by the large wheel in the diagram.

4. In the diagram, the date on the solar calendar is 13a. The combined date is 7G 13a. What will be the combined date 1 day later? 20 days later? 260 days later? 365 days later?

5. How many possible combined dates are there? Based on your results from part **c** of Step 3, do you think every possible date will occur? How many years will it be before the date 7G 13a occurs again?

Cooperative Learning

Since the activities are sequential, have each group do all three activities.

Discuss the fact that with each new day, both wheels advance in the directions shown by the arrows.

Part **c** of Step 3 is a difficult notion for most students to understand without guidance. Help them focus on finding common factors in the number of days in each cycle.

Discuss the structure of the large wheel before doing Steps 4 and 5. Students can think of the two small wheels as a single wheel with 260 dates. Encourage students to look for shortcuts in Step 4. To help them, ask "What is a shortcut for finding the date that comes 31 days after January 19?"

Discuss

Ask students what they think was the benefit of the Mayan calendar. Have students research another Mayan time-keeping system called the *long count*, a base-20 system.

A Closer Look at If-Then Statements

Prepare

Teach

W/hy   Studying the converse, inverse, and contrapositive of a conditional statement is important because of their uses in logical arguments. Conditional statements are essential for the discussion and communication of mathematical ideas.

Objectives

● Create truth tables for conditionals and for converses, inverses, and contrapositives of conditionals.

● Use if-then statements and forms of valid argument for problems involving logical reasoning.

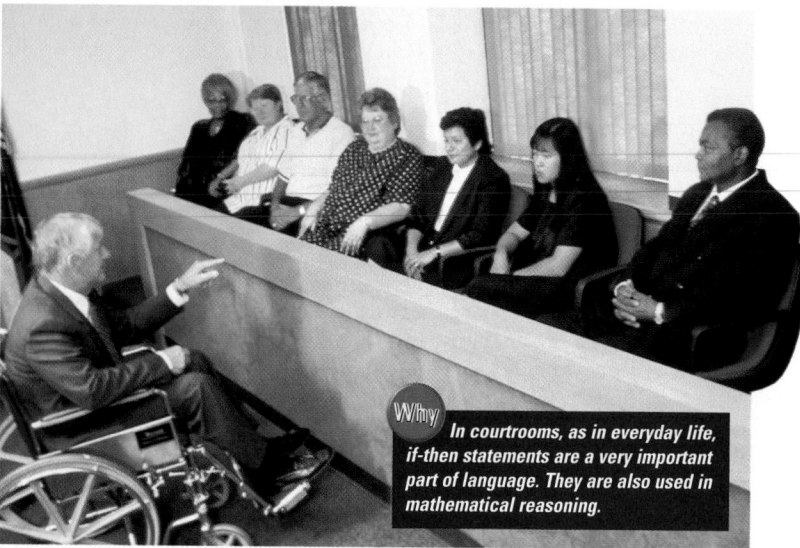

W/hy *In courtrooms, as in everyday life, if-then statements are a very important part of language. They are also used in mathematical reasoning.*

APPLICATION
CRIMINAL LAW

Lawyers use logic when presenting their cases to juries. For example, a lawyer might argue, "If the defendant committed the crime, then he could not have been at the north shopping mall between 10 A.M. and 10:30 A.M." How do you think the argument would continue?

The Truth Table for a Conditional

Various forms of if-then statements have been used throughout this book. You may recall from Lesson 2.2 that if-then statements are called conditionals. You can think of a conditional as a promise. In logic, if the "promise" is broken, the conditional is said to false. Otherwise, it is said to be true.

Suppose that your neighbor makes the following promise:

If you mow my lawn, then I will give you $10.

$$p \qquad\qquad q$$

Four possible situations can occur.

1. *You mow the lawn and your neighbor gives you $10.*
   The promise is kept. Therefore, the conditional is said to be true.

2. *You mow the lawn and your neighbor does not give you $10.*
   The promise is broken. Therefore, the conditional is said to be false.

3. *You do not mow the lawn. Your neighbor gives you $10.*
   The promise is not broken. Therefore, the conditional is said to be true.

4. *You do not mow the lawn. Your neighbor does not give you $10.*
   The promise is not broken. Therefore, the conditional is said to be true.

Alternative Teaching Strategy

TECHNOLOGY   Use the program given at right to demonstrate how if-then logic may be used by a graphics calculator. This program uses the converse of the Pythagorean Theorem to determine whether a triangle is a right triangle.

```
PROGRAM:PYTHAG
:Input "INPUT SIDE A", A
:Input "INPUT SIDE B", B
:Input "INPUT SIDE C", C
:If A^2+B^2=C^2
:Then : Disp "RIGHT"
:Else : Disp "NOT RIGHT"
:End
```

The truth table below summarizes the truth values for the conditional $p \Rightarrow q$. Recall that the logical notation for "if p then q" is $p \Rightarrow q$ (read as "p implies q").

p	q	$p \Rightarrow q$
T	T	T
T	F	F
F	T	T
F	F	T

The first two columns of the truth table list all possible combinations of T and F for the two statements p and q. The third column list the truth values of the conditional $p \Rightarrow q$. Notice that the only time $p \Rightarrow q$ is false is when p is true and q is false—that is, when the promise is broken.

The Converse of a Conditional

Recall from Lesson 2.2 that the converse of a conditional results from interchanging the statements following *if* (the hypothesis) and *then* (the conclusion). Consider the following conditional and converse:

Conditional
If Tamika lives in Montana, then she lives in the United States.

CHECKPOINT ✔ Is this statement true?

Converse of the conditional
If Tamika lives in the United States, then she lives in Montana.

CHECKPOINT ✔ Is this statement true?

The truth table below summarizes the truth values for the converse of the conditional $p \Rightarrow q$.

p	q	$q \Rightarrow p$
T	T	T
T	F	T
F	T	F
F	F	T

CRITICAL THINKING Recall that two statements are truth functionally equivalent if and only if they have the same values in their truth tables. Compare the truth values for the conditional and its converse. Are they truth functionally equivalent? Explain your reasoning.

Interdisciplinary Connection

COMPUTER PROGRAMMING One of the most powerful elements of a computer program is its ability to branch to a different part of the program. Explain to students that this ability is based on if-then logic.

The Inverse of a Conditional

The **inverse of a conditional** is formed by negating both the hypothesis and the conclusion. Below is the inverse of the conditional in the previous example.

Inverse of the conditional
If Tamika does not live in Montana, then she does not live in the United States.

CHECKPOINT ✔ Is this statement true?

The truth table below represents the inverse of the conditional $p \Rightarrow q$. Notice that extra columns are required for the negations of p and q.

p	q	$\sim p$	$\sim q$	$\sim p \Rightarrow \sim q$
T	T	F	F	T
T	F	F	T	T
F	T	T	F	F
F	F	T	T	T

CRITICAL THINKING Are a conditional and its inverse truth functionally equivalent? Is the converse of a conditional truth functionally equivalent to its inverse? Explain your reasoning.

The Contrapositive of a Conditional

The **contrapositive of a conditional** is formed by interchanging the hypothesis and the conclusion of the conditional and then negating each part. Below is the converse of the original conditional.

Contrapositive of the conditional
If Tamika does not live in the United States, then she does not live in Montana.

CHECKPOINT ✔ Is the statement true?

The truth table below summarizes the truth values for the contrapositive of the conditional $p \Rightarrow q$.

p	q	$\sim q$	$\sim p$	$\sim q \Rightarrow \sim p$
T	T	F	F	T
T	F	T	F	F
F	T	F	T	T
F	F	T	T	T

Notice that the final columns for the truth tables of the original conditional and those of its contrapositive are the same. Thus, the two statements are truth functionally equivalent. If a conditional is true, its contrapositive must also be true. Moreover, if a conditional is false, its contrapositive must also be false.

Thus, every theorem or postulate that can be written in if-then form can be rewritten in contrapositive form—which will also be true.

CHECKPOINT ✔
no

CRITICAL THINKING
A conditional and its inverse are not truth-functionally equivalent because they do not have the same truth tables. The converse of a conditional is truth-functionally equivalent to its inverse because they have the same truth tables.

CHECKPOINT ✔
yes

Cooperative Learning

Have students write three conditional statements and then form groups of three. Each student should keep one statement and give one to each of the other two members of the group. Each student should then write the converse, inverse, and contrapositive of their statements. Encourage students to share their answers.

Enrichment

Write a truth table for the following statement:

$(p \text{ AND } q) \text{ OR } (p \text{ AND } r) \rightarrow (p \text{ OR } q \text{ OR } r)$

If the statement is always true, it is called a tautology. Is this statement a tautology?

Inclusion Strategies

USING SYMBOLS Some students may be more comfortable using symbols and truth tables to find the truth value of statements and arguments. Encourage them to represent the hypothesis and the conclusion of each statement with symbols and then replace the symbols with the original statements.

Summary of Conditionals

THREE RELATED FORMS OF AN IF-THEN STATEMENT OR CONDITIONAL		
Conditional	If p then q	$p \Rightarrow q$
Converse	If q then p	$q \Rightarrow p$
Inverse	If $\sim p$ then $\sim q$	$\sim p \Rightarrow \sim q$
Contrapositive	If $\sim q$ then $\sim p$	$\sim q \Rightarrow \sim p$

TRY THIS Write the converse, inverse, and contrapositive of each conditional below. Determine whether the original statement and each new statement are true or false.

a. If a triangle is equilateral, then the triangle is isosceles.

b. If a quadrilateral is a rhombus, then the quadrilateral is a square.

Exercises

 Communicate

In Exercises 1–4, write the converse, inverse, and contrapositive of each conditional.

1. If today is February 30, then the moon is made of green cheese.

2. If all three sides of a triangle are congruent, then the triangle is equilateral.

3. If I do not go to the market, then I will not buy cereal.

4. If the car starts, then I will not be late for school.

5. Describe the circumstances that would make the statement "If a then b" false.

(See Exercise 1.)

Reteaching the Lesson

USING REVIEW Write the following conditional on the board:

If $ABCD$ is a square, then $ABCD$ is a parallelogram.

Have each student write the converse, inverse, and contrapositive of the conditional. Ask them to state the truth value of each one.

Converse: If $ABCD$ is a parallelogram, then $ABCD$ is a square. (false)

Inverse: If $ABCD$ is not a square, then $ABCD$ is not a parallelogram. (false)

Contrapositive: If $ABCD$ is not a parallelogram, then $ABCD$ is not a square. (true)

Assess

Selected Answers

Exercises 6–9, 11–31 odd

ASSIGNMENT GUIDE	
In Class	1–9
Core	10–25
Core Plus	17–27
Review	28–32
Preview	33, 34

✏ Extra Practice can be found beginning on page 818.

Error Analysis

Some students may have difficulty with exercises that use variables instead of statements. Encourage them to replace the variables with statements.

10. Conditional: True, because squares are parallelograms with four 90° angles.

Converse: If a figure is a rectangle, then it is a square. False; rectangles do not necessarily have four congruent sides.

Inverse: If a figure is not a square, then it is not a rectangle. False; the figure can still be a rectangle if it is not a square—it could still have congruent opposite sides.

Contrapositive: If a figure is not a rectangle, then it is not a square. True; if the figure is not a rectangle, it cannot be a square.

☑ internet connect

Homework Help Online

Go To: **go.hrw.com**
Keyword:
MG1 Homework Help
for Exercises 11-12, 14, and 25

APPLICATION

● *Guided Skills Practice*

6. Complete the truth table for the conditional $p \Rightarrow q$.

(TRUTH TABLE OF A CONDITIONAL)

p	q	$p \Rightarrow q$	
T	T	?	T
T	F	?	F
F	T	?	T
F	F	?	T

7. Complete the truth table for the converse of the condtional $p \Rightarrow q$.

(CONVERSE OF A CONDITIONAL)

p	q	$q \Rightarrow p$	
T	T	?	T
T	F	?	T
F	T	?	F
F	F	?	T

8. Complete the truth table for the inverse of the conditional $p \Rightarrow q$.

(INVERSE OF A CONDITIONAL)

p	q	$\sim p$	$\sim q$	$\sim p \Rightarrow \sim q$
T	T	? F	? F	? T
T	F	? F	? T	? T
F	T	? T	? F	? F
F	F	? T	? T	? T

9. Complete the truth table for the contrapositive of a conditional.

(CONTRAPOSITIVE OF A CONDITIONAL)

p	q	$\sim p$	$\sim q$	$\sim q \Rightarrow \sim p$
T	T	? F	? F	? T
T	F	? F	? T	? F
F	T	? T	? F	? T
F	F	? T	? T	? T

● *Practice and Apply*

For each conditional in Exercises 10–15, write the converse, inverse, and contrapositive. Decide whether each is true or false and explain your reasoning.

10. If a figure is a square, then it is a rectangle.

11. If $a = b$, then $a^2 = b^2$.

12. If $a < b$, then $a^2 < b^2$.

13. If three angles of a triangle are congruent to three angles of another triangle, then the triangles are congruent.

14. If p and q are even numbers, then $p + q$ is an even number.

15. **PHYSICS** If water is frozen at normal atmospheric pressure, then its temperature is less than or equal to 32°F.

---

11. Conditional: True, since squaring both sides maintains equality due to the Multiplication Property of Equality.

Converse: If $a^2 = b^2$, then $a = b$. False; if $a = -3$ and $b = 3$, then $(-3)^2 = (3)^2$ but a and b are not equal.

Inverse: If $a \neq b$, then $a^2 \neq b^2$. False; if $a = -3$ and $b = 3$, they are not equal, but $(-3)^2 = (3)^2$.

Contrapositive: If $a^2 \neq b^2$, then $a \neq b$. True; if $a^2 \neq b^2$ then taking the square root of each side will not yield the same number, so $a \neq b$.

12. Conditional: False, since squaring both sides does not necessarily maintain inequality; it can reverse it.

Converse: If $a^2 < b^2$ then $a < b$. False; if $b = -4$ and $a = 3$ then $a^2 < b^2$ but a is not less than b.

Inverse: If $a \geq b$ then $a^2 \geq b^2$ False; if $b = -4$ and $a = 3$ then $a \geq b$ but b^2 is not less than a^2.

Contrapositive: If $a^2 \geq b^2$ then $a \geq b$. False; if $a = -4$ and $b = 3$ then $a^2 \geq b^2$ but a is not greater than or equal to b.

16. Given: If p then q. Write the contrapositive of the statement. Then write the contrapositive of the contrapositive. What can you conclude about the contrapositive of the contrapositive of an if-then statement?

17. Suppose that the following statement is true: If the snow exceeds 6 in., then school will be canceled. Which of the following statements must also be true?

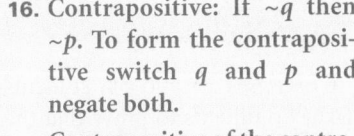

    a. If the snow does not exceed 6 in., then school will not be canceled.

    b. If school is not canceled, then the snow does not exceed 6 in.

    c. If school is canceled, then the snow exceeds 6 in.

18. State the Pythagorean Theorem and its converse, inverse, and contrapositive. Determine whether each is true or false, and explain your reasoning.

19. Choose a postulate or theorem from Chapter 3 that is written in if-then form. Write its converse, inverse, and contrapositive, decide whether each is true or false, and explain your reasoning.

20. Choose a postulate or theorem from Chapter 9 that is written in if-then form. Write its converse, inverse, and contrapositive, decide whether each is true or false, and explain your reasoning.

Some statements that are not written in if-then form can be rewritten in if-then form. For example, "Every rectangle is a parallelogram" can be rewritten as "If a figure is a rectangle, then it is a parallelogram." In Exercises 21–24, rewrite each statement in if-then form.

21. All seniors must report to the auditorium.

22. A point on the perpendicular bisector of a segment is equidistant from the endpoints of the segment.

23. She will call me if she is going to be late.

24. Doing mathematics homework every night will improve your grade in mathematics.

25. The statement "p if and only if q," written as $p \Leftrightarrow q$, is equivalent to the following two statements:

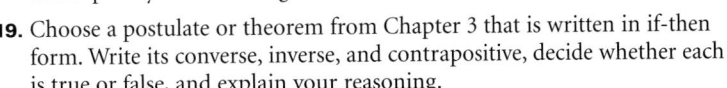

              If p then q

    and   If q then p

Suppose that the statement $r \Leftrightarrow s$ is true. Which of the following must also be true? Explain your reasoning.

    a. If r then s

    b. If s then r

    c. If $\sim r$ then $\sim s$

    d. If $\sim s$ then $\sim r$

26. Statements using "if and only if" are known as **biconditionals**. Based on your results from Exercise 25, what can you conclude about certain theorems in this book? Explain your reasoning and give two examples.

---

16. Contrapositive: If $\sim q$ then $\sim p$. To form the contrapositive switch q and p and negate both.

Contrapositive of the contrapositive: If $\sim(\sim p)$ then $\sim(\sim q)$, or if p then q. The result of taking the contrapositive of a contrapositive is the original conditional.

17. **b** is true because it is the contrapositive of the original conditional. Since the statement is true, the contrapositive is also true. The other two could be false because school could be canceled for other reasons besides snow.

18. Pythagorean Theorem: If a triangle is a right triangle, then the square of the length of the hypotenuse is equal to the sum of the squares of the lengths of the legs.

Converse: If the square of the length of the hypotenuse of a triangle is equal to the sum of the squares of the lengths of the legs then the triangle is a right triangle. True; this is proven in Chapter 5 Section 4.

Inverse: If a triangle is not a right triangle, then the square of the length of the hypotenuse is not equal to the sum of the squares of the lengths of the legs. True; if a triangle is not a right triangle then the square of the length of the hypotenuse is greater than or less than the sum of the squares of the lengths of the legs.

Contrapositive: If the square of the length of the hypotenuse of a triangle is not equal to the sum of the squares of the lengths of the legs, then the triangle is not a right triangle. True; the contrapositive is logically equivalent to the original conditional, which is the Pythagorean Theorem.

---

19. Sample answer: Theorem 3.8.2: If two nonvertical lines are parallel, then they have the same slope.

Converse: If two nonvertical lines have the same slope, then they are parallel. True; since Theorem 3.8.2 is actually an if-and-only-if statement, the converse is true.

Inverse: If two nonvertical lines are not parallel, then they do not have the same slope. True, because the inverse has the same truth value as the converse.

Contrapositive: If two nonvertical lines do not have the same slope, then they are not parallel. True, because the contrapositive has the same truth value as the conditional.

In Exercises 33 and 34, students write arguments to prove that the conclusion of a conditional is true. The strategy is to begin the argument by proving that the conditional is true.

27. a. Same meaning; implies that getting at least a B in mathematics is a requirement to make the honor roll.

b. Different meaning; this is the converse of the original statement.

c. Different meaning; this is the inverse of the original statement.

d. Same meaning; implies that getting at least a B in mathematics is a requirement to make the honor roll.

30.

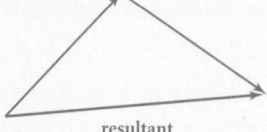

31.

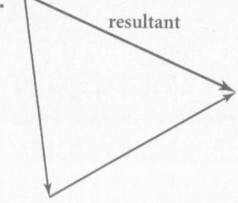

32.

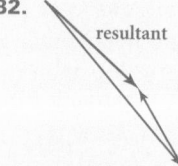

33. p is given. q is to be proved. Once $p \Rightarrow q$ is proved to be true then given p, q follows.

34. p is given. q is to be proved. Once $\sim q \Rightarrow \sim p$ is proved, then its contrapositive ($p \Rightarrow q$) is true and since p is given then q follows. However then q is true so $\sim q$ is false.

27. ACADEMICS Consider the following statement: You will make the honor roll only if you get at least a B in mathematics.

Which of the following statements appear to convey the same meaning as the original statement? Explain your reasoning.

a. If you make the honor roll, then you must have gotten at least a B in mathematics.

b. If you get at least a B in mathematics, then you will make the honor roll.

c. If you do not make the honor roll, then you did not get at least a B in mathematics.

d. If you do not get at least a B in mathematics then you will not make the honor roll.

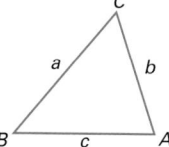

Look Back

Use the triangle below and the law of sines to find each missing measure. *(LESSON 10.4)*

28. Given $m\angle B = 37°$, $m\angle A = 50°$, and $b = 100$, find c. **165.94**

29. Given $m\angle C = 65°$, $m\angle A = 47°$, and $c = 3.45$, find b. **3.53**

Copy the vectors below and draw the resultant vector by using the head-to-tail method. You may need to translate one of the vectors. *(LESSON 10.7)*

30. 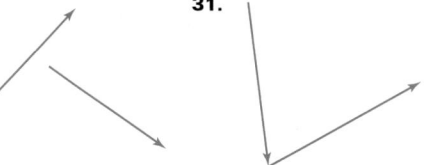 **31.** **32.**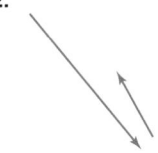

Look Beyond

33. Suppose you know that statement p is true and you want to prove that statement q is true. You begin by proving that the conditional $p \Rightarrow q$ is true, which leads immediately to the conclusion.

Write an outline of the argument in terms of what is given, what is to be proved, and the logical principles that allow you to draw your conclusion.

34. Suppose that, as in Exercise 33, you know that statement p is true and you want to prove that statement q is true. You are unable to prove directly that the conditional $p \Rightarrow q$ is true, so you prove instead that its contrapositive, $\sim q \Rightarrow \sim p$, is true. If the contrapositive is true, can $\sim q$ possibly be true? (Remember that you know p is true.)

Write an outline of the argument in terms of what is given, what is to be proved, and the logical principles that allow you to draw your conclusion. You will study this type of argument in the next lesson.

Indirect Proof

12.4

Objectives

- Develop the concept of indirect proof (*reductio ad absurdum*, or proof by contradiction).

- Use indirect proof with problems involving logical reasoning.

Why *Lewis Carroll, the author of* Alice in Wonderland *and* Through the Looking-Glass, *was a logician who was fond of absurdity as a form of entertainment. But does absurdity have any real place in logic or mathematics? In this lesson, you will see that it can, in fact, be quite useful.*

'The time has come,' the Walrus said,
'To talk of many things:
Of shoes—and ships—and sealing-wax—
Of cabbages—and kings—
And why the sea is boiling hot—
And whether pigs have wings.'
—Lewis Carroll, *Through the Looking-Glass*

Modus Tollens Revisited

You may have heard expressions such as "If he's twenty-one, then pigs have wings!"

The speaker, perhaps without realizing it, is inviting you to use the law of indirect reasoning (*modus tollens*), which you studied in Lesson 12.1. It is certainly not true that pigs have wings, so the statement in question (the hypothesis of the conditional) must be false.

If he's twenty-one, then pigs have wings.	*(If p then q)*
Pigs do not have wings.	*(~q)*
Therefore, he's not twenty-one.	*(Therefore, ~p)*

Prepare

QUICK WARM-UP

1. Write the contrapositive of the following statement: If the defendant was in his car at 10:00 P.M., then he was in an automobile accident. If the defendant was not in an automobile accident, then he was not in his car at 10:00 P.M.

2. If a triangle is a right triangle, then is it possible for it also to be an obtuse triangle? Explain. No, it is not possible because if the sum of the angles is 180°, then it cannot contain both an angle of 90° and an angle greater than 90°.

Also on Quiz Transparency 12.4

Teach

Why Indirect proof is a very common form of argument that is used when normal deduction is difficult.

Alternative Teaching Strategy

USING VISUAL MODELS Help students understand the method of proof by contradiction by having them draw a flowchart for the proof. At the top of the paper, show the statements *p* and *not p* for a given proof. From each of these statements, write conclusions that follow deductively until one branch leads to a contradiction. Reject that branch of the argument and accept the other branch.

Teaching Tip

Explain circumstances under which indirect reasoning applies to an argument. Have students write examples of indirect reasoning on the board and discuss them in class.

Cooperative Learning

Indirect Proofs

A form of argument closely related to *modus tollens* is known by its Latin name ***reductio ad absurdum***—literally, "reduction to absurdity." In this form, an assumption is shown to lead to an absurd or impossible conclusion, which means that the assumption must be rejected.

In formal logic and mathematics, certain proofs use the strategy of *reductio ad absurdum*, but with an important twist. In such proofs, you assume the opposite or, in logical terms, the negation of the statement that you want to prove. If this assumption leads to an impossible result, then it must be concluded that the assumption was false. Hence, the original statement was true. Such proofs are known as **indirect proofs** or **proofs by contradiction**.

Using a Contradiction to Prove Your Point

What is meant by an "absurd" or "impossible" result? In logic, a *contradiction* is such a result. A **contradiction** has the following form:

$$p \text{ AND } \sim p$$

That is, a contradiction asserts that a statement and its negation are both true. The following compound statement is a contradiction:

A horse is a vegetarian, and a horse is not a vegetarian.

In formal logic and in mathematics, an assumption that leads to a contradiction must be rejected. Thus, contradictions turn out to be very useful.

Proof by Contradiction

To prove a statement is true, assume that it is false and show that this leads to a contradiction.

12.4.1

An Indirect Proof

The following proof uses a contradiction to prove the converse of the Corresponding Angles Postulate. Recall that the converse of this postulate (which is itself a theorem) states the following:

If two lines are cut by a transversal in such a way that corresponding angles are congruent, then the two lines are parallel. (Theorem 3.4.1)

To prove Theorem 3.4.1, use the "if" part of the theorem as the given.

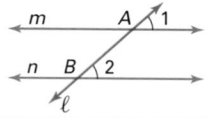

Interdisciplinary Connection

CRIMINAL LAW Detectives often use indirect reasoning to eliminate or implicate suspects in a crime. Have students research the types of courses that people studying to become detectives must take.

Enrichment

Have students simulate a defense lawyer's job by writing a short argument with conclusions drawn by deduction and indirect proof. Have students share their arguments with the class.

INDIRECT PROOF

Given: Line ℓ is a transversal that intersects lines m and n, and $\angle 1 \cong \angle 2$.

Prove: $m \| n$

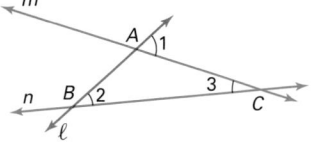

Proof: Assume that m is not parallel to n.

By assumption, m is not parallel to n, so the two lines will meet at some point, C, as shown in the redrawn figure above.

Because $\angle 1$ is an exterior angle of $\triangle ABC$, $m\angle 1 = m\angle 2 + m\angle 3$. But this means that $m\angle 1 > m\angle 2$ (because $m\angle 3 > 0°$). Therefore, $\angle 1$ is not congruent to $\angle 2$. Thus, the assumption that m is not parallel to n has led to the following contradiction:

$$(\angle 1 \cong \angle 2) \text{ AND } (\angle 1 \ncong \angle 2)$$

Therefore, the assumption must be false. The conclusion is $m \| n$.

Alibis and *Modus Tollens*

Arguments using the law of indirect reasoning (*modus tollens*) are more common than you might think. In a court of law, for example, a lawyer might want to show that a statement made by the prosecutor contradicts the accepted evidence, or the given. Arguments such as the following are quite common:

"If the defendant set the fire, then she must have been at the restaurant between 7:30 P.M. and 11:00 P.M. But three witnesses have testified that the defendant was not at the restaurant during those hours—she was in fact at a party on the other side of town. Therefore, the defendant did not set the fire."

Courtroom artists make sketches of trial scenes

The form of this argument can be represented as follows:

1. If the defendant set the fire, then she was at the restaurant between 7:30 P.M. and 11:30 P.M. (If p then q)

2. The defendant was not at the restaurant between 7:30 P.M. and 11:30 P.M. ($\sim q$)

3. Therefore, the defendant did not set the fire. (Therefore, $\sim p$)

TRY THIS   Show how the argument above could be made into a proof by contradiction.

ASSIGNMENT GUIDE

In Class	1–12
Core	13–17 odd, 18–35
Core Plus	18–38
Review	39–48
Preview	49–51

✐ Extra Practice can be found beginning on page 818.

7. (Lines ℓ and m are parallel) AND (lines ℓ and m are not parallel).

8. ($\triangle ABC$ is isosceles) AND ($\triangle ABC$ is not isosceles).

9. (All squares are rectangles) AND (all squares are not rectangles).

10. ($ABCD$ is a square) AND ($ABCD$ is not a square).

13. Yes; the argument has the *modus tollens* form: if p then q; ~q; therefore, ~p. (*Modus tollens* is the law of indirect reasoning.)

14. Yes; the argument has the *modus tollens* form: if p then q; ~q; therefore, ~p. (*Modus tollens* is the law of indirect reasoning.)

15. No; the argument has the *modus ponens* form: if p then q; p; therefore q.

16. No; the argument has the invalid form of denying the antecedent: if p then q; ~p; therefore, ~q.

17. Yes; the argument has the *modus tollens* form: if p then q; ~q; therefore, ~p. (*Modus tollens* is the law of indirect reasoning.)

Exercises

Communicate

1. What is a contradiction? Give the logical form and explain what it means.

Are the following statements contradictions? Explain your reasoning.

2. A tiger is a cat, and a tiger is a mammal.

3. A crocodile is a reptile, and a crocodile is not a reptile.

4. Some insects are butterflies, and some insects are not butterflies.

5. Some dogs are not pets, and some pets are not dogs.

6. Summarize the steps for writing an indirect proof.

Guided Skills Practice

Form a contradiction by using each statement and its negation. (*USING A CONTRADICTION*)

7. Lines ℓ and m are parallel.

8. $\triangle ABC$ is isosceles.

9. All squares are rectangles.

10. $ABCD$ is a square.

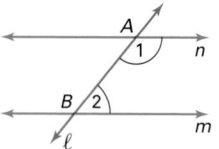

Suppose that two lines are cut by a transversal in such a way that the same-side interior angles are supplementary. Complete the following proof that the lines are parallel. (*INDIRECT PROOFS*)

Suppose that **11.** __?__ . Then lines ℓ and m must meet at some point, C, forming $\triangle ABC$. The sum of the measures of the angles of $\triangle ABC$ must be 180°. However, this is a contradiction because **12.** __?__ . Thus, lines ℓ and m must be parallel.

11. lines ℓ and m are not parallel

12. same-side interior angles are supplementary

Practice and Apply

For Exercises 13–17, determine whether the given argument is an example of indirect reasoning. Explain why or why not.

13. If it were snowing, there would be snowflakes in the air. There are no snowflakes in the air. Therefore, it is not snowing.

14. If you were not ill, then you would eat a large dinner. You did not eat a large dinner. Therefore, you must be ill.

15. If I see my shadow, the sun must be shining. I see my shadow. Therefore, the sun is shining.

16. If I am not in the United States, then I am not in New York. I am in the United States. Therefore, I am in New York.

17. If the dog knocked over the trash, then he would look guilty. The dog does not look guilty. Therefore, the dog did not knock over the trash.

The following indirect proof, based on Euclid's proof of Proposition 6 in Book I of the _Elements_, has the classical _reductio ad absurdum_ form. From a certain assumption an absurd result is produced—namely, that a larger triangle is congruent to a smaller one contained in it. No contradiction is stated formally.

Proposition 6 If in a triangle two angles equal one another, then the sides opposite the equal angles also equal one another.

INDIRECT PROOF

18. $\overline{AB} \not\cong \overline{AC}$
19. $\overline{DB} \cong \overline{AC}$
20. $\overline{BC}$
21. $\angle ACB$
22. $\triangle DCB$
23. SAS
24. $\overline{AB} \cong \overline{AC}$

Given: In $\triangle ABC$, $\angle B \cong \angle C$.

Prove: $\overline{AB} \cong \overline{AC}$

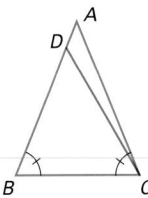

Proof: Suppose that **18.** ___?___ . Then one of the two sides $\overline{AB}$ or $\overline{AC}$ must be longer than the other. Let $\overline{AB}$ be the longer side. Then there exists a point, D, on $\overline{AB}$ such that **19.** ___?___ . Since $\overline{BC} \cong$ **20.** ___?___ , $\overline{DB} \cong \overline{AC}$, and $\angle B \cong$ **21.** ___?___ , $\triangle ABC \cong$ **22.** ___?___ by **23.** ___?___ , which is absurd. Therefore, **24.** ___?___ .

Complete the indirect proof below.

INDIRECT PROOF

25. $\overline{JK} \cong \overline{JL}$
26. isosceles
27. $\angle K \cong \angle L$
28. $\angle KJM \cong \angle LJM$
29. $\triangle KJM \cong \triangle LJM$
30. ASA
31. $\overline{KM} \cong \overline{LM}$

Given: $\overline{JM}$ bisects $\angle KJL$, and $\overline{JM}$ is not a median of $\triangle JKL$.

Prove: $\overline{JK} \not\cong \overline{JL}$

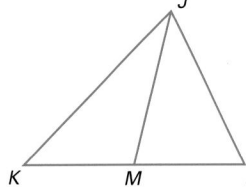

Proof: Suppose that **25.** ___?___ . Then $\triangle JKL$ is **26.** ___?___ . By the Isosceles Triangle Theorem, **27.** ___?___ . Since $\overline{JM}$ bisects $\angle KJL$, **28.** ___?___ , and so **29.** ___?___ by **30.** ___?___ . Thus, **31.** ___?___ because CPCTC. This gives a contradiction because **32.** ___?___ . Therefore, **33.** ___?___ .

---

internet connect

Homework Help Online
Go To: **go.hrw.com**
Keyword:
MG1 Homework Help
for Exercises 34-37

NUMBER THEORY **Write an indirect proof of each of the following:**

34. There is no largest integer. (Hint: If n is an integer, then $n + 1$ is an integer.)

35. There is no smallest positive real number. (Hint: $\frac{1}{2}x < x$ for all positive values of x)

36. For two integers m and n, if m^2 does not divide n^2 with no remainder, then m does not divide n with no remainder. (Hint: If m divides n with no remainder, then n can be factored into $k \cdot m$ for some integer k.)

37. If a fraction has a terminating decimal expansion, then the denominator divides some power of 10 (10, 100, 1000, etc.) with no remainder. (Hint: If a decimal terminates after n places, then multiplying it by 10^n gives an integer.)

APPLICATION

38. LAW A defense attorney begins an argument with the following words: "Suppose that the defendent is guilty of the crime. Then he was capable of carrying a heavy load up a steep hill for nearly a mile…" How might the attorney continue the argument? Explain how this would be an example of an indirect proof. State what the contradiction would be.

37. Given: a fraction $\frac{x}{y}$ with a decimal expansion that terminates after n places. Suppose the denominator y does not divide any power of 10 with no remainder. Then y does not divide 10^n. But $10^n\left(\frac{x}{y}\right) = m$, an integer, and so $10^n x = my$. Then y divides my so y divides $10^n x$. But y does not divide x, so y divides 10^n, a contradiction. Therefore y divides 10^n for some n.

38. Sample answer: This contradicts the evidence that my client had a broken leg, and thus was unable to perform that task. Therefore, the assumption that my client is guilty is false, so my client is innocent.

In Exercises 18–33, students can use geometry proof programs. Various options can be used to test different possible conclusions.

Error Analysis

The proofs in Exercises 34–37 may be difficult for students to begin on their own. It may be helpful to have students form groups to discuss how to begin each proof before working on them individually.

32. $\overline{JM}$ is not a median of $\triangle JKL$

33. $\overline{JK} \not\cong \overline{JL}$

34. Suppose that there is a largest integer m. But if m is an integer, then $m + 1$ is an integer, and $m + 1$ is larger than m. So m is not the largest integer. This is a contradiction. Therefore there is no largest integer.

35. Suppose that there is a smallest positive real number x. But $\frac{1}{2}x < x$ so x is not the smallest positive real number. This is a contradiction. Therefore there is no smallest positive real number.

36. Given: m and n are integers and m^2 does not divide n^2 with no remainder. Suppose m does divide n with no remainder (assume the opposite of what is to be proven). Then n can be factored into $k \cdot m$ for some integer k.

$$n = k \cdot m$$
$$n^2 = k^2 m^2$$
$$\frac{n^2}{m^2} = k^2$$

Thus m^2 does divide n^2 with no remainder. This is a contradiction. Therefore, for two integers m and n, if m^2 does not divide n^2 with no remainder, then m does not divide n with no remainder.

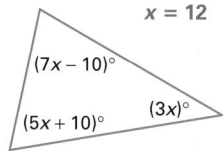
Look Back

Algebra

39. Solve for x in the figure below. *(LESSON 3.8)*

$x = 12$

$(7x - 10)°$
$(5x + 10)°$
$(3x)°$

In Exercises 40–43, state whether the given side lengths determine an acute triangle, a right triangle, an obtuse triangle, or no triangle. *(LESSON 5.4)*

40. $a = 10, b = 12, c = 14$ acute
41. $a = 7, b = 9, c = 15$ obtuse
42. $a = 4, b = 7, c = 12$ no triangle
43. $a = 8, b = 12, c = 17$ obtuse

Find the sine, cosine, and tangent of each angle below. *(LESSONS 10.1 AND 10.2)*

44. $72°$ **45.** $45°$ **46.** $140°$ **47.** $5°$

APPLICATION

48. TELECOMMUNICATIONS Suppose that you wish to measure the height of a radio tower. From point A, you measure the angle of elevation to the top of the tower to be $55°$. Then you walk 90 feet toward the tower to point B and measure the new angle of elevation, which is $68°$. Estimate the height of the tower. *(LESSON 10.4)* 304 ft

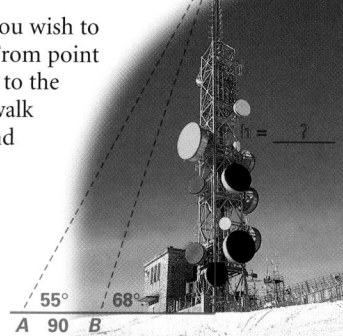

Look Beyond

Mathematical Induction: The principle of mathematical induction is useful in many proofs. The principle can be stated as follows: Suppose that S_n is a statement in terms of n, which can be any positive integer. Suppose that the statement is true for $n = 1$ and that if the statement is true for n, then it is always true for $n + 1$. Then the statement is true for all values of n.

49. Show that the statement "$5^n - 1$ is divisible by 4" is true for $n = 1$.

50. Assume that "$5^n - 1$ is divisible by 4" is true for some value of n. Show that it must be true for $n + 1$; that is, show that "$5^{n+1} - 1$ is divisible by 4." (Hint: If a number is divisible by 4, then it can be written as $4k$ for some positive integer k. Since $5^n - 1$ is divisible by 4, write 5^n in terms of some positive integer k.)

51. According to the principle of mathematical induction, what have you proved?

49. $n = 1$
$5^n - 1 = 5 - 1 = 4$ is divisible by 4.

50. Assume that $5^n - 1$ is divisible by 4. It means that $5^n - 1$ can be written as $4k$ for some positive integer k. Then
$$5^n = 4k + 1$$
$$5 \cdot 5^n = 5(4k + 1)$$
$$5^{n+1} = 20k + 5$$
$$5^{n+1} - 1 = 20k + 4$$
$$5^{n+1} - 1 = 4(5k + 1)$$
So $5^{n+1} - 1$ is divisible by 4.

51. The statement is true for $n = 1$ and if it is true for n, then it is true for $n + 1$. Then according to the Principle of Mathematical Induction, the statement is true for all values of n.

MORE ON THE EUCLIDEAN ALGORITHM

The Euclidean algorithm you studied in the Portfolio Activity in Lesson 12.1 can be proved by using an indirect proof. There are two parts to the proof—the first part is to prove that the result is a divisor of both numbers, and the second is to prove that it is the largest such number.

Examine the steps of the Euclidean algorithm on page 775. The process for finding the gcd of any two numbers a and b, where $a > b$, can be written as follows:

Portfolio Extension

Go To: **go.hrw.com**
Keyword:
MG1 Farey

$$a = b(x_1) + r_1 \qquad a \div b = x_1 \text{ remainder } r_1$$
$$b = r_1(x_2) + r_2$$
$$r_1 = r_2(x_3) + r_3$$
$$\vdots \qquad \vdots \qquad \vdots$$
$$r_{n-2} = r_{n-1}(x_n) + r_n \qquad r_n \text{ is the gcd of } a \text{ and } b.$$
$$r_{n-1} = r_n(x_{n+1}) + 0$$

Note: Since the remainders are always nonnegative integers, and the remainder gets smaller at each step, the sequence of remainders will always end in 0.

1. Use the last equation above to explain why r_n must be a divisor of r_{n-1}. Use the next-to-last equation to explain why r_n must be a divisor of r_{n-2}. (Hint: Is the right side of the equation divisible by r_n?) Use the entire sequence of equations to explain why r_n must be a divisor of both a and b.

2. Suppose that there is a number $r > r_n$ which is a divisor of both a and b. Use the first equation above to explain why r must be a divisor of r_1. Use the entire sequence of equations to explain why r must be a divisor of r_n. How does this lead to a contradiction?

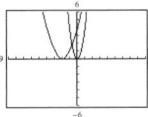

1. Give examples of electronic devices with on-off switches. **Sample: VCR, CD player, etc.**

2. Write a truth table for *p* AND *q*.

p	*q*	*p* AND *q*
T	T	T
T	F	F
F	T	F
F	F	F

3. Write a truth table for ~*p*.

p	~*p*
T	F
F	T

Also on Quiz Transparency 12.5

Teach

Why This lesson will help students make connections between computer logic, networks, and logical arguments.

Activity Notes

In this Activity, students create tables that simulate the on-off structure of computer logic gates. This Activity will help students understand that computer logic can be represented by the binary condition of *off* (0 or false) or *on* (1 or true).

For a student worksheet of this Activity and detailed Teacher Notes, see page 214 in the Lesson Activities booklet.

Computer Logic

Objectives

● Explore on-off tables, logic gates, and computer logic networks.

● Solve problems by using computer logic.

Why *Logic provides the foundation for the arithmetic and decision-making processes of many "smart" electronic devices. The fundamental units of logical circuits are logic gates, which function like logical operators such as AND, OR, and NOT.*

APPLICATION
COMPUTER ARCHITECTURE

A single computer chip in this "motherboard" contains hundreds of thousands of logic gates like the one shown greatly magnified here.

The Binary Number System

Computers use the **binary number system**. *Binary* means "having two parts," and the binary number system is based on the two numbers 1 and 0. You can think of a computer as a system of electrical switches that can be in two possible conditions, *on* and *off*. A switch that is on is represented by a 1, and a switch that is off is represented by a 0.

Activity
On-Off Tables

YOU WILL NEED
no special tools

The tables in this Activity simulate the working of three different devices. The columns of the tables represent the different states of the devices and their components. Let 1 = ON and 0 = OFF. Work in pairs to complete each table.

Part I

Pushing the power button on a TV remote control will turn the TV on if it is off and turn it off if it is on. Determine whether the TV will be on or off after pressing the power button, and fill in the blanks in the table below.

CHECKPOINT ✓

TV before pressing POWER button	TV after pressing POWER button
1	?
0	?

Alternative Teaching Strategy

INVITING PARTICIPATION Have group of students write an example of a simple on-off network such as a tape recorder. Encourage them to draw a diagram and write a description of the options available through the network. Have groups compare their examples. Ask each group to describe their network to the class.

Part II

To record on a particular video recorder, the user must press both PLAY and RECORD. Determine whether the video recorder will record or not for each combination of buttons listed in the table below, and fill in the blanks.

CHECKPOINT ✔

PLAY button	RECORD button	Video recorder
1	1	?
1	0	?
0	1	?
0	0	?

Part III

The student driver car at Dover High School has been equipped with two brake pedals so that either the student driver or the instructor can stop the car. Determine whether the car brakes will be on or off for each combination of pedals listed in the table below, and fill in the blanks.

CHECKPOINT ✔

Student pedal	Instructor pedal	Brakes
1	1	?
1	0	?
0	1	?
0	0	?

CRITICAL THINKING   How do the tables in the Activity compare with truth tables?

Logic Gates

Each table in the Activity above corresponds to a particular type of electronic circuitry called a **logic gate**. Logic gates are the building blocks of "smart" electronic devices. Each logic gate is named for a special logical function, such as NOT, AND, or OR.

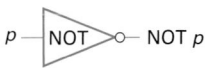

In the diagram below, p represents the electrical input. The input-output table shows the resulting output for the given input. If the input, p, of a NOT gate has a value of 1, the output, NOT p, will have a value of 0—and vice versa. Notice that the input-output table corresponds to the table in Part I of the Activity.

NOT Logic Gate

p — NOT ▷∘ — NOT p

Input   Gate   Output

Input-Output Table	
Input	**Output**
p	NOT p
1	0
0	1

CHECKPOINT ✔
Part I

TV before pressing POWER button	TV after pressing POWER button
1	0
0	1

Part II

PLAY button	RECORD button	Video recorder
1	1	1
1	0	0
0	1	0
0	0	0

Part III

Student pedal	Instructor pedal	Brakes
1	1	1
1	0	1
0	1	1
0	0	0

CRITICAL THINKING

The analogy between these tables and the truth tables can be drawn as follows:

1=T

0=F

Table in Part I is analogous to p; NOT p.

Table in Part II is analogous to p; q; p AND q.

Table in Part III is analogous to p; q; p OR q.

Interdisciplinary Connection

ELECTRONICS Electronic circuitry designs require knowledge of logic gates and on-off switches. Have students of electronics display and describe some of their network designs for the class.

Enrichment

Ask students to research the work of George Boole and Boolean algebra. Have them describe how Boolean algebra is used to solve problems in a computer system.

Cooperative Learning

Have students compare input-output tables with truth tables and with the three tables in the Activity. Have students list similarities and differences between any two tables.

Teaching Tip

Have students use chips, blocks, and other models to design a logic network. Ask students to describe how to translate the network into a physical model.

ADDITIONAL
EXAMPLE 1

Construct an input-output table for the network below.

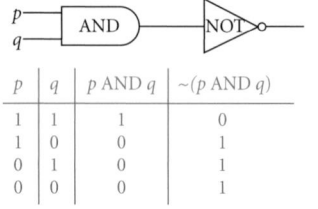

p	q	p AND q	~(p AND q)
1	1	1	0
1	0	0	1
0	1	0	1
0	0	0	1

An AND logic gate needs two inputs, which are represented by p and q in the diagram below. Notice that in order to get an output value of 1, both inputs must have a value of 1. The input-output table corresponds to the table in Part II of the Activity.

AND Logic Gate

p
q —[AND]— p AND q

Input   Gate   Output

Input-Output Table		
Input		Output
p	q	p AND q
1	1	1
1	0	0
0	1	0
0	0	0

An OR logic gate is represented below, along with its input-output table. Notice that the input-output table corresponds to the table in Part III of the Activity.

OR Logic Gate

p
q —[OR]— p OR q

Input   Gate   Output

Input-Output Table		
Input		Output
p	q	p OR q
1	1	1
1	0	1
0	1	1
0	0	0

Network of Logic Gates

Logic gates can be combined to form networks. You can use input-output tables to determine how a network operates.

EXAMPLE 1 Construct an input-output table for the network at right.

P
q —[OR]—[NOT]o—

SOLUTION

Read from left to right. The first gate is p OR q. Determine the output from this gate. Then perform the NOT operation: NOT (p OR q). To construct the input-output table, consider all possible input combinations. Fill in the values as you would for a truth table, where 1 is T and 0 is F.

Input-Output Table			
Input		Output	
p	q	p OR q	NOT (p OR q)
1	1	1	0
1	0	1	0
0	1	1	0
0	0	0	1

Inclusion Strategies

USING COGNITIVE STRATEGIES Some students may have difficulty representing electronic circuitry as logic gates. Suggest that they read this lesson carefully, take notes, and explain each diagram in terms of the model that it represents. You can use cooperative groups to discuss the diagrams.

Reteaching the Lesson

USING DISCUSSION Have each student give a simple logical expression using *AND*, *OR*, and *NOT*. Ask them to create a network for the logical expression and to describe to the class how to use the network.

EXAMPLE 2 Create a logical expression that corresponds to the network at right.

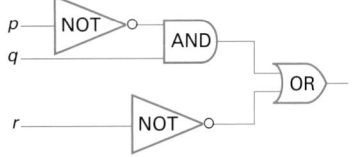

● **SOLUTION**

Read from left to right, one branch at a time. Combine the results of the branches when they flow together.

1. NOT appears first: NOT p.

2. The AND gate gives (NOT p) AND q.

3. The bottom branch gives NOT r.

4. The OR gate combines the output from Steps 2 and 3:

((NOT p) AND q) OR (NOT r)

TRY THIS How many rows and how many columns would be in an input-output table for the network in Example 2? Explain your reasoning.

Exercises

● *Communicate*

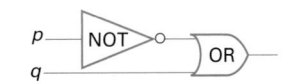

Activities Online
Go To: go.hrw.com
Keyword:
MG1 Binary

1. What is a logic gate? What do logic gates represent?

2. Does the OR logic gate represent an inclusive or exclusive *or*? How can you tell?

3. How are input-output tables for networks of logic gates related to truth tables?

4. In a network of logic gates with two inputs, there are four possible combinations of input values. How many combinations are possible with three inputs? with four inputs? Explain your reasoning.

● *Guided Skills Practice*

Complete the input-output table for each network. *(EXAMPLE 1)*

5.

Input-Output Table

Input		Output	
p	q	NOT p	NOT p OR q
1	1	? 0	? 1
1	0	? 0	? 0
0	1	? 1	? 1
0	0	? 1	? 1

6.

Input-Output Table

Input		Output	
p	q	p AND q	NOT (p AND q)
1	1	? 1	? 0
1	0	? 0	? 1
0	1	? 0	? 1
0	0	? 0	? 1

ADDITIONAL
EXAMPLE 2

Create a logical expression that corresponds to the following network:

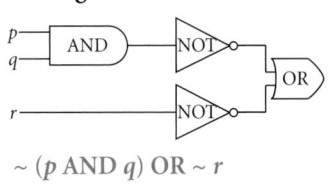

~ (p AND q) OR ~ r

TRY THIS
An input-output table would require eight rows because there are three inputs. The table would require seven columns: three for the inputs and four for the combinations.

Assess

Selected Answers
Exercises 5–8, 9–45 odd

ASSIGNMENT GUIDE

In Class	1–8
Core	9–27, 29–39 odd
Core Plus	16–27, 29–39 odd
Review	40–45
Preview	46–48

✐ Extra Practice can be found beginning on page 818.

Create a logical expression that corresponds to each network.
(EXAMPLE 2)

7.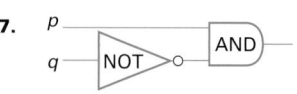

p AND (NOT *q*)

8.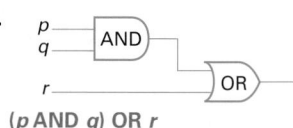

(*p* AND *q*) OR *r*

● *Practice and Apply*

Draw the symbol for the gate that corresponds to each of the following:

9. NOT   10. AND   11. OR

Use the logic gates below to answer each question.

12. If *p* = 1, what is the output? **0**

13. If *p* = 1, what is the output? **1**

14. If *p* = 1 and *q* = 0, what is the output? **0**

15. If *p* = 1 and *q* = 0, what is the output? **1**

For Exercises 16–19, complete the input-output table for the network of logic gates at right.

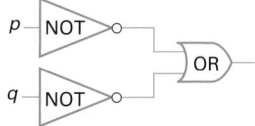

	p	*q*	NOT *p*	NOT *q*	(NOT *p*) OR (NOT *q*)
16.	1	1	? 0	? 0	? 0
17.	1	0	? 0	? 1	? 1
18.	0	1	? 1	? 0	? 1
19.	0	0	? 1	? 1	? 1

For Exercises 20–27, complete the input-output table for the network of logic gates at right.

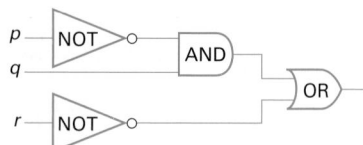

internet connect

Homework Help Online

Go To: **go.hrw.com**
Keyword:
MG1 Homework Help
for Exercises 20-27

	p	*q*	*r*	NOT *p*	(NOT *p*) AND *q*	NOT *r*	((NOT *p*) AND *q*) OR (NOT *r*)
20.	1	1	1	? 0	? 0	? 0	? 0
21.	1	1	0	? 0	? 0	? 1	? 1
22.	1	0	1	? 0	? 0	? 0	? 0
23.	1	0	0	? 0	? 0	? 1	? 1
24.	0	1	1	? 1	? 1	? 0	? 1
25.	0	1	0	? 1	? 1	? 1	? 1
26.	0	0	1	? 1	? 0	? 0	? 0
27.	0	0	0	? 1	? 0	? 1	? 1

Create a logical expression that corresponds to each network below.

28. NOT ((NOT *p*) OR *q*)

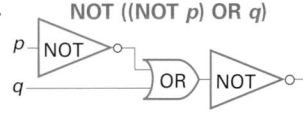

29. (NOT *p*) AND (NOT *q*)

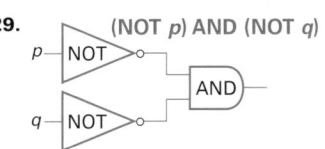

30. (*p* OR *q*) OR *r*

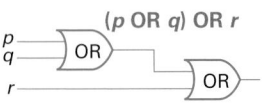

31. NOT ((NOT *p*) OR (*q* AND *r*))

Construct a network of logic gates for the following expressions:

32. *p* AND (NOT *q*)

33. NOT (*p* OR *q*)

34. (*p* AND (NOT *q*)) OR *r*

35. (*p* OR *q*) OR (NOT *r*)

Two logical expressions are **functionally equivalent** if, given the same input, they produce the same output. For Exercises 36 and 37, identify the two expressions in each list that are functionally equivalent.

36. a. NOT (*p* AND *q*)   **a and c**
   b. (NOT *p*) AND (NOT *q*)
   c. (NOT *p*) OR (NOT *q*)

37. a. *p* OR (NOT *q*)   **b and c**
   b. NOT (*p* OR *q*)
   c. (NOT *p*) AND (NOT *q*)

CHALLENGE

38. Create a network of logic gates that corresponds to the input-output table below.

Sample answer:

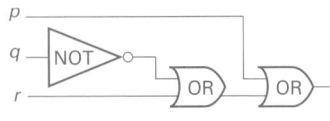

	Input		Output
p	**q**	**r**	**?**
1	1	1	1
1	1	0	1
1	0	1	1
1	0	0	1
0	1	1	1
0	1	0	0
0	0	1	1
0	0	0	1

APPLICATION

a. The arrangement corresponds to the logic function AND. Both *A* and *B* must be closed for the bulb to be on.

b. The arrangement corresponds to the logic function OR. either *A* or *B* must be closed for the bulb to be on.

39. ELECTRICITY In electrical circuit diagrams, a switch is represented as shown below. An open switch does not allow current to flow through the circuit.

open switch   closed switch

In the circuits at right, the bulb will light up if there is a closed path through the circuit. Tell what logical function is represented by each arrangement of switches. Explain how switches A and B work together to form a logic gate in each case.

battery: ⊣||⊢   bulb: ⊸⊙⊶

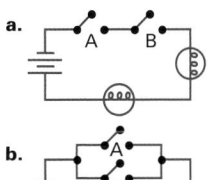

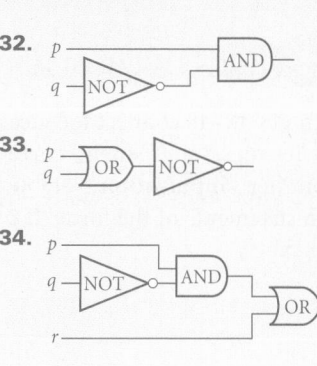

Exercises 46–48 connect the ideas of networks and their corresponding input-output tables with statements of the form "If p, then q."

43. False; example: $\frac{2}{3} = \frac{4}{6}$, but $\frac{(2+1)}{3} \neq \frac{(4+1)}{6}$ because $\frac{3}{3} \neq \frac{5}{6}$.

44. False; example: $\frac{2}{3} = \frac{4}{6}$, but $\frac{2}{3} \neq \frac{2+4+1}{3+6+1}$ because $\frac{2}{3} \neq \frac{7}{10}$.

45. $\frac{18}{25}$ or $\frac{25}{18}$ = the ratio of the perimeters; $\frac{25}{18}$ or $\frac{18}{25}$ = the ratio of the sides. The ratio of the perimeters is equal to the ratio of the sides in similar triangles.

46.

p	q	If p, then q
1	1	1
1	0	0
0	1	1
0	0	1

47. a.

p	q	$\sim p$	$\sim p$ AND q
1	1	0	0
1	0	0	0
0	1	1	1
0	0	1	0

b.

p	q	$\sim q$	p OR $\sim q$
1	1	0	1
1	0	1	1
0	1	0	0
0	0	1	1

c.

p	q	$\sim p$	$\sim p$ OR q
1	1	0	1
1	0	0	0
0	1	1	1
0	0	1	1

APPLICATION

40. Each square in the grid is about 10 feet on each side. Sample answer: About five of the grids will cover the spill.

≈ 6000 ft²

40. ENVIRONMENTAL SCIENCE The aerial photograph at right shows workers cleaning up an oil spill. Use objects in the photo to estimate the size of the squares in the yellow grid. Then estimate the area of the spill. *(LESSON 5.1)*

41. A 50-ft pipe carries water from a well to a house. Suppose that the pipe springs a leak. If the leak is equally likely to occur at any point along the pipe, what is the probability that the leak will be within 5 ft of the house? *(LESSON 5.8)* $\frac{1}{10}$ or 0.1

42. A point is chosen at random in the diagram at right. What is the probability that the point lies inside the shaded triangle? *(LESSON 5.8)* $\frac{9}{50}$ or 0.18

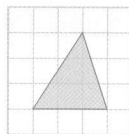

Algebra

In Exercises 43 and 44, determine whether the conditional is true for all values of the variables for which the proportions are defined. If the conditional is false, give a numerical counterexample. *(LESSON 8.2)*

43. If $\frac{x}{y} = \frac{r}{s}$, then $\frac{x+c}{y} = \frac{r+c}{s}$.

44. If $\frac{x}{y} = \frac{r}{s}$, then $\frac{x}{y} = \frac{x+r+m}{y+s+m}$.

45. If the perimeters of two similar triangles are 18 cm and 25 cm, what is the ratio of their corresponding sides? *(LESSON 8.6)*

In Exercises 46–48, you will discover a network of logic gates that is functionally equivalent to the statement "If p then q."

46. Create a truth table for the statement "If p then q." Use 1 for true statements and 0 for false statements.

47. Create an input-output table for each network below.

a. 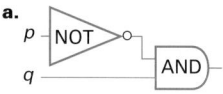 **b.** 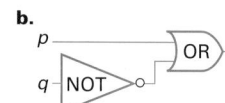 **c.**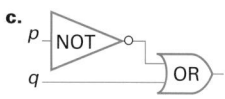

48. Which of the networks in Exercise 47 is functionally equivalent to the statement "If p then q"? Create a logical expression for this network.

48. The network whose logical expression is $\sim p$ OR q is functionally equivalent to "If p, then q."

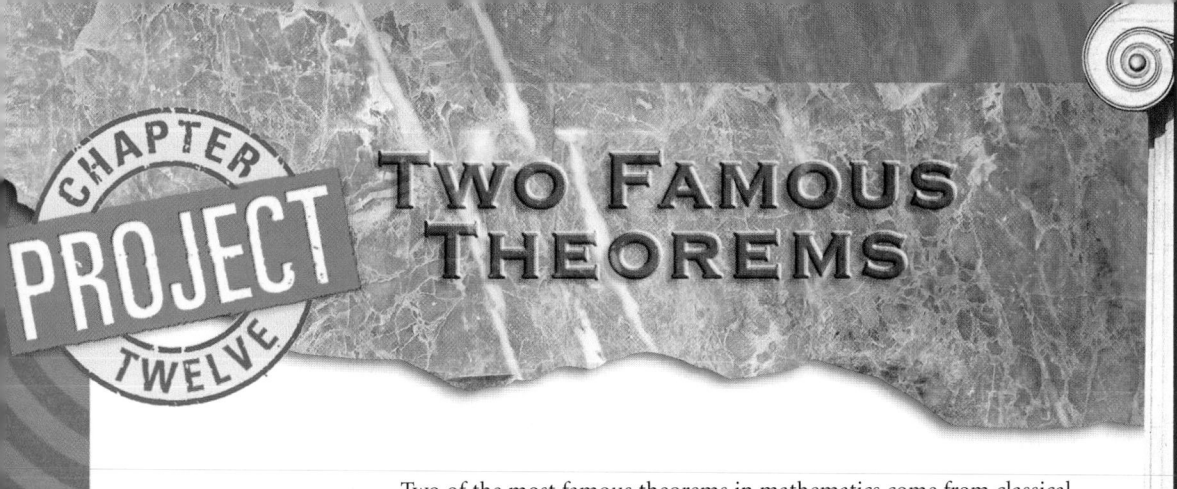

TWO FAMOUS THEOREMS

Focus

In the chapter project, students explore two famous proofs from number theory that involve indirect proof. Students are asked to explain some proofs in their own words and to prove some conjectures on their own.

Motivate

Review prime factorization with students. Remind them that either a number is prime or it has a unique prime factorization. Review the concepts of rational and irrational numbers. Ask students to think about the meaning of the word *incommensurability*. Provide a definition of the word to help students understand the incommensurability of the square root of two.

Two of the most famous theorems in mathematics come from classical times. Both of them involve indirect proof, which you studied in Lesson 12.4. Both of them also involve the prime factorization theorem, which is known as the Fundamental Theorem of Number Theory. It is not difficult to prove, but here it will be assumed true.

Every number has a unique prime factorization.

The Infinity of the Primes

Euclid's proof that there are infinitely many primes has captured the imagination of people over the ages. To help you understand the proof, which is actually very brief, a numerical example is given first.

The proof will assume that there is only a finite number of primes. Suppose, for example, that there are only 5 primes, and that 11 is the largest prime number. Then the list of all the prime numbers would read:

2, 3, 5, 7, 11

Now form a new number, m, by multiplying the prime numbers together and adding 1.

$$m = (2)(3)(5)(7)(11) + 1 = 2311$$

The new number m must be composite, according to the assumption, which states that 11 is the largest prime number. Therefore, m must be factorable into a combination of prime numbers, which (according to the assumption) range from 2 to 11. However, none of the numbers in the list of primes will divide the number exactly because you will always get a remainder of 1.

Cooperative Learning

Each of the activities can be individually assigned to a cooperative group. Have groups work on their activity as a long-term project and present their results to the class.

Discuss

Students have probably taken the number system for granted. Lead a discussion contrasting mathematics as an invented discipline with mathematics as discovered fact.

For example: $\frac{462 \text{ R } 1}{5\overline{)2311}}$

If 2311 is not divisible by one of the primes in the list, then it must be either prime, or divisible by some other prime greater than 11, which contradicts the assumption that 11 is the largest prime.

Once you understand this example, you should be able to follow the generalization of it, which is given below.

Prove: There are infinitely many primes.
Proof: Assume that there are a finite number of primes, say, n of them. Then the list of the primes would be $p_1, p_2, p_3, \ldots, p_n$. Form a new number, $m = (p_1)(p_2)(p_3) \cdots (p_n) + 1$.

By assumption, m must be composite. However, m is not divisible by any of the numbers in the list of primes, so there must be a prime larger than p, which contradicts the assumption. Thus, there are infinitely many primes.

The "Incommensurability" of the Square Root of 2

In the early history of mathematics, it seems to have been widely believed that any number could be represented as a fraction—that is, as a ratio of two integers. This was certainly believed by the early Pythagoreans. Thus, the proof that the square root of 2 *cannot* be represented by such a ratio (which means that it is irrational or *incommensurable*) came as a profound shock to them and shook the foundations of their beliefs.

Before studying the incommensurability theorem, you will need to know three simple theorems from number theory. The three theorems, given below, use the prime factorization theorem.

Prove: The square of an even number is even.
Proof: If a number is even, then 2 is one of its prime factors. When the number is squared, the number 2 will appear at least twice in the prime factorization of the square. Therefore, the square is divisible by 2, which means that it is an even number.

Prove: The square of an even number is divisible by 4.
Proof: The proof is left as an exercise for you in Activity 2.

Prove: The square of an odd number is odd.
Proof: The proof is left as an exercise for you in Activity 2.

You are now ready to tackle the incommensurability theorem.

Prove: The square root of 2 cannot be written as a ratio of two integers.
Proof: Assume that the square root of 2 can be represented as the ratio of two numbers—that is, as a fraction. If this fraction is not in lowest terms, then there is a fraction in lowest terms to which it can be reduced. Thus, the assumption can be written in the following way, which is equivalent to the original statement: Assume that the square root of 2 can be represented as a fraction in lowest terms.

Let p be the numerator and q be the denominator of the fraction that is assumed to exist.

$$\frac{p}{q} = \sqrt{2} \qquad \textit{p and q have no common factors.}$$

$$\left(\frac{p}{q}\right)^2 = (\sqrt{2})^2 = 2 \qquad \textit{Square both sides of the equation.}$$

$$p^2 = 2q^2 \qquad \textit{Multiply to clear the fraction.}$$

The last equation implies that p^2 is an even number and so p is an even number. (Why?) Therefore, p^2 is the square of an even number. This implies that p^2 is divisible by 4. But if p^2 is divisible by 4, then q^2 must be an even number. (Why?) Therefore, q is even, as well.

If p and q are both even, then they must have a common factor of 2, which contradicts the original assumption. Therefore, the assumption must be false, and the theorem is proven.

Activity ① The Infinity of the Primes

1. Repeat the numerical example in the proof for different numbers of primes. Explain why you always get a remainder of 1 when you divide m by one (or more) of the primes in your list.

2. If you subtract 1 instead of adding 1 to obtain the number m in the proof, how would the proof be affected?

3. Do some research on number theory and learn about some of the unproven conjectures such as the *Goldbach Conjecture* and the *Twin Primes Conjecture*. Explain them in your own words and give numerical illustrations for each one.

Activity ② The "Incommensurability" of the Square Root of 2

1. Prove that the square of an even number is divisible by 4.

2. Prove that the square of an odd number is odd.

3. In the incommensurability proof, there are two points at which the reader is asked to explain why a certain result occurs. Write these explanations in your own words.

Chapter Review and Assessment

Chapter Test, Form A

NAME _____ CLASS _____ DATE _____

Chapter Assessment
Chapter 12, Form A, page 1

Write the letter that best answers the question or completes the statement.

For Exercises 1–2, use the following: If p then q
                                      p
                             Therefore, q

a   1. This form of argument is called _____.
    a. *modus ponens*   b. *modus tollens*
    c. affirming the antecedent   d. denying the consequent

b   2. Statement q is called _____.
    a. an inverse   b. a conclusion
    c. a premise   d. a contrapositive

b   3. What conclusion can be drawn from the premises below?
    If an animal is a mammal, then it lives in trees.
    A cow does not live in trees.
    a. A cow is a mammal.   b. A cow is not a mammal.
    c. A cow is not an animal.   d. cannot be determined

b   4. Which is the converse of the following statement?
    If Sue plants an apple seed, a tree will grow.
    a. If Sue does not plant an apple seed, a tree will not grow.
    b. If a tree grows, Sue planted an apple seed.
    c. If a tree does not grow, Sue did not plant an apple seed.
    d. If Sue does not plant an apple seed, a tree will grow.

c   5. Which of the following results is *not* a true statement for $p \Rightarrow q$?
    a. p is true and q is true.   b. p is false and q is true.
    c. p is true and q is false.   d. p is false and q is false.

b   6. Given the statements p, "the temperature is 90°F," and q, "the apartment is hot," which is the symbolic representation of, "the temperature is 90°F and the apartment is not hot"?
    a. p AND q b. p AND $\sim q$ c. $\sim p$ AND q d. $\sim p$ AND $\sim q$

d   7. In an indirect proof, which logical form of the statement to be proven is assumed?
    a. inverse   b. converse   c. disjunction   d. negation

ARGUMENT FORMS

Lesson	Section	Argument Form
12.1	12.1.1 Argument Form: *Modus Ponens*	If p then q p Therefore, q
	12.2.2 Argument Form: *Modus Tollens*	If p then q $\sim q$ Therefore, $\sim p$
	12.1.3 Invalid Form: Affirming the Consequent	If p then q q Therefore, p
	12.1.4 Invalid Form: Denying the Antecedent	If p then q $\sim p$ Therefore, $\sim q$
12.4	12.4.1 Proof by Contradiction	To prove statement s, assume $\sim s$. Then the following argument form is valid: If $\sim s$ then (t AND $\sim t$) Therefore, s

NAME _____ CLASS _____ DATE _____

Chapter Assessment
Chapter 12, Form A, page 2

b   8. In an indirect proof for the statement "An obtuse triangle cannot contain a right triangle," the contradiction used to prove the statement is _____.
    a. the triangle is obtuse   b. the obtuse angle measures 90°
    c. the triangle is acute   d. all the angles are acute

c   9. If p is true and q is false, which of the following statements is true?
    a. p AND q b. $\sim p$ OR q c. p AND $\sim q$ d. $p \Rightarrow q$

d   10. If p is true and q is true, which of the following statements is false?
    a. $\sim q \Rightarrow p$ b. $\sim p$ OR q c. p AND q d. $p \Rightarrow \sim q$

c   11. Which is the logical expression for this network?
    a. $\sim (p$ AND $q)$
    b. $(\sim p)$ AND q
    c. $\sim [(\sim p)$ AND $q]$
    d. $(\sim p)$ AND $(\sim q)$

b   12. Which is the logical expression for this network?
    a. $[\sim (p$ OR $q)]$ AND $(\sim r)$
    b. $[p$ OR $(\sim q)]$ AND $(\sim r)$
    c. $[p$ AND $(\sim r)]$ OR $(\sim q)$
    d. $[\sim (p$ AND $r)]$ OR $(\sim q)$

c   13. If p is 1 and q is 0, which of the following outputs is equal to 1?
    a. $\sim p$ AND q b. $\sim p$ OR q c. p AND $\sim q$ d. $\sim (p$ OR $q)$

In Exercises 14–16, choose the term from the list below that best matches each definition.

> a. biconditional
> b. fallacy
> c. counterexample
> d. indirect proof
> e. binary number system

a   14. an "if and only if" statement

b   15. a common logical mistake of affirming the consequent or denying the antecedent

d   16. an argument that assumes the negation of what is to be proven

Key Skills & Exercises

LESSON 12.1

Key Skills

Determine whether an argument is valid or invalid.

Is the following argument valid or invalid?

If the Memorial Day parade passes the house, then the house is on Main Street.
The Memorial Day parade does not pass the house.

Therefore, the house is not on Main Street.

This argument is invalid. It contains the invalid argument form known as denying the antecedent.

Use the law of indirect reasoning *(modus tollens)*.

Give a valid conclusion based on the following premises:

If the oven is on, then the oven door feels warm. The oven door does not feel warm.

Using indirect reasoning (*modus tollens*), the statement "The oven is not on" is a valid conclusion.

Exercises

Determine whether each argument is valid or invalid.

1. If a polygon has 5 sides, then it is a pentagon.
   Polygon *ABCDE* has 5 sides.
   Therefore, polygon *ABCDE* is a pentagon.

2. If the car is out of gas, then the car will not start. The car will not start.
   Therefore, the car is out of gas.

Use the law of indirect reasoning to give valid conclusions based on the following premises:

3. If an animal is a rodent, then it has incisors that grow continuously.
   A cat does not have incisors that grow continuously.

4. If a number is divisible by 6, then it is divisible by 3.
   25 is not divisible by 3.

1. Valid

2. Invalid

3. Therefore, a cat is not a rodent.

4. Therefore, 25 is not divisible by 6.

5. John is my brother and 17 is prime.

LESSON 12.2

Key Skills

Write the conjuction and disjunction of two statements.

Write the conjunction and disjunction of the following statements:

The sky is green.
Your pants are on fire.

Conjunction: The sky is green, and your pants are on fire.

Disjunction: The sky is green, or your pants are on fire.

Exercises

For Exercises 5–8, refer to the statements below.

John is my brother.
17 is prime.

5. Write the conjunction of the statements above.

6. Write the disjunction of the statements above.

6. John is my brother or 17 is prime.

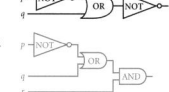
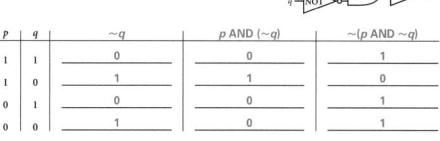

7.

p	q	p AND q
T	T	T
T	F	F
F	T	F
F	F	F

8.

p	q	p OR q
T	T	T
T	F	T
F	T	T
F	F	F

9. Converse: If the ground is wet, then it is raining. Inverse: If it is not raining, then the ground is not wet. Contrapositive: If the ground is not wet, then it is not raining.

10.

p	q	$q \Rightarrow p$
T	T	T
T	F	T
F	T	F
F	F	T

Create a truth table for a compound statement.

Create a truth table for the conjuction of the statements above.

Let p represent the statement "The sky is green" and let q represent "Your pants are on fire." Then the truth table for the conjunction p AND q is shown below.

p	q	p AND q
T	T	T
T	F	F
F	T	F
F	F	F

7. Create a truth table for the compound statement from Exercise 5.

8. Create a truth table for the compound statement from Exercise 6.

Key Skills

Create a truth table for a conditional.

Create a truth table for the following conditional: If it is 1:00 P.M., then Jill is in geometry class.

The conditional can be separated into its hypothesis, p, "It is 1:00 P.M.," and its conclusion, q, "Jill is in geometry class." The truth table is below.

p	q	$p \Rightarrow q$
T	T	T
T	F	F
F	T	T
F	F	T

Write a converse, inverse, and contrapositive of a conditional.

Write the converse, inverse, and contrapositive of the conditional above.

Converse: If Jill is in geometry class, then it is 1:00 P.M.

Inverse: If it is not 1:00 P.M., then Jill is not in geometry class.

Contrapositive: If Jill is not in geometry class, then it is not 1:00 P.M.

Exercises

For Exercises 9–12, refer to the following conditional: If it is raining, then the ground is wet.

9. Write the converse, inverse, and contrapositive of the given conditional.

10. Create a truth table for the converse of the given conditional.

11. Create a truth table for the inverse of the given conditional.

12. Create a truth table for the contrapositive of the given conditional.

11.

p	q	$\sim p$	$\sim q$	$\sim p \Rightarrow \sim q$
T	T	F	F	T
T	F	F	T	T
F	T	T	F	F
F	F	T	T	T

12.

p	q	$\sim p$	$\sim q$	$\sim q \Rightarrow \sim p$
T	T	F	F	T
T	F	F	T	F
F	T	T	F	T
F	F	T	T	T

LESSON 12.4
Key Skills

Use indirect reasoning in a proof.

In $\triangle PQR$, point S is on $\overline{PR}$ and point T is on $\overline{QR}$. $\overline{QS}$ and $\overline{PT}$ intersect at point X inside $\triangle PQR$. Prove that $\overline{QS}$ and $\overline{PT}$ do not bisect each other.

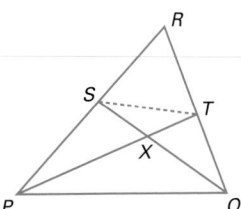

Assume that $\overline{QS}$ and $\overline{PT}$ bisect each other. Since $\overline{QS}$ and $\overline{PT}$ are diagonals of quadrilateral $PQTS$, if they bisect each other, then $PSTQ$ is a parallelogram. If $PSTQ$ is a parallelogram, then $\overline{PS}$ and $\overline{QT}$ are parallel. But $\overline{PS}$ and $\overline{QT}$ are not parallel because they lie on intersecting lines. This contradiction proves indirectly that $\overline{QS}$ and $\overline{PT}$ do not bisect each other.

Exercises

Given: $\triangle ABC$ is a scalene triangle, and $\overrightarrow{AD}$, the bisector of $\angle A$, intersects $\overline{BC}$ at point D.

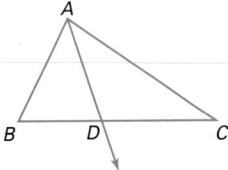

Prove: $\overrightarrow{AD}$ is not perpendicular to $\overline{BC}$.

Proof:

Assume **13.** ? . Then $\angle ADB$ and $\angle ADC$ both measure 90°. Since $\overrightarrow{AD}$ is the bisector of $\angle A$, $\angle BAD \cong \angle CAD$ and $\overline{AD} \cong \overline{AD}$, so **14.** ? by ASA. Thus, **15.** ? because CPCTC. However, this contradicts **16.** ? , thus proving indirectly that $\overrightarrow{AD}$ is not perpendicular to $\overline{BC}$.

13. $\overrightarrow{AD} \perp \overline{BC}$

14. $\triangle ADC \cong \triangle ADB$

15. $\overline{AC} \cong \overline{AB}$

16. the given fact that $\triangle ABC$ is scalene

17. p OR $\sim q$

18.

p	q	$\sim q$	p OR $\sim q$
1	1	0	1
1	0	1	1
0	1	0	0
0	0	1	1

19. (p OR q) AND $\sim r$

LESSON 12.5
Key Skills

Write a logical expression for a network.

Write a logical expression for the network below.

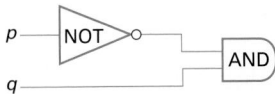

(NOT p) AND q

Create input-output tables.

Create an input-output table for the network above.

p	q	(NOT p) AND q
1	1	0
1	0	0
0	1	1
0	0	0

Exercises

For Exercises 17 and 18, refer to the diagram below.

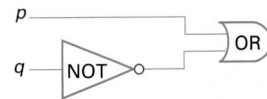

17. Write a logical expression for the given network.

18. Create an input-output table for the given network.

19. Write a logical expression for the network below.

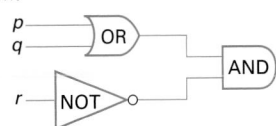

20. Create an input-output table for the network from Exercise 19.

20.

p	q	r	p OR q	NOT r	(p OR q) AND (NOT r)
1	1	1	1	0	0
1	1	0	1	1	1
1	0	1	1	0	0
1	0	0	1	1	1
0	1	1	1	0	0
0	1	0	1	1	1
0	0	1	0	0	0
0	0	0	0	1	0

21. It is a valid argument. It has *modus tollens* form.

Applications

21. LAW Determine whether the argument below is valid or invalid. Explain your reasoning.

If Mr. Smith committed the crime, then he would have been at the bank on Tuesday morning.
Mr. Smith was not at the bank on Tuesday morning.

Therefore, Mr. Smith did not commit the crime.

22. ELECTRICITY Imagine that you are planning your dream house and you want certain appliances to work in an interconnected way. Suppose that you want the living room TV to be off and the kitchen radio to be on if and only if the coffee maker and toaster are on or the kitchen light is on. Draw a network of logic gates to illustrate this situation.

22.

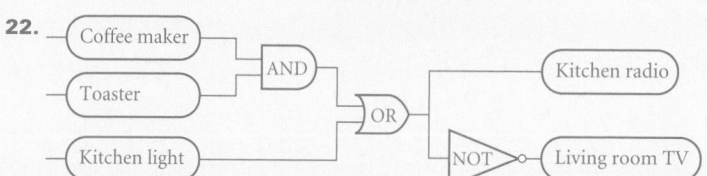

Note: Each appliance is turned on by a 1 input and off by a 0 input. Each appliance sends a 1 output if it is on and a 0 output if it is off.

12 Chapter Test

Determine whether each argument is valid or invalid.

1. If today is Tuesday, then Natalie has softball practice.
   Today is Tuesday.

   Therefore, Natalie has softball practice. **valid**

2. If the car is a Subaru, then it has four-wheel drive.
   The car has four-wheel drive.

   Therefore, the car is a Subaru. **invalid**

3. If the switch is off, then the video game will not play.
   The switch is on.

   Therefore, the video game will play. **invalid**

Use the law of indirect reasoning to give valid conclusions based on the following premises:

4. If a parallelogram is a square, then it is a rectangle.
   A rhombus is not a rectangle.

5. If a number is a multiple of 8, then it is a multiple of 4.
   The number 50 is not a multiple of 4.

For Exercises 6–10, refer to the statements below.

A dolphin is a mammal.
Circles are polygons.

6. Write the conjunction of the statements above.

7. State whether the conjunction is true or false. **false**

8. Write the disjunction of the statements above.

9. State whether the disjunction is true or false. **true**

10. Create a truth table for the disjunction p OR q.

11. Write the converse, inverse, and contrapositive of the following conditional:

    If $QRST$ is a trapezoid, then $QRST$ is a quadrilateral.

Create a truth table for the following:

12. the converse of the conditional $p \Rightarrow q$

13. the inverse of the conditional $p \Rightarrow q$

14. the contrapositive of the conditional $p \Rightarrow q$

15. **CHEMISTRY** Write the converse, inverse, and contrapositive of the conditional below. Decide whether each is true or false. Explain.

    If water boils at normal atmospheric pressure, then its temperature is greater than or equal to 100°C.

Complete the indirect proof below.

Given: $\overline{AD}$ bisects $\angle BAC$;
      $\overline{AD}$ is not a median
      of $\triangle ABC$.

Prove: $\overline{AB} \not\cong \overline{AC}$

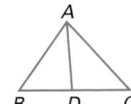

Proof: Suppose that **16.** ? . Then $\triangle ABC$ is **17.** ? . By the Isosceles Triangle Theorem, **18.** ? . Since $\overline{AD}$ bisects $\angle BAC$, **19.** ? , and so **20.** ? by **21.** ? . Thus, **22.** ? because CPCTC. This gives a contradiction because **23.** ? . Therefore, **24.** ? .

25. Write a logical expression for the network below. **~(p OR q)**

26. Create an input-output table for the network above.

27. Write a logical expression for the network below. **(p AND ~q) OR r**

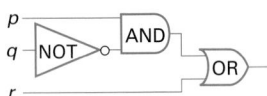

28. Create an input-output table for the network above.

---

16. $\overline{AB} \cong \overline{AC}$

17. isosceles

18. $\angle B \cong \angle C$

19. $\angle BAD \cong \angle CAD$

20. $\triangle BAD \cong \triangle CAD$

21. ASA

22. $\overline{BD} \cong \overline{CD}$

23. $\overline{AD}$ is not a median of $\triangle ABC$

24. $\overline{AB} \not\cong \overline{AC}$

---

4. Therefore, a rhombus is not a square.

5. Therefore, the number 50 is not a multiple of 8.

6. A dolphin is a mammal and circles are polygons.

8. A dolphin is a mammal, or circles are polygons.

10.

p	q	p OR q
T	T	T
T	F	T
F	T	T
F	F	F

11. Converse: If $QRST$ is a quadrilateral, then $QRST$ is a trapezoid.
    Inverse: If $QRST$ is not a trapezoid, then $QRST$ is not a quadrilateral.
    Contrapositive: If $QRST$ is not a quadrilateral, then $QRST$ is not a trapezoid.

12.

p	q	$q \Rightarrow p$
T	T	T
T	F	T
F	T	F
F	F	T

13.

p	q	$\sim p$	$\sim q$	$\sim p \Rightarrow \sim q$
T	T	F	F	T
T	F	F	T	T
F	T	T	F	F
F	F	T	T	T

14.

p	q	$\sim p$	$\sim q$	$\sim q \Rightarrow \sim p$
T	T	F	F	T
T	F	F	T	F
F	T	T	F	T
F	F	T	T	T

15. Conditional: True.
    Converse: If the water temperature is greater than or equal to 100°C, then it will boil at normal atmospheric pressure. True, by the laws of physics.
    Inverse: If water does not boil at normal atmospheric pressure, then its temperature is less than 100°C. True, by the laws of physics.
    Contrapositive: If water's temperature is less than 100°C, then it will not boil at normal atmospheric pressure. True, by the laws of physics.

26.

p	q	p OR q	$\sim(p$ OR $q)$
1	1	1	0
1	0	1	0
0	1	1	0
0	0	0	1

28.

p	q	r	$\sim q$	p AND $\sim q$	$(p$ AND $\sim q)$ OR r
1	1	1	0	0	1
1	1	0	0	0	0
1	0	1	1	1	1
1	0	0	1	1	1
0	1	1	0	0	1
0	1	0	0	0	0
0	0	0	1	0	0
0	0	1	1	0	1

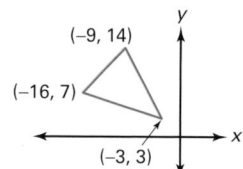
Multiple-Choice Samples

The first half of the Cumulative Assessment contains a multiple-choice section. This part of the Cumulative Assessment consists of items commonly found on standardized tests.

Free-Response Grid Samples

The second half of the Cumulative Assessment consists of a free-response section. This part requires student-produced response items like those commonly found on college entrance exams. These questions require the use of machine-scored answer grids. You may wish to have students practice answering these items in preparation for standardized tests.

1. c

2. a

3. b

4. b

5. d

MULTIPLE-CHOICE For Questions 1–8, write the letter that indicates the best answer.

1. Refer to the figure below. Which coordinates are the result of a 180° rotation of the triangle about the origin? *(LESSON 1.7)*

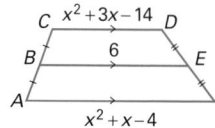

  a. $(7, -16)$, $(14, -9)$, $(3, -3)$
  b. $(-16, -7)$, $(-9, -14)$, $(-3, -3)$
  c. $(16, -7)$, $(9, -14)$, $(3, -3)$
  d. $(16, 7)$, $(9, 14)$, $(3, 3)$

2. Refer to the figure below. Find *AF*. *(LESSON 3.7)*

  a. 8
  b. 3
  c. 4
  d. 16

3. Find the approximate perimeter of the triangle below. *(LESSON 5.4)*

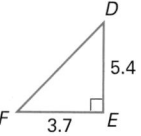

  a. 52
  b. 16
  c. 10
  d. 85

4. Which of the following statements is a definition? *(LESSON 2.3)*
  a. A monarch is an orange butterfly.
  b. A line segment is the shortest path between two points.
  c. A pile of loose rubble is a hazard.
  d. A rhombus is a parallelogram.

5. $\overline{AF}$ has endpoints at $(13, 15)$ and $(9, 20)$. The endpoints of a segment parallel to $\overline{AF}$ are at ___?___. *(LESSON 3.8)*
  a. $(0, 0)$ and $(25, 16)$
  b. $(15, 13)$ and $(20, 9)$
  c. $(4, -6)$ and $(8, -1)$
  d. $(4, -3)$ and $(12, -13)$

6. Identify the correct expression for *EC*. *(LESSON 9.5)*

a. $ED \times \dfrac{EB}{EA}$

b. $EB \times \dfrac{EA}{ED}$

c. $EB \times \dfrac{CD}{AB}$

d. $EB \times \dfrac{AB}{CD}$

7. Which statement is true in hyperbolic geometry? *(LESSON 11.5)*
  a. No lines are parallel.
  b. Two-sided polygons exist.
  c. The sum of the angles in a triangle is less than 180°.
  d. The shortest path between two points is along a great circle.

For Exercises 8–10, refer to the statements below. *(LESSONS 12.1 AND 12.3)*

p: If the fog has lifted, the boat can leave the harbor.

q: The boat cannot leave the harbor.

8. What conclusion can be drawn from statements *p* and *q*?
  a. The fog has lifted.
  b. The fog has not lifted.
  c. If the fog has not lifted, the boat cannot leave the harbor.
  d. No valid conclusion can be drawn from statements *p* and *q*.

9. Write the negation of statement *q*.

10. Write the contrapositive of statement *p*.

11. In the diagram at right, △*ABC* is equilateral. What are the coordinates of point *C*? *(LESSON 5.7)*

  a. $(0, a)$
  b. (a, a)
  c. $\left(\dfrac{a}{2}, \dfrac{a\sqrt{3}}{2}\right)$
  d. $(a, a\sqrt{3})$

FREE-RESPONSE GRID

Items 12–15 may be answered by using a free-response grid such as that commonly used by standardized-test services.

12. What is the central angle of a regular 13-gon? *(LESSON 3.1)*

13. A point is chosen at random inside the square in the coordinate plane below. What is the probability that the point lies within the shaded area? *(LESSON 5.7)*

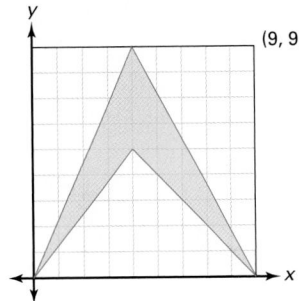

14. Find *MN*. *(LESSON 10.1)*

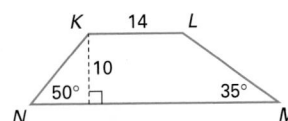

15. Find *AB*. *(LESSONS 10.5 AND 10.6)*

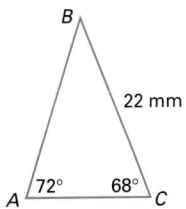

6. b

7. c

8. b

9. The boat can leave the harbor.

10. If the boat cannot leave the harbor, then the fog has not lifted.

11. c

12. 27.7°

13. 22%

14. 36.67 units

15. 21.4 mm

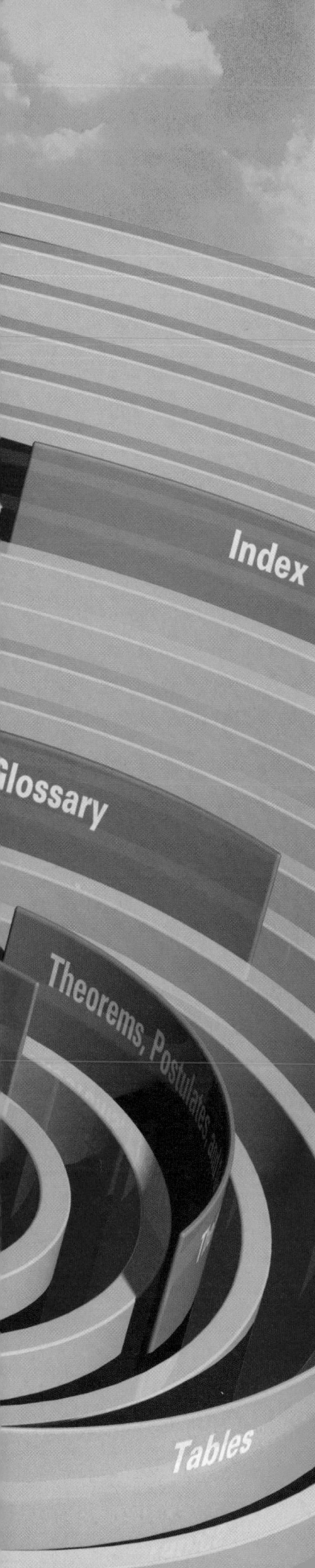

Info Bank

Lesson 1.1

3. $\overrightarrow{AB}$ and $\overrightarrow{AD}$ form angle 1, $\overrightarrow{BA}$ and $\overrightarrow{BC}$ form angle 2, $\overrightarrow{CB}$ and $\overrightarrow{CD}$ form angle 3, $\overrightarrow{DC}$ and $\overrightarrow{DA}$ form angle 4.

5. False. The two lines might not be coplanar, they could be skew lines.

6. False. Only collinear points are contained in one line.

Lesson 1.2

7. Does not make sense because $\overline{AB}$ refers to the segment, not the measure of the segment.

8. Does not make sense. Either $VW = 8$ in or $m\overline{VW} = 8$ in makes sense.

Extra Practice

CHAPTER 1

LESSON 1.1

In Exercises 1–3, refer to the rectangle at right.

1. Name all of the segments in the rectangle. $\overline{AB}$, $\overline{BC}$, $\overline{CD}$, $\overline{DA}$

2. Name the plane that contains the rectangle. plane *ABC*

3. Name the rays that form each side of the angles in the rectangle.

In Exercises 4–6, classify each statement as true or false, and explain your reasoning.

4. Two planes intersect in a line. True

5. Two lines are contained in exactly one plane.

6. Any three points are contained in exactly one line.

Refer to the figure at right for Exercises 7 and 8.

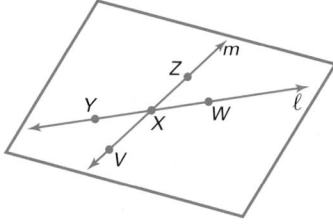

7. Name the intersection of lines ℓ and m. point *X*

8. Name an angle at the intersection of lines ℓ and m. Name the vertex of this angle and the two rays that form the sides of the angle.

$\angle VXW$, point *X*, $\overrightarrow{XV}$ and $\overrightarrow{XW}$

LESSON 1.2

In Exercises 1–3, find the length of $\overline{AB}$.

1.
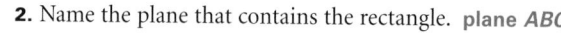
5

2.
6

3.

8

In Exercises 4 and 5, point C is between points A and B on $\overline{AB}$. Sketch the figure and find the missing lengths.

4. $AC = 12$; $CB = \underline{\quad?\quad}$; $AB = 34$ 22

5. $AC = 11$; $CB = 18$; $AB = \underline{\quad?\quad}$ 29

Find the indicated value.

6. $WY = 20$; $v = \underline{\quad?\quad}$ 4

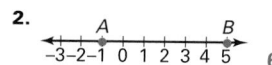

In Exercises 7 and 8, explain why each statement does or does not make sense.

7. $\overline{AB} + BC = 15$ cm

8. $mVW = 8$ in.

LESSON 1.3

Find the measure of each angle. Refer to the diagram at right.

1. m∠AVB **42°**

2. m∠AVC **99°**

3. m∠BVC **57°**

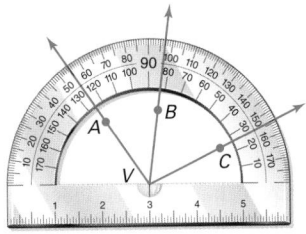

In the figure at right, m∠BVC = 55° and ∠AVB and ∠BVD form a linear pair. Find the following:

4. m∠CVD **55°**

5. m∠AVB **70°**

6. m∠AVC **125°**

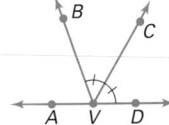

In the diagram at right, m∠WXY = (7x + 14)°. Find the value of x, and then find each indicated angle measure.

7. m∠WXZ **x = 5.5, m∠WXZ = 35°**

8. m∠ZXY **m∠ZXY = 17.5°**

9. m∠WXY **m∠WXY = 52.5°**

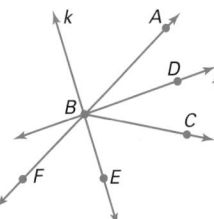

LESSON 1.4

Use folding paper to construct the figures described in Exercises 1–4. Do not use a ruler or protractor. Trace over each figure and label all relevant parts.

1. a 135° angle

2. a 112.5° angle

3. a parallelogram that is not a rectangle

4. a rhombus

For Exercises 5–7, suppose that ℓ is the angle bisector of ∠ABC, line k is perpendicular to ℓ, and m∠ABD = 35°.

5. Find m∠ABC and m∠DBC. **70°, 35°**

6. Find m∠EBC. **55°**

7. If ∠FBC and ∠CBA are supplementary angles, find m∠FBE. **55°**

Lesson 1.4

1.
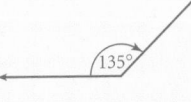

2.

3.
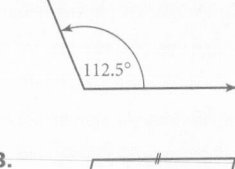

4.
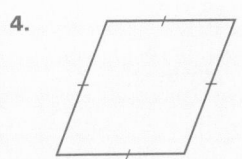

Lesson 1.5

1.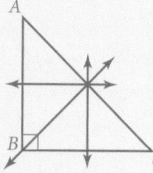

2.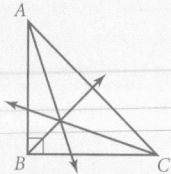

3.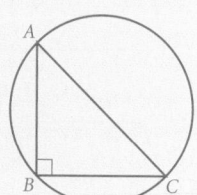

4.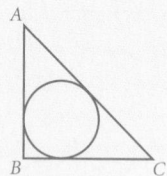

5.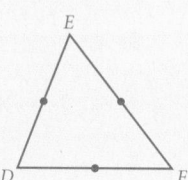

6.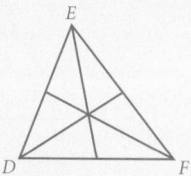

Lesson 1.6

1.

2.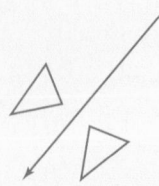

LESSON 1.5

Trace each of the triangles at right onto a separate piece of folding paper or draw them by using geometry graphics software. Triangles are named with the △ symbol and the names of the vertices, such as △*ABC*.

Find the following:

1. the perpendicular bisectors of △*ABC*

2. the angle bisectors of △*ABC*

3. the circumscribed circle of △*ABC*

4. the inscribed circle of △*ABC*

5. the midpoints of each side of ∠*DEF*

6. the medians of ∠*DEF*

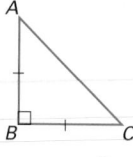

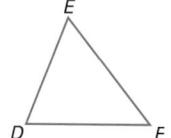

LESSON 1.6

For Exercises 1–4, trace the figures onto folding paper.

Reflect each figure across the given line.

1.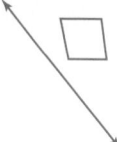

2.

Rotate each figure about the given point.

3.

4.

5. Make a list of capital letters that stay the same when reflected across a vertical line. What is the longest word you can write with the letters in your list?   A H I M O T U V W X Y,
Sample answer: YAHOO

3.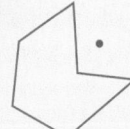

4.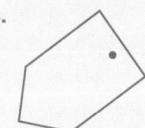

LESSON 1.7

In Exercises 1–4, use graph paper to draw the transformations of the figure as indicated.

1. Rotate the figure 180° about the origin.

2. Reflect the figure across the *y*-axis.

3. Reflect the figure across the *x*-axis.

4. Reflect the figure across the *x*-axis and then reflect the image across the *y*-axis. How does this figure relate to the transformed figure in Exercise 1?

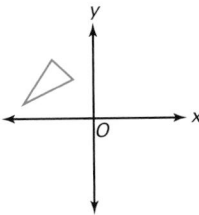

In Exercises 5–8, describe the result of applying each rule to a figure in the coordinate plane.

5. $H(x, y) = (x, y - 3)$ **vertical shift down 3 units**

6. $Z(x, y) = (x + 2, y)$ **horizontal shift right 2 units**

7. $G(x, y) = (-x, y - 5)$ **reflection across the *y*-axis, vertical shift down 5 units**

8. $F(x, y) = (y, x)$ **reflection across the line *y* = *x***

CHAPTER 2

LESSON 2.1

Use the table below to answer Exercises 1–4. The numbers in the table are perfect squares: $1^2 = 1$, $2^2 = 4$, $3^2 = 9$, $4^2 = 16$, . . .

1. Suppose you want to know which column contains the number 10,201. Ten thousand two hundred one (10,201) is the square of what number? **101**

A	B	C	D	E
1	4	9	16	25
36	49	64	81	100
⋮	⋮	⋮	⋮	⋮

2. Look at the numbers that occur in each column of the table. What is true of every number in column E? **They are all squares of multiples of 5.**

3. In what column does the number 100^2 occur? **column E**

4. Prove that 10,201 occurs in column A.

Suppose that you know the square of a positive integer *n* and you want to find the square of *n* + 2.

5. Draw a square of dots to represent the square of the number *n*.

6. Increase the side lengths of your square by 2 to represent the square of the number *n* + 2.

7. Use your diagram to prove that the square of *n* + 2 is found by adding 2*n* and 2*n* + 4 to the square of *n*. Your proof should work for all values of *n*.

5. (a square array of dots)

6. (a dot diagram showing the extended square)

7. The representation of $(n + 2)^2$ has an additional 2*n* dots directly below the original *n* columns and 2*n* + 4 dots added to the right of the original square. The total dots in the new array is then $n^2 + 2n + (2n + 4)$.

1.
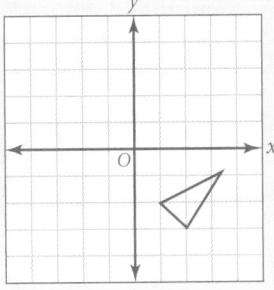

2.
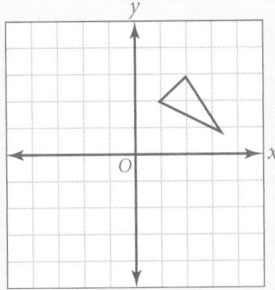

3.
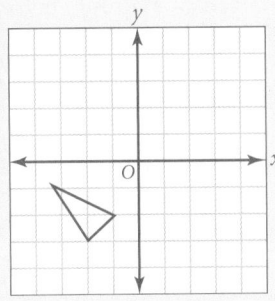

4.
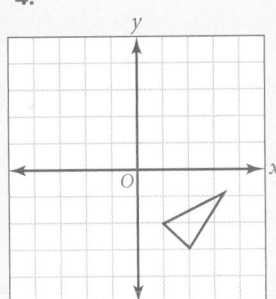

This figure is equivalent to a 180° rotation about the origin.

Lesson 2.1

4. $101 = 100 + 1$, so the square of 101 occurs in the column after the square of 100. The square of 100 is in column E. Therefore, the square of 101 is in column A.

Lesson 2.2

2. Hypothesis: An animal lives in the Everglades. **Conclusion:** It lives in Florida.

3.
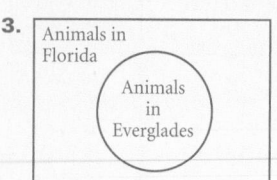

5. Hypothesis: Two angles are supplementary. **Conclusion:** The sum of their measures is 180°.
Converse: If the sum of the measures of two angles is 180°, then the angles are supplementary. Converse is true.

6. Hypothesis: Your transportation is a motorcycle.
Conclusion: You ride on two wheels.
Converse: If you ride on two wheels, then your transportation is a motorcycle.
Converse is false: You may ride on a bicycle.

7. Hypothesis: You live in Baltimore. **Conclusion:** You live in Maryland.
Converse: If you live in Maryland, then you live in Baltimore.
Converse is false: You may live in Annapolis.

LESSON 2.2

For Exercises 1–4, refer to the following statement:

All animals that live in the Everglades live in Florida.

1. Rewrite the statement as a conditional. **If an animal lives in the Everglades, then it lives in Florida.**

2. Identify the hypothesis and the conclusion of the conditional.

3. Draw an Euler diagram that illustrates the conditional.

4. Write the converse of the conditional. **If an animal lives in Florida, then it lives in the Everglades.**

For Exercises 5–8, identify the hypothesis and conclusion of each conditional. Write the converse of each conditional. If the converse is false, give a counterexample.

5. If two angles are supplementary, then the sum of their measures is 180°.

6. If your transportation is a motorcycle, then you ride on two wheels.

7. If you live in Baltimore, then you live in Maryland.

8. If you have a rose, then you have a flower.
Hypothesis: You have a rose.
Conclusion: You have a flower.
Converse: If you have a flower, then you have a rose.
Converse is false: You may have a daisy.

LESSON 2.3

In Exercises 1–5, use the following steps to determine whether the given sentence is a definition:
a. Write the sentence as a conditional statement.
b. Write the converse of the conditional.
c. Write a biconditional statement.
d. Decide whether the sentence is a definition, and explain your reasoning.

1. Mercury is the planet closest to our Sun.

2. School is the place where kids learn mathematics.

3. A multiple of ten has a ones digit of zero.

4. The set of points in a plane equidistant from a given point form a circle.

5. An equilateral triangle is a closed figure with three congruent angles.

6. The following are pentagons: The following are not pentagons:

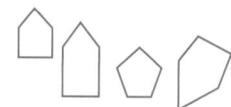 

Which of the figures at right are not pentagons? **b and d**

a. b. c. d.

Lesson 2.3

1. a. If a planet is Mercury, then it is the planet closest to our Sun.
b. If a planet is the closest to our Sun, then it is Mercury.
c. A planet is Mercury if and only if it is the closest to our Sun.
d. Definition: The biconditional is true.

2. a. If a place is a school, then it is where kids learn mathematics.
b. If kids are learning mathematics, then they are at school.
c. A place is a school if and only if it is the place where kids learn mathematics.
d. Not a definition: Kids may also learn mathematics in places other than school.

3. a. If a number is a multiple of ten, then it has a ones digit of zero.
b. If a number has a ones digit of zero, then it is a multiple of ten.
c. A number is a multiple of ten if and only if it has a ones digit of zero.
d. Definition: The biconditional is true.

LESSON 2.4

In Exercises 1 and 2, identify the Algebraic Properties of Equality that justify the steps.

1.
$$x - 7 = 15 \qquad \text{Given}$$
$$x - 7 + 7 = 15 + 7 \qquad \underline{\quad?\quad}$$
$$x = 22 \qquad \text{Addition Property of Equality}$$

2.
$$4x + 6 = 38 \qquad \text{Given} \qquad \text{Subtraction}$$
$$4x + 6 - 6 = 38 - 6 \qquad \underline{\quad?\quad} \qquad \text{Property of}$$
$$4x \div 4 = 32 \div 4 \qquad \underline{\quad?\quad} \qquad \text{Equality; Division}$$
$$x = 8 \qquad\qquad\qquad \text{Property of}$$
$$\qquad\qquad\qquad\qquad \text{Equality}$$

Refer to the diagram at right, in which
m∠ABC = m∠EBD. Use the Overlapping
Angles Theorem to complete the following:

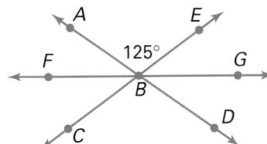

3. m∠ABC = 75° m∠ABD = 110° m∠CBD = __?__ **35°**

4. m∠ABC = (3x + 4)° m∠CBD = (x − 5)° m∠CBE = 67°
  x = __?__ **17** m∠ABD = __?__ **67°** m∠EBD = __?__ **55°**

LESSON 2.5

For Exercises 1–4, refer to the diagram at right,
which consists of three intersecting lines.

1. Which angle is congruent to ∠FBC? **∠GBE**

2. Which angle is congruent to ∠GBC? **∠FBE**

3. Which angle is congruent to ∠EBD? **∠ABC**

4. Find the measure of ∠ABC. **55°**

5. Find the measures of all the angles in
the diagram at right. **m∠ABE = m∠CBD = 137°**
m∠ABC = m∠EBD = 43°

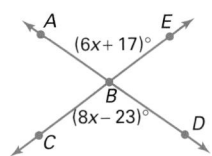

Tell whether each of the following is an example of inductive or
deductive reasoning. Is the argument a proof? Why or why not?

6. Every Saturday for the last six weeks, it has rained. **Inductive reasoning; not a proof. The**
Tomorrow is Saturday. **argument is based on an examination of**
Therefore, it will rain tomorrow. **specific examples—it is a generalization.**

7. Multiples of 5 are not prime.
The number 35 is a multiple of 5.
Therefore, 35 is not prime.
Deductive reasoning; is a proof.
The argument is true for all multiples of 5.

4. a. If a set consists of all points in a plane equidistant from a given point, then the set forms a circle.

 b. If a set forms a circle, then the set consists of all points in a plane equidistant from a given point.

 c. A set forms a circle if and only if it consists of all points in a plane equidistant from a given point.

 d. Definition: The biconditional is true.

5. a. If a closed figure is an equilateral triangle, then it has three congruent angles.

 b. If a closed figure has three congruent angles, then it is an equilateral triangle.

 c. A closed figure is an equilateral triangle if and only if it has three congruent angles.

 d. Not a definition: A pentagon may contain three congruent angles and two other angles.

1.

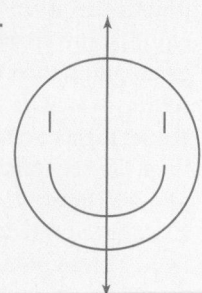

2.

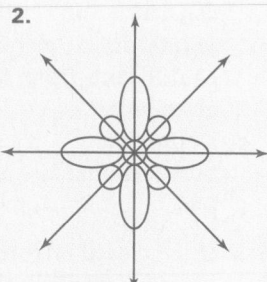

3.

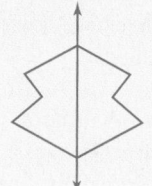

4.

5.

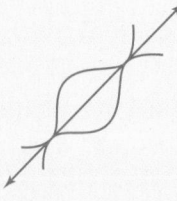

6.

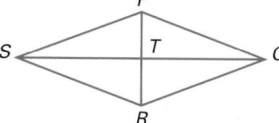

LESSON 3.1

For Exercises 1–3, copy each figure, and draw all axes of symmetry.

1.    **2.**    **3.**

Each figure below shows part of a shape with reflectional symmetry.
Copy and complete each figure.

4.    **5.**    **6.**

7. Which of the completed figures from Exercises 4–6 also has rotational symmetry?

8. Graph $y = 2(x - 3)^2 + 1$ on a graphing calculator or graph paper. Then write an equation for the axis of symmetry of the graph. $x = 3$

LESSON 3.2

For Exercises 1–8, use your conjectures from Activity 2 in Lesson 3.2 and the diagram below to find the indicated measurements. In rhombus *PQRS*, *PQ* = 5, *PR* = 6, and m∠*PQR* = 74°.

1. *QR* 5   **2.** *RS* 5

3. *PT* 3   **4.** m∠*PSR* 74°

5. m∠*QPS* 106°   **6.** m∠*PTQ* 90°

7. In rectangle *KLMN*, diagonal *KM* = *x* + 4 and diagonal *LN* = 3*x*. Find the value of *x* and the length of the diagonals of *KLMN*. *x* = 2; 6

Use the definitions of quadrilaterals and your conjectures from Activities 1–4 in Lesson 3.2 to decide whether each statement is true or false. If the statement is false, give a counterexample.

8. If a figure is not a rectangle, then it is not a square. True

9. If a figure is a rhombus, then it is a square. False; rhombus may have interior angles ≠ 90°.

10. If a figure is not a parallelogram, then it cannot be a rhombus. True

---

7. The figure in Exercise 6 has 180° rotational symmetry.

LESSON 3.3

In the figure at right, lines *k* and *ℓ* are parallel.

1. List all angles that are congruent to ∠3. ∠2, ∠6, and ∠7

2. List all angles that are congruent to ∠4. ∠1, ∠5, and ∠8

3. Are there any angles in the figure that are not congruent to ∠3 or ∠4? Explain.

4. If m∠3 = 140°, find the measure of each angle in the figure.

5. If m∠2 = 2*x*° and m∠6 = (3*x* − 50)°, find the measure of each angle in the figure.

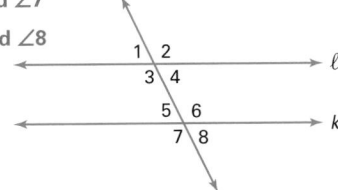

m∠1 = m∠4 = m∠5 = m∠8 = 80°
m∠2 = m∠3 = m∠6 = m∠7 = 100°

In △*FGH*, $\overline{IJ}\|\overline{HG}$ and ∠*FIJ* ≅ ∠*FJI*. Find the indicated angle measures.

6. m∠*HIG* 75°   7. m∠*FIJ* 70°

8. m∠*FJI* 70°   9. m∠*JGH* 70°

10. m∠*GIJ* 35°   11. m∠*FGI* 35°

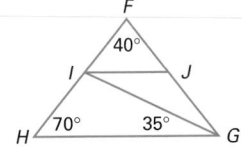

LESSON 3.4

For Exercises 1–4, refer to the diagram below, and fill in the name of the appropriate theorem or postulate.

1. If m∠*EDB* = m∠*DBC*, then $\overline{ED}\|\overline{BC}$ by the converse of the _____.

2. If m∠*AED* = m∠*ABC*, then $\overline{ED}\|\overline{BC}$ by the converse of the _____.

3. If m∠*DEB* + m∠*EBC* = 180°, then $\overline{ED}\|\overline{BC}$ by the converse of the _____.

4. If m∠*ADE* = m∠*ACB*, then $\overline{ED}\|\overline{BC}$ by the converse of the _____.

5. If m∠*DEC* = m∠*ECB*, then $\overline{ED}\|\overline{BC}$ by the converse of the _____.

6. If m∠*EDC* + m∠*DCB* = 180°, then $\overline{ED}\|\overline{BC}$ by the converse of the _____.

7. Given m∠*CBE* = 43° and m∠*BED* = 137°, write a two-column proof that $\overline{ED}\|\overline{BC}$.

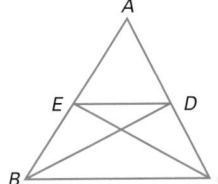

1. Alternate Interior Angles Theorem

2. Corresponding Angles Postulate

3. Same-Side Interior Angles Theorem

4. Corresponding Angles Postulate

5. Alternate Interior Angles Theorem

6. Same-Side Interior Angles Theorem

Lesson 3.3

3. No; all the angles are listed in either Exercise 1 or Exercise 2.

4. m∠1 = m∠4 = m∠8 = 40°

   m∠2 = m∠6 = m∠7 = 140°

Lesson 3.4

7.

Statements	Reasons
m∠*CBE* = 43° m∠*BED* = 137°	Given
m∠*CBE* + m∠*BED* = 180°	Angle Addition Postulate
m∠*CBE* and m∠*CBE* are supplementary	Definition of Supplementary Angles
$\overline{ED} \| \overline{BC}$	Converse of Same-Side Interior Angles Theorem

825

4. m∠HLK = m∠KJG
  = 55°
  m∠KGJ = 105°
  m∠HKJ = 160°
  m∠LKH = 20°
  m∠LHK = 105°

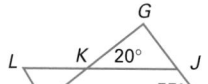

LESSON 3.5

For Exercises 1–3, two angle measurements of a triangle are given. Find the missing angle measure, or state that the triangle does not exist.

1. m∠A = 20°   m∠B = 70°   m∠C = __?__ 90°

2. m∠A = 30°   m∠B = __?__ **Does not exist**   m∠C = 160°

3. m∠A = __?__ 48°   m∠B = 66°   m∠C = 66°

4. In the figure at right, $\overline{LJ} \| \overline{HI}$, $\overline{IJ} \| \overline{HL}$, m∠GKJ = 20°, and m∠GIH = 55°. Find the measures of the other 8 angles in the figure.

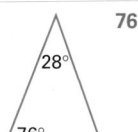

In Exercises 5–7, find the missing angle measures.

5.   55°

6.   57°

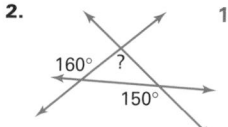

7.   76°

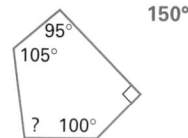

LESSON 3.6

In Exercises 1–3, find the unknown angle measures.

1.   50°

120°
?
110°
80°

2.   130°

160° ?
        150°

3.   150°

95°
105°
?   100°

In Exercises 4–6, an interior angle measure of a regular polygon is given. Find the number of sides of the polygon.

4. 108° **5**   **5.** 156° **15**   **6.** 140° **9**

In Exercises 7–9, an exterior angle measure of a regular polygon is given. Find the number of sides of the polygon.

7. 72° **5**   **8.** 45° **8**   **9.** 30° **12**

10. In quadrilateral FGHI, m∠F = (2x)°, m∠G = x°, m∠H = (x + 16)°, and m∠I = (2x + 20)°. Find the value of x and the measure of each angle in the quadrilateral. **x = 54, m∠F = 108°, m∠G = 54°, m∠H = 70°, m∠I = 128°**

LESSON 3.7

Use the conjectures you made in Activities 1–3 in Lesson 3.7 to find the indicated measures.

1. $DE =$ __?__ **30**

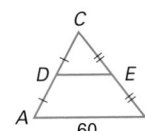

2. $GF =$ __?__ **35**

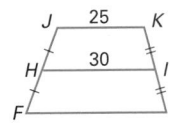

3. $NO =$ __?__ **45**

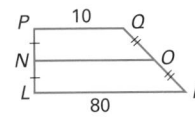

A young girl is using a 15-ft ladder to rescue her kitten from the branches of a tree. The ladder's base is 6 ft from the base of the tree.

4. Determine the distance from the ladder to the tree at a point halfway up the ladder. **3 feet**

5. Determine the distance from the ladder to the tree at a point three-quarters of the way up the ladder. **1.5 feet**

6. The little girl can reach the tree from 1 ft away or less. Estimate the percent of the ladder's height that the little girl must climb to reach the tree. **about 83%**

LESSON 3.8

In Exercises 1–4, the endpoints of a segment are given. Determine the slope and midpoint of the segment.

1. (1, 1) and (3, 2) $\frac{1}{2}$; (2, 1.5)

2. (−2, 1) and (2, 3) $\frac{1}{2}$; (0, 2)

3. (1, −2) and (−2, 5) $-\frac{7}{3}$; (−0.5, 1.5)

4. (0, −4) and (3, −1) **1**; (1.5, −2.5)

In Exercises 5–8, the endpoints of two segments are given. Determine whether the segments are parallel, perpendicular, or neither.

5. (4, 1) and (7, 6); (2, 5) and (12, −1)

6. (−1, 2) and (2, 4); (−3, 1) and (−1, −2)

7. (2, 3) and (6, 2); (0, 5) and (−1, 9)

8. (−2, −2) and (5, 2); (4, 3) and (11, 7)

For Exercises 9–12, the vertices of a triangle are given. Use slopes to determine whether each is a right triangle.

9. (1, 2), (3, 1), (2, 4)

10. (0, 1), (4, 2), (−3, 1)

11. (−3, −1), (−1, 1), (−2, −4)

12. (−1, 3), (2, 1), (1, 6)

5. Perpendicular

6. Perpendicular

7. Neither

8. Parallel

9. Right triangle

10. Not a right triangle

11. Not a right triangle

12. Right triangle

Lesson 4.1

1. Not congruent. They are different lengths.

2. Congruent. They have equal angle measures.

3. Not congruent. They have different side lengths.

4. Congruent. All corresponding parts are congruent.

5. Congruent. All corresponding parts are congruent.

6. Not congruent. Corresponding angle measures are not congruent.

CHAPTER 4

LESSON 4.1

In Exercises 1–6, determine whether the figures in each pair are congruent. Explain your reasoning.

1.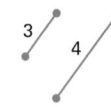

2.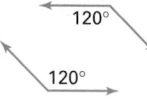

3.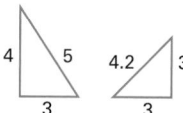

4.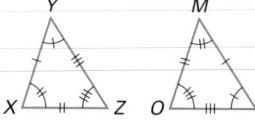

5.

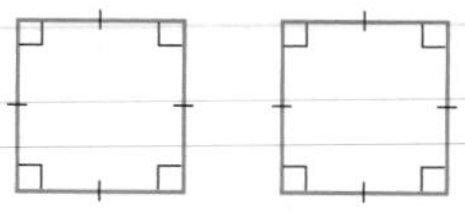

6.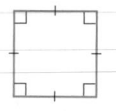

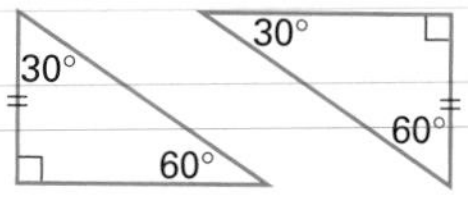

Suppose that heptagon *LKJMNOP* ≅ heptagon *AEFXWCB*.

7. Name the segment that is congruent to each segment below.

   a. $\overline{FX}$ $\overline{JM}$ **b.** $\overline{NO}$ $\overline{WC}$ **c.** $\overline{PL}$ $\overline{BA}$ **d.** $\overline{CB}$ $\overline{OP}$

8. List all pairs of congruent angles.

   ∠L ≅ ∠A, ∠K ≅ ∠E, ∠J ≅ ∠F, ∠M ≅ ∠X,

   ∠N ≅ ∠W, ∠O ≅ ∠C, and ∠P ≅ ∠B

LESSON 4.2

In Exercises 1–3, determine whether each pair of triangles can be proven congruent by using the SSS, SAS, or ASA Congruence Postulate. If so, identify which postulate is used.

1.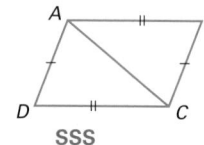

SSS

2.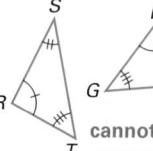

cannot be proven congruent

3.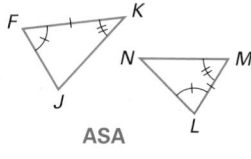

ASA

In Exercises 4–9, some measurements of a triangle are given. Is there a unique triangle that can be constructed with the given measurements? If so, identify the postulate that applies.

4. △PQR: PQ = 3, PR = 1, m∠P = 30°

5. △EFG: FG = 1, EG = 7, m∠E = 12°

6. △ABC: BC = 4, m∠B = 15°, m∠C = 100°

7. △RST: m∠R = 20°, m∠S = 100°, m∠T = 60°

8. △MNO: MN = 2, NO = 2, MO = 3

9. △JKL: JK = 6, m∠J = 15°, m∠K = 30°

4. Yes. SAS

6. Yes. ASA

8. Yes. SSS

5. No

7. No

9. Yes. ASA

LESSON 4.3

For each pair of triangles given in Exercises 1–10, is it possible to prove that the triangles are congruent? If so, write a congruence statement and name the postulate or theorem used.

1.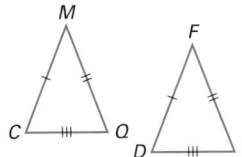

2.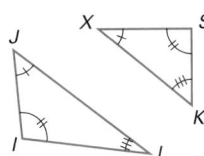

3.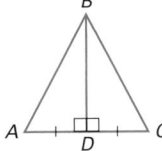

4. △ABC and △PQR with ∠A ≅ ∠P, ∠B ≅ ∠Q, and $\overline{AB} ≅ \overline{PQ}$ △ABC ≅ △PQR by ASA

5. △RST and △JKL with ∠R ≅ ∠L, $\overline{RS} ≅ \overline{JL}$, and $\overline{RT} ≅ \overline{LK}$ △RST ≅ △JKL by SAS

6. △WXY and △ERG with ∠X ≅ ∠R, ∠Y ≅ ∠G, and $\overline{WY} ≅ \overline{EG}$ △WXY ≅ △ERG by ASA or AAS

7. △ALS and △BMK with ∠A ≅ ∠B, ∠L ≅ ∠K, and ∠S ≅ ∠M No

8. △KIP and △EMU with ∠K ≅ ∠M, $\overline{IP} ≅ \overline{EU}$, and $\overline{KP} ≅ \overline{UM}$ No

9. △MNO and △GHI with $\overline{MN} ≅ \overline{GH}$, $\overline{MO} ≅ \overline{GI}$, and $\overline{NO} ≅ \overline{HI}$ △MNO ≅ △GHI by SSS

10. △RVU and △OLP with ∠R ≅ ∠L, ∠V ≅ ∠P, and $\overline{VU} ≅ \overline{OP}$ △RVU ≅ △OLP by AAS or ASA

LESSON 4.4

Given the following information about a triangle, find the missing measure.

1. △MRS: $\overline{MR} ≅ \overline{RS}$, m∠S = 40°, m∠M = __?__ 40°

2. △UVW: $\overline{VW} ≅ \overline{UV}$, m∠U = 85°, m∠W = __?__ 85°

3. △ABC: ∠A ≅ ∠B, AC = 12, CB = __?__ 12

4. △DLJ: $\overline{DJ} ≅ \overline{JL}$, m∠D = 15°, m∠J = __?__ 150°

5. △KXQ: $\overline{KX} ≅ \overline{QK}$, m∠X = 50°, m∠Q = __?__ 50°

6. △RST: ∠R ≅ ∠T, RS = 5, ST = __?__ 5

7. △DEF: $\overline{DE} ≅ \overline{DF}$, m∠E = 45°, m∠D = __?__ 90°

8. △LMN: ∠M ≅ ∠N, LM = 13, LN = __?__ 13

9. △KUD: $\overline{KU} ≅ \overline{UD}$, m∠K = (3x)°, m∠U = 90°, x = __?__ 15

10. △COL: ∠C ≅ ∠L, CO = 23, OL = __?__ 23

Lesson 4.3

1. △CMQ ≅ △FDB by SSS

2. No

3. △ABD ≅ △CBD by SAS

Lesson 4.5

7. $\triangle JKL$ is congruent to $\triangle OMN$ by ASA, so they can be put together to form a parallelogram.

8. $\triangle ABC$ could not fit together with $\triangle LMN$ to form a parallelogram because they do not have any congruent side pairs.

9. $\triangle XYZ$ is congruent to $\triangle DEF$ by SSS, so they can be put together to form a parallelogram.

Lesson 4.6

1. *WXYZ* is a parallelogram. If the diagonals of a quadrilateral bisect each other, then the quadrilateral is a parallelogram.

2. *WXYZ* is a parallelogram. If one pair of opposite sides of a quadrilateral are parallel and congruent, then the quadrilateral is a parallelogram.

3. *WXYZ* is a parallelogram. If two pairs of opposite sides of a quadrilateral are congruent, then the quadrilateral is a parallelogram.

4. *WXYZ* is not a parallelogram.

5. *ABCD* is a rhombus. If the diagonals of a parallelogram bisect the angles of the parallelogram, then the parallelogram is a rhombus.

6. *ABCD* is a rectangle. If the diagonals of a parallelogram are congruent, then the parallelogram is a rectangle.

LESSON 4.5

For Exercises 1–6, use the parallelogram shown at right. Find the indicated measures.

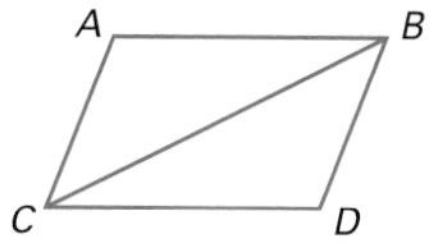

1. Given $m\angle A = (3x)°$ and $m\angle D = (x + 100)°$, find $m\angle D$. **150°**

2. Given $AB = t$ and $CD = (9 - 2t)$, find CD. **3**

3. Given $m\angle CDB = 40°$, find $m\angle A$. **40°**

4. Given $m\angle B = 50°$, $m\angle C = (2x)°$, and $BD = (x - 4)$, find BD. **21**

5. Given $m\angle ABC = 15°$ and $m\angle ACD = 45°$, find $m\angle CBD$. **30°**

6. Given $m\angle C = 80°$ and $m\angle D = x°$, find $m\angle A$. **100°**

Determine whether each pair of triangles could fit together to form a parallelogram, and justify your answer.

7. 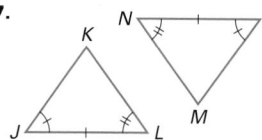 **8.** 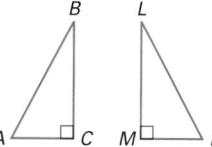 **9.**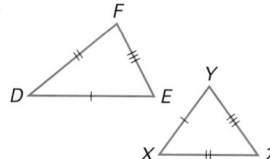

LESSON 4.6

Exercises 1–4 refer to quadrilateral *WXYZ* with diagonals $\overline{WY}$ and $\overline{XZ}$ intersecting at point *P*. For each set of conditions given, determine whether the quadrilateral is a parallelogram. If so, give the theorem that justifies your answer.

1. $\overline{WP} \cong \overline{PY}$, $\overline{XP} \cong \overline{PZ}$

2. $\overline{WX}\|\overline{ZY}$, $\overline{WX} \cong \overline{ZY}$

3. $\overline{WX} \cong \overline{ZY}$, $\overline{WZ} \cong \overline{XY}$

4. $\overline{XY}\|\overline{WZ}$, $\overline{WP} \cong \overline{PX}$

Exercises 5–10 refer to parallelogram *ABCD* with diagonals $\overline{AC}$ and $\overline{BD}$ intersecting at point *P*. For each set of conditions given, determine whether the parallelogram is a rhombus, a rectangle, or neither. Give the theorem that justifies your answer.

5. $m\angle ABP = m\angle CBP$, $m\angle BCP = m\angle DCP$

6. $\overline{AC} \cong \overline{BD}$

7. $\overline{AD} \cong \overline{AB}$

8. $m\angle ABC = 90°$

9. $\overline{AP} \cong \overline{AD}$ **neither**

10. $m\angle DPA = 90°$

---

7. *ABCD* is a rhombus. If one pair of adjacent sides of a parallelogram are congruent, then the parallelogram is a rhombus.

8. *ABCD* is a rectangle. If one angle of a parallelogram is a right angle, then the parallelogram is a rectangle.

10. *ABCD* is a rhombus. If the diagonals of a parallelogram are perpendicular, then the parallelogram is a rhombus.

LESSON 4.7

Construct a figure congruent to each figure below.

1.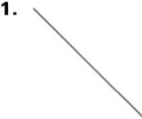

2.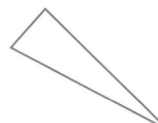

3.

Trace each triangle below, and construct the angle bisector of each angle. Using the intersection of the angle bisectors, construct the inscribed circle of each triangle.

4.

5.

6.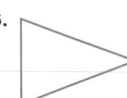

Trace each triangle below, and construct the perpendicular bisector of each side. Using the intersection of the perpendicular bisectors, construct the circumscribed circle of each triangle.

7.

8.

9.

LESSON 4.8

Trace each figure below and translate it by the direction and distance of the given translation vector.

1.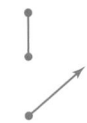

2.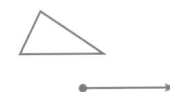

3.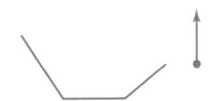

Which of the following triangles are possible? Explain your reasoning.

4. $WX = 1, XY = 2, YW = 3$

5. $MN = 14, NO = 17, OM = 15$

6. $PQ = 3, QR = 4, RP = 5$

7. $AB = 10, BC = 5, CA = 4$

8. $GH = 14, HI = 20, GI = 24$
Possible triangle. The sum of the lengths of any two sides is greater than the length of the third side.

9. $DE = 12, EF = 5, DF = 28$
Not possible since *DE + EF < DF*

1.

2.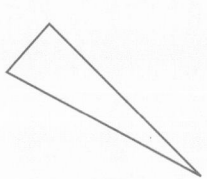

3.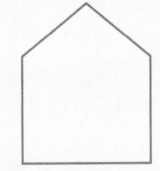

4.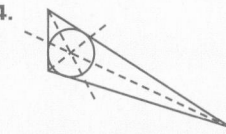

5.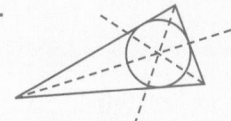

6.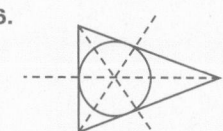

7.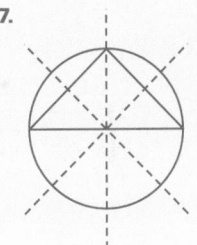

8.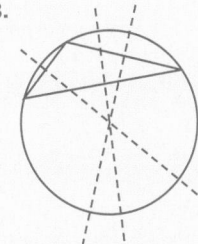

9.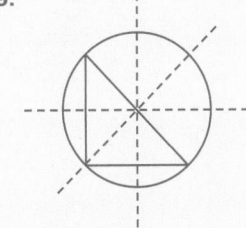

Lesson 4.8

1.

2.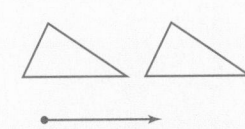

3.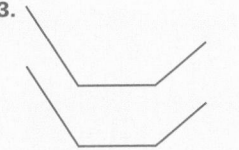

4. Not possible since $WX + XY = YW$

5. Possible triangle. The sum of the lengths of any two sides is greater than the length of the third side.

6. Possible triangle. The sum of the lengths of any two sides is greater than the length of the third side.

7. Not possible since $BC + CA < AB$

831

CHAPTER 5

LESSON 5.1

In Exercises 1–5, find the area of the rectangle with vertices at the given points. You may find it helpful to sketch a graph.

1. $(3, 0), (0, 0), (3, 4), (0, 4)$ **12**

2. $(4, 0), (4, -5), (0, 0), (0, -5)$ **20**

3. $(2, 3), (6, 4), (2, 4), (6, 3)$ **4**

4. $(-1, 2), (-1, -4), (2, 2), (2, -4)$ **18**

5. $(-7, -3), (-2, 0), (-7, 0), (-2, -3)$ **15**

6. The perimeter of a rectangle is 42 in. The length is twice the width. What are the dimensions? What is the area? $h = 14$ in., $b = 7$ in., $A = 98$ in.^{2}

7. The area of a rectangle is 36 cm^2. The width is 4 times the length. What are the dimensions? What is the perimeter? $h = 3$ cm, $b = 12$ cm, $P = 30$ cm

8. The area of a rectangle is $64x^2$. The length is 4 times the width. In terms of x, what are the dimensions? What is the perimeter? $h = 16x$, $b = 4x$, $P = 40x$

LESSON 5.2

In Exercises 1–3, find the area of each triangle.

1. $A = 24$

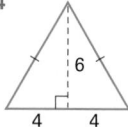

2. $A = \dfrac{3}{2}$

3. $A = 73.5$

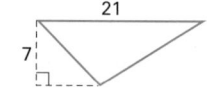

In Exercises 4–7, find the area of a parallelogram with vertices at the given points. You may find it helpful to sketch a graph.

4. $(3, 4), (8, 4), (1, 1), (6, 1)$ $A = 15$ **5.** $(-5, -1), (-3, -1), (5, 1), (3, 1)$ $A = 4$

6. $(-1, -3), (7, -7), (7, 3), (-1, -7)$ $A = 80$ **7.** $(2, -1), (0, -5), (5, -1), (3, -5)$ $A = 12$

In Exercises 8–11, find the area of a trapezoid with vertices at the given points. You may find it helpful to sketch a graph.

8. $(-3, 2), (2, 5), (6, 2), (1, 5)$ $A = 15$ **9.** $(2, 3), (6, -1), (6, 6), (2, 5)$ $A = 18$

10. $(4, 3), (2, -1), (8, -1), (6, 3)$ $A = 16$ **11.** $(4, 0), (4, 4), (7, -1), (7, 5)$ $A = 15$

LESSON 5.3

In Exercises 1–8, find the circumference and area of each circle whose radius or diameter is given. Use 3.14 for π. Round your answers to the nearest tenth.

1. $r = 3$ $C = 18.8$, $A = 28.3$ **2.** $r = 8$ $C = 50.2$, $A = 201.0$

3. $d = 14$ $C = 44.0$, $A = 153.9$ **4.** $d = 36$ $C = 113.0$, $A = 1017.4$

Use $\frac{22}{7}$ for π. Leave your answers in fractional form.

5. $r = 20$ $C = \frac{880}{7}$, $A = \frac{8800}{7}$ **6.** $d = 32$ $C = \frac{704}{7}$, $A = \frac{5632}{7}$

7. $r = 12$ $C = \frac{528}{7}$, $A = \frac{3168}{7}$ **8.** $d = 15$ $C = \frac{330}{7}$, $A = \frac{2475}{14}$

Find the radius of the circle with the given area or circumference. Give your answers both in terms of π and rounded to the nearest tenth.

9. $A = 20\pi$ $r = 2\sqrt{5} \approx 4.5$ **10.** $A = 628$ $r = \sqrt{\frac{628}{\pi}} \approx 14.1$ **11.** $C = 24$ $r = \frac{12}{\pi} \approx 3.8$

12. $C = 2\pi$ $r = 1$ **13.** $A = 117$ $r = \sqrt{\frac{117}{\pi}} \approx 6.1$ **14.** $A = 17.4$ $r = \sqrt{\frac{17.4}{\pi}} \approx 2.4$

LESSON 5.4

For Exercises 1–5, two lengths of sides of a right triangle are given. Find the missing length. Leave your answers in radical form.

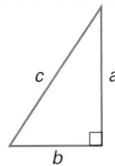

1. $a = 5$, $b = 12$, $c = \underline{\;?\;}$ 13 **2.** $a = 7$, $b = 9$, $c = \underline{\;?\;}$ $\sqrt{130}$

3. $a = 10$, $b = \underline{\;?\;}$, $c = 25$ **4.** $a = \underline{\;?\;}$, $b = 6$, $c = 10$ 8
  $5\sqrt{21}$

Each of the following triples represent the sides of a triangle. Determine whether the triangle is right, acute, or obtuse.

5. 15, 17, 3 obtuse   **6.** 6, 8, 10 right   **7.** 11, 10, 13 acute

8. 5, 12, 13 right   **9.** 18, 28, 40 obtuse   **10.** 7, 9, 11 acute

Find the area and perimeter of each triangle. Round your answers to the nearest tenth.

11. 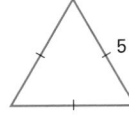   **12.** 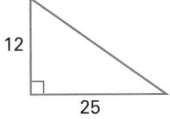   **13.**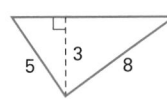

$A = 10.8$, $P = 15$ $A = 150$, $P = 64.7$ $A = 17.1$, $P = 24.4$

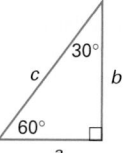

LESSON 5.5

For the given length, find the two remaining lengths. Give your answers in simplest radical form.

1. $a = 7$ $b = 7\sqrt{3}, c = 14$ **2.** $b = 3$ $a = \sqrt{3}, c = 2\sqrt{3}$

3. $c = 4$ $a = 2, b = 2\sqrt{3}$ **4.** $b = 6\sqrt{3}$ $a = 6, c = 12$

In Exercises 5–10, find the perimeter and area of each figure. Give your answers in simplest radical form.

5. an equilateral triangle with side lengths of 22 $A = 121\sqrt{3}, P = 66$

6. a square with a diagonal of 6 $A = 18, P = 12\sqrt{2}$

7. a 30-60-90 triangle with a hypotenuse of 16 $A = 32\sqrt{3}, P = 24 + 8\sqrt{3}$

8. a regular hexagon with side lengths of 10 $A = 150\sqrt{3}, P = 60$

9. a 45-45-90 triangle with leg lengths of 3 $A = 4.5, P = 6 + 3\sqrt{2}$

10. a regular octagon with side lengths of 6 $A = 72 + 72\sqrt{2}, P = 48$

LESSON 5.6

Find the distance between each pair of points. Round your answers to the nearest hundredth.

1. $(1, 3)$ and $(2, -1)$ **4.12**   **2.** $(-5, 6)$ and $(3, -2)$ **11.31**

3. $(3, 4)$ and $(0, 0)$ **5**   **4.** $(2, 2)$ and $(5, 11)$ **9.49**

5. $(6, 3)$ and $(2, 1)$ **4.47**   **6.** $(-5, -7)$ and $(-1, 7)$ **14.56**

7. $(0, 1)$ and $(4, 5)$ **5.66**   **8.** $(1, -2)$ and $(12, 3)$ **12.08**

In Exercises 9–14, use the converse of the Pythagorean Theorem to determine whether the triangle with the given vertices is a right triangle. You may wish to plot the points and draw each triangle.

9. $(1, 5), (2, -1),$ and $(7, 6)$ **yes**   **10.** $(-1, 3), (4, -2),$ and $(-4, 0)$ **yes**

11. $(2, 3), (4, -2),$ and $(-4, 0)$ **no**   **12.** $(2, 3), (8, 4),$ and $(1, 7)$ **no**

13. $(2, 10), (12, 0),$ and $(-4, 4)$ **yes**   **14.** $(-16, 18), (4, 2),$ and $(12, 8)$ **no**

LESSON 5.7

Determine the coordinates of the unknown vertex or vertices of each figure below. Use variables to represent any coordinates that are not completely determined by the given information.

1. isosceles triangle ABC with $\overline{AC} \cong \overline{BC}$
$A(0, 0), B(0, 2a), C(?, ?)$ $C(b, a)$

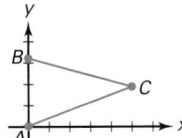

2. parallelogram $DEFG$ $E(c, b), G(a - c, 0)$
$D(0, 0), E(?, ?), F(a, b), G(?, ?)$

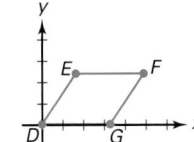

For Exercises 3–8, refer to right triangle **ABC** shown at right with vertices at **A(0, 0)**, **B(b, 0)**, and **C(0, c)**. Using the coordinates of **A, B,** and **C**, find the following lengths:

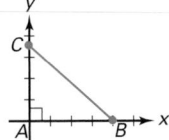

3. AB b **4.** AC c **5.** BC

6. $(AB)^2 + (AC)^2$ **7.** $(BC)^2$ $b^2 + c^2$

8. Using the results from Exercises 3–7, draw a conclusion about right triangle ABC. $(\overline{AB})^2 + (\overline{AC})^2 = (\overline{BC})^2$

5. $\sqrt{b^2 + c^2}$

6. $b^2 + c^2$

LESSON 5.8

Find the theoretical probability that a dart tossed at random onto each figure will land in the shaded area.

1.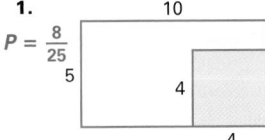
$P = \frac{8}{25}$

2. $P = \frac{1}{2}$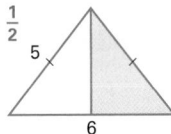

3. $P = \frac{1}{9}$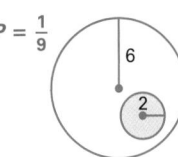

Convert the following probabilities to percents. Round your answers to the nearest tenth of a percent.

4. 0.61   61%   **5.** $\frac{4}{5}$ 80%   **6.** $\frac{3}{7}$ 42.9%   **7.** 1   100%

Convert the following percents to decimal probabilities:

8. 40%   0.4   **9.** 29%   0.29   **10.** 73%   0.73   **11.** 9%   0.09

12. 12%   0.12   **13.** 0.1%   0.001   **14.** 5.8%   0.058   **15.** 45.3%   0.453

Convert the following percents to fractional probabilities. Give your answers in lowest terms.

16. 10% $\frac{1}{10}$ **17.** 25% $\frac{1}{4}$ **18.** 50% $\frac{1}{2}$ **19.** 75% $\frac{3}{4}$

20. 2% $\frac{1}{50}$ **21.** 91% $\frac{91}{100}$ **22.** 44% $\frac{11}{25}$ **23.** 28% $\frac{7}{25}$

Lesson 6.1

3.

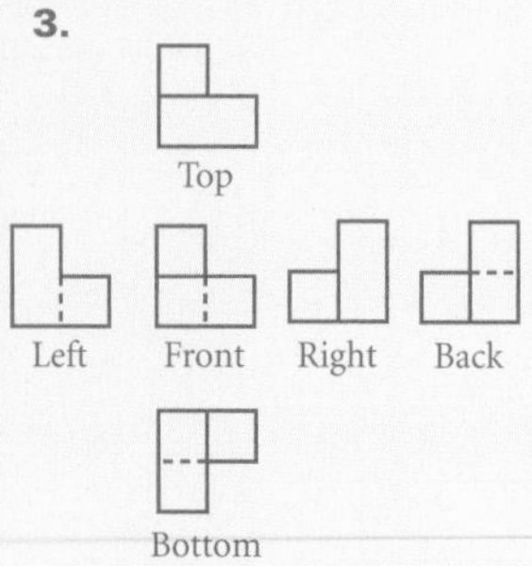

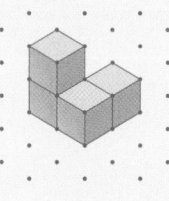

Top

Left   Front   Right   Back

Bottom

4. Sample answer:

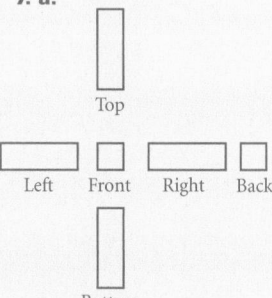

5.

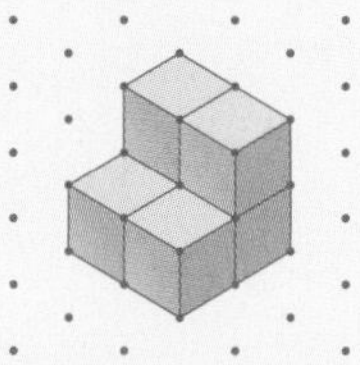

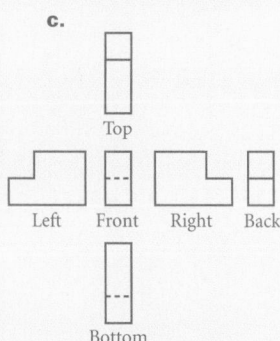

7. a.

Top

Left   Front   Right   Back

Bottom

b.

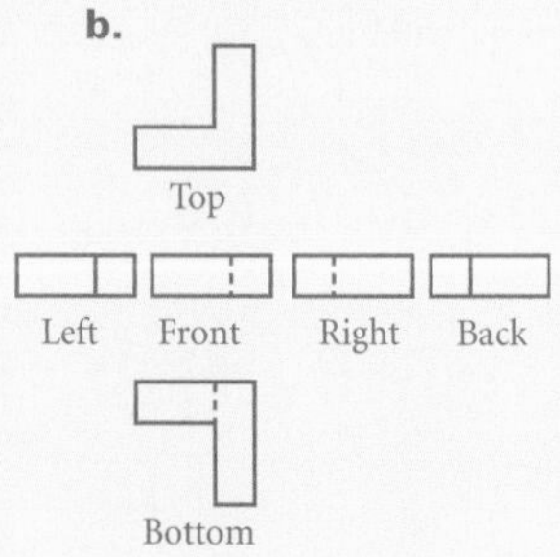

Top

Left   Front   Right   Back

Bottom

c.

Top

Left   Front   Right   Back

Bottom

CHAPTER 6

LESSON 6.1

For Exercises 1–5, refer to the isometric drawing at right.

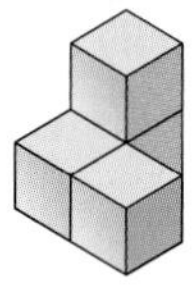

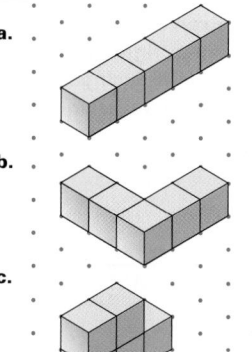

1. Give the volume in cubic units. **4 cubic units**

2. Give the surface area in square units. **18 square units**

3. Draw six orthographic projections of the solid.

4. Draw the solid on isometric dot paper from a different viewpoint.

5. Draw the solid on isometric dot paper with a cube added at each shaded face.

The three solids at right each have a volume of 5 cubic units.

6. Find the surface area of each solid in square units. Which has the least surface area?

7. Draw six orthographic projections of each solid.

6. a. 22 square units
b. 22 square units
c. 20 square units

a.
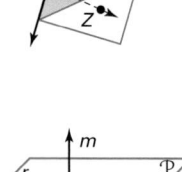

b.
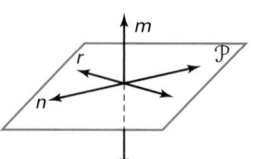

c.
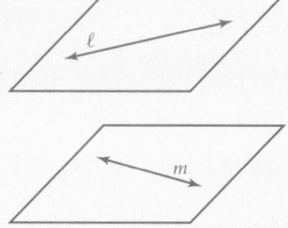

LESSON 6.2

In the figure at right, $\overline{XY}$, $\overline{ZY}$, and $\overline{AB}$ are perpendicular to ℓ, and $\overline{BC}$ is not perpendicular to ℓ.

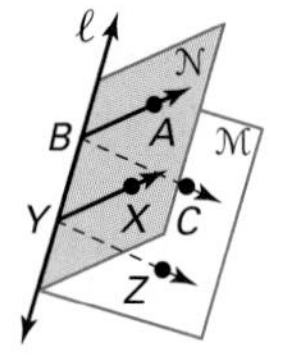

1. Is m∠ABC the measure of the dihedral angle formed by planes M and N? **no**

2. Is m∠XYZ the measure of the dihedral angle formed by planes M and N? **∠XYZ is dihedral since both rays are perpendicular to line ℓ.**

In the figure at right, line m is perpendicular to plane $\mathcal{P}$.

3. Is line m perpendicular to line n? Explain.

4. What is the relationship between line m and line r?

5. Indicate whether the statement below is true or false for a figure in space. Explain your answer by using sketches.

If two lines never intersect, then they are parallel.

Lesson 6.2

3. Yes. Since line m is perpendicular to plane $\mathcal{P}$, line m is perpendicular to all lines in $\mathcal{P}$ that it intersects.

4. Line m is perpendicular to line r.

5. False. The two lines could be skew. For example, in this sketch, the lines ℓ and m lie in parallel planes.

LESSON 6.3

Which of the figures below appear to be prisms? Give the name for each prism. If the figure is not a prism, explain why.

1.

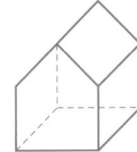

2.

3.

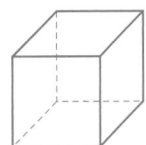

Use the oblique triangular prism at right for Exercises 4–6.

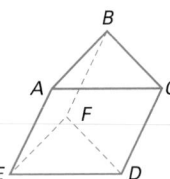

4. Which face is congruent to triangle ABC? Explain.

5. Name all segments congruent to $\overline{AE}$. $\overline{BF}$ and $\overline{CD}$

6. What type of quadrilateral is $ACDE$? parallelogram

For Exercises 7–10, find the length of a diagonal of a right rectangular prism with the given dimensions.

7. $\ell = 10, w = 10, h = 5$ $d = 15$

8. $\ell = 3, w = 4, h = 5$ $d = 5\sqrt{2}$

9. $\ell = 5, w = 6, h = 8$ $d = 5\sqrt{5}$

10. $\ell = 8, w = 24, h = 6$ $d = 26$

LESSON 6.4

Name the octant, coordinate plane, or axis where each point is located.

1. $(2, 1, 3)$ first octant

2. $(-14, 0, 0)$ x-axis

3. $(-1, 1, 8)$ top-back-right octant

4. $(53, 1.2, 0)$ xy-plane

5. $(-8, -1, -3)$ bottom-back-left octant

6. $(8, 0, -4)$ xz-plane

Find the distance between each pair of points.

7. $(1, 0, 2)$ and $(4, 1, -2)$ $\sqrt{26}$

8. $(2, 4, 0)$ and $(4, 3, 2)$ 3

9. $(-1, -2, -3)$ and $(1, 4, -1)$ $2\sqrt{11}$

10. $(4, 1, 6)$ and $(10, -4, 14)$ $5\sqrt{5}$

Lesson 6.3

1. right pentagonal prism

2. not a prism; Lateral faces are not parallelograms.

3. right rectangular prism

4. $\triangle EFD$ is congruent to $\triangle ABC$ because they are opposite faces.

Lesson 6.5

1.

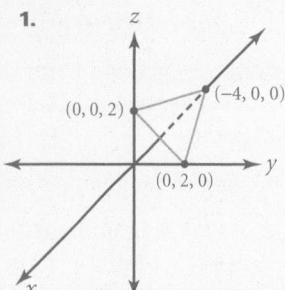

2.

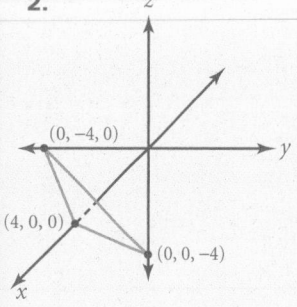

3.

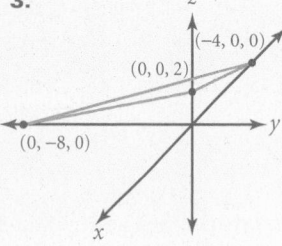

4.

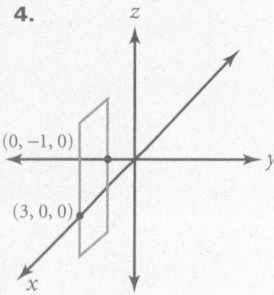

5.

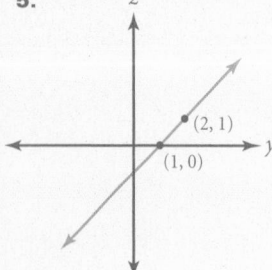

6.

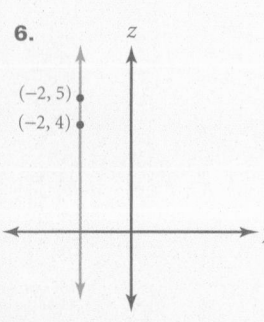

LESSON 6.5

Use intercepts to sketch a graph of the plane represented by each equation.

1. $3x - 6y - 6z = -12$

2. $2x - 2y - 2z = 8$

3. $-4x - 2y + 8z = 16$

4. $x - 3y = 3$

In a coordinate plane, plot the lines represented by each pair of parametric equations.

5. $x = t + 1$
   $y = t$

6. $x = -2$
   $y = t + 3$

Recall that a *trace* of a plane is its intersection with the *xy*-plane. Find the equation of the trace for each equation of a plane below.

7. $7x + 2y - z = 1$ $7x + 2y = 1$

8. $2x + 4y - 18z = 2$ $2x + 4y = 2$

LESSON 6.6

In Exercises 1–6, locate the vanishing point and horizon line for each figure.

1.

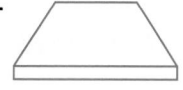

2.

3.

4.

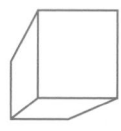

5.

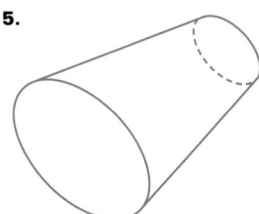

6.

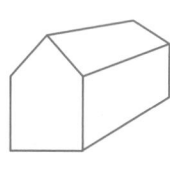

7. Make a one-point perspective drawing of a triangular prism.

Lesson 6.6

1.

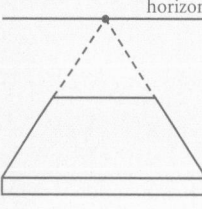

2.

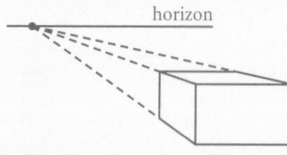

3.

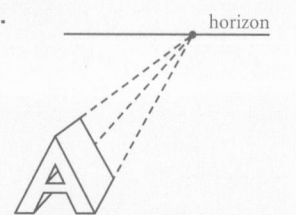

CHAPTER 7

LESSON 7.1

Determine the surface-area-to-volume ratio for a rectangular prism with the following dimensions:

1. $7 \times 1 \times 1$ **4.29**

2. $3 \times 2 \times 5$ **2.07**

3. $6 \times 6 \times 6$ **1**

4. $12 \times 12 \times 12$ **0.5**

5. $57 \times 2 \times 18$ **1.15**

6. $31 \times 94 \times 95$ **0.11**

Find the surface-area-to-volume ratio for each of the following:

7. a cube with a volume of 125 cubic units **1.2**

8. a cube with a surface area of 96 square units **1.5**

9. a rectangular prism with dimensions $1 \times n \times 2n$ $\frac{3+2n}{n}$

For each situation, determine whether you would want to maximize the volume or minimize the surface area.

10. designing a bowl out of a limited amount of clay **Maximize volume.**

11. building a carton with a fixed amount of cardboard **Maximize volume.**

LESSON 7.2

Find the volume of a prism with the given dimensions.

1. $B = 6 \text{ in.}^2, h = 4 \text{ in.}$ **24 in.^{3}**

2. $B = 14 \text{ cm}^2, h = 3 \text{ cm}$ **42 cm^3**

3. $B = 10 \text{ m}^2, h = 13 \text{ m}$ **130 m^3**

4. $B = 42 \text{ ft}^2, h = 10 \text{ ft}$ **420 ft^3**

Find the surface area and volume of a right rectangular prism with the given dimensions.

5. $\ell = 4, w = 12, h = 5$ **S = 256; V = 240**

6. $\ell = 13, w = 16, h = 3.1$ **S = 595.8; V = 644.8**

7. $\ell = 42, w = 30, h = 1$

8. $\ell = 0.05, w = 1.1, h = 2$ **S = 4.71; V = 0.11**

9. $\ell = 21, w = 21, h = 21$

10. $\ell = 4.3, w = 3.7, h = 6.8$ **S = 140.62; V = 108.19**

11. Find the height of a rectangular prism with a surface area of 2880 in.^{2} and a base measuring 5 in. by 8 in. **107.69 in.**

12. Find the width of a rectangular prism with a volume of 144 cm^3 and a base measuring 3 cm by 6 cm. **8 in.**

7. **S = 2664; V = 1260**

9. **S = 2646; V = 9261**

4.
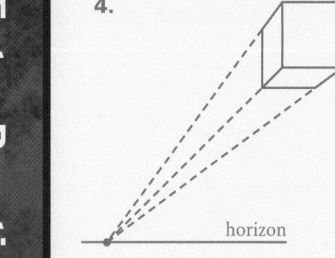

5. horizon
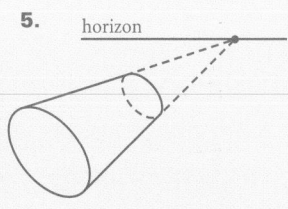

6. horizon
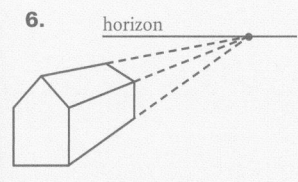

7. horizon
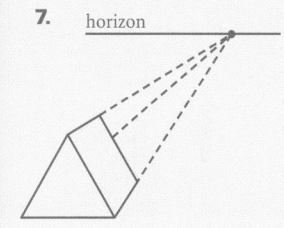

LESSON 7.3

Find the surface area of each regular pyramid with side length *s* and slant height ℓ given below. The number of sides of the base is given by *n*.

1. $n = 6, s = 4, \ell = 4$ 89.6

2. $n = 4, s = 2, \ell = 3$ 16

3. $n = 3, s = 5, \ell = 10$ 85.8

4. $n = 6, s = 14, \ell = 7$ 803.2

Find the volume of a rectangular pyramid with height *h* and base dimensions ℓ and *w*.

5. $h = 4, \ell = 7.2, w = 6.8$ 65.3

6. $h = 12, \ell = 18, w = 10$ 720

7. $h = 10, \ell = 24, w = 20$ 1600

8. $h = 17, \ell = 13, w = 5$ 368.3

9. $h = 42, \ell = 15, w = 20$ 4200

10. $h = 23, \ell = 19, w = 12$ 1748

Find the height of each pyramid described below.

11. a regular triangular pyramid with a base area of 16 square units and a volume of 48 cubic units   9 units

12. a regular square pyramid with a base length of 24 units and a volume of 14,400 cubic units   75 units

LESSON 7.4

Find the unknown value for a right cylinder with radius *r*, height *h*, and surface area *S* given below. Round your answers to the nearest tenth.

1. $r = 6, h = 3, S = \underline{\ ?\ }$ 339.3

2. $r = 14, h = 7, S = \underline{\ ?\ }$ 1846.3

3. $r = 0.5, h = 1.2, S = \underline{\ ?\ }$ 5.3

4. $r = \underline{\ ?\ }, h = 3, S = 36\pi$ 3

5. $r = 2, h = \underline{\ ?\ }, S = 72$ 3.7

6. $r = \underline{\ ?\ }, h = 3, S = 140\pi$ 7

Find the unknown value for a right cylinder with radius *r*, height *h*, and volume *V*. Round your answers to the nearest tenth.

7. $r = 4.5, h = 3.2, V = \underline{\ ?\ }$ 203.6

8. $r = \underline{\ ?\ }, h = 12, V = 150$ 2.0

9. $r = 15, h = 25, V = \underline{\ ?\ }$ 17,671.5

10. $r = 4, h = \underline{\ ?\ }, V = 24\pi$ 1.5

11. $r = \underline{\ ?\ }, h = 20, V = 180\pi$ 3

12. $r = 10, h = 8, V = \underline{\ ?\ }$ 2513.3

13. A cylinder with a diameter of 4 in. and a height of 8 in. is replaced by a cylinder that has the same volume. The new cylinder has a diameter of 6 in. What is its height? $\frac{32}{9}$ in.

LESSON 7.5

Find the surface area and volume of each right cone. Express your answers in terms of π.

1.
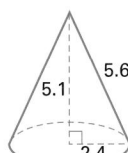
5.6
5.1
2.4
$S = 19.2\pi$
$V = 9.792\pi$

2.
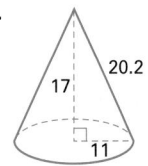
11.7
10
6
$S = 106.2\pi$
$V = 120\pi$

3.
17 20.2
11
$S = 343.2\pi$
$V = 685.67\pi$

4.
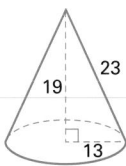
23
19
13
$S = 468\pi$
$V = 1070.33\pi$

5.

17.9
16
8
$S = 207.2\pi$
$V = 341.33\pi$

6.
56.6
38
42
$S = 4141.2\pi$
$V = 22,344\pi$

7.
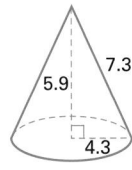
7.3
5.9
4.3
$S = 49.88\pi$
$V = 36.36\pi$

8.
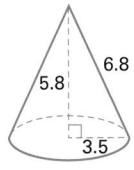
6.8
5.8
3.5
$S = 36.05\pi$
$V = 23.68\pi$

9.
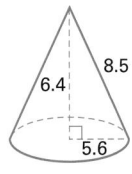
8.5
6.4
5.6
$S = 78.96\pi$
$V = 66.90\pi$

10. A right cone has a radius of 17 in. and a height of 12 in. What is the slant height of the cone? **20.81 in.**

LESSON 7.6

Find the surface area and volume of each sphere, with the given radius or diameter. Round your answers to the nearest hundredth.

1. $r = 5$ **2.** $d = 1.4$ **3.** $d = 66$

4. $r = 2.8$ **5.** $r = 2.9$ **6.** $d = 24$

7. $d = 4.02$ **8.** $r = 62$ **9.** $d = 16$

In Exercises 10–15, find the surface area and volume of each sphere with the given radius or diameter. Express your answers exactly in terms of π and a variable.

10. $r = 3x$ **11.** $d = 7y$ **12.** $d = x$

13. $r = 1.1y$ **14.** $r = 2y$ **15.** $d = \frac{x}{2}$

10. $S = 36\pi x^2$
$V = 36\pi x^3$

11. $S = 49\pi y^2$
$V = \frac{343}{6}\pi y^3$

12. $S = \pi x^2$
$V = \frac{1}{6}\pi x^3$

13. $S = 4.84\pi y^2$
$V = \frac{5.324}{3}\pi y^3$

14. $S = 16\pi y^2$
$V = \frac{32}{3}\pi y^3$

15. $S = \frac{1}{4}\pi x^2$
$V = \frac{1}{48}\pi x^3$

1. $S = 314.16$;
   $V = 523.60$

2. $S = 6.16$;
   $V = 1.44$

3. $S = 13,684.78$;
   $V = 150,532.55$

4. $S = 98.52$;
   $V = 91.95$

5. $S = 105.68$;
   $V = 102.16$

6. $S = 1,809.56$;
   $V = 7,238.23$

7. $S = 50.77$;
   $V = 34.02$

8. $S = 48,305.13$;
   $V = 998,305.99$

9. $S = 804.25$;
   $V = 2144.66$

Lesson 8.1

7. $y = 3x$;

Plugging in $O(0, 0)$ gives $0 = 0$, which is true. Thus, the origin is on this line.

8. $y = \frac{1}{4}x$;

Plugging in $O(0, 0)$ gives $0 = 0$, which is true. Thus, the origin is on this line.

9. $y = -5x$;

Plugging in $O(0, 0)$ gives $0 = 0$, which is true. Thus, the origin is on this line.

10. $y = \frac{3}{2}x$;

Plugging in $O(0, 0)$ gives $0 = 0$, which is true. Thus, the origin is on this line.

What are the coordinates of the reflection image when each point below is reflected across the *xy*-plane?

1. $(1, 1, 2)$ (1, 1, −2)   **2.** $(3.1, −2.5, −7)$ (3.1, −2.5, 7)   **3.** $(2, 1, −3)$ (2, 1, 3)

4. $(4, −4, 1.2)$ (4, −4, −1.2)   **5.** $(−2, 15, 0)$ (−2, 15, 0)   **6.** $(−12, −3, −8)$ (−12, −3, 8)

What are the coordinates of the reflection image when each point below is reflected across the *xz*-plane?

7. $(32, 32, 32)$ (32, −32, 32)   **8.** $(0, −3, −6)$ (0, 3, −6)   **9.** $(14, −4, 6.2)$ (14, 4, 6.2)

10. $(1.3, 5.2, −4)$ (1.3, −5.2, −4)   **11.** $(−12.1, −3, 6)$ (−12.1, 3, 6)   **12.** $(−53, −64, −0.2)$

(−53, 64, −0.2)

What are the coordinates of the reflection image when each point below is reflected across the *yz*-plane?

(7.8, −7.8, −10)

13. $(15, 15, 15)$ (−15, 15, 15)   **14.** $(−2, 3, 3)$ (2, 3, 3)   **15.** $(−7.8, −7.8, −10)$

16. $(3, −6, 8.9)$ (−3, −6, 8.9)   **17.** $(2, −2, −2)$ (−2, −2, −2)   **18.** $\left(\frac{1}{2}, −4, 3\right)$ $(-\frac{1}{2}, −4, 3)$

CHAPTER 8

Find the image of the point transformed by the given dilation. Plot the point and its image on a coordinate plane.

1. $(3, −1)$; $D(x, y) = (2x, 2y)$ (6, −2)   **2.** $(4, 1)$; $D(x, y) = (−4x, −4y)$ (−16, −4)

3. $(−2, −6)$; $D(x, y) = (0.5x, 0.5y)$ (−1, −3)   **4.** $(−5, 3)$; $D(x, y) = (−x, −y)$ (5, −3)

For Exercises 5 and 6, the dashed figures represent preimages of dilations and the solid figures represent images. Find the scale factor of each dilation.

5. 3

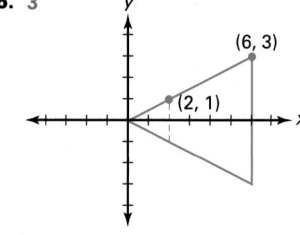

6. −2

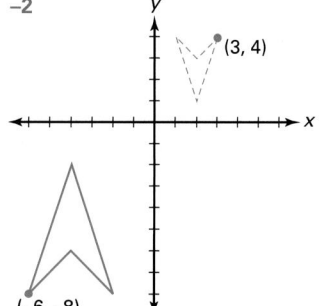

For Exercises 7–10, a point and a scale factor *n* are given. Find the line that passes through the preimage and image, and show that the origin is on this line.

7. $(−1, −3)$; $n = \frac{3}{4}$ **8.** $(4, 1)$; $n = -\frac{1}{2}$ **9.** $(1, −5)$; $n = 2$ **10.** $(2, 3)$; $n = −1$

LESSON 8.2

In Exercises 1 and 2, determine whether the polygons are similar. Explain your reasoning. If the polygons are similar, write a similarity statement.

1.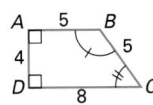

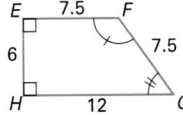

2.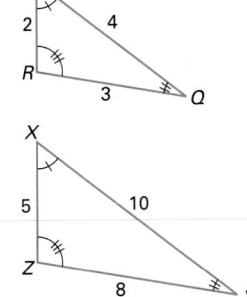

In Exercises 3–6, solve for *x*.

3. $\frac{2}{x} = \frac{6}{9}$ $x = 3$

4. $\frac{2}{x+2} = \frac{1}{x-1}$ $x = 4$

5. $\frac{\frac{1}{3}}{x} = \frac{\frac{1}{4}}{9}$ $x = 12$

6. $\frac{3.5}{x} = \frac{5.2}{15.6}$ $x = 10.5$

LESSON 8.3

Determine whether each pair of triangles can be proven similar by using the AA, SSS, or SAS Similarity Postulates and Theorems. If so, write a similarity statement and identify the postulate or theorem used.

1.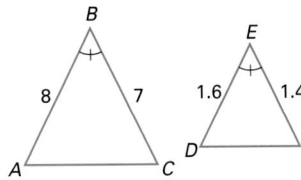

2.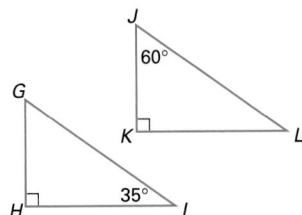

3. Indicate which figures are similar. Explain your reasoning.

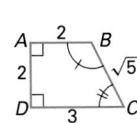 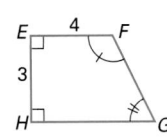 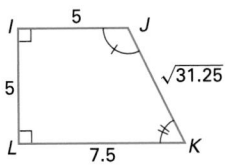

Quadrilateral *ABCD* ~ quadrilateral *IJKL* since corresponding angles are congruent and $\frac{AB}{IJ} = \frac{BC}{JK} = \frac{AD}{IL} = \frac{DC}{LK} = \frac{2}{5}$.

Quadrilateral *EFGH* is not similar to quadrilaterals *ABCD* and *IJKL* since $\frac{AB}{EF} = \frac{1}{2}$ but $\frac{AD}{EH} = \frac{2}{3} \neq \frac{1}{2}$.

Lesson 8.2

1. Yes.
$\angle A \cong \angle E$, $\angle B \cong \angle F$, $\angle D \cong \angle H$, and $\angle C \cong \angle G$.
Also, $\frac{AD}{EH} = \frac{AB}{EF} = \frac{BC}{FG} = \frac{DC}{HG} = \frac{2}{3}$.
Quadrilateral *ABCD* ~ quadrilateral *EFGH*.

2. No, since $\frac{PR}{XZ} = \frac{2}{5}$, but $\frac{RQ}{ZY} = \frac{3}{8} \neq \frac{2}{5}$.

Lesson 8.3

1. Yes, since $m\angle B = m\angle E$ and since $\frac{AB}{DE} = \frac{BC}{EF} = 5$, $\triangle ABC \sim \triangle DEF$ by SAS Similarity Theorem.

2. No. Since $m\angle G = 55°$ and $m\angle L = 30°$, the two triangles do not have congruent corresponding angles and thus are not similar.

843

LESSON 8.4

Use the Side-Splitting Theorem to find x. Some exercises may have more than one possible answer for x.

1. 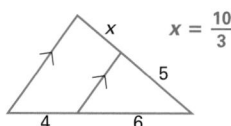 $x = \frac{10}{3}$

2. 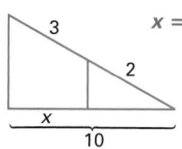 $x = 6$

3. 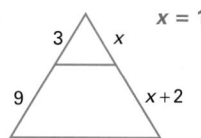 $x = 1$

Name all of the similar triangles in each figure. State the postulates or theorems that justify each similarity.

4.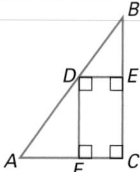

Thus $\triangle DBE \sim \triangle ABC \sim \triangle ADF$, by the AA Similarity Postulate.

5.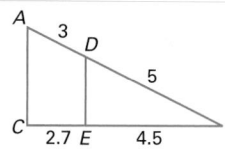

$\triangle ABC \sim \triangle DBE$ by the SAS Similarity Theorem.

6.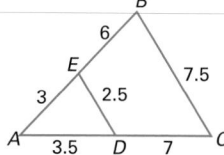

$\triangle ABC \sim \triangle AED$ by SSS Similarity Theorem.

LESSON 8.5

In Exercises 1–3, use the diagrams to find the height of each rectangle.

1. $h = 37.5$ ft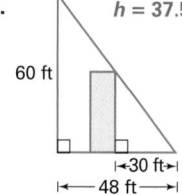

2. $h = 25$ ft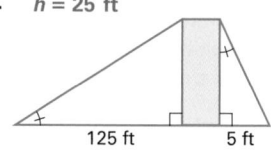

3. $h = 28.8$ ft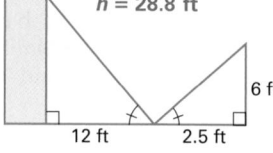

In Exercises 4–6, the triangles are similar. Find x.

4.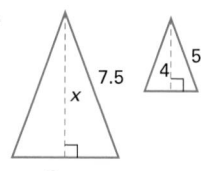

$x = 6$

5.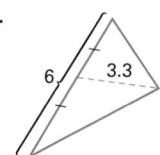

$x = 2.2$

6.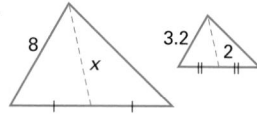

$x = 5$

LESSON 8.6

In the figure at right, $\overline{DE} \| \overline{AB}$.

1. Find the ratio of the perimeters of $\triangle ABC$ and $\triangle DEC$. $\frac{3}{2}$

2. Find the ratio of the areas of $\triangle ABC$ and $\triangle DEC$. $\frac{9}{4}$

The ratios of the sides of two similar hexagons is $\frac{2}{7}$. Find the ratio of

3. their perimeters. $\frac{2}{7}$

4. their areas. $\frac{4}{49}$

The ratios of the sides of two similar triangular prisms is $\frac{3}{4}$. Find the ratio of

5. the perimeters of their triangular faces. $\frac{3}{4}$

6. the areas of their triangular faces. $\frac{9}{16}$

7. their volumes. $\frac{27}{64}$

CHAPTER 9

LESSON 9.1

Determine the length of an arc with the given central angle measure, $m\angle P$, in a circle with radius r. Round your answers to the nearest hundredth.

1. $m\angle P = 40°; r = 6$ 4.19
2. $m\angle P = 20°; r = 8$ 2.79
3. $m\angle P = 75°; r = 20$ 26.18
4. $m\angle P = 100°; r = 16$ 27.93
5. $m\angle P = 55°; r = 13$ 12.48
6. $m\angle P = 118°; r = 30$ 61.78
7. $m\angle P = 66°; r = 40$ 46.08
8. $m\angle P = 130°; r = 61$ 138.4
9. $m\angle P = 80°; r = 39$ 54.45
10. $m\angle P = 82°; r = 5$ 7.16

Determine the degree measure of an arc with the given length, L, in a circle with radius r.

11. $L = 52; r = 14$ 212.8°
12. $L = 27; r = 5$ 309.4°
13. $L = 35; r = 11$ 182.3°
14. $L = 56; r = 30$ 107.0°
15. $L = 8; r = 2$ 229.2°
16. $L = 2.3; r = 85$ 1.6°
17. $L = 25; r = 25$ 57.3°
18. $L = 100; r = 79$ 72.5°

LESSON 9.2

For Exercises 1–6, refer to circle *J*, in which $\overline{JP} \perp \overline{KL}$ at *S*.

1. $\overline{SL} \cong$ __?__ $\overline{KS}$

2. If *JL* = 4 and *JS* = 1, what is *KS*? What is *SL*? **3.87, 3.87**

3. If *JL* = 10 and *JS* = 3, what is *KS*? What is *SL*? **9.54, 9.54**

4. If *JK* = 26 and *JS* = 11, what is *KS*? What is *SL*? **23.56, 23.56**

5. If *JK* = 60 and *JS* = 12, what is *KS*? What is *SL*? **58.79, 58.79**

6. If *JP* = 42 and *JS* = 30, what is *KS*? What is *SL*? **29.39, 29.39**

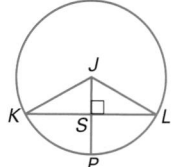

LESSON 9.3

In circle *P*, m∠*LPJ* = 30° and m∠*KMJ* = 45°. Find the following:

1. m∠*LMP* **15°**
2. m∠*JPK* **90°**
3. m∠*MJK* **45°**
4. m∠*LPM* **150°**
5. m$\widehat{MLJ}$ **180°**
6. m∠*MPK* **90°**
7. m∠*JKP* **45°**
8. m$\widehat{LJ}$ **30°**
9. m$\widehat{KM}$ **90°**
10. m∠*PLM* **15°**
11. m$\widehat{JK}$ **90°**
12. m∠*KJP* **45°**
13. m$\widehat{ML}$ **150°**
14. m∠*PKM* **45°**

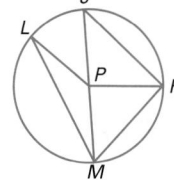

LESSON 9.4

In circle **E**, m$\widehat{BD}$ = 20°, m$\widehat{DF}$ = 180°, and m$\widehat{AF}$ = 45°. $\overrightarrow{GA}$ is tangent to circle **E** at **A**. Find each of the following:

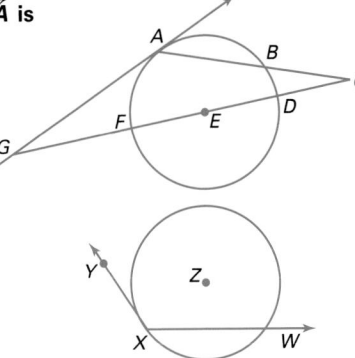

1. m∠CAG 122.5°

2. m∠GCA 12.5°

3. m∠CGA 45°

In the figure, $\overleftrightarrow{XY}$ is tangent to circle **Z** at **X**.

4. If m$\widehat{XW}$ = 95°, find m∠YXW. 132.5°

5. If m∠YXW = 100°, find m$\widehat{XW}$. 160°

6. If m$\widehat{XW}$ = x + 15, find m∠YXW. 180 − 0.5(x + 15)

LESSON 9.5

$\overleftrightarrow{XY}$ and $\overleftrightarrow{XZ}$ are tangent to circle **R**, XY = 4, and the radius of circle **R** is 2. Find the following:

1. XZ 4

2. YR 2

3. QX 2.47

4. RX 4.47

5. Name an angle congruent to ∠YXR. ∠ZXR

6. Name an angle congruent to ∠YRX. ∠ZRX

7. Name an arc congruent to $\widehat{YQ}$. $\widehat{ZQ}$

Find the value of x in each figure.

8. 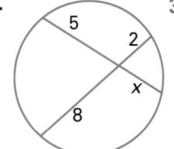 3.2

9. 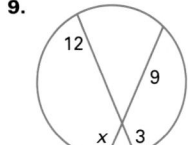 4

Lesson 9.6

1. *x*-intercepts:
$(7, 0)$, $(-7, 0)$
y-intercepts:
$(0, 7)$, $(0, -7)$

2. *x*-intercepts:
$(3.46, 0)$, $(-3.46, 0)$
y-intercepts:
$(0, 3.46)$, $(0, -3.46)$

3. *x*-intercepts:
$(5, 0)$, $(-3, 0)$
y-intercepts:
$(0, 3.87)$, $(0, -3.87)$

4. *x*-intercepts:
$(5.66, 0)$, $(-5.66, 0)$
y-intercepts:
$(0, 4)$, $(0, -8)$

5. *x*-intercepts:
$(11.06, 0)$, $(-5.06, 0)$
y-intercepts:
$(0, 4.49)$, $(0, -12.49)$

6. *x*-intercepts:
$(17.95, 0)$, $(-3.95, 0)$
y-intercepts:
$(0, 7.49)$, $(0, -9.49)$

7. center: $(0, 0)$ $r = 2$

8. center: $(0, 0)$ $r = 8$

9. center: $(-1, 0)$ $r = \sqrt{14}$

10. center: $(0, 1)$ $r = 3\sqrt{3}$

11. center: $(-3, 5)$ $r = \sqrt{31}$

12. center: $(7, -1)$ $r = 2\sqrt{3}$

Find the *x*- and *y*-intercepts for the graph of each equation below.

1. $x^2 + y^2 = 49$

2. $x^2 + y^2 = 12$

3. $(x - 1)^2 + y^2 = 16$

4. $x^2 + (y + 2)^2 = 36$

5. $(x - 3)^2 + (y + 4)^2 = 81$

6. $(x - 7)^2 + (y + 1)^2 = 121$

Find the center and radius of each circle.

7. $x^2 + y^2 = 4$

8. $x^2 + y^2 = 64$

9. $(x + 1)^2 + y^2 = 14$

10. $x^2 + (y - 1)^2 = 27$

11. $(x + 3)^2 + (y - 5)^2 = 31$

12. $(x - 7)^2 + (y + 1)^2 = 12$

Write an equation for the circle with the given center and radius.

13. center: $(1, 2)$; radius = 4
$(x - 1)^2 + (y - 2)^2 = 16$

14. center: $(-1, -6)$; radius = 12
$(x + 1)^2 + (y + 6)^2 = 144$

CHAPTER 10

Find tan *A* for each triangle below.

1.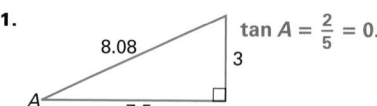
$\tan A = \frac{2}{5} = 0.4$

2.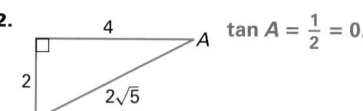
$\tan A = \frac{1}{2} = 0.5$

3.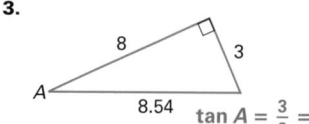
$\tan A = \frac{3}{8} = 0.375$

4.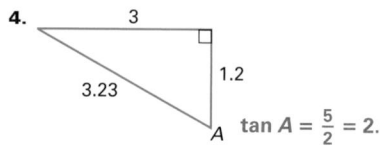
$\tan A = \frac{5}{2} = 2.5$

Use a scientific calculator to find the tangent of each angle below.
Round your answers to the nearest hundredth.

5. 15° 0.27

6. 23° 0.42

7. 47° 1.07

8. 69° 2.61

9. 42° 0.90

10. 54° 1.38

LESSON 10.2

For Exercises 1–6, refer to △*XYZ*. Find each of the following:

1. sin ___?___ = $\frac{2}{\sqrt{13}}$ *X*

2. cos ___?___ = $\frac{2}{\sqrt{13}}$ *Y*

3. cos ___?___ = $\frac{3}{\sqrt{13}}$ *X*

4. tan ___?___ = $\frac{3}{2}$ *Y*

5. tan ___?___ = $\frac{2}{3}$ *X*

6. sin ___?___ = $\frac{3}{\sqrt{13}}$ *Y*

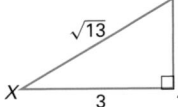

Use a scientific calculator to find the following. Round your answers to the nearest hundredth.

7. sin 15° **0.26**

8. cos 23° **0.92**

9. tan 17° **0.31**

10. cos 69° **0.36**

11. sin 42° **0.67**

12. tan 47° **1.07**

LESSON 10.3

Use a scientific calculator to find the following. Round your answers to four decimal places.

1. sin 55° **0.8192**

2. sin 169° **0.1908**

3. sin 340° **−0.3420**

4. cos 55° **0.5736**

5. cos 169° **−0.9816**

6. cos 340° **0.9397**

In Exercises 7–9, use a scientific calculator to find the sine and cosine of each angle. Use these values to give the *x*- and *y*-coordinates of a point at the given angle on the unit circle. Round your answers to four decimal places.

7. 20° **(0.9397, 0.3420)**

8. 160° **(−0.9397, 0.3420)**

9. 200° **(−0.9397, −0.3420)**

In Exercises 10–15, give two values for angle θ between 0° and 180° for the given value of sin θ. Round your answers to the nearest degree.

10. 0.2250 **13° and 167°**

11. 0.6157 **38° and 142°**

12. 0.8746 **61° and 119°**

13. 0.3907 **23° and 157°**

14. 0.3090 **18° and 162°**

15. 0.5150 **31° and 149°**

LESSON 10.4

For Exercises 1–4, find the indicated measures. Assume that all angles are acute. It may be helpful to sketch the triangles roughly to scale.

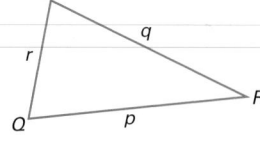

1. m∠A = 52°   m∠B = 68°   b = 4.2 cm   a = ___?___ **3.57 cm**
2. m∠A = 72°   m∠C = 32°   a = 1.4 cm   c = ___?___ **0.78 cm**
3. m∠B = 64°   a = 2.34 cm   b = 3.5 cm   m∠A = ___?___ **36.9°**
4. m∠A = 25°   m∠C = 65°   c = 5 cm   a = ___?___ **2.33 cm**

Find all unknown sides and angles in each triangle. If the triangle is ambiguous, give both possible angles. It may be helpful to sketch the triangles roughly to scale.

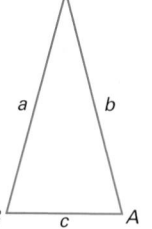

5. m∠P = 36°   m∠Q = 68°   q = 7
6. m∠R = 75°   p = 10   r = 14
7. m∠Q = 40°   m∠P = 25°   q = 12

5. m∠R = 76°   r ≈ 7.33   p ≈ 4.44
6. m∠P ≈ 43.6°   m∠Q ≈ 61.4°   q ≈ 12.73
7. m∠R = 115°   r ≈ 16.92   p ≈ 7.89

LESSON 10.5

For Exercises 1–4, find the indicated measures. It may be helpful to sketch the triangles roughly to scale. Round to scale. Round your answers to the nearest tenth.

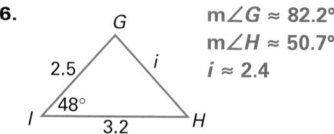

1. a = 1   b = 13   c = ___?___ **12.1** m∠C = 20°
2. a = 12   b = 5   c = ___?___ **11.1** m∠C = 68°
3. a = 4   b = 7   c = 5   m∠C = ___?___ **44.4°**
4. a = 9   b = 3   c = 8   m∠C = ___?___ **61.2°**

Solve each triangle.

5.
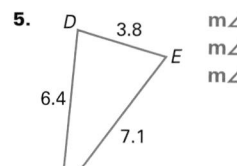

m∠D ≈ 84.1°
m∠E ≈ 63.7°
m∠F ≈ 32.2°

6.

m∠G ≈ 82.2°
m∠H ≈ 50.7°
i ≈ 2.4

LESSON 10.6

Copy each pair of vectors below and draw the vector sum, $\vec{a} + \vec{b}$, by using the head-to-tail method. You may need to translate one of the vectors.

1.

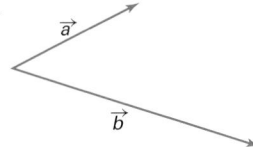

2.

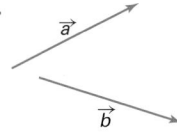

Copy each pair of vectors below and draw the vector sum, $\vec{a} + \vec{b}$, by using the parallelogram method. You may need to translate one of the vectors.

3.

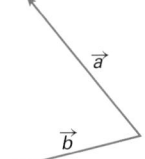

4.

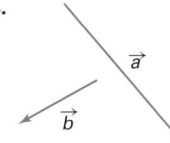

LESSON 10.7

For Exercises 1–8, a point, *P*, and an angle of rotation, *θ*, are given. Determine the coordinates of the image, *P′*. Round your answers to the nearest tenth.

1. $P(0, 5)$; $\theta = 48°$ (–3.7, 3.3)

2. $P(4, 2)$; $\theta = 270°$ (2, –4)

3. $P(1, 4)$; $\theta = 17°$ (–0.2, 4.1)

4. $P(-2, -5)$; $\theta = 65°$ (3.7, –3.9)

5. $P(2, -2)$; $\theta = 30°$ (2.7, –0.7)

6. $P(-1, 2)$; $\theta = 180°$ (1, –2)

7. $P(3, 0)$; $\theta = 90°$ (0, 3)

8. $P(-3, -2)$; $\theta = 200°$ (2.1, 2.9)

Find the rotation matrix for each angle below by filling in the sine and the cosine values. Round your answers to the nearest hundredth.

9. 135°

11. 35°

13. 40°

15. 300°

10. 215°

12. 345°

14. 180°

16. 120°

9. $\begin{bmatrix} -.71 & -.17 \\ .71 & -.71 \end{bmatrix}$

10. $\begin{bmatrix} -.82 & .57 \\ -.57 & -.82 \end{bmatrix}$

11. $\begin{bmatrix} .82 & -.57 \\ .57 & .82 \end{bmatrix}$

12. $\begin{bmatrix} .97 & .26 \\ -.26 & .97 \end{bmatrix}$

13. $\begin{bmatrix} .77 & -.64 \\ .64 & .77 \end{bmatrix}$

14. $\begin{bmatrix} -1 & 0 \\ 0 & -1 \end{bmatrix}$

15. $\begin{bmatrix} .5 & .87 \\ -.87 & .5 \end{bmatrix}$

16. $\begin{bmatrix} -.5 & -.87 \\ .87 & -.5 \end{bmatrix}$

Lesson 10.6

1.

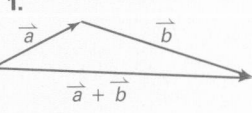

2.

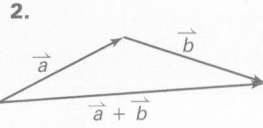

3.

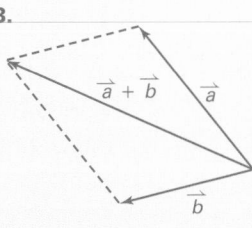

4.

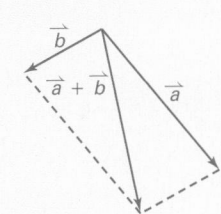

CHAPTER 11

LESSON 11.1

Determine the indicated side length of each golden rectangle. Round your answers to the nearest hundredth.

1.  2.47

?

4

2. 7.44

4.6

?

3. 6.18

?

10

4. 4.85

3

?

5. 5.07

?

8.2

6. 2.43

1.5

?

LESSON 11.2

Find the taxidistance between each pair of points.

1. $(2, 4)$ and $(1, 6)$ **3**

2. $(-5, 8)$ and $(0, -2)$ **15**

3. $(-9, -5)$ and $(-4, -7)$ **7**

4. $(-10, 10)$ and $(10, -10)$ **40**

Find the number of points on the taxicab circle with the given radius.

5. $r = 3$ **12**

6. $r = 7$ **28**

7. $r = 10$ **40**

8. $r = 8$ **32**

9. $r = 5$ **20**

10. $r = 20$ **80**

Find the circumference of the taxicab circle with the given radius.

11. $r = 3$ **24**

12. $r = 6$ **48**

13. $r = 8$ **64**

14. $r = 9$ **72**

15. $r = 11$ **88**

16. $r = 20$ **160**

LESSON 11.3

Determine whether the graphs below contain an Euler path, an Euler circuit, or neither.

1.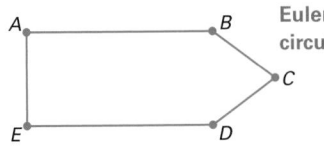

Euler circuit

2.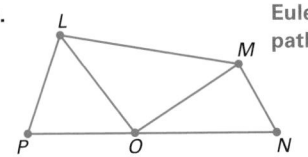

Euler path

3.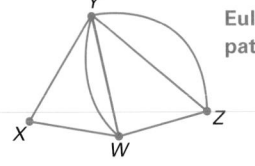

Euler path

4.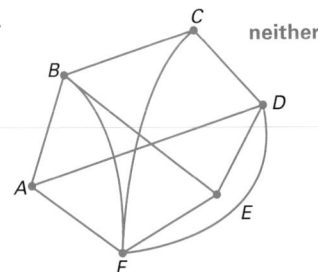

neither

LESSON 11.4

1. Which, if any, of the following figures are topologically equivalent?

a.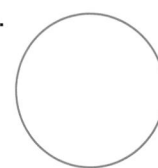

b.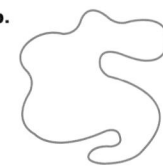

c.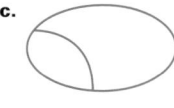

a. and b. are topologically equivalent.

Verify Euler's formula for each polyhedron below.

2. tetrahedron

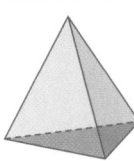

$V - E + F = 2$
$4 - 6 + 4 = 2$

3. dodecahedron

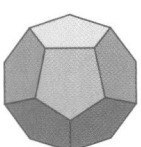

$V - E + F = 2$
$20 - 30 + 12 = 2$

1.

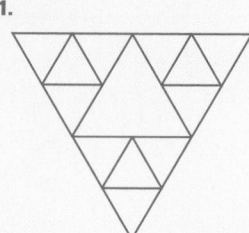

2.

3.

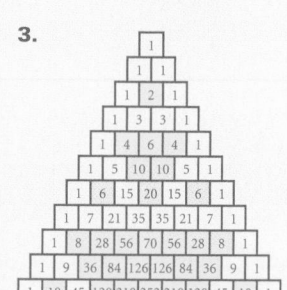

The shaded boxes in Activity 2 are unshaded in this Pascal's triangle. The unshaded boxes in Activity 2 are shaded in this Pascal's triangle.

LESSON 11.5

Refer to the diagram of a sphere at right. Determine whether the following figures are lines in spherical geometry:

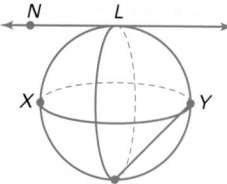

1. $\overleftrightarrow{XY}$ is a line
2. $\overleftrightarrow{NL}$ is not a line
3. $\overleftrightarrow{MY}$ is not a line
4. $\overleftrightarrow{LM}$ is a line

Refer to the diagram at right. Determine whether the following figures are lines in hyperbolic geometry:

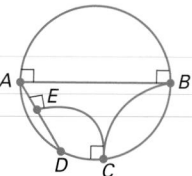

5. $\overleftrightarrow{AE}$ is not a line
6. $\overleftrightarrow{AB}$ is a line
7. $\overleftrightarrow{EC}$ is not a line
8. $\overleftrightarrow{BC}$ is a line

LESSON 11.6

Using the technique shown in the construction of the Sierpinski gasket, construct a Sierpinski gasket with the following triangles:

1.

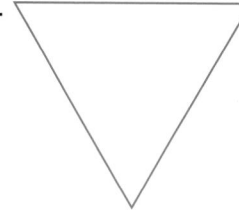

2.

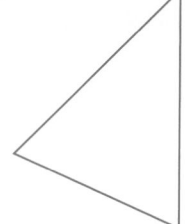

3. Build Pascal's triangle with at least 12 rows. Shade around each number that is divisible by 2, and leave the other numbers unshaded. How is the pattern different from the pattern you found in Activity 2 of Lesson 11.6? How is it the same?

LESSON 11.7

Draw a circle with a radius of 3 centered at the origin on a coordinate plane. Use the grids of parallelograms below to transform the circle.

1.

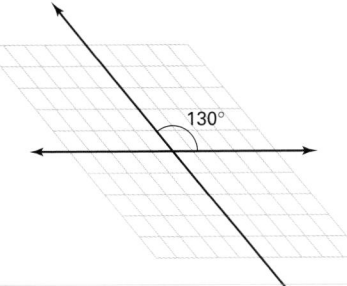

130°

2.

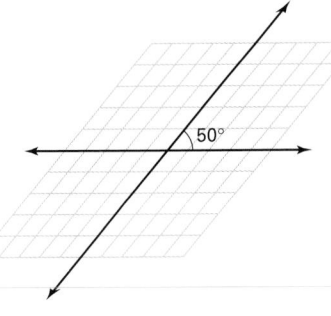

50°

Copy the diagram at right and draw each projection.

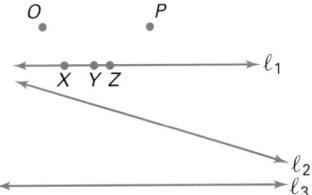

3. Using O as the center of projection, project points X, Y, and Z onto line ℓ_2. Label the projected points X', Y', and Z'.

4. Using P as the center of projection, project points X', Y', and Z' onto line ℓ_3. Label the projected points X'', Y'', and Z''.

CHAPTER 12

LESSON 12.1

In Exercises 1–4, write a valid conclusion from the given premises. Identify the form of the argument.

1. If the baby is hungry, then he cries. The baby is hungry. **Therefore, he cries. (*modus ponens*)**

2. If the baby is hungry, then he cries. The baby is not crying.

3. If the baby is not hungry, then he throws his cereal on the floor. The baby did not throw his cereal on the floor. **Therefore, he is hungry. (*modus tollens*)**

4. If the baby is not hungry, then he will play in his playpen. The baby is not hungry. **Therefore, he will play in his playpen. (*modus ponens*)**

For Exercises 5–7, refer to the following premises:

If Harry is a purple hippopotamus, then Larry is a blue dinosaur.
If Larry is not a blue dinosaur, then Barry is a pink alligator.
Larry is not a blue dinosaur.

State whether each conclusion below is valid. Explain.

5. Harry is not a purple hippopotamus. **Valid by *modus tollens*.**

6. Harry is a purple hippopotamus.

7. Barry is a pink alligator. **Valid by *modus ponens*.**

Lesson 11.7

1.

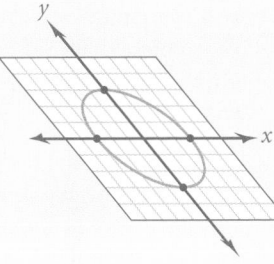

2.

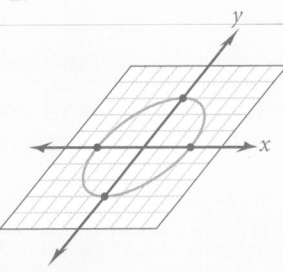

3. and 4.

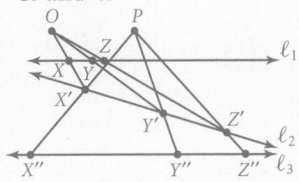

Lesson 12.1

2. Therefore, he is not hungry. (*modus tollens*)

6. Invalid. Harry is not a purple hippopotamus by *modus tollens*.

Lesson 12.2

1. Some flowers are red and cucumbers are vegetables. True since both statements are true.

2. Parallel lines intersect or all rectangles are squares. False since neither statement is true.

3. Harry is not older than Larry.

4. Larry and Barry are not twins or Harry is not younger than Barry.

5. Larry and Barry are not twins.

Lesson 12.3

1. *Conditional*: True. Twins are always the same age.
Converse: If Harry and Larry are the same age, then they are twins. False. They can be the same age without even being related.
Inverse: If Harry and Larry are not twins, then they are not the same age. False: They could be the same age without being twins.
Contrapositive: If Harry and Larry are not the same age, then they are not twins. True. To be twins, they must be the same age.

2. *Conditional*: True. All congruent triangles are similar, with scaling factor 1.
Converse: If two triangles are similar, then they are congruent. False. Triangles with sides 1, 2, 3 and 2, 4, 6 are similar by the SSS Similarity Postulate, but they are not congruent.

LESSON 12.2

1. Write a conjunction for the pair of statements below. Determine whether the conjunction is true or false.

   Some flowers are red. Cucumbers are vegetables.

2. Write a disjunction for the pair of statements below. Determine whether the disjunction is true or false.

   Parallel lines intersect. All rectangles are squares.

For Exercises 3–8, write the statement expressed by the symbols, where *p*, *q*, and *r* represent the statements shown below.

p: Harry is older than Larry.

q: Larry and Barry are twins.

r: Harry is younger than Barry.

3. ~*p*   4. ~*q* OR ~*r*   5. ~*q*
6. *p* AND ~*q*   7. ~(*p* AND *r*)   8. ~(*p* OR ~*r*)

6. Harry is older than Larry and Larry and Barry are not twins.

7. It is not the case that Harry is both older than Larry and younger than Barry.

8. It is not the case that either Harry is older than Larry or Harry is not younger than Barry.

LESSON 12.3

For each conditional in Exercises 1–3, write the converse, inverse, and contrapositive. Decide whether each is true or false and explain your reasoning.

1. If Harry and Larry are twins, then they are the same age.

2. If two triangles are congruent, then they are similar.

3. If $m \times n = 0$, then $m = 0$.

For Exercises 4–6, write each statement in if-then form.

4. All blue pigs can play the banjo. If a pig is blue, then it can play the banjo.

5. No odd numbers are divisible by 2. If a number is odd, then it is not divisible by 2.

6. I will buy you a camera if they go on sale. If cameras go on sale, I will buy you one.

Inverse: If two triangles are not congruent, then they are not similar. False: the triangles with sides 1, 2, 3 and 2, 4, 6 are not congruent, but they are similar by the SSS Similarity Postulate.
Contrapositive: If two triangles are not similar, then they are not congruent. True. Congruent triangles are always similar.

3. *Conditional*: False. It could be the case that $m \neq 0$ and $n = 0$.
Converse: If $m = 0$, then $m \times n = 0$. True. Zero multiplied by any number is zero.

Inverse: If $m \times n \neq 0$ then $m \neq 0$. True: If m were zero, $m \times n$ would be zero.
Contrapositive: If $m \neq 0$, then $m \times n \neq 0$. False. If $n = 0$ and $m \neq 0$, $m \times n$ would still equal zero.

LESSON 12.4

In Exercises 1–4, determine whether the given argument is an example of an indirect reasoning. Explain why or why not.

1. If Bobby ate some cookies from the cookie jar, then the cookie jar is not full. The cookie jar is not full. Therefore, Bobby ate some cookies from the cookie jar.

2. If Juan eats strawberries, then he breaks out in hives. Juan did not break out in hives. Therefore, he did not eat strawberries.

3. I was in an accident. If I was in an accident, then I must have been driving my car. Therefore, I was driving my car.

4. If the alarm rings, then the dogs bark. The dogs did not bark. Therefore, the alarm did not ring.

Complete the indirect proof below.

Given: $\triangle ABC$, $m\angle A < 30°$, and $m\angle B < 45°$

Prove: $\angle C$ is obtuse.

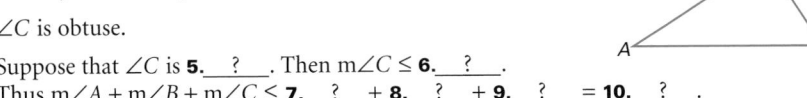

Proof: Suppose that $\angle C$ is **5.** _?_ . Then $m\angle C \leq$ **6.** _?_ .
Thus $m\angle A + m\angle B + m\angle C \leq$ **7.** _?_ + **8.** _?_ + **9.** _?_ = **10.** _?_ .
This contradicts the property of triangles that **11.** _?_ . Therefore, **12.** _?_ .

    5. not obtuse (acute or right)   6. 90°
    7. 30°   8. 45°
    9. 90°   10. 165°
    11. the sum of the measures of   12. $\angle C$ is obtuse
        the three angles is 180°

LESSON 12.5

Use the logic gates below to answer each question.

1. If $p = 0$, what is the output? **1**
2. If $p = 1$ and $q = 1$, what is the output? **0**

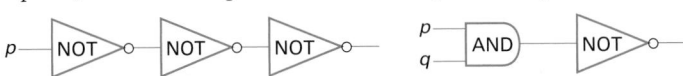

For Exercises 3–10, complete the input-output table for the network of logic gates at right.

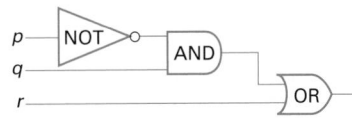

	p	q	r	$\sim p$	$\sim p$ AND q	$(\sim p$ AND $q)$ OR r
3.	1	1	1	0	0	1
4.	1	1	0	0	0	0
5.	1	0	1	0	0	1
6.	1	0	0	0	0	0
7.	0	1	1	1	1	1
8.	0	1	0	1	1	1
9.	0	0	1	1	0	1
10.	0	0	0	1	0	0

Lesson 12.4

1. No. To prove that Bobby ate some cookies by an indirect argument, it should first be assumed that Bobby did not eat some cookies from the cookie jar. Also, the argument assumes that the converse of a conditional has the same truth value as the conditional, which is not the case.

2. Yes. The proof starts by assuming the opposite of the statement to be proved. Then a contradiction is reached after logical arguments. Finally, the opposite of the assumption is stated to be true because the assumption must be false.

3. No. The proof does not start by assuming the opposite of the statement to be proved.

4. Yes. The proof starts by assuming the opposite of the statement to be proved. Then a contradiction is reached after logical arguments. Finally, the opposite of the assumption is stated to be true because the assumption must be false.

Postulates, Theorems, and Definitions

Def 1.1.1 **Segment** A **segment** is a part of a line that begins at one point and ends at another. The points are called the **endpoints** of the segment.

Def 1.1.2 **Ray** A **ray** is a part of a line that starts at a point and extends infinitely in one direction. The point is called the **endpoint** of the ray.

Def 1.1.3 **Angle** An **angle** is a figure formed by two rays with a common endpoint. The common endpoint is called the **vertex of the angle**, and the rays are the **sides of the angle**. An angle divides a plane into two regions: the **interior** and the **exterior** of the angle. If two points, one from each side of an angle, are connected by a segment, the segment passes through the interior of the angle.

Post 1.1.4 **Postulate** The intersection of two lines is a point.

Post 1.1.5 **Postulate** The intersection of two planes is a line.

Post 1.1.6 **Postulate** Through any two points there is exactly one line.

Post 1.1.7 **Postulate** Through any three noncollinear points there is exactly one plane.

Post 1.1.8 **Postulate** If two points are in a plane, then the line containing them is in the plane.

Def 1.2.1 **Length of $\overline{AB}$** Let A and B be points on a number line, with coordinates a and b. Then the measure of $\overline{AB}$, which is called its **length**, is $|a - b|$ or $|b - a|$.

$m\overline{AB}$, or $AB =$ $|a - b|$ or $|b - a|$

Post 1.2.2 **Segment Congruence Postulate** If two segments have the same length as measured by a given fair ruler, then the segments are congruent. Also, if two segments are congruent, then they have the same length as measured by a given ruler.

Post 1.2.3 **Segment Addition Postulate** If point R is between points P and Q on a line, then $PR + RQ = PQ$.

Def 1.3.1 **Measure of an Angle** Suppose that the vertex, V, of $\angle AVB$ is placed on the center point of a half-circle with coordinates from 0° to 180° so that $\overrightarrow{VA}$ and $\overrightarrow{VB}$ intersect the half-circle. Let a and b be the coordinates of the intersections. Then the **measure of the angle**, written as $m\angle AVB$, is $|a - b|$ or $|b - a|$.

Post 1.3.2 **Angle Addition Postulate** If point S is in the interior of $\angle PQR$, then $m\angle PQS + m\angle SQR = m\angle PQR$.

Post 1.3.3 **Angle Congruence Postulate** If two angles have the measure, then they are congruent. If two angles are congruent, then they have the same measure.

Def 1.3.4 **Special Angle Pairs**
Complementary angles are two angles whose measures have a sum of 90°. Each angle is called the **complement** of the other.
Supplementary angles are two angles whose measures have a sum of 180°. Each angle is called the **supplement** of the other.

Thm 1.3.5 **Linear Pair Property** If two angles form a linear pair, then they are supplementary.

Def 1.3.6 **Three Types of Angles**
A **right angle** is an angle whose measure is 90°.
An **acute angle** is an angle whose measure is less than 90°.
An **obtuse angle** is an angle whose measure is greater than 90° and less than 180°.

Def 1.4.1 **Perpendicular and Parallel Lines** **Perpendicular lines** are two lines that intersect to form a right angle. **Parallel lines** are two coplanar lines that do not intersect.

Def 1.4.2 **Bisectors and Midpoint** A **segment bisector** is a line that divides a segment into two congruent parts. The point where a bisector intersects a segment is the **midpoint** of the segment. A bisector that is perpendicular to a segment is called a **perpendicular bisector**. An **angle bisector** is a line or ray that divides an angle into two congruent angles.

Def 1.6.1 **Translation** A **translation** is a transformation in which every point of the preimage is moved the same distance in the same direction.

Def 1.6.2 **Rotation** A **rotation** is a transformation in which every point of the preimage is moved by the same angle through a circle centered at a given fixed point known as the *center of rotation*.

Def 1.6.3 **Reflection** A **reflection** is a transformation in which every point of the preimage is moved across a line known as the *mirror line* so that the mirror is the perpendicular bisector of the segment connecting the point and its image.

Def 1.7.1 **Horizontal and Vertical Coordinate Translations**
Horizontal translation of h units: $H(x, y) = (x + h, y)$
Vertical translation of v units: $V(x, y) = (x, y + v)$

Def 1.7.2 **Reflection Across the *x*- or *y*-axis**
Reflection across the x-axis: $M(x, y) = (x, -y)$
Reflection across the y-axis: $N(x, y) = (-x, y)$

Def 1.7.3 **180° Rotation About the Origin** $R(x, y) = (-x, -y)$

Post 2.2.1 **If-Then Transitive Property** Given: "If A, then B, and if B, then C." You can conclude: "If A, then C."

Def 2.3.1 **Adjacent Angles** Adjacent angles are angles in a plane that have their vertex and one side in common but have no interior points in common.

Post 2.4.1 **Addition Property** If $a = b$, then $a + c = b + c$.

Post 2.4.2 **Subtraction Property** If $a = b$, then $a - c = b - c$.

Post 2.4.3 **Multiplication Property** If $a = b$, then $ac = bc$.

Post 2.4.4 **Division Property** If $a = b$ and $c \neq 0$, then $\frac{a}{c} = \frac{b}{c}$.

Post 2.4.5 **Substitution Property** If $a = b$, you may replace a with b in any true equation containing a and the resulting equation will still be true.

Thm 2.4.6 **Overlapping Segments Theorem** Given a segment with points A, B, C, and D (in order), the following statements are true:
1. If $AB = CD$, then $AC = BD$.
2. If $AC = BD$, then $AB = CD$.

Post 2.4.7 **Reflexive Property of Equality** For any real number a, $a = a$.

Post 2.4.8 **Symmetric Property of Equality** For all real numbers a and b, if $a = b$, then $b = a$.

Post 2.4.9 **Transitive Property of Equality** For all real numbers a, b, and c, if $a = b$ and $b = c$, then $a = c$.

Post 2.4.10 **Reflexive Property of Congruence** figure $A \cong$ figure A

Post 2.4.11 **Symmetric Property of Congruence** If figure $A \cong$ figure B, then figure $B \cong$ figure A.

Post 2.4.12 **Transitive Property of Congruence** If figure $A \cong$ figure B and figure $B \cong$ figure C, then figure $A \cong$ figure C.

Thm 2.4.13 **Overlapping Angles Theorem** Given $\angle AOD$ with points B and C in its interior, the following statements are true:
1. If $m\angle AOB = m\angle COD$, then $m\angle AOC = m\angle BOD$.
2. If $m\angle AOC = m\angle BOD$, then $m\angle AOB = m\angle COD$.

Thm 2.5.1 **Vertical Angles Theorem** If two angles form a pair of vertical angles, then they are congruent.

Thm 2.5.2 **Theorem** Reflection across two parallel lines is equivalent to a translation of twice the distance between the lines and in a direction perpendicular to the lines.

Thm 2.5.3 **Theorem** Reflection across two intersecting lines is equivalent to a rotation about the point of intersection through twice the measure of the angle between the lines.

Def 3.1.1 **Polygon** A polygon is a plane figure formed from three or more segments such that each segment intersects exactly two other segments, one at each endpoint, and no two segments with a common endpoint are collinear. The segments are called the **sides of the polygon**, and the common endpoints are called the **vertices of the polygon**.

Def 3.1.2 **Reflectional Symmetry** A figure has **reflectional symmetry** if and only if its reflected image across a line coincides exactly with the preimage. The line is called an **axis of symmetry**.

Def 3.1.3 **Triangles Classified by Number of Congruent Sides**
Three congruent sides: equilateral
At least two congruent sides: isosceles
No congruent sides: scalene

Def 3.1.4 **Rotational Symmetry** A figure has **rotational symmetry** if and only if it has at least one rotation image, not counting rotation images of 0° or multiples of 360°, that coincides with the original image.

Def 3.3.1 **Transversal** A transversal is a line, ray, or segment that intersects two or more coplanar lines, rays, or segments, each at a different point.

Post 3.3.2 **Corresponding Angles Postulate** If two lines cut by a transversal are parallel, then corresponding angles are congruent.

Thm 3.3.3 **Alternate Interior Angles Theorem** If two lines cut by a transversal are parallel, then alternate interior angles are congruent.

Thm 3.3.4 **Alternate Exterior Angles Theorem** If two lines cut by a transversal are parallel, then alternate exterior angles are congruent.

Thm 3.3.5 **Same-Side Interior Angles Theorem** If two lines cut by a transversal are parallel, then same-side interior angles are supplementary.

Thm 3.4.1 **Theorem: Converse of the Corresponding Angles Postulate** If two lines are cut by a transversal in such a way that corresponding angles are congruent, then the two lines are parallel.

Thm 3.4.2 **Converse of the Same-Side Interior Angles Theorem** If two lines are cut by a transversal in such a way that same-side interior angles are supplementary, then the two lines are parallel.

Thm 3.4.3 **Converse of the Alternate Interior Angles Theorem** If two lines are cut by a transversal in such a way that alternate interior angles are congruent, then the two lines are parallel.

Thm 3.4.4 **Converse of the Alternate Exterior Angles Theorem** If two lines are cut by a transversal in such a way that alternate exterior angles are congruent, then the two lines are parallel.

Thm 3.4.5 **Theorem** If two coplanar lines are perpendicular to the same line, then the two lines are parallel.

Thm 3.4.6 **Theorem** If two lines are parallel to the same line, then the two lines are parallel.

Post 3.5.1 **The Parallel Postulate** Given a line and a point not on the line, there is one and only one line that contains the given point and is parallel to the given line.

Thm 3.5.2 **Triangle Sum Theorem** The sum of the measures of the angles of a triangle is 180°.

Thm 3.5.3 **Exterior Angle Theorem** The measure of an exterior angle of a triangle is equal to the sum of the measures of the remote interior angles.

Thm 3.6.1 **Sum of the Interior Angles of a Polygon** The sum, s, of the measures of the interior angles of a polygon with n sides is given by $s = (n-2)180°$.

Thm 3.6.2 **The Measure of an Interior Angle of a Regular Polygon** The measure, m, of an interior angle of a regular polygon with n sides is given by $m = 180° - \frac{360°}{n}$.

Thm 3.6.3 **Sum of the Exterior Angles of a Polygon** The sum of the measures of the exterior angles of a polygon is 360°.

Def 3.7.1 **Midsegment of a Triangle** A midsegment of a triangle is a segment whose endpoints are the midpoints of two sides.

Def 3.7.2 **Midsegment of a Trapezoid** A midsegment of a trapezoid is a segment whose endpoints are the midpoints of the nonparallel sides.

Def 3.8.1 **Slope** The slope of a nonvertical line that contains the points (x_1, y_1) and (x_2, y_2) is equal to the ratio $\frac{y_2 - y_1}{x_2 - x_1}$.

Thm 3.8.2 **Parallel Lines Theorem** Two nonvertical lines are parallel if and only if they have the same slope.

Thm 3.8.3 **Perpendicular Lines Theorem** Two nonvertical lines are perpendicular if and only if the product of their slopes is –1.

Post 4.1.1 **Polygon Congruence Postulate** Two polygons are congruent if and only if there is a way of setting up a correspondence between their sides and angles, in order, such that (1) all pairs of corresponding angles are congruent and (2) all pairs of corresponding sides are congruent.

Post 4.2.1 **SSS (Side-Side-Side) Postulate** If the sides of one triangle are congruent to the sides of another triangle, then the two triangles are congruent.

Post 4.2.2 **SAS (Side-Angle-Side) Postulate** If two sides and their included angle in one triangle are congruent to two sides and their included angle in another triangle, then the two triangles are congruent.

Post 4.2.3   **ASA (Angle-Side-Angle) Postulate** If two angles and their included side in one triangle are congruent to two angles and their included side in another triangle, then the two triangles are congruent.

Thm 4.3.1   **AAS (Angle-Angle-Side) Congruence Theorem** If two angles and a nonincluded side of one triangle are congruent to the corresponding angles and nonincluded side of another triangle, then the triangles are congruent.

Thm 4.3.2   **HL (Hypotenuse-Leg) Congruence Theorem** If the hypotenuse and a leg of a right triangle are congruent to the hypotenuse and corresponding leg of another right triangle, then the two triangles are congruent.

Thm 4.4.1   **Isosceles Triangle Theorem** If two sides of a triangle are congruent, then the angles opposite those sides are congruent.

Thm 4.4.2   **Converse of the Isosceles Triangle Theorem** If two angles of a triangle are congruent, then the sides opposite those angles are congruent.

Cor 4.4.3   **Corollary** The measure of each angle of an equilateral triangle is 60°.

Cor 4.4.4   **Corollary** The bisector of the vertex angle of an isosceles triangle is the perpendicular bisector of the base.

Thm 4.5.1   **Theorem** A diagonal of a parallelogram divides the parallelogram into two congruent triangles.

Thm 4.5.2   **Theorem** The opposite sides of a parallelogram are congruent.

Thm 4.5.3   **Theorem** The opposite angles of a parallelogram are congruent.

Thm 4.5.4   **Theorem** Consecutive angles of a parallelogram are supplementary.

Thm 4.5.5   **Theorem** The diagonals of a parallelogram bisect each other.

Thm 4.5.6   **Theorem** A rhombus is a parallelogram.

Thm 4.5.7   **Theorem** A rectangle is a parallelogram.

Thm 4.5.8   **Theorem** The diagonals of a rhombus are perpendicular.

Thm 4.5.9   **Theorem** The diagonals of a rectangle are congruent.

Thm 4.5.10   **Theorem** The diagonals of a kite are perpendicular.

Thm 4.5.11   **Theorem** A square is a rectangle.

Thm 4.5.12   **Theorem** A square is a rhombus.

Thm 4.5.13   **Theorem** The diagonals of a square are congruent and are the perpendicular bisectors of each other.

Thm 4.6.1   **Theorem** If two pairs of opposite sides of a quadrilateral are congruent, then the quadrilateral is a parallelogram.

Thm 4.6.2   **Theorem** If one pair of opposite sides of a quadrilateral are parallel and congruent, then the quadrilateral is a parallelogram.

Thm 4.6.3   **Theorem** If the diagonals of a quadrilateral bisect each other, then the quadrilateral is a parallelogram.

Thm 4.6.4   **Theorem** If one angle of a parallelogram is a right angle, then the parallelogram is a rectangle.

Thm 4.6.5   **Housebuilder Theorem** If the diagonals of a parallelogram are congruent, then the parallelogram is a rectangle.

Thm 4.6.6   **Theorem** If one pair of adjacent sides of a parallelogram are congruent, then the quadrilateral is a rhombus.

Thm 4.6.7   **Theorem** If the diagonals of a parallelogram bisect the angles of the parallelogram, then the parallelogram is a rhombus.

Thm 4.6.8   **Theorem** If the diagonals of a parallelogram are perpendicular, then the parallelogram is a rhombus.

Thm 4.6.9   **Triangle Midsegment Theorem** A midsegment of a triangle is parallel to a side of the triangle, and its length is equal to half the length of that side.

Post 4.8.1   **Betweenness Postulate** Given three points P, Q, and R, if $PQ + QR = PR$, then Q is between P and R on a line.

Thm 4.8.2   **Triangle Inequality Theorem** The sum of the lengths of any two sides of a triangle is greater than the length of the third side.

Def 5.1.1   **Perimeter** The **perimeter** of a closed plane figure is the distance around the figure.

Def 5.1.2   **Area** The **area** of a closed plane figure is the number of non-overlapping squares of a given size that will exactly cover the interior of the figure.

Post 5.1.3   **Postulate: Sum of Areas** If a figure is composed of non-overlapping regions A and B, then the area of the figure is the sum of the areas of regions A and B.

Thm 5.1.4   **Perimeter of a Rectangle** The perimeter of a rectangle with base b and height h is $P = 2b + 2h$.

Post 5.1.5   **Postulate: Area of a Rectangle** The area of a rectangle with base b and height h is $A = bh$.

Def 5.2.1   **Parts of a Triangle** Any side of a triangle can be called the **base of the triangle**. The **altitude of the triangle** is a perpendicular segment from a vertex to a line containing the base of the triangle. The **height of the triangle** is the length of the altitude.

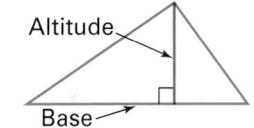

Thm 5.2.2   **Area of a Triangle** The area of a triangle with base b and height h is $A = \frac{1}{2}bh$.

Def 5.2.3   **Parts of a Parallelogram** Any side of a parallelogram can be called the **base of the parallelogram**. An **altitude of a parallelogram** is a perpendicular segment from a line containing the base to a line containing the side opposite the base. The **height of the parallelogram** is the length of the altitude.

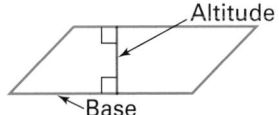

Thm 5.2.4   **Area of a Parallelogram** The area of a parallelogram with base b and height h is $A = bh$.

Def 5.2.5   **Parts of a Trapezoid** The two parallel sides of a trapezoid are known as the **bases of the trapezoid**. The two nonparallel sides are called the **legs of the trapezoid**. An **altitude of a trapezoid** is a perpendicular segment from a line containing one base to a line containing the other base. The **height of a trapezoid** is the length of an altitude.

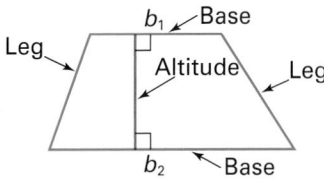

Thm 5.2.6 **Area of a Trapezoid** The area of a trapezoid with bases b_1 and b_2 and height h is $A = \frac{1}{2}(b_1 + b_2)h$.

Def 5.3.1 **Circle** A **circle** is the set of all points in a plane that are the same distance, r, from a given point in the plane known as the **center** of the circle. The distance r is known as the **radius of the circle**. The distance $d = 2r$ is known as the **diameter of the circle**.

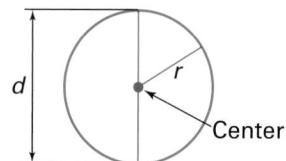

Thm 5.3.2 **Circumference of a Circle** The circumference of a circle with diameter d and radius r is $C = \pi d$ or $C = 2\pi r$.

Thm 5.3.3 **Area of a Circle** The area of a circle with radius r is $A = \pi r^2$.

Thm 5.4.1 **Pythagorean Theorem** For any right triangle, the square of the length of the hypotenuse is equal to the sum of the squares of the lengths of the legs; that is, $c^2 = a^2 + b^2$.

Thm 5.4.2 **Converse of the Pythagorean Theorem** If the square of the length of one side of a triangle equals the sum of the squares of the lengths of the other two sides, then the triangle is a right triangle.

Thm 5.4.3 **Pythagorean Inequalities** For any triangle ABC, with c as the length of the longest side:

  If $c^2 = a^2 + b^2$, then $\triangle ABC$ is a right triangle.
  If $c^2 > a^2 + b^2$, then $\triangle ABC$ is an obtuse triangle.
  If $c^2 < a^2 + b^2$, then $\triangle ABC$ is an acute triangle.

Thm 5.5.1 **45-45-90 Triangle Theorem** In any 45-45-90 triangle, the length of the hypotenuse is $\sqrt{2}$ times the length of a leg.

Thm 5.5.2 **30-60-90 Triangle Theorem** In any 30-60-90 triangle, the length of the hypotenuse is 2 times the length of the shorter leg, and the length of the longer leg is $\sqrt{3}$ times the length of the shorter leg.

Thm 5.5.3 **Area of a Regular Polygon** The area of a regular polygon with apothem a and perimeter p is $A = \frac{1}{2}ap$.

Thm 5.6.1 **Distance Formula** On a coordinate plane, the distance between two points (x_1, y_1) and (x_2, y_2) is $d = \sqrt{(x_2 - x_1)^2 + (y_2 - y_1)^2}$.

Def 6.2.1 **Polyhedron** A **polyhedron** is a closed spatial figure composed of polygons, called the **faces** of the polyhedron. The intersections of the faces are the **edges** of the polyhedron. The vertices of the faces are the **vertices** of the polyhedron.

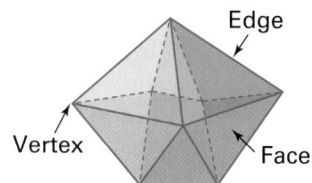

Def 6.2.2 **Parallel Planes** Two planes are parallel if and only if they do not intersect.

Def 6.2.3 **A Line Perpendicular to a Plane** A line is perpendicular to a plane at a point P if and only if it is perpendicular to every line in the plane that passes through P.

Def 6.2.4 **A Line Parallel to a Plane** A line that is not contained in a given plane is parallel to the plane if and only if it is parallel to a line contained in the plane.

Def 6.2.5 **Dihedral Angle** A **dihedral angle** is the figure formed by two half-planes with a common edge. Each half-plane is called a **face** of the angle, and the common edge of the half-planes is called the **edge** of the angle.

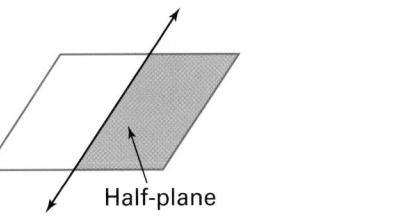
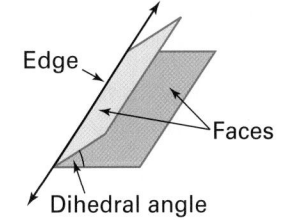

Def 6.2.6 **Measure of a Dihedral Angle** The measure of a dihedral angle is the measure of an angle formed by two rays that are on the faces and that are perpendicular to the edge.

Thm 6.3.1 **Diagonal of a Right Rectangular Prism** The length of the diagonal, d, of a right rectangular prism is given by $d = \sqrt{\ell^2 + w^2 + h^2}$.

Thm 6.4.1 **Distance Formula in Three Dimensions** The distance, d, between the points (x_1, y_1, z_1) and (x_2, y_2, z_2) in space is given by $d = \sqrt{(x_2 - x_1)^2 + (y_2 - y_1)^2 + (z_2 - z_1)^2}$.

Thm 6.4.2 **Midpoint Formula in Three Dimensions** The midpoint of a segment with endpoints at (x_1, y_1, z_1) and (x_2, y_2, z_2) in space is given by
$$\left(\frac{x_1 + x_2}{2}, \frac{y_1 + y_2}{2}, \frac{z_1 + z_2}{2} \right).$$

Thm 6.6.1 **Theorem: Sets of Parallel Lines** In a perspective drawing, all lines that are parallel to each other, but not parallel to the picture plane, will seem to meet at a single point known as the vanishing point.

Thm 6.6.2 **Theorem: Lines Parallel to the Ground** In a perspective drawing, a line that is on the plane of the ground and is not parallel to the picture plane will meet the horizon of the drawing. Any line parallel to this line will meet the horizon at the same point.

Thm 7.1.1 **Surface Area and Volume of a Right Rectangular Prism** The surface area, S, and volume, V, of a right rectangular prism with length ℓ, width w, and height h are $S = 2\ell w + 2wh + 2\ell h$ and $V = \ell wh$.

Thm 7.1.2 **Surface Area and Volume of a Cube** The surface area, S, and volume, V, of a cube with side length s are $S = 6s^2$ and $V = s^3$.

Thm 7.2.1 **Surface Area of a Right Prism** The surface area, S, of a right prism with lateral area L, base area B, perimeter p, and height h is $S = L + 2B$ or $S = hp + 2B$.

Thm 7.2.2 **Cavalieri's Principle** If two solids have equal heights and the cross sections formed by every plane parallel to the bases of both have equal areas, then the two solids have equal volumes.

Thm 7.2.3 **Volume of a Prism** The volume, V, of a prism with height h and base area B is $V = Bh$.

Thm 7.3.1 **Surface Area of a Regular Pyramid** The surface area, S, of a regular pyramid with lateral area L, base area B, perimeter of the base p, and slant height ℓ is $S = L + B$ or $S = \frac{1}{2}\ell p + B$.

Thm 7.3.2 **Volume of a Pyramid** The volume, V, of a pyramid with height h and base area B is $V = \frac{1}{3}Bh$.

Thm 7.4.1 **Surface Area of a Right Cylinder** The surface area, S, of a right cylinder with lateral area L, base area B, radius r, and height h is $S = L + 2B$ or $S = 2\pi rh + 2\pi r^2$.

Thm 7.4.2   **Volume of a Cylinder**   The volume, V, of a cylinder with radius r, height h, and base area B is $V = Bh$ or $V = \pi r^2 h$.

Thm 7.5.1   **Surface Area of a Right Cone**   The surface area, S, of a right cone with lateral area L, base of area B, radius r, and slant height ℓ is $S = L + B$ or $S = \pi r \ell + \pi r^2$.

Thm 7.5.2   **Volume of a Cone**   The volume, V, of a cone with radius r, height h, and base area B is $V = \frac{1}{3}Bh$ or $V = \frac{1}{3}\pi r^2 h$.

Thm 7.6.1   **Volume of a Sphere**   The volume, V, of a sphere with radius r is $V = \frac{4}{3}\pi r^3$.

Thm 7.6.2   **Surface Area of a Sphere**   The surface area, S, of a sphere with radius r is $S = 4\pi r^2$.

Def 8.2.1   **Similar Figures**   Two figures are similar if and only if one is congruent to the image of the other by a dilation.

Post 8.2.2   **Polygon Similarity Postulate**   Two polygons are similar if and only if there is a way of setting up a correspondence between their sides and angles such that the following conditions are met:
1. All pairs of corresponding angles are congruent.
2. All pairs of corresponding sides are proportional.

Thm 8.2.3   **Cross-Multiplication Property of Proportions**   For real numbers a, b, c, and d such that b and $d \neq 0$, if $\frac{a}{b} = \frac{c}{d}$, then $ad = bc$.

Thm 8.2.4   **Reciprocal Property of Proportions**   For real numbers a, b, c, and d such that a, b, c, and $d \neq 0$, if $\frac{a}{b} = \frac{c}{d}$, then $\frac{b}{a} = \frac{d}{c}$.

Thm 8.2.5   **Exchange Property of Proportions**   For real numbers a, b, c, and d such that a, b, c, and $d \neq 0$, if $\frac{a}{b} = \frac{c}{d}$, then $\frac{a}{c} = \frac{b}{d}$.

Thm 8.2.6   **"Add-One" Property of Proportions**   For real numbers a, b, c, and d such that b and $d \neq 0$, if $\frac{a}{b} = \frac{c}{d}$, then $\frac{a+b}{b} = \frac{c+d}{d}$.

Post 8.3.1   **AA (Angle-Angle) Similarity Postulate**   If two angles of one triangle are congruent to two angles of another triangle, then the triangles are similar.

Thm 8.3.2   **SSS (Side-Side-Side) Similarity Theorem**   If the three sides of one triangle are proprtional to the three sides of another triangle, then the triangles are similar.

Thm 8.3.3   **SAS (Side-Angle-Side) Similarity Theorem**   If two sides of one triangle are proprtional to two sides of another triangle and their included angles are congruent, then the triangles are similar.

Thm 8.4.1   **Side-Splitting Theorem**   A line parallel to one side of the triangle divides the other two sides proportionally.

Cor 8.4.2   **Two-Transversal Proportionality Corollary**   Three or more parallel lines divide two intersecting transversals proportionally.

Thm 8.5.1   **Proportional Altitudes Theorem**   If two triangles are similar, then their corresponding altitudes have the same ratio as their corresponding sides.

Thm 8.5.2   **Proportional Medians Theorem**   If two triangles are similar, then their corresponding medians have the same ratio as their corresponding sides.

Thm 8.5.3   **Proportional Angle Bisectors Theorem**   If two triangles are similar, then their corresponding angle bisectors have the same ratio as the corresponding sides.

Thm 8.5.4   **Proportional Segments Theorem**   An angle bisector of a triangle divides the opposite side into two segments that have the same ratio as the other two sides.

Def 9.1.1 **Circle** A circle is the set of all points in a plane that are equidistant from a given point in the plane known as the center of the circle. A **radius** (plural, *radii*) is a segment from the center of the circle to a point on the circle. A **chord** is a segment whose endpoints line on a circle. A **diameter** is a chord that contains the center of a circle.

Def 9.1.2 **Central Angle and Intercepted Arc** A **central angle** of a circle is an angle in the plane of a circle whose vertex is the center of the circle. An arc whose endpoints lie on the sides of the angle and whose other points lie in the interior of the angle is the **intercepted arc** of the central angle.

Def 9.1.3 **Degree Measure of Arcs** The degree measure of a minor arc is the measure of its central angle. The degree measure of a major arc is 360° minus the degree measure of its minor arc. The degree measure of a semicircle is 180°.

Thm 9.1.4 **Arc Length** If r is the radius of a circle and M is the degree measure of an arc of the circle, then the length, L, of the arc is given by the following: $L = \frac{M}{360°}(2\pi r)$.

Thm 9.1.5 **Chords and Arcs Theorem** In a circle, or in congruent circles, the arcs of congruent chords are congruent.

Thm 9.1.6 **Converse of the Chords and Arcs Theorem** In a circle, or in congruent circles, the chords of congruent arcs are congruent.

Def 9.2.1 **Secants and Tangents** A **secant** to a circle is a line that intersects the circle at two points. A **tangent** is a line in the plane of the circle that intersects the circle at exactly one point, which is known as the **point of tangency**.

Thm 9.2.2 **Tangent Theorem** If a line is tangent to a circle, then the line is perpendicular to a radius of the circle drawn to the point of tangency.

Thm 9.2.3 **Radius and Chord Theorem** A radius that is perpendicular to a chord of a circle bisects the chord.

Thm 9.2.4 **Converse of the Tangent Theorem** If a line is perpendicular to a radius of a circle at its endpoint on the circle, then the line is tangent to the circle.

Thm 9.2.5 **Theorem** The perpendicular bisector of a chord passes through the center of the circle.

Thm 9.3.1 **Inscribed Angle Theorem** The measure of an angle inscribed in a circle is equal to one-half the measure of the intercepted arc.

Cor 9.3.2 **Right Angle Corollary** If an inscribed angle intercepts a semicircle, then the angle is a right angle.

Cor 9.3.3 **Arc-Intercept Corollary** If two inscribed angles intercept the same arc, then they have the same measure.

Thm 9.4.1 **Theorem** If a tangent and a secant (or a chord) intersect on a circle at the point of tangency, then the measure of the angle formed is one-half the measure of its intercepted arc.

Thm 9.4.2 **Theorem** The measure of an angle formed by two secants or chords that intersect in the interior of a circle is one-half the sum of the measures of the arcs intercepted by the angle and its vertical angle.

Thm 9.4.3 **Theorem** The measure of an angle formed by two secants that intersect in the exterior of a circle is one-half the difference of the measures of the intercepted arcs.

Thm 9.4.4   **Theorem** The measure of a secant-tangent angle with its vertex outside the circle is one-half the difference of the measures of the intercepted arcs.

Thm 9.4.5   **Theorem** The measure of a tangent-tangent angle with its vertex outside the circle is one-half the difference of the measures of the intercepted arcs, or the measure of the major arc minus 180°.

Thm 9.5.1   **Theorem** If two segments are tangent to a circle from the same external point, then the segments are of equal length.

Thm 9.5.2   **Theorem** If two secants intersect outside a circle, then the product of the lengths of one secant segment and its external segment equals the product of the lengths of the other secant segment and its external segment. (Whole × Outside = Whole × Outside)

Thm 9.5.3   **Theorem** If a secant and a tangent intersect outside a circle, then the product of the lengths of the secant segment and its external segment equals the length of the tangent segment squared. (Whole × Outside = Tangent Squared)

Thm 9.5.4   **Theorem** If two chords intersect inside a circle, then the product of the lengths of the segments of one chord equals the product of the lengths of the segments of the other chord.

Def 10.1.1   **Tangent Ratio** For a given acute angle $\angle A$ with a measure of $\theta°$, the **tangent** of $\angle A$, or tan θ, is the ratio of the length of the leg opposite $\angle A$ to the length of the leg adjacent to $\angle A$ in any right triangle with A as one vertex, or tan $\theta = \frac{\text{opposite}}{\text{adjacent}}$.

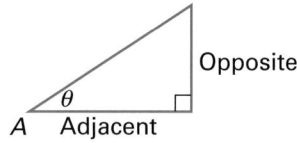

Def 10.2.1   **Sine Ratio** For a given angle $\angle A$ with a measure of $\theta°$, the **sine** of $\angle A$, or sin θ, is the ratio of the length of the leg opposite A to the length of the hypotenuse in a right triangle with A as one vertex, or sin $\theta = \frac{\text{opposite}}{\text{hypotenuse}}$.

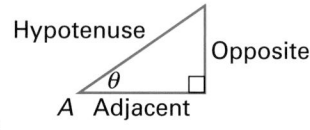

Def 10.2.2   **Cosine Ratio** The cosine of $\angle A$, or cos θ, is the ratio of the length of the leg adjacent to A to the length of the hypotenuse, or cos $\theta = \frac{\text{adjacent}}{\text{hypotenuse}}$.

Thm 10.2.3   **Identity** tan $\theta = \frac{\sin \theta}{\cos \theta}$

Thm 10.2.4   **Identity** $(\sin \theta)^2 + (\cos \theta)^2 = 1$

Def 10.3.1   **Unit Circle Definition of Sine and Cosine** Let θ be a rotation angle. Then **sin θ** is the y-coordinate of the image of point $P(1, 0)$ rotated $\theta°$ about the origin, and **cos θ** is the x-coordinate.

Thm 10.4.1   **The Law of Sines** For any triangle $\triangle ABC$ with sides a, b, and c:
$$\frac{\sin A}{a} = \frac{\sin B}{b} = \frac{\sin C}{c}$$

Thm 10.5.1   **The Law of Cosines** For any triangle $\triangle ABC$ with sides a, b, and c:
$$a^2 = b^2 + c^2 - 2bc \cos A \qquad b^2 = a^2 + c^2 - 2ac \cos B$$
$$c^2 = a^2 + b^2 - 2ab \cos C$$

Thm 11.3.1   **Theorem** A graph contains an Euler path if and only if there are at most two odd vertices.

Thm 11.4.1   **Jordan Curve Theorem** Every simple closed curve in a plane divides the plane into two distinct regions, the inside and the outside. Every curve that connects a point on the inside to a point on the outside must intersect the curve.

Thm 11.4.2   **Euler's Formula** For any polyhedron with V vertices, E edges, and F faces, $V - E + F = 2$.

Def 11.7.1 **Affine Transformation** An **affine transformation** transforms each preimage point P in a plane to an image point P' in such a way that
1. collinear points are transformed into collinear points,
2. straight lines are transformed into straight lines,
3. intersecting lines are transformed into intersecting lines, and
4. parallel lines are transformed into parallel lines.

Def 11.7.2 **Main Features of Projective Geometry**
1. Projective geometry is the study of the properties of figures that do not change under a projection.
2. There is no concept of size, measurement, or congruence.
3. Its theorems state facts about things such as the positions of points and the intersections of lines.
4. An unmarked straightedge is the only tool allowed for drawing figures.

Thm 11.7.3 **The Theorem of Pappus** If A_1, B_1, and C_1 are three distinct points on one line and A_2, B_2, and C_2 are three distinct points on a second line, then the intersections of $\overline{A_1B_2}$ and $\overline{A_2B_1}$, of $\overline{A_1C_2}$ and $\overline{A_2C_1}$, and of $\overline{B_1C_2}$ and $\overline{B_2C_1}$ are collinear.

Thm 11.7.4 **The Theorem of Desargues** If one triangle is a projection of another triangle, then the intersections of the lines containing the corresponding sides of the two triangles are collinear.

Thm 12.1.1 **Argument Form: *Modus Ponens***
If p then q
p
Therefore, q

Thm 12.1.2 **Argument Form: *Modus Tollens***
If p then q
$\sim q$
Therefore, $\sim p$

Thm 12.1.3 **Invalid Form: Affirming the Consequent**
If p then q
q
Therefore, p

Thm 12.1.4 **Invalid Form: Denying the Antecedent**
If p then q
$\sim p$
Therefore, $\sim q$

Thm 12.4.1 **Proof by Contradiction** To prove statement s, assume $\sim s$. Then the following argument form is valid:
If $\sim s$ then (t AND $\sim t$)
Therefore, s

Table of Trigonometric Ratios

Angle	sin	cos	tan	Angle	sin	cos	tan
0°	0.0000	1.0000	0.0000	45°	0.7071	0.7071	1.0000
1°	0.0175	0.9998	0.0175	46°	0.7193	0.6947	1.0355
2°	0.0349	0.9994	0.0349	47°	0.7314	0.6820	1.0724
3°	0.0523	0.9986	0.0524	48°	0.7431	0.6691	1.1106
4°	0.0698	0.9976	0.0699	49°	0.7547	0.6561	1.1504
5°	0.0872	0.9962	0.0875	50°	0.7660	0.6428	1.1918
6°	0.1045	0.9945	0.1051	51°	0.7771	0.6293	1.2349
7°	0.1219	0.9925	0.1228	52°	0.7880	0.6157	1.2799
8°	0.1392	0.9903	0.1405	53°	0.7986	0.6018	1.3270
9°	0.1564	0.9877	0.1584	54°	0.8090	0.5878	1.3764
10°	0.1736	0.9848	0.1763	55°	0.8192	0.5736	1.4281
11°	0.1908	0.9816	0.1944	56°	0.8290	0.5592	1.4826
12°	0.2079	0.9781	0.2126	57°	0.8387	0.5446	1.5399
13°	0.2250	0.9744	0.2309	58°	0.8480	0.5299	1.6003
14°	0.2419	0.9703	0.2493	59°	0.8572	0.5150	1.6643
15°	0.2588	0.9659	0.2679	60°	0.8660	0.5000	1.7321
16°	0.2756	0.9613	0.2867	61°	0.8746	0.4848	1.8040
17°	0.2924	0.9563	0.3057	62°	0.8829	0.4695	1.8807
18°	0.3090	0.9511	0.3249	63°	0.8910	0.4540	1.9626
19°	0.3256	0.9455	0.3443	64°	0.8988	0.4384	2.0503
20°	0.3420	0.9397	0.3640	65°	0.9063	0.4226	2.1445
21°	0.3584	0.9336	0.3839	66°	0.9135	0.4067	2.2460
22°	0.3746	0.9272	0.4040	67°	0.9205	0.3907	2.3559
23°	0.3907	0.9205	0.4245	68°	0.9272	0.3746	2.4751
24°	0.4067	0.9135	0.4452	69°	0.9336	0.3584	2.6051
25°	0.4226	0.9063	0.4663	70°	0.9397	0.3420	2.7475
26°	0.4384	0.8988	0.4877	71°	0.9455	0.3256	2.9042
27°	0.4540	0.8910	0.5095	72°	0.9511	0.3090	3.0777
28°	0.4695	0.8829	0.5317	73°	0.9563	0.2924	3.2709
29°	0.4848	0.8746	0.5543	74°	0.9613	0.2756	3.4874
30°	0.5000	0.8660	0.5774	75°	0.9659	0.2588	3.7321
31°	0.5150	0.8572	0.6009	76°	0.9703	0.2419	4.0108
32°	0.5299	0.8480	0.6249	77°	0.9744	0.2250	4.3315
33°	0.5446	0.8387	0.6494	78°	0.9781	0.2079	4.7046
34°	0.5592	0.8290	0.6745	79°	0.9816	0.1908	5.1446
35°	0.5736	0.8192	0.7002	80°	0.9848	0.1736	5.6713
36°	0.5878	0.8090	0.7265	81°	0.9877	0.1564	6.3138
37°	0.6018	0.7986	0.7536	82°	0.9903	0.1392	7.1154
38°	0.6157	0.7880	0.7813	83°	0.9925	0.1219	8.1443
39°	0.6293	0.7771	0.8098	84°	0.9945	0.1045	9.5144
40°	0.6428	0.7660	0.8391	85°	0.9962	0.0872	11.4301
41°	0.6561	0.7547	0.8693	86°	0.9976	0.0698	14.3007
42°	0.6691	0.7431	0.9004	87°	0.9986	0.0523	19.0811
43°	0.6820	0.7314	0.9325	88°	0.9994	0.0349	28.6363
44°	0.6947	0.7193	0.9657	89°	0.9998	0.0175	57.2900
45°	0.7071	0.7071	1.0000	90°	1.0000	0.0000	∞

Glossary

acute angle An angle whose measure is less than 90°. (29)

acute triangle A triangle with three acute angles. (654)

adjacent angles Two angles in a plane that share a common vertex and a common side but have no interior points in common. (101)

affine transformation A transformation in which all preimage points are mapped to image points in such a way that collinear points, straight lines, intersecting lines, and parallel lines are transformed as such. (747)

alternate exterior angles Two nonadjacent exterior angles that lie on opposite sides of a transversal. (156)

alternate interior angles Two nonadjacent interior angles that lie on opposite sides of a transversal. (156)

altitude of a cone A segment from the vertex perpendicular to the plane of the base. (460)

altitude of a cylinder A segment that has endpoints in the planes containing the bases and is perpendicular to both planes. (453)

altitude of a parallelogram A segment from a line containing the base to a line containing the side opposite the base that is perpendicular to both lines. (305)

altitude of a prism A segment that has endpoints in the planes containing the bases and is perpendicular to both planes. (437)

altitude of a pyramid A segment from the vertex perpendicular to the plane of the base. (445)

altitude of a trapezoid A segment from a line containing one base to the line containing the other base, which is perpendicular to both lines. (306)

altitude of a triangle A segment from a vertex perpendicular to the line containing the opposite side. (303)

angle A figure formed by two rays with a common endpoint. (11)

angle bisector A ray that divides an angle into two congruent angles. (38)

angle of rotation A figure formed by a rotating ray and a stationary reference ray. (647)

annulus The region between two circles in a plane that have the same center but different radii. (469)

apothem A segment from the center of a regular polygon perpendicular to a side of the polygon. (334)

arc An unbroken part of a circle. (566)

arc length The measure of the arc of a circle in terms of linear units, such as centimeters. (567)

arc measure The measure of an arc in a circle in terms of degrees. (567)

area The number of nonoverlapping unit squares of a given size that will exactly cover the interior of a figure. (295)

argument A sequence of statements that lead to a logical conclusion. (768)

axis of cylinder The segment joining the centers of the two bases. (453)

axis of symmetry A line that divides a planar figure into two congruent reflected halves. (139)

base angles of an isosceles triangle The angles whose vertices are the endpoints of the base of an isosceles triangle. (237)

base edge of a pyramid An edge that is part of the base of a pyramid; each lateral face has one edge in common with the base. (445)

base of a cone The circular face of the cone. (460)

base of a parallelogram Any side of a parallelogram. (305)

base of a prism The faces formed by the polygonal region and its image. (388)

base of a pyramid The polygonal face that is opposite the vertex. (445)

bases of a trapezoid The two parallel sides of a trapezoid. (306)

base of a triangle Any side of a triangle. (303)

base of isosceles triangle The side opposite the vertex angle. (237)

bases of a cylinder The faces formed by the circular region and its translated image. (453)

betweenness Given three points, A, B, and C, if $AB + BC = AC$, then A, B, and C are collinear and B is between A and C. (273)

biconditional A statement using "if and only if." (99)

binary number system A number system based on the digits 0 and 1. (798)

blocks The grid units in taxicab geometry. (706)

Cantor's dust A one-dimensional fractal created from a segment by removing the center one-third of segments at all levels of the structure. (740)

center of a circle The point inside a circle that is equidistant from all points on the circle. (314)

center of a regular polygon The point that is equidistant from all vertices of a polygon. (139)

center of dilation The point in a dilation through which every line connecting a preimage point to an image point passes. (500)

center of projection The central point in a class of transformations known as central projections. (749)

central angle of a circle An angle formed by two rays originating from the center of a circle. (566)

central angle of a regular polygon An angle whose vertex is the center of the polygon and whose sides pass through adjacent vertices. (139)

central projections A class of transformations in which every projected point of an image lies on a ray containing the center of projection and a point in the image. (749)

centroid The point where the three medians of a triangle intersect. (47)

chord A segment whose endpoints lie on a circle. (564)

circle The set of points in a plane that are equidistant from a given point known as the center of the circle. (314)

circumcenter The center of a circumscribed circle; the point where the three perpendicular bisectors of the sides of a triangle intersect; it is equidistant from the three vertices of the triangle. (45)

circumference The distance around a circle. (315)

circumscribed circle A circle that is drawn around the outside a triangle and contains all three vertices; a circle is circumscribed about a polygon if each vertex of the polygon lies on the circle. (44)

collinear Lying on the same line. (10)

complementary angles Two angles whose measures have a sum of 90°. (28)

compound statement A statement formed when two statements are connected by AND or OR. (776)

concave polygon A polygon that is not convex. (177)

conclusion The phrase following the word *then* in a conditional statement; the final statement of an argument. (90)

concurrent Literally, "running together"; of three or more lines, intersecting at a single point. (45)

conditional statement A statement that can be written in the form "If *p*, then *q*," where *p* is the hypothesis and *q* is the conclusion. (90)

cone A three-dimensional figure that consists of a circular base and a curved lateral surface that connects the base to a single point not in the plane of the base, called the vertex. (460)

congruence The relationship between figures having the same shape and same size; congruent segments are segments that match exactly. (19)

congruent polygons Two polygons are congruent if and only if there is a correspondence between their sides and angles such that each pair of corresponding angles is congruent and each pair of corresponding sides is congruent. (212)

conjecture A statement that is believed to be true. (36)

conjunction A compound statement that uses the word *AND*. (776)

contraction A dilation in which the preimage is reduced in size. (500)

contradiction A contradiction asserts that a statement and its negation are both true. (792)

converse of a conditional The statement formed by interchanging the hypothesis and conclusion of a conditional statement. (92)

convex polygon A polygon in which any line segment connecting two points of the polygon has no part outside the polygon. (177)

coordinate of a point The real number represented by a point on a number line. (17)

coordinate plane A grid formed by two or more coordinatized lines, known as *axes*, that intersect at right angles at a point known as the *origin*. (397)

contrapositive of a conditional The statement formed by interchanging the hypothesis and conclusion of a conditional statement and negating each part. (786)

coplanar Lying in the same plane. (10)

corresponding angles Two nonadjacent angles, one interior and one exterior, that lie on the same side of a transversal. (156)

corresponding angles of a polygon Angles of a polygon that are matched up with angles of another polygon with the same number of angles. (211)

corresponding sides of a polygon Sides of a polygon that are matched up with sides of another polygon with the same number of sides. (211)

corollary of a theorem A theorem that follows directly from another theorem and that can easily be proved from that theorem. (237)

cosine In a right triangle, the ratio of the length of the side adjacent to an acute angle to the length of the hypotenuse. (639)

counterexample An example that proves that a statement, often a conjecture, is false. (92)

CPCTC Abbreviation for "corresponding parts of congruent triangles are congruent." (235)

cylinder A solid that consists of a circular region and its translated image in a parallel plane with a lateral surface connecting the circles. (453)

deductive reasoning The process of drawing conclusions by using logical reasoning in an argument. (91)

degree In a graph, the number of edges at a vertex. (713)

degree measure of arcs The measure of a minor arc is the measure of its central angle. The degree measure of a major arc is 360° minus the degree measure of its central angle. (567)

diagonal of a polygon A segment that joins two nonadjacent vertices of a polygon. (390)

diagonal of a polyhedron A segment whose endpoints are vertices of two different faces of the polyhedron. (390)

diameter A chord that passes through the center of a circle; twice the length of the radius of the circle. (314)

dihedral angle An angle formed by two half-planes with a common edge. (382)

dilation A transformation in which every point P has an image point P' such that a line connecting the two points passes through a point O, known as the *center of dilation*, and $OP' = k \bullet OP$, where k is the *scale factor* of the dilation. (498)

direction of vector The orientation of a vector, generally indicated by an arrowhead. (672)

disjunction A compound statement that uses the word *OR*. (777)

displacement vector A vector that represents the change in position of an object. (674)

distance formula In a coordinate plane, the distance, d, between two points, (x_1, y_1) and (x_2, y_2), is given by the formula $d = \sqrt{(x_2 - x_1)^2 + (y_2 - y_1)^2}$. (340)

edge Segments or curves that connect vertices of a graph. (712)

edge of a dihedral angle The common edge of the half-planes of the angle. (382)

edge of a polyhedron The segment formed by the intersection of two faces of a polyhedron. (379)

endpoint A point at an end of a segment or the starting point of a ray. (10)

equiangular polygon A polygon in which all angles are congruent. (139)

equilateral polygon A polygon in which all sides are congruent. (139)

equilateral triangle A triangle in which all three sides are congruent. (196)

Equivalence Properties of Equality or Congruence The Reflexive, Symmetric, and Transitive Properties. (110)

equivalence relation Any relation that satisfies the Reflexive, Symmetric, and Transitivie Properties. (110)

Euler characteristic The Euler number $V - E + F$; an invariant of a surface. (723)

Euler circuit An Euler path that starts and ends at the same vertex and passes through all vertices of the graph. (715)

Euler path A continuous path that travels along each edge of a graph exactly once. (712)

even vertices The vertices of a graph that have an even number of edges leading to them. (713)

exclusive OR Indicating either one or the other, but not both. (777)

expansion A dilation in which the preimage is enlarged in size. (500)

exterior angle An angle formed between one side of polygon and the extension of an adjacent side. (174)

external secant segment The portion of a secant segment that lies outside the circle. (600)

face of a dihedral angle One of the half-planes of a dihedral angle. (382)

face of a polyhedron One of the polygons that form a polyhedron. (379)

face of a prism Each flat surface of a prism. (388)

fractal A structure that is self-similar; each subdivision has the same structure as the whole and the structure of the object looks the same from any view. (738)

glide reflection A combination of a translation and a reflection. (56)

golden ratio The ratio of the long side to the short side of a golden rectangle. (698)

golden rectangle A rectangle in which the length ℓ and the width w satisfy the proportion $\frac{w}{\ell - w}$. (698)

graph A diagram of vertices and edges. (712)

great circle A circle on a sphere that divides the sphere into two equal parts. (172)

half-plane The portion of a plane that lies on one side of a line in the plane and includes the line. (382)

head-to-tail method In vector addition, finding the sum of two vectors by placing the tail of one vector at the head of the other; the vector drawn from the tail of the first vector to the head of the second is the vector sum. (674)

height The length of an altitude of a polygon. (303)

hyperbolic geometry The geometry of a surface that curves like a saddle. (731)

hypotenuse The side opposite the right angle in a right triangle. (190)

hypothesis The phrase following the word *if* in a conditional statement. (90)

identity An equation that is true for all values of the variables in the equation for which the terms in the equation are defined. (641)

image A shape that results from a transformation of a figure known as the *preimage*. (50)

indirect proof A proof in which the statement that you want to prove is assumed to be false and a contradiction or other "absurdity" is shown to follow from the assumption. (792)

incenter The center of an inscribed circle; the point where the three angle bisectors of a triangle intersect; it is equidistant from the three sides of the triangle. (45)

inclusive OR Indicating either one or the other or both. (777)

inductive reasoning Forming conjectures on the basis of observations. (118)

input-output table A table that gives the outputs of a logic gate for different input combinations. (799)

inscribed angle An angle whose vertex lies on the circle and whose sides are chords of the circle. (580)

inscribed circle An inscribed circle in a triangle is inside the triangle and touches each side at one point; a circle is inscribed in a polygon if each side of the polygon is tangent to the circle. (44)

intercepted arc An arc whose endpoints lie on the sides of an inscribed angle. (566)

intercepts The points where a line in a coordinate plane passes through the x- and y-axes or where a plane in a three-dimensional space crosses the x-, y-, and z-axes. (403)

intersect To one or more points in common. (11)

intersection point A point that two or more geometric figures have in common. (11)

invariant Properties of a figure that stay the same regardless of how the figure is deformed. (723)

inverse of a conditional The statement formed by negating both the hypothesis and conclusion of a conditional statement. (786)

inverse tangent The function used to find the angle with a given tangent ratio; abbreviated $\tan^{-1}$. (633)

isometric drawing A type of three-dimensional drawing. (372)

isosceles triangle A triangle with at least two congruent sides. (237)

iteration The repetitive application of the same rule. (738)

kite A quadrilateral with exactly two pairs of adjacent congruent sides. (310)

lateral area The sum of the areas of the lateral faces of a polyhedron. (437)

lateral edge The intersection of two lateral faces of a polyhedron. (445)

lateral edges of a prism The edges of the lateral faces of a polyhedron that are not edges of either base. (388)

lateral faces The faces of a prism or pyramid that are not bases. (388)

lateral surface The curved surface of a cylinder or cone. (453)

legs of a right triangle The sides adjacent to the right angle. (229)

legs of a trapezoid The two nonparallel sides of a trapezoid. (306)

legs of an isosceles triangle The two congruent sides of an isosceles triangle. (237)

length The length of a segment is the measure of the distance from one endpoint to the other. (17)

line An undefined term in geometry, a line is understood to be perfectly straight, contain an infinite number of points, extend infinitely in two directions, and have no thickness. (9)

linear pair of angles The two angles formed by the endpoint of a ray when the endpoint falls on a line; two adjacent angles whose noncommon sides are opposite rays. (28)

logic gate An electronic circuit that represents *not*, *and*, or *or*. (799)

logical chain A series of logically linked conditional statements. (92)

logically equivalent statements Two statements, each of which can be logically derived from the other. (730)

magnitude The numerical measure of a vector. (672)

major arc An arc of a circle that is longer than a semicircle of that circle. (566)

matrix Data arranged in rows and columns and enclosed in brackets. (682)

median A segment from a vertex to the midpoint of the opposite side in a triangle. (47)

Menger sponge A three-dimensional fractal created from a cube by removing cubes from the centers of cubic structures at all levels. (738)

midpoint of a segment The point that divides a segment into two congruent segments. (349)

midsegment of a trapezoid A line connecting the midpoints of the two nonparallel segments of a trapezoid. (309)

midsegment of a triangle A segment whose endpoints are the midpoints of two sides. (183)

minor arc An arc of a circle that is shorter than a semicircle of that circle. (566)

Möbius strip A one-sided surface formed by taking a long rectangular strip and joining the ends together after giving it half a twist. (721)

modus ponens In logic, a valid argument of the following form:

      If p then q

      p

      Therefore, q (769)

modus tollens In logic, a valid argument of the following form:

      If p then q

      Not q

      Therefore Not p (770)

negation If p is a statement, then not p is its negation. (778)

non-Euclidean geometry A system of geometry in which the Parallel Postulate does not hold. (729)

nonoverlapping Having no points in common (except for common boundary points). (295)

number line A line whose points correspond with the set of real numbers. (17)

oblique cone A cone that is not a right cone. (460)

oblique cylinder A cylinder that is not a right cylinder. (453)

oblique prism A prism that has at least one nonrectangular lateral face. (389)

obtuse angle An angle whose measure is greater than 90° and less than 180°. (29)

obtuse triangle A triangle that has one obtuse angle. (654)

octant One of the eight spaces into which a three-dimensional coordinate system is divided by the xy-, yz-, and xz-planes. (397)

odd vertices The vertices of a graph that have an odd number of edges leading to them. (713)

orthogonal Pertaining to right angles. (732)

orthographic projection A view of an object in which the line of sight is perpendicular to the plane of the picture. (374)

paragraph proof A form of proof in which one's reasoning is explained in paragraph form, as opposed to flow chart or two-column proofs. (109)

parallel Of coplanar lines or any two planes in space, that they do not meet, no matter how far they might be extended. (35)

parallelogram A quadrilateral with two pairs of parallel sides. (148)

parallelogram method A method of adding vectors. Two vectors are represented as acting on a common point, and a parallelogram is formed by adding two sides to the figure. The vector sum is a vector along the diagonal of the parallelogram, starting from the common point. (674)

parametric equations Rules for the coordinates of points in a coordinate system, given in terms of a certain *parameter*, such as the time, *t*. (404)

perimeter The distance around a closed plane figure. (294)

perpendicular bisector A line that is perpendicular to a segment at its midpoint. (38)

perpendicular lines Lines that intersect to form right angles. (35)

plane An undefined term in geometry; a plane is understood to be a flat surface that extends infinitely in all directions. (10)

point An undefined term in geometry; a point can be thought of as a dot that represents a location on a plane or in space. Geometric points have no size. (9)

point of tangency The point of intersection of a circle or sphere with a tangent line or plane. (573)

polygon A closed plane figure formed from three or more segments such that each segment intersects exactly two other segments, one at each endpoint, and no two segments with a common endpoint are collinear. (138)

polyhedron A geometric solid with polygons as faces. (379)

postulate A statement that is accepted as true without proof. (11)

preimage A shape that undergoes a motion or transformation. (50)

premise A statement which is given or accepted as true in a logical argument and is used to establish a conclusion. (768)

prism A polyhedron that consists of a polygonal region and its translated image in a parallel plane, with quadrilateral faces connecting the corresponding edges. (388)

probability A number from 0 to 1 (or from 0 to 100%) that indicates how likely an event is to occur. (353)

projected points In projective geometry, the points that lie on rays containing the center of projection and the original points. (749)

projective geometry The study of the properties of figures that do not change under projection. (749)

projective rays Rays drawn from the center of a projection. (749)

proof A convincing argument that uses logic to show that a statement is true. (82)

proof by contradiction An indirect proof in which the statement that you want to prove is assumed to be false; the assumption is shown to lead to a contradiction, which indicates that the original statement must be true rather then false. (792)

proportion A statement of the equality of two ratios. (508)

proportional sides The sides of two polygons are proportional if all of the ratios of the corresponding sides are equal. (508)

pyramid A polyhedron in which all but one of the polygonal faces intersect at a single point known as the *vertex* of the pyramid. (445)

Pythagorean triple A set of positive integers a, b, and c such that $a^2 + b^2 = c^2$. (322)

quadrilateral A polygon with four sides. (148)

radius A segment that connects the center of a circle with a point on the circle; one-half the diameter of a circle. (314)

ray A part of a line that starts at a point and extends infinitely in one direction. (10)

rectangle A quadrilateral with four right angles. (148)

reductio ad absurdum A form of argument in which an assumption is shown to lead to an absurd or impossible conclusion so that the assumption must be rejected; literally "reduction to the absurd." (792)

reflection A transformation such that every point of the preimage may be connected to its image point by a segment that (a) is perpendicular to the line or plane that is the "mirror" of the reflection and (b) has its midpoint on the mirror of the reflection. (53)

reflectional symmetry A plane figure has reflectional symmetry if its reflection image across a line coincides with the preimage, the original figure. (139)

reflex angles Angles with measures greater than 180°. (29)

Reflexive Property of Equality For any real number a, $a = a$. (110)

regular polygon A polygon that is both equilateral and equiangular. (139)

regular polyhedron A polyhedron in which all faces are congruent and the same number of polygons meet at each vertex. (379)

regular pyramid A pyramid whose base is a regular polygon and whose lateral faces are congruent isosceles triangles. (445)

remote interior angle An interior angle of a triangle that is not adjacent to a given exterior angle. (174)

resultant vector The vector that represents the sum of two given vectors. (673)

rhombus A quadrilateral with four congruent sides. (148)

right angle An angle with a measure of 90°. (29)

right cone A cone in which the altitude intersects the base at its center point. (460)

right cylinder A cylinder whose axis is perpendicular to the bases. (453)

right-handed system A three-dimensional system of coordinates, named from a mnemonic device involving the fingers of the right hand. (396)

right prism A prism in which all of the lateral faces are rectangles. (389)

rigid transformation A transformation that does not change the size or shape of a figure. (50)

rise The vertical distance between two points in a coordinate plane. (190)

rotation A transformation in which every point of the preimage is rotated by a given angle about a point (in two dimensions) or a line (in three dimensions). (51)

rotation matrix A matrix used to rotate a figure about the origin through a given angle. (682)

rotational symmetry A figure has rotational symmetry if and only if it has at least one rotation image, not counting rotation images of 0° or multiples of 360°, that coincides with the original figure. (141)

run The horizontal distance between two points in a coordinate plane. (190)

same-side interior angles Interior angles that lie on the same side of a transversal. (156)

scale factor In a transformation, the number by which the distance of the preimage from the center of dilation is multiplied to determine the distance of the image point from the center. (498)

secant A line that intersects a circle at two points. (573)

secant segment A segment that contains a chord of a circle and has one endpoint exterior to the circle and the other endpoint on the circle. (600)

sector of a circle A region of a circle bounded by two radii and their intercepted arc. (316)

segment A part of a line that begins at one point and ends at another; a segment of a line has two endpoints. (10)

segment bisector A line that divides a segment into two congruent segments. (38)

self-similarity The property, possessed by fractals, that every subdivision of the fractal has a structure similar to the structure of the whole. (738)

semicircle The arc of a circle whose endpoints are the endpoints of a diameter. (566)

sides of an angle The two rays that form an angle. (11)

Sierpinski gasket A two-dimensional fractal created from a triangle by removing triangles from the centers of triangular structures at all levels. (741)

similar figures Two figures that have the same shape, but not necessarily the same size; two figures are similar if and only if one is congruent to the image of the other by a dilation. (507)

simple closed curve A shape that is topologically equivalent to a circle; a shape that does not intersect itself. (722)

sine In a right triangle, the ratio of the length of the side opposite an acute angle to the length of the hypotenuse. (639)

skew lines Lines that are not coplanar and do not intersect. (380)

slant height In a regular pyramid, the length of an altitude of a lateral face. (445)

slope The ratio of rise to run for a segment; the slope of a nonvertical line that contains the points (x_1, y_1) and (x_2, y_2) is the ratio $\frac{y_2 - y_1}{x_2 - x_1}$. (191)

solid Closed spatial figures. (379)

solid of revolution An object formed by rotating a plane figure about an axis in space. (481)

sphere The set of points in space that are equidistant from a given point known as the center of the sphere. (469)

spherical geometry Geometry on a sphere; a geometry in which line is defined as a great circle and there are no parallel lines. (731)

square A quadrilateral with four congruent sides and four right angles. (148)

straight angle An angle with a measure of 180°. (29)

supplementary angles Two angles whose measures have a sum of 180°. (28)

surface area of a prism The sum of the areas of all faces of a prism. (438)

Symmetric Property of Equality For all real numbers a and b, if $a = b$, then $b = a$. (110)

tangent In a right triangle, the ratio of the length of the side opposite an acute angle to the length of the side adjacent to it. (631)

tangent segment A segment that is contained by a line tangent to a circle and has one of its endpoints on the circle. (600)

tangent to a circle A line in the plane of a circle that intersects a circle at a exactly one point. (573)

taxicab geometry A geometry in which points are located on a special kind of map or coordinate grid whose horizontal and vertical lines represent streets and whose coordinates are always integers. (706)

taxicab radius The distance between the center of a taxicab circle and any point on the taxicab circle. (706)

taxidistance In taxicab geometry, the smallest number of grid units that must be traveled to move from one point to another point. (706)

theorem A statement that has been proven to be true deductively. (109)

topology A branch of mathematics that studies the most basic properties of figures. (721)

topologically equivalent Able to be stretched, shrunk, or otherwise distorted into another figure without cutting, tearing, or intersecting itself or compressing a segment or curve to a point. (722)

torus A three-dimensional donut-shaped surface. (723)

trace of a plane The intersection of a plane with the *xy*-plane in a three-dimensional coordinate system. (406)

transformation The movement of a figure in a plane from its original position, the preimage, to a new position, the image. (50)

transformation equations The equations $x' = x \cos \theta - y \sin \theta$ and $y' = x \sin \theta + y \sin \theta$ are known as transformation equations. (680)

Transitive Property of Equality For all real numbers a, b, and c, if $a = b$ and $b = c$, then $a = c$. (110)

translation A transformation in which every point of the preimage moves in the same direction by the same amount to form the image. (51)

transversal A line, ray, or segment that intersects two or more coplanar lines, rays, or segments, each at a different point. (155)

trapezoid A quadrilateral with one and only one pair of parallel sides. (148)

45-45-90 triangle A right triangle whose base angles have measures of 45°. (332)

30-60-90 triangle A right triangle whose acute angles have measures of 30° and 60°. (332)

triangle rigidity A property of triangles which states that if the sides of a triangle are fixed, the triangle can have only one shape. (217)

truth functionally equivalent When two logic statements have the same truth tables. (778)

truth table A table that lists all possible combinations of truth values for a given statement or combinations of statements. (776)

two-column proof A proof in which the statements are written in the left-hand column and the reasons are given in the right-hand column. (109)

unit circle A circle with a radius of 1 centered at the origin of the coordinate plane. (648)

valid argument An argument in which the premises force a conclusion; if the premises are true, then the conclusion is true. (768)

valid conclusion The conclusion in a valid argument that follows logically from the premises. (768)

vanishing point In a perspective drawing, the point at which parallel segments of a depicted object will meet if they are extended. (410)

vector A mathematical quantity that has both magnitude (a numerical measure) and direction. (672)

vector addition The process of combining two vectors to create a resultant vector. (673)

vertex angle of an isosceles triangle The angle opposite the base of the triangle. (237)

vector sum The resultant vector created by vector addition. (674)

vertex A point where the edges of a figure intersect; plural, *vertices*. (379)

vertex of a cone The point opposite the base of the cone. (460)

vertex of an angle The point in common of the two rays that form an angle. (11)

vertical angles The opposite angles formed by two intersecting lines. (117)

Lesson 1.1, pages 9–16

Activity

CHECKPOINT ✔
1. point

CHECKPOINT ✔
2. line

CHECKPOINT ✔
3. line

CHECKPOINT ✔
4. plane

CHECKPOINT ✔
5. is in the plane

Exercises
Communicate

1. Sample answer: Everything in the real world has length, width, and thickness. Geometric points have no size at all, and geometric lines and planes have no thickness.

2. Sample answers: A corner where two walls and the ceiling meet and a speck of dust represent points. The edge where a wall and the floor meet and a pencil represent lines. A wall and a desktop represent planes.

3. There are an infinite number of lines through any 1 point.

4. There are an infinite number of planes through any 2 points.

5. The first letter represents the endpoint, so the order indicates which direction the ray is going.

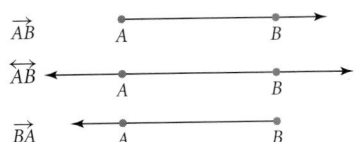

Lesson 1.2, pages 17–24

Activity

1–5. Check student constructions.

CHECKPOINT ✔
6. Answers may vary. Students should realize that different units give different measurements.

Exercises
Communicate

1. Sample answer: centimeter, meter, foot, mile

2. If the intervals are not equal, the ruler is not fair. Thus it can not be used to determine if two segments are the same length.

3. The distance between cities is a large number of centimeters. It is useful to have a large unit for greater lengths.

4. The smaller intervals make it easier to measure shorter lengths.

5. **a.** $\overline{MN}$ is not a number. The statement does not make sense.
   b. Both MN and OP are numbers. The statement makes sense.
   c. Both m$\overline{MN}$ are m$\overline{OP}$ numbers. The statement makes sense.

Lesson 1.3, pages 25–34

CRITICAL THINKING (page 25)
The starting ray $\overrightarrow{AB}$ points to the right where the bottom scale is zero, so the bottom scale must be used.

Exercises
Communicate

1. If north is 0°, east would be 90° in a clockwise direction, south 180°, and west 270°.

2. Sample answer: Complementary and corner both start with *co*. Corner is usually 90°. Supplementary and straight both start with *s* and supplementary angles form a straight line.

3. With any measure, a unit is compared to an object. When measuring length, the unit is a length such as an inch or centimeter. When measuring angles, the unit is an angle size such as a degree.

4. In the figure below, S is not in the interior of $\angle PQR$.
   m$\angle PQS$ + m$\angle SQR$ = 100° + 160° = 260°
   m$\angle PQR$ = 60°
   So, m$\angle PQS$ + m$\angle SQR$ ≠ m$\angle PQR$.

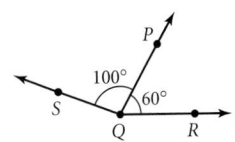

5. a. True. All right angles have the same measure, 90°.

b. False. A 60° angle and a 30° angle are both acute, but they are not congruent.

c. False. A 135° and a 150° angle are both obtuse, but they are not congruent.

6. a. The statement makes sense, because m∠A, m∠2 and 190° are all numbers.

b. The statement does not make sense, because ∠X and ∠Y are not numbers.

7. A straight angle has a measure of 180° and forms a straight line. A line segment joining a point from each side of a straight angle will lie on the angle. A reflex angle has a measure greater than 180°. A line segment joining a point from each side of a reflex angle would pass through the exterior of the angle.

Lesson 1.4, pages 35–42

Activity 1

CHECKPOINT ✔

3. 90°. Perpendicular lines.

6. the lines are parallel

Activity 2

CHECKPOINT ✔

6. Answers will vary.

Activity 3

1. perpendicular

CHECKPOINT ✔

2. $AC = BC$; equal

3. n is the angle bisector of ∠QPR.

CHECKPOINT ✔

4. The distances are equal.

Exercises

Communicate

1. The angle in the linear pairs formed on line ℓ by line m match up. If two lines intersect to form congruent linear pairs, then the lines are perpendicular.

2. 8 right angles. If two lines form 8 right angles with a third line, then the two lines are parallel.

3. The line segments lined up exactly with each other.

4. By finding the length of the perpendicular segment from S to j and from S to h.

Practice and Apply

23. Triangles that have a vertex located on the perpendicular bisector of one of the sides have at least two sides of equal length.

24. Sample answer: Trace $\overline{AB}$. Construct its perpendicular bisector m. Fold a line through A so that B falls on m. Mark this point C. △ABC will have 3 congruent sides.

Lesson 1.5, pages 43–49

Activity 1

1. They intersect at a single point.

2. They intersect at a single point.

CHECKPOINT ✔

3. intersect, point
intersect, point

Activity 2

1. Yes.

CHECKPOINT ✔

2. inscribed
circumscribed

Exercises

Communicate

1. The lines intersect, or run together, at a single point.

2. No. The inscribed circle is totally inside the triangle, so its center must be inside also.

3. Yes. The circumscribed circle is outside the triangle, so its center may be either inside or outside the triangle.

Lesson 1.6, pages 50–58

Activity 1

CHECKPOINT ✔

3. Check students' drawings.

Activity 2

CHECKPOINT ✔

3. Check students' drawings.

Activity 3

CHECKPOINT ✔ (page 52)

4. They are perpendicular, and ℓ bisects $\overline{PP'}$. Perpendicular bisector.

CHECKPOINT ✔ (page 53)

4. congruent

Exercises

Communicate

1. Translation. All points of the canoe are moving in the same direction. The direction of motion is straight ahead.

2. This is not a rotation, because the center of rotation of the ball is not a fixed point; it is moving in a straight line. It is not a translation, because the object is turning as it moves.

3. Reflection. The lake gives a mirror image of the building. The line of reflection is between the mountain and the building.

4. Rotation. All parts of each hand are moving around a point. The center of rotation is the center of the clock.

5. Rotation. The parts of the scissors move in the arc of a circle about a fixed point. The center of rotation is the pivot point of the scissors.

6. This is not a rigid transformation, because the image on the slide and the image on the screen are not congruent.

7. Sample answer: The figure on the left is the better illustration of a reflection across a line, because the line formed by the intersection of the mirror and the object is clearly the line of reflection.

Lesson 1.7, pages 59–67

Activity 1

2. Answers will vary.

3. Yes.

4. In the direction and by the amount chosen in Step 2.

CHECKPOINT ✔
5. Horizontal: $H(x, y) = (x + h, y)$
   Vertical: $V(x, y) = (x, y + v)$

Activity 2

3. Yes.

5. the x-axis and the y-axis

CHECKPOINT ✔
6. Reflection across the x-axis: $M(x, y) = (x, -y)$
   Reflection across the y-axis: $N(x, y) = (-x, y)$

Activity 3

3. Yes.

4. $(0, 0)$; $180°$

CHECKPOINT ✔
5. $R(x, y) = (-x, -y)$

Exercises

Communicate

1. To get to point A from the origin, go right 5 units and up 5 units. Thus, the coordinates are $(5, 5)$. To get to point B from the origin, go left 4 units and down 3 units. Thus, the coordinates are $(-4, -3)$.

2. The order of the coordinates matters. For $(1, 5)$, go right 1 and up 5. For $(5, 1)$, go right 5 and up 1. Coordinates are called ordered pairs because they have a pair of numbers in a significant order.

3. Algebraic rules can be used. Also, distances do not need to be measured.

4. The result is the same as a $180°$ rotation about the origin.

Practice and Apply

15. reflection across the y-axis;

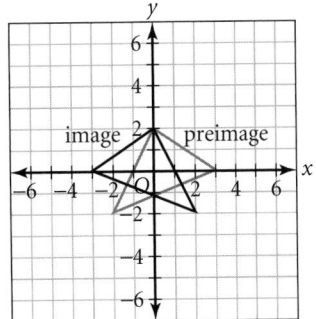

16. $180°$ rotation;

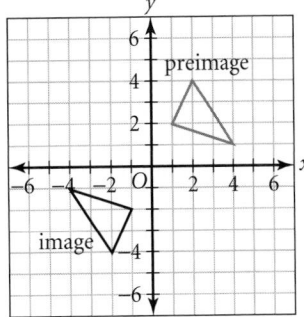

17. $180°$ rotation;

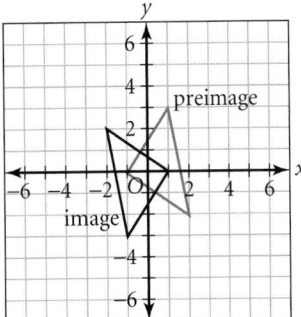

35. If K' is the result of the transformation of K under $T(x, y) = (x + h, y + k)$, then $\overleftrightarrow{KK'}$ has slope $\frac{k}{h}$.

Lesson 2.1, pages 80–89

Activity Part I

CHECKPOINT ✔ (page 81)

Sample answer: There are 64 squares on the chessboard, 32 of each color. Since one of each color is removed, there will be 31 of each color remaining. If the dominoes are lined up inside the walls of the maze, there will be no places that can't be covered. Students should convince themselves that they can always go around the corners in the maze in all situations they might encounter.

Activity Part II

n	First n odd numbers	Sum of the first n odd numbers
5	1, 3, 5, 7, 9	25
6	1, 3, 5, 7, 9, 11	36
n	1, 3, 5, 7, 9, 11, …, $2n - 1$	n^2

Notice the pattern: $1 = 1^2$, $4 = 2^2$, $9 = 3^2$, $16 = 4^2$, $25 = 5^2$, $36 = 6^2$. According to the pattern, numbers would be n^2.

CHECKPOINT ✔ (page 81)

If you represent 1 with one object, 3 with three objects, 5 with five objects, and so on, the sum of these numbers can be represented by the collection of the objects. Because the sequential collection of objects for the odd numbers can be arranged in a square, it is easy to see that the sum of the first n odd numbers is n^2.

Activity Part III

CHECKPOINT ✔ (page 81)

Answers will vary. Students should demonstrate their calculations. A proof might begin as follows: "Squares D and C have side lengths of 9 and 8, so square I must have a side lenght of 1. The side lengths of squares H and I must add up to the length of side C, so square H has a side length of 7…." When all the side lengths have been determined, the overall dimensions can be found by adding appropriately.

Exercises

Communicate

1. Sample answer: A proof is an argument used to show that a statement is true.

2. Sample answer: If the indicated squares are removed, there will be 32 light squares and 30 dark squares remaining. Each domino covers a dark and a light square so 31 dominoes could only cover 31 light squares.

3. Sample answer: A geometric solution sometimes shows the solution in a way which is easier to understand.

4. It is not a proof, because measurements are imprecise and pictures are sometimes misleading.

5. Answers may vary depending on student's earlier answer.

Lesson 2.2, pages 90–98

1. No, it does not always rain when it is cloudy.

2.

3. Sample answer: If it is cloudy, then it rains. This statement is not always true. A time when it was cloudy but not raining would fall in the rectangle but outside the circle.

4. Determine the *hypothesis* and the *conclusion*, then interchange them.

5. To disprove a conditional, find a counterexample in which the hypothesis is true but the conclusion is false.

Practice and Apply

30. If it is winter, then the days are short. If the days are short, then it is cold. If it is cold, then the birds fly south. Conclusion: If it is winter, then the birds fly south.

31. If Tim drives a car, then Tim drives too fast. If Tim drives too fast, then the police catch Tim speeding. If the police catch Tim speeding, then Tim gets a ticket. Conditional: If Tim drives a car, then Tim gets a ticket.

32. If ruskers bleer, then homblers frain. If homblers frain, then quompies plaun. If quompies plaun, then romples gleer. Conclusion: If ruskers bleer, then romples gleer.

33. If you clean your room, then you will go to the movie. If you go to the movie, then you will spend all of your money. If you spend all of your money, then you can't buy gas for your car. If you can't buy gas for your car, then you will be stranded. Conclusion: If you clean your room, then you will be stranded.

34. If a person is an independent farmer, then the person is disappearing. The conclusion does not follow, because the given statement refers to independent farmers as a group, not to individual independent farmers. In addition, as used in the given statement, "disappearing" means that the number of independent farmers is decreasing, not that they are no longer visible to the eye.

Lesson 2.3, pages 99–106

Activity I

1. Sample answer: A "house" has 5 sides, with 2 right angles which are adjacent and 3 obtuse angles.

2. Students should decide which figures fit their definition and not just the figures most similar to the original.

CHECKPOINT ✔ (Page 100)

3. Answers will vary. The answer to this will closely resemble the answer to Part 1.

TRY THIS (page 101)
Answers will vary. Definitions should accurately describe the object in clear and concise terms.

Activity 2

1. ∠1 and ∠2; ∠3 and ∠4; ∠4 and ∠5

2. Adjacent angles share a common vertex and a common side. They do not overlap.

3. Adjacent angles are angles in a plane that have their vertex and one side in common but do not overlap.

Exercises
Communicate

1. Definitions are true biconditionals, so that the conditional and the converse are both true.

2. Answers may vary. The definition should be written in the form "__ if and only if __."

3. The statement cannot be written as a true biconditional, because the converse is false. A plant with leaves is not necessarily a tree.

4. Sample answer: A figure is a blop if and only if it consists of a circle "body" with two smaller circles inside the larger circle for "eyes" and 3 small line segments for a "tail." Shapes *a* and *d* fit this definition.

5. Undefined terms are necessary, because they allow a vocabulary of "understood" terms on which to build the defined terms, or definitions.

Practice and Apply

13. a. Conditional: If an angle is a right angle, then it has a measure of 90°.
 b. Converse: If an angle has a measure of 90°, then it is a right angle.
 c. Biconditional: An angle is a right angle if and only if it has a measure of 90°.
 d. The statement is a definition, because the conditional and the converse are both true.

14. a. Conditional: If a rock is granite, then it is hard and crystalline.
 b. Converse: If a rock is hard and crystalline, then it is granite.
 c. Biconditional: A rock is granite if and only if it is hard and crystalline.
 d. The statement is not a definition because the converse is not true. A rock such as quartz is hard and crystalline, but is different from granite.

15. a. Conditional: If a substance is hydrogen, then it is the lightest of all known substances.
 b. Converse: If a substance is the lightest of all known substances, then it is hydrogen.
 c. Biconditional: A substance is hydrogen if and only if it is the lightest of all known substances.
 d. The statement is a definition because the conditional and the converse are both true.

16. a. Conditional: If an animal is an otter, then it is a small furry mammal with webbed feet that are used for swimming.
 b. Converse: If an animal is a small furry mammal with webbed feet that are used for swimming, then it is an otter.
 c. Biconditional: An animal is an otter if and only if it is a small furry mammal with webbed feet that are used for swimming.
 d. The statement is not a definition because the converse is not true. A beaver is a small furry mammal with webbed feet that are used for swimming, but it is not an otter.

17. ∠WVX and ∠XVY; ∠XVY and ∠YVZ; ∠WVY and ∠YVZ; ∠WVX and ∠XVZ

18. They do not share a common vertex or a common side.

19. They do not share a common side.

20. They do not share a common vertex.

21. They do not share a common vertex.

22. The angles overlap.

Lesson 2.4, pages 107–116
Exercises
Communicate

1. Postulates are necessary, because certain facts must be agreed on without proof. In a proof, statements must be justified in terms of things that are already known. If no statements are accepted without proof, then there would be no way to begin a proof without circular reasoning.

2. A postulate is accepted as true without proof. A theorem is a statement that has been proved.

3. Sample answer: If two equal things are added to two other equal things, then the results are also equal. For example: If $XY = YZ$ and $AB = BC$, then $XY + AB = YZ + BC$.

4. Sample answer: If two things are equal to a third thing, then they are equal to each other; Euclid's first common notion is the Transitive Property of Equality used with the Symmetric Property of Equality.

Lesson 2.5, pages 117–125

Activity 1

2. The vertical angles are equal in measure.

3. If two angles form a pair of vertical angles, then they have equal angle measures (or are congruent).

4. They are supplementary angles.

5. 180°; 180°

6. Substitution Property of Equality

CHECKPOINT ✔
7. Subtraction Property of Equality

Activity 2

1–3.

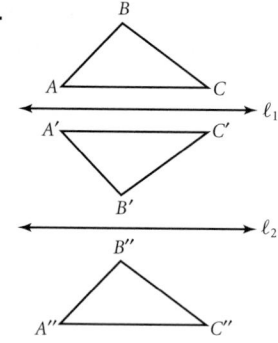

4. A vertical translation.

5. The distances are the same.

6. The distance between the lines is half the distance of the vertical translation of the vertices.

7. Yes. All of the points seemed to have moved in the same direction, because the segments connecting A and A'', B and B'', and C and C'' appear to be parallel.

CHECKPOINT ✔ (Page 117)
8. translation; twice; perpendicular

Exercises
Communicate

1. Yes. No, the statement is not true, because the integer 1,000,000,001 is greater than 1,000,000,000.

2. Inductive reasoning is based on observations. Deductive reasoning uses facts that are agreed on and logic to show the truth of statements.

3. Deductive reasoning allows you to show that a result is true for all possible cases that satisfy the hypothesis. Inductive reasoning is not acceptable in proofs because conjectures can be false.

4. A conjecture is a statement which is believed to be true based on observations. A theorem is a statement which has been proven to be true.

5. A figure can be translated by reflecting first along one line and then reflecting along another line parallel to the first line.

6. No. You cannot know a result to be true based on observations alone. The films you have seen and enjoyed are only your observations of preference and no indication that the next film will meet the same criteria.

Look Beyond

51.

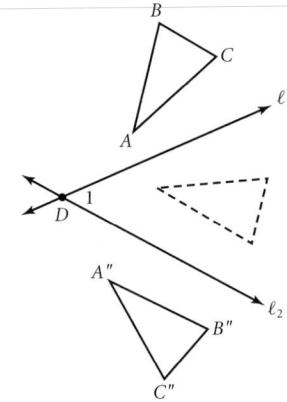

52. A rotation.

53. D is the center of rotation.

54. They are twice the measure of $\angle 1$.

55. Conjecture: The reflection of a figure across two intersecting lines is equivalent to a rotation about the point of intersection of the two lines through an angle twice the measure of the angle between the lines.

Lesson 3.1, pages 138–147

Activity 1

1. equilateral—3; isosceles—1; scalene—0

2. equilateral—all angles; isosceles—angles opposite congruent sides; scalene—none

3.

Type of triangle	Number of axes of symmetry	Number of congruent angles
equilateral	3	3
isosceles	1	2
scalene	0	0

CHECKPOINT ✔
4. segment bisector, bisector

Activity 2

3. 5

CHECKPOINT ✔
4. Yes. $\frac{360}{n}$

Exercises

Communicate

1. The picture has many axes of symmetry. One is a vertical line through the center of the Taj Mahal. Another is a horizontal line at the base of the Taj Mahal. There are also vertical axes through individual columns, windows, etc.

2. The picture shows how an image is reflected over the axes of symmetry. If the reflected top point on the Taj Mahal is the image, the preimage is the top point on the Taj Mahal. The image and preimage are equal distance from their axis of symmetry.

3. 0° and 360° rotations put a figure into its original position, no matter what the figure looks like.

4. Reflectional symmetry: 3 axes of symmetry pass through midpoints of opposite sides, and 3 axes of symmetry pass through opposite vertices. Rotational symmetry: Each central angle is 60°, so there are 60°, 120°, 180°, 240°, and 300° nontrivial rotational symmetries.

5. Each snowflake-like region has nontrivial rotational symmetries of 120° and 240°. The centers of rotation are the centers of each snowflake.

Practice and Apply

18.

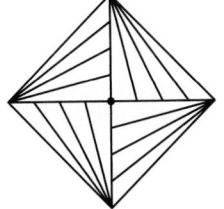

Lesson 3.2, pages 148–154

Activity 1

1. Opposite angles; opposite sides

2. The two halves of each diagonal are congruent. The conjecture does not hold if the shape is varied by dragging one of the vertices.

3. Consecutive angles are supplementary.

CHECKPOINT ✔
4. congruent; congruent; bisect each other; supplementary

Activity 2

1. Opposite angles; all sides

CHECKPOINT ✔
2. Yes. Notice that the adjacent sides of a rhombus and the diagonal connecting them form an isoceles triangle that is congruent to the isosceles triangle formed by the other two sides and the diagonal. Parallelogram.

3. perpendicular

Activity 3

1. All angles; opposite sides

CHECKPOINT ✔
2. Yes, because reflecting a rectangle about a horizontal axis of symmetry leads to the same conjectures. Parallelogram.

3. are congruent

Activity 4

1. All angles; all sides

CHECKPOINT ✔
2. All previous conjectures are true for squares as well, because reflecting a square about a horizontal axis of symmetry leads to the same conjectures. Rhombus. Rectangle. Parallelogram.

3. bisect, congruent, perpendicular

Exercises

Communicate

1. A square is a rhombus with equal angles.

2. A rectangle is a parallelogram with equal angles.

3. The drawing appears to be a cube, viewed either from below or above.

4. This definition would allow parallelograms to be considered trapezoids.

Look Beyond

52. If the diagonals of a quadrilateral are congruent, then it is a rectangle. The converse is not true.

53.

Any isosceles trapezoid has congruent diagonals, but it is not a rectangle. This shows that the converse in Exercise 52 is not true.

Lesson 3.3, pages 155–161

Activity

1. Alternate interior angles are congruent.

2. Angles 2 and 7. Alternate exterior angles are congruent.

3. Angles 4 and 6. Same-side interior angles are supplementary.

4. Angles 2 and 6, angles 3 and 7, angles 4 and 8. Corresponding angles are congruent.

CHECKPOINT ✔

5. congruent; congruent; supplementary; congruent

Exercises
Communicate

1. Sample answer: Horizontal and diagonal lines are transversals of the vertical lines. Vertical lines are transversals of the horizontal and diagonal lines.

2. ∠1 and ∠2 form a linear pair and are supplementary. Sample answers: ∠1 and ∠3, ∠3 and ∠4, ∠4 and ∠2; ∠1 and ∠4 are vertical angles and are congruent. Sample Answers: ∠2 and ∠3, ∠5 and ∠8, ∠6 and ∠7

3. Same-side exterior angles; they are supplementary. ∠2 and ∠8

4. a. No transversals, because there are only two lines.
   b. Each line is a transversal of the other two.
   c. No transversals, because these lines do not intersect.
   d. No transversals, because the three lines intersect in just one point.

Practice and Apply

34. Given: Line ℓ ∥ line m
   Line p is a transversal.
   Prove: ∠1 ≅ ∠2

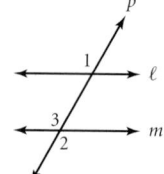

Statements	Reasons
Line ℓ ∥ line m Line p is a transversal.	Given
∠1 ≅ ∠3	Corresponding Angles Postulate
∠3 ≅ ∠2	Vertical Angles Theorem
∠1 ≅ ∠2	Transitive Property of Congruence or Substitution

35. Given: Line ℓ ∥ line m
   Line p is a transversal.
   Prove: m∠2 + m∠1 = 180°

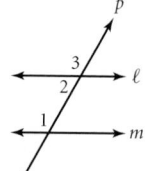

Statements	Reasons
Line ℓ ∥ line m Line p is a transversal.	Given
m∠2 + m∠3 = 180°	Linear Pair Property
m∠3 = m∠1	Corresponding Angles Postulate
m∠2 + m∠1 = 180°	Substitution Property

36. Given: Line ℓ ∥ line m
   Line p is a transversal.
   m∠1 = 90°

   Prove: m∠2 = m∠3 = m∠4 = m∠5 = m∠6 = m∠7 = m∠8 = 90°

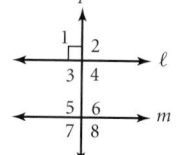

Statements	Reasons
Line ℓ ∥ line m Line p is a transversal. m∠1 = 90°	Given
m∠1 + m∠2 = 180°	Linear Pair Property
90° + m∠2 = 180°	Substitution Property
m∠2 = 90°	Subtraction Property
∠1 ≅ ∠4 and ∠2 ≅ ∠3 (m∠1 = m∠4 and m∠2 = m∠3)	Vertical Angles Theorem
∠1 ≅ ∠8 and ∠2 ≅ ∠7 (m∠1 = m∠8 and m∠2 = m∠7)	Alternate Exterior Angles Theorem
∠1 ≅ ∠5 and ∠2 ≅ ∠6 (m∠1 = m∠5 and m∠2 = m∠6)	Corresponding Angles Theorem
m∠1 = m∠4 = m∠8 = m∠5 = 90° m∠2 = m∠3 = m∠7 = m∠6 = 90°	Transitive Property of Congruence or Substitution

Lesson 3.4, pages 162–167
Exercises
Communicate

1. If corresponding angles are congruent, then the two lines cut by the transversal are parallel.

2. If same-side interior angles are supplementary, then the two lines cut by the transversal are parallel.

3. Sample answer: Since congruent corresponding angles imply parallel lines, we can draw parallel lines by constructing congruent angles with vertices at two points along a line and the angle on the same side of the line. The lines containing the rays of the angles not on the common line are parallel.

4. If the lines were parallel, then corresponding angles would be congruent. Because the indicated angles are not congruent, these lines are not parallel.

Practice and Apply

17. Given: Line $m \parallel n$
$m\angle RSU = m\angle RTU = 70°$

Prove: $RSUT$ is a parallelogram.

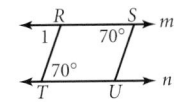

Statements	Reasons
$m \parallel n$; $m\angle RSU = m\angle RTU = 70°$	Given
$\angle RSU \cong \angle RTU$	Angle Congruence Postulate
$\angle RTU \cong \angle 1$ where $\angle 1$ is its alternate interior angle with transversal $\overline{RT}$.	Alternate Interior Angles Theorem
$\angle RSU \cong \angle 1$	Transitive Property of Congruence
$\overline{RT} \parallel \overline{SU}$	Converse of the Corresponding Angles Postulate
$RSUT$ is a parallelogram.	Definition of parallelogram

18. Given

19. Definition of supplementary angles

20. Linear Pair Property

21. Transitive Property or Substitution

22. Subtraction Property

23. Converse of the Corresponding Angles Postulate

24. Given: $\angle 1 \cong \angle 2$

Prove: $\ell_1 \parallel \ell_2$

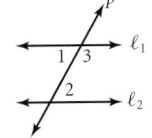

Statements	Reasons
$\angle 1 \cong \angle 2$	Given
$m\angle 1 = m\angle 2$	Angle Congruence Postulate
$m\angle 1 + m\angle 3 = 180°$	Linear Pair Property
$m\angle 2 + m\angle 3 = 180°$	Substitution Property
$\angle 2$ and $\angle 3$ are supplementary.	Definition of supplementary angles
$\ell_1 \parallel \ell_2$	Converse of the Same-side Interior Angles Theorem.

25. Given: $\angle 1 \cong \angle 2$

Prove: $\ell_1 \parallel \ell_2$

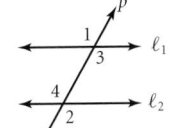

Statements	Reasons
$\angle 1 \cong \angle 2$	Given
$\angle 2 \cong \angle 4$	Vertical Angles Theorem
$\angle 1 \cong \angle 4$	Transitive Property of Congruence
$\ell_1 \parallel \ell_2$	Converse of the Corresponding Angles Postulate.

Look Beyond

49. Sample answer: Since objects at a distance appear smaller, the distance between the railroad tracks will also. "Meet at infinity" refers to the illusion of the lines meeting far away, with the full knowledge that they do not meet at any finite distance from any particular vantage point. In a way, it does *not* make sense because "infinity" is not a place at which lines could meet. On the other hand, it *does* make sense if "at infinity" is taken to mean "not any place a finite distance from here."

Lesson 3.5, pages 170–176

Activity

CHECKPOINT ✔

4. Conjecture: The sum of the angle measures equals 180°.

5. The Parallel Postulate guarantees that there is exactly one line ℓ containing C and parallel to $\overline{AB}$.

6.

$m\angle 1$	$m\angle 2$	$m\angle 3$	$m\angle 4$	$m\angle 5$	$m\angle 3 + m\angle 4 + m\angle 5$
40°	30°	110°	40°	30°	180°
20°	80°	80°	20°	80°	180°
30°	100°	50°	30°	100	180°

7. Yes. The table shows examples of 3 triangles for which the angle measures sum to 180°.

Exercises

Communicate

1. Both the torn-triangle Activity and the proof of the Triangle Sum Theorem use the idea that 3 angles fit together to form a straight line that measures 180°.

2. The torn-triangle Activity only shows that the truth of the Triangle Sum Theorem appears to be correct in a few cases. A proof is needed to show that it is true in all cases.

3. The parallel postulate allows us to assume that we can draw exactly one line ℓ which is parallel to $\overline{AB}$ and contains C.

4. Sample answer: On a globe that has a circumference of 97 cm, the distance from New York City to Bangkok is about 34 cm. The distance is estimated to be $\left(\frac{34}{97}\right)2\pi(6378) \approx 14{,}046$ km. Difficulties which may be encountered on this route include travel over mountains, ice, and water.

Lesson 3.6, pages 177–182

Activity 1

1. 180°, 180°, 180°

2. 540°

3. 540°

4.

Polygon	Number of sides	Number of triangular regions	Sum of measures of angles
triangle	3	1	180°
quadrilateral	4	2	360°
pentagon	5	3	540°
hexagon	6	4	720°
n-gon	n	$n-2$	$180°(n-2)$

CHECKPOINT ✔

5. $180°(n - 2)$

Regular Polygon	Number of sides	Sum of measures of angles	Measure of an interior angle
triangle	3	180°	60°
quadrilateral	4	360°	90°
pentagon	5	540°	108°
hexagon	6	720°	120°
n-gon	n	$180°(n-2)$	$\dfrac{180°(n-2)}{n}$

$\dfrac{180°(n - 2)}{n}$

Activity 2

1. Triangles will vary. Sum of exterior angle measures = 360°

2. 360°

3. 360°

4. 360°

5. For any n-sided polygon, the sum of exterior angle measures = 360°.

6. 540°

7. 720°

8. $(180n)°$

9.

Polygon	Number of sides	Sum of (ext. and int. angles)	Sum of interior angles	Sum of exterior angles
triangle	3	540°	180°	360°
quadrilateral	4	720°	360°	360°
pentagon	5	900°	540°	360°
hexagon	6	1080°	720°	360°
n-gon	n	$(180n)°$	$180°(n-2)$	360°

CHECKPOINT ✔

10. $(180n)° - 180°(n - 2)$
$= (180n)° - (180°n - 360°)$
$= (180n)° - (180n)° + 360° = 360°;\ 360°$

Exercises

Communicate

1. Sample answer: A triangle can't be made with 1 or 2 sides, so we get 1 triangle with 3 sides and 1 more triangle with each additional side. So there are $n - 2$ triangles.

2. Sample answer: No. The exterior angle measures must sum to 360°. If 3 angles have interior measure 60°, then the corresponding exterior angles measure 120°, and these three exterior angles alone add to 360°, leaving no room for a fourth exterior angle.

3. Sample answer: As you get farther away, the figure gets closer to a point, with the sum of the angles measuring 360°.

Lesson 3.7, pages 183–189

Activity 1

2. Measurements may vary, but $BC = 2MN$.

3. Measurements may vary, but $m\angle 1 = m\angle 2$ and $m\angle 3 = m\angle 4$. This suggests $\overline{BC} \parallel \overline{MN}$, based on the converse of the Corresponding Angles Postulate.

4. parallel; one-half

Activity 2

2. Measurements will vary.

3. $MN = \frac{1}{2}(AB + DC)$

4. Measurements will vary, but $m\angle 1 = m\angle 2$ and $m\angle 4 = m\angle 5$. This suggests $\overline{DC} \parallel \overline{MN}$.

5. Measurements may vary, but $m\angle 2 + m\angle 3 = m\angle 5 + m\angle 6 = 180°$. This suggests $\overline{AB} \parallel \overline{MN}$.

6. parallel; $\frac{1}{2}$(base 1 + base 2)

Activity 3

1. Answers will vary. The given table can be completed as follows:

DC	AB	MN
6	5	5.5
6	4	5
6	3	4.5
6	2	4
6	1	3.5
6	.5	3.25
6	.1	3.05

2. a triangle

3. $\frac{1}{2}(\text{base 1} + \text{base 2}) = \frac{1}{2}(0 + \text{base 2})$
$= \frac{1}{2}(\text{base 2})$

This is the formula for the length of a triangle midsegment length.

Exercises

Communicate

1. $\frac{1}{8}$. Sample answer: If the base of the red triangle is x, then the lower base of the yellow trapezoid is $2x$, the lower base of the green trapezoid is $4x$, and the lower base of the blue trapezoid is $8x$. Thus, the ratio is $\frac{x}{8x} = \frac{1}{8}$.

2. $\frac{1}{2}, \frac{1}{4}, \frac{1}{8}$. Each midsegment is one-half the length of the previous midsegment.

3. Sample answer: A limiting case is what happens as a feature gets closer and closer to some value that it will never actually reach.

4. This method works. Let the length of the shorter base = x and the length of the longer base = $x + y$.

length of the midsegment $= \frac{1}{2}(x + x + y)$
$= \frac{1}{2}(2x + y)$
$= x + \frac{y}{2}$

Practice and Apply

22. Midsegments: 8, 9, 13; Outer = 60, Inner = 30; Inner Triangle Perimeter = $\frac{1}{2}$(Outer Triangle Perimeter)

26.

Square. Two sides of the figure are triangle midsegments; therefore, a parallelogram is formed. Because the triangle is isosceles, the sides of the quadrilateral along the perimeter of the triangle are congruent. The sides in the interior of the triangle are midsegments of the congruent triangle sides and thus half their length. Thus, they are congruent to each other and to the sides of the quadrilateral. Consecutive interior angles are supplementary, so all angles measure 90°.

27. Midsegment lengths are $\frac{1}{2}, \frac{1}{4}, \frac{1}{8}, \frac{1}{16}, \ldots$ Looking at the bottom edge of the box, we can figure out the lengths of the segments along the bottom by the lengths of the midsegments. The first segment is $1 - \frac{1}{2} = \frac{1}{2}$. The 2nd segment is $\frac{1}{2} - \frac{1}{4} = \frac{1}{4}$. The 3rd segment is $\frac{1}{4} - \frac{1}{8} = \frac{1}{8}$ and so on. Thus, $\frac{1}{2} + \frac{1}{4} + \frac{1}{8} + \ldots$ must add up to the length of bottom edge of the square, or 1.

Lesson 3.8, pages 190–197

TRY THIS (page 193)

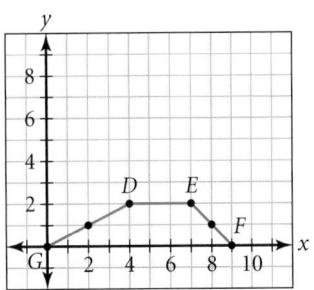

$\overline{GD}$ midpoint: (2, 1); $\overline{EF}$ midpoint: (8, 1); $DE = 3$; $GF = 9$; length of midsegment = 6. So the length of the midsegment is

$\frac{1}{2}(DE + GF) = \frac{1}{2}(3 + 9) = \frac{1}{2}(12) = 6$, confirming the trapezoid midsegment conjecture.

Exercises

Communicate

1. a. Line rises from left to right.
b. Line falls from left to right.
c. Line is horizontal.
d. Line is vertical.

2. Given two points on a line or segment, the *rise* is the vertical distance between the two points and the *run* is the horizontal distance between the two points.

3. $-\frac{1}{m}$, because the slope of a line perpendicular to ℓ, must have slope equal to the negative reciprocal of m.

4. The unit lengths on each axis are different, causing a distortion in the appearance of the graph.

5. Sample answer: It's easy to calculate lengths of bases and midsegments when they are horizontal.

Practice and Apply

31.

Parallelogram. Sides $\overline{AB}$ and $\overline{CD}$ have slope 2; thus, they are parallel. Sides $\overline{AD}$ and $\overline{BC}$ have slope $\frac{-1}{3}$; thus, they are parallel. Since it has two pairs of parallel sides, $ABCD$ is a parallelogram.

32.

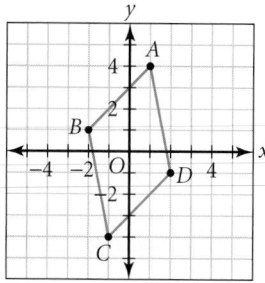

Parallelogram. Sides $\overline{AB}$ and $\overline{CD}$ have slope $\frac{2}{3}$; thus, they are parallel. Sides $\overline{AD}$ and $\overline{BC}$ have slope -4; thus, they are parallel. Since it has two pairs of parallel sides, $ABCD$ is a parallelogram.

33.

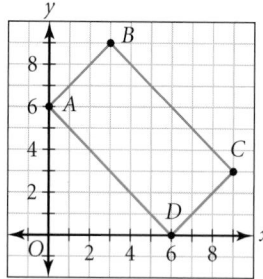

Rectangle. Sides $\overline{BC}$ and $\overline{AD}$ have slope -1; thus, they are parallel. Sides $\overline{AB}$ and $\overline{CD}$ have slope 1; thus, they are parallel. There are two sets of parallel sides, and the adjacent sides have slopes whose product is -1; therefore, the adjacent sides are perpendicular. $ABCD$ is a rectangle.

34.

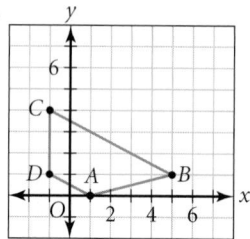

Trapezoid. Sides $\overline{AD}$ and $\overline{BC}$ have slope $\frac{-1}{2}$; thus, they are parallel. $\overline{AB}$ has slope $\frac{1}{4}$, and $\overline{CD}$ has undefined slope; thus, these sides are not parallel. Since there is exactly one pair of parallel sides, $ABCD$ is a trapezoid.

45.

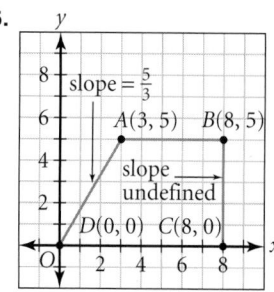

slope $\overline{AB} = 0$
slope $\overline{CD} = 0$

46.

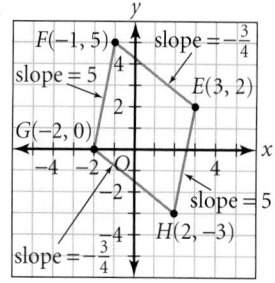

47.

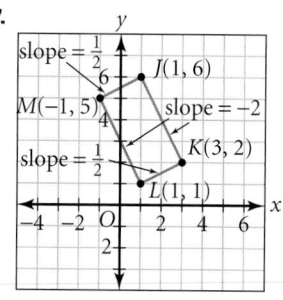

48.

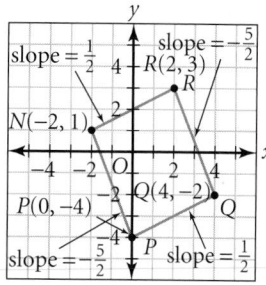

49.

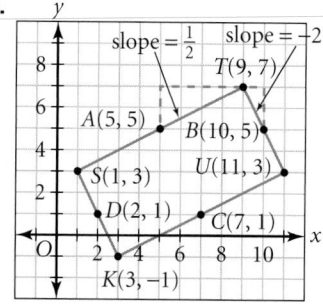

slope $\overline{SK} = -2$
slope $\overline{KU} = \frac{1}{2}$

Chapter 3 Review and Assessment

16. Given: $ABCD$ is a rectangle (quadrilateral with 4 right angles)

Prove: $\overline{AB} \parallel \overline{CD}$ and $\overline{BC} \parallel \overline{AD}$

Statements	Reasons
$\overline{AB}$ and $\overline{CD}$ are perpendicular to $\overline{BC}$	Given
$\overline{AB} \parallel \overline{CD}$	Converse of the Same-Side Interior Angles Theorem
$\overline{BC}$ and $\overline{AD}$ are perpendicular to $\overline{AB}$	Given
$\overline{BC} \parallel \overline{AD}$	Converse of the Same-Side Interior Angles Theorem

Lesson 4.1, pages 210–216

Exercises

Communicate

1.

△ABC	△BAC	△CAB
△ACB	△BCA	△CBA
△ACB	△BCA	△CBA

The order of the vertices does not matter.

2.

∠M ≅ ∠Q	$\overline{MN} ≅ \overline{QR}$
∠N ≅ ∠R	$\overline{NO} ≅ \overline{RS}$
∠O ≅ ∠S	$\overline{OP} ≅ \overline{ST}$
∠P ≅ ∠T	$\overline{PM} ≅ \overline{TQ}$

3. Two sides are congruent when they have the same length.

4. Two angles are congruent when they have the same measure.

5. $\overline{AB} ≅ \overline{CD}$ is a congruence statement about two segments. $AB = CD$ is a statement about the lengths of two segments.

6. $AB ≅ CD$ does not make sense, since a congruence refers to segments, not lengths.

Practice and Apply

30.

Statements	Reasons
∠ADB ≅ ∠ABD ≅ ∠CDB ≅ ∠CBD (m∠ADB = m∠ABD = m∠CDB = m∠CBD)	Given
m∠ADB + m∠ABD + m∠DAB = 180° m∠CDB + m∠CBD + m∠DCB = 180°	Triangle Sum Theorem
m∠ADB + m∠ABD + m∠DAB = m∠CDB + m∠CBD + m∠DCB	Substitution Property
m∠ADB + m∠ABD + m∠DAB = m∠ADB + m∠ABD + m∠DCB	Transitive Property or Substitution
m∠DAB = m∠DCB	Subtraction Property
$\overline{DB} ≅ \overline{DB}$	Reflexive Property of Congruence
$\overline{AB} ≅ \overline{BC} ≅ \overline{CD} ≅ \overline{DA}$	Definition of rhombus
△ABD ≅ △CBD	Polygon Congruence Postulate

Lesson 4.2, pages 217–225

Activity 1

2. Yes

3. No. Any two triangles with same side lengths are congruent. Two quadrilaterals with same side lengths will not necessarily be the same shape. Unlike triangles, quadrilaterals are not rigid.

4. If you know all of the side lengths, you do not need to know angle measures of a triangle to get an exact copy of it.

CHECKPOINT ✔
5. sides; sides; congruent

CRITICAL THINKING, (page 218)
No.

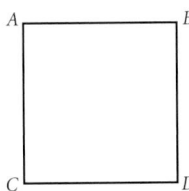 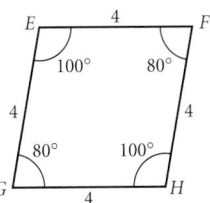

Activity 2

PART I
4. Yes   5. Yes

CHECKPOINT ✔
6. sides; included; sides; included; congruent

PART II
4. Yes   5. Yes

CHECKPOINT ✔
6. angles; side; angles; side; congruent

Exercises

Communicate

1. Sample answer: The postulates allow you to find only 3 congruences when determining if triangles are congruent. When using the Polygon Congruence Postulate, you must find 6 congruences when determining if triangles are congruent.

2. Sample answer: △ABC has sides $AB = 2$ in., $AC = 3$ in., and their included angle is ∠A.

3. Yes. The two angles must be the ones which have the given side as one side of the angle.

4. A triangle is rigid because as long as the sides do not bend or break, the shape will not change. Sample answer: When building structures, triangle braces and supports are used to give the structure rigidity.

Lesson 4.3, pages 226–234

Exercises

Communicate

1. Sample answers:
   Valid tests:
   SSS: Same Shape and Size.
   SAS: Shape And Size
   ASA: All Shapes Agree.
   AAS: All Anchors Set.
   Invalid tests:
   SSA: "Shape Size And…" doesn't make sense.
   AAA: "And, And, And" needs something more.

2. Yes. Either situation will correspond to the ASA or AAS postulate.

3. No. SSA is not a valid test for congruence because of the "swinging door."

4. The triangles will either be congruent by SAS or the third sides will be congruent by the Pythagorean Theorem, in which case the triangles will be congruent by SSS.

5. The distance between the towers is fixed. The angle from each tower to the observed object can be determined. The triangle is uniquely determined by ASA.

Practice and Apply

37. **Given:** $\angle B \cong \angle E$
    $\angle A \cong \angle D$
    $\overline{BC} \cong \overline{EF}$

    Prove: $\triangle ABC \cong \triangle DEF$ (AAS is valid)

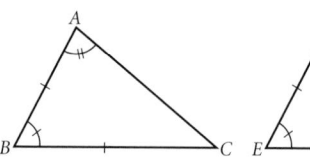

Statements	Reasons
$\angle B \cong \angle E$ ($m\angle B = m\angle E$)	Given
$\angle A \cong \angle D$ ($m\angle A = m\angle D$)	
$\overline{BC} \cong \overline{EF}$	
$m\angle F = 180° - m\angle E - m\angle D$	Triangle Sum Theorem
$m\angle C = 180° - m\angle B - m\angle A$	
$m\angle C = 180° - m\angle E - m\angle D$	Substitution Property
$m\angle C = m\angle F$ ($\angle C \cong \angle F$)	Substitution Property
$\triangle ABC \cong \triangle DEF$	ASA

TRY THIS

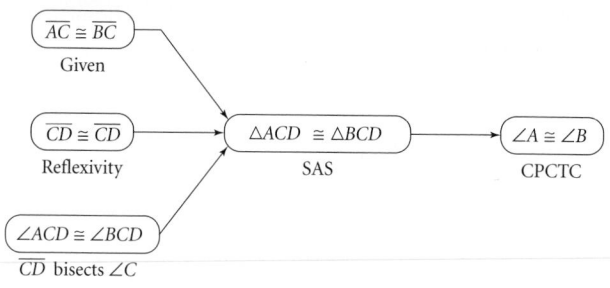

$\overline{CD}$ bisects $\angle C$

Exercises

Communicate

1. CPCTC means corresponding parts of congruent triangles are congruent. You would use it after proving that two triangles are congruent to prove that two angles or two sides of those triangles are congruent.

2. A corollary of a theorem is an additional theorem that can be easily derived from the theorem.

3. Yes, by the Converse of the Isosceles Triangle Theorem.

4. A right triangle may be isosceles (45°-45°-90°) but not equilateral because it contains a 90° angle. The measure of each angle of an equilateral triangle is 60°.

Practice and Apply

18.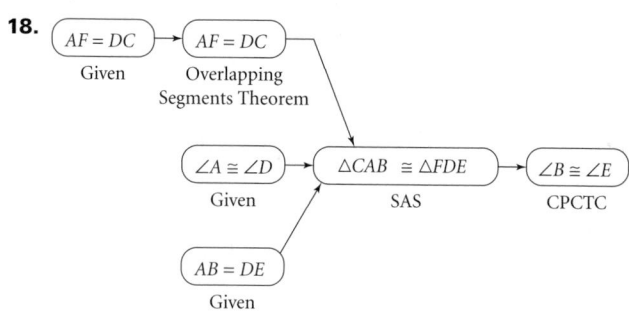

22.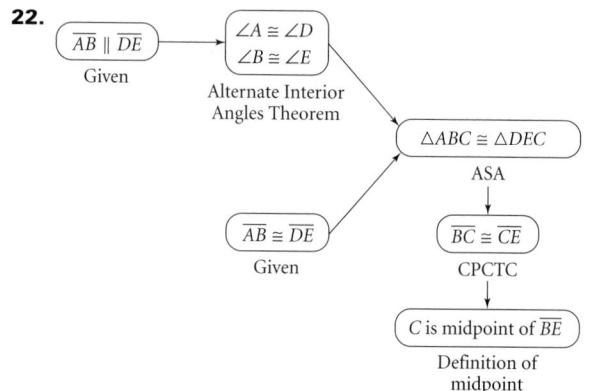

Lesson 4.4, pages 235–242

CRITICAL THINKING, (page 237)

Yes. An isosceles triangle has at least two congruent sides, and an equilateral triangle has 3.

Lesson 4.5, pages 243–252

Activity 1

CHECKPOINT ✔

3. Yes; 2 congruent triangles

CRITICAL THINKING

Since $\angle 2 \cong \angle 3$ and $\angle 1 \cong \angle 4$,
$m\angle 2 = m\angle 3$, and $m\angle 1 = m\angle 4$.
Therefore, $m\angle PLG = m\angle 1 + m\angle 2$
$\qquad\qquad\qquad = m\angle 4 + m\angle 3$
$\qquad\qquad\qquad = m\angle PMG$.

Exercises

Communicate

1. Along with 180° rotational symmetry, a rectangle has reflectional symmetry about lines connecting the midpoints of opposite sides. Along with 180° rotational symmetry, a rhombus also has reflectional symmetry about both diagonals. A square has rotational symmetry at 90°, 180°, and 270° and has reflectional symmetry about the lines connecting the midpoints of opposite sides and about the diagonals.

2. $\triangle PQS \cong \triangle RSQ$
$\triangle PSR \cong \triangle RQP$
$\triangle PQX \cong \triangle RSX$
$\triangle PSX \cong \triangle RQX$

3. Yes. Rectangles are parallelograms.

4. Yes. Rhombuses are parallelograms.

Practice and Apply

68. Given: Kite $WXYZ$ with diagonals $\overline{WY}$ and $\overline{XZ}$ intersecting at point A

Prove: $\overline{WY} \perp \overline{XZ}$

Statements	Reasons
$\overline{WX} \cong \overline{WZ}$	Given
$\overline{XY} \cong \overline{ZY}$	
$\overline{WY} \cong \overline{WY}$	Reflexive Property of Congruence
$\triangle WXY \cong \triangle WZY$	SSS
$\angle ZWY \cong \angle XWY$	CPCTC
$\overline{WA} \cong \overline{WA}$	Reflexive Property of Congruence
$\triangle WAX \cong \triangle WAZ$	SAS
$\angle WAX \cong \angle WAZ$ $(m\angle WAX = m\angle WAZ)$	CPCTC
$m\angle WAX + m\angle WAZ = 180°$	Linear Pair Property
$m\angle WAX + m\angle WAX = 180°$	Substitution Property
$m\angle WAX = 90°$	Division Property
$\overline{WY} \perp \overline{ZX}$	Definition of perpendicular

Lesson 4.6, pages 253–260

Activity 1

Part I

1. False **2.** True **3.** True
4. True **5.** True

Part II

1. False **2.** True **3.** False
4. True **5.** False

Part III

1. False **2.** True **3.** False
4. True **5.** True

Activity 2

CHECKPOINT ✔

3. If the diagonals of a parallelogram are congruent, then the parallelogram is a rectangle.

Exercises

Communicate

1. Sample answer: Yes, look for converses to the theorems in section 4–6. Another defintion: A parallelogram is a quadrilateral whose diagonals bisect each other.

2. Sample answer: An isosceles trapezoid is a counterexample to #1 in part I.

3. Sample answer: An isosceles trapezoid is a counterexample to #3 in Part II.

4. Sample answer: In Part III, a kite is a counterexample to #1.

Practice and Apply

39. Given: Quadrilateral $ABCD$, $\overline{AB} \cong \overline{CD}$, $\overline{AB} \parallel \overline{CD}$

Prove: $ABCD$ is a parallelogram.

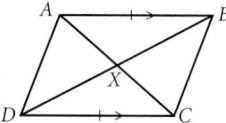

Statements	Reasons
$\overline{AB} \parallel \overline{CD}$, $\overline{AB} \cong \overline{CD}$	Given
$\angle XAB \cong \angle XCD$ $\angle XBA \cong \angle XDC$	Alternate Interior Angles Theorem
$\triangle AXB \cong \triangle CXD$	ASA
$\overline{XB} \cong \overline{XD}$ $\overline{XC} \cong \overline{XA}$	CPCTC
$\angle AXD \cong \angle BXC$	Vertical Angles Theorem
$\triangle AXD \cong \triangle CXB$	SAS
$\angle DAX \cong \angle BCX$	CPCTC
$\overline{AD} \parallel \overline{BC}$	Converse of the Alternate Interior Angles Theorem
$ABCD$ is a parallelogram.	Definition of a parallelogram

40. Given: Quadrilateral $ABCD$, $\overline{AC}$ and $\overline{BD}$ bisect each other.

Prove: $ABCD$ is a parallelogram.

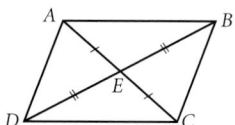

Statements	Reasons
$\overline{AC}$ bisects $\overline{BD}$. $\overline{BD}$ bisects $\overline{AC}$.	Given
$\overline{AE} \cong \overline{EC}$ $\overline{DE} \cong \overline{EB}$	Definition of bisector
$\angle AED \cong \angle BEC$	Vertical Angles Theorem
$\triangle AED \cong \triangle CEB$	SAS
$\angle EAD \cong \angle ECB$	CPCTC
$\overline{AD} \parallel \overline{BC}$	Converse of Alternate Interior Angles Theorem
$\angle AEB \cong \angle DEC$	Vertical Angles Theorem
$\triangle ABE \cong \triangle CDE$	SAS
$\angle EAB \cong \angle ECD$	CPCTC
$\overline{AB} \parallel \overline{CD}$	Converse of Alternate Interior Angles Theorem
$ABCD$ is a parallelogram.	Definition of parallelogram

41. Given: Parallelogram $ABCD$, with m$\angle A = 90°$.

Prove: $ABCD$ is a rectangle.

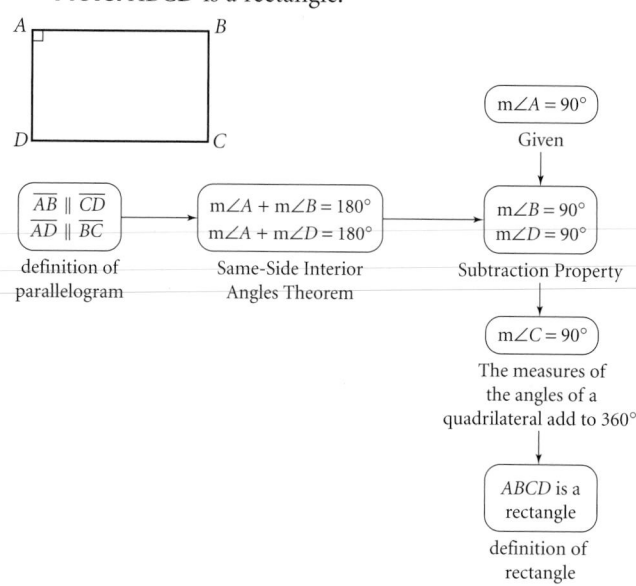

42. Given: $ABCD$ is a parallelogram, and $\overline{AB} \cong \overline{BC}$.

Prove: $ABCD$ is a rhombus.

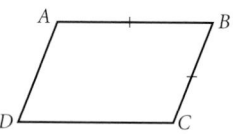

Statements	Reasons
$ABCD$ is a parallelogram.	Given
$\overline{CD} \cong \overline{AB}$ $\overline{AD} \cong \overline{BC}$	Opposite sides of a parallelogram are congruent.
$\overline{AB} \cong \overline{BC}$	Given
$\overline{AB} \cong \overline{BC} \cong \overline{CD} \cong \overline{AD}$	Transitive Property of Congruence
$ABCD$ is a rhombus.	Definition of rhombus

43. Given: $ABCD$ is a parallelogram, $\overline{AC}$ bisects $\angle A$ and $\angle C$, and $\overline{BD}$ bisects $\angle B$ and $\angle D$.

Prove: $ABCD$ is a rhombus.

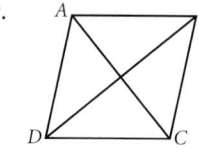

Statements	Reasons
$ABCD$ is a parallelogram.	Given
m$\angle B$ = m$\angle D$	Opposite angles of a parallelogram are congruent.
$\overline{BD}$ bisects $\angle B$ and $\angle D$	Given
m$\angle ABD = \frac{1}{2}$m$\angle B$ m$\angle ADB = \frac{1}{2}$m$\angle D$	Definition of angle bisector
m$\angle ADB = \frac{1}{2}$m$\angle B$	Substitution
m$\angle ABD$ = m$\angle ADB$	Substitution
$\overline{AD} \cong \overline{AB}$	Converse of Isosceles Triangle Theorem
$ABCD$ is a rhombus.	Theorem 4.6.6

44. Given: $ABCD$ is a parallelogram, and $\overline{AC} \perp \overline{BD}$.

Prove: $ABCD$ is a rhombus.

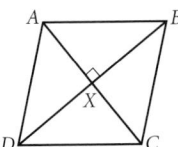

Statements	Reasons
$ABCD$ is a parallelogram.	Given
$\overline{AX} \cong \overline{XC}$	Diagonals of a parallelogram bisect each other.
$\overline{XB} \cong \overline{XB}$	Reflexive Property of Congruence
$\overline{AC} \perp \overline{BD}$	Given
$m\angle AXB = 90°$ $m\angle CXB = 90°$	Definition of perpendicular
$m\angle AXB \cong m\angle CXB$ ($\angle AXB \cong \angle CXB$)	Transitive Property of Equality
$\triangle AXB \cong \triangle CXB$	SAS
$\overline{AB} \cong \overline{CB}$	CPCTC
$ABCD$ is a rhombus.	Theorem 4.6.6

Lesson 4.7, pages 261–270

TRY THIS (page 263)

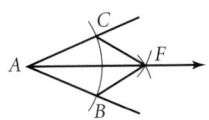

Statements	Reasons
$\overline{AC} \cong \overline{AB}$	The same compass setting is used.
$\overline{CF} \cong \overline{BF}$	The same compass setting is used.
$\overline{AF} \cong \overline{AF}$	Reflexive Property of Congruence
$\angle ACF \cong \angle ABF$	SSS
$\angle CAF \cong \angle BAF$	CPCTC

Exercises

Communicate

1. A straightedge has no measurements on it, whereas a ruler does. A ruler is not used in constructions because it is not as precise at measuring distance with a compass.

2. Assumptions #1 and 3. The compass had to be placed precisely on Point A, and the distance didn't change while the arc was being drawn.

3. Assumptions #2 and 3. The compass can be set to exact distances and doesn't change when it is set.

4. Sample answer: A computer has much more precision, but you must first know how to program it or operate the software.

Practice and Apply

14.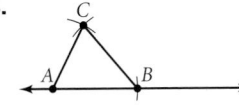

15.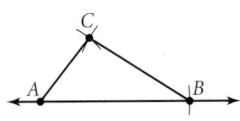

36.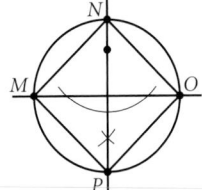

37.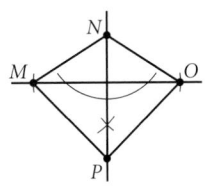

42. The same compass setting was used to create $\overline{AB}$, $\overline{AD}$, $\overline{CB}$, and $\overline{CD}$, so the segments are congruent. Thus, $ABCD$ is a rhombus by definition. By Theorem 4.5.8, the diagonals of a rhombus are perpendicular, so $\overleftrightarrow{BD} \perp \overline{AC}$. By Theorem 4.5.6, $ABCD$ is a parallelogram, so by Theorem 4.5.5, the diagonals of $ABCD$ bisect each other. Thus, $\overleftrightarrow{BD}$ bisects $\overline{AC}$.

Lesson 4.8, pages 271–280

Activity 1

1. The two lines are parallel since they have the same slope as the translation vector. Check student work. Make sure directions from the previous exercises in 4.7 were followed.

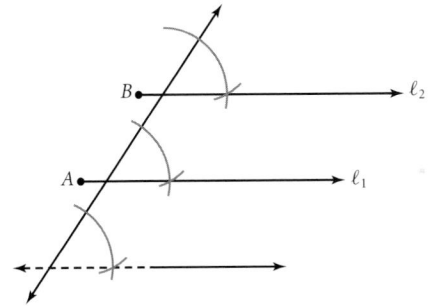

2.

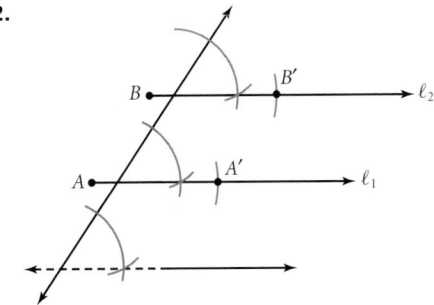

CHECKPOINT ✔

3. The segments $\overline{AB}$ and $\overline{A'B'}$ are congruent since opposite sides of a parallelogram have the same length.

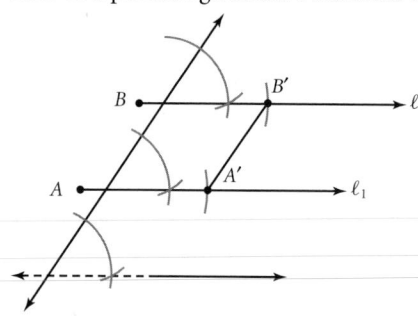

Activity 2

1. The triangles will be congruent by the SSS postulate.

2. The quadrilateral is made up of two triangles; under translation the triangles remain congruent. Therefore the quadrilaterals are congruent.

3. The translation of a polygon is congruent to the polygon.

4. The open figure can be closed by joining the two end-points with a line segment, thereby turning it into a polygon. Under translation the polygon remains congruent; therefore, the open figure is congruent to its translation.

CRITICAL THINKING (page 272)
Curves remain congruent under translation.

CRITICAL THINKING (page 273)
Translation preserves angles and distances because of CPCTC.

TRY THIS, (page 274)
a. not possible, since $14 + 8 < 25$
b. not possible, since $16 + 7 = 23$
c. possible, since $18 + 8 > 24$

Activity 3

1. $AB = A'B'$
$AX = A'X'$
$BX = B'X'$
from the results from Activity 1

2. Since $AX + XB = AB$ by the Segment Addition Postulate, $A'X' + X'B' = A'B'$ by substitution.

3. Postulate 4.8.1 applied to #2 shows that X' is between A' and B'; this proves translations preserve collinearity and betweenness.

Exercises
Communicate

1. Sample answer: In Activity 1, the line segment is translated by translating its endpoints and then drawing a line between the new endpoints. The Betweenness Postulate justifies drawing that new line, rather than translating every possible point on the original line.

2. Sample answer: "ABCD" refers to the properties preserved by rigid translation: Angles, Betweenness, Collinearity, and Distance. When transforming a triangle with a rigid transformation, the angles remain the same, the vertices remain in the same order, points on the sides remain collinear, and the lengths of the sides do not change.

3. "Iso" means "same." An isometry preserves measures of angles and segment length.

4. The minimum possible value for BC is 3 when A, B and C are collinear and B is between A and C. The maximum possible value for BC is 15, when A, B and C are collinear and A is between B and C. Otherwise, $3 < BC < 15$ by the triangle inequality.

Look Beyond

49. Check student's drawing

50. Yes. By the Triangle Midsegment Theorem, the sides of the second triangle are half the length of the first triangle, which is equilateral, so the second triangle is equilateral. The sides of the third triangle are half the length of the second triangle, so the triangle is equilateral, and so on.

51. $P_1 = 3$ units
$P_2 = \frac{3}{2}$ units
$P_3 = \frac{3}{4}$ units

52. $\frac{3}{2} + \frac{3}{4} + \frac{3}{8} + \frac{3}{16} = 2\frac{13}{16}$. The perimeter of the first triangle is 3. As the perimeters of the additional triangles are added to the sum, the total approaches the perimeter of the first triangle.

Chapter 4 Review and Assessment

19. Given: $VWXY$ is a square.

Prove: $\triangle VWX \cong \triangle WXY$

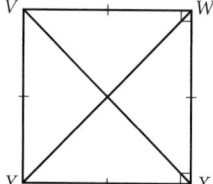

Statements	Reasons
$VWXY$ is a square.	Given
$\overline{VW} \cong \overline{WX}$	Definition of a square
$\overline{WX} \cong \overline{XY}$	
$VWXY$ is a rectangle.	Theorem 4.5.11
$m\angle W = 90°$	Definition of a rectangle
$m\angle X = 90°$	
$m\angle W = m\angle X$	Transitive Property or Substitution Property
$\triangle VWX \cong \triangle WXY$	SAS

20. Parallelogram by Theorem 4.6.2

21. Parallelogram by definition
Rhombus by Theorem 4.6.6

22. Parallelogram by Theorem 4.6.3, rhombus by Theorem 4.6.8, rectangle by Theorem 4.6.4, square by definition

23. Parallelogram by HL and Theorem 4.6.2, rectangle by Theorem 4.6.4 or Theorem 4.6.5

Lesson 5.1, pages 294–302

CRITICAL THINKING (page 294)
Sample answer: Draw a polygon with more vertices.

Activity 1

1. Sample answers: $b = 8$, $h = 4$, area = 32 sq. units; $b = 10$, $h = 2$, area = 20 sq. units; $b = 6$, $h = 6$, area = 36 sq. units.

2.
$$P = 2b + 2h$$
$24 = 2b + 2h$ Substitute 24 for P.
$12 = b + h$ Divide both sides by 2.
$12 - b = h$ Subtract b from both sides.
$h = 12 - b$ Reflexive Property

3.

b	$h = 12 - b$	$A = bh$
1	11	11
2	10	20
3	9	27
4	8	32
5	7	35
6	6	36
7	5	35
8	4	32
⋮	⋮	⋮

The area increases as the base increases to 6; then it decreases.

4.

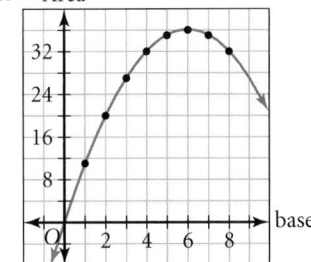
Area

5. $b = 6$, $h = 6$

CHECKPOINT ✔ (page 297)

6. The rectangle is a square. This is true for any rectangle with a fixed perimeter.

Activity 2

1.
$$A = bh$$
$3600 = bh$ Substitute 3600 for A.
$\frac{3600}{b} = h$ Divide both sides by b.
$h = \frac{3600}{b}$ Reflexive Property

2.

b	$h = 3600 \div b$	$P = 2b + 2h$
10	360	740
20	180	400
30	120	300
40	90	260
50	72	244
60	60	240
70	51.43	242.86
80	45	250
90	40	260
100	36	272

3.

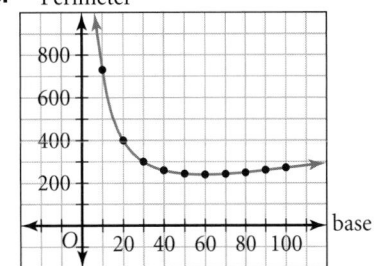
Perimeter

4. $b = 60$, $h = 60$

CHECKPOINT ✔ (page 297)

5. The rectangle is a square. This is true for any rectangle with a fixed area.

Exercises

Communicate

1. Sample answer: Find a polygon which is a good "fit" to the shape, then find the perimeter of the polygon.

2. Sample answer: Select an appropriate grid and place it over the figure. Estimate the area by counting the number of squares inside the figure. Take half the number of squares that are only partially inside the figure and add the number of squares entirely inside the figure. Another method: Divide the figure into pieces that are approximately rectangles and find the sum of the areas of the rectangles.

3. Figure C is composed of figure A and figure B, but they overlap in the middle. Thus, the areas cannot simply be added.

4. Sample answer: Count the total number of full squares which are shaded, then estimate the partial squares which are shaded.

Lesson 5.2, pages 303–311

Activity 1

PART I

1. Answers will vary. Students should draw a rectangle, measure its side lengths, and calculate its area using $A = bh$.

2. The diagonal forms two congruent triangles. The area of each triangle is half the area of the rectangle.

3. Answers will vary. Students should fit two identical triangles together to form a rectangle.

CHECKPOINT ✔ (page 304)

4. $A = \frac{1}{2}bh$

PART II

1–2.

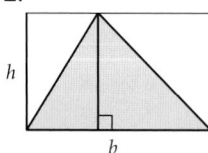

Yes. Theorem 3.4.5 can be used to show that if one side of the rectangle and the altitude are perpendicular to the same line (the base of the rectangle), then they are parallel to each other.

3. Theorem 4.5.1 says that the diagonal of a parallelogram divides the parallelogram into two congruent triangles. (Theorem 4.5.6 says a rectangle is a parallelogram.)

4. The area of the shaded triangle is half the area of the rectangle, because each of the smaller rectangles are divided into two congruent triangles: one shaded and one unshaded. Each smaller rectangle is half shaded; therefore, the larger rectangle is half shaded.

CHECKPOINT ✔ (page 304)

5. $A = \frac{1}{2}bh$

Activity 2

1.

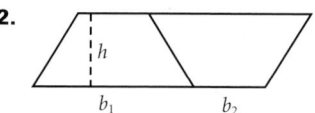

2. A rectangle is formed. The area is bh. The two figures have the same area.

CHECKPOINT ✔ (page 305)

3. $A = bh$

4. $\overline{AB} \cong \overline{DC}$ because opposite sides of a parallelogram are congruent. $\overline{BE} \cong \overline{CF}$ and $\overline{AE} \cong \overline{DF}$ because $\overline{BE} \cong \overline{AE}$ were translated to form $\overline{CF} \cong \overline{DF}$. So $\triangle AEB \cong \triangle DFC$ by the SSS congruence postulate. $\angle AEC$ is a right angle since AE is an altitude. $\angle DFC$ is a right angle, since $\triangle DFC \cong \triangle AEB$ and $\angle AEB$ is a right angle. Opposite sides are parallel, so opposite angles are congruent. Therefore, all angles are right angles and the figure is a rectangle.

Activity 3

1–2.

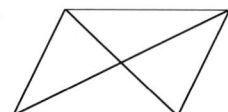

CHECKPOINT ✔ (page 306)

3. Parallelogram: $A = (b_1 + b_2)h$
Trapezoid: $A = \frac{1}{2}(b_1 + b_2)h$

Exercises

Communicate

1.

Sample answer: The two triangles formed by one diagonal are congruent. The second diagonal forms two sets of congruent triangles. Use Theorem 4.5.2 to show that opposite sides are congruent, then use Theorem 4.5.5 to show that the diagonals bisect each other. Finally, use the SSS Postulate (4.2.1) to show that each pair of triangles is congruent.

2. Sample answers:

Yes.

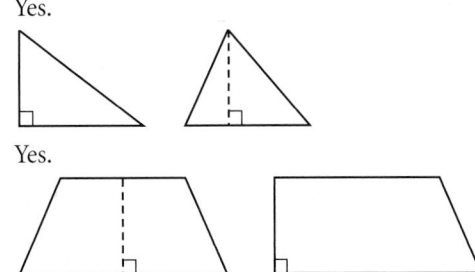

Yes.

3. AB is larger, since it is the hypotenuse of a right triangle with side h. The parallelogram has the smallest perimeter when $AB = h$, because the shortest distance from a line to a point is along the perpendicular. The parallelogram with the smallest perimeter must be a rectangle.

4. The average of the lengths of the bases is $\frac{1}{2}(b_1 + b_2)$. To find the area of the trapezoid, multiply the average of the lengths of the bases by h.

5. The rectangle and the parallelogram will have the same area, since their formulas are both base × height. The area of the trapezoid cannot be determined because it requires two bases.

Practice and Apply

49. Given a parallelogram $ABCD$ with base b and height h, let $\overline{AE}$ be a segment perpendicular to the base so that $\triangle AED$ is formed. Translate $\triangle AED$ by b units so that points D and C coincide, forming a rectangle with base b and height h. Its area is bh, so the area of the parallelogram is bh.

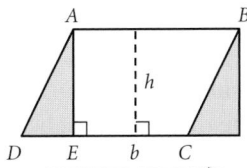

50. Given a trapezoid $ABCD$ with bases b_1 and b_2 and height h. Let $EFGH \cong ABCD$. Rotate $EFGH$ 180° and translate it so that C and F coincide and B and G coincide, forming a parallelogram $AHED$. The parallelogram has base $b_1 + b_2$ and height h, so its area is $(b_1 + b_2) \cdot h$. The area of the trapezoid is half the area of the parallelogram, so the area is $\frac{1}{2}(b_1 + b_2) \cdot h$.

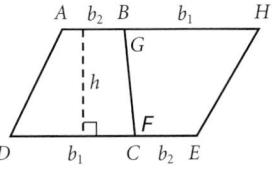

51. **Prove:** Area $ABCD = \frac{1}{2}(BD)(AC)$

Statements	Reasons
$ABCD$ is a kite with $\overline{AC} \perp \overline{BD}$	Given
Area kite $ABCD$ $= $ Area $\triangle ABD + $ Area $\triangle BCD$	Sum of Areas Postulate
Area kite $ABCD$ $= \frac{1}{2}(BD)(AX) + \frac{1}{2}(BD)(XC)$	Area of a Triangle Formula
Area kite $ABCD = \frac{1}{2}(BD)(AX + XC)$	Distributive Property
Area kite $ABCD = \frac{1}{2}(BD)(AC)$	Segment Addition Postulate

52. Rectangle $KQLN$ is divided into 2 congruent right triangles by side $\overline{KL}$. The shaded area is the area of $\triangle KLN$ minus the area of $\triangle KMN$. The area of a right triangle is $\frac{1}{2}bh$. Therefore:

Area of $\triangle KLN = \frac{1}{2}(b + x) \cdot h = \frac{1}{2}bh + \frac{1}{2}xh$

Area of $\triangle KMN = \frac{1}{2}xh$

So Area of $\triangle KLM = \left(\frac{1}{2}bh + \frac{1}{2}xh\right) - \frac{1}{2}xh = \frac{1}{2}bh$.

Lesson 5.3, pages 314–320

Activity 1

1. Answers will vary. The ratio $\frac{C}{d}$ should be approximately equal to 3.14.

2. The average of the values is about 3.14.

3. $\pi \approx 3.141592$

CHECKPOINT ✔ (page 315)
4. $\pi = \frac{C}{d}$
$\pi d = C$
$C = \pi d$
$C = \pi(2r)$
$C = 2\pi r$

Activity 2

2. A parallelogram

3. Yes.

4. A triangle. If the number of sectors increases, the height will still be approximately r.

5. The base is approximately half the circumference because it is made up of 4 sectors, which is half the circle.
Base $= \frac{1}{2}(2\pi r) = \pi r$

CHECKPOINT ✔ (page 316)
6. $A = (\text{base})(\text{height}) = (\pi r)r = \pi r^2$

Exercises

Communicate

1. Sample answer: The perimeter of a square is $4s$. If 100 feet of fencing is available, then $4s = 100$, so $s = 25$. The square pen would be 25 feet on each side, so the area of the pen would be $25^2 = 625$ ft.^{2} The circumference of a circle is $2\pi r$, so
$2\pi r = 100$ gives $r = \frac{100}{2\pi} = \frac{50}{\pi} \approx 15.92$ ft. The area of a circular pen would be $\pi r^2 = \pi(15.92)^2 \approx 796.23$ ft.^{2} The circular pen would provide the most area. Other considerations might include the shape of the yard and the "empty" space left in the corners if a circular area is fitted into a square or rectangular yard.

2. Sample answer: 3.14 is smaller than π, and $\frac{22}{7}$ is larger than π. 3.14 might be used if an underestimate is preferred to an overestimate. $\frac{22}{7}$ might be used when an overestimate is preferred. π must be estimated, because the exact value of π has an infinite number of decimal places, and so that the area or circumference can be given in terms of decimal units.

3. Sample answer: Point B moves faster, because the circumference of the tape is larger than at point A. Since a greater distance is traveled in the same amount of time, the speed is greater.

Lesson 5.4, pages 321–330

Activity

1. Answers will vary. Sample answer: $785^2 - 86^2 = 608,829$. $\sqrt{608,829} \approx 780.27$ The result is not an integer.

2. Answers may vary. Students should rarely get an integer result.

3.

a	b		b^2	Difference $b^2 - a^2$	Square Root of Difference
119	169	14161	28561	14400	120
3367	4825	11336689	23280625	11943936	3456
4601	6649	21169201	44209201	23040000	4800
12709	18541	161518681	343768681	182250000	13500
65	97	4225	9409	5184	72
319	481	101761	231361	129600	360
2291	3541	5248681	12538681	7290000	2700
799	1249	638401	1560001	921600	960
481	769	231361	591361	360000	600
4961	8161	24611521	66601921	41990400	6480
45	75	2025	5625	3600	60
1679	2929	2819041	8579041	5760000	2400
161	289	25921	83521	57600	240
1771	3229	3136441	10426441	7290000	2700
56	106	3136	11236	8100	90

An integer results every time.

4. Yes. From parts 1 and 2 of the activity, it appears that an integer result is very unlikely using random numbers. They must have known which numbers would give an integer result.

CRITICAL THINKING, (page 323)
The shape of the right triangle does not matter. It will be difficult for large squares of different sizes. The large figure is a square, because the Chinese drawing shows it on a square grid with the same number of units on each side. If the right triangles are isosceles, the square in the center will have vertices at the midpoints of the sides of the large square.

Exercises

Communicate

1. Sample answer: The sum of the squares of the two shorter sides of a right triangle equals the square of the longest side.

2. Sample answer: It could be used in construction to determine whether a corner is square by measuring the sides and the diagonal of a rectangular room. If the measurements satisfy the Converse of Pythagorean theorem, the corner has a 90° angle measure.

3. Since $3^2 + 4^2 = 5^2$, any triangular object with measurements 3, 4 and 5 has a right angle.

4. The sum of the areas of the squares on the legs of the triangle is equal to the area of the square on the hypotenuse of the triangle.

Practice and Apply

46. Sample answer: False. Any rhombus with an angle other than 90° is not a rectangle.

47. Sample answer: True. Opposite sides of a rhombus have equal lengths. Therefore, the rhombus must be a parallelogram.

48. Sample answer: False. A diagonal of a kite divides the kite, which is a quadrilateral, into two congruent triangles (by SSS), but a kite is not a parallelogram.

63. 90°

64. Definition of a right triangle.

Lesson 5.5, pages 331–338

Activity

1.

Shorter leg	Hypotenuse	Longer leg
1	2	$\sqrt{3}$
2	4	$2\sqrt{3}$
3	6	$3\sqrt{3}$

2. The length of the hypotenuse is twice the length of the shorter leg, and the longer leg is $\sqrt{3}$ times the length of the shorter leg.

3. a. $2x$

 b. $x\sqrt{3}$

CHECKPOINT ✔ (page 333)

4.

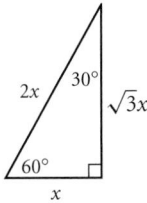

30-60-90 Triangle Theorem

$2, \sqrt{3}$

Exercises

Communicate

1. To form a square with two 45-45-90 triangles, fit them together hypotenuse to hypotenuse. To form a square with four 45-45-90 triangles, fit them together so that the four hypotenuses form the sides of the square.

2. The length of the apothem of an equilateral triangle is $\frac{1}{3}$ the height of the triangle.

3. The length of the apothem of a square is $\frac{1}{2}$ the side length of the square.

4. No. The sides of a 45-45-90 or a 30-60-90 triangle can never be a Pythagorean triple, because they can't all be integers: $\sqrt{2}$ or $\sqrt{3}$ will always be involved.

Lesson 5.6, pages 339–346

Activity

PART I: *Method A (Left-Hand Rule)*

1.

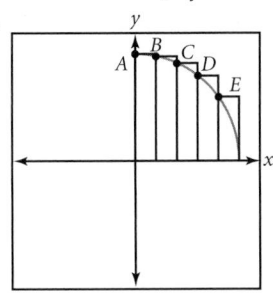

2. $A: y = \sqrt{5^2 - 0^2} = \sqrt{25} = 5$

 $B: y = \sqrt{5^2 - 1^2} = \sqrt{24}$

 $C: y = \sqrt{5^2 - 2^2} = \sqrt{21}$

 $D: y = \sqrt{5^2 - 3^2} = \sqrt{16} = 4$

 $E: y = \sqrt{5^2 - 4^2} = \sqrt{9} = 3$

3. Area of rectangle $A = 1 \cdot 5 = 5$ units^2

 Area of rectangle $B = 1 \cdot \sqrt{24} = \sqrt{24}$ units^2

 Area of rectangle $C = 1 \cdot \sqrt{21} = \sqrt{21}$ units^2

 Area of rectangle $D = 1 \cdot 4 = 4$ units^2

 Area of rectangle $E = 1 \cdot 3 = 3$ units^2

4. $\sqrt{5^2 - 0^2} + \sqrt{5^2 - 1^2} + \sqrt{5^2 - 2^2} + \sqrt{5^2 - 3^2} + \sqrt{5^2 - 4^2}$

 $= \sqrt{25} + \sqrt{24} + \sqrt{21} + \sqrt{16} + \sqrt{9} \approx 21.48$ units^2

CHECKPOINT ✔ (page 341)

5. $(21.48) \times 4 = 85.92$ units^2

 It overestimates, because each rectangle covers more area than the circle in the same 1-inch "strip."

6. $A = \pi \cdot 5^2 = 25\pi \approx 78.5398$ units^2

 $E = \dfrac{|V_e - V_t|}{V_t} \times 100 \approx \dfrac{|85.92 - 78.5398|}{78.5398} \times 100$
 ≈ 9.40 percent error

PART I: *Method B (Right-Hand Rule)*
CHECKPOINT ✔ (page 342)

 y-coordinates:

 Point $A: y = \sqrt{5^2 - 1^2} = \sqrt{24}$

 Point $B: y = \sqrt{5^2 - 2^2} = \sqrt{21}$

 Point $C: y = \sqrt{5^2 - 3^2} = \sqrt{16} = 4$

 Point $D: y = \sqrt{5^2 - 4^2} = \sqrt{9} = 3$

 Point $E: y = \sqrt{5^2 - 5^2} = \sqrt{0} = 0$

 Sum of the areas:

 $\sqrt{5^2 - 1^2} + \sqrt{5^2 - 2^2} + \sqrt{5^2 - 3^2} + \sqrt{5^2 - 4^2} + \sqrt{5^2 - 5^2}$

 $= \sqrt{24} + \sqrt{21} + \sqrt{164} + \sqrt{9} + \sqrt{0} \approx 16.48$ units^2

 $(16.48) \times 4 = 65.92$ units^2

 This result is an underestimate, since each rectangle covers less area than the circle in the same 1-inch "strip."

 Actual area $= \pi \cdot 5^2 \approx 78.5398$ units^2

 Relative error:

 $E \approx \dfrac{|65.92 - 78.5398|}{78.5398} \times 100$
 ≈ 16.07 percent error

PART III: *Combining Methods*
CHECKPOINT ✔ (page 342)

 Average $= \dfrac{85.92 + 65.92}{2} = 75.92$ units^2

 Relative error $\approx \dfrac{|75.92 - 78.5398|}{78.5398} \times 100$
 ≈ 3.34 percent error

Exercises

Communicate

1. Sample answer: The distance formula is derived from the Pythagorean Theorem, so the distance formula depends on it.

2. Sample answer: No. It does not matter which point is (x_1, y_1) and which is (x_2, y_2), because $(x_2 - x_1)^2$ is the same as $(x_1 - x_2)^2$, and $(y_2 - y_1)^2$ is the same as $(y_1 - y_2)^2$.

3. Sample answer: *Method 1*: Draw rectangles of width 1, with height even with the left endpoint of the graph in the 1-inch "strip." The sum of the areas of the rectangle is an estimate. *Method 2*: Draw rectangles of width 1, with the height even with the right endpoint of the graph in the 1-inch "strip." The sum of the areas is an estimate.

4. Sample answer: One way would be to average the areas found by the other two methods. Another way would be to use a rectangle height which is about halfway between the lowest and highest part of the graph in the 1 inch "strip." A third way would be to use rectangles which are more narrow.

Lesson 5.7, pages 347–352

Activity 1

CHECKPOINT ✔

1.

Vertices of a triangle	Coordinates of midpoints M and S		Slope		Length	
	$\overline{AB}$	$\overline{BC}$	$\overline{MS}$	$\overline{AC}$	$\overline{MS}$	$\overline{AC}$
$A(0, 0)$, $B(2, 6)$, $C(8, 0)$	$M(1, 3)$	$S(5, 3)$	0	0	4	8
$A(0, 0)$, $B(6, -8)$, $C(10, 0)$	$M(3, -4)$	$S(8, -4)$	0	0	5	10
$A(0, 0)$, $B(2_p, 2_q)$, $C(2r, 0)$	$M(p, q)$	$S(p + r, q)$	0	0	r	$2r$

2. Yes.

3. Because of the relationships between coordinates.

CHECKPOINT ✔ (page 348)

4. $\overline{MS}$ is equal to half of $\overline{AC}$, and $\overline{MS}$ and $\overline{AC}$ are parallel (they have the same slope), which proves the Midsegment Theorem.

Activity 2

1.

Three vertices of a parallelogram	Fourth Vertex	Midpoint of $\overline{BD}$	Midpoint of $\overline{AC}$
$A(0, 0)$, $B(2, 6)$, $D(10, 0)$	$C(12, 6)$	$(6, 3)$	$(6, 3)$
$A(0, 0)$, $B(2p, 2q)$, $D(2r, 0)$	$C(2p + 2r, 2q)$	$(p + r, q)$	$(p + r, q)$

CHECKPOINT ✔ (page 348)

2. The diagonals of a parallelogram bisect each other. The intersection of the diagonals is the midpoint of each diagonal.

CRITICAL THINKING (page 348)

Answers may vary. Sample answer: Given that the intersection of the two diagonals is M, you would have to prove that $\overline{AM} \cong \overline{MC}$ and $\overline{BM} \cong \overline{MD}$ by proving congruent triangles and using CPCTC.

Activity 3

1. The y-coordinate will be equal to the x-coordinate.

2. Midpoint between $P_1(a, b)$ and $P_2(b, a)$ is $M\left(\frac{a+b}{2}, \frac{b+a}{2}\right)$
Yes, point M lies on the line $y = x$ because its x-coordinate and y-coordinate are equal.

3. $A(a, a)$ $B(b, b)$
slope $= \frac{b - a}{b - a} = 1$

4. slope $= \frac{a - b}{b - a} = \frac{a - b}{-a(a - b)} = -1$

5. The lines between the two sets of points have slopes which are the negative reciprocal of each other. Therefore the lines are perpendicular to each other.

CHECKPOINT ✔, (page 349)

6. Reversing the coordinates of a point results in a line joining the two points that has a slope of -1. That line is perpendicular to the line $y = x$. In addition, the midpoint of the two points lies on the line $y = x$.

Exercises

Communicate

1. Because that does not put any restriction on the type of figure used.

2. No, because that will make it a right triangle which is a special type of triangle.

3. To prove that two lines are parallel, you need to prove they have the same slope. To prove that two lines are perpendicular, you need to prove the slope of one line is the negative reciprocal of the slope of the other.

4. To prove that two segments bisect each other, you need to prove that their intersection is the midpoint of each line segment.

Practice and Apply

25.

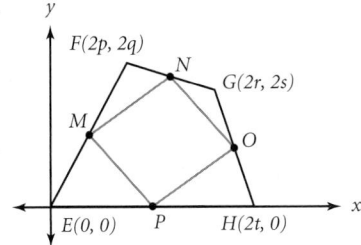

26. slope of $\overline{NO} = \dfrac{s-(q+s)}{r+t-(p+r)} = \dfrac{-q}{t-p}$

slope of $\overline{PM} = \dfrac{0-q}{t-p} = \dfrac{-q}{t-p}$

slope of $\overline{MN} = \dfrac{q+s-q}{p+r-p} = \dfrac{s}{r}$

slope of $\overline{PO} = \dfrac{0-s}{t-(r+t)} = \dfrac{-s}{-r} = \dfrac{s}{r}$

Slopes of opposite sides are equal; thus, $MNOP$ is a parallelogram.

27. Let $ABCD$ be a parallelogram formed by the following vertices: $A(0, 0)$, $B(p, q)$, $C(r, q)$, and $D(s, 0)$, where $\overline{AB}$ and $\overline{CD}$ are opposite sides and $\overline{BC}$ and $\overline{AD}$ are opposite sides. Slopes of opposite sides are equal:

$\dfrac{q}{p} = \dfrac{q}{r-s}$

$p = r - s$

$AB = \sqrt{p^2 + q^2}$

$CD = \sqrt{(s-r)^2 + q^2} = \sqrt{p^2 + q^2}$

$AB = CD$

$BC = \sqrt{(r-p)^2} = \sqrt{[r-(r-s)]^2} = \sqrt{s^2}$

$AD = \sqrt{s^2}$

$BC = AD$

28. Let $ABCD$ be a square formed by $A(0, 0)$, $B(0, p)$, $C(p, p)$, and $D(p, 0)$.

slope of diagonal $\overline{AC} = \dfrac{p-0}{p-0} = 1$

slope of diagonal $\overline{BD} = \dfrac{0-p}{p-0} = -1$

The two diagonals have slopes that are negative reciprocals of each other. Therefore, they are perpendicular to each other.

29. Let $ABCD$ be a rectangle formed by $A(0, 0)$, $B(0, p)$, $C(p, q)$, and $D(q, 0)$.

length of diagonal $\overline{AC} = \sqrt{(p-0)^2 + (q-0)^2}$
$= \sqrt{p^2 + q^2}$

length of diagonal $\overline{BD} = \sqrt{(q-0)^2 + (0-p)^2}$
$= \sqrt{p^2 + q^2}$

Thus, the diagonals of a rectangle are congruent.

30. Let $ABCD$ be a parallelogram formed by $A(0, 0)$, $B(p, q)$, $C(s, q)$, $D(r, 0)$ Since opposite sides of a parallelogram are congruent, $s = p + r$. Therefore,

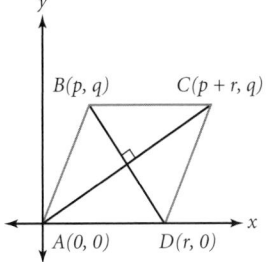

Since the diagonals of the parallelogram are perpendicular,

slope $\overline{AC}$ • slope $\overline{BD} = -1$

$\dfrac{q}{p+r} \cdot \dfrac{q}{p-r} = -1$

$q^2 = -1(p+r)(p-r)$

$q^2 = -(p^2 - r^2)$

$q^2 = r^2 - p^2$

$p^2 + q^2 = r^2$

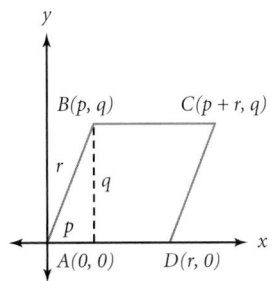

This implies that the length of side $\overline{AB}$ is r, by the Pythagorean Theorem. Since one pair of adjacent sides of the parallelogram are congruent, it is a rhombus.

31. Let $\triangle ABC$ be a triangle formed by $A(0, 0)$, $B(2p, 2q)$, and $C(2r, 0)$. Then the midpoints of the sides of the triangle are $M(p, q)$, $N(p+r, q)$, and $P(r, 0)$.

The slope of median $\overleftrightarrow{AN}$ is $m = \dfrac{q-0}{p+r-0} = \dfrac{q}{p+r}$.

The equation of $\overleftrightarrow{AN}$ is $y = \dfrac{q}{p+r}x$

The slope of median $\overleftrightarrow{CM}$ is $m = \dfrac{q-0}{p+r-0} = \dfrac{q}{p+r}$.

The equation of $\overleftrightarrow{CM}$ is $y = \dfrac{q}{p-2r}(x-2r)$

The slope of median $\overleftrightarrow{BP}$ is $m = \dfrac{2q-0}{2p-r} = \dfrac{2q}{2p-r}$.

The equation of $\overleftrightarrow{BP}$ is $y = \dfrac{2q}{2p-r}(x-r)$

Set the first two equations equal and solve for x:

$\dfrac{q}{p+r}x = \dfrac{q}{p-2r}(x-2r)$

$qx(p-2r) = q(x-2r)(p+r)$

$xp - 2rx = xp + xr - 2rp - 2r^2$

$2r^2 + 2rp = 3rx$

$x = \dfrac{2}{3}(p+r)$

Substitute into the first equation to find y:

$y = \dfrac{q}{p+r}\dfrac{2}{3}(p+r) = \dfrac{2}{3}q$

So the first two equations intersect at $\left(\dfrac{2}{3}(p+r), \dfrac{2}{3}q\right)$.

Substitute these values into the third equation to verify that the three lines intersect at a single point:

$$\frac{2}{3}q = \frac{2q}{2p-r}\left(\frac{2}{3}(p+r)-r\right)$$
$$= \frac{2q}{2p-r} \cdot \frac{2p+2r-3r}{3}$$
$$= \frac{2q}{2p-r} \cdot \frac{2p-r}{3}$$
$$= \frac{2}{3}q$$

Since this results in a true statement, the three medians intersect at a single point.

Lesson 5.8, pages 353–359

Activity

1. Answers will vary. It would be expected that about 16 of 20 tosses would result in the penny touching an intersection point.

2. Answers will vary. Sample answer:
$R = \frac{16+17+18+15+13}{20+20+20+20+20} = \frac{79}{100} = 0.79$

3. Answers will vary. Sample answer: $R = 0.79$, so $\pi \approx (0.79) \times 4 = 3.16$

4. $E = \frac{|3.16 - \pi|}{\pi} \approx 0.0059$ or 0.59 percent error

Exercises

Communicate

1. Sample answer: 0% probability represents an impossible event, so any event cannot be less likely to occur than impossible. 100% probability represents a certain event, and any event cannot more likely occur than a certain occurrence.

2. Probability 0 means the event is impossible.

3. Probability 1 means the event is certain.

4. Sample answer: never, rarely, seldom, maybe, sometimes, frequently, often, usually, always.

5. The estimate of π using the tosses of the entire class should be more accurate. To improve the estimate, increase the number of tosses.

Lesson 6.1, pages 372–378

Activity

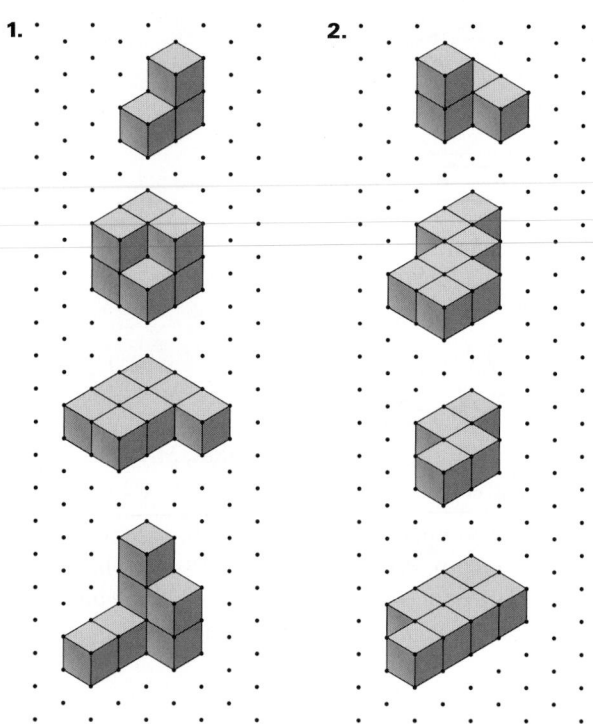

CHECKPOINT ✔ (page 373)

3. Sample answer: When you add cubes, segments that represent edges of cubes that get hidden must be erased. When you subtract cubes, segments must be added to represent edges of cubes that become visible.

Activity 2

1. The figures should be the same size and shape, but their orientation and colors may be different.

2. Answers will vary. Check that the drawings match the solid which were built. Sample answers:

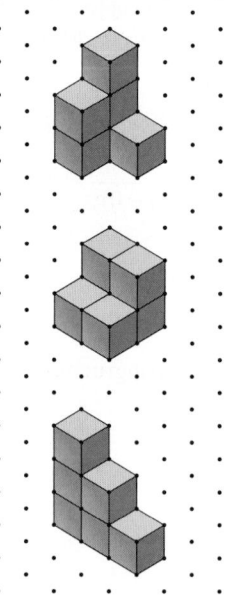

CHECKPOINT ✔ (page 373)

3. In some isometric drawings, a cube can be completely hidden behind other cubes. The drawing would look the same with or without the hidden cube. Students should indicate on their drawings where one or more cubes could be hidden.

Activity 3

1.

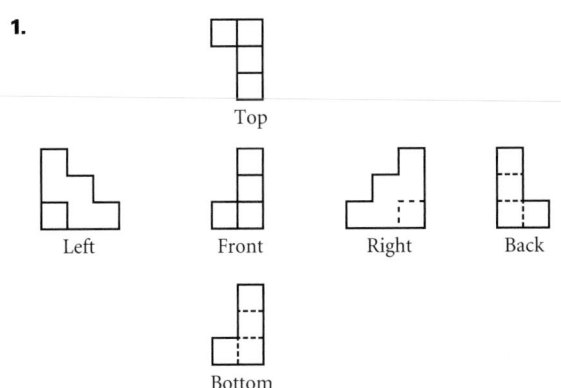

Top

Left   Front   Right   Back

Bottom

CHECKPOINT ✔ (page 374)

2. Count the number of cubes used to build the figure.

CHECKPOINT ✔ (page 374)

3. Sum the areas of the orthographic projections.

Exercises

Communicate

1. Sample answer: Tower 2 cubes high and tower 5 cubes high

2. Sample answer:

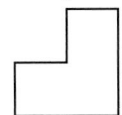

3. Sample answer: Angles of 30° from the horizontal permit a regular hexagon to be drawn, from which a representation of a cube in which all the visible edges are equal can be created.

4. Sample answer: Volume is a measure of the total space in three dimensions taken up by an object and is measured in cubic units. Surface area is the amount of space in two dimensions that is taken up by the surface of an object and is measured in square units.

Guided Skills Practice

8.

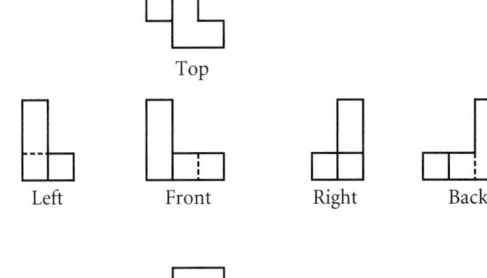

Top

Left   Front   Right   Back

Bottom

Practice and Apply

29. a.

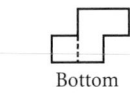

Top

Left   Front   Right   Back

Bottom

b.

Top

Left   Front   Right   Back

Bottom

c.

Top

Left   Front   Right   Back

Bottom

30. Sample answer:

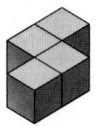 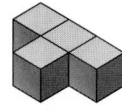

31.

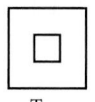

Top

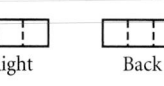

Left   Front   Right   Back

Bottom

33. Sample answer:

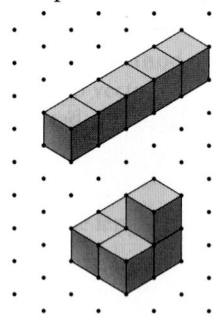

34. Sample answer:

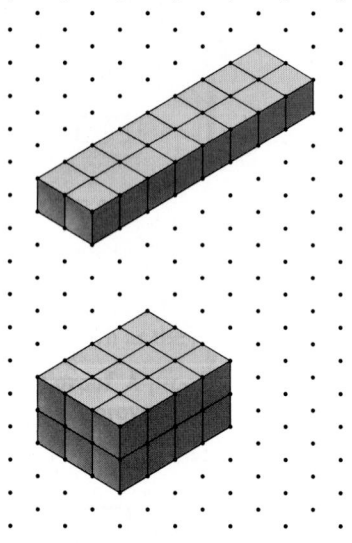

Lesson 6.2, pages 379–387

Activity 1

PART I

1. $\overline{AE}$, $\overline{DH}$, $\overline{BF}$, $\overline{CG}$
Yes; yes; no

2. Edges with the same orientation (vertical, horizontal, angled): $\overline{AE}$, $\overline{DH}$, $\overline{BF}$, and $\overline{CG}$; $\overline{EH}$, $\overline{FG}$, $\overline{AD}$, and $\overline{BC}$; $\overline{HG}$, $\overline{EF}$, $\overline{DC}$, and $\overline{AB}$.

CHECKPOINT ✔ (page 380)

3. Yes. Lines that are in parallel planes can never meet, because the planes that contain them never do.
Sample answer:
$\overline{AD}$ and $\overline{HG}$
$\overline{DC}$ and $\overline{BF}$
$\overline{BC}$ and $\overline{DH}$
$\overline{AE}$ and $\overline{GH}$

PART II

1. 6 faces
ADHE and *BCGF*
DCGH and *ABFE*
ABCD and *EFGH*

CHECKPOINT ✔ (page 380)

2. they never intersect

Activity 2

PART I

1.

Edges	Faces perpendicular to this edge
$\overline{AE}$	*ABCD*, *EFGH*
$\overline{DH}$	*ABCD*, *EFGH*
$\overline{BF}$	*ABCD*, *EFGH*
$\overline{CG}$	*ABCD*, *EFGH*
$\overline{BC}$	*ABFE*, *CDHG*
$\overline{GF}$	*ABFE*, *CDHG*
$\overline{HE}$	*ABFE*, *CDHG*
$\overline{AD}$	*ABFE*, *CDHG*
$\overline{AB}$	*ADHE*, *BCGF*
$\overline{EF}$	*ADHE*, *BCGF*
$\overline{DC}$	*ADHE*, *BCGF*
$\overline{HG}$	*ADHE*, *BCGF*

2. Sample answer: A line is perpendicular to a plane if it is perpendicular to a line in the plane and only intersects the plane in one point. Yes.

3. Yes.

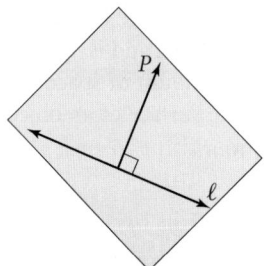

4. They are perpendicular.

5. Yes.

CHECKPOINT ✔ (page) 381

6. intersects *P*

PART II

1.

$\overline{AD}$	BCGF, EFGH
$\overline{AE}$	DCGH, BCGF
$\overline{EH}$	ABCD, BCGF
$\overline{DH}$	ABFE, BCGF
$\overline{DC}$	ABFE, EFGH
$\overline{CG}$	ABFE, ADHE
$\overline{GH}$	ABFE, ABCD
$\overline{GF}$	ABCD, ADHE
$\overline{BC}$	ADHE, EFGH
$\overline{BF}$	ADHE, DCGH
$\overline{AB}$	DCGH, EFGH
$\overline{EF}$	DCGH, ABCD

Sample answer: A segment or line is parallel to a plane if it is parallel to a line in the plane and is not in the plane.

2. Yes.

3. Yes.

CHECKPOINT ✔ (page 381)

4. a line in that plane

Activity 3

1. 4 other faces

2. Line ℓ is perpendicular to the line of the crease.

3. Student should observe an angle by opening paper slightly.

CHECKPOINT ✔ (page 382)

4. perpendicular

5. Students should construct the rays as described.

6. different

CHECKPOINT ✔ (page 383)

7. Sample answer: Theoretically, the smallest angle is only slightly bigger than 0°. The largest angle is only slightly less than 180°. By using rays perpendicular to the common edge of the two faces (the crease), a nonarbitrary and standard way of measuring dihedral angles is achieved.

Exercises

Communicate

1. Not necessarily. It is possible for the two lines to be parallel, but they could also intersect or be skew.

2. Not necessarily. A line can be perpendicular to a line on a plane but not perpendicular to the plane. For example, consider a line making a 45° angle with a line in the plane of the desk. There is another line in the plane of the desk that is perpendicular to the original line, but the original line is not perpendicular to the plane of the desk.

3. There are infinitely many ways a line can be perpendicular to a given line through a certain point on the given line, but only one way it can also be perpendicular to a second line that passes through the given point. If the line is perpendicular to both lines, then it will also be perpendicular to *every* line that passes through the given point and is in the same plane as those two lines. Thus, the line will be perpendicular to the plane of the two lines.

4. The top face of a cube is square, so the segments that connect the centers of the corner atoms of opposite edges of the top face are parallel segments.

5. Opposite faces of a cube are parallel, so the plane passing through the centers of the atoms in the top face of the cubic crystal is parallel to the plane passing through the centers of the atoms of the bottom face.

6. Adjacent faces of a cube are at right angles, so the planes that pass through the centers of the atoms in adjacent faces of the cubic crystal are at right angles.

Lesson 6.3, pages 388–395

Activity

PART I

1. a. rectangle
b. a rectangle (appears to be a square)
c. rectangle

2. $BASE \cong B'A'S'E'$, $BB'A'A \cong EE'S'S$, $BB'E'E \cong AA'S'S$

3. a. rectangle **b.** parallelogram **c.** rectangle

4. All of them are still congruent.

PART II

1. No. No face is a translation of the face opposite it.

2. No.

3. If the bases are not translations of one another, as in the figure for Part II, then lateral surfaces of the figure are not planar.

4. Yes. All of the lateral faces of a prism can be nonrectangular parallelograms. For example, if the top base of the prism is translated horizontally in the direction of a slide arrow from B to E and then in the direction of a slide arrow from B to A, the new prism will have lateral faces that are nonrectangular parallelograms.

Exercises

Communicate

1. Because each lateral face makes a 90° angle with the bases.

2. *ABFE* and *DCGH*
ADHE and *BCGF*
ABCD and *EFGH*
Yes. Each lateral face would be a parallelogram, because

opposite sides would be congruent. (Every point moves the same distance in a translation, and the length of a translated segment does not change.)

3. Yes, the bases could be any pair of opposite faces. In each pair, one face is a translation of the other.

4. Sample answer: A box of cereal, an aquarium with a hexagonal bottom (and top), a board cut (mitered) at 45° angles.

Lesson 6.4, pages 396–401

CHECKPOINT ✔ (page 397)
top-back-right: $(-, +, +)$
top-back-left: $(-, -, +)$
bottom-front-right: $(+, +, -)$
bottom-front-left: $(+, -, -)$
bottom-back-right: $(-, +, -)$
bottom-back-left: $(-, -, -)$

CHECKPOINT ✔ (page 397)
Its z-coordinate and y-coordinate are 0.
Its x-coordinate and z-coordinate are 0.
Its x-coordinate and y-coordinate are 0.

Exercises
Communicate

1. It is in a plane that contains two of the axes. It is on an axis.

2. No. The relationship of the point to the axes cannot be determined. Two coordinates of the point are needed, which can be given as numbers or shown by dashed lines on the figure.

3. No. The octant cannot be determined from the given information. But it can be determined that point P cannot be in the top-front-left, the bottom-front-left, or the bottom-front-right quadrants.

4. Because numbers are squared.

5. Sample answer: If the bottom, back, left corner of the room is the origin and the door is near the front right corner of the room, the door knob could be (18, 16, 4).

Guided Skills Practice

7. 8.
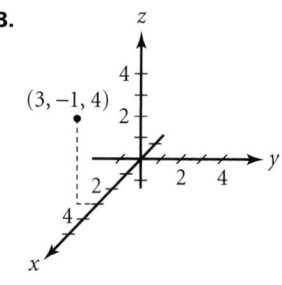

9. (0, 0, 0), (2, 0, 0), (0, 7, 0), (2, 7, 0), (0, 0, 6), (2, 0, 6), (0, 7, 6), (2, 7, 6)

Practice and Apply

13. 14.
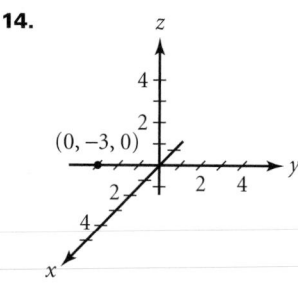

15. 16.
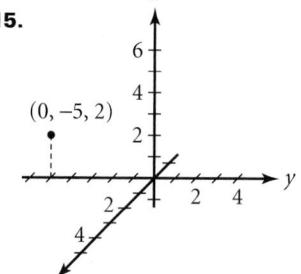
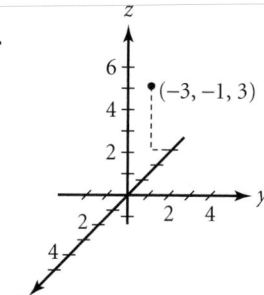

Lesson 6.5, pages 402–408
Exercises
Communicate

1. $Ax + By + Cz = D$
It resembles the equation of a line, with an extra variable for the added third dimension of space.

2. The plane is perpendicular to the yz-plane.

3. The plane is perpendicular to the xy-plane.

4. The plane is parallel to the yz-plane and perpendicular to xz-plane and xy-plane.

5. Sample answer: No. Yes. Yes. Yes. Consider a line in a two-dimensional space. It cannot pass through exactly 1 quadrant. It can pass through exactly 2 quadrants, and it can pass through exactly 3 quadrants. Take the 3-quadrant example and tilt it so it goes through a fourth octant in a three-dimensional coordinate system.

Guided Skills Practice

7.

8.

t	x	y
1	3	4
2	4	8
3	5	12

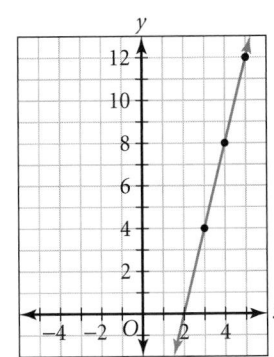

9.

t	x	y
1	0	5
2	1	7
3	2	9

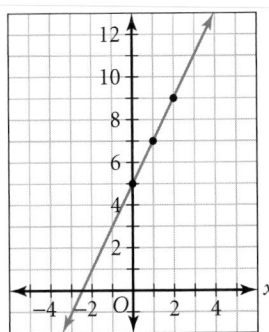

10.

t	x	y	z
1	1	2	2
2	2	4	3
3	3	6	4

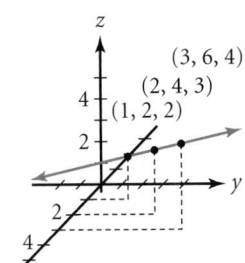

11.

t	x	y	z
1	1	7	−1
2	4	8	−2
3	7	9	−3

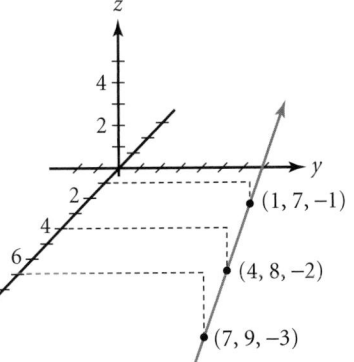

Practice and Apply

12.

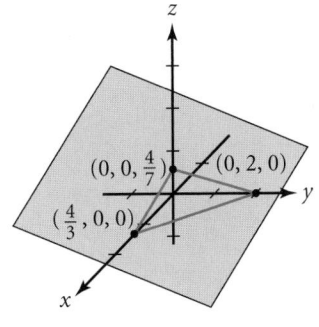

13.

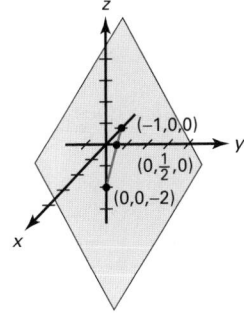

14.

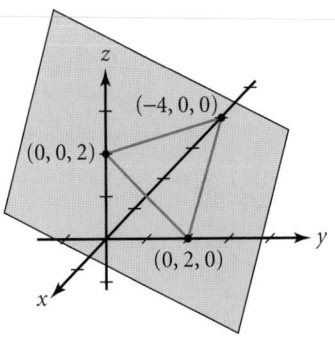

15.

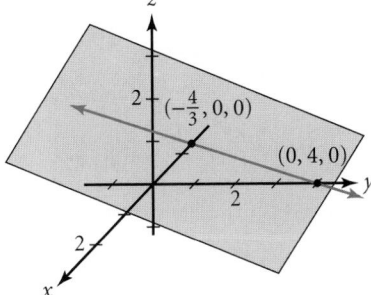

16.

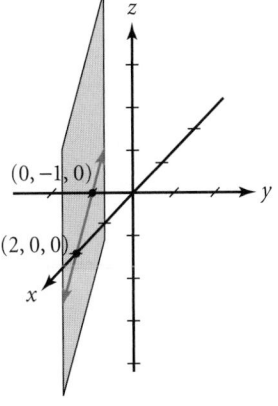

17.

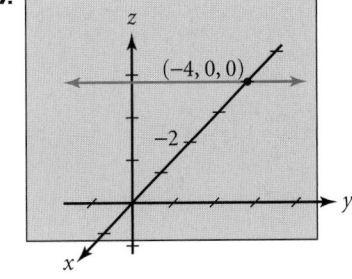

18.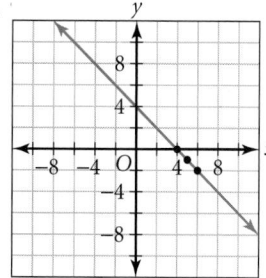

19.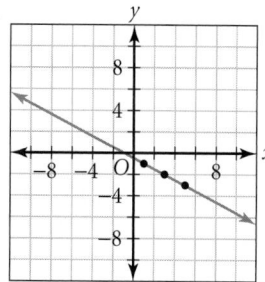

20.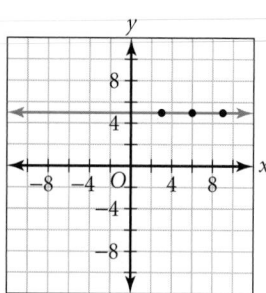

21.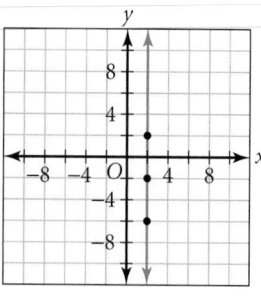

24.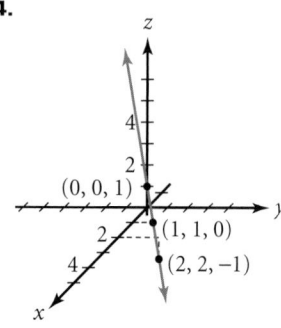

25.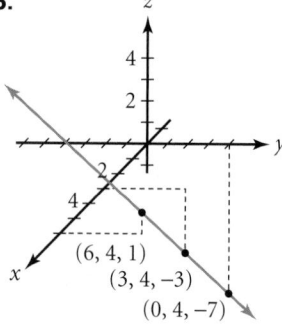

26. trace: $x + 3y = 7$

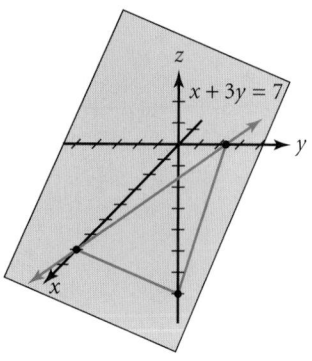

27. trace: $5x - 2y = 2$

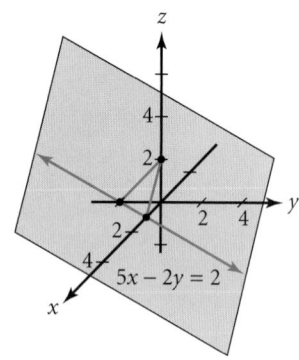

28. trace: $2x + 7y = 2$

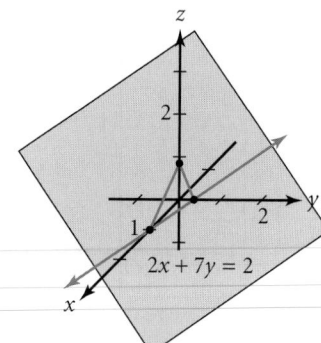

29. trace: $-4x - 2y = 1$

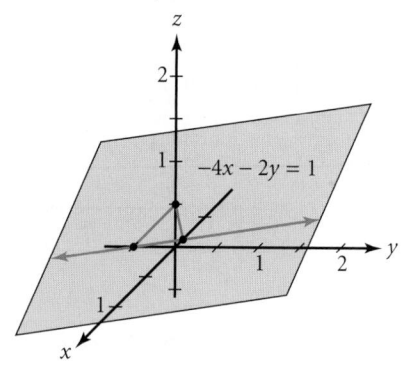

32.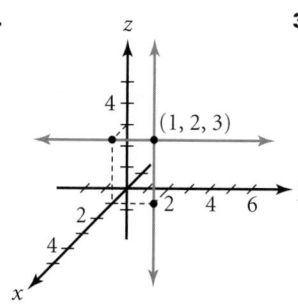

Intersecting when $t = 3$ and $s = 2$.

33.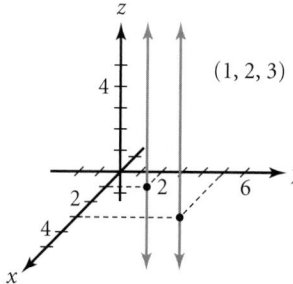

Parallel. These are vertical lines perpendicular to the xy-plane at $(1, 2, 0)$ and $(3, 5, 0)$.

34.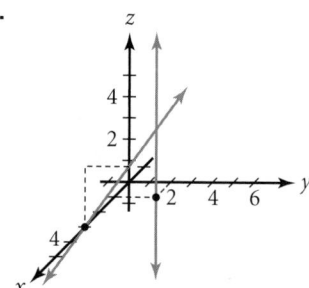

Skew. The lines never intersect because their x-values are always different.

36. $5x + 25y - z = -200$

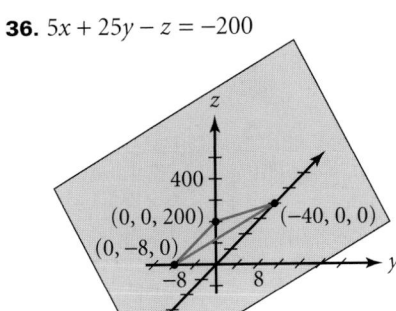

44.

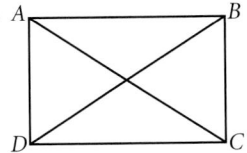

Statements	Reasons
Rectangle $ABCD$ with diagonals $\overline{AC}$ and $\overline{BD}$	Given
$ABCD$ is a parallelogram.	Theorem 4.5.7
$\overline{AD} \cong \overline{BC}$	Theorem 4.5.2
$\overline{DC} \cong \overline{DC}$	Reflexive Property of Congruence
$\overline{AC} \cong \overline{BD}$	Theorem 4.5.9
$\triangle ADC \cong \triangle BCD$	SSS

45.

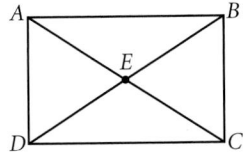

Statements	Reasons
$ABCD$ is a rectangle.	Given
$ABCD$ is a parallelogram.	Theorem 4.5.6
$\overline{AD} \cong \overline{BC}$	Theorem 4.5.2
$\overline{AE} \cong \overline{CE}$	Theorem 4.5.4
$\overline{BE} \cong \overline{DE}$	
$\triangle AED \cong \triangle CEB$	SSS

46.

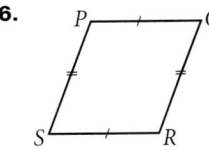

Statements	Reasons
Quadrilateral $PQRS$	Given
$\overline{PQ} \cong \overline{RS}$	
$\overline{PS} \cong \overline{QR}$	
$PQRS$ is a parallelogram.	Theorem 4.6.1

Lesson 6.6, pages 409–417

CHECKPOINT ✔ (page 410)
The string models the line of sight from a point on the object to the eye. The point where the string crosses the picture plane is where the point on the object should be drawn in the picture.

Exercises

Communicate

1. A vanishing point is the point where all lines that are parallel to each other, but not to the picture plane, seem to meet.

2. By Theorem 3.4.6, each time you add a line parallel to one of the lines, it must be parallel to both.

3. Sample answer: If a line in the plane of the ground does not meet the horizon, it would appear to either extend out into the sky or disappear into the ground.

4. Buildings and houses usually contain many sets of parallel edges that each have a single vanishing point. Also, buildings and houses are very often laid out on a rectangular grid, so that their faces are parallel to each other.

5. The line of sight of a person looking straight ahead in real life is parallel to the ground, which extends to the horizon.

6. Sample answer: Perspective drawings have a vanishing point, whereas isometric drawings do not.

CRITICAL THINKING (page 413)
7. Sample answer: One appears to be further away, and our brains believe objects futher away should appear smaller. Since the farther polar bear image is the same size, we think that the bear is actually larger.

Guided Skills Practice

8. They are parallel to each other but not to the picture plane. They are parallel to each other but not to the picture plane.

9. The lines containing the vertical segments will not meet, because they are parallel to the picture plane.

10. The lines containing the nonvertical sides of the building meet at the horizon of the drawings.

Look Beyond

38.

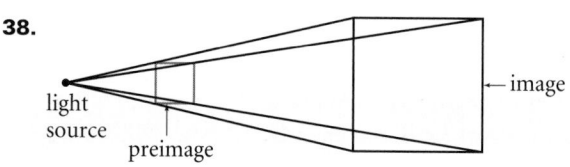

light source
preimage
← image

39. Sample answer: The slide projector projects a preimage onto a plane, like a perspective drawing. In a diagram for a perspective drawing, the eye of the observer would be at the light source point (Exercise 38), and the image and preimage would be reversed.

Lesson 7.1, pages 430–436

Activity

Ratio of Surface Area and Volume
PART I

1.

Length, ℓ	Surface area = $2\ell w + 2\ell h + 2wh$	Volume = ℓwh	Ratio of $\frac{\text{surface area}}{\text{volume}}$
1	26	1	6
2	10	2	5
3	14	3	$\frac{14}{3}$
n	$4n + 2$	n	$\frac{4n+2}{n}$

2. $\frac{402}{100}$; decreases; yes; 4

PART II

1.

Side, s	Surface area = $6s^2$	Volume = s^3	Ratio of $\frac{\text{surface area}}{\text{volume}}$
1	6	1	6
2	24	8	3
3	54	27	2
n	$6n^2$	n^3	$\frac{6}{n}$

CHECKPOINT ✔

2. $\frac{6}{100}$; decreases; The surface-area-to volume ratio is larger for smaller cubes.

Activity 2

Maximizing Volume

1.

Side of square, x	Length, ℓ	Width, w	Height, h	Volume, ℓwh
1	9	6.5	1	58.5
2	7	4.5	2	63
3	5	2.5	3	37.5
x	$11 - 2x$	$8.5 - 2x$	x	$(11 - 2x)(8.5 - 2x)x$

2. Base on the table, a side length between 1 and 2 will maximize the volume. $1 < x < 2$

3.

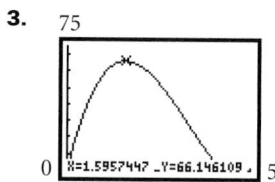

75

0   X=1.5957447 Y=66.146109   5
0

CHECKPOINT ✔ (page 432)
4. a. 1.60 inches
b. 66.15 inches^2

CRITICAL THINKING (page 433)
Sample answer: the "look" of the box the packaging of other, similar products.

Exercises

Communicate

1. Sample answer: A right rectangular prism has six faces; the faces that face each other have the same surface area. The sum of the surface area of all faces is
$\ell w + \ell h + wh + \ell w + \ell h + wh = 2\ell w + 2\ell h + 2wh$

2. Sample answer: A solid with a surface-area-to-volume ratio greater than 1 is less "bunched-up" or is smaller than a solid with surface-area-to-volume ratio less than 1.

3. Sample answer: As the value of ℓ increases, the 1×1 faces stay the same. If the value of ℓ increases by 1, the volume increases by 1, while the surface area increases by 4. Thus, the surface-area-to-volume ratio approaches $\frac{4}{1} = 4$.

4. The surface area to volume ratio of a cube could never equal zero, because this ratio is equal to $\frac{6s^2}{s^3} = \frac{6}{s}$, which is never equal to zero, because a fraction can only equal zero if its numerator is zero.

5. Sample answer: Single-celled organisms have large surface area to volume ratios. Larger animals do not have enough surface area to absorb enough oxygen and food for their larger bodies.

Lesson 7.2, pages 437–444

CRITICAL THINKING (page 438)
Use the fact that the net for the lateral area of a right prism is a rectangle whose length is the perimeter of the base and its width is the height of the prism.

Exercises
Communicate

1. Use the formula $S = hp + 2B$, where h is the height, p is the perimeter of the base, and B is the area of the base.

2. The formula for the volume of a right prism is $V = Bh$, where B is the base area and h is the height.

3. Sample answer: Cavalieri's Principle says that two solids will have equal volumes if they have the same heights and all the cross-sectional areas are equal. This means that the volume of an oblique prism is the same as the volume of a right prism having the same height and base, because their cross-sections are the same.

4. Yes, as long as the areas of the bases are the same. Cavalier's Principle only states that the areas must be the same.

Lesson 7.3, pages 445–452

Activity

2. 3 times

CHECKPOINT ✔
3. $V_{\text{pyramid}} = \frac{1}{3} V_{\text{prism}} = \frac{1}{3} Bh$

Exercises
Communicate

1. A pyramid is a polyhedron consisting of a base, which is a polygon, and three or more lateral faces. The lateral faces are triangles that share a single vertex, called the vertex of the pyramid. Yes, consider a pyramid whose base and lateral faces are all equilateral triangles.

2. To find the surface area of a regular pyramid, calculate the surface area (B) and the perimeter (p) of the base. $S = \frac{1}{2} \ell p + B$, where ℓ is the slant height.

3. To find the volume of a pyramid, calculate the area of the base (B). $V = \frac{1}{3} Bh$, where h is the height of the pyramid.

4. The slant height is larger because it is the hypotenuse of a right triangle formed by the height and another segment. The hypotenuse of a right triangle is always the longest side of the triangle.

5. Volume of pyramid = $\frac{1}{3}$ of the volume of a prism with the same base and height.

Lesson 7.4, pages 453–459

Activity

1. A right rectangular prism with $\ell = \pi r$ and $w = r$

CHECKPOINT ✔
2. $V = \pi r^2 h$

Exercises
Communicate

1. Though both an altitude and the axis of a cylinder have endpoints in the planes containing the bases, an altitude is perpendicular to both planes, and the axis joins the centers of the two bases. If a cylinder is right, the axis is an altitude of the cylinder.

2. No, a cylinder is not a polyhedron, because the lateral surface is not a polygon.

3. Calculate the lateral area ($L = 2\pi rh$), and the base area ($B = \pi r^2$). $S = L + 2B = 2\pi rh + 2\pi r^2$

4. $S = 2\pi rh + 2\pi r^2 = 2\pi r(h + r)$ Sample answer: the original form, because it is easier to remember.

5. Sample answer: Cylinders and prisms are similar because they have two congruent bases connected by something whose net is a rectangle. They are different, because all of the faces of a prism are polygons, while a cylinder has a curved surface.

Lesson 7.5, pages 460–468

Activity 1

The surface area of a right cone:
1. 6π 2. 16π 3. $\frac{3}{8} = 0.375$

4. $64\pi \approx 201.06$ units^2; $24\pi \approx 75.40$ units^2

CHECKPOINT ✔
5. 9π; $33\pi \approx 103.67$ units^2; the surface area of the cone

Activity 2

1. $S = \pi \ell^2 \cdot \frac{r}{\ell}$ 2. $S = \pi r \ell + \pi r^2$

CHECKPOINT ✔
3. The results are the same.

Exercises

Communicate

1. $S = \pi r \ell + \pi r^2$, where r is the radius of the base of the cone and ℓ is the slant height.

2. $V = \frac{1}{3}\pi r^2 h$, where r is the radius of the base of the cone and h is the height of the cone.

3. The slant height is longer than the altitude. The slant height is the hypotenuse of a right triangle formed by the altitude and a radius of the base.

4. Sample answer: There are many segments that can be drawn from the vertex to a point on the edge of the base, but since they have different lengths it is impossible to define the slant height.

5. The volume is multiplied by 4. The volume is doubled. If both are doubled, the volume is multiplied by 8.

Lesson 7.6, pages 469–475

CRITICAL THINKING (page 471)
There is no small circle inside the large circle, because the plane cuts the cylinder at the point where the cones meet; therefore, the area of the annulus is πr^2 (because $y = 0$).

Exercises

Communicate

1. The surface area of a sphere of radius r is 4 times the area of a circle with radius r.

2. 156 ft². Multiply the area of the circle by 4.

3. The area is multiplied by 4. The area is multiplied by 9.

4. The surface area is multiplied by 4. The surface area is multiplied by 9.

5. The volume is multiplied by 8. The volume of a sphere is multiplied by 27.

Lesson 7.7, pages 478–485

Activity 1

PART I

2.

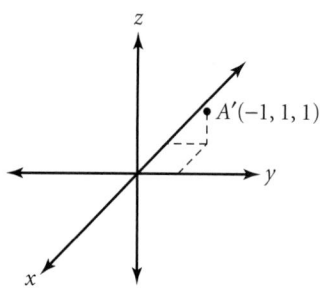

3. yz-plane; perpendicular

4. In three-dimensional coordinate space, the reflection of a point across a plane is a point that has the same coordinates on that plane and the opposite coordinate on the third axis.

5. If you multiply the x-coordinate by -1, you get the reflection of A through the yz-plane. If you multiply the y-coordinate by -1, you get the reflection of A through the xz-plane. If you multiply the z-coordinate by -1, you get the reflection of A through the xy-plane.

6. Students should experiment with the reflections of other figures and should find that these figures are reflected across the planes.

7. The reflection of a figure in three-dimensional coordinate space across a plane consists of the reflections of all the points of the figure across that plane.

PART II:

1. front-left-top, $(2, -3, 4)$; back-right-top, $(-2, 3, 4)$; back-right-bottom, $(-4, 5, -6)$; back-left-top, $(-4, -5, 6)$; front-right-top, $(4, 5, 6)$

2. a. $(x, y, -z)$
   b. $(x, -y, z)$
   c. $(-x, y, z)$

CHECKPOINT ✔ (page 480 –Top)
Each figure is symmetrical across a plane, or mirror, dividing it into two values. A figure has reflectional symmetry in space if and only if its reflected image across a plane coincides with its preimage. The plane is called a plane of symmetry.

Activity 2

2.

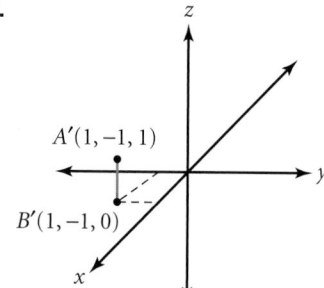

3. $(0, 0, 1)$

4. Yes. Yes. Perpendicular.

CHECKPOINT ✔

5. The rotation of a figure about an axis consists of the rotations of all the points of the figure about that axis.

CHECKPOINT ✔ (page 481)

Each figure can be rotated about a line and have its image coincide with its preimage. A figure in space has rotational symmetry if and only if it has at least one rotation image, not counting rotations of 0° or multiples of 360°, that coincides with the original image.

Exercises

Communicate

1. Sample answer: Both three-dimensional and two-dimensional reflectional symmetry require the reflected image of an object to be congruent to its preimage. Two-dimensional reflectionally symmetric objects reflect across a line, while three-dimensional reflectionally symmetric objects reflect across a plane.

2. cylinder

3. cone

4. Sample answer: pencil, lamp shade, globe, clock.

Practice and Apply

39.

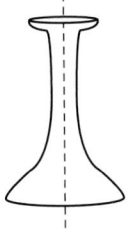

40.

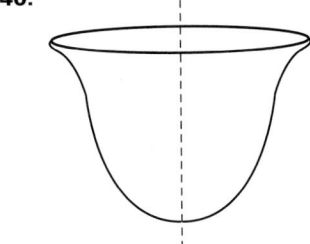

Chapter Project, pages 486–487

Activity 2

Trapezoidal pyramid (2). The missing edge lengths are (going clockwise from top center) $4\sqrt{2}, 4\sqrt{2}, 4\sqrt{3}, 4\sqrt{3}, 2\sqrt{5}, 2\sqrt{5}$, and the missing base edge is $2\sqrt{7}$. Volume = 16 cm^3.

Cube. The missing edge lengths are all 2 cm. Volume = 8 cm^3.

Trapezoidal prism. The missing edge lengths are (going clockwise from top left) $2\sqrt{2}, 2, 2, 2, 2, 2\sqrt{2}, 2\sqrt{2}$, and $2\sqrt{2}$, Volume = 12 cm^3.

Triangular prism. The missing edge lengths are (going clockwise from top left) 2, 2, 2, 2, and the missing base edges are each $2\sqrt{2}$, Volume = 8 cm^3.

Triangular prism. The missing edge lengths are (going clockwise from top left) 2, 2, 2, 2, and the missing base edges are each $2\sqrt{2}$, Volume = 4 cm^3.

EXTENSION

1. Each edge of the cube is 4 cm. Yes, the volume of the cube equals the sum of the pieces:
$2(16) + 8 + 12 + 8 + 4 = 64$
$4^3 = 64$

2. Check students' work.

Lesson 8.1, pages 498–506

Activity 1

1.

A	OA	Scale factor	Image, A'	OA'	Ratio $\frac{OA'}{OA}$
(3, 4)	5	2	(6, 8)	10	2
(3, 4)	5	0.5	(1.5, 2)	2.5	0.5
(3, 4)	5	−1	(−3, −4)	5	1
(3, 4)	5	n	(3n, 4n)	5n	n

2.

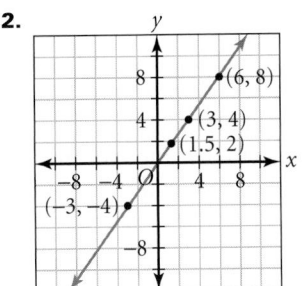

A line is the simplest geometric figure that contains all of those points.

CHECKPOINT ✔

3. n times

4.

B	AB	Scale factor	Image, B'	A'B'	Ratio $\frac{A'B'}{AB}$
(5, 6)	$2\sqrt{2}$	2	(10, 12)	$4\sqrt{2}$	2
(5, 6)	$2\sqrt{2}$	0.5	(2.5, 3)	$\sqrt{2}$	0.5
(5, 6)	$2\sqrt{2}$	−1	(−5, −6)	$2\sqrt{2}$	1
(5, 6)	$2\sqrt{2}$	n	(5n, 6n)	$2n\sqrt{2}$	n

5. 1

CHECKPOINT ✔

6. n times, equal

CRITICAL THINKING (page 500-middle)

For a scale factor of −1, the image would be the same as the one formed by a rotation of 180° about the center of dilation. For other negative scale factors, the image is also expanded or contracted.

Activity 2

CHECKPOINT ✔

4. Check student drawings.

Exercises

Communicate

1. A dilation is a transformation that is not rigid. A dilation preserves the shape of an object, but it may change its size.

2. Let (x, y) be a point of the preimage. Then $D(x, y) = (nx, ny)$ is the corresponding point of the image, where n is the scale factor. The size of the image segment is n times the size of its preimage. If n is negative, then, in addition to the latter, the image is the same as a rotation about the center of dilation of the preimage segment.

3. The dimensions of the image are twice as large as the dimensions of the preimage.

4. The dimensions of the image are half as large as the dimensions of the preimage.

5. The image is the same size as the preimage; but it is, in effect, a rotation about the center of dilation.

6. The image is the same as the preimage.

12.

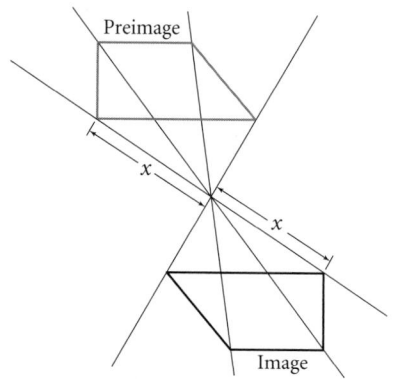

Practice and Apply

14. (2, 6), (4, 10), (8, 6)

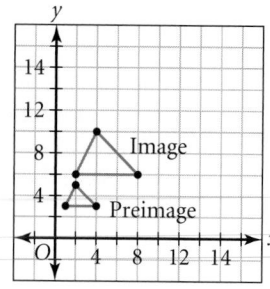

15. $\left(-1, \frac{5}{3}\right), \left(\frac{8}{3}, 3\right), \left(\frac{2}{3}, -2\right)$

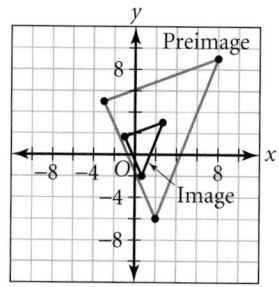

16. (0, 0), (−3, 0), (−2, −2), $\left(-1, -\frac{3}{2}\right)$

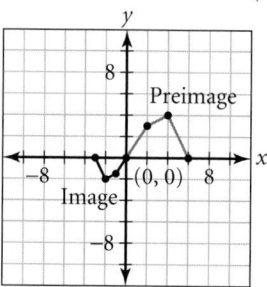

17. (1.6, 1.6), (4.8, −1.6), (−3.2, −4.8)

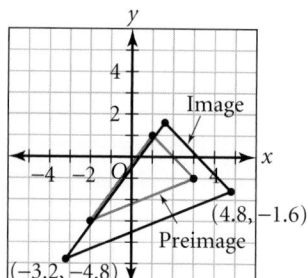

26.

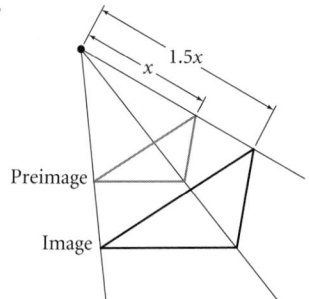

39.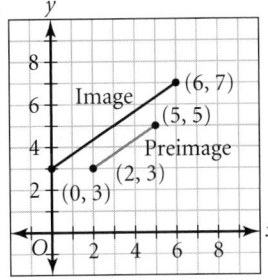

Scale factor is 2. Center of dilation is (4, 3).
$D(x, y) = (4x - 2, 4y - 1)$.

Lesson 8.2, pages 507–516

Exercises

Communicate

1. True. Corresponding angles occur in pairs. For example, if $\angle A \cong \angle D$, then $\angle D \cong \angle A$.

2. False. $\angle A \cong \angle D$ does not mean necessarily that $\angle A \cong \angle E$.

3. True. If two figures are congruent, then corresponding angles and sides are congruent and thus are proportional.

4. False. If two figures are similar, then corresponding angles and sides are proportional, but not necessarily congruent.

5. True. Two regular polygons with the same number of sides will have all the same angles, and all corresponding sides will be proportional.

Lesson 8.3, pages 517–524

Activity 1

1. 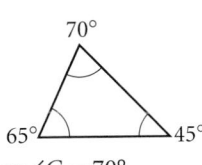 2.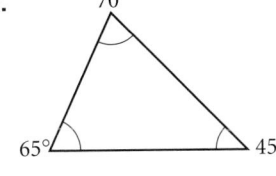

$m\angle C = 70°$

3. Sample answer:

	Sides	Angles
△ABC	AB = 1.33	m∠A = 45°
	BC = 1	m∠B = 65°
	AC = 1.282	m∠C = 70°
△DEF	DE = 3.99	m∠A = 45°
	EF = 3	m∠B = 65°
	DF = 3.846	m∠C = 70°
Ratio	$\frac{DE}{AB} = 3$ $\frac{EF}{BC} = 3$ $\frac{DF}{AC} = 3$	

4. They are similar, because corresponding sides are proportional and corresponding angles are congruent.

Activity 2

1.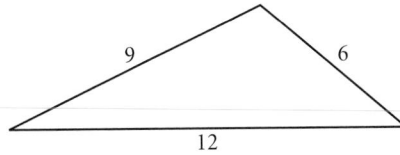

2.

2. Check students' drawings.

3.

	Sides	Angles
△ABC	AB = 2	m∠A = 47°
	BC = 3	m∠B = 104°
	AC = 4	m∠C = 29°
△DEF	DE = 6	m∠A = 47°
	EF = 9	m∠B = 104°
	DF = 12	m∠C = 29°
Ratio	$\frac{DE}{AB} = 3$ $\frac{EF}{BC} = 3$ $\frac{DF}{AC} = 3$	

4. They are similar, because corresponding sides are proportional and corresponding angles are congruent.

Activity 3

1. 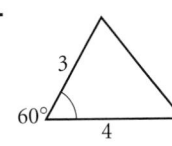 2.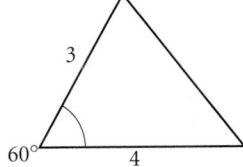

3.

	Sides	Angles
△ABC	AB = 3	m∠A = 74°
	BC = 4	m∠B = 60°
	AC = 3.6	m∠C = 46°
△DEF	DE = 6	m∠A = 74°
	EF = 8	m∠B = 60°
	DF = 7.2	m∠C = 46°
Ratio	2 2 2	

4. They are similar, because corresponding sides are proportional and corresponding angles are congruent.

5. sides, sides, included angles, similar

CRITICAL THINKING (page 519)
Because both are redundant to the AA Triangle Similarity Theorem.

Exercises

Communicate

1. We have the following labels for convenience:

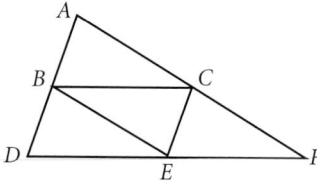

BC is a midsegment, so $\frac{AF}{AC} = \frac{AD}{AB} = 2$. Also, $\angle A$ is shared by both $\triangle ABC$ and $\triangle ADF$. Thus, by the SAS Similarity Theorem, $\triangle ABC \sim \triangle ADF$. Similarly, it follows that $\triangle BDE \sim \triangle ADF$ and $\triangle CEF \sim \triangle ADF$. Then, EB, CE, and BC are proportional to FA, AD, and DF, respectively; so, by the SSS Similarity Theorem, $\triangle ECB \sim \triangle ADF$.

2. Yes, the HL Similarity Theorem is a version of the SAS Similarity Theorem. The HL Similarity Theorem states that two right triangles are similar if there are a pair of corresponding legs and a pair of corresponding hypotenuses that are proportional.

3. Consider a square and a rectangle that is not a square. Each angle in both quadrilaterals is a right angle, so corresponding angles are congruent. But one is not congruent to a dilation of the other, so they are not similar.

4. Consider a square and a rhombus. For each shape, all sides are equal, so corresponding sides are proportional. But one is not congruent to a dilation of the other, so they are not similar.

Lesson 8.4, pages 525–532

Exercises

Communicate

1. $\frac{c}{d} = \frac{a}{b}, \frac{c}{a} = \frac{d}{b}, \frac{c}{c+d} = \frac{a}{a+b}, \frac{d}{c+d} = \frac{b}{a+b}$.

Note: Many others are possible.

2. No. Consider the isosceles 45°-45°-90° right triangle and any isosceles non-right triangle. They are not similar, because one is not congruent to a dilation of the other.

3. Yes. Each angle in an equilateral triangle is a 60° angle. So, by the AA Similarity Postulate, any two equilateral triangles are similar.

4. No. Consider a 45°-45°-90° right triangle and a 30°-60°-90° right triangle. Corresponding sides are not proportional.

5. By joining the endpoints of the "legs" of the capital letter A by a segment, a triangle with a segment parallel to the base is formed. Thus, the Side-Splitting Theorem applies to it. It would also apply if the A were slanted (italic), because the crossbar would remain horizontal and thus parallel to the base.

Practice and Apply

35. Given: $\frac{DB}{AD} = \frac{EC}{AE}$

Prove: $\overline{DE} \parallel \overline{BC}$

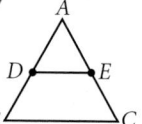

Statements	Reasons
$\frac{DB}{AD} = \frac{EC}{AE}$	Given
$1 + \frac{DB}{AD} = 1 + \frac{EC}{AE}$	Addition Property
$\frac{AD}{AD} + \frac{DB}{AD} = \frac{AE}{AE} + \frac{EC}{AE}$	Substitution Property
$\frac{AD+DB}{AD} = \frac{AE+EC}{AE}$	Addition of fractions
$AD + DB = AB$ $AE + EC = AC$	Segment Addition Postulate
$\frac{AB}{AD} = \frac{AC}{AE}$	Substitution Property
$\angle A \cong \angle A$	Reflexive Property
$\triangle ABC \sim \triangle ADE$	SAS Similarity Theorem
$\angle ADE \cong \angle B$	$\triangle ABC \sim \triangle ADE$
$\overline{DE} \parallel \overline{BC}$	Converse of the Corresponding Angles Postulate

Lesson 8.5, pages 533–542

Exercises

Communicate

1. Consider two similar triangles of which an accessible distance of a side of the smaller triangle is known and corresponds to an inaccessible distance of a side of the larger triangle that is unknown. Then, since the triangles are similar, a proportion can be set up to find the inaccessible distance.

2. No. The converse of the Proportional Altitudes Theorem is not true. Notice, for example, that the altitude of either triangle could fall outside the triangle, with its altitude and base measures remaining as shown.

3. Sample answer: Assume the object to be measured is vertical, like a flag pole or a face of a building. Set up a vertical stick. By AA Similarity, the triangle formed by the stick, its shadow, and a ray of the sun, is similar to the triangle formed by the object in question, its shadow, and another ray of the sun (The sun's rays are parallel and strike the ground at the same angle in each triangle. Also, one angle in each triangle is a right angle.) Use the measurements of the stick and its shadow and the measurement of the shadow of the object to set up a proportion and find the object's height.

4. Sample answer: Assume the object is vertical, as in Exercise 3. Set up a stick at some distance from the object. Establish a line of sight from (a) the top of the object to (b) the top of the stick to (c) a point on the ground (a mirror on the ground can be used for this purpose). Two similar triangles are formed by the object, the stick, the line of sight, and a line on the ground. Use the height of the stick and the distances along the ground from (a) the object to the stick and (b) the object to the sighting point on the ground to set up a proportion and find the object's height.

Lesson 8.6, pages 543–551

Activity 1

1. $\frac{9}{1}$ **2.** $\frac{4}{1}$ **3.** $\frac{9}{4}$ **4.** $\frac{16}{9}$

CHECKPOINT ✔

5. $\frac{a^2}{b^2}$

Activity 2

1. $\frac{27}{1}$ **2.** $\frac{27}{8}$ **3.** $\frac{8}{1}$ **4.** $\frac{8}{1}$

CHECKPOINT ✔

5. $\frac{a^3}{b^3}$

Activity 3

1. 8, 125, $\sqrt{125}$ or $5\sqrt{5}$, 1,000,000, 1000

2. Cross-sectional radius of leg bone is about 223.61 cm. Cross-sectional radius of leg bone is about 883.88 cm. Cross-sectional radius of leg bone is 2500 cm.

CHECKPOINT ✔

3. The cross-sectional radius scale factor increases faster than the height scale factor, so the legs would be proportionally thicker.

Exercises
Communicate

1. The ratio of their volumes is the cube of the ratio of their edges.

2. They are similar, because all spheres and cubes have the same shape and they have essentially only one linear dimension. No, all cylinders are not similar.

3. Find the ratio of their surface areas, take the square root, and then cube that result to get the ratio of their volumes.

4. The volume is multiplied by 4.

5. The amount of weight a structure can support is proportional to its cross-sectional area. As the height increases by a factor of n, the cross-sectional area required increases by a factor of $\sqrt{n^3}$.

Lesson 9.1, pages 564–572

Activity

CHECKPOINT ✔

4. Equilateral; 60°
The triangle is equilateral since all of its sides are the same length.

8. It is a regular hexagon since all of the sides and angles are the same measurement.

CHECKPOINT ✔

9. Each of the central angles measures 60° since each of the triangles is equilateral. Their sum is 360° since there are six triangles in the hexagon.

TRY THIS (page 568)
$L = \frac{5\pi}{2} \approx 7.85$ units

CRITICAL THINKING (page 568)
Yes. If the circles have different radii, then the lengths of arcs with the same angle measure are different.

Activity

2. $\triangle APB$ and $\triangle CPD$ are congruent by the SSS Postulate.
$\angle APB \cong \angle CPD$ so $\overset{\frown}{AB} \cong \overset{\frown}{CD}$

CHECKPOINT ✔

3. If two chords are congruent, then the corresponding arcs are congruent.

4. If two chords of congruent circles are congruent, then the corresponding arcs are congruent.

5.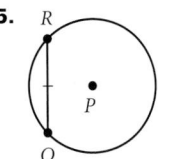

If $\odot P$ and $\odot M$ are congruent and $\overline{QR}$ and $\overline{NO}$ are congruent, then $\triangle RPQ$ is congruent to $\triangle NMO$ by the SSS Postulate. Thus $\angle RPQ \cong \angle NMO$. Hence $\overset{\frown}{RQ} \cong \overset{\frown}{NO}$ In a circle, or in congruent circles, the arcs of congruent chords are congruent.

Exercises
Communicate

1. True. A diameter is a chord that contains the center of the circle.

2. False. Only one endpoint of a radius is on the circle.

3. True. The endpoints of a chord are on the circle.

4. True. Use the triangles formed by joining the endpoints of each chord to the center of the circle.

5. True. Use the triangles formed by joining the endpoints of each arc to the center of the circle.

6. There is more than one way: show that the central angles of the arcs are congruent or show that the chords of the arcs are congruent.

7. All semicircles of a circle are congruent since the central angle of any semicircle is 180°.

Practice and Apply

44. Congruent

Proof:

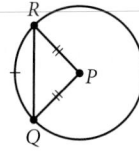 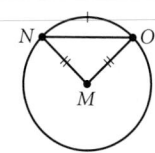

Since $\overarc{QR}$ is congruent to $\overarc{NO}$, $\angle QPR$ is congruent to $\angle NMO$. Since $\overline{RP}$ is congruent to $\overline{OM}$ and $\overline{QP}$ is congruent to $\overline{NM}$, $\triangle QPR$ is congruent to $\triangle NMO$ by the SAS Postulate. Therefore, $\overline{QR}$ is congruent to $\overline{NO}$ because CPCTC.

Lesson 9.2, pages 573–579

CRITICAL THINKING (page 573)
A tangent line touches a circle at exactly one point. A secant line "cuts" the circle into two pieces.

Activity 1

3. As point R gets closer and closer to Q, m$\angle PQR$ gets closer and closer to 90°. When $R = Q$, $\angle PQR$ is a right angle.

CHECKPOINT ✔
4. If a line is tangent to a circle, then the line is perpendicular to a radius of the circle drawn to the point of tangency.

Activity 2

3. $\overline{AX}$ and $\overline{BX}$ are congruent.

CHECKPOINT ✔
4. A radius that is perpendicular to a chord of a circle bisects the chord.

---

5. Given: $\overline{PR} \perp \overline{AB}$
Prove: $\overline{AX} \cong \overline{BX}$

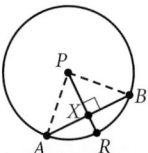

Statements	Reasons
$\overline{AP} \cong \overline{BP}$	Both $\overline{AP}$ and $\overline{BP}$ are radii.
$\overline{PX} \cong \overline{PX}$	Reflexive Property of Congruence
$\angle PXA$ and $\angle PXB$ are right angles	Definition of perpendicular
$\angle PXA \cong \angle PXB$	Angle Congruence Postulate
$\triangle PXA \cong \triangle PXB$	Hypotenuse-Leg Congruence Theorem
$\overline{AX} \cong \overline{BX}$	Polygon Congruence Postulate

Activity 3

1. In both diagrams, $\overline{PR}$ is perpendicular to $\overline{AB}$.

2. As X gets closer to R, both A and B get closer to R. When X touches R, then A, B, X, and R coincide.

CHECKPOINT ✔
If a line is perpendicular to a radius of a circle at its endpoint on the circle, then the line is tangent to the circle.

CRITICAL THINKING (page 575)
By the Pythagorean Theorem, $c^2 = a^2 + b^2$. Since a^2 is positive, $c^2 > b^2$. Since c and b are positive numbers, $c > b$. The proof that $c > a$ is similar.

Exercises
Communicate

1. A line may intersect a circle at two points, one point, or never.

2. A secant intersects a circle at two points.

3. An infinite number of lines are tangent to a circle since there are an infinite number of points on a circle.

4. Given a point on a circle, there is only one tangent line. This is a result of the Tangent Theorem.

5. The wheel is tangent to the ground where it touches.

Guided Skills Practice

9. $100 = 64 + 36$
      $10^2 = 8^2 + 6^2$
   $(AC)^2 = (BC)^2 + (AB)^2$

Thus, $\triangle ABC$ is a right triangle and $\overline{BC} \perp \overline{AB}$ Thus $\overleftrightarrow{AB}$ is tangent to $\odot C$ by the Converse of the Tangent Theorem.

Lesson 9.3, pages 580–587

CRITICAL THINKING (page 580)

$m\overset{\frown}{AVC} + m\overset{\frown}{AC} = 360°$ This result is true for all possible inscribed angles in a circle.

Activity 1

CHECKPOINT ✔

4. $m\angle AVC = \frac{1}{2}m\overset{\frown}{AC}$

Activity 2

PART I:

1. $m\angle 1 = m\angle 2$

2. $m\angle 1 + m\angle 2 = m\angle 3$

3.

m∠1	m∠2	m∠3	m$\overset{\frown}{AC}$
20°	20°	40°	40°
30°	30°	60°	60°
40°	40°	80°	80°
$x°$	$x°$	$2x°$	$2x°$

$\angle 3$ is the central angle corresponding to $\overset{\frown}{AC}$.

CHECKPOINT ✔

4. $m\angle 1 = \frac{1}{2}m\overset{\frown}{AC}$

PART II:

1. $m\angle 1 = \frac{1}{2}m\overset{\frown}{AX}$

$m\angle 4 = \frac{1}{2}m\overset{\frown}{XC}$

2. $m\overset{\frown}{AXC} = m\overset{\frown}{AX} + m\overset{\frown}{CX}$

m∠1	m$\overset{\frown}{AX}$	m∠4	m$\overset{\frown}{CX}$	m∠AVC	m$\overset{\frown}{AXC}$
20°	40°	20°	40°	40°	80°
30°	60°	20°	40°	50°	100°
40°	80°	50°	100°	90°	180°
$x°$	$2x°$	$y°$	$2y°$	$x° + y°$	$2(x + y)°$

CHECKPOINT ✔

3. $m\angle AVC = \frac{1}{2}m\overset{\frown}{AXC}$

PART III:

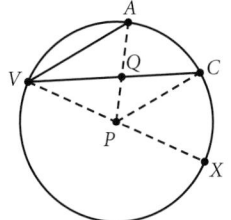

Paragraph Proof:

Let $m\angle AVC = x°$ and let $m\angle VCP = y°$. Since $\triangle VCP$ is isosceles, $m\angle CVP = y°$. Then $m\angle AVP$ is $x° + y°$ and since $\triangle AVP$ is isosceles, we have $m\angle VAP = x° + y°$.
Then $m\angle AQV = 180° - x° - (x° + y°)$.

$$= 180° - 2x° - y°$$

The $m\angle CQP = 180° - 2x° - y°$ by the Vertical Angles Theorem. Thus from $\angle QPC$,

$$m\angle APC = 180° - y° - (180° - 2x° - y°)$$

$$= 2x°.$$

Since $\angle APC$ is the central angle corresponding to $\overset{\frown}{AC}$

$m\overset{\frown}{AC} = 2x°$ so $m\angle AVC = \frac{1}{2}m\overset{\frown}{AC}$

CHECKPOINT ✔

The measure of an angle inscribed in a circle is equal to one-half the measure of the intercepted arc.

Exercises

Communicate

1. $\angle BAD$ or $\angle CBA$

2. $m\angle BAD = \frac{1}{2}m\overset{\frown}{BD}$, so $m\overset{\frown}{BD} = 2m\angle BAD = 44°$

3. $m\angle 1 = m\angle BOD$ by the Vertical Angles Theorem.
$m\angle BOD = m\overset{\frown}{BD}$, so $m\angle 1 = 44°$.

4. $m\overset{\frown}{AC} = m\angle 1 = 44°$

5. Since $\overline{OA}$ and $\overline{OB}$ are both radii, $\triangle OAB$ is isosceles, so $m\angle 2 = 22°$.

6. They are both one-half the measure of the intercepted arc.

Practice and Apply

33. 200°   **34.** 160°   **35.** 80°   **36.** 180°

37. 180°; The sum of the measures of the interior angles of a quadrilateral is 360°.

38. 80°   **39.** 200°   **40.** 100°   **41.** 180°

42. 180°; The sum of the measures of the interior angles of a quadrilateral is 360°.

43. $2x°$ **44.** $360° - 2x°$ **45.** $180° - x°$ **46.** 180°

47. 180°; The sum of the measures of the interior angles of a quadrilateral is 360°.

49. The longest sides of the triangles intersect at the center of the circle. Both right angles intercept an arc that measures 180°, so the intercepted arcs are semicircles. Thus, the longest sides of the triangles are diameters of the circles, which intersect at the center of the circle

Look Beyond

62. Each of the four areas is adjacent to the other three areas.

63. Sample answer:

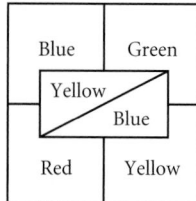

Lesson 9.4, pages 588–597

Activity 1

1. m∠AVC = 90°
m$\overset{\frown}{AV}$ = 180°

CHECKPOINT ✔
The relationship is the same.

2.

m$\overset{\frown}{AV}$	m∠1	m∠2	m∠PVC	m∠AVC
120°	120°	30°	90°	60°
100°	100°	40°	90°	50°
80°	80°	50°	90°	40°
$x°$	$x°$	$\left(90 - \frac{x}{2}\right)°$	90°	$\frac{x°}{2}$

CHECKPOINT ✔
The measure of an acute secant-tangent angle with its vertex on the circle is one-half the measure of its intercepted arc.

m$\overset{\frown}{AXV}$	m∠1	m∠2	m∠PVC	m∠AVC
200°	160°	10°	90°	100°
220°	140°	20°	90°	110°
240°	120°	30°	90°	120°
$x°$	$360° - x°$	$\frac{x°}{2} - 90°$	90°	$\frac{x°}{2}$

CHECKPOINT ✔
The measure of an obtuse secant-tangent angle with its vertex on a circle is one-half the measure of its intercepted arc.

CHECKPOINT ✔
4. If a tangent and a secant (or a chord) intersect on a circle at the point of tangency, then the measure of the angle formed is one-half the measure of its intercepted arc.

Activity 2

1. m∠AVC = m∠1 + m∠2

2.

m$\overset{\frown}{AC}$	m$\overset{\frown}{BD}$	m∠1	m∠2	m∠AVC	m∠DVB
160°	40°	80°	20°	100°	100°
180°	70°	90°	35°	125°	125°
200°	60°	100°	30°	130°	130°
$x_1°$	$x_2°$	$\frac{x_1°}{2}$	$\frac{x_2°}{2}$	$\frac{x_1° + x_2°}{2}$	$\frac{x_1° + x_2°}{2}$

CHECKPOINT ✔
3. The measure of an angle formed by two secants or chords that intersect in the interior of the circle is one-half the sum of the measures of the arcs intercepted by the angle and its vertical angle.

Activity 3

1. m∠1 = m∠AVC + m∠2

2.

m$\overset{\frown}{BD}$	m$\overset{\frown}{AC}$	m∠1	m∠2	m∠AVC
200°	40°	100°	20°	80°
250°	60°	125°	30°	95°
100°	50°	50°	25°	25°
$x_1°$	$x_2°$	$\frac{x_1°}{2}$	$\frac{x_2°}{2}$	$\frac{x_1° - x_2°}{2}$

CHECKPOINT ✔
3. The measure of an angle formed by two secants that intersect in the exterior of circle is one-half the difference of the measures of the intercepted arcs.

Exercises
Communicate

1. one-half the measure of the intercepted arc

2. one-half the measure of the intercepted arc

3. one-half the sum of the measures of the arcs intercepted by the angle and its vertical angle

4. one-half the difference of the measures of the intercepted arcs

---

Lesson 9.5, pages 600–609

Activity 1

3. $\overline{XA} \cong \overline{XB}$

4. $\overline{AP} \cong \overline{BP}$ since both are radii of ⊙P. ∠PAX and ∠PBX are both right angles by the Tangent Theorem. Hence, ∠PAX ≅ ∠PBX. $\overline{PX} \cong \overline{PX}$ by the Reflexive Property. Thus, △AXP ≅ △BXP by the SAS Postulate. $\overline{XB} \cong \overline{XA}$ by the Polygon Congruence Postulate.

CHECKPOINT ✔
5. If two segments are tangent to a circle from the same external point, then the segments are of equal length.

Activity 2

PART I:
3. △AXD, △BXC
△AXD ~ △BXC
4. $\frac{AX}{BX} = \frac{XD}{XC}$
AX · XC = BX · XD

5. By the Arc-Intercept Corollary, m∠XAD = m∠XBC, so ∠XAD ≅ ∠XBC. By the Reflexive Property, ∠AXD ≅ ∠BXC, so by the AA Similarity Postulate, △AXD ~ △BXC. Thus $\frac{AX}{BX} = \frac{XD}{XC}$, and by cross multiplying,
we get that $AX \cdot XC = BX \cdot XD$.

CHECKPOINT ✔
If two secants intersect outside a circle, the product of the lengths of one secant segment and its external segment equals the product of the lengths of the other secant segment and its external segment.

PART II:

1. When $B = D$, $XB = XD$.

2. $AX \cdot XC = BX \cdot XD = (BX)^2$

3. If a secant and a tangent intersect outside a circle, then the product of the lengths of the secant segment and its external segment equals the length of the tangent squared.

Activity 3

2. △AXD and △BXC

3. m∠DAX = m∠CBX since they intercept $\overset{\frown}{CD}$

4. m∠XCB = m∠XDA since they intercept $\overset{\frown}{AB}$
m∠AXD = m∠BXC since they are vertical angles △AXD and △BXC are similar.

5. $\frac{DX}{XA} = \frac{CX}{XB} \Rightarrow DX \cdot XB = AX \cdot XC$

6. By the Arc-Intercept Corollary, m∠DAX = m∠CBX and m∠XCB = m∠XDA, so ∠DAX ≅ ∠CBX and ∠XCB = ∠XDA. Thus by the AA Similarity Postulate, △DXA ~ △CXB so that $\frac{DX}{XA} = \frac{CX}{XB}$ or $DX \cdot XB = AX \cdot XC$.

CHECKPOINT ✔
If two chords intersect inside a circle, then the product of the lengths of the segments of one chord equals the product of the lengths of the segments of the other chord.

Exercises
Communicate

1. When two secants of a circle intersect at a point outside the circle, the segments formed on each line are related as in the given statement. Theorem 9.5.2.

2. When a secant and a tangent of a circle intersect at a point outside the circle, the segments formed on each line are related as in the given statement. Theorem 9.5.3.

3. No. In the examples shown there are no relationships between the lengths of segments. One quantity is 0 on each side of the statement given in the Theorem.

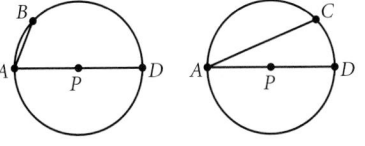

4. There is no relationship. One quantity is 0 on each side of the statement given in the Theorem.

5. Sample answer: For example, 3 points on a circle will form two chords. The intersection of the perpendicular bisectors of those chords is the center of the circle.

Lesson 9.6, pages 610–617

CRITICAL THINKING (page 611)
The radius of the circle changes.

CHECKPOINT ✔ (page 612)
If $x_1^2 + y_1^2 = r^2$, then (x_1, y_1) would be on the circle since it is the same distance from the center as all other points on the circle.

Exercises
Communicate

1. To find y-intercepts: set $x = 0$ and solve for y.
$$0^2 + y^2 = 4$$
$$y^2 = 4$$
$$y \pm 2$$
Therefore, the x-intercepts are (0, 2) and (0, −2).
To find x-intercepts, set $y = 0$ and solve for x.
$$x^2 + 0^2 = 4$$
$$x^2 = 4$$
$$x \pm 2$$
Therefore, the x-intercepts are (2, 0) and (−2, 0).

2. To find y-intercepts: set $x = 0$ and solve for y.
$$(0 - 2)^2 + (y + 2)^2 = 4$$
$$4 + (y + 2)^2 = 4$$
$$(y + 2)^2 = 0$$
$$y = -2$$
Therefore, the y-intercept is (0, −2).
To find x-intercepts: set $y = 0$ and solve for x.
$$(x - 2)^2 + (0 + 2)^2 = 4$$
$$(x - 2)^2 + 4 = 4$$
$$(x - 2)^2 = 0$$
$$x = 2$$
Therefore, the x-intercept is (2, 0).

3. Yes, the graph of a circle with no x- or y-intercepts is a circle that is contained entirely in one of the quadrants.

4. First, subtract the $(x - h)^2$ term from both sides. Next take the square root of both sides, then add k to both sides. Separating the positive and negative square roots gives two equations for y in terms of x.

5. $\sqrt{(x-h)^2 + (y-k)^2}$
The expression is the radius of the circle.

Practice and Apply

51. $(x+5)^2 + (y+4)^2 = 9$

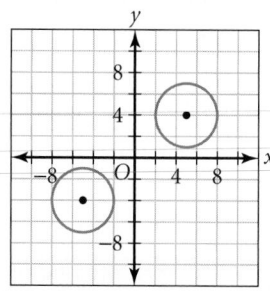

52. $y = \frac{3}{4}x + \frac{25}{2}$

53. $(x-4)^2 + (y-3)^2 = 25$

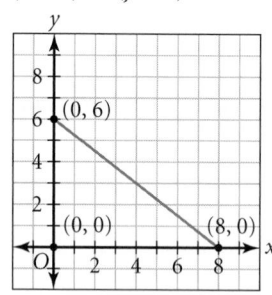

Chapter 9 Review and Assessment

21.
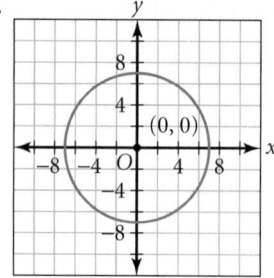

22.
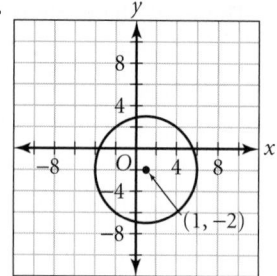

Lesson 10.1, pages 630–638

Activity 1

1–4. Answers may vary. The ratio of opposite to adjacent should be between 1.19 and 0.58, for angles between 30° and 50°.

CHECKPOINT ✔

5. Using the same angle but different side lengths produces approximately the same ratio.

Activity 2

2.

θ	$\tan\theta$
15°	0.27
30°	0.58
45°	1
60°	1.73
75°	3.73

4. $y = \tan\theta$

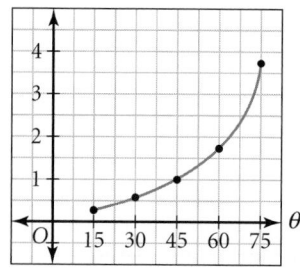

CHECKPOINT ✔

5. The graph increases at an increasing rate. The graph gets steeper as θ increases.

Activity 3

1–2. Answers may vary. The drawing should look like the one shown.

3. Sample answer: $\angle Z = 35°$ and $YZ = 1.2$ cm. Tangent of $\angle Z \approx 0.70$.

CHECKPOINT ✔

4. Sample answer: $\tan Z = \frac{XY}{YZ}$ so $0.70 = \frac{XY}{YZ}$ and $0.70 = \frac{XY}{1.2}$. So, $XY = (0.70)(1.2) \approx 0.84$

5. Sample answer: $XY = 0.84$ cm on the diagram. The actual distance is 0.84310 = 8.4 meters. Possible sources of error are small inaccuracies in measurement and rounding of the measurements to the nearest degree or centimeter.

Exercises

Communicate

1. Answers may vary. Sample answer: "Tri-gon" could mean a three-sided polygon, which is a triangle. So the name could come from the study of triangles.

2. The tangent ratio increases as the angle gets larger, when the angle is between 0° and 90°.

3. As the angle approaches 0°, the tangent ratio approaches 0. Tangent of 0° = tan 0 = 0. This answer makes sense, since if a triangle could have an angle measure of 0°, the side opposite it would have no length, so the ratio would be $\frac{0}{\text{adjacent}} = 0$.

4. As the angle approaches 90°, the tangent ratio gets very large. The calculator gives an error message when trying to calculate tan 90°. In a right triangle, the opposite side becomes very large and the adjacent side approaches zero as the angle approaches 90°.

5. It does not matter which units are used, as long as the units for the opposite side and the adjacent side are the same. The units in the ratio will cancel, leaving no units.

Lesson 10.2, pages 639–646

Activity 1

1. As θ increases, the ratio of $\frac{\text{opposite}}{\text{hypotenuse}}$ increases, so when θ gets close to 0°, the opposite side approaches 0, so $\sin\theta$ gets close to 0. When θ gets close to 90°, the length of the opposite side is close to the length of the hypotenuse, so $\sin\theta = \frac{\text{opposite}}{\text{hypotenuse}}$ approaches 1. Conjecture: $\sin 0° = 0$ and $\sin 90° = 1$.

2. As θ increases, the ratio of $\frac{\text{adjacent}}{\text{hypotenuse}}$ decreases, so the value of $\cos\theta$ decreases. When θ gets close to 0°, the length of the opposite side is close to the length of the hypotenuse, so the ratio approaches 1. When θ gets close to 90°, the hypotenuse is much larger than the adjacent side, so the ratio approaches 0. Conjecture: $\cos 0° = 1$ and $\cos 90° = 0$

CHECKPOINT ✔

3. Sample answer: The values of $\sin\theta$ increase as θ increases from 0° to 90°. The values of $\cos\theta$ decrease as θ increases from 0° to 90°. The values of both $\sin\theta$ and $\cos\theta$ are between 0 and 1 for values of θ between 0° and 90°.

4.

θ	0°	10°	20°	30°	40°	50°	60°	70°	80°	90°
$\sin\theta$	0	0.17	0.34	0.50	0.64	0.77	0.87	0.94	0.98	1
$\cos\theta$	1	0.98	0.94	0.87	0.77	0.64	0.50	0.34	0.17	0

5.

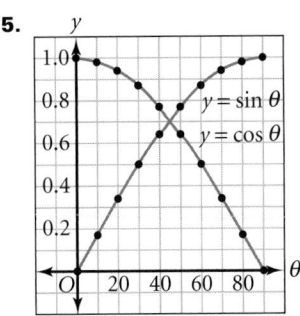

The graphs verify the conclusions made in step 3, since $\sin\theta$ increases and $\cos\theta$ decreases, as θ increases from 0° to 90°.

Activity 2

PART I:

1.

θ	$\sin\theta$	$\cos\theta$	$\frac{\sin\theta}{\cos\theta}$	$\tan\theta$
20°	0.3420	0.9397	0.3640	0.3640
40°	0.6428	0.7660	0.8391	0.8391
60°	0.8660	0.5000	1.7321	1.7321

CHECKPOINT ✔

2. The values in the tangent column are the same as the values in the $\frac{\sin\theta}{\cos\theta}$ column.
$$\tan\theta = \frac{\sin\theta}{\cos\theta}$$

3. $\dfrac{\sin\theta}{\cos\theta} = \dfrac{\frac{\text{opposite}}{\text{hypotenuse}}}{\frac{\text{adjacent}}{\text{hypotenuse}}}$

$= \dfrac{\text{opposite}}{\text{hypotenuse}} \cdot \dfrac{\text{hypotenuse}}{\text{adjacent}}$

$= \dfrac{\text{opposite}}{\text{adjacent}}$

Yes, $\tan\theta = \dfrac{\text{opposite}}{\text{adjacent}}$, so $\dfrac{\sin\theta}{\cos\theta} = \tan\theta$.

PART II:

1.

θ	$\sin\theta$	$\cos\theta$	$(\sin\theta)^2 + (\cos\theta)^2$
20°	0.3420	0.9397	0.9999 or 1.0000
40°	0.6428	0.7660	1.0000
60°	0.8660	0.5000	1.0000

CHECKPOINT ✔

2. All the values are very near 1. $(\sin\theta)^2 + (\cos\theta)^2 = 1$

3. Step 1: by the definition of sine and cosine. Step 2: by a property of exponents. Step 3: adding fractions with like denominators. By the Pythagorean theorem, $a^2 + b^2 = c^2$, so simplifying the ratio gives $\frac{c^2}{c^2} = 1$.

Exercises

Communicate

1. Sample answer: $\tan\theta = \frac{O}{A}$, $\sin\theta = \frac{O}{H}$, $\cos\theta = \frac{A}{H}$

2. $\sin\theta = \frac{3}{5} \Rightarrow \theta \sin^{-1}\left(\frac{3}{5}\right) \approx 37°$

$\cos\theta = \frac{4}{5} \Rightarrow \theta \cos^{-1}\left(\frac{4}{5}\right) \approx 37°$

$\tan\theta = \frac{3}{4} \Rightarrow \theta \tan^{-1}\left(\frac{3}{4}\right) \approx 37°$

The answers agree for each way.

3. Since $\sin\theta$ approaches 0 and $\cos\theta$ approaches 1, as θ approaches 0°, $\tan\theta = \frac{\sin\theta}{\cos\theta}$ should approach $\frac{0}{1} = 0$. Since $\sin\theta$ approaches 1 and $\cos\theta$ approaches 0, as θ approaches 90°, $\tan\theta = \frac{\sin\theta}{\cos\theta}$ should approach $\frac{1}{0}$, which is undefined. So $\tan\theta$ should be undefined at $\theta = 90°$.

4. a. Not an identity, since it has one solution, $x = 1$.
 b. This is an identity, since the equation is true for all values of n.
 c. Not an identity, since it is true only for specific values of a, b, and c.
 d. This is an identity, since it is true for all values of θ between 0° and 90°.

Look Beyond

62. a. $(90 - \theta)$ **b.** $(90 - \theta)$ **c.** $(90 - \theta)$ **d.** $(90 - \theta)$

Lesson 10.3, pages 647–653

Activity 1

1. The hypotenuse has length 1, so the x-coordinate of P' is $\frac{\sqrt{3}}{2} \approx 0.866$ and the y-coordinate is $\frac{1}{2} = 0.5$ So the coordinates of P' are $(0.866, 0.5)$.

2. The angle made from the negative side of the x-axis is $210° - 180° = 30°$. The triangle formed has a hypotenuse of length 1, so the x-coordinate of P'' is $-\frac{\sqrt{3}}{2} \approx -0.866$ and the y-coordinate is $-\frac{1}{2} = -0.5$ So the coordinates of P'' are $(-0.866, -0.5)$.

3. The angle made from the positive side of the x-axis, measured clockwise, is $360° - 330° = 30°$. The triangle formed has a hypotenuse of length 1, so the x-coordinate of P''' is $\frac{\sqrt{3}}{2} \approx 0.866$ and the y-coordinate is $-\frac{1}{2} = -0.5$ So the coordinates of P''' are $(0.866, -0.5)$.

CHECKPOINT ✔

4. a. and **b.**

Rotation Angle θ	x-coordinate of image point	y-coordinate of image point	$\cos \theta$	$\sin \theta$
30°	$\frac{\sqrt{3}}{2} \approx 0.866$	$\frac{1}{2} = 0.5$	0.866	0.5
150°	$-\frac{\sqrt{3}}{2} \approx -0.866$	$\frac{1}{2} = 0.5$	−0.866	0.5
210°	$-\frac{\sqrt{3}}{2} \approx -0.866$	$-\frac{1}{2} = -0.5$	−0.866	−0.5
330°	$\frac{\sqrt{3}}{2} \approx 0.866$	$-\frac{1}{2} = -0.5$	0.866	−0.5

5. The cosine of an angle is the x-coordinate of the image point. The sine of an angle is the y-coordinate of the image point.

CHECKPOINT ✔

6. y; x

Activity 2

1.

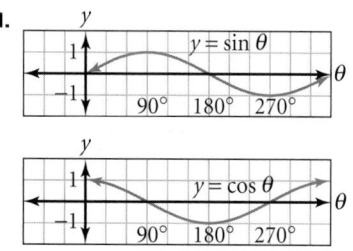

CHECKPOINT ✔

2.

Quadrant of image point	Sign of $\sin \theta$	Sign of $\cos \theta$
I(0° to 90°)	+	+
II(90° to 180°)	+	−
III(180° to 270°)	−	−
IV(270° to 360°)	−	+

TRY THIS (page 650)

$\sin^{-1}(0.5736) \approx 35°$

The second angle is $180° - 35° = 145°$

Exercises

Communicate

1. Sample answer: A rotating ray can form angles with degrees greater than 360°, since it can make an unlimited number of complete rotations around the origin.

2. Sample answer: In a unit circle, the angle can be greater than 90°, and it can be negative. The trigonometric ratios are calculated by using the angle formed from the horizontal.

3. sin 90° is the y-coordinate and cos 90° is the x-coordinate of the image point, P', obtained by rotating $P(1, 0)$ through an angle of 90°. P' would be the point on the positive y-axis which touches the unit circle. So, sin 90° = 1 and cos 90° = 0.

4. sin 180° is the y-coordiante and cos 180° is the x-coordinate of the image point, P', obtained by rotating $P(1, 0)$ through an angle of 180°. P' would be the point on the negative x-axis which touches the unit circle. So, sin 180° = 0 and cos 180° = −1.

5. Since the cosine of an angle gives the x-coordinate of the image point on the unit circle, in Quadrant III the x-coordinate will be negative, so cosine will be negative.

6. Since the sine of an angle gives the y-coordinate of the image point on the unit circle, in Quadrant II the y-coordinate will be positive, so sine will be positive.

7. As the point X rotates uniformly about the center, it's vertical motion is described by a sine function.

Lesson 10.4, pages 654–662

Activity 1

1. Answers may vary. Students should measure sides a, b, and c, and angles A, B, and C.

2. Answers may vary. The ratios $\frac{\sin A}{a}$, $\frac{\sin B}{b}$, and $\frac{\sin C}{c}$ should be equal in each row.

CHECKPOINT ✔

3. Conjecture: $\frac{\sin A}{a} = \frac{\sin B}{b} = \frac{\sin C}{c}$ for any triangle with sides a, b, and c and with angles A, B, and C opposite the given sides, respectively.

Exercises
Communicate

1. The law of sines cannot be used because no side lengths are given.

2. The law of sines cannot be used because no angle measures are given.

3. The law of sines requires one angle and the side across from the angle, along with one other side or angle. This triangle does not have the side across from the angle given.

4. The ambiguous case occurs when the SSA combination is given and the triangle is not a right triangle. In this case, the side across from the angle can "swing" to form two different triangles, one acute and one obtuse.

Lesson 10.5, pages 663–669

TRY THIS (page 665)
$b^2 = a^2 + c^2 - 2ac \cos B$
$b^2 = 4000^2 + 3600^2 - 2(4000)(3600) \cos 84°$
$b^2 = 28{,}960{,}000 - 28{,}800{,}000 \cos 84°$
$b^2 = 25{,}949{,}580.26$
$b \approx 5094.07$ ft
Since 5094.07 < 5280 ft, campers at Site 1 and 3 will be able to communicate directly with each other.

Exercises
Communicate

1. $c^2 = a^2 + b^2 - 2ab \cos C$
$c^2 = a^2 + b^2 - 2ab \cos 90°$
$c^2 = a^2 + b^2 - 2ab \cdot 0$
$c^2 = a^2 + b^2$
For a right triangle, the law of cosines gives the Pythagorean theorem.

2. Law of sines. The triangle has two angles and a side given.

3. Law of cosines. The triangle has three sides given.

4. No. If only three angles are given, then there are an infinite number of triangles possible with those angle measures.

Lesson 10.6, pages 672–679

CHECKPOINT ✔ (page 673–Top)
Windsock: the magnitude of the vector is the speed of the wind, and its direction is the direction of the wind.
Tugboat: the magnitude of the vector is the speed of the tugboat, and its direction is the direction of the motion of the tugboat. Car: the magnitude of the vector is the speed of the car, and its direction is the direction of travel along the track.

CHECKPOINT ✔ (page 673–Middle)
In each picture, a possible reference direction is North. Other answers are possible, such as the direction of the current could be used for a reference direction of the tugboat, and the orientation of the racetrack could be used as a reference direction of the car.

Exercises
Communicate

1. The magnitude of a vector its length.

2. The direction of the vector is the angle the vector makes with a reference point, usually the positive x-axis.

3. The magnitude is 175 knots and the direction is northwest.

4. The magnitude of the boat is 15 knots and the direction is upstream. The magnitude of the current is 3 knots and the direction is downstream.

5. The magnitudes of each vector are equal, and the vectors are pointing in opposite directions.

Lesson 10.7, pages 680–685

CRITICAL THINKING ✔ (page 680)
$x' = 0 \cdot \cos \theta - 0° \sin \theta = 0$
$y' = 0 \cdot \sin \theta - 0 \cos \theta = 0$
The point (0, 0) does not change under a rotation of $\theta°$ about the origin.

Activity

PART I
1–3. Sample answers:

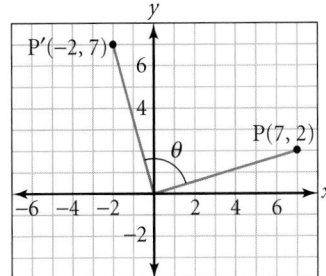

Choose point P(7, 2) and $\theta = 90°$. Then
$x' = 7 \cdot \cos 90° - 2 \cdot \sin 90° = -2$
$y' = 7 \cdot \sin 90° - 2 \cdot \cos 90° = 7$
$P' = (-2, 7)$
The angle of rotation is 90°.

PART II

1.

θ	0°	90°	180°	270°	360°
sin θ	0	1	0	−1	0
cos θ	1	0	−1	0	1

Exercises

Communicate

1. Sample answer: A counterclockwise rotation of 90° and a clockwise rotation of −270° produce the same result. A 90° rotation of point (x, y) produces:
$x' = x \cos 90° - y \sin 90° = -y$
$y' = x \sin 90° + y \cos 90° = x$
A −270° rotation of point (x, y) produces:
$x' = x \cos(-270°) - y \sin(-270°) = -y$
$y' = x \sin(-270°) + y \cos(-270°) = x$
The results are the same.

2. Sample answer:
Choose $\theta = 30°$ and 150°.
$$\begin{bmatrix} \cos 30° & -\sin 30° \\ \sin 30° & \cos 30° \end{bmatrix} = \begin{bmatrix} 0.87 & -0.5 \\ 0.5 & 0.87 \end{bmatrix}$$
$$\begin{bmatrix} \cos 150° & -\sin 150° \\ \sin 150° & \cos 150° \end{bmatrix} = \begin{bmatrix} -0.87 & -0.5 \\ 0.5 & -0.87 \end{bmatrix}$$
The sine values for two supplementary angles are the same while the cosine values change sign.
$\cos \theta = -\cos(180° - \theta)$ and $\sin \theta = \sin(180° - \theta)$.

3. Sample answer:
Choose $\theta = 20°$ and 70°.
$$\begin{bmatrix} \cos 20° & -\sin 20° \\ \sin 20° & \cos 20° \end{bmatrix} = \begin{bmatrix} 0.94 & -0.34 \\ 0.34 & 0.94 \end{bmatrix}$$
$$\begin{bmatrix} \cos 70° & -\sin 70° \\ \sin 70° & \cos 70° \end{bmatrix} = \begin{bmatrix} 0.34 & -0.94 \\ 0.94 & 0.34 \end{bmatrix}$$
$\sin \theta = \cos(90° - \theta)$ and $\cos \theta = \sin(90° - \theta)$ describe the relationship between the entries of each matrix.

4. $\theta = 0°$: $\begin{bmatrix} \cos 0° & -\sin 0° \\ \sin 0° & \cos 0° \end{bmatrix} = \begin{bmatrix} 1 & 0 \\ 0 & 1 \end{bmatrix}$

$\theta = 90°$: $\begin{bmatrix} \cos 90° & -\sin 90° \\ \sin 90° & \cos 90° \end{bmatrix} = \begin{bmatrix} 0 & -1 \\ 1 & 0 \end{bmatrix}$

$\theta = 180°$: $\begin{bmatrix} \cos 180° & -\sin 180° \\ \sin 180° & \cos 180° \end{bmatrix} = \begin{bmatrix} -1 & 0 \\ 0 & -1 \end{bmatrix}$

$\theta = 270°$: $\begin{bmatrix} \cos 270° & -\sin 270° \\ \sin 270° & \cos 270° \end{bmatrix} = \begin{bmatrix} 0 & 1 \\ -1 & 0 \end{bmatrix}$

$\theta = 360°$: $\begin{bmatrix} \cos 360° & -\sin 360° \\ \sin 360° & \cos 360° \end{bmatrix} = \begin{bmatrix} 1 & 0 \\ 0 & 1 \end{bmatrix}$

When multiplying each of the matrices shown above by $\begin{bmatrix} x \\ y \end{bmatrix}$ you get all the ordered pairs in **2.** of Part II of the Activity on page 681. For example,
$\theta = 90°$: $\begin{bmatrix} \cos 90° & -\sin 90° \\ \sin 90° & \cos 90° \end{bmatrix} = \begin{bmatrix} 0 & -1 \\ 1 & 0 \end{bmatrix}$
This is $(-y, x)$ which is the rotation of 90° of a point (x, y).

Practice and Apply

31. $\begin{bmatrix} \cos 30° & -\sin 30° \\ \sin 30° & \cos 30° \end{bmatrix} \begin{bmatrix} \cos 40° & -\sin 40° \\ \sin 40° & \cos 40° \end{bmatrix}$

$\approx \begin{bmatrix} 0.34 & -0.94 \\ 0.94 & 0.34 \end{bmatrix}$;

$\begin{bmatrix} \cos 70° & -\sin 70° \\ \sin 70° & \cos 70° \end{bmatrix} \approx \begin{bmatrix} 0.34 & -0.94 \\ 0.94 & 0.34 \end{bmatrix}$.

A rotation of 30° followed by a rotation of 40° is equivalent to a rotation of 70°.

32.

Multiply $\begin{bmatrix} 4 \\ 0 \end{bmatrix}$ repeatedly by the rotation matrix for a rotation of 72°, $\begin{bmatrix} .309 & -.951 \\ .951 & .309 \end{bmatrix}$, until all vertices are found. The other vertices are $\begin{bmatrix} 1.24 \\ 3.80 \end{bmatrix}, \begin{bmatrix} -3.24 \\ 2.35 \end{bmatrix},$
$\begin{bmatrix} -3.24 \\ -2.35 \end{bmatrix}$ and $\begin{bmatrix} 1.24 \\ -3.80 \end{bmatrix}$.

33. $[R_0] = \begin{bmatrix} \cos 0° & -\sin 0° \\ \sin 0° & \cos 0° \end{bmatrix} = \begin{bmatrix} 1 & 0 \\ 0 & 1 \end{bmatrix}$
The product of any rotation matrix and the identity matrix gives the rotation matrix .

34. The product is $[R_0] = \begin{bmatrix} 1 & 0 \\ 0 & 1 \end{bmatrix}$ because six 60° rotations give a 360° rotation, which is equivalent to a 0° rotation.

35. The product of the two matrices gives the identity matrix.
$[R_{35}] \approx \begin{bmatrix} 0.82 & -0.57 \\ 0.57 & 0.82 \end{bmatrix}$
$[R_{-35}] \approx \begin{bmatrix} 0.82 & 0.57 \\ -0.57 & 0.82 \end{bmatrix}$
$[R_{35}] \times [R_{-35}] = \begin{bmatrix} 1 & 0 \\ 0 & 1 \end{bmatrix}$

36. Let $A = (x, y)$. Since $O = (0, 0)$,
$(OA)^2 = \sqrt{(x - 0)^2 + (y - 0)^2} = x^2 + y^2$.
$A' = (x \cdot \cos \theta - y \sin \theta, x \sin \theta + y \cos \theta)$, so
$(OA')^2 = \left(\sqrt{(x \cos \theta - y \sin \theta - 0)^2 + (x \sin \theta - y \cos \theta - 0)^2}\right)^2$
$= x^2(\cos \theta)^2 - 2x \cos \theta y \sin \theta + y^2(\sin \theta)^2 + x^2(\sin \theta)^2$
$\quad + 2x \sin \theta y \cos \theta + y^2(\cos \theta)^2$
$= x^2(\cos \theta)^2 + y^2(\sin \theta)^2 + x^2(\sin \theta)^2 + y^2(\cos \theta)^2$
$= x^2[(\cos \theta)^2 + (\sin \theta)^2] + y^2[(\cos \theta)^2 + (\sin \theta)^2]$
$= x^2(1) + y^2(1)$
$= x^2 + y^2$
Since $(OA)^2 = x^2 + y^2 = (OA')^2$, the distances are equal.

37. a. (21.37, 46.98)

   b. The coordinates are $(30\cos(6t°) + 42\sin(6t°),$
      $30\sin(6t°) - 42\cos(6t°))$ after rotating for t minutes.

Lesson 11.1, pages 698–705

Activity 1

2. Sample answer:

ℓ	s	$\dfrac{\ell}{s}$	$\dfrac{s}{\ell - s}$
20	10	2	1
18	10	1.8	1.25
16	10	1.6	1.67
16.2	10	1.62	1.62

3. Check students' drawings.

CHECKPOINT ✔

4. 1.62

Activity 2

1–2. Check students' drawings.

CHECKPOINT ✔

3. Each rectangle should be a golden rectangle.

Activity 3

2. $BC = 2$; $MB = 1$; $MC = \sqrt{5}$

CHECKPOINT ✔

4. $AD = 2$; $AE = AM + ME = 1 + \sqrt{5}$; $\dfrac{AE}{AD} = \dfrac{1 + \sqrt{5}}{2} = \phi$

Exercises

Communicate

1. If you take a golden rectangle and remove from it a square with side length equal to the length of the shorter side, the result is another golden rectangle. (Also, if you take a golden rectangle and add to it a square equal to the length of the longer side, the result is another golden rectangle.)

2. Take a square with side s. Take the midpoint of one side and connect it to an opposite corner. The result is a right triangle with side lengths $\frac{s}{2}$, s, and (using the Pythagorean Theorem) $\sqrt{\frac{5s^2}{4}} = \frac{s\sqrt{5}}{2}$. Extend opposite sides of the square by a length of $\frac{s\sqrt{5}}{2} - \frac{s}{2}$. This will create a golden rectangle with length $\ell = s + \frac{s\sqrt{5}}{2} - \frac{s}{2} = s\left(\frac{1 + \sqrt{5}}{2}\right)$.

3. Sample answer: The ratio for a standard (8.5×11 in.) piece of paper is ≈ 1.29. Students could measure doors, windows, chalkboards, textbooks, or pictures.

4. Answers will vary, but should be close to 1.62.

Lesson 11.2, pages 706–711

Activity 1

PART I

1.

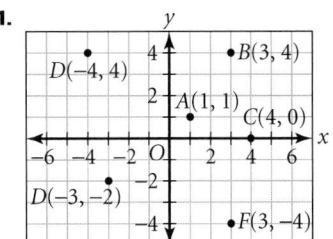

2.

Point	Coordinates	Taxidistance from O
A	$(1, 1)$	2
B	$(3, 4)$	7
C	$(4, 0)$	4
D	$(-4, 4)$	8
E	$(-3, -2)$	5
F	$(3, -4)$	7

CHECKPOINT ✔

3. $|x| + |y|$

PART II

1.

(x_1, y_1)	(x_2, y_2)	x_1	x_2	y_1	y_2	Taxidistance
A	B	1	3	1	4	5
A	D	1	-4	1	4	8
A	E	1	-3	1	-2	7
C	D	4	-4	0	4	12
D	E	-4	-3	4	-2	7
E	F	-3	3	-2	-4	8
B	F	3	3	4	-4	8

CHECKPOINT ✔

2. $|x_1 - x_2| + |y_1 - y_2|$

Activity 2

1.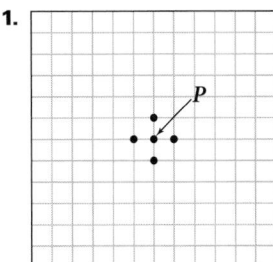

2.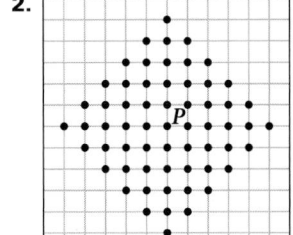

3.

Radius	Number of points on circle	Circumference
1	4	8
2	8	16
2	12	24
4	16	32
5	20	40
r	$4r$	$8r$

CHECKPOINT ✔

4. $\dfrac{\text{circumference}}{\text{diameter}} = \dfrac{8r}{2r} = 4$

Exercises

Communicate

1. You measure distance between points along a grid, the way a taxicab travels in city blocks.

2. Given two points $P(x_1, y_1)$ and $Q(x_2, y_2)$ the distance between the points is the sum of the absolute values of the difference in x-coordinates and the difference in y-coordinates, i.e. $PQ = |x_1 - x_2| + |y_1 - y_2|$.

3. π is not used in taxicab geometry since the definitions of circumference and diameter are different from that of Euclidian geometry. The taxicab equivalent for π is 4, the ratio of circumference to diameter.

4. Sample answer: Finding the shortest distance to travel for cars along roads.

5. Sample answer: Stoplights, traffic congestion, speed limit.

Practice and Apply

27.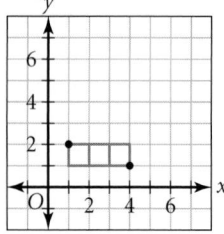

There are 4 minimum distance pathways.

28.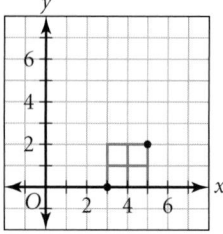

There are 6 minimum distance pathways.

31.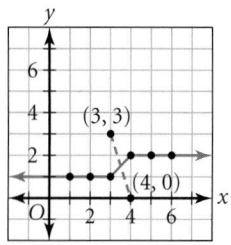

The bisector has a bend in it. However, the bend consists of only two points, rather than the three formed in Exercise 29.

32.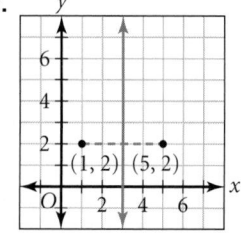

Any point with coordinates $(3, y)$ lies on the bisector. The bisector is the same line obtained when using Euclidian geometry.

33. Because the taxicab distance between $(2, 0)$ and $(3, 2)$ is odd, a perpendicular bisector cannot be formed.

34. A taxicab circle with radius r and center (h, k) has the general formula $|x - h| + |y - k| = r$.

35. The points are: $(-1, 0)$, $(-1, 1)$, $(-1, 2)$, $(0, -1)$, $(0, 3)$, $(1, -1)$, $(1, 3)$, $(2, -1)$, $(2, 3)$, $(3, 0)$, $(3, 1)$, and $(3, 2)$.

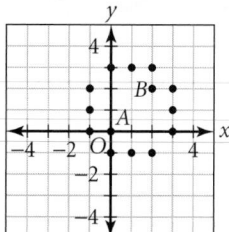

Sample answer: The points lie on the sides of a square with side length 4 that has its corner points deleted.

36. Sample answer: If one corner of the city grid is $(0, 0)$, then put call boxes at $(4m, 4n)$ and $(2 + 4m, 2 + 4n)$ where m and n are non-negative integers.

37. The set of points (x, y) will be the intersection of the circle of radius 10 centered about $(0, 0)$ and its interior and the circle of radius 8 centered about $(5, 4)$ and its interior. Jenny should look for places within the region shown, including the edge points and sides.

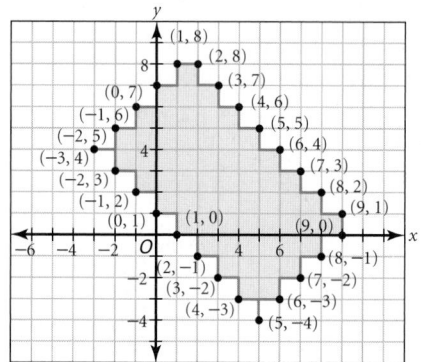

Lesson 11.3, pages 712–720

Activity

	# of vertices	# of odd vertices	# of even vertices	Euler path?
a.	4	0	4	Yes
b.	4	2	2	Yes
c.	4	4	0	No
d.	4	0	4	Yes
e.	4	4	0	No
f.	5	2	3	Yes
g.	5	4	1	No
h.	7	6	1	No

CHECKPOINT ✔
A graph contains an Euler path if and only if there are at most two odd vertices.

CRITICAL THINKING (page 714)
Start at one of the odd vertices and end at the other odd vertex.

CRITICAL THINKING (page 715)
No.

Exercises
Communicate

1. Sample answer: The Konigsberg problem is the problem of arranging a route so that a traveler would cross each of the bridges once and only once. It is impossible with the arrangement involved, because each of the four land masses has an odd number of bridges delivering the traveler to them.

2. Sample answer: The easiest way would be to build another bridge from Land Mass C to Land Mass A called h. A possible solution to this new problem would be to start at D and cross g-h-c-d-e-f-b-a.

3. An Euler path is a path through a graph which travels along each edge exactly once and has nothing to do with ending at the starting point. An Euler circuit is an Euler path that begins and ends at the same point.

4. Sample answer: The problem has to do with the positions of the bridges relative to the land masses.

Practice and Apply

34.

35. Yes. All vertices of the graph are even.

Look Back

41. If a shape is a square, then the shape is a rectangle. Converse: If a shape is a rectangle, then the shape is a square.

42. If a shape is a square, then the shape is a parallelogram with four congruent sides and four congruent angles. Converse: If a shape is a parallelogram with four congruent sides and four congruent angles, then the shape is a square.

43. If a shape is a dodecagon, then the shape is a polygon with 12 sides. Converse: If a shape is a polygon with 12 sides, then the shape is a dodecagon.

44. If θ is an angle with $\theta < 45°$, then $\sin \theta < \cos \theta$. Converse: If θ is an angle with $\sin \theta < \cos \theta$, then $\theta < 45°$.

Lesson 11.4, pages 721–728

CRITICAL THINKING
Yes. Figures 2 and 4 are toplogically equivalent. Figure 3 is not topologically equivalent to 2 and 4, because the segment in the center cannot be compressed to a point.

Exercises
Communicate

1. Two figures are topologically equivalent if one of them can be stretched, shrunk or otherwise distorted into the other without cutting, tearing or intersecting itself, or compressing a segment or curve to a point.

2. Yes. You can stretch the square into the hexagon.

3. No. The first is a simple closed curve, the second is not.

4. The Jordan Curve Theorem is true for spheres, but not for tori, so the two are not topologically equivalent. The sphere has Euler Characteristic 2, while the torus has Euler Characteristic 0, so the two are not topologically equivalent.

5. Sample answer: sometimes it is easier to recognize a difference (such as an Euler characteristic) than it is to demonstrate how one figure can be transformed into the other.

Lesson 11.5, pages 729–731

CRITICAL THINKING (page 731)
Yes, if the postulates are understood in a certain way. For example, a "line" on a sphere is a great circle. It can be extended indefinitely in the sense that one can keep traveling on it without coming to an endpoint. Any theorem that follows from just Euclid's first four postulates must be true on the surface of a sphere.

Exercises
Communicate

1. Sample answer: The fifth postulate of Euclid states that if two lines are crossed by a third line and if the sum of the internal angles on one side is less than 180°, then the two lines will meet.

2. **a.** No. There are no parallel lines in a sphere
   b. Infinitely many. Any line that does not contain P is parallel to ℓ.

3. Continually traveling on the great circle extends a line indefinitely.

4. The distance from any point to the edge of the universe in Poincaré's model is considered to be infinite, so a line can be thought of as extending indefinitely.

Practice and Apply

24.

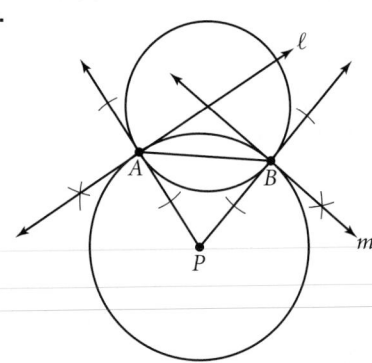

25. If A and B are endpoints of a diameter, then ℓ and m are parallel and never intersect. In this case, the line from A to B would be a straight arc of an infinitely large circle.

26.

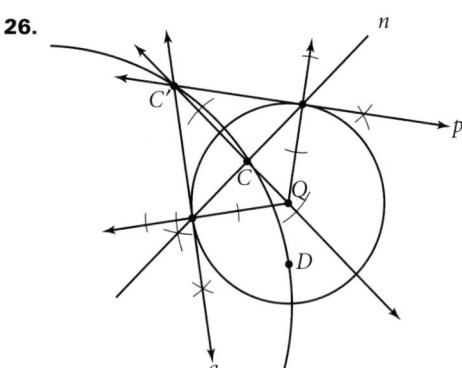

27. Infinitely many lines can be drawn through A that do not intersect ℓ. This shows that the Parallel Postulate does not hold in Poincaré's system, because there are infinitely many lines through A that are parallel to ℓ.

34. Sample answer: The model is a convex surface (or concave, depending on how it is turned).

35. Sample answer: The triangles form a flat surface.

36. Sample answer: The model will not lie flat and it is neither concave nor convex overall. The surface is "wavy," going up and down.

37. Sample answer: The models are all composed of a number of equilateral triangles about a given vertex. In the first model the angle measures of the triangles around the vertex add to less than 360°, so the model will not lie flat: it is convex. In the second model the sum of the angle measures equals 360°, so the model will lie flat. In the third model the sum of the angle measures is greater than 360, so it will not lie flat: it "bunches up."

38. Sample answer: The surface cannot be extended to infinitely many triangles; it will come back on itself to form a closed surface like a sphere. (In fact, an icosohedron will be formed, which is a polyhedral approximation of a sphere.)

39. Sample answer: A Euclidean plane will be formed. This is called a tessellation of the plane by equilateral triangles and will extend to infinitely many triangles.

40. Sample answer: A surface will be formed that has no overall tendency to be either concave or convex and will extend to infinitely many triangles.

Lesson 11.6, pages 738–746

Activity 1

5.

Interation	0	1	2	3	4	5	n
Number of segments	1	2	4	8	16	32	2^n
Combined length	27 cm	18 cm	12 cm	8 cm	$\frac{16}{3}$	$\frac{32}{9}$	$27\left(\frac{2}{3}\right)^n$

CHECKPOINT ✔

6. As the number of iterations increases, there are more segments with smaller combined lengths. The combined lengths of the segments approaches zero.

Activity 2

1. The first 12 rows of Pascal's triangle are shown below.

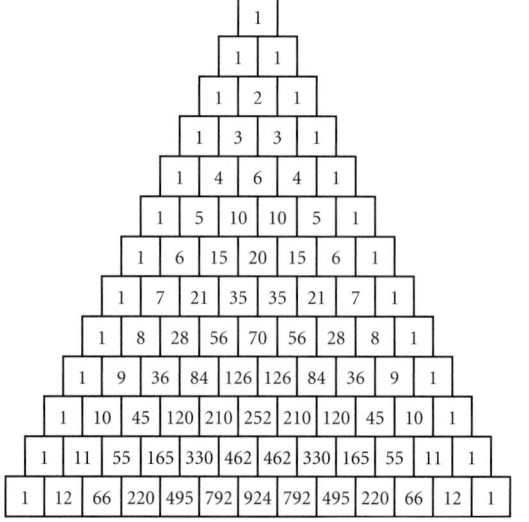

2.

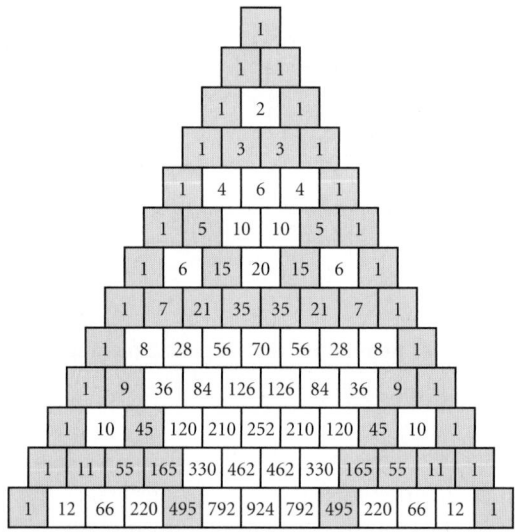

Note that the pattern that occurs in the first 8 rows continues occurring... twice in the next 8 rows and two times in the third 8.

Exercises

Communicate

1. Self-similarity means that the structure of an object looks similar when examined at any level of magnification.

2. The Cantor Dust is a self-similar object. To obtain it, remove the middle third of an interval and continue the process for all remaining intervals.

3. Arnold's coastline = 3 • 24 = 72 feet long. Roberto's coastline = (2.5)(36) = 90 feet long. Mimi's coastline = 2(47) = 94 feet long.

4. In a theoretical fractal coastline, the depth of self-similarity is endless. By taking steps of increasingly small length, the length of the fractal grows without limit.

Lesson 11.7, pages 747–755

Activity 1

PART I

1–5.

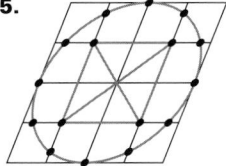

CHECKPOINT ✔

6. Note that collinear points have a collinear image, that straight lines have a linear image, that parallel lines have parallel images, and that intersecting lines have intersecting images.

PART II:

2. The result is a rectangle and its diagonals. The length of the rectangle along the x-axis is twice the side length of the square, and the height along the y-axis is three times the height of the square.

3. The result is a rectangle and its diagonals. The length of the rectangle along the x-axis is three times the side length of the square and the height along the y-axis is twice the side length of the square.

CHECKPOINT ✔

4. Note that collinear points have a collinear image, that straight lines have a linear image, that parallel lines have parallel images, and that intersecting lines have intersecting images.

Activity 2

PART I:
CHECKPOINT ✔

It appears that X, Y, and Z are collinear; collinear.

PART II:

7. It appears that X, Y, and Z are collinear; collinear.

Exercises

Communicate

1. An affine transformation is one which preserves collinearity of points, transforms straight lines to straight lines, transforms intersecting lines to intersecting lines, and preserves parallel lines. All of the transformations are special cases of affine transformations. Dilations also preserve the angles between lines. Translations, reflections, and rotations preserve angles and sizes of geometric objects.

2. A central projection projects points from one line onto another line along rays from a fixed point called the center of projection. A dilation is a central projection in which a consistent scale factor is used to determine where the image points occur. A rotation is a central projection in which the scale factor used to determine where the image points lie is −1. Neither translations or reflections are central projections.

3. There is no concept of size, measurement, or congruence in projective geometry as there is in the other geometries.

4. Near the equator, the distances from the preimage to the images are nearly the same, while near the poles the distances can vary greatly.

Chapter 11 Review and Assessment

8.

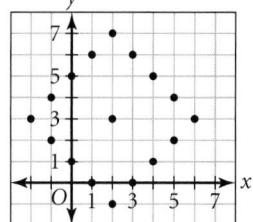

9. Euler path; no Euler circuit

10. Euler path; no Euler circuit

11. Neither

12. Euler path; no Euler circuit

Lesson 12.1, pages 768–775

CRITICAL THINKING (page 770)
Yes. Sample answer: If something is yellow, then it is a fruit. A banana is yellow. Therefore, a banana is a fruit.

Exercises

Communicate

1. Valid, because the *modus ponens* form is used, where *p* is "today is Wednesday" and *q* is "the cafeteria is serving beef stew." *p* is given, therefore *q* follows.

2. Invalid, because the argument has an invalid form (affirming the consequent). Pigs may or may not fly on February 30.

3. Invalid, because the argument has an invalid form (denying the antecedent). Not only men are mortal.

4. Valid, because the *modus tollens* form is used, where *p* is "*y* = *x*" and *q* is "*a* = *b*."

5. Yes. Validity doesn't guarantee the premises are true, so the conclusion may not be true. Validity only guarantees that the conclusion is true if all the premises are true.

Lesson 12.2, pages 776–781

Activity

1.

p	*q*	*p* AND *q*	~(*p* AND *q*)
T	T	T	F
T	F	F	T
F	T	F	T
F	F	F	T

2.

p	*q*	~*p*	~*q*	~*p* OR ~*q*
T	T	F	F	F
T	F	F	T	T
F	T	T	F	T
F	F	T	T	T

3. The last column from Step 1 and the last column from Step 2 are the same. ~ (*p* AND *q*) is equivalent to ~ *p* OR ~ *q*.

4. ~ (*p* AND *q*) is truth functionally equivalent to ~ *p* OR ~ *q*.

Exercises

Communicate

1. A conjunction is true if and only if both of its statements are true.

2. A disjunction is true if one or both of the statements is true.

3. The difference between inclusive or and exclusive or is that the inclusive or means that both statements can be true, the exclusive or means only one statement can be true, not both.

4. Kimba does not like pizza and Lin likes spaghetti. Kimba likes pizza and Lin does not like spaghetti. Kimba does not like pizza and Lin does not like spaghetti.

Practice and Apply

31.

p	*q*	*r*	*p* OR *q*	(*p* OR *q*) OR *r*
T	T	T	T	T
T	T	F	T	T
T	F	T	T	T
T	F	F	T	T
F	T	T	T	T
F	T	F	T	T
F	F	T	F	T
F	F	F	F	F

All three statements must be false in order for (*p* OR *q*) OR *r* to be false.

32.

p	*q*	*r*	*s*	*p* AND *q*	*r* AND *s*	(*p* AND *q*) OR (*r* AND *s*)
T	T	T	T	T	T	T
T	T	T	F	T	F	T
T	T	F	T	T	F	T
T	T	F	F	T	F	T
T	F	T	T	F	T	T
T	F	T	F	F	F	F
T	F	F	T	F	F	F
T	F	F	F	F	F	F
F	T	T	T	F	T	T
F	T	T	F	F	F	F
F	T	F	T	F	F	F
F	T	F	F	F	F	F
F	F	T	T	F	T	T
F	F	T	F	F	F	F
F	F	F	T	F	F	F
F	F	F	F	F	F	F

(*p* AND *q*) OR (*r* AND *s*) is false if and only if both of the statements in the disjunction are false. This happens when one or both of *p* or *q* is false and when one or both of *r* or *s* is false.

33. The sentence is true if any of the following combinations are true:
Flora will cook; Vernon will vacuum
Flora will cook; Vernon will wash the windows
Flora will wash the dishes; Vernon will vacuum
Flora will wash the dishes; Vernon will wash the windows.

34. No

35. a.

Last name	First name	State	Year of birth	Annual income
Mallo	Elizabeth	TX	1956	50,000
Brookshier	Mary	OH	1960	62,000
Lamb	Charles	TX	1951	41,000
Raemsch	Martin	OK	1965	32,000

b.

Last name	First name	State	Year of birth	Annual income
Craighead	Alicia	TX	1955	25,000
Tuggle	Lawrence	LA	1972	20,000
Mallo	Elizabeth	TX	1956	50,000
Tony	Jun	TX	1952	18,000
Lamb	Charles	TX	1951	41,000

Lesson 12.3, pages 784–790

CHECKPOINT ✔ (page 785)
Yes.

CHECKPOINT ✔ (page 785)
No.

CHECKPOINT ✔ (page 786)
No.

CHECKPOINT ✔ (page 786)
Yes.

Exercises
Communicate

1. Converse: If the moon is made of green cheese, then today is February 30. Inverse: If today is not February 30, the moon is not made of green cheese. Contrapositive: If the moon is not made of green cheese, then today is not February 30.

2. Converse: If a triangle is equilateral, then all three sides are congruent. Inverse: If all three sides of a triangle are not congruent, then the triangle is not equilateral. Contrapositive: If a triangle is not equilateral, then all three sides are not congruent.

3. Converse: If I will not buy cereal, then I do not go to the market. Inverse: If I go to the market, then I will buy cereal. Contrapositive: If I will buy cereal, then I go to the market.

4. Converse: If I am not late for school, then the car starts. Inverse: If the car does not start, then I will be late for school. Contrapositive: If I will be late for school, then the car does not start.

5. The statement is false when *a* is true and *b* is false.

Practice and Apply

13. Conditional: False, since AAA does not guarantee triangle congruence. Converse: If two triangles are congruent, then the three angles of one triangle are congruent to the three angles of the other triangle. True; given that two triangles are congruent, their corresponding angles are congruent by definition. Inverse: If the three angles of one triangle are not congruent to the three angles of another triangle, then the triangles are not congruent. True; if all the corresponding angles are not congruent, then the triangles are not congruent. Contrapositive: If two triangles are not congruent, then the three angles of one triangle are not congruent to the three angles of the other triangle. False; if the triangles are not congruent they may still have congruent corresponding angles. They could be similar.

14. Conditional: True, because the sum of two even numbers is always even. Converse: If $p + q$ is an even number, then p and q are even numbers. False; p and q can be odd and their sum even ($3 + 7 = 10$). Inverse: If p and q are not even numbers, then $p + q$ is not an even number. False; if p and q are odd then their sum is even. Contrapositive: If $p + q$ is not an even number, then p and q are not even numbers. True; if $p + q$ is odd then either p or q must be odd but not both.

15. Conditional: True, by the laws of physics. Converse: If the water temperature is less than or equal to 32°F, then it will freeze at normal atmospheric pressure. True, by the laws of physics. Inverse: If water does not freeze at normal atmospheric pressure, then its temperature is greater than 32°F. True, by the laws of physics. Contrapositive: If water's temperature is greater than 32°F, then it will not freeze at normal atmospheric pressure. True, by the laws of physics.

20. Sample answer: Theorem 9.3.3: If two inscribed angles intercept the same arc, then they have the same measure. Converse: If two inscribed angles have the same measure, then they intercept the same arc. False; inscribed angles may have the same measure and not intercept the same arc. Inverse: If two inscribed angles do not intercept the same arc, then they do not have the same measure. False; if two inscribed angles do not intercept the same arc they can still have the same measure. Contrapositive: If two inscribed angles do not have the same measure, then they do not intercept the same arc. True; if two inscribed angles do not have the same measure, then they cannot intercept the same arc.

21. If you are a senior, then you must report to the auditorium.

22. If a point is on the perpendicular bisector of a segment, then it is equidistant from the endpoints of the segment.

23. If she is going to be late, then she will call me.

24. If you do mathematics homework every night, then you will improve your grade in mathematics.

25. All of them are true. *r* if and only if *s* can be written as *r* ⇔ *s*. In other words, *r* implies *s* and *s* implies *r*. So if the statement *r* if and only if *s* is true then:

  a. If *r* is true then *s* must also be true, because *r* implies *s*.

  b. The statement is true because *s* implies *r*.

  c. The statement is true because it is the contrapositive of **a**, and since **a** is true, then this statement is also true.

  d. The statement is true because it is the contrapositive of **b**, and **b** is true.

26. Many of the theorems in the book can be rewritten in biconditional form. Examples: Theorem 4.4.1—The Isosceles Triangle Theorem: Two sides of a triangle are congruent if and only if the angles opposite those sides are congruent. The converse is true, the inverse is true, and the contrapositive is true, so an "if and only if" statement can be substituted in for the IF-THEN statement. This also works with Theorem 9.2.2, the Tangent Theorem: A line is perpendicular to a radius of a circle at its endpoint if and only if the line is tangent to the circle. Any theorem in the book will work if its converse can be proven true, because the inverse has the same truth value as the converse, and the contrapositive has the same truth value as the original statement.

Lesson 12.4, pages 791–797

Exercises

Communicate

1. A contradiction is an "absurd" or "impossible" result. It has the following logical form: *p* AND ~ *p*. That is, a contradiction asserts that a statement and its negation are both true.

2. No, a tiger is a cat and a mammal at the same time.

3. Yes, because the second statement is the negation of the first one.

4. No, because not the same insect is described as a butterfly and not a butterfly.

5. No, because the statements imply that dogs and pets are two different sets which do intersect.

6. The proof starts by assuming the negation of the statement to be proved. Then a contradiction results after logical arguments. Finally, the opposite of the assumption is stated to be true because the assumption must be false.

Guided Skills Practice

7. (Lines ℓ and *m* are parallel) AND (lines ℓ and *m* are not parallel)

8. (△*ABC* is isosceles) AND (△*ABC* is not isosceles)

9. (All squares are rectangles) AND (all squares are not rectangles)

10. (*ABCD* is a square) AND (*ABCD* is not a square)

Practice and Apply

44. $\sin 72° \approx 0.951$; $\cos 72° \approx 0.309$; $\tan 72° \approx 3.078$

45. $\sin 45° \approx 0.707$; $\cos 45° \approx 0.707$; $\tan 45° \approx 1$

46. $\sin 140° \approx 0.643$; $\cos 140° \approx -0.766$; $\tan 140° \approx -0.839$

47. $\sin 5° \approx 0.087$; $\cos 5° \approx 0.996$; $\tan 5° \approx 0.087$

Lesson 12.5, pages 798–804

Activity

PART I

TV before pressing POWER button	TV after pressing POWER button
1	0
0	1

PART II

PLAY button	RECORD button	Video recorder
1	1	1
1	0	0
0	1	0
0	0	0

PART III

Student pedal	Instructor pedal	Brakes
1	1	1
1	0	1
0	1	1
0	0	0

Exercises

Communicate

1. A logic gate is a particular type of electronic circuitry. Logic gates represent building blocks for "smart" electronic devices. Each logic gate has a special symbol, such as NOT, AND, or OR.

2. The OR logic gate represents an inclusive or. One can tell from its input-output table in which in order to get an output value of 1 only one pulse needs to have a value of 1.

3. The input-output tables for networks of logic gates are related to truth tables as follows: 1 corresponds to True 0 corresponds to False.

4. There are $2^3 = 8$ possible combinations in a network of logic gates with three inputs. There are $2^4 = 16$ possible combinations in a network of logic gates with four inputs. The first input has two possible values. For each value of the first input, there are two possible values of the second input. The total number of combinations is $2 \times 2 = 4$, etc.

Chapter 12 Review and Assessment

20.

p	q	r	p OR q	NOT r	(p OR q) AND (NOT) r
1	1	1	1	0	0
1	1	0	1	1	1
1	0	1	1	0	0
1	0	0	1	1	1
0	1	1	1	0	0
0	1	0	1	1	1
0	0	1	0	0	0
0	0	0	0	1	0

22.

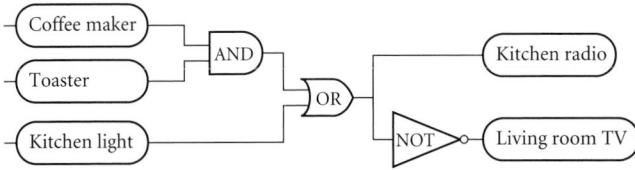

Note: Each appliance is turned on by a 1 input and off by a 0 input. Each appliance sends a 1 output if it is on and a 0 output if it is off.

INDEX

Credits

PHOTOS

Abbreviations used: (t) top, (c) center, (b) bottom, (l) left, (r) right, (bckgd) background, (bdr) border

FRONT COVER: Dale Sanders/Masterfile. **TABLE OF CONTENTS:** Page vi(tl), Michael Newman/PhotoEdit; vi(bc), Gary Vestal/Tony Stone Images(butterfly on daisy); vi(br), Tim Davis/Tony Stone Images(tail); vi(bl), Breck P. Kent/Earth Scenes; vi(bc), Alfred B. Thomas/Earth Scenes(fern); vi(br), Lester Lefkowitz/The Stock Market(Nautilus shell); vii(tl), Joe McDonald/Corbis-Bettmann; viii(tl), Gary Randall/FPG International; viii(br), John Blaustein/Liaison Agency; ix(tl), Michael P. Gadomski/Photo Researchers Inc.; ix(br), Cosmo Condina/Tony Stone Images; x(tl),(b), Christine Galida/HRW Photo; xi(tl), Don Couch/HRW Photo; xi(br), Artbase Inc.; xii(tl), M. Antman/The Image Works; xii(br), Michael Keller/Index Stock Imagery Inc.(beach and ball); xii(bl), ChromaZone Images/Index Stock Imagery Inc.(Iris); xii(br), Artbase Inc.(starfish); xii(bc), Vee Sawyer/MertzStock; xiii(tl), Peter Van Steen/HRW Photo; xiii(br), Michael Collier/Stock Boston; xiv(tl), Chad Ehlers/Tony Stone Images; xiv(br), Dugald Bremner/Tony Stone Images; xv(tl), Don Couch/HRW Photo; xv(br), Steven C. Amstrup Photo; xvi(tl), Tomas Pantin/HRW Photo; xvi(br), Don Couch/HRW Photo; xvii(tl), Tim Davis/Photo Researchers Inc.; xvii(br), Allsport USA/Al Bello; vii(br), Dennis Fagan/HRW Photo. **CHAPTER ONE:** Page 2(br), The Granger Collection, New York; 2(bc), Tim Davis/Science Photo Library/Photo Researchers Inc.; 3(tl), Artbase Inc.; 3(c), Corbis-Bettmann/Leonard de Selva; 3(bckgd), Corbis-Bettmann/Richard T. Nowitz; 3(bl), Artbase Inc.; 4(bckgd), Georgia Tblisi/Sovfoto/Eastphoto(background); 4(c), Artbase Inc.(woman); 5(tc), Gary Vestal/Tony Stone Images(butterfly on daisy); 5(tr), Tim Davis/Tony Stone Images(tail); 5(tl), Breck P. Kent/Earth Scenes; 5(tc), Alfred B. Thomas/Earth Scenes(fern); 5(tr), Lester Lefkowitz/The Stock Market(Nautilus shell); 5(cr), Herbert Bayer, *Structure with Three Squares*. 1967. Painted aluminum. 62.9 cm h x 62.9 cm w x 5.7 cm d. Gift of Jan van der Marck in loving memory of his wife Ingeborg. Photograph ©1999 The Detroit Institute of Arts.; 5(br), Miwako Ikeda/International Stock; 5(bl), Stefano Amantini/Bruce Coleman Inc.; 5(bc), Dale Knuepfer/Bruce Coleman Inc.; 6(tr), Planet Art Collection; 6(b),(br) Peter Van Steen/HRW Photo; 7(bl), Dennis Fagan/HRW Photo; 7(br), Christine Galida/HRW Photo; 7(cr), Robert Sieck Flandes, 1939, National Institute of Anthropology, Mexico; 8(cr), Erich Lessing/Art Resource; 8(cl), The Granger Collection, New York; 8(tr), Addison Geary/Stock Boston; 8(bc),(br), Artbase Inc.; 9(tr), Celestial Image Picture Co./Science Photo Library/Photo Researchers Inc.; 9(br), Artbase Inc.; 10(tr),(cr),(br), Artbase Inc.; 10(bl), Michael J. Howell/International Stock; 11(tl),(c), Artbase Inc.; 13(cr), Alan Levenson/Tony Stone Images; 14(cr), E.R. Degginger/Bruce Coleman Inc.; 16(br), Dennis Fagan/HRW Photo; 17(tc), ©2001 PhotoDisc, Inc./HRW; 17(tr), Michael Abbey/Photo Researchers Inc.; 19(tr), Peter Van Steen/HRW Photo; 24(br), Elliott Smith/International Stock; 25(tr), George Hall/Corbis-Bettmann; 29(bl), Eric Sanford/International Stock; 33(bl), Sam Dudgeon/HRW Photo; 34(br), The Granger Collection, New York; 34(tl), Peter Van Steen/HRW Photo; 35(tr),(cr), H.DE.Marcillac/GLMR/Liaison Agency; 35(c), Frederic Reglain/Gamma Liaison/Liaison Agency; 36(all) Don Couch/HRW Photo; 37(tr), Gabe Palmer/The Stock Market; 37(c),(cr),(br), 38(cr),(br), Don Couch/HRW Photo; 39(tr), Peter Van Steen/HRW Photo; 40(tr), Dennis Fagan/HRW Photo; 41(cr), Michael Newman/PhotoEdit; 42(all), Christine Galida/HRW Photo; 43(tr),(tc), Martha Cooper/Peter Arnold, Inc.; 43(br), Don Couch/HRW Photo; 44(br), Rogers Fund, 1918/The Metropolitan Museum of Art; 45(tr), John Langford/HRW Photo; 47(br), Peter Van Steen/HRW Photo; 49(bc),(br), Dennis Fagan/HRW Photo; 50(tr), Jean Y. Ruszniewski/Tony Stone Images; 50(br), Peter Van Steen/HRW Photo; 52(tr), Dennis Fagan/HRW Photo; 54(tr), Craig Aurness/Corbis-Bettmann(clock); 54(br), Bob Firth/International Stock(boy in canoe); 54(tr), Brian Vikander/Corbis-Bettmann(palace); 54(cl),(cr), Dennis Fagan/HRW Photo; 56(cr), Courtesy of Quarto Publishing, England(butterfly); 58(cr), Carl Zeiss/Bruce Coleman Inc.; 58(cr), Nuridsany Et Pérennou/Science Source/Photo Researchers Inc.; 63(br), Tom Till Photography(skyline); 67(tr), Christine Galida/HRW Photo; 68(bl), Walter Hodges/Tony Stone Images; 68-69, Dennis Fagan/HRW Photo(paper folding steps); 69(br), Artbase Inc. **CHAPTER TWO:** Page 78(bl), Hulton Getty/Liaison Agency; 78(br), Bruce Ayres/Tony Stone Images; 78(tr), The Granger Collection, New York; 78(cr), Corbis-Bettmann; 78(cr), Chris Thomaidis/Tony Stone Images(chess); 79(tr), (tl), Artbase Inc.; 79(c),80(tr),82(cr),(br), Dennis Fagan/HRW Photo; 86(c), Christine Galida/HRW Photo; 87(br), Dennis Fagan/HRW Photo; 88(bckgd), Artbase Inc.; 89(tr), Corbis-Bettmann/Bob Rowan/Progressive Image(Alaska); 89(cr),(br), Corbis-Bettmann/Joseph Sohm/ChromoSohm Inc.(Utah), (California), (Illinois), (Arizona); 90(tc), Bertram G. Murray Jr./Earth Scenes; 90(tr), Steve Solum/Bruce Coleman Inc.; 90(c), John Shaw/Bruce Coleman Inc.; 90(cr), D. Madison/Bruce Coleman Inc.; 90(cl), Bruce MacDonald/Earth Scenes; 90(tl), Robert Lubeck/Earth Scenes; 91(t), Gerald Gusthall/FPG International; 94(tr), European Space Agency/Science Photo/Photo Researchers Inc.; 95(cl), Henry Holiday; 96(br), Andy Sacks/Tony Stone Images; 97(c), Tate Gallery, London/Art Resource; 97(cr), Alexandre Calder, *Fond jaune*, 1972, Galerie Maeght, © Alexandre Calder/ARS (New-York)1999; 97(br), ©1988 Dan Lamont; 98(br), Joseph Sachs/Still Life Stock; 99(tr), Dennis Fagan/HRW Photo; 103(cl), Joe McDonald/Corbis-Bettmann; 107(tr), Corbis-Bettmann; 111(tr), Ernest Manewal/FPG International; 114(all), Christine Galida/HRW Photo; 115(br), Coco McCoy/Rainbow; 116(bl),117(tr), Dennis Fagan/HRW Photo; 117(br), Artbase Inc.; 120(bc), Everett Collection(*ET*),(*Jurassic Park*),(*Close Encounters*); 120(br), Paramount Pictures/Shooting Star; 122(br), Robert Franz/Corbis-Bettmann; 132(br), Gerard Lacz/Publiphoto. **CHAPTER THREE:** Page 136(tr), Craig D. Wood; 136(cr), Jackie Foryst/Bruce Coleman Inc.; 136(bc), Greg Hursley; 137(cr), Giraudon/Art Resource, NY; 137(tr), Paul Avis/Liaison Agency; 138(tr), John Blaustein/Liaison Agency; 138(tc), Suzanne A. Vlamis/International Stock; 140(tl),(tc),(tr), Don Couch/HRW Photo; 140(cr), Dennis Fagan/HRW Photo; 141(tr), John Shaw/Bruce Coleman Inc.; 142(tr), M. Thonig/H. Armstrong Roberts/Comstock; 142(bl), Joy Spurr/Bruce Coleman Inc.; 142(br), *Symmetry Drawing E85* by M.C. Escher ©1995 Cordon Art-Baarn-Holland. All rights reserved; 146(tc),(tr), Artbase Inc.; 146(b), MapQuest.com, Inc.; 147(tl),(tc),(tr), Copyrighted material of Meredith Corporation, used with their permission. All rights reserved.; 147(cr), The Quilt Complex; 147(br), M. Eastcott/The Image Works; 148(tr), Spencer Jones/Bruce Coleman Inc.; 148(tc), David Madison/Bruce Coleman Inc.; 153(tr), Peter Van Steen/HRW Photo; 154(b), *Symmetry Drawing E121* by M.C. Escher ©1999 Cordon Art-Baarn-Holland. All rights reserved; 155(tr), Gary Randall/FPG International; 157(cr), Milwaukee Art Museum, Gift of Frederick Layton Art League in Memory of Miss Charlotte Partridge and Miss Miriam Frink, Photographed by Richard Beauchamp 1/92; 159(tl), Peter Pearson/Tony Stone Images; 162(tc), Courtesy of NASA; 162(tr), Space Frontiers/Masterfile; 164(tl),166(cr),167(tl), Randal Alhadeff/HRW Photo; 167(br), Daryl Benson/Masterfile; 168(tl),(bl),(cl), Annette Del Zoppo; 168(c), Robin White/Fotolex Associates; 168(tc), Artbase Inc.; 170(cr), Mark E. Gibson Photography; 170(tr), Richard Pharaoh/International Stock; 170(tc), Gene Ahrens/Bruce Coleman Inc.; 171(tr), Randal Alhadeff/HRW Photo; 172(bl), Tony Craddock/Science Photo Library/Photo Researchers Inc.; 173(tr), Randal Alhadeff/HRW Photo; 175(br), Neil Rabinowitz/Corbis-Bettmann; 176(cr), Corbis-Bettmann/The Mariner's Museum; 177(tr), Tom Sanders/The Stock Market; 182(br), Brownie Harris/The Stock Market; 182(tl), Rosemary Weller/Tony Stone Images; 182(cr), Ric Ergenbright/Corbis-Bettmann; 183(tr), Paul G. Adam/Publiphoto; 184(tr), Alex S. MacLean/Landslides; 185(t), Norman Owen Tomalin/Bruce Coleman Inc.; 188(bl), Lester Lefkowitz/Tony Stone Images; 189(tr), Jeff Greenberg/Photo Researchers Inc.; 189(cr), Maresa Pryor/Animals Animals; 190(tr), Alan Thornton/Tony Stone Images; 195(br), Peter Van Steen/HRW Photo; 197(tr), *Symmetry Drawing E25* by M.C. Escher ©1999 Cordon Art-Baarn-Holland. All rights reserved; 197(br), Tomas Pantin/HRW Photo; 204(c), Copyrighted material of Meredith Corporation, used with their permission. All rights reserved; 204(br), Danny Lehman/Corbis-Bettmann. **CHAPTER FOUR:** Page 208(cr), Corbis-Bettmann(Fuller); 208(cr), Robert Fried/Stock Boston(dome); 208(bckgd), D.P.Hershkowitz/Bruce Coleman Inc.; 208(bl), Keith Gunnar/Bruce Coleman Inc.; 209(tr), Dick Dickinson/ International Stock; 210(bckgd), SEF/Art Resource, NY; 210(tc), Rick Smolan/Stock Boston; 210(c), Jean-Claude Lejeune/Stock Boston; 215(cr), John Langford/HRW Photo; 216(tr), John Langford/HRW Photo(earrings); 216(br), Dennis Fagan/HRW Photo; 216(tr), Randal Alhadeff/HRW Photo(girl); 217(tr), Hilary Wilkes/International Stock; 217(br), Randal Alhadeff/HRW Photo; 220(br), Cosmo Condina/Tony Stone Images; 223(tr), J C Carton/Bruce Coleman Inc.; 223(br), Mark E. Gibson Photography; 224(br), Sam Dudgeon/HRW Photo; 225(b), Don Couch/HRW Photo; 225(tr), Peter Van Steen/HRW Photo; 226(tc),(tr), Sullivan/Texastock; 226(cr), Daniel Brody/Stock Boston; 229(cr), Jerry Schad/Photo Researchers Inc.; 230(cr), Michael P. Gadomski/Photo Researchers Inc.; 233(cr), Cliff Hollenbeck/International Stock; 234(bl), David Madison/Bruce Coleman Inc.; 234(br), Chuck Mason/International Stock; 235(tr), *Square Limit* by M.C. Escher ©1999 Cordon Art - Baarn - Hollland. All rights reserved.; 242(tr),(cr), Artbase Inc.; 243(tr), Eric Beggs/HRW Photo; 243(bl), Randal Alhadeff/HRW Photo; 253(tr), Peter Van Steen/HRW Photo; 254(tl), Randal Alhadeff/HRW Photo; 261(tr), Scott Van Osdol/HRW Photo; 269(bl), Courtesy Big Bend National Park; (br), U. S. Geological Survey; 270(bl), Todd Gipstein/National Geographic; 271(tr), William Swartz/Index Stock Imagery Inc.; 274(all), Randal Alhadeff/HRW Photo; 279(tr), Kevin R. Morris/Corbis-Bettmann; 280(cr), 282(all), 283(all) Randal Alhadeff. **CHAPTER FIVE:** Page 292(b), James Montgomery/Bruce Coleman Inc.; 292(cr), Mark E. Gibson Photography; 293(tl), Barrett & Mackay Photography; 293(cl), J-C Carton/Bruce Coleman Inc.; 293(tr), Wolfgang Volz/Bilderberg; 294(tl), NASA/Corbis-Bettmann; 294(cr), O. Louis Mazzatenta/National Geographic Society; 294(tr), David Boyer/National Geographic Society; 297(tr), Peter Van Steen/HRW Photo; 298(tl), Steve Elmore/The Stock Market; 298(tr), Artbase Inc.; 300(br), James Amos/National Geographic Society; 301(br), Jacqui Hurst/The Garden Picture Library; 302(tl), Mark Douet/Tony Stone Images; 303(tr),(tl), Janice K. Wright/Down East Books, Camden, Maine, USA; 304(br), Christine Galida/HRW Photo; 306(br), Barrett & Mackay Photography; 311(tr), Gary Irving/Tony Stone Images; 312(b), James Montgomery/Bruce Coleman Inc.; 314(tr), W. Muller/Peter Arnold, Inc.; 314(cr), Peter Van Steen/HRW Photo; 317(cr),(tr), Randal Alhadeff/HRW Photo; 318(br), F. Gohier/Photo Researchers Inc.; 318(bl), Peter Stephenson/Masterfile; 318(c), Peter Van Steen/HRW Photo; 319(cl), W. Muller/Peter Arnold, Inc.; 320(cr), Robert & Linda Mitchell; 321(c), Columbia University Rare Book and Manuscript Library. Gift of George A. Plimpton; 321(tr), Erich Lessing/Art Resource; 322(bl), Vee Sawyer/Mertzstock; 325(tc), G. Gardner/The Image Works; 328(cl), National Portrait Gallery, Smithsonian Institution/Art Resource, NY; 328(tl), Dinodia Picture Agency; 329(cr), Frank Siteman/Monkmeyer Press Photo; 331(tr),332(b), Christine Galida/HRW Photo; 334(bl), Artbase Inc.; 337(cr), Barrett & MacKay Photography; 337(br), Christine Galida/HRW Photo; 339(tr), L. Kolvoord/The Image Works; 341(tr),342(tr), Don Couch/HRW Photo; 345(tr), Clark James Mishler/The Stock Market; 347(tc), Artbase Inc.; 347(tr), Mike Clemmer; 349(tl), Dennis Fagan/HRW Photo; 352(tl), Steve Myers Studios/International Stock; 354(bl), Peter Van Steen/HRW Photo; 358(tl), Jump Run Productions/The Image Bank (c) 2000; 359(tr), G. Schiele/Publiphoto; 360(bckgd),(tr),361(bckgd),(br),366(b), John Langford/HRW Photo. **CHAPTER SIX:** Page 370(tr), Laguna Design/Science Photo Library/Photo Researchers Inc.; 370(bl), Prof. K. Seddon & Dr. T. Evans, Queen's University Belfast/Science Photo Library/Photo Researchers Inc.; 370(br), Alfred Pasieka/Science Photo Library/Photo Researchers Inc.; 372(tr), Jon Feingersh/Stock Boston; 373(bl), Randal Alhadeff/HRW Photo; 377(cr), Scott Van Osdol/HRW Photo; 377(br), Christine Galida/HRW Photo; 379(tr), Breck P. Kent/Earth Scenes; 381(tr),(cr),385(tr),(cr), Randal Alhadeff/HRW Photo; 385(br), Bair/Monkmeyer Press Photo; 386(b),(br), John Langford/HRW Photo; 387(all), Peter Van Steen/HRW Photo; 388(tr), Charles D. Winters/Photo Researchers Inc.; 389(all),391(br), Dennis Fagan/HRW Photo; 393(cl), Michael Nicholson/Corbis-Bettmann; 393(br),(bc), Peter Van Steen/HRW Photo; 394(br), David Parker/Science Photo Library/Photo Researchers Inc.; 395(bl),(br), Scott Van Osdol/HRW Photo; 396(tr), Artbase Inc.; 396(br), Dennis Fagan/HRW Photo; 401(tl), Michael P. Gadomski/Photo Researchers Inc.; 402(tc),(cl), Roman E. Maeder; 402(cr), ©Wolfram Research Inc.; 406(bl), Peter Van Steen/HRW Photo; 407(tr), Mark E. Gibson Photography; 408(bl), Don Couch/HRW Photo(crocheted ball); 408(bl), Artbase Inc.(soccer ball); 409(tr), Scala/Art Resource; 409(tc), Erich Lessing/Art Resource; 409(bc), Dennis Fagan/HRW Photo; 409(bl), Sam Dudgeon/HRW Photo; 410(tr), Fotomas Index; 410(cr), Donovan Reese/Tony Stone Images; 410(br), Dennis Fagan/HRW Photo; 410(bc), Cindy Verheyden/HRW Photo; 411(br), H. L. Romberg; 412(t), Roger Ressmeyer/Corbis-Bettmann; 412(b), Brunelleschi, "Interno verso l'altare S. Lorenzo"/Art Resource; 413(br), Kathy Bushue/Tony Stone Images; 417(br), Scott Van Osdol/HRW Photo; 418(bl), Sam Dudgeon/HRW Photo; 418-419(bckgd),419(bl), Don Couch/HRW Photo; 419(tl), Corbis-Bettmann(representation of orbits); 419(tl), Science Photo Library/Photo Researchers Inc.(portrait); 421(tl), Randal Alhadeff/HRW Photo; 424(cr), Gary Retherford/Photo Researchers Inc. **CHAPTER SEVEN:** Page 428(br), Chris Drury; 428(bl), Courtesy Danese, New York; 428(tr), Houston Museum of Natural Science; 429(cr), Lutz Schmidt/Reuters/Corbis-Bettmann; 429(bckgd), Fabrizio Bensch REUTERS/Corbis-Bettmann; 430(tr), Renee Lynn/Photo Researchers Inc.; 430(tc),

ILLUSTRATIONS

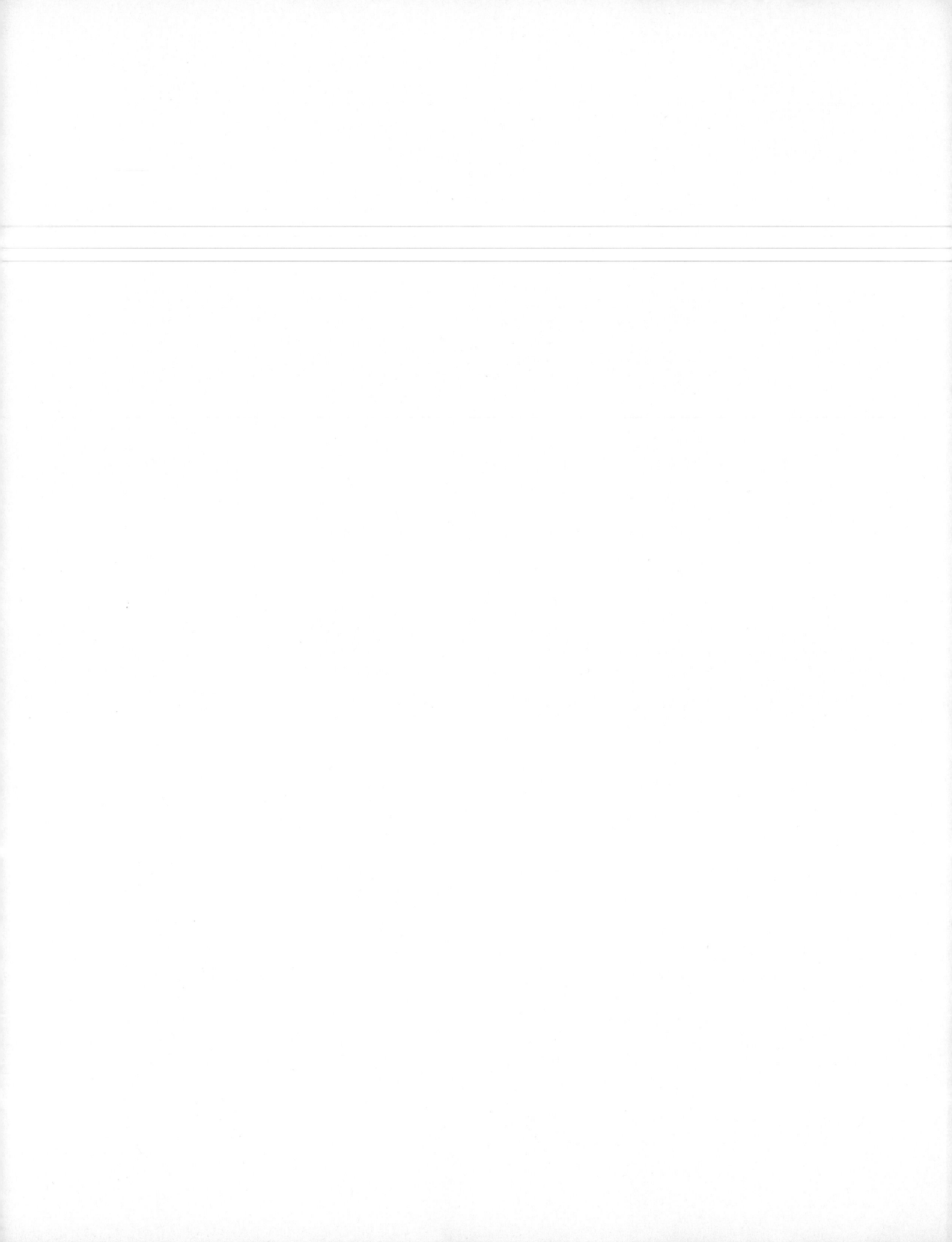

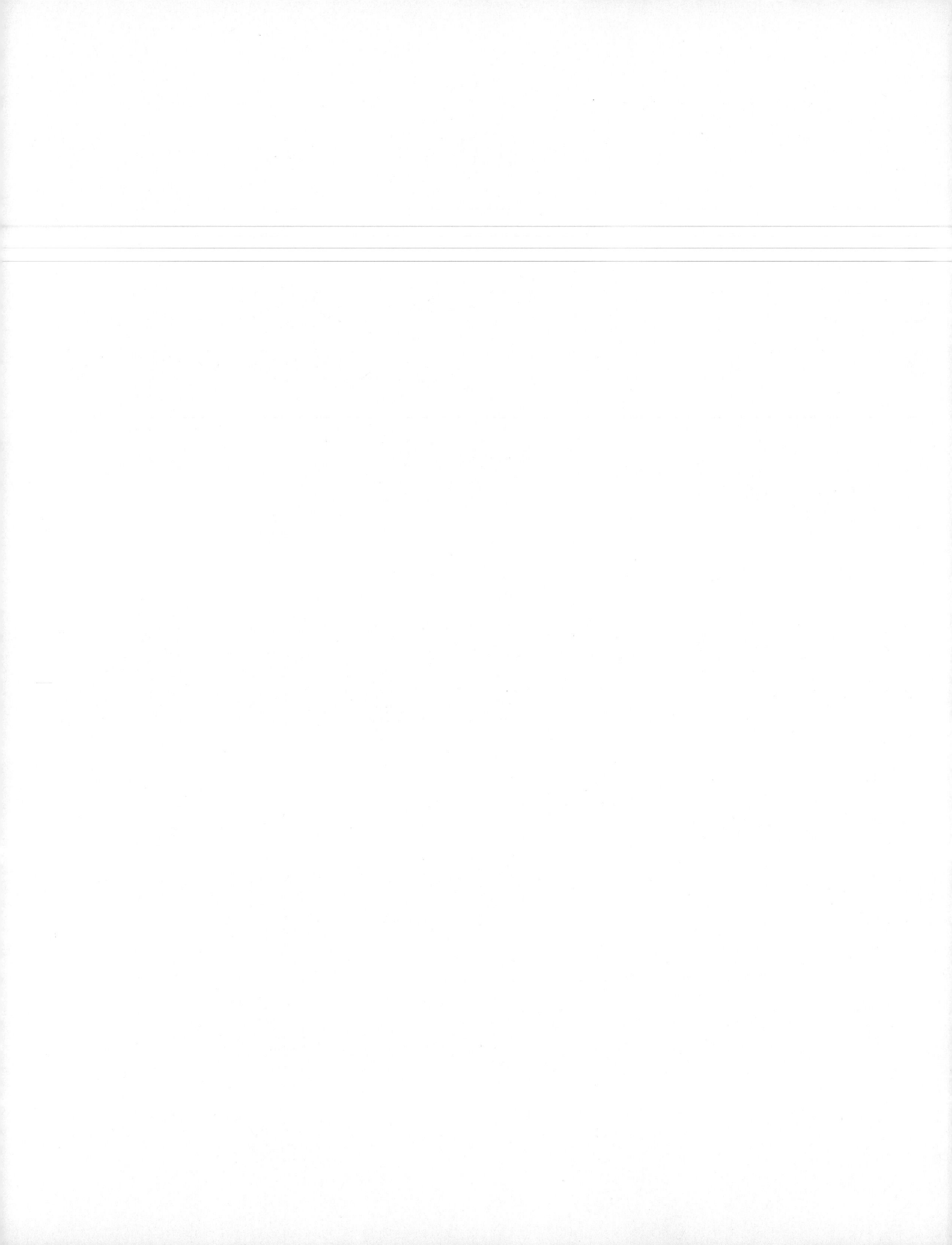